W9-BMH-488

Principles of Anatomy and Physiology

Principles of Anatomy and Physiology

Fifth Edition

Gerard J. Tortora
Bergen Community College

Nicholas P. Anagnostakos
Late, Bergen Community College

1817

HARPER & ROW, PUBLISHERS, New York
Cambridge, Philadelphia, San Francisco, Washington,
London, Mexico City, São Paulo, Singapore, Sydney

Sponsoring Editor: Claudia M. Wilson
Development Editor: Robert Ginsberg
Project Editor: Holly D. Gordon
Text Design Adaptation: Caliber Design Planning, Inc.
Cover Design: Ron Gross
Cover Art: Composite photograph of x ray of skull with brain by
 Bill Longcore/Longcore Maciel Studio. Anteroposterior projection
 of the right scapula and proximal end of the humerus. Courtesy
 of Garret J. Pinke, Sr., R.T. Scanning electron micrograph of fibrin
 threads and red cells. Courtesy of Fisher Scientific Company and
 S.T.E.M. Laboratories, Inc. Copyright 1975. The above three pho-
 tographs have been color enhanced by Bill Longcore/Longcore
 Maciel Studio.
Medical Illustrators: Leonard Dank, Charles Bridgman, Paul Melloni,
 Marsha J. Dohrmann, Helen Gee Jeung
Text Art: Vantage Art, Inc.
Photo Research: Mira Schachne
Production: Kewal K. Sharma
Compositor: Arcata Graphics/Kingsport
Printer and Binder: Arcata Graphics/Kingsport

PRINCIPLES OF ANATOMY AND PHYSIOLOGY, Fifth Edition

Library of Congress Cataloging-in-Publication Data

Tortora, Gerard J.
  Principles of anatomy and physiology.

  Bibliography: p.
  Includes index.
  1. Human physiology.   2. Anatomy, Human.
I. Anagnostakos, Nicholas Peter, 1924–1981
[DNLM: 1. Anatomy.   2. Homeostasis.   3. Physiology.
QS 4 T712p]
QP34.5.T67   1987   612   86–14843
ISBN 0–06–046669–3

Harper International Edition
ISBN 0–06–350729–3

  88 89 9 8 7 6 5 4 3

The fifth edition of *Principles of Anatomy and Physiology* is dedicated to
Lynne Marie, Gerard Joseph, Jr., and Kenneth Stephen Tortora

Contents in Brief

Preface   xix
Note to the Student   xxiii

UNIT I   ORGANIZATION OF THE HUMAN BODY   2

CHAPTER   1   An Introduction to the Human Body   5
CHAPTER   2   The Chemical Level of Organization   24
CHAPTER   3   The Cellular Level of Organization   45
CHAPTER   4   The Tissue Level of Organization   76
CHAPTER   5   The Integumentary System   100

UNIT II   PRINCIPLES OF SUPPORT AND MOVEMENT   116

CHAPTER   6   Skeletal Tissue   119
CHAPTER   7   The Skeletal System: The Axial Skeleton   132
CHAPTER   8   The Skeletal System: The Appendicular Skeleton   158
CHAPTER   9   Articulations   175
CHAPTER 10   Muscle Tissue   194
CHAPTER 11   The Muscular System   215

UNIT III   CONTROL SYSTEMS OF THE HUMAN BODY   258

CHAPTER 12   Nervous Tissue   261
CHAPTER 13   The Spinal Cord and the Spinal Nerves   280
CHAPTER 14   The Brain and the Cranial Nerves   306
CHAPTER 15   The Sensory, Motor, and Integrative Systems   338
CHAPTER 16   The Autonomic Nervous System   356
CHAPTER 17   The Special Senses   370
CHAPTER 18   The Endocrine System   397

UNIT IV   MAINTENANCE OF THE HUMAN BODY   436

CHAPTER 19   The Cardiovascular System: The Blood   439
CHAPTER 20   The Cardiovascular System: The Heart   460
CHAPTER 21   The Cardiovascular System: Vessels and Routes   487
CHAPTER 22   The Lymphatic System and Immunity   529
CHAPTER 23   The Respiratory System   554
CHAPTER 24   The Digestive System   589
CHAPTER 25   Metabolism   633
CHAPTER 26   The Urinary System   663
CHAPTER 27   Fluid, Electrolyte, and Acid–Base Dynamics   692

UNIT V   CONTINUITY   702

CHAPTER 28   The Reproductive Systems   705
CHAPTER 29   Development and Inheritance   742

APPENDIXES

APPENDIX A   Measurements   A-1
APPENDIX B   Normal Values for Selected Blood and Urine Tests   A-5
APPENDIX C   Abbreviations   A-11
APPENDIX D   Eponyms Used in This Text   A-13
Glossary of Prefixes, Suffixes, and Combining Forms   G-1
Glossary of Terms   G-5
Bibliography   B-1
Index   I-1

Contents in Detail

Preface   xix
Note to the Student   xxiii

UNIT I   ORGANIZATION OF THE HUMAN BODY   2

CHAPTER 1   An Introduction to the Human Body   5

Anatomy and Physiology Defined   6
Levels of Structural Organization   6
Life Processes   8
Structural Plan   9
Anatomical Position and Anatomical Names   9
Directional Terms   9
Planes and Sections   12
Body Cavities   13
Abdominopelvic Regions   13
Abdominopelvic Quadrants   13
Radiographic Anatomy   17
  Conventional Radiography   17
  Computed Tomography (CT) Scanning   17
  Dynamic Spatial Reconstructor (DSR)   17
Homeostasis   17
  Stress and Homeostasis   18
  Homeostasis of Blood Pressure   20
  Homeostasis of Blood Sugar Level   20
Measuring the Human Body   21

CHAPTER 2   The Chemical Level of Organization   24

Introduction to Basic Chemistry   25
  Chemical Elements   25
  Structure of Atoms   25
  Atoms and Molecules   26
    Ionic Bonds   27
    Covalent Bonds   28
    Hydrogen Bonds   30
  Radioisotopes   30
  Chemical Reactions   31
    Synthesis Reactions—Anabolism   31
    Decomposition Reactions—Catabolism   31
    Exchange Reactions   31
    Reversible Reactions   31
    Metabolism   32
    How Chemical Reactions Occur   32
    Energy and Chemical Reactions   33
Chemical Compounds and Life Processes   33
  Inorganic Compounds   33
    Water   33
    Acids, Bases, and Salts   34
    Acid–Base Balance: The Concept of pH   35
    Maintaining pH: Buffer Systems   36
  Organic Compounds   37
    Carbohydrates   37
    Lipids   38
    Proteins   39
    Nucleic Acids: Deoxyribonucleic Acid (DNA) and Ribonucleic Acid (RNA)   40
    Adenosine Triphosphate (ATP)   42
    Cyclic Adenosine-3', 5'-Monophosphate (Cyclic AMP)   42

CHAPTER 3   The Cellular Level of Organization   45

Generalized Animal Cell   46
Plasma (Cell) Membrane   46
  Chemistry and Structure   46
  Functions   47
  Movement of Materials Across Plasma Membranes   48
    Passive Processes   48
    Active Processes   51
Cytoplasm   53
Organelles   53
  Nucleus   54
  Ribosomes   55
  Endoplasmic Reticulum (ER)   56
  Golgi Complex   56
  Mitochondria   58
  Lysosomes   59
  Peroxisomes   59
  The Cytoskeleton   59
  Centrosome and Centrioles   60
  Flagella and Cilia   61
Cell Inclusions   61
Extracellular Materials   61
Gene Action   62
  Protein Synthesis   62
  Transcription   62
  Translation   62
  "SOS" Genes: DNA Repair   64
Normal Cell Division   64
  Somatic Cell Division   65
    Mitosis   65
    Cytokinesis   67
  Reproductive Cell Division   67
    Meiosis   69
Abnormal Cell Division: Cancer   69
  Definition   69
  Spread   70
  Types   70

Possible Causes   71
Treatment   71
Cells and Aging   71

Medical Terminology   72

CHAPTER 4   The Tissue Level of Organization   76

Types of Tissues   77
Epithelial Tissue   77
Covering and Lining Epithelium   77
Arrangement of Layers   77
Cell Shapes   77
Classification   78
Simple Epithelium   78
Stratified Epithelium   84
Pseudostratified Epithelium   84
Glandular Epithelium   84
Structural Classification of Exocrine Glands   84
Functional Classification of Exocrine Glands   85
Connective Tissue   85
Classification   86
Embryonic Connective Tissue   86
Adult Connective Tissue   86
Connective Tissue Proper   86
Cartilage   93
Osseous Tissue (Bone)   94
Vascular Tissue (Blood)   94
Muscle Tissue   94
Nervous Tissue   95
Membranes   96
Mucous Membranes   96
Serous Membranes   96
Cutaneous Membrane   96
Synovial Membranes   97
Tissue Repair   97
Repair Process   97
Conditions Affecting Repair   97

CHAPTER 5   The Integumentary System   100

Skin   101
Functions   101
Structure   101
Epidermis   101
Dermis   103
Skin Color   104
Epidermal Ridges and Grooves   104
Skin Wound Healing   105
Epidermal Wound Healing   105
Deep Wound Healing   105
Epidermal Derivatives   105
Hair   106
Glands   108
Sebaceous (Oil) Glands   108
Sudoriferous (Sweat) Glands   108
Ceruminous Glands   108
Nails   109
Homeostasis   109
Aging and the Integumentary System   110
Developmental Anatomy of the Integumentary System   110

Disorders: Homeostatic Imbalances   110
Medical Terminology   113

UNIT II   PRINCIPLES OF SUPPORT AND MOVEMENT   116

CHAPTER 6   Skeletal Tissue   119

Functions   120
Histology   120
Compact Bone   120
Spongy Bone   123
Ossification: Bone Formation   123
Intramembranous Ossification   123
Endochondral Ossification   123
Bone Growth   125
Homeostasis   126
Aging and the Skeletal System   126
Developmental Anatomy of the Skeletal System   126

Disorders: Homeostatic Imbalances   128
Medical Terminology   130

CHAPTER 7   The Skeletal System: The Axial Skeleton   132

Types of Bones   133
Surface Markings   133
Divisions of the Skeletal System   133
Skull   133
Sutures   134
Fontanels   136
Cranial Bones   136
Frontal Bone   136
Parietal Bones   137
Temporal Bones   137
Occipital Bone   140
Sphenoid Bone   141
Ethmoid Bone   142
Facial Bones   143
Nasal Bones   143
Maxillae   143
Paranasal Sinuses   144
Zygomatic Bones   144
Mandible   144
Lacrimal Bones   145
Palatine Bones   145
Inferior Nasal Conchae   146
Vomer   146
Foramina   146
Hyoid Bone   146
Vertebral Column   146
Divisions   146
Curves   149
Typical Vertebra   149
Cervical Region   150
Thoracic Region   150
Lumbar Region   150
Sacrum and Coccyx   150
Thorax   152
Sternum   152
Ribs   154

Disorders: Homeostatic Imbalances   155

**CHAPTER 8   The Skeletal System:
            The Appendicular Skeleton   158**

Pectoral (Shoulder) Girdle   159
  Clavicle   160
  Scapula   160
Upper Extremity   160
  Humerus   160
  Ulna and Radius   162
  Carpals, Metacarpals, and Phalanges   164
Pelvic (Hip) Girdle   165
Lower Extremity   166
  Femur   166
  Patella   166
  Tibia and Fibula   168
  Tarsals, Metatarsals, and Phalanges   170
  Arches of the Foot   171
Female and Male Skeletons   173

CHAPTER 9   Articulations   175

Classification   176
  Functional   176
  Structural   176
Fibrous Joints   176
  Suture   176
  Syndesmosis   176
  Gomphosis   176
Cartilaginous Joints   176
  Synchondrosis   176
  Symphysis   177
Synovial Joints   177
  Structure   177
  Movements   178
    Gliding   *178*
    Angular   *179*
    Rotation   *181*
    Circumduction   *181*
    Special   *181*
  Types   181
    Gliding   *181*
    Hinge   *181*
    Pivot   *182*
    Ellipsoidal   *182*
    Saddle   *182*
    Ball-and-Socket   *182*
  Summary of Joints   182
Selected Articulations of the Body   183
  Humeroscapular (Shoulder) Joint   183
  Coxal (Hip) Joint   187
  Tibiofemoral (Knee) Joint   188

Disorders: Homeostatic Imbalances   191
Medical Terminology   192

CHAPTER 10   Muscle Tissue   194

Characteristics   195
Functions   195
Types   195
Skeletal Muscle Tissue   195
  Connective Tissue Components   195
  Nerve and Blood Supply   197
  Histology   197

Contraction   199
  Sliding-Filament Theory   199
  Neuromuscular Junction (Motor End Plate)   199
  Motor Unit   200
  Physiology of Contraction   201
  Energy for Contraction   202
  Muscle Length and Force of Contraction   203
  All-or-None Principle   203
  Kinds of Contractions   203
    Twitch   *203*
    Tetanus   *204*
    Treppe   *204*
    Isotonic and Isometric   *204*
  Muscle Tone   205
  Types of Skeletal Muscle Fibers   205
Cardiac Muscle Tissue   206
Smooth Muscle Tissue   208
Homeostasis   210
  Oxygen Debt   210
  Heat Production   210
Aging and Muscle Tissue   210
Developmental Anatomy of the Muscular System   210

Disorders: Homeostatic Imbalances   211
Medical Terminology   212

CHAPTER 11   The Muscular System   215

How Skeletal Muscles Produce Movement   216
  Origin and Insertion   216
  Lever Systems and Leverage   216
  Arrangement of Fasciculi   216
  Group Actions   218
Naming Skeletal Muscles   218
Principal Skeletal Muscles   219
Intramuscular Injections   255

**UNIT III   CONTROL SYSTEMS OF THE HUMAN
BODY   258**

CHAPTER 12   Nervous Tissue   261

Organization   262
Histology   263
  Neuroglia   263
  Neurons   264
    Structure   *264*
    Structural Variation   *267*
    Classification   *267*
Physiology   268
  Nerve Impulse   268
    Membrane Potentials   *268*
    Excitability   *269*
    All-or-None Principle   *271*
    Saltatory Conduction   *271*
    Speed of Nerve Impulses   *272*
  Conduction Across Synapses   272
    Excitatory Transmission   *273*
    Inhibitory Transmission   *274*
    Integration at Synapses   *274*
  Neurotransmitters   275
  Regeneration   276
Organization of Neurons   276

**CHAPTER 13   The Spinal Cord and
the Spinal Nerves   280**

Grouping of Neural Tissue   281
Spinal Cord   281
Spinal Cord   281
  General Features   281
  Protection and Coverings   281
    Vertebral Canal   281
    Meninges   281
  Structure in Cross Section   284
  Functions   285
    Spinal Tracts   285
    Reflex Center   285
    Reflex Arc and Homeostasis   287
    Reflexes and Diagnosis   292
Spinal Nerves   292
  Names   292
  Composition and Coverings   293
  Distribution   294
    Branches   294
    Plexuses   294
    Intercostal (Thoracic) Nerves   300
  Dermatomes   302

Disorders: Homeostatic Imbalances   302

CHAPTER 14   The Brain and the Cranial Nerves   306

Brain   307
  Principal Parts   307
  Protection and Coverings   307
    Cerebrospinal Fluid (CSF)   307
  Blood Supply   310
  Brain Stem   310
    Medulla Oblongata   310
    Pons   313
    Midbrain   313
  Diencephalon   314
    Thalamus   314
    Hypothalamus   314
  Cerebrum   316
    Lobes   318
    White Matter   318
    Basal Ganglia (Cerebral Nuclei)   318
    Limbic System   319
    Functional Areas of Cerebral Cortex   321
    Electroencephalogram (EEG)   323
  Brain Lateralization (Split-Brain Concept)   323
  Cerebellum   323
Neurotransmitters in the Brain   325
Cranial Nerves   327
Aging and the Nervous System   327
Developmental Anatomy of the Nervous System   328

Disorders: Homeostatic Imbalances   332
Medical Terminology   335

**CHAPTER 15   The Sensory, Motor, and
Integrative Systems   338**

Sensations   339
  Definition   339
  Characteristics   339
  Classification of Receptors   340
    Location   340
    Stimulus Detected   340
    Simplicity or Complexity   340

General Senses   340
  Cutaneous Sensations   340
    Tactile Sensations   341
    Thermoreceptive Sensations   342
    Pain Sensations   342
  Proprioceptive Sensations   344
    Receptors   344
  Levels of Sensation   345
Sensory Pathways   345
  Somatosensory Cortex   345
  Posterior Column Pathway   346
  Spinothalamic Pathway   346
    Pain and Temperature   346
    Light Touch and Pressure   347
    *Discriminative Touch, Proprioception, and
      Vibration   347*
  Cerebellar Tracts   347
Motor Pathways   348
  Linkage of Sensory Input and Motor Responses   348
  Motor Cortex   349
  Pyramidal Pathways   349
  Extrapyramidal Pathways   350
Integrative Functions   351
  Memory   351
  Wakefulness and Sleep   352

CHAPTER 16   The Autonomic Nervous System   356

**Somatic Efferent and Autonomic Nervous
Systems   357**
Structure of the Autonomic Nervous System   358
  Visceral Efferent Pathways   358
    Preganglionic Neurons   358
    Autonomic Ganglia   358
    Postganglionic Neurons   360
  Sympathetic Division   361
  Parasympathetic Division   362
Physiology of the Autonomic Nervous System   363
  Neurotransmitters   363
  Receptors   363
  Activities   364
Visceral Autonomic Reflexes   366
Control by Higher Centers   367
  Biofeedback   367
  Meditation   368

CHAPTER 17   The Special Senses   370

Olfactory Sensations   371
  Structure of Receptors   371
  Stimulation of Receptors   371
  Adaptation and Odor Thresholds   371
  Olfactory Pathway   371
Gustatory Sensations   372
  Structure of Receptors   372
  Stimulation of Receptors   373
  Adaptation and Taste Thresholds   373
  Gustatory Pathway   373
Visual Sensations   373
  Accessory Structures of Eye   374
  Structure of Eyeball   376
    Fibrous Tunic   376
    Vascular Tunic   376
    Retina (Nervous Tunic)   378
    Lens   379
    Interior   379

Physiology of Vision   380
   Retinal Image Formation   380
   Stimulation of Photoreceptors   382
   Visual Pathway   384
Auditory Sensations and Equilibrium   386
   External or Outer Ear   386
   Middle Ear   388
   Internal or Inner Ear   389
   Sound Waves   390
   Physiology of Hearing   390
   Physiology of Equilibrium   392
   Static Equilibrium   392
   Dynamic Equilibrium   393

Disorders: Homeostatic Imbalances   394
Medical Terminology   395

CHAPTER 18   The Endocrine System   397

Endocrine Glands   399
Chemistry of Hormones   399
Mechanism of Hormonal Action   399
   Overview   399
   Receptors   400
   Interaction with Plasma Membrane Receptors   401
   Interaction with Intracellular Receptors   403
   Prostaglandins (PGs) and Hormones   403
Control of Hormonal Secretions:
   Feedback Control   403
Pituitary (Hypophysis)   404
   Adenohypophysis   404
   Growth Hormone (GH)   405
   Thyroid-Stimulating Hormone (TSH)   408
   Adrenocorticotropic Hormone (ACTH)   408
   Follicle-Stimulating Hormone (FSH)   408
   Luteinizing Hormone (LH)   408
   Prolactin (PRL)   408
   Melanocyte-Stimulating Hormone (MSH)   409
   Neurohypophysis   409
   Oxytocin (OT)   410
   Antidiuretic Hormone (ADH)   411
Thyroid   412
   Formation, Storage, and Release of Thyroid
     Hormones   414
   Function and Control of Thyroid Hormones   414
   Calcitonin (CT)   416
Parathyroids   416
   Parathyroid Hormone (PTH)   416
Adrenals (Suprarenals)   419
   Adrenal Cortex   419
   Mineralocorticoids   420
   Glucocorticoids   421
   Gonadocorticoids   423
   Adrenal Medulla   423
   Epinephrine and Norepinephrine (NE)   423
Pancreas   424
   Glucagon   424
   Insulin   424
Ovaries and Testes   427
Pineal (Epiphysis Cerebri)   428
Thymus   428
Developmental Anatomy of the Endocrine
   System   428
Other Endocrine Tissues   430
Aging and the Endocrine System   430

Stress and the General Adaptation Syndrome   430
   Stressors   430
   Alarm Reaction   430
   Resistance Reaction   432
   Exhaustion   432
   Stress and Disease   433
Medical Terminology   433

UNIT IV   MAINTENANCE OF THE HUMAN BODY   436

CHAPTER 19   The Cardiovascular System:
         The Blood   439

Physical Characteristics   440
Functions   440
Components   440
   Formed Elements   440
   Origin   441
   Erythrocytes   442
   Leucocytes   445
   Thrombocytes   447
   Plasma   448
Hemostasis   449
   Vascular Spasm   449
   Platelet Plug Formation   449
   Coagulation   449
   Extrinsic Pathway   450
   Intrinsic Pathway   450
   Retraction and Fibrinolysis   452
   Prevention   452
   Tests   453
Grouping (Typing)   453
   ABO   453
   Rh   455
Interstitial Fluid and Lymph   456

Disorders: Homeostatic Imbalances   456
Medical Terminology   457

CHAPTER 20   The Cardiovascular System:
         The Heart   460

Location   461
Pericardium   462
Heart Wall   462
Chambers of the Heart   462
Great Vessels of the Heart   463
Valves of the Heart   465
   Atrioventricular (AV) Valves   465
   Semilunar Valves   466
   Surface Projection   466
Blood Supply   468
Conduction System   469
Electrocardiogram   470
Blood Flow Through the Heart   472
Cardiac Cycle   472
   Timing   472
   Sounds   473
Cardiac Output (CO)   474
   Stroke Volume   474
   End-Diastolic Volume (EDV)   474
   End-Systolic Volume (ESV)   474

Heart Rate   475
   Autonomic Control   475
   Chemicals   477
   Temperature   477
   Emotions   477
   Sex and Age   477
Circulatory Shock and Homeostasis   477
Artificial Heart   478
Diagnosing Heart Disorders   478
Heart-Lung Machine   479
Risk Factors in Heart Disease   479
Developmental Anatomy of the Heart   480

Disorders: Homeostatic Imbalances   481
Medical Terminology   484

CHAPTER 21   The Cardiovascular System:
                   Vessels and Routes   487

Arteries   488
   Elastic (Conducting) Arteries   489
   Muscular (Distributing) Arteries   489
   Anastomoses   489
Arterioles   489
Capillaries   490
Venules   490
Veins   490
Blood Reservoirs   492
Physiology of Circulation   492
   Blood Flow and Blood Pressure   492
      Factors That Affect Arterial Blood Pressure   492
      Control of Blood Pressure   493
      Capillary Exchange   494
      Factors That Aid Venous Return   496
Checking Circulation   496
   Pulse   496
   Measurement of Blood Pressure (BP)   497
Circulatory Routes   498
   Systemic Circulation   498
   Pulmonary Circulation   520
   Hepatic Portal Circulation   520
   Fetal Circulation   521
Exercise and the Cardiovascular System   524
Aging and the Cardiovascular System   524
**Developmental Anatomy of Blood and Blood
   Vessels   524**

Disorders: Homeostatic Imbalances   525
Medical Terminology   526

CHAPTER 22   The Lymphatic System
                  and Immunity   529

Lymphatic Vessels   530
Lymphatic Tissue   531
   Lymph Nodes   531
   Tonsils   533
   Spleen   534
   Thymus Gland   536
Lymph Circulation   536
   Route   536
      Thoracic (Left Lymphatic) Duct   536
      Right Lymphatic Duct   537
   Maintenance   537

Nonspecific Resistance to Disease   537
   Skin and Mucous Membranes   538
      Mechanical Factors   538
      Chemical Factors   538
   Antimicrobial Substances   538
      Interferon (IFN)   538
      Complement   539
      Properdin   539
   Phagocytosis   539
      Kinds of Phagocytes   539
      Mechanism   539
   Inflammation   540
      Symptoms   540
      Stages   540
   Fever   542
Immunity (Specific Resistance to Disease)   542
   Antigens   542
      Definition   542
      Characteristics   542
   Antibodies   544
      Definition   544
      Structure   544
   Cellular and Humoral Immunity   544
      Formation of T Cells and B Cells   545
      T Cells, B Cells, and Macrophages   545
      T Cells and Cellular Immunity   545
      Natural Killer (NK) Cells   547
      B Cells and Humoral Immunity   547
   The Skin and Immunity   548
   Monoclonal Antibodies   549
   Immunology and Cancer   549
Developmental Anatomy of the Lymphatic System   549

Disorders: Homeostatic Imbalances   550
Medical Terminology   552

CHAPTER 23   The Respiratory System   554

Organs   556
   Nose   556
      Anatomy   556
      Physiology   556
   Pharynx   556
   Larynx   558
      Anatomy   558
      Voice Production   558
   Trachea   560
   Bronchi   560
   Lungs   561
      Gross Anatomy   562
      Lobes and Fissures   564
      Lobules   564
      Alveolar-Capillary (Respiratory) Membrane   564
      Blood Supply   565
Respiration   565
   Pulmonary Ventilation   565
      Inspiration   566
      Expiration   569
      Atelectasis (Collapsed Lung)   569
      Compliance   569
      Airway Resistance   569
   Modified Respiratory Movements   570
   Pulmonary Air Volumes and Capacities   570
      Spirometry   570
      Pulmonary Volumes   570
      Pulmonary Capacities   572

Exchange of Respiratory Gases   572
  Dalton's Law   *572*
  Henry's Law   *573*
External Respiration   573
Internal Respiration   575
Transport of Respiratory Gases   575
  Oxygen   *575*
  Hypoxia   *578*
  Carbon Dioxide   *578*
Control of Respiration   579
Nervous Control   579
  Medullary Rhythmicity Area   *579*
  Pneumotaxic Area   *581*
  Apneustic Area   *581*
Regulation of Respiratory Center Activity   581
  Cortical Influences   *581*
  Inflation Reflex   *581*
  Chemical Stimuli   *581*
  Other Influences   *582*
Aging and the Respiratory System   582
**Developmental Anatomy of the Respiratory
  System   583**

Disorders: Homeostatic Imbalances   583
Medical Terminology   586

CHAPTER 24   The Digestive System   589

Regulation of Food Intake   590
Digestive Processes   590
Organization   590
General Histology   591
  Mucosa   *591*
  Submucosa   *592*
  Muscularis   *592*
  Serosa   *592*
Peritoneum   592
Mouth (Oral Cavity)   595
Tongue   596
Salivary Glands   596
  Composition of Saliva   *596*
  Secretion of Saliva   *598*
Teeth   598
  Dentitions   *598*
Digestion in the Mouth   599
  Mechanical   *599*
  Chemical   *599*
Deglutition   600
Esophagus   600
Histology   601
Activities   601
Stomach   603
Anatomy   603
Histology   605
Digestion in the Stomach   605
  Mechanical   *605*
  Chemical   *605*
Regulation of Gastric Secretion   607
  Stimulation   *607*
  Inhibition   *607*
Regulation of Gastric Emptying   608
Absorption   609
Pancreas   609
Anatomy   609
Histology   610

Pancreatic Juice   610
Regulation of Pancreatic Secretions   611
Liver   611
Anatomy   611
Histology   611
Blood Supply   611
Bile   612
Regulation of Bile Secretion   613
Functions of the Liver   613
Gallbladder   614
Histology   614
Function   614
Emptying of the Gallbladder   614
Small Intestine   614
Anatomy   614
Histology   614
Intestinal Juice   616
Digestion in the Small Intestine   617
  Mechanical   *617*
  Chemical   *617*
Regulation of Intestinal Secretion   618
Absorption   618
  Carbohydrates   *618*
  Proteins   *619*
  Lipids   *619*
  Water   *621*
  Electrolytes   *621*
  Vitamins   *621*
Large Intestine   622
Anatomy   622
Histology   623
Digestion in the Large Intestine   624
  Mechanical   *624*
  Chemical   *624*
  Absorption and Feces Formation   *624*
  Defecation   *624*
Aging and the Digestive System   625
**Developmental Anatomy of the Digestive
  System   625**

Disorders: Homeostatic Imbalances   627
Medical Terminology   629

CHAPTER 25   Metabolism   633

Metabolism   634
Anabolism   634
Catabolism   634
Oxidation-Reduction Reactions   634
Metabolism and Enzymes   635
  Characteristics of Enzymes   *635*
  Parts of an Enzyme   *635*
  Mechanism of Enzymatic Action   *635*
Carbohydrate Metabolism   636
Fate of Carbohydrates   636
Glucose Movement into Cells   636
Glucose Catabolism   636
  Glycolysis   *636*
  Formation of Acetyl Coenzyme A   *638*
  Krebs Cycle   *639*
  Electron Transport Chain   *640*
  ATP Production   *642*
Glucose Anabolism   643
  Glucose Storage: Glycogenesis   *643*
  Glucose Release: Glycogenolysis   *643*

Formation of Glucose from Proteins and Fats:
Gluconeogenesis   643
Lipid Metabolism   644
Fate of Lipids   644
Fat Storage   644
Lipid Catabolism   644
Glycerol   645
Fatty Acids   645
Lipid Anabolism: Lipogenesis   646
Protein Metabolism   646
Fate of Proteins   646
Protein Catabolism   646
Protein Anabolism   646
Absorptive and Postabsorptive States   647
Absorptive State   647
Postabsorptive (Fasting) State   648
Regulation of Metabolism   649
Minerals   649
Vitamins   652
Metabolism and Body Heat   655
Measuring Heat   655
Production of Body Heat   655
Basal Metabolic Rate   656
Loss of Body Heat   656
Radiation   656
Conduction   656
Convection   656
Evaporation   656
Body Temperature Regulation   657
Hypothalamic Thermostat   657
Mechanisms of Heat Production   657
Mechanisms of Heat Loss   658
Body Temperature Abnormalities   658

Disorders: Homeostatic Imbalances   658

CHAPTER 26   The Urinary System   663

Kidneys   664
External Anatomy   664
Internal Anatomy   666
Nephron   666
Blood and Nerve Supply   667
Juxtaglomerular Apparatus   669
Physiology   670
Glomerular Filtration   671
Tubular Reabsorption   673
Tubular Secretion   675
Mechanism of Urine Dilution   676
Mechanism of Urine Concentration   678
Hemodialysis Therapy   679
Homeostasis   680
Other Excretory Organs   680
Urine   680
Volume   680
Physical Characteristics   681
Chemical Composition   682
Abnormal Constituents   682
Ureters   684
Structure   684
Histology   684
Physiology   685
Urinary Bladder   685
Structure   685
Histology   685
Physiology   685

Urethra   686
Histology   686
Physiology   686
Aging and the Urinary System   686
Developmental Anatomy of the Urinary System   686

Disorders: Homeostatic Imbalances   688
Medical Terminology   689

CHAPTER 27   Fluid, Electrolyte, and
Acid–Base Dynamics   692

Fluid Compartments and Fluid Balance   693
Water   693
Fluid Intake and Output   693
Regulation of Intake   693
Regulation of Output   693
Electrolytes   694
Concentration   694
Distribution   694
Functions and Regulation   694
Sodium   694
Chloride   696
Potassium   696
Calcium and Phosphate   696
Magnesium   696
Movement of Body Fluids   697
Between Plasma and Interstitial Compartments   697
Between Interstitial and Intracellular
Compartments   697
Acid–Base Balance   697
Buffer Systems   698
Carbonic Acid–Bicarbonate   698
Phosphate   698
Hemoglobin-Oxyhemoglobin   698
Protein   699
Respirations   699
Kidney Excretion   699
Acid–Base Imbalances   700

UNIT V   CONTINUITY   702

CHAPTER 28   The Reproductive Systems   705

Male Reproductive System   706
Scrotum   706
Testes   707
Spermatogenesis   708
Spermatozoa   711
Testosterone and Inhibin   711
Ducts   713
Ducts of the Testis   713
Epididymis   713
Ductus (Vas) Deferens   713
Ejaculatory Duct   714
Urethra   714
Accessory Sex Glands   714
Semen (Seminal Fluid)   715
Penis   716
Female Reproductive System   717
Ovaries   717
Oogenesis   720
Uterine (Fallopian) Tubes   720
Uterus   721

Endocrine Relations: Menstrual and Ovarian
　Cycles　722
　　Hormonal Control　722
　　Menstrual Phase (Menstruation)　723
　　Preovulatory Phase　725
　　Ovulation　725
　　Postovulatory Phase　725
　　Menarche and Menopause　726
Vagina　726
Vulva　727
Perineum　728
Mammary Glands　728
Sexual Intercourse　730
Male Sexual Act　730
　　Erection　730
　　Lubrication　730
　　Orgasm　730
Female Sexual Act　731
　　Erection　731
　　Lubrication　731
　　Orgasm (Climax)　731
Birth Control　731
Removal of Gonads and Uterus　731
Sterilization　731
Contraception　731
　　Natural　731
　　Mechanical　731
　　Chemical　732
Aging and the Reproductive Systems　732
**Developmental Anatomy of the Reproductive
Systems　733**

Disorders: Homeostatic Imbalances　735
Medical Terminology　738

CHAPTER 29　Development and Inheritance　742

Development During Pregnancy　743
Fertilization and Implantation　743
　　Fertilization　743
　　Formation of the Morula　744
　　Development of the Blastocyst　744
　　Implantation　744
External Human Fertilization　746
Embryo Transfer　746
Gamete Intrafallopian Transfer (GIFT) and Transvaginal
　Oocyte Retrieval　746
Embryonic Development　747
Beginnings of Organ Systems　747
　　Embryonic Membranes　747
　　Placenta and Umbilical Cord　750

Fetal Growth　751
Hormones of Pregnancy　752
Gestation　753
Prenatal Diagnostic Techniques　755
Amniocentesis　755
Chorionic Villi Sampling (CVS)　756
Parturition and Labor　756
Adjustments of the Infant at Birth　758
Respiratory System　758
Cardiovascular System　759
**Potential Hazards to the Developing Embryo and
Fetus　759**
Chemicals and Drugs　759
Irradiation　759
Alcohol　759
Cigarette Smoking　759
Lactation　760
Inheritance　760
Genotype and Phenotype　760
Inheritance of Sex　762
Color Blindness and X-Linked Inheritance　762

Medical Terminology　763

APPENDIXES

APPENDIX A　Measurements　A-1

**APPENDIX B　Normal Values for Selected Blood and
　　　　　Urine Tests　A-5**

APPENDIX C　Abbreviations　A-11

APPENDIX D　Eponyms Used in This Text　A-13

**Glossary of Prefixes, Suffixes, and
Combining Forms　G-1**

Glossary of Terms　G-5

Bibliography　B-1

Index　I-1

Preface

Principles of Anatomy and Physiology, Fifth Edition, is designed for use in an introductory course in anatomy and physiology and assumes no previous study of the human body. The text is geared to students in health-oriented, medical, and biological programs who are aiming for careers as nurses, medical assistants, medical laboratory technologists, radiologic technologists, inhalation therapists, dental hygienists, physical therapists, morticians, and medical records keepers. Because of its scope, the text is also useful for students in the biological sciences, science technology, liberal arts, physical education, and in premedical, predental, and prechiropractic programs.

OBJECTIVES

The objectives of the fifth edition are the same as those of previous editions: (1) to provide a basic understanding and working knowledge of the human body and (2) to present this essential material at a level that average students can handle.

Throughout, our goal has been to eliminate barriers to a ready comprehension of the structure and function of the human body. We recognize, however, that some technical vocabulary and difficult concepts are vital to the course. Such material is developed in step-by-step, easy-to-understand explanations that avoid needlessly difficult nontechnical vocabulary and syntax.

THEMES

Two major themes still dominate the book—**homeostasis** and **pathology.** Throughout, the book shows students how dynamic counterbalancing forces maintain normal anatomy and physiology. Pathology is viewed as a disruption in homeostasis. Accordingly, we present a number of clinical topics and contrast them with specific normal processes.

ORGANIZATION

The book follows the same unit and topic sequence as its four earlier editions. It is divided into five principal areas of concentration. Unit I, Organization of the Human Body, provides an understanding of the structural and functional levels of the body, from molecules to organ-systems. Unit II, Principles of Support and Movement, analyzes the anatomy and physiology of the skeletal system, articulations, and the muscular system. Unit III, Control Systems of the Human Body, emphasizes the importance of the nerve impulse in the immediate mainte-

nance of homeostasis, the role of receptors in providing information about the internal and external environment, and the significance of hormones in maintaining long-range homeostasis. Unit IV, Maintenance of the Human Body, illustrates how the body maintains itself on a day-to-day basis through the mechanisms of circulation, respiration, digestion, cellular metabolism, urine production, and buffer systems. Unit V, Continuity, covers the anatomy and physiology of the reproductive systems, sexual development, and the basic concepts of genetics.

GENERAL CHANGES

1. Developmental anatomy is no longer covered exclusively in a single chapter (Chapter 29). The developmental anatomy of each organ-system is covered in its respective chapter.
2. The effects of aging on body systems are no longer considered in a single exhibit in Chapter 3 but are discussed in each respective chapter.
3. The general histological characteristics of bone, blood, muscle, and nervous tissue are now treated in Chapter 4, along with epithelium and other connective tissues.
4. Definitions of medical specialties have been added and placed in the chapter on the specific organ-system to which each pertains. For example, hematology is included in the chapter on blood, cardiology in the chapter on the heart, dermatology in the chapter on the skin, and so on.
5. Current terms are used in place of eponyms throughout the text and in the labels on the figures. Where a current term appears, the eponym is indicated in parentheses following the term. For example, "auditory (Eustachian) tube." In addition, a new appendix (D) has been added, consisting of an alphabetical listing of eponyms and their corresponding current terms. In the index, all eponyms are cross-referenced to current terms.
6. The line drawings for most of the bones in Chapter 8 have been replaced by color photos.
7. A new appendix on normal values for selected components of blood and urine has been added.
8. Discussions of drugs associated with various body systems have been moved to the *Instructor's Manual.* They have been updated and revised for the fifth edition.

In the fifth edition, the strengths of the previous editions have been maintained. Revisions made for the new edition have focused on updating certain topics and strengthening the coverage of physiology. Among the specific changes made in topic coverage in the fifth edition are the following.

UNIT I. ORGANIZATION OF THE HUMAN BODY

Chapter 1 has been expanded to include sections dealing with life processes, autopsy, and cine CT scanning. In Chapter 2, a new section has been added on the collision theory of chemical reactions, and the detailed discussion of enzymes has been moved to Chapter 25. Chapter 3 includes an updated discussion of cancer. New to Chapter 3 are discussions on microtomography and receptor-mediated endocytosis. In terms of reorganization of Chapter 3, gene action is now discussed before cell division, and meiosis has been moved from Chapter 29 to Chapter 3, where it is contrasted with mitosis. In Chapter 4, there is a new section on suction lipectomy, and the general histological characteristics of bone, blood, muscle, and nervous tissue are described. The detailed discussion of inflammation has been moved to Chapter 22, along with other components of nonspecific resistance. Changes in Chapter 5 consist of revised discussions of the histology of the epidermis, skin cancer, and skin grafts, and new material on chemical exfoliation, skin wound healing, and development of the skin.

UNIT II. PRINCIPLES OF SUPPORT AND MOVEMENT

New to Chapter 6 are sections on osteogenic sarcoma and development of bone tissue. The discussion of osteoporosis has been revised. In Chapter 7, there are new sections on gravity inversion, caudal anesthesia, and fractured ribs. New to Chapter 8 is a section on shinsplints. Chapter 9 has been expanded to include discussions of dislocation of the shoulder joint, runner's knee, swollen knee, and dislocated knee. In Chapter 10, there are new sections on muscle-building anabolic steroids, muscle length and force of contraction, and the development of muscle tissue. Revised discussions relate to the histology of muscle tissue, the neuromuscular junction, the physiology of skeletal muscle contraction, and types of skeletal muscle fibers. Chapter 11 includes a new exhibit that describes and illustrates the muscles of the floor of the oral cavity.

UNIT III. CONTROL SYSTEMS OF THE HUMAN BODY

Chapter 12 has been updated to include revised sections on membrane potentials, generation of nerve impulses, impulse conduction across synapses, and neurotransmitters. In Chapter 13, the discussions of the stretch reflex, spinal cord injury, and shingles have been revised, and there are new sections on tendon reflexes and plexuses. In Chapter 14, discussions of cerebrospinal fluid (CSF), neurotransmitters, Parkinson's disease, and Alzheimer's disease (AD) have been revised, and new sections have been added on brain death, brain tumors, and development of the nervous system. Chapter 15 now contains a revised discussion of adaptation, acupuncture, proprioceptive receptors, memory, and NREM sleep. New

sections on the linkage of sensory input and motor responses and narcolepsy have been added. New to Chapter 16 are sections on the comparisons between the somatic efferent and autonomic nervous system, and nicotinic, muscarinic, alpha, and beta receptors. In Chapter 17, coverage has been expanded to include radial keratotomy, the artificial ear, and vertigo. The treatment of the histology of the olfactory epithelium, taste buds, and retina, the rhodopsin cycle, and the physiology of hearing has been revised. Chapter 18 contains revised material on hormone chemistry, the mechanism of hormone action, the histology of the thyroid gland, diabetes mellitus, and stress. New to Chapter 18 are sections on receptors, stress and disease, and development of the endocrine system.

UNIT IV. MAINTENANCE OF THE HUMAN BODY

Chapter 19 contains a revised discussion of erythropoiesis, clotting time, bleeding time, blood clotting, blood transfusions, and sickle-cell anemia, as well as new sections on multiple myeloma, bone marrow transplantation, apheresis, and tissue plasminogen activator (t-PA). In Chapter 20, there are revised discussions of the pericardium, cardiac cycle, Starling's law of the heart, the artificial heart, and atherosclerosis. New material has been added on ventricular filling, isovolumetric contraction and relaxation, syncope, diagnosing heart disorders, coronary artery bypass surgery, coarctation of the aorta, and development of the heart. Coronary artery disease, atherosclerosis, coronary artery spasm, and cor pulmonale have been moved from Chapter 21 to Chapter 20. New topics in Chapter 21 are deep-venous thrombosis (DVT) and development of blood vessels. Several new flow diagrams of arterial distribution and venous drainage have also been added. The discussions of capillary exchange, exercise and the cardiovascular system, aneurysm, and hypertension have been revised. Chapter 22 contains new material on the organization of the lymphatic system, the histology of lymph capillaries, types of T cells, the skin and immunity, hypersensitivity (allergy), severe combined immunodeficiency disease (SCID), Hodgkin's disease, and development of the lymphatic system. Among the revised topics are interferon (IFN), T cells, B cells and macrophages, monoclonal antibodies, and acquired immune deficiency syndrome (AIDS). Chapter 23 contains new sections on sleep apnea, hypoxia, respiratory failure, pulmonary embolism (PE), pulmonary edema, and development of the respiratory system. In Chapter 24, new material has been added on the regulation of food intake, small intestinal digestion, endoscopy, and development of the digestive system. Chapter 25 contains expanded and reorganized discussions of anabolism, catabolism, oxidation-reduction reactions, metabolism and enzymes, production of body heat, basal metabolic rate (BMR), and fever. The chapter also includes new material on abnormal conditions associated with megadoses of vitamins, obesity, and starvation. The detailed discussion of enzymes has been moved from Chapter 2 to Chapter 25. Chapter 26 provides new material on mechanisms of urine dilution and concentration, cystoscopy, renal failure, urinary tract infections (UTIs), and development of the urinary system.

The discussion of the juxtaglomerular apparatus and tubular reabsorption has been revised.

UNIT V. CONTINUITY

New topics in Chapter 28 include the blood-testis barrier, semen analysis, artificial insemination, and development of the reproductive systems. Spermatogenesis, oogenesis, sexual intercourse, and birth control have been moved from Chapter 29 to Chapter 28. In Chapter 29, the sections on fertilization and gestation have been expanded. There are also new sections on embryo transfer, gamete intrafallopian transfer (GIFT), transvaginal oocyte retrieval, chorionic villi sampling (CVS), and premature infants.

SPECIAL FEATURES

As in previous editions, the book contains numerous learning aids. Users of the book have cited the pedagogical aids as one of the book's many strengths. All of the tested and successful learning aids of previous editions have been retained in the fifth edition, and several new ones have been added. These special features are:

1. **Student Objectives.** Each chapter opens with a comprehensive list of Student Objectives. Each objecive describes a knowledge or skill students should acquire while studying the chapter. (See **Note to the Student** for an explanation of how the objectives may be used.)
2. **Study Outline.** A Study Outline at the end of each chapter provides a brief summary of major topics. This section consolidates the essential points covered in the chapter so that students can recall and relate the points to one another. Page numbers given beside major headings in the outlines make it easy to refer to topics within chapters.
3. **Review Questions.** Review Questions at the end of each chapter provide a check to see if the objectives stated at the beginning of the chapter have been met. After answering the questions, students should reread the objectives to determine whether they have met the goals.
4. **Exhibits.** Health-science students are generally expected to learn a great deal about the anatomy of certain organ-systems, specifically, skeletal muscles, articulations, blood vessels, and nerves. To avoid interrupting the discussion of concepts and to organize the data, anatomical details have been presented in tabular form in exhibits, most of which are accompanied by illustrations. New summary exhibits have been added to the fifth edition, especially in relation to physiological principles.
5. **Disorders: Homeostatic Imbalances.** Abnormalities of structure or function are grouped at the end of appropriate chapters in sections titled "Disorders: Homeostatic Imbalances." These sections provide a review of normal body processes and demonstrate the importance of the study of anatomy and physiology to a career in any of the health fields. All disorders have been updated and several new ones have been added.
6. **Phonetic Pronunciations.** Throughout the text, phonetic pronunciations are provided in parentheses for selected anatomical and physiological terms. These pronun-

ciations are given at the point where the terms are introduced and are repeated in the Glossary of Terms. The **Note to the Student** explains the pronunciation key.
7. **Medical Terminology.** Glossaries of selected medical terms appear at the end of appropriate chapters; these listings are entitled "Medical Terminology." All of the glossaries have been revised for the fifth edition.
8. **Clinical Applications.** Throughout the text, Clinical Applications are boxed off for greater emphasis. Many new ones have been added.
9. **Line Art.** The line drawings in the book are large so that details are easily seen. In the fifth edition, some line art has been redrawn and many new illustrations have been added. Full color is used throughout to differentiate structures and regions.
10. **Photographs.** The photographs amplify the narrative and the line drawings. Numerous photomicrographs (in full color), scanning electron micrographs, and transmission electron micrographs enhance the histological discussions. Color photographs of specimens and regional dissections clarify gross anatomy discussions.
11. **Appendixes.** Appendix A, Measurements, summarizes U.S., metric, and apothecary units of length, mass, volume, and time. Appendix B, Normal Values for Selected Blood and Urine Tests, contains a listing of normal values for the principal constituents of these fluids. Appendix C, Abbreviations, is an alphabetical list of commonly encountered medical abbreviations that has been expanded for the fifth edition. Appendix D, Eponyms Used in This Text, is an alphabetical list of commonly encountered eponyms and the corresponding current terminology. Eponyms are cited in the text in parentheses immediately following the preferred current terms.
12. **Glossaries.** Two glossaries appear at the end of the book. The first deals with prefixes, suffixes, and combining forms. The second is a comprehensive glossary of terms that has been updated and expanded.
13. **Bibliography.** The extensive bibliography lists current suggested readings that correspond to the unit organization of the text.

SUPPLEMENTS

The following supplementary items are available to accompany the fifth edition of *Principles of Anatomy and Physiology:*

1. **Learning Guide.** By Kathleen S. Prezbindowski and Gerard J. Tortora, the *Learning Guide* is designed to help students *learn* anatomy and physiology. Rather than asking for mere repetition of the text material, the *Learning Guide* requires students to perform activities such as labeling a drawing or coloring parts of a diagram, or to role-play the components of the major physiological systems, as well as to answer multiple-choice, matching, and fill-in questions. The book's emphasis is on *using* new information in order to learn it. To enhance the effectiveness of the exercises, answers are provided for key-concept exercises so that the student has immediate feedback. A Mastery Test at the end of each chapter provides the student with a means of evaluating his or her learning of the chapter material and also gives practice for classroom testing situations.

2. **Test Bank.** The complimentary *Test Bank,* which contains over 1500 true-false questions, has been updated and greatly expanded for the fifth edition. The test bank is available in standard printed format as well as on the PAR Testing System (a microcomputer-based test generator).

3. **Instructor's Manual.** Each chapter in the complimentary *Instructor's Manual,* prepared by Professor Tortora, consists of a chapter overview, a list of instructional concepts relating to the chapter, suggested problem-solving essay questions, and lists of audiovisual materials relating to the chapter topic. Lists of drugs associated with various body systems have been moved from the text to the *Instructor's Manual.*

4. **Transparencies.** One hundred twenty-five full-color transparencies will be available. Illustrations from the text that are often shown and discussed in class are the subjects for the transparencies, and special care has been taken to make them clear and usable as overhead projections.

ACKNOWLEDGMENTS

I wish to thank the following reviewers for their outstanding contributions to the fifth edition of *Principles of Anatomy and Physiology.* They have shared their extensive knowledge, teaching experience, and concern for students with me, and their feedback has been invaluable.

Wayne W. Carley
Lamar University

Anthony N. C. Chee
Houston Community College

Victor P. Eroschenko
University of Idaho

Richard T. Fraga
Lane Community College

Bonnie K. Gordon
Memphis State University

Gayle Dranch Insler
Adelphi University

Elden W. Martin
Bowling Green State University

Robert E. Nabors
Tarrant County Junior College

Richard E. Welton
Southern Oregon State College

In addition to the individuals who reviewed the entire manuscript for this edition, many people have been kind enough to take time to offer their comments and suggestions about the previous edition of my book. Most have written to me directly and a few have corresponded with me several times. I wish to publicly acknowledge their efforts and to thank them for encouraging me to write a better fifth edition. Among those to whom I express my deepest gratitude are: Norman L. Abel, Shelton, CT; William R. Belzer, Clarion University of Pennsylvania; Bob Boettcher, Lane Community College; Steve Borecky, Carlow College; Joyce Bork, Central Oregon Community College; Ray D. Burkett, Shelby State Community College; Maureen Carney, Oakton Community College; Kenneth Carpenter, Shelby State Community College; David S. Castelan, R.N., San Gabriel, CA; Anthony N. C. Chee, Houston Community College; Charles D. Cornell, West Liberty State College; John G. Deaton, University of Texas at Austin; Charles J. Ellis, Iowa State University; John H. Emes, British Columbia Institute of Technology; Daniel S. Fertig, East Los Angeles College; Carol Gerding, Cuyahoga Community College; Norman Goldstein, California State University (Hayward); John Gwinn, University of Akron; Rebecca Halyard, Clayton Jr. College; Martin L. Heath, North Dakota State School of Science; Abraham A. Held, Herbert H. Lehman College; Julius J. Ivanus, University of Belgrade; Charles Jeffrey, Harrisburg Area Community College; Victor R. Johnson, Madison Area Technical College; John Martin, Clark College; Molly F. Mastrangelo, Allegany Community College; Denis J. Meerdink, University of Arizona; Dean Milligan, Anchorage Community College; Robert R. Montgomery, Oakland Community College; George Newman, Hardin-Simmons University; Patricia Palanker, Middlesex County College; Izak Paul, Mount Royal College; Mike Postula, Parkland College; Louis Renaud, Prince George's Community College; Jack Segurson, Tucson, Arizona; Eddie J. Shellman, Central Florida Community College; Bette C. Slutsky, Truman College; Judith Carroll Stanton, Georgia State University; Russell Stullken, Augustana College; Robert B. Tallitsch, Augustana College; Patricia B. Taylor, R.N., St. Francis Hospital and Medical Center; Robert Tiplady, Brainerd Community College; Robert C. Wall, Lake-Sumter Community College; Paul Weaver, Muhlenberg College; Richard E. Welton, Southern Oregon State College; Margurite Zemek, College of Du Page.

Continuing thanks go to those people who worked on the previous editions. Gratitude is also extended for the contributions of those people whose names appear with the photographs in the text. Special thanks to Dr. Michael H. Ross of the University of Florida for providing the excellent color photographs of bones in Chapter 8, Dr. Michael C. Kennedy of Hahnemann Medical College for his help in reviewing the illustration program, and Professor John LoRusso of Bergen Community College for researching and providing the data for the normal values in Appendix B. Finally, for typing drafts of the manuscript and numerous other clerical duties associated with the task of putting together a textbook, thanks to Geraldine C. Tortora.

As the acknowledgments indicate, the participation of many individuals of diverse talent and expertise is required in the production of a textbook of this scope and complexity. For this reason, readers and users of the fifth edition are invited to send their reactions and suggestions to me so that plans can be formulated for subsequent editions.

Gerard J. Tortora
Natural Sciences and Mathematics, S229
Bergen Community College
400 Paramus Road
Paramus, NJ 07652

Note to the Student

At the beginning of each chapter is a listing of **Student Objectives.** Before you read the chapter, please read the objectives carefully. Each objective is a statement of a skill or knowledge that you should acquire. To meet these objectives, you will have to perform several activities. Obviously, you must read the chapter carefully. If there are sections of the chapter that you do not understand after one reading, you should reread those sections before continuing. In conjunction with your reading, pay particular attention to the figures and exhibits; they have been carefully coordinated with the textual narrative.

At the end of each chapter are two and sometimes three other learning guides that you may find useful. The first, **Study Outline,** is a concise summary of important topics discussed in the chapter. This section is designed to consolidate the essential points covered in the chapter, so that you may recall and relate them to one another. The second guide, **Review Questions,** is a series of questions designed specifically to help you master the objectives. A third aid, **Medical Terminology,** appears in some chapters. This is a listing of terms designed to build your medical vocabulary. After you have answered the review questions, you should return to the beginning of the chapter and reread the objectives to determine whether you have achieved the goals.

As a further aid, we have included pronunciations for many terms that may be new to you. These appear in parentheses immediately following the new words, and they are repeated in the Glossary of Terms at the back of the book. (Of course, since there will always be some conflict among medical personnel and dictionaries about pronunciation, you will come across variations in different sources.) Look at the words carefully and say them out loud several times. Learning to pronounce a new word will help you remember it and make it a useful part of your medical vocabulary. Take a few minutes now to read the following pronunciation key, so it will be familiar as you encounter new words. The key is repeated at the beginning of the Glossary of Terms.

PRONUNCIATION KEY

1. The strongest accented syllable appears in capital letters, for example, bilateral (bī-LAT-er-al) and diagnosis (dī-ag-NŌ-sis).
2. If there is a secondary accent, it is noted by a single quote mark ('), for example, constitution (kon'-sti-TOO-shun) and physiology (fiz'-ē-OL-ō-jē). Any additional secondary accents are also noted by a single quote mark, for example, decarboxylation (dē'-kar-bok'-si-LĀ-shun).
3. Vowels marked with a line above the letter are pronounced with the long sound as in the following common words.

    ā as in *māke*
    ē as in *bē*
    ī as in *īvy*
    ō as in *pōle*

4. Vowels not so marked are pronounced with the short sound as in the following words.

    e as in *bet*
    i as in *sip*
    o as in *not*
    u as in *bud*

5. Other phonetic symbols are used to indicate the following sounds:

    a as in *above*
    oo as in *sue*
    yoo as in *cute*
    oy as in *oil*

Principles of Anatomy and Physiology

U N I T   I

Organization of the Human Body

This unit is designed to show you how your body is organized at different levels. After you study the various regions and parts of your body, you will discover the importance of the chemicals that make it up. You will then find out how your cells, tissues, and organs form the systems that keep you alive and healthy.

1 / An Introduction to the Human Body

2 / The Chemical Level of Organization

3 / The Cellular Level of Organization

4 / The Tissue Level of Organization

5 / The Integumentary System

An Introduction to the Human Body

1

■ **STUDENT OBJECTIVES**

- Define anatomy, with its subdivisions, and physiology.
- Define each of the following levels of structural organization that make up the human body: chemical, cellular, tissue, organ, system, and organismic.
- Identify the principal systems of the human body, list the representative organs of each system, and describe the function of each system.
- List and define several important life processes of humans.
- Describe the general anatomical characteristics of the structural plan of the human body.
- Define the anatomical position and compare common and anatomical terms used to describe various regions of the human body.
- Define several directional terms used in association with the human body.
- Define the common anatomical planes that may be passed through the human body and distinguish a cross section, frontal section, and midsagittal section.
- List by name and location the principal body cavities and the organs contained within them.
- Explain how the abdominopelvic cavity is divided into nine regions and into quadrants.
- Contrast the principles employed in conventional radiography and in computed tomography (CT) scanning.
- Explain how the dynamic spatial reconstructor (DSR) operates and describe its diagnostic importance.
- Define homeostasis and explain why homeostasis is a state that results in normal body activities and why the inability to achieve homeostasis leads to disorders.
- Define a stress and identify the effects of stress on homeostasis.
- Describe the interrelationships of body systems in maintaining homeostasis.
- Contrast the homeostasis of blood pressure (BP) through nervous control and of blood sugar (BS) level through hormonal control.
- Define a feedback system and explain its role in homeostasis.

You are about to begin a study of the human body in order to learn how your body is organized and how it functions. The study of the human body involves many branches of science. Each contributes to a comprehensive understanding of how your body normally works and what happens when it is injured, diseased, or placed under stress.

ANATOMY AND PHYSIOLOGY DEFINED

Two branches of science that will help you understand your body parts and functions are anatomy and physiology. **Anatomy** (*anatome* = to dissect) refers to the study of *structure* and the relationships among structures. Anatomy is a broad science, and the study of structure becomes more meaningful when specific aspects of the science are considered. Several subdivisions of anatomy are described in Exhibit 1-1.

Whereas anatomy and its branches deal with structures of the body, **physiology** (fiz'-ē-OL-ō-jē) deals with *functions* of the body parts, that is, how the body parts work. Since physiology cannot be completely separated from anatomy, you will learn about the human body by study-

ing its structures and functions together, and you will see how each structure of the body is custom-modeled to carry out a particular function. The structure of a part often determines the functions it will perform. In turn, body functions often influence the size, shape, and health of the structures.

LEVELS OF STRUCTURAL ORGANIZATION

The human body consists of several levels of structural organization that are associated with one another in various ways. Here we will consider the principal levels that will help you to understand how your body is organized (Figure 1-1). The lowest level of organization, the **chemical level,** includes all chemical substances essential for maintaining life. All these chemicals are made up of atoms joined together in various ways.

The chemicals, in turn, are put together to form the next higher level of organization: the **cellular level. Cells** are the basic structural and functional units of an organism. Among the many kinds of cells in your body are muscle cells, nerve cells, and blood cells. Figure 1-1 shows several isolated cells from the lining of the stomach. Each has a different structure, and each performs a different function.

The next higher level of structural organization is the **tissue level. Tissues** are groups of similar cells and their intercellular material (substance between cells) that have a similar embryological origin and perform special functions. When the isolated cells shown in Figure 1-1 are joined together, they form a tissue called epithelium, which lines the stomach. Each cell in the tissue has a specific function. Mucous cells produce mucus, a secretion that lubricates food as it passes through the stomach. Parietal cells produce acid in the stomach. Chief cells produce enzymes needed to digest proteins. Other examples of tissues in your body are muscle tissue, connective tissue, and nervous tissue.

In many places in the body, different kinds of tissues are joined together to form an even higher level of organization: the **organ level. Organs** are structures of definite form and function composed of two or more different tissues. Organs usually have a recognizable shape. Examples of organs are the heart, liver, lungs, brain, and stomach. Figure 1-1 shows several tissues that make up the stomach. The serosa is a layer of connective tissue and epithelium around the outside that protects the stomach and reduces friction when the stomach moves and rubs against other organs. The muscle tissue layers of the stomach contract to mix food and pass it on to the next digestive organ. The epithelial tissue layer lining the stomach produces mucus, acid, and enzymes.

The next higher level of structural organization in the body is the **system level. A system** consists of an association of organs that have a common function. The digestive system, which functions in the breakdown of food, is composed of the mouth, saliva-producing glands called salivary glands, pharynx (throat), esophagus (gullet),

EXHIBIT 1-1 SUBDIVISIONS OF ANATOMY

SUBDIVISION	DESCRIPTION
Surface anatomy	Study of the form (morphology) and markings of the surface of the body.
Gross (macroscopic) anatomy	Study of structures that can be examined without the use of a microscope.
Systemic (systematic) anatomy	Study of specific systems of the body such as the nervous system or respiratory system.
Regional anatomy	Study of a specific region of the body such as the head or chest.
Developmental anatomy	Study of development from the fertilized egg to adult form.
Embryology	Study of development from the fertilized egg through the eighth week in utero.
Pathological (*patho* = disease) **anatomy**	Study of structural changes associated with disease.
Histology (*histio* = tissue)	Microscopic study of the structure of tissues.
Cytology (*cyto* = cell)	Microscopic study of the structure of cells.
Radiographic anatomy	Study of the structure of the body using radiographic (x-ray) techniques.

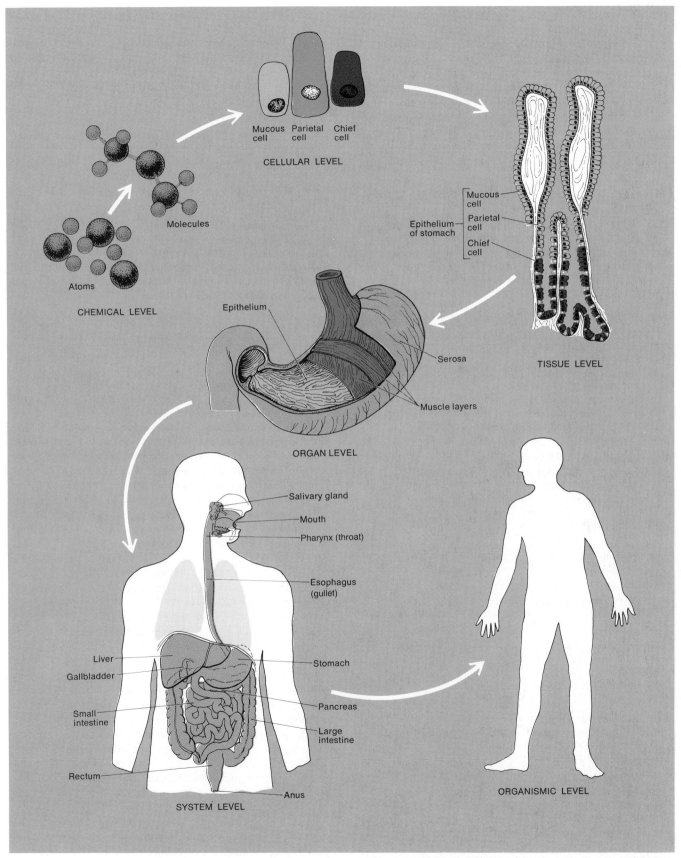

FIGURE 1-1 Levels of structural organization that compose the human body.

stomach, small intestine, large intestine, rectum, liver, gallbladder, and pancreas.

The highest level is the **organismic level.** All the parts of the body functioning with one another constitute the total **organism**—one living individual.

In the chapters that follow, you will examine the anatomy and physiology of the major body systems. Exhibit 1-2 describes these systems in terms of their representative organs and their general functions. The systems are presented in the exhibit in the order in which they are discussed in later chapters.

LIFE PROCESSES

All living forms have certain characteristics that distinguish them from nonliving things. Following are several of the more important life processes of humans:

1. Metabolism. Metabolism is the sum of all the chemical processes that occur in the body. One phase of metabolism, called **catabolism,** provides us with energy needed to sustain life. The other phase of metabolism, called **anabolism,** uses the energy to make various substances that form the body's structural and functional components. Among the many processes contributing to metabolism are: *ingestion,* the taking in of foods; *digestion,* the breaking down of foods into simpler forms that can be used by cells; *absorption,* the uptake of substances by cells; *assimilation,* the buildup of absorbed substances into different substances required by cells; *respiration,* the generation of energy, usually in the presence of oxygen with the release of carbon dioxide; *secretion,* the production and release of a useful substance by cells; and *excretion,* the elimination of wastes produced as a result of metabolism.

2. Excitability. Excitability is our ability to sense changes within and around us. We do this by responding to stimuli (changes in the environment) such as light, pressure, heat, noises, chemicals, and pain. We constantly respond to our environment to make adjustments that maintain health.

EXHIBIT 1-2 PRINCIPAL SYSTEMS OF HUMAN BODY, REPRESENTATIVE ORGANS, AND FUNCTIONS

1. Integumentary
Definition: The skin and structures derived from it, such as hair, nails, and sweat and oil glands.
Function: Helps regulate body temperature, protects the body, eliminates wastes, synthesizes vitamin D, and receives certain stimuli such as temperature, pressure, and pain.
Reference: See Figure 5-1.

2. Skeletal
Definition: All the bones of the body, their associated cartilages, and the joints of the body.
Function: Supports and protects the body, provides leverage, produces blood cells, and stores minerals.
Reference: See Figure 7-1.

3. Muscular
Definition: All the muscle tissue of the body, including skeletal, visceral, and cardiac.
Function: Participates in bringing about movement, maintains posture, and produces heat.
Reference: See Figure 11-3.

4. Nervous
Definition: Brain, spinal cord, nerves, and sense organs, such as the eye and ear.
Function: Regulates body activities through nerve impulses.
Reference: See Figures 13-1 and 14-1.

5. Endocrine
Definition: All glands that produce hormones.
Function: Regulates body activities through hormones transported by the cardiovascular system.
Reference: See Figure 18-1.

6. Cardiovascular
Definition: Blood, heart, and blood vessels.
Function: Distributes oxygen and nutrients to cells, carries carbon dioxide and wastes from cells, maintains the acid–base balance of the body, protects against disease, prevents hemorrhage by forming blood clots, and helps regulate body temperature.
Reference: See Figures 20-1, 21-12, and 21-17.

7. Lymphatic
Definition: Lymph, lymph vessels, and structures or organs containing lymphatic tissue (large numbers of white blood cells called lymphocytes), such as the spleen, thymus gland, lymph nodes, and tonsils.
Function: Returns proteins and plasma to the cardiovascular system, transports fats from the digestive system to the cardiovascular system, filters body fluid, produces white blood cells, and protects against disease.
Reference: See Figure 22-2.

8. Respiratory
Definition: The lungs and a series of passageways leading into and out of them.
Function: Supplies oxygen, eliminates carbon dioxide, and helps regulate the acid–base balance of the body.
Reference: See Figure 23-1.

9. Digestive
Definition: A long tube and associated organs such as the salivary glands, liver, gallbladder, and pancreas.
Function: Performs the physical and chemical breakdown of food for use by cells and eliminates solid and other wastes.
Reference: See Figure 24-1.

10. Urinary
Definition: Organs that produce, collect, and eliminate urine.
Function: Regulates the chemical composition of blood, eliminates wastes, regulates fluid and electrolyte balance and volume, and helps maintain the acid–base balance of the body.
Reference: See Figure 26-1.

11. Reproductive
Definition: Organs (testes and ovaries) that produce reproductive cells (sperm and ova) and organs that transport and store reproductive cells.
Function: Reproduces the organism.
Reference: See Figures 28-1 and 28-11.

3. Conductivity. Conductivity refers to the ability to carry the effect of a stimulus from one part of a cell to another. This characteristic is highly developed in nerve cells and is developed to a great extent in muscle cells.

4. Contractility. Contractility is the capacity of cells to undergo shortening and change form. Muscle cells exhibit a high degree of contractility.

5. Growth. Growth refers to an increase in size. It may involve an increase in the number of cells or an increase in the size of existing cells.

6. Differentiation. Differentiation is the basic mechanism whereby unspecialized cells change to specialized cells. Specialized cells have structural and functional characteristics that differ from cells from which they originated. Through differentiation, a fertilized egg normally develops into an embryo, fetus, infant, child, and adult, each of which consists of a variety of diversified cells.

7. Reproduction. Reproduction refers to either the formation of new cells for growth, repair, or replacement, or the production of a new individual. Through reproduction, life is transmitted from one generation to the next.

STRUCTURAL PLAN

The human body has certain general **anatomical characteristics** that will help you to understand its overall structural plan. For example, humans have a **backbone (vertebral column),** a characteristic that places them in a large group of organisms called *vertebrates.* Another characteristic is the body's **tube within a tube** construction. The outer tube is formed by the body wall; the inner tube is the digestive tract. Moreover, humans are for the most part **bilaterally symmetrical,** that is, essentially the left and right sides of the body are mirror images.

DIRECTIONAL TERMS

In order to explain exactly where various body structures are located in relationship to each other, anatomists use certain **directional terms.** If you want to point out the sternum (breastbone) to someone who knows where the clavicle (collarbone) is, you can say that the sternum is inferior (farther away from the head) and medial (toward the middle of the body) to the clavicle. As you can see, using the terms *inferior* and *medial* avoids a great deal of complicated description. Many directional terms are defined in Exhibit 1-3, and the parts of the body referred to in the examples are labeled in Figure 1-2. Studying the exhibit and the figure together should make clear to you many of the directional relationships among various body parts.

ANATOMICAL POSITION AND ANATOMICAL NAMES

In all anatomical texts and charts, descriptions of any region or part of the human body assume that the body is in a specific position, called the **anatomical position,** so that directional terms are clear and any part can be related to any other part. In the anatomical position, the subject is standing erect facing the observer, the upper extremities are placed at the sides, and the palms of the hands are turned forward (Figure 1-3). Once the body is in the anatomical position, it is easier to visualize and understand how it is organized into various regions. The common and anatomical terms of the principal body regions are also presented in Figure 1-3.

EXHIBIT 1-3 DIRECTIONAL TERMS

TERM	DEFINITION	EXAMPLE
Superior (cephalad or **craniad)**	Toward the head or the upper part of a structure; generally refers to structures in the trunk.	The heart is superior to the liver.
Inferior (caudad)	Away from the head or toward the lower part of a structure; generally refers to structures in the trunk.	The stomach is inferior to the lungs.
Anterior (ventral)	Nearer to or at the front of the body.	The sternum is anterior to the heart.
Posterior (dorsal)	Nearer to or at the back of the body.	The esophagus is posterior to the trachea.
Medial	Nearer to the midline of the body or a structure.	The ulna is on the medial side of the forearm.
Lateral	Farther from the midline of the body or a structure.	The ascending colon of the large intestine is lateral to the urinary bladder.
Intermediate	Between two structures.	The ring finger is intermediate between the little and middle fingers.
Ipsilateral	On the same side of the body.	The gallbladder and ascending colon of the large intestine are ipsilateral.
Contralateral	On the opposite side of the body.	The ascending and descending colons of the large intestine are contralateral.
Proximal	Nearer to the attachment of an extremity to the trunk or a structure; nearer to the point of origin.	The humerus is proximal to the radius.

EXHIBIT 1-3 (*Continued*)

TERM	DEFINITION	EXAMPLE
Distal	Farther from the attachment of an extremity to the trunk or a structure; farther from the point of origin.	The phalanges are distal to the carpals (wrist bones).
Superficial	Toward or on the surface of the body.	The muscles of the thoracic wall are superficial to the viscera in the thoracic cavity. (See Figure 1-7.)
Deep	Away from the surface of the body.	The muscles of the arm are deep to the skin of the arm.
Parietal	Pertaining to or forming the outer wall of a body cavity.	The parietal pleura forms the outer layer of the pleural sacs that surround the lungs. (See Figure 1-7.)
Visceral	Pertaining to the covering of an organ (viscus).	The visceral pleura forms the inner layer of the pleural sacs and covers the external surface of the lungs. (See Figure 1-7.)

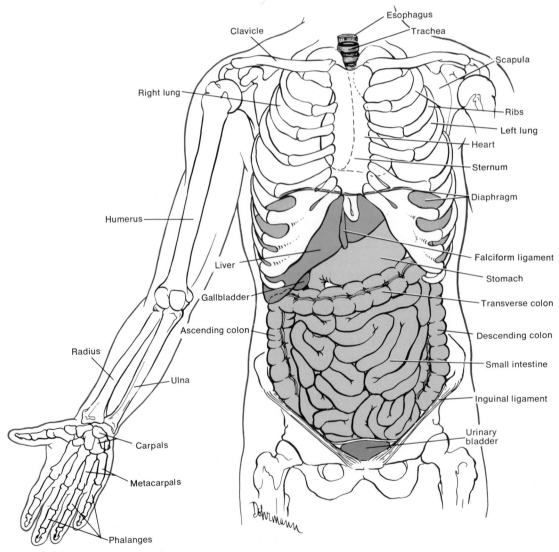

FIGURE 1-2 Anatomical and directional terms. By studying Exhibit 1-3 with this figure, you should gain an understanding of the meanings of the terms *superior, inferior, anterior, posterior, medial, lateral, intermediate, ipsilateral, contralateral, proximal,* and *distal.*

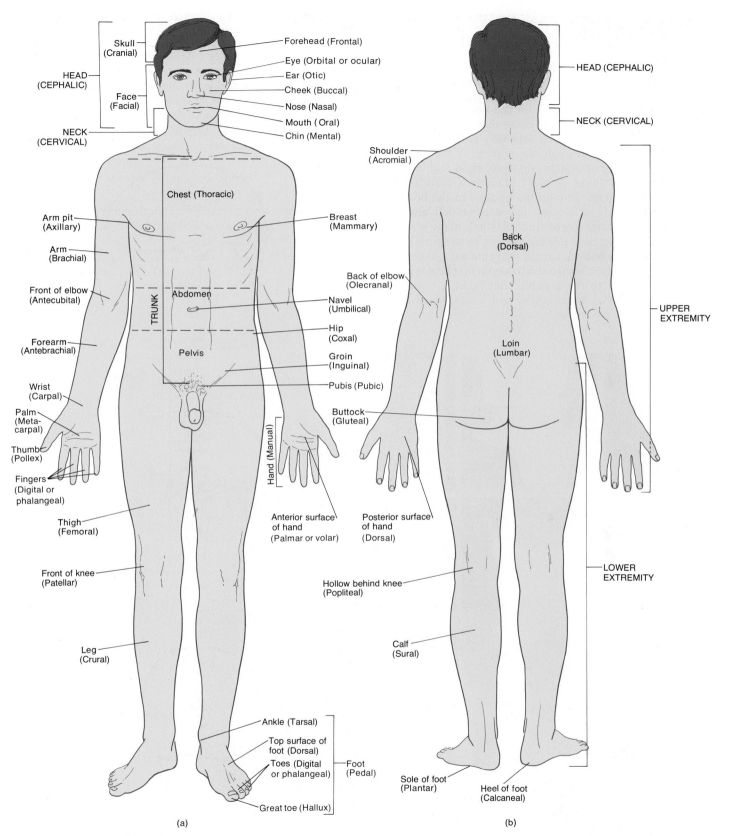

FIGURE 1-3 Anatomical position. The common names and anatomical terms, in parentheses, are indicated for many of the regions of the body. (a) Anterior view. (b) Posterior view.

PLANES AND SECTIONS

The structural plan of the human body may also be discussed with respect to **planes** (imaginary flat surfaces) that pass through it. Several of the commonly used planes are illustrated in Figure 1-4. A **sagittal** (SAJ-i-tal) **plane** is a vertical plane that divides the body or organs into right and left sides. Such a plane may be midsagittal or parasagittal. A **midsagittal (median) plane** is a vertical plane that passes through the midline of the body and divides the body or organs into *equal* right and left sides. A **parasagittal** (*para* = near) **plane** is a vertical plane that does not pass through the midline of the body and divides the body or organs into *unequal* left and right portions. A **frontal (coronal; kō-RŌ-nal) plane** is a plane at a right angle to a midsagittal (or parasagittal) plane that divides the body or organs into anterior and posterior portions. Finally, a **horizontal (transverse) plane** is a plane that is parallel to the ground, that is, at a right angle to the midsagittal, parasagittal, and frontal planes. It di-

vides the body or organs into superior and inferior portions.

When you study a body structure, you will often view it in section, meaning that you look at the flat surface resulting from a cut made through the three-dimensional structure. It is important to know the plane of the section so that you can understand the anatomical relationship of one part to another. Figure 1-5 indicates how three

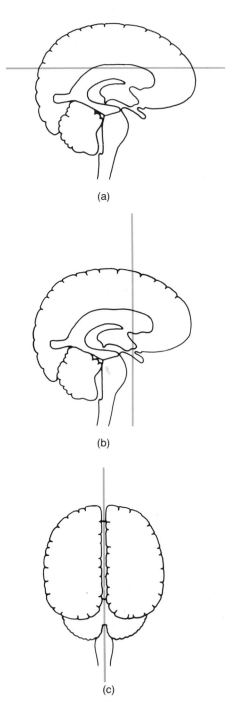

(a)

(b)

(c)

FIGURE 1-5 Sections through different parts of the brain. (a) Cross (horizontal or transverse) section; see also Figure 14-7d. (b) Frontal section; see also Figure 14-6a. (c) Midsaggittal section; see also Figure 14-1b.

MIDSAGITTAL PLANE

HORIZONTAL PLANE

FRONTAL PLANE

PARASAGITTAL PLANE

FIGURE 1-4 Planes of the human body.

different sections—a *cross section,* a *frontal section,* and a *midsagittal section*—are made through different parts of the brain.

BODY CAVITIES

Spaces within the body that contain internal organs are called **body cavities.** Specific cavities may be distinguished if the body is divided into right and left halves. Figure 1-6 shows the two principal body cavities. The **dorsal body cavity** is located near the dorsal (posterior) surface of the body. It is further subdivided into a **cranial cavity,** which is a bony cavity formed by the cranial (skull) bones and contains the brain, and a **vertebral (spinal) canal,** which is a bony cavity formed by the vertebrae of the backbone and contains the spinal cord and the beginnings of spinal nerves.

The other principal body cavity is the **ventral body cavity.** This cavity is located on the ventral (anterior) aspect of the body. Its walls are composed of skin, connective tissue, bone, muscles, and a membrane called a serous membrane. The organs inside the ventral body cavity are called **viscera** (VIS-er-a). Like the dorsal body cavity, the ventral body cavity has two principal subdivisions—an upper portion, called the **thoracic** (thō-RAS-ik) **cavity** (or chest cavity), and a lower portion, called the **abdominopelvic** (ab-dom'-i-nō-PEL-vik) **cavity.** The anatomical landmark that divides the ventral body cavity into the thoracic and abdominopelvic cavities is the muscular diaphragm.

The thoracic cavity contains several divisions. There are two **pleural cavities,** each containing a small amount of fluid (Figure 1-7). Each is a small potential space (not a natural space) between the parietal pleura, a membrane that lines the pleural cavities, and the visceral pleura, a membrane that covers the lungs. The **mediastinum** (mē'-dē-as-TĪ-num), is a broad, median partition, actually a mass of tissues, between the lungs that extends from the sternum to the vertebral column (Figure 1-7). The mediastinum includes all of the contents of the thoracic cavity, except the lungs themselves. Among the structures in the mediastinum are the heart, thymus gland, esophagus, and many large blood and lymphatic vessels. The **pericardial** (per'-i-KAR-dē-al; *peri* = around, *cardi* = heart) **cavity** is a small potential space between the visceral pericardium and parietal pericardium, the membranes covering the heart (Figure 1-7). It also contains a fluid.

The abdominopelvic cavity, as the name suggests, is divided into two portions, although no wall separates them (see Figure 1-6). The upper portion, the **abdominal cavity,** contains the stomach, spleen, liver, gallbladder, pancreas, small intestine, and most of the large intestine. The lower portion, the **pelvic cavity,** contains the urinary bladder, sigmoid colon, rectum, and the internal male or female reproductive organs. One way to mark the division between the abdominal and pelvic cavities is to draw an imaginary line from the symphysis pubis (anterior joint between hipbones) to the superior border of the sacrum (sacral promontory).

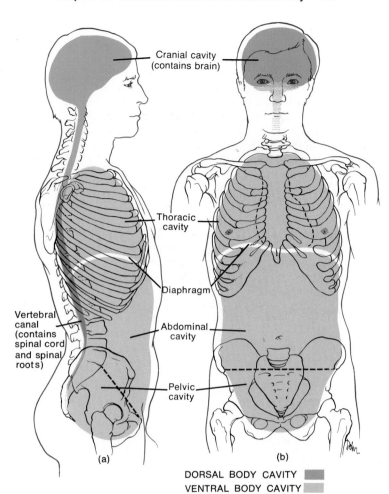

FIGURE 1-6 Body cavities. (a) Location of the dorsal and ventral body cavities. (b) Subdivisions of the ventral body cavity.

ABDOMINOPELVIC REGIONS

To describe the location of organs easily, the abdominopelvic cavity may be divided into the **nine regions** shown in Figure 1-8a. Note which organs and parts of organs are in the different regions by carefully examining Figure 1-8b–d. Although some parts of the body in the illustrations may be unfamiliar to you at this point, they will be discussed in detail in later chapters.

ABDOMINOPELVIC QUADRANTS

The abdominopelvic cavity may be divided more simply into **quadrants** (*quad* = four). These are shown in Figure 1-9. In this method, frequently used by clinicians, a horizontal line and a vertical line are passed through the umbilicus. These two lines divide the abdomen into a **right upper quadrant (RUQ), left upper quadrant (LUQ), right lower quadrant (RLQ),** and **left lower quadrant (LLQ).** Whereas the nine-region designation is more widely used for anatomical studies, the quadrant designation is better suited for locating the site of an abdominopelvic pain, tumor, or other abnormality.

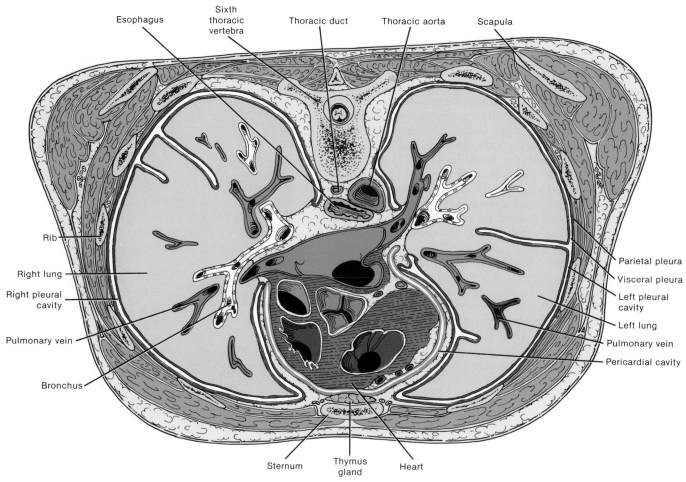

FIGURE 1-7 Mediastinum seen in a cross section of the thorax. Some of the structures shown and labeled may be unfamiliar to you now, but they will be discussed in detail in later chapters.

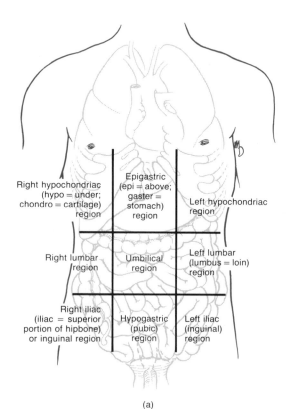

FIGURE 1-8 Abdominopelvic cavity. (a) The nine regions. The top horizontal (transpyloric) line is drawn just inferior to the bottom of the rib cage, across the lower portion (pylorus) of the stomach. The bottom horizontal (transtubercular) line is drawn just inferior to the tops of the hipbones. The two vertical (left and right lateral) lines are drawn just medial to the nipples. The horizontal and vertical lines divide the area into a larger middle section and smaller left and right sections.

(a)

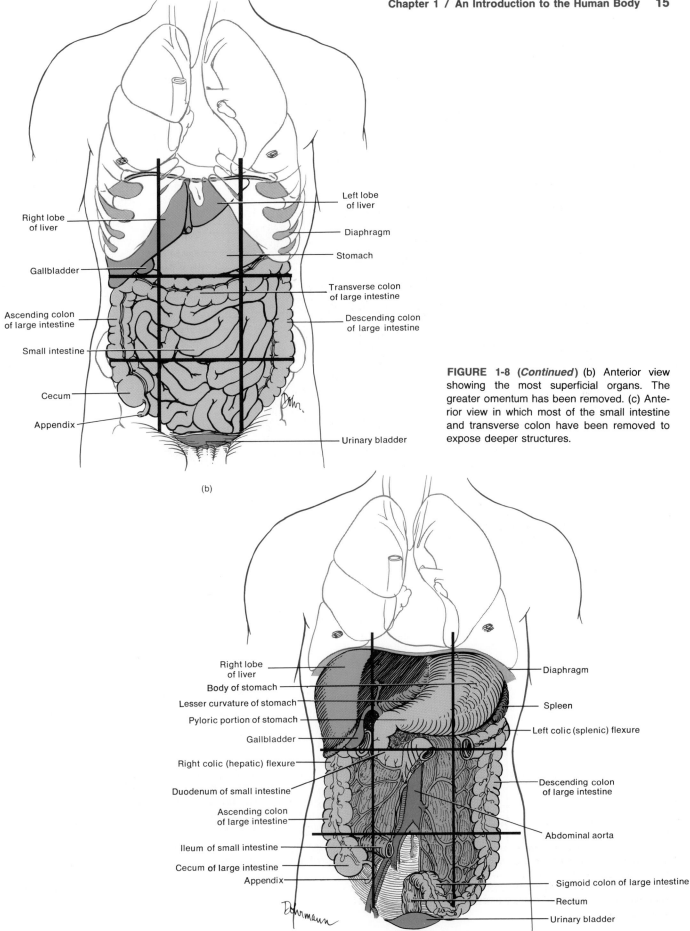

Right lobe
of liver

Gallbladder

Ascending colon
of large intestine

Small intestine

Cecum

Appendix

Left lobe
of liver

Diaphragm

Stomach

Transverse colon
of large intestine

Descending colon
of large intestine

Urinary bladder

(b)

FIGURE 1-8 (*Continued*) (b) Anterior view showing the most superficial organs. The greater omentum has been removed. (c) Anterior view in which most of the small intestine and transverse colon have been removed to expose deeper structures.

Right lobe
of liver

Body of stomach

Lesser curvature of stomach

Pyloric portion of stomach

Gallbladder

Right colic (hepatic) flexure

Duodenum of small intestine

Ascending colon
of large intestine

Ileum of small intestine

Cecum of large intestine

Appendix

Diaphragm

Spleen

Left colic (splenic) flexure

Descending colon
of large intestine

Abdominal aorta

Sigmoid colon of large intestine

Rectum

Urinary bladder

(c)

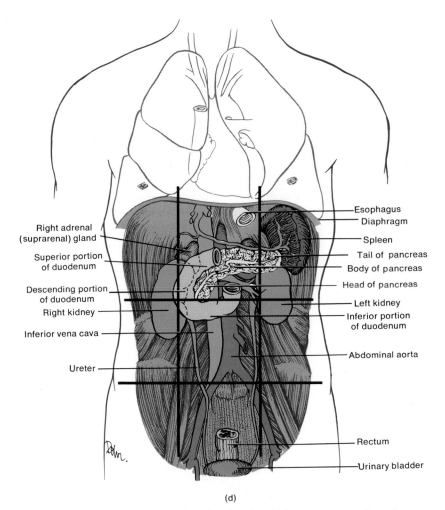

Right adrenal (suprarenal) gland

Superior portion of duodenum

Descending portion of duodenum

Right kidney

Inferior vena cava

Ureter

Esophagus

Diaphragm

Spleen

Tail of pancreas

Body of pancreas

Head of pancreas

Left kidney

Inferior portion of duodenum

Abdominal aorta

Rectum

Urinary bladder

(d)

FIGURE 1-8 (*Continued*) (d) Anterior view in which many organs have been removed, exposing the posterior structures.

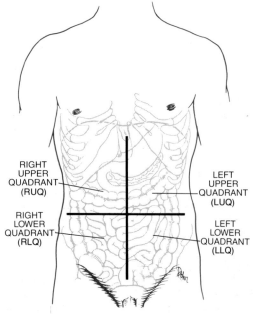

RIGHT UPPER QUADRANT (RUQ)

LEFT UPPER QUADRANT (LUQ)

RIGHT LOWER QUADRANT (RLQ)

LEFT LOWER QUADRANT (LLQ)

FIGURE 1-9 Quadrants of the abdominopelvic cavity. The two lines intersect at right angles at the umbilicus.

CLINICAL APPLICATION

To determine the cause of death accurately, it is necessary to perform an **autopsy** (AW-top-sē; *auto* = self; *opsis* = to view), that is, a postmortem examination of the body. In addition, an autopsy can also be used to uncover the existence of diseases not detected during life, support the accuracy of diagnostic tests, determine the effectiveness and side effects of drugs, analyze the effects of environmental influences on the body, and educate medical students.

A typical autopsy consists of three principal phases of examination. The first phase is examination of the exterior of the body for the presence of wounds, scars, tumors, or other abnormalities. The second phase includes the dissection and gross examination of the major body organs. The third phase of an autopsy consists of microscopic examination of tissues from organs to ascertain any pathology. Depending on the circumstances, techniques may also be used to detect and recover microbes and determine the presence of foreign substances in the body.

A complete autopsy itself begins with a Y-shaped incision to expose the thoracic, abdominal, and pelvic viscera. The upper parts of the Y begin in front of each shoulder, extend inferior to the nipples, and join just inferior to the sternum. The incision is then continued down the middle of the abdominopelvic wall to the symphysis pubis. The sternum is removed by cutting the ribs to expose the thoracic viscera, and the tissues of the abdominopelvic wall are folded back to expose the contents of the abdominopelvic cavity. Unless the information is of special clinical significance, the face and extremities are not usually dissected.

An autopsy usually lasts two to four hours, depending on its comprehensiveness. When the procedure is completed, the organs are returned to the body, except for those that are donated or used to extract a particular pharmacologic substance, and all incisions are sewn up.

RADIOGRAPHIC ANATOMY

A specialized branch of anatomy that is essential for the diagnosis of many disorders is **radiographic** (*radius* = ray) **anatomy,** or **radiology,** which includes the use of several radiographic techniques (x-rays).

CONVENTIONAL RADIOGRAPHY

The most common and familiar type of radiographic anatomy employs the use of a single barrage of x-rays. The x-rays pass through the body and expose an x-ray film, producing a photographic image called a **roentgenogram** (RENT-gen-ō-gram), a two-dimensional shadow image of the interior of the body (Figure 1-10a). As valuable as they are in diagnosis, conventional x-rays compress the body image onto a flat sheet of film, often resulting in an overlap of organs and tissues that could make diagnosis difficult. Moreover, x-rays do not always differentiate between subtle differences in tissue density.

COMPUTED TOMOGRAPHY (CT) SCANNING

The diagnostic difficulties described for conventional radiography have been largely eliminated by the use of an x-ray technique called **computed tomography (CT) scanning** or **computerized axial tomography (CAT) scanning.** First introduced in 1971, CT scanning combines the principles of x-ray and advanced computer technologies. An x-ray source moves in an arc around the part of the body being scanned and repeatedly sends out x-ray beams. As the beams pass through the body, the tissues absorb small amounts of radiation, depending on their densities. Once the beams pass through, they are converted by light-sensitive crystal detectors to electronic signals that are transmitted to the scanner's computer. The computer then projects an image, called a **CT scan,** onto a television

screen (physician's console). The CT scan provides a very accurate cross-sectional picture of any area of the body (Figure 1-10b). A series of scans permits a physician to examine layer after layer of a patient's tissues.

The CT scan permits a significant differentiation of body parts that was never possible with conventional x-rays. The entire CT scanning process takes only seconds, it is completely painless, and the x-ray dose is equal to or less than that of many other diagnostic procedures.

A recently devised variation of CT scanning is called **cine CT.** It has the ability to take stop-action photos of a beating heart, measure blood flow through grafted coronary arteries, measure the weight of different parts of the heart, identify defects in children with congenital heart diseases prior to surgery, and identify obstructions in respiratory airways.

DYNAMIC SPATIAL RECONSTRUCTOR (DSR)

A recent development in radiographic anatomy is a highly sophisticated x-ray machine called the **dynamic spatial reconstructor (DSR).** It has the ability to construct *moving, three-dimensional,* life-size images of all or part of an internal organ from any view desired. The image produced can be rotated and tipped and, like an electronic knife, the DSR can pictorially slice open an organ and expose its interior. The instrument also provides enlargements, stop-action, replay, high-speed, and slow-motion viewing. The level of radiation exposure of the DSR is about double that for a chest roentgenogram.

The DSR has been designed to provide three-dimensional imaging of the heart, lungs, and circulation. It can be used to measure the volumes and movements of the heart and lungs and other internal organs, to detect cancer and heart defects, and to measure tissue damage following a heart attack, stroke, or other disease.

HOMEOSTASIS

Having considered some of the principal anatomical features of the body, we will now turn our attention to an important physiological feature of the body. This is homeostasis, one of the major themes of this textbook.

Homeostasis (hō'-mē-ō-STĀ-sis) is the condition in which the body's internal environment remains relatively constant, within limits (*homeo* = same; *stasis* = standing still). For the body's cells to survive, the composition of the surrounding fluids must be precisely maintained at all times. Fluid outside body cells is called **extracellular** (*extra* = outside) **fluid (ECF)** and is found in two principal places. The fluid filling the microscopic spaces between the cells of tissues is called *intercellular fluid, interstitial* (in'-ter-STISH-al) *fluid* (*inter* = between), or *tissue fluid.* The extracellular fluid in blood vessels is termed *plasma* (Figure 1-11). Fluid within cells is called **intracellular** (*intra* = within, inside) **fluid (ICF).** Among the substances in extracellular fluid are gases, nutrients, and electrically charged chemical particles called ions—all needed for the maintenance of life. Extracellular fluid circulates through the blood and lymph vessels and from there

Left lobe of liver —
Body of pancreas —
Gallbladder —
Right lobe of liver —
Right kidney —
Stomach —
Colon of large intestine —
Rib —
Left kidney —
Vertebra —

(a)

Left lobe of liver —
Body of pancreas —
Gallbladder —
Right lobe of liver —
Right kidney —
Stomach —
Colon of large intestine —
Rib —
Left kidney —
Vertebra —

(b)

FIGURE 1-10 Radiographic anatomy. (a) Roentgenogram of the abdomen. (b) CT scan of the abdomen. Note the greater detail in the CT scan. (Courtesy of Stephen A. Kieffer and E. Robert Heitzman, *An Atlas of Cross-Sectional Anatomy,* Harper & Row, Publishers, Inc., Hagerstown, MD, 1979.)

moves into the spaces between the tissue cells. Thus it is in constant motion throughout the body. Essentially, all body cells are surrounded by the same fluid environment. For this reason, extracellular fluid is often called the body's internal environment.

An organism is said to be in homeostasis when its internal environment (1) contains the optimum concentrations of gases, nutrients, ions, and water, (2) has an optimal temperature, and (3) has an optimal pressure for the health of the cells, When homeostasis is disturbed, ill health may result. If the body fluids are not eventually brought back into balance, death may occur.

STRESS AND HOMEOSTASIS

Homeostasis in all organisms is continually disturbed by **stress,** which is any stimulus that creates an imbalance in the internal environment. The stress may come from the external environment in the form of stimuli such as heat, cold, loud noises, or lack of oxygen. Or the stress may originate within the body in the form of stimuli such as high blood pressure, pain, tumors, or unpleasant thoughts. Most stresses are mild and routine. Poisoning, overexposure, severe infection, and surgical operations are examples of extreme stress.

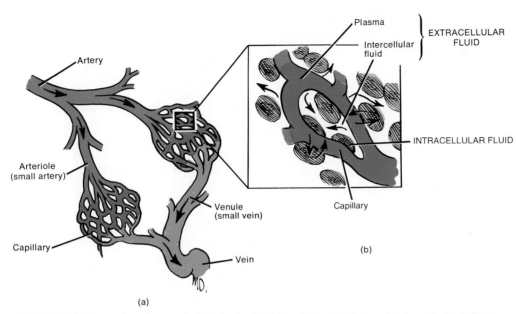

FIGURE 1-11 Internal environment of the body. (a) Extracellular fluid is found in two principal places: in blood vessels as plasma and between cells as interstitial fluid. Plasma circulates through arteries and arterioles and then into microscopic blood vessels called capillaries. From there it moves into the spaces between body cells, where it is called interstitial fluid. This fluid then returns to capillaries as plasma and passes through the venules, and then into the veins. (b) Enlarged detail.

Fortunately, the body has many regulating (homeostatic) devices that oppose the forces of stress and bring the internal environment back into balance. High resistance to stress is a striking feature of all organisms. Some people live in deserts where the daytime temperatures easily reach 49°C (120°F). Others work outside all day in subzero weather. Yet everyone's internal body temperature remains near 37°C (98.6°F). Mountain climbers exercise strenuously at high altitudes, where the oxygen content of the air is low. But once they adjust to the new altitude, they do not suffer from oxygen shortage. The extremes in temperature and in oxygen content of the air are external stresses, and the exercise performed is an internal stress, yet the body compensates and remains in homeostasis. Walter B. Cannon (1871–1945), an American physiologist who coined the term *homeostasis*, noted that the heat produced by the muscles during strenuous exercise would curdle and inactivate the body's proteins if the body did not dissipate heat quickly. Muscles that are being exercised also produce, in addition to heat, a great deal of lactic acid. If the body did not have a homeostatic mechanism for reducing the amount of the acid, the extracellular fluid would become acidic and destroy the cells.

Every body structure, from the cellular to the system level, contributes in some way to keeping the internal environment within normal limits. One homeostatic function of the cardiovascular system, for example, is to keep the fluids of all parts of the body constantly moving. When we are at rest, fresh blood is circulated throughout the entire body about once every minute. But when we are active and our muscles need nutrients rapidly, the heart quickens its pace and pumps fresh blood to the organs much more frequently. In this way, the cardiovascular system helps compensate for the stress of increased activity.

The respiratory system offers another example of a homeostatic mechanism in the body. The cells use more oxygen and produce more carbon dioxide when they are very active. Therefore, during periods of activity the respiratory system must work faster to keep the oxygen in the extracellular fluid from falling below normal limits and prevent excessive amounts of carbon dioxide from accumulating.

The digestive system and related organs help maintain the homeostasis by providing nutrients and removing wastes. As circulating blood passes through the organs of digestion, the products of digestion are transported to the body fluids so they can be used as nutrients by the cells. The liver, kidneys, endocrine glands, and other organs help to alter or store the products of digestion in various ways. The kidneys also help to remove the wastes produced by cells after the cells have utilized the nutrients.

The homeostatic mechanisms of the body, such as those performed by the cardiovascular, respiratory, and digestive systems, are themselves subject to control by the nervous system and the endocrine system. The nervous system regulates homeostasis by detecting when the body deviates from its balanced state and by sending messages to the proper organs to counteract the stress. For instance, when muscle cells are active, they take a great deal of oxygen from the blood. They also give off carbon dioxide, which is picked up by the blood. Certain nerve cells detect the chemical changes occurring in the blood and send a message to the brain. The brain then sends a message to the heart to pump blood more quickly to the lungs so the blood can give up its excess carbon dioxide and take on more oxygen. Simultaneously, the brain sends a message to the muscles that control breathing to contract faster. As a result, carbon dioxide can be exhaled and more oxygen can be inhaled.

Homeostasis is also controlled by the endocrine system—a series of glands that secrete chemical regulators, called hormones, into the blood. Whereas nerve impulses coordinate homeostasis rapidly, hormones work slowly. Both means of control are directed toward the same end.

HOMEOSTASIS OF BLOOD PRESSURE

Blood pressure (BP) is the force exerted by blood as it presses against and attempts to stretch the walls of the blood vessels, especially the arteries. It is determined primarily by three factors: the rate and strength of the heartbeat, the amount of blood, and the resistance offered by the arteries as blood passes through them. The resistance of the arteries results from the chemical properties of the blood and the size of the arteries.

If some stress, either internal or external, causes the heartbeat to speed up, the following sequence occurs (Figure 1-12). As the heart pumps faster, it pushes more blood into the arteries per minute, increasing pressure in the arteries. The higher pressure is detected by pressure-sensitive nerve cells in the walls of certain arteries, which send nerve impulses to the brain. The brain interprets the message and responds by sending impulses to the heart to slow the heart rate, thus decreasing blood pressure. The continual monitoring of blood pressure by the nervous system is an attempt to maintain a normal blood pressure and employs what is called a feedback system.

A **feedback system** is any circular situation in which information about the status of something is continually reported (fed back) to a central control region. The nervous control that results in a normal constant blood pressure is an example. In the case of regulating blood pressure, the **input (stimulus)** is the information picked up by the pressure-sensitive nerve cells (high blood pressure), and the **output (response)** is the return toward normal

blood pressure due to decreased heartbeat. Figure 1-12 shows that the system runs in a circle. The pressure-sensitive nerve cells continue to monitor pressure and to feed this information back to the brain, even after the return to homeostasis has begun and blood pressure has begun to normalize. In other words, the pressure-sensitive nerve cells send impulses to the brain about the changed blood pressure, and if the pressure is still too high, the brain continues to send out impulses to slow the heartbeat.

This type of feedback system reverses the direction of the initial condition from a rising to a falling blood pressure and is called a **negative feedback system.** The reaction of the body (output) counteracts the stress (input) in order to restore homeostasis. Such a system is therefore a stimulatory-inhibitory one. If, instead, the brain had signaled the heart to beat even faster and the blood pressure had kept on rising, the system would have been a **positive feedback system.** In a positive feedback system the output **intensifies** the input. This system is therefore a stimulatory-stimulatory one. Most positive feedback systems are destructive and result in various disorders. Most of the feedback systems of the body are negative.

Figure 1-12 also shows that a second negative feedback control is involved in maintaining normal blood pressure. Small arteries, called arterioles, have muscular walls that can constrict or relax upon receiving an appropriate signal from the brain. When the blood pressure increases, pressure-sensitive nerve cells in certain arteries send messages to the brain. The brain interprets the messages and responds by sending impulses to the arterioles, causing them to relax (dilate). Thus the blood flowing through the arterioles is offered less resistance and blood pressure drops back to normal.

HOMEOSTASIS OF BLOOD SUGAR LEVEL

An example of homeostatic regulation by hormones is the maintenance of blood sugar level. Glucose, the sugar found in blood, is one of the body's principal sources of energy. Under normal circumstances, the concentration of sugar in the blood averages about 90 milligrams/100

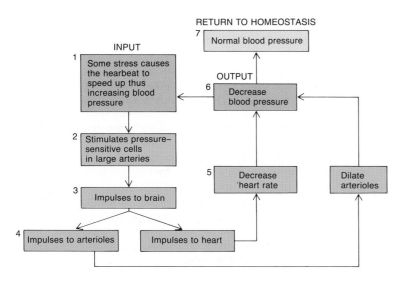

FIGURE 1-12 Homeostasis of blood pressure. Note that the output is fed back into the system, and the system continues to lower blood pressure until there is a return to homeostasis.

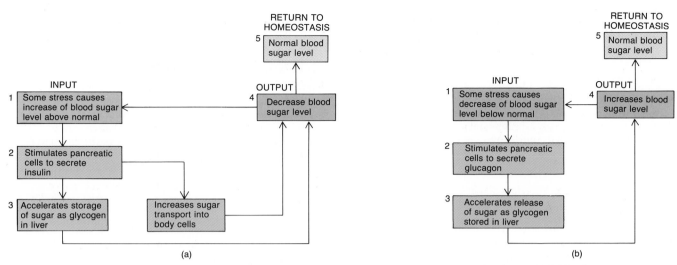

FIGURE 1-13 Homeostasis of blood sugar level. (a) Homeostatic mechanism that lowers high blood sugar level down to normal. (b) Homeostatic mechanism that raises low blood sugar level up to normal.

milliliters of blood.* This level is maintained primarily by two hormones secreted by the pancreas: insulin and glucagon. Suppose you have just eaten some candy. The sugar in the candy is broken down by the organs of digestion and moves from the digestive tract into the blood. The sugar then becomes a stress because it raises the blood sugar level above normal. In response to this stress, the cells of the pancreas are stimulated to secrete insulin (Figure 1-13a). Once insulin enters the blood, it has two principal effects. First, it increases sugar uptake by cells. The blood sugar level is lowered by this action since the sugar moves from the blood into body cells. Second, insulin accelerates the process by which sugar is stored in the liver and muscles. Thus even more sugar is removed from the blood. In essence, insulin decreases blood sugar concentration until it returns to normal.

The other hormone produced by the pancreas, glucagon, has the opposite effect of insulin. Suppose you have not eaten for several hours and your blood sugar level is steadily decreasing. Lack of sugar is now the stress, and under this condition other cells of the pancreas are

stimulated to secrete glucagon (Figure 1-13b). This hormone accelerates the process by which sugar stored in the liver is sent back into the bloodstream. The blood sugar level is thus increased until it returns to normal.

MEASURING THE HUMAN BODY

An important aspect of describing the body and understanding how it works is **measurement.** Examples of measurements involve the size of an organ, the weight of an organ, the time it takes for a physiological reaction to occur, and the amount of a medication to be administered. Measurements involving time, weight, temperature, size, length, and volume are routine to a medical science program. The metric system of measurement is standardly used in sciences. In this book, measurements are given in metric units followed by the approximate U.S. equivalent in parentheses, where convenient. If you are not familiar with U.S.–metric conversions, consult Appendix A.

___

STUDY OUTLINE

Anatomy and Physiology Defined (p. 6)
1. Anatomy is the study of structure and the relationship among structures.
2. Subdivisions of anatomy include surface anatomy (form and markings of surface features), gross anatomy (macroscopic), systemic or systematic anatomy (systems), regional anatomy (regions), developmental anatomy (development from fertilization to adulthood), embryology (development from fertilized egg through eighth week in utero), pathological anatomy (disease), histology (tissues), cytology (cells), and radiographic anatomy (x-rays).
3. Physiology is the study of how body structures function.

* One milligram (mg) = 0.001 gram (g), 1 g = 0.035 ounce (oz), and 100 milliliters (ml) = 3.38 oz.

Levels of Structural Organization (p. 6)
1. The human body consists of several levels of structural organization; among these are the chemical, cellular, tissue, organ, system, and organismic levels.
2. Cells are the basic structural and functional units of an organism.
3. Tissues consist of groups of similarly specialized cells and their intercellular material that perform certain special functions.
4. Organs are structures of definite form and function composed of two or more different tissues.
5. Systems consist of associations of organs that have a common function.
6. The human organism is a collection of structurally and functionally integrated systems.

7. The systems of the human body are the integumentary, skeletal, muscular, nervous, endocrine, cardiovascular, lymphatic, respiratory, digestive, urinary, and reproductive (see Exhibit 1-2).

Life Processes (p. 8)

1. All living forms have certain characteristics that distinguish them from nonliving things.
2. Among the life processes in humans are metabolism, excitability, conductivity, contractility, growth, differentiation, and reproduction.

Structural Plan (p. 9)

1. The human body has certain general characteristics.
2. Among the characteristics are a backbone, a tube within a tube organization, and bilateral symmetry.

Directional Terms (p. 9)

1. Directional terms indicate the relationship of one part of the body to another.
2. Commonly used directional terms are superior (toward the head or upper part of a structure), inferior (away from the head or toward the lower part of a structure), anterior (near or at the front of the body), posterior (near or at the back of the body), medial (nearer the midline of the body or a structure), intermediate (between a medial and lateral structure), ipsilateral (on the same side of the body), contralateral (on the opposite side of the body), proximal (nearer the attachment of an extremity to the trunk or a structure), distal (farther from the attachment of an extremity to the trunk or a structure), superficial (toward or on the surface of the body), deep (away from the surface of the body), parietal (pertaining to the outer wall of a body cavity), and visceral (pertaining to the covering of an organ).

Anatomical Position and Anatomical Names (p. 9)

1. When in the anatomical position, the subject stands erect facing the observer, the upper extremities are placed at the sides, and the palms of the hands are turned forward.
2. Regional names are terms given to specific regions of the body for reference. Examples of regional names include cranial (skull), thoracic (chest), brachial (arm), patellar (knee), cephalic (head), and gluteal (buttock).

Planes and Sections (p. 12)

1. Planes are imaginary flat surfaces that are used to divide the body or organs into definite areas. A midsagittal (median) plane is a vertical plane through the midline of the body that divides the body or organs into equal right and left sides; a parasagittal plane is a plane that does not pass through the midline of the body and divides the body or organs into unequal right and left sides; a frontal (coronal) plane is a plane at a right angle to a midsagittal (or parasagittal) plane that divides the body or organs into anterior and posterior portions; and a horizontal (transverse) plane is a plane parallel to the ground and at a right angle to the midsagittal, parasagittal, and frontal planes that divides the body or organs into superior and inferior portions.
2. Sections are flat surfaces resulting from cuts through body structures. They are named according to the plane on which the cut is made and include cross sections, frontal sections, and midsagittal sections.

Body Cavities (p. 13)

1. Spaces in the body that contain internal organs are called cavities.
2. The dorsal and ventral cavities are the two principal body cavities. The dorsal cavity contains the brain and spinal cord.

The organs of the ventral cavity are collectively called viscera.
3. The dorsal cavity is subdivided into the cranial cavity, which contains the brain, and the vertebral (spinal) canal, which contains the spinal cord and beginnings of spinal nerves.
4. The ventral body cavity is subdivided by the diaphragm into an upper thoracic cavity and a lower abdominopelvic cavity.
5. The thoracic cavity contains two pleural cavities and a mediastinum, which includes the pericardial cavity.
6. The mediastinum is a broad, median partition, actually a mass of tissue between the lungs that extends from the sternum to the vertebral column; it contains all contents of the thoracic cavity, except the lungs.
7. The abdominopelvic cavity is divided into a superior abdominal and an inferior pelvic cavity by an imaginary line extending from the symphysis pubis to the sacral promontory.
8. Viscera of the abdominal cavity include the stomach, spleen, pancreas, liver, gallbladder, small intestine, and most of the large intestine.
9. Viscera of the pelvic cavity include the urinary bladder, sigmoid colon, rectum, and internal female and male reproductive structures.

Abdominopelvic Regions (p. 13)

1. To describe the location of organs easily, the abdominopelvic cavity may be divided into nine regions by drawing four imaginary lines (left lateral, right lateral, transpyloric, and transtubercular).
2. The names of the nine abdominopelvic regions are epigastric, right hypochondriac, left hypochondriac, umbilical, right lumbar, left lumbar, hypogastric (pubic), right iliac (inguinal), and left iliac (inguinal).

Abdominopelvic Quadrants (p. 13)

1. To locate the site of an abdominopelvic abnormality in clinical studies, the abdominopelvic cavity may be divided into quadrants by passing imaginary horizontal and vertical lines through the umbilicus.
2. The names of the abdominopelvic quadrants are right upper quadrant (RUQ), left upper quadrant (LUQ), right lower quadrant (RLQ), and left lower quadrant (LLQ).

Radiographic Anatomy (p. 17)

1. Radiographic anatomy is a specialized branch of anatomy that makes use of x-rays.
2. The value of radiographic anatomy is its application in the diagnosis of disease.

Conventional Radiography (p. 17)

1. Conventional radiography uses a single barrage of x-rays.
2. The photographic two-dimensional image produced is called a roentgenogram.
3. Conventional radiography has several diagnostic drawbacks including overlapping of organs and tissues and inability to differentiate subtle differences in tissue density.

Computed Tomography (CT) Scanning (p. 17)

1. CT scanning combines the principles of x-ray and advanced computer technology.
2. The image produced, called a CT scan, provides a very accurate cross-sectional picture of any area of the body.

Dynamic Spatial Reconstructor (DSR) (p. 17)

1. The DSR is a highly sophisticated x-ray machine that can produce moving, three-dimensional images of different organs of the body.
2. The DSR has been designed to provide imaging of the heart, lungs, and circulation.

Homeostasis (p. 17)

1. Homeostasis is a condition in which the internal environment (extracellular fluid) of the body remains relatively constant in terms of chemical composition, temperature, and pressure.
2. All body systems attempt to maintain homeostasis.
3. Homeostasis is controlled mainly by the nervous and endocrine systems.

Stress and Homeostasis (p. 18)

1. Stress is any external or internal stimulus that creates a change in the internal environment.
2. If a stress acts on the body, homeostatic mechanisms attempt to counteract the effects of the stress and bring the condition back to normal.

Homeostasis of Blood Pressure (p. 20)

1. Blood pressure (BP) is the force exerted by blood as it presses against and attempts to stretch the walls of arteries. It is determined by the rate and force of the heartbeat, the amount of blood, and arterial resistance.
2. If a stress causes the heartbeat to increase, blood pressure also increases; pressure-sensitive nerve cells in certain arteries inform the brain, and the brain responds by sending impulses that decrease heartbeat, thus decreasing blood pressure back to normal; a rise in blood pressure also causes the brain to send impulses that dilate arterioles, thereby also helping to decrease blood pressure back to normal.
3. Any circular situation in which information about the status of something is continually fed back to a control region is called a feedback system.
4. A negative feedback system is one in which the reaction of the body (output) counteracts the stress (input) in order to maintain homeostasis; most feedback systems of the body are negative. In a positive feedback system, the output intensifies the input; the system is usually destructive.

Homeostasis of Blood Sugar Level (p. 20)

1. A normal blood sugar level is maintained by the actions of two different pancreatic hormones: insulin and glucagon.
2. Insulin lowers blood sugar level by increasing sugar uptake by cells and accelerating sugar storage as glycogen in the liver and skeletal muscles.
3. Glucagon raises blood sugar level by accelerating the rate of sugar released from glycogen by the liver.

Measuring the Human Body (p. 21)

1. Various kinds of measurements are important in understanding the human body.
2. Examples of such measurements include organ dimensions and weight, physiological response time, and amount of medication to be administered.
3. Measurements in this book are given in metric units followed by the approximate U.S. equivalents in parentheses, where convenient.
4. The principles and applications of the various metric and U.S. units of measurement are discussed in Appendix A.

REVIEW QUESTIONS

1. Define anatomy. List and define the various subdivisions of anatomy. Define physiology.
2. Give several examples of how structure and function are related.
3. Define each of the following terms: cell, tissue, organ, system, and organism.
4. Using Exhibit 1-2 as a guide, outline the functions of each system of the body, and list several organs that compose each system.
5. List and define the life processes of humans.
6. Define the anatomical position. Why is the anatomical position used?
7. Review Figure 1-2. See if you can locate each region on your own body, and name each by its common and anatomical term.
8. What is a directional term? Why are these terms important? Use each of the directional terms listed in Exhibit 1-3 in a complete sentence.
9. Define the various planes that may be passed through the body. Explain how each plane divides the body. Describe the meaning of cross section, frontal section, and midsagittal section.
10. Define a body cavity. List the body cavities discussed, and tell which major organs are located in each. What landmarks separate the various body cavities from one another? What is the mediastinum?
11. Describe how the abdominopelvic area is subdivided into nine regions. Name and locate each region and list the organs, or parts of organs, in each.
12. Describe how the abdominopelvic cavity is divided into quadrants, and name each quadrant.
13. Why is an autopsy performed? Describe the basic procedure.
14. Explain the principle of computed tomography (CT) scan-
ning. Contrast it with conventional radiography in terms of principle and diagnostic value.
15. Explain the principle and clinical application of the dynamic spatial reconstructor (DSR).
16. Define homeostasis. What is extracellular fluid (ECF)? Why is it called the internal environment of the body?
17. Under what conditions are the internal environment said to be in homeostasis?
18. What is a stress? Give several examples. How is stress related to homeostasis?
19. How is homeostasis related to normal and abnormal conditions in the body?
20. Substantiate this statement: "Homeostasis is a cooperative effort of all body parts."
21. What systems of the body control homeostasis? Explain. Discuss briefly how the regulation of blood pressure and blood glucose level are examples of homeostasis.
22. Define a feedback system. Distinguish between a negative and a positive feedback system.
23. Describe several situations that involve measurement of the human body.
24. Using Appendix A as a guide, solve the following problems.
    a. A bacterial cell measures 100 μm in length. Give its length in nanometers.
    b. A person's arm measures 2 ft in length. How many centimeters is this?
    c. The indicated dosage of a drug is 50 μg. How many milligrams is this?
    d. A person weighs 110 lb. Give the weight in kilograms.
    e. If you excrete 1,300 ml of urine in a day, how many liters is this?
    f. If you remove 15 cm^3 of blood from a patient, how many milliliters have you removed?

The Chemical Level of Organization

2

- Identify by name and symbol the principal chemical elements of the human body.
- Explain, by diagraming, the structure of an atom.
- Describe the principle of magnetic resonance imaging (MRI) and its diagnostic importance.
- Define a chemical reaction as a function of electrons in incomplete outer energy levels.
- Describe ionic bond formation in a molecule of sodium chloride (NaCl).
- Discuss covalent bond formation as the sharing of outer energy level electrons.
- Explain the nature and importance of hydrogen bonding.
- Define a radioisotope and explain the principle of positron emission tomography (PET) scanning and its diagnostic importance.
- Explain the basic differences among synthesis, decomposition, exchange, and reversible chemical reactions.
- Discuss how chemical reactions are related to metabolism.
- Define and distinguish between inorganic and organic compounds.
- Discuss the functions of water as a solvent, suspending medium, chemical reactant, heat absorber, and lubricant.
- List and compare the properties of acids, bases, and salts.
- Define pH as the degree of acidity or alkalinity of a solution.
- Explain the role of a buffer system as a homeostatic mechanism that maintains the pH of a body fluid.
- Compare the structure and functions of carbohydrates, lipids, and proteins.
- Contrast the structure and functions of deoxyribonucleic acid (DNA) and ribonucleic acid (RNA).
- Identify the function and importance of adenosine triphosphate (ATP) and cyclic adenosine-3′,5′-monophosphate (cyclic AMP).

Many of the common substances we eat and drink—water, sugar, table salt, cooking oil—play vital roles in keeping us alive. In this chapter you will learn something about how these substances function in your body. Fundamental to this study is a knowledge of basic chemistry and chemical processes, since your body is composed of chemicals and all body activities are chemical in nature. To understand the nature of the matter you are made from and the changes this matter goes through in your body, you will need to know which chemical elements are present in the human organism and how they interact.

INTRODUCTION TO BASIC CHEMISTRY

CHEMICAL ELEMENTS

All living and nonliving things consist of **matter,** which is anything that occupies space and has mass. Matter may exist in a solid, liquid, or gaseous state. All forms of matter are made up of a limited number of building units called **chemical elements,** substances that cannot be decomposed into simpler substances by ordinary chemical reactions. At present scientists recognize 106 different elements, of which 92 occur naturally. Elements are designated by letter abbreviations, usually derived from the first or first and second letters of the Latin or English name for the element. Such letter abbreviations are called **chemical symbols.** Examples of chemical symbols are H (hydrogen), C (carbon), O (oxygen), N (nitrogen), Na (sodium), K (potassium), Fe (iron), and Ca (calcium).

Approximately 26 elements are found in the human organism. Oxygen, carbon, hydrogen, and nitrogen make up about 96 percent of the body's weight. These 4 elements together with calcium and phosphorous constitute approximately 99 percent of the total body weight. Twenty other chemical elements, called trace elements, are found in low concentrations and compose the remaining 1 percent.

STRUCTURE OF ATOMS

Each element is made up of units of matter called **atoms,** the smallest units of matter that enter into chemical reactions. An element is simply a quantity of matter composed of atoms all of the same type. A handful of the element carbon, such as pure coal, contains only carbon atoms. A tank of oxygen contains only oxygen atoms. Measurements indicate that the smallest atoms are less than 0.00000001 cm (1/250,000,000 inch) in diameter, and the largest atoms are 0.00000005 cm (1/50,000,000 inch) in diameter. In other words, if 50 million of the largest atoms were placed end to end, they would measure approximately 2.5 cm (1 inch) in length.

An atom consists of two basic parts: the nucleus and the electrons (Figure 2-1). The centrally located **nucleus** comprises most of the atomic mass and contains positively charged particles called **protons** (p⁺) and uncharged (neutral) particles called **neutrons** (n⁰). Because each proton has one positive charge, the nucleus itself is positively charged. Together, protons and neutrons are referred to as particles called **nucleons.** Various coordinated movements of nuclear particles cause the nucleus to vibrate in several distinctive patterns. **Electrons** (e⁻) are negatively charged particles that move around the nucleus. (For reasons far beyond the scope of this textbook, the diagrams of atoms are very simplified, resembling planetary models of the solar system. In actuality, electrons do not follow fixed paths around the nucleus, but move about in probable locations.) The number of electrons in an atom of an element always equals the number of protons. Since each electron carries one negative charge, the negatively charged electrons and the positively charged protons balance each other, and the atom is electrically neutral.

What makes the atoms of one element different from those of another? The answer lies in the number of protons. Figure 2-2 shows that the hydrogen atom contains one proton. The helium atom contains two. The carbon atom has six, and so on. Each different kind of atom has a different number of protons in its nucleus. The number of protons in an atom is called the atom's **atomic number.** Therefore we can say that each kind of atom, or element, has a different atomic number. The total number of protons and neutrons in an atom is its **atomic weight.**

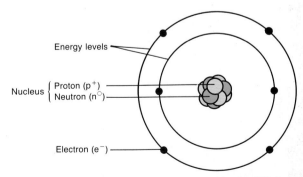

FIGURE 2-1 Structure of an atom. In this highly simplified version of a carbon atom, note the centrally located nucleus. The nucleus contains six neutrons and six protons, although all are not visible in this view since some are behind others. The six electrons move about the nucleus at varying distances from its center.

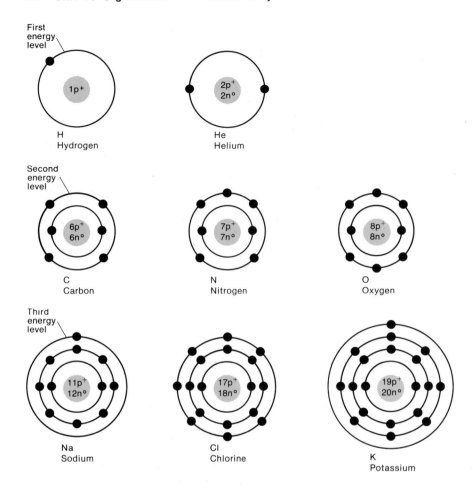

First
energy
level

1p+

H
Hydrogen

2p+
2n°

He
Helium

Second
energy
level

6p+
6n°

C
Carbon

7p+
7n°

N
Nitrogen

8p+
8n°

O
Oxygen

Third
energy
level

11p+
12n°

Na
Sodium

17p+
18n°

Cl
Chlorine

19p+
20n°

K
Potassium

FIGURE 2-2 Atomic structures of some representative atoms.

field (Figure 2-3a). The image produced shows the density and energy loss (voltage) of the nuclei of a particular element and somewhat resembles a CT scan (Figure 2-3b). The images can be displayed in color and as two- or three-dimensional images, and indicate a biochemical blueprint of cellular activity.

The diagnostic advantage of MRI is that in addition to providing images of diseased organs and tissues, it also provides information about what chemicals are present in the organ and tissue. Such an analysis can indicate a particular disease is in progress, even before symptoms occur. MRI also offers the advantages of being noninvasive (not involving puncture or incision of the skin or insertion of an instrument or foreign material into the body), not utilizing radiation, and gathering biochemical information without time-consuming chemical analyses.

In clinical and preclinical trials, MRI studies have confirmed the findings of other diagnostic techniques such as CT scans and, in some cases, have added more detail. Thus, MRI has proved useful in identifying existing pathologies. Scientists hope that MRI can also be used to perform a "biopsy" on tumors without an operation, assess mental disorders, measure blood flow, study the evolution of hematomas and infarctions in the brain following a stroke, identify the potential for developing a stroke, assess treatment for conditions such as heart disease and stroke, monitor the progress and treatment of a disease, study the effects of toxic drugs on tissues, measure intracellular pH, study metabolism, and determine how donor organs from cadavers may function after transplantation.

ATOMS AND MOLECULES

When atoms combine with or break apart from other atoms, a **chemical reaction** occurs. In the process, new products with different properties are formed. Chemical reactions are the foundation of all life processes.

The electrons of the atom actively participate in chemical reactions. The electrons move around the nucleus in regions, shown in Figure 2-2 as concentric circles lying at varying distances from the nucleus. We call these regions **energy levels.** Each energy level has a maximum number of electrons it can hold. For instance, the energy level nearest the nucleus never holds more than two electrons, no matter what the element. This energy level can be referred to as the first energy level. The second energy level holds a maximum of 8 electrons. The third level of atoms whose atomic number is less than 20 also can hold a maximum of 8 electrons. The third level of more complex atoms can hold a maximum of 18 electrons.

(a)

(b)

FIGURE 2-3 Magnetic resonance imaging (MRI). (a) MRI scanner. (b) Image of the human brain. (Courtesy of Technicare Corporation, Cleveland, Ohio.)

An atom always attempts to have its outermost energy level with the maximum number of electrons it can hold. To do this, the atom may give up an electron, take on an electron, or share an electron with another atom—whichever is easiest. The **valence** (combining capacity) is the number of extra or deficient electrons in the outermost energy level. Take a look at the chlorine atom. Its outermost energy level, which happens to be the third level, has seven electrons. Since the third level of an atom can hold a maximum of eight electrons, chlorine can be described as having a shortage of one electron. In fact, chlorine usually does try to pick up an extra electron. Sodium, by contrast, has only one electron in its outer level. This again happens to be the third energy level. It is much easier for sodium to get rid of the one electron than to fill the third level by taking on seven more electrons. Atoms of a few elements, like helium, have completely filled outer energy levels and do not need to gain or lose electrons. These are called **inert** elements.

Atoms with incompletely filled outer energy levels, like sodium and chlorine, tend to combine with other atoms

in a chemical reaction. During the reaction, the atoms can trade off or share electrons and, thereby, fill their outer energy levels. Atoms that already have filled outer levels generally do not participate in chemical reactions for the simple reason that they do not need to gain or lose electrons. When two or more atoms combine in a chemical reaction, the resulting combination is called a **molecule** (MOL-e-kyool). A molecule may contain two atoms of the same kind, as in the hydrogen molecule: H_2. The subscript 2 indicates that there are two hydrogen atoms in the molecule. Molecules may also be formed by the reaction of two or more different kinds of atoms, as in the hydrochloric acid molecule: HCl. Here an atom of hydrogen is attached to an atom of chlorine. A **compound** is a substance that can be broken down into two or more other substances by chemical means. The molecules of a compound always contain atoms of two or more different elements. Hydrochloric acid, which is present in the digestive juices of the stomach, is a compound. A molecule of hydrogen is not.

The atoms in a molecule are held together by forces of attraction called **chemical bonds.** Here we will consider ionic bonds, covalent bonds, and hydrogen bonds.

Ionic Bonds

Atoms are electrically neutral because the number of positively charged protons equals the number of negatively charged electrons. But when an atom gains or loses electrons, this balance is upset. If the atom gains electrons, it acquires an overall negative charge. If the atom loses electrons, it acquires an overall positive charge. Such a negatively or positively charged particle is called an **ion** (Ī-on).

Consider the sodium ion (Figure 2-4a). The sodium atom (Na) has 11 protons and 11 electrons, with 1 electron in its outer energy level. When sodium gives up the single electron in its outer level, it is left with 11 protons and only 10 electrons. It is an *electron donor*. The atom now has an overall positive charge of one (+1). This positively charged sodium atom is called a sodium ion (written Na^+).

Another example is the formation of the chloride ion (Figure 2-4b). Chlorine has a total of 17 electrons, 7 of them in the outer energy level. Since this energy level can hold 8 electrons, chlorine tends to pick up an electron that has been lost by another atom. Chlorine is an *electron acceptor*. By accepting an electron, chlorine acquires a total of 18 electrons. However, it still has only 17 protons in its nucleus. The chloride ion therefore has a negative charge of one (−1) and is written as Cl^-.

The positively charged sodium ion (Na^+) and the negatively charged chloride ion (Cl^-) attract each other—unlike charges attract each other. The attraction, called an **ionic bond,** holds the two atoms together, and a molecule is formed (Figure 2-4c). The formation of this molecule, sodium chloride (NaCl) or table salt, is one of the most common examples of ionic bonding. Thus an ionic bond is an attraction between atoms in which one atom loses electrons and another atom gains electrons. Generally, atoms whose outer energy level is less than half-filled lose electrons and form positively charged ions called

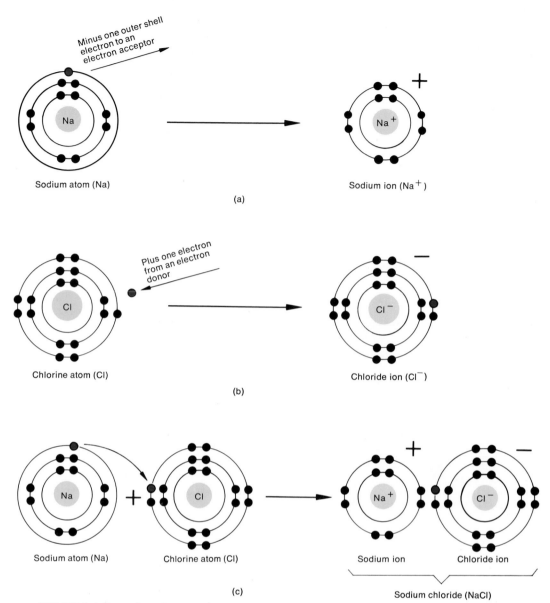

Minus one outer shell electron to an electron acceptor

Sodium atom (Na)

Sodium ion (Na$^+$)

(a)

Plus one electron from an electron donor

Chlorine atom (Cl)

Chloride ion (Cl$^-$)

(b)

Sodium atom (Na)   +   Chlorine atom (Cl)

Sodium ion   Chloride ion

Sodium chloride (NaCl)

(c)

FIGURE 2-4 Formation of an ionic bond. (a) An atom of sodium attains stability by passing a single electron to an electron acceptor. The loss of this single electron results in the formation of a sodium ion (Na$^+$). (b) An atom of chlorine attains stability by accepting a single electron from an electron donor. The gain of this single electron results in the formation of a chloride ion (Cl$^-$). (c) When the Na$^+$ and the Cl$^-$ ions are combined, they are held together by the attraction of opposite charges, which is known as an ionic bond, and a molecule of sodium chloride (NaCl) is formed.

cations (KAT-ī-ons). Examples of cations are potassium ion (K$^+$), calcium ion (Ca^{2+}), iron ion (Fe^{2+}), and sodium ion (Na$^+$). By contrast, atoms whose outer energy level is more than half-filled tend to gain electrons and form negatively charged ions called **anions** (AN-ī-ons). Examples of anions include iodine ion (I$^-$), chloride ion (Cl$^-$), and sulfur ion (S^{2-}).

Notice that an ion is always symbolized by writing the chemical abbreviation followed by the number of positive ($+$) or negative ($-$) charges the ion acquires.

Hydrogen is an example of an atom whose outer level is exactly half-filled. The first energy level can hold two electrons, but in hydrogen atoms it contains only one. Hydrogen may lose its electron and become a positive

ion (H$^+$). This is precisely what happens when hydrogen combines with chlorine to form hydrochloric acid (H$^+$Cl$^-$). However, hydrogen is equally capable of forming another kind of bond altogether: a covalent bond.

Covalent Bonds

The second chemical bond to be considered is the **covalent bond.** This bond is far more common in organisms than an ionic bond and is more stable. When a covalent bond is formed, neither of the combining atoms loses or gains an electron. Instead, the two atoms share one, two, or three electron pairs. Look at the hydrogen atom again. One way a hydrogen atom can fill its outer energy level

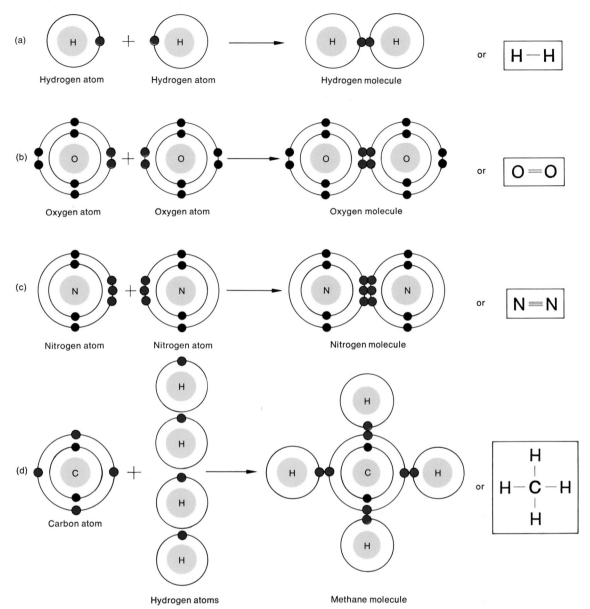

FIGURE 2-5 Formation of a covalent bond between atoms of the same element and between atoms of different elements. (a) A single covalent bond between two hydrogen atoms. (b) A double covalent bond between two oxygen atoms. (c) A triple covalent bond between two nitrogen atoms. (d) Single covalent bonds between a carbon atom and four hydrogen atoms. In the symbols on the right, each covalent bond is represented by a straight line between atoms.

is to combine with another hydrogen atom to form the molecule H_2 (Figure 2-5a). In the H_2 molecule, the two atoms share a pair of electrons. Each hydrogen atom has its own electron plus one electron from the other atom. The two electrons actually circle the nuclei of both atoms, spending their time equally between both atoms. Therefore the outer energy levels of both atoms are filled half the time. When one pair of electrons is shared between atoms, as in the H_2 molecule, a *single covalent bond* is formed. A single covalent bond is expressed as a single line between the atoms (H—H). When two pairs of electrons are shared between two atoms, a *double covalent bond* is formed, which is expressed as two parallel

lines (=) (Figure 2-5b). A *triple covalent bond,* expressed by three parallel lines (≡), occurs when three pairs of electrons are shared (Figure 2-5c).

The same principles that apply to covalent bonding between atoms of the same element also apply to atoms of different elements. Methane (CH_4), also know as marsh gas, is an example of covalent bonding between atoms of different elements (Figure 2-5d). The outer energy level of the carbon atom can hold eight electrons, but has only four of its own. Each hydrogen atom can hold two electrons, but has only one of its own. In the methane molecule the carbon atom shares four pairs of electrons. One pair is shared with each hydrogen atom. Each of the four carbon electrons orbits around both the carbon nucleus and a hydrogen nucleus. Each hydrogen electron circles around its own nucleus and the carbon nucleus.

Elements whose outer energy levels are half-filled, such as hydrogen and carbon, form covalent bonds quite easily. In fact, carbon always forms covalent bonds. It never becomes an ion. However, many atoms whose outer energy levels are more than half-filled also form covalent bonds. An example is oxygen. We won't go into the reasons why some atoms tend to form covalent bonds rather than ionic bonds.

Hydrogen Bonds

A **hydrogen bond** consists of a hydrogen atom covalently bonded to one oxygen atom or one nitrogen atom, but attracted to another oxygen or nitrogen atom. Because hydrogen bonds are weak, only about 5 percent as strong as covalent bonds, they do not bind atoms into molecules. However, they do serve as bridges between different molecules or between various parts of the same molecule. The weak bonds may be formed and broken fairly easily. It is this property that accounts for the temporary bonding between certain atoms within large complex molecules such as proteins and nucleic acids. It should be noted that even though hydrogen bonds are relatively weak, such large molecules may contain several hundred of these bonds, resulting in considerable strength and stability.

Chemical reactions are nothing more than the making or breaking of bonds between atoms. And these reactions occur continually in all the cells of your body. As you will see again and again, chemical reactions are the processes by which body structures are built and body functions carried out.

RADIOISOTOPES

Atoms of an element, although chemically alike, may have different nuclear masses because of one or more extra neutrons in some of the atoms, so the atomic weight assigned to an element is only an average. Each of the chemically identical atoms of an element with a particular nuclear mass is an **isotope** of that element. All isotopes of an element have the same number of protons in their nuclei, but their atomic weights differ because of the difference in the number of neutrons. In a sample of oxygen, for example, most atoms have 8 neutrons, but a few have 9 or 10, even though all have 8 protons. The isotopes of oxygen are designated as ^{16}O, ^{17}O, and ^{18}O. The number indicates their atomic weights.

Certain isotopes called **radioisotopes** are unstable—they "decay" or change their nuclear structure to a more stable configuration. And in decaying they emit radiation that can be detected by instruments. These instruments estimate the amount of radioisotope present in a part of the body or in a sample of material and form an image of its distribution.

Radioisotopes of iodine were among the first discovered, and their specificity for the thyroid gland made them the cornerstone for the study of thyroid physiology. Nuclear medicine now uses ^{32}P to treat leukemia and

^{59}Fe for the study of red blood cell production. Moreover, short-lived agents such as technetium-99m pertechnetate (^{99m}Tc) have improved the quality of the images and reduced the patient's radiation dose.

CLINICAL APPLICATION

In Chapter 1 you learned that conventional radiographic techniques, CT scanners, and the DSR use radiation in the form of x-rays to produce images that are invaluable in the diagnosis of disease. In **radioisotopic scanning** the chemical activity of various tissues of the body can be recorded. Physicians can diagnose certain diseases by charting the course of an injected radioisotope through the body and by noting how much of the substance concentrates and where. Radioisotope scans, like the DSR, can provide early warning of disease. The branch of medicine that is concerned with the use of radioisotopes in the diagnosis of disease and therapy is called **nuclear medicine.**

Although radioisotope scans have been used in diagnosis since the 1950s, their value was limited by distorted images of the distribution of a radioisotope. Within the past few years more precise localization of radioisotopes in the body has been made possible by new computer imaging techniques similar to those used in CT scanning. The result is a sophisticated version of radioisotope scanning called **positron emission tomography (PET).**

The principle behind PET scanning is as follows. Short-lived radioisotopes such as ^{11}C, ^{13}N, or ^{15}O are produced and incorporated into a solution that can be injected into the body. As the radioisotope circulates through the body, it emits positively charged electron-like particles called *positrons*. Positrons collide with negatively charged electrons in body tissues, causing their annihilation and the release of gamma rays. Gamma rays are bundles of energy similar to high-energy x-rays. Once released, they travel in opposite directions and are detected and recorded by PET receptors. A computer then takes the information and constructs a colored **PET scan** that shows where the radioisotopes are being used in the body (Figure 2-6).

Although PET may not become a clinical tool for several years, it has already provided information that

(a)   (b)   (c)

FIGURE 2-6 Positron emission tomography (PET) scans. (a) Oxygen metabolism in the brain. (b) Blood volume in the brain. (c) Blood flow through the brain. (Courtesy of Dr. Michel M. Ter-Rogossian, Washington University, School of Medicine.)

cannot be obtained by any other technique by sensing function rather than structure. Using PET, physicians can study the effects of drugs in body organs, measure blood flow through organs such as the brain and heart, diagnose coronary artery disease, identify the extent of damage as a result of strokes or heart attacks, and detect cancers and measure the effects of treatment. PET studies with schizophrenic and manic-depressive patients have revealed that schizophrenics tend to use less glucose in certain brain areas, whereas manic-depressive patients use more glucose during manic phases. PET technology is also now being used to study the chemical changes that occur during epileptic seizures and in persons suffering from senile dementia. It is hoped that PET could eventually become a routine part of psychiatric examinations, helping to diagnose patients who show symptoms of more than one kind of mental illness. As impressive as PET is in the diagnosis of disease, it is also being used by scientists to probe the healthy brain. For example, by detecting and recording changes in glucose consumption in the brain, scientists can identify which specific areas of the brain are involved in specific sensory and motor activities.

CHEMICAL REACTIONS

As we said earlier, **chemical reactions** involve the making or breaking of bonds between atoms. After a chemical reaction, the total number of atoms remains the same, but because they are rearranged, there are new molecules with new properties. In this section, we will look at the basic chemical reactions common to all living cells. Once you have learned them, you will be able to understand the chemical reactions discussed later.

Synthesis Reactions—Anabolism

When 2 or more atoms, ions, or molecules combine to form new and larger molecules, the process is called a **synthesis reaction.** The word *synthesis* means "combination," and synthesis reactions involve the *forming of new bonds.* Synthesis reactions can be expressed in the following way:

$$A \quad + \quad B \xrightarrow{\text{combine to form}} AB$$

Atom, ion,   Atom, ion,   New
or molecule A   or molecule B   molecule AB

The combining substances, A and B, are called the **reactants;** the substance formed by the combination is the **end product.** The arrow indicates the direction in which the reaction is proceeding. An example of a synthesis reaction is:

$$N \quad + \quad 3H \quad \rightarrow \quad NH_3$$

Nitrogen   Hydrogen   Ammonia
atom   atoms   molecule

All the synthesis reactions that occur in your body are collectively called anabolic reactions, or simply **anabolism** (a-NAB-ō-lizm). Combining glucose molecules to form glycogen and combining amino acids to form proteins

are two examples of anabolism. The importance of anabolism is considered in detail in Chapter 25.

Decomposition Reactions—Catabolism

The reverse of a synthesis reaction is a **decomposition reaction.** The word *decompose* means to break down into smaller parts. In a decomposition reaction, the *bonds are broken.* Large molecules are broken down into smaller molecules, ions, or atoms. A decomposition reaction occurs in this way:

$$AB \xrightarrow{\text{breaks down into}} A \quad + \quad B$$

Molecule AB   Atom, ion, or   Atom, ion,
              molecule A   or molecule B

Under the proper conditions, methane can decompose into carbon and hydrogen:

$$CH_4 \quad \rightarrow \quad C \quad + \quad 4H$$

Methane   Carbon   Hydrogen
molecule   atom   atoms

The subscript 4 on the left-hand side of the reaction equation indicates that four atoms of hydrogen are bonded to one carbon atom in the methane molecule. The number 4 on the right-hand side of the equation shows that four single hydrogen atoms have been set free.

All the decomposition reactions that occur in your body are collectively called catabolic reactions, or simply **catabolism** (ka-TAB-ō-lizm). The digestion and oxidation of food molecules are examples. The importance of catabolism is also considered in detail in Chapter 25.

Exchange Reactions

All chemical reactions are based on synthesis or decomposition processes. In other words, chemical reactions are simply the making and/or breaking of ionic or covalent bonds. Many reactions, such as **exchange reactions,** are partly synthesis and partly decomposition. An exchange reaction works like this:

$$AB + CD \rightarrow AD + BC \quad \text{or} \quad AC + BD$$

The bonds between A and B and between C and D are broken in a decomposition process. New bonds are then formed between A and D and between B and C or between A and C and between B and D in a synthesis process.

Reversible Reactions

When chemical reactions are reversible, the end product can revert to the original combining molecules. A **reversible reaction** is indicated by two arrows:

$$A + B \underset{\text{breaks down to}}{\overset{\text{combines with}}{\rightleftarrows}} AB$$

Some reversible reactions occur because neither the reactants nor the end products are stable. Other reactions reverse themselves only under special conditions:

$$A + B \underset{\text{water}}{\overset{\text{heat}}{\rightleftarrows}} AB$$

Whatever is written above or below the arrows indicates the special condition under which the reaction occurs. In this case, A and B react to produce AB only when heat is applied, and AB breaks down into A and B only when water is added. Figure 2-7 summarizes the basic chemical reactions that can occur.

Metabolism

The word **metabolism** (me-TAB-ō-lizm) represents the sum of all the synthesis and decomposition reactions occurring in the body—the sum of all anabolic and catabolic reactions that go on inside you. When we say that a person has a high metabolism that is due to a disruption of homeostasis, we mean that the chemical reactions in the body are proceeding at a faster rate than normal. The decomposition reactions are occurring so quickly that foods are broken down completely before the body has a chance to store them. Consequently, people with a high metabolism can usually eat a great deal without gaining weight. Because rapid decomposition reactions generate large amounts of energy, including heat, such people often complain that they are too hot.

Chemical reactions in the bodies of people with low metabolism that is due to a disruption of homeostasis proceed more slowly that normal. Food is broken down slowly. Much of it is only partially broken down and then stored. Such people tend to gain weight easily, have little energy, and are often cold. Because their synthesis reactions are also slowed down, their bodies build up new structures very slowly. Wounds, for instance, often take a long time to heal.

How Chemical Reactions Occur

The **collision theory** explains how chemical reactions occur and how certain factors affect the rates of those reactions. According to this theory, all atoms, ions, and molecules are continuously moving and colliding with one another. The energy transferred by the particles in the collision might disrupt their electron structures enough that chemical bonds are broken or new ones are formed.

Several factors determine whether a collision will actually cause a chemical reaction. Among these are the velocities of the colliding particles, their energy, and their specific chemical configurations. Up to a point, the higher the particles' velocities, the greater the probability that their collision will result in a reaction. Also, each chemical reaction requires a specific level of energy. The collision energy required for a chemical reaction is its activation

GENERAL REACTION   SPECIFIC EXAMPLE

(a) Synthetic or anabolic reaction : A + B ⟶ AB

Glucose molecules   Glucose molecules   Glycogen molecule

(b) Decomposition or catabolic reaction : AB ⟶ A + B

Fat molecule   Glycerol   Fatty acids

(c) Exchange reaction: AB + CD ⟶ AD + BC

Hydrochloric acid   Sodium bicarbonate   Sodium chloride   Carbonic acid

(d) Reversible reaction: A + B ⇌ AB

ADP   Phosphate   ATP

FIGURE 2-7 Kinds of chemical reactions. (a) Synthetic or anabolic reaction. When linked together as shown, molecules of glucose form a molecule of glycogen. Glucose is a sugar that is the primary source of energy. Glycogen is a stored form of that sugar found in the liver and skeletal muscles. (b) Decomposition or catabolic reaction. The example shows a molecule of fat breaking down into glycerol and fatty acids. This reaction occurs whenever a food that contains fat is digested. (c) Exchange reaction. In this reaction, atoms of different molecules are exchanged with each other. Shown is a buffer reaction in which the body eliminates strong acids to help maintain homeostasis. (d) Reversible reaction. ATP (adenosine triphosphate) is an important source of stored energy. When such energy is needed, the ATP breaks down into ADP (adenosine diphosphate) and PO_4^{3-} (phosphate group), releasing energy in the reaction. The phosphate group is symbolized as P. The cells of the body reconstruct ATP by using the energy of foods to attach ADP to PO_4^{3-}.

energy, which is the amount of energy needed to disrupt the stable electronic configuration of a specific molecule so that the electrons can be rearranged. (The relationship of enzymes to activation energy is considered in Chapter 25.) But, even if colliding particles possess the minimum energy needed for reaction, no reaction will take place unless the particles are properly oriented toward each other.

Energy and Chemical Reactions

Energy is the capacity to do work. The two principal kinds of energy are **potential** (inactive or stored) and **kinetic** (energy of motion). Energy, whether potential or kinetic, exists in a number of different forms.

Chemical energy is the energy released or absorbed in the breaking or forming of chemical bonds. When a chemical bond is formed, energy is required. When a bond is broken, energy is released. This means that synthesis reactions need energy in order to occur, whereas decomposition reactions give off energy. The building processes of the body—the construction of bones, the growth of hair and nails, the replacement of injured cells—occur basically through synthesis reactions. The breakdown of foods, on the other hand, occurs through decomposition reactions. When foods are decomposed, they release energy that can be used by the body for its building processes.

Mechanical energy is energy directly involved in moving a part of the body. For example, when skeletal muscles contract, they pull on bones to bring about movement.

Radiant energy, such as heat and light, travels in waves. Some of the energy released during decomposition reactions is heat energy, which is used to help maintain normal body temperature.

Electrical energy is the result of the flow of charges, electrons, or charged particles called ions. As you will see later, electrical energy is essential for the conduction of nerve impulses by nerve cells.

The various forms of energy can be transduced (transformed) from one form into another.

CHEMICAL COMPOUNDS AND LIFE PROCESSES

Most of the chemicals in the body exist in the form of compounds. Biologists and chemists divide these compounds into two principal classes: inorganic compounds and organic compounds. **Inorganic compounds** usually lack carbon. They are usually small, ionically bonded molecules that are vital to body functions. They include water, many salts, acids, and bases. **Organic compounds** always contain carbon and hydrogen. Carbon is a unique element in the chemistry of life. Because carbon has four electrons in its outer shell, it can combine with a variety of atoms, including other carbon atoms, to form straight or branched chains and rings. Carbon chains are the backbone for many substances of living cells. Organic compounds are held together mostly or entirely by covalent bonds. Organic compounds present in the body include carbohydrates, lipids, proteins, nucleic acids, and adenosine triphosphate (ATP).

INORGANIC COMPOUNDS

Water

One of the most important, as well as the most abundant, inorganic substances in the human organism is **water.** In fact, with a few exceptions, such as tooth enamel and bone tissue, water is by far the most abundant material in tissues. About 60 percent of red blood cells, 75 percent of muscle tissue, and 92 percent of blood plasma is water. The following functions of water explain why it is such a vital compound in living systems:

1. Water is an excellent solvent and suspending medium. A *solvent* is a liquid or gas in which some other material (solid, liquid, or gas), called a *solute,* has been dissolved. The combination of solvent plus solute is called a **solution,** and one common example of a solution is salt water. A solute, such as salt in water, typically does not settle out of its solution. The solute can be retrieved through a chemical reaction or, in some cases, by boiling off the solvent. In a **suspension,** by contrast, the suspended material mixes with the liquid or suspending medium, but it will eventually settle out of the mixture. An example of a suspension is cornstarch and water. If the two materials are shaken together, a milky mixture forms. After the mixture sits for a while, however, the water clears at the top and the cornstarch settles to the bottom.

The solvating property of water is essential to health and survival. For example, water in the blood forms a solution with a small portion of the oxygen you inhale, allowing the oxygen to be carried to your body cells. Water in the blood also dissolves a small portion of the carbon dioxide that is carried from the cells to the lungs to be exhaled. Furthermore, if the surfaces of the air sacs in your lungs are not moist, oxygen cannot dissolve and, therefore, cannot move into your blood to be distributed throughout your body. Water, moreover, is the solvent that carries nutrients into and wastes out of your body cells.

As a suspending medium, water is also vital to your survival. Many large organic molecules are suspended in the water of your body cells. These molecules are consequently able to come in contact with other chemicals, allowing various essential chemical reactions to occur.

2. Water can participate in chemical reactions. During digestion, for example, water can be added to large nutrient molecules in order to break them down into smaller molecules. This kind of breakdown is necessary if the body is to utilize the energy in nutrients. Water molecules are also used in synthesis reactions. Such reactions occur in the production of hormones and enzymes.

3. Water absorbs and releases heat very slowly. In comparison to other substances, water requires a large amount of heat to increase its temperature and a great loss of heat to decrease its temperature. Thus the presence of a large amount of water moderates the effects of fluctuations in environmental temperature and thereby helps to maintain a homeostatic body temperature.

4. Water requires a large amount of heat to change from a liquid to a gas. When water (perspiration) evaporates from the skin, it takes with it large quantities of heat and provides an excellent cooling mechanism.

5. Water serves as a lubricant in various regions of the body. It is a major part of mucus and other lubricating fluids. Lubrication is especially necessary in the chest and abdomen, where internal organs touch and slide over each other. It is also needed at joints, where bones, ligaments, and tendons rub against each other. In the digestive tract, water moistens foods to ensure their smooth passage.

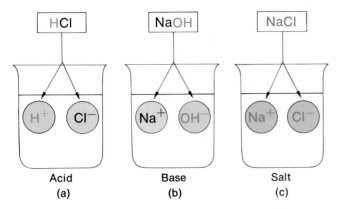

FIGURE 2-8 Ionization of acids, bases, and salts. (a) When placed in water, hydrochloric acid (HCl) dissociates into H^+ ions and Cl^- ions. Acids are proton donors. (b) When the base sodium hydroxide (NaOH) is placed in water, it dissociates into OH^- ions and Na^+ ions. Bases are proton acceptors. (c) When table salt (NaCl) is placed in water, it dissociates into positive and negative ions (Na^+ and Cl^-), neither of which is H^+ or OH^-.

Acids, Bases, and Salts

When molecules of inorganic acids, bases, or salts are dissolved in water in the body cells, they undergo **ionization** (ī'-on-i-ZĀ-shun) or **dissociation** (dis'-sō-sē-Ā-shun), that is, they break apart into ions. Such particles are also called **electrolytes** (ē-LEK-trō-līts) because the solution will conduct an electric current (the chemistry and importance of electrolytes are discussed in detail in Chapter 27). An **acid** may be defined as a substance that dissociates into one or more *hydrogen ions* (H^+) and one or more negative ions (anions). An acid may also be defined

as a proton (H^+) donor. A **base,** by contrast, dissociates into one or more *hydroxyl ions* (OH^-) and one or more positive ions (cations). A base may also be viewed as a proton acceptor. Hydroxyl ions, as well as some other negative ions, have a strong attraction for protons. A **salt,** when dissolved in water, dissociates into cations and anions, neither of which is H^+ or OH^- (Figure 2-8). Acids and bases react with one another to form salts. For example, the combination of hydrochloric acid (HCl), an acid, and sodium hydroxide (NaOH), a base, produces sodium chloride (NaCl), a salt, and water (H_2O).

Many salts are found in the body. Some are in cells, whereas others are in the body fluids, such as lymph, blood, and the extracellular fluid of tissues. The ions of salts are the source of many essential chemical elements. Exhibit 2-1 shows how salts dissociate into ions that pro-

---

EXHIBIT 2-1   DISSOCIATION OF REPRESENTATIVE SALTS INTO IONS THAT PROVIDE ESSENTIAL CHEMICAL ELEMENTS FOR THE BODY

SALT	DISSOCIATES INTO	CATION		ANION
NaCl Sodium chloride	⟶	Na^+ Sodium ion	+	Cl^- Chloride ion
KCl Potassium chloride	⟶	K^+ Potassium ion	+	Cl^- Chloride ion
$CaCl_2$ Calcium chloride	⟶	Ca^{2+} Calcium ion	+	$2Cl^-$ Chloride ions
$MgCl_2$ Magnesium chloride	⟶	Mg^{2+} Magnesium ion	+	$2Cl^-$ Chloride ions
$CaCo_3$ Calcium carbonate	⟶	Ca^{2+} Calcium ion	+	CO_3^{2-} Carbonate ion
$Ca_3(PO_4)_2$ Calcium phosphate	⟶	$3Ca^{2+}$ Calcium ions	+	$2PO_4^{3-}$ Phosphate ions
Na_2SO_4 Sodium sulfate	⟶	$2Na^+$ Sodium ions	+	SO_4^{2-} Sulfate ion

---

EXHIBIT 2-2   REPRESENTATIVE CHEMICAL ELEMENTS FOUND IN THE BODY

CHEMICAL ELEMENT	COMMENT
Oxygen (O)	Constituent of water and organic molecules; functions in cellular respiration.
Carbon (C)	Found in every organic molecule.
Hydrogen (H)	Constituent of water, all foods, and most organic molecules.
Nitrogen (N)	Component of all protein molecules and nucleic acid molecules.
Calcium (Ca)	Constituent of bone and teeth; required for blood clotting, intake (endocytosis) and output (exocytosis) of substances through plasma membranes, motility of cells, movement of chromosomes prior to cell division, glycogen metabolism, synthesis and release of neurotransmitters, and contraction of muscle.
Phosphorus (P)	Component of many proteins, nucleic acids, ATP, and cyclic AMP; required for normal bone and tooth structure; found in nerve tissue.
Chlorine (Cl)	Cl^- is an anion of NaCl, a salt important in water movement between cells.
Sulfur (S)	Component of many proteins, especially the contractile proteins of muscle.
Potassium (K)	Required for growth and important in conduction of nerve impulses and muscle contraction.
Sodium (Na)	Na^+ is a cation of NaCl; structural component of bone; essential in blood to maintain water balance; needed for conduction of nerve impulses.
Magnesium (Mg)	Component of many enzymes.
Iodine (I)	Vital to functioning of thyroid gland.
Iron (Fe)	Essential component of hemoglobin and respiratory enzymes.

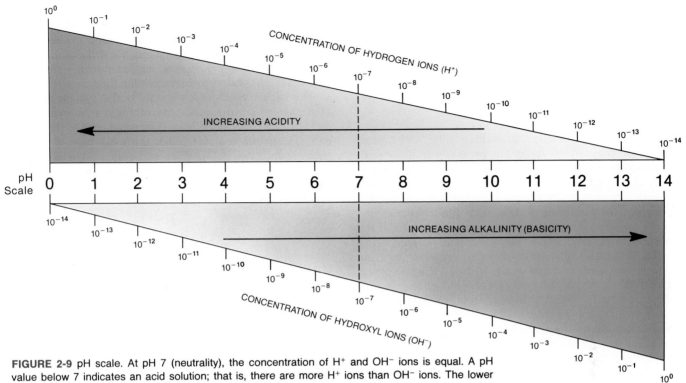

FIGURE 2-9 pH scale. At pH 7 (neutrality), the concentration of H^+ and OH^- ions is equal. A pH value below 7 indicates an acid solution; that is, there are more H^+ ions than OH^- ions. The lower the numerical value of the pH, the more acid the solution is because the H^+ ion concentration becomes progressively greater. A pH value above 7 indicates an alkaline (basic) solution; that is, there are more OH^- ions than H^+ ions. The higher the numerical value of the pH, the more alkaline the solution is because the OH^- ion concentration becomes progressively greater. A change in one whole number on the pH scale represents a 10-fold change from the previous concentration ($10^0 = 1.0$, $10^{-1} = 0.1$, $10^{-2} = 0.01$, $10^{-3} = 0.001$, and so on).

vide these elements. Chemical analyses reveal that sodium and chloride ions are present in higher concentrations than other ions in extracellular body fluids. Inside the cells, phosphate and potassium ions are more abundant than other ions. Chemical elements such as sodium, phosphorus, potassium, or iodine are present in the body only in chemical combination with other elements or as ions. Their presence as free, un-ionized atoms could be instantly fatal. Exhibit 2-2 lists representative elements found in the body.

Acid–Base Balance: The Concept of pH

The fluids of your body must maintain a fairly constant balance of acids and bases. In solutions such as those found in body cells or in extracellular fluids, acids dissociate into hydrogen ions (H^+) and anions. Bases, on the other hand, dissociate into hydroxyl ions (OH^-) and cations. The more hydrogen ions that exist in a solution, the more acid the solution; conversely, the more hydroxyl ions, the more basic (alkaline) the solution. The term **pH** is used to describe the degree of *acidity* or *alkalinity* (*basicity*) of a solution.

Biochemical reactions—reactions that occur in living systems—are extremely sensitive to even small changes in the acidity or alkalinity of the environment in which they occur. In fact, H^+ and OH^- ions are involved in practically all biochemical processes, and the functions of cells are modified greatly by any departure from narrow limits of normal H^+ and OH^- concentrations. For this reason, the acids and bases that are constantly formed in the body must be kept in balance.

A solution's acidity or alkalinity is expressed on a **pH scale** that runs from 0 to 14 (Figure 2-9). The pH scale is based on the number of H^+ ions in a solution expressed in chemical units called moles per liter. A pH of 7 means that a solution contains one ten-millionth (0.0000001) of a mole* of H^+ ions per liter. The number 0.0000001 is written 10^{-7} in exponential form. To convert this value to pH, the negative exponent (-7) is converted into the positive number 7. A solution with a concentration of 0.0001 (10^{-4}) of H^+ ions per liter has a pH of 4, a solution with a concentration of 0.000000001 (10^{-9}) has a pH of 9, and so on.

A solution that is zero on the pH scale has many H^+ ions and few OH^- ions. A solution that rates 14, by contrast, has many OH^- ions and few H^+ ions. The midpoint in the scale is 7, where the concentration of H^+ and OH^- ions is equal. A substance with a pH of 7, such as pure water, is neutral. A solution that has more H^+ ions than OH^- ions is an *acid solution* and has a pH below 7. A solution that has more OH^- ions than H^+ ions is a *basic (alkaline) solution* and has a

---

* A **mole** of any substance is the weight, in grams, of the combined atomic weights of the atoms that make up a molecule of the substance. Example: A mole of H_2O weighs 18 grams (2 for the 2 hydrogen atoms + 16 for the oxygen atom).

pH above 7. A change of one whole number on the pH scale represents a 10-fold change from the previous concentration. That is, a pH of 2 indicates 10 times fewer H^+ ions than a pH of 1. A pH of 3 indicates 10 times fewer H^+ ions than a pH of 2 and 100 times fewer H^+ ions than a pH of 1.

Maintaining pH: Buffer Systems

Although the pH of body fluids may differ, the normal limits for the various fluids are generally quite specific and narrow. Exhibit 2-3 shows the pH values for certain body fluids compared with common substances. Even though strong acids and bases are continually taken into the body, the pH levels of these body fluids remain relatively constant. The mechanisms that maintain these homeostatic pH values in the body are called **buffer systems.**

The essential function of a buffer system is to react with strong acids or bases in the body and replace them with weak acids or bases that can change the normal pH values only slightly. Strong acids (or bases) ionize easily and contribute many H^+ (or OH^-) ions to a solution. They therefore change the pH drastically. Weak acids (or bases) do not ionize so easily. They contribute

EXHIBIT 2-3   NORMAL pH VALUES OF REPRESENTATIVE SUBSTANCES

SUBSTANCE	pH VALUE
Gastric juice (digestive juice of the stomach)	1.2–3.0
Lemon juice	2.2–2.4
Grapefruit juice, vinegar, beer, wine	3.0
Cider	2.8–3.3
Pineapple juice, orange juice	3.5
Tomato juice	4.2
Coffee	5.0
Clam chowder	5.7
Urine	4.6–8.0
Saliva	6.35–6.85
Milk	6.6–6.9
Pure (distilled) water	7.0
Blood	7.35–7.45
Semen (fluid containing sperm)	7.20–7.60
Cerebrospinal fluid (fluid associated with nervous system)	7.4
Pancreatic juice (digestive juice of the pancreas)	7.1–8.2
Eggs	7.6–8.0
Bile (liver secretion that aids in fat digestion)	7.6–8.6
Milk of magnesia	10.0–11.0
Limewater	12.3

fewer H^+ (or OH^-) ions and have little effect on the pH. The chemicals that replace strong acids or bases with weak ones are called **buffers** and are found in the body's fluids. Of the several buffer systems, we will examine only the *carbonic acid–bicarbonate buffer system*—the most important one found in extracellular fluid.

The carbonic acid–bicarbonate buffer system consists of a *pair* of compounds. One of them is a *weak acid,* and the other is a *weak base.* The weak acid of the buffer pair is *carbonic acid* (H_2CO_3); the weak base is *sodium bicarbonate* ($NaHCO_3$). The carbonic acid is a proton (H^+) donor and the bicarbonate ion of the sodium bicarbonate is a proton acceptor. In solution, the members of this buffer pair dissociate as follows:

Acidic component:

$$H_2CO_3 \rightleftharpoons H^+ + HCO_3^-$$

Carbonic acid   Hydrogen ion   Bicarbonate ion

Basic component:

$$NaHCO_3 \rightleftharpoons Na^+ + HCO_3^-$$

Sodium bicarbonate   Sodium ion   Bicarbonate ion

Each member of the buffer pair has a specific role in helping the body maintain a constant pH. If the body's pH is threatened by the presence of a strong acid, the weak base of the buffer pair goes into operation. If the body's pH is threatened by a strong base, the weak acid goes into play.

Consider the following situation. If a strong acid, such as HCl, is added to extracellular fluid, the weak base of the buffer system goes to work and the following acid-buffering reaction occurs:

$$HCl + NaHCO_3 \rightleftharpoons NaCl + H_2CO_3$$

Hydrochloric acid (strong acid)   Sodium bicarbonate (weak base of buffer system)   Sodium chloride (salt)   Carbonic acid (weak acid)

The chloride ion of HCl and the sodium ion of sodium bicarbonate combine to form NaCl, a substance that has no effect on pH. The hydrogen ion of the HCl could greatly lower pH by making the solution more acid, but this H^+ ion combines with the bicarbonate ion (HCO_3^-) of sodium bicarbonate to form carbonic acid, a weak acid that lowers pH only slightly. In other words, because of the action of the weak base of the buffer system, the strong acid (HCl) has been replaced by a weak acid and a salt, and the pH remains relatively constant.

Now suppose a strong base, such as sodium hydroxide (NaOH), is added to the extracellular fluid. In this instance, the weak acid of the buffer system goes to work and the following base-buffering reaction take place:

$$NaOH + H_2CO_3 \rightleftharpoons H_2O + NaHCO_3$$

Sodium hydroxide (strong base)   Carbonic acid (weak acid of buffer system)   Water   Sodium bicarbonate (weak base)

In this reaction, the OH^- ion of sodium hydroxide could greatly raise the pH of the solution by making it more alkaline. However, the OH^- ion combines with an H^+ ion of carbonic acid and forms water, a substance that has no effect on pH. In addition, the Na^+ ion of sodium hydroxide combines with the bicarbonate ion (HCO_3^-) to form sodium bicarbonate, a weak base that has little effect on pH. Thus, because of the action of the buffer system, the strong base is replaced by water and a weak base, and the pH remains relatively constant.

Whenever a buffering reaction occurs, the concentration of one member of the buffer pair is increased while the concentration of the other decreases. When a strong acid is buffered, for example, the concentration of carbonic acid is increased, but the concentration of sodium bicarbonate is decreased. This happens because carbonic acid is produced and sodium bicarbonate is used up in the acid-buffering reaction. When a strong base is buffered, the concentration of sodium bicarbonate is increased, but the concentration of carbonic acid is decreased because sodium bicarbonate is produced and carbonic acid is used up in the base-buffering reaction. When the buffered substances, HCl and NaOH in this case, are removed from the body via the lungs or kidneys, the carbonic acid and sodium bicarbonate formed as products of the reactions function again as components of the buffer pair. Do you now understand why buffers are sometimes called "chemical sponges"?

ORGANIC COMPOUNDS

In addition to carbon, the most frequently found elements in organic compounds are hydrogen (which can form one bond), oxygen (two bonds), and nitrogen (three bonds). Sulfur (two bonds) and phosphorus (five bonds) appear less often. Other elements are found, but only in a relatively few organic compounds. Carbon has several properties that make it particularly useful to living organisms. For one thing, it can react with one to several hundred other carbon atoms to form large molecules of many different shapes. This means that the body can build many compounds out of carbon, hydrogen, and oxygen. Each compound can be especially suited for a particular structure or function. The relatively large size of most carbon-containing molecules and the fact that they do not dissolve easily in water make them useful materials for building body structures. Carbon compounds are mostly or entirely held together by covalent bonds and tend to decompose easily. This means that organic compounds are also a good source of energy. Ionic compounds are not good energy sources because they form new ionic bonds as soon as the old ones are broken.

Carbohydrates

A large and diverse group of organic compounds found in the body is the **carbohydrates,** also known as sugars and starches. The carbohydrates perform a number of major functions in living systems. A few even form structural units. For instance, one type of sugar (deoxyribose) is a building block of DNA, the molecule that carries hereditary information. Some carbohydrates are converted to proteins and to fats or fatlike substances, which are used to build structures and provide an emergency source of energy. Other carbohydrates function as food reserves. One example is glycogen, which is stored in the liver and skeletal muscles. The principal function of carbohydrates, however, is to provide the most readily available source of energy to sustain life.

Carbon, hydrogen, and oxygen are the elements found in carbohydrates. The ratio of hydrogen to oxygen atoms is typically 2 to 1, the same as in water. This ratio can be seen in the formulas for carbohydrates such as ribose ($C_5H_{10}O_5$), glucose ($C_6H_{12}O_6$), and sucrose ($C_{12}H_{22}O_{11}$). Although there are exceptions, the general formula for carbohydrates is $(CH_2O)_n$, where n symbolizes three or more CH_2O units. Carbohydrates can be divided into three major groups on the basis of size: monosaccharides, disaccharides, and polysaccharides.

1. Monosaccharides. Monosaccharides (mon-ō-SAK-a-rīds), or simple sugars, are compounds containing from three to seven carbon atoms. Simple sugars with three carbons in the molecule are called trioses. The number of carbon atoms in the molecule is indicated by the prefix *tri.* There are also tetroses (four-carbon sugars), pentoses (five-carbon sugars), hexoses (six-carbon sugars), and heptoses (seven-carbon sugars). Pentoses and hexoses are exceedingly important to the human organism. The pentose called deoxyribose is a component of genes. The hexose called glucose is the main energy-supplying molecule of the body.

2. Disaccharides. A second group of carbohydrates, the **disaccharides** (dī-SAK-a-rīds), consists of two monosaccharides joined chemically. In the process of disaccharide formation, two monosaccharides combine to form a disaccharide molecule and a molecule of water is lost. This reaction is known as **dehydration synthesis** (*dehydration* = loss of water). The following reaction shows disaccharide formation. Molecules of the monosaccharides glucose and fructose combine to form a molecule of the disaccharide sucrose (table sugar):

$$C_6H_{12}O_6 \ + \ C_6H_{12}O_6 \ \rightarrow C_{12}H_{22}O_{11} + H_2O$$
Glucose　　　　Fructose　　　　Sucrose　　Water
(monosaccharide)　(monosaccharide)　(disaccharide)

You may be puzzled to see that glucose and fructose have the same chemical formulas. Actually, they are different monosaccharides, since the relative positions of the oxygens and carbons vary in the two different molecules (see Figure 2-10). The formula for sucrose is $C_{12}H_{22}O_{11}$ and not $C_{12}H_{24}O_{12}$, since a molecule of H_2O is lost in the process of disaccharide formation. In every dehydration synthesis a molecule of water is lost. Along with this water loss, there is the synthesis of two small molecules, such as glucose and fructose, into one large, more complex molecule, such as sucrose (Figure 2-10). Similarly, the dehydration synthesis of the two monosaccharides glucose and galactose forms the disaccharide lactose (milk sugar).

Disaccharides can also be broken down into smaller, simpler molecules by adding water. This reverse chemical reaction is called **digestion (hydrolysis),** which means to split by using water. A molecule of sucrose, for example, may be digested into its components of glucose and fructose by the addition of water. The mechanism of this reaction also is represented in Figure 2-10.

CLINICAL APPLICATION

In recent years, **artificial sweeteners** have been used in soft drinks and foods and for table use. Aspartame (as-PAR-tām) is now used in many such products as a replacement for saccharin. Aspartame (NutraSweet

Glucose   Fructose   Sucrose   Water

FIGURE 2-10 Dehydration synthesis and hydrolysis of a molecule of sucrose. In the dehydration synthesis reaction (read from left to right), the two smaller molecules, glucose and fructose, are joined to form a larger molecule of sucrose. Note the loss of a water molecule. In hydrolysis (read from right to left), the larger sucrose molecule is broken down into the two smaller molecules, glucose and fructose. Here, a molecule of water is added to sucrose for the reaction to occur.

or Equal) is about 180 times sweeter than sucrose and reportedly lacks the bitter aftertaste of saccharin. Because of the small amounts needed, aspartame adds very few calories to drinks and foods, less than 0.5 calories for the equivalent of 1 teaspoon of sugar. It also poses no risk for the development of dental caries, but some evidence suggests it can cause an allergic reaction (hives) in some people.

3. Polysaccharides. The third major group of carbohydrates, the **polysaccharides** (pol'-ē-SAK-a-rīds), consists of several monosaccharides joined together through dehydration synthesis. Polysaccharides have the formula $(C_6H_{10}O_5)_n$. Like disaccharides, polysaccharides can be broken into their constituent sugars through hydrolysis reactions. Unlike monosaccharides or disaccharides, however, they usually lack the characteristic sweetness of sugars like fructose or sucrose and are usually not soluble in water. One of the chief polysaccharides is glycogen.

Lipids

A second group of organic compounds that is vital to the human organism is the **lipids.** Like carbohydrates, lipids are composed of carbon, hydrogen, and oxygen, but they do not have a 2:1 ratio of hydrogen to oxygen. In addition, the amount of oxygen in fats is usually less than that in carbohydrates. Most lipids are insoluble in water, but they readily dissolve in solvents such as alcohol, chloroform, and ether. Among the groups of lipids are fats, phospholipids (lipids that contain phosphorous), steroids, carotenes, vitamins E and K, and prostaglandins. A listing of the various types of lipids is shown in Exhibit 2-4, along with their relationships to the human organism. Since lipids are a large and diverse group of compounds, we will discuss only two types in detail at this point: fats and prostaglandins.

A molecule of **fat** (triglyceride) consists of two basic components: **glycerol** and **fatty acids** (Figure 2-11). A single molecule of fat is formed when a molecule of glycerol combines with three molecules of fatty acids. This reaction, like the one described for disaccharide formation, is a dehydration synthesis reaction. During hy-

drolysis, a single molecule of fat is broken down into fatty acids and glycerol.

In subsequent chapters, we will be talking about saturated, unsaturated, and polyunsaturated fats. These terms have the following meanings. A *saturated fat* contains no double bonds between any of its carbon atoms. It contains only single covalent bonds between its carbon atoms, and all the carbon atoms are bonded to the maximum number of hydrogen atoms; thus, such a fat is saturated with hydrogen atoms. Saturated fats (and some cholesterol) occur mostly in animal foods such as beef, pork, butter, whole milk, eggs, and cheese. They also occur in some plant products such as cocoa butter, palm oil, and coconut oil. The liver uses some breakdown products of saturated fats to produce cholesterol. An *unsaturated fat* contains one or more double covalent bonds between its carbon atoms; it is not completely saturated with hydrogen atoms. Examples are olive oil and peanut oil and they have no significant effect on cholesterol level. A *polyunsaturated fat* contains two or more double covalent bonds between its carbon atoms. Corn oil, safflower oil, sunflower oil, cottonseed oil, sesame oil, and soybean oil are examples of polyunsaturated fats that researchers believe help to reduce cholesterol in the blood.

Fats represent the body's most highly concentrated source of energy. They provide more than twice as much energy per weight as either carbohydrates or proteins. In general, however, fats are about 10 to 12 percent less efficient as body fuels than are carbohydrates. A great amount of the fat calorie is wasted and thus not available for the body to use.

Prostaglandins (pros'-ta-GLAN-dins), also called **PG**, are a large group of membrane-associated lipids composed of 20-carbon fatty acids containing 5 carbon atoms joined to form a ring (cyclopentane ring). Prostaglandins are produced in all nucleated cells in the body and are able to influence the functioning of any type of cell.

Prostaglandins are produced in cell membranes and are rapidly decomposed by catabolic enzymes. Although synthesized in minute quantities, they are potent substances and exhibit a wide variety of effects on the body. Basically, prostaglandins mimic hormones. They are involved in the modulation of many hormonal responses (Chapter 18), inducing menstruation, and inducing sec-

EXHIBIT 2-4   RELATIONSHIPS OF REPRESENTATIVE LIPIDS TO THE HUMAN ORGANISM

LIPIDS	RELATIONSHIP
FATS	Protection, insulation, source of energy.
PHOSPHOLIPIDS	
Lecithin	Major lipid component of cell membranes; constituent of plasma.
Cephalin and sphingomyelin	Found in high concentrations in nerves and brain tissue.
STEROIDS	
Cholesterol	Constituent of all animal cells, blood, and nervous tissue; suspected relationship to heart disease and atherosclerosis; precursor of bile salts, vitamin D, and steroid hormones.
Bile salts	Substances that emulsify or suspend fats before their digestion and absorption; needed for absorption of fat-soluble vitamins (A, D, E, K).
Vitamin D	Produced in skin on exposure to ultraviolet radiation; necessary for bone growth, development, and repair.
Estrogens	Sex hormones produced in large quantities by females.
Androgens	Sex hormones produced in large quantities by males.
OTHER LIPOID SUBSTANCES	
Carotenes	Pigment in egg yolk, carrots, and tomatoes; vitamin A is formed from carotenes; retinene, formed from vitamin A, is a photoreceptor in the retina of the eye.
Vitamin E	May promote wound healing, prevent scarring, and contribute to the normal structure and functioning of the nervous system; deficiency causes sterility in rats and muscular dystrophy in monkeys; deficiency in humans believed to cause the oxidation of certain fats (unsaturated) resulting in abnormal structure and function of certain parts of cells (mitochondria, lysosomes, and plasma membranes); may reduce the severity of visual loss associated with retrolental fibroplasia (eye disease in premature infants caused by too much oxygen in incubators) by functioning as an antioxidant.
Vitamin K	Promotes blood clotting to prevent excessive bleeding.
Prostaglandins	Membrane-associated lipids that stimulate uterine contractions, induce labor and abortions, regulate blood pressure, transmit nerve impulses, regulate metabolism, regulate stomach secretions, inhibit lipid breakdown, and regulate muscular contractions of the gastrointestinal tract; mediate hormones.

ond-trimester abortions. They are also involved in the inflammatory response (Chapter 22), preventing peptic ulcers, opening bronchial and nasal passages, platelet aggregation and inhibition of aggregation, and regulating body temperature.

Proteins

A third principal group of organic compounds is **proteins.** These compounds are much more complex in structure than the carbohydrates or lipids. They are also responsible

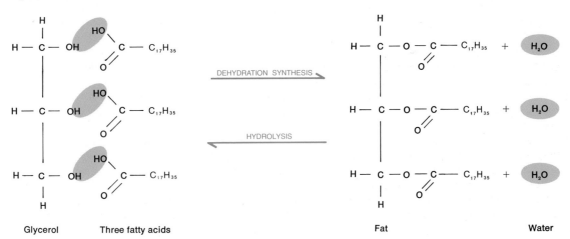

FIGURE 2-11 Dehydration synthesis and hydrolysis of a fat. In the dehydration synthesis reaction (read from left to right), one molecule of glycerol combines with three fatty acid molecules, resulting in a loss of three molecules of water. In hydrolysis (read from right to left), a molecule of fat is broken down into a single molecule of glycerol and three fatty acid molecules when three molecules of water are added. The fatty acid shown is stearic acid, a component of corn oil, coconut oil, beef fat, and pork fat.

for much of the structure of body cells and are related to many physiological activities. For example, proteins in the form of enzymes speed up many essential biochemical reactions. This is described in detail in Chapter 25. Other proteins assume a necessary role in muscular contraction. Antibodies are proteins that provide the human organism with defenses against invading microbes. And some hormones that regulate body functions are also proteins. A classification of proteins on the basis of function is shown in Exhibit 2-5.

Chemically, proteins always contain carbon, hydrogen, oxygen, and nitrogen. Many proteins also contain sulfur and phosphorus. Just as monosaccharides are the building units of sugars, and fatty acids and glycerol are the building units of fats, **amino acids** are the building blocks of proteins. In protein formation, amino acids combine to form more complex molecules, while water molecules are lost. The process is a dehydration synthesis reaction, and the bonds formed between amino acids are called *peptide bonds* (Figure 2-12).

When two amino acids combine, a **dipeptide** results. Adding another amino acid to a dipeptide produces a **tripeptide.** Further additions of amino acids result in the formation of **polypeptides,** which are large protein molecules. At least 20 different amino acids are found in proteins. Although all amino acids have the same basic composition, each is distinguished on the basis of additional atoms or groups of atoms arranged in a specific way. A great variety of proteins is possible because each variation in the number or sequence of amino acids can produce a different protein. The situation is similar to using an alphabet of 20 letters to form words. Each letter could be compared to a different amino acid, and each word would be a different protein.

Proteins have four levels of structural organization. The *primary structure* is the sequence of amino acids making up the protein. An alteration in primary structure can have serious consequences. For example, a single substitution of an amino acid in a blood protein can result in a deformed hemoglobin molecule that produces sickle-cell anemia. The *secondary structure* of a protein is its coiling or zigzag arrangement along two dimensions, such as clockwise spirals and pleated sheets. The *tertiary structure* refers to the bending or folding of a protein into a three-dimensional shape. The *quaternary structure* of a protein refers to two or more tertiary patterns bonded to each other.

EXHIBIT 2-5 CLASSIFICATION OF PROTEINS BY FUNCTION

TYPE OF PROTEIN	DESCRIPTION
Structural	Form the structural framework of various parts of the body. Examples: keratin in the skin, hair, and fingernails and collagen in connective tissue (Chapter 5).
Regulatory	Function as hormones and regulate various physiological processes. Example: insulin, which regulates blood sugar level (Chapter 18).
Contractile	Serve as contractile elements in muscle tissue. Examples: myosin and actin (Chapter 10).
Immunological	Serve as antibodies to protect the body against invading microbes. Example: gamma globulin (Chapter 22).
Transport	Transport vital substances throughout the body. Example: hemoglobin, which transports oxygen and carbon dioxide in the blood (Chapter 19).
Catalytic	Act as enzymes and function in regulating biochemical reactions. Examples: salivary amylase, lipase, and lactase (Chapters 24 and 25).

Nucleic Acids: Deoxyribonucleic Acid (DNA) and Ribonucleic Acid (RNA)

Nucleic (noo-KLĒ-ic) **acids,** compounds first discovered in the nuclei of cells, are exceedingly large organic molecules containing carbon, hydrogen, oxygen, nitrogen, and phosphorus. They are divided into two principal kinds: **deoxyribonucleic** (dē-ok'-sē-rī'-bō-noo-KLĒ-ik) **acid (DNA)** and **ribonucleic acid (RNA).**

Whereas the basic structural units of proteins are amino acids, the basic units of nucleic acids are **nucleotides.** A molecule of DNA is a chain composed of repeating nucleotide units. Each nucleotide of DNA consists of three basic parts (Figure 2-13a):

FIGURE 2-12 Protein formation. When two or more amino acids are chemically united, the resulting bond between them is called a peptide bond. In the example, the amino acids glycine and alanine are joined to form the dipeptide glycylalanine. The peptide bond is formed at the point where water is lost.

1. It contains one of four possible *nitrogen bases,* which are ring-shaped structures containing atoms of C, H, O, and N. The nitrogen bases found in DNA are adenine, thymine, cytosine, and guanine. Adenine and guanine are double-ring structures, collectively referred to as *purines* (see Figure 2-15). Thymine and cytosine are smaller, single-ring structures, called *pyrimidines.*

2. It contains a pentose sugar called *deoxyribose.*

3. It also contains *phosphate groups.*

The nucleotides are named according to the nitrogen base that is present. Thus a nucleotide containing thymine is called a *thymine nucleotide.* One containing adenine is called an *adenine nucleotide,* and so on.

The chemical components of the DNA molecule were known before 1900, but it was not until 1953 that a model of the organization of the chemicals was constructed. This model was proposed by J. D. Watson and F. H. C. Crick on the basis of data from many investigations. Figure 2-13b shows the following structural characteristics of the DNA molecule.

1. The molecule consists of two strands with crossbars. The strands twist about each other in the form of a *double helix* so that the shape resembles a twisted ladder. For many years it was assumed that all DNA was in the form of a double helix that twisted smoothly to the right (right-handed DNA). There are at least four varieties of right-handed DNA designated as A-DNA, B-DNA, C-DNA, and D-DNA. Watson and Crick's model, the one most commonly found in cells, is B-DNA. It was discovered, however, that DNA can also twist jaggedly to the left (Z-DNA or left-handed DNA). This alternate form of DNA is located near the ends of genes and may help to explain how genes turn on and off and how some cells may become malignant.

2. The uprights of the DNA ladder consist of alternating phosphate groups and the deoxyribose portions of the nucleotides.

3. The rungs of the ladder contain paired nitrogen bases. As shown, adenine always pairs with thymine and cytosine always pairs with guanine.

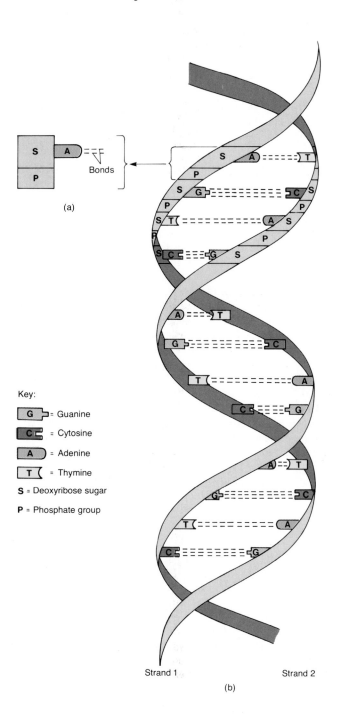

Key:

G = Guanine

C = Cytosine

A = Adenine

T = Thymine

S = Deoxyribose sugar

P = Phosphate group

Strand 1   Strand 2

(b)

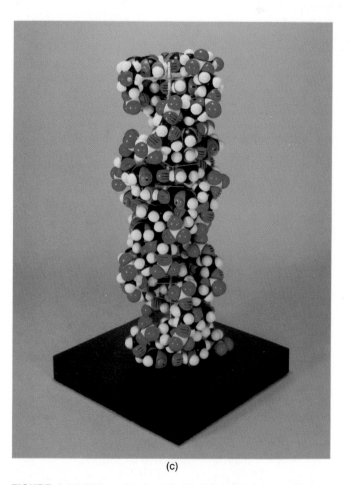

(c)

FIGURE 2-13 DNA molecule. (a) Adenine nucleotide. (b) Portion of an assembled DNA molecule. (c) Space-filling (three-dimensional) model to show relative size and location of atoms. (Courtesy of Ealing Corporation.)

ADENOSINE DIPHOSPHATE (ADP)

ADENOSINE TRIPHOSPHATE (ATP)

FIGURE 2-14 Structure of ATP and ADP. The high-energy bonds are indicated by $\sim$.

Cells contain hereditary material called *genes,* each of which is a segment of a DNA molecule. Our genes determine which traits we inherit, and they control all the activities that take place in our cells throughout a lifetime. When a cell divides, its hereditary information is passed on to the next generation of cells. The passing of information is possible because of DNA's unique structure.

RNA, the second principal kind of nucleic acid, differs from DNA in several respects. RNA is single-stranded; DNA is double-stranded. The sugar in the RNA nucleotide is the pentose ribose. And RNA does not contain the nitrogen base thymine. Instead of thymine, RNA has the nitrogen base uracil. At least three different kinds of RNA have been identified in cells. Each type has a specific role to perform with DNA in protein synthesis reactions (Chapter 3).

Adenosine Triphosphate (ATP)

A molecule that is indispensable to the life of the cell is **adenosine** (a-DEN-ō-sēn) **triphosphate (ATP).** This substance is found universally in living systems and performs the essential function of storing energy for various cellular activities. Structurally, ATP consists of three phosphate groups (PO_4^{3-}) and an adenosine unit composed of adenine and the five-carbon sugar ribose (Figure 2-14). ATP is regarded as a high-energy molecule because of the total amount of usable energy it releases when it is broken down by the addition of a water molecule (hydrolysis).

When the terminal phosphate group (here symbolized by P) is hydrolyzed, the reaction liberates a great deal of energy. This energy is used by the cell to perform its basic activities. Removal of the terminal phosphate group leaves a molecule called **adenosine diphosphate (ADP).** This reaction may be represented as follows.

$$\text{ATP} \quad \rightleftharpoons \quad \text{ADP} \quad + \quad \text{P} \quad + \quad \text{E}$$

Adenosine Adenosine Phosphate Energy
triphosphate diphosphate

The energy supplied by the catabolism of ATP into ADP is constantly being used by the cell. Since the supply of

ATP at any given time is limited, a mechanism exists to replenish it—a phosphate group is added to ADP to manufacture more ATP. The reaction may be represented as follows:

$$\text{ADP} \quad + \quad \text{P} \quad + \quad \text{E} \quad \rightleftharpoons \quad \text{ATP}$$

Adenosine Phosphate Energy Adenosine
diphosphate   triphosphate

Logically, energy is required to manufacture ATP. The energy required to attach a phosphate group to ADP is supplied by various decomposition reactions taking place in the cell, particularly by the decomposition of glucose. Glucose is completely metabolized in cells into carbon dioxide and water, and the energy released in this process is used to attach a phosphate group to ADP to resynthesize ATP. ATP can be stored in every cell, where it provides potential energy that is not released until needed.

Cyclic Adenosine-3′,5′-Monophosphate (Cyclic AMP)

A chemical substance closely related to ATP is **cyclic adenosine-3′,5′-monophosphate (cyclic AMP).** Essentially it is a molecule of adenosine monophosphate with the

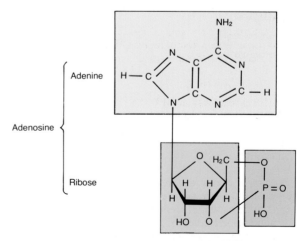

FIGURE 2-15 Structure of cyclic AMP.

phosphate attached to the ribose sugar at two places (Figure 2-15). The attachment forms a ring-shaped structure and thus the name cyclic AMP.

Cyclic AMP is formed from ATP by the action of a special enzyme, called *adenylate cyclase,* located in the cell membrane. Although cyclic AMP was discovered in 1958, only recently has its function in cells become clear. One function is related to the action of hormones, a topic we explore in detail in Chapter 18.

STUDY OUTLINE

Introduction to Basic Chemistry (p. 25)

Chemical Elements (*p. 25*)
1. Matter is anything that occupies space and has mass. It is made up of building units called chemical elements.
2. Oxygen, carbon, hydrogen, and nitrogen make up 96 percent of body weight. These elements together with calcium and phosphorus make up 99 percent of total body weight.

Structure of Atoms (*p. 25*)
1. Units of matter of all chemical elements are called atoms.
2. Atoms consist of a nucleus, which contains protons and neutrons (nucleons), and electrons that move about the nucleus in energy levels.
3. The total number of protons in an atom is its atomic number. This number is equal to the number of electrons in the atom.
4. Magnetic resonance imaging (MRI) is based on the reaction of atomic nuclei to magnetism.
5. MRI can identify existing pathologies, evaluate drug therapy, measure metabolism, and assess the potential for certain diseases.

Atoms and Molecules (*p. 26*)
1. The electrons are the part of an atom that actively participate in chemical reactions.
2. A molecule is the smallest unit of two or more combined atoms. A molecule containing two or more different kinds of atoms is a compound.
3. In an ionic bond, outer-energy-level electrons are transferred from one atom to another. The transfer forms ions, whose unlike charges attract each other and form ionic bonds.
4. In a covalent bond, there is a sharing of pairs of outer-energy-level electrons.
5. Hydrogen bonding provides temporary bonding between certain atoms within large complex molecules such as proteins and nucleic acids.

Radioisotopes (*p. 30*)
1. Radioisotopes emit radiation that can be detected by instruments.
2. A sophisticated version of radioisotopic scanning is called positron emission tomography (PET).
3. PET scans provide images of chemical activity in tissues.

Chemical Reactions (*p. 31*)
1. Synthesis reactions involve the combination of reactants to produce a new molecule. The reactions are anabolic: bonds are formed.
2. In decomposition reactions, a substance breaks down into other substances. The reactions are catabolic: bonds are broken.
3. Exchange reactions involve the replacement of one atom or atoms by another atom or atoms.
4. In reversible reactions, end products can revert to the original combining molecules.
5. The sum of all synthetic and decomposition reactions that occur within an organism is referred to as metabolism.

6. When chemical bonds are formed, energy is needed. When bonds are broken, energy is released. This is known as chemical bond energy.
7. Other forms of energy include mechanical, radiant, and electrical.

Chemical Compounds and Life Processes (p. 33)
1. Inorganic substances usually lack carbon, contain ionic bonds, resist decomposition, and dissolve readily in water.
2. Organic substances always contain carbon and hydrogen. Most organic substances contain covalent bonds and many are insoluble in water.

Inorganic Compounds (*p. 33*)
1. Water is the most abundant substance in the body. It is an excellent solvent and suspending medium, participates in chemical reactions, absorbs and releases heat slowly, and lubricates.
2. Acids, bases, and salts dissociate into ions in water. An acid ionizes into H^+ ions; a base ionizes into OH^- ions. A salt ionizes into neither H^+ nor OH^- ions. Cations are positively charged ions; anions are negatively charged ions.
3. The pH of different parts of the body must remain fairly constant for the body to remain healthy. On the pH scale, 7 represents neutrality. Values below 7 indicate acid solutions, and values above 7 indicate alkaline solutions.
4. The pH values of different parts of the body are maintained by buffer systems, which usually consist of a weak acid and a weak base. Buffer systems eliminate excess H^+ ions and excess OH^- ions in order to maintain pH homeostasis.

Organic Compounds (*p. 37*)
1. Carbohydrates are sugars or starches that provide most of the energy needed for life. They may be monosaccharides, disaccharides, or polysaccharides. Carbohydrates, and other organic molecules, are joined together to form larger molecules with the loss of water by a process called dehydration synthesis. In the reverse process, called digestion (hydrolysis), large molecules are broken down into smaller ones upon the addition of water.
2. Lipids are a diverse group of compounds that includes fats, phospholipids, steroids, carotenes, vitamins E and K, and prostaglandins (PG). Fats protect, insulate, provide energy, and are stored. Prostaglandins mimic the effects of hormones and are involved in the inflammatory response and the modulation of hormonal responses.
3. Proteins are constructed from amino acids. They give structure to the body, regulate processes, provide protection, help muscles to contract, transport substances, and serve as enzymes.
4. Deoxyribonucleic acid (DNA) and ribonucleic acid (RNA) are nucleic acids consisting of nitrogen bases, sugar, and phosphate groups. DNA is a double helix and is the primary chemical in genes. RNA differs in structure and chemical composition from DNA and is mainly concerned with protein synthesis reactions.
5. The principal energy-storing molecule in the body is adeno-

sine triphosphate (ATP). When its energy is liberated, it is decomposed to adenosine diphosphate (ADP). ATP is manufactured from ADP using the energy supplied by various decomposition reactions, particularly of glucose.

6. Cyclic AMP is closely related to ATP and assumes a function in certain hormonal reactions.

REVIEW QUESTIONS

1. What is the relationship of matter to the body?
2. Define a chemical element. List the chemical symbols for 10 different chemical elements. Which chemical elements make up the bulk of the human organism?
3. What is an atom? Diagram the positions of the nucleus, protons, neutrons, and electrons in an atom of oxygen and an atom of nitrogen. What is atomic number?
4. Explain the principle of magnetic resonance imaging (MRI). How is MRI useful in the diagnosis of disease?
5. What is meant by an energy level?
6. How are chemical bonds formed? Distinguish between an ionic bond and a covalent bond. Give at least one example of each.
7. Can you determine how a molecule of $MgCl_2$ is ionically bonded? Magnesium has two electrons in its outer energy level. Construct a diagram to verify your answer.
8. Refer to Figure 2-5b and c. See if you can determine why there is a double covalent bond between atoms in an oxygen molecule (O_2) and a triple covalent bond between atoms in a nitrogen molecule (N_2).
9. Define a hydrogen bond. Why are hydrogen bonds important?
10. Define a radioisotope. Explain the principle of positron emission tomography (PET) scanning. What is the diagnostic importance of PET scanning?
11. What are the four principal kinds of chemical reactions? How are anabolism and catabolism related to synthesis and decomposition reactions, respectively? How is energy related to chemical reactions?
12. Identify what kind of reaction each of the following represents:
    a. $H_2 + Cl_2 \rightarrow 2HCl$
    b. $3\ NaOH + H_3PO_4 \rightarrow Na_3PO_4 + 3\ H_2O$
    c. $CaCO_3 + CO_2 + H_2O \rightarrow Ca(HCO_3)_2$
    d. $HNO_3 \rightarrow H^+ + NO_3^-$
    e. $NH_3 + H_2O \rightleftharpoons NH_4^+ + OH^-$
13. How do inorganic compounds differ from organic compounds? List and define the principal inorganic and organic compounds that are important to the human body.
14. What are the essential functions of water in the body? Distinguish between a solution and a suspension.
15. Define an acid, a base, and a salt. How does the body acquire some of these substances? List some functions of the chemical elements furnished as ions of salts.
16. What is pH? Why is it important to maintain a relatively constant pH? What is the pH scale? If there are 100 OH^- ions at a pH of 8.5, how many OH^- ions are there at a pH of 9.5?
17. List the normal pH values of some common fluids, biological solutions, and foods. Refer to Exhibit 2-3, and select the two substances whose pH values are closest to neutrality. Is the pH of milk or of cerebrospinal fluid closer to 7? Is the pH of bile or of urine farther from neutrality?
18. What are the components of a buffer system? What is the function of a buffer pair? Diagram and explain how the carbonic acid–bicarbonate buffer system of extracellular fluid maintains a constant pH even in the presence of a strong acid or strong base. How is this an example of homeostasis?
19. Why are the reactions of buffer pairs more important with strong acids and bases than with weak acids and bases?
20. Define a carbohydrate. Why are carbohydrates essential to the body? How are carbohydrates classified?
21. Compare dehydration synthesis and hydrolysis. Why are they significant?
22. How do lipids differ from carbohydrates? What are some relationships of lipids to the body?
23. Define a prostaglandin (PG). List some physiological effects of prostaglandins.
24. Define a protein. What is a peptide bond? Discuss the classification of proteins on the basis of function.
25. What is a nucleic acid? How do deoxyribonucleic acid (DNA) and ribonucleic acid (RNA) differ with regard to chemical composition, structure, and function?
26. What is adenosine triphosphate (ATP)? What is the essential function of ATP in the human body? How is this function accomplished?
27. How is cyclic AMP related to ATP? What is the function of cyclic AMP?

The Cellular
Level of Organization

3

- Define a cell and list its generalized parts.
- Explain the chemistry, structure, and functions of the plasma membrane.
- Describe how materials move across plasma membranes by diffusion, facilitated diffusion, osmosis, filtration, dialysis, active transport, and endocytosis.
- Describe the chemical composition and functions of cytoplasm.
- Describe the structure and functions of a cell nucleus.
- Define the structure and function of ribosomes.
- Distinguish between agranular and granular endoplasmic reticulum (ER) with regard to structure and function.
- Describe the structure and functions of the Golgi complex.
- Discuss the structure and function of mitochondria as ''powerhouses of the cell.''
- Explain the structure and function of lysosomes.
- Discuss the structure and role of peroxisomes.
- Distinguish between the structure and function of microfilaments, microtubules, and intermediate filaments as components of the cytoskeleton.
- Discuss the structure and function of centrioles in cellular reproduction.
- Differentiate between cilia and flagella.
- Define a cell inclusion and give several examples.
- Define extracellular material and give several examples.
- Define a gene and explain the sequence of events involved in protein synthesis.
- Describe the principle and importance of genetic engineering.
- Describe the principal events of interphase, the stage between cell divisions.
- Discuss the stages, events, and significance of somatic and reproductive cell division.
- Describe cancer as a homeostatic imbalance of cells.
- Explain the relationship of aging to cells.
- Define medical terminology associated with cells.

The study of the body at the cellular level of organization is important because activities essential to life occur in cells and disease processes originate there. A **cell** may be defined as the basic, living, structural and functional unit of the body and, in fact, of all organisms. **Cytology** is the branch of science concerned with the study of cells. This chapter concentrates on the structure, functions, and reproduction of cells.

A series of illustrations accompanies each cell structure that you study. A diagram of a generalized animal cell shows the location of the structure within the cell. An electron micrograph shows the actual appearance of the structure.* A diagram of the electron micrograph clarifies some of the small details by exaggerating their outlines. Finally, an enlarged diagram of the structure shows its details.

CLINICAL APPLICATION

Despite the enormous advantages provided by electron microscopy in helping scientists understand detailed cellular structure, one drawback is the fact that specimens must be killed, since they are placed in a vacuum. This problem is now overcome by a procedure called **microtomography** (mī-krō-tō-MOG-ra-fē), in which the principles of electron microscopy and computed tomography (CT) scanning are combined. Microtomography produces high-magnification, three-dimensional images of *living* cells.

The potential applications of microtomography include studying how normal and cancerous cells move, grow, and reproduce. It will also be used to follow the changes that occur as an embryo progresses through its developmental sequences. The technique will also enable scientists to observe the microscopic effects of drugs on living cells and how cancer-causing substances alter living cells.

GENERALIZED ANIMAL CELL

A **generalized animal cell** is a composite of many different cells in the body. Examine the generalized cell illustrated in Figure 3-1, but keep in mind that no such single cell actually exists.

For convenience, we can divide the generalized cell into four principal parts:

1. **Plasma (cell) membrane.** The outer, limiting membrane separating the cell's internal parts from the extracellular fluid and external environment.
2. **Cytoplasm.** The substance between the nucleus and the plasma membrane that surrounds organelles.
3. **Organelles.** Permanent structures with characteristic morphology that are highly specialized for specific cellular activities.
4. **Inclusions.** The secretions and storage products of cells.

* An **electron micrograph (EM)** is a photograph taken with an electron microscope. Some electron microscopes can magnify objects up to 1,000,000 times. In comparison, the light miscroscope that you probably use in your laboratory magnifies objects up to 1,000 times their size.

Extracellular materials, which are substances external to the cell surface, will also be examined in connection with cells.

PLASMA (CELL) MEMBRANE

The exceedingly thin structure that separates one cell from other cells and from the external environment is called the **plasma (cell) membrane** (Figure 3-2a). The membrane ranges from 6.5 to 10 nm (formerly Å) in thickness.

CHEMISTRY AND STRUCTURE

Plasma membranes consist primarily of phospholipids and proteins. Other constituents in lesser amounts include cholesterol, glycolipids (combinations of carbohydrates and lipids), water, carbohydrates, and ions. Studies of membrane structure suggest a concept regarding the arrangement of the various molecules referred to as the **fluid mosaic model** (Figure 3-2b).

The phospholipid molecules are arranged in two parallel rows, forming a **phospholipid bilayer.** A phospholipid molecule consists of a polar, phosphate-containing "head," that mixes with water and nonpolar fatty acid "tails" that do not mix with water. The molecules are oriented in the bilayer so that "heads" face outward on either side and the "tails" face each other in the membrane's interior. The phospholipid bilayer forms the basic framework of plasma membranes.

The membrane proteins are classified into two categories: integral and peripheral. **Integral proteins** are embedded in the phospholipid bilayer among the fatty acid "tails." Some of the integral proteins lie at or near the inner and outer membrane surfaces; others penetrate the membrane completely. Since the phospholipid bilayer is somewhat fluid and flexible and the integral proteins have been observed moving from one location to another in the membrane, the relationship has been compared to icebergs (proteins) floating in the sea (phospholipid bilayer). The subunits of some integral proteins form minute channels through which substances can be transported into and out of the cell (described shortly). Other integral proteins carry branching chains of carbohydrates. Such combinations of carbohydrate and protein, called *glycoproteins,* provide receptor sites that enable a cell to recognize other cells of its own kind so that they can associate to form a tissue, to recognize and respond to foreign cells that might be potentially dangerous, and to recognize and attach to hormones, nutrients, and other chemicals. Red blood cells also have glycoprotein receptors that prevent them from clumping and producing unwanted blood clots. Diabetes mellitus is a disease believed to be caused by faulty cell receptor sites.

Peripheral proteins are loosely bound to the membrane surface and easily separated from it. Far less is known about them than integral proteins, and their functions are not yet completely understood. For example, some peripheral proteins are believed to serve as enzymes that catalyze cellular reactions. An example is cytochrome

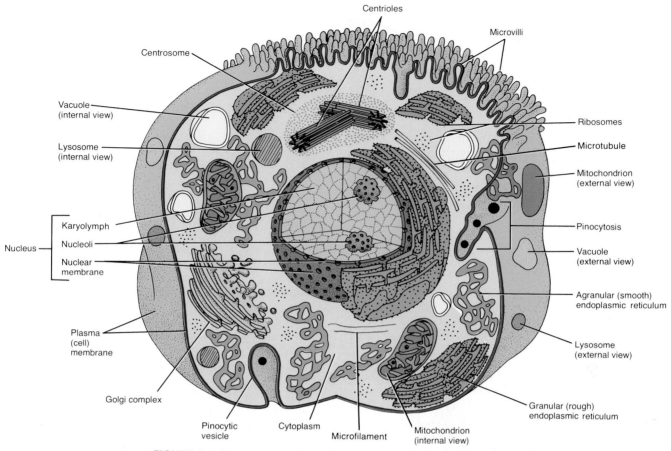

FIGURE 3-1 Generalized animal cell based on electron microscope studies.

c, involved in cellular respiration. Other peripheral proteins, such as spectrin of red blood cells, are believed to have a mechanical function by serving as a scaffolding to support the plasma membrane. It is also believed that peripheral proteins may assume a role in changes in membrane shape during such processes as cell division, locomotion, and ingestion.

FUNCTIONS

Based upon the discussion of the chemistry and structure of the plasma membrane, we can now describe its several important functions. First, the plasma membrane provides a flexible boundary that encloses the cellular contents and separates them from the external environment. Second, the membrane facilitates contact with other body cells or with foreign cells or substances. Third, the membrane provides receptors for chemicals such as hormones, neurotransmitters, enzymes, nutrients, and antibodies. Fourth, the plasma membrane mediates the entrance and exit of materials. The ability of a plasma membrane to permit certain substances to enter and exit, but to restrict the passage of others, is called **selective permeability.** Let us look at this mechanism in a bit more detail.

A membrane is *permeable* to a substance if it allows the free passage of that substance. Although plasma membranes are not actually freely permeable to any substance, they do permit some substances to pass more readily than others. For example, water passes more readily than most other substances. The permeability of a plasma membrane appears to be a function of several factors.

1. Size of molecules. Large molecules cannot pass through the plasma membrane. Water and amino acids are small molecules and can enter and exit the cell easily. However, most proteins, which consist of many amino acids linked together, seem to be too large to pass through the membrane. Many scientists believe that the giant-sized molecules do not enter the cell because they are larger than the minute channels in the integral proteins.

2. Solubility in lipids. Substances that dissolve easily in lipids pass through the membrane more readily than other substances, since a major part of the plasma membrane consists of lipid molecules. Examples of lipid-soluble substances are oxygen, carbon dioxide, and steroid hormones.

3. Charge on ions. The charge of an ion attempting to cross the plasma membrane can determine how easily the ion enters or leaves the cell. The protein portion of the membrane is capable of ionization. If an ion has a charge opposite that of the membrane, it is attracted to the membrane and passes through more readily. If the ion attempting to cross the membrane has the same charge as the membrane, it is repelled by the membrane and its passage is restricted. This phenomenon conforms to the rule of physics that opposite charges attract, whereas like charges repel each other.

4. Presence of carrier molecules. Some integral proteins called carriers are capable of attracting and transporting substances across the membrane regardless of size, ability to dissolve

Plasma membrane
Cytoplasm
Intercellular space
Cytoplasm
Plasma membrane

(a)

Carbohydrates
Channels
Separation of phospholipid bilayer
Peripheral proteins
Integral proteins
Cholesterol
Phospholipid bilayer

(b)

FIGURE 3-2 Plasma membrane. (a) Electron micrograph of portions of two plasma membranes separated by an intercellular space at a magnification of 200,000×. (Copyright © Dr. Donald Fawcett, Science Source/Photo Researchers.) (b) Enlargement of the plasma membrane showing the relationship of the phospholipid bilayer and protein molecules. The separation of the bilayer is for illustrative purposes only.

in lipids, or membrane charge. Their purpose is to modify membrane permeability. The mechanism by which carriers do this will be described shortly.

MOVEMENT OF MATERIALS ACROSS PLASMA MEMBRANES

The mechanisms whereby substances move across a plasma membrane are essential to the life of the cell. Certain substances, for example, must move into the cell to support life, whereas waste materials or harmful substances must be moved out. Plasma membranes mediate the movements of such materials. The processes involved in these movements may be classified as either passive or active. In **passive processes,** substances move across plasma membranes without assistance from the cell. Their movement involves the kinetic energy of individual molecules. The substances move on their own down a concentration gradient from an area where their concentration is high to an area where their concentration is low. The substances may also be forced across the plasma membrane by pressure from an area where the pressure is high to an area where it is low. In **active processes,** the cell contributes energy (ATP) in moving the substance across the membrane since the substance moves against a concentration gradient.

Passive Processes

• *Diffusion*   A passive process called **diffusion** occurs when there is a *net* or greater movement of molecules or ions from a region of higher concentration to a region of lower concentration. The movement from high to low concentration continues until the molecules are evenly distributed. At this point, they move in both directions at an equal rate. This point of even distribution is called *equilibrium.* The difference between high and low concentrations is called the *concentration gradient.* Molecules moving from the high-concentration area to the low-concentration area are said to move *down* or *with* the concentration gradient.

If a dye pellet is placed in a beaker filled with water, the color of the dye is seen immediately around the pellet. At increasing distances from the pellet, the color becomes lighter (Figure 3-3). Later, however, the water solution will be a uniform color. The dye molecules possess kinetic energy (energy of motion) and move about at random. The dye molecules move down the concentration gradient from an area of high dye concentration to an area of low dye concentration. The water molecules also move from their high-concentration to their low-concentration area. When dye molecules and water molecules are evenly distributed among themselves, equilibrium is reached and diffusion ceases, even though molecular movements continue. As another example of diffusion, consider what would happen if you opened a bottle of perfume in a room. The perfume molecules would diffuse until an equilibrium is reached between the perfume molecules and the air molecules in the room.

In the examples cited, no membranes were involved. Diffusion may occur, however, through selectively permeable membranes in the body. Large and small lipid-soluble molecules pass through the phospholipid bilayer of the membrane. A good example of diffusion in the body is the movement of oxygen from the blood into the cells and the movement of carbon dioxide from the cells back into the blood. This is essential in order for body cells to maintain homeostasis. It ensures that cells receive adequate amounts of oxygen and eliminate carbon dioxide as part of their normal metabolism. Small molecules that are not lipid-soluble, such as certain ions (sodium, potassium, chloride), are able to diffuse through channels formed by integral proteins in the membrane.

- *Facilitated Diffusion* Another type of diffusion through a selectively permeable membrane occurs by a process called **facilitated diffusion.** This process is accomplished with the assistance of integral proteins in the membrane that serve as carriers. Although some chemical substances are large molecules and insoluble in lipids, they can still pass through the plasma membrane. Among these are different sugars, especially glucose. In the process of facilitated diffusion, it is believed that glucose is picked up by a carrier. The combined glucose-carrier is soluble in the phospholipid bilayer of the membrane, and the carrier moves the glucose to the inside of the membrane and then inside the cell. The carrier makes the glucose soluble in the phospholipid bilayer of the membrane so it can pass through the membrane. By itself, glucose is insoluble and cannot penetrate the membrane. In the process of facilitated diffusion, the cell does not expend energy, and the substance moves from a region of its higher concentration to a region of its lower concentration.

The rate of facilitated diffusion is considerably faster than that of simple diffusion and depends on (1) the difference in concentration of the substance on either side of the membrane, (2) the amount of carrier available to transport the substance, and (3) how quickly the carrier and substance combine. The process for glucose is greatly accelerated by insulin, a hormone produced by the pancreas. One of insulin's functions is to lower the blood glucose level by accelerating the transportation of glucose from the blood into body cells. This transportation, as we have just seen, is by facilitated diffusion.

- *Osmosis* Another passive process by which materials move across membranes is **osmosis.** It is the net movement of water molecules through a selectively permeable membrane from an area of higher water concentration to an area of lower water concentration. The water molecules pass through channels in integral proteins in the membrane. Once again, a simple apparatus may be used to demonstrate the process. The apparatus shown in Figure 3-4 consists of a tube constructed from cellophane, a selectively permeable membrane. The cellophane tube is filled with a colored, 20 percent sugar (sucrose) solution. Such a solution consists of 20 parts sugar and 80 parts water. The upper portion of the cellophane tube is plugged with a rubber stopper through which a glass tube is fitted. The cellophane tube is placed into a beaker containing distilled (pure) water. Pure water consists of 0 parts sugar and 100 parts water. Initially, the concentrations of water on either side of the selectively permeable membrane are different. There is a lower concentration of water inside the cellophane tube than outside. Because of this difference, water moves from the beaker into the cellophane tube. The passage of water through a selectively permeable membrane generates an osmotic pressure. **Osmotic pressure** is the pressure required to prevent the movement of pure water into a solution containing solutes when the solutions are separated by a selectively permeable membrane. The greater the solute concentration of the solution, the greater its osmotic pressure. There is no movement of sugar from the cellophane tube inside the beaker, since the cellophane is impermeable to molecules of sugar—sugar molecules are too large to go through the pores of the membrane. As water moves into the

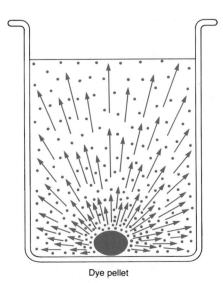

Dye pellet

FIGURE 3-3 Principle of diffusion. Molecules of dye (solute) in a beaker of water (solvent) move down the concentration gradient from a region of high concentration to a region of low concentration.

Glass tube
Rubber stopper
Sugar (sucrose) molecules
Selectively permeable membrane
Water molecules

FH (osmotic pressure)

(a)   (b)

FIGURE 3-4 Principle of osmosis. (a) Apparatus at the start of the experiment. (b) Apparatus at equilibrium. In (a), the cellophane tube contains a 20 percent sugar solution and is immersed in a beaker of distilled water. The 20 percent sugar solution contains 20 parts sugar and 80 parts water, whereas the distilled water contains 0 parts sugar and 100 parts water. The arrows indicate that water molecules can pass freely into the tube, but that sugar (sucrose) molecules are held back by the selectively permeable membrane. As water moves into the tube by osmosis, the sugar solution is diluted and the volume of the solution in the cellophane tube increases. This increased volume is shown in (b), with the sugar solution moving up the glass tubing. The final height reached (FH) occurs at equilibrium and represents the osmotic pressure. At this point, the number of water molecules leaving the cellophane tube is equal to the number of water molecules entering the tube.

cellophane tube, the sugar solution becomes increasingly diluted and the increased volume forces the mixture up the glass tubing. In time, the weight of the water that has accumulated in the cellophane tube and the glass tube will be equal to the osmotic pressure, and this keeps the water volume from increasing in the tube. When water molecules leave and enter the cellophane tube at the same rate, equilibrium is reached. Osmotic pressure is an important force in the movement of water between various compartments of the body.

● *Isotonic, Hypotonic, and Hypertonic Solutions* Osmosis may also be understood by considering the effects of different water concentrations on red blood cells. If the normal shape of a red blood cell is to be maintained, the cell must be placed in an **isotonic solution.** This is a solution in which the total concentrations of water molecules and solute molecules are the same on both sides of the selectively permeable cell membrane. The concentrations of water and solute in the extracellular fluid outside the red blood cell must be the same as the concentration of the intracellular fluid. Under ordinary circumstances, a 0.85 percent NaCl solution is isotonic for red blood cells. In this condition, water molecules enter and exit the cell at the same rate, allowing the cell to maintain its normal shape.

A different situation results if red blood cells are placed in a solution that has a lower concentration of solutes and, therefore, a higher concentration of water. This is called a **hypotonic solution.** In this condition, water molecules enter the cells faster than they can leave, causing the red blood cells to swell and eventually burst. The rupture of red blood cells in this manner is called **hemolysis** (hē-MOL-i-sis) or **laking.** Distilled water is a strongly hypotonic solution.

A **hypertonic solution** has a higher concentration of solutes and a lower concentration of water than the red blood cells. One example of a hypertonic solution is a 10 percent NaCl solution. In such a solution, water molecules move out of the cells faster than they can enter.

This situation causes the cells to shrink. The shrinkage of red blood cells in this manner is called **crenation** (krē-NĀ-shun). Red blood cells may be greatly impaired or destroyed if placed in solutions that deviate significantly from the isotonic state.

● *Filtration* A third passive process involved in moving materials in and out of cells is **filtration.** This process involves the movement of solvents such as water and dissolved substances such as sugar across a selectively permeable membrane by gravity or mechanical pressure, usually hydrostatic (water) pressure. Such a movement is always from an area of higher pressure to an area of lower pressure and continues as long as a pressure difference exists. Most small- to medium-sized molecules can be forced through a cell membrane.

An example of filtration occurs in the kidneys, where the blood pressure supplied by the heart forces water and small molecules like urea through thin cell membranes of tiny blood vessels and into the kidney tubules. In this basic process, protein molecules are retained by the body since they are too large to be forced through the cell membranes of the blood vessels. The molecules of harmful substances such as urea are small enough to be forced through and eliminated, however.

● *Dialysis* The final passive process to be considered is **dialysis.** Dialysis is the diffusion of solute particles across a selectively permeable membrane and involves the separation of small molecules from large molecules. Suppose a solution containing molecules of various sizes is placed in a sac that is permeable only to the smaller molecules. The sac is then placed in a beaker of distilled water. Eventually, the smaller molecules will move from the sac into the water in the beaker and the larger molecules will be left behind.

The principle of dialysis is employed in artificial kidney machines. It probably does not occur within the human body. The patient's blood is passed across a dialysis membrane outside the body. The dialysis membrane takes the place of the kidneys. As the blood moves across the membrane, small-particle waste products pass from the blood into a solution surrounding the dialysis membrane. At the same time, certain nutrients can be passed from the solution into the blood. The blood is then returned to the body.

Active Processes

When cells actively participate in moving substances across membranes, they must expend energy. Cells can even move substances against a concentration gradient. The active processes considered here are active transport and endocytosis (phagocytosis, pinocytosis, and receptor-mediated endocytosis).

• *Active Transport*   The process by which substances are transported across plasma membranes typically from an area of low concentration to an area of high concentration is called **active transport.** In order to transport a substance against a concentration gradient, the membrane uses energy supplied by ATP. In fact, a typical body cell probably expends up to 40 percent of its ATP for active transport. Although the molecular events involved in active transport are not completely understood, integral proteins in the plasma membrane do assume a role.

As just noted, glucose can be transported across cell membranes via facilitated diffusion from areas of high to low concentration. Glucose can also be moved by the cells lining the digestive tract from the cavity of the tract into the blood, even though blood concentration of glucose is higher. This movement involves active transport. One proposed mechanism is that glucose (or other substance) enters a channel in an integral membrane protein (Figure 3-5). When the glucose molecule makes contact with an active site in the channel, the energy from ATP induces a change in the membrane protein that expels the glucose on the opposite side of the membrane. As you will see later, kidney cells also have the ability to actively transport glucose back into the blood so that it is not lost in the urine.

Active transport is also an important process in maintaining the concentrations of some ions inside body cells and other ions outside body cells. For example, before a nerve cell can conduct an impulse, the concentration of potassium (K^+) ions must be considerably higher inside the cell than outside, even when the intracellular K^+ concentration is higher than the extracellular concentration. This is accomplished by a "potassium pump," which consists of an integral protein serving as the "pump," powered by ATP (see Figure 12-7a). Nerve impulse conduction also requires that nerve cells have a higher concentration of sodium (Na^+) ions outside than inside, even when the extracellular Na^+ concentration is higher than the intracellular concentration. This is accomplished by a "sodium pump," whose operation is coupled with that of the "potassium pump." The "sodium pump" is also an integral protein powered by ATP (see Figure 12-7a).

• *Endocytosis*   Large molecules and particles pass through plasma membranes by a process called **endocyto-**

FIGURE 3-5 Proposed mechanism of active transport. The number ① indicates an earlier stage of the process; the number ② indicates a later stage.

sis, in which a segment of the plasma membrane surrounds the substance, encloses it, and brings it into the cell. The export of substances from the cell by the reverse process is called **exocytosis,** a very important mechanism for secretory cells. There are three basic kinds of endocytosis: phagocytosis, pinocytosis, and receptor-mediated endocytosis.

In **phagocytosis** (fag'-ō-sī-TŌ-sis) or "cell eating," projections of cytoplasm, called *pseudopodia* (soo'-dō-PŌ-dē-a), engulf large solid particles external to the cell (Figure 3-6a,b). Once the particle is surrounded, the membrane folds inwardly, forming a membrane sac around the particle. This newly formed sac, called a *phagocytic vesicle,* breaks off from the outer cell membrane, and the solid material inside the vesicle, is digested. Indigestible particles and cell products are removed from the cell by a reverse phagocytosis. Phagocytosis is important because molecules and particles of material that would normally be restricted from crossing the plasma membrane because of their large size can be brought into or removed from the cell. The phagocytic white blood cells of the body constitute a vital defense mechanism. Through phagocytosis, the white blood cells engulf and destroy bacteria and other foreign substances.

In **pinocytosis** (pi'-nō-sī-TŌ-sis), or "cell drinking," the engulfed material consists of an extracellular liquid rather than a solid (Figure 3-6c). Moreover, no cytoplasmic projections are formed. Instead, a minute droplet of liquid is attracted to the surface of the membrane. The membrane folds inwardly, forms a pinocytic vesicle that surrounds the liquid, and detaches from the rest of the intact membrane. Few cells are capable of phagocytosis, but many cells carry on pinocytosis. Examples include cells in the kidneys and urinary bladder.

Receptor-mediated endocytosis is a highly selective process in which cells can take up large molecules or particles. This process is accomplished as follows. The

(a)

Pseudopodium   White blood cell

Pseudopodium   Microbe

(b)

Small particle in liquid droplet

Plasma membrane

Plasma membrane

Pinocytic vesicle

(c)

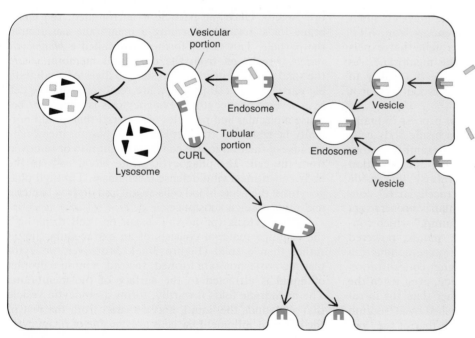

Vesicular portion

Endosome

Vesicle

Endosome

Vesicle

Lysosome

CURL

Tubular portion

Key:

▭ Intact ligand
⊔ Protein receptor
▶ Digestive enzyme
▢ Degraded ligand

(d)

FIGURE 3-6 Endocytosis. (a) Diagram of phagocytosis. (b) Color-enhanced photomicrographs of phagocytosis. The photomicrograph on the left shows a human white blood cell (neutrophil) engulfing a microbe, and the photomicrograph on the right shows a later stage of phagocytosis in which the engulfed microbe is being destroyed. (Courtesy of Abbott Laboratories.) (c) Two variations of pinocytosis. In the variation on the left, the ingested substance enters a channel formed by the plasma membrane and becomes enclosed in a pinocytic vesicle at the base of the channel. In the variation on the right, the ingested substance becomes enclosed in a pinocytic vesicle that forms at the surface of the cell and detaches. (d) Receptor-mediated endocytosis.

plasma membrane contains protein receptors that have binding sites for extracellular large molecules or particles. Such extracellular materials are known as *ligands* (Figure 3-6d). The binding between receptor and ligand causes the plasma membrane to invaginate, forming a membrane-bounded *vesicle* around the ligand. As vesicles move inward from the plasma membrane, they fuse with each other and form larger structures called *endosomes*. Each endosome gives rise to an even larger structure called a *CURL* (compartment of uncoupling of receptor and ligand). Within the CURL, the ligands separate from the receptors. The ligands are distributed into the vesicular portion of the CURL and the receptors accumulate in the tubular portion of the CURL. Following this segregation, the vesicular portion fuses with a lysosome where the ligands are broken down by powerful digestive enzymes. The tubular portion recycles receptors to the plasma membrane for reuse.

The various passive and active processes by which substances move across plasma membranes are summarized in Exhibit 3-1.

CYTOPLASM

The substance inside the cell's plasma membrane and external to the nucleus is called **cytoplasm** (Figure 3-7a,b). It is the matrix, or ground substance, in which various cellular components are found. Physically, cytoplasm may be described as a thick, semitransparent, elastic fluid containing suspended particles and a series of minute tubules and filaments that form a cytoskeleton (described shortly).

The cytoskeleton provides support and shape, and is involved in the movement of structures in the cytoplasm and even the entire cell, as occurs in phagocytosis. Chemically, cytoplasm is 75 to 90 percent water plus solid components. Proteins, carbohydrates, lipids, and inorganic substances compose the bulk of the solid components. The inorganic substances and most carbohydrates are soluble in water and are present as a true solution. The majority of organic compounds, however, are found as colloids—particles that remain suspended in the surrounding ground substance. Since the particles of a colloid bear electrical charges that repel each other, they remain suspended and separated from each other.

Functionally, cytoplasm is the substance in which some chemical reactions occur. The cytoplasm receives raw materials from the external environment and converts them into usable energy by decomposition reactions. Cytoplasm is also the site where new substances are synthesized for cellular use. It packages chemicals for transport to other parts of the cell or other cells of the body and facilitates the excretion of waste materials.

ORGANELLES

Despite the myriad chemical activities occurring simultaneously in the cell, there is little interference of one reaction with another. The cell has a system of compartmentalization provided by structures called **organelles.** These structures are specialized portions of the cell with characteristic morphology that assume specific roles in growth, maintenance, repair, and control.

EXHIBIT 3-1 SUMMARY OF PROCESSES BY WHICH SUBSTANCES MOVE ACROSS MEMBRANES

PROCESS	DESCRIPTION
PASSIVE PROCESSES	Substances move on their own down a concentration gradient from an area of higher to lower concentration or pressure; cell does not expend energy.
Diffusion	Net movement of molecules or ions due to their kinetic energy from an area of higher to lower concentration until an equilibrium is reached.
Facilitated diffusion	Diffusion of larger molecules across a selectively permeable membrane with the assistance of integral proteins in the membrane that serve as carriers.
Osmosis	Net movement of water molecules due to kinetic energy across a selectively permeable membrane from an area of higher to lower concentration of water until an equilibrium is reached.
Filtration	Movement of solvents (such as water) and solutes (such as glucose) across a selectively permeable membrane as a result of gravity or hydrostatic (water) pressure from an area of higher to lower pressure.
Dialysis	Diffusion of solute particles across a selectively permeable membrane in which small molecules are separated from large ones.
ACTIVE PROCESSES	Substances move against a concentration gradient from an area of lower to higher concentration; cell must expend energy.
Active transport	Movement of substances, usually ions, across a selectively permeable membrane from a region of lower to higher concentration by an interaction with integral proteins in the membrane; the process requires energy expenditure in the form of ATP.
Endocytosis	Movement of large molecules and particles through plasma membranes in which the membrane surrounds the substance, encloses it, and brings it into the cell. Examples include phagocytosis ("cell eating"), pinocytosis ("cell drinking"), and receptor-mediated endocytosis.
Exocytosis	Export of substances from the cell by reverse endocytosis.

NUCLEUS

The **nucleus** is generally a spherical or oval organelle and is the largest structure in the cell (Figure 3-7a–c). It contains hereditary factors of the cell, called genes, which control cellular structure and direct many cellular activities. Most body cells contain a single nucleus, although some, such as mature red blood cells, do not. Skeletal muscle cells and a few other cells contain several nuclei.

The nucleus is separated from the cytoplasm by a double membrane called the *nuclear membrane* or *envelope* (Figure 3-7c). Between the two layers of the nuclear membrane is a space called the *perinuclear cisterna.* Each of the nuclear membranes resembles the structure of the plasma membrane. Minute *pores* in the nuclear membrane allow the nucleus to communicate with a membranous network in the cytoplasm called the endoplasmic reticulum. Substances entering and leaving the nucleus are believed to pass through the tiny pores. At many points the two nuclear membranes fuse with each other and, at the points of fusion, the membranes are permeable

to practically all dissolved or suspended substances, including newly synthesized ribosomes.

Three prominent structures are visible internal to the nuclear membrane. The first of these is a gel-like fluid that fills the nucleus called *karyolymph (nucleoplasm).* One or more spherical bodies called the *nucleoli* may also be present. These structures do not contain a membrane and are composed of protein, DNA, and RNA. Nucleoli disperse and disappear during cell division and re-form once new cells are formed. Nucleoli are the sites of the synthesis of a type of RNA called ribosomal RNA. The RNA is stored in nucleoli and, as you will see shortly, assumes a function in protein synthesis. Finally, there is the *genetic material* consisting principally of DNA. When the cell is not reproducing, the genetic material appears as a threadlike mass called *chromatin.* Prior to cellular reproduction the chromatin shortens and coils into rod-shaped bodies called *chromosomes.*

Chromosomes consist of DNA and proteins called *histones* organized in a very specific way. Elementary subunits of chromosomal structure are called **nucleosomes.** A nucleosome consists of different histone proteins associ-

(a)

(b)

(c)

FIGURE 3-7 Cytoplasm and nucleus. (a) Electron micrograph of cytoplasm and the nucleus at a magnification of 31,600×. (Copyright © Dr. Myron C. Ledbetter, Biophoto Associates/Photo Researchers.) (b) Diagram of the electron micrograph. (c) Diagram of a nucleus with two nucleoli.

ated with a relatively fixed length of DNA (about 200 nitrogenous base pairs). Present evidence indicates that the DNA is wrapped around the histones in some way. One type of histone maintains adjacent nucleosomes in a helical coil (*solenoid*). Nucleosomes are separated by stretches of DNA only so that the arrangement resembles beads on a string. The functional significance of nucleosome structure is still speculative. It is believed that histones may facilitate changes in chromosomal structure that expose activated genes (DNA) to perform a specific task in the cell.

RIBOSOMES

Ribosomes are tiny granules, 25 nm at their largest dimension, that are composed of a type of RNA called ribosomal RNA (rRNA) and a number of specific ribosomal proteins. The rRNA is manufactured by DNA in the nucleolus. Ribosomes were so named because of their high content of rRNA. Structurally, a ribosome consists of two subunits, one about half the size of the other. Recently scientists have provided three-dimensional models of the structure of a ribosome (Figure 3-8d). Functionally, ribo-

(a)

(b)

(c)

Ribosomes

Cisternae of granular (rough) endoplasmic reticulum

Large subunit   +   Small subunit   →   Complete functional ribosome

(d)

FIGURE 3-8 Endoplasmic reticulum and ribosomes. (a) Electron micrograph of the endoplasmic reticulum and ribosomes at a magnification of 76,000×. (Copyright © Dr. Myron C. Ledbetter, Biophoto Associates/Photo Researchers.) (b) Diagram of the electron micrograph. (c) Diagram of the endoplasmic reticulum and ribosomes. See if you can find the agranular (smooth) endoplasmic reticulum in Figure 3-9a. (d) Diagram of the three-dimensional structure of ribosomes.

somes are the sites of protein synthesis: they receive genetic instructions and use them to produce protein. Amino acids are joined one at a time on ribosomes into a protein chain. The completed chain then folds into a protein molecule that can serve as part of the cell's structure or as an enzyme, for example. The mechanism of ribosomal function is discussed later in the chapter.

Some ribosomes, called *free ribosomes,* are scattered in the cytoplasm; they have no attachments to other parts of the cell. The free ribosomes occur singly or in clusters, and they are primarily concerned with synthesizing proteins for use inside the cell. Other ribosomes are attached to a cellular structure called the endoplasmic reticulum (ER). These ribosomes are concerned with the synthesis of proteins for export from the cell.

ENDOPLASMIC RETICULUM (ER)

Within the cytoplasm, there is a system of pairs of parallel membranes enclosing narrow cavities of varying shapes. This system is known as the **endoplasmic reticulum,** or **ER** (Figure 3-8). The ER, in other words, is a network of channels (*cisternae*) running through the cytoplasm. These channels are continuous with the nuclear membrane.

On the basis of its association with ribosomes, ER is distinguished into two types. **Granular (rough) ER** is studded with ribosomes; **agranular (smooth) ER** is free of ribosomes. Recent studies using radioactive tracers suggest that agranular ER is synthesized from granular ER.

Numerous functions related to homeostasis are attributed to the ER. It contributes to the mechanical support and distribution of the cytoplasm. The ER is also involved in the intracellular exchange of materials with the cytoplasm and provides a surface area for chemical reactions. Various products are transported from one portion of the cell to another via the ER, so the ER is considered an intracellular circulatory system. The ER also serves as a storage area for synthesized molecules. And, together with a cellular structure called the Golgi complex, the ER assumes a role in the synthesis and packaging of molecules.

GOLGI COMPLEX

A cytoplasmic structure called the **Golgi** (GOL-jē) **complex** is generally near the nucleus. This structure consists of four to eight flattened membranous sacs, stacked upon each other like a pile of dishes with expanded areas at their ends. Like those of the ER, the stacked elements are called *cisternae* (Figure 3-9). On the basis of function, cisternae are designated as cis, medial, and trans (described shortly).

The principal function of the Golgi complex is to process, sort, and deliver proteins to various parts of the cell. Proteins synthesized at ribosomes associated with granular ER are transported into the granular ER (Figure 3-10). While still in the granular ER, sugar molecules are added to the proteins if needed (glycoproteins). Next, the proteins become surrounded by a vesicle formed by

FIGURE 3-9 Golgi complex. (a) Electron micrograph of the two Golgi complexes at a magnification of 78,000×. (Copyright © Dr. Myron C. Ledbetter, Biophoto Associates/Photo Researchers.) (b) Diagram of the electron micrograph. (c) Diagram of a Golgi complex.

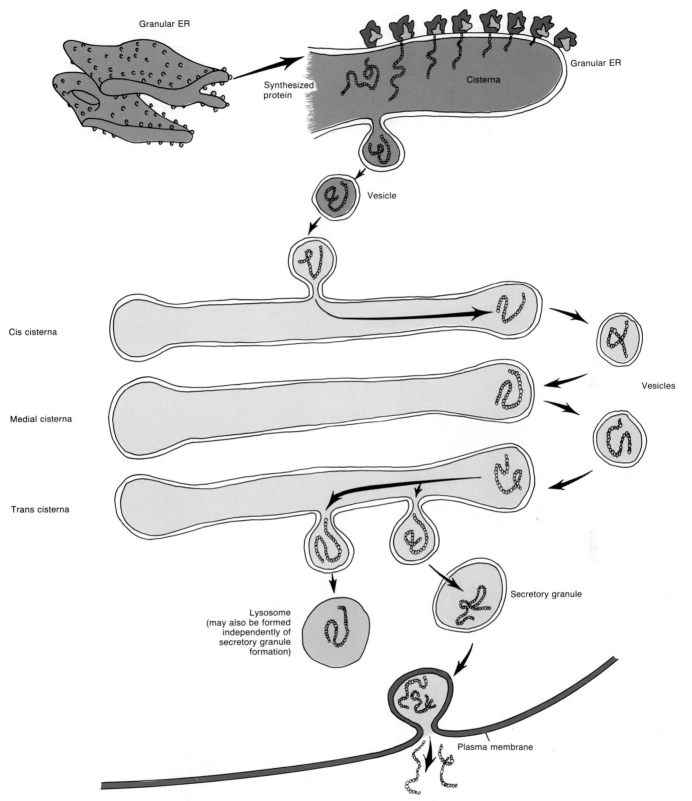

FIGURE 3-10 Packaging of synthesized protein for export from the cell.

a piece of the granular ER membrane, the vesicle buds off and fuses with the cis cisterna of the Golgi complex. The cis cisternae are the ones closest to the agranular ER. As a result of this fusion, the proteins enter the Golgi complex. Once inside the Golgi complex, the pro-teins are transported by other vesicles formed by the Golgi complex from the cis cisternae to the medial cisternae to the trans cisternae, the ones farthest from the agranular ER. As the proteins pass in succession through the Golgi cisternae, they become modified in various ways depending on their function and destination. The proteins are sorted and packaged (in vesicles) in the trans cisternae. Some vesicles become *secretory granules,* which move toward the surface of the cell where the protein is released. The contents of the granules are discharged into the extracellular space, and the granule membrane is incorporated into the plasma membrane. Cells of the digestive tract that secrete protein enzymes utilize this mechanism. The secretory granule prevents "digestion" of the cytoplasm of the cells by the enzymes as it moves toward the cell surface. Other vesicles that pinch off from the Golgi complex are loaded with special digestive enzymes and remain within the cell. They become cellular structures called lysosomes.

The Golgi complex is also associated with lipid secretion. Lipids synthesized by the agranular ER pass through the ER into the Golgi complex. Packaged lipids are discharged at the surface of the cell. In the course of moving through the cytoplasm, the vesicle may release lipids into the cytoplasm before being discharged from the cell. These appear in the cytoplasm as lipid droplets. Among the lipids secreted in this manner are steroids (see Exhibit 2-4).

MITOCHONDRIA

Small, spherical, rod-shaped, or filamentous structures called **mitochondria** (mī'-tō-KON-drē-a) appear throughout the cytoplasm. When sectioned and viewed under an electron microscope, each reveals an elaborate internal organization (Figure 3-11). A mitochondrion consists of two membranes, each of which is similar in structure to the plasma membrane. The outer mitochondrial membrane is smooth, but the inner membrane is arranged in a series of folds called *cristae.* The center of a mitochondrion is called the *matrix.*

Because of the nature and arrangement of the cristae, the inner membrane provides an enormous surface area for chemical reactions. Enzymes involved in energy-releasing reactions that form ATP are located on the cristae. Mitochondria are frequently called the "powerhouses of the cell" because they are the sites for the production of ATP. Active cells, such as muscle, liver, and kidney tubule cells, have a large number of mitochondria because of their high energy expenditure.

Mitochondria are self-replicative, that is, they can divide to form new ones. The replication process is con-

(a)   (b)

- Double mitochondrial membrane
- Matrix
- Cristae

- Inner mitochondrial membrane
- Outer mitochondrial membrane
- Cristae
- Matrix

(c)

FIGURE 3-11 Mitochondria. (a) Electron micrograph of an entire mitochondrion (above) and a portion of another (below). (Courtesy of Lester V. Bergman & Associates, Inc.) (b) Diagram of the electron micrograph. (c) Diagram of a mitochondrion.

FIGURE 3-12 Lysosome. (a) Electron micrograph of a lysosome at a magnification of 55,000×. (Courtesy of F. Van Hoof, Université Catholique de Louvain.) (b) Diagram of the electron micrograph.

trolled by DNA that is incorporated into the mitochondrial structure. Self-replication usually occurs in response to increased cellular need for ATP.

LYSOSOMES

When viewed under the electron microscope, **lysosomes** (*lysis* = dissolution; *soma* = body) appear as membrane-enclosed spheres (Figure 3-12). They are formed from Golgi complexes, have a single membrane, and lack detailed structure. They contain powerful digestive enzymes capable of breaking down many kinds of molecules. You will see in Chapter 14 that Tay-Sachs disease results from a deficiency of a lysosomal enzyme. These enzymes are also capable of digesting bacteria that enter the cell (see Figure 22-8). White blood cells, which ingest bacteria by phagocytosis, contain large numbers of lysosomes.

Scientists have wondered why these powerful enzymes do not also destroy their own cells. Perhaps the lysosomal membrane in a healthy cell is impermeable to enzymes so they cannot move out into the cytoplasm. However, when a cell is injured, the lysosomes release their enzymes. The enzymes then promote reactions that break the cell down into its chemical constituents. This process of self-destruction by cells is known as **autolysis.** The chemical remains are either reused by the body or excreted. Because of this function, lysosomes have been called "suicide packets."

Lysosomes are crucial in the removal of cell parts, whole cells, and even extracellular material, which may be the process underlying bone removal. In bone reshaping, especially during the growth process, special bone-destroying cells, called osteoclasts, secrete extracellular enzymes that dissolve bone. Bone tissue cultures given excess amounts of vitamin A remove bone apparently through an activation process involving lysosomes.

CLINICAL APPLICATION

Animals overfed on vitamin A interestingly develop spontaneous fractures, suggesting greatly increased lysosomal activity. On the other hand, cortisone and hydrocortisone, steroid hormones produced by the adrenal gland, have a stabilizing effect on lysosomal membranes. The steroid hormones are well known for their antiinflammatory properties, which suggests that they reduce destructive cellular activity by lysosomes.

PEROXISOMES

Organelles similar in structure to lysosomes, but smaller, are called **peroxisomes** (pe-ROKS-i-sōms). They are abundant in liver cells and contain several enzymes related to the metabolism of hydrogen peroxide (H_2O_2), a substance that is toxic to body cells. One of the enzymes in peroxisomes, called *catalase,* immediately breaks down H_2O_2 into water and oxygen:

$$H_2O_2 \xrightarrow{\text{catalase}} H_2O + O_2$$

Hydrogen   Water   Oxygen
peroxide

THE CYTOSKELETON

In recent years, electron microscopy has revealed that the cytoplasm of cells is more than just a structureless medium in which the various cellular components are suspended. It has been shown that cytoplasm actually has a complex internal structure, consisting of a series of exceedingly small microfilaments, microtubules, and intermediate filaments, together referred to as the **cytoskeleton** (see Figure 3-1).

Microfilaments are rodlike structures ranging from 3 to 12 nm in diameter. They are of variable length and may occur in bundles, randomly scattered throughout the cytoplasm, or arranged in a meshwork, depending on the type of cell in which they are found. Some microfilaments consist of a protein called *actin;* others consist of a protein called *myosin.* In muscle tissue, actin microfilaments (thin myofilaments) and myosin microfilaments (thick myofilaments) are involved in the contraction of muscle cells. This mechanism is described in Chapter 10. In nonmuscle cells, microfilaments help to provide support and shape and assist in the movement of entire cells (phagocytes and cells of developing embryos) and movements within cells (secretion, phagocytosis, pinocytosis).

Microtubles are relatively straight, slender, cylindrical structures that range in diameter from 18 to 30 nm, usually averaging about 24 nm. They consist of a protein called *tubulin*. Microtubules dispersed in the cytoplasm, together with microfilaments, help to provide support and shape for cells. Some evidence suggests that microtubules may form conducting channels through which various substances can move throughout the cytoplasm. This mechanism has been studied extensively in nerve cells. Microtubules also assist in the movement of pseudopodia that are characteristic of phagocytes. As you will see shortly, microtubules form the structure of flagella and cilia (cellular appendages involved in motility), centrioles (organelles that may direct the assembly of microtubules), and the mitotic spindle (structures that are involved in cell division).

Intermediate filaments range between 7 to 11 nm in diameter. Their proteins vary among cell types. In epithelial cells, for example, they consist of keratin. Intermediate filaments are usually dispersed throughout cytoplasm. In some cells, they are individual filaments; in others, they are bundled together. Although the functions of intermediate filaments are not completely understood, it appears that they provide structural reinforcement in some cells.

CENTROSOME AND CENTRIOLES

A dense area of cytoplasm, generally spherical and located near the nucleus, is called the **centrosome (centrosphere).** Within the centrosome is a pair of cylindrical structures: the **centrioles** (Figure 3-13). Each centriole is composed of nine triplet clusters of microtubules arranged in a circular pattern. Centrioles lack the two central single microtubules found in flagella and cilia. The two centrioles are situated so that the long axis of one is at right angles to the long axis of the other. Centrioles assume a role in cell reproduction by serving as centers about which microtubules involved in chromosome movement are organized. This role will be described shortly as part of cell division. Certain cells, such as most mature nerve cells, do not have a centrosome and so do not reproduce. This is why they cannot be replaced if they are destroyed. Like mitochondria, centrioles contain DNA that controls their self-replication.

(a)

(b)

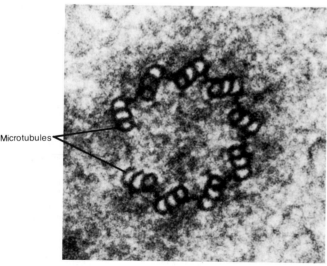

(c)

FIGURE 3-13 Centrosome and centrioles. (a) Diagram of a centriole in longitudinal view. (b) Diagram of a centriole in cross section. (c) Electron micrograph of a centriole in cross section at a magnification of 67,000×. (Courtesy of Biophoto Associates/Photo Researchers.)

FLAGELLA AND CILIA

Some body cells possess projections for moving the entire cell or for moving substances along the surface of the cell. These projections contain cytoplasm and are bounded by the plasma membrane. If the projections are few and long in proportion to the size of the cell, they are called **flagella.** The only example of a flagellum in the human body is the tail of a sperm cell, used for locomotion (see Figure 28-5). If the projections are numerous and short, resembling many hairs, they are called **cilia.** In humans, ciliated cells of the respiratory tract move mucus that has trapped foreign particles over the surface of the tissue (see Figure 23-4a). Electron microscopy has revealed no fundamental structural difference between cilia and flagella. Both consist of nine pairs of microtubules that form a ring around two single microtubules in the center.

CELL INCLUSIONS

Cell inclusions are a large and diverse group of chemical substances, some of which have recognizable shapes. These products are principally organic and may appear or disappear at various times in the life of the cell. *Melanin* is a pigment stored in certain cells of the skin, hair, and eyes. It protects the body by screening out harmful ultraviolet rays from the sun. *Glycogen* is a polysaccharide that is stored in the liver, skeletal muscle cells, and the vaginal mucosa. When the body requires quick energy, liver cells can break down the glycogen into glucose and release it. *Lipids,* which are stored in adipocytes (fat cells), may be decomposed for producing energy. A final example of an inclusion is *mucus,* which is produced by cells that line organs. Its function is to provide lubrication and protection.

The major parts of the cell and their functions are summarized in Exhibit 3-2.

EXTRACELLULAR MATERIALS

The substances that lie outside cells are called **extracellular materials.** They include body fluids, which provide a medium for dissolving, mixing, and transporting substances. Among the body fluids are interstitial fluid, the fluid that fills the microscopic spaces (interstitial spaces), and plasma, the liquid portion of blood. Extracellular materials also include secreted inclusions like mucus and special substances that form the matrix in which some cells are embedded.

The matrix materials are produced by certain cells and deposited outside their plasma membranes. The matrix supports the cells, binds them together, and gives strength and elasticity to the tissue. Some matrix materials are *amorphous* (a = without, *morpho* = shape); they have no specific shape. These include hyaluronic acid and chondroitin sulfate. **Hyaluronic** (hī'-a-loo-RON-ik) **acid** is a viscous, fluidlike substance that binds cells together, lubricates joints, and maintains the shape of the eyeballs. **Chondroitin** (kon-DROY-tin) **sulfate** is a jelly-like substance that provides support and adhesiveness in

EXHIBIT 3-2 CELL PARTS AND THEIR FUNCTIONS

PART	FUNCTIONS
PLASMA MEMBRANE	Protects cellular contents; makes contact with other cells; provides receptors for hormones, enzymes, and antibodies; mediates the entrance and exit of materials.
CYTOPLASM	Serves as the ground substance in which chemical reactions occur.
ORGANELLES	
Nucleus	Contains genes and controls cellular activities.
Ribosomes	Sites of protein synthesis.
Endoplasmic reticulum (ER)	Contributes to mechanical support; conducts intracellular nerve impulses in muscle cells; facilitates intracellular exchange of materials with cytoplasm; provides a surface area for chemical reactions; provides a pathway for transporting chemicals; serves as a storage area; together with Golgi complex synthesizes and packages molecules for export.
Golgi complex	Packages synthesized proteins for secretion in conjunction with endoplasmic reticulum; forms lysosomes; secretes lipids; synthesizes carbohydrates; combines carbohydrates with proteins to form glycoproteins for secretion.
Mitochondria	Sites for production of ATP.
Lysosomes	Digest substances and foreign microbes; may be involved in bone removal.
Peroxisomes	Contain several enzymes, such as catalase, related to hydrogen peroxide metabolism.
Microfilaments	Form part of cytoskeleton; involved in muscle cell contraction; provide support and shape; assist in cellular and intracellular movement.
Microtubules	Form part of cytoskeleton; provide support and shape; form intracellular conducting channels; assist in cellular movement; form the structure of flagella, cilia, centrioles, and spindle fibers.
Intermediate filaments	Form part of cytoskeleton; probably provide structural reinforcement in some cells.
Centrioles	Help organize mitotic spindle during cell division.
Flagella and cilia	Allow movement of entire cell (flagella) or movement of particles along surface of cell (cilia).
INCLUSIONS	Melanin (pigment in skin, hair, eyes) screens out ultraviolet rays; glycogen (stored glucose) can be decomposed to provide energy; lipids (stored in fat cells) can be decomposed to produce energy; mucus provides lubrication and protection.

cartilage, bone, heart valves, the cornea of the eye, and the umbilical cord.

Other matrix materials are *fibrous,* or threadlike. Fibrous materials provide strength and support for tissues. Among these are **collagenous** (*kolla* = glue) **fibers** consisting of the protein *collagen.* These fibers are found in all types of connective tissue, especially in bones, cartilage, tendons, and ligaments. **Reticular** (*rete* = net) **fibers** consisting of the protein collagen and a coating of glycoprotein, form a network around fat cells, nerve fibers, muscle cells, and blood vessels. They also form the framework or stroma for many soft organs of the body such as the spleen. **Elastic fibers,** consisting of the protein *elastin,* give elasticity to skin and to tissues forming the walls of blood vessels.

GENE ACTION

PROTEIN SYNTHESIS

Although cells synthesize numerous chemicals in order to maintain homeostasis, much of the cellular machinery is concerned with producing proteins. Some of the proteins are structural, helping to form plasma membranes, microfilaments, microtubules, centrioles, flagella, cilia, the mitotic spindle, and other parts of cells. Other proteins serve as contractile elements in muscle tissue, hormones, and antibodies. Still other proteins serve as enzymes that regulate the myriad chemical reactions that occur in cells. Basically, cells are protein factories that constantly synthesize large numbers of diverse proteins that determine the physical and chemical characteristics of cells and, therefore, of organisms.

The genetic instructions for making proteins are found in DNA. Cells make proteins by translating the genetic information encoded in DNA into specific proteins. In the process, the genetic information in a region of DNA is copied to produce a specific molecule of RNA. Through a complex series of reactions, the information contained in RNA is translated into a corresponding specific sequence of amino acids in a newly produced protein molecule. Let us take a look at how DNA directs protein synthesis by considering the two principal steps in protein synthesis: transcription and translation.

TRANSCRIPTION

Transcription is the process by which genetic information encoded in DNA is copied by a strand of RNA called **messenger RNA (mRNA).** It is called transcription because it resembles the transcription of a sequence of words from one tape to another. By using a specific portion of the cell's DNA as a template, the genetic information stored in the sequence of nitrogen bases of DNA is rewritten so that the same information appears in the nitrogen bases of mRNA. As in DNA replication (see Figure 3-15), a cytosine (C) in the DNA template dictates a guanine (G) in the mRNA strand being made; a G in the DNA template dictates a C in the mRNA strand; and a thymine (T) in the DNA template dictates an adenine (A) in the mRNA. Since RNA contains uracil (U) instead of T, an A in the DNA template dictates a U in the mRNA. As an example, if the template portion of DNA has the base sequence ATGCAT, the transcribed mRNA strand would have the complementary base sequence UACGUA. Only one of the two DNA strands serves as a template for RNA synthesis. This strand is referred to as the *sense strand.* The other strand, the one not transcribed, is the complement of the sense strand and is called the *antisense strand.*

In addition to serving as the template for the synthesis of mRNA, DNA also synthesizes two other kinds of RNA. One is called **ribosomal RNA (rRNA)** which, together with ribosomal proteins, makes up ribosomes. The other is called **transfer RNA (tRNA).** Once synthesized, mRNA, rRNA, and tRNA leave the nucleus of the cell. In the cytoplasm, they participate in the next principal step in protein synthesis—translation.

TRANSLATION

The process by which information in the nitrogen base sequence of mRNA is used to specify the amino acid sequence of a protein is called **translation.** The key events involved in translation are as follows (Figure 3-14).

1. In the cytoplasm, the small ribosomal subunit binds one end of the mRNA molecule (Figure 3-14a). In this respect, ribosomes are the sites of protein synthesis.

2. There are 20 different amino acids in the cytoplasm that may participate in protein synthesis. Whichever amino acids participate to form a particular protein are picked up by tRNAs (Figure 3-14b). For each different amino acid there is a different type of tRNA. One end of a tRNA molecule couples with a specific amino acid. This amino acid activation requires energy from ATP breakdown. The other end has a specific sequence of three nitrogen bases (triplet) known as an *anticodon.* The complementary triplet on an mRNA strand is called a *codon.*

3. By base pairing, the anticodon of a specific tRNA recognizes the corresponding codon of mRNA and attaches to it. If the tRNA anticodon is UAC, the mRNA codon would be AUG (Figure 3-14c). In the process, the tRNA also brings along the specific amino acid. The pairing of anticodon and codon occurs only where mRNA is attached to a ribosome.

4. Once the first tRNA attaches to mRNA, the ribosome moves along the mRNA, and the next tRNA with its amino acid moves into position (Figure 3-14d).

5. The two amino acids are joined by a peptide bond and the first tRNA detaches itself from the mRNA strand. The larger ribosomal subunit contains the enzymes that join the amino acids together (Figure 3-14e). The released tRNA can now pick up another similar amino acid if necessary.

6. As the proper amino acids are brought into line, one by one, peptide bonds are formed between them and the protein gets progressively longer (Figure 3-14f).

7. When the specified protein is completed, further synthesis is stopped by a special *termination codon.* The assembled protein is released from the ribosome and the ribosome comes apart into its component subunits (Figure 3-14g,h).

As each ribosome moves along the mRNA strand, it "reads" the information coded in mRNA and synthesizes a protein according to that information; the ribosome synthesizes the protein by translating the codon sequences into an amino acid sequence.

FIGURE 3-14 Translation. Note that during protein synthesis the ribosomal subunits join together. When not involved in protein synthesis, the two subunits are separate entities.

Protein synthesis progresses at the rate of about 15 amino acids per second. As the ribosome moves along the mRNA and before it completes translation of that gene, another ribosome may attach and begin translation of the same mRNA strand, so that several ribosomes may be attached to the same mRNA strand. Such an mRNA strand with its several ribosomes attached is called a **polyribosome.** Several ribosomes moving simultaneously in tandem along the same mRNA molecule permit the translation of a single mRNA strand into several identical proteins simultaneously.

Based upon the description of protein synthesis just presented, we can appropriately define a **gene** as a group of nucleotides on a DNA molecule that serves as the master mold for manufacturing a specific protein. Genes average about 1,000 pairs of nucleotides, which appear in a specific sequence on the DNA molecule. No two genes have exactly the same sequence of nucleotides, and this is the key to heredity.

Remember that the base sequence of the gene determines the sequence of the bases in the mRNA. The sequence of the bases in the mRNA then determines the order and kind of amino acids that will form the protein. Thus each gene is responsible for making a particular protein as follows:

$$DNA \xrightarrow{\text{transcription}} RNA \xrightarrow{\text{translation}} protein$$

"SOS" GENES: DNA REPAIR

The normal daily activities of a cell and the continuity of life from one generation to another depend on the stable and precise replication of the DNA molecule in which hereditary information is encoded. Unfortunately, the architecture of DNA is such that it is vulnerable to damage by harmful radiations and various chemical agents. If the damage results in the incorporation of an incorrect or altered nitrogen base or other structural change that distorts the DNA molecule, its ability to replicate and synthesize proteins is impaired. If the damage is not repaired, the cell functions abnormally and could die. The damage seems sometimes to initiate a sequence of events leading to cancer.

Extensive damage to DNA can be repaired by certain enzymes. From what can be learned, it appears that damage to DNA switches on an **"SOS response."** As part of this response, certain genes produce increased quantities of proteins (enzymes) that operate together to repair the genetic damage. Scientists are now trying to learn the mechanisms of different repair systems and under what conditions they operate. It is hoped that such information will provide clues to the nature of cancer and other diseases and the aging process. [i]

CLINICAL APPLICATION

Different kinds of cells make different proteins following instructions encoded in the DNA of their genes. Since 1973 scientists have been able to alter those instructions in bacterial cells by adding genes from other organisms to the bacterial genes. This causes the bacterial cells to produce proteins that they normally do not synthesize. Bacteria so altered are called **recombinants,** and their DNA, a combination of DNA from different sources, is called **recombinant DNA.** When recombinant DNA is introduced into a bacterium, the bacterium will synthesize the proteins of whatever new

gene it has acquired. The new technology that has arisen from manipulating the genetic material is called **genetic engineering.** The bacterium that currently plays the most important role is *Escherichia coli,* a common inhabitant of the human intestine. Yeast cells are also being used in recombinant DNA research.

An important goal of recombinant DNA research is to increase our understanding of how genes are organized, how they function, and how they are regulated. This information will enable scientists to identify each of the thousands of genes in the human cell and might be used to find a way to replace defective genes that are responsible for hemophilia, sickle-cell anemia, and other diseases. Recombinant DNA technology might also help scientists understand why normal cells turn malignant.

The practical applications of recombinant DNA technology are staggering. Strains of recombinant bacteria are presently producing several important therapeutic substances. These include *growth hormone (GH)*, required for growth during childhood; *somatostatin,* a brain hormone that helps regulate growth; *insulin,* a hormone that helps regulate blood sugar level and is used by diabetics; *human chorionic gonadotropin (HCG)*, a hormone required to maintain pregnancy; *interferon (INF)*, an antiviral (and possibly anticancer) substance; *surfactant,* a phospholipid that lowers surface tension in the lungs; *factor VIII,* a clotting factor missing in people with hemophilia A, the main form of the inherited blood disorder; *tissue-plasminogen activator* (t-PA), a substance used to dissolve blood clots in coronary arteries; and *beta-endorphin,* a neuropeptide that has morphinelike properties that suppress pain. Scientists are also using recombinant DNA techniques to develop vaccines against several viruses, including those that cause herpes, hepatitis B, influenza, and malaria. In addition to the medical benefits, recombinant DNA technology could be applied to producing proteins to help alleviate food shortages, to increase the alcohol yield from corn, to consume hazardous oil spills in the ocean, and to extract scarce minerals from soil.

NORMAL CELL DIVISION

Most of the cell activities mentioned thus far maintain the life of the cell on a day-to-day basis. However, cells become damaged, diseased, or worn out and then die. New cells must be produced as replacements and for growth. In addition, sperm and egg cells must be produced by cell division.

Cell division is the process by which cells reproduce themselves. It consists of a nuclear division and a cytoplasmic division (cytokinesis). Because nuclear division can be of two types, two kinds of cell division are recognized.

In the first kind of division, often called **somatic cell division,** a single parent cell duplicates itself. This process consists of a nuclear division called **mitosis** and cytokinesis. The process ensures that each new daughter cell has the same *number* and *kind* of chromosomes as the original parent cell. After the process is complete, the two daughter cells have the same hereditary material and genetic potential as the parent cell. This kind of cell division results in an increase in the number of body cells. In a 24-hour period, the average adult loses billions of cells from different parts of the body. Obviously, these cells must be replaced. Cells that have a short life span—the cells of the outer layer of skin, the cornea of the eye, the digestive tract—are continually being replaced. Mitosis and cytokinesis are the means by which dead or injured cells are replaced and new cells are added for body growth.

The second type of cell division is the mechanism by which sperm and egg cells are produced, preliminary to the formation of a new organism. The process consists of a nuclear division called **meiosis** plus cytokinesis. It is often called **reproductive cell division.** We will first discuss somatic cell division.

SOMATIC CELL DIVISION

When a cell reproduces, it must replicate its chromosomes so its hereditary traits may be passed on to succeeding generations of cells. A **chromosome** is a highly coiled DNA molecule that is partly covered by protein. The protein causes changes in the length and thickness of the chromosome. Hereditary information is contained in the DNA portion of the chromosome in units called **genes.**

When a cell is between divisions it is said to be in **interphase (metabolic phase).** It is during this stage that the replication (synthesis) of chromosomes occurs and the RNA and protein needed to produce structures required for doubling all cellular components are manufactured. When DNA replicates, its helical structure partially uncoils (Figure 3-15). Those portions of DNA that remain coiled stain darker than the uncoiled portions. This unequal distribution of stain causes the DNA to appear as a granular mass called **chromatin** (see Figure 3-16a). During uncoiling, DNA separates at the points where the nitrogen bases are connected. Each exposed nitrogen base then picks up a complementary nitrogen base (with associated sugar and phosphate group) from the cytoplasm of the cell. This uncoiling and complementary base pairing continues until each of the two original DNA strands is matched and joined with two newly formed DNA strands. The original DNA molecule has become two DNA molecules.

A microscopic view of a cell during interphase shows a clearly defined nuclear membrane, nucleoli, karyolymph, chromatin, and a pair of centrioles. Once a cell completes its replication of DNA and its production of RNA and proteins during interphase, mitosis begins.

Mitosis

The succession of events that takes place during mitosis and cytokinesis is plainly visible under a microscope after the cells have been stained in the laboratory.

The process called **mitosis** is the distribution of the two sets of chromosomes into two separate and equal

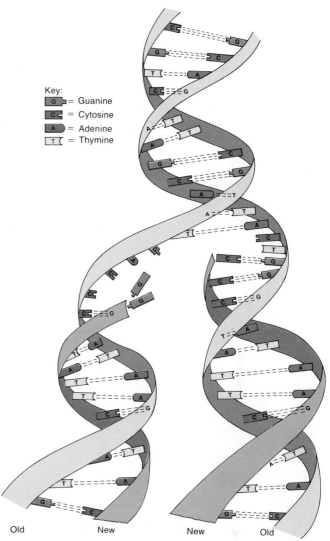

Key:
- G ▭ = Guanine
- C ▭ = Cytosine
- A ▭ = Adenine
- T ▭ = Thymine

Old New   New Old

FIGURE 3-15 Replication of DNA. The two strands of the double helix separate by breaking the bonds between nucleotides. New nucleotides attach at the proper sites, and a new strand of DNA is paired off with each of the original strands. After replication, the two DNA molecules, each consisting of a new and an old strand, return to their helical structure.

nuclei following the replication of the chromosomes of the parent nucleus. For convenience, biologists divide the process into four stages: prophase, metaphase, anaphase, and telophase. These are arbitrary classifications. Mitosis is actually a continuous process, one stage merging imperceptibly into the next.

• *Prophase* During **prophase** (Figure 3-16b), the first stage of mitosis, chromatin shortens and coils into chromosomes. The nucleoli become less distinct and the nuclear membrane disappears. Each prophase "chromosome" is actually composed of a pair of structures called **chromatids.** Each chromatid is a complete chromosome consisting of a double-stranded DNA molecule and each is attached to its chromatid pair by a small spherical body called a **centromere.** During prophase, the chromatid pairs assemble near the equatorial plane region (equator) of the cell.

FIGURE 3-16 Cell division: mitosis and cytokinesis. Photomicrographs and diagrammatic representations of the various stages of cell division in whitefish eggs. Read the sequence starting at (a), and move clockwise until you complete the cycle. (Photographs by Carolina Biological Supply Company.)

Also during prophase, the paired centrioles separate and each pair moves to an opposite pole (end) of the cell. Between the centrioles, a series of microtubules is organized into two groups of fibers. The *continuous (interpolar) microtubules* originate from the vicinity of each pair of centrioles and grow toward each other. Thus, they extend from one pole of the cell to another. As they grow toward each other, the second group of microtubules develops. These are called *chromosomal microtubules* and grow out of the centromeres; they extend from a centromere to a pole of the cell. Together, the continuous and chromosomal microtubules constitute the **mitotic spindle** and, with the centrioles, are referred to as the **mitotic apparatus.**

• *Metaphase* During **metaphase** (Figure 3-16c), the second stage of mitosis, the centromeres of the chromatid pairs line up on the equatorial plane of the cell. The centromeres of each chromatid pair form a chromosomal microtubule that attaches the centromere to a pole of the cell.

• *Anaphase* The third stage of mitosis, **anaphase** (Figure 3-16d), is characterized by the division of the centromeres and the movement of complete identical sets of chromatids, now called chromosomes, to opposite poles of the cell. During this movement, the centromeres attached to the chromosomal microtubules seem to drag the trailing parts of the chromosomes toward opposite poles. Although several theories have been proposed, the mechanism by which the chromosomes move to opposite poles is not completely understood.

• *Telophase* Telophase (Figure 3-16e), the final stage of mitosis, consists of a series of events nearly the reverse of prophase. By this time, two identical sets of chromosomes have reached opposite poles. As telophase progresses, new nuclear membranes begin to enclose the chromosomes, the chromosomes start to assume their chromatin form, nucleoli reappear, and the mitotic spindles disappear. The centrioles also replicate so that each cell has two centriole pairs. The formation of two nuclei identical to those of cells in interphase terminates telophase. A mitotic cycle has thus been completed (Figure 3-16f).

• *Time Required* The time required for mitosis varies with the kind of cell, its location, and the influence of factors such as temperature. Furthermore, the different stages of mitosis are not equal in duration. Prophase is usually the longest stage, lasting from one to several hours. Metaphase is considerably shorter, ranging from 5 to 15 minutes. Anaphase is the shortest stage, lasting from 2 to 10 minutes. Telophase lasts from 10 to 30 minutes. These lengths of time vary considerably with cell type. Overall, mitosis represents only a small part of the life cycle of the cell.

Cytokinesis

Division of the cytoplasm, a process called **cytokinesis** (sī'-tō-ki-NĒ-sis), often begins during late anaphase and terminates at the same time as telophase. Cytokinesis begins with the formation of a *cleavage furrow* that extends around the cell's equator. The furrow progresses inward, resembling a constricting ring, and cuts completely through the cell to form two separate portions of cytoplasm (Figure 3-16d–f).

A summary of the events that occur during interphase and somatic cell division is presented in Exhibit 3-3.

REPRODUCTIVE CELL DIVISION

In sexual reproduction, each new organism is produced by the union and fusion of two different sex cells, one

EXHIBIT 3-3 SUMMARY OF EVENTS ASSOCIATED WITH INTERPHASE AND SOMATIC CELL DIVISION

PERIOD OR STAGE	ACTIVITY
Interphase	Cell is between divisions. Cell engages in growth, metabolism, and production of substances required for division; chromosomal replication occurs.
Cell Division	Single parent cell produces two identical daughter cells.
Prophase	Chromatin shortens and coils into chromosomes (chromatids), nucleoli and nuclear membrane become less distinct, centrioles separate and move to opposite poles of cell, and mitotic spindle forms.
Metaphase	Centromeres of chromatid pairs line up on equatorial plane of cell and form chromosomal microtubules that attach centromeres to poles of cell.
Anaphase	Centromeres divide and identical sets of chromosomes move to opposite poles of cell.
Telophase	Nuclear membrane reappears and encloses chromosomes, chromosomes resume chromatin form, nucleoli reappear, mitotic spindle disappears, and centrioles duplicate.
Cytokinesis	Cleavage furrow forms around equatorial plane of cell, progresses inward, and separates cytoplasm into two separate and equal portions.

produced by each parent. The sex cells, called **gametes,** are the ovum produced in the female gonads (ovaries) and the sperm produced in the male gonads (testes). The union and fusion of gametes is called fertilization and the cell thus produced is known as a zygote. The zygote contains a mixture of chromosomes (DNA) from the two parents and, through its repeated mitotic division, develops into a new organism.

Gametes differ from all other body cells (somatic cells) with respect to the number of chromosomes in their nuclei. Somatic cells, such as brain cells, stomach cells, kidney cells, and all other uninucleated somatic cells, contain 46 chromosomes in their nuclei. Some somatic cells, such as skeletal muscle cells, are multinucleated and thus contain more than 46 chromosomes. However, since most somatic cells are uninucleated, these are the cells to which we will refer in the following discussion. Of the 46 chromosomes, 23 are a complete set that contain all the genes necessary for carrying out the activities of the cell. In a sense, the other 23 chromosomes are a duplicate set. The symbol n is used to designate the number of different chromosomes within the nucleus. Since somatic cells contain two sets of chromosomes, they are referred to as **diploid** (DIP-loyd) **cells** (*di* = two), symbolized as $2n$. In a diploid cell, two chromosomes that belong to a pair

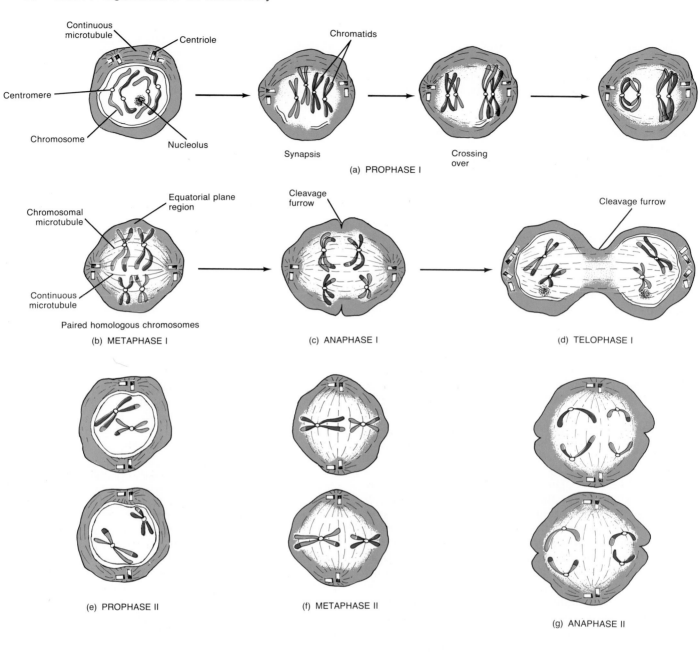

Continuous microtubule

Centriole

Chromatids

Centromere

Chromosome

Nucleolus

Synapsis

Crossing over

(a) PROPHASE I

Equatorial plane region

Chromosomal microtubule

Cleavage furrow

Cleavage furrow

Continuous microtubule

Paired homologous chromosomes

(b) METAPHASE I

(c) ANAPHASE I

(d) TELOPHASE I

(e) PROPHASE II

(f) METAPHASE II

(g) ANAPHASE II

(h) TELOPHASE II

FIGURE 3-17 Meiosis. See text for details.

are called **homologous** (hō-MOL-ō-gus) **chromosomes,** or **homologues.**

If gametes had the same number of chromosomes as somatic cells, the zygote formed from their fusion would have double the number. The somatic cells of the resulting individual would have twice the number of chromosomes ($4n$) as the somatic cells of the parents, and with every succeeding generation, the number of chromosomes would double. The chromosome number does not double with each generation because of a special nuclear division called **meiosis.** Meiosis occurs only in the production of gametes. It causes a developing sperm or ovum to relinquish its duplicate set of chromosomes so that the mature gamete has only 23. Thus gametes are **haploid** (HAP-loyd) **cells,** meaning "one-half," and are symbolized as n.

Meiosis

The formation of haploid sperm cells in the testes of the male by meiosis is called spermatogenesis, while the formation of haploid ova (eggs) in the ovaries of the female by meiosis is referred to as oogenesis. Both spermatogenesis and oogenesis are discussed in detail in Chapter 28. At this point, we will examine the essentials of meiosis.

Meiosis occurs in two successive nuclear divisions referred to as **reduction division (meiosis I)** and **equatorial division (meiosis II).** During the interphase that precedes reduction division of meiosis, the chromosomes replicate themselves. This replication is similar to that in interphase preceding the mitosis of somatic cell division. Once chromosomal replication is complete, reduction division begins. It consists of four phases referred to as prophase I, metaphase I, anaphase I, and telophase I (Figure 3-17).

Prophase I is an extended phase in which the chromosomes shorten and thicken, the nuclear membrane and nucleoli disappear, the centrioles replicate, and the spindle appears. Unlike the prophase of mitosis, however, a unique event occurs in prophase I of meiosis. The chromosomes line up in the nuclear region in homologous pairs. The pairing is called **synapsis.** The four chromatids of each homologous pair are referred to as a **tetrad.** Another unique event of meiosis occurs within a tetrad. Portions of one chromatid may be exchanged with portions of another, a process called **crossing-over** (Figure 3-18). This process, among others, permits an exchange of genes among chromatids so that subsequent cells produced are unlike each other genetically and unlike the cell that produced them. This phenomenon accounts for the great genetic variation among humans and other organisms that form gametes by meiosis. In metaphase I, the paired chromosomes line up along the equatorial plane of the cell, with one member of each pair on either side. Recall that there is no pairing of homologous chromosomes during the metaphase of mitosis. The centromeres of each chromatid pair also form chromosomal microtubules that attach the centromeres to opposite poles of the cell. Anaphase I is characterized by separation of the members of each pair, with one member of each pair moving to an opposite pole of the cell. During anaphase I, unlike mitotic anaphase, the centromeres do not split and the paired chromatids, held by a centromere, remain together. Telophase I and cytokinesis are similar to telophase and cytokinesis of mitosis. The net effect of reduction division is that each resulting daughter cell contains the haploid number of chromosomes; each cell contains only one member of each pair of the original homologous chromosomes in the starting parent cell.

The interphase between reduction division and equatorial division is either brief or lacking altogether. It does, however, differ from the interphase preceding reduction division or mitosis in that there is replication of DNA between the reduction and equatorial divisions.

The equatorial division of meiosis consists of four phases referred to as prophase II, metaphase II, anaphase II, and telophase II. These phases are essentially similar to those that occur during mitosis since the centromeres

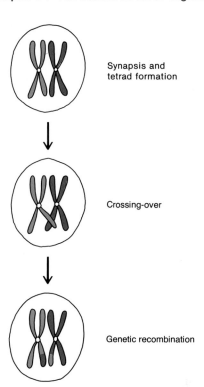

FIGURE 3-18 Crossing-over within a tetrad resulting in genetic recombination.

divide and chromatids separate and move toward opposite poles of the cell.

In reviewing the overall process, note that during reduction division we start with a parent cell with the diploid number and end up with two daughter cells, each with the haploid number. During equatorial division, each haploid cell formed during reduction division divides and the net result is four haploid cells. As you will see later, all four haploid cells develop into sperm cells in the testes of the male, but only one of the haploid cells develops into an ovum in the ovaries in the female. The other three become polar bodies which do not function as gametes.

A simplified comparison of mitosis and meiosis is illustrated in Figure 3-19.

ABNORMAL CELL DIVISION: CANCER

DEFINITION

When cells in some area of the body duplicate without control, the excess of tissue that develops is called a **tumor, growth,** or **neoplasm.** The study of tumors is called **oncology** (*onco* = swelling or mass; *logos* = study of), and a physician who specializes in this field is called an **oncologist.** Tumors may be cancerous and sometimes fatal or they may be quite harmless. A cancerous growth is called a **malignant tumor,** or **malignancy.** A noncancerous growth is called a **benign growth.** Benign tumors are composed of cells that do not spread to other parts of the

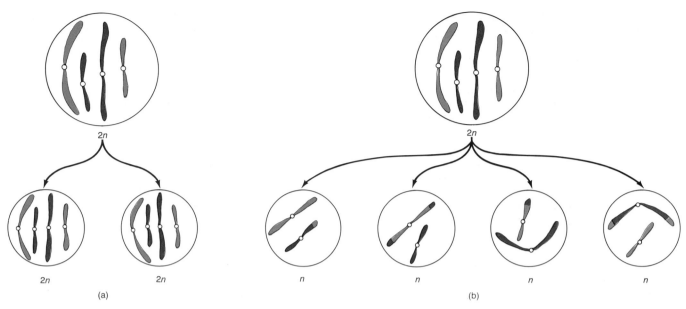

FIGURE 3-19 Very simplified comparison between (a) mitosis and (b) meiosis.

body, and they may be removed if they interfere with a normal body function or are disfiguring.

SPREAD

Cells of malignant growths duplicate continuously and very often quickly and without control. The majority of cancer patients are not killed by the **primary tumor** that develops but by secondary infections of bacteria and viruses due to lowered resistance as a result of **metastasis** (me-TAS-ta-sis), the spread of the disease to other parts of the body. One of the unique properties of a malignant tumor is its ability to metastasize. Metastatic groups of cells are harder to detect and eliminate than primary tumors.

In the process of metastasis, there is an initial invasion of the malignant cells into surrounding tissues. As the cancer grows, it expands and begins to compete with normal tissues for space and nutrients. Eventually, the normal tissue atrophies and dies. The invasiveness of the malignant cells may be related to mechanical pressure of the growing tumor, motility of the malignant cells, and enzymes produced by the malignant cells. Also, malignant cells lack what is called *contact inhibition*. When nonmalignant cells of the body divide and migrate (for example, skin cells that multiply to heal a superficial cut), their further migration is inhibited by contact on all sides with other skin cells. Unfortunately, malignant cells do not conform to the rules of contact inhibition; they have the ability to invade healthy body tissues with very few restrictions.

Following invasion, some of the malignant cells may detach from the primary tumor and invade a body cavity (abdominal or thoracic) or enter the blood or lymph. This latter condition can lead to widespread metastasis. In the next step in metastasis, those malignant cells that survive in the blood or lymph invade adjacent body tissues

and establish **secondary tumors.** It is believed that some of the invading cells involved in metastasis have properties different from those of the primary tumor that enhance metastasis. These include appropriate mechanical, enzymatic, and surface properties. In the final stage of metastasis, the secondary tumors become vascularized, that is, they take on new networks of blood vessels that provide nutrients for their further growth. Any new tissue, whether it results from repair, normal growth, or tumors requires a blood supply. A protein that serves as a chemical trigger for blood vessel growth is called *angiogenin,* which has been isolated from human colon tumors. In all stages of metastasis, the malignant cells resist the antitumor defenses of the body. The pain associated with cancer develops when the growth puts pressure on nerves or blocks a passageway so that secretions build up pressure.

TYPES

At present, cancers are classified by their microscopic appearance and the body site from which they arise. At least 100 different cancers have been identified in this way. If finer details of appearance are taken into consideration, the number can be increased to 200 or more. The name of the cancer is derived from the type of tissue in which it develops. **Carcinoma** (*carc* = cancer; *oma* = tumor) refers to a malignant tumor consisting of epithelial cells. A tumor that develops from a gland is called an **adenosarcoma** (*adeno* = gland). **Sarcoma** is a general term for any cancer arising from connective tissue. **Osteogenic sarcomas** (*osteo* = bone; *genic* = origin), the most frequent type of childhood cancer, destroy normal bone tissue and eventually spread to other areas of the body. **Myelomas** (*myelos* = marrow) are malignant tumors, occurring in middle-aged and older people, that interfere with the blood-cell-producing function of bone marrow and cause anemia. **Chondrosarcomas** are cancerous growths of cartilage (*chondro* = cartilage).

POSSIBLE CAUSES

What triggers a perfectly normal cell to lose control and become abnormal? Scientists are uncertain. First, there are environmental agents: substances in the air we breathe, the water we drink, the food we eat. A chemical or other environmental agent that produces cancer is called a **carcinogen.** The World Health Organization estimates that carcinogens may be associated with 60 to 90 percent of all human cancer. Examples of carcinogens are the hydrocarbons found in cigarette tar. Ninety percent of all lung cancer patients are smokers. Another environmental factor is radiation. Ultraviolet (UV) light from the sun, for example, may cause genetic mutations in exposed skin cells and lead to cancer, especially among light-skinned people.

Viruses are a second cause of cancer, at least in animals. These agents are tiny packages of nucleic acids, either DNA or RNA, that are capable of infecting cells and converting them to virus-producers. With over 100 separate viruses identified as carcinogens in many species and tissues of animals, it is also probable that at least some cancers in humans are due to viruses. For example, *human T-cell leukemia-lymphoma virus-1* (*HTLV*-1) is strongly associated with *leukemia,* a malignant disease of blood-forming tissues, and *lymphoma,* a cancer of lymphoid tissue. A variant of HTLV-I, known as HTLV-III, is the causative agent of acquired immune deficiency syndrome (AIDS). The *Epstein-Barr virus* (*EBV*), the causative agent of infectious mononucleosis, has recently been linked as the causative agent of several human cancers—*Burkitt's lymphoma* (a cancer of white blood cells called B cells), *nasopharyngeal carcinoma* (common in Chinese males), and *Hodgkin's disease* (a cancer of the lymphatic system). The *hepatitis B virus* (*HBV*) has been associated with cancer of the liver. Also, *type 2 herpes simplex virus,* the causative agent of genital herpes, has been implicated in cancer of the cervix of the uterus, and the *papilloma virus,* a virus that causes warts, has been associated with cancer of the cervix and cancer of the colon.

A great deal of cancer research is now directed toward studying **oncogenes,** genes that have the ability to transform a normal cell into a cancerous cell. Oncogenes are slightly altered versions of their normal genes, referred to as **proto-oncogenes.** Every human cell contains oncogenes. In fact, oncogenes apparently carry out normal cellular functions until a malignant change occurs. It appears that some proto-oncogenes are activated to oncogenes by various types of mutations in which the DNA of the proto-oncogenes is altered. Such mutations are induced by carcinogens. Other proto-oncogenes are activated by viruses. Some oncogenes can also be activated by a rearrangement of a cell's chromosomes in which segments of DNA are exchanged. This rearrangement is sufficient to activate oncogenes by placing them near genes that enhance their activity. Such a mechanism occurs in Burkitt's lymphoma.

Researchers have also begun to uncover new evidence that some cancers are not caused by oncogenes, but may be caused by the absence of a normal gene. The prototype for such cancers is a rare, inherited childhood cancer of the eye called retinoblastoma.

Currently, scientists are also trying to establish a relationship between stress and cancer. Some believe that stress may play a role not only in the development but also in the metastasis of cancer.

TREATMENT

Treating cancer is difficult because it is not a single disease and because all the cells in a single population (tumor) do not behave in the same way. The same cancer may contain a diverse population of cells by the time it reaches a clinically detectable size. Although tumor cells look alike when stained and viewed under the microscope, they do not necessarily behave in the same manner in the body. For example, some metastasize and others do not. Some divide and others do not. Some are sensitive to drugs and some are resistant. As a consequence of differences in drug resistance, a single chemotherapeutic drug may destroy susceptible cells, but permit resistant cells to proliferate. This is probably one of the reasons that combination chemotherapy is usually more successful. In addition to chemotherapy, radiation therapy, surgery, and hyperthermia (abnormally high temperatures) may be used alone or in combination.

Scientists are moving closer to developing a vaccine for cancer. What happens in cancer is that the immune system fails to protect the body. Accordingly, the goal of a cancer vaccine is to stimulate the immune system into marshaling a successful attack against the cancer cells.

In recent years, there has been considerable debate over the use of **Laetrile** (LĀ-e-tril) in the treatment of human cancer. Laetrile is a naturally derived substance prepared from apricot pits. In response to public pressure, the National Cancer Institute (NCI) and Food and Drug Administration (FDA) conducted a clinical trial to determine the effectiveness of Laetrile in the treatment of advanced cancer. The results of the trial were published in 1982 and indicated that Laetrile does not work and is a toxic drug.

CELLS AND AGING

Aging is a progressive failure of the body's homeostatic adaptive responses. It is a general response that produces observable changes in structure and function and increased vulnerability to environmental stress and disease. Disease and aging probably accelerate each other. The specialized branch of medicine that deals with the medical problems and care of elderly persons is called **geriatrics** (jer'-ē-AT-riks; *geras* = old age; *iatrike* = surgery, medicine).

The obvious characteristics of aging are well known: graying and loss of hair, loss of teeth, wrinkling of skin, decreased muscle mass, and increased fat deposits. The physiological signs of aging are gradual deterioration in function and capacity to respond to environmental stress. Thus basic kidney and digestive metabolic rates decrease, as does the ability to respond effectively to changes in temperature, diet, and oxygen supply in order to maintain a constant internal environment. These manifestations of aging are related to a net decrease in the number of cells in the body (thousands of brain cells are lost each day) and to the dysfunctioning of the cells that remain.

The extracellular components of tissues also change with age. Collagen fibers, responsible for the strength in tendons, increase in number and change in quality with aging. These changes in the collagen of arterial walls are as much responsible for their loss of extensibility as are the deposits associated with atherosclerosis, the deposition of fatty materials in arterial walls. Elastin, another extracellular component, is responsible for the elasticity of blood vessels and skin. It thickens, fragments, and acquires a greater affinity for calcium with age—changes that may also be associated with the development of atherosclerosis.

Several kinds of cells in the body—heart cells, skeletal muscle cells, nerve cells—are incapable of replacement. Experiments have proved that many other cell types are limited when it comes to cell division. Cells grown outside the body divide only a certain number of times and then stop. The number of divisions correlates with the donor's age and with the normal life span of the different species from which the cells are obtained—strong evidence for the hypothesis that cessation of mitosis is a normal, genetically programmed event. According to this view, an "aging" gene is part of the genetic blueprint at birth, and it turns on at a preprogrammed time, slowing down or halting processes vital to life.

Another theory of aging is the free radical theory. Free radicals, or oxygen radicals, are oxygen molecules that bear free electrons, are highly reactive, and can easily tie up and weaken proteins. As a result, cells grow rigid as nutrients are excluded and wastes are locked in. Such effects are exhibited as wrinkled skin, stiff joints, and hardened arteries. Free radicals may also cause damage to DNA. Among the factors that produce free radicals are pollution, radiation, and certain foods we eat. Other substances in the diet such as vitamin E, vitamin C, beta-carotene, and selenium, are antioxidants and inhibit free radical formation.

Whereas some theories of aging explain the process at the cellular level, others concentrate on regulatory mechanisms operating within the entire organism. For example, one such theory holds that the immune system, which manufactures antibodies against foreign invaders, turns on its own cells. This autoimmune response might be caused by changes in the surfaces of cells, causing antibodies to attack the body's own cells. As surface changes in cells increase, the autoimmune response intensifies, producing the well-known signs of aging. Another organismic theory suggests that aging is programmed in the pituitary gland, a hormone-producing gland attached to the undersurface of the brain. Supposedly, at a set time in life, the gland releases a hormone that triggers age-associated disruptions.

The effects of aging on the various body systems are discussed in their respective chapters.

MEDICAL TERMINOLOGY

NOTE TO THE STUDENT

Each chapter in this text that discusses a major system of the body is followed by a glossary of **medical terminology**. Both normal and pathological conditions of the system are included in these glossaries. You should familiarize yourself with the terms, since they will play an essential role in your medical vocabulary.

Some of these disorders, as well as disorders discussed in the text, are referred to as local or systemic. A **local disease** is one that affects one part or a limited area of the body. A **systemic disease** affects either the entire body or several parts.

The science that deals with why, when, and where diseases occur and how they are transmitted in a human community is known as **epidemiology** (ep'-i-dē'-mē-OL-ō-jē; *epidemios* = prevalent; *logos* = study of). The science that deals with the effects and uses of drugs in the treatment of disease is called **pharmacology** (far'-ma-KOL-ō-jē; *pharmakon* = medicine; *logos* = study of).

Atrophy (*a* = without; *tropho* = nourish)   A decrease in the size of cells with subsequent decrease in the size of the affected tissue or organ; wasting away.

Biopsy (*bio* = life; *opsis* = vision)   The removal and microscopic examination of tissue from the living body for diagnosis.

Deterioration (*deterior* = worse or poorer)   The process or state of growing worse; disintegration or wearing away.

Dysplasia (*dys* = abnormal; *plas* = to grow)   Alteration in the size, shape, and organization of cells due to chronic irritation or inflammation; may progress to neoplasia or revert to normal if the stress is removed.

Hyperplasia (*hyper* = over)   Increase in the number of cells due to an increase in the frequency of cell division.

Hypertrophy   Increase in the size of cells without cell division.

Insidious   Hidden, not apparent, as a disease that does not exhibit distinct symptoms of its arrival.

Metaplasia (*meta* = change)   The transformation of one cell into another.

Metastasis (*stasis* = standing still)   The transfer of disease from one part of the body to another that is not directly connected with it.

Necrosis (*necros* = death; *osis* = condition)   Death of a group of cells.

Neoplasm (*neo* = new)   Any abnormal formation or growth, usually a malignant tumor.

Progeny (*progignere* = to bring forth)   Offspring or descendants.

Senescence   The process of growing old.

STUDY OUTLINE

Generalized Animal Cell (p. 46)

1. A cell is the basic, living, structural and functional unit of the body.
2. A generalized cell is a composite that represents various cells of the body.
3. Cytology is the science concerned with the study of cells.
4. The principal parts of a cell are the plasma (cell) membrane, cytoplasm, organelles, and inclusions. Extracellular materials are manufactured by the cell and deposited outside the plasma membrane.

Plasma (Cell) Membrane (p. 46)

Chemistry and Structure (p. 46)

1. The plasma (cell) membrane, surrounds the cell and separates it from other cells and the external environment.
2. It is composed primarily of proteins and phospholipids. According to the fluid mosaic model, the membrane consists of a phospholipid bilayer with integral and peripheral proteins.

Functions (p. 47)

1. Functionally, the plasma membrane facilitates contact with other cells, provides receptors, and mediates the passage of materials.
2. The membrane's selectively permeable nature restricts the passage of certain substances. Substances can pass through the membrane depending on their molecular size, lipid solubility, electrical charges, and the presence of carriers.

Movement of Materials Across Plasma Membranes (p. 48)

1. Passive processes involve the kinetic energy of individual molecules.
2. Diffusion is the net movement of molecules or ions from an area of higher concentration to an area of lower concentration until an equilibrium is reached.
3. In facilitated diffusion, certain molecules, such as glucose, combine with a carrier to become soluble in the phospholipid portion of the membrane.
4. Osmosis is the movement of water through a selectively permeable membrane from an area of higher water concentration to an area of lower water concentration.
5. In an isotonic solution, red blood cells maintain their normal shape; in a hypotonic solution, they undergo hemolysis; in a hypertonic solution, they undergo crenation.
6. Filtration is the movement of water and dissolved substances across a selectively permeable membrane by pressure.
7. Dialysis is the diffusion of solute particles across a selectively permeable membrane and involves the separation of small molecules from large molecules.
8. Active processes involve the use of ATP by the cell.
9. Active transport is the movement of ions across a cell membrane from lower to higher concentration.
10. Endocytosis is the movement of substances through plasma membranes in which the membrane surrounds the substance, encloses it, and brings it into the cell.
11. Phagocytosis is the ingestion of solid particles by pseudopodia. It is an important process used by white blood cells to destroy bacteria that enter the body.
12. Pinocytosis is the ingestion of a liquid by the plasma membrane. In this process, the liquid becomes surrounded by a vacuole.
13. Receptor-mediated endocytosis is the selective uptake of large molecules and particles by cells.

Cytoplasm (p. 53)

1. Cytoplasm is the substance inside the cell between the plasma membrane and nucleus that contains organelles and inclusions.
2. It is composed mostly of water plus proteins, carbohydrates, lipids, and inorganic substances. The chemicals in cytoplasm are either in solution or in a colloid (suspended) form.
3. Functionally, cytoplasm is the medium in which chemical reactions occur.

Organelles (p. 53)

1. Organelles are specialized portions of the cell with characteristic morphology that carry on specific activities.
2. They assume specific roles in cellular growth, maintenance, repair, and control.

Nucleus (p. 54)

1. Usually the largest organelle, the nucleus controls cellular activities and contains the genetic information.
2. Cells without nuclei, such as mature red blood cells, do not grow or reproduce.
3. The parts of the nucleus include the nuclear membrane, karyolymph, nucleoli, and genetic material (DNA), comprising the chromosomes.
4. Chromosomes consist of DNA and histones and of subunits called nucleosomes.

Ribosomes (p. 55)

1. Ribosomes are granular structures consisting of ribosomal RNA and ribosomal proteins.
2. They occur free (singly or in clusters) or in conjunction with endoplasmic reticulum.
3. Functionally, ribosomes are the sites of protein synthesis.

Endoplasmic Reticulum (ER) (p. 56)

1. The ER is a network of parallel membranes continuous with the plasma membrane and nuclear membrane.
2. Granular or rough ER has ribosomes attached to it. Agranular or smooth ER does not contain ribosomes.
3. The ER provides mechanical support, conducts intracellular nerve impulses in muscle cells, exchanges materials with cytoplasm, transports substances intracellularly, stores synthesized molecules, and helps export chemicals from the cell.

Golgi Complex (p. 56)

1. The Golgi complex consists of four to eight stacked, flattened membranous sacs (cisternae) referred to as cis, medial, and trans.
2. The principal function of the Golgi complex is to process, sort, and deliver proteins within the cell. It also secretes proteins.

Mitochondria (p. 58)

1. Mitochondria consist of a smooth outer membrane and a folded inner membrane surrounding the interior matrix. The inner folds are called cristae.
2. The mitochondria are called "powerhouses of the cell" because ATP is produced in them.

Lysosomes (p. 59)

1. Lysosomes are spherical structures that contain digestive enzymes. They are formed from Golgi complexes.
2. They are found in large numbers in white blood cells, which carry on phagocytosis.

3. If the cell is injured, lysosomes release enzymes and digest the cell. Thus they are called "suicide packets."
4. Lysosomes may be involved in bone removal.

Peroxisomes (p. 59)

1. Peroxisomes are similar to lysosomes, but smaller.
2. They contain enzymes (e.g., catalase) involved in the metabolism of hydrogen peroxide.

The Cytoskeleton (p. 59)

1. Together microfilaments, microtubules, and intermediate filaments form the cytoskeleton.
2. Microfilaments are rodlike structures consisting of the protein actin or myosin. They are involved in muscular contraction, support, and movement.
3. Microtubules are cylindrical structures consisting of the protein tubulin. They support, provide movement, and form the structure of flagella, cilia, centrioles, and the mitotic spindle.
4. Intermediate filaments appear to provide structural reinforcement in some cells.

Centrosome and Centrioles (p. 60)

1. The dense area of cytoplasm containing the centrioles is called a centrosome.
2. Centrioles are paired cylinders arranged at right angles to one another. They assume an important role in cell reproduction.

Flagella and Cilia (p. 61)

1. These cellular projections have the same basic structure and are used in movement.
2. If projections are few and long, they are called flagella. If they are numerous and hairlike, they are called cilia.
3. The flagellum on a sperm cell moves the entire cell. The cilia on cells of the respiratory tract move foreign matter trapped in mucus along the cell surfaces toward the throat for elimination.

Cell Inclusions (p. 61)

1. Cell inclusions are chemical substances produced by cells. They are usually organic and may have recognizable shapes.
2. Examples are melanin, glycogen, lipids, and mucus.

Extracellular Materials (p. 61)

1. These are substances that lie outside the plasma membrane.
2. They provide support and a medium for the diffusion of nutrients and wastes.
3. Some, like hyaluronic acid and chondroitin sulfate, are amorphous. Others, like collagenous, reticular, and elastic fibers, are fibrous.

Gene Action (p. 62)

Protein Synthesis (p. 62)

1. Most of the cellular machinery is concerned with synthesizing proteins.
2. Cells make proteins by translating the genetic information encoded in DNA into specific proteins. This involves transcription and translation.
3. In transcription, genetic information encoded in DNA is copied by a strand of messenger RNA (mRNA); the DNA strand that serves as the template is called the sense strand.
4. DNA also synthesizes ribosomal RNA (rRNA) and transfer RNA (tRNA).
5. The process of using the information in the nitrogen base sequence of mRNA to dictate the amino acid sequence of a protein is known as translation.
6. mRNA associates with ribosomes, which consist of rRNA and protein.
7. Specific amino acids are attached to molecules of tRNA. Another portion of the tRNA has a triplet of bases called an anticodon; a codon is a segment of three bases of mRNA.
8. tRNA delivers a specific amino acid to the codon; the ribosome moves along an mRNA strand as amino acids are joined to form a growing polypeptide.

"SOS" Genes: DNA Repair (p. 64)

1. The structure of DNA is vulnerable to damage by harmful radiations and various chemicals.
2. Damage could lead to cellular malfunction that might lead to cancer.
3. In response to DNA damage, an "SOS response" occurs—certain genes produce enzymes that repair genetic damage.

Normal Cell Division (p. 64)

1. Cell division is the process by which cells reproduce themselves. It consists of nuclear division and cytoplasmic division (cytokinesis).
2. Cell division that results in an increase in body cells is called somatic cell division and involves a nuclear division called mitosis and cytokinesis.
3. Cell division that results in the production of sperm and eggs is called reproductive cell division and consists of a nuclear division called meiosis and cytokinesis.

Somatic Cell Division (p. 65)

1. Prior to mitosis and cytokinesis, the DNA molecules, or chromosomes, replicate themselves so the same chromosomal complement can be passed on to future generations of cells.
2. A cell carrying on every life process except division is said to be in interphase (metabolic phase).
3. Mitosis is the distribution of two sets of chromosomes into separate and equal nuclei following their replication.
4. It consists of prophase, metaphase, anaphase, and telophase.
5. Cytokinesis usually begins in late anaphase and terminates in telophase.
6. A cleavage furrow forms at the cell's equator and progresses inward, cutting through the cell to form two separate portions of cytoplasms.

Reproductive Cell Division (p. 67)

1. Gametes contain the haploid chromosome number and uninucleated somatic cells contain the diploid chromosome number.
2. Meiosis is the process that produces haploid gametes. It consists of two successive nuclear divisions called reduction division and equatorial division.
3. During reduction division, homologous chromosomes undergo synapsis and crossing-over; the net result is two haploid daughter cells.
4. During equatorial division, the two haploid daughter cells undergo mitosis and the net result is four haploid cells.

Abnormal Cell Division: Cancer (p. 69)

1. Cancerous tumors are referred to as malignant; noncancerous tumors are called benign; the study of tumors is called oncology.
2. The spread of cancer from its primary site is called metastasis.
3. Carcinogens include environmental agents and viruses.
4. Treating cancer is difficult because all the cells in a single population do not behave the same way.

Cells and Aging (p. 71)
1. Aging is a progressive failure of the body's homeostatic adaptive responses.
2. Many theories of aging have been proposed, including genetically programmed cessation of cell division and excessive

immune responses, but none successfully answers all the experimental objections.
3. All of the various body systems exhibit definitive and sometimes extensive changes with aging.

REVIEW QUESTIONS

1. Define a cell. What are the four principal portions of a cell? What is meant by a generalized cell?
2. Discuss the chemistry and structure of the plasma membrane with respect to the fluid mosaic model.
3. How do integral and peripheral membrane proteins differ in function?
4. Describe the various functions of the plasma membrane. What determines selective permeability?
5. What are the major differences between active processes and passive processes in moving substances across plasma membranes?
6. Define and give an example of each of the following: diffusion, facilitated diffusion, osmosis, filtration, dialysis, active transport, phagocytosis, pinocytosis, and receptor-mediated endocytosis.
7. Compare the effect on red blood cells of an isotonic, hypertonic, and hypotonic solution. What is osmotic pressure?
8. Discuss the chemical composition and physical nature of cytoplasm. What is its function?
9. What is an organelle? By means of a labeled diagram, indicate the parts of a generalized animal cell.
10. Describe the structure and functions of the nucleus of a cell. What are nucleosomes?
11. Discuss the distribution of ribosomes. What is their function?
12. Distinguish between granular (rough) and agranular (smooth) endoplasmic reticulum (ER). What are the functions of ER?
13. Describe the structure and functions of the Golgi complex.
14. Why are mitochondria referred to as "powerhouses of the cell"?
15. List and describe the various functions of lysosomes.
16. What is the importance of peroxisomes?
17. Contrast the structure and functions of microfilaments, microtubules, and intermediate filaments.
18. Describe the structure and function of centrioles.
19. How are cilia and flagella distinguished on the basis of structure and function?
20. Define a cell inclusion. Provide examples and indicate their functions.
21. What are extracellular materials? Give examples and the functions of each.
22. Summarize the steps involving gene action in protein synthesis.
23. How is the repair of DNA accomplished?
24. What is recombinant DNA? What is its clinical usefulness?
25. Distinguish between the two types of cell division. Why is each important?
26. Define interphase.
27. How does DNA replicate itself?
28. Describe the principal events of each stage of mitosis.
29. Distinguish between haploid and diploid cells.
30. Define meiosis and contrast the principal events of reduction division and equatorial division.
31. What is a tumor? Distinguish between malignant and benign tumors. Describe the principal types of malignant tumors.
32. Define metastasis. What factors contribute to metastasis?
33. Discuss several possible causes of cancer.
34. What are some of the problems with respect to treating cancer?
35. What is aging? List some of the characteristics of aging.
36. Briefly describe the various theories regarding aging.
37. Refer to the glossary of medical terminology associated with cells. Be sure that you can define each term.

The Tissue Level of Organization

- Define a tissue.
- Classify the tissues of the body into four major types and define each type.
- Discuss the distinguishing characteristics of epithelial tissue.
- Contrast the structural and functional differences between covering and lining epithelium and glandular epithelium.
- Compare the layering arrangements and cell shapes of covering and lining epithelium.
- List the structure, location, and function for the following types of epithelium: simple squamous, simple cuboidal, simple columnar (nonciliated and ciliated), stratified squamous, stratified cuboidal, stratified columnar, transitional, and pseudostratified.
- Define a gland and distinguish between exocrine and endocrine glands.
- Classify exocrine glands according to structural complexity and function and give an example of each.
- Identify the distinguishing characteristics of connective tissue.
- Contrast the structural and functional differences between embryonic and adult connective tissues.
- Discuss the intercellular substance, fibers, and cells that constitute connective tissue.
- List the structure, function, and location of loose (areolar) connective tissue, adipose tissue, and dense, elastic, and reticular connective tissue.
- List the structure, function, and location of the three types of cartilage.
- Distinguish between interstitial and appositional growth of cartilage.
- Describe the structure and functions of osseous tissue (bone) and vascular tissue (blood).
- Contrast the three types of muscle tissue with regard to structure and location.
- Describe the structural features and functions of nervous tissue.
- Define an epithelial membrane.
- List the location and function of mucous, serous, cutaneous, and synovial membranes.
- Describe the conditions necessary for tissue repair.
- Explain the importance of nutrition, adequate circulation, and age to tissue repair.

Cells are highly organized units, but they do not function in isolation. They work together in a group of similar cells called a tissue.

TYPES OF TISSUES

A **tissue** is a group of similar cells and their intercellular substance functioning together to perform a specialized activity. The science that deals with the study of tissues is called **histology** (*histio* = tissue; *logos* = study of). Certain tissues function to move body parts. Others move food through body organs. Some tissues protect and support the body. Others produce chemicals such as enzymes and hormones. The various tissues of the body are classified into four principal types according to their function and structure:

1. **Epithelial** (ep'-i-THĒ-lē-al) **tissue**, which covers body surfaces or tissues, lines body cavities, and forms glands.
2. **Connective tissue**, which protects and supports the body and its organs and binds organs together.
3. **Muscular tissue**, which is responsible for movement.
4. **Nervous tissue**, which initiates and transmits nerve impulses that coordinate body activities.

Epithelial tissue and connective tissue, except for bone and blood, will be discussed in detail in this chapter. The general features of bone tissue and blood will be introduced here, but their detailed discussion occurs later in the book. Similarly, the detailed discussion of muscle tissue and nervous tissue will be postponed.

EPITHELIAL TISSUE

The tissues in this principal type perform many activities in the body, ranging from protection of underlying tissues against microbial invasion, drying out, and harmful environmental factors to secretion. **Epithelial tissue,** or more simply **epithelium,** may be divided into two subtypes: (1) *covering and lining epithelium* and (2) *glandular epithelium.* Covering and lining epithelium forms the outer covering of external body surfaces and the outer covering of some internal organs. It lines the body cavities and the interiors of the respiratory and digestive tracts, blood vessels, and ducts. It makes up, along with nervous tissue, the parts of the sense organs that are sensitive to the stimuli that produce smell and hearing sensations. And, it is the tissue from which gametes (sperm and eggs) develop. Glandular epithelium constitutes the secreting portion of glands.

Both types of epithelium consist largely or entirely of closely packed cells with little or no intercellular material between adjacent cells. (Such intercellular material is also called the matrix.) The points of attachment between adjacent plasma membranes of epithelial cells are called **cell junctions.** They not only provide cell-to-cell attachments but also inhibit the movement of materials into certain cells and provide channels for communication between other cells. Epithelial cells are arranged in continuous sheets that may be either single or multilayered. Nerves may extend through these sheets, but blood vessels

do not. They are *avascular* (*a* = without, *vascular* = blood vessels). The vessels that supply nutrients and remove wastes are located in underlying connective tissue.

Both types of epithelium overlie and adhere firmly to the connective tissue, which holds the epithelium in position and prevents it from being torn. The surface of attachment between the epithelium and the connective tissue is a thin extracellular layer called the **basement membrane.** With few exceptions, epithelial cells secrete along their basal surfaces a material consisting of a special type of collagen and glycoproteins. This structure averages 50 to 80 nm in thickness and is referred to as the **basal lamina.** Frequently, the basal lamina is reinforced by an underlying **reticular lamina** consisting of reticular fibers and glycoproteins. This lamina is produced by cells in the underlying connective tissue. The combination of the basal lamina and reticular lamina constitutes the basement membrane.

Some tissues of the body are so highly differentiated (specialized) that they have lost their capacity for mitosis. Examples are muscle tissue and nervous tissue. Other tissues, such as epithelium, which are subjected to a certain amount of wear and tear and injury, have a continuous capacity for renewal because they contain stem cells that are capable of renewing the tissue. Examining cells that are sloughed off provides the basis for the Pap smear, a test for precancer and cancer diagnosis of the uterus (Chapter 28).

COVERING AND LINING EPITHELIUM

Arrangement of Layers

Covering and lining epithelium is arranged in several different ways related to location and function. If the epithelium is specialized for absorption or filtration and is in an area that has minimal wear and tear, the cells of the tissue are arranged in a single layer. Such an arrangement is called **simple epithelium.** If the epithelium is not specialized for absorption or filtration and is found in an area with a high degree of wear and tear, then the cells are stacked in several layers. This tissue is referred to as **stratified epithelium.** A third, less common arrangement of epithelium is called **pseudostratified.** Like simple epithelium, pseudostratified epithelium has only one layer of cells. However, some of the cells do not reach the surface—an arrangement that gives the tissue a multilayered, or stratified, appearance. The pseudostratified cells that do reach the surface either secrete mucus or contain cilia that move mucus and foreign particles for eventual elimination from the body.

Cell Shapes

In addition to classifying covering and lining epithelium according to the number of its layers, we may also categorize it by cell shape. The cells may be flat, cubelike, or columnar or may resemble a cross between shapes. **Squamous** (SKWĀ-mus) cells are flattened and scalelike. They

EXHIBIT 4-1   EPITHELIAL TISSUES

COVERING AND LINING EPITHELIUM

Simple squamous (250×)

Description: Single layer of flat, scalelike cells; large, centrally located nuclei.

Location: Lines air sacs of lungs, glomerular (Bowman's) capsule of kidneys, and inner surfaces of the membranous labyrinth and tympanic membrane of ear. Called endothelium when it lines heart, blood and lymphatic vessels, and forms capillaries. Called mesothelium when it lines the ventral body cavity and covers viscera as part of a serous membrane.

Function: Filtration, absorption, and secretion in serous membranes.

Cell boundary

Cytoplasm

Nuclei of simple squamous cells

are attached to each other like a mosaic. **Cuboidal** cells are usually cube-shaped in cross section. They sometimes appear as hexagons. **Columnar** cells are tall and cylindrical, appearing as rectangles set on end. **Transitional** cells often have a combination of shapes and are found where there is a great degree of distention or expansion in the body. Transitional cells in the bottom layer of an epithelial tissue may range in shape from cuboidal to columnar. In the intermediate layer, they may be cuboidal or polyhedral. In the superficial layer, they may range from cuboidal to squamous, depending on how much they are pulled out of shape during certain body functions.

Classification

Considering layers and cell type in combination, we may classify covering and lining epithelium as follows:

Simple
1. Squamous
2. Cuboidal
3. Columnar

Stratified
1. Squamous
2. Cuboidal
3. Columnar
4. Transitional

Pseudostratified

Each of the epithelial tissues described in the following sections is illustrated in Exhibit 4-1.

Simple Epithelium

● *Simple Squamous Epithelium*   This type of simple epithelium consists of a single layer of flat, scalelike cells. Its surface resembles a tiled floor. The nucleus of each cell is centrally located and oval or spherical. Since simple squamous epithelium has only one layer of cells, it is highly adapted to diffusion, osmosis, and filtration. Thus it lines the air sacs of the lungs, where oxygen is exchanged with carbon dioxide. It is present in the part of the kidney that filters blood. It also lines the inner surfaces of the membranous labyrinth and tympanic membrane of the ear. Simple squamous epithelium is found in body parts that have little wear and tear.

Simple squamous epithelium that lines the heart, blood vessels, and lymph vessels and forms the walls of capillaries is known as **endothelium.** Simple squamous epithelium that forms the epithelial layer of serous membranes is called **mesothelium.** This tissue lines the thoracic and abdominopelvic cavities and covers the viscera within them.

● *Simple Cuboidal Epithelium*   Viewed from above, the cells of simple cuboidal epithelium appear as closely fitted polygons. The cuboidal nature of the cells is obvious only when the tissue is sectioned at right angles. Like simple squamous epithelium, these cells possess a central nucleus. Simple cuboidal epithelium covers the surface of the ovaries, lines the anterior surface of the capsule of the lens of the eye, and forms the pigmented epithelium

EXHIBIT 4-1 (*Continued*)

Simple cuboidal (450×)
Description: Single layer of cube-shaped cells; centrally located nuclei.
Location: Covers surface of ovary, lines anterior surface of capsule of the lens of eyes, forms pigmented epithelium of retina of eye, and lines kidney tubules and smaller ducts of many glands.
Function: Secretion and absorption.

Simple columnar (nonciliated) (600×)
Description: Single layer of nonciliated rectangular cells; contains goblet cells; nuclei at bases of cells.
Location: Lines the digestive tract from the cardia of the stomach to the anus, excretory ducts of many glands, and gallbladder.
Function: Secretion and absorption.

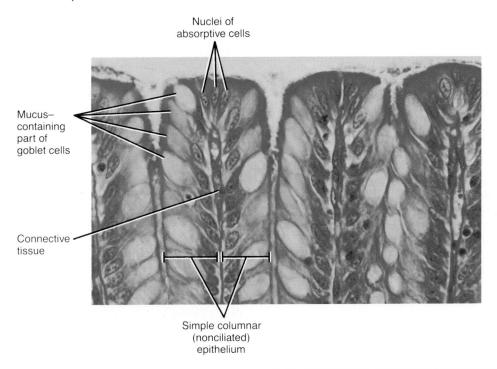

of the retina of the eye. In the kidneys, where it forms the kidney tubules and contains microvilli, it functions in water reabsorption. It also lines the smaller ducts of some glands and the secreting units of glands, such as the thyroid.

This tissue performs the functions of secretion and absorption. **Secretion,** usually a function of epithelium, is the production and release by cells of a fluid that may contain a variety of substances such as mucus, perspiration, or enzymes. **Absorption** is the intake of fluids or other substances by cells of the skin or mucous membranes.

EXHIBIT 4-1 (*Continued*)

Simple columnar (ciliated) (400×)
Description: Single layer of ciliated columnar cells; contains goblet cells; nuclei at bases of cells.
Location: Lines a few portions of upper respiratory tract, uterine (Fallopian) tubes, uterus, some paranasal sinuses, and central canal of spinal cord.
Function: Moves mucus by ciliary action.

Stratified squamous (185×)
Description: Several layers of cells; cuboidal to columnar shape in deep layers; scalelike shape in superficial layers; basal cells replace surface cells as they are lost.
Location: Nonkeratinized variety lines wet surfaces such as lining of the mouth, tongue, esophagus, part of epiglottis, and vagina; keratinized variety forms outer layer of skin.
Function: Protection.

EXHIBIT 4-1 (*Continued*)

Stratified cuboidal (185×)
Description: Two or more layers of cells in which the surface cells are cube-shaped.
Location: Ducts of adult sweat glands, fornix of conjunctiva of eye, cavernous urethra of male urogenital system, pharynx, and epiglottis.
Function: Protection.

Ducts of sweat glands   Lumen of duct

Nuclei of stratified cuboidal cells

Connective tissue

Secretory portion of sweat gland

Stratified cuboidal epithelium

Stratified columnar (600×)
Description: Several layers of polyhedral cells; columnar cells only in superficial layer.
Location: Lines part of male urethra, large excretory ducts of some glands, and small areas in anal mucous membrane.
Function: Protection and secretion.

Connective tissue

Lumen of urethra

Stratified columnar epithelium

Nuclei of stratified columnar cells

EXHIBIT 4-1 (Continued)

Stratified transitional (185×)
Description: Resembles nonkeratinized stratified squamous tissue, except that superficial cells are larger and more rounded.
Location: Lines urinary bladder.
Function: Permits distention.

Lumen of urinary bladder

Stratified transitional epithelium

Nuclei of transitional cells

Connective tissue

Pseudostratified (500×)
Description: Not a true stratified tissue; nuclei of cells at different levels; all cells attached to basement membrane, but not all reach surface.
Location: Lines larger excretory ducts of many large glands, male urethra, and auditory (Eustachian) tubes; ciliated variety with goblet cells lines most of the upper respiratory tract and some ducts of male reproductive system.
Function: Secretion and movement of mucus by ciliary action.

Lumen of duct

Nucleus of columnar cell

Nucleus of basal cell

Connective tissue

Pseudostratified epithelium

• *Simple Columnar Epithelium* The surface view of simple columnar epithelium is similar to that of simple cuboidal tissue. When sectioned at right angles to the surface, however, the cells appear somewhat rectangular. The nuclei are located near the bases of the cells.

The luminal surfaces (surfaces adjacent to the lumen, or cavity of a hollow organ, vessel, or duct) of simple columnar epithelial cells are modified in several ways,

depending on location and function. Simple columnar epithelium lines the digestive tract from the cardia of the stomach to the anus, the gallbladder, and excretory ducts of many glands. In such sites, the cells protect the underlying tissues. Many of them are also modified to aid in food-related activities. In the small intestine especially, the plasma membranes of the cells are folded into **microvilli** (see Figure 3-1). The microvilli arrange-

EXHIBIT 4-1 (*Continued*)

GLANDULAR EPITHELIUM

Exocrine gland (300×)
Description: Secretes products into ducts.
Location: Sweat, oil, wax, and mammary glands of the skin; digestive glands such as salivary glands that secrete into mouth cavity.
Function: Produces mucus, perspiration, oil, wax, milk, or digestive enzymes.

Pancreatic
duct

Pancreatic
secretory
units
(epithelial
cells)

Endocrine gland (180×)
Description: Secretes hormones into blood.
Location: Pituitary at base of brain, thyroid and parathyroids near larynx, adrenals above kidneys, pancreas below stomach, ovaries in pelvic cavity, testes in scrotum, pineal at base of brain, and thymus between lungs.
Function: Produces hormones that regulate various body activities.

Thyroid
follicle

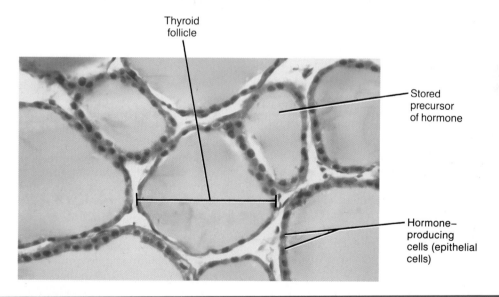

Stored
precursor
of hormone

Hormone-
producing
cells (epithelial
cells)

Photomicrographs Copyright © 1983 by Michael H. Ross. Used by permission.

ment increases the surface area of the plasma membrane and thereby allows larger amounts of digested nutrients and fluids to be absorbed into the body.

Interspersed among the typical columnar cells of the intestine are other modified columnar cells called **goblet cells.** These cells, which secrete mucus, are so named because the mucus accumulates in the upper half of the cell, causing the area to bulge out. The whole cell resem-

bles a goblet or wine glass. The secreted mucus serves as a lubricant between the food and the walls of the digestive tract.

A third modification of columnar epithelium is found in cells with hairlike processes called **cilia.** In a few portions of the upper respiratory tract, ciliated columnar cells are interspersed with goblet cells. Mucus secreted by the goblet cells forms a film over the respiratory surface. This film traps foreign particles that are inhaled. The cilia wave in unison and move the mucus, with any foreign particles, toward the throat, where it can be swallowed or eliminated. Air is filtered by this process before entering the lungs. Ciliated columnar epithelium is also found in the uterus and uterine (Fallopian) tubes of the female reproductive system, some paranasal sinuses, and the central canal of the spinal cord.

Stratified Epithelium

In contrast to simple epithelium, stratified epithelium consists of at least two layers of cells. Thus it is durable and can protect underlying tissues from the external environment and from wear and tear. Some stratified epithelium cells also produce secretions. The name of the specific kind of stratified epithelium depends on the shape of the *surface* cells.

● *Stratified Squamous Epithelium* In the more superficial layers of this type of epithelium, the cells are flat, whereas in the deep layers, cells vary in shape from cuboidal to columnar. The basal, or bottom, cells continually multiply by cell division. As new cells grow, they compress the cells on the surface and push them outward. The basal cells continually shift upward and outward. As they move farther from the deep layer and their blood supply, they become dehydrated, shrink, and become harder. At the surface, the cells are rubbed off. New cells continually emerge, are sloughed off, and replaced.

One form of stratified squamous epithelium is called **nonkeratinized stratified squamous epithelium.** This tissue is found on wet surfaces that are subjected to considerable wear and tear and does not perform the function of absorption—the lining of the mouth, the tongue, the esophagus, and the vagina. Another form of stratified squamous epithelium is called **keratinized stratified squamous epithelium.** The surface cells of this type are modified into a tough layer of material containing keratin. **Keratin** is a protein that is waterproof and resistant to friction and helps to resist bacterial invasion. The outer layer of skin consists of keratinized tissue.

● *Stratified Cuboidal Epithelium* This relatively rare type of epithelium is found in the ducts of the sweat glands of adults, fornix of the conjunctiva of the eye, cavernous urethra of the male urogenital system, pharynx, and epiglottis. It sometimes consists of more than two layers of cells. Its function is mainly protective.

● *Stratified Columnar Epithelium* Like stratified cuboidal epithelium, this type of tissue is also uncommon in the body. Usually the basal layer or layers consist of shortened, irregularly polyhedral cells. Only the superficial cells are columnar in form. This kind of epithelium lines part of the male urethra, some larger excretory ducts such as lactiferous (milk) ducts in the mammary glands, and small areas in the anal mucous membrane. It functions in protection and secretion.

● *Transitional Epithelium* This kind of epithelium is very much like nonkeratinized stratified squamous epithe-

lium. The distinction is that cells of the outer layer in transitional epithelium tend to be large and rounded rather than flat. This feature allows the tissue to be stretched (distended) without the outer cells breaking apart from one another. When stretched, they are drawn out into squamouslike cells. Because of this arrangement, transitional epithelium lines hollow structures that are subjected to expansion from within, such as the urinary bladder. Its function is to help prevent a rupture of the organ.

Pseudostratified Epithelium

The third category of covering and lining epithelium consists of columnar cells, and is called pseudostratified epithelium. The nuclei of the cells are at varying depths. Even though all the cells are attached to the basement membrane in a single layer, some do not reach the surface. This feature gives the impression of a multilayered tissue, the reason for the designation *pseudo*stratified epithelium. It lines the larger excretory ducts of many glands, parts of the male urethra, and the auditory (Eustachian) tubes, the tubes that connect the middle ear cavity and upper part of the throat. Pseudostratified epithelium may be ciliated and may contain goblet cells. In this form, it lines most of the upper respiratory tract and certain ducts of the male reproductive system.

GLANDULAR EPITHELIUM

The function of glandular epithelium is secretion, accomplished by glandular cells that lie in clusters deep to the covering and lining epithelium. A **gland** may consist of one cell or a group of highly specialized epithelial cells that secrete substances into ducts, onto a surface, or into the blood. The production of such substances always requires active work by the glandular cells and results in an expenditure of energy.

All glands of the body are classified as exocrine or endocrine according to whether they secrete substances into ducts (or directly onto a free surface) or into the blood. **Exocrine glands** secrete their products into ducts (tubes) that empty at the surface of covering and lining epithelium or directly onto a free surface. The product of an exocrine gland may be released at the skin surface or into the lumen of a hollow organ. The secretions of exocrine glands include mucus, perspiration, oil, wax, and digestive enzymes. Examples of exocrine glands are sweat glands, which eliminate perspiration to cool the skin, salivary glands, which secrete a digestive enzyme, and goblet cells, which produce mucus. **Endocrine glands** are ductless and secrete their products into the blood. The secretions of endocrine glands are always hormones, chemicals that regulate various physiological activities. The pituitary, thyroid, and adrenal glands are endocrine glands.

Structural Classification of Exocrine Glands

Exocrine glands are classified into two structural types: unicellular and multicellular. **Unicellular glands** are single-celled. A good example of a unicellular gland is a goblet cell (see Exhibit 4-1, simple columnar, nonciliated). Goblet cells are found in the epithelial lining of the digestive, respiratory, urinary, and reproductive systems. They produce mucus to lubricate the free surfaces of these membranes.

Multicellular glands occur in several different forms (Figure 4-1). If the secretory portions of a gland are tubular, it is referred to as a **tubular gland.** If they are flasklike, it is called an **acinar (AS-i-nar) gland.** If the gland contains both tubular and flasklike secretory portions, it is called a **tubuloacinar gland.** Further, if the duct of the gland does not branch, it is referred to as a **simple gland;** if the duct does branch, it is called a **compound gland.** By combining the shape of the secretory portion with the degree of branching of the duct, we arrive at the following structural classification for exocrine glands:

I. **Unicellular.** Single-celled gland that secretes mucus. Example: goblet cell of the digestive and respiratory systems.
II. **Multicellular.** Many-celled glands.
  A. *Simple.* Single, nonbranched duct.
   1. **Tubular.** The secretory portion is straight and tubular. Example: intestinal glands.
   2. **Branched tubular.** The secretory portion is branched and tubular. Examples: gastric and uterine glands.
   3. **Coiled tubular.** The secretory portion is coiled. Example: sudoriferous (sweat) glands.
   4. **Acinar.** The secretory portion is flasklike. Example: seminal vesicle glands.
   5. **Branched acinar.** The secretory portion is branched and flasklike. Example: sebaceous (oil) glands.
  B. *Compound.* Branched duct.
   1. **Tubular.** The secretory portion is tubular. Examples: bulbourethral (Cowper's) glands, testes, and liver.

2. **Acinar.** The secretory portion is flasklike. Examples: salivary glands (sublingual and submandibular).
3. **Tubuloacinar.** The secretory portion is both tubular and flasklike. Examples: salivary glands (parotid) and pancreas.

Functional Classification of Exocrine Glands

The functional classification of exocrine glands is based on how the gland releases its secretion. The three recognized categories are holocrine, merocrine, and apocrine glands. **Holocrine glands** accumulate a secretory product in their cytoplasm. The cell then dies and is discharged with its contents as the glandular secretion (Figure 4-2a). The discharged cell is replaced by a new cell. One example of a holocrine gland is a sebaceous (oil) gland of the skin. **Merocrine glands** simply form the secretory product and discharge it from the cell (Figure 4-2b). Examples of merocrine glands are the salivary glands and pancreas. **Apocrine glands** accumulate their secretory product at the apical (outer) margin of the secreting cell. That portion of the cell pinches off from the rest of the cell to form the secretion (Figure 4-2c). The remaining part of the cell repairs itself and repeats the process. An example of an apocrine gland is the mammary gland.

CONNECTIVE TISSUE

The most abundant tissue in the body is **connective tissue.** This binding and supporting tissue usually is highly vascular and thus has a rich blood supply. An exception is cartilage, which is avascular. The cells are widely scat-

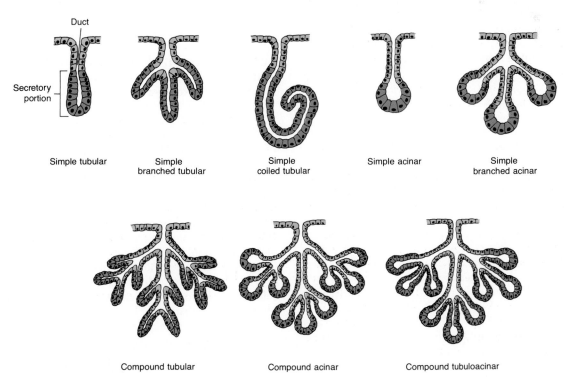

FIGURE 4-1 Structural types of multicellular exocrine glands. The secretory portions of the glands are indicated in purple. The blue areas represent the ducts of the glands.

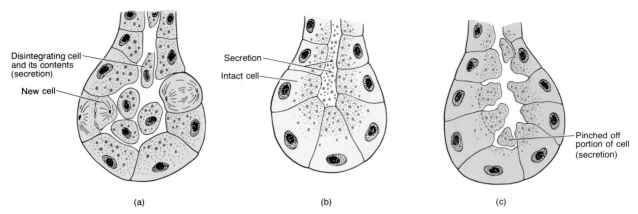

FIGURE 4-2 Functional classification of multicellular exocrine glands. (a) Holocrine gland. (b) Merocrine gland. (c) Apocrine gland.

tered, rather than closely packed, and there is considerable intercellular material, or matrix. In contrast to epithelium, connective tissues do not occur on free surfaces, such as the surfaces of a body cavity or the external surface of the body. The general functions of connective tissues are protection, support, and the binding together of various organs.

The intercellular substance in a connective tissue largely determines the tissue's qualities. This substance is nonliving and may consist of fluid, semifluid, mucoid (mucuslike), or fibrous material. In cartilage, the intercellular material is firm but pliable. In bone, it is considerably harder and not pliable. The cells of connective tissue produce the intercellular substances. The cells may also store fat, ingest bacteria and cell debris, form anticoagulants, or give rise to antibodies that protect against disease.

CLASSIFICATION

Connective tissues may be classified in several ways. We will classify them as follows:

I. **Embryonic connective tissue**
  A. Mesenchyme
  B. Mucous connective tissue
II. **Adult connective tissue**
  A. Connective tissue proper
    1. Loose (areolar) connective tissue
    2. Adipose tissue
    3. Dense (collagenous) connective tissue
    4. Elastic connective tissue
    5. Reticular connective tissue
  B. Cartilage
    1. Hyaline cartilage
    2. Fibrocartilage
    3. Elastic cartilage
  C. Osseous (bone) tissue
  D. Vascular (blood) tissue

Each of the connective tissues described in the following sections is illustrated in Exhibit 4-2.

EMBRYONIC CONNECTIVE TISSUE

Connective tissue that is present primarily in the embryo or fetus is called **embryonic connective tissue.** The term *embryo* refers to a developing human from fertilization through the first two months of pregnancy; a *fetus* refers to a developing human from the third month of pregnancy to birth.

One example of embryonic connective tissue found almost exclusively in the embryo is **mesenchyme (MEZ-en-kīm)**—the tissue from which all other connective tissues eventually arise. Mesenchyme may be observed beneath the skin and along the developing bones of the embryo. Some mesenchymal cells are scattered irregularly throughout adult connective tissue, most frequently around blood vessels. Here mesenchymal cells differentiate into fibroblasts that assist in wound healing.

Another kind of embryonic connective tissue is **mucous connective tissue (Wharton's jelly),** found primarily in the fetus. This tissue is located in the umbilical cord of the fetus, where it supports the wall of the cord.

ADULT CONNECTIVE TISSUE

Adult connective tissue is connective tissue that exists in the newborn and that does not change after birth. It is subdivided into several kinds.

Connective Tissue Proper

Connective tissue that has a more or less fluid intercellular material and a fibroblast as the typical cell is termed **connective tissue proper.** Five examples of such tissues may be distinguished.

• *Loose (Areolar) Connective Tissue* Loose or areolar (a-RĒ-ō-lar) connective tissue is one of the most widely distributed connective tissues in the body. Structurally, it consists of fibers and several kinds of cells embedded in a semifluid intercellular substance. The term "loose" refers to the loosely woven arrangement of fibers in the intercellular substance. The fibers are neither abundant nor arranged to prevent stretching.

The intercellular substance consists of a viscous material called **hyaluronic acid.** Hyaluronic acid normally facilitates the passage of nutrients from the blood vessels of the connective tissue into adjacent cells and tissues,

EXHIBIT 4-2   CONNECTIVE TISSUES

EMBRYONIC

Mesenchymal (180×)

Description: Consists of highly branched mesenchymal cells embedded in a fluid substance.

Location: Under skin and along developing bones of embryo; some mesenchymal cells found in adult connective tissue, especially along blood vessels.

Function: Forms all other kinds of connective tissue.

Blood vessels

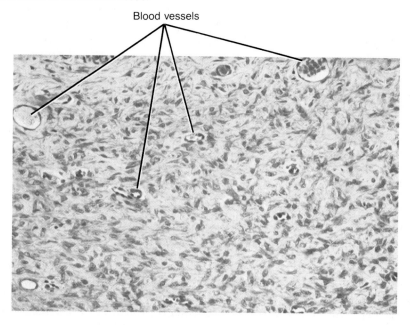

Mucous (320×)

Description: Consists of flattened or spindle-shaped cells embedded in a mucuslike substance containing fine collagenous fibers.

Location: Umbilical cord of fetus.

Function: Support.

Epithelial surface cells of umbilical cord

Collagenous fibers

Nuclei of fiber-forming cells

EXHIBIT 4-2 (*Continued*)

ADULT

Loose or areolar (180×)

Description: Consists of fibers (collagenous, elastic, and reticular) and several kinds of cells (fibroblasts, macrophages, plasma cells, and mast cells) embedded in a semifluid ground substance.

Location: Subcutaneous layer of skin, mucous membranes, blood vessels, nerves, and body organs.

Function: Strength, elasticity, and support.

Collagenous fibers

Elastic fibers

Adipose (250×)

Description: Consists of adipocytes, "signet ring"-shaped cells with peripheral nuclei, that are specialized for fat storage.

Location: Subcutaneous layer of skin, around heart and kidneys, marrow of long bones, and padding around joints.

Function: Reduces heat loss through skin, serves as an energy reserve, supports, and protects.

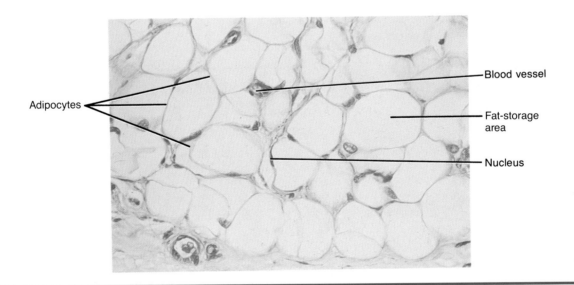

Adipocytes

Blood vessel

Fat-storage area

Nucleus

although the thick consistency of this acid may impede the movement of some drugs. But if an enzyme called **hyaluronidase** is injected into the tissue, hyaluronic acid changes to a watery consistency. This feature is of clinical importance because the reduced viscosity hastens the absorption and diffusion of injected drugs and fluids through the tissue and thus can lessen tension and pain. Some bacteria, white blood cells, and sperm cells produce hyaluronidase.

The three types of fibers embedded between the cells of loose connective tissue are collagenous, elastic, and reticular fibers. **Collagenous (white) fibers** are very tough and resistant to a pulling force, yet are somewhat flexible because they are usually wavy. These fibers often occur in bundles. They are composed of many minute fibers called fibrils lying parallel to one another. The bundle arrangement affords a great deal of strength. Chemically, collagenous fibers consist of the protein collagen. **Elastic**

EXHIBIT 4-2 (*Continued*)

Dense or collagenous (250×)
Description: Consists of predominately collagenous (white) fibers arranged in bundles; fibroblasts present in rows between bundles.
Location: Forms tendons, ligaments, aponeuroses, membranes around various organs, and fasciae.
Function: Provides strong attachment between various structures.

Fibroblasts

Collagenous fibers

Elastic (180×)
Description: Consists of predominately freely branching elastic (yellow) fibers; fibroblasts present in spaces between fibers.
Location: Lung tissue, cartilage of larynx, wall of arteries, trachea, bronchial tubes, true vocal cords, and ligamenta flava of vertebrae.
Function: Allows stretching of various organs.

Elastic fibers

Collagenous fiber

(yellow) fibers, by contrast, are smaller than collagenous fibers and freely branch and rejoin one another. Elastic fibers consist of a protein called elastin. These fibers also provide strength and have great elasticity, up to 50 percent of their length. **Reticular fibers** also consist of collagen, plus some glycoprotein. They are very thin fibers that branch extensively and are not as strong as collagenous fibers. Some authorities believe that reticular fibers are immature collagenous fibers. Like collagenous fibers, retic-ular fibers provide support and strength and form the *stroma* (framework) of many soft organs.

The cells in loose connective tissue are numerous and varied. Most are **fibroblasts**—large, flat cells with branching processes. If the tissue is injured, the fibroblasts are believed to form collagenous fibers, elastic fibers, and the viscous ground substance. Mature fibroblasts are referred to as **fibrocytes.** The basic distinction between the two is that fibroblasts (or *"blasts"* of any form of cell) are

EXHIBIT 4-2 (Continued)

Reticular (250×)
Description: Consists of a network of interlacing reticular fibers with thin, flat cells wrapped around fibers.
Location: Liver, spleen, and lymph nodes.
Function: Forms stroma of organs; binds together smooth muscle tissue cells.

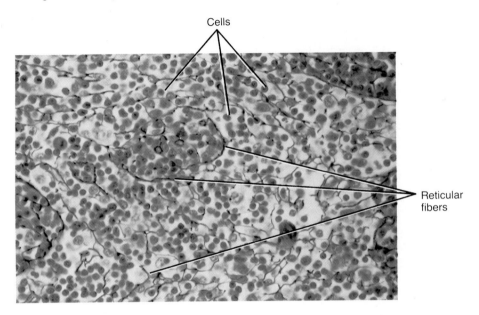

Cells

Reticular fibers

Hyaline cartilage (350×)
Description: Also called gristle; appears as a bluish white, glossy mass; contains numerous chondrocytes; is the most abundant type of cartilage.
Location: Ends of long bones, ends of ribs, nose, parts of larynx, trachea, bronchi, bronchial tubes, and embryonic skeleton.
Function: Provides movement at joints, flexibility, and support.

Lacunae containing chondrocytes

Intercellular substance (matrix)

Nuclei of chondrocytes

involved in the formation of immature tissue or repair of mature tissue, and fibrocytes (or "*cytes*" of any form of cell) are inactive cells that no longer produce fibers or matrix.

Other cells found in loose connective tissue are called **macrophages** (MAK-rō-fā-jez; *macro* = large; *phagein* = to eat), which are derived from white blood cells called monocytes. They are irregular in form with short branch-ing projections and are capable of engulfing bacteria and cellular debris by phagocytosis. Thus they provide a vital defense for the body. The types of macrophages are dis-cussed in Chapter 22 in relation to tissue inflammation.

A third kind of cell in loose connective tissue is the **plasma cell.** These cells are small and either round or irregular. Plasma cells develop from a type of white blood cell called a lymphocyte (B cell). They give rise to anti-

EXHIBIT 4-2 (*Continued*)

Fibrocartilage (180×)
Description: Consists of chondrocytes scattered among bundles of collagenous fibers.
Location: Symphysis pubis, intervertebral discs, and menisci of knee.
Function: Support and fusion.

Collagenous fibers

Intercellular substance (matrix)

Chondrocyte

Elastic cartilage (350×)
Description: Consists of chondrocytes located in a threadlike network of elastic fibers.
Location: Epiglottis of larynx, external ear, and auditory (Eustachian) tubes.
Function: Gives support and maintains shape.

Chondrocytes

Elastic fibers in intercellular substance (matrix)

Elastic fibers in perichondrium

bodies and, accordingly, provide a defensive mechanism through immunity. Plasma cells are found in many places of the body, but most are found in connective tissue, especially that of the digestive tract and the mammary glands.

Another cell in loose connective tissue is the **mast cell.** It may develop from another type of white blood cell called a basophil. The mast cell is somewhat larger than a basophil and is found in abundance along blood vessels. It forms heparin, an anticoagulant that prevents blood from clotting in the vessels. Mast cells are also believed to produce histamine and serotonin, chemicals that dilate small blood vessels.

Other cells in loose connective tissue include **melanocytes (pigment cells), adipocytes (fat cells),** and **leucocytes (white blood cells).**

EXHIBIT 4-2 (*Continued*)

Osseous (bone) (150×)

Description: Compact bone consists of osteons (Haversian systems) that contain lamellae, lacunae, osteocytes, canaliculi, and central (Haversian) canals.

Location: Both compact and spongy bone comprise the various bones of the body.

Function: Support, protection, movement, storage, blood cell production.

Vascular (blood) (1800×)

Description: Consists of plasma (intercellular substance) and formed elements (erythrocytes, leucocytes, and thrombocytes).

Location: Within blood vessels (arteries, arterioles, capillaries, venules, and veins).

Function: Erythrocytes transport oxygen and carbon dioxide, leucocytes carry on phagocytosis and are involved in allergic reactions and immunity, and thrombocytes are essential to the clotting of blood.

Loose connective tissue is continuous throughout the body. It is present in all mucous membranes and around all blood vessels and nerves. And it occurs around body organs and in the papillary region of the dermis of the skin. Combined with adipose tissue, it forms the **subcutaneous** (sub'-kyoo-TĀ-nē-us; *sub* = under, *cut* = skin) **layer**—the layer of tissue that attaches the skin to underlying tissues and organs. The subcutaneous layer is also referred to as the **superficial fascia** (FASH-ē-a) or **hypodermis.**

• *Adipose Tissue*   Adipose tissue is basically a form of loose connective tissue in which the cells, called **adipocytes,** are specialized for fat storage. Adipocytes are derived from fibroblasts and the cells have the shape of a "signet ring" because the cytoplasm and nucleus are pushed to the edge of the cell by a large droplet of fat. Adipose tissue is found wherever loose connective tissue is located. Specifically, it is in the subcutaneous layer below the skin, around the kidneys, at the base and on the surface of the heart, in the marrow of long bones, as a padding around joints, and behind the eyeball in the orbit. Adipose tissue is a poor conductor of heat and therefore reduces heat loss through the skin. It is also a major energy reserve and generally supports and protects various organs.

CLINICAL APPLICATION

A surgical procedure, called **suction lipectomy** (*lipo* = fat; *ectomy* = removal of) involves suctioning out small amounts of fat from certain areas of the body. The technique can be used in a number of areas of the body in which fat accumulates, such as the inner and outer thighs, buttocks, area behind the knee, underside of the arms, breasts, and abdomen. However, suction lipectomy is not a treatment for obesity, since it will not result in a permanent reduction in fat in the area.

• *Dense (Collagenous) Connective Tissue*   Dense (collagenous) connective tissue is characterized by a close packing of fibers and less intercellular substance than in loose connective tissue. The fibers can be either irregularly arranged or regularly arranged. In areas of the body where tensions are exerted in various directions, the fiber bundles are interwoven and without regular orientation. Such a dense connective tissue is referred to as **irregularly arranged** and occurs in sheets. It forms most fasciae, the reticular region of the dermis of the skin, the periosteum of bone, the perichondrium of cartilage, and the *membrane (fibrous) capsules* around organs, such as the kidneys, liver, testes, and lymph nodes.

In other areas of the body, dense connective tissue is adapted for tension in one direction, and the fibers have an orderly, parallel arrangement. Such a dense connective tissue is known as **regularly arranged.** The most common variety of dense regularly arranged connective tissue has a predominance of collagenous (white) fibers arranged in bundles. Fibroblasts are placed in rows between the bundles. The tissue is silvery white, tough, yet somewhat pliable. Because of its great strength, it is the principal component of *tendons,* which attach muscles to bones; many *ligaments* (collagenous ligaments), which hold bones together at joints; and *aponeuroses* (ap'-ō-noo-RŌ-sēz), which are flat bands connecting one muscle with another or with bone.

CLINICAL APPLICATION

Scientists have developed a carbon fiber implant for **reconstructing severely torn ligaments and tendons.** The implant consists of carbon fibers coated with a plastic called polylactic acid. The coated fibers are sewn in and around torn ligaments and tendons to reinforce them and to provide a scaffolding around which the body's own collagenous fibers grow. Within 2 weeks, the polylactic acid is absorbed by the body and the carbon fibers eventually fracture. By this time, the fibers are completely clad in collagen produced by fibroblasts.

• *Elastic Connective Tissue*   Unlike collagenous connective tissue, elastic connective tissue has a predominance of freely branching elastic fibers. These fibers give the tissue a yellowish color. Fibroblasts are present only in the spaces between the fibers. Elastic connective tissue can be stretched and will snap back into shape. It is a component of the cartilages of the larynx, the walls of elastic arteries, the trachea, the bronchial tubes to the lungs, and the lungs themselves. Elastic connective tissue provides stretch and strength, allowing structures to perform their functions efficiently. Yellow elastic ligaments, as contrasted with collagenous ligaments, are composed mostly of elastic fibers; they form the ligamenta flava of the vertebrae (ligaments between successive vertebrae), the suspensory ligament of the penis, and the true vocal cords.

• *Reticular Connective Tissue*   Reticular connective tissue consists of interlacing reticular fibers. It helps to form a delicate supporting stroma for many organs, including the liver, spleen, and lymph nodes. Reticular connective tissue also helps to bind together the cells of smooth muscle tissue.

Cartilage

One type of connective tissue, called cartilage, is capable of enduring considerably more stress than the tissues just discussed. Unlike other connective tissues, cartilage has no blood vessels (except for the perichondrium) or nerves of its own. **Cartilage** consists of a dense network of collagenous fibers and elastic fibers firmly embedded in chondroitin sulfate, a jellylike substance. Whereas the strength of cartilage is due to its collagenous fibers, its resilience (ability to assume its original shape after deformation) is due to chondroitin sulfate. The cells of mature cartilage, called **chondrocytes** (KON-drō-sīts), occur singly or in groups within spaces called **lacunae** (la-KOO-nē) in the intercellular substance. The surface of cartilage is surrounded by irregularly arranged dense connective tissue called the **perichondrium** (per'-i-KON-drē-um; *peri* = around, *chondro* = cartilage). Three kinds of cartilage are recognized: hyaline cartilage, fibrocartilage, and elastic cartilage (Exhibit 4-2).

• *Growth of Cartilage*   The growth of cartilage follows two basic patterns. In **interstitial (endogenous) growth,** the cartilage increases rapidly in size through the division of existing chondrocytes and continuous deposition of increasing amounts of intercellular matrix by the chondrocytes. The formation of new chondrocytes and their production of new intercellular matrix causes the cartilage to expand from within, thus the term *interstitial growth.* This growth pattern occurs while the cartilage is young and pliable—during childhood and adolescence.

In **appositional (exogenous) growth,** the growth of cartilage occurs because of the activity of the inner chondrogenic layer of the perichondrium. The deeper cells of the perichondrium, the fibroblasts, divide. Some differentiate into chondroblasts (immature cells that develop into specialized cells) and then into chondrocytes. As differentiation occurs, the chondroblasts become surrounded with intercellular matrix and become chondrocytes. As a result, the matrix is deposited on the surface of the cartilage, increasing its size. The new layer of cartilage is added beneath the perichondrium on the surface of the cartilage, causing it to grow in width. Appositional growth starts later than interstitial growth and continues throughout life.

• *Hyaline Cartilage*   This cartilage, also called gristle, appears as a bluish-white, glossy, homogeneous mass. The collagenous fibers, although present, are not visible with ordinary staining techniques, and the prominent chondrocytes are found in lacunae. Hyaline cartilage is the most abundant kind of cartilage in the body. It is found at joints over the ends of the long bones (where it is called *articular cartilage*) and forms the *costal cartilages* at the ventral ends of the ribs. Hyaline cartilage also helps to form the nose, larynx, trachea, bronchi, and bronchial tubes leading to the lungs. Most of the embryonic skeleton consists of hyaline cartilage. The role of hyaline cartilage in bone formation is discussed in Chapter 6. Hyaline cartilage affords flexibility and support.

• *Fibrocartilage*   Chondrocytes scattered through many bundles of visible collagenous fibers are found in this type of cartilage. Fibrocartilage is found at the symphysis pubis, the point where the coxal (hip) bones fuse anteriorly at the midline. It is also found in the intervertebral discs between vertebrae and the menisci of the knee. This tissue combines strength and rigidity.

• *Elastic Cartilage*   In this tissue, chondrocytes are located in a threadlike network of elastic fibers. Elastic cartilage provides strength and maintains the shape of certain organs—the epiglottis of the larynx, the external part of the ear (the pinna), and the auditory (Eustachian) tubes.

Osseous Tissue (Bone)

Together, cartilage and **osseous** (OS-ē-us) **tissue (bone)** comprise the skeletal system. Like other connective tissues, osseous tissue contains abundant intercellular substance surrounding widely separated cells. Mature bone cells are called **osteocytes.** The intercellular substances consist of mineral salts, primarily calcium phosphate and calcium carbonate, and collagenous fibers. The salts are responsible for the hardness of bone.

Bone tissue is classified as either compact (dense) or spongy (cancellous), depending on how the intercellular substance and cells are organized. At this point, we will discuss compact bone only. The basic unit of compact bone is called an **osteon (Haversian system).** Each osteon consists of **lamellae,** concentric rings of hard, intercellular substance; **lacunae,** small spaces between lamellae that contain osteocytes; **canaliculi,** radiating minute canals that provide numerous routes so that nutrients can reach osteocytes and wastes can be removed from them; and a **central (Haversian) canal** that contains blood vessels and nerves. Notice that whereas osseous tissue is vascular, cartilage is not. Also, the lacunae of osseous tissue are interconnected by canaliculi; those of cartilage are not.

Functionally, the skeletal system supports soft tissues, protects delicate structures, works with skeletal muscles to produce movement, stores calcium and phosphorus, and houses red marrow, which produces several kinds of blood cells.

The details of compact and spongy bone are discussed in Chapter 6.

Vascular Tissue (Blood)

Vascular tissue (blood) is a liquid connective tissue that consists of an intercellular substance called plasma and formed elements (cells and cell-like structures). **Plasma** is a straw-colored liquid that consists mostly of water plus some dissolved substances (nutrients, enzymes, hormones, respiratory gases, and ions). The formed elements are erythrocytes (red blood cells), leucocytes (white blood cells), and thrombocytes (platelets).

Erythrocytes function in transporting oxygen to body cells and removing carbon dioxide from them. **Leucocytes** are involved in phagocytosis, immunity, and allergic reactions. **Thrombocytes** function in blood clotting.

The details of blood are considered in Chapter 19.

MUSCLE TISSUE

Muscle tissue consists of highly specialized cells that are modified for contraction. As a result of this characteristic, muscle tissue provides motion, maintenance of posture, and heat production. On the basis of certain structural and functional characteristics, muscle tissue is classified into three types: skeletal, cardiac, and smooth (Exhibit 4-3).

Skeletal muscle tissue is named for its location—attached to bones. It is also *striated,* that is, the cells contain alternating light and dark bands (striations) that are perpendicular to the long axes of the cells. The striations are visible under a microscope. Skeletal muscle tissue is also *voluntary* because it can be made to contract by conscious control. A single skeletal muscle cell (fiber) is cylindrical, and the cells are parallel to each other in a tissue. Each muscle fiber contains a plasma membrane, the *sarcolemma,* surrounding the cytoplasm, or *sarcoplasm.* Skeletal muscle fibers are multinucleate (they have more than one nucleus), and the nuclei lie close to the sarcolemma. The contractile elements of skeletal muscle fibers are proteins called *myofibrils.* They contain wide, transverse, dark bands and narrow light ones that give the fibers the striated appearances.

Cardiac muscle tissue forms the bulk of the wall of the heart. Like skeletal muscle tissue, it is striated. However, unlike skeletal muscle tissue, it is usually involuntary; its contraction is usually not under conscious control. Cardiac muscle fibers are roughly quadrangular and branch to form networks throughout the tissue. The fibers usually have only one nucleus that is centrally located. Cardiac muscle fibers are separated from each other by transverse thickenings of the sarcolemma called *intercalated discs.* These are unique to cardiac muscle and serve to strengthen the tissue and aid nerve impulse conduction.

Smooth muscle tissue is located in the walls of hollow internal structures such as blood vessels, the stomach, intestines, and urinary bladder. Smooth muscle fibers are usually involuntary and they are *nonstriated (smooth).* Each smooth muscle cell is spindle-shaped, with either end tapering to a point, and contains a single, centrally located nucleus.

A more detailed discussion of muscle tissue is considered in Chapter 10.

NERVOUS TISSUE

Despite the tremendous complexity of the nervous system, it consists of only two principal kinds of cells: neurons and neuroglia. **Neurons,** or nerve cells, are highly specialized cells that are capable of picking up stimuli; converting stimuli to nerve impulses; and conducting nerve impulses to other neurons, muscle fibers, or glands. Neurons are the structural and functional units of the nervous system. Each consists of three basic portions: cell body and two kinds of processes called dendrites and axons (Exhibit 4-4). The *cell body (perikaryon)* contains the nucleus and other organelles. *Dendrites* are highly branched processes of the cell body that conduct nerve impulses toward the cell body. *Axons* are single, long processes of the cell body that conduct nerve impulses away from the cell body.

Neuroglia are cells that protect and support neurons. They are of clinical interest because they are frequently the sites of tumors of the nervous system.

The detailed structure and function of neurons and neuroglia are considered in Chapter 12.

EXHIBIT 4-3   MUSCLE TISSUE

Skeletal muscle tissue (800×)
Description: Cylindrical, striated fibers with several peripheral nuclei; voluntary.
Location: Attached to bones.
Function: Motion, posture, heat production.

Cardiac muscle tissue (400×)
Description: Quadrangular, branching, striated fibers with one centrally located nucleus; contains intercalated discs; usually involuntary.
Location: Heart wall.
Function: Motion (contraction of heart).

EXHIBIT 4-3 (Continued)

Smooth muscle tissue (840×)

Description: Spindle-shaped, nonstriated fibers with one centrally located nucleus; usually involuntary.

Location: Walls of hollow internal structures such as blood vessels, stomach, intestines, and urinary bladder.

Function: Motion (constriction of blood vessels, propulsion of foods through gastrointestinal tract, contraction of gallbladder).

Muscle fibers   Nuclei of muscle fibers

Sarcolemma

Photomicrographs Copyright © 1983 by Michael H. Ross. Used by permission.

MEMBRANES

The combination of an epithelial layer and an underlying connective tissue layer constitutes an **epithelial membrane.** The principal epithelial membranes of the body are mucous membranes, serous membranes, and the cutaneous membrane, or skin. Another kind of membrane, a **synovial membrane,** does not contain epithelium.

MUCOUS MEMBRANES

A **mucous membrane,** or **mucosa,** lines a body cavity that opens directly to the exterior. Mucous membranes line the entire digestive, respiratory, excretory, and reproductive tracts (see Figure 24-2). The surface tissue of a mucous membrane may vary in type—it is stratified squamous epithelium in the esophagus and simple columnar epithelium in the intestine.

The epithelial layer of a mucous membrane secretes mucus, which prevents the cavities from drying out. It also traps dust in the respiratory passageways and lubricates food as it moves through the digestive tract. In

EXHIBIT 4-4   NERVOUS TISSUE (640×)

Description: Neurons (nerve cells) consist of a cell body and processes extending from the cell body called dendrites or axons.

Location: Nervous system.

Function: Pick up stimuli, convert stimuli to nerve impulses, and conduct nerve impulses to other neurons, muscle fibers, or glands.

Processes

Cell body or perikaryon

Photomicrograph courtesy of Biophoto Associates/Photo Researchers.

addition, the epithelial layer is responsible for the secretion of digestive enzymes and the absorption of food.

The connective tissue layer of a mucous membrane is called the *lamina propria.* It binds the epithelium to the underlying structures and allows some flexibility of the membrane. It also holds the blood vessels in place, protects underlying muscles from abrasion or puncture, provides the epithelium covering it with oxygen and nutrients, and removes wastes.

SEROUS MEMBRANES

A **serous membrane,** or **serosa,** lines a body cavity that does not open directly to the exterior, and it covers the organs that lie within the cavity. Serous membranes consist of thin layers of loose connective tissue covered by a layer of mesothelium, and they are in the form of invaginated double-walled sacs. The part attached to the cavity wall is called the *parietal* (pa-RĪ-e-tal) *portion.* The part that covers the organs inside these cavities is the *visceral portion.* The serous membrane lining the thoracic cavity and covering the lungs is called the *pleura* (see Figure 1-7). The membrane lining the heart cavity and covering the heart is the *pericardium* (*cardio* = heart). The serous membrane lining the abdominal cavity and covering the abdominal organs and some pelvic organs is called the *peritoneum.*

The epithelial layer of a serous membrane secretes a lubricating fluid that allows the organs to glide easily against one another or against the walls of the cavities. The connective tissue layer of the serous membrane consists of a relatively thin layer of loose connective tissue.

CUTANEOUS MEMBRANE

The **cutaneous membrane,** or skin, constitutes an organ of the integumentary system and is discussed in the next chapter.

SYNOVIAL MEMBRANES

Synovial membranes line the cavities of the joints (see Figure 9-1a). Like serous membranes, they line structures that do not open to the exterior. Unlike mucous, serous, and cutaneous membranes, they do not contain epithelium and are therefore not epithelial membranes. They are composed of loose connective tissue with elastic fibers and varying amounts of fat. Synovial membranes secrete *synovial fluid,* which lubricates the ends of bones as they move at joints and nourishes the articular cartilage covering the bones that form the joints.

TISSUE REPAIR

Tissue repair is the process by which new tissues replace dead or damaged cells. New cells originate by cell duplication from the *stroma,* the supporting connective tissue, or from the *parenchyma,* cells which form the organ's functioning part. The epithelial cells that secrete and absorb are the parenchymal cells of the intestine, for example. The restoration of an injured organ or tissue to normal structure and function depends entirely on which type of cell—parenchymal or stromal—is active in the repair. If only parenchymal elements accomplish the repair, a perfect or near-perfect reconstruction of the injured tissue may occur. However, if the fibroblast cells of the stroma are active in the repair, the tissue will be replaced with new connective tissue called *scar tissue.* In this process, fibroblasts synthesize collagen and protein polysaccharides that aggregate to form scar tissue. The process of scar tissue formation is known as **fibrosis.** Since scar tissue is not specialized to perform the functions of the parenchymal tissue, the function of the tissue is impaired. If the rate of collagen breakdown in a scar exceeds production, the scar becomes softer and less bulky. If, on the other hand, the rate of collagen production exceeds breakdown, a **keloid (hypertrophic) scar** develops. Such a scar is sharply elevated and irregularly shaped.

CLINICAL APPLICATION

The scar tissue formed by fibrosis can cause the abnormal joining of tissues called **adhesions.** They are common in the abdomen and may occur around a site of previous inflammation such as an inflamed appendix, or they can follow surgery. Such adhesions make subsequent surgery more difficult. Surgery may be required to remove adhesions.

The cardinal factor in tissue repair lies in the capacity of parenchymal tissue to regenerate. This capacity, in turn, depends on the ability of the parenchymal cells to replicate quickly.

REPAIR PROCESS

If injury to a tissue is slight, repair may sometimes be accomplished with the drainage and reabsorption of pus, followed by parenchymal regeneration. When the area of skin loss is great, fluid moves out of the capillaries and the area becomes dry. Fibrin seals the open tissue by hardening into a **scab.**

When tissue and cell damage are extensive and severe, as in large, open wounds, both the connective tissue stroma and the parenchymal cells are active in repair. This repair involves the rapid cell division of many fibroblasts, the manufacture of new collagenous fibers to provide strength, and an increase by cell division of the number of small blood vessels in the area. All these processes create an actively growing, connective tissue called **granulation tissue.** This new granulation tissue forms across a wound or surgical incision to provide a framework (stroma). The framework supports the epithelial cells that migrate into the open area and fill it. The newly formed granulation tissue also secretes a fluid that kills bacteria.

CONDITIONS AFFECTING REPAIR

Three factors affect tissue repair: nutrition, blood circulation, and age. Nutrition is vital in the healing process since a great demand is placed on the body's store of nutrients. Protein-rich diets are important since most of the cell structure is made from proteins. Vitamins also play a direct role in wound healing. Among the vitamins involved and their roles in wound healing are the following:

1. Vitamin A is essential in the replacement of epithelial tissues, especially in the respiratory tract.
2. The B vitamins—thiamine, nicotinic acid, riboflavin—are needed by many enzyme systems in cells. They are needed especially for the enzymes involved in decomposing glucose to CO_2 and H_2O, which is crucial to both heart and nervous tissue. These vitamins may relieve pain in some cases.
3. Vitamin C directly affects the normal production and maintenance of intercellular substances. It is required for the manufacture of cementing elements of connective tissues, especially collagen. Vitamin C also strengthens and promotes the formation of new blood vessels. With vitamin C deficiency, even superficial wounds fail to heal, and the walls of the blood vessels become fragile and are easily ruptured.
4. Vitamin D is necessary for the proper absorption of calcium from the intestine. Calcium gives bones their hardness and is necessary for the healing of fractures.
5. Vitamin E is believed to promote healing of injured tissues and may prevent scarring.
6. Vitamin K assists in the clotting of blood and thus prevents the injured person from bleeding to death.

In tissue repair, proper blood circulation is indispensable. It is the blood that transports oxygen, nutrients, antibodies, and many defensive cells to the site of injury. The blood also plays an important role in the removal of tissue fluid, blood cells that have been depleted of oxygen, bacteria, foreign bodies, and debris. The elements would otherwise interfere with healing.

Generally, tissues heal faster and leave less obvious scars in the young than in the aged. The young body is generally in a much better nutritional state, its tissues have a better blood supply, and the cells of younger people have a faster metabolic rate. Thus cells can duplicate their materials and divide more quickly.

STUDY OUTLINE

Types of Tissues (p. 77)

1. A tissue is a group of similar cells and their intercellular substance that are specialized for a particular function.
2. Depending on their function and structure, the various tissues of the body are classified into four principal types: epithelial, connective, muscular, and nervous.

Epithelial Tissue (p. 77)

1. Epithelium has many cells, little intercellular material, and no blood vessels (avascular). It is attached to connective tissue by a basement membrane. It can replace itself.
2. The subtypes of epithelium include covering and lining epithelium and glandular epithelium.

Covering and Lining Epithelium (p. 77)

1. Layers are arranged as simple (one layer), stratified (several layers), and pseudostratified (one layer that appears as several); cell shapes include squamous (flat), cuboidal (cubelike), columnar (rectangular), and transitional (variable).
2. Simple squamous epithelium is adapted for diffusion and filtration and is found in lungs and kidneys. Endothelium lines the heart and blood vessels. Mesothelium lines the thoracic and abdominopelvic cavities and covers the organs within them.
3. Simple cuboidal epithelium is adapted for secretion and absorption. It is found covering ovaries, in kidneys and eyes, and lining some glandular ducts.
4. Nonciliated simple columnar epithelium lines most of the digestive tract. Specialized cells containing microvilli perform absorption. Goblet cells secrete mucus. In a few portions of the respiratory tract, the cells are ciliated to move foreign particles trapped in mucus out of the body.
5. Stratified squamous epithelium is protective. It lines the upper digestive tract and vagina and forms the outer layer of skin.
6. Stratified cuboidal epithelium is found in adult sweat glands, pharynx, epiglottis, and portions of the urethra.
7. Stratified columnar epithelium protects and secretes. It is found in the male urethra and large excretory ducts.
8. Transitional epithelium lines the urinary bladder and is capable of stretching.
9. Pseudostratified epithelium has only one layer but gives the appearance of many. It lines larger excretory ducts, parts of urethra, auditory (Eustachian) tubes, and most upper respiratory structures, where it protects and secretes.

Glandular Epithelium (p. 84)

1. A gland is a single cell or a mass of epithelial cells adapted for secretion.
2. Exocrine glands (sweat, oil, and digestive glands) secrete into ducts or directly onto a free surface.
3. Structural classification includes unicellular and multicellular glands; multicellular glands are further classified as tubular, acinar, tubuloacinar, simple, and compound.
4. Functional classification includes holocrine, merocrine, and apocrine glands.
5. Endocrine glands secrete hormones into the blood.

Connective Tissue (p. 85)

1. Connective tissue is the most abundant body tissue. It has few cells, an extensive intercellular substance, and a rich blood supply (vascular), except for cartilage. It does not occur on free surfaces.
2. The intercellular substance determines the tissue's qualities.

3. Connective tissue protects, supports, and binds organs together.
4. Connective tissue is classified into two principal types: embryonic and adult.

Embryonic Connective Tissue (p. 86)

1. Mesenchyme forms all other connective tissues.
2. Mucous connective tissue is found in the umbilical cord of the fetus, where it gives support.

Adult Connective Tissue (p. 86)

1. Adult connective tissue is connective tissue that exists in the newborn and does not change after birth. It is subdivided into several kinds: connective tissue proper, cartilage, bone tissue, and vascular tissue.
2. Connective tissue proper has a more or less fluid intercellular material, and a typical cell is the fibroblast. Five examples of such tissues may be distinguished.
3. Loose (areolar) connective tissue is one of the most widely distributed connective tissues in the body. Its intercellular substance (hyaluronic acid) contains fibers (collagenous, elastic, and reticular) and various cells (fibroblasts, macrophages, plasma, mast, and melanocytes). Loose connective tissue is found in all mucous membranes, around body organs, and in the subcutaneous layer.
4. Adipose tissue is a form of loose connective tissue in which the cells, called adipocytes, are specialized for fat storage. It is found in the subcutaneous layer and around various organs.
5. Dense (collagenous) connective tissue has a close packing of fibers (regularly or irregularly arranged). It is found as a component of fascia, membranes of organs, tendons, ligaments, and aponeuroses.
6. Elastic connective tissue has a predominance of freely branching elastic fibers that give it a yellow color. It is found in the cartilages of the larynx, elastic arteries, trachea, bronchial tubes, and true vocal cords.
7. Reticular connective tissue consists of interlacing reticular fibers and forms the stroma of the liver, spleen, and lymph nodes.
8. Cartilage has a jellylike matrix containing collagenous and elastic fibers and chondrocytes.
9. The growth of cartilage is accomplished by interstitial growth (from within) and appositional growth (from without).
10. Hyaline cartilage is found in the embryonic skeleton, at the ends of bones, in the nose, and in respiratory structures. It is flexible, allows movement, and provides support.
11. Fibrocartilage connects the pelvic bones and the vertebrae. It provides strength.
12. Elastic cartilage maintains the shape of organs such as the larynx, auditory (Eustachian) tubes, and external ear.
13. Osseous tissue (bone) consists of mineral salts and collagenous fibers that contribute to the hardness of bone and cells called osteocytes. It supports, protects, helps provide movement, stores minerals, and houses red marrow.
14. Vascular tissue (blood) consists of plasma and formed elements (erythrocytes, leucocytes, and thrombocytes). Functionally, its cells transport, carry on phagocytosis, participate in allergic reactions, provide immunity, and bring about blood clotting.

Muscle Tissue (p. 94)

1. Muscle tissue is modified for contraction and thus provides motion, maintenance of posture, and heat production.

2. Skeletal muscle tissue is attached to bones, is striated, and is voluntary.
3. Cardiac muscle tissue forms most of the heart wall, is striated, and is usually involuntary.
4. Smooth muscle tissue is found in the walls of hollow internal structures (blood vessels and viscera), is nonstriated, and is usually involuntary.

Nervous Tissue (p. 95)

1. The nervous system is composed of neurons (nerve cells) and neuroglia (protective and supporting cells).
2. Neurons consist of a cell body and two types of processes called dendrites and axons.
3. Neurons are specialized to pick up stimuli, convert stimuli into nerve impulses, and conduct nerve impulses.

Membranes (p. 96)

1. An epithelial membrane is an epithelial layer overlying a connective tissue layer. Examples are mucous, serous, and cutaneous membranes.
2. Mucous membranes line cavities that open to the exterior, such as the digestive tract.
3. Serous membranes (pleura, pericardium, peritoneum) line closed cavities and cover the organs in the cavities. These membranes consist of parietal and visceral portions.
4. The cutaneous membrane is the skin.
5. Synovial membranes line joint cavities and do not contain epithelium.

Tissue Repair (p. 97)

1. Tissue repair is the replacement of damaged or destroyed cells by healthy ones.
2. It begins during the active phase of inflammation and is not completed until after harmful substances in the inflamed area have been neutralized or removed.

Repair Process (p. 97)

1. If the injury is superficial, tissue repair involves pus removal (if pus is present), scab formation, and parenchymal regeneration.
2. If damage is extensive, granulation tissue is involved.

Conditions Affecting Repair (p. 97)

1. Nutrition is important to tissue repair. Various vitamins (A, some B, D, C, E, and K) and a protein-rich diet are needed.
2. Adequate circulation of blood is needed.
3. The tissues of young people repair rapidly and efficiently; the process slows down with aging.

REVIEW QUESTIONS

1. Define a tissue. What are the four basic kinds of human tissue?
2. Distinguish covering and lining epithelium from glandular epithelium. What characteristics are common to all epithelium?
3. Describe the origin and composition of the basement membrane.
4. Describe the various layering arrangements and cell shapes of epithelium.
5. How is epithelium classified? List the various types.
6. For each of the following kinds of epithelium, briefly describe the microscopic appearance, location in the body, and functions: simple squamous, simple cuboidal, simple columnar (nonciliated and ciliated), stratified squamous, stratified cuboidal, stratified columnar, transitional, and pseudostratified.
7. Define the following terms: endothelium, mesothelium, secretion, absorption, goblet cell, and keratin.
8. What is a gland? Distinguish between endocrine and exocrine glands.
9. Describe the classification of exocrine glands according to structure and function and give at least one example of each.
10. Enumerate the ways in which connective tissue differs from epithelium.
11. How are connective tissues classified? List the various types.
12. How are embryonic connective tissue and adult connective tissue distinguished?
13. Describe the following connective tissues with regard to microscopic appearance, location in the body, and function: loose (areolar), adipose, dense (collagenous), elastic, reticular, hyaline cartilage, fibrocartilage, elastic cartilage, osseous tissue (bone), and vascular tissue (blood).
14. Distinguish between the interstitial and appositional growth of cartilage.
15. Define the following terms: hyaluronic acid, collagenous fiber, elastic fiber, reticular fiber, fibroblast, macrophage, plasma cell, mast cell, melanocyte, adipocyte, chondrocyte, lacuna, osteocyte, lamella, and canaliculi.
16. Describe the histology of muscle tissue. How is it classified? What are its functions?
17. Distinguish between neurons and neuroglia. Describe the structure and function of neurons.
18. Define the following kinds of membranes: mucous, serous, cutaneous, and synovial. Where is each located in the body? What are their functions?
19. Following are some descriptions of various tissues of the body. For each description, name the tissue described.
    a. An epithelium that permits distention (stretching).
    b. A single layer of flat cells concerned with filtration and absorption.
    c. Forms all other kinds of connective tissue.
    d. Specialized for fat storage.
    e. An epithelium with waterproofing qualities.
    f. Forms the framework of many organs.
    g. Produces perspiration, wax, oil, or digestive enzymes.
    h. Cartilage that shapes the external ear.
    i. Contains goblet cells and lines the intestine.
    j. Most widely distributed connective tissue.
    k. Forms tendons, ligaments, and aponeuroses.
    l. Specialized for the secretion of hormones.
    m. Provides support in the umbilical cord.
    n. Lines kidney tubules and is specialized for absorption and secretion.
    o. Permits extensibility of lung tissue.
    p. Stores red marrow, protects, supports.
    q. Nonstriated, usually involuntary muscle tissue.
    r. Consists of granulocytes and agranulocytes.
    s. Composed of a cell body, dendrites, and axons.
20. What is meant by tissue repair? Distinguish between stromal and parenchymal repair.
21. What is the importance of granulation tissue?
22. What conditions affect tissue repair?

5

The Integumentary System

- Define the integumentary system.
- Describe the various functions of the skin.
- List the various layers of the epidermis and describe their structure and functions.
- Describe the composition and function of the dermis.
- Explain the basis for skin color.
- Explain the basic pattern of epidermal ridges and grooves.
- Outline the steps involved in epidermal wound healing and deep wound healing.
- Describe the development, distribution, and structure of hair.
- Compare the structure, distribution, and functions of sebaceous (oil), sudoriferous (sweat), and ceruminous glands.
- List the parts of a nail and describe their composition.
- Explain the role of the skin in helping to maintain the homeostasis of normal body temperature.
- Describe the effects of aging on the integumentary system.
- Describe the development of the epidermis, its derivatives, and the dermis.
- Describe the causes and effects for the following skin disorders: acne, systemic lupus erythematosus (SLE), psoriasis, decubitus ulcers, sunburn, and skin cancer.
- Define a burn and list the systemic effects of a burn.
- Classify burns into first, second, and third degrees and describe how to estimate the extent of a burn.
- Define medical terminology associated with the integumentary system.

An aggregation of tissues that performs a specific function is an **organ**. The next higher level of organization is a **system**—a group of organs operating together to perform specialized functions. The skin and its derivatives, such as hair, nails, glands, and several specialized receptors, constitute the **integumentary** (in'-teg-yoo-MEN-tar-ē) **system** of the body.

The developmental anatomy of the integumentary system is considered at the end of the chapter.

SKIN

The **skin** is an organ because it consists of tissues structurally joined together to perform specific activities. It is one of the larger organs of the body in terms of surface area. For the average adult, the skin occupies a surface area of approximately 19,355 sq cm (3,000 sq inches). The skin is not just a simple thin covering that keeps the body together and gives it protection. The skin is quite complex in structure and performs several functions essential for survival. **Dermatology** (*dermato* = skin; *logos* = study of) is the medical specialty that deals with the diagnosis and treatment of skin disorders.

FUNCTIONS

The numerous functions of the skin may be grouped as follows:

1. Maintenance of body temperature. In response to high environmental temperature or strenuous exercise, the production of perspiration by sweat glands helps to lower body temperature back to normal. This is described in detail later in the chapter.

2. Protection. The skin covers the body and provides a physical barrier that protects underlying tissues from physical abrasion, bacterial invasion, dehydration, and ultraviolet (UV) radiation.

3. Perception of stimuli. The skin contains numerous nerve endings and receptors that detect stimuli related to temperature, touch, pressure, and pain (Chapter 15).

4. Excretion. Not only does perspiration assume a role in helping to maintain normal body temperature, it also assists in the excretion of small amounts of water, salts, and several organic compounds.

5. Synthesis of vitamin D. The term *vitamin D* actually refers to a group of closely related compounds synthesized naturally from a precursor molecule present in the skin upon exposure to ultraviolet (UV) radiation. In the skin, the precursor substance, 7-dehydrocholesterol, is converted to cholecalciferol (vitamin D_3) in the presence of UV radiation. In the liver, cholecalciferol is converted to 25-hydroxycholecalciferol. Then, in the kidneys, this substance is changed into 1,25-dihydroxycalciferol (calcitriol), the most active form of vitamin D that stimulates the absorption of calcium and phosphorous from dietary foods. In subsequent chapters, when we speak of vitamin D, we are really referring to 1,25-dihydroxycalciferol. Vitamin D is actually a hormone, since it is produced in one location in the body, transported by the blood, and then exerts its effect in another location.

6. Immunity. As you will see in Chapter 22, certain cells of the epidermis assume a role in bolstering immunity.

STRUCTURE

Structurally, the skin consists of two principal parts (Figure 5-1). The outer, thinner portion, which is composed of epithelium, is called the **epidermis**. The epidermis is cemented to the inner, thicker, connective tissue part called the **dermis**. Thick skin has a relatively thick epidermis, whereas thin skin has a relatively thin epidermis. Beneath the dermis is a **subcutaneous layer.** This layer, also called the **superficial fascia** or **hypodermis,** consists of areolar and adipose tissues. Fibers from the dermis extend down into the superficial fascia and anchor the skin to the subcutaneous layer. The superficial fascia, in turn, is firmly attached to underlying tissues and organs.

EPIDERMIS

The **epidermis** is composed of stratified squamous epithelium and contains four distinct types of cells. The most numerous is known as a *keratinocyte,* a cell that undergoes keratinization. This process will be described shortly. The functions of these cells are to produce keratin, which helps waterproof and protect the skin and underlying tissues, and to participate in immunity. A second type of cell in the epidermis is called a *melanocyte.* Its role is to produce melanin, one of the pigments responsible for skin color. The third type of cell is referred to as a *Langerhan's cell,* which arises from bone marrow and invades the epidermis and other stratified squamous epithelial sites of the body where it functions in immunity (Chapter 22). The *Granstein cell,* the fourth principal type of cell, also functions in immunity.

The epidermis is organized into four or five cell layers, depending on its location in the body (see Figures 5-1 and 5-2). Where exposure to friction is greatest, such as in the palms and soles, the epidermis has five layers. In all other parts it has four layers. The names of the five layers from the deepest to the most superficial are as follows:

1. Stratum basale. This single layer of cuboidal to columnar cells is capable of continued cell division. As these cells multiply, they push up toward the surface and become part of the layers to be described next. Their nuclei degenerate, and the cells die. Eventually, the cells are shed from the top layer of the epidermis. The stratum basale is sometimes referred to as the **stratum germinativum** (jer'-mi-na-TĒ-vum) to indicate its role in germinating new cells.

2. Stratum spinosum. This layer of the epidermis contains eight to ten rows of polyhedral (many-sided) cells that fit closely together. The surfaces of these cells may assume a prickly appearance when prepared for microscope examination (*spinosum* = prickly). The deeper layers of the epidermis of hairless skin contain nerve endings sensitive to touch called *tactile (Merkel's) discs* (see Figure 15-1).

3. Stratum granulosum. The third layer of the epidermis consists of three to five rows of flattened cells that contain darkly staining granules of a substance called *keratohyalin* (ker'-a-tō-HĪ-a-lin). This compound is involved in the first step of keratin formation. *Keratin* is a waterproofing protein found in the top layer of the epidermis. The nuclei of the cells in the stratum

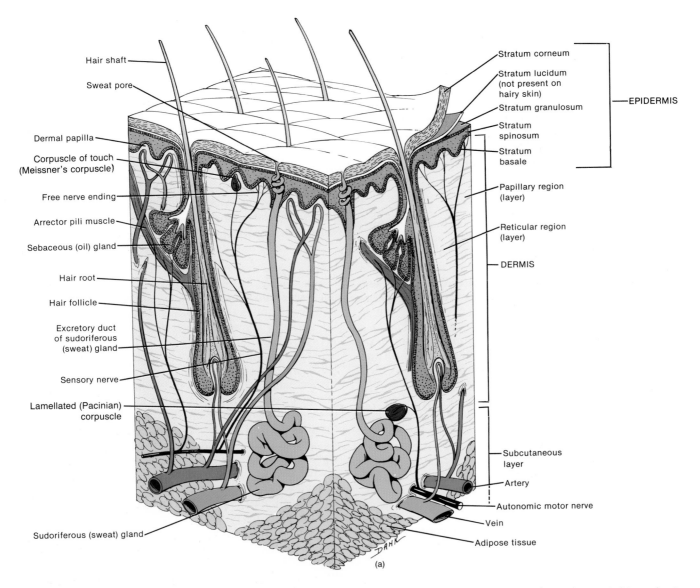

Hair shaft

Sweat pore

Dermal papilla

Corpuscle of touch
(Meissner's corpuscle)

Free nerve ending

Arrector pili muscle

Sebaceous (oil) gland

Hair root

Hair follicle

Excretory duct
of sudoriferous
(sweat) gland

Sensory nerve

Lamellated (Pacinian)
corpuscle

Sudoriferous (sweat) gland

Stratum corneum

Stratum lucidum
(not present on
hairy skin)

Stratum granulosum

Stratum
spinosum

Stratum
basale

EPIDERMIS

Papillary region
(layer)

Reticular region
(layer)

DERMIS

Subcutaneous
layer

Artery

Autonomic motor nerve

Vein

Adipose tissue

(a)

Hair shafts

EPIDERMIS

DERMIS

Blood
vessels

(b)

FIGURE 5-1 Skin. (a) Structure of the skin and underlying subcutaneous layer. (b) Scanning electron micrograph of the skin and several hairs at a magnification of 260×. (From *Tissues and Organs: A Text-Atlas of Scanning Electron Microscopy* by Richard G. Kessel and Randy H. Kardon. W. H. Freeman and Company. Copyright © 1979.)

granulosum are in various stages of degeneration. As these nuclei break down, the cells are no longer capable of carrying out vital metabolic reactions and die.

4. Stratum lucidum. This layer is normally found only in the thick skin of the palms and soles and is absent in thin skin. It consists of several rows of clear, flat, dead cells that contain droplets of a substance called *eleidin* (el-Ē-i-din). The layer is so named because eleidin is translucent (*lucidum* = clear). Eleidin is formed from keratohyalin and is eventually transformed to keratin.

5. Stratum corneum. This layer consists of 25 to 30 rows of flat, dead cells completely filled with keratin. These cells are continuously shed and replaced. The stratum corneum serves as an effective barrier against light and heat waves, bacteria, and many chemicals.

FIGURE 5-2 Photomicrograph of thick skin at a magnification of 50×. The stratum lucidum is not shown here, but is illustrated in Figure 5-1a. (Copyright © 1983 by Michael H. Ross. Used by permission.)

In the process of **keratinization,** newly formed cells produced in the basal layers are pushed up to more superficial layers. As the cells move toward the surface, the cytoplasm, nucleus, and other organelles are replaced by keratin and the cells die. Eventually, the keratinized cells are sloughed off and replaced by underlying cells that, in turn, become keratinized. The whole process by which a cell forms in the basal layers, rises to the surface, becomes keratinized, and sloughs off takes about 2 weeks.

DERMIS

The second principal part of the skin, the **dermis,** is composed of connective tissue containing collagenous and elastic fibers (see Figure 5-1). The dermis is very thick in the palms and soles and very thin in the eyelids, penis, and scrotum. It also tends to be thicker on the dorsal aspects of the body than the ventral and thicker on the lateral aspects of extremities than medial aspects. Numerous blood vessels, nerves, glands, and hair follicles are embedded in the dermis.

The upper region of the dermis, about one-fifth of the thickness of the total layer, is named the **papillary region** or **layer.** Its surface area is greatly increased by small, fingerlike projections called **dermal papillae** (pa-PIL-ē). These structures project into the concavities between ridges in the deep surface of the epidermis, and many contain loops of capillaries. Some dermal papillae contain *corpuscles of touch,* also called *Meissner's* (MĪS-nerz) *corpuscles.* As the name implies, these nerve endings are sensitive to touch. The papillary region consists of loose connective tissue containing fine elastic fibers.

The remaining portion of the dermis is called the **reticular region** or **layer.** It consists of dense, irregularly arranged connective tissue containing interlacing bundles of collagenous and coarse elastic fibers. It is named the reticular region because the bundles of collagenous fibers interlace in a netlike manner. Spaces between the fibers

are occupied by a small quantity of adipose tissue, hair follicles, nerves, oil glands, and the ducts of sweat glands. Varying thicknesses of the reticular region, among other factors, are responsible for differences in the thickness of the skin.

The combination of collagenous and elastic fibers in the reticular region provides the skin with strength, extensibility, and elasticity. (Extensibility is the ability to stretch; elasticity is the ability to return to original shape after extension or contraction.) The ability of the skin to stretch can readily be seen during conditions of pregnancy, obesity, and edema. The small tears that occur during extreme stretching are initially red and remain visible afterward as silvery white streaks called *striae* (STRĪ-ē).

The reticular region is attached to underlying organs, such as bone and muscle, by the subcutaneous layer. The subcutaneous layer also contains nerve endings called *lamellated* or *Pacinian* (pa-SIN-ē-an) *corpuscles* that are sensitive to pressure (see Figure 15-1).

CLINICAL APPLICATION

The collagenous fibers in the dermis run in all directions, but in particular regions of the body they tend to run more in one direction than another. The predominant direction of the underlying collagenous fibers is indicated in the skin by **lines of cleavage (tension lines).** The lines are especially evident on the palmar surfaces of the fingers where they run parallel to the long axis of the digit. Lines of cleavage are of particular interest to a surgeon because an incision running parallel to the collagen fibers will heal with only a fine scar. An incision made across the rows of fibers disrupts the collagen, and the wound tends to gape open and heal in a broad, thick scar.

SKIN COLOR

The color of the skin is due to melanin, a pigment in the epidermis; carotene, a pigment mostly in the dermis; and blood in capillaries in the dermis. The amount of **melanin** varies the skin color from pale yellow to black. This pigment is found primarily in the basale and spinosum layers. Melanin is synthesized in cells called **melanocytes** (MEL-a-nō-sīts), located either just beneath or between cells of the stratum basale. Since the number of melanocytes is about the same in all races, differences in skin color are due to the amount of pigment the melanocytes produce and disperse. An inherited inability of an individual in any race to produce melanin results in **albinism** (AL-bin-izm). The pigment is absent in the hair and eyes as well as the skin. An individual affected with albinism is called an **albino** (al-BĪ-no). The partial or complete loss of melanocytes from areas of skin produces patchy, white spots and the condition is called **vitiligo** (vit-i-LĪ-gō). In some people, melanin tends to form in patches called **freckles.**

Melanocytes synthesize melanin from the amino acid *tyrosine* in the presence of an enzyme called *tyrosinase*. Exposure to ultraviolet radiation increases the enzymatic activity of melanocytes and leads to increased melanin production. The cell bodies of melanocytes send out long processes between epidermal cells. Upon contact with the processes, epidermal cells take up the melanin by phagocytosis. When the skin is again exposed to ultraviolet radiation, both the amount and the darkness of melanin increase, tanning and further protecting the body against radiation. Thus melanin serves a vital protective function. Melanocyte-stimulating hormone (MSH) produced by the anterior pituitary gland causes increased melanin synthesis and dispersion through the epidermis.

CLINICAL APPLICATION

Overexposure of the skin to the ultraviolet light of the sun may lead to skin cancer. Among the most malignant and lethal skin cancers is **melanoma** (*melano* = dark-colored; *oma* = tumor), cancer of the melanocytes. Fortunately, most skin cancers involve basal and squamous cells and are curable.

Another pigment, called **carotene** (KAR-o-tēn), is found in the stratum corneum and fatty areas of the dermis in people of Asian origin. Together carotene and melanin account for the yellowish hue of their skin.

The pink color of Caucasian skin is due to blood in capillaries in the dermis. The redness of the vessels is not heavily masked by pigment. The epidermis has no blood vessels, a characteristic of all epithelia.

EPIDERMAL RIDGES AND GROOVES

The outer surface of the skin of the palms and fingers and soles and toes is marked by a series of ridges and grooves that appear either as fairly straight lines or as a pattern of loops and whorls, as on the tips of the digits.

The **epidermal ridges** develop during the third and fourth fetal months as the epidermis conforms to the contours of the underlying dermal papillae (see Figure 5-1a). Since the ducts of sweat glands open on the summits of the epidermal ridges, fingerprints (or footprints) are left when a smooth object is touched. The ridge pattern, which is genetically determined, is unique for each individual. Normally, it does not change throughout life, except to enlarge, and thus can serve as the basis for identification through fingerprints or footprints. The function of the ridges is to increase the grip of the hand or foot by increasing friction.

Epidermal grooves on other parts of the skin divide the surface into a number of diamond-shaped areas. Examine the dorsum of the hand as an example. Note that hairs typically emerge at the points of intersection of the grooves. Note also that the grooves increase in frequency and depth as regions of free joint movement are approached.

CLINICAL APPLICATION

It is now possible to improve in many cases the appearance of scars from deep acne, chickenpox, burns, cleft lip, and other disorders and to make age-related wrinkles disappear. The procedure that makes this possible is called a **collagen implant.** Collagen is the body's principal structural material. It is found in skin, bone, cartilage, tendons, ligaments, and various viscera and accounts for almost one-third of the total protein of the body.

The collagen implant is prepared from cattle collagen that is suspended in a saline solution containing lidocaine, a local anesthetic. Once injected into the skin, the solution disappears and the collagen, under the influence of body temperature, becomes a stationary fleshlike substance that is incorporated into the skin. The collagen implant becomes colonized by blood vessels and cells and acts as a natural structural framework in the skin.

Another procedure, called **chemical exfoliation (skin peel),** is used for more superficial problems such as frown lines, wrinkles, age spots, and superficial scars. It is performed under sedation or with light anesthesia. In this procedure, the natural oils in the skin are removed with solvents. Then, keratolytic agents (keratin-dissolving chemicals) are mixed with surgical soap or croton oil, applied thickly, and the area is wrapped in bandages for a few days. When the bandages are removed, the keratin-containing layers of the epidermis are peeled away. When the new surface layers grow back, the superficial problems are either greatly reduced or disappear completely.

SKIN WOUND HEALING

Here we will discuss the basic mechanisms by which skin wounds are repaired. First we will consider wounds that affect primarily the epidermis. Then we will describe wounds that extend to tissues deep to the epithelium such as the dermis and subcutaneous layer.

EPIDERMAL WOUND HEALING

The exposed location of the skin (and mucous membranes) causes it to be subjected to physical and chemical trauma. One common type of epidermal wound is an abrasion such as might be experienced in the form of a skinned knee or elbow. Another type is a first-degree or second-degree burn. In an epidermal wound, the central portion of the wound usually extends deep down to the dermis, whereas the edges of the wound usually involve only superficial damage to the epidermal cells.

In response to injury, basal epidermal cells in the area of the wound break their contacts with the basement membrane. These cells then enlarge and migrate across the wound (Figure 5-3). The cells appear to migrate as a sheet until advancing cells from opposite sides of the wound meet. When epidermal cells encounter each other, their continued migration is stopped by *contact inhibition.* According to this phenomenon, when one epidermal cell encounters another, its direction of movement is changed until it encounters another like cell, and so on. Continued migration of the epidermal cell is inhibited when it is finally in contact on all sides with other epidermal cells. Contact inhibition appears to occur only among like cells; in other words, contact inhibition does not occur between epidermal cells and other types of cells. Recall from Chapter 3 that malignant cells do not conform to the rules of contact inhibition and so they have the ability to invade body tissues with few restrictions.

Simultaneous with the migration of some basal epidermal cells, stationary basal epidermal cells divide to replace the migrated cells. Migration continues until resurfacing of the wound is accomplished. Following this, the migrated cells themselves divide to form new strata, thus thickening the new epidermis. Once coverage of the wound surface beneath a scab is sufficient, the scab sloughs off and the surface epidermal cells become keratinized (see Figure 5-4b). The various events involved in epidermal wound healing occur within 24 to 48 hours after wounding.

DEEP WOUND HEALING

When an injury extends to tissues deep to the epidermis, due to accidental lacerations or surgical incisions, the repair process is more complex than epidermal healing and scar formation results. The first step in deep wound healing involves inflammation (Chapter 22). Inflammation is a vascular and cellular response that serves to dispose of microbes, foreign material, and dying tissue in preparation for repair. During the **inflammatory phase,** a blood clot forms in the wound and loosely unites the wound edges, epithelial cells begin migrating across the wound (see Figure 5-3), vasodilation and increased permeability of blood vessels deliver neutrophils and monocytes that phagocytize microbes, and mesenchymal cells develop into fibroblasts (Figure 5-4a).

In the next phase, the **migratory phase,** the clot becomes a scab and epithelial cells migrate beneath the scab to bridge the wound, fibroblasts migrate along fibrin threads and begin synthesizing scar tissue (collagenous fibers and protein polysaccharides), and damaged blood vessels begin to regrow. During this phase, tissue filling the wound is called *granulation tissue.*

The **proliferative phase** is characterized by an extensive growth of epithelial cells beneath the scab, the deposition of collagenous fibers in random patterns by fibroblasts, and the continued growth of blood vessels.

In the final phase, the **maturation phase,** the scab sloughs off once the epidermis is restored to normal thickness, collagenous fibers become more organized, fibroblasts begin to disappear, and blood vessels are restored to normal (Figure 5-4b).

EPIDERMAL DERIVATIVES

Structures developed from the embryonic epidermis—hair, glands, nails—perform functions that are necessary

FIGURE 5-3 Epidermal wound healing. (a) Division of basal cells and migration across wound. (b) Resurfacing of wound.

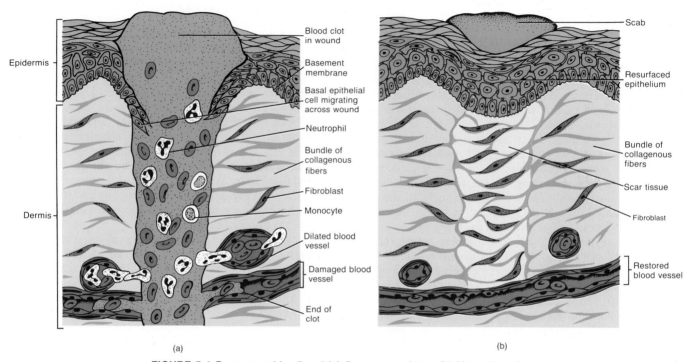

FIGURE 5-4 Deep wound healing. (a) Inflammatory phase. (b) Maturation phase.

and sometimes vital. Hair and nails protect the body. The sweat glands help regulate body temperature.

HAIR

Growths of the epidermis variously distributed over the body are **hair (pili).** The primary function of hair is protection. Though the protection is limited, hair guards the scalp from injury and the sun's rays. Eyebrows and eyelashes protect the eyes from foreign particles. Hair in the nostrils and external ear canal protects these structures from insects and dust.

Each hair consists of a shaft and a root (Figure 5-5a). The **shaft** is the superficial portion, most of which projects above the surface of the skin. The shaft of coarse hairs consists of three principal parts. The inner **medulla** is composed of rows of polyhedral cells containing granules of eleidin and air spaces. The middle **cortex** forms the major part of the shaft and consists of elongated cells that contain pigment granules in dark hair, but mostly air in white hair. The **cuticle of the hair,** the outermost layer, consists of a single layer of thin, flat, scalelike cells that are the most heavily keratinized. They are arranged like shingles on the side of a house, but the free edges of the cuticle cells point upward rather than downward like shingles (Figure 5-5c).

The **root** is the portion below the surface that penetrates into the dermis and even into the subcutaneous layer and, like the shaft, contains a medulla, cortex, and cuticle (Figure 5-5a).

Surrounding the root is the **hair follicle,** which is made up of an external root sheath and an internal root sheath. The **external root sheath** is a downward continuation of the basale and spinosum layers of the epidermis. Near the surface it contains all the epidermal layers. At the bottom of the hair follicle, the external root sheath contains only the stratum basale. The **internal root sheath** is formed from proliferating cells of the matrix (described shortly) and takes the form of a cellular tubular sheath deep to the external root sheath.

The base of each follicle is enlarged into an onion-shaped structure, the **bulb.** This structure contains an indentation, the **papilla of the hair,** filled with loose connective tissue. The papilla of the hair contains many blood vessels and provides nourishment for the growing hair. The bulb also contains a region of cells called the **matrix,** a germinal layer. The cells of the matrix produce new hairs by cell division when older hairs are shed. This replacement occurs within the same follicle.

Normal hair loss in an adult scalp is about 70 to 100 hairs per day. Both the rate of growth and the replacement cycle may be altered by illness, diet, and other factors.

Medulla
Cortex
Cuticle of
the hair
Hair
root
Internal
root
sheath

External root
sheath

Connective
tissue
sheath

Matrix

Blood
vessels

Papilla of
hair in bulb

(a)

EPIDERMIS

DERMIS

Opening of hair follicle
containing sebum from
sebaceous (oil) gland

Hair follicle

Site of sebaceous
(oil) gland opening
into follicle

Portion of
hair shaft

Hair
follicle

Secretory portions
of sebaceous (oil) glands

(b)

(c)

FIGURE 5-5 Hair. (a) Principal parts of a hair root and associated structures seen in longitudinal section. The relationship of a hair to the epidermis, sebaceous (oil) glands, and arrector pili muscle should be noted in Figure 5-1a. (b) Photomicrograph of thin skin of the face showing the location of sebaceous glands at a magnification of 60×. (Copyright © 1983 by Michael H. Ross. Used by permission). (c) Scanning electron micrograph of the surface of a hair shaft showing the shinglelike cuticular scales at a magnification of 1,000×. (Courtesy of Fisher Scientific Company and S.T.E.M. Laboratories, Inc., Copyright © 1975.)

For example, high fever, major illness, major surgery, blood loss, or severe emotional stress may increase the rate of shedding. Rapid weight-loss diets involving severe restriction of calories or protein also increase hair loss. An increase in the rate of shedding can also occur for 3 to 4 months after childbirth. Certain drugs and radiation therapy are also factors in increasing hair loss.

A substance that removes superfluous hair is called a **depilatory.** It dissolves the protein in the hair shaft, turning it into a gelatinous mass that can be wiped away. Since the hair root is not affected, regrowth of the hair occurs. In **electrolysis,** the hair bulb is destroyed by an electric current so that the hair cannot regrow.

Sebaceous glands and a bundle of smooth muscle are also associated with hair. Details of the sebaceous glands will be discussed shortly. The smooth muscle is called **arrector pili;** it extends from the dermis of the skin to the side of the hair follicle (see Figure 5-1a). In its normal position hair is arranged at an angle to the surface of the skin. The arrectores pilorum muscles contract under stresses of fright and cold and pull the hairs into a vertical position. This contraction results in "goosebumps" or "gooseflesh" because the skin around the shaft forms slight elevations.

Around each hair follicle are nerve endings, called *root hair plexuses,* that are sensitive to touch (see Figure 15-1). They respond if a hair shaft is moved.

CLINICAL APPLICATION

The age of onset, degree of thinning, and ultimate hair pattern associated with **common baldness (male-pattern baldness)** are determined by male hormones, called androgens, and heredity. Androgens are involved in promoting normal sexual development. In recent years, minoxidil, a drug used to treat high blood pressure, has been applied topically to stimulate hair regrowth in some persons with common baldness.

GLANDS

Three kinds of glands associated with the skin are sebaceous glands, sudoriferous glands, and ceruminous glands.

Sebaceous (Oil) Glands

Sebaceous (se-BĀ-shus) or **oil glands,** with few exceptions, are connected to hair follicles (Figure 5-5b). The secreting portions of the glands lie in the dermis, and those glands associated with hairs open into the necks of hair follicles. Sebaceous glands not associated with hair follicles open directly onto the surface of the skin (lips, glans penis, labia minora, and tarsal glands of the eyelids). Sebaceous glands are simple branched acinar glands. Absent in the palms and soles, they vary in size and shape in other regions of the body. For example, they are small in most areas of the trunk and extremities, but large in the skin of the breasts, face, neck, and upper chest.

The sebaceous glands secrete an oily substance called **sebum** (SĒ-bum), a mixture of fats, cholesterol, proteins, and inorganic salts. Sebum helps keep hair from drying and becoming brittle, forms a protective film that prevents excessive evaporation of water from the skin, keeps the skin soft and pliable, and inhibits the growth of certain bacteria.

CLINICAL APPLICATION

When sebaceous glands of the face become enlarged because of accumulated sebum, **blackheads** develop. Since sebum is nutritive to certain bacteria, **pimples** or **boils** often result. The color of blackheads is due to melanin and oxidized oil, not dirt.

Sudoriferous (Sweat) Glands

Sudoriferous (soo'-dor-IF-er-us; *sudor* = sweat; *ferre* = to bear) or **sweat glands** are divided into two principal

types on the basis of structure and location. **Apocrine sweat glands** are simple, branched tubular glands. Their distribution is limited primarily to the skin of the axilla, pubic region, and pigmented areas (areolae) of the breasts. The secretory portion of apocrine sweat glands is located in the dermis or subcutaneous layer and the excretory duct opens into hair follicles. Apocrine sweat glands begin to function at puberty and produce a more viscous secretion than eccrine sweat glands.

Eccrine sweat glands are much more common than apocrine sweat glands and are simple, coiled tubular glands. They are distributed throughout the skin except for the margins of the lips, nail beds of the fingers and toes, glans penis, glans clitoris, labia minora, and eardrums. Eccrine sweat glands are most numerous in the skin of the palms and the soles; their density can be as high as 3,000 per square inch in the palms. The secretory portion of eccrine sweat glands is located in the subcutaneous layer and the excretory duct projects upward through the dermis and epidermis to terminate at a pore at the surface of the epidermis (see Figure 5-1). Eccrine sweat glands function throughout life and produce a secretion that is more watery than that of apocrine sweat glands.

Perspiration, or **sweat,** is the substance produced by sudoriferous glands. It is a mixture of water, salts (mostly NaCl), urea, uric acid, amino acids, ammonia, sugar, lactic acid, and ascorbic acid. Its principal function is to help maintain body temperature. It also helps to eliminate wastes.

Since the mammary glands are actually modified sudoriferous glands, they could be discussed here. However, because of their relationship to the reproductive system, they will be considered in Chapter 28.

Ceruminous Glands

In certain parts of the skin, sudoriferous glands are modified as **ceruminous** (se-ROO-mi-nus) **glands.** Such modified glands are simple, coiled tubular glands present in the external auditory meatus (canal). The secretory portions of ceruminous glands lie in the submucosa, deep to sebaceous glands, and the excretory ducts open either directly onto the surface of the external auditory meatus or into sebaceous ducts. The combined secretion of the ceruminous and sebaceous glands is called **cerumen** (*cera* = wax). Cerumen, together with hairs in the external auditory meatus, provides a sticky barrier that prevents the entrance of foreign bodies.

CLINICAL APPLICATION

Some people produce an abnormal amount of cerumen, or earwax, in the external auditory canal. It then becomes impacted and prevents sound waves from reaching the tympanic membrane. The treatment for **impacted cerumen** is usually periodic ear irrigation or removal of wax with a blunt instrument by trained medical personnel.

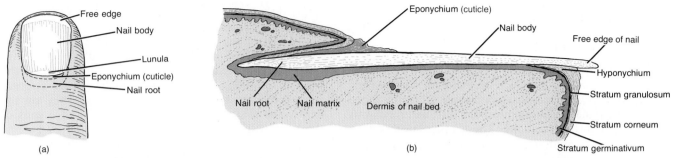

FIGURE 5-6 Structure of nails. (a) Fingernail viewed from above. (b) Sagittal section of a fingernail and nail bed.

NAILS

Hard, keratinized cells of the epidermis are referred to as **nails.** The cells form a clear, solid covering over the dorsal surfaces of the terminal portions of the fingers and toes. Each nail (Figure 5-6) consists of a nail body, a free edge, and a nail root. The **nail body** is the portion of the nail that is visible, the **free edge** is the part that projects beyond the distal end of the digit, and the **nail root** is the portion that is hidden in the proximal nail groove. Most of the nail body is pink because of the underlying vascular tissue. The whitish semilunar area of the proximal end of the body is called the **lunula** (LOO-nyoo-la). It appears whitish because the vascular tissue underneath does not show through.

The fold of skin that extends around the proximal and lateral borders of the nail is known as the **nail fold,** and the epidermis beneath the nail constitutes the **nail bed.** The furrow between the two is the **nail groove.**

The **eponychium** (ep'-ō-NIK-ē-um) or **cuticle** is a narrow band of epidermis that extends from the margin of the nail wall (lateral border), adhering to it. It occupies the proximal border of the nail and consists of stratum corneum. The thickened area of stratum corneum below the free edge of the nail is referred to as the **hyponychium** (hī'-pō-NIK-ē-um).

The epithelium of the proximal part of the nail bed is known as the **nail matrix.** Its function is to bring about the growth of nails. Essentially, growth occurs by the transformation of superficial cells of the matrix into nail cells. In the process, the outer, harder layer is pushed forward over the stratum germinativum. The average growth in the length of fingernails is about 1 mm (0.04 inch) per week. The growth rate is somewhat slower in toenails. For some reason, the longer the digit, the faster the nail grows. Adding supplements such as gelatin to an otherwise healthy diet has no effect on making nails grow faster or stronger.

Functionally, nails help us to grasp and manipulate small objects in various ways.

HOMEOSTASIS

One of the best examples of homeostasis in humans is the maintenance of body temperature by the skin. Hu-

mans, like other mammals, are *homeotherms*—warm-blooded organisms. This means that we are able to maintain a remarkably constant body temperature of 37°C (98.6°F) even though the environmental temperature may vary over a broad range.

Suppose you are in an environment where the temperature is 37.8°C (100°F). A sequence of events is set into operation to counteract this above-normal temperature, which may be considered a stress. Sensing structures in the skin called receptors pick up the stimulus—in this case, heat—and activate nerves that send the message to your brain. A temperature-regulating area of the brain then sends nerve impulses to the sudoriferous glands, which produce more perspiration. As the perspiration evaporates from the surface of your skin, it is cooled and your body temperature is maintained. This sequence of events is shown in Figure 5-7.

Note that temperature regulation by the skin involves a feedback system because the output (cooling of the skin) is fed back to the skin receptors and becomes part of a new stimulus-response cycle. In other words, after the sudoriferous glands are activated, the skin receptors keep the brain informed about the external temperature. The brain, in turn, continues to send messages to the sudoriferous glands until the temperature returns to 37°C (98.6°F).

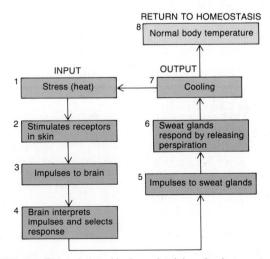

FIGURE 5-7 Role of the skin in maintaining the homeostasis of body temperature.

Like most of the body's feedback systems, temperature regulation is a negative feedback system—the output, cooling, is the opposite of the original condition, overheating.

Temperature maintenance by perspiration represents only one mechanism by which we maintain normal body temperature. Other mechanisms include adjusting blood flow to the skin, regulating metabolic rate, and regulating skeletal muscle contractions. These other mechanisms, as well as increased temperature resulting from fever, are discussed in detail in Chapter 25.

AGING AND THE INTEGUMENTARY SYSTEM

Although skin is constantly aging, the pronounced effects do not occur until a person reaches the late forties. Collagen fibers stiffen, break apart, and form into a shapeless, matted tangle. Elastic fibers thicken into clumps and fray, and as a result, the skin wrinkles. Fibroblasts, which produce both collagenous and elastic fibers, decrease in number, and macrophages become less efficient phagocytes. In addition, there is a loss of subcutaneous fat; atrophy of sebaceous (oil) glands, producing dry and broken skin that is susceptible to infection; a decrease in the number of functioning melanocytes, resulting in gray hair and atypical skin pigmentation; and an increase in the size of some melanocytes that produces pigmented blotching. Aged skin also becomes susceptible to pathological conditions such as senile pruritis (itching), decubitus ulcers (bedsores), and herpes zoster (shingles). Prolonged exposure to the ultraviolet (UV) rays of sunlight accelerates the aging of skin.

DEVELOPMENTAL ANATOMY OF THE INTEGUMENTARY SYSTEM

In this and subsequent chapters, the developmental anatomy of the systems of the body will be discussed at the end of each appropriate chapter. Since the principal features of embryonic development are not treated in detail until Chapter 29, it will be necessary to explain a few terms at this point so that you can follow the development of organ systems.

As part of the early development of a fertilized egg, a portion of the developing embryo differentiates into three layers of tissue called primary germ layers. On the basis of position, the primary germ layers are referred to as **ectoderm, mesoderm,** and **endoderm.** They are the embryonic tissues from which all tissues and organs of the body will eventually develop (see Exhibit 29-1).

The *epidermis* is derived from the **ectoderm.** At the beginning of the second month, the ectoderm consists of simple cuboidal epithelium. These cells become flattened and are known as the **periderm.** By the fourth month, all layers of the epidermis are formed and each layer assumes its characteristic structure.

Nails are developed during the third month. Initially, they consist of a thick layer of epithelium called the **primary nail field.** The nail itself is keratinized epithelium and grows forward from its base. It is not until the ninth month that the nails actually reach the tips of the digits.

Hair follicles develop between the third and fourth months as downgrowths of the stratum basale of the epidermis into the dermis below. The downgrowths soon differentiate into the bulb, papilla of the hair, beginnings of the epithelial portions of sebaceous glands, and other structures associated with hair follicles. By the fifth or sixth month, the follicles produce **lanugo** (delicate fetal hair), first on the head and then on other parts of the body. The lanugo is usually shed prior to birth.

The epithelial (secretory) portions of *sebaceous (oil) glands* develop from the sides of the hair follicles and remain connected to the follicles.

The epithelial portions of *sudoriferous (sweat) glands* are also derived from downgrowths of the stratum basale of the epidermis into the dermis. They appear during the fourth month on the palms and soles and a little later in other regions. The connective tissue and blood vessels associated with the glands develop from **mesoderm.**

The *dermis* is derived from wandering **mesenchymal (mesodermal) cells.** The mesenchyme becomes arranged in a zone beneath the ectoderm and there undergoes changes into the connective tissues that form the dermis.

DISORDERS: HOMEOSTATIC IMBALANCES

Acne

Acne is an inflammation of sebaceous (oil) glands and usually begins at puberty. At puberty, the sebaceous glands, under the influence of androgens (male hormones) grow in size and increase production of sebum. Although testosterone, a male hormone, appears to be the most potent circulating androgen for sebaceous cell stimulation, adrenal and ovarian androgens can stimulate sebaceous secretions as well.

Acne occurs predominantly in sebaceous follicles. The four basic types of acne lesions in order of increasing severity are comedones, papules, pustules, and cysts. The sebaceous follicles are rapidly colonized by bacteria that thrive in the lipid-rich sebum of the follicles. When this occurs, the cyst or sac of connective tissue cells can destroy and displace epidermal cells,

resulting in permanent scarring. This type of acne, called *cystic acne,* may be treated successfully with a drug called Accutane, a synthetic form of vitamin A. Care must be taken to avoid squeezing, pinching, or scratching the lesions.

Systemic Lupus Erythematosus (SLE)

Systemic lupus erythematosus (er-i'-them-a-TŌ-sus), **SLE,** or **lupus** is an autoimmune, inflammatory disease, occurring mostly in young women in their reproductive years. An *autoimmune disease* is one in which the body attacks its own tissues, failing to differentiate between what is foreign and what is not. In SLE, damage to blood vessel walls results in the release of chemicals that mediate the inflammatory response. The blood vessel damage can be associated with virtually every body system.

The cause of SLE is not known and its onset may be abrupt or gradual. It is not contagious, and is thought to be hereditary. There seems to be a strong incidence of other connective tissue disorders—especially rheumatoid arthritis (RA) and rheumatic fever—in relatives of SLE victims. The disease may be triggered by medication, such as penicillin, sulfa, or tetracycline, exposure to excessive sunlight, injury, emotional upset, infection, or other stress. These triggering factors, once recognized, are to be avoided by the patient in the future.

Symptoms include painful joints, low-grade fever, fatigue, photosensitivity, rapid loss of large amounts of scalp hair, and sometimes an eruption across the bridge of the nose and cheeks called a "butterfly rash." Other skin lesions may occur with blistering and ulceration. The erosive nature of some of the SLE skin lesions was thought to resemble the damage inflicted by the bite of a wolf, thus the term *lupus*. The most serious complications of the disease involve inflammation of the kidneys, liver, spleen, lungs, heart, and the central nervous system.

Psoriasis

Psoriasis (sō-RĪ-a-sis) is a chronic, occasionally acute, noncontagious, relapsing skin disease. It is characterized by distinct, reddish, slightly raised plaques or papules (small, round skin elevations) covered with scales. Itching is seldom severe, and the lesions heal without scarring. Psoriasis ordinarily involves the scalp, the elbows and knees, the back, and the buttocks. Occasionally the disease is generalized.

Studies have traced the complex cause of psoriasis to an abnormally high rate of mitosis in epidermal cells that may be related to a substance carried in the blood, a defect in the immune system, or a virus. Triggering factors such as trauma, infections, seasonal and hormonal changes, and emotional stress can initiate and intensify the skin eruptions.

Psoriasis is treated by a number of methods including steroid ointments and creams, natural sunlight, tar preparations, retinoids (chemicals similar to vitamin A), and PUVA, a therapy that combines psoralen (a chemical that increases the skin's reaction to light) and artificial ultraviolet (UV) light.

Decubitus Ulcers

Decubitus (dē-KYOO-bi-tus) **ulcers,** also known as **bedsores, pressure sores,** or **trophic ulcers** are caused by a constant deficiency of blood to tissues overlying a bony projection that has been subjected to prolonged pressure against an object such as a bed, cast, or splint. The deficiency results in tissue ulceration. Small breaks in the epidermis become infected, and the sensitive subcutaneous and deeper tissues are damaged. Eventually the tissue is destroyed.

Decubitus ulcers are seen most frequently in patients who are bedridden for long periods of time. The most common areas involved are the skin over the sacrum, heels, ankles, buttocks, and other large bony projections. The chief causes are pressure from infrequent turning of the patient, trauma and maceration of the skin, and malnutrition. Maceration of the skin often follows soaking of bed and clothing by perspiration, urine, or feces.

Sunburn

Sunburn is injury to the skin as a result of acute, prolonged exposure to the ultraviolet (UV) rays of sunlight. The damage to skin cells caused by sunburn is due to inhibition of DNA and RNA synthesis which leads to cell death. There can also be damage to blood vessels as well as other structures in the dermis. Overexposure over a period of years results in a leathery skin texture, wrinkles, skin folds, sagging skin, warty growths called keratoses, freckling, a yellow discoloration due to abnormal elastic tissue, premature aging of the skin, and skin cancer.

Skin Cancer

Excessive sun exposure can result in **skin cancer.** Everyone, regardless of skin pigmentation, is a potential victim of skin cancer if exposure to sunlight is sufficiently intense and continuous. Natural skin pigment can never give complete protection. As noted earlier, **melanoma** is a skin cancer that arises from melanocytes. It usually develops from normal-looking skin, most commonly between ages 35 to 50. It occurs about six times more frequently in Caucasians than in blacks and Asians, with over 14,000 new cases being diagnosed each year. Ultraviolet (UV) light is the principal cause of melanoma. The thickness of the melanoma and the depth of invasion into the skin are the principal factors that determine the likelihood of cure.

If you must be in direct sunlight for long periods of time, use a suitable sunscreen. One of the best agents for protection against overexposure to ultraviolet rays of the sun is para-amino-benzoic acid (PABA). The alcohol preparations of PABA are best because the active ingredient binds to the stratum corneum of the skin.

Burns

Tissues may be damaged by thermal (heat), electrical, radioactive, or chemical agents, all of which can destroy the proteins in the exposed cells and cause cell injury or death. Such damage is a **burn.** The injury to tissues directly or indirectly in contact with the damaging agent, such as the skin or the linings of the respiratory and digestive tracts, is the local effect of a burn. Generally, however, the systemic effects of a burn are a greater threat to life than the local effects. The systemic effects of a burn may include (1) a large loss of water, plasma, and plasma proteins, which causes shock; (2) bacterial infection; (3) reduced circulation of blood; and (4) decreased production of urine.

Classification

A **first-degree burn** involves only the surface epithelium. It is characterized by mild pain, erythema (redness), dry skin, slight edema, and no blisters. Skin functions remain intact. Generally, a first-degree burn will heal in about 2 to 3 days and may be accompanied by flaking or peeling. A typical sunburn is an example of a first-degree burn.

A **second-degree burn** involves the deeper layers of the epidermis or the upper levels of the dermis and skin functions are lost. In a superficial second-degree burn, the deeper layers of the epidermis are injured and there is characteristic erythema, blister formation, marked edema, and pain. Blisters beneath or within the epidermis are called *bullae* (BYOOL-ē), meaning bubbles. Such an injury usually heals within 7 to 10 days with only mild scarring. In a deep second-degree burn, there is destruction of the epidermis as well as the upper levels of the dermis. Epidermal derivatives, such as hair follicles, sebaceous glands, and sweat glands are usually not injured. If there is no infection, deep second-degree burns heal without grafting in about 3 to 4 weeks. Scarring may result.

First- and second-degree burns are collectively referred to as **partial-thickness burns. A third-degree burn** or **full-thickness burn** involves destruction of the epidermis, dermis, and the epidermal derivatives and skin functions are lost. Such burns vary in appearance from marble-white to mahogany colored to charred, dry wounds. There is marked edema and such a burn is usually not painful to the touch due to destruction of nerve endings. Regeneration is slow and much granulation tissue forms before being covered by epithelium. Even if skin grafting is quickly begun, third-degree burns quickly contract and produce scarring.

A fairly accurate method for estimating the amount of surface area affected by a burn is to apply the *Lund-Browder*

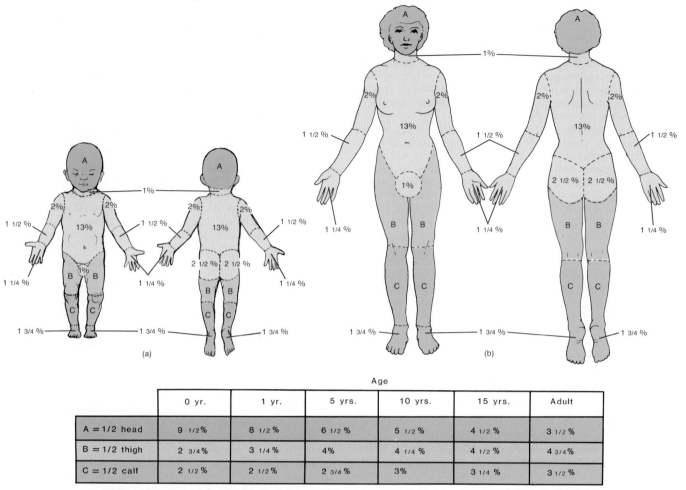

	Age					
	0 yr.	1 yr.	5 yrs.	10 yrs.	15 yrs.	Adult
A = 1/2 head	9 1/2 %	8 1/2 %	6 1/2 %	5 1/2 %	4 1/2 %	3 1/2 %
B = 1/2 thigh	2 3/4 %	3 1/4 %	4%	4 1/4 %	4 1/2 %	4 3/4 %
C = 1/2 calf	2 1/2 %	2 1/2 %	2 3/4 %	3%	3 1/4 %	3 1/2 %

FIGURE 5-8 The Lund-Browder method for estimating the extent of burns. Relative proportions of various body regions in (a) a young child compared to (b) an adult.

method. This method estimates the extent by measuring the areas affected against the percentage of total surface area for body parts shown in Figure 5-8. For example, if the anterior of the head and neck of an adult is affected, the burn covers 4½ percent of the body surface. Because the proportions of the body change with growth, the percentages vary for different ages. Thus, the extent of burn damage can be made fairly accurately for any age group.

Treatment

A severely burned individual should be moved as quickly as possible to a hospital. Treatment may then include:

1. Cleansing the burn wounds thoroughly.
2. Removing all dead tissue (debridement) so antibacterial agents can directly contact the wound surface and thereby prevent infection.

3. Replacing lost body fluids and electrolytes (ions).
4. Covering wounds with temporary protection as soon as possible.
5. Removing a thin layer of skin from another part of the burn victim's body and transplanting it to the injured area (skin graft). When a victim is so extensively burned that there is not enough healthy skin left intact for removal and grafting onto burned areas, a new source of skin grafts is needed. Just recently, a new procedure makes this possible. Human skin cells taken from a small skin-biopsy sample of the burn victim can be grown in the laboratory to produce thin sheets of skin that can be used to cover burn areas. Once the sheets adhere to burn wounds, they generate a permanent new skin to cover the burned areas. The obvious advantage to this procedure is that there is no limit to the amount of new epithelium that can be grown from a very small skin sample.

MEDICAL TERMINOLOGY

Anhidrosis (*an* = without; *hidr* = sweating; *osis* = condition)   A rare genetic condition characterized by inability to sweat.

Antiperspirant (*anti* = against; *perspirare* = to breathe through)   An agent that inhibits or prevents perspiration and usually contains aluminum compounds as the active ingredient.

Athlete's foot   A superficial fungus infection of the skin of the foot.

Callus (callosity)   An area of hardened and thickened skin that is usually seen in palms and soles and is due to pressure and friction.

Carbuncle (*carbunculus* = little coal)   A hard, round, deep, painful inflammation of the subcutaneous tissue that causes necrosis (death) and pus formation (abscess).

Cold sore   Lesion, usually in oral mucous membrane, caused by type 1 herpes simplex virus (HSV), transmitted by oral or respiratory routes. Triggering factors include ultraviolet (UV) radiation, hormonal changes, and emotional stress. Also called a **fever blister.**

Comedo (*comedo* = to eat up)   A collection of sebaceous material and dead cells in the hair follicle and excretory duct of the sebaceous gland. Usually found over the face, chest, and back, and more commonly during adolescence. Also called **blackhead** or **whitehead.**

Corn   A painful conical thickening of the skin found principally over toe joints and between the toes. It may be hard or soft, depending on the location. Hard corns are usually found over toe joints, and soft corns are usually found between the fourth and fifth toes.

Cyst (*cyst* = sac containing fluid)   A sac with a distinct connective tissue wall, containing a fluid or other material.

Deodorant (*de* = from; *odorare* = perfume)   An agent that masks offensive odors; usually contains aluminum compounds as the active ingredient. Body odor is not caused by perspiration itself as much as by the activities of bacteria that grow in perspiration.

Dermabrasion (*derm* = skin)   Removal of acne, scars, tattoos, or moles by sandpaper or a high-speed brush.

Dermatome (*tomy* = cut)   An instrument for excising areas of skin to be used for grafting.

Detritus (*deterere* = to rub away)   Particulate matter produced by or remaining after the wearing away or disintegration of a substance or tissue; scales, crusts, and loosened skin.

Eczema (*ekzein* = to boil out)   An acute or chronic superficial inflammation of the skin, characterized by redness, oozing, crusting, and scaling. Also called **chronic dermatitis.**

Erythema (*erythema* = redness)   Redness of the skin caused by engorgement of capillaries in lower layers of the skin. Erythema occurs with any skin injury, infection, or inflammation.

Furuncle   A boil; an abscess resulting from infection of a hair follicle.

Hypodermic (*hypo* = under)   Relating to the area beneath the skin. Also called **subcutaneous.**

Impetigo   Superficial skin infection caused by staphylococci or streptococci; most common in children.

Intradermal (*intra* = within)   Within the skin. Also called **intracutaneous.**

Keratosis (*kera* = horn)   Formation of a hardened growth of tissue.

Nevus   A round, pigmented, flat, or raised skin area that may be present at birth or develop later. Varying in color from yellow-brown to black. Also called a **mole** or **birthmark.**

Nodule (*nodulus* = little knot)   A large cluster of cells raised above the skin but extending deep into the tissues.

Papule   A small, round skin elevation varying in size from a pinpoint to that of a split pea. One example is a pimple.

Polyp   A tumor on a stem found especially on mucous membranes.

Pruritus (*pruire* = to itch)   Itching, one of the most common dermatological disorders. It may be caused by skin disorders (infections), systemic disorders (cancer, kidney failure), or psychogenic factors (emotional stress).

Pustule   A small, round elevation of the skin containing pus.

Subcutaneous (*sub* = under)   Beneath the skin; also called **hypodermis.**

Topical   Pertaining to a definite area; local. Also, of a medication, applied to the surface rather than ingested or injected.

Wart   Mass produced by uncontrolled growth of epithelial skin cells; caused by a virus (papovavirus). Most warts are noncancerous.

STUDY OUTLINE

Skin (p. 101)

1. The skin and its derivatives (hair, glands, and nails) constitute the integumentary system.
2. The skin is one of the larger organs of the body. It performs the functions of maintaining body temperature; protection; receiving stimuli; excretion of water, salts, and several organic compounds; and synthesis of vitamin D.
3. The principal parts of the skin are the outer epidermis and inner dermis. The dermis overlies the subcutaneous layer.
4. The epidermal layers, from deepest to most superficial, are the stratum basale, spinosum, granulosum, lucidum, and corneum. The basale undergoes continuous cell division and produces all other layers.
5. The dermis consists of a papillary region and a reticular region. The papillary region is loose connective tissue containing blood vessels, nerves, hair follicles, dermal papillae, and corpuscles of touch (Meissner's). The reticular region is dense, irregularly arranged connective tissue containing adipose tissue, hair follicles, nerves, sebaceous (oil) glands, and ducts of sudoriferous (sweat) glands.
6. Lines of cleavage indicate the direction of collagenous fiber bundles in the dermis and are considered during surgery.
7. The color of skin is due to melanin, carotene, and blood in capillaries in the dermis.
8. Epidermal ridges increase friction for better grasping ability and provide the basis for fingerprints and footprints.

Skin Wound Healing (p. 105)

Epidermal Wound Healing (p. 105)

1. In an epidermal wound, the central portion of the wound usually extends deep down to the dermis, whereas the wound edges usually involve only superficial damage to the epidermal cells.
2. Epidermal wounds are repaired by enlargement and migration of basal cells, contact inhibition, and division of migrating and stationary basal cells.

Deep Wound Healing (p. 105)

1. During the inflammatory phase, a blood clot unites the wound edges, epithelial cells migrate across the wound, vasodilation and increased permeability of blood vessels deliver phagocytes, and fibroblasts form.
2. During the migratory phase, epithelial cells beneath the scab bridge the wound, fibroblasts begin to synthesize scar tissue, and damaged blood vessels begin to regrow.
3. During the proliferative phase, the events of the migratory phase intensify and the open wound tissue is called granulation tissue.
4. During the maturation phase, the scab sloughs off, the epidermis is restored to normal thickness, collagenous fibers become more organized, fibroblasts begin to disappear, and blood vessels are restored to normal.

Epidermal Derivatives (p. 105)

1. Epidermal derivatives are structures developed from the embryonic epidermis.
2. Among the epidermal derivatives are hair, skin glands (sebaceous, sudoriferous, and ceruminous), and nails.

Hair (p. 106)

1. Hairs are epidermal growths that function in protection.
2. Hair consists of a shaft above the surface, a root that penetrates the dermis and subcutaneous layer, and a hair follicle.
3. Associated with hairs are sebaceous (oil) glands, arrectores pilorum muscles, and root hair plexuses.
4. New hairs develop from cell division of the matrix in the bulb; hair replacement and growth occurs in a cyclic pattern. "Male-pattern" baldness is caused by androgens and heredity.

Glands (p. 108)

1. Sebaceous (oil) glands are usually connected to hair follicles; they are absent in the palms and soles. Sebaceous glands produce sebum, which moistens hairs and waterproofs the skin. Enlarged sebaceous glands may produce blackheads, pimples, and boils.
2. Sudoriferous (sweat) glands are divided into apocrine and eccrine. Apocrine sweat glands are limited in distribution to the skin of the axilla, pubis, and areolae; their ducts open into hair follicles. Eccrine sweat glands have an extensive distribution; their ducts terminate at pores at the surface of the epidermis. Sudoriferous glands produce perspiration, which carries small amounts of wastes to the surface and

assists in maintaining body temperature.
3. Ceruminous glands are modified sudoriferous glands that secrete cerumen. They are found in the external auditory meatus.

Nails (p. 109)

1. Nails are hard, keratinized epidermal cells over the dorsal surfaces of the terminal portions of the fingers and toes.
2. The principal parts of a nail are the body, free edge, root, lunula, eponychium, hyponychium, and matrix. Cell division of the matrix cells produces new nails.

Homeostasis (p. 109)

1. One of the functions of the skin is the maintenance of a normal body temperature of 37°C (98.6°F).
2. If environmental temperature is high, skin receptors sense the stimulus (heat) and generate impulses that are transmitted to the brain. The brain then causes the sweat glands to produce perspiration. As the perspiration evaporates, the skin is cooled.
3. The skin-cooling response is a negative feedback mechanism.
4. Temperature maintenance is also accomplished by adjusting blood flow to the skin, regulating metabolic rate, and regulating skeletal muscle contractions.

Aging and the Integumentary System (p. 110)

1. Most effects of aging occur when an individual reaches the late forties.
2. Among the effects of aging are wrinkling, loss of subcutaneous fat, atrophy of sebaceous glands, and decrease in the number of melanocytes.

Developmental Anatomy of the Integumentary System (p. 110)

1. The epidermis is derived from ectoderm. Hair, nails, and skin glands are epidermal derivatives.
2. The dermis is derived from wandering mesodermal cells.

Disorders: Homeostatic Imbalances (p. 110)

1. Acne is an inflammation of sebaceous glands.
2. Systemic lupus erythematosus (SLE) is an autoimmune disease of connective tissue.
3. Psoriasis is a chronic skin disease characterized by reddish, raised plaques or papules.
4. Decubitus ulcers are caused by a chronic deficiency of blood to tissues subjected to prolonged pressure.
5. Sunburn is a skin injury resulting from prolonged exposure to the ultraviolet (UV) rays of sunlight.
6. Skin cancer can be caused by excessive exposure to sunlight.
7. Tissue damage that destroys protein is called a burn. Depending on the depth of damage, skin burns are classified as first-degree and second-degree (partial-thickness) and third-degree (full-thickness). One method employed for determining the extent of a burn is the Lund-Browder method. Burn treatment may include cleansing the wound, removing dead tissue, replacing lost body fluids, covering wounds with temporary protection, and skin grafting.

REVIEW QUESTIONS

1. What is the integumentary system?
2. List the principal functions of the skin.
3. Compare the structure of epidermis and dermis. What is the subcutaneous layer?
4. List and describe the epidermal layers from the deepest outward. What is the importance of each layer?
5. Contrast the structural differences between the papillary and reticular regions of the dermis.

6. What are lines of cleavage? What is their importance during surgery?
7. Explain the factors that produce skin color. What is an albino?
8. Describe how melanin is synthesized and distributed to epidermal cells.
9. How are epidermal ridges formed? Why are they important?
10. List the receptors in the epidermis, dermis, and subcutaneous layer, and indicate the location and role of each.
11. Outline the steps involved in epidermal wound healing and deep wound healing.
12. Describe the development and distribution of hair.
13. Describe the structure of a hair. How are hairs moistened? What produces "goosebumps" or "gooseflesh"?
14. Contrast the locations and functions of sebaceous (oil) glands, sudoriferous (sweat) glands, and ceruminous glands. What are the names and chemical components of the secretions of each?
15. Distinguish between apocrine and eccrine sweat glands.
16. From what layer of the skin do nails form? Describe the principal parts of a nail.
17. Explain with a labeled diagram how the skin helps maintain normal body temperature.
18. Describe the effects of aging on the integumentary system.
19. Describe the origin of the epidermis, its derivatives, and the dermis.
20. Define each of the following disorders of the integumentary system: acne, systemic lupus erythematosus (SLE), psoriasis, decubitus ulcers, sunburn, and skin cancer.
21. Define a burn. Classify burns according to degree.
22. Explain how the Lund-Browder method is used to estimate the extent of a burn. How are burns treated?
23. Refer to the glossary of medical terminology associated with the integumentary system. Be sure that you can define each term.

U N I T   II

Principles of Support and Movement

This unit considers two primary themes—support and movement. You will study the various ways in which the body is supported and the different movements it can perform. Both support and movement are made possible by the cooperative effort of bones, joints, and muscles.

6 / Skeletal Tissue

7 / The Skeletal System: The Axial Skeleton

8 / The Skeletal System: The Appendicular Skeleton

9 / Articulations

10 / Muscle Tissue

11 / The Muscular System

Skeletal Tissue

- Discuss the components and functions of the skeletal system.
- List and describe the gross features of a long bone.
- Describe the histological features of compact and spongy bone tissue.
- Contrast the steps involved in intramembranous and endochondral ossification.
- Identify the zones and growth pattern of the epiphyseal plate.
- Describe the processes of bone construction and destruction involved in the homeostasis of bone remodeling.
- Describe the conditions necessary for normal bone growth and replacement.
- Explain the effects of aging on the skeletal system.
- Describe the development of the skeletal system.
- Define rickets and osteomalacia as vitamin deficiency disorders.
- Contrast the causes and clinical symptoms associated with osteoporosis, Paget's disease, and osteomyelitis.
- Define a fracture and describe several common kinds of fractures.
- Describe the sequence of events involved in fracture repair.
- Define medical terminology associated with the skeletal system.

Without the skeletal system we would be unable to perform movements, such as walking or grasping. The slightest jar to the head or chest could damage the brain or heart. It would even be impossible to chew food. The framework of bones and cartilage that protects our organs and allows us to move is called the **skeletal system.** The specialized branch of medicine that deals with the preservation and restoration of the skeletal system, articulations (joints), and associated structures is called **orthopedics** (*ortho* = correct or straighten; *pais* = child).

The developmental anatomy of the skeletal system is considered at the end of the chapter.

FUNCTIONS

The skeletal system performs several basic functions.

1. Support. The skeleton provides a framework for the body and, as such, it supports soft tissues and provides a point of attachment for many muscles.

2. Protection. Many internal organs are protected from injury by the skeleton. For example, the brain is protected by the cranial bones, the spinal cord by the vertebrae, the heart and lungs by the rib cage, and internal reproductive organs by the pelvic bones.

3. Movement. Bones serve as levers to which muscles are attached. When the muscles contract, the bones acting as levers produce movement.

4. Mineral storage. Bones store several minerals that can be distributed to other parts of the body upon demand. The principal stored minerals are calcium and phosphorus.

5. Blood cell production. Red marrow in certain bones is capable of producing blood cells, a process called **hematopoiesis** (hēm'-a-tō-poy-Ē-sis) or **hemopoiesis.** Red marrow consists of blood cells in immature stages, fat cells, and macrophages. Red marrow produces red blood cells, some white blood cells, and platelets.

HISTOLOGY

Structurally, the skeletal system consists of two types of connective tissue: cartilage and bone. We described the microscopic structure of cartilage in Chapter 4. Here, our attention will be directed to a detailed discussion of the microscopic structure of bone tissue.

Like other connective tissues, **bone,** or **osseous** (OS-ē-us) **tissue,** contains a great deal of intercellular substance surrounding widely separated cells. Mature bone cells are called **osteocytes.** Unlike other connective tissues, the intercellular substance of bone contains abundant mineral salts, primarily calcium phosphate ($Ca_3(PO_4)_2 \cdot (OH)_2$) and some calcium carbonate ($CaCO_3$). Together these salts are referred to as *hydroxyapatites*. As these salts are deposited in the framework formed by the collagenous fibers of the intercellular substance, the tissue hardens, that is, becomes ossified. The hydroxyapatites compose 67 percent of the weight of bone, and the collagenous fibers make up the remaining 33 percent.

The microscopic structure of bone may be analyzed by first considering the antamy of a long bone such as the humerus (arm bone) shown in Figure 6-1a. A typical long bone consists of the following parts.

1. Diaphysis (dī-AF-i-sis). The shaft or long, main portion of the bone.

2. Epiphyses (ē-PIF-i-sēz). The extremities or ends of the bone.

3. Metaphysis (me-TAF-i-sis). The region in a mature bone where the diaphysis joins the epiphysis. In a growing bone, it is the region (epiphyseal plate) where calcified cartilage is reinforced and then replaced by bone (described later in the chapter).

4. Articular cartilage. A thin layer of hyaline cartilage covering the epiphysis where the bone forms a joint with another bone.

5. Periosteum (per'-ē-OS-tē-um). A dense, white, fibrous covering around the remaining surface of the bone. The periosteum (*peri* = around; *osteo* = bone) consists of two layers. The outer **fibrous layer** is composed of connective tissue containing blood vessels, lymphatic vessels, and nerves that pass into the bone. The inner **osteogenic** (os'-tē-ō-JEN-ik) **layer** contains elastic fibers, blood vessels, and cells capable of developing into **osteoblasts**—cells responsible for forming new bone during growth and repair. The word *blast* means a germ or bud. It denotes an immature cell or tissue that later develops into a specialized form. The periosteum is essential for bone growth, repair, and nutrition. It also serves as a point of attachment for ligaments and tendons.

6. Medullary (MED-yoo-lar'-ē) or **marrow cavity.** The space within the diaphysis that contains the fatty **yellow marrow** in adults. Yellow marrow consists primarily of fat cells and a few scattered blood cells.

7. Endosteum (end-OS-tē-um). A layer of osteoblasts that lines the medullary cavity and contains scattered osteoclasts (cells that may assume a role in the removal of bone).

Bone is not a completely solid, homogeneous substance. In fact, all bone has some spaces between its hard components. The spaces provide channels for blood vessels that supply bone cells with nutrients. The spaces also make bones lighter. Depending on the size and distribution of the spaces, the regions of a bone may be categorized as spongy or compact (Figure 6-2; see Figure 6-1a,b also).

Spongy (cancellous) bone tissue contains many large spaces filled with red marrow. It makes up most of the bone tissue of short, flat, and irregularly shaped bones and most of the epiphyses of long bones. Spongy bone tissue also provides a storage area for marrow. **Compact (dense)** bone tissue, by contrast, contains few spaces. It is deposited in a layer over the spongy bone tissue. The layer of compact bone is thicker in the diaphysis than the epiphyses. Compact bone tissue provides protection and support and helps the long bones resist the stress of weight placed on them.

COMPACT BONE

We can compare the differences between spongy and compact bone tissue by looking at the highly magnified sections in Figure 6-3. One main difference is that adult compact bone has a concentric-ring structure, whereas spongy bone does not. Blood vessels and nerves from the periosteum penetrate the compact bone through **perforating (Volkmann's) canals.** The blood vessels of these canals connect with blood vessels and nerves of the medullary cavity and those of the **central (Haversian) canals.**

FIGURE 6-1 Osseous tissue. (a) Macroscopic appearance of a long bone that has been partially sectioned. (b) Histological structure of bone.

The central canals run longitudinally through the bone. Around them are **concentric lamellae** (la-MEL-ē)—rings of hard, calcified, intercellular substance. Between the lamellae are small spaces called **lacunae** (la-KOO-nē), which contain osteocytes. **Osteocytes** are mature osteoblasts that no longer produce new bone tissue and function to support daily cellular activities of bone tissue. Radiating in all directions from the lacunae are minute canals called **canaliculi** (kan'-a-LIK-yoo-lī), which contain slender processes of osteocytes. The canaliculi connect with other lacunae and, eventually, with the central canals. Thus an intricate network is formed throughout the bone. This branching network of canaliculi provides numerous

FIGURE 6-2 Spongy and compact bone. Photograph of a section through the femur to illustrate the positional and structural differences between spongy and compact bone. (Courtesy of Lester V. Bergman & Associates, Inc.)

FIGURE 6-3 Histology of osseous tissue. (a) Enlarged aspect of several osteons (Haversian systems) in compact bone. (b) Photomicrograph of portions of several osteons with interstitial lamellae between systems at a magnification of 150×. (Courtesy of Biophoto Associates/Photo Researchers.)

routes so that nutrients can reach the osteocytes and wastes can be removed. Each central canal, with its surrounding lamellae, lacunae, osteocytes, and canaliculi, is called an **osteon (Haversian system).** Osteons are charac-

teristic of adult bone. The areas between osteons contain **interstitial lamellae.** These also possess lacunae with osteocytes and canaliculi, but their lamellae are usually not connected to the osteons. Interstitial lamellae are frag-

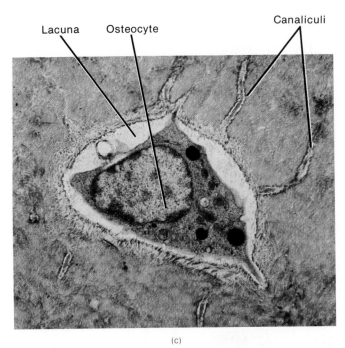

FIGURE 6-3 (*Continued*) (c) Electron micrograph of an osteocyte at a magnification of 10,000×. (Courtesy of Biophoto Associates/Photo Researchers.)

ments of older osteons that have been partially destroyed during bone remodeling.

SPONGY BONE

In contrast to compact bone, spongy bone does not contain true osteons. It consists of an irregular latticework of thin plates of bone called **trabeculae** (tra-BEK-yoo-lē). See Figure 6-2. The spaces between the trabeculae of some bones are filled with red marrow. The cells of red marrow are responsible for producing blood cells. Within the trabeculae lie lacunae, which contain the osteocytes. Blood vessels from the periosteum penetrate through to the spongy bone, and osteocytes in the trabeculae are nourished directly from the blood circulating through the marrow cavities.

Most people think of all bone as a very hard, rigid material. Yet the bones of an infant are not hard at all, and a child's bones are generally more pliable than those of an adult. The final shape and hardness of adult bones require many years to develop and depend on a complex series of chemical changes. Let us now see how bones are formed and how they grow.

OSSIFICATION: BONE FORMATION

The process by which bone forms in the body is called **ossification (osteogenesis).** The "skeleton" of a human embryo is composed of fibrous membranes and hyaline cartilage. Both are shaped like bones and provide the medium for ossification. Ossification begins around the sixth or seventh week of embryonic life and continues throughout adulthood. Two kinds of bone formation occur. The first is called intramembranous (in'-tra-MEM-bra-nus) ossification. This term refers to the formation of bone directly on or within the fibrous membranes (*intra* = within; *membranous* = membrane). The second kind, endochondral (en'-dō-KON-dral) ossification, refers to the formation of bone in cartilage (*endo* = within; *chondro* = cartilage). These two kinds of ossification do *not* lead to differences in the structure of mature bones. They simply indicate different methods of bone formation. Both mechanisms involve the replacement of a preexisting connective tissue with bone.

The first stage in the development of bone is the migration of mesenchymal cells (embryonic connective tissue cells) into the area where bone formation is about to begin. These cells increase in number and size. In some skeletal structures where capillaries are lacking they become chondroblasts; in others where capillaries are present some become osteoblasts. The **chondroblasts** will be responsible for cartilage formation. The osteoblasts will form bone tissue by intramembranous or endochondral ossification.

INTRAMEMBRANOUS OSSIFICATION

Of the two types of bone formation, the simpler and more direct is **intramembranous ossification.** Most of the surface skull bones and the clavicles (collarbones) are formed in this way. The essentials of this process are as follows.

Osteoblasts formed from mesenchymal cells cluster in the fibrous membrane. The site of such a cluster is called a *center of ossification.* The osteoblasts then secrete intercellular substances partly composed of collagenous fibers that form a framework, or matrix, in which calcium salts are quickly deposited. The deposition of calcium salts is called *calcification.* When a cluster of osteoblasts is completely surrounded by the calcified matrix, it is called a *trabecula.* As trabeculae form in nearby ossification centers, they fuse into the open latticework characteristic of spongy bone. With the formation of successive layers of bone, some osteoblasts become trapped in the lacunae. The entrapped osteoblasts lose their ability to form bone and are called osteocytes. The spaces between the trabeculae fill with red marrow. The original connective tissue that surrounds the growing mass of bone then becomes the periosteum. The ossified area has now become true spongy bone. Eventually, the surface layers of the spongy bone will be reconstructed into compact bone. Much of this newly formed bone will be destroyed and reformed so the bone may reach its final adult size and shape.

ENDOCHONDRAL OSSIFICATION

The replacement of cartilage by bone is called **endochondral (intracartilaginous) ossification.** Most bones of the body, including parts of the skull, are formed in this way. Since this type of ossification is best observed in a long bone, we will investigate the tibia, or shinbone (Figure 6-4).

Early in embryonic life, a cartilage model or template of the future bone is laid down. This model is covered by a membrane called the *perichondrium* (per-i-KON-drē-um). Midway along the shaft of this model, a blood vessel penetrates the perichondrium, stimulating the cells

FIGURE 6-4 Endochondral ossification of the tibia. (a) Cartilage model. (b) Collar formation. (c) Development of primary ossification center. (d) Entrance of blood vessels. (e) Marrow-cavity formation. (f) Thickening and lengthening of collar. (g) Formation of secondary ossification centers. (h) Remains of cartilage as articular cartilage and epiphyseal plate. (i) Formation of epiphyseal lines.

in the internal layer of the perichondrium to enlarge and become osteoblasts. The cells begin to form a collar of compact bone around the middle of the diaphysis of the cartilage model. Once the perichondrium starts to form bone, it is called the *periosteum.* Simultaneously with the appearance of the bone collar and the penetration of blood vessels, changes occur in the cartilage in the center of the diaphysis. In this area, called the **primary ossification center,** cartilage cells hypertrophy (increase in size)—probably because they accumulate glycogen for energy and produce enzymes to catalyze future chemical

reactions. These cells burst, resulting in a change in extracellular pH to a more alkaline pH causing the intercellular substance to become *calcified,* that is, minerals are deposited within it. Once the cartilage becomes calcified, nutritive materials required by the cartilage cells can no longer diffuse through the intercellular substance, and this may cause the cartilage cells to die. Then the intercellular substance begins to degenerate, leaving large cavities in the cartilage model. The blood vessels grow along the spaces where cartilage cells were previously located and enlarge the cavities further. Gradually, these spaces in the middle of the shaft join with each other, and the marrow cavity is formed.

As these developmental changes are occurring, the osteoblasts of the periosteum deposit successive layers of bone on the outer surface so that the collar thickens, becoming thickest in the diaphysis. The cartilage model continues to grow at its ends, steadily increasing in length. Eventually, blood vessels enter the epiphyses, and **secondary ossification centers** appear in the epiphyses and lay down spongy bone. In the tibia, one secondary ossification center develops in the proximal epiphysis soon after birth. The other center develops in the distal epiphysis during the child's second year.

After the two secondary ossification centers have formed, bone tissue completely replaces cartilage, except in two regions. Cartilage continues to cover the articular surfaces of the epiphyses, where it is called **articular cartilage.** It also remains as a plate between the epiphysis and diaphysis, where it is called the **epiphyseal plate** (metaphysis of a growing bone).

BONE GROWTH

In order to understand how a bone grows in length, you will need to know some of the details of the structure of the epiphyseal plate.

The epiphyseal (ep'-i-FIZ-ē-al) plate consists of four zones (Figure 6-5). The *zone of reserve cartilage* is adjacent to the epiphysis and consists of small chondrocytes that are scattered irregularly throughout the intercellular matrix. The cells of this zone do not function in bone growth; they anchor the epiphyseal plate to the bone of the epiphysis.

The *zone of proliferating cartilage* consists of slightly larger chondrocytes arranged like stacks of coins. The function of this zone is to make new chondrocytes by cell division to replace those that die at the diaphyseal surface of the epiphyseal plate.

The *zone of hypertrophic* (hī-per-TRŌF-ik) *cartilage* consists of even larger chondrocytes that are also arranged

in columns, with the more mature cells closer to the diaphysis. The lengthwise expansion of the epiphyseal plate is the result of cellular proliferation of the zone of proliferating cartilage and maturation of the cells in the zone of hypertrophic cartilage.

The *zone of calcified matrix* is only a few cells thick and consists mostly of dead cells because the intercellular matrix around them has calcified. The calcified matrix is taken up by osteoclasts, and the area is invaded by osteoblasts and capillaries from the bone in the diaphysis. These cells lay down bone on the calcified cartilage that persists. As a result, the diaphyseal border of the epiphyseal plate is firmly cemented to the bone of the diaphysis.

The region between the diaphysis and epiphysis of a bone where the calcified matrix is replaced by bone is called the *metaphysis* (me-TAF-i-sis). The activity of the epiphyseal plate is the only mechanism by which the diaphysis can increase in length. Unlike cartilage, which can grow by both interstitial and appositional growth, bone can grow only by appositional growth.

The epiphyseal plate allows the diaphysis of the bone to increase in length until early adulthood. The rate of growth is controlled by hormones such as growth hormone (GH) produced by the pituitary gland. As the child grows, cartilage cells are produced by mitosis on the epiphyseal side of the plate. Cartilage cells are then destroyed and the cartilage is replaced by bone on the diaphyseal side of the plate. In this way, the thickness of the epiphyseal plate remains fairly constant, but the bone on the diaphyseal side increases in length.

Growth in diameter occurs along with growth in length. In this process, the bone lining the marrow cavity is destroyed so that the cavity increases in diameter. At the same time, osteoblasts from the periosteum add new osseous tissue around the outer surface of the bone. Initially, diaphyseal and epiphyseal ossification produce only spongy bone. Later, by reconstruction, the outer region of spongy bone is reorganized into compact bone.

The epiphyseal cartilage cells stop dividing and the cartilage is replaced by bone at about age 18 in females and age 20 in males. The newly formed bony structure is called the **epiphyseal line,** a remnant of the once active epiphyseal plate. With the appearance of the epiphyseal line, appositional bone growth stops. The clavicle is the last bone to stop growing. Ossification of most bones is usually completed by age 25.

CLINICAL APPLICATION

Osteogenic sarcoma is a malignant bone tumor that primarily affects osteoblasts. It predominantly affects young people between the ages of 10 and 25 years, and occurs slightly more often in males than females. Tumors often occur in the metaphyses of long bones such as the femur (thighbone), tibia (shinbone), and humerus (arm bone). Left untreated, osteogenic sarcoma metastasizes and leads to death quite rapidly. Metastases occur most frequently in the lungs. Treatment consists of multidrug chemotherapy following amputation or resection.

Epiphyseal side

Zone of reserve cartilage

Zone of proliferating cartilage

Zone of hypertropic cartilage

Zone of calcified matrix

Bone

Marrow space

Diaphyseal side

FIGURE 6-5 Epiphyseal plate. Photomicrograph of the epiphysis of a long bone showing the various zones of the epiphyseal plate at a magnification of 160×. (Copyright © 1983 by Michael H. Ross. Used by permission.)

HOMEOSTASIS

Bones undergoing either intramembranous or endochondral ossification are continually remodeled from the time that initial calcification occurs until the final structure appears. **Remodeling** is the replacement of old bone tissue by new bone tissue. Compact bone is formed by the transformation of spongy bone. The diameter of a long bone is increased by the destruction of the bone closest to the marrow cavity and the construction of new bone around the outside of the diaphysis. However, even after bones have reached their adult shapes and sizes, old bone is perpetually destroyed and new osseous tissue is formed in its place.

Bone shares with skin the feature of replacing itself throughout adult life. Remodeling takes place at different rates in various body regions. The distal portion of the femur (thighbone) is replaced about every 4 months. By contrast, bone in certain areas of the shaft will not be completely replaced during the individual's life. Remodeling allows worn or injured bone to be removed and replaced with new tissue. It also allows bone to serve as the body's storage area for calcium. Many other tissues in the body need calcium in order to perform their functions. For example, nerve cells need calcium for nerve impulse conduction, muscle needs calcium to contract, and blood needs calcium to clot. The blood continually trades off calcium with the bones, removing calcium when it and other tissues are not receiving enough of this element and resupplying the bones with dietary calcium to keep them from losing too much bone mass.

The cells believed to be responsible for the resorption (loss of a substance through a physiological or pathological process) of bone tissue are called **osteoclasts** (*clast* = break). In the healthy adult, a delicate homeostasis is maintained between the action of the osteoclasts in removing calcium and the action of the bone-making osteoblasts in depositing calcium. Should too much new tissue be formed, the bones become abnormally thick and heavy. If too much calcium is deposited in the bone, the surplus may form thick bumps, or spurs, on the bone that interfere with movement at joints. A loss of too much tissue or calcium weakens the bones and allows them to break easily or to become very flexible. As you will see later, a greatly accelerated remodeling process results in a condition called Paget's disease.

In the process of resorption, it is believed that osteoclasts send out projections that secrete proteolytic enzymes released from lysosomes and several acids (lactic and citric). The enzymes may function by digesting the collagen and other organic substances, while the acids may cause the bone salts (minerals) to dissolve in solution. It is also presumed that the osteoclastic projections may phagocytose whole fragments of collagen and bone salts.

Normal bone growth in the young and bone replacement in the adult depend on several factors. First, sufficient quantities of calcium and phosphorus, components of the primary salt that makes bone hard, must be included in the diet.

Second, the individual must obtain sufficient amounts of vitamins, particularly vitamin D, which participates in the absorption of calcium from the digestive tract into the blood, calcium removal from bone, and kidney reabsorption of calcium that might otherwise be lost in urine.

Third, the body must manufacture the proper amounts of the hormones responsible for bone tissue activity (Chapter 18). Growth hormone (GH), secreted by the pituitary gland, is responsible for the general growth of bones. Too much or too little of these hormones during childhood makes the adult abnormally tall or short. Other hormones specialize in regulating the osteoclasts. Calcitonin (CT), produced by the thyroid gland, inhibits osteoclastic activity and accelerates calcium absorption by bones, while parathormone (PTH), synthesized by the parathyroid glands, increases the number and activity of osteoclasts. PTH also releases calcium and phosphate from bones into blood, transports calcium from urine into blood, and transports phosphate from blood into urine. And still others, especially the sex hormones, aid osteoblastic activity and thus promote the growth of new bone. The sex hormones act as a double-edged sword. They aid in the growth of new bone, but they also bring about the degeneration of all the cartilage cells in epiphyseal plates. Because of the sex hormones, the typical adolescent experiences a spurt of growth during puberty, when sex hormone levels start to increase. The individual then quickly completes the growth process as the epiphyseal cartilage disappears. Premature puberty can actually prevent one from reaching an average adult height because of the simultaneous premature degeneration of the plates.

AGING AND THE SKELETAL SYSTEM

There are two principal effects of aging on the skeletal system. The first effect is the loss of calcium from bones. This loss usually begins after age 40 in females and continues thereafter until as much as 30 percent of the calcium in bones is lost by age 70. In males, calcium loss typically does not begin until after age 60. The loss of calcium from bones is one of the factors related to a condition called osteoporosis, which will be described shortly.

The second principal effect of aging on the skeletal system is a decrease in the rate of protein formation that results in a decreased ability to produce the organic portion of bone matrix. As a consequence, bone matrix accumulates a lesser proportion of organic matrix and a greater proportion of inorganic matrix. In some elderly individuals, this process can cause their bones to become quite brittle and more susceptible to fracture.

DEVELOPMENTAL ANATOMY OF THE SKELETAL SYSTEM

Bone forms about the sixth or seventh week by either of two processes, **intramembranous ossification** or **endochondral ossification.** Both processes begin when **mesenchymal (mesodermal) cells** migrate into the area where bone formation will occur. In some skeletal structures, the mesenchymal cells develop into **chondroblasts** that form *cartilage.* In other skeletal structures, the mesenchymal cells develop into **osteoblasts** that form *bone tissue* by intramembranous or endochondral ossification. (The details of ossification have already been discussed earlier in the chapter).

Discussion of the development of the skeletal system provides us with an excellent opportunity to trace the development of the extremities. The *extremities* make their appearance about the fifth week as small elevations at the sides of the trunk called **limb buds** (Figure 6-6). They consist of masses of general **mesoderm** surrounded by **ectoderm.** At this point, a mesenchymal skeleton exists in the limbs; some of the masses of mesoderm surrounding the developing bones will develop into the skeletal muscles of the extremities. By the sixth week, the limb buds develop a constriction around the middle portion. The constriction produces distal segments of the upper buds called **hand plates** and distal segments of the lower buds called **foot plates.** These plates represent the beginnings of the *hands* and *feet,* respectively. At this stage of limb development, a cartilaginous skeleton is present. By the seventh week, the *arm, forearm,* and *hand* are evident in the upper limb bud, and the *thigh, leg,* and *foot* appear in the lower limb bud. Endochondral ossification has begun. By the eighth week, the upper limb bud is appropriately called the *upper extremity* as the *shoulder, elbow,* and *wrist* areas become apparent, and the lower limb bud is

FIGURE 6-6 External features of a developing embryo at various stages of development. (a) Fifth week. (b) Sixth week. (c) Seventh week. (d) Eighth week. Many of the labeled structures are discussed in later chapters.

referred to as the *lower extremity* with the appearance of the *knee* and *ankle* areas.

The **notochord** is a flexible rod of tissue that lies in a position where the future vertebral column will develop (see Figure 10-13b). As the vertebrae develop, the noto-chord becomes surrounded by the developing vertebral bodies and the notochord eventually disappears, except for remnants that persist as the *nucleus pulposus* of the intervertebral discs (see Figure 7-19).

DISORDERS: HOMEOSTATIC IMBALANCES

Many bone disorders result from deficiencies in vitamins or minerals or from too much or too little of the hormones that regulate bone homeostasis. Infection and tumors are also responsible for certain bone disorders.

Vitamin Deficiencies

Rickets

A deficiency of vitamin D in children results in a condition called **rickets.** It is characterized by an inability of the body to transport calcium and phosphorus from the digestive tract into the blood for utilization by bones. As a result, epiphyseal cartilage cells cease to degenerate and new cartilage continues to be produced. Epiphyseal cartilage thus becomes wider than normal. At the same time, the soft matrix laid down by the osteoblasts in the diaphysis fails to calcify. As a result, the bones stay soft. When the child walks, the weight of the body causes the bones in the legs to bow. Malformations of the head, chest, and pelvis also occur.

The cure and prevention of rickets consist of adding generous amounts of calcium, phosphorus, and vitamin D to the diet. Exposing the skin to the ultraviolet rays of sunlight also aids the body in manufacturing additional vitamin D.

Osteomalacia

Deficiency of vitamin D in the adult causes the bones to give up excessive amounts of calcium and phosphorus. This loss, called *demineralization,* is especially heavy in the bones of the pelvis, legs, and spine. Demineralization caused by vitamin D deficiency is called **osteomalacia** (os'-tē-ō-ma-LĀ-shē-a; *malacia* = softness). After the bones demineralize, the weight of the body produces bowing of the leg bones, shortening of the backbone, and flattening of the pelvic bones. Osteomalacia mainly affects women who live on nutritionally poor cereal diets devoid of milk, are seldom exposed to the sun, and have repeated pregnancies that deplete the body of calcium. The condition responds to the same treatment as rickets; if the disease is severe enough and threatens life, large doses of vitamin D are given.

Osteomalacia may also result from failure to absorb fats (steatorrhea) because vitamin D is soluble in fats and calcium combines with fats. As a result, vitamin D and calcium remain with the unabsorbed fat and are lost in the feces.

Osteoporosis

Osteoporosis (os'-tē-ō-pō-RŌ-sis) is an age-related disorder characterized by decreased bone mass and increased susceptibility to fractures. It primarily affects middle-aged and elderly people—women more than men and whites more than blacks. Between puberty and the middle years, sex hormones maintain osseous tissue by stimulating the osteoblasts to form new bone. Women produce smaller amounts of sex hormones (estrogens) after menopause, and both men and women produce smaller amounts during old age. As a result, osteoblasts become less active and there is a decrease in bone mass. Osteoporosis can also occur during nursing and pregnancy, and in individuals exposed to prolonged treatment with cortisone. Osteoporosis affects the entire skeletal system, but especially the spine, hips, legs, and feet. The disorder is responsible for shrinkage of the backbone and height loss, hunched backs, hip fractures, and considerable pain.

Among the factors implicated in bone loss are a decrease in levels of estrogens, calcium deficiency and malabsorption, vitamin D deficiency, loss of muscle mass, and inactivity. Estrogen replacement therapy (ERT), calcium supplements, and exercise are prescribed to treat osteoporosis.

Paget's Disease

Paget's disease is characterized by a greatly accelerated remodeling process in which osteoclastic resorption is massive and new bone formation by osteoblasts is extensive. As a result, there is an irregular thickening and softening of the bones and greatly increased vascularity, especially in bones of the skull, pelvis, and extremities. It rarely occurs in individuals under 50. The cause, or **etiology** (ē'-tē-OL-ō-jē), of the disease is unknown.

Osteomyelitis

The term **osteomyelitis** (os'-tē-ō-mī-i-LĪ-tis) includes all infectious diseases of bone. These diseases may be localized or widespread and may also involve the periosteum, marrow, and cartilage. Various microorganisms may give rise to bone infection, but the most frequent are bacteria known as *Staphylococcus aureus,* commonly called "staph." These bacteria may reach the bone by various means: the bloodstream, an injury such as a fracture, or an infection such as a sinus infection or a tooth abscess. Antibiotics may be effective in treating the disease and in preventing it from spreading through extensive areas of bone, but complete recovery is usually a prolonged and painful process.

Fractures

Very simply, a **fracture** is any break in a bone. Usually, the fracture is restored to normal position by manipulation without surgery. This procedure of setting a fracture is called *closed reduction.* In other cases, the fracture must be exposed by surgery before the break is rejoined. This procedure is known as *open reduction.*

Types

Although fractures may be classified in several different ways, the following scheme is useful (Figure 6-7):

1. Partial (incomplete). A fracture in which the break across the bone is incomplete.

2. Complete. A fracture in which the break across the bone is complete, so that the bone is broken into two pieces.

3. Closed (simple). A fracture in which the bone does not break through the skin.

4. Open (compound). A fracture in which the broken ends of the bone protrude through the skin.

5. Comminuted (KOM-i-nyoo'-ted). A fracture in which the bone is splintered at the site of impact and smaller fragments of bone are found between the two main fragments.

6. Greenstick. A partial fracture in which one side of the bone is broken and the other side bends; occurs only in children.

7. Spiral. A fracture in which the bone is usually twisted apart.

FIGURE 6-7 Types of fractures. (a) Comminuted. (b) Colles'. (c) Impacted. (d) Pott's. (e) Greenstick. (f) Compound.

8. Transverse. A fracture at right angles to the long axis of the bone.

9. Impacted. A fracture in which one fragment is firmly driven into the other.

10. Pott's. A fracture of the distal end of the fibula, with serious injury of the distal tibial articulation.

11. Colles' (KOL-ēz). A fracture of the distal end of the radius in which the distal fragment is displaced posteriorly.

12. Displaced. A fracture in which the anatomical alignment of the bone fragments is not preserved.

13. Nondisplaced. A fracture in which the anatomical alignment of the bone fragments is preserved.

14. Stress. A partial fracture resulting from inability to withstand repeated stress due to a change in training, harder surfaces, longer distances, and greater speed. About 25 percent of all stress fractures involve the fibula, typically the distal third.

15. Pathologic. A fracture due to weakening of a bone caused by disease processes such as neoplasia, osteomyelitis, osteoporosis, or osteomalacia.

Fracture Repair

Unlike the skin, which may repair itself within days, or muscle, which may mend in weeks, a bone sometimes requires months to heal. A fractured femur, for example, may take six months to heal. Sufficient calcium to strengthen and harden new bone is deposited only gradually. Bone cells also grow and reproduce slowly. Moreover, the blood supply to bone is decreased, which helps to explain the difficulty in the healing of an infected bone.

The following steps occur in the repair of a fracture (Figure 6-8):

1. As a result of the fracture, blood vessels crossing the fracture line are broken. These vessels are found in the periosteum, osteons (Haversian systems), and marrow cavity. As blood pours from the torn ends of the vessels, it coagulates and forms a clot in and about the site of the fracture. This clot, called a **fracture hematoma** (hē'-ma-TŌ-ma), usually occurs 6 to 8 hours

(a)   (b)   (c)

FIGURE 6-8 Fracture repair. (a) Formation of fracture hematoma. (b) Formation of external and internal calli. (c) Completely healed fracture.

after the injury. Since the circulation of blood ceases when the fracture hematoma forms, bone cells and periosteal cells at the fracture line die. The hematoma serves as a focus for subsequent cellular invasion.

2. A growth of new bone tissue—a **callus**—develops in and around the fractured area. It forms a bridge between separated areas of bone. The part of the callus that forms from the osteogenic cells of the torn periosteum and develops around the outside of the fracture is called an *external callus*. The part of the callus that forms from the osteogenic cells of the endosteum and develops between the two ends of bone fragments and between the two marrow cavities is called the *internal callus*.

Approximately 48 hours after a fracture occurs, the cells that ultimately repair the fracture become actively mitotic. These cells come from the osteogenic layer of the periosteum, the endosteum of the marrow cavity, and the bone marrow. As a result of their accelerated mitotic activity, the cells of the three regions grow toward the fracture. During the first week following the fracture, the cells of the endosteum and bone marrow form new trabeculae in the marrow cavity near the line of fracture. This is the internal callus. During the next few days, osteogenic cells of the periosteum form a collar around each bone fragment. The collar, or external callus, is replaced by trabeculae. The trabeculae of the calli are joined to living and dead portions of the original bone fragments.

3. The final phase of fracture repair is the **remodeling** of the calli. Dead portions of the original fragments are gradually resorbed by osteoclasts. Compact bone replaces spongy bone around the periphery of the fracture. In some cases, the healing is so complete that the fracture line is undetectable, even by x-ray. However, a thickened area on the surface of the bone usually remains as evidence of the fracture site.

CLINICAL APPLICATION

In the past when a fracture failed to unite, the patient could either wait, hoping that nature and time would solve the problem, or choose surgery. Today there is another alternative. The new procedure, called **pulsating electromagnetic fields (PEMFs),** involves electrotherapy to stimulate bone repair.

Essentially, the fracture is exposed to a weak electrical current generated from coils that are fastened around the cast. Osteoblasts adjacent to the fracture site become more active metabolically in response to the stimulation. The increased activity apparently causes an acceleration of calcification, vascularization, and endochondral ossification, resulting in acceleration of fracture repair. It was noted earlier that parathyroid hormone (PTH) increases osteoclastic activity and thus stimulates bone destruction. One hypothesis suggests that electricity also heals fractures by keeping PTH from acting on osteoclasts, thus increasing bone formation and repair.

Although the original application of PEMFs was to treat improperly healing fractures, research is continuing to determine if electrical stimulation can be used to regenerate limbs, stop the growth of tumor cells in humans, reduce postoperative swelling, and improve the healing of tendons and ligaments.

MEDICAL TERMINOLOGY

Achondroplasia (*a* = without; *chondro* = cartilage; *plasia* = growth) Imperfect ossification within cartilage of long bones during fetal life; also called **fetal rickets.**

Craniotomy (*cranium* = skull; *tome* = a cutting) Any surgery that requires cutting through the bones surrounding the brain.

Necrosis (*necros* = death; *osis* = condition) Death of tissues or organs; in the case of bone, results from deprivation of blood supply resulting from fracture, extensive removal of periosteum in surgery, exposure to radioactive substances, or other causes.

Osteitis (*osteo* = bone) Inflammation or infection of bone.

Osteoarthritis (*arthro* = joint) A degenerative condition of bone and also the joint.

Osteoblastoma (*oma* = tumor) A benign tumor of the osteoblasts.

Osteochondroma (*chondro* = cartilage) A benign tumor of the bone and cartilage.

Osteoma A benign bone tumor.

Osteosarcoma (*sarcoma* = connective tissue tumor) A malignant tumor composed of osseous tissue.

Pott's disease Inflammation of the backbone, caused by the microorganism that produces tuberculosis.

STUDY OUTLINE

Functions (p. 120)

1. The skeletal system consists of all bones attached at joints and cartilage between joints.

2. The functions of the skeletal system include support, protection, leverage, mineral storage, and blood cell production.

Histology (p. 120)

1. Osseous tissue consists of widely separated cells surrounded by large amounts of intercellular substance. The intercellular substance contains collagenous fibers and abundant hydroxyapatites (mineral salts).
2. Parts of a typical long bone are the diaphysis (shaft), epiphyses (ends), metaphysis, articular cartilage, periosteum, medullary (marrow) cavity, and endosteum.
3. Compact (dense) bone consists of osteons (Haversian systems) with little space between them. Compact bone lies over spongy bone and composes most of the bone tissue of the diaphysis. Functionally, compact bone protects, supports, and resists stress.
4. Spongy (cancellous) bone consists of trabeculae surrounding many red marrow-filled spaces. It forms most of the structure of short, flat, and irregular bones, and the epiphyses of long bones. Functionally, spongy bone stores marrow and provides some support.

Ossification: Bone Formation (p. 123)

1. Bone forms by a process called ossification (osteogenesis), which begins when mesenchymal cells become transformed into osteoblasts.
2. The process begins during the sixth or seventh week of embryonic life and continues throughout adulthood. The two types of ossification, intramembranous and endochondral, involve the replacement of a preexisting connective tissue with bone.
3. Intramembranous ossification occurs within fibrous membranes of the embryo and the adult.
4. Endochondral ossification occurs within a cartilage model. The primary ossification center of a long bone is in the diaphysis. Cartilage degenerates, leaving cavities that merge to form the marrow cavity. Osteoblasts lay down bone. Next, ossification occurs in the epiphyses, where bone replaces cartilage, except for the epiphyseal plate.

Bone Growth (p. 125)

1. The anatomical zones of the epiphyseal plate are the zones of reserve cartilage, proliferating cartilage, hypertrophic cartilage, and calcified matrix.
2. Because of the activity of the epiphyseal plate, the diaphysis of a bone increases in length by appositional growth.
3. Bone grows in diameter as a result of the addition of new bone tissue by periosteal osteoblasts around the outer surface of the bone.

Homeostasis (p. 126)

1. The homeostasis of bone growth and development depends on a balance between bone formation and resorption.
2. Old bone is constantly destroyed by osteoclasts, while new bone is constructed by osteoblasts. This process is called remodeling.
3. Normal growth depends on calcium, phosphorus, and vitamins, especially vitamin D, and is controlled by hormones that are responsible for bone mineralization and resorption.

Aging and the Skeletal System (p. 126)

1. The principal effect of aging is a loss of calcium from bones, which may result in osteoporosis.
2. Another effect is a decreased production of organic matrix, which makes bones more susceptible to fracture.

Developmental Anatomy of the Skeletal System (p. 126)

1. Bone forms from mesoderm by intramembranous or endochondral ossification.
2. Extremities develop from limb buds, which consist of mesoderm and ectoderm.

Disorders: Homeostatic Imbalances (p. 128)

1. Rickets is a vitamin D deficiency in children in which the body does not absorb calcium and phosphorus. The bones soften and bend under the body's weight.
2. Demineralization caused by vitamin D deficiency in adults results in osteomalacia.
3. Osteoporosis is a decrease in the amount and strength of bone tissue due to decreases in hormone output.
4. Paget's disease is the irregular thickening and softening of bones, related to a greatly accelerated remodeling process.
5. Osteomyelitis is a term for the infectious diseases of bones, marrow, and periosteum. It is frequently caused by "staph" bacteria.
6. A fracture is any break in a bone.
7. The types of fractures include: partial, complete, simple, compound, comminuted, greenstick, spiral, transverse, impacted, Pott's, Colles', displaced, nondisplaced, and stress.
8. Fracture repair consists of forming a fracture hematoma, forming a callus, and remodeling.
9. Treatment by pulsating electromagnetic fields (PEMFs) has provided dramatic results in healing fractures that would otherwise not have mended properly. Its application for limb regeneration and stopping the growth of tumor cells is being investigated.

REVIEW QUESTIONS

1. Define the skeletal system. What are its five principal functions?
2. Why is osseous tissue considered a connective tissue? What is its composition?
3. Diagram the parts of a long bone, and list the functions of each part.
4. Distinguish between spongy and compact bone in terms of microscopic appearance, location, and function.
5. Diagram the microscopic appearance of compact bone and indicate the functions of the various components.
6. What is meant by ossification? Describe the initial events of ossification.
7. Outline the major events involved in intramembranous and endochondral ossification and explain the main differences.
8. Describe the histology of the various zones of the epiphyseal plate. How does the plate grow? What is the significance of the epiphyseal line?
9. Define remodeling. How does balance between osteoblast activity and osteoclast activity demonstrate the homeostasis of bone?
10. List the primary factors involved in bone growth and replacement.
11. Explain the effects of aging on the skeletal system.
12. Describe the development of the skeletal system.
13. Distinguish between rickets and osteomalacia. What do the two diseases have in common?
14. What are the principal symptoms of osteoporosis, Paget's disease, and osteomyelitis? What is the etiology of each?
15. What is a fracture? Distinguish several principal kinds. Outline the three basic steps involved in fracture repair.
16. Explain the principal of pulsating electromagnetic fields (PEMFs) in the repair of fractures.
17. Refer to the glossary of medical terminology associated with the skeletal system. Be sure that you can define each term.

The Skeletal System: The Axial Skeleton

7

STUDENT OBJECTIVES

- Define the four principal types of bones.
- Describe the various markings on the surfaces of bones.
- Relate the structure of a surface marking to its function.
- List the components of the axial and appendicular skeleton.
- Identify the bones of the skull and the major markings associated with each.
- Identify the principal sutures and fontanels of the skull.
- Identify the paranasal sinuses of the skull.
- Identify the principal foramina of the skull.
- Identify the bones of the vertebral column and their principal markings.
- List the defining characteristics and curves of each region of the vertebral column.
- Identify the bones of the thorax and their principal markings.
- Contrast herniated (slipped) disc, curvatures, spina bifida, and fractures of the vertebral column as disorders associated with the skeletal system.

The skeletal system forms the framework of the body. For this reason, a familiarity with the names, shapes, and positions of individual bones will help you understand some of the other organ-systems. For example, movements such as throwing a ball, typing, and walking require the coordinated use of bones and muscles. To understand how muscles produce different movements, you need to learn the parts of the bones to which the muscles attach. The respiratory system is also highly dependent on bone structure. The bones in the nasal cavity form a series of passageways that help clean, moisten, and warm inhaled air. Furthermore, the bones of the thorax are specially shaped and positioned so the chest can expand during inhalation. Many bones also serve as landmarks to students of anatomy as well as to surgeons. As you will see, bony landmarks can be used to locate the outlines of the lungs and heart, abdominal and pelvic viscera, and structures within the skull. Blood vessels and nerves often run parallel to bones. These structures can be located more easily if the bone is identified first.

We will study bones by examining the various regions of the body. We will look at the skull first and see how the bones of the skull relate to each other. We will then move on to the vertebral column and the chest. This regional approach will allow you to see how all the many bones of the body relate to each other.

TYPES OF BONES

Almost all the bones of the body may be classified into four principal types on the basis of shape: long, short, flat, and irregular. **Long bones** have greater length than width and consist of a diaphysis and two epiphyses. They are slightly curved for strength. A curved bone is structurally designed to absorb the stress of the body weight at several different points so the stress is evenly distributed. If such bones were straight, the weight of the body would be unevenly distributed and the bone would easily fracture. Examples of long bones include bones of the thighs, legs, toes, arms, forearms, and fingers. Figure 6-1a shows the parts of a long bone.

Short bones are somewhat cube-shaped and nearly equal in length and width. Their texture is spongy except at the surface, where there is a thin layer of compact bone. Examples of short bones are the wrist and ankle bones.

Flat bones are generally thin and composed of two more or less parallel plates of compact bone enclosing a layer of spongy bone. Flat bones afford considerable protection and provide extensive areas for muscle attachment. Examples of flat bones include the cranial bones (which protect the brain), the sternum and ribs (which protect organs in the thorax), and the scapulas.

Irregular bones have complex shapes and cannot be grouped into any of the three categories just described. They also vary in the amount of spongy and compact bone present. Such bones include the vertebrae and certain facial bones.

There are two additional types of bones that are not included in this classification by shape. **Sutural** (SOO-chur-al) or **Wormian bones** are small bones between the joints of certain cranial bones (see Figure 7-2e). Their number varies greatly from person to person. **Sesamoid bones** are small bones in tendons where considerable pressure develops, for instance, in the wrist. These, like sutural bones, are also variable in number. Two sesamoid bones, the patellas, or kneecaps, are present in all individuals.

SURFACE MARKINGS

The surfaces of bones reveal various structural features adapted to specific functions. These features are called **surface markings.** Long bones that bear a great deal of weight have large, rounded ends that can form sturdy joints. Other bones have depressions that receive the rounded ends. Rough areas serve as points of attachment for muscles, tendons, and ligaments. Grooves in the surfaces of bones provide for the passage of blood vessels. Openings occur where blood vessels and nerves pass through the bone. Exhibit 7-1 describes the different markings and their functions.

DIVISIONS OF THE SKELETAL SYSTEM

The adult human skeleton usually consists of 206 bones grouped in two principal divisions: the **axial skeleton** and the **appendicular skeleton.** The longitudinal *axis,* or center, of the human body is a straight line that runs vertically along the body's center of gravity. This imaginary line extends through the head and down to the space between the feet. The midsagittal plane is drawn through this line. The axial division of the skeleton consists of the bones that lie around the axis: ribs, breastbone, bones of the skull, and backbone.

The appendicular division contains the bones of the free *appendages,* which are the upper and lower extremities, plus the bones called *girdles,* which connect the free appendages to the axial skeleton.

The 80 bones of the axial division and the 126 bones of the appendicular division are typically grouped as shown in Exhibit 7-2.

Now that you understand how the skeleton is organized into axial and appendicular divisions, refer to Figure 7-1 to see how the two divisions are joined to form the complete skeleton. The bones of the axial skeleton are shown in gold. Be certain to locate the following regions of the skeleton: skull, cranium, face, hyoid bones, vertebral column, thorax, shoulder girdle, upper extremity, pelvic girdle, and lower extremity.

SKULL

The **skull,** which contains 22 bones, rests on the superior end of the vertebral column and is composed of two sets of bones: cranial bones and facial bones. The **cranial bones** enclose and protect the brain. The 8 cranial bones are the frontal bone, parietal bones (2), temporal bones (2), occipital bone, sphenoid bone, and ethmoid bone. There are 14 **facial bones:** nasal bones (2), maxillae (2), zygomatic bones (2), mandible, lacrimal bones (2), palatine

EXHIBIT 7-1   BONE MARKINGS

MARKING	DESCRIPTION	EXAMPLE
DEPRESSIONS AND OPENINGS		
Fissure (FISH-ur)	A narrow, cleftlike opening between adjacent parts of bones through which blood vessels or nerves pass.	Superior orbital fissure of the sphenoid bone (Figure 7-2).
Foramen (fō-RĀ-men; *foramen* = hole)	An opening through which blood vessels, nerves, or ligaments pass.	Infraorbital foramen of the maxilla (Figure 7-2).
Meatus (mē-Ā-tus; *meatus* = canal)	A tubelike passageway running within a bone.	External auditory meatus of the temporal bone (Figure 7-2).
Paranasal sinus (*sin* = cavity)	An air-filled cavity within a bone connected to the nasal cavity.	Frontal sinus of the frontal bone (Figure 7-8).
Groove or **sulcus** (*sulcus* = ditchlike groove)	A furrow or depression that accommodates a soft structure such as a blood vessel, nerve, or tendon.	Intertubercular sulcus of the humerus (Figure 8-4).
Fossa (*fossa* = basinlike depression)	A depression in or on a bone.	Mandibular fossa of the temporal bone (Figure 7-4).
PROCESSES	Any prominent projection.	Mastoid process of the temporal bone (Figure 7-2).
Processes that form joints		
Condyle (KON-dīl; *condylus* = knucklelike process)	A large, rounded articular prominence.	Medial condyle of the femur (Figure 8-10).
Head	A rounded, articular projection supported on the constricted portion (neck) of a bone.	Head of the femur (Figure 8-10).
Facet	A smooth, flat surface.	Articular facet for the tubercle of rib on a vertebra (Figure 7-18).
Processes to which tendons, ligaments, and other connective tissues attach		
Tubercle (TOO-ber-kul; *tube* = knob)	A small, rounded process.	Greater tubercle of the humerus (Figure 8-4).
Tuberosity	A large, rounded, usually roughened process.	Ischial tuberosity of the hipbone (Figure 8-8).
Trochanter (trō-KAN-ter)	A large, blunt projection found only on the femur.	Greater trochanter of the femur (Figure 8-10).
Crest	A prominent border or ridge.	Iliac crest of the hipbone (Figure 8-7).
Line	A less prominent ridge than a crest.	Linea aspera of the femur (figure 8-10).
Spinous process (spine)	A sharp, slender process.	Spinous process of a vertebra (Figure 7-12).
Epicondyle (*epi* = above)	A prominence above a condyle.	Medial epicondyle of the femur (Figure 8-10).

bones (2), inferior nasal conchae (2), and vomer. Be sure you can locate all the skull bones in the anterior, lateral, median, and posterior views of the skull (Figure 7-2).

SUTURES

A **suture** (SOO-chur), meaning seam or stitch, is an immovable joint found only between skull bones. Very little connective tissue is found between the bones of the suture. Four prominent skull sutures are:

1. **Coronal suture** between the frontal bone and the two parietal bones.
2. **Sagittal suture** between the two parietal bones.
3. **Lambdoidal** (lam-DOY-dal) **suture** between the parietal bones and the occipital bone.
4. **Squamosal** (skwa-MŌ-sal) **suture** between the parietal bones and the temporal bones.

Refer to Figures 7-2 and 7-3 for the locations of these sutures. Several other sutures are also shown. Their names are descriptive of the bones they connect. For example, the frontonasal suture is between the frontal bone and

FIGURE 7-1 Divisions of the skeletal system. The axial skeleton is indicated in gold. (a) Anterior view. (b) Posterior view.

EXHIBIT 7-2 DIVISIONS OF THE SKELETAL SYSTEM

REGIONS OF THE SKELETON	NUMBER OF BONES
AXIAL SKELETON	
Skull	
Cranium	8
Face	14
Hyoid	1
Auditory ossicles* (3 in each ear)	6
Vertebral column	26
Thorax	
Sternum	1
Ribs	24
	80
APPENDICULAR SKELETON	
Pectoral (shoulder) girdles	
Clavicle	2
Scapula	2
Upper extremities	
Humerus	2
Ulna	2
Radius	2
Carpals	16
Metacarpals	10
Phalanges	28
Pelvic (hip) girdle	
Coxal, pelvic, or hip bone	2
Lower extremities	
Femur	2
Fibula	2
Tibia	2
Patella	2
Tarsals	14
Metatarsals	10
Phalanges	28
	126
	Total = 206

* Although the auditory ossicles are not considered part of the axial or appendicular skeleton, but rather as a separate group of bones, they are placed with the axial skeleton for convenience. The auditory ossicles are exceedingly small bones named for their shapes. Their names are the malleus, incus, and stapes, commonly called the hammer, anvil, and stirrup, respectively. The middle portion of each ear contains three auditory ossicles held together by a series of ligaments. The auditory ossicles vibrate in response to sound waves that strike the eardrum and assume a key function in the mechanism involved in hearing. This is described in detail in Chapter 17.

the nasal bones. These sutures are indicated in Figures 7-2 to 7-5.

FONTANELS

The "skeleton" of a newly formed embryo consists of cartilage or fibrous membrane structures shaped like bones. Gradually the cartilage or fibrous membrane is replaced by bone. At birth, membrane-filled spaces called **fontanels** (fon'-ta-NELZ; = little fountains) are found between cranial bones (Figure 7-3). These "soft spots" are areas where intramembranous ossification will eventually replace the membrane-filled spaces with bone. They allow the skull to be compressed during birth. Physicians find the fontanels helpful in determining the position of the infant's head prior to delivery. Although an infant may have many fontanels at birth, the form and location of six are fairly constant.

The **anterior (frontal) fontanel** is located between the angles of the two parietal bones and the two segments of the frontal bone. This fontanel is roughly diamond-shaped, and it is the largest of the six fontanels. It usually closes 18 to 24 months after birth.

The **posterior (occipital;** ok-SIP-i-tal**) fontanel** is situated between the two parietal bones and the occipital bones. This diamond-shaped fontanel is considerably smaller than the anterior fontanel. It generally closes about 2 months after birth.

The **anterolateral (sphenoidal;** sfē-NOY-dal**) fontanels** are paired. One is located on each side of the skull at the junction of the frontal, parietal, temporal, and sphenoid bones. These fontanels are quite small and irregular in shape. They normally close about 3 months after birth.

The **posterolateral (mastoid) fontanels** are also paired. One is situated on each side of the skull at the junction of the parietal, occipital, and temporal bones. These fontanels are irregularly shaped. They begin to close 1 or 2 months after birth, but closure is not generally complete until 12 months.

CRANIAL BONES

Frontal Bone

The **frontal bone** forms the forehead (the anterior part of the cranium), the roofs of the *orbits* (eye sockets), and most of the anterior part of the cranial floor. Soon after birth, the left and right parts of the frontal bone are united by a suture. The suture usually disappears by age 6. If, however, the suture persists throughout life, it is referred to as the **metopic suture.**

If you examine the anterior and lateral views of the skull in Figure 7-2, you will note the **frontal squama** (SKWĀ-ma), or **vertical plate** (*squam* = scale). This scalelike plate, which corresponds to the forehead, gradually slopes down from the coronal suture, then turns abruptly downward.

A thickening of the frontal bone is called the **supraorbital margin.** From this margin the frontal bone extends posteriorly to form the roof of the orbit and part of the floor of the cranial cavity.

Within the supraorbital margin, slightly medial to its midpoint, is a hole called the **supraorbital foramen.** The supraorbital nerve and artery pass through this foramen. The **frontal sinuses** lie deep to the frontal squama. These mucus-lined cavities act as sound chambers that give the voice resonance.

Parietal Bones

The two **parietal bones** (pa-RĪ-e-tal; *paries* = wall) form the greater portion of the sides and roof of the cranial cavity (see Figure 7-2). The internal surfaces of the bones contain many eminences and depressions that accommo-

date the blood vessels supplying the outer meninx (covering) of the brain called the *dura mater.*

Temporal Bones

The two **temporal bones** form the inferior sides of the cranium and part of the cranial floor. The term *tempora* pertains to the temples.

In the lateral view of the skull in Figure 7-2, notice the **temporal squama**—a thin, large, expanded area that forms the anterior and superior part of the temple. Projecting from the inferior portion of the temporal squama is the **zygomatic process,** which articulates with the temporal process of the zygomatic bone. The zygomatic process of the temporal bone and the temporal process of the zygomatic bone constitute the **zygomatic arch.**

At the floor of the cranial cavity, shown in Figure 7-5, is the **petrous portion** of the temporal bone. This portion is triangular and located at the base of the skull between the sphenoid and occipital bones. The petrous portion contains the internal ear, the essential part of the organ of hearing. It also contains the **carotid foramen (canal)** through which the internal carotid artery passes (see Figure 7-4). Posterior to the carotid foramen and anterior to the occipital bone is the **jugular foramen (fossa)** through which the internal jugular vein and the

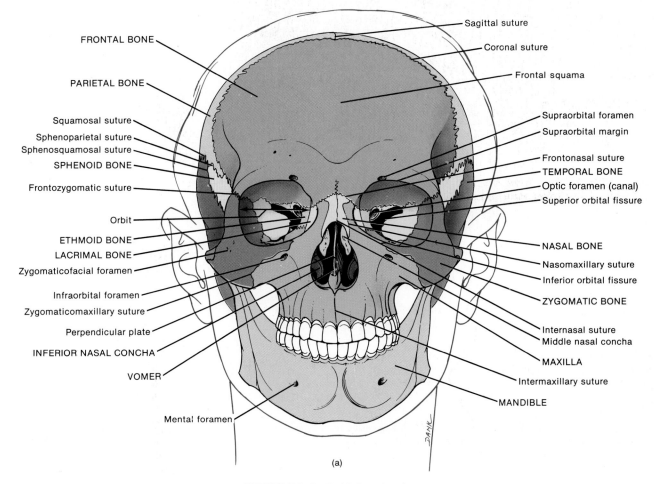

(a)

FIGURE 7-2 Skull. (a) Anterior view.

(b)

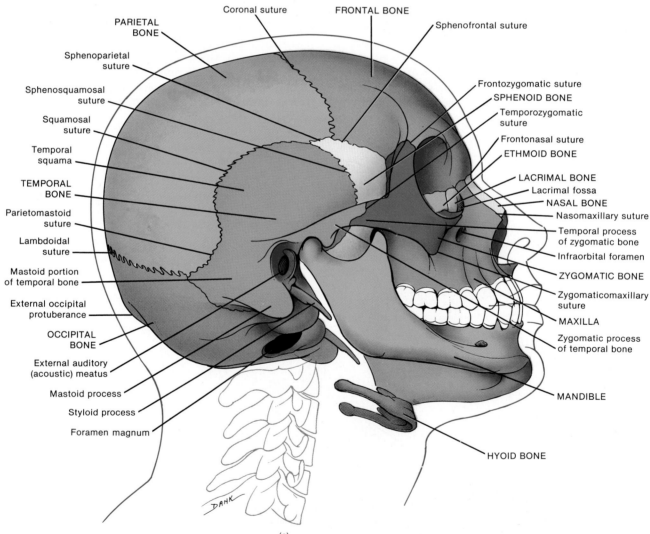

(c)

FIGURE 7-2 (*Continued*) (b) Details of the right orbit in anterior view. (c) Right lateral view.

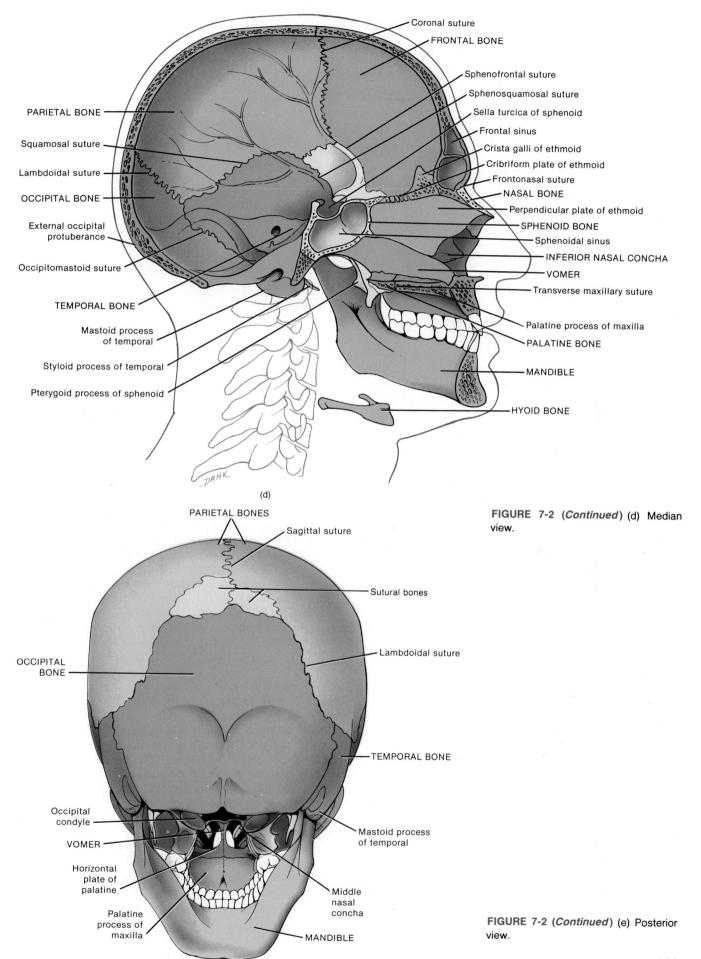

Coronal suture
FRONTAL BONE
Sphenofrontal suture
Sphenosquamosal suture
Sella turcica of sphenoid
Frontal sinus
Crista galli of ethmoid
Cribriform plate of ethmoid
Frontonasal suture
NASAL BONE
Perpendicular plate of ethmoid
SPHENOID BONE
Sphenoidal sinus
INFERIOR NASAL CONCHA
VOMER
Transverse maxillary suture
Palatine process of maxilla
PALATINE BONE
MANDIBLE
HYOID BONE

PARIETAL BONE
Squamosal suture
Lambdoidal suture
OCCIPITAL BONE
External occipital protuberance
Occipitomastoid suture
TEMPORAL BONE
Mastoid process of temporal
Styloid process of temporal
Pterygoid process of sphenoid

DANK

(d)

PARIETAL BONES
Sagittal suture
Sutural bones
Lambdoidal suture
TEMPORAL BONE
Mastoid process of temporal
Middle nasal concha
MANDIBLE

OCCIPITAL BONE
Occipital condyle
VOMER
Horizontal plate of palatine
Palatine process of maxilla

(e)

FIGURE 7-2 (*Continued*) (d) Median view.

FIGURE 7-2 (*Continued*) (e) Posterior view.

139

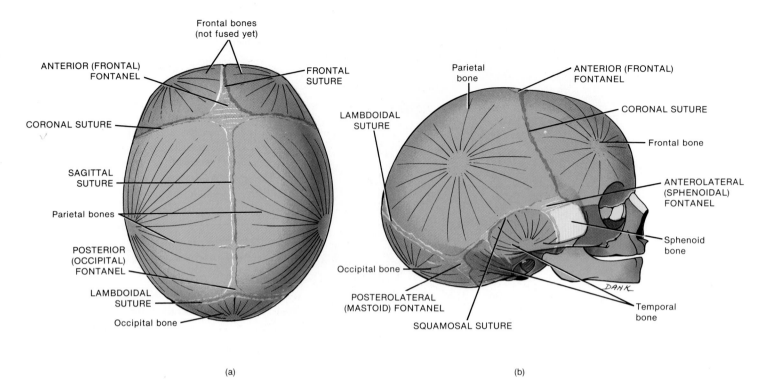

FIGURE 7-3 Fontanels of the skull at birth. (a) Superior view. (b) Right lateral view.

glossopharyngeal (IX) nerve, vagus (X) nerve, and accessory (XI) nerve pass. (As you will see later, the Roman numerals associated with cranial nerves indicate the order in which the nerves arise from the brain, from front to back.)

Between the squamous and petrous portions is a socket called the **mandibular fossa.** Anterior to the mandibular fossa is a rounded eminence, the **articular tubercle.** The mandibular fossa and articular tubercle articulate with the condylar process of the mandible (lower jawbone) to form the temporomandibular (TM) joint. The mandibular fossa and articular tubercle are seen best in Figure 7-4.

In the lateral view of the skull in Figure 7-2, you will see the **mastoid portion** of the temporal bone, located posterior and inferior to the external auditory meatus, or ear canal. In the adult, this portion of the bone contains a number of **mastoid air "cells."** These air spaces are separated from the brain only by thin bony partitions.

CLINICAL APPLICATION

If **mastoiditis,** the inflammation of these bony cells, occurs, the infection may spread to the brain or its outer covering. The mastoid air cells do not drain as do the paranasal sinuses.

The **mastoid process** is a rounded projection of the temporal bone posterior to the external auditory meatus. It serves as a point of attachment for several neck muscles.

Near the posterior border of the mastoid process is the **mastoid foramen** through which a vein (emissary) to the transverse sinus and a small branch of the occipital artery to the dura mater pass. The **external auditory (acoustic) meatus** is the canal in the temporal bone that leads to the middle ear. The **internal auditory (acoustic) meatus** is superior to the jugular foramen (see Figure 7-5a). It transmits the facial (VII) and vestibulocochlear (VIII) nerves and the internal auditory artery. The **styloid process** projects downward from the undersurface of the temporal bone and serves as a point of attachment for muscles and ligaments of the tongue and neck.

Occipital Bone

The **occipital** (ok-SIP-i-tal) **bone** forms the posterior part and a prominent portion of the base of the cranium (Figure 7-4).

The **foramen magnum** is a large hole in the inferior part of the bone through which the medulla oblongata and its membranes, the spinal portion of the accessory (XI) nerve, and the vertebral and spinal arteries pass.

The **occipital condyles** are oval processes with convex surfaces, one on either side of the foramen magnum, that articulate (form a joint) with depressions on the first cervical vertebra.

The **external occipital protuberance** is a prominent projection on the posterior surface of the bone just superior to the foramen magnum. You can feel this structure as a definite bump on the back of your head, just above your neck. The protuberance is also visible in Figure 7-2d.

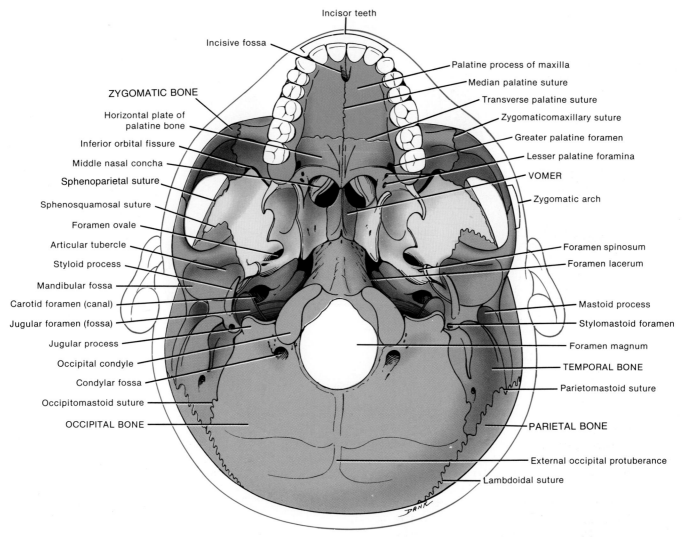

FIGURE 7-4 Skull in inferior view.

Sphenoid Bone

The **sphenoid** (SFĒ-noyd) **bone** is situated at the middle part of the base of the skull (Figure 7-5). The combining form *spheno* means wedge. This bone is referred to as the keystone of the cranial floor because it articulates with all the other cranial bones. If you view the floor of the cranium from above, you will note that the sphenoid articulates with the temporal bones anteriorly and the occipital bone posteriorly. It lies posterior and slightly superior to the nasal cavities and forms part of the floor and sidewalls of the orbit (eye socket). The shape of the sphenoid is frequently described as a bat with outstretched wings.

The **body** of the sphenoid is the cubelike central portion between the ethmoid and occipital bones. It contains the **sphenoidal sinuses,** which drain into the nasal cavity (see Figure 7-8). On the superior surface of the sphenoid body is a depression called the **sella turcica** (SEL-a TUR-si-ka = Turk's saddle). This depression houses the pituitary gland.

The **greater wings** of the sphenoid are lateral projections from the body and form the anterolateral floor of the cranium. The greater wings also form part of the lateral wall of the skull just anterior to the temporal bone. The **lesser wings** are anterior and superior to the greater wings. They form part of the floor of the cranium and the posterior part of the orbit.

Between the body and lesser wing, you can locate the **optic foramen (canal)** through which the optic (II) nerve and ophthalmic artery pass. Lateral to the body between the greater and lesser wings is a somewhat triangular slit called the **superior orbital fissure.** It is an opening for the oculomotor (III) nerve, trochlear (IV) nerve, ophthalmic branch of the trigeminal (V) nerve, and abducens (VI) nerve. This fissure may also be seen in the anterior view of the skull in Figure 7-2.

On the inferior part of the sphenoid bone you can see the **pterygoid** (TER-i-goyd) **processes.** These structures project inferiorly from the points where the body and greater wings unite. The pterygoid processes form part of the lateral walls of the nasal cavities.

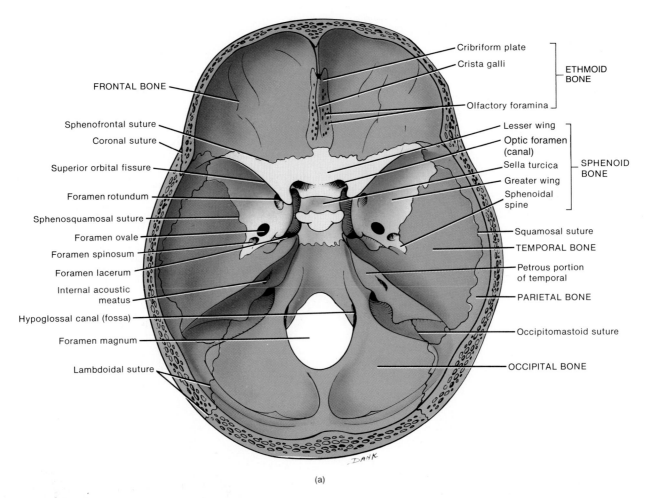

FRONTAL BONE

Sphenofrontal suture

Coronal suture

Superior orbital fissure

Foramen rotundum

Sphenosquamosal suture

Foramen ovale

Foramen spinosum

Foramen lacerum

Internal acoustic meatus

Hypoglossal canal (fossa)

Foramen magnum

Lambdoidal suture

Cribriform plate

Crista galli

ETHMOID BONE

Olfactory foramina

Lesser wing

Optic foramen (canal)

Sella turcica

SPHENOID BONE

Greater wing

Sphenoidal spine

Squamosal suture

TEMPORAL BONE

Petrous portion of temporal

PARIETAL BONE

Occipitomastoid suture

OCCIPITAL BONE

(a)

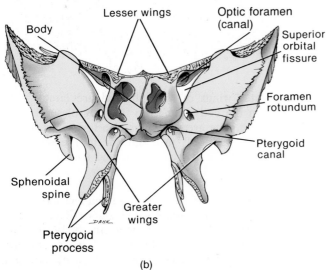

Body

Lesser wings

Optic foramen (canal)

Superior orbital fissure

Foramen rotundum

Pterygoid canal

Sphenoidal spine

Greater wings

Pterygoid process

(b)

FIGURE 7-5 Sphenoid bone. (a) Viewed in the floor of the cranium from above. (b) Anterior view.

Ethmoid Bone

The **ethmoid bone** is a light, spongy bone located in the anterior part of the floor of the cranium between the orbits. It is anterior to the sphenoid and posterior to the nasal bones (Figure 7-6). The ethmoid bone forms part of the anterior portion of the cranial floor, the medial wall of the orbits, the superior portions of the nasal septum, or partition, and most of the sidewalls of the nasal roof. The ethmoid is the principal supporting structure of the nasal cavities.

Its **lateral masses (labyrinths)** compose most of the wall between the nasal cavities and the orbits. They contain several air spaces, or "cells," ranging in number from 3 to 18. It is from these "cells" that the bone derives its name (*ethmos* = sieve). The ethmoid "cells" together form the **ethmoidal sinuses.** The sinuses are shown in Figure 7-8. The **perpendicular plate** forms the superior portion of the nasal septum (see Figure 7-7). The **cribriform (horizontal) plate,** lies in the anterior floor of the cranium and forms the roof of the nasal cavity. Projecting upward from the horizontal plate is a triangular process called the **crista galli** (= cock's comb). This structure serves as a point of attachment for the membranes (meninges) that cover the brain.

The labyrinths contain two thin, scroll-shaped bones on either side of the nasal septum. These are called the **superior nasal concha** (KONG-ka; *concha* = shell) and the **middle nasal concha.** The conchae allow for the efficient circulation and filtration of inhaled air before it passes into the trachea, bronchi, and lungs.

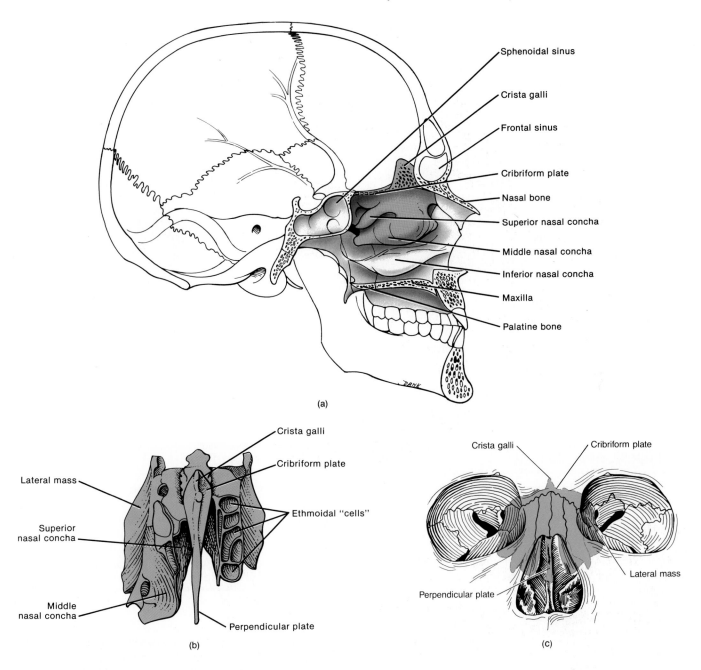

FIGURE 7-6 Ethmoid bone. (a) Median view showing the ethmoid bone on the inner aspect of the left part of the skull. (b) Anterior view. A frontal section has been made through the left side of the bone to expose the ethmoidal "cells." (c) Highly diagrammatic representation showing, in anterior view, the approximate position of the ethmoid bone in the skull.

FACIAL BONES

Nasal Bones

The paired **nasal bones** are small, oblong bones that meet at the middle and superior part of the face (see Figures 7-2 and 7-7). Their fusion forms part of the bridge of the nose. The inferior portion of the nose, indeed the major portion, consists of cartilage.

Maxillae

The paired maxillary bones unite to form the upper jaw-bone (Figure 7-7). The **maxillae** (mak-SIL-ē) articulate with every bone of the face except the mandible, or lower jawbone. They form part of the floors of the orbits, part of the roof of the mouth (most of the hard palate), and part of the lateral walls and floor of the nasal cavity.

Each maxillary bone contains a **maxillary sinus** that empties into the nasal cavity (see Figure 7-8). The **alveolar** (al-VĒ-ō-lar) **process** (*alveolus* = hollow) contains the **alveoli** (bony sockets) into which the maxillary (upper) teeth are set. The **palatine process** is a horizontal projection of the maxilla that forms the anterior three-fourths of the hard palate, or anterior portion of the roof of the oral cavity. The two portions of the maxillary bones unite, and the fusion is normally completed before birth.

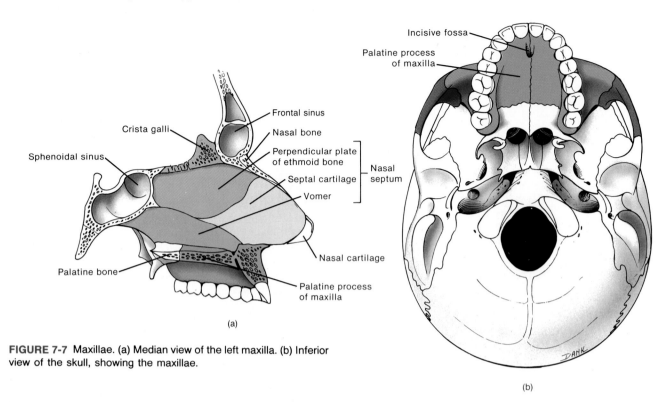

FIGURE 7-7 Maxillae. (a) Median view of the left maxilla. (b) Inferior view of the skull, showing the maxillae.

CLINICAL APPLICATION

If the palatine processes of the maxillary bones do not unite before birth, a condition called **cleft palate** results. The condition may also involve incomplete fusion of the horizontal plates of the palatine bones (see Figure 7-4). Another form of this condition, called **cleft lip,** involves a split in the upper lip. Cleft lip is often associated with cleft palate. Depending on the extent and position of the cleft, speech and swallowing may be affected. A surgical procedure can sometimes improve the condition.

A fissure associated with the maxilla and sphenoid bone is the **inferior orbital fissure.** It is located between the greater wing of the sphenoid and the maxilla (see Figure 7-4). It transmits the maxillary branch of the trigeminal (V) nerve, the infraorbital vessels, and the zygomatic nerve.

Paranasal Sinuses

Paired cavities, called **paranasal sinuses,** are located in certain bones near the nasal cavity (Figure 7-8). The paranasal sinuses are lined with mucous membranes that are continuous with the lining of the nasal cavity. Cranial bones containing paranasal sinuses are the frontal, sphenoid, ethmoid, and maxillae. (The paranasal sinuses were described in the discussion of each of these bones.) Besides producing mucus, the paranasal sinuses lighten the skull bones and serve as resonant chambers for sound.

CLINICAL APPLICATION

Secretions produced by the mucous membranes of the paranasal sinuses drain into the nasal cavity. An inflammation of the membranes due to an allergic reaction or infection is called **sinusitis.** If the membranes swell enough to block drainage into the nasal cavity, fluid pressure builds up in the paranasal sinuses and a sinus headache results.

Zygomatic Bones

The two **zygomatic bones (malars),** commonly referred to as the cheekbones, form the prominences of the cheeks and part of the outer wall and floor of the orbits (Figure 7-2b).

The **temporal process** of the zygomatic bone projects posteriorly and articulates with the zygomatic process of the temporal bone. These two processes form the **zygomatic arch** (Figure 7-4).

Mandible

The **mandible,** or lower jawbone, is the largest, strongest facial bone (Figure 7-9). It is the only movable skull bone.

In the lateral view you can see that the mandible consists of a curved, horizontal portion called the **body** and two perpendicular portions called the **rami.** The **angle** of the mandible is the area where each ramus meets the body. Each ramus has a **condylar** (KON-di-lar) **process** that articulates with the mandibular fossa and articular

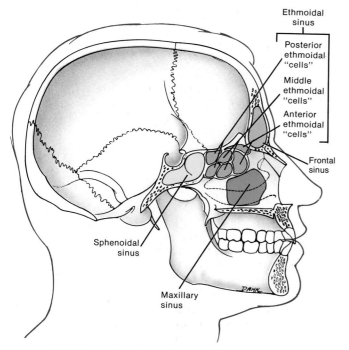

FIGURE 7-8 Paranasal sinuses seen in median view.

the mandible is the **mandibular foramen** on the medial surface of the ramus, another site frequently used by dentists to inject anesthetics. It transmits the inferior alveolar nerve and vessels. The mandibular foramen is the beginning of the **mandibular canal,** which runs forward in the ramus deep to the roots of the teeth. The canal carries branches of the inferior alveolar nerve and vessels to the teeth. Parts of these nerves and vessels emerge through the mental foramen.

CLINICAL APPLICATION

Opening of the mouth very wide, as in a very large yawn, may cause displacement of the condylar process of the mandible from the mandibular fossa of temporal bone, thereby producing a **dislocation of the jaw.** The displacement is usually anterior and bilateral and the patient is unable to close the mouth.

Lacrimal Bones

The paired **lacrimal bones** (LAK-ri-mal; *lacrima* = tear) are thin bones roughly resembling a fingernail in size and shape. They are the smallest bones of the face. These bones are posterior and lateral to the nasal bones in the medial wall of the orbit. They can be seen in the anterior and lateral views of the skull in Figure 7-2. The lacrimal bones form a part of the medial wall of the orbit.

Palatine Bones

The two **palatine** (PAL-a-tīn) **bones** are L-shaped and form the posterior portion of the hard palate, part of the floor and lateral wall of the nasal cavity, and a small portion of the floors of the orbits. The posterior portion of the hard palate, which separates the nasal cavity from the oral cavity, is formed by the **horizontal plates** of the palatine bones. These can be seen in Figure 7-4.

tubercle of the temporal bone to form the temporomandibular (TM) joint. It also has a **coronoid** (KOR-ō-noyd) **process** to which the temporalis muscle attaches. The depression between the coronoid and condylar processes is called the **mandibular notch.** The **alveolar process** is an arch containing the **alveoli** (sockets) for the mandibular (lower) teeth.

The **mental foramen** (*mentum* = chin) is approximately below the first molar tooth. The mental nerve and vessels pass through this opening. It is through this foramen that dentists sometimes reach the nerve when injecting anesthetics. Another foramen associated with

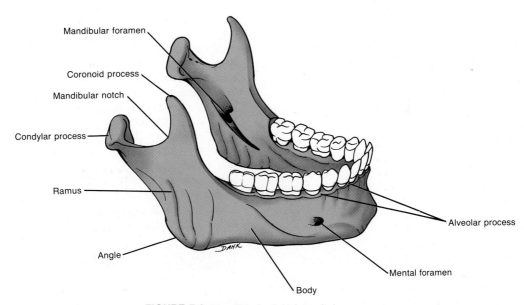

FIGURE 7-9 Mandible in right lateral view.

Inferior Nasal Conchae

Refer to the views of the skull in Figures 7-2a and 7-6a. The two **inferior nasal conchae** are scroll-like bones that form a part of the lateral wall of the nasal cavity and project into the nasal cavity inferior to the superior and middle nasal conchae of the ethmoid bone. They serve the same function as the superior and middle nasal conchae, that is, they allow for the circulation and filtration of air before it passes into the lungs. The inferior nasal conchae are separate bones and not part of the ethmoid.

Vomer

The **vomer** (= plowshare) is a roughly triangular bone that forms the inferior and posterior part of the nasal septum. It is clearly seen in the anterior view of the skull in Figure 7-2a and the inferior view in Figure 7-4.

The inferior border of the vomer articulates with the cartilage septum that divides the nose into a right and left nostril. Its superior border articulates with the perpendicular plate of the ethmoid bone. Thus the structures that form the **nasal septum,** or partition, are the perpendicular plate of the ethmoid, the septal cartilage, and the vomer (Figure 7-7a).

CLINICAL APPLICATION

A **deviated nasal septum** is deflected laterally from the midline of the nose. The deviation usually occurs at the junction of bone with the septal cartilage. If the deviation is severe, it may entirely block the nasal passageway. Even though the blockage may not be complete, infection and inflammation develop and cause nasal congestion, blockage of the paranasal sinus openings, and chronic sinusitis.

FORAMINA

Some foramina (singular = foramen) of the skull were mentioned along with the descriptions of the cranial and facial bones with which they are associated. As preparation for studying other systems of the body, especially the nervous and cardiovascular systems, these foramina, as well as additional ones, and the structures passing through them are listed in Exhibit 7-3. For your convenience and for future reference, the foramina are listed alphabetically.

HYOID BONE

The single **hyoid bone** (*hyoedes* = U-shaped) is a unique component of the axial skeleton because it does not articulate with any other bone. Rather, it is suspended from the styloid process of the temporal bone by ligaments and muscles. The hyoid is located in the neck between the mandible and larynx. It supports the tongue and provides attachment for some of its muscles. Refer to the

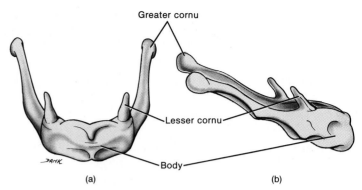

FIGURE 7-10 Hyoid bone. (a) Anterior view. (b) Right lateral view.

median and lateral views of the skull in Figure 7-2c,d to see the position of the hyoid bone.

The hyoid consists of a horizontal **body** and paired projections called the **lesser cornu** (*cornu* = horn) and the **greater cornu** (Figure 7-10). Muscles and ligaments attach to these paired projections.

The hyoid bone is frequently fractured during strangulation. As a result, it is carefully examined in an autopsy when strangulation is suspected.

VERTEBRAL COLUMN

DIVISIONS

The **vertebral column,** or **spine,** together with the sternum and ribs, constitutes the skeleton of the **trunk** of the body. The vertebral column is composed of a series of bones called **vertebrae.** In the average adult, the column measures about 71 cm (28 inches) in length. In effect, the vertebral column is a strong, flexible rod that moves anteriorly, posteriorly, and laterally. It encloses and protects the spinal cord, supports the head, and serves as a point of attachment for the ribs and the muscles of the back. Between the vertebrae are openings called **intervertebral foramina.** The nerves that connect the spinal cord to various parts of the body pass through these openings.

The adult vertebral column typically contains 26 vertebrae (Figure 7-11a,b). These are distributed as follows: 7 **cervical vertebrae** (*cervix* = neck) in the neck region; 12 **thoracic vertebrae** posterior to the thoracic cavity; 5 **lumbar vertebrae** (*lumbus* = loin) supporting the lower back; 5 **sacral vertebrae** fused into one bone called the **sacrum;** and usually 4 **coccygeal** (kok-SIJ-ē-al) **vertebrae** fused into one or two bones called the **coccyx** (KOK-six). Prior to the fusion of the sacral and coccygeal vertebrae, the total number of vertebrae is 33.

Between adjacent vertebrae from the first vertebra (axis) to the sacrum are fibrocartilaginous **intervertebral discs.** Each disc is composed of an outer fibrous ring consisting of fibrocartilage called the *annulus fibrosus* and an inner soft, pulpy, highly elastic structure called the *nucleus pulposus* (see Figure 7-19). The discs form strong joints, permit various movements of the vertebral column, and absorb vertical shock. Under compression, they flat-

EXHIBIT 7-3   SUMMARY OF FORAMINA OF THE SKULL

FORAMEN	LOCATION	STRUCTURES PASSING THROUGH
Carotid (Figure 7-4)	Petrous portion of temporal.	Internal carotid artery.
Greater palatine (Figure 7-4)	Posterior angle of hard palate.	Greater palatine nerve and greater palatine vessels.
Hypoglossal (Figure 7-5)	Superior to base of occipital condyles.	Hypoglossal (XII) nerve and branch of ascending pharyngeal artery.
Incisive (Figure 7-7b)	Posterior to incisor teeth.	Branches of descending palatine vessels and naso-palatine nerve.
Inferior orbital (Figure 7-4)	Between greater wing of sphenoid and maxilla.	Maxillary branch of trigeminal (V) nerve, zygomatic nerve, and infraorbital vessels.
Infraorbital (Figure 7-2a)	Inferior to orbit in maxilla.	Infraorbital nerve and artery.
Jugular (Figure 7-4)	Posterior to carotid canal between petrous portion of temporal and occipital.	Internal jugular vein, glossopharyngeal (IX) nerve, vagus (X) nerve, accessory (XI) nerve, and sigmoid sinus.
Lacerum (Figure 7-5)	Bounded anteriorly by sphenoid, posteriorly by petrous portion of temporal, and medially by the sphenoid and occipital.	Internal carotid artery and branch of ascending pharyngeal artery.
Lacrimal (Figure 7-2c)	Lacrimal bone.	Lacrimal (tear) duct.
Lesser palatine (Figure 7-4)	Posterior to greater palatine foramen.	Lesser palatine nerves and artery.
Magnum (Figure 7-4)	Occipital bone.	Medulla oblongata and its membranes, the accessory (XI) nerve, and the vertebral and spinal arteries and meninges.
Mandibular (Figure 7–9)	Medial surface of ramus of mandible.	Inferior alveolar nerve and vessels.
Mastoid	Posterior border of mastoid process of temporal bone.	Emissary vein to transverse sinus and branch of occipital artery to dura mater.
Mental (Figure 7-9)	Inferior to second premolar tooth in mandible.	Mental nerve and vessels.
Olfactory (Figure 7-5)	Cribriform plate of ethmoid.	Olfactory (I) nerve.
Optic (Figure 7-5)	Between upper and lower portions of small wing of sphenoid.	Optic (II) nerve and ophthalmic artery.
Ovale (Figure 7-5)	Greater wing of sphenoid.	Mandibular branch of trigeminal (V) nerve.
Rotundum (Figure 7-5)	Junction of anterior and medial parts of sphenoid.	Maxillary branch of trigeminal (V) nerve.
Spinosum (Figure 7-5)	Posterior angle of sphenoid.	Middle meningeal vessels.
Stylomastoid (Figure 7-4)	Between styloid and mastoid processes of temporal.	Facial (VII) nerve and stylomastoid artery.
Superior orbital (Figure 7-5)	Between greater and lesser wings of sphenoid.	Oculomotor (III) nerve, trochlear (IV) nerve, ophthalmic branch of trigeminal (V) nerve, and abducens (VI) nerve.
Supraorbital (Figure 7-2a)	Supraorbital margin of orbit.	Supraorbital nerve and artery.
Zygomaticofacial (Figure 7-2a)	Zygomatic bone.	Zygomaticofacial nerve and vessels.

FIGURE 7-11 Vertebral column. (a) Anterior view. (b) Right lateral view. (c) Intervertebral disc in its normal position (left) and under compression (right). The relative size of the disc has been enlarged for emphasis. A "window" has been cut in the annulus fibrosus so that the nucleus pulposus can be seen.

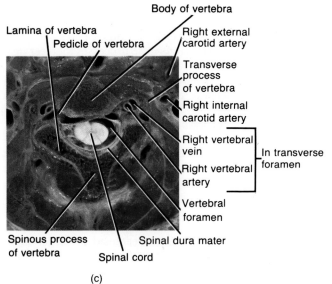

FIGURE 7-12 Typical vertebra. (a) Superior view. (b) Right lateral view. (c) Photograph of a cross section through the upper cervical region of the vertebral column. (Courtesy of Stephen A. Kieffer and E. Robert Heitzman, *An Atlas of Cross-Sectional Anatomy*, Harper & Row, Publishers, Inc., Hagerstown, MD, 1979.)

ten, broaden, and bulge from their intervertebral spaces (Figure 7-11c).

CURVES

When viewed from the side with the subject facing to the right, the vertebral column shows four **curves** (Figure 7-11b), two of which are convex ()), and two of which are concave ((). The curves of the column, like the curves in a long bone, are important because they increase its strength, help maintain balance in the upright position, absorb shocks from walking, and help protect the column from fracture.

In the fetus, there is only a single anteriorly concave curve. At approximately the third postnatal month, when an infant begins to hold its head erect, the **cervical curve** develops. Later, when the child stands and walks, the **lumbar curve** develops. The cervical and lumbar curves

are anteriorly convex. Because they are modifications of the fetal positions, they are called **secondary curves.** The other two curves, the **thoracic curve** and the **sacral curve,** are anteriorly concave. Since they retain the anterior concavity of the fetus, they are referred to as **primary curves.**

CLINICAL APPLICATION

In recent years, **gravity inversion** has been used by some people to decompress the backbone by using gravity and the body's own weight. This may be accomplished by hanging upside down by the ankles in gravity inversion boots suspended from a horizontal bar. Despite the popularity of this method of traction and exercise, there are some potentially harmful effects. For example, it has been found that some healthy young individuals may experience increases in general blood pressure, pulse rate, intraocular pressure, and blood pressure in the eyes. Moreover, gravity inversion is not recommended for people with glaucoma, high blood pressure, heart disease, hiatal hernias, or disorders of the vertebral column.

TYPICAL VERTEBRA

Although there are variations in size, shape, and detail in the vertebrae in different regions of the column, all the vertebrae are basically similar in structure (Figure 7-12). A typical vertebra consists of the following components.

1. The **body** is the thick, disc-shaped anterior portion that is the weight-bearing part of a vertebra. Its superior and inferior surfaces are roughened for the attachment of intervertebral discs. The anterior and lateral surfaces contain nutrient foramina for blood vessels.

2. The **vertebral (neural) arch** extends posteriorly from the body of the vertebra. With the body of the vertebra, it surrounds the spinal cord. It is formed by two short, thick processes, the **pedicles** (PED-i-kuls), which project posteriorly from the body to unite with the laminae. The **laminae** (LAM-i-nē) are the flat parts that join to form the posterior portion of the vertebral arch. The space that lies between the vertebral arch and body contains the spinal cord. This space is known as the **vertebral foramen.** The vertebral foramina of all vertebrae together form the **vertebral (spinal) canal.** The pedicles are notched superiorly and inferiorly in such a way that, when they are arranged in the column, there is an opening between vertebrae on each side of the column. This opening, the **intervertebral foramen,** permits the passage of a single spinal nerve.

3. Seven **processes** arise from the vertebral arch. At the point where a lamina and pedicle join, a **transverse process** extends laterally on each side. A single **spinous process (spine)** projects posteriorly and inferiorly from the junction of the laminae. These three processes serve as points of muscular attachment. The remaining four processes form joints with other vertebrae. The two **superior articular processes** of a vertebra articulate with the vertebra immediately superior to them. The two **inferior articular processes** of a vertebra articulate with the vertebra inferior to them. The articulating surfaces of the articular processes are referred to as **facets.**

CERVICAL REGION

When viewed from above, it can be seen that the bodies of **cervical vertebrae** are smaller than those of thoracic vertebrae (Figure 7-13). The arches, however, are larger. The spinous processes of the second through sixth cervical vertebrae are often *bifid,* that is, with a cleft. All cervical vertebrae have three foramina: the vertebral foramen and two transverse foramina. Each cervical transverse process contains a **transverse foramen** through which the vertebral artery and its accompanying vein and nerve fibers pass.

The first two cervical vertebrae differ considerably from the others. The first cervical vertebra (C1), the **atlas,** is named for its support of the head. Essentially, the atlas is a ring of bone with **anterior** and **posterior arches** and large **lateral masses.** It lacks a body and a spinous process. The superior surfaces of the lateral masses, called **superior articular facets,** are concave and articulate with the occipital condyles of the occipital bone. This articulation permits the movement seen when nodding the head. The inferior surfaces of the lateral masses, the **inferior articular facets,** articulate with the second cervical vertebra. The transverse processes and transverse foramina of the atlas are quite large.

The second cervical vertebra (C2), the **axis,** does have a body. A peglike process called the **dens (odontoid process)** projects up through the ring of the atlas. The dens makes a pivot on which the atlas and head rotate. This arrangement permits side-to-side rotation of the head.

The third through sixth cervical vertebrae (C3–C6) correspond to the structural pattern of the typical cervical vertebra previously described.

The seventh cervical vertebra (C7), called the **vertebra prominens,** is somewhat different. It is marked by a large, nonbifid spinous process that may be seen and felt at the base of the neck.

THORACIC REGION

Viewing a typical **thoracic vertebra** from above, you can see that it is considerably larger and stronger than a vertebra of the cervical region (Figure 7-14). In addition, the spinous process on each vertebra is long, laterally flattened, and directed inferiorly. Thoracic vertebrae also have longer and heavier transverse processes than cervical vertebrae.

Except for the eleventh and twelfth thoracic vertebrae, the transverse processes have **facets** for articulating with the tubercles of the ribs. The bodies of thoracic vertebrae also have whole **facets** or half-facets, called **demifacets,** for articulation with the heads of the ribs. The first thoracic vertebra (T1) has, on either side of its body, a superior whole facet and an inferior demifacet. The superior facet articulates with the first rib, and the inferior demifacet, together with the superior demifacet of the second thoracic vertebra (T2), forms a facet for articulation with the second rib. The second through eighth thoracic vertebrae (T2–T8) have two demifacets on each side, a larger superior demifacet and a smaller inferior demifacet. When the vertebrae are articulated, they form whole facets for the heads of the ribs. The ninth thoracic vertebra (T9) has a single superior demifacet on either side of its body. The tenth through twelfth thoracic vertebrae (T10–T12) have whole facets on either side of their bodies.

LUMBAR REGION

The **lumbar vertebrae** (L1–L5) are the largest and strongest in the column (Figure 7-15). Their various projections are short and thick. The superior articular processes are directed medially instead of superiorly. The inferior articular processes are directed laterally instead of inferiorly. The spinous process is quadrilateral in shape, thick and broad, and projects nearly straight posteriorly. The spinous process is well adapted for the attachment of the large back muscles.

SACRUM AND COCCYX

The **sacrum** is a triangular bone formed by the union of five sacral vertebrae. These are indicated in Figure

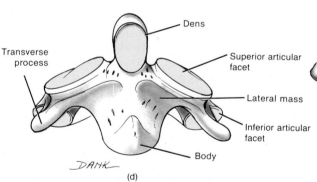

FIGURE 7-13 Cervical vertebrae. (a) Superior view of a cervical vertebrae. (b) Superior view of the atlas. (c) Superior view of the axis. (d) Anterior view of the axis. (e) Articulated in posterior view.

7-16 as S1 through S5. The sacrum serves as a strong foundation for the pelvic girdle. It is positioned at the posterior portion of the pelvic cavity between the two hipbones.

The concave anterior side of the sacrum faces the pelvic cavity. It is smooth and contains four **transverse lines (ridges)** that mark the joining of the sacral vertebral bodies. At the ends of these lines are four pairs of **anterior sacral (pelvic) foramina.**

The convex, posterior surface of the sacrum is irregular. It contains a **median sacral crest,** the fused spinous processes of the upper sacral vertebrae; a **lateral sacral crest,** the transverse processes of the sacral vertebrae;

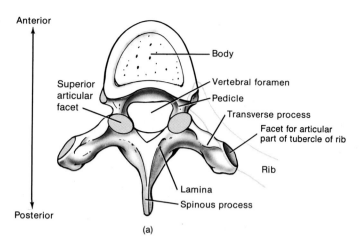

Anterior

Posterior

Body

Superior articular facet

Vertebral foramen

Pedicle

Transverse process

Facet for articular part of tubercle of rib

Rib

Lamina

Spinous process

(a)

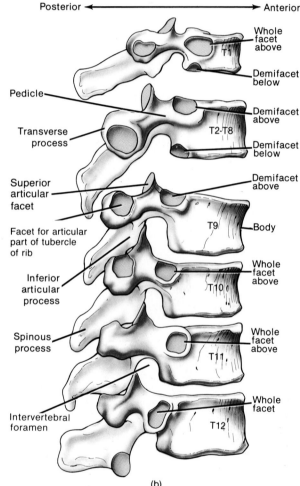

Posterior ←————————→ Anterior

Pedicle

Transverse process

Superior articular facet

Facet for articular part of tubercle of rib

Inferior articular process

Spinous process

Intervertebral foramen

Whole facet above

Demifacet below

T1

Demifacet above

T2-T8

Demifacet below

Demifacet above

T9

Body

Whole facet above

T10

Whole facet above

T11

Whole facet

T12

(b)

FIGURE 7-14 Thoracic vertebrae. (a) Superior view. (b) Articulated in right lateral view.

and four pairs of **posterior sacral (dorsal) foramina.** These foramina communicate with the anterior sacral foramina through which nerves and blood vessels pass. The **sacral canal** is a continuation of the vertebral canal. The laminae of the fifth sacral vertebra, and sometimes the fourth, fail to meet. This leaves an inferior entrance to the vertebral canal called the **sacral hiatus** (hī-Ā-tus). On either side of the sacral hiatus are the **sacral cornua,** the inferior

articular processes of the fifth sacral vertebra. They are connected by ligaments to the coccygeal cornua of the coccyx.

CLINICAL APPLICATION

Anesthetic agents that act on the sacral and coccygeal nerves are sometimes injected through the sacral hiatus, a procedure called **caudal anesthesia.** Since the sacral hiatus is between the sacral cornua, the cornua are important bony landmarks for locating the hiatus. Anesthetic agents may also be injected through the posterior sacral (dorsal) foramina.

The superior border of the sacrum exhibits an anteriorly projecting border, the **sacral promontory** (PROM-on-tō'-rē). It is an obstetrical landmark for measurements of the pelvis. An imaginary line running from the superior surface of the symphysis pubis to the sacral promontory separates the abdominal and pelvic cavities. Laterally, the sacrum has a large **auricular surface** for articulating with the ilium of the hipbone. This is the sacroiliac joint. The **superior articular processes** of the sacrum articulates with the fifth lumbar vertebra.

The **coccyx** is also triangular in shape and is formed by the fusion of the coccygeal vertebrae, usually the last four. These are indicated in Figure 7-16 as Co1 through Co4. The dorsal surface of the body of the coccyx contains two long **coccygeal cornua** that are connected by ligaments to the sacral cornua. The coccygeal cornua are the pedicles and superior articular processes of the first coccygeal vertebra. On the lateral surfaces of the body of the coccyx are a series of **transverse processes,** the first pair being the largest. The coccyx articulates superiorly with the sacrum.

THORAX

Anatomically, the term **thorax** refers to the chest. The skeletal portion of the thorax is a bony cage formed by the sternum, costal cartilage, ribs, and the bodies of the thoracic vertebrae (Figure 7-17).

The thoracic cage is roughly cone-shaped, the narrow portion being superior and the broad portion inferior. It is flattened from front to back. The thoracic cage encloses and protects the organs in the thoracic cavity. It also provides support for the bones of the shoulder girdle and upper extremities.

STERNUM

The **sternum,** or breastbone, is a flat, narrow bone measuring about 15 cm (6 inches) in length. It is located in the median line of the anterior thoracic wall.

The sternum (see Figure 7-17) consists of three basic portions: the **manubrium** (ma-NOO-brē-um), the triangular, superior portion; the **body,** the middle, largest portion; and the **xiphoid** (ZĪ-foyd) **process,** the inferior, smallest portion. The manubrium has a depression on its superior

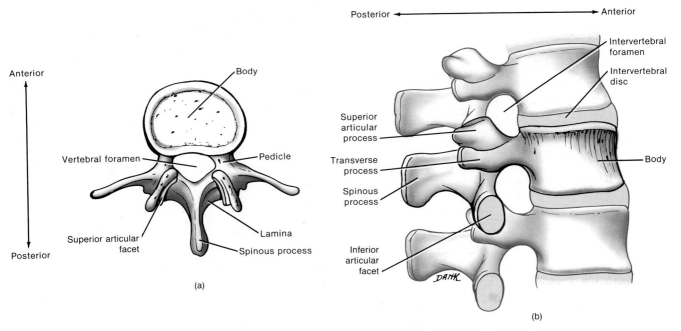

FIGURE 7-15 Lumbar vertebrae. (a) Superior view. (b) Articulated in right lateral view.

surface called the **jugular (suprasternal) notch.** On each side of the jugular notch are **clavicular notches** that articulate with the medial ends of the clavicles. The manubrium also articulates with the first and second ribs. The body of the sternum articulates directly or indirectly with the second through tenth ribs. The xiphoid process has no ribs attached to it but provides attachment for some abdominal muscles. The xiphoid process consists of hyaline cartilage during infancy and childhood and does not ossify completely until about age 40. If the hands of a rescuer are mispositioned during cardiopulmonary resuscitation (CPR), there is the danger of fracturing the ossified xiphoid process, separating it from the body, and driving it into the liver.

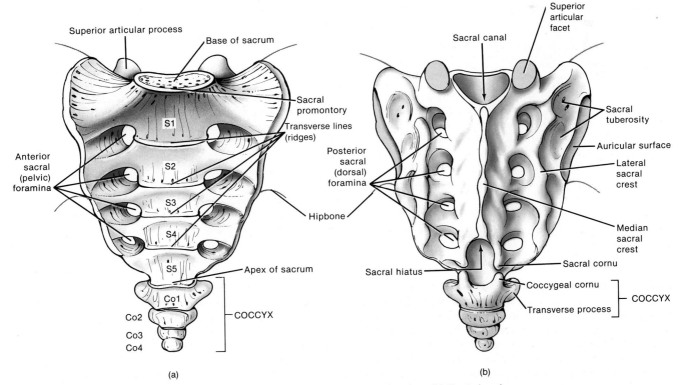

FIGURE 7-16 Sacrum and coccyx. (a) Anterior view. (b) Posterior view.

FIGURE 7-17 Skeleton of the thorax in anterior view.

CLINICAL APPLICATION

Since the sternum possesses red bone marrow throughout life and because it is readily accessible, it is a common site for **marrow biopsy.** Under a local anesthetic, a wide-bore needle is introduced into the marrow cavity of the sternum for aspiration of a sample of red bone marrow. This procedure is called a **sternal puncture.**

The sternum may also be split in the midsagittal plane to allow surgeons access to mediastinal structures such as the thymus gland, heart, and great vessels of the heart.

RIBS

Twelve pairs of **ribs** make up the sides of the thoracic cavity (see Figure 7-17). The ribs increase in length from the first through seventh. Then they decrease in length to the twelfth rib. Each rib articulates posteriorly with its corresponding thoracic vertebra.

The first through seventh ribs have a direct anterior attachment to the sternum by a strip of hyaline cartilage,

called **costal cartilage** (*costa* = rib). These ribs are called **true (vertebrosternal) ribs.** The remaining five pairs of ribs are referred to as **false ribs** because their costal cartilages do not attach directly to the sternum. The cartilages of the eighth, ninth, and tenth ribs attach to each other and then to the cartilage of the seventh rib. These false ribs are called **vertebrochondral ribs.** The eleventh and twelfth false ribs are designated as **floating (vertebral) ribs** because their anterior ends do not attach even indirectly to the sternum. They attach only posteriorly to the thoracic vertebrae.

Although there is some variation in rib structure, we will examine the parts of a typical (third through ninth) rib when viewed from the right side and from behind (Figure 7-18). The **head** of a typical rib is a projection at the posterior end of the rib. It is wedge-shaped and consists of one or two **facets** that articulate with facets on the bodies of adjacent thoracic vertebrae. The facets on the head of a rib are separated by a horizontal **interarticular crest.** The inferior facet on the head of a rib is larger than the superior facet. The **neck** is a constricted portion just lateral to the head. A knoblike structure on the posterior surface where the neck joins the body is called a **tubercle.** It consists of a **nonarticular part,** which affords attachment to the ligament of the tubercle and an **articular part,** which articulates with the facet of a transverse process of the inferior of the two vertebrae to which the head of the rib is connected. The **body (shaft)** is the main part of the rib. A short distance beyond the tubercle, there is an abrupt change in the curvature of the shaft. This point is called the **costal angle.** The inner surface of the rib has a **costal groove** that protects blood vessels and a small nerve.

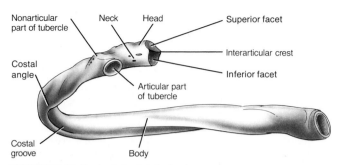

FIGURE 7-18 Typical rib. A left rib viewed from below and behind.

CLINICAL APPLICATION

Rib fractures usually result from direct blows, most commonly from steering wheel impact, falls, and crushing injuries to the chest. Ribs tend to break at the weakest point, that is, the site of greatest curvature just anterior to the costal angle. In children, the ribs are highly elastic and fractures are less frequent than in adults. Since the first two ribs are protected by the clavicle and pectoralis major muscle, and the last two ribs are mobile, they are the least commonly injured. The middle ribs are the ones most commonly fractured. In some cases, fractured ribs may cause damage to the heart, great vessels of the heart, lungs, trachea, bronchi, esophagus, spleen, and liver.

The posterior portion of the rib is connected to a thoracic vertebra by its head and articular part of a tubercle. The facet of the head fits into a facet on the body of a vertebra, and the articular part of the tubercle articulates with the facet of the transverse process of the vertebra. Each of the second through ninth ribs articulates with the bodies of two adjacent vertebrae. The first, tenth, eleventh, and twelfth ribs articulate with only one vertebra each. On the eleventh and twelfth ribs, there is no articulation between the tubercles and the transverse processes of their corresponding vertebrae.

Spaces between ribs, called **intercostal spaces,** are occupied by intercostal muscles, blood vessels, and nerves.

CLINICAL APPLICATION

Surgical access to the lungs or structures in the mediastinum is commonly undertaken through an intercostal space. Special rib retractors are used to create a wide separation between ribs. The costal cartilages are sufficiently elastic to permit considerable bending.

DISORDERS: HOMEOSTATIC IMBALANCES

Herniated (Slipped) Disc

In their function as shock absorbers, intervertebral discs are subject to compressional forces. The discs between the fourth and fifth lumbar vertebrae and between the fifth lumbar vertebra and sacrum usually are subject to more forces than other discs. If the anterior and posterior ligaments of the discs become injured or weakened, the pressure developed in the nucleus pulposus may be great enough to rupture the surrounding fibrocartilage. If this occurs, the nucleus pulposus may protrude posteriorly or into one of the adjacent vertebral bodies (herniate). This condition is called a **herniated (slipped) disc.**

Most often the nucleus pulposus slips posteriorly toward the spinal cord and spinal nerves (Figure 7-19). This movement exerts pressure on the spinal nerves, causing considerable, sometimes very acute, pain. If the roots of the sciatic nerve, which passes from the spinal cord to the foot, are pressured, the pain radiates down the back of the thigh, through the calf, and occasionally into the foot. If pressure is exerted on the spinal cord itself, nervous tissue may be destroyed.

Traction, bed rest, and analgesia usually relieve the pain. If such treatment is ineffective, surgical decompression of the spinal nerves or removal of some of the nucleus pulposus may be necessary to relieve pain. Removal may involve conventional surgery or *chemonucleolysis,* the use of a proteolytic enzyme called chymopapain (Chymodiactin). This enzyme is injected

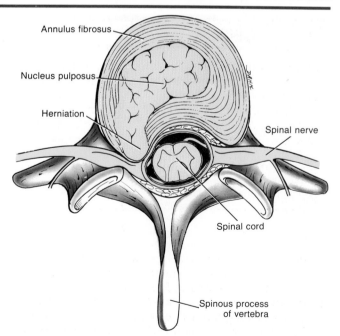

FIGURE 7-19 Superior view of a herniated (slipped) disc.

into a herniated disc where it dissolves the nucleus pulposus, thus relieving pressure on spinal nerves and pain.

Curvatures

As a result of various conditions, the normal curves of the vertebral column may become exaggerated or the column may acquire a lateral bend. Such conditions are called **curvatures** of the spine.

Scoliosis (skō'-lē-Ō-sis; *scolio* = bent) is a lateral bending of the vertebral column, usually in the thoracic region. This is the most common of the curvatures. It may be congenital, due to an absence of the lateral half of a vertebra (hemivertebra). It may also be acquired from a persistent severe sciatica. Poliomyelitis may cause scoliosis by the paralysis of muscles on one side of the body, which produces a lateral deviation of the trunk toward the unaffected side. Poor posture is also a contributing factor. Scoliosis can also result from one leg being shorter than the other.

Until recently, a child with progressive scoliosis faced either years of treatment with a brace or major corrective surgery. Several research teams are experimenting with using electrical stimulation of muscles to limit the progression of scoliosis and even to reduce curvatures in some cases. The skeletal muscles on the convex side of the spinal curve are electrically stimulated.

Kyphosis (kī-FŌ-sis; *kypho* = hunchback) is an exaggeration of the thoracic curve of the vertebral column. In tuberculosis of the spine, vertebral bodies may partially collapse, causing an acute angular bending of the vertebral column. In the elderly, degeneration of the intervertebral discs leads to kyphosis. Kyphosis may also be caused by rickets and poor posture. The term "round-shouldered" is an expression for mild kyphosis. Electrical muscle stimulation is being studied to assess its effects on kyphosis as well as on scoliosis.

Lordosis (lor-DŌ-sis; *lordo* = swayback) is an exaggeration of the lumbar curve of the vertebral column. It may result from increased weight of abdominal contents as in pregnancy or extreme obesity. Other causes include poor posture, rickets, and tuberculosis of the spine.

Spina Bifida

Spina bifida (SPĪ-na BIF-i-da) is a congenital defect of the vertebral column in which laminae fail to unite at the midline. The lumbar vertebrae are involved in about 50 percent of all cases. In less serious cases, the defect is small and the area is covered with skin. Only a dimple or tuft of hair may mark the site. Symptoms are mild, with perhaps intermittent urinary problems. An operation is not ordinarily required. Larger defects in the vertebral arches with protrusion of the membranes around the spinal cord or the spinal cord itself produce serious problems, such as partial or complete paralysis, partial or complete loss of urinary bladder control, and the absence of reflexes. Spina bifida may be diagnosed prenatally by a test of the mother's blood, sonography, and/or amniocentesis.

Fractures of the Vertebral Column

Fractures of the vertebral column most commonly involve T12, L1, and L2. They usually result from a flexion-compression type of injury such as might be sustained in landing on the feet or buttocks after a fall from a height or having a heavy weight fall on the shoulders. The forceful compression wedges the involved vertebrae. If, in addition to compression, there is forceful forward movement, one vertebra may displace forward on its adjacent vertebra below, with either dislocation or fracture of the articular facets between the two (fracture dislocation) and with rupture of the interspinous ligaments.

The cervical vertebrae may be fractured or, more commonly, dislodged by a fall on the head with acute flexion of the neck, as might happen on diving into shallow water. Dislocation may even result from the sudden forward jerk which may occur when an automobile or airplane crashes ("whiplash"). The relatively horizontal intervertebral facets of the cervical vertebrae allow dislocation to take place without their being fractured, whereas the relatively vertical thoracic and lumbar intervertebral facets nearly always fracture in forward dislocation of the thoracolumbar region. Spinal nerve damage may occur as a result of fractures of the vertebral column.

STUDY OUTLINE

Types of Bones (p. 133)
1. On the basis of shape, bones are classified as long, short, flat, or irregular.
2. Sutural (Wormian) bones are found between the sutures of certain cranial bones. Sesamoid bones develop in tendons or ligaments.

Surface Markings (p. 133)
1. Markings are areas on the surfaces of bones.
2. Each marking is structured for a specific function—joint formation, muscle attachment, or passage of nerves and blood vessels.
3. Terms that describe markings include fissure, foramen, meatus, fossa, process, condyle, head, facet, tuberosity, crest, and spine.

Divisions of the Skeletal System (p. 133)
1. The axial skeleton consists of bones arranged along the longitudinal axis. The parts of the axial skeleton are the skull, hyoid bone, auditory ossicles, vertebral column, sternum, and ribs.
2. The appendicular skeleton consists of the bones of the girdles and the upper and lower extremities. The parts of the appendicular skeleton are the pectoral (shoulder) girdles, bones of the upper extremities, pelvic (hip) girdle, and bones of the lower extremities.

Skull (p. 133)
1. The skull consists of the cranium and the face. It is composed of 22 bones.
2. Sutures are immovable joints between bones of the skull. Examples are coronal, sagittal, lambdoidal, and squamosal sutures.
3. Fontanels are membrane-filled spaces between the cranial bones of fetuses and infants. The major fontanels are the anterior, posterior, anterolaterals, and posterolaterals.
4. The 8 cranial bones include the frontal, parietal (2), temporal (2), occipital, sphenoid, and ethmoid.
5. The 14 facial bones are the nasal (2), maxillae (2), zygomatic (2), mandible, lacrimal (2), palatine (2), inferior nasal conchae (2), and vomer.
6. Paranasal sinuses are cavities in bones of the skull that communicate with the nasal cavity. They are lined by mucous membranes. The cranial bones containing the paranasal sinuses are the frontal, sphenoid, ethmoid, and maxillae.

7. The foramina of the skull bones provide passages for nerves and blood vessels.

Hyoid Bone (p. 146)
1. The hyoid bone is a U-shaped bone that does not articulate with any other bone.
2. It supports the tongue and provides attachment for some of its muscles.

Vertebral Column (p. 146)
1. The vertebral column, sternum, and ribs constitute the skeleton of the trunk.
2. The bones of the adult vertebral column are the cervical vertebrae (7), thoracic vertebrae (12), lumbar vertebrae (5), the sacrum (5, fused), and the coccyx (4, fused).
3. The vertebral column contains primary curves (thoracic and sacral) and secondary curves (cervical and lumbar). These curves give strength, support, and balance.
4. The vertebrae are similar in structure, each consisting of a body, vertebral arch, and seven processes. Vertebrae in the different regions of the column vary in size, shape, and detail.

Thorax (p. 152)
1. The thoracic skeleton consists of the sternum, ribs and costal cartilages, and thoracic vertebrae.
2. The thorax protects vital organs in the chest area.

Disorders: Homeostatic Imbalances (p. 155)
1. Protrusion of the nucleus pulposus of an intervertebral disc posteriorly or into an adjacent vertebral body is called a herniated (slipped) disc.
2. Exaggeration of a normal curve of the vertebral column is called a curvature. Examples include scoliosis, kyphosis, and lordosis.
3. The imperfect union of the vertebral laminae at the midline, a congenital defect, is referred to as spina bifida.
4. Fractures of the vertebral column most often involve T12, L1, and L2.

REVIEW QUESTIONS

1. What are the four principal types of bones? Give an example of each. Distinguish between a sutural and a sesamoid bone.
2. What are surface markings? Describe and give an example of each.
3. Distinguish between the axial and appendicular skeletons. What subdivisions and bones are contained in each?
4. What are the bones that compose the skull? The cranium? The face?
5. Define a suture. What are the four prominent sutures of the skull? Where are they located?
6. What is a fontanel? Describe the location of the six fairly constant fontanels.
7. What is a paranasal sinus? What cranial bones contain paranasal sinuses?
8. Define the following: mastoiditis, cleft palate, cleft lip, sinusitis, dislocation of the jaw, and deviated nasal septum.
9. What is the hyoid bone? In what respect is it unique? What is its function?
10. What bones form the skeleton of the trunk? Distinguish between the number of nonfused vertebrae found in the adult vertebral column and that of a child.
11. What are the normal curves in the vertebral column? How are primary and secondary curves differentiated? What are the functions of the curves?
12. What are the principal distinguishing characteristics of the bones of the various regions of the vertebral column?
13. What bones form the skeleton of the thorax? What are the functions of the thoracic skeleton?
14. How are ribs classified on the basis of their attachment to the sternum?
15. What is a herniated (slipped) disc? Why does it cause pain? How is it treated?
16. What is a curvature? Describe the symptoms and causes of scoliosis, kyphosis, and lordosis.
17. Define spina bifida.
18. What are some common causes of fractures of the vertebral column?

The Skeletal System: The Appendicular Skeleton

8

STUDENT OBJECTIVES

- Identify the bones of the pectoral (shoulder) girdle and their major markings.
- Identify the upper extremity, its component bones, and their markings.
- Identify the components of the pelvic (hip) girdle and their principal markings.
- Identify the lower extremity, its component bones, and their markings.
- Define the structural features and importance of the arches of the foot.
- Compare the principal structural differences between female and male skeletons, especially those that pertain to the pelvis.

This chapter discusses the bones of the appendicular skeleton, that is, the bones of the pectoral (shoulder) and pelvis (hip) girdles and extremities. The differences between female and male skeletons are also compared.

PECTORAL (SHOULDER) GIRDLE

The **pectoral** (PEK-tō-ral) or **shoulder girdles** attach the bones of the upper extremities to the axial skeleton (Figure 8-1). Each of the two pectoral girdles consists of two

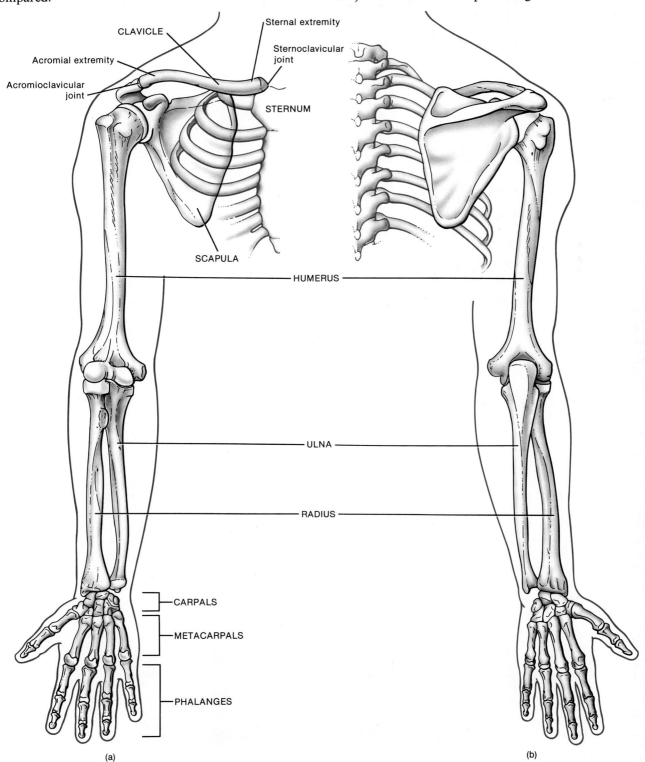

CLAVICLE

Sternal extremity

Acromial extremity

Sternoclavicular joint

Acromioclavicular joint

STERNUM

SCAPULA

HUMERUS

ULNA

RADIUS

CARPALS

METACARPALS

PHALANGES

(a)

(b)

FIGURE 8-1 Right pectoral (shoulder) girdle and upper extremity. (a) Anterior view. (b) Posterior view.

bones: a clavicle and a scapula. The clavicle is the anterior component and articulates with the sternum at the sterno-clavicular joint. The posterior component, the scapula, which is positioned freely by complex muscle attachments, articulates with the clavicle and humerus. The pectoral girdles have no articulation with the vertebral column. Although the shoulder joints are not very stable, they are freely movable and thus allow movement in many directions.

CLAVICLE

The **clavicles** (KLAV-i-kuls), or collarbones, are long, slender bones with a double curvature (Figure 8-2). The two bones lie horizontally in the superior and anterior part of the thorax superior to the first rib.

The medial end of the clavicle, the **sternal extremity,** is rounded and articulates with the sternum. The broad, flat, lateral end, the **acromial** (a-KRŌ-mē-al) **extremity,** articulates with the acromion of the scapula. This joint is called the **acromioclavicular joint.** (Refer to Figure 8-1 for a view of these articulations.) The **conoid tubercle** on the inferior surface of the lateral end of the bone serves as a point of attachment for a ligament. The **costal tuberosity** on the inferior surface of the medial end also serves as a point of attachment for a ligament.

CLINICAL APPLICATION

Because of its position, the clavicle transmits forces from the upper extremity to the trunk. If such forces are excessive, as in falling on one's outstretched arm, a **fractured clavicle** may result. In fact, it is the most frequently broken bone in the body.
   Separation of the shoulder simply means dislocation of the acromioclavicular joint.

SCAPULA

The **scapulae** (SCAP-yoo-lē), or shoulder blades, are large, triangular, flat bones situated in the dorsal part of the thorax between the levels of the second and seventh ribs (Figure 8-3). Their medial borders are located about 5 cm (2 inches) from the vertebral column.

A sharp ridge, the **spine,** runs diagonally across the dorsal surface of the flattened, triangular **body.** The end of the spine projects as a flattened, expanded process called the **acromion** (a-KRŌ-mē-on). This process articulates with the clavicle. Inferior to the acromion is a depression called the **glenoid cavity.** This cavity articulates with the head of the humerus to form the shoulder joint.

The thin edge of the body near the vertebral column is the **medial (vertebral) border.** The thick edge closer to the arm is the **lateral (axillary) border.** The medial and lateral borders join at the **inferior angle.** The superior edge of the scapular body, called the **superior border,** joins the vertebral border at the **superior angle.** The **scapular notch** is a prominent indentation along the superior border near the coracoid process; the notch permits passage of the suprascapular nerve.

At the lateral end of the superior border is a projection of the anterior surface called the **coracoid** (KOR-a-koyd) **process** to which muscles attach. Above and below the spine are two fossae: the **supraspinous** (soo'-pra-SPĪ-nus) **fossa** and the **infraspinous fossa,** respectively. Both serve as surfaces of attachment for shoulder muscles. On the ventral (costal) surface is a lightly hollowed-out area called the **subscapular fossa,** also a surface of attachment for shoulder muscles.

UPPER EXTREMITY

The **upper extremities** consist of 60 bones. The skeleton of the right upper extremity is shown in Figure 8-1. Each upper extremity includes a humerus in the arm, ulna and radius in the forearm, carpals (wrist bones), metacarpals (palm bones), and phalanges in the fingers of the hand.

HUMERUS

The **humerus** (HYOO-mer-us), or arm bone, is the longest and largest bone of the upper extremity (Figure 8-4). It articulates proximally with the scapula and distally at the elbow with both ulna and radius.

The proximal end of the humerus consists of a **head** that articulates with the glenoid cavity of the scapula. It also has an **anatomical neck,** which is an oblique groove just distal to the head. The **greater tubercle** is a lateral projection distal to the neck. The **lesser tubercle** is an anterior projection. Between these tubercles runs an **inter-tubercular sulcus (bicipital groove).** The **surgical neck** is a constricted portion just distal to the tubercles. It is so named because of its liability to fracture.

Acromial extremity

Sternal extremity

Conoid tubercle

Costal tuberosity

FIGURE 8-2 Right clavicle viewed from below. (Copyright © 1987 by Michael H. Ross. Reprinted by permission.)

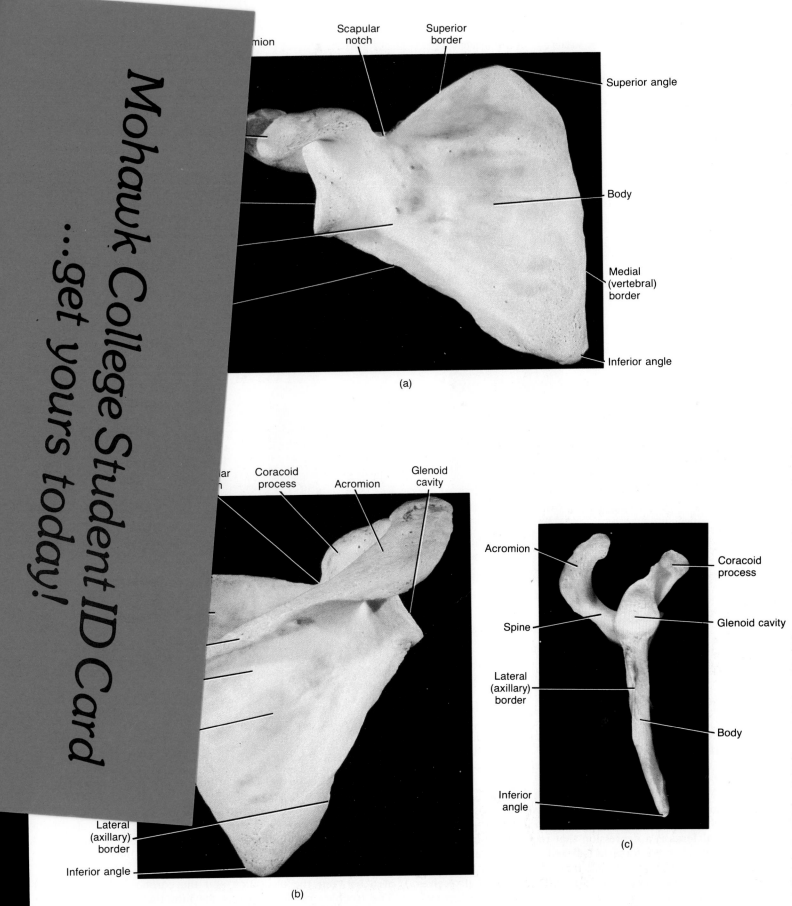

Scapular notch

Superior border

Acromion

Superior angle

Body

Medial (vertebral) border

Inferior angle

(a)

Coracoid process

Acromion

Glenoid cavity

Scapular notch

Acromion

Coracoid process

Spine

Glenoid cavity

Lateral (axillary) border

Body

Inferior angle

(c)

Lateral (axillary) border

Inferior angle

(b)

FIGURE 8-3 Right scapula. (a) Anterior view. (b) Posterior view. (c) Lateral border view. (Copyright © 1987 by Michael H. Ross. Reprinted by permission.)

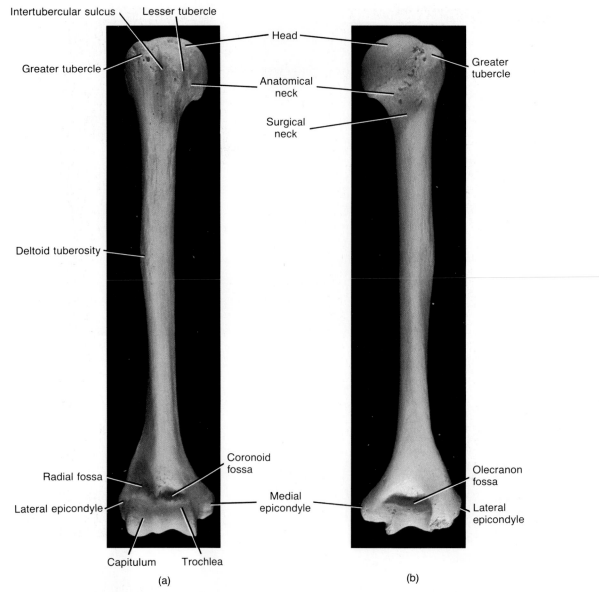

Intertubercular sulcus Lesser tubercle
Head
Greater tubercle
Anatomical neck
Surgical neck
Greater tubercle
Deltoid tuberosity
Coronoid fossa
Radial fossa
Olecranon fossa
Medial epicondyle
Lateral epicondyle
Lateral epicondyle
Capitulum   Trochlea
(a)
(b)

FIGURE 8-4 Right humerus. (a) Anterior view. (b) Posterior view. (Copyright © 1987 by Michael H. Ross. Reprinted by permission.)

The **body (shaft)** of the humerus is cylindrical at its proximal end. It gradually becomes triangular and is flattened and broad at its distal end. Along the middle portion of the shaft, there is a roughened, V-shaped area called the **deltoid tuberosity.** This area serves as a point of attachment for the deltoid muscle.

The following parts are found at the distal end of the humerus. The **capitulum** (ka-PIT-yoo-lum) is a rounded knob that articulates with the head of the radius. The **radial fossa** is a depression that receives the head of the radius when the forearm is flexed. The **trochlea** (TRŌK-lē-a) is a pulleylike surface that articulates with the ulna. The **coronoid fossa** is an anterior depression that receives part of the ulna when the forearm is flexed. The **olecranon** (ō-LEK-ra-non) **fossa** is a posterior depression that receives the olecranon of the ulna when the forearm is

extended. The **medial epicondyle** and **lateral epicondyle** are rough projections on either side of the distal end.

ULNA AND RADIUS

The **ulna** is the medial bone of the forearm (Figure 8-5). In other words, it is located at the little finger side. The proximal end of the ulna presents an **olecranon (olecranon process),** which forms the prominence of the elbow. The **coronoid process** is an anterior projection that, together with the olecranon, receives the trochlea of the humerus. The **trochlear (semilunar) notch** is a curved area between the olecranon and the coronoid process. The trochlea of the humerus fits into this notch. The **radial notch** is a depression located laterally and inferiorly to the trochlear notch. It receives the head of the radius.

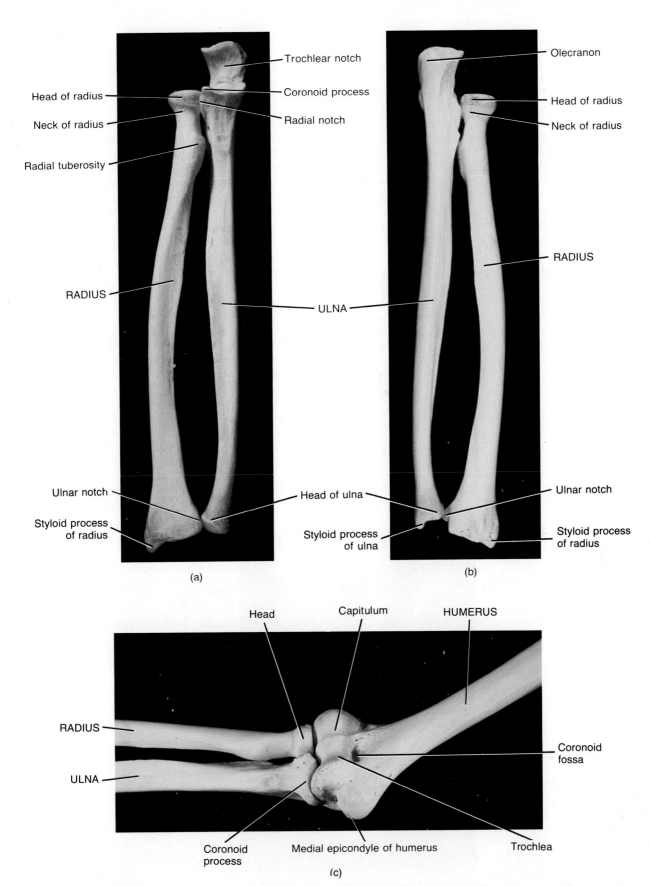

FIGURE 8-5 Right ulna and radius. (a) Anterior view. (b) Posterior view. (c) Medial view of the right elbow. (Copyright © 1987 by Michael H. Ross. Reprinted by permission.)

The distal end of the ulna consists of a **head** that is separated from the wrist by a fibrocartilage disc. A **styloid process** is on the posterior side of the distal end.

The **radius** is the lateral bone of the forearm, that is, it is situated on the thumb side. The proximal end of the radius has a disc-shaped **head** that articulates with the capitulum of the humerus and radial notch of the ulna. It also has a raised, roughened area on the medial side called the **radial tuberosity.** This is a point of attachment for the biceps muscle. The shaft of the radius widens distally to form a concave inferior surface that articulates with two bones of the wrist called the lunate and scaphoid bones. Also at the distal end is a **styloid process** on the lateral side and a medial, concave **ulnar notch** for articulation with the distal end of the ulna.

CLINICAL APPLICATION

When one falls on the outstretched arm, the radius bears the brunt of forces transmitted through the hand. If a fracture occurs in such a fall, it is usually a transverse break about 3 cm (1 inch) from the distal end of the bone. In this type of injury, called a **Colles' fracture,** the hand is displaced backward and upward (see Figure 6-7b).

CARPALS, METACARPALS, AND PHALANGES

The **carpus** (or wrist) consists of eight small bones, the **carpals,** united to each other by ligaments (Figure 8-6).

The bones are arranged in two transverse rows, with four bones in each row. In the anatomical position, the proximal row of carpals, from the lateral to medial position, consists of the **scaphoid, lunate, triquetral,** and **pisiform.** In about 70 percent of cases involving carpal fractures, only the scaphoid is involved. The distal row of carpals, from the lateral to medial position, consists of the **trapezium, trapezoid, capitate,** and **hamate.**

The five bones of the **metacarpus** constitute the palm of the hand. Each metacarpal bone consists of a proximal **base,** a **shaft,** and a distal **head.** The metacarpal bones are numbered I to V, starting with the lateral bone. The bases articulate with the distal row of carpal bones and with one another. The heads articulate with the proximal phalanges of the fingers. The heads of the metacarpals are commonly called the "knuckles" and are readily visible when the fist is clenched.

The **phalanges** (fa-LAN-jēz), or bones of the fingers, number 14 in each hand. A single bone of the finger (or toe) is referred to as a **phalanx** (FĀ-lanks). Each phalanx consists of a proximal **base,** a **shaft,** and a distal **head.** There are two phalanges in the first digit, called the thumb (*pollex*), and three phalanges in each of the remaining four digits. These digits, moving medially from the thumb, are commonly referred to as the index finger, middle finger, ring finger, and little finger. The first row of phalanges, the **proximal row,** articulates with the metacarpal bones and second row of phalanges. The second row of phalanges, the **middle row,** articulates with the proximal row and the third row. The third row of phalanges, the **distal row,** articulates with the middle row. The thumb has no middle phalanx.

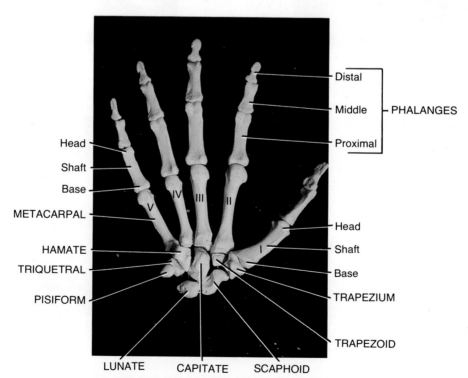

FIGURE 8-6 Right wrist and hand in anterior view. (Courtesy of J. A. Gosling et al., *Atlas of Human Anatomy with Integrated Text,* Copyright © 1985 by Gower Medical Publishing Ltd.)

PELVIC (HIP) GIRDLE

The **pelvic (hip) girdle** consists of the two **coxal (KOK-sal) bones,** commonly called the pelvic, innominate, or hipbones (Figure 8-7). The pelvic girdle provides a strong and stable support for the lower extremities on which the weight of the body is carried. The coxal bones are united to each other anteriorly at the symphysis (SIM-fi-sis) pubis. They unite posteriorly to the sacrum.

Together with the sacrum and coccyx, the two bones of the pelvic girdle form the basinlike structure called the **pelvis.** The pelvis is divided into a greater pelvis and a lesser pelvis by an oblique plane that passes through sacral promontory (posterior), iliopectineal lines (laterally), and symphysis pubis (anteriorly). The circumference of this oblique plane is called the **brim of the pelvis.** The **greater (false) pelvis** represents the expanded portion situated superior to the brim of the pelvis. The greater pelvis consists laterally of the superior portions of the ilia and posteriorly of the superior portion of the sacrum. There is no bony component in the anterior aspect of the greater pelvis. Rather, the front is formed by the walls of the abdomen.

The **lesser (true) pelvis** is inferior and posterior to the brim of the pelvis. It is formed by the inferior portions of the ilia and sacrum, the coccyx, and the pubes. The lesser pelvis contains a superior opening called the **pelvic inlet** and an inferior opening called the **pelvic outlet.**

CLINICAL APPLICATION

Pelvimetry is the measurement of the size of the inlet and outlet of the birth canal. Measurement of the pelvic cavity is important to the physician, because the fetus must pass through the narrower opening of the lesser pelvis at birth.

Each of the two coxal bones of a newborn consists of three components: a superior **ilium,** an inferior and anterior **pubis,** and an inferior and posterior **ischium** (Figure 8-8). Eventually, the three separate bones fuse into one. The area of fusion is a deep, lateral fossa called the **acetabulum** (as'-e-TAB-yoo-lum). Although the adult coxae are both single bones, it is common to discuss the bones as if they still consisted of three portions.

The ilium is the largest of the three subdivisions of the coxal bone. Its superior border, the **iliac crest,** ends anteriorly in the **anterior superior iliac spine.** Posteriorly, the iliac crest ends in the **posterior superior iliac spine.** The spines serve as points of attachment for muscles of the abdominal wall. Inferior to the posterior superior iliac spine is the **greater sciatic (sī-AT-ik) notch.** The internal surface of the ilium seen from the medial side is the **iliac fossa.** It is a concavity where the iliacus muscle attaches. Posterior to this fossa is the **auricular surface,** which articulates with the sacrum.

The ischium is the inferior, posterior portion of the coxal bone. It contains a prominent **ischial spine,** a **lesser sciatic notch** below the spine, and an **ischial tuberosity.** The rest of the ischium, the **ramus,** joins with the pubis and together they surround the **obturator (OB-too-rā'-ter) foramen.**

The pubis is the anterior and inferior part of the coxal bone. It consists of a **superior ramus,** an **inferior ramus,** and a **body** that contributes to the formation of the symphysis pubix.

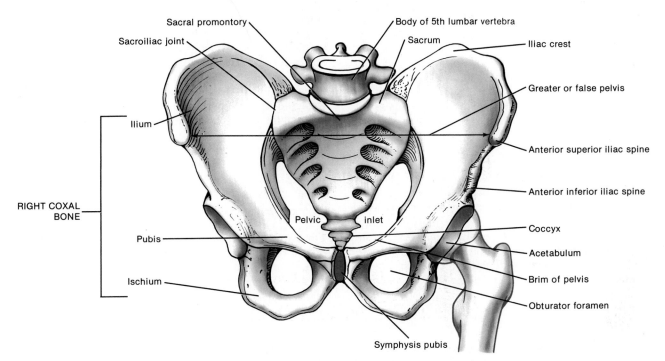

FIGURE 8-7 Pelvic (hip) girdle of a female in anterior view.

FIGURE 8-8 Right coxal bone. (a) Lateral view. (b) Medial view. The lines of fusion of the ilium, ischium, and pubis are not actually visible in an adult bone. (Copyright © 1987 by Michael H. Ross. Reprinted by permission.)

The **symphysis pubis** is the joint between the two coxal bones (see Figure 8-7). It consists of fibrocartilage. The **acetabulum** is the fossa formed by the ilium, ischium, and pubis. It is the socket for the head of the femur. Two-fifths of the acetabulum is formed by the ilium, two-fifths by the ischium, and one-fifth by the pubis.

LOWER EXTREMITY

The **lower extremities** are composed of 60 bones (Figure 8-9). Each extremity includes the femur in the thigh, patella (kneecap), fibula and tibia in the leg, tarsals (ankle bones), metatarsals, and phalanges in the toes.

FEMUR

The **femur,** or thighbone, is the longest and heaviest bone in the body (Figure 8-10). Its proximal end articulates with the coxal bone. Its distal end articulates with the tibia. The shaft of the femur bows medially so that it approaches the femur of the opposite thigh. As a result of this convergence, the knee joints are brought nearer to the body's line of gravity. The degree of convergence is greater in the female because the female pelvis is broader.

The proximal end of the femur consists of a rounded **head** that articulates with the acetabulum of the coxal bone. The **neck** of the femur is a constricted region distal to the head. A fairly common fracture in the elderly occurs at the neck of the femur. Apparently the neck becomes so weak that it fails to support the body. The **greater trochanter** (trō-KAN-ter) and **lesser trochanter** are projections that serve as points of attachment for some of the thigh and buttock muscles.

The shaft of the femur contains a rough vertical ridge on its posterior surface called the **linea aspera.** This ridge serves for the attachment of several thigh muscles.

The distal end of the femur is expanded and includes the **medial condyle** and **lateral condyle.** These articulate with the tibia. Superior to the condyles are the **medial epicondyle** and **lateral epicondyle.** A depressed area between the condyles on the posterior surface is called the **intercondylar** (in'-ter-KON-di-lar) **fossa.** The **patellar surface** is located between the condyles on the anterior surface.

PATELLA

The **patella,** or kneecap, is a small, triangular bone anterior to the knee joint (Figure 8-11). It is a sesamoid bone that develops in the tendon of the quadriceps femoris muscle. The broad superior end of the patella is called the **base.** The pointed inferior end is the **apex.** The posterior surface contains two **articular facets,** one for the medial condyle and the other for the lateral condyle of the femur.

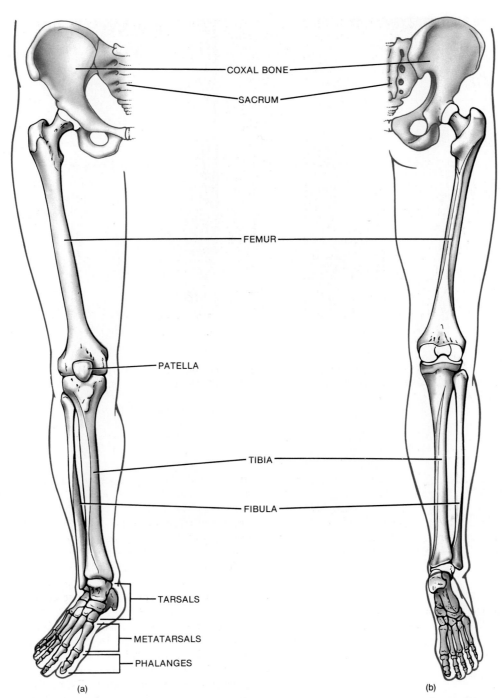

COXAL BONE

SACRUM

FEMUR

PATELLA

TIBIA

FIBULA

TARSALS

METATARSALS

PHALANGES

(a)

(b)

FIGURE 8-9 Right pelvic (hip) girdle and lower extremity. (a) Anterior view. (b) Posterior view.

Greater trochanter
Head
Neck
Lesser trochanter
Greater trochanter
Gluteal tuberosity
Linea aspera

Nutrient foramen
Lateral epicondyle
Lateral condyle
Medial epicondyle
Medial condyle
Patellar surface
Lateral epicondyle
Lateral condyle
Intercondylar fossa

(a)

(b)

FIGURE 8-10 Right femur. (a) Anterior view. (b) Posterior view. (Copyright © 1987 by Michael H. Ross. Reprinted by permission.)

TIBIA AND FIBULA

The **tibia,** or shinbone, is the larger, medial bone of the leg (Figure 8-12). It bears the major portion of the weight of the leg. The tibia articulates at its proximal end with the femur and fibula, and at its distal end with the fibula of the leg and talus of the ankle.

The proximal end of the tibia is expanded into a **lateral condyle** and a **medial condyle.** These articulate with the condyles of the femur. The inferior surface of the lateral condyle articulates with the head of the fibula. The slightly concave condyles are separated by an upward projection called the **intercondylar eminence.** The **tibial tuberosity** on the anterior surface is a point of attachment for the patellar ligament.

The medial surface of the distal end of the tibia forms the **medial malleolus** (mal-LĒ-ō-lus). This structure articulates with the talus bone of the ankle and forms the prominence that can be felt on the medial surface of your ankle. The **fibular notch** articulates with the fibula.

CLINICAL APPLICATION

Shinsplints (tibia stress syndrome) refers to soreness or pain along the tibia, probably caused by inflammation of the periosteum (periostitis) brought on by the repeated tugging of the muscles and tendons attached to the periosteum. The condition may occur as a result

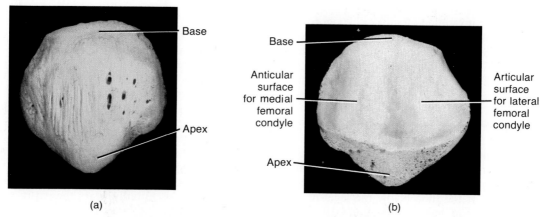

(a)   (b)

FIGURE 8-11 Right patella. (a) Anterior view. (b) Posterior view. (Copyright © 1987 by Michael H. Ross. Reprinted by permission.)

(a)   (b)

FIGURE 8-12 Right tibia and fibula. (a) Anterior view. (b) Posterior view. (Copyright © 1987 by Michael H. Ross. Reprinted by permission.)

of walking or running up and down hills or by vigorous activity of the legs following a period of relative inactivity. Rest usually alleviates the pain. Patients who do not respond to rest may be given local injections of cortisonelike steroid drugs or may have to undergo minor surgery to release pressure in the soft tissues around the bone.

The **fibula** is parallel and lateral to the tibia. It is considerably smaller than the tibia. The **head** of the fibula, the proximal end, articulates with the inferior surface of the lateral condyle of the tibia below the level of the knee joint. The distal end has a projection called the **lateral malleolus,** that articulates with the talus bone of the ankle. This forms the prominence on the lateral surface of the ankle. The inferior portion of the fibula also articulates with the tibia at the fibular notch. A fracture of the lower end of the fibula with injury to the tibial articulation is called a **Pott's fracture.**

TARSALS, METATARSALS, AND PHALANGES

The **tarsus** is a collective designation for the seven bones of the ankle called **tarsals** (Figure 8-13). The term *tarsos* pertains to a broad, flat surface. The **talus** and **calcaneus** (kal-KĀ-nē-us) are located on the posterior part of the foot. The anterior part contains the **cuboid, navicular,** and three **cuneiform bones** called the **first (medial), second (intermediate),** and **third (lateral) cuneiform.** The talus, the uppermost tarsal bone, is the only bone of the foot that articulates with the fibula and tibia. It is surrounded on one side by the medial malleolus of the tibia and on the other side by the lateral malleolus of the fibula. During walking, the talus initially bears the entire weight of the body. About half the weight is then transmitted to the calcaneus. The remainder is transmitted to the other tarsal bones. The calcaneus, or heel bone, is the largest and strongest tarsal bone.

The **metatarsus** consists of five metatarsal bones numbered I to V from the medial to lateral position. Like the metacarpals of the palm of the hand, each metatarsal

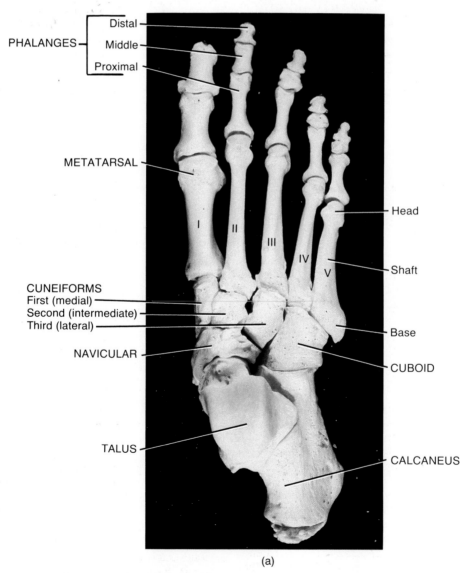

(a)

FIGURE 8-13 Right foot. (a) Superior view.

TIBIA

Medial
malleolus

TALUS

CALCANEUS

NAVICULAR

Distal   Proximal   METATARSAL

PHALANGES

FIRST (MEDIAL)
CUNEIFORM

(b)

FIGURE 8-13 (Continued) (b) Medial view. (Copyright © 1987 by Michael H. Ross. Reprinted by permission.)

consists of a proximal **base,** a **shaft,** with a distal **head.** The metatarsals articulate proximally with the first, second, and third cuneiform bones and with the cuboid. Distally, they articulate with the proximal row of phalanges. The first metatarsal is thicker than the others because it bears more weight.

The **phalanges** of the foot resemble those of the hand both in number and arrangment. Each also consists of a proximal **base,** a middle **shaft,** and a distal **head.** The great (big) toe, or *hallux,* has two large, heavy phalanges called proximal and distal phalanges. The other four toes each have three phalanges—proximal, middle, and distal.

ARCHES OF THE FOOT

The bones of the foot are arranged in two **arches** (Figure 8-14). These arches enable the foot to support the weight of the body and provide leverage while walking. The arches are not rigid. They yield as weight is applied and spring back when the weight is lifted.

The **longitudinal arch** has two parts. Both consist of tarsal and metatarsal bones arranged to form an arch from the anterior to the posterior part of the foot. The **medial,** or inner, part of the longitudinal arch originates at the calcaneus. It rises to the talus and descends through the navicular, the three cuneiforms, and the three medial metatarsals. The talus is the keystone of this arch. The **lateral,** or outer, part of the longitudinal arch also begins

at the calcaneus. It rises at the cuboid and descends to the two lateral metatarsals. The cuboid is the keystone of this arch.

The **transverse arch** is formed by the calcaneus, navicular, cuboid, and the posterior parts of the five metatarsals.

CLINICAL APPLICATION

The bones composing the arches are held in position by ligaments and tendons. If these ligaments and tendons are weakened, the height of the medial longitudinal arch may decrease or "fall." The result is **flatfoot.**

Clawfoot is a condition in which the medial longitudinal arch is abnormally elevated. It is frequently caused by muscle imbalance, such as may result from poliomyelitis.

A **bunion (hallux valgus,** *valgus* = bent outward) is a deformity of the great toe. Although the condition may be inherited, it is typically caused by wearing tightly fitting shoes and is characterized by lateral deviation of the great toe, with partial dislocation at its junction with metatarsal I. As a result, the head of metatarsal I is displaced medially. The condition produces inflammation of bursae (fluid-filled sacs at the joint), bone spurs, and calluses.

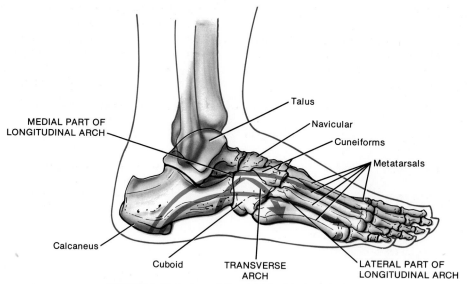

MEDIAL PART OF
LONGITUDINAL ARCH

Talus

Navicular

Cuneiforms

Metatarsals

Calcaneus

Cuboid

TRANSVERSE
ARCH

LATERAL PART OF
LONGITUDINAL ARCH

FIGURE 8-14 Arches of the right foot in lateral view.

EXHIBIT 8-1   COMPARISON OF TYPICAL FEMALE AND MALE PELVIS

POINT OF COMPARISON	FEMALE	MALE
General structure	Light and thin.	Heavy and thick.
Joint surfaces	Small.	Large.
Muscle attachments	Rather indistinct.	Well marked.
Greater pelvis	Shallow.	Deep.
Pelvic inlet	Larger and more oval.	Heart-shaped.
Pelvic outlet	Comparatively large.	Comparatively small.
First piece of sacrum	Superior surface of the body spans about a third the width of sacrum.	Superior surface of the body spans nearly half the width of the sacrum.
Sacrum	Short, wide, flat, curving forward in lower part.	Long, narrow, with smooth concavity.
Auricular surface	Extends only to upper border of third piece of the sacrum.	Extends well down the third piece of the sacrum.
Pubic arch	Greater than a 90° angle.	Less than a 90° angle.
Inferior ramus of pubis	Everted surface not present.	Presents strong everted surface for attachment of the crus of the penis.
Symphysis pubis	Less deep.	More deep.
Ischial spine	Turned inward less.	Turned inward more.
Ischial tuberosity	Turned outward.	Turned inward.
Ilium	Less vertical.	More vertical.
Iliac fossa	Shallow.	Deep.
Iliac crest	Less curved.	More curved.
Anterior superior iliac spine	Wider apart.	Closer.
Acetabulum	Small.	Large.
Obturator foramen	Oval.	Round.
Greater sciatic notch	Wide.	Narrow.

FEMALE AND MALE SKELETONS

The bones of the male are generally larger and heavier than those of the female. The articular ends are thicker in relation to the shafts. In addition, since certain muscles of the male are larger than those of the female, the points of attachment—tuberosities, lines, ridges—are larger in the male skeleton.

Many significant structural differences between female and male skeletons are noted in the pelvis; most are related to pregnancy and childbirth. The typical differences are listed in Exhibit 8-1 and illustrated in Figure 8-15.

STUDY OUTLINE

Pectoral (Shoulder) Girdle (p. 159)
1. Each pectoral (shoulder) girdle consists of a clavicle and scapula.
2. Each attaches an upper extremity to the trunk.

Upper Extremity (p. 160)
The bones of each upper extremity include the humerus, ulna, radius, carpals, metacarpals, and phalanges.

Pelvic (Hip) Girdle (p. 164)
1. The pelvic (hip) girdle consists of two coxal bones or hip-bones.
2. It attaches the lower extremities to the trunk at the sacrum.
3. Each coxal bone consists of three fused components—ilium, pubis, and ischium.

(a)

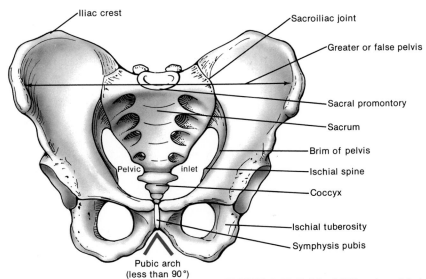

(b)

FIGURE 8-15 Pelvis. (a) Female pelvis in anterior view. (b) Male pelvis in anterior view.

Lower Extremity (p. 166)

1. The bones of each lower extremity include the femur, tibia, fibula, tarsals, metatarsals, and phalanges.
2. The bones of the foot are arranged in two arches, the longitudinal arch and the transverse arch, to provide support and leverage.

Female and Male Skeletons (p. 173)

1. The female pelvis is adapted for pregnancy and childbirth. Differences in pelvic structure are listed in Exhibit 8-1.
2. Male bones are generally larger and heavier than female bones and have more prominent markings for muscle attachment.

REVIEW QUESTIONS

1. What is the pectoral (shoulder) girdle? Why is it important?
2. What are the bones of the upper extremity? What is a Colles' fracture?
3. What is the pelvic (hip) girdle? Why is it important?
4. What are the bones of the lower extremity? What is a Pott's fracture?
5. What is pelvimetry? What is its clinical importance?
6. In what ways do the upper extremity and lower extremity differ structurally?
7. Describe the structure of the longitudinal and transverse arches of the foot. What is the function of an arch?
8. How do flatfeet, clawfeet, and bunions arise?
9. What are the principal structural differences between typical female and male skeletons? Use Exhibit 8-1 as a guide in formulating your response.

Articulations

9

STUDENT OBJECTIVES

- Define an articulation and identify the factors that determine the degree of movement at a joint.
- Contrast the structure, kind of movement, and location of fibrous, cartilaginous, and synovial joints.
- Explain the principle of arthroscopy and its clinical importance.
- Discuss and compare the movements possible at various synovial joints.
- Describe selected articulations of the body with respect to the bones that enter into their formation, structural classification, and anatomical components.
- Describe the causes and symptoms of common joint disorders, including rheumatism, rheumatoid arthritis (RA), osteoarthritis, gouty arthritis, bursitis, dislocation, and sprain.
- Define medical terminology associated with articulations.

Bones are too rigid to bend without damage. Fortunately, the skeletal system consists of many separate bones, which are held together at joints by flexible connective tissue. All movements that change the positions of the bony parts of the body occur at joints. You can understand the importance of joints if you imagine how a cast over the knee joint prevents flexing the leg or how a splint on a finger limits the ability to manipulate small objects.

An **articulation (joint)** is a point of contact between bones or between cartilage and bones. The scientific study of joints is referred to as **arthrology** (*arthro* = joint; *logos* = study of). The joint's structure determines how it functions. Some joints permit no movement, others permit slight movement, and still others afford considerable movement. In general, the closer the fit at the point of contact, the stronger the joint. At tightly fitted joints, however, movement is restricted. The looser the fit, the greater the movement. Unfortunately, loosely fitted joints are prone to dislocation. Movement at joints is also determined by the flexibility of the connective tissue that binds the bones together and by the position of ligaments, muscles, and tendons.

CLASSIFICATION

FUNCTIONAL

The functional classification of joints takes into account the degree of movement they permit. Functionally, joints are classified as **synarthroses** (sin'-ar-THRŌ-sēz), which are immovable joints; **amphiarthroses** (am'-fē-ar-THRŌ-sēz), which are slightly movable joints; and **diarthroses** (dī'-ar-THRŌ-sēz), which are freely movable joints.

STRUCTURAL

The structural classification of joints is based on the presence or absence of a joint cavity (a space between the articulating bones) and the kind of connective tissue that binds the bones together. Structurally, joints are classified as **fibrous,** in which there is no joint cavity and the bones are held together by fibrous connective tissue; **cartilaginous,** in which there is no joint cavity and the bones are held together by cartilage; and **synovial,** in which there is a joint cavity and the bones forming the joint are united by a surrounding articular capsule and frequently by accessory ligaments (described in detail later). We will discuss the joints of the body based upon their structural classification, but with reference to their functional classification as well.

FIBROUS JOINTS

Fibrous joints lack a joint cavity, and the articulating bones are held very closely together by fibrous connective tissue. They permit little or no movement. The three types of fibrous joints are (1) sutures, (2) syndesmoses, and (3) gomphoses.

SUTURE

Sutures are found between bones of the skull. In a suture, the bones are united by a thin layer of dense fibrous connective tissue. The irregular structure of sutures gives them added strength and decreases their chance of fractures. Since sutures are immovable, they are functionally classified as synarthroses. Some sutures, present during growth, are replaced by bone in the adult. In this case they are called **synostoses** (sin'-os-TŌ-sēz), or bony joints—joints in which there is a complete fusion of bone across the suture line. An example is the frontal suture between the left and right sides of the frontal bone which begins to fuse during infancy (see Figure 7-3a). Synostoses are also functionally classified as synarthroses.

SYNDESMOSIS

A **syndesmosis** (sin'-dez-MŌ-sis) is a fibrous joint in which the uniting fibrous connective tissue is present in a much greater amount than in a suture, but the fit between the bones is not quite as tight. The fibrous connective tissue forms an interosseous membrane or ligament. A syndesmosis is slightly movable because the bones are separated more than in a suture and some flexibility is permitted by the interosseous membrane or ligament. Thus, syndesmoses are functionally classified as amphiarthrotic. Examples of syndesmoses include the distal articulation of the tibia and fibula (see Figure 8-12) and the articulations between the shafts of the ulna and radius (see Figure 8-5).

GOMPHOSIS

A **gomphosis** (gom-FŌ-sis) is a type of fibrous joint in which a cone-shaped peg fits into a socket. The intervening substance is the periodontal ligament. A gomphosis is functionally classified as synarthrotic. Examples are the articulations of the roots of the teeth with the alveoli (sockets) of the maxillae and mandible.

CARTILAGINOUS JOINTS

Another joint that has no joint cavity is a **cartilaginous joint.** Here the articulating bones are tightly connected by cartilage. Like fibrous joints, they allow little or no movement. The two types of cartilaginous joints are (1) synchondroses and (2) symphyses.

SYNCHONDROSIS

A **synchondrosis** (sin'-kon-DRŌ-sis) is a cartilaginous joint in which the connecting material is hyaline cartilage. The most common type of synchondrosis is the epiphyseal plate (see Figure 6-5). Such a joint is found between the epiphysis and diaphysis of a growing bone and is immovable. Thus it is synarthrotic. Since the hyaline cartilage is eventually replaced by bone when growth ceases, the joint is temporary. It is replaced by a synostosis. Another example of a synchondrosis is the joint between the first rib and the sternum. The cartilage in this joint undergoes ossification during adult life.

SYMPHYSIS

A **symphysis** (SIM-fi-sis) is a cartilaginous joint in which the connecting material is a broad, flat disc of fibrocartilage. This joint is found between bodies of vertebrae (see Figure 7-11). A portion of the intervertebral disc is cartilaginous material. The symphysis pubis between the anterior surfaces of the coxal bones is another example (see Figure 8-15). These joints are slightly movable, or amphiarthrotic.

SYNOVIAL JOINTS

STRUCTURE

A joint in which there is a space between articulating bones is called a **synovial** (si-NŌ-vē-al) **joint.** The space is called the **synovial (joint) cavity** (Figure 9-1). Because of this cavity and because of the arrangement of the articular capsule and accessory ligaments, synovial joints are freely movable. Thus, synovial joints are functionally classified as diarthrotic.

Synovial joints are also characterized by the presence of **articular cartilage.** Articular cartilage covers the surfaces of the articulating bones, but does not bind the bones together. The articular cartilage of synovial joints is hyaline cartilage.

Synovial joints are surrounded by a sleevelike **articular capsule** that encloses the synovial cavity and unites the articulating bones. The articular capsule is composed of two layers. The outer layer, the **fibrous capsule,** consists of dense connective (collagenous) tissue. It is attached to the periosteum of the articulating bones at a variable distance from the edge of the articular cartilage. The flexi-

bility of the fibrous capsule permits movement at a joint, whereas its great tensile strength resists dislocation. The fibers of some fibrous capsules are arranged in parallel bundles and are therefore highly adapted to resist recurrent strain. Such fibers are called *ligaments* and are given special names. The strength of the ligaments is one of the principal factors in holding bone to bone.

The inner layer of the articular capsule is formed by a **synovial membrane.** The synovial membrane is composed of loose connective tissue with elastic fibers and a variable amount of adipose tissue. It secretes *synovial fluid,* which lubricates the joint and provides nourishment for the articular cartilage. Synovial fluid also contains phagocytic cells that remove microbes and debris resulting from wear and tear in the joint. Synovial fluid consists of hyaluronic acid and an interstitial fluid formed from blood plasma and is similar in appearance and consistency to egg white. When there is no joint movement, the fluid is viscous, but as movement increases, the fluid becomes less viscous. The amount of synovial fluid varies in different joints of the body, ranging from a thin, viscous layer to about 3.5 ml (about 0.1 oz) of free fluid in a large joint such as the knee. The amount present in each joint is sufficient only to form a thin film over the surfaces within an articular capsule.

FIGURE 9-1 Synovial joint. (a) Diagram of a generalized synovial joint in frontal section. (b) Photograph of the internal structure of the right knee joint in frontal section. (Courtesy of C. Yokochi and J. W. Rohen, *Photographic Anatomy of the Human Body,* 2nd ed., 1978, IGAKU-SHOIN, Ltd., Tokyo, New York.)

(a)

(b)

Many synovial joints also contain **accessory ligaments,** which are called extracapsular ligaments and intracapsular ligaments. *Extracapsular ligaments* are outside of the articular capsule. An example is the fibular collateral ligament of the knee joint (see Figure 9-8e). *Intracapsular ligaments* occur within the articular capsule, but are excluded from the synovial cavity by reflections of the synovial membrane. Examples are the cruciate ligaments of the knee joint (see Figure 9-8e).

Inside some synovial joints, there are pads of fibrocartilage that lie between the articular surfaces of the bones and are attached by their margins to the fibrous capsule. These pads are called **articular discs (menisci).** The discs usually subdivide the synovial cavity into two separate spaces. Articular discs allow two bones of different shapes to fit tightly; they modify the shape of the joint surfaces of the articulating bones. Articular discs also help to maintain the stability of the joint and direct the flow of synovial fluid to areas of greatest friction.

CLINICAL APPLICATION

A tearing of articular discs in the knee, commonly called **torn cartilage,** occurs frequently among athletes. Such damaged cartilage requires surgical removal (meniscectomy) or it will begin to wear and cause arthritis. Until recently, knee joint surgery for torn cartilage necessitated cutting through layers of healthy tissue and removing much, if not all, of the cartilage. This procedure is usually painful and expensive and does not always provide full recovery.

These problems have been overcome by **arthroscopy (arthroscopic surgery).** The arthroscope is a device, several inches long, that resembles a pencil and is designed like a telescope with a series of lenses aligned one above the other. Its optical fibers give off light, like a lamp on a miner's hard hat. It is inserted into the knee joint through an incision as small as a quarter-inch. A second small incision is made to insert a tube through which a salt solution is injected into the joint. Another small incision is used for insertion of an instrument that shaves off and reshapes the damaged cartilage and then suctions out the shaved cartilage along with the salt solution. Some orthopedic surgeons attach a lightweight television camera to the arthroscope so that the image from inside the knee can be projected onto a screen. Since arthroscopy requires only small incisions, recovery is usually speedy and there is very little pain or discomfort. Although arthroscopy is used mostly to remove torn cartilage, it can also be used for other types of knee surgery, for surgery on other joints of the body to obtain tissue samples, for diagnosing certain pathologies, deciding whether surgery is needed, and for planning surgical procedures.

The various movements of the body create friction between moving parts. To reduce this friction, saclike structures called **bursae** are situated in the body tissues.

These sacs resemble joints in that their walls consist of connective tissue lined by a synovial membrane. They are also filled with a fluid similar to synovial fluid. Bursae are located between the skin and bone in places where skin rubs over bone. They are also found between tendons and bones, muscles and bones, and ligaments and bones. As fluid-filled sacs, they cushion the movement of one part of the body over another. An inflammation of a bursa is called **bursitis.**

The articular surfaces of synovial joints are kept in contact with each other by several factors. One factor is the fit of the articulating bones. This interlocking is very obvious at the hip joint, where the head of the femur articulates with the acetabulum of the coxal bone. Another factor is the strength of the joint ligaments. This is especially important in the hip joint. A third factor is the tension of the muscles around the joint. For example, the fibrous capsule of the knee joint is formed principally from tendinous expansions by muscles acting on the joint.

MOVEMENTS

The movements permitted at synovial joints are limited by several factors. One is the **apposition of soft parts.** For example, during bending of the elbow, the anterior surface of the forearm is pressed against the anterior surface of the arm. This apposition limits movement. A second factor is the **tension of ligaments.** The different components of a fibrous capsule are tense only when the joint is in certain positions. Tense ligaments not only restrict the range of movement but also direct the movement of the articulating bones with respect to each other. In the knee joint, for example, the major ligaments are lax when the knee is bent, but tense when the knee is straightened. Also, when the knee is straightened, the surfaces of the articulating bones are in fullest contact with each other. A third factor that restricts movement at a synovial joint is **muscle tension,** which reinforces the restraint placed on a joint by ligaments. A good example of the effect of muscle tension on a joint is seen at the hip joint. When the thigh is raised with the knee straight, the movement is restricted by the tension of the hamstring muscles on the posterior surface of the thigh. But if the knee is bent, the tension on the hamstring muscles is lessened and the thigh can be raised further. Movements at synovial joints are also limited by the **structure of the articulating bones.**

Following is a description of the specific movements that occur at synovial joints.

Gliding

A **gliding movement** is the simplest kind that can occur at a joint. One surface moves back and forth and from side to side over another surface without angular or rotary motion. Some joints that glide are those between the carpals and between the tarsals. The heads and tubercles of ribs glide on the bodies and transverse processes of vertebrae.

Angular

Angular movements increase or decrease the angle between bones. Among the angular movments are flexion, extension, abduction, and adduction (Figure 9-2). **Flexion** usually involves a decrease in the angle between the anterior surfaces of the articulating bones. An exception to this definition is flexion of the knee and the toe joints in which there is a decrease in the angle between the posterior surfaces of the articulating bones. Examples of flexion include bending the head forward (the joint is between the occipital bone and the atlas), bending the elbow, and bending the knee.

Extension involves an increase in the angle between the anterior surfaces of the articulating bones, with the same exceptions of knee and toe joints. Extension restores a body part to its anatomical position after it has been flexed. Examples of extension are returning the head to the anatomical position after flexion, straightening the arm after flexion, and straightening the leg after flexion. Continuation of extension beyond the anatomical position, as in bending the head backward, is called **hyperextension.**

Abduction usually means movement of a bone *away from* the midline of the body. An example of abduction is moving the arm upward and away from the body until it is held straight out at right angles to the chest. With the fingers and toes, however, the midline of the body is not used as the line of reference. Abduction of the fingers is a movement away from an imaginary line drawn through the middle finger; in other words, it is spreading the fingers. Abduction of the toes is relative to an imaginary line drawn through the second toe.

(a)

(b)

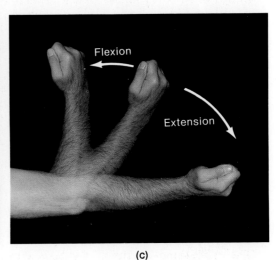

(c)

FIGURE 9-2 Angular movements at synovial joints. (Copyright © 1983 by Gerard J. Tortora. Courtesy of Lynne Tortora and James Borghesi.)

(d)

(e)

(f)

(g)

(h)

(i)

(j)

Adduction is usually movement of a part *toward* the midline of the body. An example of adduction is returning the arm to the side after abduction. As in abduction, adduction of the fingers is relative to the middle finger, and adduction of the toes is relative to the second toe.

Rotation

Rotation is the movement of a bone around its own longitudinal axis. During rotation, no other motion is permitted. In *medial rotation,* the anterior surface of a bone or extremity moves toward the midline. In *lateral rotation,* the anterior surface moves away from the midline. We rotate the atlas around the odontoid process of the axis when we shake the head from side to side. Moving from the shoulder and turning the forearm up, then palm down, and then palm up again is an example of slight lateral and medial rotation of the humerus (Figure 9-3a).

Circumduction

Circumduction is a movement in which the distal end of a bone moves in a circle while the proximal end remains stable. The bone describes a cone in the air. Circumduction typically involves flexion, abduction, adduction, extension, and rotation. It involves a 360° rotation. An example is moving the outstretched arm in a circle to wind up to pitch a ball (Figure 9-3b).

Special

Special movements are those found only at the joints indicated in Figure 9-4. **Inversion** is the movement of the sole of the foot inward (medially) at the ankle joint. **Eversion** is the movement of the sole outward (laterally) at the ankle joint. Flexion of the foot at the ankle joint is called **dorsiflexion.** Extension of the foot at the ankle joint is **plantar flexion.**

   Protraction is the movement of the mandible or clavicle forward on a plane parallel to the ground. Thrusting the jaw outward is protraction of the mandible. Bringing your arms forward until the elbows touch requires protraction of the clavicle. **Retraction** is the movement of a protracted part of the body backward on a plane parallel to the ground. Pulling the lower jaw back in line with the upper jaw is retraction of the mandible.

   Supination is a movement of the forearm in which the palm of the hand is turned anterior or superior. To demonstrate supination, flex your arm at the elbow to prevent rotation of the humerus in the shoulder joint. **Pronation** is a movement of the flexed forearm in which the palm is turned posterior or inferior.

   Elevation is an upward movement of a part of the body. You elevate your mandible when you close your mouth. **Depression** is a downward movement of a part of the body. You depress your mandible when you open your mouth. The shoulders can also be elevated and depressed.

   A summary of movements that occur at synovial joints is presented in Exhibit 9-1.

TYPES

Though all synovial joints are similar in structure, variations exist in the shape of the articulating surfaces. Accordingly, synovial joints are divided into six subtypes: gliding, hinge, pivot, ellipsoidal, saddle, and ball-and-socket joints.

Gliding

The articulating surfaces of bones in **gliding joints** or **arthrodia** (ar-THRŌ-dē-a) are usually flat. Only side-to-side and back-and-forth movements are permitted (Figure 9-5a). Twisting and rotation are inhibited at gliding joints, generally because ligaments or adjacent bones restrict the range of movement. Since gliding joints do not move around an axis, they are referred to as *nonaxial.* Examples are the joints between carpal bones, tarsal bones, the sternum and clavicle, and the scapula and clavicle.

Hinge

A **hinge** or **ginglymus** (JIN-gli-mus) **joint** is one in which the convex surface of one bone fits into the concave surface of another bone. Movement is primarily in a single plane,

(a)   (b)

FIGURE 9-3 Rotation and circumduction. (a) Rotation at the atlantoaxial joint (left) and rotation of the humerus (right). (b) Circumduction of the humerus at the shoulder joint. (Copyright © 1983 by Gerard J. Tortora.)

EXHIBIT 9-1 SUMMARY OF MOVEMENTS AT SYNOVIAL JOINTS

MOVEMENT	DEFINITION
GLIDING	One surface moves back and forth and from side to side over another surface without angular or rotary motion.
ANGULAR	There is an increase or decrease at the angle between bones.
Flexion	Usually involves a decrease in the angle between the anterior surfaces of articulating bones.
Extension	Usually involves an increase in the angle between the anterior surfaces of articulating bones.
Hyperextension	Continuation of extension beyond the anatomical position.
Abduction	Movement of a bone away from the midline.
Adduction	Movement of a bone toward the midline.
ROTATION	Movement of a bone around its longitudinal axis; may be medial or lateral.
CIRCUMDUCTION	A movement in which the distal end of a bone moves in a circle while the proximal end remains stable.
SPECIAL	Occurs at specific joints.
Inversion	Movement of the sole of the foot inward at the ankle joint.
Eversion	Movement of the sole of the foot outward at the ankle joint.
Dorsiflexion	Flexion of the foot at the ankle joint.
Plantar flexion	Extension of the foot at the ankle joint.
Protraction	Movement of the mandible or clavicle forward on a plane parallel to the ground.
Retraction	Movement of a protracted part backward on a plane parallel to the ground.
Supination	Movement of the forearm in which the palm is turned anterior or superior.
Pronation	Movement of the flexed forearm in which the palm is turned posterior or inferior.
Elevation	Movement of a part of the body upward.
Depression	Movement of a part of the body downward.

and the joint is therefore known as *monaxial,* or *uniaxial* (Figure 9-5b). The motion is similar to that of a hinged door. Movement is usually flexion and extension. Examples of hinge joints are the elbow, ankle, and interphalangeal joints. The movement allowed by a hinge joint is illustrated by flexion and extension at the elbow (see Figure 9-2c).

Pivot

In a **pivot** or **trochoid** (TRŌ-koyd) **joint,** a rounded, pointed, or conical surface of one bone articulates within a ring formed partly by bone and partly by a ligament. The primary movement permitted is rotation, and the joint is therefore *monaxial* (Figure 9-5c). Examples include the joint between the atlas and axis (atlantoaxial) and between the proximal ends of the radius and ulna. Movement at a pivot joint is illustrated by supination and pronation of the palms and rotation of the head from side to side (see Figure 9-3a).

Ellipsoidal

In an **ellipsoidal** or **condyloid** (KON-di-loyd) **joint,** an oval-shaped condyle of one bone fits into an elliptical cavity of another bone. Since the joint permits side-to-side and back-and-forth movements, it is *biaxial* (Figure 9-5d). The joint at the wrist between the radius and carpals is ellipsoidal. The movement permitted by such a joint is illustrated when you flex and extend (see Figure 9-2d) and abduct and adduct the wrist.

Saddle

In a **saddle** or **sellaris** (sel-A-ris) **joint,** the articular surfaces of both bones are saddle-shaped, that is, concave in one direction and convex in the other. Essentially, the saddle joint is a modified ellipsoidal joint in which the movement is somewhat freer. Movements at a saddle joint are side to side and back and forth. Thus the joint is *biaxial* (Figure 9-5e). The joint between the trapezium of the carpus and metacarpal of the thumb is an example of a saddle joint.

Ball-and-Socket

A **ball-and-socket** or **spheroid** (SFĒ-royd) **joint** consists of a ball-like surface of one bone fitted into a cuplike depression of another bone. Such a joint permits *triaxial* movement, or movement in three planes of motion: flexion-extension, abduction-adduction, and rotation (Figure 9-5f). Examples of ball-and-socket joints are the shoulder joint and hip joint. The range of movement at a ball-and-socket joint is illustrated by circumduction of the arm (Figure 9-3b).

SUMMARY OF JOINTS

The summary of joints presented in Exhibit 9-2 is based on the anatomy of the joints. If we rearrange the types

FIGURE 9-4 Special movements. (a) Inversion. (b) Eversion. (c) Dorsiflexion. (d) Plantar flexion. (e) Retraction. (f) Protraction. (g) Pronation. (h) Supination. (i) Elevation. (j) Depression. (Copyright © 1983 by Gerard J. Tortora. Courtesy of Lynne Tortora and James Borghesi.)

of joints into a classification based on movement, we arrive at the following:

Synarthroses: immovable joints
1. **Suture**
2. **Synchondrosis**
3. **Gomphosis**

Amphiarthroses: slightly movable joints
1. **Syndesmosis**
2. **Symphysis**

Diarthroses: freely movable joints
1. **Gliding**
2. **Hinge**
3. **Pivot**
4. **Ellipsoidal**
5. **Saddle**
6. **Ball-and-socket**

SELECTED ARTICULATIONS OF THE BODY

We will now examine in some detail the structure of three articulations of the body: humeroscapular (shoulder), coxal (hip), and tibiofemoral (knee) joint.

HUMEROSCAPULAR (SHOULDER) JOINT

The **humeroscapular (shoulder) joint** is formed by the head of the humerus and the glenoid cavity of the scapula (Figure 9-6). It is a ball-and-socket (spheroid) joint.

Its anatomical components are as follows:

1. **Articular capsule.** Loose sac that completely envelops the joint, extending from the circumference of the glenoid cavity to the anatomical neck of the humerus.

2. **Coracohumeral ligament.** Strong, broad ligament that extends from the coracoid process of the scapula to the greater tubercle of the humerus.

Navicular

Second (intermediate) cuneiform

Third (lateral) cuneiform

(a)

Trochlea of humerus

Trochlear notch of ulna

(b)

Radial notch

Head of radius

(c)

FIGURE 9-5 Subtypes of synovial joints. For each subtype shown, there is a simplified diagram and a photograph of the actual joint. (a) Gliding joint between the navicular and second and third cuneiforms of the tarsus. (b) Hinge joint at the elbow between the trochlea of the humerus and trochlear notch of the ulna. (c) Pivot joint between the head of the radius and the radial notch of

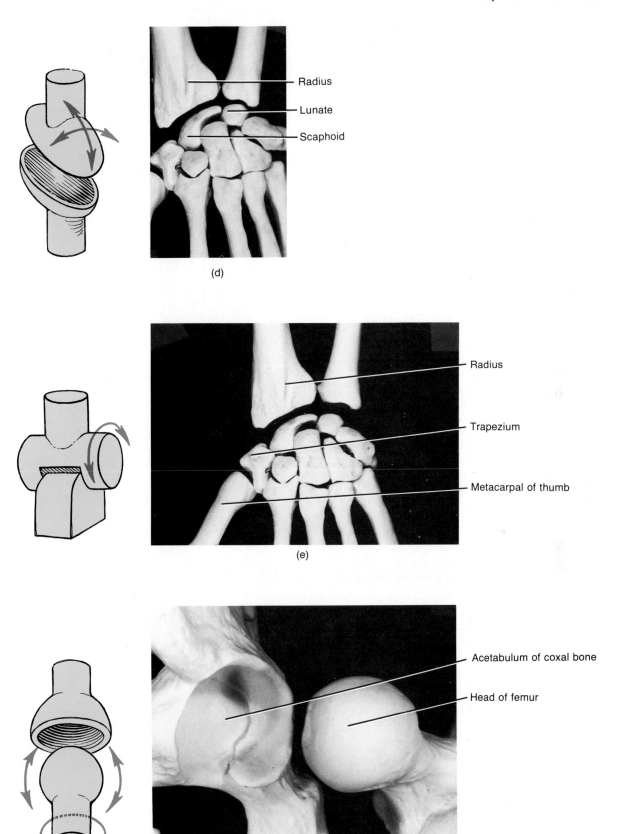

the ulna. (d) Ellipsoidal joint at the wrist between the distal end of the radius and the scaphoid and lunate bones of the carpus. (e) Saddle joint between the trapezium of the carpus and the metacarpal of the thumb. (f) Ball-and-socket joint between the head of the femur and the acetabulum of the coxal bone. (Copyright © 1985 by Michael H. Ross. Used by permission.)

EXHIBIT 9-2 SUMMARY OF JOINTS

TYPE	DESCRIPTION	MOVEMENT	EXAMPLES
FIBROUS	No joint cavity; bones held together by a thin layer of fibrous tissue or dense fibrous tissue.		
Suture	Found only between bones of the skull; articulating bones separated by a thin layer of fibrous tissue.	None (synarthrotic).	Lambdoidal suture between occipital and parietal bones.
Syndesmosis	Articulating bones united by dense fibrous tissue.	Slight (amphiarthrotic).	Distal ends of tibia and fibula.
Gomphosis	Cone-shaped peg fits into a socket; articulating bones separated by periodontal ligament.	None (synarthrotic).	Roots of teeth in alveoli (sockets).
CARTILAGINOUS	No joint cavity, articulating bones united by cartilage.		
Synchondrosis	Connecting material is hyaline cartilage.	None (synarthrotic).	Temporary joint between the diaphysis and epiphyses of a long bone and permanent joint between first rib and sternum.
Symphysis	Connecting material is a broad, flat disc of fibrocartilage.	Slight (amphiarthrotic).	Intervertebral joints and symphysis pubis.
SYNOVIAL	Joint cavity and articular cartilage present; articular capsule composed of an outer fibrous capsule and an inner synovial membrane; may contain accessory ligaments, articular discs (menisci), and bursae.	Freely movable (diarthrotic).	
Gliding	Articulating surfaces usually flat.	Nonaxial.	Intercarpal and intertarsal joints.
Hinge	Spoollike surface fits into a concave surface.	Monaxial (flexion-extension).	Elbow, ankle, and interphalangeal joints.
Pivot	Rounded, pointed, or concave surface fits into a ring formed partly by bone and partly by a ligament.	Monaxial (rotation).	Atlantoaxial and radioulnar joints.
Ellipsoidal	Oval-shaped condyle fits into an elliptical cavity.	Biaxial (flexion-extension, abduction-adduction).	Radiocarpal joint.
Saddle	Articular surfaces concave in one direction and convex in opposite direction.	Biaxial (flexion-extension, abduction-adduction).	Carpometacarpal joint of thumb.
Ball-and-socket	Ball-like surface fits into a cuplike depression.	Triaxial (flexion-extension, abduction-adduction, rotation).	Shoulder and hip joints.

3. **Glenohumeral ligaments.** Three thickenings of the articular capsule over the ventral surface of the joint.

4. **Transverse humeral ligament.** Narrow sheet extending from the greater tubercle to the lesser tubercle of the humerus.

5. **Glenoid labrum.** Narrow rim of fibrocartilage around the edge of the glenoid cavity.

6. Among the **bursae** associated with the shoulder joint are:
   a. **Subscapular bursa** between the tendon of the subscapularis muscle and the underlying joint capsule.
   b. **Subdeltoid bursa** between the deltoid muscle and joint capsule.
   c. **Subacromial bursa** between the acromion and joint capsule.

d. **Subcoracoid bursa** either lies between the coracoid process and joint capsule or appears as an extension from the subacromial bursa.

CLINICAL APPLICATION

The strength and stability of the shoulder joint are not provided by the shape of the articulating bones or its ligaments. Instead, deep muscles of the shoulder and their tendons—subscapularis, supraspinatus, in-

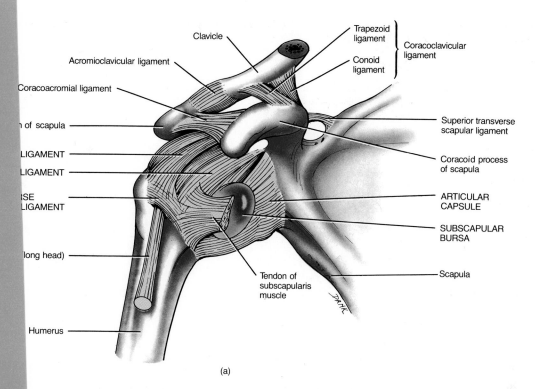

Clavicle

Acromioclavicular ligament

Coracoacromial ligament

...n of scapula

...LIGAMENT

...LIGAMENT

...SE ...LIGAMENT

...(long head)

Humerus

Trapezoid ligament

Conoid ligament

Coracoclavicular ligament

Superior transverse scapular ligament

Coracoid process of scapula

ARTICULAR CAPSULE

SUBSCAPULAR BURSA

Scapula

Tendon of subscapularis muscle

(a)

...lar (shoulder) joint. (a) Diagram of ante-
...anterior view. (Courtesy of C. Yokochi
...phic Anatomy of the Human Body, 2nd
...td., Tokyo, New York.)

Glenoid cavity   Scapula

(b)

fraspinatus, and teres minor—strengthen and stabilize the shoulder joint (see Figure 11-16). The muscles and their tendons are so arranged as to form a nearly complete encirclement of the joint. This arrangement is referred to as the **rotator (musculotendinous) cuff** and is a common site of injury to baseball pitchers, especially tearing of the supraspinatus muscle.

Dislocations of the shoulder joint are based on the relationship of the displaced head of the humerus to the glenoid cavity. Anterior displacement of the humeral head accounts for almost all shoulder dislocations.

COXAL (HIP) JOINT

The **coxal (hip) joint** is formed by the head of the femur and the acetabulum of the coxal bone (Figure 9-7). It is a ball-and-socket (spheroid) joint.

Its anatomical components are as follows:

1. **Articular capsule.** Extends from the rim of the acetabulum to the neck of the femur. One of the strongest ligaments of the body, the capsule consists of circular and longitudinal fibers. The circular fibers, called the **zona orbicularis,** form a collar around the neck of the femur. The longitudinal fibers are reinforced by accessory ligaments known as the iliofemoral ligament, pubofemoral ligament, and ischiofemoral ligament.

2. **Iliofemoral ligament.** Thickened portion of the articular capsule that extends from the anterior inferior iliac spine of the coxal bone to the intertrochanteric line of the femur.

3. **Pubofemoral ligament.** Thickened portion of the articular capsule that extends from the pubic part of the rim of the acetabulum to the neck of the femur.

4. **Ischiofemoral ligament.** Thickened portion of the articular capsule that extends from the ischial wall of the acetabulum to the neck of the femur.

(a)

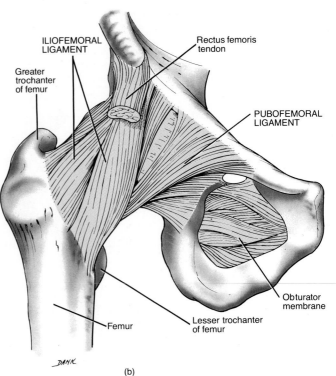

(b)

FIGURE 9-7 Coxal (hip) joint. (a) Frontal section. (b) Anterior view. (c) Posterior view.

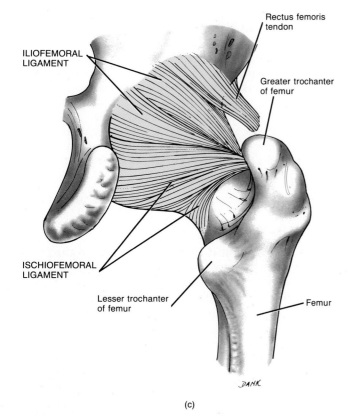

(c)

5. Ligament of the head of the femur (capitate ligament). Flat, triangular band that extends from the fossa of the acetabulum to the head of the femur.

6. Acetabular labrum. Fibrocartilage rim attached to the margin of the acetabulum.

7. Transverse ligament of the acetabulum. Strong ligament that crosses over the acetabular notch, converting it to a fora-

men. It supports part of the acetabular labrum and is connected with the ligament of the head of the femur and the articular capsule.

CLINICAL APPLICATION

Dislocation of the hip among adults is quite rare because of (1) the stability of the ball-and-socket joint, (2) the strong, tough, articular capsule, (3) the strength of the intracapsular ligaments, and (4) the extensive musculature over the joint.

TIBIOFEMORAL (KNEE) JOINT

The **tibiofemoral (knee) joint** is the largest joint of the body, actually consisting of three joints: (1) an intermediate patellofemoral joint between the patella and the patellar surface of the femur, (2) a lateral tibiofemoral joint between the lateral condyle of the femur, lateral meniscus, and lateral condyle of the tibia, and (3) a medial tibiofemoral joint between the medial condyle of the femur, medial meniscus, and medial condyle of the tibia (Figure 9-8). See Figure 9-1b also. The patellofemoral joint is a gliding (arthrodial) joint and the lateral and medial tibiofemoral joints are hinge (ginglymus) joints.

The anatomical components of the tibiofemoral joint are as follows:

1. Articular capsule. No complete, independent capsule unites the bones. The ligamentous sheath surrounding the joint consists mostly of muscle tendons or expansions of them. There

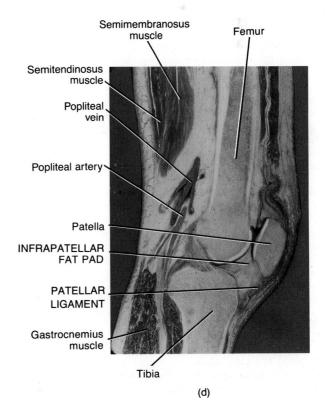

FIGURE 9-8 Tibiofemoral (knee) joint. (a) Diagram of anterior view. (b) Diagram of posterior view. (c) Diagram of sagittal section. (d) Photograph of sagittal section. (Courtesy of C. Yokochi and J. W. Rohen, *Photographic Anatomy of the Human Body,* 1st ed., 1969, IGAKU-SHOIN, Ltd., Tokyo, New York.)

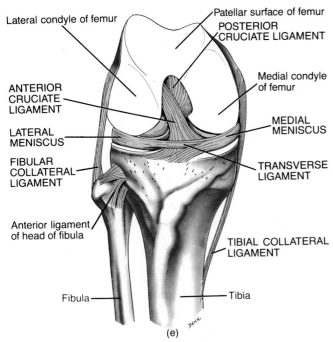

Lateral condyle of femur

Patellar surface of femur

POSTERIOR CRUCIATE LIGAMENT

ANTERIOR CRUCIATE LIGAMENT

Medial condyle of femur

LATERAL MENISCUS

MEDIAL MENISCUS

FIBULAR COLLATERAL LIGAMENT

TRANSVERSE LIGAMENT

Anterior ligament of head of fibula

TIBIAL COLLATERAL LIGAMENT

Fibula

Tibia

(e)

FIGURE 9-8 (Continued) (e) Diagram of anterior view (flexed).

are, however, some capsular fibers connecting the articulating bones.

2. Medial and lateral patellar retinacula. Fused tendons of insertion of the quadriceps femoris muscle and the fascia lata that strengthen the anterior surface of the joint.

3. Patellar ligament. Central portion of the common tendon of insertion of the quadriceps femoris muscle that extends from the patella to the tibial tuberosity. This ligament also strengthens the anterior surface of the joint. The posterior surface of the ligament is separated from the synovial membrane of the joint by an **infrapatellar fat pad.**

4. Oblique popliteal ligament. Broad, flat ligament that connects the intercondylar fossa of the femur to the head of the tibia. The tendon of the semimembranosus muscle is superficial to the ligament and passes from the medial condyle of the tibia to the lateral condyle of the femur. The ligament and tendon afford strength for the posterior surface of the joint.

5. Arcuate popliteal ligament. Extends from the lateral condyle of the femur to the styloid process of the head of the fibula. It strengthens the lower lateral part of the posterior surface of the joint.

6. Tibial collateral ligament. Broad, flat ligament on the medial surface of the joint that extends from the medial condyle of the femur to the medial condyle of the tibia. The ligament is crossed by tendons of the sartorius, gracilis, and semitendinosus muscles, all of which strengthen the medial aspect of the joint.

7. Fibular collateral ligament. Strong, rounded ligament on the lateral surface of the joint that extends from the lateral condyle of the femur to the lateral side of the head of the fibula. The ligament is covered by the tendon of the biceps femoris muscle. The tendon of the popliteal muscle is deep to the tendon.

8. Intraarticular ligaments. Ligaments within the capsule that connect the tibia and femur.

    a. Anterior cruciate ligament. Extends posteriorly and laterally from the area anterior to the intercondylar eminence of the tibia to the posterior part of the medial surface of the lateral condyle of the femur. This

ligament is stretched or torn in about 70 percent of all serious knee injuries. Both O. J. Simpson and Gale Sayers had their careers ended by torn anterior cruciate ligaments.

    b. Posterior cruciate ligament. Extends anteriorly and medially from the posterior intercondylar fossa of the tibia and lateral meniscus to the anterior part of the medial surface of the medial condyle of the femur.

9. Articular discs. Fibrocartilage discs between the tibial and femoral condyles. They help to compensate for the incongruence of the articulating bones.

    a. Medial meniscus. Semicircular piece of fibrocartilage. Its anterior end is attached to the anterior intercondylar fossa of the tibia, in front of the anterior cruciate ligament. Its posterior end is attached to the posterior intercondylar fossa of the tibia between the attachments of the posterior cruciate ligament and lateral meniscus.

    b. Lateral meniscus. Circular piece of fibrocartilage. Its anterior end is attached anterior to the intercondylar eminence of the tibia and lateral and posterior to the anterior cruciate ligament. Its posterior end is attached posterior to the intercondylar eminence of the tibia and anterior to the posterior end of the medial meniscus. The medial and lateral menisci are connected to each other by the **transverse ligament** and to the margins of the head of the tibia by the **coronary ligaments.**

10. The principal **bursae** of the knee include:

    a. Anterior bursae: (1) between the patella and skin (**prepatellar bursa**), (2) between upper part of tibia and patellar ligament (**infrapatellar bursa**), (3) between lower part of tibial tuberosity and skin, and (4) between lower part of femur and deep surface of quadriceps femoris muscle (**suprapatellar bursa**).

    b. Medial bursae: (1) between medial head of gastrocnemius muscle and the articular capsule, (2) superficial to the tibial collateral ligament between the ligament and tendons of the sartorius, gracilis, and semitendinosus muscles, (3) deep to the tibial collateral ligament between the ligament and the tendon of the semimembranosus muscle, (4) between the tendon of the semimembranosus muscle and the head of the tibia, and (5) between the tendons of the semimembranosus and semitendinosus muscles.

    c. Lateral bursae: (1) between the lateral head of the gastrocnemius muscle and articular capsule, (2) between the tendon of the biceps femoris muscle and fibular collateral ligament, (3) between the tendon of the popliteal muscle and fibular collateral ligament, and (4) between the lateral condyle of the femur and the popliteal muscle.

CLINICAL APPLICATION

The most common type of **knee injury** in football is rupture of the tibial collateral ligament, often associated with tearing of the anterior cruciate ligament and medial meniscus (torn cartilage). It is caused by a blow to the lateral side of the knee. When a knee is examined for such an injury, the three C's are kept in mind: collateral ligament, cruciate ligament, and cartilage.

Runner's knee (patello-femoral syndrome) refers to an aching pain around the front of a runner's knee that may result from excessive pronation of the foot or tight hamstrings. The pain typically occurs after a person has been sitting for a while, especially after exercise. A common cause of runner's knee is constantly walking, running, or jogging on the same side of the road. Since roads are high in the middle and slope down on the sides, the slope stresses the knee that is closer to the center of the road.

A **swollen knee** may occur immediately or be delayed. The immediate swelling is due to escape of blood from rupture of the anterior cruciate ligament, torn menisci, fractures, or collateral ligament sprains. Delayed swelling is due to an excessive production of synovial fluid as a result of conditions that irritate the synovial membrane.

A **dislocated knee** refers to the displacement of the tibia relative to the femur. Accordingly, such dislocations are classified as anterior, posterior, medial, lateral, or rotatory. The most common type is anterior dislocation, resulting from hyperextension of the knee. A frequent consequence of a dislocated knee is damage to the popliteal artery.

DISORDERS: HOMEOSTATIC IMBALANCES

Rheumatism

Rheumatism (*rheumat* = subject to flux) refers to any painful state of the supporting structures of the body—its bones, ligaments, joints, tendons, or muscles. Arthritis is a form of rheumatism in which the joints have become inflamed.

Arthritis

The term **arthritis** refers to many different diseases, the most common of which are rheumatoid arthritis (RA), osteoarthritis, and gouty arthritis. All are characterized by inflammation in one or more joints. Inflammation, pain, and stiffness may also be present in adjacent parts of the body, such as the muscles near the joint. The chronic pain that accompanies various arthritic conditions may be related to the patient's inability to produce endorphins, naturally produced painkillers (Chapter 14). The causes of arthritis are unknown.

Rheumatoid Arthritis (RA)

Rheumatoid (ROO-ma-toyd) **arthritis (RA)** is the most common inflammatory form of arthritis. It involves inflammation of the joint, swelling, pain, and a loss of function. Usually this form occurs bilaterally—if your left knee is affected, your right knee may also be affected, although usually not to the same degree.

The primary symptom of rheumatoid arthritis is inflammation of the synovial membrane. If untreated, the following sequential pathology may occur. The membrane thickens and synovial fluid accumulates. The resulting pressure causes pain and tenderness. The membrane then produces an abnormal tissue called *pannus,* which adheres to the surface of the articular cartilage. The pannus formation sometimes erodes the cartilage completely. When the cartilage is destroyed, fibrous tissue joins the exposed bone ends. The tissue ossifies and fuses the joint so that it is immovable—the ultimate crippling effect of rheumatoid arthritis. Most cases do not progress to this stage, but the range of motion of the joint is greatly inhibited by the severe inflammation and swelling.

Damaged joints may be surgically replaced, either partly or entirely, with artificial joints. The artificial parts are inserted after removal of the diseased portion of the articulating bone and its cartilage. The new metal or plastic joint is fixed in place with a special acrylic cement. When freshly mixed in the operating room, it hardens as strong as bone in minutes. These new parts function nearly as well as a normal joint—and much better than a diseased joint.

Osteoarthritis

A degenerative joint disease far more common than rheumatoid arthritis, and usually less damaging, is **osteoarthritis** (os'-tē-ō-ar-THRĪ-tis). It apparently results from a combination of aging, irritation of the joints, and wear and abrasion.

Degenerative joint disease is a noninflammatory, progressive disorder of movable joints, particularly weight-bearing joints. It is characterized by the deterioration of articular cartilage and by formation of new bone in the subchondral areas and at the margins of the joint. The cartilage slowly degenerates, and as the bone ends become exposed, small bumps, or *spurs,* of new osseous tissue are deposited on them. These spurs decrease the space of the joint cavity and restrict joint movement. Unlike rheumatoid arthritis, osteoarthritis usually affects only the articular cartilage. The synovial membrane is rarely destroyed, and other tissues are unaffected.

Gouty Arthritis

Uric acid is a waste product produced during the metabolism of nucleic acids. Normally, all the acid is quickly excreted in the urine. In fact, it gives urine its name. The person who suffers from *gout* either produces excessive amounts of uric acid or is not able to excrete normal amounts. The result is a buildup of uric acid in the blood. This excess acid then reacts with sodium to form a salt called sodium urate. Crystals of this salt are deposited in soft tissues. Typical sites are the kidneys and the cartilage of the ears and joints.

In **gouty** (GOW-tē) **arthritis,** sodium urate crystals are deposited in the soft tissues of the joints. The crystals irritate the cartilage, causing inflammation, swelling, and acute pain. Eventually, the crystals destroy all the joint tissues. If the disorder is not treated, the ends of the articulating bones fuse and the joint becomes immovable.

Gouty arthritis occurs primarily in males of any age. It is believed to be the cause of 2 to 5 percent of all chronic joint diseases. Numerous studies indicate that gouty arthritis is sometimes caused by an abnormal gene. As a result of this gene, the body manufactures unusually large amounts of uric acid. Diet and environmental factors such as stress and climate are also suspected causes of gouty arthritis.

Although other forms of arthritis cannot be treated with complete success, the treatment of gouty arthritis with the use of various drugs has been quite effective. A chemical called colchicine has been utilized periodically since the sixth century to relieve pain, swelling, and tissue destruction that occur during attacks of gouty arthritis. This chemical is derived from the variety of crocus plant from which the spice saffron is obtained. Other drugs, which either inhibit uric acid production or assist in the elimination of excess uric acid by the kidneys, are used to prevent further attacks. The drug allopurinal is used for the treatment of gouty arthritis because it prevents the formation of uric acid without interfering with nucleic acid synthesis.

Bursitis

An acute chronic inflammation of a bursa is called **bursitis.**

The condition may be caused by trauma, by an acute or chronic infection (including syphilis and tuberculosis), or by rheumatoid arthritis. Repeated excessive friction often results in a bursitis with local inflammation and the accumulation of fluid. Bunions are frequently associated with a friction bursitis over the head of the first metatarsal bone. Symptoms include pain, swelling, tenderness, and the limitation of motion involving the inflamed bursa. The prepatellar or subcutaneous infrapatellar bursa may become inflamed in individuals who spend a great deal of time kneeling. This bursitis is usually called **"housemaid's knee"** **("carpet layer's knee").**

Aspiration of the knee joint may be necessary to relieve pressure, to evacuate blood, or to obtain a fluid sample for laboratory studies. In this procedure, the needle is inserted somewhat proximal and lateral to the patella through the tendinous part of the vastus lateralis muscle and is directed toward the middle of the joint. Through the same route, the joint cavity may be anesthetized or irrigated, and steroids may be administered in the treatment of knee pathologies.

Dislocation

A **dislocation** or **luxation** (luks-Ā-shun) is the displacement of a bone from a joint with tearing of ligaments, tendons, and articular capsules. A partial or incomplete dislocation is called a **subluxation.** The most common dislocations are those involving a finger, thumb, or shoulder. Those of the mandible, elbow, knee, or hip are less common. Symptoms include loss of motion, temporary paralysis of the involved joint, pain, swelling, and occasional shock. A dislocation is usually caused by a blow or fall, although unusual physical effort may lead to this condition.

Sprain and Strain

A **sprain** is the forcible wrenching or twisting of a joint with partial rupture or other injury to its attachments without luxation. It occurs when the attachments are stressed beyond their normal capacity. There may be damage to the associated blood vessels, muscles, tendons, ligaments, or nerves. A sprain is more serious than a **strain,** which is the overstretching of a muscle. Severe sprains may be so painful that the joint cannot be moved. There is considerable swelling, with reddish to blue discoloration due to hemorrhage from ruptured blood vessels. The ankle joint is most often sprained; the low back area is another frequent location for sprains.

MEDICAL TERMINOLOGY

Ankylosis (*ankyle* = stiff joint; *osis* = condition)   Severe or complete loss of a movement at a joint.
Arthralgia (*arth* = joint; *algia* = pain)   Pain in a joint.
Arthrosis   Refers to an articulation; also a disease of a joint.
Bursectomy (*ectomy* = removal of)   Removal of a bursa.

Chondritis (*chondro* = cartilage)   Inflammation of cartilage.
Rheumatology (*rheumat* = subject to flux)   The medical specialty devoted to arthritis.
Synovitis (*synoy* = joint)   Inflammation of a synovial membrane in a joint.

STUDY OUTLINE

Classification (p. 176)
1. An articulation (joint) is a point of contact between two or more bones.
2. Functional classification of joints is based on the degree of movement permitted. Joints may be synarthroses, amphiarthroses, or diarthroses.
3. Structural classification is based on the presence of a synovial (joint) cavity and type of connecting tissue. Structurally, joints are classified as fibrous, cartilaginous, or synovial.

Fibrous Joints (p. 176)
1. Bones held by fibrous connective tissue, with no synovial (joint) cavity, are fibrous joints.
2. These joints include immovable sutures (found in the skull), slightly movable syndesmoses (such as the tibiofibular articulation), and immovable gomphoses (roots of teeth in alveoli of mandible and maxilla).

Cartilaginous Joints (p. 176)
1. Bones held together by cartilage, with no synovial (joint) cavity, are cartilaginous joints.
2. These joints include immovable synchondroses united by hyaline cartilage (temporary cartilage between diaphysis and epiphyses) and partially movable symphyses united by fibrocartilage (the symphysis pubis).

Synovial Joints (p. 177)
1. Synovial joints contain a synovial (joint) cavity, articular cartilage, and a synovial membrane; some also contain ligaments, articular discs, and bursae.
2. All synovial joints are freely movable.
3. Movements at synovial joints are limited by the apposition of soft parts, tension of ligaments, and muscle tension.
4. Types of movements at synovial joints include gliding movements, angular movements, rotation, circumduction, and special movements.
5. Types of synovial joints include gliding joints (wrist bones), hinge joints (elbow), pivot joints (radioulnar), ellipsoidal joints (radiocarpal), saddle joints (carpometacarpal), and ball-and-socket joints (shoulder and hip).
6. A joint may be described according to the number of planes of movement it allows as nonaxial, monaxial, biaxial, or triaxial.

Selected Articulations of the Body (p. 183)
1. The humeroscapular (shoulder joint) is formed by the humerus and scapula.
2. The coxal (hip) joint is formed by the femur and coxal bone.
3. The tibiofemoral (knee) joint is formed by the patella and femur and by the tibia and femur.

Disorders: Homeostatic Imbalances (p. 191)

1. Rheumatism is a painful state of supporting body structures such as bones, ligaments, tendons, joints, and muscles.
2. Arthritis refers to several disorders characterized by inflammation of joints, often accompanied by stiffness of adjacent structures.
3. Rheumatoid arthritis (RA) refers to inflammation of a joint accompanied by pain, swelling, and loss of function.
4. Osteoarthritis is a degenerative joint disease characterized by deterioration of articular cartilage and spur formation.
5. Gouty arthritis is a condition in which sodium urate crystals are deposited in the soft tissues of joints and eventually destroy the tissues.
6. Bursitis is an acute or chronic inflammation of bursae.
7. A dislocation (luxation) is a displacement of a bone from its joint; a partial dislocation is called subluxation.
8. A sprain is the forcible wrenching or twisting of a joint with partial rupture to its attachments without dislocation, while a strain is the stretching of a muscle.

REVIEW QUESTIONS

1. Define an articulation. What factors determine the degree of movement at joints?
2. Distinguish among the three kinds of joints on the basis of structure and function. List the subtypes. Be sure to include degree of movement and specific examples.
3. Explain the components of a synovial joint. Indicate the relationship of ligaments and tendons to the strength of the joint and restrictions on movement.
4. Explain how the articulating bones in a synovial joint are held together.
5. What is an accessory ligament? Define the two principal types.
6. What is an articular disc? Why are they important?
7. Describe the principle and importance of anthroscopy.
8. What are bursae? What is their function?
9. Define the following principal movements: gliding, angular, rotation, circumduction, and special. Name a joint where each occurs.
10. Have another person assume the anatomical position and execute for you each of the movements at joints discussed in the text. Reverse roles, and see if you can execute the same movements.
11. Contrast nonaxial, monaxial, biaxial, and triaxial planes of movement. Give examples of each, and name a joint at which each occurs.
12. Be sure that you can name the bones that form the joint, identify the joint by type, and list the anatomical components of the joint for each of the following: humeroscapular (shoulder), coxal (hip), and tibiofemoral (knee) joints.
13. Distinguish between rheumatoid arthritis (RA), osteoarthritis, and gouty arthritis with respect to causes and symptoms.
14. Define bursitis. How is it caused?
15. Define dislocation. What are the symptoms of dislocation?
16. Distinguish between a sprain and a strain.
17. Refer to the glossary of medical terminology associated with articulation at the end of the chapter. Be sure that you can define each term.

Muscle Tissue

- List the characteristics and functions of muscle tissue.
- Compare the location, microscopic appearance, nervous control, and functions of the three kinds of muscle tissue.
- Define fascia, epimysium, perimysium, endomysium, tendons, and aponeuroses and list their modes of attachment to muscles.
- Explain the relationship of blood vessels and nerves to skeletal muscles.
- Describe the principal events associated with the sliding-filament theory.
- Describe the structure and importance of a neuromuscular junction (motor end plate) and a motor unit.
- Identify the source of energy for muscular contraction.
- Define the all-or-none principle of muscular contraction.
- Describe different types of normal contractions performed by skeletal muscles.
- Describe the phases of contraction in a typical myogram of a twitch contraction.
- Compare the structure and function of the three types of skeletal muscle fibers.
- Compare oxygen debt, fatigue, and heat production as examples of muscle homeostasis.
- Explain the effects of aging on muscle tissue.
- Describe the development of the muscular system.
- Define such common muscular disorders as fibrosis, fibrositis, "charleyhorse," muscular dystrophy, and myasthenia gravis (MG).
- Compare spasms, cramps, convulsions, fibrillation, and tics as abnormal muscular contractions.
- Define medical terminology associated with the muscular system.

Although bones and joints provide leverage and form the framework of the body, they are not capable of moving the body by themselves. Motion is an essential body function that results from the contraction and relaxation of muscles.

Muscle tissue constitutes about 40 to 50 percent of the total body weight and is composed of highly specialized cells. The scientific study of muscles is known as **myology** (*myo* = muscle; *logos* = study of).

The developmental anatomy of the muscular system is considered at the end of the chapter.

CHARACTERISTICS

Muscle tissue has four principal characteristics that assume key roles in maintaining homeostasis.

1. Excitability is the ability of muscle tissue to receive and respond to stimuli. A stimulus is a change in the internal or external environment strong enough to initiate a nerve impulse (action potential).

2. Contractility is the ability to shorten and thicken, or contract, when a sufficient stimulus is received.

3. Extensibility is the ability of muscle tissue to be stretched (extended). Many skeletal muscles are arranged in opposing pairs. While one is contracting, the other is relaxed and is undergoing extension.

4. Elasticity is the ability of muscle to return to its original shape after contraction or extension.

FUNCTIONS

Through contraction, muscle performs three important functions:

1. Motion.
2. Maintenance of posture.
3. Heat production.

Motion is obvious in movements involving the whole body, such as walking and running, and in localized movements, such as grasping a pencil or nodding the head. All of these movements rely on the integrated functioning of the bones, joints, and muscles attached to the bones. Less noticeable kinds of motion produced by muscles are the beating of the heart, churning of food in the stomach, pushing of food through the intestines, contraction of the gallbladder to release bile, and contraction of the urinary bladder to expel urine.

In addition to the movement function, muscle tissue also enables the body to maintain posture. The contraction of skeletal muscles holds the body in stationary positions, such as standing and sitting.

The third function of muscle tissue is heat production. Skeletal muscle contractions produce heat and are thereby important in maintaining normal body temperature. It has been estimated that as much as 85 percent of all body heat is generated by muscle contractions.

TYPES

Types of muscle tissue are categorized by location, microscopic structure, and nervous control.

Skeletal muscle tissue, which is named for its location, is attached to bones and moves parts of the skeleton. It is **striated** muscle tissue because striations, or bandlike structures, are visible when the tissue is examined under a microscope. It is a **voluntary** muscle tissue because it can be made to contract by conscious control.

Cardiac muscle tissue forms the bulk of the wall of the heart. It is **striated** and **involuntary,** that is, its contraction is usually not under conscious control.

Smooth muscle tissue is involved with processes related to maintaining the internal environment. It is located in the walls of hollow internal structures, such as blood vessels, the stomach, and the intestines. It is referred to as **nonstriated** because it lacks striations. It is **involuntary** muscle tissue.

Thus all muscle tissues are classified in the following way: (1) skeletal, striated, voluntary muscle tissue, (2) cardiac, striated, involuntary muscle tissue, and (3) smooth, nonstriated, involuntary muscle tissue.

SKELETAL MUSCLE TISSUE

To understand the fundamental mechanisms of muscle movement, you will need some knowledge of its connective tissue components, nerve and blood supply, and histology.

CONNECTIVE TISSUE COMPONENTS

The term **fascia** (FASH-ē-a) is applied to a sheet or broad band of fibrous connective tissue beneath the skin or around muscles and other organs of the body. **Superficial fascia (subcutaneous layer)** is immediately deep to the skin. It is composed of adipose tissue and loose connective tissue and has a number of important functions. (1) It serves as a storehouse for water and particularly for fat. Much of the fat of an overweight person is in the superficial fascia. (2) It forms a layer of insulation protecting the body from loss of heat. (3) It provides mechanical protection from blows. (4) It provides a pathway for nerves and vessels.

Deep fascia is a dense connective tissue that lines the body wall and extremities and holds muscles together, separating them into functioning groups. Functionally, deep fascia allows free movement of muscles, carries nerves and blood vessels, fills spaces between muscles, and sometimes provides the origin for muscles.

CLINICAL APPLICATION

Lines of fusion of fascial sheets are essentially avascular and for this reason they are often sites for **surgical incisions.** Surgeons also prefer fascial junctional areas for anchoring of sutures because of the strength of fasciae and the strong union that results from wound healing.

Skeletal muscles are further protected, strengthened, and attached to other structures by several other connective tissue coverings (Figure 10-1). The entire muscle is usually wrapped with a substantial quantity of fibrous connective tissue called the **epimysium** (ep'-i-MĪZ-ē-um). The epimysium is an extension of deep fascia. When the muscle is cut in cross section, invaginations of the epimysium are seen to divide the muscle into bundles of fibers (cells) called **fasciculi** (fa-SIK-yoo-lī) or **fascicles** (FAS-i-kuls). These invaginations of the epimysium are called the **perimysium** (per'-i-MĪZ-ē-um). Perimysium, like epimysium, is an extension of deep fascia. In turn, invaginations of the perimysium, called **endomysium** (en'-dō-MĪZ-ē-um), penetrate into the interior of each fascicle and separate the muscle fibers. Endomysium is also an extension of deep fascia.

Epimysium, perimysium, and endomysium are all continuous with the connective tissue that attaches the muscle to another structure, such as bone or other muscle. All three elements may be extended beyond the muscle cells as a **tendon**—a cord of connective tissue that attaches a muscle to the periosteum of a bone. When the connective tissue elements extend as a broad, flat layer, the tendon is called an **aponeurosis.** This structure also attaches to the coverings of a bone or another muscle. An example of an aponeurosis is the galea aponeurotica (See Figure 11-4). When a muscle contracts, the tendon and its corresponding bone or muscle are pulled toward the contract-ing muscle. In this way skeletal muscles produce movement. Certain tendons, especially those of the wrist and ankle, are enclosed by tubes of fibrous connective tissue called **tendon sheaths.** They are similar in structure to bursae. The inner layer of a tendon sheath, the visceral layer, is applied to the surface of the tendon. The outer layer is known as the parietal layer. Between the layers is a cavity that contains a film of synovial fluid. Tendon sheaths permit tendons to slide easily and also prevent tendons from slipping out of place.

CLINICAL APPLICATION

Tendinitis or **tenosynovitis** (ten'-ō-sin-ō-VĪ-tis) frequently occurs as inflammation involving the tendon sheaths and synovial membrane surrounding certain joints. The wrists, shoulders, elbows (tennis elbow), finger joints (trigger finger), ankles, and associated tendons are most often affected. The affected sheaths may become visibly swollen because of fluid accumulation or they may remain dry. Local tenderness is variable, and there may be disabling pain with movement of the body part. The condition often follows some form of trauma, strain, or excessive exercise.

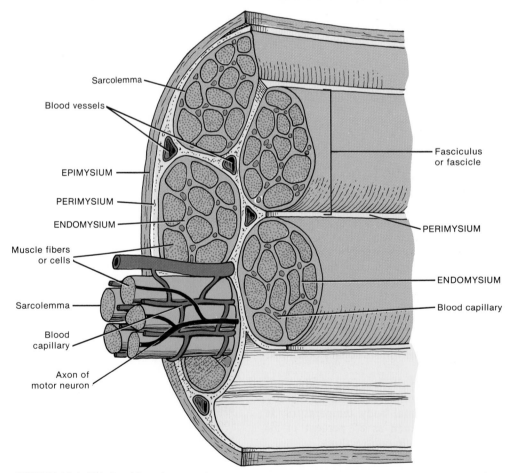

FIGURE 10-1 Relationships of connective tissue to skeletal muscle. Shown is a cross section and longitudinal section of a skeletal muscle indicating the relative positions of the epimysium, perimysium, and endomysium.

NERVE AND BLOOD SUPPLY

Skeletal muscles are well supplied with nerves and blood vessels. This innervation and vascularization is directly related to contraction, the chief characteristic of muscle. For a skeletal muscle fiber (cell) to contract, it must first be stimulated by an impulse from a nerve cell. Muscle contraction also requires a good deal of energy and therefore large amounts of nutrients and oxygen. Moreoover, the waste products of these energy-producing reactions must be eliminated. Thus prolonged muscle action depends on a rich blood supply to deliver nutrients and oxygen and remove wastes.

Generally, an artery and one or two veins accompany each nerve that penetrates a skeletal muscle. The larger branches of the blood vessels accompany the nerve branches through the connective tissue of the muscle. Microscopic blood vessels called capillaries are distributed within the endomysium. Each muscle cell is thus in close contact with one or more capillaries. Each skeletal muscle fiber usually makes contact with a portion of a nerve cell called a synaptic end bulb.

HISTOLOGY

When a typical skeletal muscle is teased apart and viewed microscopically, it can be seen to consist of thousands of elongated, cylindrical cells called **muscle fibers** or **myofibers** (see Exhibit 4-3). These fibers lie parallel to one another and range from 10 to 100 μm in diameter. Some fibers may reach lengths of 30 cm (12 inches) or more. Each muscle fiber is enveloped by a plasma membrane called the **sarcolemma** (*sarco* = flesh; *lemma* = sheath). The sarcolemma surrounds a quantity of cytoplasm called **sarcoplasm.** Within the sarcoplasm of a muscle fiber and lying close to the sarcolemma are many nuclei. Skeletal muscle fibers are thus multinucleate. The sarcoplasm also contains myofibrils (described shortly), special high-energy molecules, enzymes, and **sarcoplasmic reticulum** (sar'-kō-PLAZ-mik re-TIK-yoo-lum), a network of membrane-enclosed tubules comparable to smooth endoplasmic reticulum (Figure 10-2a). Running transversely through the fiber and perpendicularly to the sarcoplasmic reticulum are **transverse tubules (T tubules).** The tubules are extensions of the sarcolemma that open to the outside

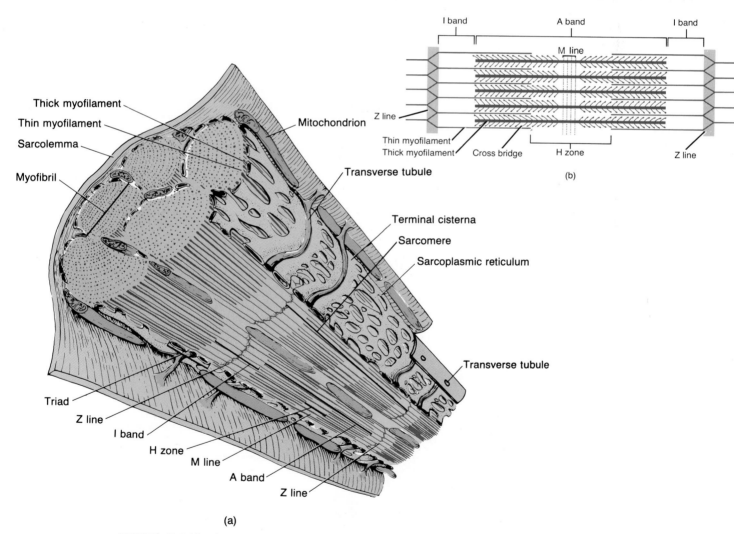

FIGURE 10-2 Histology of skeletal muscle tissue. (a) Enlarged aspect of several myofibrils of a muscle fiber based on an electron micrograph. (b) Enlarged aspect of a sarcomere showing thin and thick myofilaments.

of the fiber. A **triad** consists of a transverse tubule and the segments of sarcoplasmic reticulum on either side.

A highly magnified view of skeletal muscle fibers reveals cylindrical structures, about 1 or 2 μm in diameter, called **myofibrils** (Figure 10-2a,b). Myofibrils, ranging in number from several hundred to several thousand, run longitudinally through the muscle fiber and consist of two kinds of even smaller structures called **myofilaments.** The **thin myofilaments** are about 6 nm in diameter. The **thick myofilaments** are about 16 nm in diameter.

The myofilaments of a myofibril do not extend the entire length of a muscle fiber—they are arranged in compartments called **sarcomeres.** Sarcomeres are separated from one another by narrow zones of dense material called **Z lines.** Within a sarcomere, certain areas can be distinguished (Figure 10-2b). A dark, dense area, called the **anisotropic** or **A band,** represents the length of thick myofilaments. The sides of the A band are darkened by the overlapping of thick and thin myofilaments. The length of darkening depends on the extent of overlapping. As you will see later, the greater the degree of contraction, the greater the overlapping of thick and thin myofilaments. A light-colored, less dense area called the **isotropic band,** or **I band,** is composed of thin myofilaments only. This combination of alternating dark A bands and light I bands gives the muscle fiber its striated (striped) appearance. A narrow **H zone** contains thick myofilaments only. In the center of the H zone is the **M line,** a series of fine threads that appear to connect the middle parts of adjacent thick myofilaments.

Thin myofilaments are anchored in the Z lines and project in both directions. They are composed mostly of the protein **actin.** The actin molecules are arranged in two single strands that entwine helically and give the thin myofilaments their characteristic shape (Figure 10-3a). Each actin molecule contains a *myosin-binding site* that interacts with a cross bridge of a myosin molecule (described shortly). Besides actin, the thin myofilaments contain two other protein molecules, **tropomyosin** and **troponin,** that are involved in the regulation of muscle contractions. Tropomyosin is arranged in strands that are loosely attached to the actin helices. Troponin is located at regular intervals on the surface of tropomyosin and is made up of three subunits: troponin I which binds to actin, troponin C which binds to calcium ions, and troponin T which binds to tropomyosin. Together tropomyosin and troponin are referred to as the **tropomyosin-troponin complex.**

Thick myofilaments overlap the free ends of the thin myofilaments and occupy the A band region of a sarcomere. These myofilaments are composed mostly of the protein **myosin.** A myosin molecule is shaped like a golf club. The tails (handles of the golf club) are arranged parallel to each other forming the shaft of the thick myofilament. The heads of the golf clubs project outward from the shaft and are arranged spirally on the surface of the shaft. The projecting heads are referred to as **cross bridges** and contain an *actin-binding site* and an *ATP-binding site* (Figure 10-3b).

CLINICAL APPLICATION

Steroids are chemical substances derived from cholesterol. Most steroids are hormones. At the IX Pan

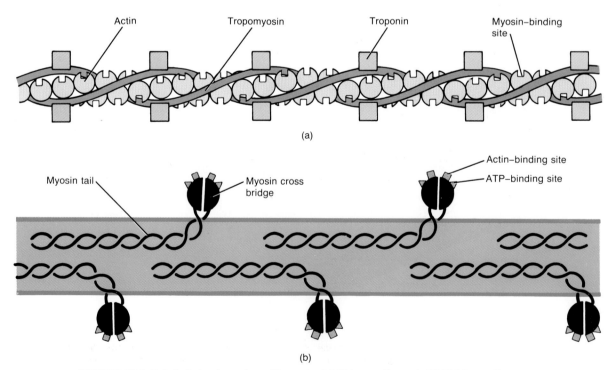

FIGURE 10-3 Detailed structure of myofilaments. (a) Thin myofilament. (b) Thick myofilament.

American Games in Caracas, Venezuela, held in 1983, worldwide attention was focused on the use of **muscle-building anabolic steroids** by amateur athletes. These steroids, a variant of the hormone testosterone, were used by the athletes supposedly to build muscle proteins and therefore increase strength and endurance during athletic events. However, physicians point out that use of anabolic steroids may cause a number of side effects, including liver cancer, kidney damage, increased risk of heart disease, increased irritability and aggressive behavior, sterility, the development of facial hair and deepening of the voice in females, and excessive development of breast glands and diminished hormone secretion and sperm production by the testes in males.

CONTRACTION

SLIDING-FILAMENT THEORY

During muscle contraction, thin myofilaments slide inward toward the H zone. The sarcomere shortens, but the lengths of the thin and thick myofilaments do not change. The myosin cross bridges of the thick myofilaments connect with portions of actin of the thin myofilaments. The myosin cross bridges move like the oars of a boat on the surface of the thin myofilaments, and the thin and thick myofilaments slide past each other. As the thin myofilaments move past the thick myofilaments, the H zone narrows and even disappears when the thin myofilaments meet at the center of the sarcomere (Figure

10-4). In fact, the myosin cross bridges may pull the thin myofilaments of each sarcomere so far inward that their ends overlap. As the thin myofilaments slide inward, the Z lines are drawn toward each other and the sarcomere is shortened. The sliding of myofilaments and shortening of sarcomeres causes the shortening of the muscle fibers. All these events associated with the movement of myofilaments are known as the **sliding-filament theory** of muscle contraction.

NEUROMUSCULAR JUNCTION (MOTOR END PLATE)

For a skeletal muscle fiber to contract, a stimulus must be applied to it. The stimulus is delivered by a nerve cell, or **neuron.** A neuron has a threadlike process called a fiber, or axon, that may run 91 cm (3 ft) or more to a muscle. A bundle of such fibers from many different neurons composes a nerve. A neuron that stimulates muscle tissue is called a **motor neuron.**

On entering a skeletal muscle, the axon of a motor neuron branches into axon terminals (telodendria) that come into close approximation with a portion of the sarcolemma of a muscle fiber. The term **neuromuscular (myoneural) junction** or **motor end plate** refers to the axon terminal of a motor neuron together with the portion of the sarcolemma of a muscle fiber in close approximation with the axon terminal (Figure 10-5). Close examination of a neuromuscular junction reveals that the distal ends of the axon terminals are expanded into bulblike structures called **synaptic end bulbs** (see Figure 12-3a). The bulbs contain membrane-enclosed sacs, the **synaptic vesicles,** that store chemicals called **neurotransmitters.** These

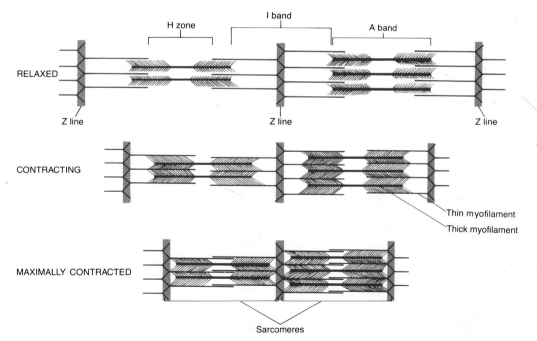

FIGURE 10-4 Sliding-filament theory of muscle contraction. Shown are the positions of the various parts of two sarcomeres in relaxed, contracting, and maximally contracted states. Note the movement of the thin myofilaments and the relative size of the H zone.

Motor nerve

Muscle fibers

Axon terminals (telodendria)

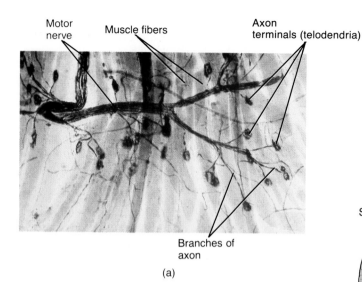

Branches of axon

(a)

Axon terminal (telodendrium)

Synaptic vesicle

Sarcolemma

Synaptic end bulb

Subneural cleft

Sarcoplasm

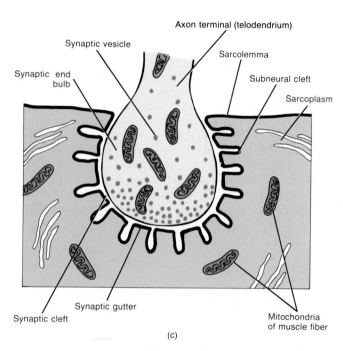

Synaptic cleft

Synaptic gutter

Mitochondria of muscle fiber

(c)

Myelin sheath surrounding axon of motor neuron

Axon terminal (telodendrium)

Sarcolemma

Nuclei of muscle fiber

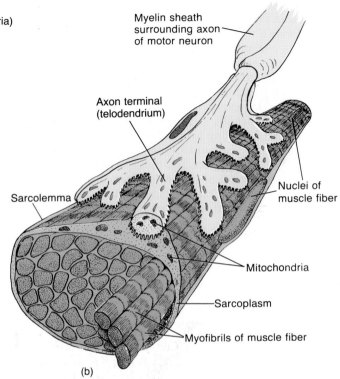

Mitochondria

Sarcoplasm

Myofibrils of muscle fiber

(b)

FIGURE 10-5 Neuromuscular junction (motor end plate). (a) Photomicrograph at a magnification of 400×. (Courtesy of D. E. Kelly, from *Introduction to the Musculoskeletal System* by Cornelius Rosse and D. Kay Clawson, Harper & Row, Publishers, Inc., New York, 1970.) (b) Diagram based on a photomicrograph. (c) Enlarged aspect based on an electron micrograph.

in the development of a nerve impulse that travels along the sarcolemma, thus initiating the events leading to contraction. (The details of nerve impulse generation are discussed in Chapter 12.)

MOTOR UNIT

A motor neuron, together with all the muscle fibers it stimulates, is referred to as a **motor unit.** A single motor neuron may innervate about 150 muscle fibers, depending on the region of the body. This means that stimulation of one neuron will tend to cause the simultaneous contraction of about 150 muscle fibers. In addition, all the muscle fibers of a motor unit that are sufficiently stimulated will contract and relax together. Muscles that control precise movements, such as the extrinsic eye muscles, have fewer than 10 muscle fibers to each motor unit. Muscles of the body that are responsible for gross movements, such as the biceps brachii and gastrocnemius, may have as many as 500 muscle fibers in each motor unit.

Stimulation of a motor neuron produces a contraction in all the muscle fibers in a particular motor unit. Accordingly, the total tension in a muscle can be varied by adjusting the number of motor units that are activated. The process of increasing the number of active motor units is called **recruitment** and is determined by the needs of the body at a given time. The various motor neurons to a given muscle fire asynchronously, that is, while some are excited, others are inhibited. This means that while

chemicals determine whether a nerve impulse is passed on to a muscle, gland, or another nerve cell. The invaginated area of the sarcolemma under the axon terminal is referred to as a **synaptic gutter (trough),** and the space between the axon terminal and sarcolemma is known as a **synaptic cleft.** There are numerous folds of the sarcolemma along the synaptic gutter, called **subneural clefts,** which greatly increase the surface area of the synaptic gutter.

When a nerve impulse (action potential) reaches an axon terminal, it initiates a sequence that liberates neurotransmitter molecules from synaptic vesicles (and possibly the cytoplasm as well). The neurotransmitter released at neuromuscular junctions is **acetylcholine** (as'-ē-til-KŌ-lēn), or **ACh.** Upon its release, ACh diffuses across the synaptic cleft and combines with receptor sites on the sarcolemma of the muscle fiber. This combination alters the permeability of the sarcolemma and ultimately results some motor units are active, others are inactive; all the motor units are not contracting at the same time. Asynchronous firing of motor neurons prevents fatigue while maintaining contraction by allowing a brief rest for the inactive units. The alternating motor units relieve one another so smoothly that the contraction can be sustained for long periods. It also helps to maintain a state of partial contraction in a muscle, a phenomenon called muscle tone (described later in the chapter). And, recruitment is one factor responsible for producing smooth movements during a muscle contraction, rather than a series of jerky movements.

PHYSIOLOGY OF CONTRACTION

When a muscle fiber is relaxed (not contracting), the concentration of calcium ions (Ca^{2+}) in the sarcoplasm is low; the ions are stored in the sarcoplasmic reticulum. Also, in a relaxed muscle fiber, ATP concentration is high and ATP is attached to the ATP-binding sites of the myosin cross bridges. The myosin cross bridges are prevented from combining with actin of the thin myofilaments by the binding of the tropomyosin-troponin complex to actin and the binding of ATP to the myosin cross bridges. In other words, the muscle fiber remains relaxed as long as there are few calcium ions in the sarcoplasm, the tropomyosin-troponin complex is attached to actin, and ATP is attached to the ATP-binding site of the myosin cross bridges (Figure 10-6a).

When a nerve impulse reaches an axon terminal, a small amount of calcium enters the terminal, causing the synaptic vesicles (and possibly the cytoplasm as well) to release acetylcholine. It diffuses across the synaptic cleft and combines with receptor sites on the sarcolemma of the muscle fiber. Acetylcholine alters the sarcolemma and this alteration ultimately results in the development of a nerve impulse that spreads over the surface of the sarcolemma and into the transverse tubules. When the impulse is conveyed from the transverse tubules to the sarcoplasmic reticulum, the reticulum releases the calcium ions from storage into the sarcoplasm surrounding the myofilaments.

Calcium ions combine with troponin, causing it to undergo a structural change. This change in some way causes the troponin to pull on the tropomyosin strand and the tropomyosin-troponin complex moves into a groove between actin strands, thus exposing the myosin-binding sites on actin (Figure 10-6b).

Muscle contraction requires energy, as well as calcium ions. This energy is supplied by ATP. ATP is found in muscle fibers attached to ATP-binding sites on myosin cross bridges. When a nerve impulse stimulates a muscle fiber, an ATPase enzyme in the myosin cross bridge splits ATP into ADP + P. As a result, energy is released and it activates the myosin cross bridges.

Next, the activated myosin cross bridges combine with myosin-binding sites on actin (Figure 10-6c). Then, the energy stored in the myosin cross bridges is discharged, causing the myosin cross bridges to move toward the H zone in the center of a sarcomere. This movement of the myosin cross bridges, called the *power stroke,* causes

FIGURE 10-6 Mechanism of muscle contraction. (a) Relaxed state of a muscle fiber in which the tropomyosin-troponin complex covers the myosin-binding site on actin and the ATP-binding site of the myosin cross bridge is occupied. (b) Calcium ions combine with troponin, and the tropomyosin-troponin complex moves, exposing the myosin-binding site. (c) The energy from the splitting of ATP → ADP + P activates the myosin cross bridge, the cross bridge combines with the myosin-binding site, and the power stroke slides the actin myofilament past the myosin myofilament.

the thin actin myofilaments to slide past the thick myosin myofilaments. Once the power stroke is complete, ATP combines with the ATP-binding sites on the myosin cross bridges, resulting in the detachment of the myosin cross bridge from actin. A given myosin cross bridge then combines with another myosin-binding site further along the actin strand. Again, ATP is split and the cycle repeats itself. The myosin cross bridges keep moving back and forth like the cogs of a ratchet with each power stroke, moving the thin actin myofilaments toward the H zone. This continual movement causes the Z lines of a sarcomere to be drawn toward each other and the sarcomere shortens. The muscle fibers thus contract and the muscle itself contracts. During maximal muscle contractions, the distance between Z lines can be shortened to 50 percent of the resting length.

What happens when a muscle fiber goes from a contracted state back to a relaxed state? Acetylcholine is rapidly destroyed by an enzyme called *acetylcholinesterase* (*AChE*), which is found on the surfaces of the subneural clefts of the sarcolemma of muscle fibers. The absence of ACh terminates nerve impulse conduction from the axon terminals to the muscle fiber sarcolemma. After the nerve impulse ends, the calcium ions are actively transported from the sarcoplasm back into the sarcoplasmic reticulum for storage. This is accomplished by two proteins called *calsequestrin* and *calcium-ATPase* and involves an expenditure of some ATP. With the removal of calcium ions from the sarcoplasm, the tropomyosin-troponin complex is reattached to the actin strands, so that the myosin-binding sites of actin become covered and myosin cross bridges separate from actin. ATP is also required to detach the cross bridges. Since the myosin cross bridges are broken, the thin myofilaments slip back to their relaxed position. After the nerve impulse ends, ADP is resynthesized into ATP, which again attaches to the ATP-binding site of the myosin cross bridge. The sarcomeres are thereby returned to their resting lengths and the muscle fiber resumes its resting state. During extensibility, sarcomere length can be increased by about 20 percent of the resting length.

CLINICAL APPLICATION

Following death, certain chemical changes occur in muscle tissue that affect the status of the muscles. Due to a lack of ATP, myosin cross bridges remain attached to the actin myofilaments, thus preventing relaxation. The resulting condition, in which muscles are in a state of partial contraction, is called **rigor mortis** (rigidity of death). The time elapsing between death and the onset of rigor mortis varies greatly among individuals. Those who have had long, wasting illnesses undergo rigor mortis more quickly.

A summary of the events associated with the contraction and relaxation of a muscle fiber is presented in Exhibit 10-1.

EXHIBIT 10-1   SUMMARY OF EVENTS INVOLVED IN CONTRACTION AND RELAXATION OF A SKELETAL MUSCLE FIBER

1. A nerve impulse causes synaptic vesicles in motor axon terminals to release acetylcholine (ACh).
2. Acetylcholine diffuses across the synaptic cleft within the neuromuscular junction and initiates an impulse that spreads over the surface of the sarcolemma.
3. The impulse enters the transverse tubules and sarcoplasmic reticulum and stimulates the sarcoplasmic reticulum to release calcium ions from storage into the sarcoplasm.
4. Calcium ions combine with troponin, causing the tropomyosin-troponin complex to move, thus exposing the myosin-binding sites on actin.
5. The nerve impulse also causes ATP → ADP + P. The released energy activates myosin cross bridges, which combine with the exposed myosin-binding sites on actin, and causes the myosin cross bridges to move toward the H zone. This movement results in the sliding of the thin myofilaments past thick myofilaments.
6. The sliding draws the Z lines toward each other, the sarcomere shortens, the muscle fibers contract, and the muscle contracts.
7. Acetylcholine is inactivated by acetylcholinesterase (AChE), thus terminating nerve impulse conduction within the neuromuscular junction.
8. Once the nerve impulse is terminated, calcium ions are actively transported back into the sarcoplasmic reticulum by calsequestrin and calcium ATPase, using energy from ATP breakdown.
9. The low calcium concentration in the sarcoplasm permits the tropomyosin-troponin complex to reattach to actin. As a result, myosin-binding sites of actin become covered, myosin cross bridges separate from actin, ADP is resynthesized into ATP (which reattaches to the ATP-binding site of the myosin cross bridge), and the thin myofilaments return to their relaxed position.
10. Sarcomeres return to their resting lengths, muscle fibers relax, and the muscle relaxes.

ENERGY FOR CONTRACTION

Contraction of a muscle requires energy. When a nerve impulse stimulates a muscle fiber, ATP, in the presence of ATPase, breaks down into ADP + P, and energy is released. ATP is the immediate, direct source of energy for muscle contraction.

Like other cells of the body, muscle fibers synthesize ATP as follows:

$$ADP + P + Energy \rightarrow ATP$$

Unlike most other cells of the body, muscle fibers function in a discontinuous manner, that is, they alternate between virtual inactivity and great activity. The stores of ATP in muscle fibers are used up in a matter of seconds during bursts of maximal muscle contraction. Muscle fibers have several basic mechanisms for generating ATP continuously. The first involves a high-energy molecule called **phosphocreatine** (fos'-fō-KRĒ-a-tin) found in muscle fibers in concentrations about five times that of ATP.

Phosphocreatine can decompose to creatine and phosphate and, in the process, large amounts of energy are released:

Phosphocreatine $\rightarrow$ Creatine + Phosphate + Energy

The released energy is used to convert ADP to ATP. The transfer of energy from phosphocreatine to ATP takes place in a fraction of a second. Together phosphocreatine and ATP provide only enough energy for muscles to contract maximally for about 15 seconds.

When muscle activity is continued so that even the supply of phosphocreatine is depleted, then the source of energy is derived from the breakdown of glycogen. Glycogen, which is stored glucose, is always present in skeletal muscles and the liver. Its breakdown results in the resynthesis of tremendous quantities of ATP. In fact, the breakdown of glycogen provides sufficient ATP for maximal muscle contraction for several minutes. If exercise continues to the point where even most of the glycogen is catabolized, then muscle fibers can break down fats to provide the energy to resynthesize ATP. The supply of energy from fats to resynthesize ATP is almost inexhaustible since fats need only to be replaced at meals.

MUSCLE LENGTH AND FORCE OF CONTRACTION

Having already considered the sliding-filament theory of muscle contraction, we can now examine the relationship between muscle length and force of contraction (tension). As you already know, a skeletal muscle fiber contracts when myosin cross bridges of thick myofilaments connect with portions of thin myofilaments within a sarcomere. As it turns out, a muscle fiber develops its greatest tension when there is maximum overlap between thick and thin myofilaments. At this length, the optimal length, the maximum number of myosin cross bridges make contact with thin myofilaments to bring about the greatest force of contraction. As a muscle fiber is stretched, fewer and fewer myosin cross bridges make contact with thin myofilaments and the force of contraction progressively decreases. In fact, if a muscle fiber is stretched to 175 percent its optimal length, no myosin cross bridges attach to thin myofilaments and no contraction occurs. At lengths less than the optimum length, the force of contraction also decreases. This is because extreme shortening of sarcomeres causes thin myofilaments to overlap and thick myofilaments to crumple as they run into Z lines, resulting in fewer myosin cross bridge contacts with thin myofilaments. In general, changes in resting muscle fiber length above or below the optimum length rarely exceed 30 percent.

ALL-OR-NONE PRINCIPLE

The weakest stimulus from a neuron that can initiate a contraction is called a **threshold (liminal) stimulus.** A stimulus of lesser intensity, or one that cannot initiate contraction, is referred to as a **subthreshold (subliminal) stimulus.** According to the **all-or-none principle,** once a stimulus is applied, individual muscle fibers of a motor unit will contract to their fullest extent or will not contract at all, provided conditions remain constant. In other words, *muscle fibers* do not partly contract. The principle does not mean that the entire muscle must be either fully relaxed or fully contracted because, of the many motor units that comprise the entire muscle, some are contracting and some are relaxing. Thus, the muscle as a whole can have graded contractions. The strength of contraction may be decreased by fatigue, lack of nutrients, or lack of oxygen.

KINDS OF CONTRACTIONS

The various skeletal muscles are capable of producing different kinds of contractions, depending on the stimulation frequency.

Twitch

The **twitch contraction** is a rapid, jerky response to a single stimulus. Twitch contractions can be artificially brought about in muscles of an animal. A recording of a twitch contraction is the classic way to illustrate the different phases of one single contraction. Figure 10-7 is a graph of a twitch contraction. The record of a muscle contraction is called a **myogram.** Note that a brief period exists between application of the stimulus and the beginning of contraction: the **latent period.** During this time it is believed that Ca^{2+} ions are released from their reservoirs and the onset of myosin cross-bridge activity occurs. In frog muscle, it lasts about 10 milliseconds (msec) (10 msec = 0.01 sec). The second phase, the **contraction period,** lasts about 40 msec (0.04 sec) and is indicated by the upward tracing. The third phase, the **relaxation period,** lasts about 50 msec (0.05 sec) and is indicated by the downward tracing. The duration of these periods varies with the muscle involved. The latent period, contraction period, and relaxation period for muscles that move the eyes are very short. For muscles that move the leg, longer periods are required.

If additional stimuli are applied to the muscle after the initial stimulus, other responses may be noted. For example, if two stimuli are applied one immediately after

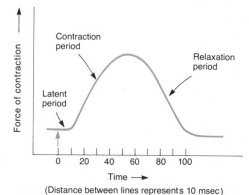

FIGURE 10-7 Myogram of a twitch contraction. The red arrow indicates the point at which the stimulus is applied.

the other, the muscle will respond to the first stimulus but not to the second. When a muscle fiber receives enough stimulation to contract, it temporarily loses its irritability and cannot contract again until its responsiveness is regained. This period of lost irritability is the **refractory period.** Its duration also varies with the muscle involved. Skeletal muscle has a short refractory period of 5 msec (0.005 sec). Cardiac muscle has a long refractory period of 300 msec (0.30 sec).

Tetanus

When two stimuli are applied and the second is delayed until the refractory period is over, the skeletal muscle will respond to both stimuli. In fact, if the second stimulus is applied after the refractory period, but before the muscle has finished relaxing, the second contraction will be stronger than the first. This phenomenon is called **wave summation** (Figure 10-8).

If a frog muscle is stimulated at a rate of 20 to 30 stimuli per second, the muscle can only partly relax between stimuli. As a result, the muscle maintains a sustained contraction called **incomplete (unfused) tetanus** (Figure 10-9a). Stimulation at an increased rate (35 to 50 stimuli per second) results in **complete (fused) tetanus,** a sustained contraction that lacks even partial relaxation (Figure 10-9b). Essentially, both kinds of tetanus result from the addition of calcium ions released from the sarcoplasmic reticulum by the second stimulus to the calcium ions still in the sarcoplasm from the first stimulus. This causes the rapid succession of separate twitches. Relaxation is either partial or does not occur at all. Voluntary contractions, such as contraction of the biceps brachii muscle in order to flex the forearm, are tetanic contractions. In fact, most of the work we do involves short-term tetanic contractions.

Treppe

Treppe is the condition in which a skeletal muscle contracts more forcefully in response to the same strength

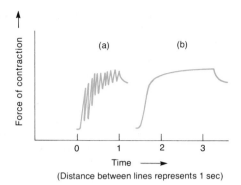

FIGURE 10-9 Myograms of (a) incomplete and (b) complete tetanus.

of stimulus after it has contracted several times. It is demonstrated by stimulating an isolated muscle with a series of stimuli at the same frequency and intensity, but not at a rate fast enough to produce tetanus. Suppose a series of liminal stimuli are introduced into a muscle. Time must be allowed for the muscle to undergo its latent period, contract, and relax, but this takes only 0.1 sec in frog muscle. If stimuli are repeated at intervals of 0.5 sec, the first few tracings on the myogram will show an increasing height with each contraction. This is treppe—the staircase phenomenon (Figure 10-10). It is the principle athletes use when warming up. After the first few stimuli, the muscle reaches its peak of performance and undergoes its strongest contraction. Treppe is thought to result from increased availability of calcium ions that relieve tropomyosin-troponin inhibition after several contractions.

Isotonic and Isometric

Isotonic (*iso* = equal; *tonos* = tension) **contractions** are probably familiar to you. As the contraction occurs, the muscle shortens and pulls on another structure, such as a bone, to produce movement. During such a contraction, the tension remains constant and energy is expended.

In an **isometric contraction,** there is a minimal shortening of the muscle. It remains nearly the same length,

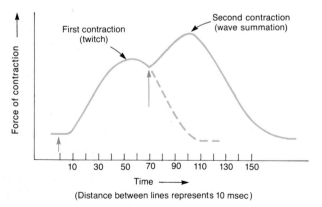

FIGURE 10-8 Myogram of wave summation. The second stimulus (long red arrow) is applied before the muscle has finished relaxing. The second contraction is even stronger than the first. The broken line represents the continuation of a twitch contraction.

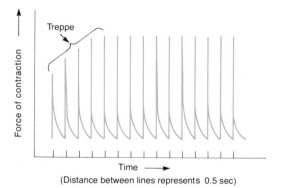

FIGURE 10-10 Myogram of treppe.

but the *tension* on the muscle increases greatly. Although isometric contractions do not result in body movement, energy is still expended. You can demonstrate such a contraction by carrying your books with your arm extended. The weight of the books pulls the arm downward, stretching the shoulder and arm muscles. The isometric contraction of the shoulder and arm muscles counteracts the stretch. The two forces—contraction and stretching—applied in opposite directions create the tension. The tension developed in a muscle for performing any kind of action depends on the total number of muscle fibers contracting at a time and the amount of tension each muscle fiber generates.

Although both isotonic and isometric training methods are able to increase muscular strength in relatively short periods, studies where direct comparisons are made tend to favor isotonic methods. The greatest advantage of isotonic exercise is that it works all the involved muscles over the entire range of a particular movement. Isometric exercise would require several separate and different maneuvers to work all the same muscles.

There is also the psychological advantage of performing isotonic exercises. The performer has some satisfaction in seeing the movements being accomplished, whereas isometrics are less satisfying and often considered boring since they are static.

Some experiments indicate that greater muscle enlargement (hypertrophy) and endurance result from isotonic exercise.

The main advantages of isometric exercises are their ease of implementation and time-saving features. Very little equipment is needed, and therefore larger groups of participants work out in a shorter period of time.

One caution needs to be added. It is well documented that the systolic and diastolic blood pressures increase considerably during isometric maneuvers. Therefore, they are a potentially dangerous form of exercise for rehabilitating cardiac patients and for older adults.

MUSCLE TONE

A muscle may be in a state of partial contraction even though the muscle fibers operate on an all-or-none basis. At any given time, some fibers in a muscle are contracted while others are relaxed. This contraction tightens a muscle, but there may not be enough fibers contracting at the time to produce movement. Asynchronous firing allows the contraction to be sustained for long periods.

A sustained partial contraction of portions of a skeletal muscle in response to activation of stretch receptors results in **muscle tone.** Tone is essential for maintaining posture. For example, when the muscles in the back of the neck are in tonic contraction, they keep the head in the anatomical position and prevent it from slumping forward onto the chest, but they do not apply enough force to pull the head back into hyperextension. The degree of tone in a skeletal muscle is monitored by receptors in the muscle called **muscle spindles.** They provide feedback information on tone to the brain so that adjustments can be made (Chapter 15).

CLINICAL APPLICATION

The term **flaccid** (FLAK-sid) is applied to muscles with less than normal tone. Such a loss of tone may be the result of damage or disease of the nerve that conducts a constant flow of impulses to the muscle. If the muscle does not receive impulses for an extended period of time, it may progress from flaccidity to **atrophy** (AT-rō-fē), which is a state of wasting away. Individual muscle cells decrease in size due to a progressive loss of myofibrils. Muscles may also become flaccid and atrophied if they are not used. Bedridden individuals and people with casts may experience atrophy because the flow of impulses to the inactive muscle is greatly reduced. If the nerve supply to a muscle is cut, it will undergo complete atrophy. In about 6 months to 2 years, the muscle will be one-quarter its original size and the muscle fibers will be replaced by fibrous tissue. The transition to fibrous tissue, when complete, cannot be reversed. Battery-operated transcutaneous muscle stimulations (TMSs) are used to maintain the strength of atrophied muscles immobilized by a cast by stimulating the muscles to contract. TMSs are also used to prevent atrophy of muscles whose motor nerves have been temporarily damaged by trauma or stroke. Some athletes even use TMSs to strengthen certain muscles.

Muscular hypertrophy (hī-PER-trō-fē) is the reverse of atrophy. It refers to an increase in the diameters of muscle fibers due to the production of more myofibrils, mitochondria, sarcoplasmic reticulum, nutrients (glycogen and triglycerides), and energy-supplying molecules (ATP and phosphocreatine). Hypertrophic muscles are capable of more forceful contractions. Weak muscular activity does not produce significant hypertrophy. It results from very forceful muscular activity or repetitive muscular activity at moderate levels.

TYPES OF SKELETAL MUSCLE FIBERS

All skeletal muscle fibers are not alike in structure or function. For example, skeletal muscle fibers vary in color depending on their content of **myoglobin,** a reddish pigment similar to hemoglobin in blood. Myoglobin stores oxygen until needed by mitochondria, the organelles in which ATP generation occurs. Skeletal muscle fibers that have a high myoglobin content are referred to as **red muscle fibers.** Conversely, skeletal muscle fibers that have a low content of myoglobin are called **white muscle fibers.** Red muscle fibers are smaller in diameter than white muscle fibers, and red muscle fibers have more mitochondria and more blood capillaries. White muscle fibers have a more extensive sarcoplasmic reticulum than red muscle fibers.

Skeletal muscle fibers contract with different velocities, depending on their ability to split ATP. Faster contracting fibers have greater ability to split ATP. In addition, skeletal muscle fibers vary with respect to the metabolic processes they use to generate ATP. They also differ in terms of the onset of fatigue, the mechanism of which is discussed at the end of the chapter.

On the basis of various structural and functional characteristics, skeletal muscle fibers are classified into three types:

1. Slow-twitch red fibers. These fibers contain large amounts of myoglobin, many mitochondria, and many blood capillaries. Slow-twitch red fibers have a high capacity to generate ATP by oxidative metabolic processes (Chapter 25). Such fibers also split ATP at a slow rate and, as a result, contraction velocity is slow. Slow-twitch red fibers are very resistant to fatigue.

2. Fast-twitch red fibers. These fibers contain very large amounts of myoglobin, very many mitochondria, and very many blood capillaries. Fast-twitch red fibers have a very high capacity for generating ATP by oxidative metabolic processes. Such fibers also split ATP at a very rapid rate and, as a result, contraction velocity is fast. Fast-twitch red fibers are resistant to fatigue, but not quite as much as slow-twitch red fibers.

3. Fast-twitch white fibers. These fibers have a low content of myoglobin, relatively few mitochondria, and relatively few blood capillaries. They do, however, contain large amounts of glycogen. Fast-twitch white fibers are geared to generating ATP by anaerobic metabolic processes, and the processes are not able to supply skeletal muscle fibers continuously with sufficient ATP. Accordingly, these fibers fatigue easily, but they split ATP at a fast rate so that contraction velocity is fast.

Most skeletal muscles of the body are a mixture of all three types of skeletal muscle fibers, but their proportion varies depending on the usual action of the muscle. For example, postural muscles of the neck, back, and legs have a higher proportion of slow-twitch red fibers. Muscles of the shoulders and arms are not constantly active but are used intermittently, usually for short periods of time, to produce large amounts of tension such as in lifting and throwing. These muscles have a higher proportion of fast-twitch white fibers. Leg muscles not only support the body but are also used for walking and running. Such muscles have higher proportions of slow-twitch red fibers and fast-twitch white fibers.

Even though most skeletal muscles are a mixture of all three types of skeletal muscle fibers, all the skeletal muscle fibers of any one motor unit are all the same. In addition, the different skeletal muscle fibers in a muscle may be used in various ways, depending on need. For example, if only a weak contraction is needed to perform a task, only slow-twitch red fibers are activated by their motor units. If a stronger contraction is needed, the motor units of fast-twitch red fibers are activated. And, if a maximal contraction is required, motor units of fast-twitch white fibers are activated as well. Activation of various motor units is determined in the brain and spinal cord.

Although the number of the different skeletal muscle fibers does not change, the characteristics of those present can be altered. Various types of exercises can bring about changes in the fibers in a skeletal muscle. Endurance-type exercises, such as running or swimming, cause a gradual transformation of fast-twitch white fibers into fast-twitch red fibers. The transformed muscle fibers show a slight increase in diameter, mitochondria, blood capillaries, and strength. Endurance exercises result in cardiovascular and respiratory changes that cause skeletal muscles to receive better supplies of oxygen and carbohydrates but do not contribute to muscle mass. On the other hand, exercises that require great strength for short periods of time, such as weight lifting, produce an increase in the size and strength of fast-twitch white fibers. The increase in size is due to increased synthesis of thin and thick myofilaments. The overall result is that the person develops large muscles.

CARDIAC MUSCLE TISSUE

The principal constituent of the heart wall is **cardiac muscle tissue.** Although it is striated in appearance like skeletal muscle, it is involuntary. The fibers of cardiac muscle tissue are roughly quadrangular and usually have only a single centrally located nucleus (see Exhibit 4-3). Skeletal muscle fibers contain several nuclei that are peripherally located. The thin sarcolemma of cardiac muscle fibers is similar to that of skeletal muscle, but the sarcoplasm is more abundant and the mitochondria are larger and more numerous. Cardiac muscle fibers have the same arrangement of actin and myosin and the same bands, zones, and lines as skeletal muscle fibers (Figure 10-11). Myofilaments in cardiac muscle fibers are not arranged in discrete myofibrils as in skeletal muscle. The transverse tubules of mammalian cardiac muscle are larger than those of skeletal muscle and are located at the Z lines rather than at the A-I band junctions as in skeletal muscle fibers. The sarcoplasmic reticulum of cardiac muscle is less well-developed than that in skeletal muscle.

Note in Exhibit 4-3 that cardiac muscle fibers branch and interconnect with each other. Recall that skeletal muscle fibers are arranged in parallel fashion. Cardiac muscle fibers are also physically connected to each other by cell junctions called *gap junctions,* which permit a direct conduction of nerve impulses from one fiber to another. Cardiac muscle fibers form two separate networks. The muscular walls and partition of the upper chambers of the heart (atria) compose one network. The muscular walls and partition of the lower chambers of the heart (ventricles) compose the other network. Each fiber in a network is separated from the next fiber by an irregular transverse thickening of the sarcolemma called an **intercalated** (in-TER-ka-lāt-ed) **disc.** These discs strengthen cardiac muscle tissue and aid in impulse conduction from one muscle fiber to another. When a single fiber of either network is stimulated, all the fibers in the network become stimulated as well. Thus each network contracts as a functional unit. As you will see in Chapter 20, when the fibers of the atria contract as a unit, blood moves into the ventricles. Then, when the ventricular fibers contract as a unit, blood is pumped into arteries.

Under normal conditions, cardiac muscle tissue contracts and relaxes rapidly, continuously, and rhythmically about 75 times a minute without stopping while a person is at rest. This is a major physiological difference between cardiac and skeletal muscle tissue. Accordingly, cardiac muscle tissue requires a constant supply of oxygen. En-

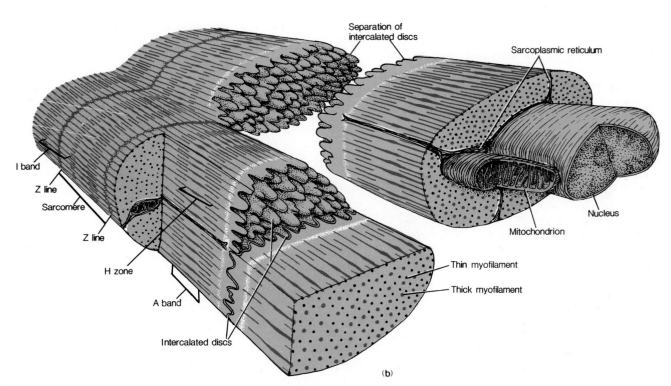

FIGURE 10-11 Histology of cardiac muscle tissue. (a) Diagram based on an electron micrograph showing several myofibrils. (b) Diagram based on an electron micrograph showing intercalated discs and related structures.

ergy generation occurs in large, numerous mitochondria. Another difference is the source of stimulation. Skeletal muscle tissue ordinarily contracts only when stimulated by a nerve impulse. In contrast, cardiac muscle tissue can contract without extrinsic (outside) nerve stimulation. Its source of stimulation is a conducting tissue of specialized intrinsic (internal) muscle within the heart. Nerve stimulation merely causes the conducting tissue to increase or decrease its rate of discharge. Some types of smooth muscle cells and nerve cells in the brain and spinal cord also possess spontaneous, rhythmical, self-excitation, a phenomenon referred to as autorhythmicity. It is discussed in detail in Chapter 20.

Another difference between cardiac and skeletal muscle tissue is that cardiac muscle tissue remains contracted (depolarized) 10 to 15 times longer than skeletal muscle tissue. This is because there is a prolonged delivery of Ca^{2+} ions into the sarcoplasm. In cardiac muscle fibers, Ca^{2+} ions are derived from sarcoplasmic reticulum (as in skeletal muscle fibers) and from extracellular fluids. Although the release of Ca^{2+} ions from the sarcoplasmic reticulum is fast, passage of Ca^{2+} ions from extracellular fluids through the sarcolemma is much slower, thus accounting for the prolonged contraction of cardiac muscle fibers.

Cardiac muscle tissue also has an extra-long refractory period, lasting several tenths of a second, which allows time for the heart to relax between beats. The long refractory period permits heart rate to be increased significantly, but prevents the heart itself from undergoing incomplete or complete tetanus. Tetanus of heart muscle would stop blood flow through the body and result in death.

SMOOTH MUSCLE TISSUE

Like cardiac muscle tissue, **smooth muscle tissue** is usually involuntary. However, it is nonstriated. Smooth muscle

fibers are considerably smaller than skeletal muscle fibers. A single fiber of smooth muscle tissue is about 5 to 10 μm in diameter and 30 to 200 μm long. It is spindle-shaped, and within the fiber is a single, oval, centrally located nucleus (Figure 10-12 and see Exhibit 4-3). The sarcoplasm of smooth muscle fibers contains thick myofilaments (longer than those of skeletal muscle fibers) and thin myofilaments, but not arranged as orderly as in striated muscle. In smooth muscle fibers, there are 10 to 15 thin myofilaments for each thick myofilament in the regions of myofilament overlap; in skeletal muscle fibers, the ratio is 2 to 1. Smooth muscle fibers also contain *intermediate filaments*. Since the various myofilaments have no regular pattern of organization and since there are no A or I bands of sarcomeres, smooth muscle fibers have no characteristic striations. Thus, the name smooth.

Intermediate filaments are attached to structures called *dense bodies*, which have characteristics similar to Z lines in striated muscle fibers. Some dense bodies are dispersed throughout the cytoplasm; others are attached to the sarcolemma. Bundles of intermediate filaments stretch from one dense body to another (Figure 10-12b). The sliding filament mechanism involving thick and thin myofilaments during contraction generates tension that is transmitted to intermediate filaments. These, in turn, pull on the dense bodies attached to the sarcolemma and there is a lengthwise shortening of the muscle fiber. Note in Figure 10-12b that shortening of the muscle fiber produces a bubblelike expansion of the sarcolemma.

Smooth muscle fibers also possess a less well-developed sarcoplasmic reticulum than skeletal muscle fibers. Small subsurface vesicles termed *caveolae* open onto the surface of the smooth muscle fiber and are thought to function in a way similar to the transverse tubules of striated fibers, namely, to carry nerve impulses into the fibers. Tubules of the sarcoplasmic reticulum are associated with longitudinal rows of caveolae.

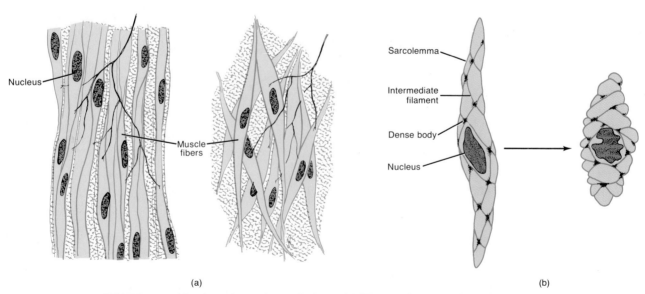

(a)   (b)

FIGURE 10-12 Histology of smooth muscle tissue. (a) Diagram of visceral (single-unit) smooth muscle tissue (left) and multiunit smooth muscle tissue (right). (b) Details of a fiber before contraction (left) and after contraction (right).

Two kinds of smooth muscle tissue, visceral and multiunit, are recognized (Figure 10-12a). The more common type is called **visceral (single-unit) muscle tissue.** It is found in wraparound sheets that form part of the walls of small arteries and veins and hollow viscera such as the stomach, intestines, uterus, and urinary bladder. The terms *smooth muscle tissue* and *visceral muscle tissue* are sometimes used interchangeably. The fibers in visceral muscle tissue are tightly bound together to form a continuous network. They contain gap junctions to facilitate nerve impulse conduction between fibers. When a neuron stimulates one fiber, the impulse travels over the other fibers so that contraction occurs in a wave over many adjacent fibers. Whereas skeletal muscle cells contract as individual units, visceral muscle cells contract in sequence as the impulse spreads from one cell to another.

The second kind of smooth muscle tissue, **multiunit smooth muscle tissue,** consists of individual fibers each with its own motor-nerve endings. Whereas stimulation of a single visceral muscle fiber causes contraction of many adjacent fibers, stimulation of a single multiunit fiber causes contraction of only that fiber. In this respect, multiunit muscle tissue is like skeletal muscle tissue. Multiunit smooth muscle tissue is found in the walls of large arteries, large airways to the lungs, in the arrector pili muscles that attach to hair follicles, and in the intrinsic muscles of the eye, such as the iris.

Although the principles of contraction are essentially the same in smooth and striated muscle tissue, smooth muscle tissue exhibits several important physiological differences. First of all, the duration of contraction of smooth muscle fibers is about 5 to 500 times longer than in skeletal muscle fibers. Smooth muscle fibers derive their Ca^{2+} ions from sarcoplasmic reticulum and extracellular fluids. Since there are no transverse tubules in smooth muscle fibers, it takes longer for Ca^{2+} ions to reach the deep filaments in the center of the fiber to trigger the contractile process, and this accounts, in part, for the slow activation and prolonged contraction of smooth muscle. Moreover, smooth muscle fibers do not contain calcium-binding troponin to facilitate myosin cross-bridge attachment to thin myofilaments. They use a different mechanism that takes more time and thus contributes to prolonged contraction.

No only do Ca^{2+} ions move slowly into deep filaments, they also move slowly out of the muscle fiber following contraction. This delays relaxation, and the prolonged stay of Ca^{2+} ions in the fibers provides for tone, the state of continued partial contraction. Smooth muscle tissue can undergo sustained, long-term tone, which is important in the gastrointestinal tract where the walls of the tract maintain a steady pressure on the contents of the tract, in the walls of blood vessels called arterioles that maintain a steady pressure on blood, and in the wall of the urinary bladder that maintains a steady pressure on urine.

Some smooth muscle fibers contract in response to nerve impulses from the autonomic nervous system. Thus, smooth muscle is normally not under voluntary control. Other smooth muscle fibers contract in response to hormones or local factors such as pH, oxygen and carbon dioxide levels, temperature, and ion concentrations.

Finally, unlike skeletal muscle fibers, smooth muscle fibers can stretch considerably without developing tension. For example, when smooth muscle fibers are stretched, they initially develop increased tension. However, there is an almost immediate decrease in the tension. This phenomenon is referred to as *stress-relaxation* and is important because it permits smooth muscle to accommodate great changes in size while still retaining the ability to contract effectively. Thus, the smooth muscle in the wall of hollow organs such as the stomach, intestines, and urinary bladder can stretch as the viscera distend, while the pressure within them remains the same.

A summary of the principal characteristics of the three types of muscle tissue is presented in Exhibit 10-2.

EXHIBIT 10-2   SUMMARY OF THE PRINCIPAL CHARACTERISTICS OF MUSCLE TISSUE

CHARACTERISTIC	SKELETAL MUSCLE	CARDIAC MUSCLE	SMOOTH MUSCLE
Location	Attached to bones.	Heart.	Walls of hollow viscera, blood vessels, iris, arrector pili.
Microscopic appearance	Striated, multinucleated, unbranched fibers.	Striated uninucleated, branched fibers with intercalated discs.	Nonstriated (smooth) uninucleated, spindle-shaped fibers.
Nervous control	Voluntary.	Involuntary.	Involuntary.
Sarcomeres	Yes.	Yes.	No.
Transverse tubules	Yes.	Yes.	No.
Gap junctions	Yes.	Yes.	Yes, in visceral smooth muscle.
Cell size	Large.	Large.	Small.
Source of calcium	Sarcoplasmic reticulum.	Sarcoplasmic reticulum and extracellular fluids.	Sarcoplasmic reticulum and extracellular fluids.
Speed of contraction	Fast.	Moderate.	Slow.

HOMEOSTASIS

Muscle tissue has a vital role in maintaining the body's homeostasis. Two examples are the relationship of muscle tissue to oxygen and to heat production.

OXYGEN DEBT

The energy required to convert ADP into ATP is produced by the breakdown of digested foods. The primary nutrient that usually supplies this energy is the sugar glucose. The reaction proceeds as follows:

Glucose $\rightarrow$ pyruvic acid + Energy (small amount)

When skeletal muscle is at rest, the breakdown of glucose is slow enough for blood to supply sufficient oxygen to participate in the complete catabolism of pyruvic acid. The waste products of this reaction are carbon dioxide and water. Since the catabolism of pyruvic acid requires oxygen, it is referred to as *aerobic*.

Pyruvic acid + $O_2 \rightarrow CO_2 + H_2O$ + Energy

When a skeletal muscle is contracting vigorously, the breakdown of glucose to pyruvic acid occurs too rapidly for the blood to supply sufficient oxygen for the pyruvic acid to be completely catabolized to carbon dioxide and water. Most of the pyruvic acid is catabolized to lactic acid. The conversion of pyruvic acid to lactic acid proceeds *anaerobically* (without oxygen). About 80 percent of the lactic acid diffuses from the skeletal muscles and is transported to the liver. There it is eventually synthesized to glycogen or glucose, which can be reused later. Some of the lactic acid accumulates in muscle tissue, however. Physiologists suspect that this acid is responsible for the feeling of muscle fatigue. Ultimately, the lactic acid in skeletal muscles, as well as that which has diffused into the blood, must be catabolized to CO_2 and H_2O. And for this conversion additional oxygen is needed— the **oxygen debt.** When vigorous activity is over, labored breathing continues in order to pay back the debt. Thus, the accumulation of lactic acid causes hard breathing and sufficient discomfort to curtail muscle activity until homeostasis is restored.

If a skeletal muscle or group of skeletal muscles is continuously stimulated for an extended period of time, the strength of contraction becomes progressively weaker until the muscle no longer responds. This condition is called **fatigue.** It results partly from the diminished availability of oxygen and partly from the toxic effects of lactic acid and carbon dioxide accumulated during exercise. The significant factors that contribute to muscle fatigue are:

1. Excessive activity, resulting in the accumulation of toxic products.
2. Malnutrition, resulting in insufficient supplies of glucose and, therefore, ATP.
3. Cardiovascular disturbances that impair the delivery of useful substances to muscles and the removal of waste products from muscles.
4. Respiratory disturbances that interfere with the oxygen supply and increase the oxygen debt.

HEAT PRODUCTION

The production of heat by skeletal muscles is an important homeostatic mechanism for maintaining normal body temperature. Of the total energy released during muscular contraction, only a small amount is used for mechanical work (contraction). As much as 85 percent can be released as heat, which is utilized to help maintain a normal body temperature.

Heat production by muscles may be divided into two phases: (1) **initial heat,** which is produced by the contraction and relaxation of a muscle, and (2) **recovery heat,** which is produced after relaxation. Initial heat is independent of O_2 and is associated with ATP breakdown. Recovery heat is associated with ATP restoration. It includes the anaerobic breakdown of glucose to pyruvic acid and pyruvic acid to lactic acid. It also includes the aerobic breakdown of pyruvic acid to CO_2 and H_2O and the aerobic conversion of lactic acid to CO_2 and H_2O.

AGING AND MUSCLE TISSUE

Beginning at about 30 years of age, there is a progressive loss of skeletal muscle mass that is largely replaced by fat. Accompanying the loss of muscle mass, there is a decrease in maximal strength and a diminishing of muscle reflexes.

DEVELOPMENTAL ANATOMY OF THE MUSCULAR SYSTEM

In this brief discussion of the development of the muscular system, we will concentrate mostly on skeletal muscles. Except for the muscles of the iris of the eyes and the arrector pili muscles attached to hairs, all muscles of the body are derived from **mesoderm.** As the mesoderm develops, a portion of it becomes arranged in dense columns on either side of the developing nervous system. These columns of mesoderm undergo segmentation into a series of blocks of cells called **somites** (Figure 10-13a). The first pair of somites appears on the twentieth day. Eventually, 44 pairs of somites are formed by the thirtieth day.

With the exception of the skeletal muscles of the head and extremities, *skeletal muscles* develop from the **mesoderm of somites.** Since there are very few somites in the head region of the embryo, most of the skeletal muscles there develop from the **general mesoderm** in the head region. The skeletal muscles of the limbs develop from masses of general mesoderm around developing bones in embryonic limb buds (origins of future extremities: see Figure 6-6a).

The cells of a somite are differentiated into three regions: (1) **myotome,** which forms most of the skeletal muscles; (2) **dermatome,** which forms the connective tissues, including the dermis, under the epidermis; and (3) **sclerotome,** which gives rise to the vertebrae (Figure 10-13b).

In the development of a skeletal muscle from a myotome of a somite, certain patterns emerge. For example, a myotome may split longitudinally into two or more portions. This represents the manner in which the trape-

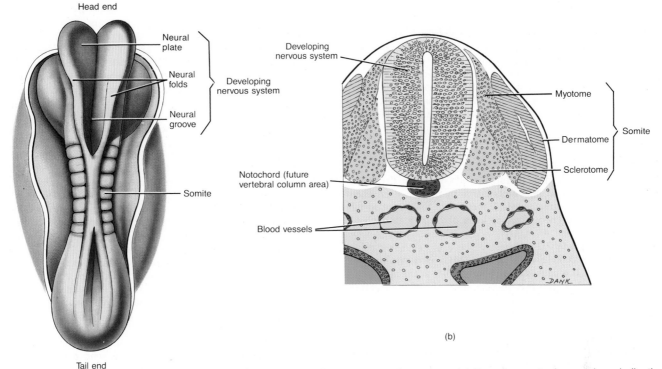

Neural plate

Neural folds

Neural groove

Developing nervous system

Somite

Tail end

(a)

Developing nervous system

Developing nervous system

Myotome

Dermatome

Sclerotome

Somite

Notochord (future vertebral column area)

Blood vessels

(b)

FIGURE 10-13 Development of the muscular system. (a) Dorsal aspect of an embryo indicating the location of somites. (b) Cross section of a portion of an embryo showing a somite.

zius muscle forms. Other myotomes split into two or more layers. Such a pattern represents how the external oblique, internal oblique, and transversus abdominis muscles develop. In other instances, several myotomes fuse to form a single muscle. Such an example is the rectus abdominis muscle. Muscles may also migrate, wholly or in part, from their sites of origin. The latissimus dorsi, for example, originates from the cervical myotomes, but extends all the way down to the thoracic and lumbar vertebrae and hip bone. Another trend that may be observed is a change in the direction of muscle fibers. Initially, muscle fibers in a myotome are parallel to the long axis of the embryo. However, nearly all developed skeletal muscles do not have fibers that are parallel to the long axis. One example is the external oblique. Finally, *fasciae, ligaments,* and *aponeuroses* may form as a result of degeneration of all or parts of myotomes.

Smooth muscle develops from **mesodermal cells** that migrate to and envelop the developing gastrointestinal tract and viscera.

Cardiac muscle develops from **mesodermal cells** that migrate to and envelop the developing heart while it is still in the form of primitive heart tubes (see Figure 20-14).

DISORDERS: HOMEOSTATIC IMBALANCES

Disorders of the muscular system are related to disruptions of homeostasis. The disorders may involve a lack of nutrients, accumulation of toxic products, disease, injury, disuse, or faulty nervous connections (innervations).

Fibrosis

The formation of fibrous (containing fibers) connective tissue in locations where it normally does not exist is called **fibrosis.** Skeletal and cardiac muscle fibers cannot undergo mitosis, and dead muscle fibers are normally replaced with fibrous connective tissue. Fibrosis, then, is often a consequence of muscle injury or degeneration.

Fibrositis

Fibrositis is an inflammation of fibrous tissue. If it occurs in the lumbar region, it is termed **lumbago** (lum-BĀ-gō). Fibrositis is a common condition characterized by pain, stiffness, or sore-

ness of fibrous tissue, especially in the muscle coverings. It is not destructive or progressive. It may persist for years or spontaneously disappear. Attacks of fibrositis may follow an injury, repeated muscular strain, or prolonged muscular tension.

"Charleyhorse"

Fibromyositis refers to a group of symptoms that include pain, tenderness, and stiffness of the joints, muscles, or adjacent structures. These symptoms ordinarily occur in various combinations. When the quadriceps femoris muscle of the thigh is involved, it is usually called **"charleyhorse."** It results from a contusion (bruising) and tearing of muscle fibers that produce a hematoma (collection of blood). "Charleyhorse" is characterized by the sudden onset of pain and is aggravated by motion. In some cases, local muscle spasms are noted. The condition is relieved with heat, massage, and rest and completely disappears, although occasionally it may become chronic or recur at frequent intervals.

Muscular Dystrophy

The term **muscular dystrophy** (*dystrophy* = degeneration) applies to a number of muscle-destroying diseases. The disease is characterized by degeneration of individual muscle fibers, which leads to a progressive atrophy of the skeletal muscle. Usually the voluntary skeletal muscles are weakened equally on both sides of the body, whereas the internal muscles, such as the diaphragm, are not affected. Histologically, the changes that occur include the variation in muscle fiber size, degeneration of fibers, and deposition of fat.

The cause of muscular dystrophy has been variously attributed to a genetic defect, faulty metabolism of potassium, protein deficiency, and inability of the body to utilize creatine.

There is no specific drug therapy for this disorder. Treatment involves attempts to prolong ambulation by muscle-strengthening exercises, corrective surgical measures, and appropriate braces. Patients are encouraged to keep physically active.

Myasthenia Gravis (MG)

Myasthenia (mī-as-THĒ-nē-a) **gravis (MG)** is a weakness of skeletal muscles. It is caused by an abnormality at the neuromuscular junction that prevents muscle fibers from contracting. Recall that motor neurons stimulate skeletal muscle fibers to contract by releasing acetylcholine. Myasthenia gravis is an autoimmune disorder caused by antibodies directed against ACh receptors of the muscle fiber sarcolemma. The antibodies bind to the receptors and hinder the attachment of ACh to the receptors (see Chapter 14). As the disease progresses, more neuromuscular junctions become affected. The muscle becomes increasingly weaker and may eventually cease to function altogether.

Myasthenia gravis is more common in females, occurring most frequently between the ages of 20 and 50. The muscles of the face and neck are most apt to be involved. Initial symptoms include a weakness of the eye muscles and difficulty in swallowing. Later, the individual has difficulty chewing and talking. Eventually, the muscles of the limbs may become involved. Death may result from paralysis of the respiratory muscles, but usually the disorder does not progress to this stage.

Anticholinesterase drugs such as neostigmine and pyridostigmine, derivatives of physostigmine, have been the primary treatment for the disease. They act as inhibitors of acetylcholinesterase, thus raising the level of ACh to bind with available receptors. More recently, steroid drugs, such as prednisone, have been used with great success to reduce antibody levels. Immunosuppressant drugs are also used to decrease the production of antibodies that interfere with normal muscle contraction. Another treatment involves *plasmapheresis,* a procedure that separates blood cells from the plasma that contains the unwanted antibodies. The blood cells are then mixed with a plasma substitute and pumped back into the individual. In some individuals, surgical removal of the thymus gland is indicated (thymectomy).

Abnormal Contractions

One kind of abnormal contraction of a muscle is **spasm,** a sudden, involuntary contraction of short duration. A **cramp** is a painful spasmodic contraction of a muscle. It is an involuntary, complete tetanic contraction. **Convulsions** are violent, involuntary tetanic contractions of an entire group of muscles. Convulsions occur when motor neurons are stimulated by fever, poisons, hysteria, or changes in body chemistry due to withdrawal of certain drugs. The stimulated neurons send many bursts of seemingly disordered impulses to muscle fibers. **Fibrillation** is the uncoordinated contraction of individual muscle fibers preventing the smooth contraction of the muscle. A **tic** is a spasmodic twitching made involuntarily by muscles that are ordinarily under voluntary control. Twitching of the eyelid and face muscles are examples. In general, tics are of psychological origin.

MEDICAL TERMINOLOGY

Electromyography or **EMG** (*electro* = electricity; *myo* = muscle; *graph* = to write) The recording and study of electrical changes that occur in muscle tissue.

Gangrene (*gangraena* = an eating sore) Death of a soft tissue, such as muscle, that results from interruption of its blood supply. It is caused by various species of *Clostridium,* bacteria that live anaerobically in the soil.

Myalgia (*algia* = painful condition) Pain in or associated with muscles.

Myoma (*oma* = tumor) A tumor consisting of muscle tissue.

Myomalacia (*malaco* = soft) Softening of a muscle.

Myopathy (*pathos* = disease) Any disease of muscle tissue.

Myosclerosis (*scler* = hard) Hardening of a muscle.

Myositis (*itis* = inflammation of) Inflammation of muscle fibers (cells).

Myospasm Spasm of a muscle.

Myotonia (*tonia* = tension) Increased muscular excitability and contractility with decreased power of relaxation; tonic spasm of the muscle.

Paralysis (*para* = beyond; *lyein* = to loosen) Loss or impairment of motor (muscular) function resulting from a lesion of nervous or muscular origin.

Trichinosis A myositis caused by the parasitic worm *Trichinella spiralis,* which may be found in the muscles of humans, rats, and pigs. People contract the disease by eating insufficiently cooked infected pork.

Volkmann's contracture (*contra* = against) Permanent contraction of a muscle due to replacement of destroyed muscle cells with fibrous tissue that lacks ability to stretch. Destruction of muscle cells may occur from interference with circulation caused by a tight bandage, a piece of elastic, or a cast.

Wryneck or **torticollis** (*tortus* = twisted; *collum* = neck) Spasmodic contraction of several superficial and deep muscles of the neck; produces twisting of the neck and an unnatural position of the head.

STUDY OUTLINE

Characteristics (p. 195)

1. Excitability is the property of receiving and responding to stimuli.
2. Contractility is the ability to shorten and thicken (contract).
3. Extensibility is the ability to be stretched (extended).
4. Elasticity is the ability to return to original shape after contraction or extension.

Functions (p. 195)

1. Through contraction, muscle tissue performs three important functions.
2. These functions are motion, maintenance of posture, and heat production.

Types (p. 195)

1. Skeletal muscle tissue is attached to bones. It is striated and voluntary.
2. Cardiac muscle tissue forms the wall of the heart. It is striated and involuntary.
3. Visceral muscle tissue is located in viscera. It is nonstriated (smooth) and involuntary.

Skeletal Muscle Tissue (p. 195)

Connective Tissue Components (p. 195)

1. The term fascia is applied to a sheet or broad band of fibrous connective tissue underneath the skin or around muscles and organs of the body.
2. Other connective tissue components are epimysium, covering the entire muscle; perimysium, covering fasiculi; and endomysium, covering fibers; all are extensions of deep fascia.
3. Tendons and aponeuroses are extensions of connective tissue beyond muscle cells that attach the muscle to bone or other muscle.

Nerve and Blood Supply (p. 197)

1. Nerves convey impulses for muscular contraction.
2. Blood provides nutrients and oxygen for contraction.

Histology (p. 197)

1. Skeletal muscle consists of fibers (cells) covered by a sarcolemma. The fibers contain sarcoplasm, nuclei, sarcoplasmic reticulum, and transverse tubules.
2. Each fiber contains myofibrils that consist of thin and thick myofilaments. The myofilaments are compartmentalized into sarcomeres.
3. Thin myofilaments are composed of actin, tropomyosin, and troponin; thick myofilaments consist mostly of myosin.
4. Projecting myosin heads are called cross bridges and contain actin- and ATP-binding sites.

Contraction (p. 199)

Sliding-Filament Theory (p. 199)

1. A nerve impulse travels over the sarcolemma and enters the transverse tubules and sarcoplasmic reticulum.
2. The nerve impulse leads to the release of calcium ions from the sarcoplasmic reticulum, triggering the contractile process.
3. Actual contraction is brought about when the thin myofilaments of a sarcomere slide toward each other.

Neuromuscular Junction (Motor End Plate) (p. 199)

1. A motor neuron transmits a nerve impulse to a skeletal muscle for contraction.
2. A neuromuscular junction (motor end plate) refers to an axon terminal of a motor neuron and the portion of the muscle fiber sarcolemma in close approximation with it.

Motor Unit (p. 200)

1. A motor neuron and the muscle fibers it stimulates form a motor unit.
2. A single motor unit may innervate up to 500 muscle fibers.

Physiology of Contraction (p. 201)

1. When a nerve impulse reaches an axon terminal, the synaptic vesicles of the terminal release acetylcholine (ACh), which initiates a nerve impulse in the muscle fiber sarcolemma. The impulse then travels into the transverse tubules and sarcoplasmic reticulum.
2. The transmitted impulse releases calcium ions that combine with troponin, causing it to pull on tropomyosin, thus exposing myosin-binding sites on actin.
3. The energy released from the breakdown of ATP causes myosin cross bridges to attach to actin, and their movement results in the sliding of thin myofilaments.

Energy for Contraction (p. 202)

1. The immediate, direct source of energy for muscle contraction is ATP.
2. Muscle fibers generate ATP continuously. This involves phosphocreatine and the metabolism of glycogen and fats.

Muscle Length and Force of Contraction (p. 203)

1. A muscle fiber develops its greatest tension when there is maximum overlap between thick and thin myofilaments (optimum length).
2. As a fiber is stretched or shortened, force of contraction continuously decreases.

All-or-None Principle (p. 203)

1. The weakest stimulus capable of causing contraction is a liminal (threshold) stimulus.
2. A stimulus not capable of inducing contraction is a subliminal (subthreshold) stimulus.
3. Muscle fibers of a motor unit contract to their fullest extent or not at all.

Kinds of Contractions (p. 203)

1. The various kinds of contractions are twitch, tetanus, treppe, isotonic, and isometric.
2. A record of a contraction is called a myogram. The refractory period is the time when a muscle has temporarily lost excitability. Skeletal muscles have a short refractory period. Cardiac muscle has a long refractory period.
3. Wave summation is the increased strength of a contraction resulting from the application of a second stimulus before the muscle has completely relaxed after a previous stimulus.

Muscle Tone (p. 205)

1. A sustained partial contraction of portions of a skeletal muscle results in muscle tone.
2. Tone is essential for maintaining posture.
3. Flaccidity is a condition of less than normal tone. Atrophy is a wasting away or decrease in size; hypertrophy is an enlargement or overgrowth.

Types of Skeletal Muscle Fibers (p. 205)

1. On the basis of structure and function, skeletal muscle fibers are classified as slow-twitch red, fast-twitch red, and fast-twitch white.
2. Most skeletal muscles contain a mixture of all three fiber types, their proportions varying with the usual action of the muscle.
3. Various exercises can modify the types of skeletal muscle fibers.

Cardiac Muscle Tissue (p. 206)

1. This muscle is found only in the heart. It is striated and involuntary.
2. The cells are quadrangular and usually contain a single centrally placed nucleus.
3. Compared to skeletal muscle tissue, cardiac muscle tissue has more sarcoplasm, more mitochondria, less well-developed

sarcoplasmic reticulum, and larger transverse tubules located at Z lines rather than at A-I band junctions. Myofilaments are not arranged in discrete myofibrils.

4. The fibers branch freely and are connected via gap junctions.
5. Intercalated discs provide strength and aid impulse conduction.
6. Unlike skeletal muscle tissue, cardiac muscle tissue contracts and relaxes rapidly, continuously, and rhythmically. Energy is supplied by glycogen and fat in large, numerous mitochondria.
7. Cardiac muscle tissue can contract without extrinsic stimulation and can remain contracted longer than skeletal muscle tissue.
8. Cardiac muscle tissue has a long refractory period which prevents tetanus.

Smooth Muscle Tissue (p. 208)

1. Smooth muscle is nonstriated and involuntary.
2. Smooth muscle fibers contain thick and thin myofilaments in a greater ratio than in skeletal muscle fibers; smooth muscle fibers also contain intermediate filaments, dense bodies (function as Z lines), and caveolae (function as transverse tubules).
3. Visceral (single-unit) smooth muscle is found in the walls of viscera. The fibers are arranged in a network.
4. Multiunit smooth muscle is found in blood vessels and the eye. The fibers operate singly rather than as a unit.
5. The duration of contraction and relaxation of smooth muscle is longer than in skeletal muscle.
6. Smooth muscle fibers contract in response to nerve impulses, hormones, and local factors.
7. Smooth muscle fibers can stretch considerably without developing tension.

Homeostasis (p. 210)

1. Oxygen debt is the amount of O_2 needed to convert accumulated lactic acid into CO_2 and H_2O. It occurs during strenuous exercise and is paid back by continuing to breathe rapidly after exercising. Until it is paid back, the homeostasis between muscular activity and oxygen requirements is not restored.
2. Muscle fatigue results from diminished availability of oxygen and toxic effects of carbon dioxide and lactic acid built up during exercise.
3. The heat given off during muscular contraction maintains the homeostasis of body temperature.

Aging and Muscle Tissue (p. 210)

1. At about 30 years of age, there is a progressive loss of skeletal muscle, which is replaced by fat.
2. There is also a decrease in muscle strength and diminished muscle reflexes.

Developmental Anatomy of the Muscular System (p. 210)

1. With few exceptions, muscles develop from mesoderm.
2. Skeletal muscles of the head and extremities develop from general mesoderm; the remainder of the skeletal muscles develop from the mesoderm of somites.

Disorders: Homeostatic Imbalances (p. 211)

1. Fibrosis is the formation of fibrous tissue where it normally does not exist; it frequently occurs in damaged muscle tissue.
2. Fibrositis is an inflammation of fibrous tissue. If it occurs in the lumbar region, it is called lumbago.
3. "Charleyhorse" refers to pain, tenderness, and stiffness of joints, muscles, and related structures in the thigh.
4. Muscular dystrophy is a hereditary disease of muscles characterized by degeneration of individual muscle cells.
5. Myasthenia gravis (MG) is a disease characterized by great muscular weakness and fatigability resulting from improper neuromuscular transmission.
6. Abnormal contractions include spasms, cramps, convulsions, fibrillations, and tics.

REVIEW QUESTIONS

1. How is the skeletal system related to the muscular system? What are the four characteristics of muscle tissue?
2. What are the three basic functions of the muscular system?
3. How can the three types of muscle tissue be distinguished on the basis of location, microscopic appearance, and nervous control?
4. What is fascia? Distinguish between superficial and deep fascia.
5. Define epimysium, perimysium, endomysium, tendon, and aponeurosis. Describe the nerve and blood supply to a skeletal muscle.
6. Describe the microscopic structure of skeletal muscle tissue.
7. What are some problems associated with muscle-building anabolic steroids?
8. In considering the contraction of skeletal muscle tissue, describe the following: neuromuscular junction (motor end plate), motor unit, role of calcium, sources of energy, and sliding-filament theory.
9. How does muscle length relate to force of contraction?
10. What is the all-or-none principle? Relate it to a threshold and subthreshold stimulus.
11. Define each of the following contractions and state the importance of each: twitch, tetanus, treppe, isotonic, and isometric.
12. What is a myogram? Describe the latent period, contraction period, and relaxation period of muscle contraction. Construct a diagram to illustrate your answer.
13. Define the refractory period. How does it differ between skeletal and cardiac muscle? What is wave summation?
14. What is muscle tone? Why is it important? Distinguish between atrophy and hypertrophy.
15. Compare slow-twitch red, fast-twitch red, and fast-twitch white fibers with respect to structure and function.
16. Compare skeletal, cardiac, and smooth muscle with regard to differences in structure and physiology.
17. Discuss each of the following as examples of muscle homeostasis: oxygen debt, fatigue, and heat production.
18. What do you think might be the relationship between shivering (uncontrolled muscular contractions) and body temperature? Can you relate sweating (cooling of the skin) after strenuous exercise to the homeostasis of body temperature?
19. Describe the effects of aging on muscle tissue.
20. Describe how the muscular system develops.
21. Define fibrosis, fibrositis, and "charleyhorse."
22. What is muscular dystrophy?
23. What is myasthenia gravis (MG)? In this disease, why do the muscles not contract normally?
24. Define each of the following abnormal muscular contractions: spasm, cramp, convulsion, fibrillation, and tic.
25. Refer to the glossary of medical terminology associated with the muscular system. Be sure that you can define each term.

The Muscular System

STUDENT OBJECTIVES

- Describe the relationship between bones and skeletal muscles in producing body movements.
- Define a lever and fulcrum and compare the three classes of levers on the basis of placement of the fulcrum, effort, and resistance.
- Identify the various arrangements of muscle fibers in a skeletal muscle and relate the arrangements to the strength of contraction and range of movement.
- Discuss most body movements as activities of groups of muscles by explaining the roles of the prime mover, antagonist, synergist, and fixator.
- Define the criteria employed in naming skeletal muscles.
- Identify the principal skeletal muscles in different regions of the body by name, origin, insertion, action, and innervation.
- Discuss the administration of drugs by intramuscular injection.

The term **muscle tissue** refers to all the contractile tissues of the body: skeletal, cardiac, and smooth muscle. The **muscular system,** however, refers to the *skeletal* muscle system: the skeletal muscle tissue and connective tissues that make up individual muscle organs, such as the biceps brachii muscle. Cardiac muscle tissue is located in the heart and is therefore considered part of the cardiovascular system. Smooth muscle tissue of the intestine is part of the digestive system, whereas smooth muscle tissue of the urinary bladder is part of the urinary system. In this chapter, we discuss only the muscular system. We will see how skeletal muscles produce movement, and we will describe the principal skeletal muscles.

HOW SKELETAL MUSCLES PRODUCE MOVEMENT

ORIGIN AND INSERTION

Skeletal muscles produce movements by exerting force on tendons, which in turn pull on bones. Most muscles cross at least one joint and are attached to the articulating bones that form the joint (Figure 11-1). When such a muscle contracts, it draws one articulating bone toward the other. The two articulating bones usually do not move equally in response to the contraction. One is held nearly in its original position because other muscles contract to pull it in the opposite direction or because its structure makes it less movable. Ordinarily, the attachment of a muscle tendon to the stationary bone is called the **origin.** The attachment of the other muscle tendon to the movable bone is the **insertion.** A good analogy is a spring on a door. The part of the spring attached to the door represents the insertion; the part attached to the frame is the origin. The fleshy portion of the muscle between the tendons of the origin and insertion is called the **belly (gaster).** The origin is usually proximal and the insertion distal, especially in the extremities. In addition, muscles that move a body part generally do not cover the moving part. Figure 11-1a shows that, although contraction of the biceps brachii muscle moves the forearm, the belly of the muscle lies over the humerus.

LEVER SYSTEMS AND LEVERAGE

In producing a body movement, bones act as levers and joints function as fulcrums of these levers. A **lever** may be defined as a rigid rod that moves about on some fixed point called a **fulcrum.** A fulcrum may be symbolized as △. A lever is acted on at two different points by two different forces: the *resistance* Ⓡ and the *effort* (E). The resistance may be regarded as a force to be overcome, whereas the effort is the force exerted to overcome the resistance. The resistance may be the weight of the body part that is to be moved. The muscular effort (contraction) is applied to the bone at the insertion of the muscle and produces motion. Consider the biceps brachii flexing the forearm at the elbow as a weight is lifted (Figure 11-

1b). When the forearm is raised, the elbow is the fulcrum. The weight of the forearm plus the weight in the hand is the resistance. The shortening of the biceps brachii pulling the forearm up is the effort.

Levers are categorized into three types according to the positions of the fulcrum, the effort, and the resistance.

1. In **first-class levers,** the fulcrum is between the effort and resistance (Figure 11-2a). An example of a first-class lever is a seesaw. There are not many first-class levers in the body. One example is the head resting on the vertebral column. When the head is raised, the facial portion of the skull is the resistance. The joint between the atlas and occipital bone (atlantooccipital joint) is the fulcrum. The contraction of the muscles of the back is the effort.

2. Second-class levers have the fulcrum at one end, the effort at the opposite end, and the resistance between them (Figure 11-2b). They operate like a wheelbarrow. Most authorities agree that there are very few examples of second-class levers in the body. One example is raising the body on the toes. The body is the resistance, the ball of the foot is the fulcrum, and the contraction of the calf muscles to pull the heel upward is the effort.

3. Third-class levers consist of the fulcrum at one end, the resistance at the opposite end, and the effort between them (Figure 11-2c). They are the most common levers in the body. An example is flexing the forearm at the elbow. As we have seen, the weight of the forearm is the resistance, the contraction of the biceps brachii is the effort, and the elbow joint is the fulcrum.

Leverage, the mechanical advantage gained by a lever, is largely responsible for a muscle's strength and range of movement. Consider strength first. Suppose we have two muscles of the same strength crossing and acting on a joint. Assume also that one is attached farther from the joint and one is nearer. The muscle attached farther will produce the more powerful movement. Thus strength of movement depends on the placement of muscle attachments.

In considering range of movement, again assume that we have two muscles of the same strength crossing and acting on a joint and that one is attached farther from the joint than the other. The muscle inserting closer to the joint will produce the greater range of movement. Thus range of movement also depends on the placement of muscle attachments. Since strength increases with distance from the joint and range of movement decreases, maximal strength and maximal range are incompatible; strength and range vary inversely.

ARRANGEMENT OF FASCICULI

Recall from Chapter 10 that skeletal muscle fibers are arranged within the muscle in bundles called fasciculi (fascicles). The muscle fibers are arranged in a parallel fashion within each bundle, but the arrangement of the fasciculi with respect to the tendons may take one of four characteristic patterns.

The first pattern is called **parallel.** The fasciculi are parallel with the longitudinal axis and terminate at either end in flat tendons. The muscle is typically quadrilateral

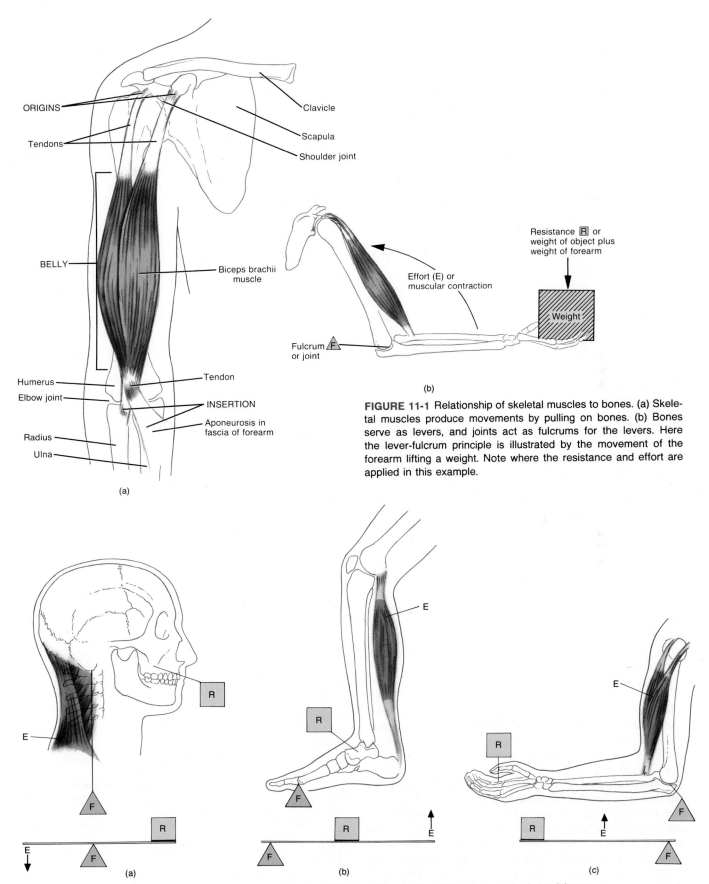

FIGURE 11-1 Relationship of skeletal muscles to bones. (a) Skeletal muscles produce movements by pulling on bones. (b) Bones serve as levers, and joints act as fulcrums for the levers. Here the lever-fulcrum principle is illustrated by the movement of the forearm lifting a weight. Note where the resistance and effort are applied in this example.

FIGURE 11-2 Classes of levers. Each is defined on the basis of the placement of the fulcrum, effort, and resistance. (a) First-class lever. (b) Second-class lever. (c) Third-class lever.

in shape. An example is the stylohyoid muscle (see Figure 11-7). In a modification of the parallel arrangement, called *fusiform,* the fasciculi are nearly parallel with the longitudinal axis and terminate at either end in flat tendons, but the muscle tapers toward the tendons, where the diameter is less than that of the belly. An example is the biceps brachii muscle (see Figure 11-17).

The second distinct pattern is called **convergent.** A broad origin of fasciculi converges to a narrow, restricted insertion. Such a pattern gives the muscle a triangular shape. An example is the deltoid muscle (see Figure 11-16).

The third distinct pattern is referred to as **pennate.** The fasciculi are short in relation to the entire length of the muscle and the tendon extends nearly the entire length of the muscle. The fasciculi are directed obliquely toward the tendon like the plumes of a feather. If the fasciculi are arranged on only one side of a tendon, as in the extensor digitorum longus muscle, the muscle is referred to as *unipennate* (see Figure 11-22). If the fasciculi are are arranged on both sides of a centrally positioned tendon, as in the rectus femoris muscle, the muscle is referred to as *bipennate* (see Figure 11-21).

The final distinct pattern is referred to as **circular.** The fasciculi are arranged in a circular pattern and enclose an orifice. An example is the orbicularis oris muscle (see Figure 11-4).

Fascicular arrangement is correlated with the power of a muscle and range of movement. When a muscle fiber contracts, it shortens to a length just slightly greater than half of its resting length. Thus, the longer the fibers in a muscle, the greater the range of movement it can produce. By contrast, the strength of a muscle depends on the total number of fibers it contains, since a short fiber can contract as forcefully as a long one. Because a given muscle can contain either a small number of long fibers or a large number of short fibers, fascicular arrangement represents a compromise between power and range of movement. Pennate muscles, for example, have a large number of fasciculi distributed over their tendons, giving them greater power, but a smaller range of movement. Parallel muscles, on the other hand, have comparatively few fasciculi that extend the length of the muscle. Thus, they have a greater range of movement but less power.

GROUP ACTIONS

Most movements are coordinated by several skeletal muscles acting in groups rather than individually and most skeletal muscles are arranged in opposing pairs at joints, that is, flexors–extensors, abductors–abbductors, and so on. Consider flexing the forearm at the elbow, for example. A muscle that causes a desired action is referred to as the **prime mover (agonist).** In this instance, the biceps brachii is the prime mover (see Figure 11-17). Simultaneously with the contraction of the biceps brachii, another muscle, called the **antagonist,** is relaxing. In this movement, the triceps brachii serves as the antagonist (see Figure 11-17). The antagonist has an effect opposite to that of the prime mover, that is, the antagonist relaxes and yields to the movement of the prime mover. You

should not assume, however, that the biceps brachii is always the prime mover and the triceps brachii is always the antagonist. For example, when extending the forearm at the elbow, the triceps brachii serves as the prime mover and the biceps brachii functions as the antagonist; their roles are reversed. Note that if the prime mover and antagonist contracted simultaneously with equal force, there would be no movement.

In addition to prime movers and antagonists, most movements also involve muscles called **synergists,** which serve to steady a movement, thus preventing unwanted movements and helping the prime mover function more efficiently. For example, flex your hand at the wrist and then make a fist. Note how difficult this is to do. Now, extend your hand at the wrist and then make a fist. Note how much easier it is to clench your fist. In this case, the extensor muscles of the wrist act as synergists in cooperation with the flexor muscles of the fingers acting as prime movers. The extensor muscles of the fingers serve as antagonists (see Figure 11-18).

Some muscles in a group also act as **fixators,** which stabilize the origin of the prime mover so that the prime mover can act more efficiently. For example, the scapula is a freely movable bone in the pectoral (shoulder) girdle that serves as an origin for several muscles that move the arm. However, for the scapula to serve as a firm origin for muscles that move the arm, it must be held steady. This is accomplished by fixator muscles that hold the scapula firmly against the back of the chest. In abduction of the arm, the deltoid muscle serves as the prime mover, whereas fixators (pectoralis minor, rhomboids, subclavius, and serratus anterior) hold the scapula firmly (see Figure 11-15). These fixators stabilize the origin of the deltoid muscle on the scapula while the insertion of the muscle pulls on the humerus to abduct the arm. Under different conditions and depending on the movement, many muscles act, at various times, as prime movers, antagonists, synergists, or fixators.

NAMING SKELETAL MUSCLES

The names of most of the nearly 700 skeletal muscles are based on several types of characteristics. Learning the terms used to indicate specific characteristics will help you remember the names of muscles.

1. Muscle names may indicate the **direction of the muscle fibers.** *Rectus* fibers usually run parallel to the midline of the body. *Transverse* fibers run perpendicular to the midline. *Oblique* fibers are diagonal to the midline. Muscles named according to directions of fibers include the rectus abdominis, transversus abdominis, and external oblique.

2. A muscle may be named according to **location.** The temporalis is near the temporal bone. The tibialis anterior is near the tibia.

3. Size is another characteristic. The term *maximus* means largest, *minimus* smallest, *longus* long, and *brevis* short. Examples include the gluteus maximus, gluteus minimus, adductor longus, and peroneus brevis.

4. Some muscles are named for their **number of origins.** The biceps brachii has two origins, the triceps brachii three, and the quadriceps femoris four.

5. Other muscles are named on the basis of **shape.** Common examples include the deltoid (meaning triangular) and trapezius (meaning trapezoid).

6. Muscles may be named after their **origin** and **insertion.** The sternocleiodomastoid originates on the sterum and clavicle and inserts at the mastoid process of the temporal bone; the stylohyoid originates on the styloid process of the temporal bone and inserts at the hyoid bone.

7. Still another characteristic of muscles used for naming is **action.** Exhibit 11-1 lists the principal actions of muscles, their definitions, and examples of muscles that perform the actions. For convenience, the actions are grouped as antagonistic pairs where possible.

PRINCIPAL SKELETAL MUSCLES

Exhibits 11-2 through 11-21 list the principal muscles of the body with their origins, insertions, actions, and innervations. (By no means have all the muscles of the body been included.) Refer to Chapters 7 and 8 to review bone markings, since they serve as points of origin and insertion for muscles. The muscles are divided into groups according to the part of the body on which they act. If you have mastered the naming of the muscles, their actions will have more meaning. Figure 11-3 shows general anterior and posterior views of the muscular system. Do not try to memorize all these muscles yet. As you study groups of muscles in the following exhibits, refer to Figure 11-3 to see how each group is related to all others.

The figures that accompany the exhibits contain superficial and deep, anterior and posterior, or medial and lateral views to show each muscle's position as clearly as possible. An attempt has been made to show the relationship of the muscles under consideration to other muscles in the area you are studying.

EXHIBIT 11-1   PRINCIPAL ACTIONS OF MUSCLES

ACTION	DEFINITION	EXAMPLE
Flexor	Usually decreases the anterior angle at a joint; some decrease the posterior angle.	Flexor carpi radialis.
Extensor	Usually increases the anterior angle at a joint; some increase the posterior angle.	Extensor carpi ulnaris.
Abductor	Moves a bone away from the midline.	Abductor hallucis longus.
Adductor	Moves a bone closer to the midline.	Adductor longus.
Levator	Produces an upward movement.	Levator scapulae.
Depressor	Produces a downward movement.	Depressor labii inferioris.
Supinator	Turns the palm upward or anteriorly.	Supinator.
Pronator	Turns the palm downward or posteriorly.	Pronator teres.
Sphincter	Decreases the size of an opening.	External anal sphincter.
Tensor	Makes a body part more rigid.	Tensor fasciae latae.
Rotator	Moves a bone around its longitudinal axis.	Obturator.

Orbicularis oculi
Frontalis
Masseter
Buccinator
Sternocleidomastoid
Orbicularis oris
Platysma
Trapezius
Deltoid
Pectoralis major
Latissimus dorsi
Serratus anterior
Biceps brachii
External oblique
Brachioradialis
Extensor carpi radialis longus
Pronator teres
Extensor digitorum communis
Extensor carpi ulnaris

Flexor digitorum superficialis
Flexor carpi radialis
Flexor carpi ulnaris
Brachioradialis
Brachialis
Triceps brachii
Coracobrachialis
Teres major
Serratus anterior
Latissimus dorsi

Rectus abdominis
Iliacus
Psoas major
Pectineus
Tensor fasciae latae
Adductor longus
Sartorius
Gracilis
Rectus femoris
Vastus lateralis
Vastus medialis

Tibialis anterior
Peroneus longus
Extensor digitorum longus

Gastrocnemius
Soleus
Flexor digitorum longus
Tibialis posterior

(a)

FIGURE 11-3 Principal superficial skeletal muscles. (a) Anterior view.

Occipitalis

Splenius capitis

Trapezius

Deltoid

Infraspinatus

Teres minor

Teres major

Rhomboideus major

Triceps brachii

Brachioradialis

Extensor carpi radialis longus

Flexor carpi ulnaris

Extensor digitorum communis

Extensor carpi ulnaris

Extensor carpi ulnaris

Extensor carpi radialis longus

Brachialis

Biceps brachii

Triceps brachii

Deltoid

Latissimus dorsi

External oblique

Gluteus medius

Gluteus maximus

Iliotibial tract

Vastus lateralis

Biceps femoris

Adductor magnus

Gracilis

Semitendinosus

Semimembranosus

Gastrocnemius

Soleus

Flexor digitorum longus

Peroneus longus

Peroneus brevis

(b)

FIGURE 11-3 (*Continued*) (b) Posterior view.

EXHIBIT 11-2   MUSCLES OF FACIAL EXPRESSION (Figure 11-4)

MUSCLE	ORIGIN	INSERTION	ACTION	INNERVATION
Epicranius (*epi* = over; *crani* = skull)	This muscle is divisible into two portions: the frontalis over the frontal bone and the occipitalis over the occipital bone. The two muscles are united by a strong aponeurosis, the galea aponeurotica, which covers the superior and lateral surfaces of the skull.			
Frontalis (*front* = forehead)	Galea aponeurotica.	Skin superior to supraorbital line.	Draws scalp forward, raises eyebrows, and wrinkles skin of forehead horizontally.	Facial (VII) nerve.
Occipitalis (*occipito* = base of skull)	Occipital bone and mastoid process of temporal bone.	Galea aponeurotica.	Draws scalp backward.	Facial (VII) nerve.
Orbicularis oris (*orb* = circular; *or* = mouth)	Muscle fibers surrounding opening of mouth.	Skin at corner of mouth.	Closes lips, compresses lips against teeth, protrudes lips, and shapes lips during speech.	Facial (VII) nerve.
Zygomaticus major (*zygomatic* = cheek bone; *major* = greater)	Zygomatic bone.	Skin at angle of mouth and orbicularis oris.	Draws angle of mouth upward and outward as in smiling or laughing.	Facial (VII) nerve.
Levator labii superioris (*levator* = raises or elevates; *labii* = lip; *superioris* = upper)	Superior to infraorbital foramen of maxilla.	Skin at angle of mouth and orbicularis oris.	Elevates (raises) upper lip.	Facial (VII) nerve.
Depressor labii inferioris (*depressor* = depresses or lowers; *inferioris* = lower)	Mandible.	Skin of lower lip.	Depresses (lowers) lower lip.	Facial (VII) nerve.

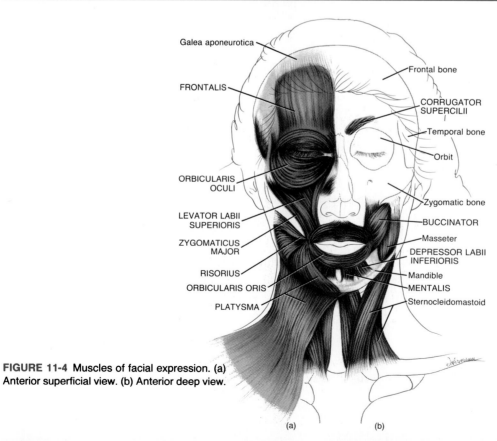

FIGURE 11-4 Muscles of facial expression. (a) Anterior superficial view. (b) Anterior deep view.

EXHIBIT 11-2 (Continued)

MUSCLE	ORIGIN	INSERTION	ACTION	INNERVATION
Buccinator (*bucc* = cheek)	Alveolar processes of maxilla and mandible and pterygomandibular raphe (fibrous band extending from the pterygoid hamulus to the mandible).	Orbicularis oris.	Major cheek muscle; compresses cheek as in blowing air out of mouth and causes cheeks to cave in, producing the action of sucking.	Facial (VII) nerve.
Mentalis (*mentum* = chin)	Mandible.	Skin of chin.	Elevates and protrudes lower lip and pulls skin of chin up as in pouting.	Facial (VII) nerve.
Platysma (*platy* = flat, broad)	Fascia over deltoid and pectoralis major muscles.	Mandible, muscles around angle of mouth, and skin of lower face.	Draws outer part of lower lip downward and backward as in pouting; depresses mandible.	Facial (VII) nerve.
Risorius (*risor* = laughter)	Fascia over parotid (salivary) gland.	Skin at angle of mouth.	Draws angle of mouth laterally as in tenseness.	Facial (VII) nerve.
Orbicularis oculi (*ocul* = eye)	Medial wall of orbit.	Circular path around orbit.	Closes eye.	Facial (VII) nerve.
Corrugator supercilii (*corrugo* = to wrinkle; *supercilium* = eyebrow)	Medial end of superciliary arch of frontal bone.	Skin of eyebrow.	Draws eyebrow downward as in frowning.	Facial (VII) nerve.
Levator palpebrae superioris (*palpebrae* = eyelids) (see Figure 11-6b)	Roof of orbit (lesser wing of sphenoid bone).	Skin of upper eyelid.	Elevates upper eyelid.	Oculomotor (III) nerve.

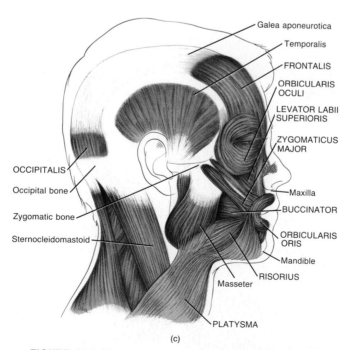

(c)

FIGURE 11-4 (*Continued*) (c) Right lateral superficial view.

EXHIBIT 11-3   MUSCLES THAT MOVE THE LOWER JAW (Figure 11-5)

MUSCLE	ORIGIN	INSERTION	ACTION	INNERVATION
Masseter (*maseter* = chewer)	Maxilla and zygomatic arch.	Angle and ramus of mandible.	Elevates mandible as in closing mouth and protracts (protrudes) mandible.	Mandibular branch of trigeminal (V) nerve.
Temporalis (*tempora* = temples)	Temporal bone.	Coronoid process of mandible.	Elevates and retracts mandible.	Temporal nerve from mandibular division of trigeminal (V) nerve.
Medial pterygoid (*medial* = closer to midline; *pterygoid* = like a wing; pterygoid plate of sphenoid bone)	Medial surface of lateral pterygoid plate of sphenoid; maxilla.	Angle and ramus of mandible.	Elevates and protracts mandible and moves mandible from side to side.	Mandibular branch of trigeminal (V) nerve.
Lateral pterygoid (*lateral* = farther from midline)	Greater wing and lateral surface of lateral pterygoid plate of sphenoid.	Condyle of mandible; temporomandibular articulation.	Protracts mandible, opens mouth, and moves mandible from side to side.	Mandibular branch of trigeminal (V) nerve.

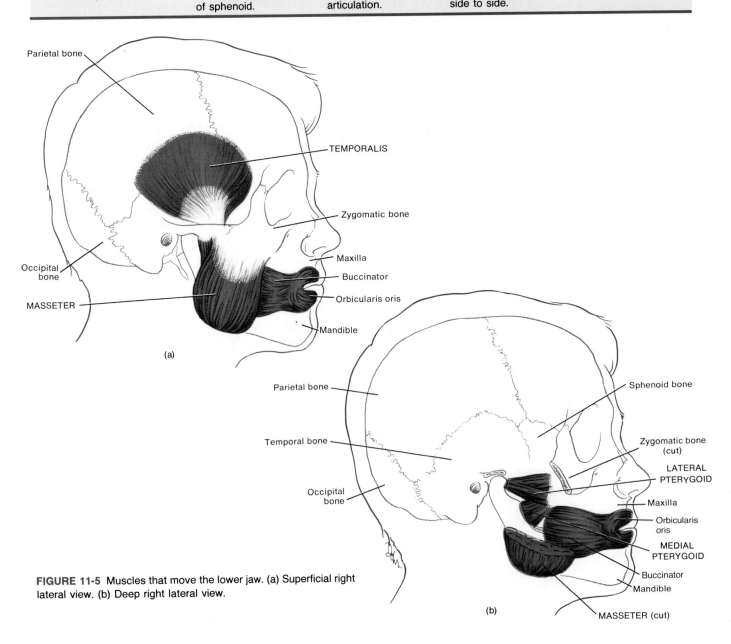

FIGURE 11-5 Muscles that move the lower jaw. (a) Superficial right lateral view. (b) Deep right lateral view.

EXHIBIT 11-4   MUSCLES THAT MOVE THE EYEBALLS—EXTRINSIC MUSCLES* (Figure 11-6)

MUSCLE	ORIGIN	INSERTION	ACTION	INNERVATION
Superior rectus (*superior* = above; *rectus* = in this case, muscle fibers running parallel to long axis of eyeball)	Tendinous ring attached to bony orbit around optic foramen.	Superior and central part of eyeball.	Rolls eyeball upward.	Oculomotor (III) nerve.
Inferior rectus (*inferior* = below)	Same as above.	Inferior and central part of eyeball.	Rolls eyeball downward.	Oculomotor (III) nerve.
Lateral rectus	Same as above.	Lateral side of eyeball.	Rolls eyeball laterally.	Abducens (VI) nerve.
Medial rectus	Same as above.	Medial side of eyeball.	Rolls eyeball medially.	Oculomotor (III) nerve.
Superior oblique (*oblique* = in this case, muscle fibers running diagonally to long axis of eyeball)	Same as above.	Eyeball between superior and lateral recti.	Rotates eyeball on its axis; directs cornea downward and laterally; note that it moves through a ring of fibrocartilaginous tissue called the trochlea (*trochlea* = pulley).	Trochlear (IV) nerve.
Inferior oblique	Maxilla (front of orbital cavity).	Eyeball between inferior and lateral recti.	Rotates eyeball on its axis; directs cornea upward and laterally.	Oculomotor (III) nerve.

* Muscles situated on the outside of the eyeball.

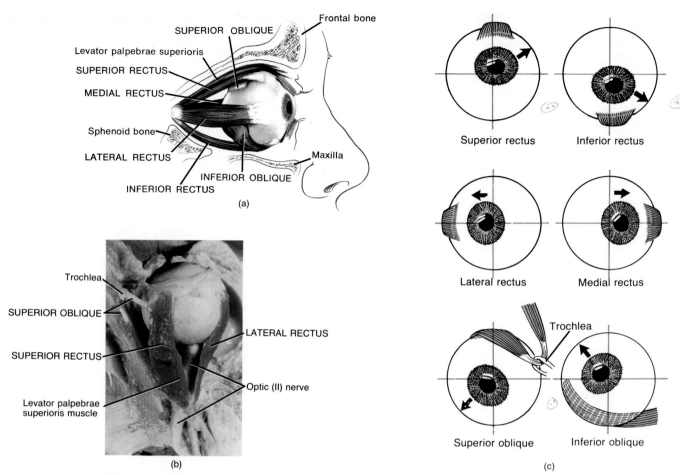

FIGURE 11-6 Extrinsic muscles of the eyeball. (a) Diagram of lateral view of the right eyeball. (b) Photograph of superior view of the right eyeball. (Courtesy of C. Yokochi and J. W. Rohen, *Photographic Anatomy of the Human Body*, 2nd ed., 1978, IGAKU-SHOIN, Ltd., Tokyo, New York.) (c) Movements of the right eyeball in response to contraction of the various extrinsic muscles.

EXHIBIT 11-5   MUSCLES THAT MOVE THE TONGUE (Figure 11-7)

MUSCLE	ORIGIN	INSERTION	ACTION	INNERVATION
Genioglossus (*geneion* = chin; *glossus* = tongue)	Mandible.	Undersurface of tongue and hyoid bone.	Depresses tongue and thrusts it forward (protraction).	Hypoglossal (XII) nerve.
Styloglossus (*stylo* = stake or pole; styloid process of temporal bone)	Styloid process of temporal bone.	Side and undersurface of tongue.	Elevates tongue and draws it backward (retraction).	Hypoglossal (XII) nerve.
Palatoglossus* (*palato* = palate)	Anterior surface of soft palate.	Side of tongue.	Elevates tongue and draws soft palate down on tongue.	Hypoglossal (XII) nerve.
Hyoglossus	Body of hyoid bone.	Side of tongue.	Depresses tongue and draws down its sides.	Hypoglossal (XII) nerve.

* Not illustrated.

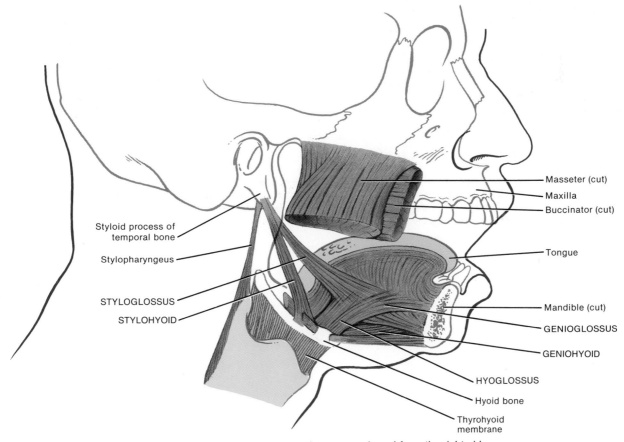

FIGURE 11-7 Muscles that move the tongue viewed from the right side.

EXHIBIT 11-6   MUSCLES OF THE PHARYNX (Figure 11-8)

MUSCLE	ORIGIN	INSERTION	ACTION	INNERVATION
Inferior constrictor (*inferior* = below; *constrictor* = decreases diameter of a lumen)	Cricoid and thyroid cartilages of larynx.	Posterior median raphe of pharynx.	Constricts inferior portion of pharynx to propel a bolus into esophagus.	Pharyngeal plexus.
Middle constrictor	Greater and lesser cornu of hyoid bone and stylohyoid ligament.	Posterior median raphe of pharynx.	Constricts middle portion of pharynx to propel a bolus into esophagus.	Pharyngeal plexus.
Superior constrictor (*superior* = above)	Pterygoid process, pterygomandibular raphe, and mylohyoid line of mandible.	Posterior median raphe of pharynx.	Constricts superior portion of pharynx to propel a bolus into esophagus.	Pharyngeal plexus.
Stylopharyngeus (*stylo* = stake or pole; styloid process of temporal bone; *pharyngo* = pharynx) (see also Figure 11-7)	Medial side of base of styloid process.	Lateral aspects of pharynx and thyroid cartilage.	Elevates larynx and dilates pharynx to help bolus descend.	Glossopharyngeal (IX) nerve.
Salpingopharyngeus (*salping* = pertaining to the auditory or uterine tube)	Inferior portion of auditory (Eustachian) tube.	Posterior fibers of palatopharyngeus muscle.	Elevates superior portion of lateral wall of pharynx during swallowing and opens orifice of auditory (Eustachian) tube.	Pharyngeal plexus.
Palatopharyngeus (*palato* = palate)	Soft palate.	Posterior border of thyroid cartilage and lateral and posterior wall of pharynx.	Elevates larynx and pharynx and helps close nasopharynx during swallowing.	Pharyngeal plexus.

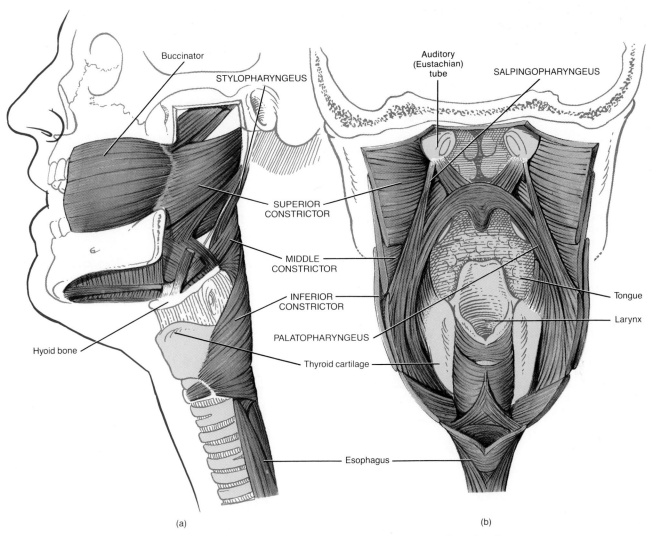

FIGURE 11-8 Muscles of the pharynx. (a) Left lateral view. (b) Posterior view.

EXHIBIT 11-7 MUSCLES OF THE FLOOR OF THE ORAL CAVITY* (Figure 11-9)

MUSCLE	ORIGIN	INSERTION	ACTION	INNERVATION
Digastric (*di* = two; *gaster* = belly)	Anterior belly from inner side of lower border of mandible; posterior belly from mastoid process of temporal bone.	Body of hyoid bone via an intermediate tendon.	Elevates hyoid bone and depresses mandible as in opening the mouth.	Anterior belly from mandibular branch of trigeminal (V) nerve; posterior belly from facial (VII) nerve.
Stylohyoid (*stylo* = stake or pole, styloid process of temporal bone; *hyoedes* = U-shaped, pertaining to hyoid bone; see also Figure 11-7)	Styloid process of temporal bone.	Body of hyoid bone.	Elevates hyoid bone and draws it posteriorly.	Facial (VII) nerve.
Mylohyoid	Inner surface of mandible.	Body of hyoid bone.	Elevates hyoid bone and floor of mouth and depresses mandible.	Mandibular branch of trigeminal (V) nerve.
Geniohyoid (*geneion* = chin) (see Figure 11-7)	Inner surface of mandible.	Body of hyoid bone.	Elevates hyoid bone, draws hyoid bone and tongue anteriorly, depresses mandible.	Hypoglossal (XII) nerve.

* These muscles together are referred to as **suprahyoid muscles.**

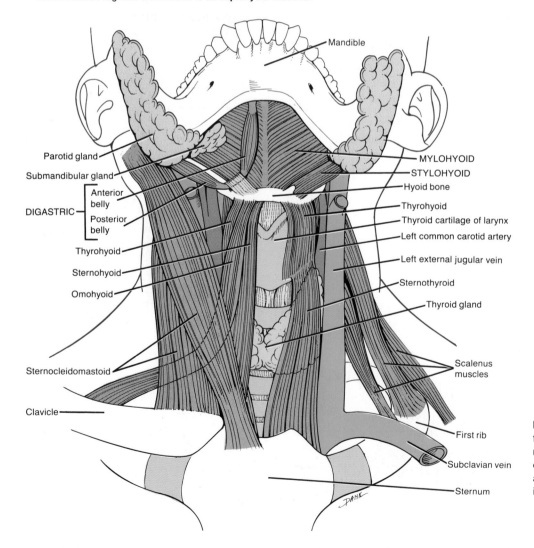

FIGURE 11-9 Muscles of the floor of the oral cavity. Superficial muscles are shown on the left side of the illustration, deep muscles are shown on the right side of the illustration.

EXHIBIT 11-8   MUSCLES OF THE LARYNX (Figure 11-10)

MUSCLE	ORIGIN	INSERTION	ACTION	INNERVATION
EXTRINSIC				
Omohyoid* (*omo* = relationship to the shoulder; *hyoedes* = U-shaped; pertaining to hyoid bone)	Superior border of scapula and superior transverse ligament.	Body of hyoid bone.	Depresses hyoid bone.	Branches of ansa cervicalis nerve (C1–C3).
Sternohyoid* (*sterno* = sternum)	Medial end of clavicle and manubrium of sternum.	Body of hyoid bone.	Depresses hyoid bone.	Branches of ansa cervicalis nerve (C1–C3).
Sternothyroid* (*thyro* = thyroid gland)	Manubrium of sternum.	Thyroid cartilage of larynx.	Depresses thyroid cartilage.	Branches of ansa cervicalis nerve (C1–C3).
Thyrohyoid*	Thyroid cartilage of larynx.	Greater cornu of hyoid bone.	Elevates thyroid cartilage and depresses hyoid bone.	Cervical nerves C1–C2 and descending hypoglossal (XII) nerve.
Stylopharyngeus	See Exhibit 11-6.			
Palatopharyngeus	See Exhibit 11-6.			
Inferior constrictor	See Exhibit 11-6.			
Middle constrictor	See Exhibit 11-6.			
INTRINSIC				
Cricothyroid (*crico* = cricoid cartilage of larynx)	Anterior and lateral portion of cricoid cartilage of larynx.	Anterior border of inferior cornu of thyroid cartilage of larynx and posterior part of inferior border of lamina of thyroid cartilage.	Produces tension and elongation of vocal folds.	External laryngeal branch of vagus (X) nerve.

* These muscles together are referred to as **infrahyoid (strap) muscles.**

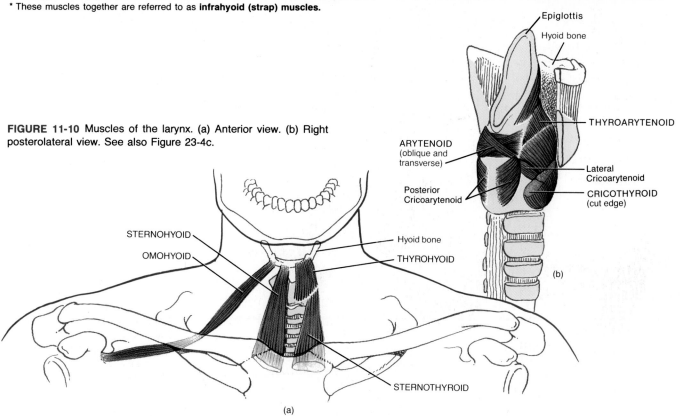

FIGURE 11-10 Muscles of the larynx. (a) Anterior view. (b) Right posterolateral view. See also Figure 23-4c.

EXHIBIT 11-9   MUSCLES THAT MOVE THE HEAD

MUSCLE	ORIGIN	INSERTION	ACTION	INNERVATION
Sternocleidomastoid (*sternum* = breastbone; *cleido* = clavicle; *mastoid* = mastoid process of temporal bone) (see Figure 11-15)	Sternum and clavicle.	Mastoid process of temporal bone.	Contraction of both muscles flex the cervical part of the vertebral column, draw the head forward, and elevate chin; contraction of one muscle rotates face toward side opposite contracting muscle.	Accessory (XI) nerve; cervical nerves C2–C3.
Semispinalis capitis (*semi* = half; *spine* = spinous process; *caput* = head) (see Figure 11-19)	Articular process of seventh cervical vertebra and transverse processes of first six thoracic vertebrae.	Occipital bone.	Both muscles extend head; contraction of one muscle rotates face toward same side as contracting muscle.	Dorsal rami of spinal nerves.
Splenius capitis (*splenion* = bandage) (see Figure 11-19)	Ligamentum nuchae and spines of seventh cervical vertebra and first four thoracic vertebrae.	Occipital bone and mastoid process of temporal bone.	Both muscles extend head; contraction of one rotates it to same side as contacting muscle.	Dorsal rami of middle and lower cervical nerves.
Longissimus capitis (*longissimus* = longest) (see Figure 11-19)	Transverse processes of last four cervical vertebrae.	Mastoid process of temporal bone.	Extends head and rotates face toward side opposite contracting muscle.	Dorsal rami of middle and lower cervical nerves.

EXHIBIT 11-10   MUSCLES THAT ACT ON THE ANTERIOR ABDOMINAL WALL (Figure 11-11)

MUSCLE	ORIGIN	INSERTION	ACTION	INNERVATION
Rectus abdominis (*rectus* = fibers parallel to midline; *abdomino* = abdomen)	Pubic crest and symphysis pubis.	Cartilage of fifth to seventh ribs and xiphoid process.	Flexes vertebral column; compresses abdomen.	Branches of thoracic nerves T7–T12.
External oblique (*external* = closer to surface; *oblique* = fibers diagonal to midline)	Lower eight ribs.	Iliac crest and linea alba (midline aponeurosis).	Contraction of both compresses abdomen; contraction of one side alone bends vertebral column laterally.	Branches of thoracic nerves T7–T12 and iliohypogastric nerve.
Internal oblique (*internal* = farther from surface)	Iliac crest, inguinal ligament, and thoracolumbar fascia.	Cartilage of last three or four ribs.	Compresses abdomen; contraction of one side alone bends vertebral column laterally.	Branches of thoracic nerves T8–T12, iliohypogastric, and ilioinguinal nerves.
Transversus abdominis (*transverse* = fibers perpendicular to midline)	Iliac crest, inguinal ligament, lumbar fascia, and cartilages of last six ribs.	Xiphoid process, linea alba, and pubis.	Compresses abdomen.	Branches of thoracic nerves T8–T12, iliohypogastric, and ilioinguinal nerves.

Deltoid

Pectoralis major

Sternum

Clavicle

Biceps brachii

Latissimus dorsi

Serratus anterior

EXTERNAL OBLIQUE

RECTUS ABDOMINIS

Linea alba

EXTERNAL OBLIQUE (cut)

Tendinous intersection

INTERNAL OBLIQUE

TRANSVERSUS ABDOMINIS

(a)   (b)

FIGURE 11-11 Muscles of the anterior abdominal wall. (a) Superficial view. (b) Deep view.

EXHIBIT 11-11   MUSCLES USED IN BREATHING (Figure 11-12)

MUSCLE	ORIGIN	INSERTION	ACTION	INNERVATION
Diaphragm (*dia* = across; *phragma* = wall)	Xiphoid process, costal cartilages of last six ribs, and lumbar vertebrae.	Central tendon.	Forms floor of thoracic cavity; pulls central tendon downward during inspiration and thus increases vertical length of thorax.	Phrenic nerve.
External intercostals (*external* = closer to surface; *inter* = between; *costa* = rib)	Inferior border of rib above.	Superior border of rib below.	Elevate ribs during inspiration and thus increase lateral and anteroposterior dimensions of thorax.	Intercostal nerves.
Internal intercostals (*internal* = farther from surface)	Superior border of rib below.	Inferior border of rib above.	Draw adjacent ribs together during forced expiration and thus decrease lateral and anteroposterior dimensions of thorax.	Intercostal nerves.

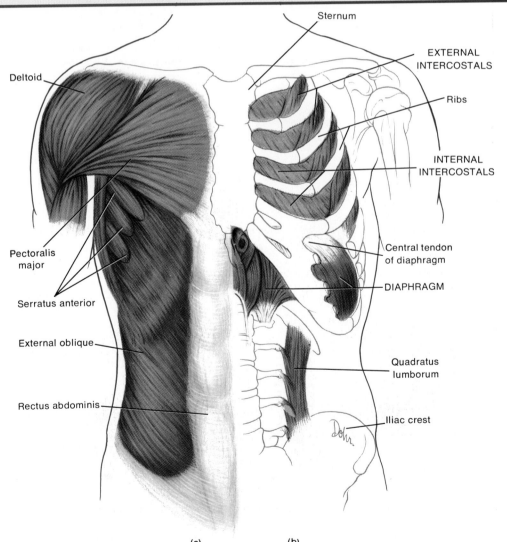

(a)   (b)

FIGURE 11-12 Muscles used in breathing. (a) Diagram of superficial view. (b) Diagram of deep view.

EXHIBIT 11-12   MUSCLES OF THE PELVIC FLOOR* (Figure 11-13; see Figure 11-14 also)

MUSCLE	ORIGIN	INSERTION	ACTION	INNERVATION
Levator ani (*levator* = raises; *ani* = anus)	This muscle is divisible into two parts, the pubococcygeus muscle and the iliococcygeus muscle. It forms the funnel-shaped floor of the pelvic cavity and supports the pelvic structures. It contains openings for the anal canal and urethra in both sexes and the vagina in the female.			
Pubococcygeus (*pubo* = pubis; *coccygeus* = coccyx)	Pubis.	Coccyx, urethra, anal canal, and central tendon of perineum.	Supports and slightly raises pelvic floor, resists increased intra-abdominal pressure, and draws anus toward pubis and constricts it.	Sacral nerves S3–S4 or S4 and perineal branch of pudendal nerve.
Iliococcygeus (*ilio* = ilium)	Ischial spine.	Coccyx.	Supports and slightly raises peivic floor, resists increased intra-abdominal pressure, and draws anus toward pubis and constricts it.	Sacral nerves S3–S4 or S4 and perineal branch of pudendal nerve.
Coccygeus	Ischial spine.	Lower sacrum and upper coccyx.	Supports and slightly raises pelvic floor, resists intraabdominal pressure, and pulls coccyx forward following defecation or parturition.	Sacral nerve S3 or S4.

* The muscles of the pelvic floor, together with the fasciae covering their internal and external surfaces, are collectively referred to as the **pelvic diaphragm.**

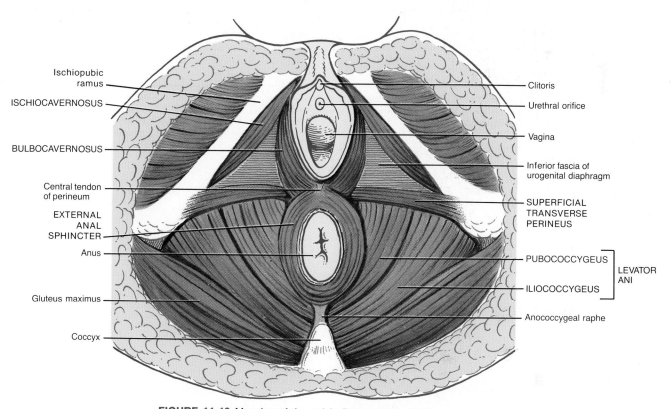

FIGURE 11-13 Muscles of the pelvic floor seen in the female perineum.

EXHIBIT 11-13   MUSCLES OF THE PERINEUM* (Figure 11-14; see Figure 11-13 also)

MUSCLE	ORIGIN	INSERTION	ACTION	INNERVATION
Superficial transverse perineus (*superficial* = closer to surface; *transverse* = across, *perineus* = perineum)	Ischial tuberosity.	Central tendon of perineum.	Helps to stabilize the central tendon of the perineum.	Perineal branch of pudendal nerve.
Bulbocavernosus (*bulbus* = bulb; *caverna* = hollow place)	Central tendon of perineum.	Inferior fascia of urogenital diaphragm, corpus spongiosum of penis, and deep fascia on dorsum of penis in male; pubic arch and root and dorsum of clitoris in female.	Helps expel last drops of urine during micturition, helps propel semen along urethra, and assists in erection of the penis in male; decreases vaginal orifice and assists in erection of clitoris in female.	Perineal branch of pudendal nerve.
Ischiocavernosus (*ischion* = hip)	Ischial tuberosity and ischial and pubic rami.	Corpus cavernosum of penis in male and clitoris in female	Maintains erection of penis in male and clitoris in female.	Perineal branch of pudendal nerve.
Deep transverse perineus† (*deep* = farther from surface)	Ischial rami.	Central tendon of perineum.	Helps eject last drops of urine and semen in male and urine in female.	Perineal branch of pudendal nerve.
Urethral sphincter† (*sphincter* = circular muscle that decreases size of an opening; *urethrae* = urethra)	Ischial and pubic rami.	Median raphe in male and vaginal wall in female.	Helps eject last drops of urine and semen in male and urine in female.	Perineal branch of pudendal nerve.
External anal sphincter	Anococcygeal raphe.	Central tendon of perineum.	Keeps anal canal and orifice closed.	Sacral nerve S4 and inferior rectal branch of pudendal nerve.

* The **perineum** is the entire outlet of the pelvis. It is a diamond-shaped area at the lower end of the trunk between the thighs and buttocks. It is bordered anteriorly by the symphysis pubis, laterally by the ischial tuberosities, and posteriorly by the coccyx. A transverse line drawn between the ischial tuberosities divides the perineum into an anterior **urogenital triangle** that contains the extenal genitals and a posterior **anal triangle** that contains the anus.
† The deep transverse perineus, the urethral sphincter, and a fibrous membrane constitute the **urogenital diaphragm.** It surrounds the urogenital ducts and helps to strengthen the pelvic floor.

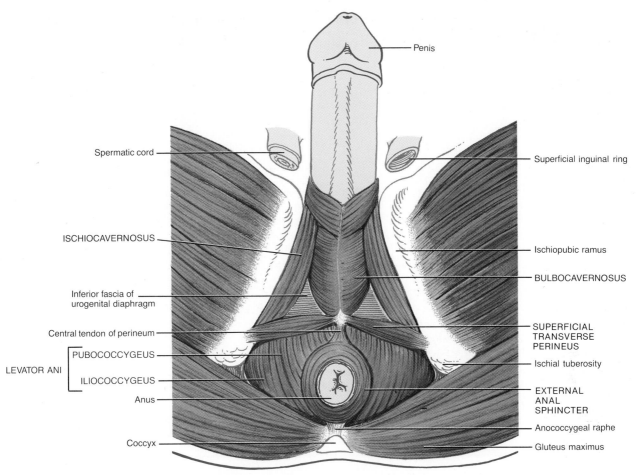

FIGURE 11-14 Muscles of the male perineum.

MUSCLE	ORIGIN	INSERTION	ACTION	INNERVATION
Subclavius (*sub* = under, *clavius* = clavicle)	First rib.	Clavicle.	Depresses clavicle.	Nerve to subclavius.
Pectoralis minor (*pectus* = breast, chest, thorax; *minor* = lesser)	Third through fifth ribs.	Coracoid process of scapula.	Depresses scapula, rotates shoulder joint anteriorly, and elevates third through fifth ribs during forced inspiration when scapula is fixed.	Medial pectoral nerve.
Serratus anterior (*serratus* = saw-toothed; *anterior* = front)	Upper eight or nine ribs.	Vertebral border and inferior angle of scapula.	Rotates scapula laterally and elevates ribs when scapula is fixed.	Long thoracic nerve.
Trapezius (*trapezoides* = trapezoid-shaped)	Occipital bone, ligamentum nuchae, and spines of seventh cervical and all thoracic vertebrae.	Clavicle and acromion and spine of scapula.	Elevates clavicle, adducts scapula, elevates or depresses scapula, and extends head.	Accessory (XI) nerve and cervical nerves C3–C4.
Levator scapulae (*levator* = raises; *scapulae* = scapula)	Upper four or five cervical vertebrae.	Vertebral border of scapula.	Elevates scapula.	Dorsal scapula nerve and cervical nerves C3–C5.
Rhomboideus major (*rhomboides* = rhomboid or diamond-shaped)	Spines of second to fifth thoracic vertebrae.	Vertebral border of scapula.	Adducts scapula and slightly rotates it upward.	Dorsal scapular nerve.
Rhomboideus minor	Spines of seventh cervical and first thoracic vertebrae.	Superior angle of scapula.	Adducts scapula.	Dorsal scapular nerve.

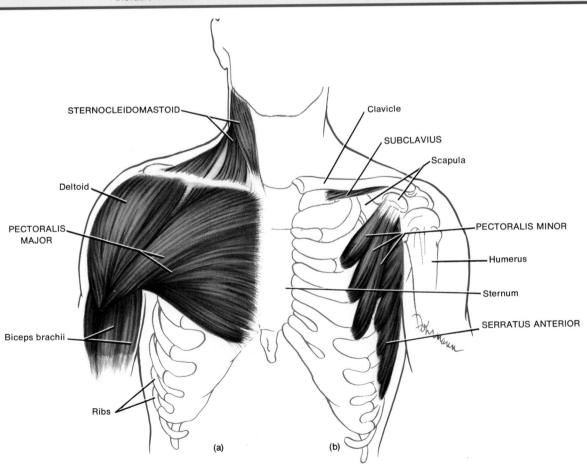

FIGURE 11-15 Muscles that move the pectoral (shoulder) girdle. (a) Anterior superficial view. (b) Anterior deep view.

FIGURE 11-15 (*Continued*) (c) Posterior superficial view. (d) Posterior deep view.

EXHIBIT 11-15   MUSCLES THAT MOVE THE ARM (Figure 11-16)

MUSCLE	ORIGIN	INSERTION	ACTION	INNERVATION
Pectoralis major (see Figure 11-15a)	Clavicle, sternum, cartilages of second to sixth ribs.	Greater tubercle of humerus.	Flexes, adducts, and rotates arm medially.	Medial and lateral pectoral nerves.
Deltoid (*delta* = triangular)	Clavicle and acromion and spine of scapula.	Deltoid tuberosity of humerus.	Abducts, flexes, and extends arm.	Axillary nerve.
Subscapularis (*sub* = below; *scapularis* = scapula)	Subscapular fossa of scapula.	Lesser tubercle of humerus.	Rotates arm medially.	Upper and lower subscapular nerves.
Supraspinatus (*supra* = above; *spinatus* = spine of scapula)	Fossa superior to spine of scapula.	Greater tubercle of humerus.	Assists deltoid muscle in abducting arm.	Suprascapular nerve.
Infraspinatus (*infra* = below)	Fossa inferior to spine of scapula.	Greater tubercle of humerus.	Rotates arm laterally.	Suprascapular nerve.
Latissimus dorsi (*latissimus* = widest; *dorsum* = back)	Spines of lower six thoracic vertebrae, lumbar vertebrae, crests of sacrum and ilium, lower four ribs.	Intertubercular groove of humerus.	Extends, adducts, and rotates arm medially; draws shoulder downward and backward.	Thoracodorsal nerve.
Teres major (*teres* = long and round)	Inferior angle of scapula.	Distal to lesser tubercle of humerus.	Extends arm and draws it down; assists in adduction and medial rotation of arm.	Lower subscapular nerve.
Teres minor	Lateral border of scapula.	Greater tubercle of humerus.	Rotates arm laterally.	Axillary nerve.

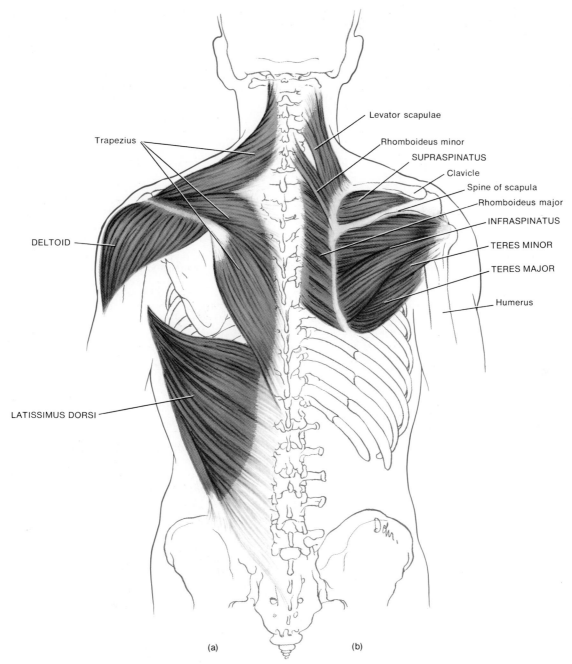

FIGURE 11-16 Muscles that move the arm. (a) Posterior superficial view. (b) Posterior deep view.

EXHIBIT 11-16   MUSCLES THAT MOVE THE FOREARM (Figure 11-17)

MUSCLE	ORIGIN	INSERTION	ACTION	INNERVATION
Biceps brachii (*biceps* = two heads of origin; *brachion* = arm)	Long head originates from tubercle above glenoid cavity; short head originates from coracoid process of scapula.	Radial tuberosity and bicipital aponeurosis.	Flexes and supinates forearm.	Musculocutaneous nerve.
Brachialis	Distal, anterior surface of humerus.	Tuberosity and coronoid process of ulna.	Flexes forearm.	Musculocutaneous and radial nerves.
Brachioradialis (*radialis* = radius) (see also Figure 11-18)	Supracondyloid ridge of humerus.	Superior to styloid process of radius.	Flexes forearm.	Radial nerve.
Triceps brachii (*triceps* = three heads of origin)	Long head originates from infraglenoid tuberosity of scapula; lateral head originates from lateral and posterior surface of humerus superior to radial groove; medial head originates from posterior surface of humerus inferior to radial groove.	Olecranon of ulna.	Extends forearm.	Radial nerve.
Coracobrachialis (*coraco* = coracoid process)	Coracoid process of scapula.	Middle of medial surface of shaft of humerus.	Flexes and adducts arm.	Musculocutaneous nerve.
Anconeus (*anconeal* = pertaining to elbow) (see Figure 11-18)	Lateral epicondyle of humerus.	Olecranon and superior portion of shaft of ulna.	Extends forearm.	Radial nerve.
Supinator (*supination* = turning palm upward or anteriorly)	Lateral epicondyle of humerus and ridge on ulna.	Oblique line of radius.	Supinates forearm.	Deep radial nerve.
Pronator teres (*pronation* = turning palm downward or posteriorly)	Medial epicondyle of humerus and coronoid process of ulna.	Midlateral surface of radius.	Pronates forearm.	Median nerve.
Pronator quadratus (*quadratus* = squared, four-sided)	Distal portion of shaft of ulna.	Distal portion of shaft of radius.	Pronates and rotates forearm.	Median nerve.

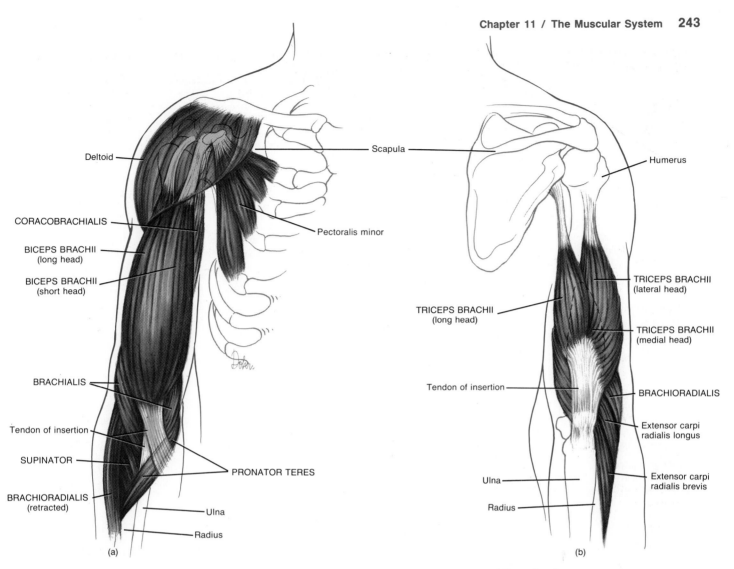

Deltoid

CORACOBRACHIALIS

BICEPS BRACHII
(long head)

BICEPS BRACHII
(short head)

BRACHIALIS

Tendon of insertion

SUPINATOR

BRACHIORADIALIS
(retracted)

Scapula

Pectoralis minor

PRONATOR TERES

Ulna

Radius

(a)

Humerus

TRICEPS BRACHII
(long head)

TRICEPS BRACHII
(lateral head)

TRICEPS BRACHII
(medial head)

Tendon of insertion

BRACHIORADIALIS

Extensor carpi
radialis longus

Ulna

Radius

Extensor carpi
radialis brevis

(b)

FIGURE 11-17 Muscles that move the forearm. (a) Anterior view. (b) Posterior view.

EXHIBIT 11-17   MUSCLES THAT MOVE THE WRIST AND FINGERS (Figure 11-18)

MUSCLE	ORIGIN	INSERTION	ACTION	INNERVATION
Flexor carpi radialis (*flexor* = decreases angle at joint; *carpus* = wrist; *radialis* = radius)	Medial epicondyle of humerus.	Second and third metacarpals.	Flexes and abducts wrist.	Median nerve.
Flexor carpi ulnaris (*ulnaris* = ulna)	Medial epicondyle of humerus and upper dorsal border of ulna.	Pisiform, hamate, and fifth metacarpal.	Flexes and adducts wrist.	Ulnar nerve.
Palmaris longus (*palma* = palm)	Medial epicondyle of humerus.	Transverse carpal ligament and palmar aponeurosis.	Flexes wrist and tenses palmar aponeurosis.	Median nerve.
Extensor carpi radialis longus (*extensor* = increases angle at joint; *longus* = long)	Lateral epicondyle of humerus.	Second metacarpal.	Extends and abducts wrist.	Radial nerve.
Extensor carpi ulnaris	Lateral epicondyle of humerus and dorsal border of ulna.	Fifth metacarpal.	Extends and adducts wrist.	Deep radial nerve.
Flexor digitorum profundus (*digit* = finger or toe; *profundus* = deep)	Anterior medial surface of body of ulna.	Bases of distal phalanges.	Flexes distal phalanges of each finger.	Median and ulnar nerves.
Flexor digitorum superficialis (*superficialis* = closer to surface)	Medial epicondyle of humerus, coronoid process of ulna, and oblique line of radius.	Middle phalanges.	Flexes middle phalanges of each finger.	Median nerve.
Extensor digitorum	Lateral epicondyle of humerus.	Middle and distal phalanges of each finger.	Extends phalanges.	Deep radial nerve.
Extensor indicis (*indicis* = index)	Dorsal surface of ulna.	Tendon of extensor digitorum of index finger.	Extends index finger.	Deep radial nerve.

FIGURE 11-18 Muscles that move the wrist and fingers. (a) Anterior view. (b) Posterior view.

EXHIBIT 11-18   MUSCLES THAT MOVE THE VERTEBRAL COLUMN (Figure 11-19)

MUSCLE	ORIGIN	INSERTION	ACTION	INNERVATION
Rectus abdominis (see Figure 11-11)	Pubic crest and symphysis pubis.	Cartilages of fifth through seventh ribs and xiphoid process.	Flexes vertebral column at lumbar spine and compresses abdomen.	Intercostal (thoracic) nerves T7–T12.
Quadratus lumborum (*quadratus* = squared, four-sided; *lumb* = lumbar region)	Iliac crest.	Twelfth rib and upper four lumbar vertebrae.	Flexes vertebral column laterally.	Intercostal nerve T12 and lumbar nerve L1.
Sacrospinalis (erector spinae)	This posterior muscle consists of three groupings: iliocostalis, longissimus, and spinalis. These groups, in turn, consist of a series of overlapping muscles. The iliocostalis group is laterally placed, the longissimus group is intermediate in placement, and the spinalis is medially placed.			

LATERAL

MUSCLE	ORIGIN	INSERTION	ACTION	INNERVATION
Iliocostalis lumborum (*ilium* = flank; *costa* = rib)	Iliac crest.	Lower six ribs.	Extends lumbar region of vertebral column.	Dorsal rami of lumbar nerves.
Iliocostalis thoracis (*thorax* = chest)	Lower six ribs.	Upper six ribs.	Maintains erect position of spine.	Dorsal rami of thoracic (intercostal) nerves.
Iliocostalis cervicis (*cervix* = neck)	First six ribs.	Transverse processes of fourth to sixth cervical vertebrae.	Extends cervical region of vertebral column.	Dorsal rami of cervical nerves.

INTERMEDIATE

MUSCLE	ORIGIN	INSERTION	ACTION	INNERVATION
Longissimus thoracis	Transverse processes of lumbar vertebrae.	Transverse processes of all thoracic and upper lumbar vertebrae and ninth and tenth ribs.	Extends thoracic region of vertebral column.	Dorsal rami of spinal nerves.
Longissimus cervicis	Transverse processes of fourth and fifth thoracic vertebrae.	Transverse processes of second to sixth cervical vertebrae.	Extends cervical region of vertebral column.	Dorsal rami of spinal nerves.
Longissimus capitis (*caput* = head)	Transverse processes of upper four thoracic vertebrae.	Mastoid process of temporal bone.	Extends head and rotates it to opposite side.	Dorsal rami of middle and lower cervical nerves.

MEDIAL

MUSCLE	ORIGIN	INSERTION	ACTION	INNERVATION
Spinalis thoracis	Spines of upper lumbar and lower thoracic vertebrae.	Spines of upper thoracic vertebrae.	Extends vertebral column.	Dorsal rami of spinal nerves.

Occipital bone

SPLENIUS CAPITIS

T1

SEMISPINALIS CAPITIS

LONGISSIMUS CAPITIS

C7

ILIOCOSTALIS CERVICIS

LONGISSIMUS CERVICIS

SPINALIS THORACIS

LONGISSIMUS THORACIS

ILIOCOSTALIS THORACIS

QUADRATUS LUMBORUM

L1

ILIOCOSTALIS LUMBORUM

External oblique

Coxal bone

FIGURE 11-19 Muscles that move the vertebral column.

EXHIBIT 11-19   MUSCLES THAT MOVE THE THIGH (Figure 11-20)

MUSCLE	ORIGIN	INSERTION	ACTION	INNERVATION
Psoas major* (*psoa* = muscle of loin)	Transverse processes and bodies of lumbar vertebrae.	Lesser trochanter of femur.	Flexes and rotates thigh laterally; flexes vertebral column.	Lumbar nerves L2–L3.
Iliacus* (*iliac* = ilium)	Iliac fossa.	Tendon of psoas major.	Flexes and rotates thigh laterally; flexes vertebral column slightly.	Femoral nerve.
Gluteus maximus (*glutos* = buttock; *maximus* = largest)	Iliac crest, sacrum, coccyx, and aponeurosis of sacrospinalis.	Iliotibial tract of fascia lata and gluteal tuberosity of femur.	Extends and rotates thigh laterally.	Inferior gluteal nerve.
Gluteus medius (*media* = middle)	Ilium.	Greater trochanter of femur.	Abducts and rotates thigh medially.	Superior gluteal nerve.
Gluteus minimus (*minimus* = smallest)	Ilium.	Greater trochanter of femur.	Abducts and rotates thigh laterally.	Superior gluteal nerve.
Tensor fasciae latae (*tensor* = makes tense; *fascia* = band; *latus* = wide)	Iliac crest.	Tibia by way of the iliotibial tract.	Flexes and abducts thigh.	Superior gluteal nerve.
Adductor longus (*adductor* = moves part closer to midline; *longus* = long)	Pubic crest and symphysis pubis.	Linea aspera of femur.	Adducts, rotates, and flexes thigh.	Obturator nerve.
Adductor brevis (*brevis* = short)	Inferior ramus of pubis.	Linea aspera of femur.	Adducts, rotates, and flexes thigh.	Obturator nerve.
Adductor magnus (*magnus* = large)	Inferior ramus of pubis and ischium to ischial tuberosity.	Linea aspera of femur.	Adducts, flexes, and extends thigh (anterior part flexes, posterior part extends).	Obturator and sciatic nerves.
Piriformis (*pirum* = pear; *forma* = shape)	Sacrum.	Greater trochanter of femur.	Rotates thigh laterally and abducts it.	Sacral nerves S2 or S1–S2.
Obturator internus (*obturator* = obturator foramen; *internus* = inside)	Margin of obturator foramen, pubis, and ischium.	Greater trochanter of femur.	Rotates thigh laterally and abducts it.	Nerve to obturator internus.
Pectineus (*pecten* = comb-shaped)	Fascia of pubis.	Pectineal line of femur.	Flexes, adducts, and rotates thigh laterally.	Femoral nerve.

CLINICAL APPLICATION

A common sports injury is a **"pulled groin."** Very simply this means there is a strain, stretching, and probably some tearing away of the tendinous origins of the adductor muscles that move the thigh.

* Together the psoas major and iliacus are sometimes termed the **iliopsoas muscle.**

PSOAS MAJOR

Coxal bone

ILIACUS

Anterior superior
iliac spine

Anterior inferior
iliac spine

Inguinal
ligament

TENSOR
FASCIAE
LATAE

Vastus intermedius

Rectus femoris
(cut)

Vastus lateralis

PECTINEUS (cut)

ADDUCTOR LONGUS

ADDUCTOR BREVIS

ADDUCTOR MAGNUS

GLUTEUS MAXIMUS
(cut)

GLUTEUS MEDIUS
(cut)

GLUTEUS MINIMUS

PIRIFORMIS

OBTURATOR INTERNUS

Ischial
tuberosity

Quadratus femoris

Sciatic nerve

ADDUCTOR MAGNUS

(a)

(b)

FIGURE 11-20 Muscles that move the thigh. (a) Anterior view. (b) Posterior view.

EXHIBIT 11-20   MUSCLES THAT ACT ON THE LEG (Figure 11-21)

MUSCLE	ORIGIN	INSERTION	ACTION	INNERVATION
Quadriceps femoris (*quadriceps* = four heads of origin; *femoris* = femur)	This composite muscle includes four distinct parts, usually described as four separate muscles. The common tendon that includes the patella and attaches to the tibial tuberosity is known as the patellar ligament.			
Rectus femoris (*rectus* = fibers parallel to midline)	Anterior inferior iliac spine.	Upper border of patella.	All four heads extend leg; rectus portion alone also flexes thigh.	Femoral nerve.
Vastus lateralis (*vastus* = large; *lateralis* = lateral)	Greater trochanter and linea aspera of femur.	Upper border and sides of patella; tibial tuberosity through patellar ligament (tendon of quadriceps).		Femoral nerve.
Vastus medialis (*medialis* = medial)	Linea aspera of femur.			Femoral nerve.
Vastus intermedius (*intermedius* = middle)	Anterior and lateral surfaces of body of femur.			Femoral nerve.
Hamstrings	A collective designation for three separate muscles.			
Biceps femoris (*biceps* = two heads of origin)	Long head arises from ischial tuberosity; short head arises from linea aspera of femur.	Head of fibula and lateral condyle of tibia.	Flexes leg and extends thigh.	Tibial nerve from sciatic nerve.
Semitendinosus (*semi* = half; *tendo* = tendon)	Ischial tuberosity.	Proximal part of medial surface of body of tibia.	Flexes leg and extends thigh.	Tibial nerve from sciatic nerve.
Semimembranosus (*membran* = membrane)	Ischial tuberosity.	Medial condyle of tibia.	Flexes leg and extends thigh.	Tibial nerve from sciatic nerve.
Gracilis (*gracilis* = slender)	Symphysis pubis and pubic arch.	Medial surface of body of tibia.	Flexes leg and adducts thigh.	Obturator nerve.
Sartorius (*sartor* = tailor; refers to cross-legged position of tailors)	Anterior superior spine of ilium.	Medial surface of body of tibia.	Flexes leg; flexes thigh and rotates it laterally, thus crossing leg.	Femoral nerve.

CLINICAL APPLICATION

"Pulled hamstrings" are common sports injuries in individuals who run very hard. Sometimes the violent muscular exertion required to perform a feat tears off part of the tendinous origins of the hamstrings from the ischial tuberosity. This is usually accompanied by a contusion (bruising) and tearing of some of the muscle fibers and rupture of blood vessels producing a hematoma (collection of blood).

Psoas major

Iliacus

Tensor fasciae latae

Iliotibial tract

VASTUS INTERMEDIUS

RECTUS FEMORIS

VASTUS LATERALIS

VASTUS MEDIALIS

Tendon of quadriceps femoris

Pectineus

Adductor longus

GRACILIS

SARTORIUS

(a)

Gluteus medius

Gluteus maximus

Adductor magnus

Iliotibial tract

BICEPS FEMORIS (long head)

SEMITENDINOSUS

SEMIMEMBRANOSUS

GRACILIS

SARTORIUS

Gastrocnemius

(b)

FIGURE 11-21 Muscles that act on the leg. (a) Anterior view. (b) Posterior view.

EXHIBIT 11-21   MUSCLES THAT MOVE THE FOOT AND TOES (Figure 11-22)

MUSCLE	ORIGIN	INSERTION	ACTION	INNERVATION
Gastrocnemius (*gaster* = belly; *kneme* = leg)	Lateral and medial condyles of femur and capsule of knee.	Calcaneus by way of calcaneal (Achilles) tendon.	Plantar flexes foot and flexes knee.	Tibial nerve.
Soleus (*soleus* = sole of foot)	Head of fibula and medial border of tibia.	Calcaneus by way of calcaneal (Achilles) tendon.	Plantar flexes foot.	Tibial nerve.
Peroneus longus (*perone* = fibula; *longus* = long)	Head and body of fibula and lateral condyle of tibia.	First metatarsal and first cuneiform.	Plantar flexes and everts foot.	Superficial peroneal nerve.
Peroneus brevis (*brevis* = short)	Body of fibula.	Fifth metatarsal.	Plantar flexes and everts foot.	Superficial peroneal nerve.
Peroneus tertius (*tertius* = third)	Distal third of fibula.	Fifth metatarsal.	Dorsiflexes and everts foot.	Deep peroneal nerve.
Tibialis anterior (*tibialis* = tibia; *anterior* = front)	Lateral condyle and body of tibia.	First metatarsal and first cuneiform.	Dorsiflexes and inverts foot.	Deep peroneal nerve.
Tibialis posterior (*posterior* = back)	Interosseus membrane between tibia and fibula.	Second, third, and fourth metatarsals; navicular; third cuneiform; and cuboid.	Plantar flexes and inverts foot.	Tibial nerve.
Flexor digitorum longus (*flexor* = decreases angle at joint; *digitorum* = finger or toe)	Tibia.	Distal phalanges of four outer toes.	Flexes toes and plantar flexes and inverts foot.	Tibial nerve.
Extensor digitorum longus (*extensor* = increases angle at joint)	Lateral condyle of tibia and anterior surface of fibula.	Middle and distal phalanges of four outer toes.	Extends toes and dorsiflexes and everts foot.	Deep peroneal nerve.

CLINICAL APPLICATION

The calcaneal tendon, the **strongest tendon of the body,** is able to withstand a 1,000-pound force without tearing. Despite this, however, the calcaneal tendon ruptures more frequently than any other tendon.

Semimembranosus

Gracilis

Semitendinosus

Biceps femoris

Femur

GASTROCNEMIUS

SOLEUS

FLEXOR
DIGITORUM
LONGUS

Tibia

Fibula

PERONEUS BREVIS

PERONEUS LONGUS

Calcaneal
tendon

Calcaneus

(a)

Biceps femoris

Semitendinosus

Gracilis

Semimembranosus

GASTROCNEMIUS
(cut)

SOLEUS
(cut)

TIBIALIS POSTERIOR

FLEXOR
DIGITORUM
LONGUS

PERONEUS
LONGUS

Flexor
hallucis
longus

Tibia

Fibula

PERONEUS
BREVIS

(b)

FIGURE 11-22 Muscles that move the foot and toes. (a) Superficial posterior view. (b) Deep posterior
view.

FIGURE 11-22 (Continued) (c) Superficial anterior view. (d) Superficial lateral view.

INTRAMUSCULAR INJECTIONS

An **intramuscular injection** penetrates the skin and subcutaneous tissue to enter the muscle itself. Intramuscular injections are preferred when prompt absorption is desired, when larger doses than can be given cutaneously are indicated, or when the drug is too irritating to give subcutaneously. The common sites for intramuscular injections include the buttock, lateral side of the thigh, and the deltoid region of the arm. Muscles in these areas, especially the gluteal muscles in the buttock, are fairly thick. Because of the large number of muscle fibers and extensive fascia, the drug has a large surface area for absorption. Absorption is further promoted by the extensive blood supply to muscles. Ideally, intramuscular injections should be given deep within the muscle and away from major nerves and blood vessels.

For many intramuscular injections, the preferred site is the gluteus medius muscle of the buttock (Figure 11-23a). The buttock is divided into quadrants, and the upper outer quadrant is used as the injection site. The iliac crest serves as a landmark for this quadrant. The spot for injection is usually about 5 to 7½ cm (2 to 3 inches) below the iliac crest. The upper outer quadrant is chosen because the muscle in this area is quite thick and has few nerves. Injection in this area thus reduces the chance of injury to the sciatic nerve, which can cause paralysis of the lower extremity. The probability of injecting the drug into a blood vessel is also remote in this area. After the needle is inserted into the gluteus medius muscle, the plunger is pulled up for a few seconds. If the syringe fills with blood, the needle is in a blood vessel and a different injection site on the opposite buttock is chosen.

Injections given in the lateral side of the thigh are inserted into the midportion of the vastus lateralis muscle (Figure 11-23b). This site is determined by using the knee and greater trochanter of the femur as landmarks. The midportion of the muscle is located by measuring a handbreadth above the knee and a handbreadth below the greater trochanter.

The deltoid injection is given in the midportion of the muscle about two to three fingerbreadths below the acromion of the scapula and lateral to the axilla (Figure 11-23c).

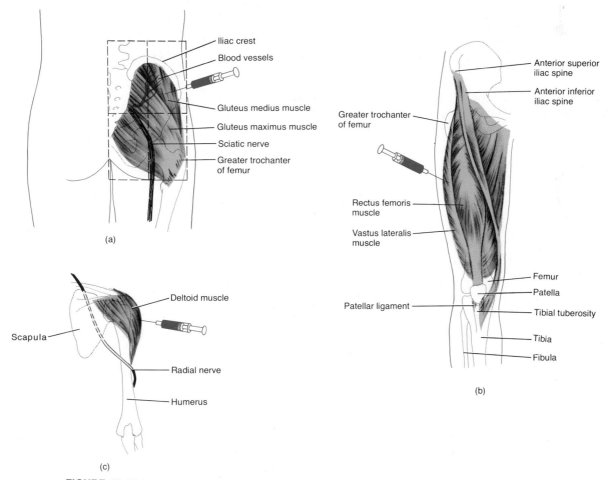

FIGURE 11-23 Intramuscular injections. Shown are the three common sites for intramuscular injections. (a) Buttock. (b) Lateral surface of the thigh. (c) Deltoid region of the arm.

STUDY OUTLINE

How Skeletal Muscles Produce Movement (p. 216)

1. Skeletal muscles produce movement by pulling on bones.
2. The attachment to the stationary bone is the origin. The attachment to the movable bone is the insertion.
3. Bones serve as levers and joints serve as fulcrums. The lever is acted on by two different forces: resistance and effort.
4. Levers are categorized into three types—first-class, second-class, and third-class—according to the position of the fulcrum, effort, and resistance on the lever.
5. Fascicular arrangements include parallel, convergent, pennate, and circular. Fascicular arrangement is correlated with the power of a muscle and the range of movement.
6. The prime mover produces the desired action. The antagonist produces an opposite action. The synergist assists the prime mover by reducing unnecessary movement. The fixator stabilizes the origin of the prime mover so that it can act more efficiently.

Naming Skeletal Muscles (p. 218)

Skeletal muscles are named on the basis of distinctive criteria: direction of fibers, location, size, number of origins (or heads), shape, origin and insertion, and action.

Principal Skeletal Muscles (p. 219)

The principal skeletal muscles of the body are grouped according to region in Exhibits 11-2 through 11-21.

Intramuscular Injections (p. 255)

1. Advantages of intramuscular injections are prompt absorption, use of larger doses than can be given cutaneously, and minimal irritation.
2. Common sites for intramuscular injections are the buttock, lateral side of the thigh, and deltoid region of the arm.

REVIEW QUESTIONS

1. Using the terms origin, insertion, and belly in your discussion, describe how skeletal muscles produce body movements by pulling on bones.
2. What is a lever? Fulcrum? Apply these terms to the body and indicate the nature of the forces that act on levers. Describe the three classes of levers and provide one example of each in the body.
3. Describe the various arrangements of fasciculi. How is fascicular arrangement correlated with the strength of a muscle and its range of movement?
4. Define the role of the prime mover, antagonist, synergist, and fixator in producing body movements.
5. Select at random several muscles presented in Exhibits 11-2 through 11-21 and see if you can determine the criterion or criteria employed for naming each. In addition, refer to the prefixes, suffixes, roots, and definitions in each exhibit as a guide. Select as many muscles as you wish, as long as you feel you understand the concept involved.
6. What muscles would you use to do the following: (a) frown, (b) pout, (c) show surprise, (d) show your upper teeth, (e) pucker your lips, (f) squint, (g) blow up a balloon, (h) smile?
7. What would happen if you lost tone in the masseter and temporalis muscles?
8. What muscles move the eyeball? In what direction does each muscle move the eyeball?
9. What tongue, facial, and mandibular muscles would you use when chewing a piece of gum?
10. What muscles constrict the pharynx? What muscle dilates the pharynx?
11. What muscles would you use to signify "yes" and "no" by moving your head?
12. What muscles accomplish compression of the anterior abdominal wall?
13. What are the principal muscles involved in breathing? What are their actions?
14. Describe the actions of the muscles of the pelvic floor and perineum.
15. What muscles are used to (a) raise your shoulders, (b) lower your shoulders, (c) join your hands behind your back, (d) join your hands in front of your chest?
16. What muscles move the arm? In which directions do these movements occur?
17. What muscles move the forearm and what actions are used when striking a match?
18. How many muscles and actions of the wrist and fingers used when writing can you list?
19. Can you perform an exercise that would involve the use of each of the muscles listed in Exhibit 11-18?
20. Review the various movements involved in your favorite kind of dancing. What muscles listed in Exhibit 11-20 would you be using and what actions would you be performing?
21. Determine the muscles and their actions listed in Exhibit 11-21 that you would use in climbing a ladder to a diving board, diving into the water, swimming the length of a pool, and then sitting at pool side.
22. Discuss the muscles that plantar flex, evert, pronate, dorsiflex, and supinate the foot.
23. What are the advantages of intramuscular injections?
24. Describe how you would locate the sites for an intramuscular injection in the buttock, lateral side of the thigh, and deltoid region of the arm.

U N I T III

Control Systems of the Human Body

This unit will show you the significance of the nerve impulse in making rapid adjustments for maintaining homeostasis. You will learn how the nervous system detects changes in the environment, decides on a course of action, and responds to the changes. We will also investigate the role of hormones in maintaining long-term homeostasis.

12 / Nervous Tissue

13 / The Spinal Cord and the Spinal Nerves

14 / The Brain and the Cranial Nerves

15 / The Sensory, Motor, and Integrative Systems

16 / The Autonomic Nervous System

17 / The Special Senses

18 / The Endocrine System

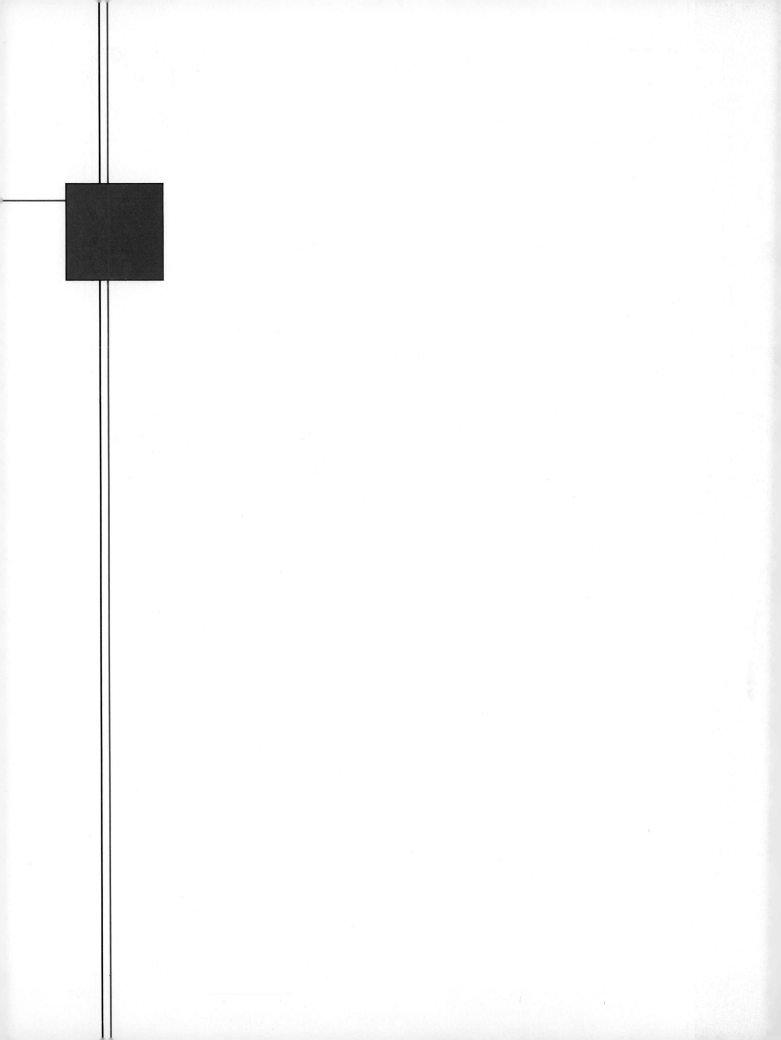

Nervous Tissue

12

STUDENT OBJECTIVES

- Identify the three basic functions of the nervous system in maintaining homeostasis.
- Classify the organs of the nervous system into central and peripheral divisions.
- Contrast the histological characteristics and functions of neuroglia and neurons.
- Classify neurons by structure and function.
- Describe the conditions that contribute to the resting state of a neuron.
- List the sequence of events involved in the generation and conduction of a nerve impulse.
- Define the all-or-none principle of nerve impulse transmission.
- Discuss the factors that determine the speed of impulse conduction.
- Define a synapse and list the factors involved in the conduction of a nerve impulse across a synapse.
- Compare the functions of excitatory transmitter-receptor interactions and inhibitory transmitter-receptor interactions in helping to maintain homeostasis.
- Describe integration at synapses.
- Define the roles of neurotransmitters in conducting a nerve impulse across a synapse.
- List the factors that may inhibit or block nerve impulses.
- List the necessary conditions for the regeneration of nervous tissue.
- Explain the organization of neurons in the nervous system.

The **nervous system** is the body's control center and communications network. In humans, the nervous system serves three broad functions: sensory, integrative, and motor. First, it senses changes within the body and in the outside environment; this is its sensory function. Second, it interprets the changes; this is its integrative function. Third, it responds to the interpretation by initiating action in the form of muscular contractions or glandular secretions; this is its motor function.

Through sensation, integration, and response the nervous system represents the body's most rapid means of maintaining homeostasis. Its split-second reactions, carried out by nerve impulses, can normally make the adjustments necessary to keep the body functioning efficiently. As you will see later, the nervous system shares the maintenance of homeostasis with the endocrine system. Although the adjustments made by hormones secreted by endocrine glands are slower than those made by nerve impulses, they are no less effective.

The branch of medical science that deals with the normal functioning and disorders of the nervous system is called **neurology** (*neuro* = nerve or nervous system; *logos* = study of).

The developmental anatomy of the nervous system is considered in Chapter 14.

ORGANIZATION

The nervous system may be divided into two principal divisions, the central nervous system and the peripheral nervous system, and several subdivisions (Figure 12-1).

The **central nervous system (CNS)** is the control center for the entire system and consists of the brain and spinal cord. All body sensations must be relayed from receptors to the central nervous system if they are to be interpreted and acted on. All the nerve impulses that stimulate muscles to contract and glands to secrete must also originate in the central nervous system.

The various nerve processes that connect the brain and spinal cord with receptors, muscles, and glands constitute the **peripheral (pe-RIF-er-al) nervous system (PNS).** The peripheral nervous system may be divided into an afferent system and an efferent system. The **afferent (AF-er-ent) system** consists of nerve cells that convey information from receptors in the periphery of the body to the central nervous system. These nerve cells, called *afferent (sensory) neurons,* are the first cells to pick up incoming information. The **efferent (EF-er-ent) system** consists of nerve cells that convey information from the central nervous system to muscles and glands. These nerve cells are called *efferent (motor) neurons.*

The efferent system is subdivided into a somatic nervous system and an autonomic nervous system. The **somatic nervous system** or **SNS** (*soma* = body) consists of efferent neurons that conduct impulses from the central nervous system to skeletal muscle tissue. Since the somatic nervous system produces movement only in skeletal muscle tissue, it is under conscious control and therefore voluntary. The **autonomic nervous system** or **ANS** (*auto* = self; *nomos* = law), by contrast, contains efferent neurons that convey impulses from the central nervous system to smooth muscle tissue, cardiac muscle tissue, and glands. Since it produces responses only in involuntary muscles and glands, it is usually considered to be involuntary.

With few exceptions, the viscera receive nerve fibers from the two divisions of the autonomic nervous system: the **sympathetic division** and the **parasympathetic divi-**

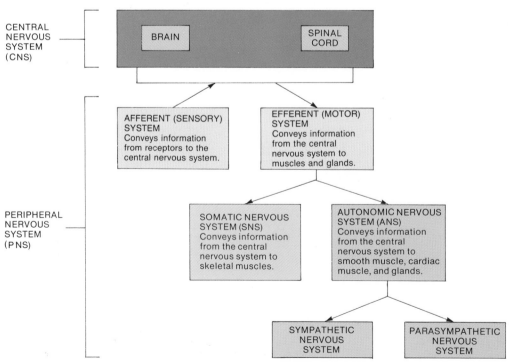

FIGURE 12-1 Organization of the nervous system.

sion. In general, the fibers of one division stimulate or increase an organ's activity, while the fibers from the other inhibit or decrease activity (see Chapter 16).

HISTOLOGY

Despite the organizational complexity of the nervous system, it consists of only two principal kinds of cells: neurons and neuroglia. Neurons make up the nervous tissue that forms the structural and functional portion of the system. They are highly specialized for impulse conduction and for all special functions attributed to the nervous system: thinking, controlling muscle activity, regulating glands. Neuroglia serve as a special supporting and protective component of the nervous system.

NEUROGLIA

The cells of the nervous system that perform the functions of support and protection are called **neuroglia** (noo-ROG-lē-a; *neuro* = nerve; *glia* = glue) or **glial cells** (Figure 12-2). Neuroglia are generally smaller than neurons and outnumber them by 5 to 10 times. Many of the glial cells form a supporting network by twining around nerve

(a)

(b)   (c)   (d)

FIGURE 12-2 Histology of neuroglia. (a) Fibrous astrocyte (left) and protoplasmic astrocyte (right) associated with a blood vessel. (b) Oligodendrocyte associated with an axon of a neuron of the central nervous system. (c) Microglial cell. (d) Ependyma. Note the cilia.

cells or lining certain structures in the brain and spinal cord. Others bind nervous tissue to supporting structures and attach the neurons to their blood vessels. A few types of glial cells also serve specialized functions. For example, some produce a phospholipid covering, called a myelin sheath, around nerve fibers in the central nervous system, which increases the speed of nerve impulse conduction and insulates the fibers. Certain small glial cells are phagocytic; they protect the central nervous system from disease by engulfing invading microbes and clearing away debris. Neuroglia are of clinical interest because they are a common source of tumors (gliomas) of the nervous system. It is estimated that gliomas account for 40 to 45 percent of intracranial tumors. Unfortunately, gliomas are very invasive.

Exhibit 12-1 lists the neuroglial cells and summarizes their functions.

NEURONS

Nerve cells, called **neurons,** are responsible for conducting impulses from one part of the body to another. They are the structural and functional units of the nervous system.

Structure

A neuron consists of three distinct portions: (1) cell body, (2) dendrites, and (3) axon (Figure 12-3a). The **cell body, soma,** or **perikaryon** (per'-i-KAR-ē-on), contains a well-defined nucleus and nucleolus surrounded by a granular cytoplasm. Within the cytoplasm are typical organelles such as lysosomes, mitochondria, and Golgi complexes. Many neurons also contain cytoplasmic inclusions such as *lipofuscin* pigment that occurs as clumps of yellowish-brown granules. Lipofuscin may be a by-product of lysosomal activity. Although its significance is unknown, lipo-

fusin is related to aging; the amount of pigment increases with age. Also located in the cytoplasm are structures characteristic of neurons: chromatophilic substance and neurofibrils. The *chromatophilic substance* (*Nissl bodies*) is an orderly arrangement of granular (rough) endoplasmic reticulum whose function is protein synthesis. Newly synthesized proteins pass from the perikaryon into the neuronal processes, mainly the axon, at the rate of about 1 mm (0.04 inch) per day. These proteins replace those lost during metabolism and are used for growth of neurons and regeneration of peripheral nerve fibers. *Neurofibrils* are long, thin fibrils composed of microtubules. They may assume a function in support and the transportation of nutrients. Mature neurons do not contain a mitotic apparatus. The significance of this absence will be noted shortly.

Neurons have two kinds of cytoplasmic processes: dendrites and axons. **Dendrites** (*dendro* = tree) are highly branched, thick extensions of the cytoplasm of the cell body. They typically contain chromatophilic substance, mitochondria, and other cytoplasmic organelles. A neuron usually has several main dendrites. Their function is to conduct an impulse toward the cell body.

The second type of cytoplasmic process, called an **axon (axis cylinder),** is a single, highly specialized, long, thin process that conducts impulses away from the cell body to another neuron or tissue. It usually originates from the cell body as a small conical elevation called the *axon hillock.* An axon contains mitochondria and neurofibrils but no chromatophilic substance; thus it does not carry on protein synthesis. Its cytoplasm, called *axoplasm,* is surrounded by a plasma membrane known as the *axolemma* (*lemma* = sheath or husk). Axons vary in length from a few millimeters (1 mm = 0.04 inch) in the brain to a meter (3.28 ft) or more between the spinal cord and toes. Along the course of an axon, there may be

EXHIBIT 12-1   NEUROGLIA OF CENTRAL NERVOUS SYSTEM

TYPE	DESCRIPTION	FUNCTION
Astrocytes (*astro* = star; *cyte* = cell)	Star-shaped cells with numerous processes. *Protoplasmic astrocytes* are found in the gray matter of the CNS, and *fibrous astrocytes* are found in the white matter of the CNS.	Twine around nerve cells to form supporting network in brain and spinal cord; attach neurons to their blood vessels.
Oligodendrocytes (*oligo* = few; *dendro* = tree)	Resemble astrocytes in some ways, but processes are fewer and shorter.	Give support by forming semirigid connective tissue rows between neurons in brain and spinal cord; produce a phospholipid myelin sheath around axons of neurons of central nervous system.
Microglia (*micro* = small; *glia* = glue)	Small cells with few processes; derived from monocytes; normally stationary, but may migrate to site of injury; also called brain macrophages.	Engulf and destroy microbes and cellular debris; may migrate to area of injured nervous tissue and function as small macrophages.
Ependyma (ependymocytes) (*ependyma* = upper garment)	Epithelial cells arranged in a single layer and ranging in shape from squamous to columnar; many are ciliated.	Form a continuous epithelial lining for the ventricles of the brain (spaces that form and circulate cerebrospinal fluid) and the central canal of the spinal cord; probably assist in the circulation of cerebrospinal fluid (CSF).

FIGURE 12-3 Structure of a typical neuron as exemplified by an efferent (motor) neuron. (a) An entire efferent neuron. Arrows indicate the direction in which nerve impulses travel. (b) Sectional planes through a myelinated fiber. (c) Photomicrograph of an efferent neuron at a magnification of 640×. (Courtesy of Biophoto Associates/Photo Researchers.)

FIGURE 12-4 Comparison between myelinated and unmyelinated axons. (a) Stages in the formation of a myelin sheath by a neurolemmocyte (Schwann cell). (b) Unmyelinated axon.

CLINICAL APPLICATION

The route taken by materials back to the cell body by axonal transport is the route by which the **herpes virus** and **rabies virus** make their way back to nerve cell bodies, where they multiply and cause their damage. The toxin produced by the **tetanus bacterium** uses the same route to reach the central nervous system. In fact, the time delay between the release of the toxin and the first appearance of symptoms is in part due to the time required for movement of the toxin by axonal transport.

side branches called *axon collaterals.* The axon and its collaterals terminate by branching into many fine filaments called *axon terminals* (*telodendria*). The distal ends of axon terminals are expanded into bulblike structures called *synaptic end bulbs,* which are important in nerve impulse conduction from one neuron to another and from a neuron to muscle or glandular tissue. They contain membrane-enclosed sacs called *synaptic vesicles* that store chemicals called neurotransmitters that determine whether impulse conduction occurs or not.

The cell body of a neuron is essential for the synthesis of many substances that sustain the life of the nerve cell. Neurons have two types of intracellular systems for transporting synthesized materials from the cell body. The slower one, called **axoplasmic flow,** conveys axoplasm in one direction only—from the cell body toward axon terminals. This mechanism may occur by protoplasmic streaming and supplies new axoplasm for developing or regenerating axons and renews axoplasm in growing and mature axons. The faster type of intracellular transport is called **axonal transport.** It conveys materials in both directions—away from the cell body and toward the cell body—possibly along tracks formed by microtubules and filaments. Axonal transport moves various organelles and materials that form the membranes of the axolemma, synaptic end bulbs, and synaptic vesicles. Materials returning to the cell body are degraded or recycled.

The term **nerve fiber** may be applied to any process projecting from the cell body. More commonly, it refers to an axon and its sheaths. Figure 12-3b shows two sectional planes of a nerve fiber of the peripheral nervous system. Many axons, especially large, peripheral axons, are surrounded by a multilayered, white, phospholipid, segmented covering called the **myelin sheath.** Axons containing such a covering are *myelinated,* while those without it are *unmyelinated* (see Figure 12-4). The function of the myelin sheath is to increase the speed of nerve impulse conduction and to insulate and maintain the axon. Myelin is responsible for the color of the white matter in the nerves, brain, and spinal cord.

The myelin sheath of axons of the peripheral nervous system is produced by flattened cells, called **neurolemmocytes (Schwann cells),** located along the axons. In the formation of a sheath, a developing neurolemmocyte encircles the axon until its ends meet and overlap (Figure 12-4). The cell then winds around the axon many times and, as it does so, the cytoplasm and nucleus are pushed to the outside layer. The inner portion, consisting of up to 20–30 layers of neurolemmocyte membrane, is the myelin shealth. The peripheral nucleated cytoplasmic layer of the neurolemmocyte (the outer layer that encloses the sheath) is called the **neurolemma (sheath of Schwann).**

The neurolemma is found only around fibers of the peripheral nervous system. Its function is to assist in the regeneration of injured axons. Between the segments of the myelin sheath are unmyelinated gaps, called **neurofibral nodes (nodes of Ranvier)** (ron-VĒ-ā). Unmyelinated fibers are also enclosed by neurolemmocytes, but without multiple wrappings.

Nerve fibers of the central nervous system may also be myelinated or unmyelinated. Myelination of central nervous system axons is accomplished by oligodendrocytes in somewhat the same manner that neurolemmocytes myelinate peripheral nervous system axons (see Figure 12-2b). Myelinated axons of the central nervous system also contain neurofibral nodes, but they are not so numerous.

Myelin sheaths are first laid down during the later part of fetal development and during the first year of life. The amount of myelin increases from birth to maturity. Since myelination is still in progress during infancy, an infant's responses to stimuli are not as rapid or coordinated as those of an older child or an adult.

Structural Variation

Although all neurons conform to the general plan described, there are considerable differences in structure. For example, cell bodies range in diameter from 5 μm for the smallest cells to 135 μm for large motor neurons. The pattern of dendritic branching is varied and distinctive for neurons in different parts of the body. The axons of very small neurons are only a fraction of a millimeter in length and lack a myelin sheath, whereas axons of large neurons are over a meter long and are usually enclosed in a myelin sheath.

Classification

The different neurons in the body may be classified by structure and function.

The structural classification is based on the number of processes extending from the cell body. **Multipolar neurons** have several dendrites and one axon (see Figure 12-3a). Most neurons in the brain and spinal cord are of this type. **Bipolar neurons** have one dendrite and one axon, and are found in the retina of the eye, the inner ear, and the olfactory area. **Unipolar neurons** have only one process extending from the cell body. The single process divides into a central branch, which functions as an axon, and a peripheral branch, which functions as a dendrite. Unipolar neurons originate in the embryo as bipolar neurons and, during development, the axon and dendrite fuse into a single process. Unipolar neurons are found in posterior (sensory) root ganglia of spinal nerves.

The functional classification of neurons is based on the direction in which they transmit impulses. **Sensory neurons,** called **afferent neurons,** transmit impulses from receptors in the skin, sense organs, and viscera to the brain and spinal cord. They are usually unipolar (Figure 12-5). **Motor neurons,** called **efferent neurons,** convey impulses from the brain and spinal cord to effectors, which may be either muscles or glands (see Figure 12-3a). Other neurons, called **association** (**connecting** or **interneuron**) **neurons,** carry impulses from sensory neurons to motor

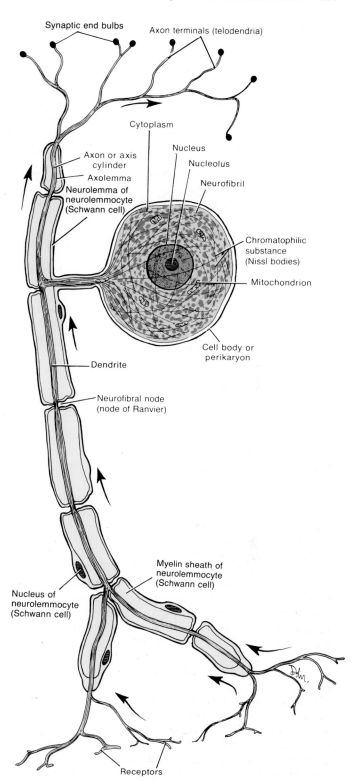

FIGURE 12-5 Structure of a typical afferent (sensory) neuron. Arrows indicate the direction in which the nerve impulse travels.

neurons and are located in the brain and spinal cord. Examples of association neurons are a *stellate cell,* a *cell of Martinotti* (mar'-ti-NOT-ē), a *horizontal cell of Cajal* (kā-HAL), and a *pyramidal cell* (pi-RAM-i-dal) cell. All are found in the cerebral cortex, the outer layer of the cerebrum. The *granule cell* and *Purkinje* (pur-KIN-jē) *cell* are association neurons in the cortex of the cerebellum. Most neurons in the body, perhaps 90 percent, are association neurons. Several association neurons are shown in Figure 12-6.

The processes of afferent and efferent neurons are arranged into bundles called *nerves*. Since nerves lie outside the central nervous system, they belong to the peripheral nervous system. The functional components of nerves are the nerve fibers, which may be grouped according to the following scheme.

   1. **General somatic afferent fibers** conduct impulses from the skin, skeletal muscles, and joints to the central nervous system.
   2. **General somatic efferent fibers** conduct impulses from the central nervous system to skeletal muscles. Impulses over these fibers cause the contraction of skeletal muscles.
   3. **General visceral afferent fibers** convey impulses from the viscera and blood vessels to the central nervous system.
   4. **General visceral efferent fibers** belong to the autonomic nervous system and are also called *autonomic fibers*. They convey impulses from the central nervous system to cause contractions of smooth and cardiac muscle and secretion by glands.

In addition to being grouped as nerves, neural tissue is also organized into other structures such as ganglia, tracts, nuclei, and horns. These are described in Chapter 13.

PHYSIOLOGY

Two striking features of nervous tissue are (1) its highly developed ability to generate and conduct electrical messages called nerve impulses, and (2) its limited ability to regenerate.

NERVE IMPULSE

At this point, we will consider the nerve impulse—your body's quickest way of controlling and maintaining homeostasis.

Membrane Potentials

Studies of cell membranes, especially in nerve and muscle cells, indicate that when a cell is at rest there is a consider-

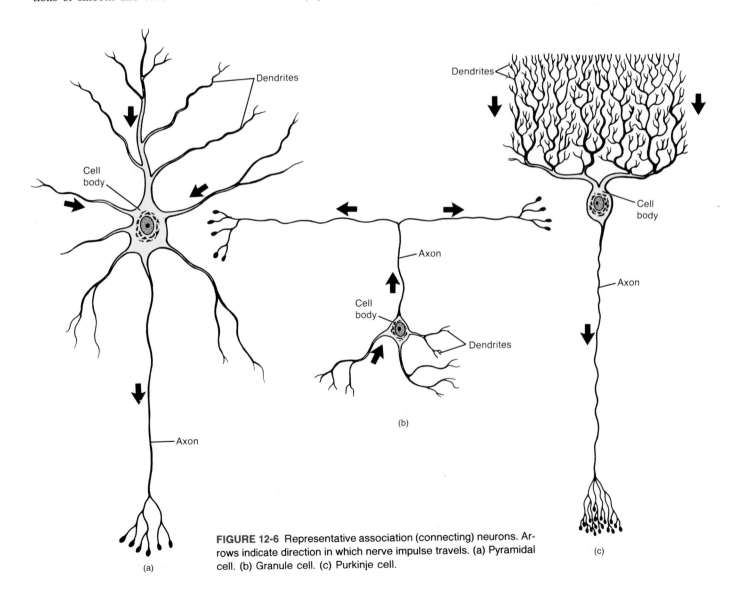

FIGURE 12-6 Representative association (connecting) neurons. Arrows indicate direction in which nerve impulse travels. (a) Pyramidal cell. (b) Granule cell. (c) Purkinje cell.

able difference between the ion concentration outside and inside the plasma membrane. In a resting neuron (one that is not conducting an impulse), there is a difference in electrical charges on either side of the membrane. This difference is partly the result of an unequal distribution of potassium (K^+) ions and sodium (Na^+) ions on either side of the membrane. In resting neurons, the K^+ ion concentration inside the cell is about 28 to 30 times greater than it is outside. The Na^+ ion concentration is about 14 times greater outside than inside. Another significant factor is the presence of large, nondiffusible negatively charged ions trapped in the cell. Most of them are proteins. What causes the outside of the nerve cell membrane to differ from the inside?

Even when a nerve cell is not conducting an impulse, it is actively transporting ions across its membrane. (The mechanism of active transport may be reviewed in Figure 3-5.) Na^+ ions are actively transported out and K^+ ions are actively transported in. The cellular system by which Na^+ and K^+ ions are actively transported simultaneously is called the **sodium-potassium pump** (Figure 12-7a). The operation of the pump requires the expenditure of ATP. The active transport of Na^+ and K^+ ions is unequal, that is, three Na^+ ions are actively transported out for every two K^+ ions actively transported in.

Neurons also contain a large number of negative ions, mostly protein anions, on the inside that cannot diffuse

FIGURE 12-7 Development of the membrane (resting) potential. (a) Schematic representation of the sodium-potassium pump and distribution of ions. Note that the large number of Na^+ ions outside the membrane results in an external positive charge. Although there are more K^+ ions inside the cell membrane than outside, there are many more negative ions inside the membrane. This results in a net internal negative charge. (b) Simplified representation of a polarized membrane.

outside or diffuse very poorly. Since Na^+ ions are positive and are actively transported outside the cell by the sodium-potassium pump, a positive charge develops outside the membrane. Even though K^+ ions are also positive and are actively transported to the inside of the cell by the sodium-potassium pump, there are insufficient K^+ ions to equalize the even larger number of nondiffusible negative ions trapped in the cell.

In addition, as a result of the operation of the sodium-potassium pump, there is a concentration and electrical gradient for Na^+ and K^+ ions. As a result, K^+ ions tend to diffuse (leak) out of the cell and Na^+ ions tend to diffuse into the cell. This diffusion occurs through membrane proteins called *potassium* and *sodium channels.* But, since membrane permeability to K^+ ions is 100 times greater than that to Na^+ ions, the number of K^+ ions that leave the cell is 100 times greater than the number of Na^+ ions that enter. The sodium-potassium pump not only actively transports Na^+ and K^+ ions but also establishes concentration gradients for the ions. The result is that there is a difference in charge on either side of the membrane—positive outside and negative inside. This difference in charge on either side of the membrane of a resting neuron is the **membrane (resting) potential.** Such a membrane is said to be **polarized** (Figure 12-7b). Since the sodium-potassium pump transports unequal numbers of Na^+ and K^+ ions that contribute to the membrane potential, the pump is referred to as **electrogenic.**

Electrical measurements of a polarized membrane indicate a voltage of about 70 millivolts (mV). This means that the inside of the membrane is 70 mV less than the outside, that is, the membrane potential is -70 mV. In subsequent discussions of membrane potentials, we will use the mV value that refers to the inside of the membrane, -70 mV. The mV value established is the result of the active separation of charges on either side of the membrane by the sodium-potassium pump.

Excitability

The events associated with the generation of nerve impulses will now be examined. The ability of nerve cells to respond to stimuli and convert them into nerve impulses is called **excitability.** A **stimulus** is any condition in the environment capable of altering the membrane potential.

If a stimulus of adequate strength (threshold stimulus) is applied to a polarized membrane, the membrane's permeability to Na^+ ions greatly increases at the point of stimulation (Figure 12-8a–d). As a result, sodium channels open and permit the influx of Na^+ ions by diffusion. The movement of Na^+ ions inward also results from the attraction of Na^+ ions to the negative ions on the inside of the membrane, and since there are more Na^+ ions entering than leaving, the electrical potential of the membrane begins to change. At first, the potential inside the membrane shifts from -70 mV toward 0, and then to a positive value. This loss of polarization is called **depolarization.** Depolarization begins at -69 mV, and from this point on the membrane is said to be **depolarized.** Throughout depolarization, the Na^+ ions continue to rush inside

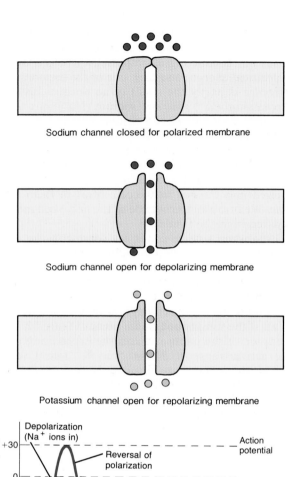

FIGURE 12-8 Initiation and conduction of a nerve impulse. (a–d) The stippled area containing the arrow represents the region of the membrane that has initiated and is conducting the nerve impulse. (e) Record of potential changes of a nerve impulse.

until the membrane potential is *reversed*—the inside of the membrane becomes positive and the outside negative. Electrical measurements indicate that the inside of the membrane is now +30 mV with respect to the outside. Thus the potential inside the membrane changes from −70 mV to 0 mV to +30 mV.

The inward movement of Na⁺ ions during depolarization is an example of a positive feedback system that is essential for homeostasis. As Na⁺ ions continue to move inward, depolarization increases. This in turn, causes more sodium channels to open. As more channels open, more Na⁺ ions move inward, and this leads to more depolarization, and so on. The continued movement of Na⁺ ions is independent of the initial stimulus; it occurs as part of a positive feedback cycle.

Once the events of depolarization have occurred, we say that an **action potential (nerve impulse)** is initiated. It lasts about 1 msec. The stimulated, negatively charged

point on the outside of the membrane sends out an electrical current to the positive point (still polarized) adjacent to it. This local current causes the adjacent inner part of the membrane to reverse its potential from −70 mV to +30 mV. The reversal repeats itself over and over until the nerve impulse is conducted the length of the neuron. As you can see, a nerve impulse is actually a wave of negativity that self-propagates along the outside surface of the membrane of a neuron. Depolarization and reversal of potential require only about 0.5 msec. Of all the cells of the body, only muscle and nerve cells produce action potentials. Their ability to do this is called excitability.

By the time the impulse has traveled from one point on the membrane to the next, the previous point becomes **repolarized**—its resting potential is restored. Repolarization results from a new series of changes in membrane permeability. The membrane now becomes more permeable to K$^+$ ions than it was at its resting potential level and is relatively impermeable again to Na$^+$ ions. The outward movement of K$^+$ ions occurs through potassium channels because of the high concentration of K$^+$ ions inside the membrane and the electrical gradient of the K$^+$ ions. As the K$^+$ ions move through the open potassium channels, the outer surface of the membrane becomes electrically positive. The heavy loss of positive ions leaves the inner surface of the membrane negative again. Eventually any ions that have moved into or out of the nerve cell are restored to their original sites.

The repolarization period returns the cell to its resting potential, from +30 mV to −70 mV. The neuron is now prepared to receive another stimulus and conduct it in the same manner. In fact, until repolarization occurs, the neuron cannot conduct another impulse. The period of time during which the neuron cannot generate another action potential is called the **refractory period.** The **absolute refractory period** refers to the period of time during which a second action potential cannot be initiated, even with a very strong stimulus. It corresponds roughly with the period of sodium permeability changes. Large fibers repolarize in about 1/2,500 second. Their absolute refractory period is 0.4 msec. Thus a second nerve impulse can be transmitted 1/2,500 second after the first—a total of up to 2,500 impulses per second. Small fibers, on the other hand, require as much as 1/250 second to repolarize. Their absolute refractory period is 4 msec. Thus they can transmit only 250 impulses per second. Under normal body conditions, the frequency of impulses conducted over nerve fibers may range between 10 and 500 impulses per second. The **relative refractory period** refers to the period of time during which a second action potential can be initiated, but only by a stronger-than-normal stimulus. It corresponds roughly to the period of increased potassium permeability.

K$^+$ ions that have leaked to the outside of the membrane and Na$^+$ ions that have leaked to the inside of the membrane must be restored to their original sides of the membrane. Ultimately, this is accomplished by the sodium-potassium pump, by which extra K$^+$ ions on the outside are pumped back in and extra Na$^+$ ions on the inside are pumped back out. However, the pump is not required for the initial repolarization of the membrane after *each* nerve impulse. This is accomplished by the outward diffusion of K$^+$ ions through the potassium channels. In fact, each time a single nerve impulse is conducted, only a minute amount of Na$^+$ ions enters the neuron and an equal amount of K$^+$ ions leaves the neuron. It literally takes thousands and thousands of nerve impulses to produce a massive shift in Na$^+$ and K$^+$ ions that results in cessation of nerve impulse conduction. The sodium-potassium pump goes into operation only after thousands of nerve impulses have caused a massive shift of ions.

A record of the electrical changes associated with a nerve impulse is illustrated in Figure 12-8e.

A sensory neuron is generally stimulated at its distal end by a receptor, a structure sensitive to changes in the environment. Association and motor neurons are usually stimulated at their dendrites or cell bodies by another neuron. However, if a neuron's plasma membrane is depolarized at some point other than the usual one, the impulse will travel in both directions over the cell membrane of the entire neuron.

All-or-None Principle

Any stimulus strong enough to initiate an impulse is referred to as a **threshold (liminal) stimulus.** When a stimulus is of threshold strength, we say the neuron has reached its threshold of stimulation. A single nerve cell, just like a single muscle fiber, transmits an impulse according to the **all-or-none principle:** If a stimulus is strong enough to generate an action potential, the impulse is conducted along the entire neuron at a constant and maximum strength for the existing conditions. The conduction is independent of any further intensity of the stimulus. However, conduction may be altered by conditions such as toxic materials in cells, fatigue, and malaise. An analogy may help in understanding this principle. If a long trail of gun powder were spilled along the ground, it could be ignited at one end to send a blazing signal down the entire length of the trail. It would not matter how tiny or great the triggering spark or flame or explosion was; the blazing signal moving along the trail of gunpowder would be just the same, a maximal one (or none at all if it never started).

Any stimulus weaker than a threshold stimulus is termed a **subthreshold** or **(subliminal) stimulus.** Such a stimulus is incapable of initiating a nerve impulse. If, however, a second stimulus or a series of subthreshold stimuli is quickly applied to the neuron, the cumulative effect may be sufficient to initiate an impulse. This phenomenon is called summation and is discussed shortly.

Saltatory Conduction

Thus far we have considered nerve impulse conduction via unmyelinated fibers. The step-by-step depolarization described is called **continuous conduction.** In myelinated fibers, conduction is somewhat different. The myelin sheath surrounding a fiber contains myelin, a phospholipid that does not conduct an electric current. It thus forms an insulating layer around the fiber and virtually inhibits the movement of ions. The myelin sheath is interrupted at various intervals called neurofibral nodes (nodes of Ranvier). At the nodes, membrane depolarization can occur and action potentials can be generated. But beneath the myelin sheath, depolarization is impossible since ionic movement is inhibited. When an impulse is conducted along a myelinated fiber, it moves from one node to another through the surrounding extracellular fluids and through the axoplasm. Thus the impulse jumps from node to node. This type of impulse conduction, characteristic of myelinated fibers, is called **saltatory conduction** (*saltare* = leaping).

Saltatory conduction is a valuable asset to your homeostasis. Since the impulse jumps long intervals as it moves from one node to the next, the speed of conduction is greatly increased. The impulse travels much faster than in the step-by-step depolarization process involved in an unmyelinated fiber of equal diameter. Another advantage of saltatory conduction is greater efficiency as a result of energy expenditure. Saltatory conduction prevents depolarization of large areas of the plasma membrane of the fiber, thus preventing leakage of large amounts of Na^+ ions to the inside of the membrane and K^+ ions to the outside each time a nerve impulse is transmitted. This results in a lower expenditure of energy by the sodium-potassium pump.

Speed of Nerve Impulses

The speed of a nerve impulse is independent of stimulus strength. Once a neuron reaches its threshold of stimulation, the speed of the nerve impulse is normally determined by temperature, the diameter of the fiber, and the presence or absence of myelin.

When warmed, nerve fibers conduct impulses at higher speeds; when cooled, they conduct impulses at lower speeds. Localized cooling of a nerve can block nerve conduction. Pain resulting from injured tissue can be reduced by the application of cold because the nerve fibers carrying the pain sensations are partially blocked.

Fibers with large diameters conduct impulses faster than those with small ones. Fibers with the largest diameter are called **A fibers** and are all myelinated. The A fibers have a brief absolute refractory period and are capable of saltatory conduction. They transmit impulses at speeds up to 130 m/sec. The A fibers are located in the axons of large sensory nerves that relay impulses associated with touch, pressure, position of joints, heat, and cold. They are also found in all motor nerves that convey impulses to the skeletal muscles. Sensory A fibers generally connect the brain and spinal cord with sensors that detect danger in the outside environment. Motor A fibers innervate the muscles that can do something about the situation. If you touch a hot object, information about the heat passes over sensory A fibers to the spinal cord. There it is relayed to motor A fibers that stimulate the muscles of the hand to withdraw instantaneously. The A fibers are located where split-second reaction may mean survival.

Other fibers, called B and C fibers, conduct impulses more slowly and are generally found where instantaneous response is not a life-and-death matter. **B fibers** have a middle-sized diameter and a somewhat longer absolute refractory period than A fibers. They are also myelinated and therefore capable of saltatory conduction. They conduct impulses at speeds of about 10 m/sec. B fibers are found in nerves that transmit impulses from the skin and viscera to the brain and spinal cord. They also constitute all the axons of the visceral efferent neurons located in the motor nerves that leave the lower part of the brain and spinal cord and terminate in relay stations called *ganglia* (GANG-lē-a). The ganglia ultimately link with other fibers that stimulate the smooth muscle and glands of the viscera.

C fibers have the smallest diameter and the longest absolute refractory periods. They conduct impulses at the rate of about 0.5 m/sec. C fibers are unmyelinated and incapable of saltatory conduction. They are located in nerves that conduct impulses from the skin and in visceral nerves. These fibers conduct impulses for pain and perhaps impulses for touch, pressure, heat, and cold from the skin and pain receptors from the viscera. C fibers are located in all motor nerves that lead from the ganglia and stimulate the smooth muscle and glands of the viscera. Examples of the motor functions of B and C fibers are constricting and dilating the pupils, increasing and decreasing the heart rate, and contracting and relaxing the urinary bladder—functions of the autonomic nervous system.

CONDUCTION ACROSS SYNAPSES

An impulse is conducted not only along the length of a neuron but also from one neuron to another or to an effector such as a muscle or gland. Impulses are conducted from a neuron to a muscle cell across an area of contact called a **neuromuscular (myoneural) junction** or **motor end plate,** which was discussed in detail in Chapter 10. The area of contact between a neuron and glandular cells is known as a **neuroglandular junction.** Together, neuromuscular and neuroglandular junctions are known as **neuroeffector junctions.**

Impulses are conducted from one neuron to another across a **synapse**—a junction between two neurons. The term *synapsis* means connection. The synapse is essential for homeostasis because of its ability to transmit certain impulses and inhibit others. Much of an organism's ability to learn will probably be explained in terms of synapses. Moreover, most diseases of the brain and many psychiatric disorders result from a disruption of synaptic communication. Synapses are also the sites of action for most drugs that affect the brain, including therapeutic and addictive substances. Figure 12-9 shows that within a synapse is a minute gap, about 20 nm across, called the **synaptic cleft. A presynaptic neuron** is a neuron located before a synapse. A **postsynaptic neuron** is located after a synapse.

The axon terminals of neurons terminate in bulblike structures referred to as **synaptic end bulbs.** The synaptic end bulbs of a presynaptic neuron commonly synapse with the dendrites, cell body, or axon hillock of a postsynaptic neuron. Accordingly, synapses may be classified as *axodendritic, axosomatic,* and *axoaxonic.* The synaptic end bulbs from a single presynaptic neuron may synapse with several postsynaptic neurons. Such an arrangement, called **divergence,** permits a single presynaptic neuron to influence several postsynaptic neurons or several muscle or gland cells at the same time (see Figure 12-13a). In another arrangement, called **convergence,** the synaptic end bulbs of several presynaptic neurons synapse with a single postsynaptic neuron (see Figure 12-13b). This arrangement permits stimulation or inhibition of the postsynaptic neuron.

At a synapse there is only **one-way inpulse conduction**—from a presynaptic axon to a postsynaptic dendrite, cell body, or axon hillock. Impulses must move forward over their pathways. They cannot back up into another

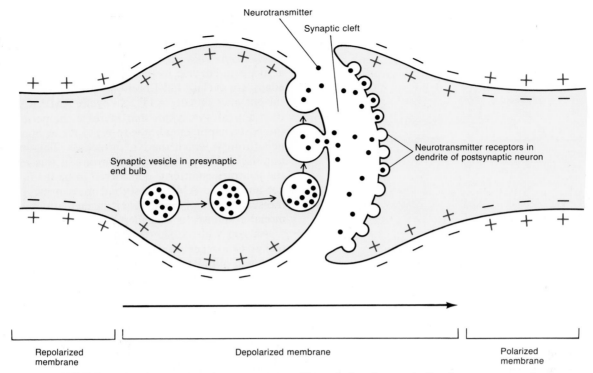

Neurotransmitter

Synaptic cleft

Synaptic vesicle in presynaptic end bulb

Neurotransmitter receptors in dendrite of postsynaptic neuron

Repolarized membrane

Depolarized membrane

Polarized membrane

FIGURE 12-9 Impulse conduction at synapses. Shown is impulse conduction from a presynaptic end bulb across a synapse to a postsynaptic dendrite in which synaptic vesicles fuse with the presynaptic membrane and discharge neurotransmitter into the synaptic cleft.

presynaptic neuron. Such a mechanism is crucial in preventing impulse conduction along improper pathways.

Whether an impulse is conducted across a synapse, neuromuscular junction, or neuroglandular junction depends on the presence of chemicals called **neurotransmitters (transmitter substances).** These chemicals are made by the neuron, usually from amino acids. Following its production and transportation to the synaptic end bulbs, the neurotransmitter is stored in the end bulbs in small membrane-enclosed sacs called **synaptic vesicles** (Figure 12-9). Each of the thousands of synaptic vesicles present may contain between 10,000 and 100,000 neurotransmitter molecules.

When a nerve impulse arrives at a synaptic end bulb of a presynaptic neuron, it is believed that a small amount of calcium ions enters the end bulb from interstitial fluid, attracts synaptic vesicles to the plasma membrane, and helps to liberate the neurotransmitter molecules from the vesicles. Some scientists believe that the synaptic vesicles fuse with the plasma membrane of the presynaptic neuron, form openings, and release the neurotransmitter through the openings into the synaptic cleft. Others think that neurotransmitter stored in cytoplasm rather than in synaptic vesicles leaves through small protein channels formed under the influence of calcium ions and enters the synaptic cleft. In either case, the neurotransmitter enters the synaptic cleft and, depending on the chemical nature of the neurotransmitter and the interaction of the neurotransmitter with receptors of the postsynaptic plasma membrane, several things can happen.

Excitatory Transmission

An **excitatory transmitter-receptor interaction** is one that can lower (make less negative) the postsynaptic neuron's membrane potential so that a new impulse can be generated across the synapse. If the potential is lowered enough, the membrane becomes depolarized, the potential inside the membrane becomes positive and the potential outside becomes negative, and a nerve impulse is initiated. Generally, the release of a neurotransmitter by a single presynaptic end bulb is not sufficient to develop an action potential in a postsynaptic neuron. However, its release does bring the resting membrane potential closer to threshold level as Na$^+$ ions move into the cell. This change from the resting membrane potential level in the direction of the threshold level is called the **excitatory postsynaptic potential (EPSP)** (Figure 12-10a). The EPSP is always less negative than the resting membrane potential of the neuron, but more negative than its threshold level. Even though the release of a neurotransmitter by a single presynaptic end bulb is not sufficient to initiate an action potential in a postsynaptic neuron, the postsynaptic neuron does become more excitable to impulses from presynaptic neurons, so that the possibility for generating an impulse is greater. This effect is called **facilitation,** that is, near excitation so that subsequent stimuli can generate an impulse. In other words, the postsynaptic neuron is prepared for subsequent stimuli that can trigger an impulse.

The EPSP lasts only a few milliseconds. If several pre-

FIGURE 12-10 Comparison between (a) EPSP and (b) IPSP.

synaptic end bulbs release their neurotransmitter at about the same time, however, the combined effect may initiate a nerve impulse—an effect known as **summation.** If hundreds of presynaptic end bulbs release their neurotransmitter simultaneously, they increase the chance of initiating a nerve impulse in the postsynaptic neuron. The greater the summation, the greater the probability an impulse will be initiated. When the summation is the result of the accumulation of neurotransmitter from several presynaptic end bulbs, it is called **spatial summation.** When summation is the result of the accumulation of neurotransmitter from a single presynaptic end bulb firing two or more times in rapid succession, it is called **temporal summation.** Since the EPSP lasts only about 15 msec, the second firing must follow quickly if temporal summation is to occur. The time required for the impulse to actually cross a synaptic cleft—the **synaptic delay**—is about 0.5 msec. This delay is caused by the liberation of neurotransmitter, its passage across the synapse, stimulation of the postsynaptic neuron to become more permeable to Na$^+$ ions, and the inward movement of Na$^+$ ions that initiates the nerve impulse in the postsynaptic neuron.

Excitatory transmitter-receptor interactions are thought to occur by two different mechanisms. In the first mechanism, neurotransmitters bind to proteins on the postsynaptic plasma membrane, called **neurotransmitter receptors** (see Figures 12-9 and 12-11). This binding causes sodium channels in the plasma membrane to be-

come more permeable to Na$^+$ ions. The inward movement of Na$^+$ ions results in depolarization and initiation of a nerve impulse. Other neurotransmitters use a different mechanism. They bind to neurotransmitter receptors and activate an enzyme in the postsynaptic membrane called **adenylate cyclase** (ad-EN-nil-āt SĪ-klās). Once activated, the enzyme converts ATP into **cyclic AMP.** Cyclic AMP then activates enzymes that increase the permeability of the postsynaptic membrane to Na$^+$ ions by opening channels through which the Na$^+$ ions pass. This initiates the impulse in the postsynaptic neuron. In this mechanism, the neurotransmitter is considered to be the first messenger and cyclic AMP the second messenger. The second messenger role of cyclic AMP in mediating certain hormonal responses is considered in detail in Chapter 18.

As you will see shortly, once a neurotransmitter attaches to a receptor site, it must be rapidly inactivated or it will stimulate the postsynaptic neuron, muscle, or gland indefinitely.

Inhibitory Transmission

An **inhibitory transmitter-receptor interaction** is one that can inhibit impulse generation at a synapse. Whereas excitatory transmitter-receptor interactions make the postsynaptic neuron's resting membrane potential less negative, inhibitory transmitter-receptor interactions make the postsynaptic neuron's resting membrane potential more negative. This is referred to as **hyperpolarization.** When it is at rest, the cell interior becomes even more negative in comparison to the outside, making it even more difficult for the neuron to generate an action potential. The alteration of the postsynaptic membrane in which the resting membrane potential is made more negative is called the **inhibitory postsynaptic potential (IPSP)** (Figure 12-10b). When the neurotransmitter attaches to the receptor site, the membrane becomes less permeable to Na$^+$ ions, but more permeable to the movement of K$^+$ ions to the exterior of the membrane and Cl$^-$ ions to the interior. As a result, there is an increase in internal negativity, or hyperpolarization. In contrast, the EPSP is less negative than the resting membrane potential of a neuron, and the IPSP is more negative than the resting membrane potential.

Integration at Synapses

A single postsynaptic neuron synapses with many presynaptic neurons. Some presynaptic end bulbs produce excitation and some produce inhibition. The sum of all the effects, excitatory and inhibitory, determines the effect on the postsynaptic neuron. Thus the postsynaptic neuron is an **integrator.** It receives signals, integrates them, and then responds accordingly. The postsynaptic neuron may respond in the following ways:

1. If the excitatory effect is greater than the inhibitory effect, but less than the threshold level of stimulation, the result is facilitation.

2. If the excitatory effect is greater than the inhibitory effect, but equal to or higher than the threshold level of stimulation, the result is generation of an impulse.

3. If the inhibitory effect is greater than the excitatory effect, the result is inhibition of the impulse.

NEUROTRANSMITTERS

Perhaps the best-studied neurotransmitter is **acetylcholine** (as'-ē-til-KŌ-lēn) or **ACh.** It is a neurotransmitter released by many neurons outside the brain and spinal cord and by some neurons inside the brain and cord. This neurotransmitter was discussed in Chapter 10 in relation to impulse conduction from a motor neuron to a muscle fiber across the neuromuscular junction (motor end plate). Following the arrival of a nerve impulse at an axon terminal, calcium ions enter an axon terminal and cause the release of ACh from synaptic vesicles or the cytoplasm of the terminal. ACh is released into the synaptic cleft in multiple units called *quanta.* One quantum contains between 1,000 to 10,000 molecules. At neuromuscular junctions, ACh binds to neurotransmitter receptors on the muscle fiber membrane and increases its permeability to Na^+ and K^+ ions. The neurotransmitter (ACh) receptor is an integral protein in the plasma membrane of the muscle fiber (Figure 12-11). When ACh molecules bind to the receptor, a change occurs in the receptor, causing its channel to open. As a result, there is an inward movement of Na^+ ions. In the absence of ACh, the receptor channel remains closed. The inward movement of Na^+ ions depolarizes the membrane, and ultimately a nerve impulse is generated, causing the muscle fiber to contract.

As long as ACh is present in the neuromuscular junction, it can stimulate a muscle fiber almost indefinitely. The transmission of a continuous succession of impulses by ACh is normally prevented by an enzyme called **acetylcholinesterase (AChE),** or simply **cholinesterase** (kō'-lin-ES-ter-ās). AChE is found on the outside surface of the subneural clefts of the membrane of muscle fibers. Within 1/500 sec, AChE inactivates ACh by breaking ACh down into its components, acetate and choline. This action permits the membrane of the muscle fiber to repolarize almost immediately so that another impulse may be generated. When the next impulse comes through, the synaptic vesicles release more ACh, the impulse is conducted, and AChE again inactivates ACh. This cycle is repeated over and over again. Eventually, acetate and choline reenter the axon terminal, where they are synthesized into new ACh in the presence of the enzyme choline acetyltransferase. Then, new synaptic vesicles form and ACh is distributed into vesicles and cytoplasm of the axon terminal.

ACh is released at some neuromuscular junctions that contain cardiac and smooth muscle, as well as those that contain skeletal muscle. It is also released at some neuroglandular junctions. In the brain and spinal cord, it appears that the effects of ACh are mediated through cyclic AMP. Although ACh leads to excitation in many parts of the body, it is inhibitory with respect to the heart (vagus nerve).

There is considerable evidence that a substance called **gamma aminobutyric** (GAM-ma am-i-nō-byoo-TĒR-ik) **acid (GABA)** leads to inhibition in the central nervous system. It probably exerts its effect by hyperpolarizing (making more negative) the postsynaptic membrane according to the mechanism previously described. The amino acid **glycine** leads to inhibition in the spinal cord.

Neurotransmitters in the central nervous system will be discussed in more detail in Chapter 14.

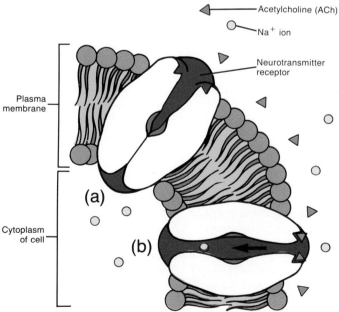

FIGURE 12-11 Neurotransmitter receptors. Such receptors are integral proteins in the plasma membrane. (a) In the absence of ACh, the channel remains closed. (b) When two molecules of ACh bind to the receptor, a change occurs permitting positive ions, such as Na^+ ions, to pass through.

CLINICAL APPLICATION

There are many ways that **synaptic conduction can be altered** by disease, drugs, and pressure. In Chapter 10 it was noted that *myasthenia gravis* results from antibodies directed against acetylcholine receptors on muscle fiber membranes at neuromuscular junctions, causing dysfunctions in muscular contractions. *Alkalosis,* an increase in pH above 7.45 up to about 8.0, results in increased excitability of neurons that can cause cerebral convulsions, while *acidosis,* a decrease in pH below 7.35 down to about 6.80, results in a depression of neuronal activity and can produce a comatose state.

Curare, the deadly South American Indian arrowhead poison, also competes for acetylcholine receptor sites. Since the muscles of respiration depend on neuromuscular transmission to initiate their contraction, death results from asphyxiation. *Neostigmine* and *physostigmine* are anticholinesterase agents that combine with acetylcholinesterase to inactivate it for several hours. As a result, acetylcholine accumulates and produces repetitive stimulation of muscles. This causes muscular spasm, and death due to laryngeal spasm can ensue. *Diisopropyl fluorophosphate* is a very powerful nerve gas that inactivates acetylcholinesterase for up to several weeks, making it a particularly lethal drug. The *botulism toxin* inhibits the release of acetylcholine, thus inhibiting muscle contraction. The toxin is exceedingly potent in even small amounts (less than 0.0001 mg) and is the substance involved in one type of food poisoning. *Hypnotics, tranquilizers,* and *anesthetics* depress synaptic conduction by increasing the threshold for excitation of neurons, while *caffeine, benzedrine,* and *nicotine* reduce the threshold for excitation of neurons and result in facilitation.

Pressure has an effect on impulse conduction also. If excessive or prolonged pressure is applied to a nerve, impulse conduction is interrupted and part of the body may "go to sleep," producing a tingling sensation. This sensation is caused by an accumulation of waste products and a depressed circulation of blood.

REGENERATION

Unlike the cells of epithelial tissue, neurons have only limited powers for **regeneration,** that is, a natural ability to renew themselves. Around 6 months of age, the cell bodies of most developing nerve cells lose their mitotic apparatus (centrioles and mitotic spindles) and their ability to reproduce. Thus, when a neuron is damaged or destroyed, it cannot be replaced by the daughter cells of other neurons. A neuron destroyed is permanently lost, and only some types of damage may be repaired.

Damage to some types of myelinated axons often can be repaired if the cell body remains intact and if the cell that performs the myelination remains active. Axons in the peripheral nervous system are myelinated by neurolemmocytes (Schwann cells). They proliferate following axonal damage and their neurolemmas form a tube that assists in regeneration (Chapter 13). Axons in the brain and spinal cord (central nervous system) are myelinated by oligodendroglial cells. These cells do not form neurolemmas to assist in regeneration and do not survive following axonal damage. An added complication in the central nervous system is that following axonal damage, the affected region is rapidly converted into a special form of scar tissue by astroglial proliferation. The scar tissue forms a barrier to regeneration. Thus, an injury to the brain or spinal cord is also permanent because axonal regeneration is blocked by rapid scar tissue formation. An injury to a nerve in the arm (peripheral nervous system) may repair itself before scar tissue forms, and so some nerve function may be restored.

ORGANIZATION OF NEURONS

In subsequent chapters, we will be concerned with the structure and physiology of the nervous system. To give you a better understanding of how your nervous system helps maintain homeostasis, we now examine how synapses are organized into functional units.

The central nervous system contains millions of neurons. Their arrangement is not haphazard. They are organized into definite patterns called **neuronal pools.** Each pool differs from all others and has its own role in regulating homeostasis.

A neuronal pool may contain thousands or even millions of neurons. To illustrate the composition of a neuronal pool, a simplified version is given in Figure 12-12. This example contains only five postsynaptic neurons and two incoming presynaptic neurons. The postsynaptic neurons are subject to stimulation by the incoming presynaptic neurons. The postsynaptic neurons in the pool may be stimulated by one or several presynaptic end bulbs. Moreover, the incoming presynaptic end bulbs may produce facilitation, excitation, or inhibition. A principal feature of a neuronal pool is the location of the presynaptic neurons in relation to the postsynaptic neurons. Compare the location of presynaptic axon 1 with that of postsynaptic neuron B. Since they are aligned, more presynaptic end bulbs of axon number 1 synapse with postsynaptic neuron B than with postsynaptic neurons A or C. We say that postsynaptic neuron B is in the center of the field of presynaptic axon 1. Consequently, postsynaptic neuron B usually receives sufficient presynaptic end bulbs from axon 1 to generate an impulse. This region where the neuron in the pool fires is called the **discharge zone.**

Now look outside the field of presynaptic axon 1. Note its relation to postsynaptic neuron A. Here postsynaptic neuron A in the pool is receiving few presynaptic end bulbs from the axon supplying postsynaptic neuron B. Thus there are insufficient presynaptic end bulbs to fire postsynaptic neuron A, but enough to cause facilitation. We therefore call this area the **facilitated zone.** Presynaptic axon 1, when stimulated, will cause excitation of postsynaptic neuron B and facilitation of postsynaptic neuron A.

Neuronal pools in the central nervous system are arranged in patterns over which the impulses are conducted. These are termed **circuits. Simple series circuits** are arranged so that a presynaptic neuron stimulates a single neuron in a pool. The single neuron then stimulates another and so on. In other words, the impulse is relayed from one neuron to another in succession as a new impulse is generated at each synapse.

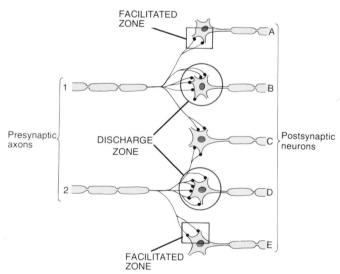

FIGURE 12-12 Relative positions of discharge and facilitated zones in a very simplified version of a neuronal pool.

Most circuits, however, are more complex. In a **diverging circuit,** the impulse from a single presynaptic neuron causes the stimulation of increasing numbers of cells along the circuit (Figure 12-13a). An example of such a circuit is a single motor neuron in the brain stimulating numerous other motor neurons in the spinal cord that, in turn, leave the spinal cord where each stimulates many skeletal muscle fibers. Thus a single impulse may result in the contraction of several skeletal muscle fibers. In another kind of diverging circuit, impulses from one pathway are relayed to other pathways so the same information travels in various directions at the same time. This circuit is common along sensory pathways of the nervous system.

Another kind of circuit is called a **converging circuit** (Figure 12-13b). In one pattern of convergence, the postsynaptic neuron receives impulses from several fibers of the same source. Here there is the possibility of strong excitation or inhibition. In a second pattern, the postsynaptic neuron receives impulses from several different sources. Here there is a possibility of reacting the same way to different stimuli. Suppose your reaction to vomit is distinctly unpleasant. The smell of vomit (one kind of stimulus), the sight of vomit (another kind), or just reading about vomit (still another kind) might all have the same effect on you—an unpleasant one.

Some circuits in your body are constructed so that once the presynaptic cell is stimulated, it will cause the postsynaptic cell to transmit a series of impulses. One such circuit is called a **reverberating (oscillatory) circuit** (Figure 12-13c). In this pattern, the incoming impulse stimulates the first neuron, which stimulates the second, which stimulates the third, and so on. Branches from the second and third neurons synapse with the first, however, sending the impulse back through the circuit again and again. A central feature of the reverberating circuit is that once fired, the output signal may last from a few seconds to many hours. The duration depends on the number and arrangement of neurons in the circuit. Among the body responses thought to be the result of output signals from reverberating circuits are the rate of breathing, coordinated muscular activities, waking up, and sleeping (when reverberation stops). Some scientists think reverberating circuits are related to short-term memory. One form of epilepsy (grand mal) is probably caused by abnormal reverberating circuits.

A final circuit worth consideration is the **parallel after-discharge circuit** (Figure 12-13d). Like a reverberating circuit, a parallel after-discharge circuit is constructed so the postsynaptic cell transmits a series of impulses. In a parallel after-discharge circuit, a single presynaptic

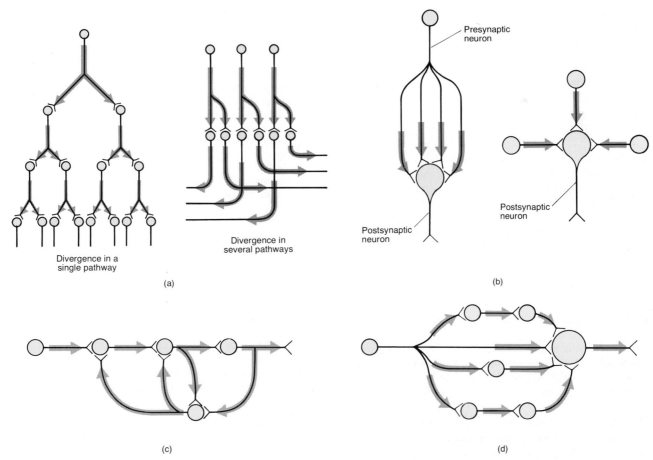

Divergence in a
single pathway

Divergence in
several pathways

(a)

Presynaptic
neuron

Postsynaptic
neuron

(b)

Postsynaptic
neuron

(c)

(d)

FIGURE 12-13 Circuits of neuronal pools. (a) Diverging. (b) Converging. (c) Reverberating. (d) Parallel after-discharge.

cell stimulates a group of neurons, each of which synapses with a common postsynaptic cell. The advantage of this circuit is that the postsynaptic neuron can send out a stream of impulses in succession as they are received. The impulses leave the postsynaptic neuron once every 1/2000 second. This circuit has no feedback system. Once all the neurons in the circuit have transmitted their impulses to the postsynaptic neuron, the circuit is broken. It is thought that the parallel after-discharge circuit is employed for precise activities like mathematical calculations.

STUDY OUTLINE

Organization (p. 262)

1. The nervous system controls and integrates all body activities by sensing changes (sensory), interpreting them (integrative), and reacting to them (motor).
2. The central nervous system (CNS) consists of the brain and spinal cord.
3. The peripheral nervous system (PNS) is classified into an afferent system and an efferent system.
4. The efferent system is subdivided into a somatic nervous system and an autonomic nervous system.
5. The somatic nervous system (SNS) consists of efferent neurons that conduct impulses from the central nervous system to skeletal muscle tissue.
6. The autonomic nervous system (ANS) contains efferent neurons that convey impulses from the central nervous system to smooth muscle tissue, cardiac muscle tissue, and glands.

Histology (p. 263)

Neuroglia (p. 263)

1. Neuroglia are specialized tissue cells that support neurons, attach neurons to blood vessels, produce the myelin sheath, and carry out phagocytosis.
2. Neuroglial cells include astrocytes, oligodendrocytes, microglia, and ependyma.

Neurons (p. 264)

1. Neurons, or nerve cells, consist of a cell body (perikaryon), dendrites that pick up stimuli and convey impulses to the cell body, and usually a single axon. The axon conducts impulses from the neuron to the dendrites or cell body of another neuron or to an effector organ of the body.
2. On the basis of structure, neurons are multipolar, bipolar, and unipolar.
3. On the basis of function, sensory (afferent) neurons conduct impulses to the central nervous system; association (connecting) neurons conduct impulses to other neurons, including motor neurons; and motor (efferent) neurons conduct impulses to effectors.

Physiology (p. 268)

Nerve Impulse (p. 268)

1. The nerve impulse is the body's quickest way of controlling and maintaining homeostasis.
2. The membrane of a nonconducting neuron is positive outside and negative inside due to the operation of the sodium-potassium pump, and the membrane is said to be polarized.
3. When a stimulus causes the inside of the cell membrane to become positive and the outside negative, the membrane is said to have an action potential, which travels from point to point along the membrane. The traveling action potential is a nerve impulse. The ability of a neuron to respond to a stimulus and convert it into a nerve impulse is called excitability.

4. Restoration of the resting potential is called repolarization. The period of time during which the membrane recovers and cannot initiate another action potential is called the refractory period.
5. According to the all-or-none principle, if a stimulus is strong enough to generate an action potential, the impulse travels at a constant and maximum strength for the existing conditions.
6. Nerve impulse conduction in which the impulse jumps from neurofibral node to node is called saltatory conduction.
7. Fibers with larger diameters conduct impulses faster than those with smaller diameters; myelinated fibers conduct impulses faster than unmyelinated fibers.

Conduction Across Synapses (p. 272)

1. Impulse conduction can occur from one neuron to another or from a neuron to an effector.
2. The junction between neurons is called a synapse.
3. At a synapse there is only one-way impulse conduction from a presynaptic axon to a postsynaptic dendrite, cell body, or axon hillock.
4. An excitatory transmitter-receptor interaction is one that can lower (make less negative) the postsynaptic neuron's membrane potential, so that a new impulse can be generated across the synapse.
5. Facilitation refers to near excitation, so that subsequent stimuli can generate an impulse.
6. If several presynaptic end bulbs release their neurotransmitter at about the same time, the combined effect may generate a nerve impulse, resulting in the phenomenon referred to as summation. Summation may be spatial or temporal.
7. The time required for an impulse to cross a synaptic cleft, about 0.5 msec, is called synaptic delay.
8. An inhibitory transmitter-receptor interaction is one that can raise (make more negative) the postsynaptic neuron's membrane potential and thus inhibit an impulse at a synapse.
9. The postsynaptic neuron is an integrator. It receives signals, integrates them, and then responds accordingly.

Neurotransmitters (p. 275)

1. It is thought that the neurotransmitter that causes excitation in a major portion of the central nervous system is acetylcholine (ACh). An enzyme called acetylcholinesterase (AChE) inactivates ACh.
2. Neurotransmitters that are probably inhibitory are gamma aminobutyric acid (GABA) and glycine.

Regeneration (p. 276)

1. Around 6 months of age, the cell body loses its mitotic apparatus and is no longer able to divide.
2. Nerve fibers (axis cylinders) that have a neurolemma are capable of regeneration.

Organization of Neurons (p. 276)

1. Neurons in the central nervous system are organized into definite patterns called neuronal pools. Each pool differs from all others and has its own role in regulating homeostasis.

2. Neuronal pools are organized into circuits. These include simple series, diverging, converging, reverberating (oscillatory), and parallel after-discharge circuits.

REVIEW QUESTIONS

1. Describe the three basic functions of the nervous system that are necessary to maintain homeostasis.
2. Distinguish between the central and peripheral nervous systems, and describe the functions of each subdivision.
3. Relate the terms *voluntary* and *involuntary* to the nervous system.
4. What are neuroglia? List the principal types and their functions. Why are they important clinically?
5. Define a neuron. Diagram and label a neuron. Next to each part list its function.
6. Distinguish between axoplasmic flow and axonal transport. Why is axonal transport important clinically?
7. What is a myelin sheath? How is it formed?
8. Define the neurolemma. Why is it important?
9. Discuss the structural classification of neurons. Give an example of each.
10. What are the structural differences between a typical afferent and efferent neuron? Give several examples of association neurons.
11. Describe the functional classification of neurons.
12. Distinguish among the following kinds of fibers: general somatic afferent, general somatic efferent, general visceral afferent, and general visceral efferent.
13. Define excitability.
14. Outline the principal steps in the origin and conduction of a nerve impulse.
15. Define the following: membrane (resting) potential, polarized membrane, action potential, depolarized membrane, repolarized membrane, and refractory period.
16. What is the all-or-none principle? Relate it to threshold stimulus, subthreshold stimulus, and summation.
17. What is saltatory conduction? Why is it important?
18. What factors determine the speed of nerve impulses?
19. What events are involved in the conduction of a nerve impulse across a synapse?
20. Distinguish between excitatory and inhibitory transmission.
21. Why does one-way impulse conduction occur?
22. How are nerve impulses inhibited? Of what advantage is this to the body?
23. Why is the postsynaptic neuron called an integrator?
24. What is a neurotransmitter? Describe, in detail, the release, action, and inactivation of acetylcholine (ACh) at a neuromuscular junction.
25. List several probable neurotransmitters, indicate their locations in the nervous system, and whether or not they lead to excitation or inhibition.
26. How do disease, drugs, and pressure affect synaptic transmission?
27. What factors determine neuron regeneration?
28. Distinguish between the discharge zone and facilitated zone and neuronal pool.
29. What is a neuron circuit? Distinguish among simple series, diverging, converging, reverberating (oscillatory), and parallel after-discharge circuits.

The Spinal Cord and the Spinal Nerves

13

- Describe how neural tissue is grouped.
- Describe the gross anatomical features of the spinal cord.
- Explain how the spinal cord is protected.
- Describe the structure and location of the spinal meninges.
- Discuss the location, general technique, purpose, and significance of a spinal puncture.
- Describe the structure of the spinal cord.
- Explain the functions of the spinal cord as a conduction pathway and a reflex center.
- List the location, origin, termination, and function of the principal ascending and descending tracts of the spinal cord.
- Describe the components of a reflex arc and its relationship to homeostasis.
- Compare the functional anatomy of a stretch reflex, tendon reflex, flexor reflex, and crossed extensor reflex.
- List and describe several clinically important reflexes.
- Name the 31 pairs of spinal nerves.
- Describe the composition and coverings of a spinal nerve.
- Explain how a spinal nerve branches upon leaving an intervertebral foramen.
- Define a plexus and an intercostal nerve.
- Define a dermatome and its clinical importance.
- Describe spinal cord injury and list the immediate and long-range effects.
- Identify the effects of peripheral nerve damage and conditions necessary for its regeneration.
- Explain the causes and symptoms of neuritis, sciatica, and shingles.

In this chapter, our main concern will be to study the structure and function of the spinal cord and the nerves that originate from it. Keep in mind, however, that the spinal cord is continuous with the brain, and that together they constitute the central nervous system.

GROUPING OF NEURAL TISSUE

The term **white matter** refers to aggregations of myelinated axons from many neurons supported by neuroglia. The lipid substance myelin has a whitish color that gives white matter its name. The **gray matter** of the nervous system contains either nerve cell bodies and dendrites or bundles of unmyelinated axons and neuroglia. The absence of myelin in these areas accounts for their gray color.

A **nerve** is a bundle of fibers located outside the central nervous system. Since the dendrites of somatic afferent neurons and axons of somatic efferent neurons of the peripheral nervous system are myelinated, most nerves are white matter. Nerve cell bodies that lie outside the central nervous system are generally grouped with other nerve cell bodies to form **ganglia** (GANG-lē-a; *ganglion* = knot). Ganglia, since they are made up principally of nerve cell bodies, are masses of gray matter.

A **tract** is a bundle of fibers in the central nervous system. Tracts may run long distances up and down the spinal cord. Tracts also exist in the brain and connect parts of the brain with each other and with the spinal cord. The chief spinal tracts that conduct impulses up the cord are concerned with sensory impulses and are called *ascending tracts.* Spinal tracts that carry impulses down the cord are motor tracts and are called *descending tracts.* The major tracts consist of myelinated fibers and are therefore white matter. A **nucleus** is a mass of nerve cell bodies and dendrites in the central nervous system, all of which have similar functions. It forms gray matter. **Horns (columns)** are the chief areas of gray matter in the spinal cord. The term *horn* describes the two-dimensional appearance of the organization of gray matter in the spinal cord as seen in cross section. The term *column* describes the three-dimensional appearance of the gray matter in longitudinal columns. Since the white matter of the spinal cord is also arranged in columns, we will refer to the gray matter as being arranged in horns.

SPINAL CORD

GENERAL FEATURES

The **spinal cord** is a cylindrical structure that is slightly flattened anteriorly and posteriorly. It begins as a continuation of the medulla oblongata, the inferior part of the brain stem, and extends from the foramen magnum of the occipital bone to the level of the second lumbar vertebra (Figure 13-1). The length of the adult spinal cord ranges from 42 to 45 cm (16 to 18 inches). The diameter of the cord is about 2.54 cm (1 inch) in the midthoracic region, but is somewhat larger in the lower cervical and midlumbar regions.

When the cord is viewed externally, two conspicuous enlargements can be seen. The superior enlargement, the **cervical enlargement,** extends from the fourth cervical to the first thoracic vertebra. Nerves that supply the upper extremities arise from the cervical enlargement. The inferior enlargement, called the **lumbar enlargement,** extends from the ninth to the twelfth thoracic vertebra. Nerves that supply the lower extremities arise from the lumbar enlargement.

Below the lumbar enlargement, the spinal cord tapers to a conical portion known as the **conus medullaris** (KŌ-nus med-yoo-LAR-is). The conus medullaris ends at the level of the intervertebral disc between the first and second lumbar vertebra. Arising from the conus medullaris is the **filum terminale** (FĪ-lum ter-mi-NAL-ē), a nonnervous fibrous tissue of the spinal cord that extends inferiorly to attach to the coccyx. The filum terminale consists mostly of pia mater, the innermost of three membranes that cover and protect the spinal cord and brain. Some nerves that arise from the lower portion of the cord do not leave the vertebral column immediately. They angle inferiorly in the vertebral canal like wisps of coarse hair flowing from the end of the cord. They are appropriately named the **cauda equina** (KAW-da ē-KWĪ-na), meaning horse's tail.

The spinal cord is a series of 31 segments, each giving rise to a pair of spinal nerves. **Spinal segment** refers to a region of the spinal cord from which a pair of spinal nerves arises. The cord is divided into right and left sides by two grooves (see Figure 13-3). The **anterior median fissure** is a deep, wide groove on the anterior (ventral) surface, and the **posterior median sulcus** is a shallower, narrow groove on the posterior (dorsal) surface.

PROTECTION AND COVERINGS

Vertebral Canal

The spinal cord is located in the vertebral canal of the vertebral column. The canal is formed by the vertebral foramina of all the vertebrae arranged on top of each other. Since the wall of the vertebral canal is essentially a ring of bone surrounding the spinal cord, the cord is well protected. A certain degree of protection is also provided by the meninges, cerebrospinal fluid, and the vertebral ligaments.

Meninges

The **meninges** (me-NIN-jēz) are coverings that run continuously around the spinal cord and brain. Those associated specifically with the cord are known as **spinal meninges** (see Figure 14-2). The outer spinal meninx is called the **dura mater** (DYOO-ra MĀ-ter), meaning tough mother. It forms a tube from the level of the second

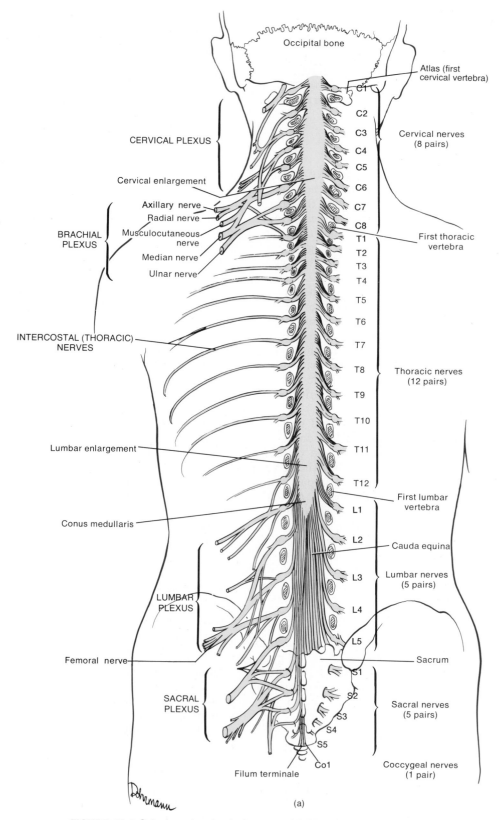

FIGURE 13-1 Spinal cord and spinal nerves. (a) Diagram of posterior view.

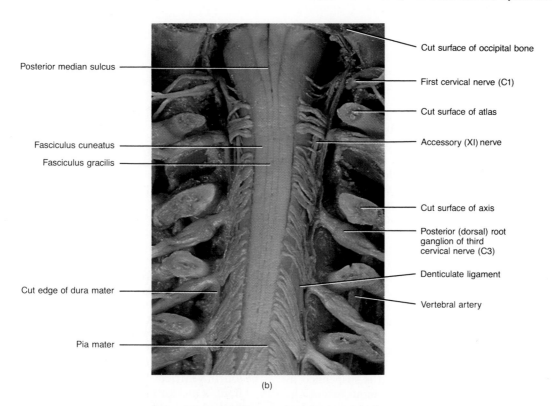

Posterior median sulcus

Fasciculus cuneatus

Fasciculus gracilis

Cut edge of dura mater

Pia mater

Cut surface of occipital bone

First cervical nerve (C1)

Cut surface of atlas

Accessory (XI) nerve

Cut surface of axis

Posterior (dorsal) root ganglion of third cervical nerve (C3)

Denticulate ligament

Vertebral artery

(b)

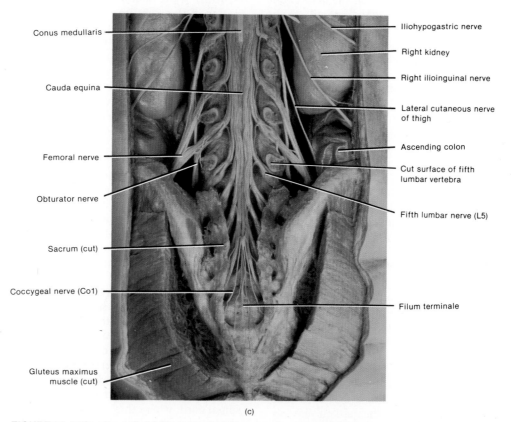

Conus medullaris

Cauda equina

Femoral nerve

Obturator nerve

Sacrum (cut)

Coccygeal nerve (Co1)

Gluteus maximus muscle (cut)

Iliohypogastric nerve

Right kidney

Right ilioinguinal nerve

Lateral cutaneous nerve of thigh

Ascending colon

Cut surface of fifth lumbar vertebra

Fifth lumbar nerve (L5)

Filum terminale

(c)

FIGURE 13-1 (*Continued*) (b) Photograph of the posterior aspect of the inferior portion of the medulla and the superior six segments of the cervical portion of the spinal cord. (c) Photograph of the posterior aspect of the conus medullaris and cauda equina. (Photographs courtesy of N. Gluhbegovic and T. H. Williams, *The Human Brain: A Photographic Guide,* Harper & Row, Publishers, Inc., Hagerstown, MD, 1980.)

sacral vertebra, where it is fused with the filum terminale, to the foramen magnum, where it is continuous with the dura mater of the brain. It is composed of dense, fibrous connective tissue. Between the dura mater and the wall of the vertebral canal is the *epidural space,* which is filled with fat, connective tissue, and blood vessels. It serves as padding around the cord. The epidural space inferior to the second lumbar vertebra is the site for the injection of anesthetics, such as a saddleblock for childbirth.

The middle spinal meninx is called the **arachnoid** (a-RAK-noyd), or spider layer. It is a delicate connective tissue membrane that forms a tube inside the dura mater. It is also continuous with the arachnoid of the brain. Between the dura mater and the arachnoid is the *subdural space,* which contains serous fluid.

The inner meninx is known as the **pia mater** (PĒ-a MĀ-ter), or delicate mother. It is a transparent fibrous membrane that forms a tube around and adheres to the surface of the spinal cord and brain. It contains numerous blood vessels. Between the arachnoid and the pia mater is the *subarachnoid space,* where the cerebrospinal fluid circulates.

All three spinal meninges cover the spinal nerves up to the point of exit from the spinal column through the intervertebral foramina. The spinal cord is suspended in the middle of its dural sheath by membranous extensions of the pia mater. These extensions, called the *denticulate* (den-TIK-yoo-lāt) *ligaments,* are attached laterally to the dura mater along the length of the cord between the ventral and dorsal nerve roots on either side. The ligaments protect the spinal cord against shock and sudden displacement. Essentially, the spinal cord is fixed in its position in the vertebral canal, since it is anchored to the coccyx inferiorly by the filum terminale, laterally to the dura mater by the denticulate ligaments, and superiorly to the brain.

CLINICAL APPLICATION

Inflammation of the meninges is known as **meningitis.** If only the dura mater becomes inflamed, the condition is *pachymeningitis.* Inflammation of the arachnoid and pia mater is *leptomeningitis,* the most common form of meningitis.

Cerebrospinal fluid is removed from the subarachnoid space in the inferior lumbar region of the spinal cord by a **spinal (lumbar) puncture (tap).** The procedure is normally performed between the third and fourth or fourth and fifth lumbar vertebrae. The spinous process of the fourth lumbar vertebra is easily located. A line drawn across the highest points of the iliac crests will pass through the spinous process. A lumbar puncture is made below the spinal cord and thus poses little danger to it. If the patient lies on one side, drawing the knees and chest together, the vertebrae separate slightly so that a needle can be conveniently inserted. In its course, the needle pierces the skin, superficial fascia, supraspinous ligament, interspi-

nous ligament, epidural space, and arachnoid, to enter the subarachnoid space. Lumbar punctures are used to withdraw fluid for diagnostic purposes and to introduce antibiotics (as in the case of meningitis) and contrast media. For example, in a radiologic technique, called **myelography** (mī-e-LOG-ra-fē), a contrast medium (usually metrizamide) is introduced into the spinal subarachnoid space to determine or exlude the presence of lesions within and around the spinal cord.

STRUCTURE IN CROSS SECTION

The spinal cord consists of both gray and white matter. Figure 13-2 shows that the gray matter forms an H-shaped area within the white matter. The gray matter consists primarily of nerve cell bodies and unmyelinated axons and dendrites of association and motor neurons. The white matter consists of bundles of myelinated axons of motor and sensory neurons.

The cross bar of the H is formed by the **gray commissure** (KOM-mi-shur). In the center of the gray commissure is a small space called the **central canal.** This canal runs the length of the spinal cord and is continuous with the fourth ventricle of the medulla. It contains cerebrospinal fluid. Anterior to the gray commissure is the **anterior (ventral) white commissure,** which connects the white matter of the right and left sides of the spinal cord.

The upright portions of the H are further subdivided into regions. Those closer to the front of the cord are called **anterior (ventral) gray horns.** They represent the motor part of the gray matter. The regions closer to the back of the cord are referred to as **posterior (dorsal) gray horns.** They represent the sensory part of the gray matter. The regions between the anterior and posterior gray horns are intermediate **lateral gray horns.** The lateral gray horns are present in the thoracic, upper lumbar, and sacral segments of the cord.

The gray matter of the cord also contains several nuclei that serve as relay stations for impulses and origins for certain nerves. Nuclei are clusters of nerve cell bodies and dendrites in the spinal cord and brain.

The white matter, like the gray matter, is organized into regions. The anterior and posterior gray horns divide the white matter on each side into three broad areas: **anterior (ventral) white columns, posterior (dorsal) white columns,** and **lateral white columns.** Each column (or *funiculus*) in turn consists of distinct bundles of myelinated fibers that run the length of the cord. These bundles are called **tracts** or **fasciculi** (fa-SIK-yoo-lī). The longer **ascending tracts** consist of sensory axons that conduct impulses that enter the spinal cord upward to the brain. The longer **descending tracts** consist of motor axons that conduct impulses from the brain downward into the spinal cord, where they synapse with other neurons whose axons pass out to muscles and glands. Thus the ascending tracts are sensory tracts and the descending tracts are motor tracts. Still other short tracts contain ascending or descending axons that convey impulses from one level of the cord to another.

(a)

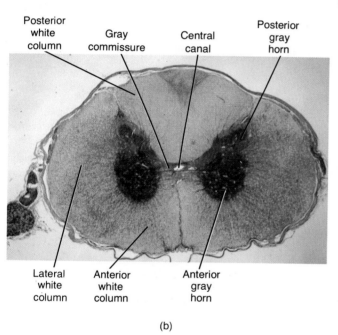

(b)

FIGURE 13-2 Spinal cord. (a) The organization of gray and white matter in the spinal cord as seen in cross section. The front of the figure has been sectioned at a lower level than the back so that you can see what is inside the posterior root ganglion, posterior root of the spinal nerve, anterior root of the spinal nerve, and the spinal nerve. In this and other illustrations of cross sections of the spinal cord, dendrites are not shown in relation to cell bodies of motor or association neurons for purposes of simplicity. (b) Photograph of the spinal cord at the seventh cervical segment. (Courtesy of Victor B. Eichler, Ph.D., Wichita, Kansas.)

tract travels, where the cell bodies of the tract originate, and where the axons of the tract terminate. Since the origin and termination are specified, the direction of impulse conduction is also indicated by the name. For example, the anterior spinothalamic tract is located in the *anterior* white column, it originates in the *spinal* cord, and it terminates in the *thalamus* (a region of the brain). It is an ascending (sensory) tract since it conveys impulses from the cord upward to the brain.

The principal ascending and descending tracts are listed in Exhibit 13-1 and shown in Figure 13-3.

FUNCTIONS

A major function of the spinal cord is to convey sensory impulses from the periphery to the brain and to conduct motor impulses from the brain to the periphery. This is accomplished within spinal tracts. A second principal function is to provide a means of integrating reflexes. Both functions are essential to maintaining homeostasis.

Spinal Tracts

The vital function of conveying sensory and motor information to and from the brain is carried out by the ascending and descending tracts of the cord. The names of the tracts indicate the white column (funiculus) in which the

Reflex Center

The second principal function of the spinal cord is to serve as a center for reflex actions. Spinal nerves are the paths of communication between the spinal cord tracts and the periphery. Figure 13-3 reveals that each pair of spinal nerves is connected to a segment of the cord by two points of attachment called roots. The **posterior** or **dorsal (sensory) root** contains sensory nerve fibers only and conducts impulses from the periphery to the spinal cord. These fibers extend into the posterior (dorsal) gray horn. Each dorsal root also has a swelling, the **posterior** or **dorsal (sensory) root ganglion,** which contains the cell bodies of the sensory neurons from the periphery. The other point of attachment of a spinal nerve to the cord

EXHIBIT 13-1   SELECTED ASCENDING AND DESCENDING TRACTS OF SPINAL CORD

TRACT	LOCATION (WHITE COLUMN)	ORIGIN	TERMINATION	FUNCTION
ASCENDING TRACTS				
Anterior (ventral) spinothalamic	Anterior (ventral) column.	Posterior (dorsal) gray horn on one side of cord, but crosses to opposite side of brain.	Thalamus; impulses eventually conveyed to cerebral cortex.	Conveys sensations for touch and pressure from one side of body to opposite side of thalamus. Eventually sensations reach cerebral cortex.
Lateral spinothalamic	Lateral column.	Posterior (dorsal) gray horn on one side of cord, but crosses to opposite side of brain.	Thalamus; impulses eventually conveyed to cerebral cortex.	Conveys sensations for pain and temperature from one side of body to opposite side of thalamus. Eventually sensations reach cerebral cortex.
Fasciculus gracilis and **fasciculus cuneatus**	Posterior (dorsal) column.	Axons of afferent neurons from periphery that enter posterior (dorsal) column on one side of cord and rise to same side of brain.	Nucleus gracilis and nucleus cuneatus of medulla; impulses eventually conveyed to cerebral cortex.	Convey sensations from one side of body to same side of medulla for touch; two-point discrimination (ability to distinguish that two points on skin are touched even though close together); proprioception (awareness of precise position of body parts and their direction of movement); stereognosis (ability to recognize size, shape, and texture of object); weight discrimination (ability to assess weight of an object); and vibration. Eventually sensations may reach cerebral cortex.
Posterior (dorsal) spinocerebellar	Posterior (dorsal) portion of lateral column.	Posterior (dorsal) gray horn on one side of cord and rises to same side of brain.	Cerebellum.	Conveys sensations from one side of body to same side of cerebellum for subconscious proprioception.
Anterior (ventral) spinocerebellar	Anterior (ventral) portion of lateral column.	Posterior (dorsal) gray horn on one side of cord; contains both crossed and uncrossed fibers.	Cerebellum.	Conveys sensations from both sides of body to cerebellum for subconscious proprioception.

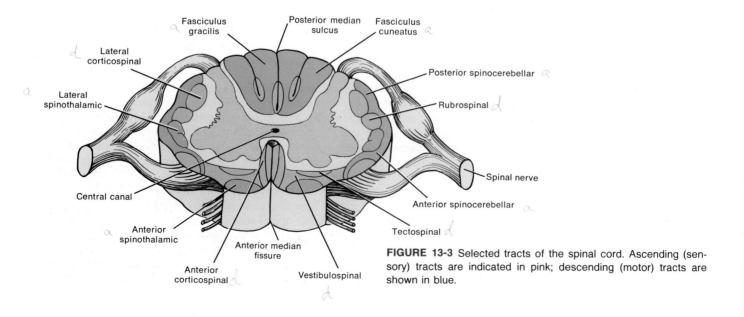

FIGURE 13-3 Selected tracts of the spinal cord. Ascending (sensory) tracts are indicated in pink; descending (motor) tracts are shown in blue.

EXHIBIT 13-1 *(Continued)*

TRACT	LOCATION (WHITE COLUMN)	ORIGIN	TERMINATION	FUNCTION
DESCENDING TRACTS				
Lateral corticospinal	Lateral column.	Cerebral cortex on one side of brain, but crosses in base of medulla to opposite side of cord.	Anterior (ventral) gray horn.	Conveys motor impulses from one side of cortex to anterior gray horn of opposite side. Eventually impulses reach skeletal muscles on opposite side of body that coordinate precise, discrete movements.
Anterior (ventral) corticospinal	Anterior (ventral) column.	Cerebral cortex on one side of brain, uncrossed in medulla, but crosses to opposite side of cord.	Anterior (ventral) gray horn.	Conveys motor impulses from one side of cortex to anterior gray horn of same side. Impulses cross to opposite side in spinal cord and reach skeletal muscles that coordinate precise, discrete movements.
Rubrospinal	Lateral column.	Midbrain (red nucleus) on one side of brain, but crosses to opposite side of cord.	Anterior (ventral) gray horn.	Conveys motor impulses from one side of midbrain to skeletal muscles on opposite side of body that are concerned with muscle tone and posture.
Tectospinal	Anterior (ventral) column.	Midbrain on one side of brain, but crosses to opposite side of cord.	Anterior (ventral) gray horn.	Conveys motor impulses from one side of midbrain to skeletal muscles on opposite side of body that control movements of head in response to auditory, visual, and cutaneous stimuli.
Vestibulospinal	Anterior (ventral) column.	Medulla on one side of brain and descends to same side of cord.	Anterior (ventral) gray horn.	Conveys motor impulses from one side of medulla to skeletal muscles on same side of body that regulate body tone in response to movements of head (equilibrium).

is the **anterior** or **ventral (motor) root.** It contains motor nerve fibers only and conducts impulses from the spinal cord to the periphery.

The cell bodies of the motor neurons are located in the gray matter of the cord. If the motor impulse supplies a skeletal muscle, the cell bodies are located in the anterior (ventral) gray horn. If, however, the impulse supplies smooth muscle, cardiac muscle, or a gland through the autonomic nervous system, the cell bodies are located in the lateral gray horn.

Reflex Arc and Homeostasis

The path an impulse follows from its origin in the dendrites or cell body of a neuron in one part of the body to its termination elsewhere in the body is called a **conduction pathway.** All conduction pathways consist of circuits of neurons. One pathway is known as a **reflex arc,** the

functional unit of the nervous system. A reflex arc contains two or more neurons over which impulses are conducted from a receptor to the brain or spinal cord and then to an effector. The basic components of a reflex arc are as follows (Figure 13-4).

1. Receptor. The distal end of a dendrite or a sensory structure associated with the distal end of a dendrite. Its role in the reflex arc is to respond to a change in the internal or external environment by initiating a nerve impulse in a sensory neuron.

2. Sensory neuron. Passes the impulse from the receptor to its axonal termination in the central nervous system.

3. Center. A region in the central nervous system where an incoming sensory impulse generates an outgoing motor impulse. In the center, the impulse may be inhibited, transmitted, or rerouted. In the center of some reflex arcs, the sensory neuron directly generates the impulse in the motor neuron. The center may also contain an association neuron between the sensory neuron and the motor neuron leading to a muscle or a gland.

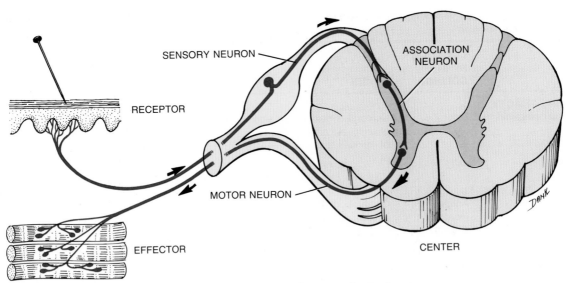

FIGURE 13-4 Components of a generalized reflex arc.

4. Motor neuron. Transmits the impulse generated by the sensory or association neuron in the center to the organ of the body that will respond.

5. Effector. The organ of the body, either a muscle or gland, that responds to the motor impulse. This response is called a reflex.

Reflexes are fast responses to changes in the internal or external environment that allow the body to maintain homeostasis. Reflexes are associated not only with skeletal muscle contraction but also with body functions such as heart rate, respiration, digestion, urination, and defecation. Reflexes carried out by the spinal cord alone are called **spinal reflexes.** Reflexes that result in the contraction of skeletal muscles are known as **somatic reflexes.**

Those that cause the contraction of smooth or cardiac muscle or secretion by glands are **visceral (autonomic) reflexes.** Our concern at this point is to examine a few somatic spinal reflexes: the stretch reflex, tendon reflex, flexor reflex, and crossed extensor reflex.

• *Stretch Reflex*   The **stretch reflex** is based on a two-neuron or *monosynaptic reflex arc.* Only two neurons are involved, and there is only one synapse in the pathway (Figure 13-5). This reflex results in the contraction of a muscle when it is stretched. Slight stretching of a muscle stimulates receptors in the muscle called *muscle spindles* (see Figure 15-3a). The spindles monitor changes in the length of the muscle by responding to the rate and degree

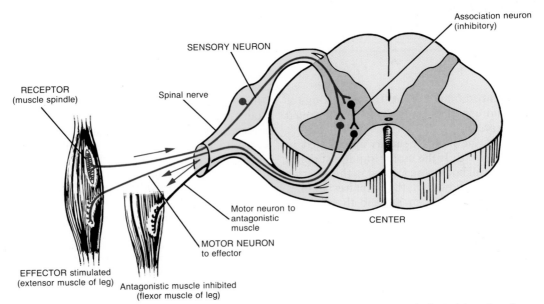

FIGURE 13-5 Stretch reflex. The neurons of the stretch reflex pathway are indicated in color. A stretch reflex is monosynaptic since only one synapse and two neurons—sensory neuron and motor neuron to effector—are involved in the pathway. The synapse is between the sensory neuron and motor neuron to the effector. This reflex is also ipsilateral. Why?

of change in length. Once the spindle is stimulated, an impulse is sent along a sensory neuron through the posterior root of the spinal nerves to the spinal cord. The sensory neuron synapses with a motor neuron in the anterior gray horn. The sensory neuron generates an impulse at the synapse that is transmitted along the motor neuron. The motor neuron in the anterior root of the spinal nerve terminates in a skeletal muscle. Once the impulse reaches the stretched muscle, it contracts. Thus the stretch is counteracted by contraction.

Since the sensory impulse enters the spinal cord on the same side that the motor impulse leaves the spinal cord, the reflex arc is called an *ipsilateral* (ip'-si-LAT-er-al) *reflex arc*. All monosynaptic reflex arcs are ipsilateral and are deep tendon reflexes (discussed shortly).

The stretch reflex is essential in maintaining muscle tone and is important for muscle functions during exercise. It also helps prevent injury from overstretching of muscles. Moreover, it is the basis for several tests used in neurological examinations. One such reflex is the *knee jerk* (*patellar reflex*). This reflex is tested by tapping the patellar ligament (stimulus). Muscle spindles in the quadriceps femoris muscle (extensor muscle of leg) attached to the ligament send the sensory impulse to the spinal cord, and the returning motor impulse causes contraction of the muscle. The response is extension of the leg at the knee, or a knee jerk.

Although the stretch reflex pathway contains only two neurons, other neurons are involved in it in various ways. For example, the sensory neuron from the muscle spindle that enters the spinal cord also synapses with an inhibitory association neuron that, in turn, synapses with a motor neuron that controls antagonistic muscles. Thus, during the stretch reflex when the stretched muscle is counteracted by contraction, antagonistic muscles that oppose the contraction are inhibited. This phenomenon, by which

impulses stimulate contraction of one muscle and simultaneously inhibit contraction of antagonistic muscles, is called **reciprocal innervation.** It avoids conflict between prime movers and antagonists, and is vital in coordinating body movements. You may recall from Chapter 11 that skeletal muscles act in groups rather than alone and that each muscle in the group has a specific role in bringing about the movement.

The sensory neuron from the muscle spindle also synapses with neurons that relay impulses to the brain. In this way, the brain is provided with information about the state of stretch or contraction of skeletal muscles that enables it to coordinate muscular movements and posture.

• *Tendon Reflex*   Reflexes other than the stretch reflex involve association neurons in addition to sensory and motor neurons. Since more than two neurons are involved, there is more than one synapse, and thus these reflexes are *polysynaptic reflex arcs*. One example of a reflex based on a polysynaptic reflex arc is the **tendon reflex.** The tendon reflex, like the stretch reflex, is also ipsilateral.

The tendon reflex is designed to protect tendons and their associated muscles from damage that might be brought about by excessive tension. The receptors for this reflex are called *tendon organs* or *Golgi tendon organs* (see Figure 15-3b). Tendon organs lie within muscle tendons near the junction of a tendon and a muscle. Whereas muscle spindles are sensitive to changes in muscle length, tendon organs detect and respond to changes in muscle tension caused by passive stretch or muscular contraction.

When an increase in tension is applied to a tendon, the tendon organ is stimulated and nerve impulses are generated and transmitted to the spinal cord via a sensory neuron (Figure 13-6). Within the spinal cord, the sensory neuron synapses with an inhibitory association neuron which, in turn, synapses with a motor neuron that inner-

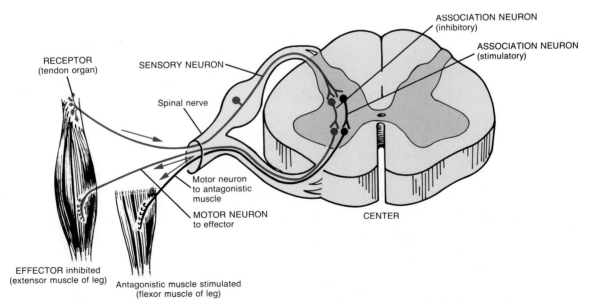

FIGURE 13-6 Tendon reflex. The neurons of the tendon reflex pathway are indicated in color. This reflex arc is polysynaptic since there is more than one synapse and more than two neurons involved in the pathway. There is one synapse between the sensory neuron and association neuron (inhibitory), and another synapse between the association neuron (inhibitory) and motor neuron to the effector.

vates the muscle associated with the tendon organ. Thus, as tension on the tendon organ increases, the frequency of inhibitory impulses increases, and the inhibition causes relaxation of the muscle. In this way, the tendon reflex protects the tendon and muscle from damage from excessive tension. Thus, the tendon reflex is protective.

The sensory neuron from the tendon organ also synapses with a stimulatory association neuron in the spinal cord. The stimulatory association neuron, in turn, synapses with motor neurons controlling antagonistic muscles. Thus, whereas the tendon reflex brings about relaxation of the muscle containing the tendon organ, it also brings about contraction of the antagonistic muscles. This is another example of reciprocal innervation. The sensory neuron also synapses with neurons that relay impulses to the brain, thus informing the brain about the state of muscle tension throughout the body.

• *Flexor Reflex and Crossed Extensor Reflex*   Another example of a reflex based on a polysynaptic reflex arc

is the **flexor reflex,** or **withdrawal reflex** (Figure 13-7). Suppose you step on a tack. As a result of the painful stimulus, you immediately withdraw your foot. What has happened? A sensory neuron transmits an impulse from the receptor to the spinal cord. A second impulse is generated in an association neuron, which generates a third impulse in a motor neuron. The motor neuron stimulates the muscles of your foot and you withdraw it. Thus a flexor reflex is protective in that it results in the movement of an extremity to avoid pain.

The flexor reflex, like the stretch reflex, is also ipsilateral. The incoming and outgoing impulses are on the same side of the spinal cord. The flexor reflex also illustrates another feature of polysynaptic reflex arcs. In the monosynaptic stretch reflex, the returning motor impulse affects only the quadriceps muscle of the thigh. When you withdraw your entire lower or upper extremity from a noxious stimulus, more than one muscle is involved, and several motor neurons are simultaneously returning impulses to several upper and lower extremity muscles

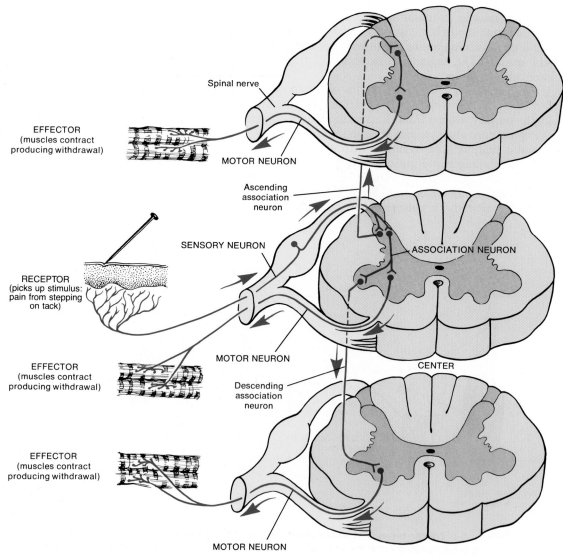

FIGURE 13-7 Flexor (withdrawal) reflex. This reflex arc is polysynaptic and ipsilateral. Why is the reflex arc shown also an intersegmental reflex arc?

at the same time. Thus a single sensory impulse causes several motor responses. This kind of reflex arc, in which a single sensory neuron splits into ascending and descending branches, each forming a synapse with association neurons at different segments of the cord, is called an *intersegmental reflex arc.* Because of intersegmental reflex arcs, a single sensory neuron can activate several motor neurons and thereby cause stimulation of more than one effector.

Something else may happen when you step on a tack. You may lose your balance as your body weight shifts to the other foot. Then you do whatever you can to regain your balance so you do not fall. This means motor impulses are also sent to your unstimulated foot and both upper extremities. The motor impulses cause extension at the knee so you can place your entire body weight

on the foot. These impulses cross the spinal cord as shown in Figure 13-8. The incoming sensory impulse not only initiates the flexor reflex that causes you to withdraw, it also initiates an extensor reflex. The incoming sensory impulse crosses to the opposite side of the spinal cord through association neurons at that level and several levels above and below the point of sensory stimulation. From these levels, the motor neurons cause extension of the knee, thus maintaining balance. Unlike the flexor reflex, which passes over an ipsilateral reflex arc, the extensor reflex passes over a *contralateral* (kon'-tra-LAT-er-al) *reflex arc*—the impulse enters one side of the spinal cord and exits on the opposite side. The reflex just described, in which extension of the muscles in one limb occurs as a result of contraction of the muscles of the opposite limb, is called a **crossed extensor reflex.**

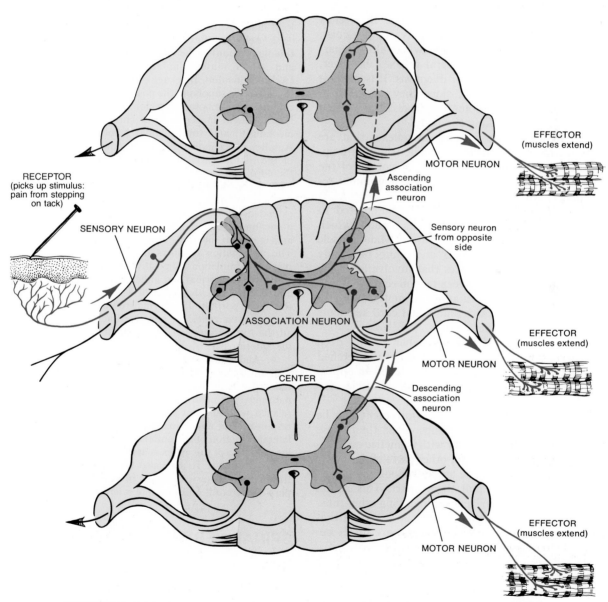

FIGURE 13-8 Crossed extensor reflex. Although the flexor reflex is shown on the left of the diagram so that you can correlate it with the crossed extensor reflex on the right, concentrate your attention on the crossed extensor reflex. Why is the crossed extensor reflex classified as a contralateral reflex arc?

In the flexor reflex, when the flexor muscles of your lower extremity are contracting, the extensor muscles of the same extremity are extending. If both sets of muscles contracted at the same time, you would not be able to flex your limb because both sets of muscles would pull on the limb bones. But because of reciprocal innervation, one set of muscles contracts while the other is being extended.

In the crossed extensor reflex, reciprocal innervation also occurs. While you are flexing the muscles of the limb that has been stimulated by the tack, the muscles of your other limb are producing extension to help maintain balance.

Reflexes and Diagnosis

Reflexes are often used for diagnosing disorders of the nervous system and locating injured tissue. If a reflex ceases to function or functions abnormally, the physician may suspect that the damage lies somewhere along a particular conduction pathway. Visceral reflexes, however, are usually not practical tools for diagnosis. It is difficult to stimulate visceral receptors, since they are deep in the body. In contrast, many somatic reflexes can be tested simply by tapping or stroking the body.

Superficial reflexes are withdrawal reflexes that are elicited by tactile or noxious stimuli. The stimuli are applied to the skin, mucous membranes, or cornea of the eye. Examples include the corneal reflex, abdominal reflex, and pharyngeal reflex.

Any skeletal muscle can normally be stimulated to contract by a slight sudden stretch of its tendon, which can be created by administering a light tap. Many muscle tendons are deeply buried, however, and cannot be readily tapped through the skin. Reflexes that involve a stretch stimulus to a tendon are called **deep tendon reflexes.** Deep tendon reflexes provide information about the integrity and function of the reflex arcs and spinal cord segments without involving the higher centers.

To obtain a substantial response when testing deep tendon reflexes, the muscle must be slightly stretched before the tap is administered. If it is stretched an appropriate amount, tapping the tendon elicits a muscle contraction.

If reflexes are weak or absent, *reinforcement* can be used. In this method, muscle groups other than those being tested are tensed voluntarily with isometric contractions to increase reflex activity in other parts of the body. For example, the person can be asked to hook the fingers together and then try to pull them apart. This action may increase the strength of reflexes involving other muscles. If the reflex can be demonstrated, it is certain that the sensory and motor nerve connections are intact between muscle and spinal cord.

Muscle reflexes can help determine the spinal cord's excitability. When a large number of facilitatory impulses are transmitted from the brain to the spinal cord, the muscle reflexes become so sensitive that simply tapping the knee tendon with the tip of one's finger may cause the leg to jump a considerable distance. On the other hand, the cord may be so intensely inhibited by other impulses from the brain that almost no degree of pounding on the muscles or tendons can elicit a response.

Neurological impairment can be evaluated by using a stopwatch to time the reflex response. Sensitivity of sensory end organs in a muscle is demonstrated by stretching it by as little as 0.05 mm and for as short a duration as $\frac{1}{20}$ sec.

Among the reflexes of clinical significance are:

1. Patellar reflex (knee jerk). This reflex involves extension of the leg by contraction of the quadriceps femoris muscle in response to tapping the patellar ligament (see Figure 13-5). The reflex is blocked by damaged afferent or efferent nerves to the muscle or reflex centers in the second, third, or fourth lumbar segments of the spinal cord. This reflex is also absent in people with chronic diabetes and neurosyphilis. The reflex is exaggerated in disease or injury involving the corticospinal tracts descending from the cortex to the spinal cord. This reflex may also be exaggerated by applying a second stimulus (a sudden loud noise) while tapping the patellar tendon.

2. Achilles reflex (ankle jerk). This reflex involves extension (plantar flexion) of the foot by contraction of the gastrocnemius and soleus muscles in response to tapping the calcaneal (Achilles) tendon. Blockage of the ankle jerk indicates damage to the nerves supplying the posterior leg muscles or to the nerve cells in the lumbosacral region of the spinal cord. This reflex is also absent in people with chronic diabetes, neurosyphilis, alcoholism, and subarachnoid hemorrhages. An exaggerated Achilles reflex indicates cervical cord compression or a lesion of the motor tracts of the first or second sacral segments of the cord.

3. Babinski sign. This reflex results from light stimulation to the outer margin of the sole of the foot. The great toe is extended, with or without fanning of the other toes. This phenomenon occurs in normal children under 1½ years of age and is due to incomplete development of the nervous system. The myelination of fibers in the corticospinal tract has not reached completion. A positive Babinski sign after age 1½ is considered abnormal and indicates an interruption of the corticospinal tract as the result of a lesion of the tract, usually in the upper portion. The normal response after 1½ years of age is the **plantar reflex,** or negative Babinski—a curling under of all the toes accompanied by a slight turning in and flexion of the anterior part of the foot.

4. Abdominal reflex. This reflex compresses the abdominal wall in response to stroking the side of the abdomen. Two separate reflexes, the upper abdominal reflex and the lower abdominal reflex, are involved. The patient should be lying down and relaxed with arms at the sides and knees slightly flexed. The response is an abdominal muscle contraction that results in a lateral deviation of the umbilicus to the side opposite the stimulus. Absence of this reflex is associated with lesions of the corticospinal system. It may also be absent because of lesions of the peripheral nerves, lesions of reflex centers in the thoracic part of the cord, and multiple sclerosis.

SPINAL NERVES

NAMES

The 31 pairs of spinal nerves are named and numbered according to the region and level of the spinal cord from which they emerge (see Figure 13-1). The first cervical pair emerges between the atlas and the occipital bone. All other spinal nerves leave the vertebral column from

FIGURE 13-9 Coverings of a spinal nerve. (a) Diagram. (b) Scanning electron micrograph at a magnification of 900×. (Courtesy of Richard G. Kessel and Randy H. Kardon, from *Tissues and Organs: A Text-Atlas of Scanning Electron Microscopy.* Copyright © 1979 by Scientific American, Inc.)

the intervertebral foramina between adjoining vertebrae. There are 8 pairs of cervical nerves, 12 pairs of thoracic nerves, 5 pairs of lumbar nerves, 5 pairs of sacral nerves, and 1 pair of coccygeal nerves.

During fetal life, the spinal cord and vertebral column grow at different rates, the cord growing more slowly. Thus not all the spinal cord segments are in line with their corresponding vertebrae. Remember that the spinal cord terminates near the level of the first or second lumbar vertebra. Thus the lower lumbar, sacral, and coccygeal nerves must descend more and more to reach their foramina before emerging from the vertebral column. This arrangement constitutes the cauda equina.

COMPOSITION AND COVERINGS

A **spinal nerve** has two points of attachment to the cord: a posterior root and an anterior root. The posterior and anterior roots unite to form a spinal nerve at the intervertebral foramen. Since the posterior root contains sensory fibers and the anterior root contains motor fibers, a spinal nerve is a *mixed nerve.* The posterior (dorsal) root ganglion contains cell bodies of sensory neurons.

In Figure 13-9, you can see that a spinal nerve contains many fibers surrounded by different coverings. The individual fibers, whether myelinated or unmyelinated, are wrapped in a connective tissue called the **endoneurium**

(en'-dō-NYOO-rē-um). Groups of fibers with their endoneurium are arranged in bundles called *fascicles,* and each bundle is wrapped in connective tissue called the **perineurium** (per'-i-NYOO-rē-um). The outermost covering around the entire nerve is the **epineurium** (ep'-i-NYOO-rē-um). The spinal meninges fuse with the epineurium as the nerve exits from the vertebral canal.

DISTRIBUTION

Branches

Shortly after a spinal nerve leaves its intervertebral foramen, it divides into several branches (Figure 13-10). These branches are known as **rami** (RĀ-mē). The **dorsal ramus** (RĀ-mus) innervates the deep muscles and skin of the dorsal surface of the back. The **ventral ramus** of a spinal nerve innervates the superficial back muscles and all the structures of the extremities and the lateral and ventral trunk. In addition to dorsal and ventral rami, spinal nerves also give off a **meningeal branch.** This branch reenters the spinal canal through the intervertebral foramen and supplies the vertebrae, vertebral ligaments, blood vessels of the spinal cord, and the meninges. Other branches of a spinal nerve are the **rami communicantes** (kō-myoo-ni-KAN-tēz), components of the autonomic nervous system whose structure and function are discussed in Chapter 16.

Plexuses

The ventral rami of spinal nerves, except for thoracic nerves T2–T11, do not go directly to the structures of the body they supply. Instead, they form networks by joining with adjacent nerves on either side of the body. Such a network is called a **plexus** (*plexus* = braid). The principal plexuses are the cervical plexus, the brachial plexus, the lumbar plexus, and the sacral plexus (see Figure 13-1). Emerging from the plexuses are nerves bearing names that are often descriptive of the general regions they supply or the course they take. Each of the nerves, in turn, may have several branches named for the specific structures they innervate.

• *Cervical Plexus* The **cervical plexus** is formed by the ventral rami of the first four cervical nerves (C1–C4) with contributions from C5. There is one on each side of the neck alongside the first four cervical vertebrae (Figure 13-11). The *roots* of the plexus indicated in the diagram are the ventral rami. The cervical plexus supplies the skin and muscles of the head, neck, and upper part of the shoulders. Branches of the cervical plexus also connect with the accessory (XI) and hypoglossal (XII) cranial nerves. The phrenic nerves are a major pair of nerves arising from the cervical plexuses that supply the motor fibers to the diaphragm. Damage to the spinal cord above the origin of the phrenic nerves results in paralysis of the diaphragm, since the phrenic nerves no longer send impulses to the diaphragm. Contractions of the diaphragm are essential for normal breathing.

Exhibit 13-2 summarizes the nerves and distributions of the cervical plexus. The relationship of the cervical plexus to the other plexuses is shown in Figure 13-1a.

FIGURE 13-10 Branches of a typical spinal nerve.

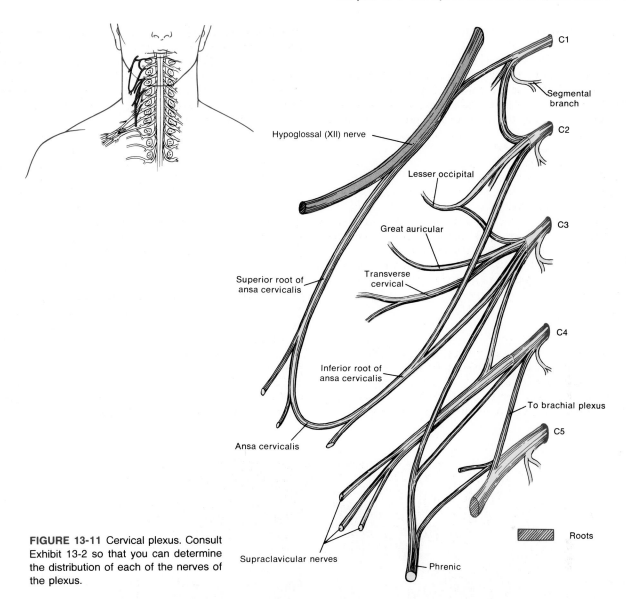

FIGURE 13-11 Cervical plexus. Consult Exhibit 13-2 so that you can determine the distribution of each of the nerves of the plexus.

EXHIBIT 13-2	CERVICAL PLEXUS	
NERVE	**ORIGIN**	**DISTRIBUTION**
SUPERFICIAL OR CUTANEOUS BRANCHES		
Lesser occipital	C2–C3.	Skin of scalp behind and above ear.
Greater auricular	C2–C3.	Skin in front, below, and over ear and over parotid glands.
Transverse cervical	C2–C3.	Skin over anterior aspect of neck.
Supraclavicular	C3–C4.	Skin over upper portion of chest and shoulder.
DEEP OR LARGELY MOTOR BRANCHES		
Ansa cervicalis	This nerve is divided into a superior root and an inferior root.	
Superior root	C1–C2.	Infrahyoid, thyrohyoid, and geniohyoid muscles of neck.
Inferior root	C3–C4.	Omohyoid, sternohyoid, and sternothyroid muscles of neck.
Phrenic	C3–C5.	Diaphragm between thorax and abdomen.
Segmental branches	C1–C5.	Prevertebral (deep) muscles of neck, levator scapulae, and middle scalene muscles.

• *Brachial Plexus*   The **brachial plexus** is formed by the ventral rami of spinal nerves C5–C8 and T1. On either side of the last four cervical and first thoracic vertebrae, the brachial plexus extends downward and laterally, passes over the first rib behind the clavicle, and then enters the axilla (Figure 13-12). The brachial plexus constitutes the entire nerve supply for the upper extremities and shoulder region.

The *roots* of the brachial plexus, like those of the cervical plexus, are the ventral rami of the spinal nerves. The roots of C5 and C6 unite to form the *superior trunk,* C7 becomes the *middle trunk,* and C8 and T1 form the *inferior trunk.* Each trunk, in turn, divides into an *anterior division* and a *posterior division.* The divisions then unite to form cords. The *posterior cord* is formed by the union of the posterior divisions of the superior, middle, and

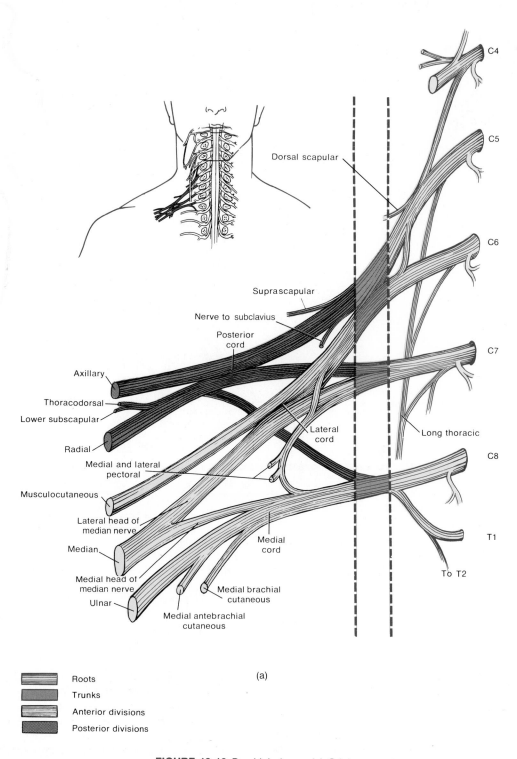

(a)

| Roots |
| Trunks |
| Anterior divisions |
| Posterior divisions |

FIGURE 13-12 Brachial plexus. (a) Origin.

inferior trunks. The *medial cord* is formed as a continuation of the anterior divison of the inferior trunk. The *lateral cord* is formed by the union of the anterior divisions of the superior and middle trunk. The peripheral nerves arise from the cords. Thus the brachial plexus begins as roots that unite to form trunks, the trunks branch into divisions, the divisions form cords, and the cords give rise to the peripheral nerves.

FIGURE 13-12 (*Continued*) (b) Distribution. Consult Exhibit 13-3 so that you can determine the distribution of each of the nerves of the plexus.

Three important nerves arising from the brachial plexus are the radial, median, and ulnar. The radial nerve supplies the muscles on the posterior aspect of the arm and forearm. The median nerve supplies most of the muscles of the anterior forearm and some of the muscles in the palm. The ulnar nerve supplies the anteromedial muscles of the forearm and most of the muscles of the palm.

CLINICAL APPLICATION

Prolonged use of a crutch that presses into the axilla may result in injury to a portion of the brachial plexus. The usual **crutch palsy** involves the posterior cord of the brachial plexus or, more frequently, just the radial nerve, which, in general, supplies extensors.

Radial nerve damage is indicated by wrist drop: inability to extend the hand at the wrist. Care must be taken not to injure the radial nerve when deltoid intramuscular injections are given. **Median nerve damage** is indicated by numbness, tingling, and pain in the palm and fingers; weak thumb movements; and inability to pronate the forearm and flex the wrist properly. Compression of the median nerve inside the carpal tunnel, formed anteriorly by the transverse carpal ligament and posteriorly by the carpal bones, is known as **carpal tunnel syndrome.** It may be caused by any condition that aggravates compression of the contents of the carpal tunnel, such as trauma, edema, and flexion of the wrist. **Ulnar nerve damage** is indicated by an inability to flex and adduct the wrist and difficulty in spreading the fingers.

A summary of the nerves and distributions of the brachial plexus is given in Exhibit 13-3. The relationship of the brachial plexus to the other plexuses is shown in Figure 13-1a.

• *Lumbar Plexus* The **lumbar plexus** is formed by the ventral rami of spinal nerves L1–L4. It differs from the brachial plexus in that there is no intricate interlacing of fibers. It also consists of *roots* and an *anterior* and *posterior division.* On either side of the first four lumbar vertebrae, the lumbar plexus passes obliquely outward behind the psoas major muscle (posterior division) and anterior to the quadratus lumborum muscle (anterior division), and then gives rise to its peripheral nerves (Figure 13-13). The lumbar plexus supplies the anterolateral abdominal wall, external genitals, and part of the lower extremity. The largest nerve arising from the lumbar plexus is the femoral nerve.

CLINICAL APPLICATION

Injury to the femoral nerve is indicated by an inability to extend the leg and by loss of sensation in the skin over the anteromedial aspect of the thigh.

EXHIBIT 13-3   BRACHIAL PLEXUS

NERVE	ORIGIN	DISTRIBUTION
ROOT NERVES		
Dorsal scapular	C5.	Levator scapulae, rhomboideus major, and rhomboideus minor muscles.
Long thoracic	C5–C7.	Serratus anterior muscle.
TRUNK NERVES		
Nerve to subclavius	C5–C6.	Subclavius muscle.
Suprascapular	C5–C6.	Supraspinatus and infraspinatus muscles.
LATERAL CORD NERVES		
Musculocutaneous	C5–C7.	Coracobrachialis, biceps brachii, and brachialis muscles.
Median (lateral head)	C5–C7.	See distribution for Median (medial head) in this exhibit.
Lateral pectoral	C5–C7.	Pectoralis major muscle.
POSTERIOR CORD NERVES		
Upper subscapular	C5–C6.	Subscapularis muscle.
Thoracodorsal	C6–C8.	Latissimus dorsi muscle.
Lower subscapular	C5–C6.	Subscapularis and teres major muscles.
Axillary (circumflex)	C5–C6.	Deltoid and teres minor muscles; skin over deltoid and upper posterior aspect of arm.
Radial	C5–C8 and T1.	Extensor muscles of arm and forearm (triceps brachii, brachioradialis, extensor carpi radialis longus, extensor digitorum, extensor carpi ulnaris, extensor indicis); skin of posterior arm and forearm, lateral two-thirds of dorsum of hand, and fingers over proximal and middle phalanges.
MEDIAL CORD NERVES		
Medial pectoral	C8–T1.	Pectoralis major and pectoralis minor muscles.
Medial brachial cutaneous	C8–T1.	Skin of medial and posterior aspects of lower third of arm.
Medial antebrachial cutaneous	C8–T1.	Skin of medial and posterior aspects of forearm.
Median (medial head)	C5–C8 and T1.	Medial and lateral heads of median nerve form median nerve. Distributed to flexors of forearm (pronator teres, flexor carpi radialis, flexor digitorum superficialis), except flexor carpi ulnaris; skin of lateral two-thirds of palm of hand and fingers.
Ulnar	C8–T1.	Flexor carpi ulnaris and flexor digitorum profundus muscles; skin of medial side of hand, little finger, and medial half of ring finger.
OTHER CUTANEOUS DISTRIBUTIONS		
Intercostobrachial	Second intercoastal nerve.	Skin over medial side of arm.
Upper lateral brachial cutaneous	Axillary.	Skin over deltoid muscle and down to elbow.
Posterior brachial cutaneous	Radial.	Skin over posterior aspect of arm.
Lower lateral brachial cutaneous	Radial.	Skin over lateral aspect of elbow.
Lateral antebrachial cutaneous	Musculocutaneous.	Skin over lateral aspect of forearm.
Posterior antebrachial cutaneous	Radial.	Skin over posterior aspect of forearm.

A summary of the nerves and distributions of the lumbar plexus is presented in Exhibit 13-4. The relationship of the lumbar plexus to the other plexuses is shown in Figure 13-1a.

• *Sacral Plexus*   The **sacral plexus** is formed by the ventral rami of spinal nerves L4–L5 and S1–S4. It is situated largely in front of the sacrum (Figure 13-14). Like the lumbar plexus, it contains *roots* and an *anterior* and *posterior division*. The sacral plexus supplies the buttocks, perineum, and lower extremities. The largest nerve arising from the sacral plexus—and, in fact, the largest nerve in the body—is the sciatic nerve. The sciatic nerve supplies the entire musculature of the leg and foot.

FIGURE 13-13 Lumbar plexus. (a) Origin. (b) Distribution of nerves of the lumbar and sacral plexuses in anterior view (left) and posterior view (right). Consult Exhibit 13-4 so that you can determine the distribution of the nerves of the lumbar plexus.

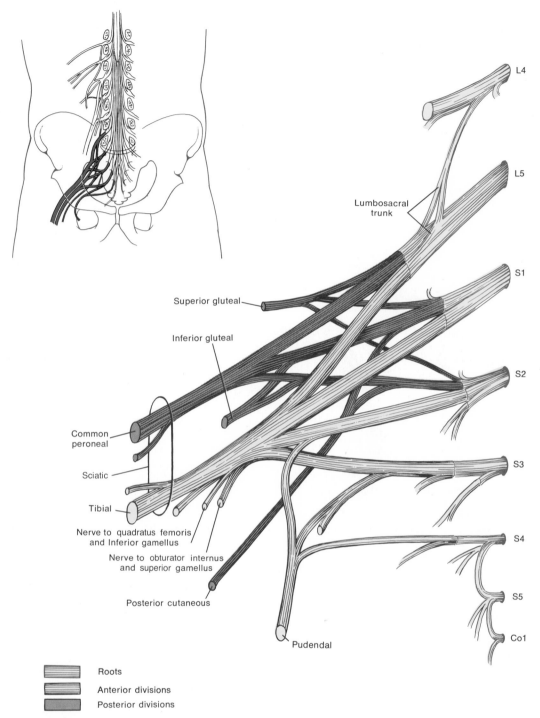

Lumbosacral trunk

L4

L5

S1

S2

S3

S4

S5

Co1

Superior gluteal

Inferior gluteal

Common peroneal

Sciatic

Tibial

Nerve to quadratus femoris and Inferior gamellus

Nerve to obturator internus and superior gamellus

Posterior cutaneous

Pudendal

Roots

Anterior divisions

Posterior divisions

FIGURE 13-14 Sacral plexus. Refer to Figure 13-13b for the distribution of nerves of the sacral plexus. Consult Exhibit 13-5 so that you can determine the distribution of each of the nerves of the plexus.

CLINICAL APPLICATION

Damage to the sciatic nerve and its branches results in foot drop, an inability to dorsiflex the foot. This nerve may be injured because of a slipped disc, dislocated hip, pressure from the uterus during pregnancy, or an improperly given gluteal intramuscular injection.

A summary of the nerves and distributions of the sacral plexus is given in Exhibit 13-5. The relationship of the sacral plexus to the other plexuses is shown in Figure 13-1a.

Intercostal (Thoracic) Nerves

Spinal nerves T2–T11 do not enter into the formation of plexuses. Nerves T2–T11 are known as **intercostal (tho-**

EXHIBIT 13-4   LUMBAR PLEXUS

NERVE	ORIGIN	DISTRIBUTION
Iliohypogastric	T12–L1.	Muscles of anterolateral abdominal wall (external oblique, internal oblique, transversus abdominis); skin of lower abdomen and buttock.
Ilioinguinal	L1.	Muscles of anterolateral abdominal wall as indicated above; skin of upper medial aspect of thigh, root of penis and scrotum in male, and labia majora and mons pubis in female.
Genitofemoral	L1–L2.	Cremaster muscle; skin over middle anterior surface of thigh, scrotum in male, and labia majora in female.
Lateral femoral cutaneous	L2–L3.	Skin over lateral, anterior, and posterior aspects of thigh.
Femoral	L2–L4.	Flexor muscles of thigh (iliacus, psoas major, pectineus, rectus femoris, sartorius); extensor muscles of leg (rectus femoris, vastus lateralis, vastus medialis, vastus intermedius); skin on front and over medial aspect of thigh and medial side of leg and foot.
Obturator	L2–L4.	Adductor muscles of leg (obturator externus, pectineus, adductor longus, adductor brevis, adductor magnus, gracilis); skin over medial aspect of thigh.

EXHIBIT 13-5   SACRAL PLEXUS

NERVE	ORIGIN	DISTRIBUTION
Superior gluteal	L4–L5 and S1.	Gluteus minimus and gluteus medius muscles and tensor fasciae latae.
Inferior gluteal	L5–S2.	Gluteus maximus muscle.
Nerve to piriformis	S1–S2.	Piriformis muscle.
Nerve to quadratus femoris	L4–L5 and S1.	Inferior gemellus and quadratus femoris muscles.
Nerve to obturator internus	L5–S2.	Superior gemellus and obturator internus muscles.
Perforating cutaneous	S2–S3.	Skin over lower medial aspect of buttock.
Posterior cutaneous	S1–S3.	Skin over anal region, lower lateral aspect of buttock, upper posterior aspect of thigh, upper part of calf, scrotum in male, and labia majora in female.
Sciatic	L4–S3.	Actually two nerves: tibial and common peroneal, bound together by common sheath of connective tissue. It splits into its two divisions, usually at knee. (See below for distributions.) As sciatic nerve descends through thigh, it sends branches to hamstring muscles (biceps femoris, semitendinosus, semimembranosus) and adductor magnus.
Tibial (medial popliteal)	L4–S3.	Gastrocnemius, plantaris, soleus, popliteus, tibialis posterior, flexor digitorum, and hallucis muscles. Branches of tibial nerve in foot are medial plantar nerve and lateral plantar nerve.
Medial plantar		Abductor hallucis, flexor digitorum brevis, and flexor hallucis muscles; skin over medial two-thirds of plantar surface of foot.
Lateral plantar		Remaining muscles of foot not supplied by medial plantar nerve; skin over lateral third of plantar surface of foot.
Common peroneal (lateral popliteal)	L4–S2.	Divides into a superficial peroneal and a deep peroneal branch.
Superficial peroneal		Peroneus longus and peroneus brevis muscles; skin over distal third of anterior aspect of leg and dorsum of foot.
Deep peroneal		Tibialis anterior, extensor hallucis longus, peroneus tertius, and extensor digitorum brevis muscles; skin over great and second toes.
Pudendal	S2–S4.	Muscles of perineum; skin of penis and scrotum in male and clitoris, labia majora, labia minora, and lower vagina in female.

racic) **nerves** and are distributed directly to the structures they supply in intercostal spaces (see Figure 13-1). After leaving its intervertebral foramen, the ventral ramus of nerve T2 supplies the intercostal muscles of the second intercostal space and the skin of the axilla and posteromedial aspect of the arm. Nerves T3 and T6 pass in the costal grooves of the ribs and are distributed to the intercostal muscles and skin of the anterior and lateral chest wall. Nerves T7–T11 supply the intercostal muscles and the abdominal muscles and overlying skin. The dorsal rami of the intercostal nerves supply the deep back muscles and skin of the dorsal aspect of the thorax.

DERMATOMES

The skin over the entire body is supplied segmentally by spinal nerves. This means that the spinal nerves innervate specific, constant segments of the skin. All spinal nerves except C1 supply branches to the skin. The skin segment supplied by the dorsal root of a spinal nerve is a **dermatome.**

In the neck and trunk, the dermatomes form consecutive bands of skin. In the trunk, there is an overlap of adjacent dermatome nerve supply. Thus there is little loss of sensation if only a single nerve supply to a dermatome is interrupted. Most of the skin of the face and scalp is supplied by the trigeminal (V) cranial nerve.

Since physicians know which spinal nerves are associated with each dermatome, it is possible to determine which segment of the spinal cord or spinal nerve is malfunctioning. If a dermatome is stimulated and the sensation is not perceived, it can be assumed that the nerves supplying the dermatome are involved.

DISORDERS: HOMEOSTATIC IMBALANCES

Spinal Cord Injury

The spinal cord may be damaged by expanding tumors, blood clots, degenerative and demyelinating disorders, fracture or dislocation of the vertebrae enclosing it, penetrating wounds caused by projectile metal fragments, or other traumatic events such as automobile accidents. Depending on the location and extent of the injury, paralysis may occur. The various types of paralysis may be classified as follows: **monoplegia** (*mono* = one; *plege* = stroke), paralysis of one extremity only; **diplegia** (*di* = two), paralysis of both upper extremities or both lower extremities; **paraplegia** (*para* = beyond), paralysis of both lower extremities; **hemiplegia** (*hemi* = half), paralysis of the upper extremity, trunk, and lower extremity on one side of the body; and **quadriplegia** (*quad* = four), paralysis of the two upper and two lower extremities.

Complete transection of the spinal cord means that the cord is cut transversely and severed from one side to the other. It results in a loss of all sensations and voluntary movement below the level of the transection. If the upper cervical cord is transected, quadriplegia results; if the transection is between the cervical and lumbar enlargements, paraplegia results. **Hemisection** of the spinal cord refers to a partial transection. It is characterized, below the hemisection, by a loss of proprioception, tactile discrimination, and feeling of vibration on the same side as the injury; paralysis on the same side; and loss of feelings of pain and temperature on the opposite side. If the hemisection is of the upper cervical cord, hemiplegia results; if the hemisection is of the thoracic cord, paralysis of one lower extremity results (monoplegia).

Following transection, there is an initial period of **spinal shock** that lasts from a few days to several weeks. During this period, all reflex activity is abolished, a condition called *areflexia* (a'-rē-FLEK-sē-a). In time, however, there is a return of reflex activity. The first reflex to return is the knee jerk. Its reappearance may take several days. Next the flexion reflexes return, over a period of up to several months. Then the crossed extensor reflexes return. Visceral reflexes such as erection and ejaculation are also affected by transection. Moreover, urinary bladder and bowel functions are no longer under voluntary control.

Until recently, severe damage resulting from transection was thought to be irreversible. However, a team of researchers has developed a technique for regenerating severed spinal cords in animals. The technique, called *delayed nerve grafting*, involves cutting away the crushed or injured section of the spinal cord and bridging the gap with nerve segments from the arm or leg. The original severed axons in the cord can then grow through the bridge. Delayed nerve grafting offers hope for paraplegics who have lost voluntary movements of the lower extremities, as well as urinary, bowel, and sexual functions.

Peripheral Nerve Damage and Repair

As we have seen, axons that have a neurolemma can be repaired as long as the cell body is intact, fibers are in association with neurolemmocytes (Schwann cells), and scar tissue formation does not occur too rapidly. Most nerves that lie outside the brain and spinal cord consist of axons that are covered with a neurolemma. A person who injures a nerve in the upper extremity, for example, has a good chance of regaining nerve function. Axons in the brain and spinal cord do not have a neurolemma. Injury there is permanent.

When there is damage to an axon (or to dendrites of somatic afferent neurons), there are usually changes in the cell body and always changes in the portions of the nerve processes distal to the site of damage. The changes associated with the cell body are referred to as the **axon reaction (retrograde degeneration).** Those associated with the distal portion of the cut fiber are called **Wallerian** (wal-LE-rē-an) **degeneration.** The axon reaction occurs in essentially the same way, whether the damaged fiber is in the central or peripheral nervous system. The Wallerian degeneration reaction, however, depends on whether the fiber is central or peripheral.

Axon Reaction

When there is damage to an axon of a central or peripheral neuron, certain structural changes occur in the cell body. One of the most significant features of the axon reaction occurs 24 to 48 hours after damage. The chromatophilic substance (Nissl bodies), arranged in an orderly fashion in an uninjured cell body, breaks down into finely granular masses. This alteration is called *chromatolysis* (krō'-ma-TOL-i-sis; *chromo* = color, *lysis* = dissolution). It begins between the axon hillock and nucleus, but spreads throughout the cell body. As a result of chromatolysis, the cell body swells and the swelling reaches its maximum between 10 and 20 days after injury (Figure 13-15b). Chromatolysis results in a loss of ribosomes by the rough endoplasmic reticulum and an increase in the number of free ribosomes.

FIGURE 13-15 Peripheral nerve damage and repair. (a) Normal neuron. (b) Chromatolysis. (c) Wallerian degeneration. (d) Regeneration.

Another sign of the axon reaction is the off-center position of the nucleus in the cell body. This change makes it possible to identify the cell bodies of damaged fibers through a microscope.

Wallerian Degeneration

The part of the axon distal to the damage becomes slightly swollen and then breaks up into fragments by the third to fifth day. The myelin sheath around the axon also undergoes degeneration (Figure 13-15c). Degeneration of the distal portion of the axon and myelin sheath is called Wallerian degeneration. Following degeneration, there is phagocytosis of the remains by macrophages.

Even though there is degeneration of the axon and myelin sheath, the neurolemma of the neurolemmocytes remains. The neurolemmocytes on either side of the site of injury multiply by mitosis and grow toward each other until they form a tube across the injured area. The tube provides a means for new axons to grow from the proximal area across the injured area into the distal area previously occupied by the original nerve fiber (Figure 13-15d). The growth of new axons will not occur if the gap at the site of injury is too large or if the gap becomes filled with dense collagenous fibers.

Regeneration

Following chromatolysis, there are signs of recovery in the cell body. There is an acceleration of RNA and protein synthesis, which favors regeneration of the axon. Recovery often takes several months and involves the restoration of normal levels of RNA, proteins, and the chromatophilic substance to their usual, uninjured patterns.

Accelerated protein synthesis is required for repair of the damaged axon. The proteins synthesized in the cell body pass into the axon by axoplasmic flow at about the rate of 1 mm (0.04 inch)/day. The proteins assist in regenerating the damaged axon. During the first few days following damage, buds of regenerating axons begin to invade the tube formed by the neurolemmocytes. Axons from the proximal area grow at the rate of about 1.5 mm (0.06 inch)/day across the area of damage, find

their way into the distal neurolemmal tubes, and grow toward the distally located receptors and effectors. Thus sensory and motor connections are reestablished. In time, a new myelin sheath is also produced by the neurolemmocytes. However, function is never completely restored after a nerve is severed.

Neuritis

Neuritis is inflammation of a single nerve, two or more nerves in separate areas, or many nerves simultaneously. It may result from irritation to the nerve produced by direct blows, bone fractures, contusions, or penetrating injuries. Additional causes include vitamin deficiency (usually thiamine) and poisons such as carbon monoxide, carbon tetrachloride, heavy metals, and some drugs.

Sciatica

Sciatica (sī-AT-i-ka) is a type of neuritis characterized by severe pain along the path of the sciatic nerve or its branches. The term is commonly applied to a number of disorders affecting this nerve. Because of its length and size, the sciatic nerve is exposed to many kinds of injury. Inflammation of or injury to the nerve causes pain that passes from the back or thigh down its length into the leg, foot, and toes.

Probably the most common cause of sciatica is a herniated (slipped) disc. Other causes include irritation from osteoarthritis, back injuries, or pressure on the nerve from certain types of exertion. Sciatica may be associated with diabetes mellitus, gout, or vitamin deficiencies. Other cases are idiopathic (unknown).

Shingles

Shingles is an acute infection of the peripheral nervous system and is frequently a relapse of the same case of chickenpox first acquired during childhood. Shingles is caused by a virus called herpes zoster (HER-pēz ZOS-ter), the chickenpox virus. Following recovery from chickenpox, the virus retreats to dorsal root ganglia where it resides. If the virus becomes activated, the immune system usually prevents it from spreading. However, from time to time the activated virus overcomes the immune system. In this case, the virus leaves the ganglion and travels down sensory neurons, causing the pain, and invades the skin where the neurons end, causing a characteristic line of skin blisters and discoloration of the skin. The line of blisters has a shape corresponding to the distribution of a particular nerve. The intercostal nerves and thoracic spinal nerves in the waist area are most commonly affected.

STUDY OUTLINE

Grouping of Neural Tissue (p. 281)
1. White matter is an aggregation of myelinated axons and associated neuroglia.
2. Gray matter is a collection of nerve cell bodies and dendrites or unmyelinated axons along with associated neuroglia.
3. A nerve is a bundle of nerve fibers outside the central nervous system.
4. A ganglion is a collection of cell bodies outside the central nervous system.
5. A tract is a bundle of fibers of similar function in the central nervous system.
6. A nucleus is a mass of nerve cell bodies and dendrites in the gray matter of the brain and spinal cord.
7. A horn (column) is an area of gray matter in the spinal cord.

Spinal Cord (p. 281)

General Features (p. 281)
1. The spinal cord begins as a continuation of the medulla oblongata and terminates at about the second lumbar vertebra.
2. It contains cervical and lumbar enlargements that serve as points of origin for nerves to the extremities.
3. The tapered portion of the spinal cord is the conus medullaris, from which arise the filum terminale and cauda equina.
4. The spinal cord is partially divided into right and left sides by the anterior median fissure and posterior median sulcus.
5. The gray matter in the spinal cord is divided into horns and the white matter into funiculi (columns).
6. In the center of the spinal cord is the central canal, which runs the length of the spinal cord and contains cerebrospinal fluid.
7. There are ascending (sensory) tracts and descending (motor) tracts.

Protection and Coverings (p. 281)
1. The spinal cord is protected by the vertebral canal, meninges, cerebrospinal fluid, and vertebral ligaments.
2. The meninges are three coverings that run continuously around the spinal cord and brain: dura mater, arachnoid, and pia mater.
3. Removal of cerebrospinal fluid from the subarachnoid space is called a spinal (lumbar) puncture. The procedure is used to diagnose pathologies and to introduce antibiotics or contrast media.

Structure in Cross Section (p. 284)
1. Parts of the spinal cord observed in cross section are the gray commissure; central canal; anterior, posterior, and lateral gray horns; anterior, posterior, and lateral white columns; and ascending and descending tracts.
2. The spinal cord conveys sensory and motor information by way of the ascending and descending tracts, respectively.

Functions (p. 285)
1. A major function of the spinal cord is to convey sensory impulses from the periphery to the brain and to conduct motor impulses from the brain to the periphery.
2. Another function is to serve as a reflex center. The posterior root, posterior root ganglion, and anterior root are involved in conveying an impulse.
3. A reflex arc is the shortest route that can be taken by an impulse from a receptor to an effector. Its basic components are a receptor, a sensory neuron, a center, a motor neuron, and an effector.
4. A reflex is a quick, involuntary response to a stimulus that passes along a reflex arc. Reflexes represent the body's principal mechanisms for responding to changes in the internal and external environment.
5. Somatic spinal reflexes include the stretch reflex, tendon reflex, flexor reflex, and crossed extensor reflex; all exhibit reciprocal innervation.
6. A two-neuron or monosynaptic reflex arc contains one sensory and one motor neuron. A stretch reflex, such as the patellar reflex, is an example.
7. The stretch reflex is ipsilateral and is important in maintaining muscle tone and muscle coordination during exercise.
8. A polysynaptic reflex arc contains a sensory, association, and motor neuron. The tendon reflex, flexor reflex, and crossed extensor reflexes are examples.
9. The tendon reflex is ipsilateral and prevents damage to muscles and tendons as a result of stretching; the flexor reflex is ipsilateral and is a withdrawal reflex; the crossed extensor reflex is contralateral.
10. Among clinically important somatic reflexes are the patellar reflex, the Achilles reflex, the Babinski sign, and the abdominal reflex.

Spinal Nerves (p. 292)

Names (p. 292)
1. The 31 pairs of spinal nerves are named and numbered according to the region and level of the spinal cord from which they emerge.
2. There are 8 pairs of cervical, 12 paris of thoracic, 5 pairs of lumbar, 5 pairs of sacral, and 1 pair of coccygeal nerves.

Composition and Coverings (p. 293)
1. Spinal nerves are attached to the spinal cord by means of a posterior root and an anterior root. All spinal nerves are mixed.
2. Spinal nerves are covered by endoneurium, perineurium, and epineurium.

Distribution (p. 294)
1. Branches of a spinal nerve include the dorsal ramus, ventral ramus, meningeal branch, and rami communicantes.
2. The ventral rami of spinal nerves, except for T2–T11, form networks of nerves called plexuses.
3. Emerging from the plexuses are nerves bearing names that are often descriptive of the general regions they supply or the course they take.

4. The cervical plexus supplies the skin and muscles of the head, neck and upper part of the shoulders, connects with some cranial nerves, and supplies the diaphragm.
5. The brachial plexus constitutes the nerve supply for the upper extremities and a number of neck and shoulder muscles.
6. The lumbar plexus supplies the anterolateral abdominal wall, external genitals, and part of the lower extremities.
7. The sacral plexus supplies the buttocks, perineum, and lower extremities.
8. Nerves T2–T11 do not form plexuses and are called intercostal (thoracic) nerves. They are distributed directly to the structures they supply in intercostal spaces.

Dermatomes (p. 302)
1. All spinal nerves except C1 innervate specific, constant segments of the skin. The skin segments are called dermatomes.

2. Knowledge of dermatomes helps a physician to determine which segment of the spinal cord or a spinal nerve is malfunctioning.

Disorders: Homeostatic Imbalances (p. 302)
1. Spinal cord injury may result in paralysis, which may be classified as monoplegia, diplegia, paraplegia, hemiplegia, or quadriplegia.
2. Transection is followed by a period of loss of reflex activity called areflexia.
3. Following peripheral nerve damage, repair is accomplished by an axon reaction, Wallerian degeneration, and regeneration.
4. Inflammation of nerves is known as neuritis.
5. Neuritis of the sciatic nerve and its branches is called sciatica.
6. Shingles is an acute infection of peripheral nerves.

REVIEW QUESTIONS

1. Define the following groupings of neural tissue: white matter, gray matter, nerve, ganglion, tract, nucleus, and horn (column).
2. Describe the location of the spinal cord. What are the cervical and lumbar enlargements?
3. Define conus medullaris, filum terminale, and cauda equina. What is a spinal segment? How is the spinal cord partially divided into a right and left side?
4. Describe the bony covering of the spinal cord.
5. Explain the location and composition of the spinal meninges. Describe the location of the epidural, subdural, and subarachnoid spaces. Define meningitis.
6. What is a spinal (lumbar) puncture? Why is it performed?
7. Based upon your knowledge of the structure of the spinal cord in cross section, define the following: gray commissure, central canal, anterior gray horn, lateral gray horn, posterior gray horn, anterior white column, lateral white column, posterior white column, ascending tract, and descending tract.
8. Describe the function of the spinal cord as a conduction pathway.
9. Using Exhibit 13-1 as a guide, be sure that you can list the location, origin, termination, and function of the principal ascending and descending tracts.
10. Describe how the spinal cord serves as a reflex center.
11. What is a reflex arc? List and define the components of a reflex arc.
12. Define a reflex. How are reflexes related to the maintenance of homeostasis?
13. Describe the mechanism and function of a stretch reflex,

tendon reflex, and crossed extensor reflex.
14. Define the following terms relating to reflex arcs: monosynaptic, ipsilateral, polysynaptic, intersegmental, contralateral, and reciprocal innervation.
15. Why are reflexes important in diagnosis? Indicate the clinical importance of the following reflexes: patellar, Achilles, Babinski sign, and abdominal.
16. Define a spinal nerve. Why are all spinal nerves classified as mixed nerves?
17. Describe how a spinal nerve is attached to the spinal cord.
18. Explain how a spinal nerve is enveloped by its several different coverings.
19. How are spinal nerves named and numbered?
20. Describe the branches and innervations of a typical spinal nerve.
21. What is a plexus? Describe the principal plexuses and the regions they supply.
22. What are intercostal (thoracic) nerves?
23. Define a dermatome. Why is a knowledge of dermatomes important?
24. Distinguish the following types of paralysis: monoplegia, diplegia, paraplegia, hemiplegia, and quadriplegia.
25. Define complete transection and hemisection. What are the consequences of each? What is spinal shock?
26. Outline the principal events that occur as part of the axon reaction, Wallerian degeneration, and regeneration following peripheral nerve damage.
27. Distinguish between sciatica and neuritis.
28. What is shingles?

14

The Brain and the Cranial Nerves

ow we will consider the principal parts of the brain, how the brain is protected, and how it is related to the spinal cord and to the 12 pairs of cranial nerves.

The developmental anatomy of the brain is explained in detail at the end of the chapter.

BRAIN

PRINCIPAL PARTS

The **brain** of an average adult is one of the largest organs of the body, weighing about 1,300 g (3 lb). Figure 14-1 shows that the brain is mushroom-shaped and divided into four principal parts: brain stem, diencephalon, cerebrum, and cerebellum. In some cases, embryological names are retained when distinguishing the various parts of the brain. The **brain stem,** the stalk of the mushroom, consists of the medulla oblongata, pons, and midbrain or mesencephalon (mes-en-SEF-a-lon). The lower end of the brain stem is a continuation of the spinal cord. Above the brain stem is the **diencephalon** (dī-en-SEF-a-lon), consisting primarily of the thalamus and hypothalamus. The **cerebrum** spreads over the diencephalon. The cerebrum

constitutes about seven-eighths of the total weight of the brain and occupies most of the cranium. Inferior to the cerebrum and posterior to the brain stem is the **cerebellum.**

PROTECTION AND COVERINGS

The brain is protected by the cranial bones (see Figure 7-2). Like the spinal cord, the brain is also protected by meninges. The **cranial meninges** surround the brain, are continuous with the spinal meninges, and have the same basic structure and bear the same names as the spinal meninges: the outermost **dura mater,** middle **arachnoid,** and innermost **pia mater** (Figure 14-2).

The cranial dura mater consists of two layers. The thicker, outer layer (periosteal layer) tightly adheres to the cranial bones and serves as periosteum. The thinner, inner layer (meningeal layer) includes a mesothelial layer on its smooth surface. The spinal dura mater corresponds to the meningeal layer of the cranial dura mater.

CEREBROSPINAL FLUID (CSF)

The brain, as well as the rest of the central nervous system, is further protected against injury by **cerebrospinal fluid**

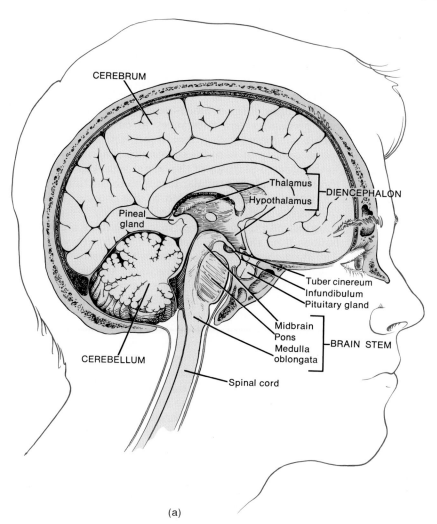

CEREBRUM

Thalamus
Hypothalamus —DIENCEPHALON

Pineal
gland

Tuber cinereum
Infundibulum
Pituitary gland

Midbrain
Pons —BRAIN STEM
Medulla
oblongata

CEREBELLUM

Spinal cord

(a)

FIGURE 14-1 Brain. (a) Principal parts of the medial aspect of the brain seen in sagittal section. The infundibulum and pituitary gland are discussed in conjunction with the endocrine system in Chapter 18.

CEREBRUM Pineal gland

Thalamus ⎤
Hypothalamus ⎦ DIENCEPHALON

Tuber cinereum

Infundibulum

Midbrain ⎤
Pons ⎥ BRAIN STEM
Medulla oblongata ⎦

Central canal

Spinal cord

CEREBELLUM

(b)

FIGURE 14-1 (*Continued*) (b) Photograph of the medial aspect of the brain seen in sagittal section. (Courtesy of C. Yokochi and J. W. Rohen, *Photographic Anatomy of the Human Body,* 2nd ed., 1978, IGAKU-SHOIN, Ltd., Tokyo, New York.)

(CSF). This fluid circulates through the subarachnoid space around the brain and spinal cord and through the ventricles of the brain. The subarachnoid space is the area between the arachnoid and pia mater.

The **ventricles** (VEN-tri-kuls) are cavities in the brain that communicate with each other, with the central canal of the spinal cord, and with the subarachnoid space (Figure 14-2). Each of the two **lateral ventricles** is located in a hemisphere (side) of the cerebrum under the corpus callosum. The **third ventricle** is a vertical slit between and inferior to the right and left halves of the thalamus and between the lateral ventricles. Each lateral ventricle communicates with the third ventricle by a narrow, oval opening, the **interventricular foramen.** The **fourth ventricle** lies between the inferior brain stem and the cerebellum. It communicates with the third ventricle via the **cerebral aqueduct,** which passes through the midbrain. The roof of the fourth ventricle has three openings: a **median aperture** and two **lateral apertures.** Through these openings, the fourth ventricle also communicates with the subarachnoid space of the brain and cord.

The entire central nervous system contains between 80 and 150 ml (3 to 5 oz) of cerebrospinal fluid. It is a clear, colorless fluid of watery consistency. Chemically, it contains proteins, glucose, urea, and salts. It also contains some lymphocytes. Cerebrospinal fluid has two principal functions related to homeostasis: protection and circulation. The fluid serves as a shock-absorbing medium to protect the brain and spinal cord from jolts that would otherwise cause them to crash against the bony walls of the cranial and vertebral cavities. The fluid also buoys the brain so that it "floats" in the cranial cavity. With regard to its circulatory function, cerebrospinal fluid delivers nutritive substances filtered from blood to the brain and spinal cord, and removes wastes and toxic substances produced by brain and spinal cord cells.

Cerebrospinal fluid is formed primarily by filtration and secretion from networks of capillaries, called **choroid** (KŌ-royd; *chorion* = delicate) **plexuses,** located in the ventricles (Figure 14-2a). Various components of the choroid plexuses form a **blood–cerebrospinal fluid barrier** that permits certain substances to enter the fluid but prohibits others. Such a barrier protects the brain and spinal cord from harmful substances. The fluid formed in the choroid plexuses of the lateral ventricles circulates through the interventricular foramina to the third ventricle, where more fluid is added by the choroid plexus of the third ventricle. It then flows through the cerebral aqueduct into the fourth ventricle. Here there are contributions from the choroid plexus of the fourth ventricle. The fluid then circulates through the apertures of the fourth ventricle into the subarachnoid space around the back of the brain. It also passes downward to the subarachnoid space around the posterior surface of the spinal cord, up the anterior surface of the spinal cord, and around the anterior part of the brain. From there it is gradually reabsorbed into veins. Some cerebrospinal fluid may be formed by ependymal (neuroglial) cells lining the central canal of the spinal cord. This small quantity of fluid ascends to reach the fourth ventricle. Most of the fluid is absorbed into the superior sagittal sinus. The absorption actually occurs through **arachnoid villi**—fingerlike projections of the arachnoid that push into the dural venous sinuses, especially the superior sagittal sinus (Figure 14-2b). Normally, cerebrospinal fluid is absorbed as rapidly as it is formed.

The formation, circulation, and absorption of cerebrospinal fluid are summarized in Figure 14-3.

Superior cerebral vein

Arachnoid villus

Cerebrum

Subarachnoid space of brain

Superior sagittal sinus

Intermediate mass

Corpus callosum

Choroid plexus of lateral ventricle

Lateral ventricle

Great cerebral vein

Interventricular foramen

Cerebellum

Pons

Pia mater

Arachnoid   } Cranial meninges

Dura mater

Third ventricle

Choroid plexus of third ventricle

Lateral aperture

Straight sinus

Medulla oblongata

Cerebral aqueduct

Choroid plexus of fourth ventricle

Spinal cord

Fourth ventricle

Pia mater

Arachnoid   } Spinal meninges

Dura mater

Median aperture

Central canal

Subarachnoid space of spinal cord

Filum terminale

(a)

FIGURE 14-2 Meninges. (a) Brain, spinal cord, and meninges seen in sagittal section. Arrows indicate the direction of flow of cerebrospinal fluid.

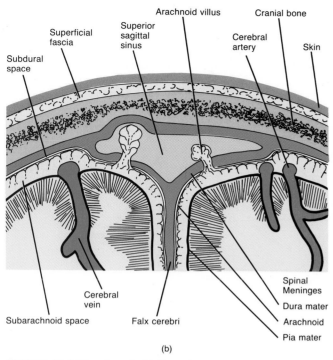

(b)

FIGURE 14-2 (*Continued*) (b) Frontal section through the superior portion of the brain showing the relationship of the superior sagittal sinus to the arachnoid villi.

CLINICAL APPLICATION

If an obstruction, such as a tumor or a congenital blockage, arises in the brain and interferes with the drainage of fluid from the ventricles into the sub-arachnoid space, large amounts of fluid accumulate in the ventricles. Fluid pressure inside the brain increases and, if the fontanels have not yet closed, the head bulges to relieve the pressure. This condition is called **internal hydrocephalus** (*hydro* = water; *enkephalos* = brain). If an obstruction interferes with drainage somewhere in the subarachnoid space and cerebrospinal fluid accumulates inside the space, the condition is termed **external hydrocephalus.**

BLOOD SUPPLY

The brain is well supplied with oxygen and nutrients by blood vessels that form the cerebral arterial circle (circle of Willis). Cerebral circulation is outlined in Exhibits 21-4 and 21-9 and Figures 21-14c and 21-18. Blood vessels that enter brain tissue pass along the surface of the brain and, as they penetrate inward, they are surrounded by a loose-fitting layer of pia mater. The space between the penetrating blood vessel and pia mater is called a *perivascular space.*

Although the brain composes only about 2 percent of total body weight, it utilizes about 20 percent of the oxygen used by the entire body. The brain is one of the most metabolically active organs of the body, and the amount of oxygen it uses varies with the degree of mental activity. If the blood flow to the brain is interrupted even briefly, unconsciousness may result. A 1- or 2-minute interruption may weaken the brain cells by starving them of oxygen, and if the cells are totally deprived of oxygen for 4 minutes, many are permanently injured. Lysosomes of brain cells are sensitive to decreased oxygen concentration. If the condition persists long enough, lysosomes break open and release enzymes that bring about self-destruction of brain cells. Occasionally during childbirth, the oxygen supply from the mother's blood is interrupted before the baby leaves the birth canal and can breathe. Often such babies are stillborn or suffer permanent brain damage that may result in mental retardation, epilepsy, and paralysis.

Blood supplying the brain also contains glucose, the principal source of energy for brain cells. Because carbohydrate storage in the brain is limited, the supply of glucose must be continuous. If blood entering the brain has a low glucose level, mental confusion, dizziness, convulsions, and loss of consciousness may occur.

Both carbon dioxide and oxygen have potent effects on cerebral blood flow. Carbon dioxide increases cerebral blood flow by combining with water to form carbonic acid (H_2CO_3), which breaks down into hydrogen ions (H^+) and bicarbonate ions (HCO_3^-). The H^+ ions then cause vasodilation of cerebral vessels and increased blood flow. Dilation is almost directly proportional to an increase in H^+ ion concentration. A decrease in oxygen in the blood also causes vasodilation and increased cerebral blood flow.

Glucose, oxygen, and certain ions pass rapidly from the circulating blood into brain cells. Other substances, such as creatinine, urea, chloride, insulin, and sucrose, enter quite slowly. Still other substances—proteins and most antibiotics—do not pass at all from the blood into brain cells. The differential rates of passage of certain materials from the blood into most parts of the brain are based upon a concept called the **blood–brain barrier (BBB).** The barrier is either absent or less selective in the hypothalamus and roof of the fourth ventricle. Electron micrograph studies of the capillaries of the brain reveal that they differ structurally from other capillaries. Brain capillaries are constructed of more densely packed cells and are surrounded by large numbers of neuroglial cells and a continuous basement membrane. These features form a barrier to the passage of certain materials. Substances that cross the barrier are either very small molecules or require the assistance of a carrier molecule to cross by active transport The blood–brain barrier functions as a selective barrier to protect brain cells from harmful substances. An injury to the brain due to trauma, inflammation, or toxins causes a breakdown of the blood–brain barrier, permitting the passage of normally restricted substances into brain tissue.

BRAIN STEM

Medulla Oblongata

The **medulla oblongata** (me-DULL-la ob'-long-GA-ta), or simply **medulla,** is a continuation of the upper portion of the spinal cord and forms the inferior part of the brain

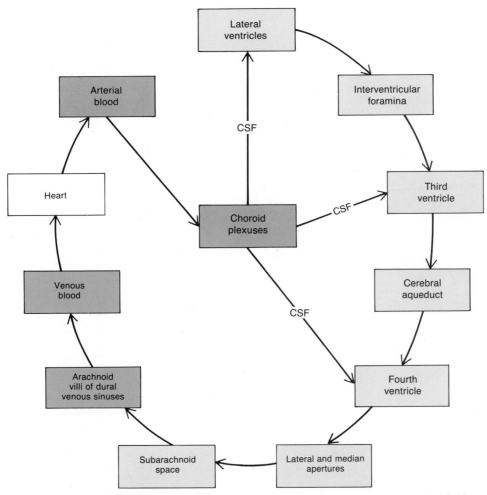

FIGURE 14-3 Summary of the formation, circulation, and absorption of cerebrospinal fluid.

stem (Figure 14-4). Its position in relation to the other parts of the brain may be noted in Figure 14-1. It lies just superior to the level of the foramen magnum and extends upward to the inferior portion of the pons. The medulla measures 3 cm (about 1 inch) in length.

The medulla contains all ascending and descending tracts that communicate between the spinal cord and various parts of the brain. These tracts constitute the white matter of the medulla. Some tracts cross as they pass through the medulla. Let us see how this crossing occurs and what it means.

On the ventral side of the medulla are two roughly triangular structures called **pyramids** (Figures 14-4 and 14-5). The pyramids are composed of the largest motor tracts that pass from the outer region of the cerebrum (cerebral cortex) to the spinal cord. Just above the junction of the medulla with the spinal cord, most of the fibers in the left pyramid cross to the right side, and most of the fibers in the right pyramid cross to the left. This crossing is called the **decussation** (dē'-ku-SĀ-shun) **of pyramids.** The adaptive value, if any, of this phenomenon is unknown. The principal motor fibers that undergo decussation belong to the lateral corticospinal tracts. These tracts originate in the cerebral cortex and pass inferiorly to the medulla. The fibers cross in the pyramids and descend in the lateral columns of the spinal cord, terminating in the anterior gray horns. Here synapses occur with motor neurons that terminate in skeletal muscles. As a result of the crossing, fibers that originate in the left cerebral cortex activate muscles on the right side of the body, and fibers that originate in the right cerebral cortex activate muscles on the left side. Decussation explains why motor areas of one side of the cerebral cortex control muscular movements on the opposite side of the body.

The dorsal side of the medulla contains two pairs of prominent nuclei: the right and left **nucleus gracilis** (gras-I-lis; *gracilis* = slender) and **nucleus cuneatus** (kyoo-nē-Ā-tus; *cuneus* = wedge). These nuclei receive sensory fibers from ascending tracts (right and left fasciculus gracilis and fasciculus cuneatus) of the spinal cord and relay the sensory information to the opposite side of the medulla (see Figure 15-7). The information is conveyed to the thalamus and then to the sensory areas of the cerebral cortex. Nearly all sensory impulses received on one side of the body cross in the medulla or spinal cord and are perceived in the opposite side of the cerebral cortex.

In addition to its function as a conduction pathway for motor and sensory impulses between the brain and spinal cord, the medulla also contains an area of dispersed

FIGURE 14-4 Brain stem. Diagram of the ventral surface of the brain showing the structure of the brain stem in relation to the cranial nerves and associated structures.

gray matter containing some white fibers. This region is called the **reticular formation.** Actually, portions of the reticular formation are also located in the spinal cord, pons, midbrain, and diencephalon. The reticular formation functions in consciousness and arousal.

CLINICAL APPLICATION

The most common knockout blow is one that makes contact with the mandible. Such a blow twists and distorts the brain stem and overwhelms the reticular activating system (RAS) by sending a sudden volley of nerve impulses to the brain, resulting in **unconsciousness.**

Within the medulla are three vital reflex centers of the reticular system. The **cardiac center** regulates heartbeat and force of contraction (see Figure 20-9), the **medullary rhythmicity area** adjusts the basic rhythm of breathing (see Figure 23-20), and the **vasomotor (vasoconstric-**

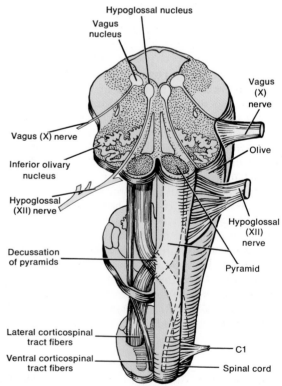

FIGURE 14-5 Details of the medulla showing the decussation of pyramids.

tor) **center** regulates the diameter of blood vessels. Other centers in the medulla are considered nonvital and coordinate swallowing, vomiting, coughing, sneezing, and hiccuping.

CLINICAL APPLICATION

Neurosurgeons are now using an ultrasonic device called a **cavitron ultrasonic surgical aspirator (CUSA)** that can shatter certain kinds of brain tumors. High-frequency sound waves stimulate the slender tip of the instrument to vibrate 23,000 times per sound. This vibratory action disintegrates the tumors. The instrument also delivers an irrigating saline solution that aspirates the fragmented particles. One major advantage of CUSA is that it decreases the probability of damaging either adjacent normal tissues such as the brain stem, which could cause abnormal heart rhythms, or large blood vessels, which could cause hemorrhage.

The medulla also contains the nuclei of origin for several pairs of cranial nerves (Figure 14-4 and 14-5). These are the cochlear and vestibular branches of the vestibulocochlear (VIII) nerves, which are concerned with hearing and equilibrium (there is also a nucleus for the vestibular branches in the pons); the glossopharyngeal (IX) nerves, which relay impulses related to swallowing, salivation, and taste; the vagus (X) nerves, which relay impulses to and from many thoracic and abdominal viscera; the cranial portions of the accessory (XI) nerves, which convey impulses related to head and shoulder movements (a part of this nerve also arises from the first five segments of the spinal cord); and the hypoglossal (XII) nerves, which convey impulses that involve tongue movements.

On each lateral surface of the medulla is an oval projection called the **olive** (Figure 14-4), which contains an inferior olivary nucleus and two accessory olivary nuclei. The nuclei are connected to the cerebellum by fibers.

Also associated with the medulla is the greater part of the **vestibular nuclear complex.** This nuclear group consists of the *lateral, medial,* and *inferior vestibular nuclei* in the medulla and the *superior vestibular nucleus* in the pons. As you will see later (Chapter 17), the vestibular nuclei assume an important role in helping the body maintain its sense of equilibrium.

In view of the many vital activities controlled by the medulla, it is not surprising that a hard blow to the base of the skull can be fatal. Nonfatal medullary injury may be indicated by cranial nerve malfunctions on the same side of the body as the area of medullary injury, paralysis and loss of sensation on the opposite side of the body, and irregularities in respiratory control.

Pons

The relationship of the **pons** to other parts of the brain can be seen in Figures 14-1 and 14-4. The pons, which means bridge, lies directly above the medulla and anterior

to the cerebellum. It measures about 2.5 cm (1 inch) in length. Like the medulla, the pons consists of white fibers scattered throughout with nuclei. As the name implies, the pons is a bridge connecting the spinal cord with the brain and parts of the brain with each other. These connections are provided by fibers that run in two principal directions. The transverse fibers connect with the cerebellum through the *middle cerebellar peduncles.* The longitudinal fibers of the pons belong to the motor and sensory tracts that connect the spinal cord or medulla with the upper parts of the brain stem.

The nuclei for certain paired cranial nerves are also contained in the pons (Figure 14-4). These include the trigeminal (V) nerves, which relay impulses for chewing and for sensations of the head and face; the abducens (VI) nerves, which regulate certain eyeball movements; the facial (VII) nerves, which conduct impulses related to taste, salivation, and facial expression; and the vestibular branches of the vestibulocochlear (VIII) nerves, which are concerned with equilibrium.

Other important nuclei in the reticular formation of the pons are the **pneumotaxic** (noo-mō-TAK-sik) **area** and the **apneustic** (ap-NOO-stik) **area.** Together with the medullary rhythmicity area in the medulla, they help control respiration.

Midbrain

The **midbrain,** or **mesencephalon** (*meso* = middle; *enkephalos* = brain), extends from the pons to the lower portion of the diencephalon (Figures 14-1 and 14-4). It is about 2.5 cm (1 inch) in length. The cerebral aqueduct passes through the midbrain and connects the third ventricle above with the fourth ventricle below.

The ventral portion of the midbrain contains a pair of fiber bundles referred to as **cerebral peduncles** (pe-DUNG-kulz). The cerebral peduncles contain many motor fibers that convey impulses from the cerebral cortex to the pons and spinal cord. They also contain sensory fibers that pass from the spinal cord to the thalamus. The cerebral peduncles constitute the main connection for tracts between upper parts of the brain and lower parts of the brain and the spinal cord.

The dorsal portion of the midbrain is called the **tectum** (*tectum* = roof) and contains four rounded eminences: the **corpora quadrigemina** (KOR-po-ra kwad-ri-JEM-in-a). Two of the eminences are known as the **superior colliculi** (ko-LIK-yoo-lī). These serve as reflex centers for movements of the eyeballs and head in response to visual and other stimuli. The other two eminences, the **inferior colliculi,** serve as reflex centers for movements of the head and trunk in response to auditory stimuli. The midbrain also contains the **substantia nigra** (sub-STAN-shē-a NĪ-gra), a large, heavily pigmented nucleus near the cerebral peduncles.

A major nucleus in the reticular formation of the midbrain is the **red nucleus.** Fibers from the cerebellum and cerebral cortex terminate in the red nucleus. The red nucleus is also the origin of cell bodies of the descending rubrospinal tract. Other nuclei in the midbrain are associated with cranial nerves (see Figure 14-4). These include the oculomotor (III) nerves, which mediate some movements of the eyeballs and changes in pupil size and lens shape, and the trochlear (IV) nerves, which conduct impulses that move the eyeballs.

A structure called the **medial lemniscus** (*lemniskos* = ribbon or band) is common to the medulla, pons, and midbrain. The medial lemniscus is a band of white fibers containing axons that convey impulses for fine touch, proprioception, and vibrations from the medulla to the thalamus.

DIENCEPHALON

The **diencephalon** (*dia* = through; *enkephalos* = brain) consists principally of the thalamus and hypothalamus. The relationship of these structures to the rest of the brain is shown in Figure 14-1.

Thalamus

The **thalamus** (THAL-a-mus; *thalamos* = inner chamber) is an oval structure above the midbrain that measures about 3 cm (1 inch) in length and constitutes four-fifths of the diencephalon. It consists of two oval masses of mostly gray matter organized into nuclei that form the lateral walls of the third ventricle (Figure 14-6). The masses are joined by a bridge of gray matter called the **intermediate mass.** Each mass is deeply embedded in a cerebral hemisphere and is bounded laterally by the **internal capsule.**

Although the thalamic masses are primarily gray matter, some portions are white matter. Among the white matter portions are the **stratum zonale,** which covers the dorsal surface; the **external medullary lamina,** covering the lateral surface; and the **internal medullary lamina,** which divides the gray matter masses into an anterior

nuclear group, a medial nuclear group, and a lateral nuclear group.

Within each group are nuclei that assume various roles. Some nuclei in the thalamus serve as relay stations for all sensory impulses, except smell, to the cerebral cortex. These include the **medial geniculate** (je-NIK-yoo-lāt) **nuclei** (hearing), the **lateral geniculate nuclei** (vision), and the **ventral posterior nuclei** (general sensations and taste). Other nuclei are centers for synapses in the somatic motor system. These include the **ventral lateral nuclei** (voluntary motor actions) and **ventral anterior nuclei** (voluntary motor actions and arousal). The thalamus is the principal relay station for sensory impulses that reach the cerebral cortex from the spinal cord, brain stem, cerebellum, and parts of the cerebrum.

The thalamus also functions as an interpretation center for some sensory impulses, such as pain, temperature, light touch, and pressure. The thalamus also contains a **reticular nucleus** in its reticular formation, which in some way seems to modify neuronal activity in the thalamus, and an **anterior nucleus** in the floor of the lateral ventricle, which is concerned with certain emotions and memory.

Hypothalamus

The **hypothalamus** (*hypo* = under) is a small portion of the diencephalon. Its relationship to other parts of the brain is shown in Figure 14-1 and 14-6a. The hypothalamus forms the floor and part of the lateral walls of the third ventricle. It is partially protected by the sella turcica of the sphenoid bone.

Information from the external environment comes to the hypothalamus via afferent pathways originating in the peripheral sense organs. Sound, taste, smell, and somatic sensations all come to the hypothalamus. Afferent

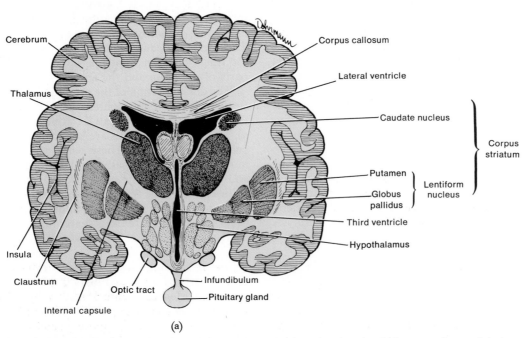

(a)

FIGURE 14-6 Thalamus. (a) Diagram of a frontal section showing the thalamus and associated structures.

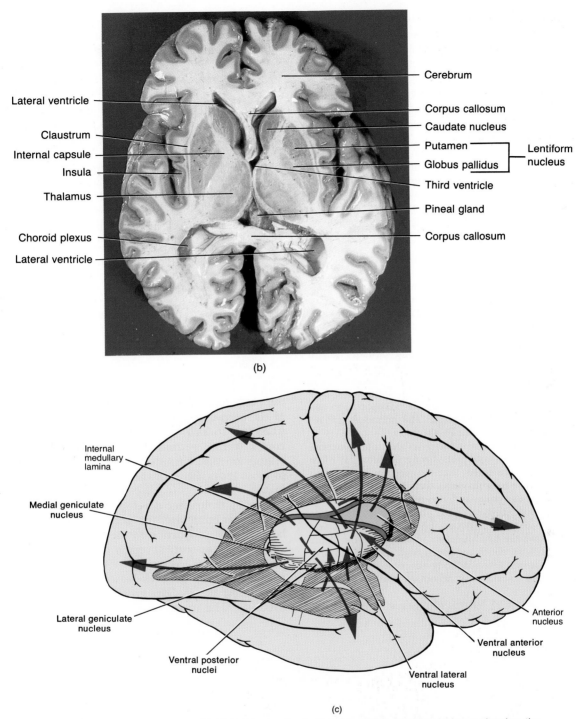

Cerebrum

Lateral ventricle

Corpus callosum

Caudate nucleus

Claustrum

Putamen ⎤

Internal capsule

Lentiform nucleus

Globus pallidus ⎦

Insula

Third ventricle

Thalamus

Choroid plexus

Pineal gland

Lateral ventricle

Corpus callosum

(b)

Internal medullary lamina

Medial geniculate nucleus

Lateral geniculate nucleus

Anterior nucleus

Ventral anterior nucleus

Ventral posterior nuclei

Ventral lateral nucleus

(c)

FIGURE 14-6 (*Continued*) (b) Photograph of a horizontal section of the cerebrum showing the thalamus and associated structures. (Courtesy of C. Yokochi and J. W. Rohen, *Photographic Anatomy of the Human Body,* 2nd ed., 1978, IGAKU-SHOIN, Ltd., Tokyo, New York.) (c) Diagram of right lateral view of the thalamic nuclei. The arrows indicate some of the connections between the thalamus and cerebral cortex.

impulses, monitoring the internal environment, arise from the internal viscera and reach the hypothalamus. Other parts of the hypothalamus itself continually monitor the water level, hormone concentrations, and temperature of blood. And, as you will see shortly, the hypothalamus has several very important connections with the pituitary gland.

Despite its small size, nuclei in the hypothalamus control many body activities, most of them related to homeostasis. Although differentiation of the hypothalamic nuclei is far from precise, it is possible to identify certain nuclei. Some of these nuclei are more readily identified in lower animals and are more distinct in fetuses than adults. Also, within a given nucleus there may be several kinds of cells that can be differentiated histologically. The localization of function, with a few exceptions, is not specific to the individual nuclei; certain functions tend to overlap nuclear boundaries. For this reason, functions are attributed to regions rather than specific nuclei.

The chief functions of the hypothalamus are as follows.

1. It controls and integrates the autonomic nervous system, which stimulates smooth muscle, regulates the rate of contraction of cardiac muscle, and controls the secretions of many glands. This is accomplished by axons of neurons whose dendrites and cell bodies are in hypothalamic nuclei. The axons form tracts from the hypothalamus to sympathetic and parasympathetic nuclei in the brain stem and spinal cord. Through the autonomic nervous system, the hypothalamus is the main regulator of visceral activities. It regulates heart rate, movement of food through the digestive tract, and contraction of the urinary bladder.

2. It is involved in the reception and integration of sensory impulses from the viscera.

3. It is the principal intermediary between the nervous system and the endocrine system—the two major control systems of the body. The hypothalamus lies just above the pituitary, the main endocrine gland. When the hypothalamus detects certain changes in the body, it releases chemicals called regulating factors that stimulate or inhibit the anterior pituitary gland. The anterior pituitary then releases or holds back hormones that regulate various physiological activities of the body. The hypothalamus also produces two hormones, antidiuretic hormone (ADH) and oxytocin (OT), which are transported to and stored in the posterior pituitary gland. The hormones are released from storage when needed by the body.

4. It is the center for the mind-over-body phenomenon. When the cerebral cortex interprets strong emotions, it often sends impulses along the tracts that connect the cortex with the hypothalamus. The hypothalamus then directs impulses via the autonomic nervous system and also releases chemicals that stimulate the anterior pituitary gland. The result can be a wide range of changes in body activities. For instance, when you panic, impulses leave the hypothalamus to stimulate your heart to beat faster. Likewise, continued psychological stress can produce long-term abnormalities in body function that result in serious illness. These so-called psychosomatic disorders are definitely real.

5. It is associated with feelings of rage and aggression.

6. It controls normal body temperature. Certain cells of the hypothalamus serve as a thermostat. If blood flowing through the hypothalamus is above normal temperature, the hypothalamus directs impulses along the autonomic nervous system to stimulate activities that promote heat loss. Heat can be lost through relaxation of the smooth muscle in the blood vessels, causing vasodilation of cutaneous vessels and increased heat loss from the skin. Heat loss also occurs by sweating. Conversely, if the temperature of the blood is below normal, the hypothalamus generates impulses that promote heat retention. Heat can be retained through the constriction of cutaneous blood vessels, cessation of sweating, and by shivering.

7. It regulates food intake through two centers. The **feeding (hunger) center** is stimulated by hunger sensations from an empty stomach. When sufficient food has been ingested, the **satiety** (sa-TĪ-e-tē) **center** is stimulated and sends out impulses that inhibit the feeding center.

8. It contains a **thirst center.** Certain cells in the hypothalamus are stimulated when the extracellular fluid volume is reduced. The stimulated cells produce the sensation of thirst.

9. It is one of the centers that maintains the waking state and sleep patterns.

10. It exhibits properties of a self-sustained oscillator and, as such, acts as a pacemaker to drive many biological rhythms.

CEREBRUM

Supported on the brain stem and forming the bulk of the brain is the **cerebrum** (see Figure 14-1). The surface of the cerebrum is composed of gray matter 2 to 4 mm (0.08 to 0.16 inch) thick and is referred to as the **cerebral cortex** (*cortex* = rind or bark). The cortex, containing billions of cells, consists of six layers of nerve cell bodies in most areas. Beneath the cortex lies the cerebral white matter.

During embryonic development, when there is a rapid increase in brain size, the gray matter of the cortex enlarges out of proportion to the underlying white matter. As a result, the cortical region rolls and folds upon itself. The folds are called **gyri** (JĪ-rī) or **convolutions** (Figure 14-7). The deep grooves between folds are referred to as **fissures;** the shallow grooves between folds are **sulci** (SUL-sī). The most prominent fissure, the **longitudinal**

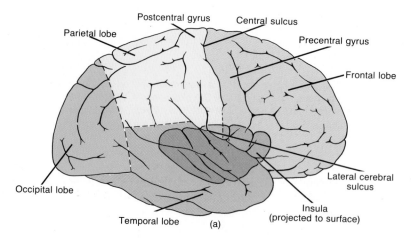

FIGURE 14-7 Lobes and fissures of the cerebrum. (a) Diagram of right lateral view. Since the insula cannot be seen externally, it has been projected to the surface. It can be seen in Figure 14-6a.

(b)

(c)

FIGURE 14-7 (*Continued*) (b) Photograph of right lateral view showing the insula after removal of a portion of the cerebrum. (Courtesy of N. Gluhbegovic and T. H. Williams, *The Human Brain: A Photographic Guide,* Harper & Row, Publishers, Inc., Hagerstown, MD, 1980). (c) Diagram of superior view. The insert to the left indicates the relative differences among a gyrus, sulcus, and fissure.

Falx cerebri

Choroid plexus of
lateral ventricle

Lateral ventricle

Cerebral cortex

Cerebral white matter

Falx cerebri

Superior sagittal sinus

Frontal lobe of cerebrum

Parietal lobe of cerebrum

Corpus callosum

Temporal lobe of cerebrum

Occipital lobe of cerebrum

Superior sagittal sinus

(d)

FIGURE 14-7 (*Continued*) (d) Photograph of a cross section through the cerebrum. (Courtesy of Stephen A. Kieffer and E. Robert Heitzman, *An Atlas of Cross-Sectional Anatomy,* Harper & Row, Publishers, Inc., Hagerstown, MD, 1979.)

fissure, nearly separates the cerebrum into right and left halves, or **hemispheres.** The hemispheres, however, are connected internally by a large bundle of transverse fibers composed of white matter called the **corpus callosum** (kal-LŌ-sum; *corpus* = body; *callosus* = hard). Between the hemispheres is an extension of the cranial dura mater called the **falx (FALKS) cerebri (cerebral fold).** It encloses the superior and inferior sagittal sinuses.

Lobes

Each cerebral hemisphere is further subdivided into four lobes by deep sulci or fissures. The **central sulcus** separates the **frontal lobe** from the **parietal lobe.** A major gyrus, the **precentral gyrus,** is located immediately anterior to the central sulcus. The gyrus is a landmark for the primary motor area of the cerebral cortex. Another major gyrus, the **postcentral gyrus,** is located immediately posterior to the central sulcus. This gyrus is a landmark for the general sensory area of the cerebral cortex. The **lateral cerebral sulcus** separates the **frontal lobe** from the **temporal lobe.** The **parietooccipital sulcus** separates the **parietal lobe** from the **occipital lobe.** Another prominent fissure, the **transverse fissure,** separates the cerebrum from the cerebellum. The frontal lobe, parietal lobe, temporal lobe, and occipital lobe are named after the bones that cover them. A fifth part of the cerebrum, the **insula,** lies deep within the lateral cerebral fissure, under the parietal, frontal, and temporal lobes. It cannot be seen in an external view of the brain (Figure 14-7a, b).

As you will see later, the olfactory (I) and optic (II) nerves are associated with specific lobes of the cerebrum.

White Matter

The white matter underlying the cortex consists of myelinated axons running in three principal directions (Figure 14-8).

1. Association fibers connect and transmit impulses between gyri in the same hemisphere.

2. Commissural fibers transmit impulses from the gyri in one cerebral hemisphere to the corresponding gyri in the opposite cerebral hemisphere. Three important groups of commissural fibers are the *corpus callosum, anterior commissure,* and *posterior commissure.*

3. Projection fibers form ascending and descending tracts that transmit impulses from the cerebrum to other parts of the brain and spinal cord. The internal capsule is an example.

Basal Ganglia (Cerebral Nuclei)

The **basal ganglia (cerebral nuclei)** are paired masses of gray matter in each cerebral hemisphere (Figures 14-6 and 14-9). The largest of the basal ganglia of each hemisphere is the **corpus striatum** (strī-Ā-tum; *corpus* = body; *striatus* = striped). It consists of the **caudate** (*cauda* = tail) **nucleus** and the **lentiform** (*lenticula* = shaped like a lentil or lens) **nucleus.** The lentiform nucleus, in turn, is subdivided into a lateral portion called the **putamen** (pu-TĀ-men; *putamen* = shell) and a medial portion called the **globus pallidus** (*globus* = ball; *pallid* = pale).

The portion of the *internal capsule* passing between the lentiform nucleus and the caudate nucleus and between the lentiform nucleus and thalamus is sometimes considered part of the corpus striatum. The internal cap-

FIGURE 14-8 White matter tracts of the left cerebral hemisphere seen in sagittal section. (Courtesy of N. Gluhbegovic and T. H. Williams, *The Human Brain: A Photographic Guide,* Harper & Row, Publishers, Inc., Hagerstown, MD, 1980.)

sule is made up of a group of sensory and motor white matter tracts that connect the cerebral cortex with the brain stem and spinal cord.

Other structures frequently considered part of the basal ganglia are the claustrum and amygdaloid nucleus. The **claustrum** (KLAWS-trum) is a thin sheet of gray matter lateral to the putamen. The **amygdaloid** (a-MIG-da-loyd; *amygda* = almond) **nucleus** is located at the tail end of the caudate nucleus. Some authorities also consider the **substantia nigra,** the **subthalamic nucleus,** and the **red nucleus** to be part of the basal ganglia. The substantia nigra is a large nucleus in the midbrain whose axons terminate in the caudate nucleus and putamen. The subthalamic nucleus lies against the internal capsule. Its major connection is with the globus pallidus.

The basal ganglia are interconnected by many fibers. They are also connected to the cerebral cortex, thalamus, and hypothalamus. The caudate nucleus and the putamen control large subconscious movements of the skeletal muscles, such as swinging the arms while walking. Such gross movements are also consciously controlled by the cerebral cortex. The globus pallidus is concerned with the regulation of muscle tone required for specific body movements.

CLINICAL APPLICATION

Damage to the basal ganglia results in abnormal body movements, such as uncontrollable shaking, called **tremor,** and **involuntary movements of skeletal muscles.** Moreover, destruction of a substantial portion of the caudate nucleus almost totally **paralyzes** the side of the body opposite to the damage.. The caudate nucleus is an area often affected by a stroke.

A lesion in the subthalamic nucleus results in a motor disturbance on the opposite side of the body called **hemiballismus** (*hemi* = half; *ballismos* = jumping), which is characterized by involuntary movements occurring suddenly with great force and rapidity. The movements are purposeless and generally of the withdrawal type, although they may be jerky. The spontaneous movements affect the proximal portions of the extremities most severely, especially the arms.

Limbic System

Certain components of the cerebral hemispheres and diencephalon constitute the **limbic** (*limbus* = border) **system.** Among its components are the following regions of gray matter.

1. **Limbic lobe.** Formed by two gyri of the cerebral hemisphere: the cingulate gyrus and the hippocampal gyrus.
2. **Hippocampus.** An extension of the hippocampal gyrus that extends into the floor of the lateral ventricle.
3. **Amygdaloid nucleus.** Located at the tail end of the caudate nucleus.
4. **Mammillary bodies of the hypothalamus.** Two round masses close to the midline near the cerebral peduncles.
5. **Anterior nucleus of the thalamus.** Located in the floor of the lateral ventricle.

The limbic system is a wishbone-shaped group of structures that encircles the brain stem and functions in the emotional aspects of behavior related to survival. The hippocampus, together with portions of the cerebrum, also functions in memory. Memory impairment results from lesions in the limbic system. People with such damage forget recent events and cannot commit anything to memory. How the limbic system functions in memory

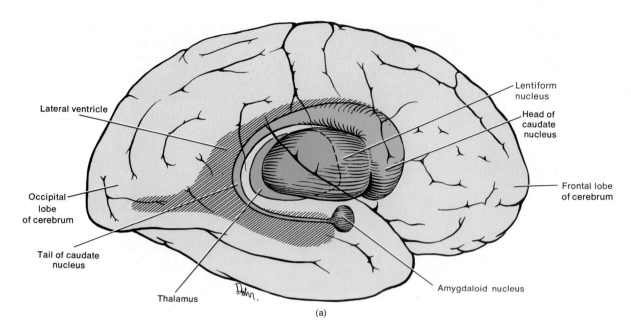

Lateral ventricle

Lentiform nucleus

Head of caudate nucleus

Frontal lobe of cerebrum

Amygdaloid nucleus

Occipital lobe of cerebrum

Tail of caudate nucleus

Thalamus

(a)

Body of caudate nucleus

Internal capsule

Tail of caudate nucleus

Occipital lobe of cerebrum

Frontal lobe of cerebrum

Head of caudate nucleus

Amygdaloid nucleus

(b)

FIGURE 14-9 Basal ganglia. (a) In this diagram of the right lateral view of the cerebrum, the basal ganglia have been projected to the surface. Refer to Figure 14-6b for the positions of the basal ganglia in the horizontal section of the cerebrum. (b) Photograph of the medial surface of the left cerebral hemisphere showing portions of the basal ganglia. (Courtesy of N. Gluhbegovic and T. H. Williams, *The Human Brain: A Photographic Guide,* Harper & Row, Publishers, Inc., Hagerstown, MD, 1980.)

is not clear. Although behavior is a function of the entire nervous system, the limbic system controls most of its involuntary aspects. Experiments on the limbic system of monkeys and other animals indicate that the amygdaloid nucleus assumes a major role in controlling the overall pattern of behavior.

Other experiments have shown that the limbic system is associated with pleasure and pain. When certain areas of the limbic system of the hypothalamus, thalamus, and midbrain are stimulated in animals, their reactions indicate they are experiencing intense punishment. When other areas are stimulated, the animals' reactions indicate

they are experiencing extreme pleasure. In still other studies, stimulation of the perifornical nuclei of the hypothalamus results in a behavioral pattern called *rage.* The animal assumes a defensive posture—extending its claws, raising its tail, hissing, spitting, growling, and opening its eyes wide. Stimulating other areas of the limbic system results in an opposite behavioral pattern: docility, tameness, and affection. Because the limbic system assumes a primary function in emotions such as pain, pleasure, anger, rage, fear, sorrow, sexual feelings, docility, and affection, it is sometimes called the "visceral" or "emotional" brain.

CLINICAL APPLICATION

Brain injuries are commonly associated with head injuries and result from displacement and distortion of neuronal tissue at the moment of impact. The various degrees of brain injury are described by the following terms.

1. Concussion. An abrupt but temporary loss of consciousness following a blow to the head or a sudden stopping of a moving head. A concussion produces no visible bruising of the brain.

2. Contusion. A visible bruising of the brain due to trauma and blood leaking from microscopic vessels. The pia mater is stripped from the brain over the injured area and may be torn, allowing blood to enter the subarachnoid space. A contusion usually results in an extended loss of consciousness, ranging from several minutes to many hours.

3. Laceration. Tearing of the brain, usually from a skull fracture or gunshot wound. A laceration results in rupture of large blood vessels with bleeding into the brain and subarachnoid space. Consequences include cerebral hematoma, edema, and increased intracranial pressure.

Functional Areas of Cerebral Cortex

The functions of the cerebrum are numerous and complex. In a general way, the cerebral cortex is divided into sensory, motor, and association areas. The **sensory areas** interpret sensory impulses, the **motor areas** control muscular movement, and the **association areas** are concerned with emotional and intellectual processes.

• *Sensory Areas* The **primary somesthetic** (sō-mes-THET-ik; *soma* = body; *aisthesis* = perception) **area** or **general sensory area** is located directly posterior to the central sulcus of the cerebrum in the postcentral gyrus of the parietal lobe. It extends from the longitudinal fissure on the top of the cerebrum to the lateral cerebral sulcus. In Figure 14-10, the general sensory area is designated by the areas numbered 1, 2, and 3.*

The primary somesthetic area receives sensations from cutaneous, muscular, and visceral receptors in various parts of the body. Each point of the area receives sensations from specific parts of the body, and essentially the entire body is spatially represented in it. The size of the portion of the sensory area receiving stimuli from body parts is not dependent on the size of the part but on the number of receptors the part contains. For example, a larger portion of the sensory area receives impulses from the lips than from the thorax (see Figure 15-4). The major function of the primary somesthetic area is to localize exactly the points of the body where the sensations originate. The thalamus is capable of localizing sensations in a general way, that is, it receives sensations from large areas of the body, but cannot distinguish be-

* These numbers, as well as most of the others shown, are based on K. Brodmann's cytoarchitectural map of the cerebral cortex. His map, first published in 1909, attempts to correlate structure and function.

FIGURE 14-10 Functional areas of the cerebrum. This lateral view indicates the sensory and motor areas of the right hemisphere. Although Broca's area is in the left hemisphere of most people, it is shown here to indicate its location.

tween specific areas of stimulation. This ability is reserved for the primary somesthetic area of the cortex.

The **secondary somesthetic area** is a small region in the posterior wall of the lateral sulcus in line with the postcentral gyrus. It is involved mainly in less discriminative aspects of sensation.

Posterior to the primary somesthetic area is the **somesthetic association area.** It corresponds to the areas numbered 5 and 7 in Figure 14-10. The somesthetic association area receives input from the thalamus, other lower portions of the brain, and the primary somesthetic area. Its role is to integrate and interpret sensations. This area permits you to determine the exact shape and texture of an object without looking at it, to determine the orientation of one object to another as they are felt, and to sense the relationship of one body part to another. Another role of the somesthetic association area is the storage of memories of past sensory experiences. Thus you can compare sensations with previous experiences.

Other sensory areas of the cortex include:

1. **Primary visual area** (area 17). Located on the medial surface of the occipital lobe and occasionally extends around to the lateral surface. It receives sensory impulses from the eyes and interprets shape, color, and movement.

2. **Visual association area** (areas 18 and 19). Located in the occipital lobe. It receives sensory signals from the primary visual area and the thalamus. It relates present to past visual experiences with recognition and evaluation of what is seen.

3. **Primary auditory area** (areas 41 and 42). Located in the superior part of the temporal lobe near the lateral cerebral sulcus. It interprets the basic characteristics of sound such as pitch and rhythm. Whereas the anterolateral portion of the auditory area responds to low pitches, the posterolateral portion responds to high pitches.

4. **Auditory association area** (area 22). Inferior to the primary auditory area in the temporal cortex. It determines if a sound is speech, music, or noise. It also interprets the meaning of speech by translating words into thoughts.

5. **Primary gustatory area** (area 43). Located at the base of the postcentral gyrus above the lateral cerebral sulcus in the parietal cortex. It interprets sensations related to taste.

6. **Primary olfactory area.** Located in the temporal lobe on the medial aspect. It interprets sensations related to smell.

7. **Gnostic** (NOS-tik; *gnosis* = knowledge) **area** (areas 5, 7, 39, and 40). This *common integrative area* is located among the somesthetic, visual, and auditory association areas. The gnostic area receives impulses from these areas, as well as from the taste and smell areas, the thalamus, and lower portions of the brain stem. It integrates sensory interpretations from the association areas and impulses from other areas so that a common thought can be formed from the various sensory inputs. It then transmits signals to other parts of the brain to cause the appropriate response to the sensory signal.

In Chapter 2, we discussed the principle and clinical applications of a recently developed type of radioisotope scanning called **positron emission tomography (PET).** At that point it was indicated that PET is being used in the diagnosis of an assortment of diseases. It is also being used by scientists to probe the healthy brain (see Figure 2-6). By detecting and recording changes in glucose metabolism (consumption), it is possible to identify which specific areas of the brain are involved in specific sensory and motor activities.

• *Motor Areas* The **primary motor area** (area 4) is located in the precentral gyrus of the frontal lobe (Figure 14-10). Like the primary somesthetic area, the primary motor area consists of regions that control specific muscles or groups of muscles (see Figure 15-8). Stimulation of a specific point of the primary motor area results in a muscular contraction, usually on the opposite side of the body.

The **premotor area** (area 6) is anterior to the primary motor area. It is concerned with learned motor activities of a complex and sequential nature. It generates impulses that cause a specific group of muscles to contract in a specific sequence, for example, writing. Thus the premotor area controls skilled movements.

The **frontal eye field area** (area 8) in the frontal cortex is sometimes included in the premotor area. This area controls voluntary scanning movements of the eyes—searching for a word in a dictionary, for instance.

The **language areas** are also significant parts of the motor cortex. The translation of speech or written words into thoughts involves sensory areas—primary auditory, auditory association, primary visual, visual association, and gnostic—as we just described. The translation of throughts into speech involves the **motor speech area** (area 44) or **Broca's** (BRŌ-kaz) **area,** located in the frontal lobe just superior to the lateral cerebral sulcus. From this area, a sequence of signals is sent to the premotor regions that control the muscles of the larynx, pharynx, and mouth. The impulses from the premotor area to the muscles result in specific, coordinated contractions that enable you to speak. Simultaneously, impulses are sent from Broca's area to the primary motor area. From here, impulses reach your breathing muscles to regulate the proper flow of air past the vocal cords. The coordinated contractions of your speech and breathing muscles enable you to translate your thoughts into speech.

CLINICAL APPLICATION

Broca's area and other language areas are located in the left cerebral hemisphere of most individuals regardless of whether they are left-handed or right-handed. Injury to the sensory or motor speech areas results in **aphasia** (a-FĀ-zē-a; *a* = without; *phasis* = speech), an inability to speak; **agraphia** (*a* = without; *graph* = write), an inability to write; **word deafness,** an inability to understand spoken words; or **word blindness,** an inability to understand written words.

• *Association Areas* The **association areas** of the cerebrum are made up of association tracts that connect motor and sensory areas (see Figure 14-8). The association region of the cortex occupies the greater portion of the lateral surfaces of the occipital, parietal, and temporal lobes, and the frontal lobes anterior to the motor areas. The association areas are concerned with memory, emotions, reasoning, will, judgment, personality traits, and intelligence.

Electroencephalogram (EEG)

Brain cells can generate electrical activity as a result of literally millions of action potentials of individual neurons. These electrical potentials are called **brain waves** and indicate activity of the cerebral cortex. Brain waves pass through the skull easily and can be detected by sensors called electrodes. A record of such waves is called an **electroencephalogram (EEG)**. An EEG is obtained by placing electrodes on the head and amplifying the waves with an electroencephalograph. As indicated in Figure 14-11, four kinds of waves are produced by normal individuals.

1. **Alpha waves.** These rhythmic waves occur at a frequency of about 8 to 13 cycles per second. (The unit commonly used to express frequency is the Hertz; 1 Hz = 1 cycle per second.) They are found in the EEGs of nearly all normal individuals when awake and in the resting state with their eyes closed. These waves disappear entirely during sleep.
2. **Beta waves.** The frequency of these waves is between 14 to 30 Hz. Beta waves generally appear when the nervous system is active, that is, during periods of sensory input and mental activity.
3. **Theta waves.** These waves have frequencies of 4 to 7 Hz. Theta waves normally occur in children and in adults experiencing emotional stress. They also occur in many disorders of the brain.
4. **Delta waves.** The frequency of these waves is 1 t 5 Hz. Delta waves occur during deep sleep. They are normal in an awake infant. When produced by an awake adult, they indicate brain damage.

Distinct EEG patterns appear in certain abnormalities. In fact, the EEG is used clinically in the diagnosis of epilepsy, infectious diseases, tumors, trauma, and hematomas. Electroencephalograms also furnish information regarding sleep and wakefulness.

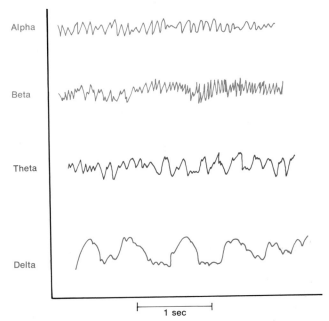

FIGURE 14-11 Kinds of waves recorded in an electroencephalogram (EEG).

BRAIN LATERALIZATION (SPLIT-BRAIN CONCEPT)

Gross examination of the brain would suggest that it is bilaterally symmetric. However, detailed examination by the use of CT scans reveals certain anatomical differences between the two hemispheres. For example, in left-handed people the parietal and occipital lobes of the right hemisphere are usually narrower than the corresponding lobes of the left hemisphere. In addition, the frontal lobe of the left hemisphere of such individuals is typically narrower than that of the right hemisphere.

In addition to the structural differences between both sides of the brain, there are also several important functional differences. It has been shown that the left hemisphere is more important for right-hand control, spoken and written language, numerical and scientific skills, and reasoning in most people. Conversely, it has been shown that the right hemisphere is more important for left-hand control; musical and artistic awareness; space and pattern perception; insight; imagination; and generating mental images of sight, sound, touch, taste, and smell in order to compare relationships.

CEREBELLUM

The **cerebellum** is the second-largest portion of the brain (almost one-eighth of the brain's mass) and occupies the inferior and posterior aspects of the cranial cavity. Specifically, it is posterior to the medulla and pons and below the occipital lobes of the cerebrum (see Figure 14-1). It is separated from the cerebrum by the **transverse fissure** and by an extension of the cranial dura mater called the **tentorium** (*tentorium* = tent) **cerebelli.** The tentorium cerebelli partially encloses the transverse sinuses and supports the occipital lobes of the cerebral hemispheres.

The cerebellum is shaped somewhat like a butterfly. The central constricted area is the **vermis,** which means worm-shaped, and the lateral "wings" or lobes are referred to as **hemispheres** (Figure 14-12). Each hemisphere consists of lobes that are separated by deep and distinct fissures. The **anterior lobe** and **posterior lobe** are concerned with subconscious movements of skeletal muscles. The **flocculonodular lobe** is concerned with the sense of equilibrium (see Chapter 17). Between the hemispheres is another extension of the cranial dura mater: the **falx cerebelli.** It passes only a short distance between the cerebellar hemispheres and contains the occipital sinus.

The surface of the cerebellum, called the **cortex,** consists of gray matter in a seies of slender, parallel ridges

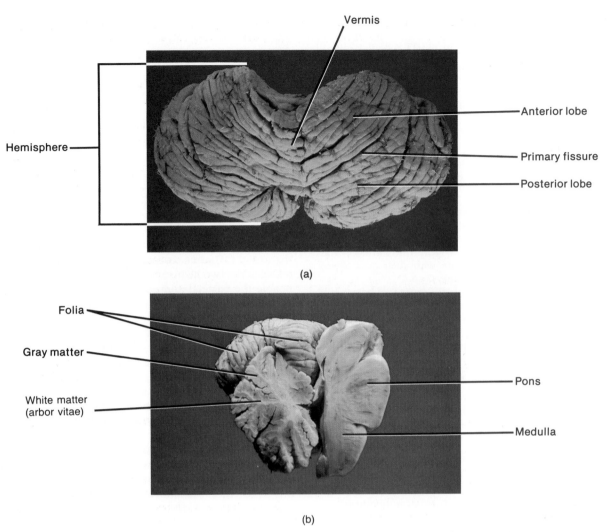

Hemisphere —

Anterior lobe

Primary fissure

Posterior lobe

(a)

Folia

Gray matter

White matter
(arbor vitae)

Pons

Medulla

(b)

FIGURE 14-12 Cerebellum. (a) Superior view. (b) Viewed in sagittal section. The arbor vitae can also be seen in Figure 14-1b. (Courtesy of Lester V. Bergman & Associates, Inc.)

called **folia.** They are less prominent than the gyri of the cerebral cortex. Beneath the gray matter are **white matter tracts (arbor vitae)** that resemble branches of a tree. Deep within the white matter are masses of gray matter, the **cerebellar nuclei.**

The cerebellum is attached to the brain stem by three paired bundles of fibers called **cerebellar peduncles** (see Figure 14-4). **Inferior cerebellar peduncles** connect the cerebellum with the medulla at the base of the brain stem and with the spinal cord. **Middle cerebellar peduncles** connect the cerebellum with the pons. **Superior cerebellar peduncles** connect the cerebellum with the midbrain.

The cerebellum is a motor area of the brain concerned with certain subconscious movements in the skeletal muscles. These movements are required for coordination, maintenance of posture, and balance. The cerebellar peduncles are the fiber tracts that allow the cerebellum to perform its functions.

Let us now see how the cerebellum produces coordinated movement. Although other parts of the brain, such as basal ganglia, assume a role in coordinating move-

ments, we will be concerned only with the role of the cerebellum at this point. Motor areas of the cerebral cortex voluntarily initiate muscle contraction. Once the movement has begun, the sensory areas of the cortex receive impulses from nerves in the joints. The impulses provide information about the extent of muscle contraction and the amount of joint movement. The body's awareness of the position of one body part relative to another is called *proprioception.* The cerebral cortex uses the proprioceptive sensations to determine which mucles are required to contract next and with what strength they must contract to continue moving in the desired direction. Then a pattern of impulses is generated by the cerebral cortex along tracts to the pons and midbrain, which relay the impulses over the middle and superior cerebellar peduncles to the cerebellum. The cerebellum then delays various impulses for fractions of a second in order to coordinate a sequence of muscle contractions. Impulses from the cerebellum pass along the inferior cerebellar peduncles to the medulla, downward along the spinal cord, and out the nerves that stimulate the prime movers and synergists to contract and that inhibit the contraction of the antagonists. The result is a smooth, coordinated movement. A well-functioning cerebellum is essential for delicate movements such as playing the piano.

The cerebellum also transmits impulses that control postural muscles, that is, it maintains normal muscle tone. The cerebellum also maintains body equilibrium. The inner ear contains structures that sense balance. Information such as whether the body is leaning to the left or right is transmitted from the inner ear to the cerebellum. The cerebellum then discharges impulses that cause the contraction of the muscles necessary for maintaining equilibrium.

There is some evidence that the cerebellum may play a role in a person's emotional development, modulating sensations of anger and pleasure.

CLINICAL APPLICATION

Damage to the cerebellum through trauma or disease is characterized by certain symptoms involving skeletal muscles on the same side of the body as the damage. The effects are ipsilateral because of a double crossing of tracts within the cerebellum. There may be lack of muscle coordination, called *ataxia* (*a* = without; *taxis* = order). Blindfolded people with ataxia cannot touch the tip of their nose with a finger because they cannot coordinate movement with their sense of where a body part is located. Another sign of ataxia is a change in the speech pattern due to a lack of coordination of speech muscles. Cerebellar damage may also result in *disturbances of gait,* in which the subject staggers or cannot coordinate normal walking movements, and *severe dizziness.*

A summary of the functions of the various parts of the brain is presented in Exhibit 14-2.

NEUROTRANSMITTERS IN THE BRAIN

There are numerous substances that are either known or suspected neurotransmitters in the brain. These substances can facilitate, excite, or inhibit postsynaptic neurons. They establish the lines of communication between brain cells and are localized in certain parts of the brain. In Chapter 12, we noted that **acetylcholine (ACh)** is released at some neuromuscular junctions, at some neuroglandular junctions, and at synapses between certain brain and spinal cord cells. In most parts of the body, ACh leads to excitation. ACh is inactivated to acetate and choline by acetylcholinesterase (AChE). The amino acids **glutamic acid** and **aspartic acid** are also believed to lead to excitation in the brain.

Norepinephrine (NE) is a neurotransmitter that is released at some neuromuscular junctions and some neuroglandular junctions. NE probably causes inhibition in most instances, but it may be involved in excitation. It is concentrated in a group of neurons in the brain stem

near the fourth ventricle called the *locus coeruleus* (LŌ-kus sē-ROO-lē-us). This area, which literally means blue place, projects axons into the hypothalamus, cerebellum, cerebral cortex, and spinal cord. In these places, NE has been implicated in maintaining arousal (awakening from deep sleep), dreaming, and the regulation of mood. Inactivation of NE is different from that of ACh. After NE is released from its synaptic vesicles, it is rapidly pumped back into the synaptic end bulbs. Here it is either destroyed by the enzymes **catechol-O-methyltransferase** (kat'-e-kōl-ō-meth-il-TRANS-fer-ās) or **COMT** and **monoamine oxidase** (mōn-ō-AM-ēn OK-si-dās) or **MAO,** or recycled back into the synaptic vesicles.

Neurons containing the neurotransmitter **dopamine (DA)** are clustered in the midbrain in the substantia nigra. DA usually leads to inhibition. Some axons projecting from the substantia nigra terminate in the cerebral cortex, where DA is thought to be involved in emotional responses. Other axons project to the corpus striatum (basal ganglia), where the DA is involved in gross subconscious movements of skeletal muscles. The degeneration of axons in the basal ganglia results in Parkinson's disease, which is discussed at the end of this chapter.

Serotonin (5-HT) is a neurotransmitter that leads to excitation and is concentrated in the neurons in a part of the brain stem called the *raphe nucleus.* Axons projecting from the nucleus terminate in the hypothalamus, thalamus, other parts of the brain, and spinal cord. Serotonin is thought to be involved in inducing sleep, sensory perception, temperature regulation, and control of mood.

The most common neurotransmitter in the brain leading to inhibition is **gamma aminobutyric acid (GABA).** It is most highly concentrated in the superior and inferior colliculi, thalamus, hypothalamus, and occipital lobes of the cerebrum. It has been implicated as a likely target for antianxiety drugs such as diazepan (Valium), which enhance the action of GABA. A neurotransmitter that leads to inhibition in the spinal cord is **glycine.**

In recent years, another group of chemical messengers in the brain has been identified. These are known as **neuropeptides,** which consist of chains of 2 to about 40 amino acids that occur naturally in the brain. Although some neuropeptides function as true neurotransmitters, most act primarily to modulate the response of or the response to a neurotransmitter. In 1975 the first neuropeptides, referred to as **enkephalins** (en-KEF-a-lins; *en* = without; *keph* = head), were discovered. These chemicals are similar in structure to morphine, the painkiller derived from the opium poppy. Enkephalins are concentrated in the thalamus, in parts of the limbic system, and in those spinal cord pathways that relay impulses for pain. It has been suggested that enkephalins are the body's natural painkillers. They do this by inhibiting impulses in the pain pathway and by binding to the same receptors in the brain as morphine.

Other naturally occurring neuropeptides, called **endorphins** (en-DOR-fins), have subsequently been isolated from the pituitary gland. Like the enkephalins, they have morphinelike properties that suppress pain. Endorphins have also been linked to memory and learning; sexual activity; control of body temperature; regulation of hor-

EXHIBIT 14-2 SUMMARY OF FUNCTIONS OF PRINCIPAL PARTS OF BRAIN

PART	FUNCTION
BRAIN STEM	
Medulla	Relays motor and sensory impulses between other parts of the brain and the spinal cord.
	Recticular formation (also in pons, midbrain, and diencephalon) functions in consciousness and arousal.
	Vital reflex centers regulate heartbeat, breathing (together with pons), and blood vessel diameter.
	Nonvital reflex centers coordinate swallowing, vomiting, coughing, sneezing, and hiccuping.
	Contains nuclei of origin for cranial nerves VIII, IX, X, XI, and XII.
	Vestibular nuclear complex helps maintain equilibrium.
Pons	Relays impulses within the brain and between parts of the brain and spinal cord.
	Contains nuclei of origin for cranial nerves V, VI, VII, and VIII.
	Pneumotaxic area and apneustic area, together with the medulla, help control breathing.
Midbrain	Relays motor impulses from the cerebral cortex to the pons and spinal cord, and relays sensory impulses from the spinal cord to the thalamus.
	Superior colliculi coordinate movements of the eyeballs in response to visual and other stimuli, and the inferior colliculi coordinate movements of the head and trunk in response to auditory stimuli.
	Contains nuclei of origin for cranial nerves III and IV.
DIENCEPHALON	
Thalamus	Several nuclei serve as relay stations for all sensory impulses, except smell, to the cerebral cortex.
	Relays motor impulses from the cerebral cortex to the spinal cord.
	Interprets pain, temperature, light touch, and pressure sensations.
	Anterior nucleus functions in emotions and memory.
Hypothalamus	Controls and integrates the autonomic nervous system.
	Receives sensory impulses from viscera.
	Articulates with the pituitary gland.
	Center for mind-over-body phenomena.
	Functions in rage and aggression.
	Controls normal body temperature, food intake, and thirst.
	Helps maintain the waking state and sleep.
	Functions as a self-sustained oscillator that drives many biological rhythms.
CEREBRUM	Sensory areas interpret sensory impulses, motor areas control muscular movement, and association areas function in emotional and intellectual processes.
	Basal ganglia control gross muscle movements and regulate muscle tone.
	Limbic system functions in emotional aspects of behavior related to survival.
CEREBELLUM	Controls subconscious skeletal muscle contractions required for coordination, posture, and balance.
	Assumes a role in emotional development, modulating sensations of anger and pleasure.

mones that affect the onset of puberty, sexual drive, and reproduction; and mental illnesses such as depression and schizophrenia. One of the better studied endorphins is **beta-endorphin (β-endorphin).** A neuropeptide that is believed to work with endorphins is known as **substance P.** It is found in sensory nerves, spinal cord pathways, and parts of the brain associated with pain. When substance P is released by neurons, it conducts pain-related nerve impulses from peripheral pain receptors into the central nervous system. It is now suspected that endorphins may exert their analgesic effects by suppressing the release of substance P. Substance P has also been shown to counter the effects of certain nerve-damaging chemicals, prompting speculation that it might prove useful as a treatment for nerve degeneration.

A neuropeptide called **dynorphin** (*dynamis* = power), discovered in 1979, is 200 times more powerful in action than morphine and 50 times more powerful than beta-endorphin. Its exact functions have yet to be determined.

However, the chemical might be a factor in controlling pain and registering emotions.

As research into neuropeptides continues, a long list of substances is being drawn up. Interestingly enough, many of these neuropeptides are also found in other parts of the body, where they serve as hormones or other regulators of physiological responses. Although we will be discussing many of them in subsequent chapters, a few will be mentioned here to give you some idea of their diversity.

1. **Angiotensin.** This is a substance that can raise blood pressure and is found in the bloodstream as well. It is produced by a kidney enzyme called renin. Some evidence suggests that angiotensin in the brain may be part of the brain's mechanism for regulating its own blood pressure.

2. **Cholecystokinin** (ko'-lē-sis'-tō-KĪN-in). This substance (also called pancreozymin) is produced by the lining of the small intestine and causes the pancreas to release digestive juice and the gallbladder to release bile. Although the significance of cholecystokinin in the brain is unclear, there are some sugges-

tions that it may be related to the control of feeding. Other neuropeptides associated with the gastrointestinal tract include neurotensin, vasoactive intestinal peptide (VIP), gastrin, and secretin.

3. **Regulating factors.** These are chemicals produced by the hypothalamus that regulate the release of hormones by the pituitary gland. Examples include thyrotropin releasing factor (TRF), luteinizing hormone releasing factor (LHRF), and growth hormone inhibiting factor (GHIF).

CRANIAL NERVES

Of the 12 pairs of **cranial nerves,** 10 originate from the brain stem, but all leave the skull through foramina of the skull (see Figure 14-4). The cranial nerves are designated with Roman numerals and with names. The Roman numerals indicate the order in which the nerves arise from the brain (front to back). The names indicate the distribution or function.

Some cranial nerves contain only sensory fibers and thus are called *sensory nerves.* The remainder contain both sensory and motor fibers and are referred to as *mixed nerves.* At one time, it was believed that some cranial nerves (oculomotor, trochlear, abducens, accessory, and hypoglossal) were entirely motor. However, it is now known that these cranial nerves also contain some sensory fibers from proprioceptors in muscles they innervate. Although these nerves are mixed, they are primarily motor in function, serving to stimulate skeletal muscle contraction. The cell bodies of sensory fibers are found outside the brain, whereas the cell bodies of motor fibers lie in nuclei within the brain.

Some motor fibers control subconscious movements, yet the somatic nervous system has been defined as a *conscious* system. The reason for this apparent contradiction is that some fibers of the autonomic nervous system leave the brain, bundled together with somatic fibers of the cranial nerves, as is the case for spinal nerves. Therefore, subconscious functions transmitted by the autonomic fibers are described along with the conscious functions of the somatic fibers of the cranial nerves.

A summary of cranial nerves and clinical applications related to dysfunction is presented in Exhibit 14-3.

AGING AND THE NERVOUS SYSTEM

One of the effects of aging on the nervous system is that nerve cells are lost. Associated with this decline, there is a decreased capacity for sending nerve impulses to and from the brain. Conduction velocity decreases, voluntary motor movements slow down, and the reflex time for skeletal muscles increases. Parkinson's disease is the most common movement disorder involving the central nervous system. Degenerative changes and disease states involving the sense organs can alter vision, hearing, taste, smell, and touch. The disorders that represent the most common visual problems and may be responsible for serious loss of vision are presbyopia (inability to focus on nearby objects), cataracts (cloudiness of the lens), and glaucoma (excessive fluid pressure in the eyeball). Impaired hearing associated with aging, known as presbycusis, is usually the result of changes in important structures of the inner ear.

EXHIBIT 14-3   SUMMARY OF CRANIAL NERVES

NERVE (TYPE)	LOCATION	FUNCTION AND CLINICAL APPLICATION
Olfactory (I) (sensory) *[handwritten: cribiform plate]*	Arises in olfactory mucosa, passes through olfactory bulb and olfactory tract, and terminates in primary olfactory areas of cerebral cortex.	Smell. Loss of the sense of smell, called *anosmia,* may result from head injuries in which the cribiform plate of the ethmoid bone is fractured and from lesions along the olfactory pathway.
Optic (II) (sensory) *[handwritten: optic foramen]*	Arises in retina of the eye, forms optic chiasma, passes through optic tracts, lateral geniculate nucleus in thalamus, and terminates in visual areas of cerebral cortex.	Vision. Fractures in the orbit, lesions along the visual pathway, and diseases of the nervous system may result in visual field defects and loss of visual acuity. A defect of vision is called *anopsia.*
Oculomotor (III) (mixed, primarily motor) *[handwritten: superior orbital fissure midbrain]*	Motor portion: originates in midbrain and is distributed to levator palpebrae superioris of upper eyelid and four extrinsic eyeball muscles (superior rectus, medial rectus, inferior rectus, and inferior oblique); parasympathetic innervation to ciliary muscle of eyeball and sphincter muscle of iris. Sensory portion: consists of afferent fibers from proprioceptors in eyeball muscles and terminates in midbrain.	Motor: movement of eyelid and eyeball, accommodation of lens for near vision, and constriction of pupil. Sensory: muscle sense (proprioception). A lesion in the nerve causes *strabismus* (squinting) *ptosis* (drooping) of the upper eyelid, pupil dilation, the movement of the eyeball downward and outward on the damaged side, a loss of accommodation for near vision, and double vision (*diplopia*).

EXHIBIT 14-3   *(Continued)*

NERVE (TYPE)	LOCATION	FUNCTION AND CLINICAL APPLICATION
Trochlear (IV) (mixed, primarily motor)	Motor portion: originates in midbrain and is distributed to superior oblique muscle, an extrinsic eyeball muscle. Sensory portion: consists of afferent fibers from proprioceptors in superior oblique muscles and terminates in midbrain.	Motor: movement of eyeball. Sensory: muscle sense (proprioception). In trochlear nerve paralysis, the head is tilted to the affected side and diplopia and strabismus occur.
Trigeminal (V) (mixed)	Motor portion: originates in pons and terminates in muscles of mastication. Sensory portion: consists of three branches: *ophthalmic*—contains sensory fibers from skin over upper eyelid, eyeball, lacrimal glands, nasal cavity, side of nose, forehead, and anterior half of scalp; *maxillary*—contains sensory fibers from mucosa of nose, palate, parts of pharynx, upper teeth, upper lip, cheek, and lower eyelid; *mandibular*—contains sensory fibers from anterior two-thirds of tongue, lower teeth, skin over mandible, and side of head in front of ear. The three branches terminate in pons. Sensory portion also consists of afferent fibers from proprioceptors in muscles of mastication.	Motor: chewing. Sensory: conveys sensations for touch, pain, and temperature from structures supplied; muscle sense (proprioception). Injury results in paralysis of the muscles of mastication and a loss of sensation of touch and temperature. *Neuralgia* (pain) of one or more branches of trigeminal nerve is called *trigeminal neuralgia* (*tic douloureux*).
Abducens (VI) (mixed, primarily motor)	Motor portion: originates in pons and is distributed to lateral rectus muscle, an extrinsic eyeball muscle. Sensory portion: consists of afferent fibers from proprioceptors in lateral rectus muscle and terminates in pons.	Motor: movement of eyeball. Sensory: muscle sense (proprioception). With damage to this nerve, the affected eyeball cannot move laterally beyond the midpoint and the eye is usually directed medially.
Facial (VII) (mixed)	Motor portion: originates in pons and is distributed to facial, scalp, and neck muscles, parasympathetic distribution to lacrimal, sublingual, submandibular, nasal, and palatine glands. Sensory portion: arises from taste buds on anterior two-thirds of tongue, passes through geniculate ganglion, a nucleus in pons that sends fibers to thalamus for relay to gustatory areas of cerebral cortex. Also consists of afferent fibers from proprioceptors in muscles of face and scalp.	Motor: facial expression and secretion of saliva and tears. Sensory: taste; muscle sense (proprioception). Injury produces paralysis of the facial muscles, called *Bell's palsy,* loss of taste, and the eyes remain open, even during sleep.
Vestibulocochlear (VIII) (sensory)	Cochlear branch: arises in spiral organ (organ of Corti), forms spiral ganglion, passes through a nucleus in the medulla, and terminates in thalamus. Fibers synapse with neurons that relay impulses to auditory areas of cerebral cortex.	Cochlear branch: conveys impulses associated with hearing. Vestibular branch: conveys impulses associated with equilibrium.

DEVELOPMENTAL ANATOMY OF THE NERVOUS SYSTEM

The development of the nervous system begins early in the third week of development and begins with a thickening of the **ectoderm** called the **neural plate** (Figure 14-13a–c). The plate folds inward and forms a longitudinal groove, the **neural groove.** The raised edges of the neural plate are called **neural folds.** As development continues the neural folds increase in height, meet, and form a tube, the **neural tube.**

The cells of the wall that encloses the neural tube differentiate into three kinds. The outer or **marginal layer** develops into the *white matter* of the nervous system; the middle or **mantle layer** develops into the *gray matter* of the system; and the inner or **ependymal layer** eventually forms the *lining of the ventricles* of the central nervous system.

The **neural crest** is a mass of tissue between the neural tube and the ectoderm (Figure 14-13). It becomes differentiated and eventually forms the *posterior (dorsal) root ganglia of spinal nerves, spinal nerves, ganglia of cranial*

EXHIBIT 14-3 *(Continued)*

NERVE (TYPE)	LOCATION	FUNCTION AND CLINICAL APPLICATION
Vestibulocochlear (VIII) (sensory) *(Continued)*	Vestibular branch: arises in semicircular canals, saccule, and utricle and forms vestibular ganglion; fibers pass through medulla and pons and terminate in thalamus.	Injury to the cochlear branch may cause *tinnitus* (ringing) or deafness. Injury to the vestibular branch may cause *vertigo* (a subjective feeling of rotation), *ataxia,* and *nystagmus* (in-voluntary rapid movement of the eyeball).
Glossopharyngeal (IX) (mixed)	Motor portion: originates in medulla and is distributed to swallowing muscles of pharynx; parasympathetic distribution to parotid gland. Sensory portion: arises from taste buds on posterior one-third of tongue and from carotid sinus and terminates in thalamus. Also consists of afferent fibers from proprioceptors in swallowing muscles supplied.	Motor: swallowing movements and secretion of saliva. Sensory: taste and regulation of blood pressure; muscle sense (proprioception). Injury results in pain during swallowing, reduced secretion of saliva, loss of sensation in the throat, and loss of taste.
Vagus (X) (mixed)	Motor portion: originates in medulla and terminates in muscles of pharynx, larynx, respiratory passageways, lungs, esophagus, heart, stomach, small intestine, most of large intestine, and gallbladder; parasympathetic fibers innervate involuntary muscles and glands of the gastrointestinal (GI) tract. Sensory portion: arises from essentially same structures supplied by motor fibers and terminates in medulla and pons. Also consists of afferent fibers from proprioceptors in muscles supplied.	Motor: visceral muscle movement and swallowing movements. Sensory: sensations from organs supplied; muscle sense (proprioception). Severing of both nerves in the upper body interferes with swallowing, paralyzes vocal cords, and interrupts sensations from many organs. Injury to both nerves in the abdominal area has little effect, since the abdominal organs are also supplied by autonomic fibers from the spinal cord.
Accessory (XI) (mixed, primarily motor)	Motor portion: consists of a bulbar portion and a spinal portion. Bulbar portion originates from medulla and supplies voluntary muscles of pharynx, larynx, and soft palate. Spinal portion originates from anterior gray horn of first five cervical segments of spinal cord and supplies sternocleiodomastoid and trapezius muscles. Sensory portion: consists of afferent fibers from proprioceptors in muscles supplied.	Motor: bulbar portion mediates swallowing movements; spinal portion mediates movements of head. Sensory: muscle sense (proprioception). If damaged, the sternocleidomastoid and trapezius muscles become paralyzed, with resulting inability to turn the head or raise the shoulders.
Hypoglossal (XII) (mixed, primarily motor)	Motor portion: originates in medulla and supplies muscles of tongue. Sensory portion: consists of fibers from proprioceptors in tongue muscles that terminate in medulla.	Motor: movement of tongue during speech and swallowing. Sensory: muscle sense (proprioception). Injury results in difficulty in chewing, speaking, and swallowing. The tongue, when protruded, curls toward the affected side and the affected side becomes atrophied, shrunken, and deeply furrowed.

nerves, cranial nerves, ganglia of the automatic nervous system, and the *adrenal medulla.*

When the neural tube is formed from the neural plate, the anterior portion of the neural tube develops into three enlarged areas called vesicles: (1) **forebrain vesicle (prosencephalon), (2) midbrain vesicle (mesencephalon),** and (3) **hindbrain vesicle (rhombencephalon)** (Figure 14-14). The vesicles are fluid-filled enlargements that develop by the fourth week of gestation. Since they are the first vesicles to form, they are called **primary vesicles.** As development progresses, the vesicular region undergoes several

flexures (bends), resulting in subdivision of the three primary vesicles, so that by the fifth week of development the embryonic brain consists of five **secondary vesicles.** The forebrain vesicle (prosencephalon) divides into an anterior **telencephalon** and a posterior **diencephalon;** the midbrain vesicle (mesencephalon) remains unchanged; the hindbrain vesicle (rhombencephalon) divides into an anterior **metencephalon** and a posterior **myelencephalon.**

Ultimately, the telencephalon develops into the *cerebral hemispheres* and *basal ganglia;* the diencephalon develops into the *thalamus, hypothalamus,* and *pineal gland;*

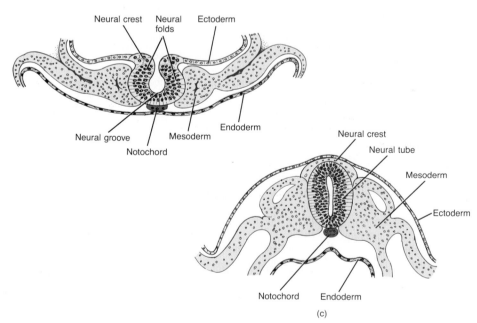

FIGURE 14-13 Origin of the nervous system. (a) Dorsal view of an embryo with three pairs of somites showing the neural plate. (b) Dorsal view of an embryo with seven pairs of somites in which the neural folds have united just medial to the somites, forming the early neural tube. (c) Cross section through the embryo showing the formation of the neural tube.

(a)

(b)

FIGURE 14-14 Development of the brain and spinal cord. (a) Primary vesicles of the neural tube seen in frontal section (left) and right lateral view (right) at about 3 to 4 weeks. (b) Secondary vesicles seen in frontal section (left) and right lateral view (right) at about 5 weeks.

the midbrain vesicle (mesencephalon) develops into the *midbrain;* the metencephalon develops into the *pons* and *cerebellum;* and the myelencephalon develops into the *medulla oblongata.* The cavities within the vesicles de-velop into the *ventricles* of the brain, whereas the fluid within them is *cerebrospinal fluid.* The area of the neural tube posterior to the myelencephalon gives rise to the *spinal cord.*

DISORDERS: HOMEOSTATIC IMBALANCES

Many disorders can affect the central nervous system. Some are caused by viruses or bacteria. Others are caused by damage to the nervous system during birth. The origins of many conditions, however, are unknown. Here we discuss the origins and symptoms of some common central nervous system disorders.

Brain Tumors

A **brain tumor** refers to any benign or malignant growth within the cranium. Tumors may arise from neuroglial cells in the cerebrum, brain stem, and cerebellum, or from supporting or neighboring structures such as cranial nerve coverings, meninges, and the pituitary gland. Intracranial pressure from a growing tumor or edema associated with the tumor produce the characteristic signs and symptoms of a brain tumor. Among these are headache, altered consciousness, edema of the optic disc of the eyeball, and vomiting. Brain tumors can also result in seizures, visual problems, cranial nerve abnormalities, hormonal syndromes, personality changes, dementia, and sensory or motor deficits. Treatment of brain tumors involves surgery, radiation therapy, and chemotherapy.

Poliomyelitis

Poliomyelitis (infantile paralysis), or simply **polio,** is most common during childhood and is caused by a virus called poliovirus. The onset of the disease is marked by fever, severe headache, a stiff neck and back, deep muscle pain and weakness, and loss of certain somatic reflexes. The virus can be ingested in drinking water contaminated with feces containing the virus. It may affect almost all parts of the body. In its most serious form, called **bulbar polio,** the virus spreads via blood to the central nervous system, where it destroys the motor nerve cell bodies, specifically those in the anterior horns of the spinal cord and in the nuclei of the cranial nerves. Injury to the spinal gray matter is the basis for the name of this disease (*polio* = gray matter; *myel* = spinal cord). Destruction of the anterior horns produces paralysis. The first sign of bulbar polio is difficulty in swallowing, breathing, and speaking. Poliomyelitis can cause death from respiratory or heart failure if the virus invades the brain cells of the vital medullary centers. The incidence of polio in the United States has decreased markedly since the availability of polio vaccines (Salk vaccine and more recently Sabin vaccine).

Cerebral Palsy

The term **cerebral palsy** refers to a group of motor disorders caused by damage to the motor areas of the brain during fetal life, birth, or infancy. One cause is infection of the mother with German measles during the first three months of pregnancy. During early pregnancy, certain cells in the fetus are dividing and differentiating in order to lay down the basic structures of the brain. These cells can be abnormally changed by toxin from the measles virus. Radiation during fetal life, temporary oxygen starvation during birth, and hydrocephalus during infancy may also damage brain cells. Cases of cerebral palsy are categorized into three groups, depending on whether the cortex, the basal ganglia of the cerebrum, or the cerebellum is affected most severely. About 70 percent of cerebral palsy

victims appear to be mentally retarded. The apparent mental slowness, however, is often due to the person's inability to speak or hear well. Such individuals are often more mentally acute than they appear. Cerebral palsy is not a progressive disease; it does not worsen as time elapses. Once the damage is done, however, it is irreversible.

Parkinson's Disease

Parkinson's disease or **parkinsonism** is a progressive disorder of the central nervous system that begins inconspicuously and typically affects its victims around age 60. The cause is unknown. The disease is related to pathological changes in the substantia nigra and basal ganglia. The substantia nigra contains cell bodies of neurons that produce the neurotransmitter dopamine (DA) in their axon terminals. The axon terminals release DA in the basal ganglia of the cerebrum. Recall that basal ganglia regulate subconscious contractions of skeletal muscles that aid activities also consciously controlled by the motor areas of the cerebral cortex—swinging the arms when walking, for example. In Parkinson's disease there is a degeneration of DA-producing neurons in the substantia nigra, and the severe reduction of DA in the basal ganglia brings about most of the symptoms of Parkinson's disease.

Diminished levels of DA cause unnecessary skeletal muscle movements that often interfere with voluntary movement. For instance, the muscles of the upper extremity may alternately contract and relax, causing the hand to shake. This shaking is called *tremor,* the most common symptom of Parkinson's disease. The tremor may spread to the ipsilateral lower extremity and then to the contralateral extremities. Motor performance is also impaired by *bradykinesia* (*brady* = slow; *kinesis* = motion), in which activities such as shaving, cutting food, and buttoning a shirt take longer and become increasingly more difficult. Muscular movements are performed not only slowly but with decreasing range of motion (*hypokinesia*). For example, as handwriting continues, letters get smaller, become poorly formed, and eventually become illegible. Some muscles may contract continuously, causing *rigidity* of the involved body part. Rigidity of the facial muscles gives the face a masklike appearance. The expression is characterized by a wide-eyed, unblinking stare and a slightly open mouth with uncontrolled drooling. Decreased DA production also results in impaired walking, in which steps become shorter and shuffling and arm swing diminishes. There are also changes that lead to stooped posture and loss of postural reflexes, autonomic dysfunction (constipation, retention), sensory complaints (pain, numbness, tingling), and sustained muscle spasms. Vision, hearing, and intelligence are unaffected by the disorder, indicating that Parkinson's disease does not attack the cerebral cortex.

Treatment of the symptoms of Parkinson's disease is directed toward increasing levels of DA. Although people with Parkinson's disease do not manufacture enough DA, injections of it are useless; the blood–brain barrier stops it. However, symptoms are somewhat relieved by a drug developed in the 1960s called levodopa. Administered by itself, levodopa may elevate brain levels of DA, causing undesirable side effects such as low blood pressure, nausea, mental changes, and liver dysfunction. Levodopa has been combined with carbidopa, which inhibits the formation of DA outside the brain. The combined drugs diminish the undesirable side effects. Currently, methods are being devised to help drugs pass through the blood–brain barrier by combining them with fat-soluble chemicals. Once the complex passes through the barrier, the fat-soluble chemical breaks down and is excreted, leaving the drug in the brain where it exerts its effect.

Multiple Sclerosis (MS)

Multiple sclerosis (MS) is often called the "great imitator." It is the progressive destruction of the myelin sheaths of neurons in the central nervous system accompanied by disappearance of oligodendrocytes and the proliferation of astrocytes. The sheaths deteriorate to *scleroses,* which are hardened scars or plaques, in multiple regions. The destruction of myelin sheaths interferes with the transmission of impulses from one neuron to another, literally short-circuiting conduction pathways. Usually the first symptoms occur between the ages of 20 and 40. Early symptoms are generally produced by the formation of a few plaques and are, consequently, mild. Plaque formation in the cerebellum may produce lack of coordination in one hand. The patient's handwriting becomes strained and irregular. A short-circuiting of pathways in the corticospinal tract may partially paralyze the leg muscles so that the patient drags a foot when walking. Other early symptoms include double vision and urinary tract infections. Following a period of remission during which the symptoms temporarily disappear, a new series of plaques develop and the victim suffers a second attack. One attack follows another over the years, usually every year or two. Each time the plaques form, some neurons are damaged by the hardening of their sheaths, while others are uninjured by their plaques. The result is a progressive loss of function interspersed with remission periods during which the undamaged neurons regain their ability to transmit impulses.

The symptoms of MS depend on the areas of the central nervous system most heavily laden with plaques. Sclerosis of the white matter of the spinal cord is common. As the sheaths of the neurons in the corticospinal tract deteriorate, the patient loses the ability to contract skeletal muscles. Damage to the ascending tracts produces numbness and short-circuits impulses related to position of body parts and flexion of joints. Damage to either set of tracts also destroys spinal cord reflexes.

As the disease progresses, most voluntary motor control is eventually lost and the patient becomes bedridden. Death occurs anywhere from 7 to 30 years after the first symptoms appear. The usual cause of death is a severe infection resulting from the loss of motor activity. Without the constricting action of the urinary bladder wall, for example, the bladder never totally empties and stagnant urine provides an environment for bacterial growth. Bladder infection may then spread to the kidney, damaging kidney cells.

Although the etiology of MS is unclear, there is increasing evidence that it might result from a viral infection that precipitates an autoimmune response. Viruses may trigger the destruction of oligodendrocytes by the antibodies and killer cells of the body's immune system. Like other demyelinating diseases, MS is incurable. However, in view of the evidence that it might be an autoimmune disease, immunosuppressive therapy is widely used (glucocorticoids such as prednisone). A recent treatment consists of administering cyclophosphamide (a powerful anticancer drug that suppresses the immune system) in combination with ACTH (a pituitary gland hormone that stimulates secretion of hormones containing cortisone) to patients with active MS. Treatment is also directed at management of complications such as spasticity, facial neuralgia and twitching, urinary bladder problems, and constipation. Electrical stimulation of the spinal cord can also improve function in certain patients. Improvement has also been shown in patients who are administered pure oxygen while in a pressure chamber (hyperbaric oxygen). This supports the idea that the destruction of myelin occurs preferentially in parts of the brain that are relatively low in oxygen.

Epilepsy

Epilepsy is the second most common neurological disorder after stroke. It is characterized by short, recurrent, periodic attacks of motor, sensory, or psychological malfunction. The attacks, called *epileptic seizures,* are initiated by abnormal and irregular discharges of electricity from millions of neurons in the brain, probably resulting from abnormal reverberating circuits. The discharges stimulate many of the neurons to send impulses over their conduction pathways. As a result, a person undergoing an attack may contract skeletal muscles involuntarily. Lights, noise, or smells may be sensed when the eyes, ears, and nose actually have not been stimulated. The electrical discharges may also inhibit certain brain centers. For instance, the waking center in the brain may be depressed so that the person loses consciousness.

The causes of epilepsy are varied. Many conditions can cause nerve cells to produce periodic bursts of impulses. These causes include metabolic disturbances (hypoglycemia, hypocalcemia, uremia, hypoxia), infections (encephalitis or meningitis), toxins (alcohol, tranquilizers, hallucinogens), vascular disturbances (hemorrhage, hypotension), head injuries, and tumors and abscesses of the brain. Most epileptic seizures, however, are *idiopathic,* that is, they have no demonstrable cause. It should be noted that epilepsy almost never affects intelligence. If frequent severe seizures are allowed to occur over a long period of time, however, some cerebral damage may occasionally result. Damage can be prevented by controlling the seizures with drug therapy.

Epileptic seizures can be eliminated or alleviated by drugs that make neurons more difficult to stimulate. Many of these drugs change the permeability of the neuron cell membrane so that it does not depolarize as easily. One such drug is valproic acid, which increases the quantity of the inhibitory neurotransmitter GABA.

Cerebrovascular Accidents (CVAs)

The most common brain disorder is a **cerebrovascular accident (CVA),** also called a **stroke** or **cerebral apoplexy.** A CVA is characterized by a relatively abrupt onset of persisting neurological symptoms due to the destruction of brain tissue (infarction), resulting from disorders in the blood vessels that supply the brain. Common causes of CVAs are intracerebral hemorrhage from aneurysms, embolism, and atherosclerosis of the cerebral arteries. An *intracerebral hemorrhage* is a rupture of a vessel in the pia mater or brain. Blood seeps into the brain and damages neurons by increasing intracranial fluid pressure. An *embolus* is a blood clot, air bubble, or bit of foreign material, most often debris from an inflammation, that becomes lodged in an artery and blocks circulation. *Atherosclerosis* is the formation of plaques in the artery walls. The plaques may slow down circulation by constricting the vessel. Both emboli and atherosclerosis cause brain damage by reducing the supply of oxygen and glucose needed by brain cells.

Dyslexia

Dyslexia (dis-LEK-sē-a; *dys* = difficulty; *lexis* = words) is unrelated to basic intellectual capacity, but it causes a mysterious difficulty in handling words and symbols. Apparently some peculiarity in the brain's organizational pattern distorts the ability to read, write, and count. Letters in words seem transposed, reversed, or upside down—*dog* becomes *god; b* changes identity with *d;* a sign saying "OIL" inverts into "710." Many dyslexics cannot orient themselves in the three dimensions of space and may show bodily awkwardness.

The exact cause of dyslexia is unknown, since it is unaccompanied by outward scars of detectable neurological damage and its symptoms vary from victim to victim. It occurs three times as often among boys as among girls. It has been variously attributed to defective vision, brain damage, lead in the air, physical trauma, or oxygen deprivation during birth. A recent theory holds that it might be related to dysfunction of the vestibular apparatus and semicircular canals of the ear (see Chapter 17).

Now available to physicians is a new technique used primarily to diagnose dyslexia called **brain electrical activity mapping (BEAM)**. It is a noninvasive procedure that measures and displays the electrical activity of the brain on a color television screen and compares the image produced with a normal image. BEAM may have other diagnostic applications for learning disabilities, schizophrenia, depression, dementia, epilepsy, and early tumor occurrence and recurrence.

Tay-Sachs Disease

Tay-Sachs disease is a central nervous system affliction that brings death before age 5. The Tay-Sachs gene is carried mostly by individuals descended from the Ashkenazi Jews of Eastern Europe. Approximately 1 in 3,600 of their offspring will be afflicted with Tay-Sachs disease. The disease involves the neuronal degeneration of the central nervous system because of excessive amounts of a lipid called ganglioside in the nerve cells of the brain. The substance accumulates because of a deficient lysosomal enzyme. The afflicted child develops normally until the age of 4 to 8 months. Then the symptoms follow a course of progressive degeneration: paralysis, blindness, inability to eat, decubitus, and death from infection. There is no known cure.

Headache

One of the most common human afflictions is **headache** or **cephalgia** (*enkephalos* = brain, *algia* = painful condition). Based on origin, two general types are distinguished: intracranial and extracranial. Serious headaches of intracranial origin are caused by brain tumors, blood vessel abnormalities, inflammation of the brain or meninges, decrease in oxygen supply to the brain, and damage to brain cells. Extracranial headaches are related to infections of the eyes, ears, nose, and sinuses, and are commonly felt as headaches because of the location of these structures.

Most headaches require no special treatment. Analgesic and tranquilizing compounds are generally effective for tension headaches, but not for migraine headaches. Drugs that constrict the blood vessels can be helpful for migraine. Biofeedback training and dietary changes may also be of some help. Taking a careful and complete history is an important procedure in determining the cause or causes of a patient's headaches.

Trigeminal Neuralgia

As noted earlier, pain arising from irritation of the trigeminal (V) nerve is known as **trigeminal neuralgia** or **tic douloureux** (doo-loo-ROO). The disorder is characterized by brief but extreme pain in the face and forehead on the affected side. The characteristic pain consists of red-hot, needlelike jabs lasting a few seconds and building to a searing pain like a hot poker being dragged or stuck into the face, lasting 10 to 15 seconds. Many patients describe sensitive regions around the mouth and nose that can cause an attack when touched. Eating, drinking, washing the face, and exposure to cold may also bring on an attack.

Treatment may be palliative (relieving symptoms without curing the disease) or surgical. One technique involves alcohol injections directly into the semilunar (Gasserian) ganglion, which controls the trigeminal (V) nerve. This method is superior to open surgery because it is safer. Moreover, it can bring lasting pain relief with preservation of touch sensation in the face and release of the patient 24 hours after treatment. Acupuncture can also provide relief in some patients.

Reye's Syndrome (RS)

Reye's syndrome (RS), first described in 1963 by the Australian pathologist R. Douglas Reye, seems to occur following a viral infection, particularly chickenpox or influenza. Aspirin at normal doses is believed to be a risk factor in the development of RS. The majority of persons affected are children or teenagers. The disease is characterized by vomiting and brain dysfunction (disorientation, lethargy, and personality changes), and may progress to coma. Also, the liver becomes infiltrated with small lipid droplets and loses some of its ability to detoxify ammonia. The disease runs its course in just a few days.

Brain dysfunction and death are typically caused by swelling of brain cells. The pressure not only kills the cells directly but also results in hypoxia that kills them indirectly. The mortality rate is about 40 percent. Swelling may result in irreversible brain damage, including mental retardation, in children who survive. Therapy is directed at controlling the swelling.

Alzheimer's Disease (AD)

Alzheimer's disease (ALTZ-hī-merz) or **AD** is a disabling neurological disorder that afflicts about 5 percent (about 1 million people) of the population over age 65. Its causes are unknown, its effects are irreversible, and it has no cure. The disease claims over 120,000 lives a year, making it the fourth leading cause of death among the elderly following heart disease, cancer, and stroke.

Victims of AD initially have trouble remembering recent events. Next they become more confused and forgetful, often repeating questions or getting lost while traveling to previously familiar places. Disorientation grows, memories of past events disappear, and there may be episodes of paranoia, hallucination, or violent changes in mood. As their minds continue to deteriorate, they lose their ability to read, write, talk, eat, walk, or take care of themselves. Finally, the disease culminates in dementia, the loss of reason. A person with AD usually dies of some complication that affects bedridden patients, such as pneumonia.

At present, there is no diagnostic test for AD. Autopsy findings, however, clearly show several characteristic pathologies in the brains of Alzheimer's victims. Although there is normally gradual loss of neurons in the brain associated with aging, the rate of loss in persons with AD is greater than normal, especially in regions of the brain that are important for memory and other intellectual processes, such as the cerebral cortex and hippocampus. Another pathological finding is the presence of **neurofibrillary tangles,** bundles of fibrous proteins, in the cell bodies of neurons in the cerebral cortex, hippocampus, and brain stem. Also present is **amyloid,** pathological protein-rich accumulations. In some cases, amyloid surrounds and invades cerebral blood vessels. This form is called **cerebrovascular amyloid,** which collects in the middle muscular layer of the blood vessels and eventually results in hemorrhage. Another form of amyloid is a component of **neuritic (senile) plaques,** which consist of abnormal axons and axon terminals that surround amyloid. Neuritic plaques are abundant in the cerebral cortex, hippocampus, and amygdala.

One hypothesis concerning the possible cause of AD centers on the observation that the enzyme choline acetyltransferase (CAT) is found in very low concentrations in the brains of AD victims. CAT is needed for the synthesis of the neurotransmitter acetylcholine (ACh) in axon terminals. Levels of CAT are quite low in the cerebral cortex and hippocampus of AD patients due to the loss of axon terminals in these regions, and thus acetylcholine levels are also low. Decreased levels of

acetylcholine are correlated with reduced transmission of nerve impulses, impulses required for memory.

Other hypotheses suggest that AD may be caused by the loss of neurons due to the inheritance of faulty genes, an abnormal accumulation of proteins in the brain, infectious agents called prions (protein particles), environmental toxins such as aluminum, and decreased blood flow to the brain resulting in inadequate amounts of oxygen and glucose.

MEDICAL TERMINOLOGY

Agnosia (*a* = without: *gnosis* = knowledge)   Inability to recognize the significance of sensory stimuli such as auditory, visual, olfactory, gustatory, and tactile.

Analgesia (*an* = without; *algia* = painful condition)   Insensibility to pain.

Anesthesia (*esthesia* = feeling)   Loss of feeling.

Apraxia (*pratto* = to do)   Inability to carry out purposeful movements in the absence of paralysis.

Coma   Abnormally deep unconsciousness with an absence of voluntary response to stimuli and with varying degrees of reflex activity. It may be due to illness or to an injury.

Electroconvulsive therapy (ECT)   A form of shock therapy in which convulsions are induced by the passage of an electric current through the brain. A patient undergoing ECT is properly anesthetized and given a muscle relaxant to minimize the convulsions. ECT is regarded as an important therapeutic option in the treatment of severe depression.

Huntington's chorea (*choreia* = dance)   A rare hereditary disease characterized by involuntary jerky movements

and mental deterioration that terminates in dementia.

Lethargy   A condition of functional torpor or sluggishness.

Nerve block   Loss of sensation in a region, such as in local dental anesthesia.

Neuralgia (*neur* = nerve)   Attacks of pain along the entire course or branch of a peripheral sensory nerve.

Paralysis   Diminished or total loss of motor function resulting from damage to nervous tissue or a muscle.

Spastic (*spas* = draw or pull)   Resembling spasms or convulsions.

Stupor   Condition of unconsciousness, torpor, or lethargy with suppression of sense or feeling.

Torpor   Abnormal inactivity or lack of response to normal stimuli.

Viral encephalitis   An acute inflammation of the brain caused by a direct attack by various viruses or by an allergic reaction to any of the many viruses that are normally harmless to the central nervous system. If the virus affects the spinal cord as well, it is called **encephalomyelitis**.

STUDY OUTLINE

Brain (p. 307)

Principal Parts (p. 307)

1. During embryological development, brain vesicles are formed and serve as forerunners of various parts of the brain.
2. The diencephalon develops into the thalamus and hypothalamus, the telencephalon forms the cerebrum, the mesencephalon develops into the midbrain, the myelencephalon forms the medulla, and the metencephalon develops into the pons and cerebellum.
3. The principal parts of the brain are the brain stem, diencephalon, cerebrum, and cerebellum.
4. The brain is protected by the cranial bones, cranial meninges, and cerebrospinal fluid.

Protection and Coverings (p. 307)

1. The brain is protected by cranial bones, meninges, and cerebrospinal fluid.
2. The cranial meninges are continuous with the spinal meninges and are named dura mater, arachnoid, and pia mater.

Cerebrospinal Fluid (CSF) (p. 307)

1. Cerebrospinal fluid is formed in the choroid plexuses and circulates through the subarachnoid space, ventricles, and central canal. Most of the fluid is absorbed by the arachnoid villi of the superior sagittal sinus.

2. Cerebrospinal fluid protects by serving as a shock absorber. It also delivers nutritive substances from the blood and removes wastes.
3. The accumulation of cerebrospinal fluid in the head is called hydrocephalus. If the fluid accumulates in the ventricles, it is called internal hydrocephalus. If it accumulates in the subarachnoid space, it is called external hydrocephalus.

Blood Supply (p. 310)

1. The blood supply to the brain is via the cerebral arterial circle (circle of Willis).
2. Any interruption of the oxygen supply to the brain can result in weakening, permanent damage, or death of brain cells. Interruption of the mother's blood supply to a child during childbirth before it can breathe may result in paralysis, mental retardation, epilepsy, or death.
3. Glucose deficiency may produce dizziness, convulsions, and unconsciousness.
4. The blood–brain barrier (BBB) is a concept that explains the differential rates of passage of certain material from the blood into the brain.

Brain Stem (p. 310)

1. The medulla oblongata is continuous with the upper part of the spinal cord. It contains nuclei that are reflex centers for regulation of heart rate, respiratory rate, vasoconstriction,

swallowing, coughing, vomiting, sneezing, and hiccuping. It also contains the nuclei of origin for cranial nerves VIII (cochlear and vestibular branches) through XII.

2. The pons is superior to the medulla. It connects the spinal cord with the brain and links parts of the brain with one another. It relays impulses related to voluntary skeletal movements from the cerebral cortex to the cerebellum. It contains the nuclei for cranial nerves V through VII and the vestibular branch of VIII. The reticular formation of the pons contains the pneumotaxic center, which helps control respiration.

3. The midbrain connects the pons and diencephalon. It conveys motor impulses from the cerebrum to the cerebellum and cord, sensory impulses from cord to thalamus, and regulates auditory and visual reflexes. It also contains the nuclei of origin for cranial nerves III and IV.

Diencephalon (p. 314)

1. The diencephalon consists primarily of the thalamus and hypothalamus.

2. The thalamus is superior to the midbrain and contains nuclei that serve as relay stations for all sensory impulses, except smell, to the cerebral cortex. It also registers conscious recognition of pain and temperature and some awareness of light touch and pressure.

3. The hypothalamus is inferior to the thalamus. It controls and integrates the autonomic nervous system, receives sensory impulses from viscera, connects the nervous and endocrine systems, coordinates mind-over-body phenomena, functions in rage and aggression, controls body temperature, regulates food and fluid intake, maintains the waking state and sleep patterns, and acts as a self-sustained oscillator that drives biological rhythms.

Cerebrum (p. 316)

1. The cerebrum is the largest part of the brain. Its cortex contains convolutions, fissures, and sulci.

2. The cerebral lobes are named the frontal, parietal, temporal, and occipital.

3. The white matter is under the cortex and consists of myelinated axons running in three principal directions.

4. The basal ganglia (cerebral nuclei) are paired masses of gray matter in the cerebral hemispheres. They help to control muscular movements.

5. The limbic system is found in the cerebral hemispheres and diencephalon. It functions in emotional aspects of behavior and memory.

6. The sensory areas of the cerebral cortex are concerned with the interpretation of sensory impulses. The motor areas are the regions that govern muscular movement. The association areas are concerned with emotional and intellectual processes.

7. Positron emission tomography (PET) helps identify which parts of the brain are involved in specific sensory and motor activities.

8. Brain waves generated by the cerebral cortex are recorded as an electroencephalogram (EEG). It may be used to diagnose epilepsy, infections, and tumors.

Brain Lateralization (Split-Brain Concept) (p. 323)

1. Recent research indicates that the two hemispheres of the brain are not bilaterally symmetrical, either anatomically or functionally.

2. The left hemisphere is more important for right-handed control, spoken and written language, numerical and scientific skills, and reasoning.

3. The right hemisphere is more important for left-handed control, musical and artistic awareness, space and pattern perception, insight, imagination, and generating mental images of sight, sound, touch, taste, and smell.

Cerebellum (p. 323)

1. The cerebellum occupies the inferior and posterior aspects of the cranial cavity. It consists of two hemispheres and a central, constricted vermis.

2. It is attached to the brain stem by three pairs of cerebellar peduncles.

3. The cerebellum functions in the coordination of skeletal muscles and the maintenance of normal muscle tone and body equilibrium.

Neurotransmitters in the Brain (p. 325)

1. There are numerous substances that are either known or suspected neurotransmitters that can facilitate, excite, or inhibit postsynaptic neurons.

2. Examples of neurotransmitters include acetylcholine (ACh), glutamic acid, aspartic acid, norepinephrine (NE), dopamine (DA), serotonin (5-HT), gamma aminobutyric acid (GABA), and glycine.

3. Neuropeptides that act as natural painkillers in the body are enkephalins, endorphins, and dynorphin.

4. Other neuropeptides serve as hormones or other regulators of physiological responses. Examples include angiotensin, cholecystokinin, and regulating factors produced by the hypothalamus.

Cranial Nerves (p. 327)

1. Twelve pairs of cranial nerves originate from the brain.

2. The pairs are named primarily on the basis of distribution and numbered by order of attachment to the brain. (See Exhibit 14-3 for summary of cranial nerves.)

Aging and the Nervous System (p. 327)

1. Age-related effects involve loss of neurons and decreased capacity for sending nerve impulses.

2. Degenerative changes also affect the sense organs.

Developmental Anatomy of the Nervous System (p. 328)

1. The development of the nervous system begins with a thickening of ectoderm called the neural plate.

2. The parts of the brain develop from primary and secondary vesicles.

Disorders: Homeostatic Imbalances (p. 332)

1. Brain tumors are neoplasms within the cranium.

2. Poliomyelitis is a viral infection that results in paralysis.

3. Cerebral palsy refers to a group of motor disorders caused by damage to motor centers of the cerebral cortex, cerebellum, or basal ganglia during fetal development, childbirth, or early infancy.

4. Parkinson's disease is a progressive degeneration of the dopamine (DA)-producing neurons in the substantia nigra resulting in insufficient DA in the basal ganglia.

5. Multiple sclerosis (MS) is the destruction of meylin sheaths of the neurons of the central nervous system. Impulse transmission is interrupted.

6. Epilepsy results from irregular electrical discharges of brain cells and may be diagnosed by an EEG. Depending on the form of the disease, the victim experiences degrees of motor, sensory, or psychological malfunction.

7. Cerebrovascular accidents (CVAs), also called strokes, involve brain tissue destruction due to hemorrhage, thrombosis, or atherosclerosis.

8. Dyslexia involves an inability of an individual to comprehend written language. It may be diagnosed by using brain electrical activity mapping (BEAM).
9. Tay-Sachs disease is an inherited disorder that involves neurological degeneration of the CNS because of excessive amounts of ganglioside.
10. Headaches are of two types: intracranial and extracranial.
11. Irritation of the trigeminal (V) nerve is known as trigeminal neuralgia.
12. Reye's syndrome (RS) is characterized by vomiting, brain dysfunction, and liver damage.
13. Alzheimer's disease (AD) is a disabling neurological disorder of the elderly that involves widespread intellectual impairment, personality changes, and sometimes delirium.

REVIEW QUESTIONS

1. Identify the four principal parts of the brain and the components of each, where applicable. What is the origin of each of the parts?
2. Describe the location of the cranial meninges. What is an extradural hemorrhage?
3. Where is cerebrospinal fluid (CSF) formed? Describe its circulation. Where is CSF absorbed?
4. Distinguish between internal and external hydrocephalus.
5. Explain the importance of oxygen and glucose to brain cells.
6. What is the blood–brain barrier (BBB)? Describe the passage of several drugs with respect to the BBB.
7. Describe the location and structure of the medulla. Define decussation of pyramids. Why is it important? List the principal functions of the medulla.
8. Describe the location and structure of the pons. What are its functions?
9. Describe the location and structure of the midbrain. What are its functions?
10. Describe the location and structure of the thalamus. List its functions.
11. Where is the hypothalamus located? Explain some of its major functions.
12. Where is the cerebrum located? Describe the cortex, convolutions, fissures, and sulci of the cerebrum.
13. List and locate the lobes of the cerebrum. How are they separated from one another? What is the insula?
14. Describe the organization of cerebral white matter. Be sure to indicate the function of each group of fibers.
15. What are basal ganglia? Name the important basal ganglia and list the function of each. Describe the effects of damage on the basal ganglia.
16. Define the limbic system. Explain several of its functions.
17. What is meant by a sensory area of the cerebral cortex? List, locate, and give the function of each sensory area.
18. Describe the principle of positron emission tomography (PET). What are its clinical applications?
19. What is meant by a motor area of the cerebral cortex? List, locate, and give the function of each motor area.
20. What conditions may result from damage to sensory or motor speech areas?
21. What is an association area of the cerebral cortex? What are its functions?
22. Define an electroencephalogram (EEG). List the principal waves recorded and the importance of each. What is the diagnostic value of an EEG?
23. Describe brain lateralization (split-brain concept).
24. Describe the location of the cerebellum. List the principal parts of the cerebellum.
25. What are cerebellar peduncles? List and explain the function of each.
26. Explain the functions of the cerebellum. Describe some effects of cerebellar damage.
27. What are neurotransmitters? How do they affect postsynaptic neurons?
28. Give some examples of neurotransmitters and indicate their functions.
29. Describe the neuropeptides that act as natural painkillers in the body. Give their particular locations in the body and their functions.
30. Define a cranial nerve. How are cranial nerves named and numbered? Distinguish between a mixed and a sensory cranial nerve.
31. For each of the 12 pairs of cranial nerves, list (a) its name, number, and type; (b) its location; and (c) its function. In addition, list the effects of damage, where applicable.
32. Describe the effects of aging on the nervous system.
33. Describe the development of the nervous system.
34. Define each of the following: brain tumors, poliomyelitis, cerebral palsy, Parkinson's disease, multiple sclerosis (MS), epilepsy, cerebrovascular accidents (CVAs), dyslexia, Tay-Sachs disease, headache, trigeminal neuralgia, Reye's syndrome (RS), and Alzheimer's disease (AD).
35. What is brain electrical activity mapping (BEAM)? What are its clinical applications?
36. Refer to the glossary of medical terminology associated with the nervous system. Be sure that you can define each term.

15

The Sensory, Motor, and Integrative Systems

- Define a sensation and list the four prerequisites necessary for its transmission.
- Describe the characteristics of sensations.
- Classify receptors on the basis of location, stimulus detected, and simplicity or complexity.
- List the location and function of the receptors for tactile sensations (touch, pressure, vibration), thermoreceptive sensations (heat and cold), and pain.
- Distinguish somatic, visceral, referred, and phantom pain, and describe the various methods used to relieve pain.
- Identify the proprioceptive receptors and indicate their functions.
- Describe the composition of the somatosensory cortex.
- Discuss the origin, neuronal components, and destination of the posterior column and spinothalamic pathways.
- Explain the neural pathways for pain and temperature; light touch and pressure; and discriminative touch, proprioception, and vibration.
- Contrast the roles of the cerebellar tracts in conveying sensations.
- Describe how sensory input and motor responses are linked in the central nervous system.
- Describe the composition of the motor cortex.
- Compare the course of the pyramidal and extrapyramidal motor pathways.
- List the functions of the lateral corticospinal, anterior corticospinal, corticobulbar, rubrospinal, tectospinal, and vestibulospinal tracts.
- Compare integrative functions such as memory, wakefulness, and sleep.

Having examined the structure of the nervous system and its activities, we will now see how its different parts cooperate in performing its three essential functions: (1) receiving sensory information; (2) transmitting motor impulses that result in movement or secretion; and (3) integration, an activity that deals with memory, sleep, and emotions.

SENSATIONS

The central nervous system requires a continual flow of information to regulate homeostasis and initiate appropriate responses to changes in the internal and external environments. At any given time, our brains receive and respond to many varieties of information. However, we are aware only of the information that we consciously focus upon. The central nervous system selects only those bits of information that are important for the moment, and it is only those bits of information that are brought to our conscious level. There is no question that we would collapse into nervous wrecks if our consciousness was forced to deal with all the information arriving at once. The conscious mind is turned off to protect itself from overstimulation.

Your ability to sense stimuli is vital to your survival. If pain could not be sensed, burns would be common. An inflamed appendix or stomach ulcer would progress unnoticed. A lack of sight would increase the risk of injury from unseen obstacles, a loss of smell would allow a harmful gas to be inhaled, a loss of hearing would prevent recognition of automobile horns, and a lack of taste would allow toxic substances to be ingested. In short, if you could not "sense" your environment and make the necessary homeostatic adjustments, you could not survive on your own.

DEFINITION

In its broadest context, **sensation** refers to a state of awareness of external or internal conditions of the body. **Perception** refers to the conscious registration of a sensory stimulus. For a sensation to occur, four prerequisites must be satisfied.

1. A **stimulus,** or change in the environment, capable of initiating a response by the nervous system must be present.

2. A **receptor** or *sense organ* must pick up the stimulus and transduce (convert) it to a nerve impulse. A receptor or sense organ may be viewed as specialized nervous tissue that is extremely sensitive to internal or external conditions.

3. The impulse must be **conducted** along a neural pathway from the receptor or sense organ to the brain.

4. A region of the brain must **translate** the impulse into a sensation.

Receptors are capable of converting a specific stimulus into a nerve impulse. The stimulus may be light, heat, pressure, mechanical energy, or chemical energy. Each stimulus is capable of causing the membrane of the receptor to deplorize. This depolarization is called a **generator (receptor) potential.**

The generator potential is a graded response; within limits, the magnitude increases with stimulus strength and frequency. When the generator potential reaches the threshold level, it initiates an action potential (nerve impulse). Once initiated, the action potential is propagated along the nerve fiber. Whereas a generator potential is a local, graded response, an action potential obeys the all-or-none principle. The function of a generator potential is to initiate an action potential by transducing a stimulus into a nerve impulse.

A receptor may be quite simple. It may consist of the dendrites of a single neuron in the skin that are sensitive to pain stimuli; or it may be contained in a complex organ such as the eye. Regardless of complexity, all sense receptors contain the dendrites of sensory neurons. The dendrites occur either alone or in close association with specialized cells of other tissues.

Receptors are at the same time very excitable and very specialized. Except for pain receptors, each has a low threshold of response to its specific stimulus and a high threshold to all others.

Once a stimulus is received by a receptor and converted into an impulse, the impulse is conducted along an afferent pathway that enters either the spinal cord or the brain. Many sensory impulses are conducted to the sensory areas of the cerebral cortex. It is in this region that stimuli produce conscious sensations. Sensory impulses that terminate in the spinal cord or brain stem can initiate motor activities, but typically do not produce conscious sensations. The thalamus detects pain sensations but cannot distinguish the intensity or location from which they arise. This is a function of the cerebrum.

CHARACTERISTICS

Most conscious sensations or perceptions occur in the cortical regions of the brain. In other words, you see, hear, and feel in the brain. You seem to see with your eyes, hear with your ears, and feel pain in an injured part of your body only because the cortex interprets the sensation as coming from the stimulated sense receptor. The term **projection** describes this process by which the brain refers sensations to their point of stimulation.

A second characteristic of many sensations is **adaptation,** that is, a decrease in sensitivity to continued stimuli. In fact, the perception of a sensation may actually disappear, even though the stimulus is still being applied. For, example, when you first get into a tub of hot water, you probably feel a burning sensation, but soon the sensation decreases to one of comfortable warmth, even though the stimulus (hot water) is still present. In time, the sensation of warmth disappears completely. Other examples of adaptation include placing a ring on your finger, putting on your shoes or hat, sitting on a chair, and pushing your glasses up onto the top of your head. Adaptation results from a change in a receptor, a change in a structure associated with a receptor, or inhibitory feedback from the brain. Receptors vary in their ability to adapt. *Rapidly adapting (phasic) receptors,* such as those associated with pressure, touch, and smell, adapt very quickly. Such receptors play a major role in signaling changes in a particular sensation. *Slowly adapting (tonic) receptors,* such as those associated with pain, body position, and detecting chemicals in blood, adapt slowly. These receptors are important in signaling information regarding steady states of the body.

Sensations may also be characterized by **afterimages,** that is, some sensations persist even though the stimulus has been removed. This phenomenon is the reverse of adaptation. One common example of afterimage occurs when you look at a bright light and then look away or close your eyes. You still see the light for several seconds or minutes afterward.

Another characteristic of sensations is **modality:** the specific type of sensation felt. The sensation may be one of pain, pressure, touch, body position, equilibrium, hearing, vision, smell, or taste. In other words, the distinct property by which one sensation may be distinguished from another is its modality.

CLASSIFICATION OF RECEPTORS

Location

One convenient method of classifying receptors is by their location. **Exteroceptors** (eks'-ter-ō-SEP-tors) provide information about the external environment. They are sensitive to stimuli outside the body and transmit sensations of hearing, sight, smell, taste, touch, pressure, temperature, and pain. Exteroceptors are located near the surface of the body.

Visceroceptors (vis'-er-ō-SEP-tors), or **enteroceptors,** provide information about the internal environment. These sensations arise from within the body and may be felt as pain, pressure, fatigue, hunger, thirst, and nausea. Visceroceptors are located in blood vessels and viscera.

Proprioceptors (prō'-prē-ō-SEP-tors) provide information about body position and movement. Such sensations give us information about muscle tension, the position and tension of our joints, and equilibrium. These receptors are located in muscles, tendons, joints, and the internal ear.

Stimulus Detected

Another method of classifying receptors is by the type of stimuli they detect. **Mechanoreceptors** detect mechanical deformation of the receptor itself or in adjacent cells. Stimuli so detected include those related to touch, pressure, vibration, proprioception, hearing, equilibrium, and blood pressure. **Thermoreceptors** detect changes in temperature. **Nociceptors** detect pain, usually as a result of physical or chemical damage to tissues. **Electromagnetic (photo) receptors** detect light on the retina of the eye. **Chemoreceptors** detect taste in the mouth, smell in the nose, and chemicals in body fluids, such as oxygen, carbon dioxide, water, and glucose.

Simplicity or Complexity

As will be described shortly, receptors may also be classified according to the simplicity or complexity of their structure and the neural pathway involved. **Simple receptors** and neural pathways are associated with **general senses.** The receptors for general sensations are numerous and widespread. Examples include cutaneous sensations such as touch, pressure, vibration, heat, cold, and pain. **Complex receptors** and neural pathways are associated with **special senses.** The receptors for each special sense are found in only one or two specific areas of the body. Among the special senses are smell, taste, sight, and hearing.

GENERAL SENSES

CUTANEOUS SENSATIONS

Cutaneous sensations include tactile sensations (touch, pressure, vibration), thermoreceptive sensations (cold and heat), and pain. The receptors for these sensations are in the skin, connective tissue, and the ends of the gastrointestinal tract.

The cutaneous receptors are distributed over the body surface in such a way that certain parts of the body are densely populated with receptors and other parts contain only a few. Areas of the body that have few cutaneous receptors are insensitive; those containing many are very sensitive.

This type of distribution can be demonstrated in the skin by using the *two-point discrimination test* for touch. A compass is applied to the skin, and the distance in millimeters between the two points of the compass is varied. The subject then indicates when two points are felt and when only one is felt. The compass may be placed on the tip of the tongue, an area where receptors are very densely packed. At a distance of 1.4 mm (0.06 inch), the points are able to stimulate two different receptors, and the subject feels touched by two objects. If the distance is less than 1.4 mm, the subject feels only one point, even though both points are touching the tongue, because the points are so close together that they reach only one receptor. If the compass is placed on the back of the neck, the subject feels two distinctly different points only if the distance between them is 36.2 mm (1.43 inches) or greater, because the receptors are few in number and far apart. The results of this test indicate that the more sensitive the area, the closer the compass points may be placed and still be felt separately. The following order for these receptors, from greatest sensitivity to least, has been established: tip of tongue, tip of finger, side of nose, back of hand, and back of neck.

Cutaneous receptors have simple structures. They consist of the dendrites of sensory neurons that may or may not be enclosed in a capsule of epithelial or connective tissue. Impulses generated by cutaneous receptors pass along somatic afferent neurons in spinal and cranial nerves, through the thalamus, to the general sensory area of the parietal lobe of the cortex.

Tactile Sensations

Even though the **tactile sensations** are divided into separate sensations of touch, pressure, and vibration, they are all detected by the same types of receptors, that is, receptors subject to deformation.

•   *Touch*   Touch sensations generally result from stimulation of tactile receptors in the skin or tissues immediately beneath the skin. The term *discriminative touch* refers to the ability to recognize exactly what point of the body is touched. *Light touch* refers to the ability to perceive that something has touched the skin, although its exact location, shape, size, or texture cannot be determined.

Tactile receptors for touch include root hair plexuses, free nerve endings, tactile discs, corpuscles of touch, and type II cutaneous mechanoreceptors (Figure 15-1). **Root hair plexuses** are dendrites arranged in networks around the roots of hairs. They are not surrounded by supportive or protective structures. If a hair shaft is moved, the dendrites are stimulated. Root hair plexuses detect movements mainly on the surface of the body.

Other receptors that are not surrounded by supportive or protective structures are called **free (naked) nerve endings.** Free nerve endings are found everywhere in the skin and many other tissues.

Tactile, or **Merkel's** (MER-kelz) **discs,** are receptors for touch that consist of disclike formations of dendrites attached to deeper layers of epidermal cells. They are distributed in many of the same locations as corpuscles of touch.

Corpuscles of touch, or **Meissner's** (MĪS-nerz) **corpuscles,** are egg-shaped receptors containing a mass of dendrites enclosed by connective tissue. They are located in the dermal papillae of the skin and are most numerous in the fingertips, palms of the hands, and soles of the feet. They are also abundant in the eyelids, tip of the tongue, lips, nipples, clitoris, and tip of penis.

Type II cutaneous mechanoreceptors, or **end organs of Ruffini,** are embedded deeply in the dermis and in deeper tissues of the body. They detect heavy and continuous touch sensations.

•   *Pressure*   **Pressure sensations** generally result from stimulation of tactile receptors in deeper tissues and are longer lasting and have less variation in intensity than touch sensations. Moreover, pressure is felt over a larger area than touch.

Pressure receptors are free nerve endings, type II cuta-

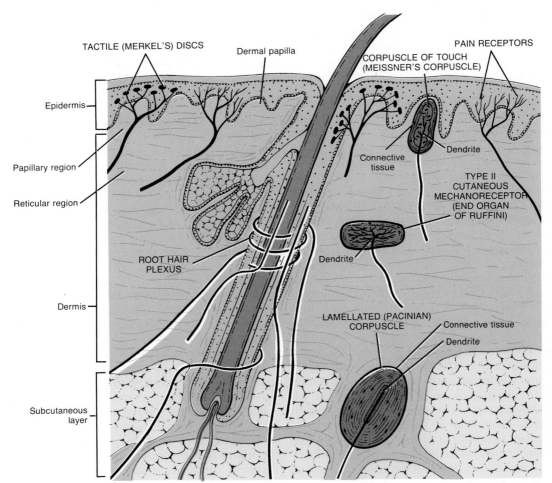

FIGURE 15-1 Structure and location of cutaneous receptors.

neous mechanoreceptors, and lamellated corpuscles. **Lamellated,** or **Pacinian** (pa-SIN-ē-an) **corpuscles** (Figure 15-1), are oval structures composed of a capsule resembling an onion that consist of connective tissue layers enclosing dendrites. Lamellated corpuscles are located in the subcutaneous tissue under the skin, the deep subcutaneous tissues that lie under mucous membranes, in serous membranes, around joints and tendons, in the perimysium of muscles, in the mammary glands, in the external genitalia of both sexes, and in certain viscera.

* *Vibration* **Vibration sensations** result from rapidly repetitive sensory signals from tactile receptors.

The receptors for vibration sensations are corpuscles of touch and lamellated corpuscles. Whereas corpuscles of touch detect low-frequency vibration, lamellated corpuscles detect higher-frequency vibration.

Thermoreceptive Sensations

The **thermoreceptive sensations** are heat and cold. The exact nature of **thermoreceptive receptors** is not known, but they might be free nerve endings.

Pain Sensations

Pain is indispensable for a normal life. It provides us with information about tissue-damaging stimuli, and thus often enables us to protect ourselves from greater damage. It is pain that initiates our search for medical assistance, and it is our subjective description and indication of the location of the pain that helps to pinpoint the underlying cause of disease.

The receptors for **pain,** called **nociceptors,** are simply the branching ends of the dendrites of certain sensory neurons (Figure 15-1). Pain receptors are found in practically every tissue of the body. They may respond to any type of stimulus. When stimuli for other sensations, such as touch, pressure, heat, and cold, reach a certain threshold, they stimulate the sensation of pain as well. Excessive stimulation of a sense organ causes pain. Additional stimuli for pain receptors include excessive distension or dilation of a structure, prolonged muscular contractions, muscle spasms, inadequate blood flow to an organ, or the presence of certain chemical substances. Pain receptors, because of their sensitivity to all stimuli, perform a protective function by identifying changes that may endanger the body. Pain receptors adapt only slightly or not at all. Adaptation is the decrease or disappearance of the perception of a sensation even though the stimulus is still present. If there were adaptation to pain, it would cease to be sensed and irreparable damage could result.

Sensory impulses for pain are conducted to the central nervous system along spinal and cranial nerves. The lateral spinothalamic tracts of the spinal cord relay impulses to the thalamus. From here the impulses may be relayed to the postcentral gyrus of the parietal lobe. Recognition of the kind and intensity of most pain is ultimately localized in the cerebral cortex. Some awareness of pain occurs at subcortical levels.

Pain may be divided into two types: somatic and visceral. **Somatic pain** arises from stimulation of receptors in the skin, in which case it is called *superficial somatic pain,* or from stimulation of receptors in skeletal muscles, joints, tendons, and fascia, then called *deep somatic pain.* **Visceral pain** results from stimulation of receptors in the viscera.

The ability of the cerebral cortex to locate the origin of pain is related to past experience. In most instances of somatic pain and in some instances of visceral pain, the cortex accurately projects the pain back to the stimulated area. If you burn your finger, you feel the pain in your finger. If the lining of your pleural cavity is inflamed, you experience pain there. In most instances of visceral pain, however, the sensation is not projected back to the point of stimulation. Rather, the pain may be felt in or just under the skin that overlies the stimulated organ. The pain may also be felt in a surface area far from the stimulated organ. This phenomenon is called **referred pain.** In general, the area to which the pain is referred and the visceral organ involved receive their innervation from the same segment of the spinal cord. Consider the following example. Afferent fibers from the heart as well as from the skin over the heart and along the medial aspect of the left upper extremity enter spinal cord segments T1–T4. Thus the pain of a heart attack is typically felt in the skin over the heart and along the left arm. Figure 15-2 illustrates cutaneous regions to which visceral pain may be referred.

CLINICAL APPLICATION

A kind of pain frequently experienced by patients who have had a limb amputated is called **phantom pain.** They still experience pain or other sensations in the extremity as if the limb were still there. This probably occurs because the remaining proximal portions of the sensory nerves that previously received impulses from the limb are being stimulated by the trauma of the amputation. Stimuli from these nerves are interpreted by the brain as coming from the nonexistent (phantom) limb.

Pain sensations may be controlled by interrupting the pain impulse between the receptors and the interpretation centers of the brain. This may be done chemically, surgically, or by other means. Most pain sensations respond to pain-reducing drugs, which, in general, act to inhibit nerve impulse conduction at synapses.

Occasionally, however, pain may be controlled only by surgery. The purpose of surgical treatment is to interrupt the pain impulse somewhere between the receptors and the interpretation centers of the brain by severing the sensory nerve, its spinal root, or certain tracts in the spinal cord or brain. *Sympathectomy* is excision of portions of the neural tissue from the autonomic nervous system; *cordotomy* is severing a spinal cord tract, usually the lateral spinothalamic; *rhizotomy* is the cutting of sen-

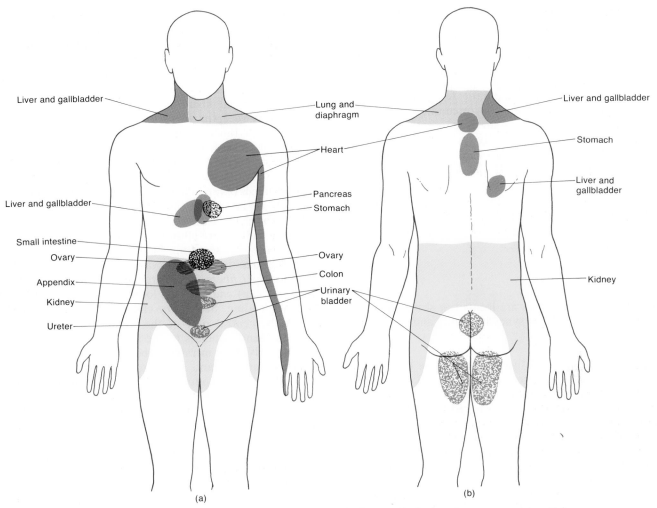

Liver and gallbladder

Lung and diaphragm

Heart

Liver and gallbladder

Pancreas

Stomach

Small intestine

Ovary

Ovary

Appendix

Colon

Kidney

Urinary bladder

Ureter

Liver and gallbladder

Stomach

Liver and gallbladder

Kidney

(a)

(b)

FIGURE 15-2 Referred pain. The colored parts of the diagrams indicate cutaneous areas to which visceral pain is referred. (a) Anterior view. (b) Posterior view.

sory nerve roots; *prefrontal lobotomy* is the destruction of the tracts that connect the thalamus with the prefrontal and frontal lobes of the cerebral cortex. In each instance, the pathway for pain is severed so that pain impulses are no longer conducted to the cortex.

Another method of inhibiting pain impulses is **acupuncture** (*acus* = needle; *pungere* = sting). Needles are inserted through selected areas of the skin and then twirled by the acupuncturist or by a mechanical device. After 20 to 30 minutes, pain is deadened for 6 to 8 hours. The location of needle insertion depends on the part of the body the acupuncturist wishes to anesthetize. To pull a tooth, a needle is inserted in the web between thumb and index finger. For a tonsillectomy, a needle is inserted approximately 5 cm (2 inches) above the wrist. For removal of a lung, a needle is placed in the forearm midway between wrist and elbow.

Acupuncture probably works by taking advantage of the body's natural inhibitory influences that can normally block pain pathways. For example, it has been established that sensory pain fibers in posterior root ganglia release

a neurotransmitter called substance P. This substance produces excitatory postsynaptic potentials (EPSP's) on the next neuron on the way to the neurons in the tract that conveys pain impulses (see Figure 15-5). Small enkephalin-containing neurons impinge on the synaptic end bulb of the sensory path fiber. The enkephalin released by these small neurons blocks the release of substance P from the nerve terminals of the sensory pain fibers and, therefore, inhibits pain transmission to the brain. Acupuncture enhances the release of these inhibitory substances, such as enkephalin, from various locations, and these substances can then be carried by the circulatory system to the pain fibers. The onset of the effect of acupuncture is delayed until the enkephalin level rises to inhibitory levels. Similarly, the effect of acupuncture lingers after the twirling or vibration of the needles stops.

Currently, acupuncture is used in the United States mostly for childbirth, tic douloureux, arthritis, and other nonsurgical conditions. It is also being used in spinal injury rehabilitation to activate useful motion in limbs, provided that there is still some intact neural tissue.

PROPRIOCEPTIVE SENSATIONS

An awareness of the activities of muscles, tendons, and joints and of equilibrium is provided by the **proprioceptive,** or **kinesthetic** (kin'-es-THET-ik), **sense.** It informs us of the degree to which muscles are contracted, the amount of tension created in the tendons, the change of position of a joint, and the orientation of the head relative to the ground and in response to movements (equilibrium). The proprioceptive sense enables us to recognize the location and rate of movement of one body part in relation to others. It also allows us to estimate weight and determine the muscular work necessary to perform a task. With the proprioceptive sense, we can judge the position and movements of our limbs without using our eyes when we walk, type, or dress in the dark.

Receptors

Proprioceptive receptors are located in skeletal muscles, tendons in and around synovial joints, and the internal ear.

• *Muscle Spindles* **Muscle spindles** are delicate proprioceptive receptors interspersed among skeletal muscle fibers and oriented parallel to the fibers (Figure 15-3a). The ends of the spindles are anchored to the endomysium and perimysium. Muscle spindles consist of three to ten specialized muscle fibers called *intrafusal fibers,* which are partially enclosed in a connective tissue capsule that is filled with lymph. The spindles are surrounded by skeletal

muscle fibers of the muscle called *extrafusal fibers.* The central region of each intrafusal fiber has few or no actin and myosin myofilaments and an accumulation of nuclei. In some intrafusal fibers the nuclei bunch at the center (*nuclear bag fibers*); in others, the nuclei form a chain at the center (*nuclear chain fibers*). The central region of the intrafusal fibers cannot contract and represents the sensory receptor area for a spindle.

Although the central receptor area cannot contract because it lacks myofilaments, it does contain two types of sensory fibers. A large sensory fiber, called a *type Ia fiber,* innervates the exact center of the intrafusal fibers. The branches of the Ia fiber, called *primary (annulospiral) endings,* wrap around the center of the intrafusal fibers. When the central part of the spindle is stretched, the primary endings are stimulated and send impulses to the spinal cord at exceedingly great velocities. The central receptor area is also innervated by two sensory fibers called *type II fibers.* Their branches, known as *secondary (flower spray) endings,* are located on either side of the primary ending. Secondary endings are also stimulated when the central part of the spindle is stretched, and they too send impulses to the spinal cord.

The ends of the intrafusal fibers contain actin and myosin myofilaments and represent the contractile portions of the fibers. The ends of the fibers contract when stimulated by *gamma efferent neurons.* These neurons are small motor neurons located in the anterior gray horn of the spinal cord. The fibers of the gamma efferent neurons terminate as motor end plates on the ends of the intrafusal fibers. Extrafusal fibers are innervated by large motor neurons called *alpha efferent neurons.* These neurons are

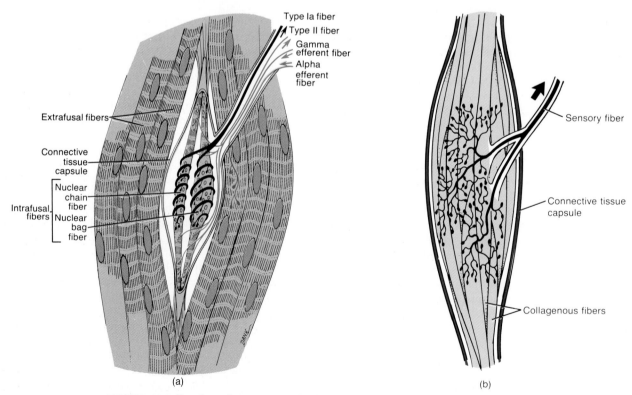

Type Ia fiber
Type II fiber
Gamma efferent fiber
Alpha efferent fiber

Extrafusal fibers

Connective tissue capsule

Nuclear chain fiber
Intrafusal fibers
Nuclear bag fiber

Sensory fiber

Connective tissue capsule

Collagenous fibers

(a)

(b)

FIGURE 15-3 Proprioceptive receptors. (a) Muscle spindle. (b) Tendon organ.

also located in the anterior gray horn of the spinal cord near gamma efferent neurons.

Muscle spindles are stimulated in response to both sudden and maintained stretch on the central areas of the intrafusal fibers. The muscle spindles monitor changes in the length of a skeletal muscle by responding to the rate and degree of change in length. This information is relayed to the central nervous system to assist in the coordination and efficiency of muscle contraction. The neural pathway involved in transmitting the information to the central nervous system, processing by the central nervous system, and eliciting a motor response is shown in Figure 15-7.

• *Tendon Organs* **Tendon organs (Golgi tendon organs)** are proprioceptive receptors found at the junction of a tendon with a muscle. They help protect tendons and their associated muscles from damage resulting from excessive tension. Each consists of a thin capsule of connective tissue that encloses a few collagenous fibers (Figure 15-3b). The capsule is penetrated by one or more sensory neurons whose terminal branches entwine among and around the collagenous fibers. When tension is applied to a tendon, tendon organs are stimulated and the information is relayed to the central nervous system. The neural pathway involved in transmitting the information to the central nervous system, processing by the central nervous system, and eliciting a motor response is shown in Figure 15-7.

• *Joint Kinesthetic Receptors* There are several types of **joint kinesthetic receptors** within and around the articular capsules of synovial joints. Encapsulated receptors, similar to type II cutaneous mechanoreceptors (end organs of Ruffini), are present in the capsules of joints and respond to pressure. Small lamellated (Pacinian) corpuscles in the connective tissue outside articular capsules are receptors that respond to acceleration and deceleration. Articular ligaments contain receptors similar to tendon organs that mediate reflex inhibition of the adjacent muscles when excessive strain is placed on the joint.

• *Maculae and Cristae* The proprioceptors in the internal ear are the macula of the saccule and the utricle and cristae in the semicircular ducts. Their function in equilibrium is discussed in Chapter 17.

Proprioceptors adapt only slightly. This feature is advantageous since the brain must be apprised of the status of different parts of the body at all times so that adjustments can be made to ensure coordination.

The afferent pathway for muscle sense consists of impulses generated by proprioceptors via cranial and spinal nerves to the central nervous system. Impulses for conscious proprioception pass along ascending tracts in the cord, where they are relayed to the thalamus and cerebral cortex. The sensation is registered in the general sensory area in the parietal lobe of the cerebral cortex posterior to the central sulcus. Proprioceptive impulses that have resulted in reflex action pass to the cerebellum along spinocerebellar tracts.

LEVELS OF SENSATION

As we have said, a receptor converts a stimulus into a nerve impulse, and only after that impulse has been conducted to a region of the spinal cord or brain can it be translated into a sensation. The nature of the sensation and the type of reaction generated vary with the level of the central nervous system at which the sensation is translated.

Sensory fibers terminating in the spinal cord can generate spinal reflexes without immediate action by the brain. Sensory fibers terminating in the lower brain stem bring about far more complex motor reactions than simple spinal reflexes. When sensory impulses reach the lower brain stem, they cause subconscious motor reactions. Sensory impulses that reach the thalamus can be localized crudely in the body. At the thalamic levels sensations are sorted by modality, that is, identified as the *specific* sensation of touch, pressure, pain, position, hearing, vision, smell, or taste. When sensory information reaches the cerebral cortex, we experience precise localization. It is at this level that memories of previous sensory information are stored and the perception of sensation occurs on the basis of past experience.

SENSORY PATHWAYS

SOMATOSENSORY CORTEX

Sensory information from receptors on one side of the body crosses over to the opposite side in the spinal cord or brain stem and then to the **somatosensory cortex (primary somesthetic** or **general sensory area)** of the cerebral cortex where conscious sensations are produced (see Figure 14-10). Areas of the somatosensory cortex have been mapped out that represent the termination of sensory information from all parts of the body. Figure 15-4 shows the location and areas of representation of the somatosensory cortex of the right cerebral hemisphere. The left cerebral hemisphere has a duplicate somatosensory cortex.

Note that some parts of the body are represented by large areas in the somatosensory cortex. These include the lips, face, and thumb. Other parts of the body, such as the trunk and lower extremities, are represented by relatively small areas. The relative sizes of the somatosensory cortex are directly proportional to the number of specialized sensory receptors in each respective part of the body. Thus, there are numerous receptors in the skin of the lips, but relatively few in the skin of the trunk. Essentially, the size of the area for a particular part of the body is determined by the functional importance of the part and its need for sensitivity.

Let us now examine how sensory information is transmitted from receptors to the central nervous system. You will find it helpful to review the principal ascending and descending tracts of the spinal cord (see Exhibit 13-1 and Figure 13-3). Sensory information transmitted from the spinal cord to the brain is conducted along two general

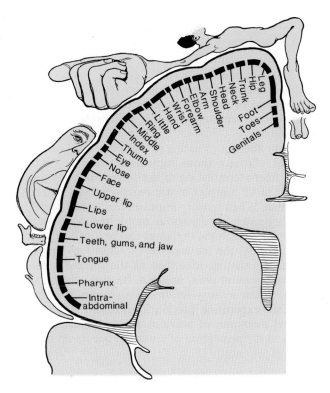

FIGURE 15-4 Somatosensory cortex of the right cerebral hemisphere.

pathways: the posterior column pathway and the spinothalamic pathway.

POSTERIOR COLUMN PATHWAY

In the **posterior column pathway (fasciculus gracilis** and **fasciculus cuneatus)** to the cerebral cortex there are three separate sensory neurons. The **first-order neuron** connects the receptor with the spinal cord and medulla on the same side of the body. The cell body of the first-order neuron is in the posterior root ganglion of a spinal nerve. The first-order neuron synapses with a **second-order neuron,** which passes from the medulla upward to the thalamus. The cell body of the second-order neuron is located in the nuclei cuneatus or gracilis of the medulla. Before passing into the thalamus, the second-order neuron crosses to the opposite side of the medulla and enters the medial lemniscus, a projection tract that terminates at the thalamus. In the thalamus, the second-order neuron synapses with a **third-order neuron,** which terminates in the somesthetic sensory area of the cerebral cortex.

The posterior column pathway conducts impulses related to proprioception, discriminative touch, two-point discrimination, and vibrations.

SPINOTHALAMIC PATHWAY

The **spinothalamic pathway** is also composed of three orders of sensory neurons. The first-order neuron connects a receptor of the neck, trunk, and extremities with the spinal cord. The cell body of the first-order neuron is in the posterior root ganglion also. The first-order neuron synapses with the second-order neuron, which has its cell

body in the posterior gray horn of the spinal cord. The fiber of the second-order neuron crosses to the opposite side of the spinal cord and passes upward to the brain stem in the lateral spinothalamic tract or ventral spinothalamic tract. The fibers from the second-order neuron terminate in the thalamus. There the second-order neuron synapses with a third-order neuron. The third-order neuron terminates in the somesthetic sensory area of the cerebral cortex. The spinothalamic pathway conveys sensory impulses for pain and temperature as well as light touch and pressure.

Now we can examine the anatomy of specific sensory pathways—for pain and temperature; for light touch and pressure; and for discriminative touch, proprioception, and vibration.

Pain and Temperature

The sensory pathway for pain and temperature is called the **lateral spinothalamic pathway** (Figure 15-5). The first-

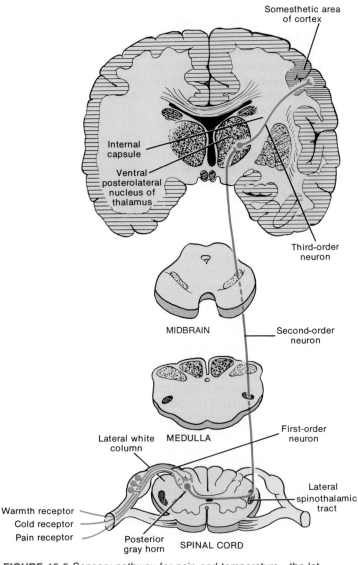

FIGURE 15-5 Sensory pathway for pain and temperature—the lateral spinothalamic pathway.

order neuron conveys the impulse for pain or temperature from the appropriate receptor to the posterior gray horn on the same side of the spinal cord. In the horn, the first-order neuron synapses with a second-order neuron. The axon of the second-order neuron crosses to the opposite side of the cord. Here it becomes a component of the *lateral spinothalamic tract* in the lateral white column. The second-order neuron passes upward in the tract through the brain stem to a nucleus in the thalamus called the ventral posterolateral nucleus. In the thalamus, conscious recognition of pain and temperature occurs. The sensory impulse is then conveyed from the thalamus through the internal capsule to the somesthetic area of the cerebral cortex by a third-order neuron. The cortex analyzes the sensory information for the precise source, severity, and quality of the pain and heat stimuli.

Light Touch and Pressure

The neural pathway that conducts impulses for light touch and pressure is the **anterior (ventral) spinothalamic pathway** (Figure 15-6). The first-order neuron conveys the impulse from a light touch or pressure receptor to the posterior gray horn on the same side of the spinal cord. In the horn, the first-order neuron synapses with a second-order neuron. The axon of the second-order neuron crosses to the opposite side of the cord and becomes a component of the **anterior spinothalamic tract** in the anterior white column. The second-order neuron passes upward in the tract through the brain stem to the ventral posterolateral nucleus of the thalamus. The sensory impulse is then relayed from the thalamus through the internal capsule to the somesthetic area of the cerebral cortex by a third-order neuron. Although there is some awareness of light touch and pressure at the thalamic level, it is not fully perceived until the impulses reach the cortex.

Discriminative Touch, Proprioception, and Vibration

The neural pathway for discriminative touch, proprioception, and vibration is called the **posterior column pathway** (Figure 15-7). This pathway conducts impulses that give rise to several discriminating senses.

1. Discriminative touch: the ability to recognize the exact location of stimulation and to make two-point discriminations.
2. Stereognosis: the ability to recognize by "feel" the size, shape, and texture of an object.
3. Proprioception: the awareness of the precise position of body parts and directions of movement.
4. Weight discrimination: the ability to assess the weight of an object.
5. The ability to sense vibrations.

First-order neurons for the discriminating senses just noted follow a pathway different from those for pain and temperature and light touch and pressure. Instead of terminating in the posterior gray horn, the first-order neurons from appropriate receptors pass upward in the fasciculus gracilis or fasciculus cuneatus in the posterior white column of the cord. From here the first-order neurons enter either the nucleus gracilis or nucleus cuneatus in

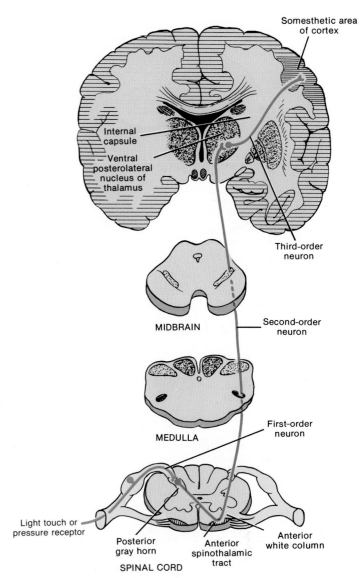

FIGURE 15-6 Sensory pathway for light touch and pressure—the anterior spinothalamic pathway.

the medulla, where they synapse with second-order neurons. The axons of the second-order neurons cross to the opposite side of the medulla and ascend to the thalamus through the medial lemniscus, a projection tract of white fibers passing through the medulla, pons, and midbrain. The second-order neuron axons synapse with third-order neurons in the ventral posterior nucleus in the thalamus. In the thalamus, there is no conscious awareness of the discriminating senses, except for a possible crude awareness of vibrations. The third-order neurons convey the sensory impulses to the somesthetic area of the cerebral cortex. It is here that you perceive your sense of position and movement and light touch.

CEREBELLAR TRACTS

The *posterior spinocerebellar tract* is an uncrossed tract that conveys impulses concerned with subconscious muscle sense and thus assumes a role in reflex adjustments

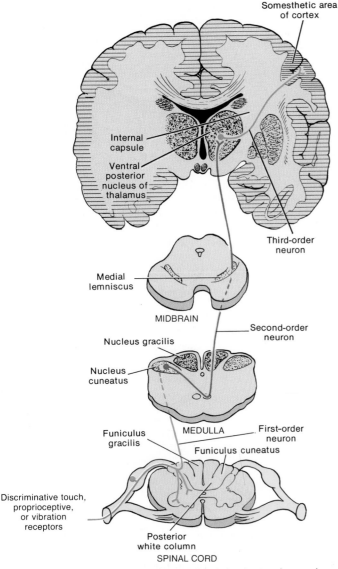

FIGURE 15-7 Sensory pathway for discriminative touch, proprioception, and vibration—the posterior column pathway.

Labels in figure:
- Somesthetic area of cortex
- Internal capsule
- Ventral posterior nucleus of thalamus
- Third-order neuron
- Medial lemniscus
- MIDBRAIN
- Second-order neuron
- Nucleus gracilis
- Nucleus cuneatus
- Funiculus gracilis
- MEDULLA
- First-order neuron
- Funiculus cuneatus
- Discriminative touch, proprioceptive, or vibration receptors
- Posterior white column
- SPINAL CORD

for posture and muscle tone. The nerve impulses originate in neurons that run between proprioceptors in muscles, tendons, and joints and the posterior gray horn of the spinal cord. Here the neurons synapse with afferent neurons that pass to the ipsilateral lateral white column of the cord to enter the posterior cerebellar tract. The tract enters the inferior cerebellar peduncles from the medulla and ends at the cerebellar cortex. In the cerebellum, synapses are made that ultimately result in the transmission of impulses back to the spinal cord to the anterior gray horn to synapse with the lower motor neurons leading to skeletal muscles.

The **anterior spinocerebellar tract** also conveys impulses for subconscious muscle sense. It, however, is made up of both crossed and uncrossed nerve fibers. Sensory neurons deliver impulses from proprioceptors to the posterior gray horn of the spinal cord. Here a synapse occurs with neurons that make up the anterior spinocerebellar

tracts. Some fibers cross to the opposite side of the spinal cord in the anterior white commissure. Others pass laterally to the ipsilateral anterior spinocerebellar tract and move upward, through the brain stem, to the pons to enter the cerebellum through the superior cerebellar peduncles. Here again the impulses for subconscious muscle sense are registered.

MOTOR PATHWAYS

After receiving and interpreting sensations, the central nervous system generates impulses to direct responses to that sensory input. We have already considered somatic reflex arcs and visceral efferent pathways. Now we will look at the transmission of motor impulses that result in the movement of skeletal muscles.

LINKAGE OF SENSORY INPUT AND MOTOR RESPONSES

Sensory systems function to keep the central nervous system aware of the external and internal environment. Active responses to this information are mediated by the motor system, which enable us to move about and change our relationship to the world around us. As sensory information is conveyed into the central nervous system, it becomes part of a large pool of sensory input. We do not respond actively to every bit of input the central nervous system receives. Rather, the incoming information is integrated with other information arriving from all other operating sensory receptors. The integration process occurs not just once but at many stations along the pathways of the central nervous system. It occurs within the spinal cord, brain stem, cerebellum, and cerebral motor cortex. As a result, a motor response to make a muscle contract or a gland secrete can be initiated at any of these levels. The cerebral motor cortex assumes the major role for controlling precise, discrete muscular movements. The basal ganglia largely integrate semivoluntary movements like walking, swimming, and laughing. The cerebellum, although not a control center, assists the motor cortex and basal ganglia by making body movements smooth and coordinated.

The design of the motor response system is very precise. Each skeletal muscle is connected to its own group of nerve fibers, and these fibers innervate only fibers of that muscle. This arrangement ensures that, when the central nervous system initiates a response, only the desired muscle contracts. Recall the description of the motor unit in Chapter 10.

Muscle actions can vary greatly in complexity, and the pattern of activation of motor units therefore also varies in complexity. Motor units can respond simultaneously to produce a sudden, forceful movement, or they can be orchestrated to respond asynchronously in a time sequence so that the force of contraction builds up more gradually. When more than one joint is involved, the sequential patterns of motor unit activation involving all the joints can be very complicated indeed. Yet, when complicated movements are performed by an Olympic gymnast, for example, the beauty of the act belies the underlying complexity of the myriad of neuromuscular interactions.

Some muscular responses are simple and occur subconsciously. The quick withdrawal of a hand when it touches a hot object or the knee jerk that results when a percussion hammer taps the patellar tendon are simple responses. Playing a piano is a much more complex response. Simple or very complex, all muscle responses involve both contraction and relaxation of muscle groups. The innervation to antagonistic (opposing) muscles always acts in a reciprocal fashion. One nerve fiber set causes the contraction of the prime movers, and the other set of fibers causes relaxation of the antagonists. The complexity of the muscular movements involved in piano playing can be astonishing. Sensory input is rapidly and continually changing as the eyes scan the music score. Each bit of information initiates and modifies the reciprocal pattern of muscle innervation to the fingers, within milliseconds. Additional information from sensory receptors relating to touch, position, pressure, speed, and sound also affect the reciprocal innervation program. This almost instantaneous analysis of the total sensory input generates the incredible array of motor responses to produce beautiful music. The linking of the sensory input and motor response occurs at several major levels in the central nervous system, and each level can add richness and variety to the repertoire of movements.

As the location of the sensory-motor linkage climbs to higher levels in the central nervous system—from spinal cord to brain stem to cerebellum to basal ganglia to cerebral motor cortex—additional contributions are introduced that enrich the growing inventory of motor responses. All information about each sensory modality arises from the peripheral nervous system and is directed into the spinal cord. At the level of entry, a synapse is made that directs the input of information to higher centers in the central nervous system. When information reaches the highest level, the cerebral cortex, conscious sensation results. Most conscious sensations originate from fibers connected to receptors on the exterior of the body.

Other inputs to the central nervous system never give rise to conscious sensation. These operate to coordinate and modulate the contractions of muscle groups. Nerve fibers that connect to proprioceptive receptors within muscles and tendons constantly deliver input to the central nervous system about the present state of the muscle length and its speed and strength of contraction. These fibers enter the spinal cord, synapse, and project their input up the cord to the cerebellum.

The number of interconnections between the regions of the central nervous system is so great that it is impossible to define the final site of sensory input. Some sensory input arrives first at the cerebral cortex and some is delivered to the cerebellum. In other cases, the sensory input is delivered to the cerebellum, which then relays some of the input to the cerebral cortex.

When the input reaches the highest center, the process of sensory-motor integration occurs. This involves not only the utilization of information contained within that center but also the information impinging on that center

that is delivered from other centers within the central nervous system. After integration occurs, the output of the center is sent down the spinal cord in two major descending motor pathways: the pyramidal pathways and extrapyramidal pathways.

MOTOR CORTEX

Just as the somatosensory cortex has been mapped to indicate the termination of sensory information from all parts of the body, the **motor cortex** has been mapped to indicate which groups of muscles are controlled by its specific areas (Figure 15-8). The right cerebral hemisphere has a duplicate of the motor cortex in the left cerebral hemisphere. Note that the different muscle groups are not represented equally in the motor cortex. In general, the degree of representation is proportional to the preciseness of movement required of a particular part of the body. For example, the thumb, fingers, lips, tongue, and vocal cords have large representations. The trunk has a relatively small representation. By comparing Figures 15-4 and 15-8, you will see that somatosensory and motor representations are not identical for the same part of the body.

PYRAMIDAL PATHWAYS

Voluntary motor impulses are conveyed from the motor areas of the brain to somatic efferent neurons leading to skeletal muscles via the **pyramidal (pi-RAM-i-dal) pathways.** Most pyramidal fibers originate from cell bodies in the precentral gyrus. They descend through the internal capsule of the cerebrum and most cross to the opposite side of the brain. They terminate in nuclei of

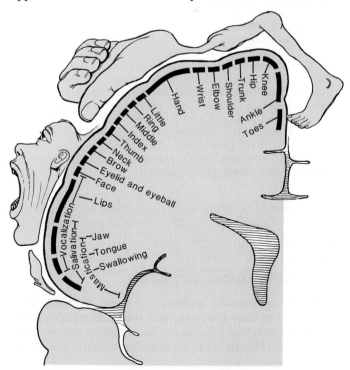

FIGURE 15-8 Motor cortex of the right cerebral hemisphere.

cranial nerves that innervate voluntary muscles or in the anterior gray horn of the spinal cord. A short connecting neuron probably completes the connection of the pyramidal fibers with the motor neurons that activate voluntary muscles.

The pathways over which the impulses travel from the motor cortex to skeletal muscles have two components: **upper motor neurons (pyramidal fibers)** in the brain and **lower motor neurons (peripheral fibers)** in the spinal cord. Here we consider three tracts of the pyramidal system.

1. **Lateral corticospinal tract (pyramidal tract proper).** This tract begins in the motor cortex and descends through the internal capsule of the cerebrum, the cerebral peduncle of the midbrain, and then to the pons on the same side as the point of origin (Figure 15-9). About 85 percent of the upper motor neurons from the motor cortex cross in the medulla. After decussation (crossing), these fibers descend through the spinal cord in the lateral white column in the lateral corticospinal tract. Thus the motor cortex of the right side of the brain controls muscles on the left side of the body and vice versa. Most upper motor neuron fibers of the lateral corticospinal tract synapse with short association neurons in the anterior gray horn of the cord. These then synapse in the anterior gray horn with lower motor neurons that exit all levels of the cord via the ventral roots of spinal nerves. The lower motor neurons terminate in skeletal muscles.

2. **Anterior corticospinal tract.** About 15 percent of the upper motor neurons from the motor cortex do not cross in the medulla. These pass through the medulla and continue to descend on the same side to the anterior white column to become part of the anterior (straight or uncrossed) corticospinal tract. The fibers of these upper motor neurons decussate and synapse with association neurons in the anterior gray horn of the spinal cord on the side opposite the origin of the anterior corticospinal tract. The association neurons in the horn synapse with lower motor neurons that exit the cervical and upper thoracic segments of the cord via the ventral roots of spinal nerves. The lower motor neurons terminate in skeletal muscles that control muscles of the neck and part of the trunk.

3. **Corticobulbar tract.** The fibers of this tract arise from upper motor neurons in the motor cortex. They accompany the corticospinal tracts through the internal capsule to the brain stem, where they decussate and terminate in the nuclei of cranial nerves in the pons and medulla. These cranial nerves include the oculomotor (III), trochlear (IV), trigeminal (V), abducens (VI), facial (VII), glossopharyngeal (IX), vagus (X), accessory (XI), and hypoglossal (XII). The corticobulbar tract conveys impulses that largely control voluntary movements of the head and neck.

The various tracts of the pyramidal system convey impulses from the cortex that result in precise muscular movements.

EXTRAPYRAMIDAL PATHWAYS

The **extrapyramidal pathways** include all descending tracts other than the pyramidal tracts. Generally, these include tracts that begin in the basal ganglia and reticular formation. The main extrapyramidal tracts are as follows.

1. **Rubrospinal tract.** This tract originates in the red nucleus of the midbrain (after receiving fibers from the cerebellum), crosses over to descend in the lateral white column of the oppo-

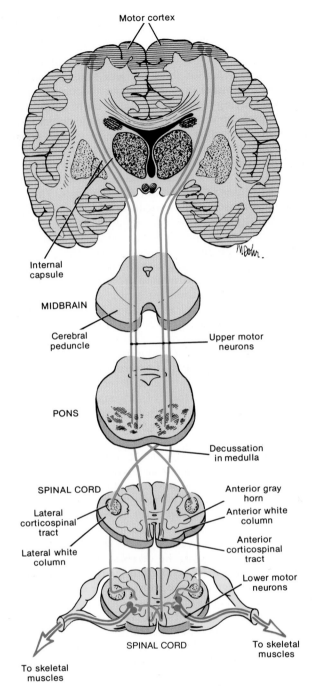

FIGURE 15-9 Pyramidal pathways.

site side, and extends through the entire length of the spinal cord. The tract transmits impulses to skeletal muscles concerned with tone and posture.

2. **Tectospinal tract.** This tract originates in the superior colliculus of the midbrain, crosses to the opposite side, descends in the anterior white column, and enters the anterior gray horns in the cervical segments of the cord. Its function is to transmit impulses that control movements of the head in response to visual stimuli.

3. **Vestibulospinal tract.** This tract originates in the vestibular nucleus of the medulla, descends on the same side of the cord in the anterior white column, and terminates in the anterior gray horns, mostly in the cervical and lumbosacral segments

of the cord. It conveys impulses that regulate muscle tone in response to movements of the head. This tract, therefore, plays a major role in equilibrium.

Only one motor neuron carries the impulse from the cerebral cortex to the cranial nerve nuclei or spinal cord: an **upper motor neuron.** Only one motor neuron in the pathway actually terminates in a skeletal muscle: the **lower motor neuron.** This neuron, a somatic efferent neuron, always extends from the central nervous system to the skeletal muscle. Since it is the final transmitting neuron in the pathway, it is also called the **final common pathway.**

CLINICAL APPLICATION

The lower motor and upper motor neurons are important clinically. If the lower motor neuron is damaged or diseased, there is neither voluntary nor reflex action of the muscle it innervates, and the muscle remains in a relaxed state, a condition called **flaccid paralysis.** Injury or disease of upper motor neurons in a motor pathway is characterized by varying degrees of continued contraction of the muscle, a condition called **spasticity,** and exaggerated reflexes. Another characteristic is the **Babinski sign,** which is dorsiflexion of the great toe accompanied by fanning of the lateral toes in response to stroking the plantar surface along the outer border of the foot. As we have seen, this response is normal only in infants; the normal adult response is plantar flexion of the toes.

Lower motor neurons are subjected to stimulation by many other presynaptic neurons. Some signals are excitatory; others are inhibitory. The algebraic sum of the opposing signals determines the final response of the lower motor neuron. It is not just a simple matter of the brain sending an impulse and the muscle always contracting.

Association neurons are of considerable importance in the motor pathways. Impulses from the brain are conveyed to association neurons before being received by lower motor neurons. These association neurons integrate the pattern of muscle contraction.

The basal ganglia have many connections with other parts of the brain. Through these connections, they help to control subconscious movements. The caudate nucleus controls gross intentional movements. The caudate nucleus and putamen, together with the cerebral cortex, control patterns of movement. The globus pallidus controls positioning of the body for performing a complex movement. The subthalamic nucleus is thought to control walking and possibly rhythmic movements. Many potential functions of the basal ganglia are held in check by the cerebrum. Thus if the cerebral cortex is damaged early in life, a person can still perform many gross muscular movements.

The role of the cerebellum is significant also. The cerebellum is connected to other parts of the brain that are concerned with movement. The vestibulocerebellar tract transmits impulses from the equilibrium apparatus in the ear to the cerebellum. The olivocerebellar tract transmits impulses from the basal ganglia to the cerebellum. The corticopontocerebellar tract conveys impulses from the cerebrum to the cerebellum. The spinocerebellar tracts relay proprioceptive information to the cerebellum. Thus the cerebellum receives considerable information regarding the overall physical status of the body. Using this information, the cerebellum generates impulses that integrate body responses.

Take tennis, for example. To make a good serve, you must bring your racket forward just far enough to make solid contact. How do you stop at the exact point without swinging too far? This is where the cerebellum comes in. It receives information about your body status while you are serving. Before you even hit the ball, the cerebellum had already sent information to the cerebral cortex and basal ganglia informing them that your swing must stop at an exact point. In response to cerebellar stimulation, the cortex and basal ganglia transmit motor impulses to your opposing body muscles to stop the swing. The cerebellar function of stopping overshoot when you want to zero in on a target is called its **damping function.** The cerebellum also helps you to coordinate different body parts while walking, running, and swimming. Finally, the cerebellum helps you maintain equilibrium.

INTEGRATIVE FUNCTIONS

We turn now to a fascinating, though poorly understood, function of the cerebrum: integration. The **integrative functions** include cerebral activities such as memory, sleep and wakefulness, and emotional response. The role of the limbic system in emotional behavior was discussed in Chapter 14.

MEMORY

Without a memory we would repeat mistakes and be unable to learn. Similarly, we would not be able to repeat our successes or accomplishments, except by chance. Although both memory and learning have been studied by scientists for many years, there is still no satisfactory explanation for how we recall information or how we remember. Some things, however, are known about how information is acquired and stored.

Learning is the ability to acquire knowledge or a skill through instruction or experience. Learning is very closely associated with rewards and punishments. **Memory** is the ability to recall thoughts. For an experience to become part of memory, it must produce changes in the CNS that represent the experience. Such a memory trace in the brain is called an **engram.** The portions of the brain thought to be associated with memory include the association cortex of the frontal, parietal, occipital, and temporal lobes; parts of the limbic system, especially the hippocampus and amygdaloid nucleus; and thalamus.

Memory may generally be classified into two kinds based on how long the memory lasts: activated and long-term memory. **Activated (short-term) memory** lasts only seconds or hours and is the ability to recall bits of information. One example is finding an unfamiliar telephone number in a telephone book and then dialing it. If the number has no special significance, it is usually forgotten within a few seconds. **Long-term memory,** on the other hand, lasts from days to years. For example, if you frequently use a telephone number for any period of time, such as your home telephone number, it becomes part of long-term memory. When the number is in long-term memory it can be retrieved for use whenever needed for quite a long period of time. Such reinforcement due to the frequent use of the telephone number is called **memory consolidation.**

As stated earlier in the chapter, the human brain receives stimuli from many places, but only a small percentage ever comes to our conscious realization. It has been estimated that, of all the information that comes to our conscious attention, only about 1 percent goes into long-term memory, and much of what goes into long-term memory is forgotten. It is fortunate that our brains select only a small percentage of our thoughts and lose much of what we stored, otherwise our brain would be overwhelmed with information.

It is a feature of memory that we can remember short lists easier than long ones. This is to say the obvious, but human memory does not record everything like an endless magnetic tape. It cannot remember or record long lists of details. Another feature of memory is that, even when details are lost, the concept or main idea is retained. Then, interestingly, we can often explain the idea or concept—not like replaying a tape—but with our own selection of words and ways of explanation.

Despite several decades of research, it is still impossible to explain the mechanisms of memory. However, by integrating several clinical and experimental observations, scientists have developed several theories to help us understand the mechanisms.

One theory of activated memory states that memories may be caused by reverberating neuronal circuits—an incoming impulse stimulates the first neuron, which stimulates the second, which stimulates the third, and so on (see Figure 12-13c). Branches from the second and third neurons synapse with the first, sending the impulse back through the circuit again and again. Thus the output neuron generates continuous impulses. Once fired, the output signal may last from a few seconds to many hours, depending on the arrangement of neurons in the circuit. If this pattern is applied to activated memory, an incoming thought—the phone number—continues in the brain even after the initial stimulus is gone. Thus you can recall the thought only for as long as the reverberation continues.

There is some evidence to support the concept that activated memory is related to electrical and chemical events rather than structural changes in the brain. For example, several conditions may inhibit the electrical activity of the brain. These include anesthesia, coma, electroconvulsive shock, and ischemia (reduced blood supply) of the brain. Although such states do interfere with the retention of recently acquired information, they do not usually interfere with long-term memory laid down prior to the interference with electrical activity. It is a common experience that individuals who suffer retrograde amnesia cannot remember any events that have occurred for about

30 minutes before the amnesia developed. But, as the person recovers from the amnesia, the most recent memories return last; loss of consciousness in no way affects memories learned before the amnesia.

Most research on long-term memory focuses on anatomical or biochemical changes at synapses that might enhance facilitation at synapses. Anatomical changes occur in neurons when they are either stimulated or made inactive. For example, electron micrographic studies of presynaptic neurons that have been subjected to prolonged, intense activity exhibit several anatomical changes. These include an increase in the number of presynaptic terminals, enlargement of synaptic end bulbs, and an increase in the branching patterns and conductance of dendrites. There is also an increase in the number of neuroglia. Moreover, neurons grow new synaptic end bulbs with increasing age, presumably as a result of increased utilization. These changes, which are correlated with faster learning, suggest enhancement of facilitation at synapses. Such changes do not occur when neurons are inactive. In fact, in animals that have lost their eyesight, there is thinning of the cerebral cortex in the visual area.

Any of the events in the synaptic transmission could be responsible for enhanced communication between neurons. There is experimental evidence that repetitive use of a synapse causes the presynaptic membrane to hyperpolarize so that, when the action potential passes over the synaptic end bulb, its amplitude is much greater. The greater the amplitude, the greater the amount of neurotransmitter released. To further enhance this possibility, the neurotransmitter availability in the presynaptic cell increases greatly and, therefore, can be released with each action potential. Other suggestions include increase in the number of receptor molecules in the postsynaptic cell membrane or a decrease in the rate of removal of neurotransmitter substance.

Finally, there is considerable interest in the possible involvement of nucleic acids in long-term memory. The molecules DNA and RNA store information, and these molecules, especially DNA, tend to persist for the lifetime of the cell. Studies have shown an increase in the RNA content of activated neurons. Conversely, there is some evidence that shows that long-term memory will not occur to any significant extent when RNA formation is inhibited.

WAKEFULNESS AND SLEEP

Humans sleep and awaken in a fairly constant 24-hour rhythm called a **circadian** (ser-KĀ-dē-an) **rhythm.** Since neuronal fatigue precedes sleep and the signs of fatigue disappear after sleep, fatigue is apparently one cause of sleep. Moreover, EEG recordings indicate that during wakefulness the cerebral cortex is very active, sending impulses continuously through the body. During sleep, however, fewer impulses are transmitted by the cerebral cortex. The activity of the cerebral cortex is thought to be related to the reticular formation.

The reticular formation has numerous connections with the cerebral cortex. Stimulation of portions of the reticular formation results in increased cortical activity.

Thus the reticular formation is also known as the **reticular activating system (RAS).** One part of the system, the mesencephalic part, is composed of areas of gray matter of the pons and midbrain. When this area is stimulated, many impulses pass upward into the thalamus and disperse to widespread areas of the cerebral cortex. The effect is a generalized increase in cortical activity. The other part of the RAS, the thalamic part, consists of gray matter in the thalamus. When the thalamic part is stimulated, signals from specific parts of the thalamus cause activity in specific parts of the cerebral cortex. Apparently the mesencephalic part of the RAS causes general wakefulness (consciousness) and the thalamic part causes **arousal,** that is, awakening from deep sleep.

For arousal to occur, the RAS must be stimulated by input signals. Almost any sensory input can activate the RAS: pain stimuli, proprioceptive signals, bright light, or an alarm clock. Once the RAS is activated, the cerebral cortex is also activated and you experience arousal. Signals from the cerebral cortex can also stimulate the RAS. Such signals may originate in the somesthetic cortex, the motor cortex, or the limbic system. When the signals activate the RAS, the RAS activates the cerebral cortex and arousal occurs.

Following arousal, the RAS and cerebral cortex continue to activate each other through a feedback system consisting of many circuits. The RAS also has a feedback system with the spinal cord that is composed of many circuits. Impulses from the activated RAS are transmitted down the spinal cord and then to skeletal muscles. Muscle activation causes proprioceptors to return impulses that activate the RAS. The two feedback systems maintain activation of the RAS, which in turn maintains activation of the cerebral cortex. The result is a state of wakefulness called **consciousness.** The RAS is the physical basis of consciousness, the brain's chief watchguard. It continuously sifts and selects, forwarding only the essential, unusual, or dangerous to the conscious mind. Since humans experience different levels of consciousness (alertness, attentiveness, relaxation, inattentiveness), it is assumed that the level of consciousness depends on the number of feedback circuits operating at the time. During resting wakefulness alpha waves appear on an EEG.

CLINICAL APPLICATION

Consciousness may be altered by various factors. Amphetamines probably activate the RAS to produce a state of wakefulness and alertness. Meditation produces a lack of consciousness. Anesthetics produce a state of consciousness called anesthesia. Damage to the nervous system, as well as disease, can produce a lack of consciousness called coma. Drugs such as LSD can also alter consciousness.

If this theory of wakefulness is accepted, how then does sleep occur if the activating feedback systems are in continual operation? One explanation is that the feedback system slows tremendously or is inhibited. Inactivation of the RAS produces a state known as **sleep.**

Just as there are different levels of consciousness, there are different levels of sleep. Normal sleep consists of two types: nonrapid eye movement sleep (NREM) and rapid eye movement sleep (REM).

NREM sleep, or **slow wave sleep,** consists of four stages, each of which gradually merges into the next. Each stage has been identified by EEG recordings:

Stage 1. This is a transition stage between waking and sleep that normally lasts from 1 to 7 minutes. The person is relaxing with eyes closed. During this time, respirations are regular, pulse is even, and the person has fleeting thoughts. If awakened, the person will frequently say he has not been sleeping. Alpha waves diminish and theta waves appear on the EEG.

Stage 2. This is the first stage of true sleep, even though the person experiences only light sleep. It is a little harder to awaken the person. Fragments of dreams may be experienced, and the eyes may slowly roll from side to side. The EEG shows *sleep spindles*—sudden, short bursts of sharply pointed waves that occur at 12 to 14 Hz (cycles) per second.

Stage 3. This is a period of moderately deep sleep. The person is very relaxed. Body temperature begins to fall and blood pressure decreases. It is difficult to awaken the person and the EEG shows a mixture of sleep spindles and delta waves. This stage occurs about 20 minutes after falling asleep.

Stage 4. Deep sleep occurs. The person is very relaxed and responds slowly if awakened. Bed-wetting and sleep-walking may occur during this stage. The EEG is dominated by delta waves.

In a typical 7- or 8-hour sleep period, a person goes from stage 1 to 4 of NREM sleep. Then the person ascends to stage 3 and 2 and then to REM sleep within 50 to 90 minutes.

In **REM sleep,** also called **paradoxical sleep,** the EEG readings are similar to those of stage 1 of NREM sleep. There are significant physiological differences, however. During REM sleep, muscle tone is depressed (except for rapid movements of the eyes), and respirations and pulse rate increase and are irregular. Blood pressure also fluctuates considerably. It is during REM sleep that most dreaming occurs. Following REM sleep, the person descends again to stages 3 and 4 of NREM sleep.

REM and NREM sleep alternate throughout the night with approximately 90-minute intervals between REM periods. This cycle repeats itself from 3 to 5 times during the entire sleep period. The REM periods start out lasting from 5 to 10 minutes and gradually lengthen until the final one lasts about 50 minutes. In a normal night's sleep, REM totals 90 to 120 minutes. As much as 50 percent of an infant's sleep is REM, as contrasted with 20 percent for adults. Most sedatives significantly decrease REM sleep.

As a person ages, the average time spent sleeping during the night decreases. In addition, the percentage of REM sleep decreases. It has been suggested that the high percentage of REM sleep in infants and children reflects increased neuronal activity that is important for the maturation of the brain. Infants apparently need this internal stimulation since the available external stimuli are restricted. Support for this idea comes from the fact that dreams, a particular kind of conscious activity in the brain, are most frequent during REM-type sleep.

Recent studies with animals suggest that two specific neural centers in the brain stem determine the occurrence of NREM and REM sleep. The NREM sleep center is found in the raphe nuclei. Its neurons contain large amounts of the neurotransmitter serotonin (5-HT). When the supply of serotonin is exhausted, the result is severe insomnia and a reduction in both NREM and REM sleep. The insomnia can be alleviated by the administration of a precursor of serotonin (5-hydroxytryptophan). Serotonin itself cannot cross the blood–brain barrier. The REM sleep center is found in the locus coeruleus. Its neurons contain large amounts of the neurotransmitter norepinephrine (NE). Destruction of the loci coerulei results in a complete disappearance of REM sleep, but has no influence on NREM sleep. The administration of reserpine, a drug that exhausts the supply of both serotonin and norepinephrine, results in the elimination of both NREM and REM sleep. All of these observations suggest that serotonin is important for NREM sleep, that norepinephrine is important for REM sleep, and that normally REM sleep is possible only if preceded by NREM sleep.

A **polysomograph** (*poly* = many; *somnus* = sleep; *graph* = to write) is an instrument that uses electrodes to record several physiological variables during sleep. Among these variables are brain activity recorded as an electroencephalogram (EEG), eye movements recorded as an electrooculogram (EOG), and muscle tonus recorded as an electromyogram (EMG). These recordings indicate precisely when patients fall asleep, how many wake periods they experience, and the quality and duration of sleep.

CLINICAL APPLICATION

At least three sleep disorders are clinically significant. **Narcolepsy** (NAR-kō-lep-sē; *narke* = numbness, *lepsis* = seizure) is a condition of involuntary attacks of sleep that last about 15 minutes and may occur at almost any time of the day. It is an inability, in the waking state, to inhibit REM sleep. **Insomnia** (in-SOM-nē-a; *in* = not, *somnus* = sleep) consists of difficulty in falling asleep and, usually, frequent awakening. It may be related to specific disorders or secondary factors, both medical and psychiatric. **Hypersomnia** refers to too much sleep, in which individuals may sleep as many as 15 hours or more per day.

STUDY OUTLINE

Sensations (p. 339)

Definition (*p. 339*)
1. Sensation is a state of awareness of external and internal conditions of the body.
2. The prerequisites for a sensation to occur are reception of a stimulus, conversion of the stimulus into a nerve impulse by a receptor, conduction of the impulse to the brain, and translation of the impulse into a sensation by a region of the brain.
3. Each stimulus is capable of causing the membrane of a receptor to depolarize. This is called the generator potential.

Characteristics (*p. 339*)
1. Projection occurs when the brain refers a sensation to the point of stimulation.
2. Adaptation is the loss of sensation even though the stimulus is still applied.
3. An afterimage is the persistence of a sensation even though the stimulus is removed.
4. Modality is the property by which one sensation is distinguished from another.

Classification of Receptors (*p. 340*)
1. According to location, receptors are classified as exteroceptors, visceroceptors, and proprioceptors.
2. On the basis of type of stimulus detected, receptors are classified as mechanoreceptors, thermoreceptors, nociceptors, electromagnetic receptors, and chemoreceptors.
3. In terms of simplicity or complexity, simple receptors are associated with general senses and complex receptors are associated with special senses.

General Senses (p. 340)

Cutaneous Sensations (*p. 340*)
1. Cutaneous sensations include tactile sensations (touch, pressure, vibration), thermoreceptive sensations (heat and cold), and pain. Receptors for these sensations are located in the skin, connective tissues, and the ends of the gastrointestinal tract.
2. Receptors for touch are root hair plexuses, free nerve endings, tactile (Merkel's) discs, corpuscles of touch (Meissner's corpuscles), and type II cutaneous mechanoreceptors (end organs of Ruffini). Receptors for pressure are free nerve endings, type II cutaneous mechanoreceptors, and lamellated (Pacinian) corpuscles. Receptors for vibration are corpuscles of touch and lamellated corpuscles.
3. Pain receptors (nociceptors) are located in nearly every body tissue.
4. Two kinds of pain recognized in the parietal lobe of the cortex are somatic and visceral.
5. Referred pain is felt in the skin near or away from the organ sending pain impulses.
6. Phantom pain is the sensation of pain in a limb that has been amputated.
7. Pain impulses may be inhibited by drugs, surgery, and acupuncture.
8. In acupuncture, enkephalins block the release of substance P and inhibit pain transmission in the brain.

Proprioceptive Sensations (*p. 344*)
1. Receptors located in skeletal muscles, tendons, in and around joints, and the internal ear convey impulses related to muscle tone, movement of body parts, and body position.
2. The receptors include muscle spindles, tendon organs (Golgi tendon organs), joint kinesthetic receptors, and the maculae and cristae.

Levels of Sensation (*p. 345*)
1. Sensory fibers terminating in the lower brain stem bring about far more complex motor reactions than simple spinal reflexes.
2. When sensory impulses reach the lower brain stem, they cause subconscious motor reactions.
3. Sensory impulses that reach the thalamus can be localized crudely in the body.

4. When sensory impulses reach the cerebral cortex, we experience precise localization.

Sensory Pathways (p. 345)

1. Sensory information from all parts of the body terminates in a specific area of the somatosensory cortex.
2. In the posterior column pathway and the spinothalamic pathway there are first-order, second-order, and third-order neurons.
3. The neural pathway for pain and temperature is the lateral spinothalamic pathway.
4. The neural pathway for light touch and pressure is the anterior spinothalamic pathway.
5. The neural pathway for discriminative touch, proprioception, and vibration is the posterior column pathway.
6. The pathways to the cerebellum are the anterior and posterior spinocerebellar tracts.

Motor Pathways (p. 348)

1. Incoming sensory information is added to, subtracted from, or integrated with other information arriving from all other operating sensory receptors.
2. The integration process occurs at many stations along the pathways of the central nervous system such as within the spinal cord, brain stem, cerebellum, and cerebral cortex.
3. A motor response to make a muscle contract or a gland secrete can be initiated at any of these stations or levels.
4. As the location of the sensory-motor linkage climbs to higher levels in the central nervous system, additional contributions are introduced that enrich the growing inventory of motor responses.
5. When the input reaches the highest center, the process of sensory-motor integration occurs. This involves not only the utilization of information contained within that center but also the information impinging on that center that is delivered from other centers within the central nervous system.
6. The muscles of all parts of the body are controlled by a specific area of the motor cortex.
7. Voluntary motor impulses are conveyed from the brain through the spinal cord along the pyramidal pathways and the extrapyramidal pathways.
8. Pyramidal pathways include the lateral corticospinal, anterior corticospinal, and corticobulbar tracts.
9. Major extrapyramidal tracts are the rubrospinal, tectospinal, and vestibulospinal tracts.

Integrative Functions (p. 351)

1. Memory is the ability to recall thoughts and is generally classified into two kinds: activated (short-term) and long-term memory.
2. A memory trace in the brain is called an engram.
3. Activated memory is related to electrical and chemical events; long-term memory is related to anatomical and biochemical changes at synapses.
4. Sleep and wakefulness are integrative functions that are controlled by the reticular activating system (RAS).
5. Nonrapid eye movement (NREM) sleep consists of four stages identified by EEG recordings.
6. Most dreaming occurs during rapid eye movement (REM) sleep.
7. The neurotransmitters that affect sleep are serotonin (5-HT) and norepinephrine (NE).
8. A polysomograph is an instrument that measures several physiological variables during sleep.

REVIEW QUESTIONS

1. Define a sensation and a sense receptor. What prerequisites are necessary for the perception of a sensation?
2. Describe the following characteristics of a sensation: projection, adaptation, afterimage, modality.
3. Name some examples of adaptation not discussed in the text.
4. Classify receptors on the basis of location, stimulus detected, and simplicity or complexity.
5. Distinguish between a general sense and a special sense.
6. What is a cutaneous sensation? Distinguish tactile, thermoreceptive, and pain sensations.
7. How are cutaneous receptors distributed over the body? Relate your response to the two-point discrimination test.
8. For each of the following cutaneous sensations, describe the receptor involved in terms of structure, function, and location: touch, pressure, vibration, and pain.
9. How do cutaneous sensations help maintain homeostasis?
10. Why are pain receptors important? Differentiate somatic pain, visceral pain, referred pain, and phantom pain.
11. Why is the concept of referred pain useful to the physician in diagnosing internal disorders?
12. What is acupuncture? Describe how acupuncture is believed to relieve pain.
13. How is acupuncture currently being used in the United States?
14. What is the proprioceptive sense? Where are the receptors for this sense located?
15. Describe the structure of muscle spindles, tendon organs (Golgi tendon organs), and joint kinesthetic receptors.
16. Relate proprioception to the maintenance of homeostasis.
17. Describe the various levels of sensation in the central nervous system.
18. Describe how various parts of the body are represented in the somatosensory cortex.
19. Distinguish between the posterior column and spinothalamic pathways.
20. What is the sensory pathway for pain and temperature? How does it function?
21. Which pathway controls light touch and pressure? How does it function?
22. Which pathway is responsible for discriminative touch, proprioception, and vibration? How does it function?
23. Describe how sensory input and motor responses are linked in the central nervous system.
24. Describe how various parts of the body are represented in the motor cortex.
25. Which pathways control voluntary motor impulses from the brain through the spinal cord? How are they distinguished?
26. Define memory. What are the two kinds of memory?
27. Define an engram and memory consolidation.
28. Describe the proposed mechanism of memory.
29. Describe how sleep and wakefulness are related to the reticular activating system (RAS).
30. What are the four stages of nonrapid eye movement (NREM) sleep? How is NREM sleep distinguished from rapid eye movement (REM) sleep?
31. Explain the roles of serotonin (5-HT) and nonrepinephrine (NE) in sleep.
32. Define narcolepsy, insomnia, and hypersomnia.

The Autonomic Nervous System

STUDENT OBJECTIVES

- Compare the structural and functional differences between the somatic efferent and autonomic portions of the nervous system.
- Identify the principal structural features of the autonomic nervous system.
- Compare the sympathetic and parasympathetic divisions of the autonomic nervous system in terms of structure, physiology, and neurotransmitters released.
- Describe the various postsynaptic receptors involved in autonomic responses.
- Describe a visceral autonomic reflex and its components.
- Explain the role of the hypothalamus and its relationship to the sympathetic and parasympathetic divisions.
- Explain the relationship between biofeedback and the autonomic nervous system.
- Describe the relationship between meditation and the autonomic nervous system.

The portion of the nervous system that regulates the activities of smooth muscle, cardiac muscle, and glands is the **autonomic nervous system (ANS).** Structurally, the system consists of visceral efferent neurons organized into nerves, ganglia, and plexuses. Functionally, it usually operates without conscious control. The system was originally named *autonomic* because physiologists thought it functioned with no control from the central nervous system, that it was autonomous or self-governing. It is now known that the autonomic system is neither structurally nor functionally independent of the central nervous system. It is regulated by centers in the brain, in particular by the cerebral cortex, hypothalamus, and medulla oblongata. However, the old terminology has been retained, and since the autonomic nervous system does differ from the somatic nervous system in some ways, the two are separated for convenience of study.

SOMATIC EFFERENT AND AUTONOMIC NERVOUS SYSTEMS

Whereas the somatic efferent nervous system produces conscious movement in skeletal muscles, the autonomic nervous system (visceral efferent nervous system) regulates visceral activities, and it generally does so involuntarily and automatically. Examples of visceral activities regulated by the autonomic nervous system are changes in the size of the pupil, accommodation for near vision, dilation and constriction of blood vessels, adjustment of the rate and force of the heartbeat, movements of the gastrointestinal tract, and secretion by most glands. These activities usually lie beyond conscious control. They are automatic.

The autonomic nervous system is generally considered to be entirely motor. All its axons are efferent fibers, which transmit impulses from the central nervous system to visceral effectors. Autonomic fibers are called **visceral efferent fibers. Visceral effectors** include cardiac muscle, smooth muscle, and glandular epithelium. This does not mean there are no afferent (sensory) impulses from visceral effectors, however. Impulses that give rise to visceral sensations pass over visceral afferent neurons that have cell bodies located in the posterior (dorsal) root ganglia of spinal nerves. Some functions of these afferent neurons were described in discussing the cranial and spinal nerves. The hypothalamus, which largely controls the autonomic nervous system, also receives impulses from the visceral sensory fibers.

In the motor portion of the neural pathway of the somatic efferent system, an efferent neuron runs from the central nervous system and synapses directly on a skeletal muscle. In the neural pathway of the autonomic nervous system, there are two efferent neurons and a ganglion between them. The first neuron runs from the central nervous system to a ganglion, where it synapses with the second efferent neuron. It is this neuron that ultimately synapses on a visceral effector. Also, whereas fibers of somatic efferent neurons release acetylcholine (ACh) as their neurotransmitter, fibers of autonomic efferent neurons release either ACh or norepinephrine (NE).

The autonomic nervous system consists of two principal divisions: the **sympathetic** and the **parasympathetic.** Many organs innervated by the autonomic nervous system receive visceral efferent neurons from both components of the autonomic system—one set from the sympathetic division, another from the parasympathetic division. In general, impulses transmitted by the fibers of one division stimulate the organ to start or increase activity, whereas impulses from the other division decrease the organ's activity. Organs that receive impulses from both sympathetic and parasympathetic fibers are said to have *dual innervation.* Thus autonomic innervation may be excitatory or inhibitory. In the somatic efferent nervous system, only one kind of motor neuron innervates an organ, which is always a skeletal muscle. Moreover, innervation is always excitatory. When a somatic neuron stimulates a skeletal muscle, the muscle becomes active. When the neuron ceases to stimulate the muscle, contraction stops altogether.

A summary of the principal differences between the somatic efferent and autonomic nervous systems is presented in Exhibit 16-1.

EXHIBIT 16-1 COMPARISON OF SOMATIC EFFERENT AND AUTONOMIC NERVOUS SYSTEMS

	SOMATIC EFFERENT	AUTONOMIC
Effectors	Skeletal muscles.	Cardiac muscle, smooth muscle, glandular epithelium.
Type of control	Voluntary.	Involuntary.
Neural pathway	One efferent neuron extends from CNS and synapses directly on a skeletal muscle.	One efferent neuron extends from the CNS and synapses with another efferent neuron in a ganglion; the second neuron synapses on a visceral effector.
Action on effector	Always excitatory.	May be excitatory or inhibitory, depending on whether stimulation is sympathetic or parasympathetic.
Neurotransmitters	Acetylcholine (ACh).	Acetylcholine (ACh) or norepinephrine (NE).

STRUCTURE OF THE AUTONOMIC NERVOUS SYSTEM

VISCERAL EFFERENT PATHWAYS

Autonomic visceral efferent pathways always consist of two neurons. One extends from the central nervous system to a ganglion. The other extends directly from the ganglion to the effector (muscle or gland).

The first of the visceral efferent neurons in an autonomic pathway is called a **preganglionic neuron** (Figure 16-1). Its cell body is in the brain or spinal cord. Its myelinated axon, called a **preganglionic fiber,** passes out of the central nervous system as part of a cranial or spinal nerve. At some point, the fiber separates from the nerve and courses to an autonomic ganglion, where it synapses with the dendrites or cell body of the postganglionic neuron, the second neuron in the visceral efferent pathway.

The **postganglionic neuron** lies entirely outside the central nervous system. Its cell body and dendrites (if it has dendrites) are located in the autonomic ganglion, where the synapse with one or more preganglionic fibers occurs. The axon of a postganglionic neuron, called a **postganglionic fiber,** is unmyelinated and terminates in a visceral effector.

Thus preganglionic neurons convey efferent impulses from the central nervous system to autonomic ganglia. Postganglionic neurons relay the impulses from autonomic ganglia to visceral effectors.

Preganglionic Neurons

In the sympathetic division, the preganglionic neurons have their cell bodies in the lateral gray horns of the twelve thoracic segments and first two or three lumbar segments of the spinal cord (Figure 16-2). It is for this reason that the sympathetic division is also called the **thoracolumbar** (thō'-ra-kō-LUM-bar) **division** and the fibers of the sympathetic preganglionic neurons are known as the **thoracolumbar outflow.**

The cell bodies of the preganglionic neurons of the parasympathetic division are located in the nuclei of cranial nerves III, VII, IX, and X in the brain stem and in the lateral gray horns of the second through fourth sacral segments of the spinal cord. Hence the parasympathetic division is also known as the **craniosacral division,** and the fibers of the parasympathetic preganglionic neurons are referred to as the **craniosacral outflow.**

Autonomic Ganglia

Autonomic pathways always include **autonomic ganglia,** where synapses between visceral efferent neurons occur. Autonomic ganglia differ from posterior root ganglia. The latter contain cell bodies of sensory neurons and no synapses occur in them.

The autonomic ganglia may be divided into three general groups. The **sympathetic trunk (vertebral chain) ganglia** are a series of ganglia that lie in a vertical row on either side of the vertebral column, extending from the base of the skull to the coccyx (Figure 16-3). They are also known as **paravertebral (lateral) ganglia.** They receive preganglionic fibers only from the thoracolumbar (sympathetic) division (Figure 16-2). Because of this, sympathetic preganglionic fibers tend to be short.

The second kind of autonomic ganglion also belongs to the sympathetic division. It is called a **prevertebral (collateral) ganglion** (Figure 16-3). The ganglia of this group lie anterior to the spinal column and close to the large abdominal arteries from which their names are derived. Examples of prevertebral ganglia so named are the celiac ganglion, on either side of the celiac artery just below the diaphragm; the superior mesenteric ganglion, near the beginning of the superior mensentric artery in the upper abdomen; and the inferior mesenteric ganglion, located near the beginning of the inferior mesenteric artery in the middle of the abdomen (Figure 16-2). Prevertebral ganglia receive preganglionic fibers from the thoracolumbar (sympathetic) division.

The third kind of autonomic ganglion belongs to the parasympathetic division and is called a **terminal (intramural) ganglion.** The ganglia of this group are located at the end of a visceral efferent pathway very close to visceral effectors or actually within the walls of visceral

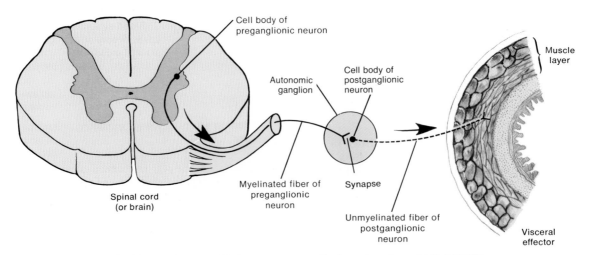

FIGURE 16-1 Relationship between preganglionic and postganglionic neurons.

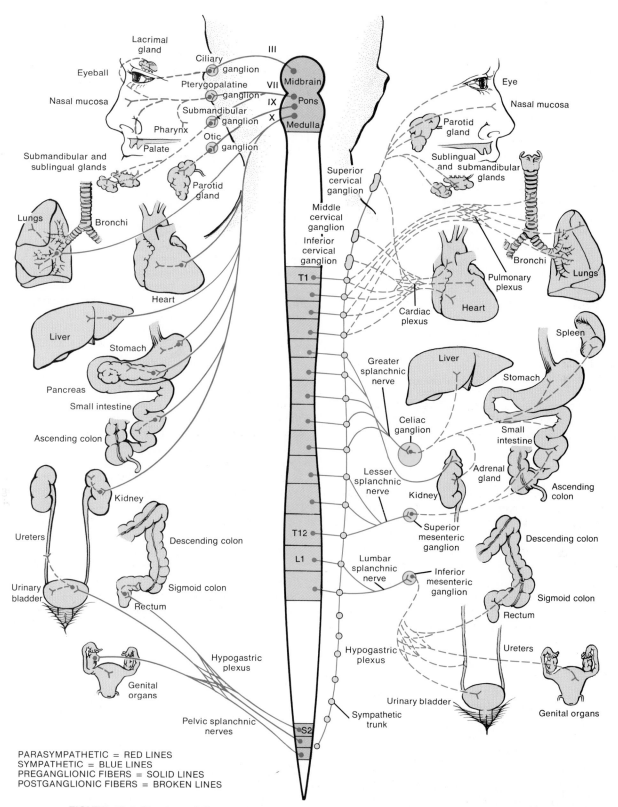

PARASYMPATHETIC = RED LINES
SYMPATHETIC = BLUE LINES
PREGANGLIONIC FIBERS = SOLID LINES
POSTGANGLIONIC FIBERS = BROKEN LINES

FIGURE 16-2 Structure of the autonomic nervous system. Although the parasympathetic division is shown only on the left side of the figure and the sympathetic division is shown only on the right side, keep in mind that each division is actually on both sides of the body (bilateral symmetry).

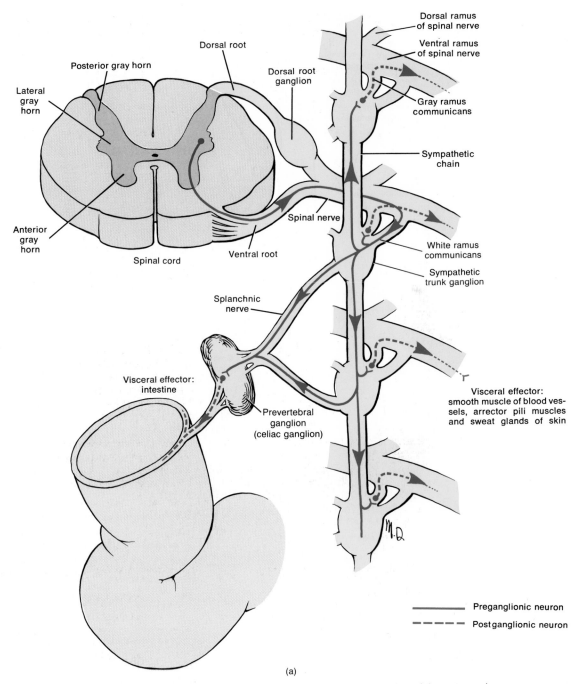

FIGURE 16-3 Ganglia and rami communicantes of the sympathetic division of the autonomic nervous system. (a) Diagram.

effectors. Terminal ganglia receive preganglionic fibers from the craniosacral (parasympathetic) division. The preganglionic fibers do not pass through sympathetic trunk ganglia (Figure 16-2). Because of this, parasympathetic preganglionic fibers tend to be long.

In addition to autonomic ganglia, the autonomic nervous system contains **autonomic plexuses.** Slender nerve fibers from ganglia containing postganglionic nerve cell bodies arranged in a branching network constitute an autonomic plexus.

Postganglionic Neurons

Axons from preganglionic neurons of the sympathetic division pass to ganglia of the sympathetic trunk. They can either synapse in the sympathetic chain ganglia with postganglionic neurons or they can continue, without synapsing, through the chain ganglia to end at a perverte-bral ganglion where synapses with the postganglionic neurons can take place. Each sympathetic preganglionic fiber synapses with several postganglionic fibers in the gan-

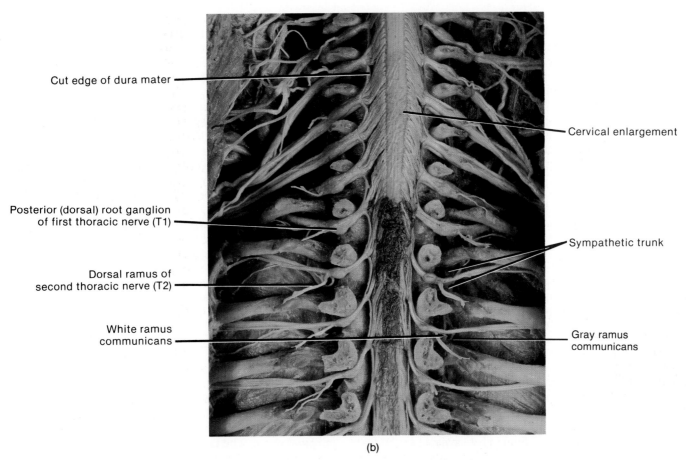

Cut edge of dura mater

Cervical enlargement

Posterior (dorsal) root ganglion of first thoracic nerve (T1)

Sympathetic trunk

Dorsal ramus of second thoracic nerve (T2)

White ramus communicans

Gray ramus communicans

(b)

FIGURE 16-3 (*Continued*) (b) Photograph of the posterior aspect of the lower cervical and upper thoracic segments of the spinal cord. (Courtesy of N. Gluhbegovic and T. H. Williams, *The Human Brain: A Photographic Guide,* Harper & Row, Publishers, Inc., Hagerstown, MD, 1980.)

glion, and the postganglionic fibers pass to several visceral effectors. After exiting their ganglia, the postsynaptic fibers innervate their visceral effectors.

Axons from preganglionic neurons of the parasympathetic division pass to terminal ganglia near or within a visceral effector. In the ganglion, the presynaptic neuron usually synapses with only four or five postsynaptic neurons to a single visceral effector. After exiting their ganglia, the postsynaptic fibers supply their visceral effectors.

With this background in mind, we can now examine some specific structural features of the sympathetic and parasympathetic divisions of the autonomic nervous system.

SYMPATHETIC DIVISION

The preganglionic fibers of the sympathetic division have their cell bodies located in the lateral gray horns of all the thoracic segments and first two or three lumbar segments of the spinal cord (Figure 16-2). The preganglionic fibers are myelinated and leave the spinal cord through the ventral root of a spinal nerve along with the somatic efferent fibers at the same segmental levels. After exiting through the intervertebral foramina, the preganglionic sympathetic fibers enter a white ramus to pass to the nearest sympathetic trunk ganglion on the same side. Collectively, the white rami are called the **white rami communicantes** (kō-myoo-ne-KAN-tēz). Their name indicates that they contain myelinated fibers. Only thoracic and upper lumbar nerves have white rami communicantes. The white rami communicantes connect the ventral ramus of the spinal nerve with the ganglia of the sympathetic trunk.

The paired sympathetic trunks are situated anterolaterally to the spinal cord, one on either side. Each consists of a series of ganglia arranged more or less segmentally. The divisions of the sympathetic trunk are named on the basis of location. Typically, there are 22 ganglia in each chain: 3 cervical, 11 thoracic, 4 lumbar, and 4 sacral. Although the trunk extends downward from the neck, thorax, and abdomen to the coccyx, it receives preganglionic fibers only from the thoracic and lumbar segments of the spinal cord (Figure 16-2).

The cervical portion of each sympathetic trunk is located in the neck anterior to the prevertebral muscles. It is subdivided into a superior, middle, and inferior ganglion (Figure 16-2). The *superior cervical ganglion* is posterior to the internal carotid artery and anterior to the transverse processes of the second cervical vertebra. Postganglionic fibers leaving the ganglion serve the head, where they are distributed to the sweat glands, the smooth muscle of the eye and blood vessels of the face, the nasal mucosa, and the submandibular, sublingual, and parotid salivary glands. Gray rami communicantes (described shortly) from the ganglion also pass to the upper two to four cervical spinal nerves. The *middle cervical ganglion* is situated near the sixth cervical vertebra at the level of the cricoid cartilage. Postganglionic fibers from it innervate the heart. The *inferior cervical ganglion* is located near the first rib, anterior to the transverse processes of the seventh cervical vertebra. Its postganglionic fibers also supply the heart.

The thoracic portion of each sympathetic trunk usually consists of 11 segmentally arranged ganglia, lying ventral to the necks of the corresponding ribs. This portion of the sympathetic trunk receives most of the sympathetic preganglionic fibers. Postganglionic fibers from the thoracic sympathetic trunk innervate the heart, lungs, bronchi, and other thoracic viscera.

The lumbar portion of each sympathetic trunk is found on either side of the corresponding lumbar vertebrae. The sacral portion of the sympathetic trunk lies in the pelvic cavity on the medial side of the sacral foramina. Postganglionic fibers from the lumbar and sacral sympathetic chain ganglia are distributed with the respective spinal nerves via gray rami, or they may join the hypogastric plexus via direct visceral branches.

When a preganglionic fiber of a white ramus communicans enters the sympathetic trunk, it may terminate (synapse) in several ways. Some fibers synapse in the first ganglion at the level of entry. Others pass up or down the sympathetic trunk for a variable distance to form the fibers on which the ganglia are strung. These fibers, known as **sympathetic chains** (Figure 16-3), may not synapse until they reach a ganglion in the cervical or sacral area. Most rejoin the spinal nerves before supplying peripheral visceral effectors such as sweat glands and the smooth muscle in blood vessels and around hair follicles in the extremities. The **gray ramus communicans** (kō-MYOO-ni-kanz) is the structure containing the postganglionic fibers that connect the ganglion of the sympathetic trunk to the spinal nerve (Figure 16-3). The fibers are unmyelinated. All spinal nerves have gray rami communicantes. Gray rami communicantes outnumber the white rami, since there is a gray ramus leading to each of the 31 pairs of spinal nerves.

In most cases, a sympathetic preganglionic fiber terminates by synapsing with a large number, usually 20 or more, of postganglionic cell bodies in a ganglion. Often the postganglionic fibers then terminate in widely separated organs of the body. Thus an impulse that starts in a single preganglionic neuron may reach several visceral effectors. For this reason, most sympathetic responses have widespread effects on the body.

Some preganglionic fibers pass through the sympathetic trunk without terminating in the trunk. Beyond the trunk, they form nerves known as **splanchnic** (SPLANK-nik) **nerves** (Figure 16-2). After passing through the trunk of ganglia, the splanchnic nerves from the thoracic area terminate in the *celiac* (SĒ-lē-ak) or *solar plexus*. In the plexus, the preganglionic fibers synapse in ganglia with postganglionic cell bodies. These ganglia are prevertebral ganglia. The greater splanchnic nerve passes to the celiac ganglion of the celiac plexus. From here, postganglionic fibers are distributed to the stomach, spleen, liver, kidney, and small intestine. The lesser splanchnic nerve passes through the celiac plexus to the superior mesenteric ganglion of the superior mesenteric plexus. Postganglionic fibers from this ganglion innervate the small intestine and colon. The lowest splanchnic nerve, not always present, enters the renal plexus. Postganglionics supply the renal artery and ureter. The lumbar splanchnic nerve enters the inferior mesenteric plexus. In the plexus, the preganglionic fibers synapse with postganglionic fibers in the inferior mesenteric ganglion. These fibers pass through the hypogastric plexus and supply the distal colon and rectum, urinary bladder, and genital organs. As noted earlier, the postganglionic fibers leaving the prevertebral ganglia follow the course of various arteries to abdominal and pelvic visceral effectors.

PARASYMPATHETIC DIVISION

The preganglionic cell bodies of the parasympathetic division are found in nuclei in the brain stem and the lateral gray horn of the second through fourth sacral segments of the spinal cord (Figure 16-2). Their fibers emerge as part of a cranial nerve or as part of the ventral root of a spinal nerve. The **cranial parasympathetic outflow** consists of preganglionic fibers that leave the brain stem by way of the oculomotor (III) nerves, facial (VII) nerves, glossopharyngeal (IX) nerves, and vagus (X) nerves. The **sacral parasympathetic outflow** consists of preganglionic fibers that leave the ventral roots of the second through fourth sacral nerves. The preganglionic fibers of both the cranial and sacral outflows end in terminal ganglia, where they synapse with postganglionic neurons. We will first look at the cranial outflow.

The cranial outflow has five components: four pairs of ganglia and the plexuses associated with the vagus nerve. The four pairs of cranial parasympathetic ganglia innervate structures in the head and are located close to the organs they innervate. The *ciliary ganglion* is near the back of an orbit lateral to each optic (II) nerve. Preganglionic fibers pass with the oculomotor (III) nerve to the ciliary ganglion. Postganglionic fibers from the ganglion innervate smooth muscle cells in the eyeball. Each *pterygopalatine* (ter'-i-gō-PAL-a-tin) *ganglion* is situated lateral to a sphenopalatine foramen. It receives preganglionic fibers from the facial (VII) nerve and transmits postganglionic fibers to the nasal mucosa, palate, pharynx, and lacrimal gland. Each *submandibular ganglion* is found near the duct of a submandibular salivary gland. It receives preganglionic fibers from the facial (VII) nerve and transmits postganglionic fibers that innervate the submandibular and sublingual salivary glands. The *otic ganglia* are situated just below each foramen ovale. The otic ganglion receives preganglionic fibers from the glossopharyngeal (IX) nerve and transmits postganglionic fibers that innervate the parotid salivary gland. Ganglia associated with the cranial outflow are classifed as terminal ganglia. Since the terminal ganglia are close to their visceral effectors, postganglionic parasympathetic fibers are short. Postganglionic sympathetic fibers are relatively long.

The last component of the cranial outflow consists of the preganglionic fibers that leave the brain via the vagus (X) nerves. This component has the most extensive distribution of the parasympathetic fibers, providing about 80 percent of the craniosacral outflow. Each vagus (X) nerve enters into the formation of several plexuses in the thorax and abdomen. As it passes through the thorax, it sends fibers to the *superficial cardiac plexus* in the arch of the aorta and the *deep cardiac plexus* anterior to the branching of the trachea. These plexuses contain terminal ganglia, and the postganglionic parasympathetic fibers emerging from them supply the heart. Also in the thorax is the *pulmonary plexus,* in front of and behind the roots of the lungs and within the lungs themselves. It receives preganglionic fibers from the vagus and transmits postganglionic parasympathetic fibers to the lungs and bronchi. Other plexuses associated with the vagus (X) nerve are described in later chapters in conjunction with the appropriate thoracic, abdominal, and pelvic viscera. Postganglionic fibers from these plexuses innervate viscera such as the liver, pancreas, stomach, kidneys, small intestine, and part of the colon.

The sacral parasympathetic outflow consists of preganglionic fibers from the ventral roots of the second through fourth sacral nerves. Collectively, they form the *pelvic splanchnic nerves.* They pass into the hypogastric plexus. From ganglia in the plexus, parasympathetic postganglionic fibers are distributed to the colon, ureters, urinary bladder, and reproductive organs.

The salient structural features of the sympathetic and parasympathetic divisions are compared in Exhibit 16-2.

EXHIBIT 16-2   STRUCTURAL FEATURES OF SYMPATHETIC AND PARASYMPATHETIC DIVISIONS

SYMPATHETIC	PARASYMPATHETIC
Forms thoracolumbar outflow.	Forms craniosacral outflow.
Contains sympathetic trunk and prevertebral ganglia.	Contains terminal ganglia.
Ganglia are close to the CNS and distant from visceral effectors.	Ganglia are near or within visceral effectors.
Each preganglionic fiber synapses with many postganglionic neurons that pass to many visceral effectors.	Each preganglionic fiber usually synapses with four or five postganglionic neurons that pass to a single visceral effector.
Distributed throughout, the body, including the skin.	Distribution limited primarily to head and viscera of thorax, abdomen, and pelvis.

PHYSIOLOGY OF THE AUTONOMIC NERVOUS SYSTEM

NEUROTRANSMITTERS

Autonomic fibers, like other axons of the nervous system, release neurotransmitters at synapses as well as at points of contact with visceral effectors. These latter points are called **neuroeffector junctions.** Neuroeffector junctions may be either neuromuscular or neuroglandular junctions. On the basis of the neurotransmitter produced, autonomic fibers may be classified as either cholinergic or adrenergic (Figure 16-4).

Cholinergic (kō'-lin-ER-jik) **fibers** release **acetylcholine (ACh)** and include the following: (1) all sympathetic and parasympathetic preganglionic axons, (2) all parasympathetic postganglionic axons, and (3) a few sympathetic postganglionic axons. The cholinergic sympathetic postganglionic axons include those to sweat glands and blood vessels in skeletal muscles, to the skin, and to the external genitalia. Since acetylcholine is quickly inactivated by the enzyme **acetylcholinesterase (AChE),** the effects of cholinergic fibers are short-lived and local.

Adrenergic (ad'-ren-ER-jik) **fibers** produce **norepinephrine (NE).** Most sympathetic postganglionic axons are adrenergic. Since norepinephrine is inactivated much more slowly by **catechol-O-methyltransferase (COMT)** or **monoamine oxidase (MAO)** than acetylcholine is by acetylcholinesterase, and since norepinephrine may enter the blood stream, the effects of sympathetic stimulation are longer lasting and more widespread than parasympathetic stimulation. The action of NE produced by sympathetic postganglionic axons is augmented by both NE and epinephrine secreted by the adrenal medulla. Since both substances are secreted into the blood, their effects are more sustained than NE released by axons. NE and epinephrine released by the adrenal medulla are ultimately destroyed by enzymes in the liver once they have acted on their visceral effectors.

RECEPTORS

ACh is synthesized and stored in an inactive form in synaptic vesicles in the axon terminals of the cholinergic fibers. The release of ACh from the axon terminals requires Ca^{2+} ions and comes in bursts of thousands of molecules. ACh then diffuses the short distance across the synaptic cleft to bind with a receptor molecule on the postsynaptic membrane. The membrane is quickly depolarized and the ACh is quickly inactivated by acetylcholinesterase (AChE), an enzyme that can be inhibited by a variety of drugs such as physostigmine and neostigmine.

The actual effect produced by ACh is determined by the type of postsynaptic receptor with which it interacts (Figure 16-4). The two types of ACh postsynaptic receptors are known as nicotinic receptors and muscarinic receptors. **Nicotinic receptors** are found on both sympathetic and parasympathetic postganglionic neurons. These receptors are so named because the actions of ACh on

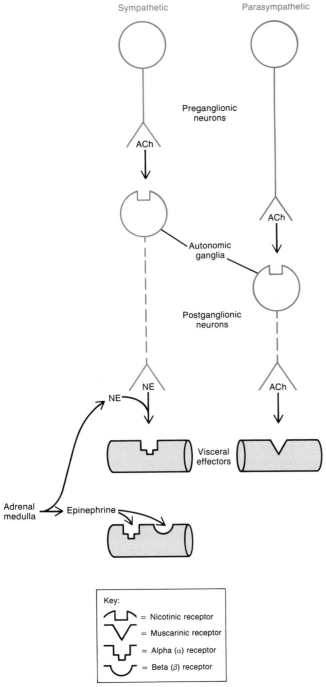

FIGURE 16-4 Neurotransmitters and receptors associated with the autonomic nervous system.

an action potential reaches the axon terminal, NE is rapidly released into the synaptic cleft. The molecules diffuse across the cleft and combine with specific receptor molecules on the postsynaptic membrane to elicit a specific effector response.

The effects of NE and epinephrine, like those of ACh, are also determined by the type of postsynaptic receptor with which they interact. Such receptors are found on visceral effectors innervated by most sympathetic postganglionic axons and are referred to as **alpha (α) receptors** and **beta (β) receptors** (Figure 16-4). These receptors are distinguished by the specific responses they elicit and by their selective combination with drugs that excite or inhibit them. In general, stimulation of an alpha receptor results in an increase in permeability of a postsynaptic membrane to Na⁺, followed by depolarization. For this reason, alpha receptors are generally excitatory. Conversely, stimulation of beta receptors generally results in an increase in permeability of a postsynaptic membrane to the outward movement of K⁺ and the inward movement of Cl⁻. This brings about hyperpolarization of a postsynaptic membrane. Thus, beta receptors are generally inhibitory.

Although cells of most effectors contain either alpha or beta receptors, some visceral effector cells contain both. NE, in general, stimulates alpha receptors to a greater extent than beta receptors, and epinephrine, in general, stimulates both alpha and beta receptors.

Exhibit 16-3 shows the types of receptors present on the cells of visceral effectors and their response to autonomic stimulation. Note that there are some very important (and perplexing) exceptions to the general rule describing alpha and beta receptors. For example, heart muscle has beta receptors that operate as excitatory receptors and cause the heart muscle to contract more forcibly. Other visceral effector cells possess both types of receptors. In these situations, the majority rules. The beta receptors greatly outnumber the alpha receptors in the smooth muscle cells of the blood vessels that flow through skeletal muscles. Therefore, an injection of NE, which binds only to alpha receptors, attaches onto the surface of such smooth muscle cells and causes a vasoconstriction. An injection of epinephrine, however, would result in a vasodilation because epinephrine acts equally on both types of receptors. But since there is a preponderance of beta receptors, the effect is that of the beta receptor, that is, vasodilation due to relaxation of smooth muscle cells.

ACTIVITIES

Most visceral effectors have dual innervation, that is, they receive fibers from both the sympathetic and the parasympathetic divisions. In these cases, impulses from one division stimulate the organ's activities, whereas impulses from the other division inhibit the organ's activities. The stimulating division may be either the sympathetic or the parasympathetic, depending on the organ. For example, sympathetic impulses increase heart activity, whereas parasympathetic impulses decrease it. On the other hand, parasympathetic impulses increase digestive activities,

such receptors are similar to those produced by nicotine. **Muscarinic receptors** are found on all effectors innervated by parasympathetic postganglionic axons and some effectors innervated by sympathetic postganglionic axons. These postsynaptic receptors are so named because the actions of ACh on such receptors are similar to those produced by muscarine, a toxin produced by a mushroom.

NE is released from most postganglionic sympathetic fibers. It is synthesized and stored in synaptic vesicles located in the axon terminals of adrenergic fibers. When

EXHIBIT 16-3   ACTIVITIES OF AUTONOMIC NERVOUS SYSTEM

VISCERAL EFFECTOR	RECEPTOR	EFFECT OF SYMPATHETIC STIMULATION	EFFECT OF PARASYMPATHETIC STIMULATION
Eye			
Radial muscle of iris	α	Contraction that results in dilation of pupil (mydriasis).	No known functional innervation.
Sphincter muscle of iris	α	No known functional innervation.	Contraction that results in constriction of pupil (miosis).
Ciliary muscle	β	Relaxation that results in far vision.	Contraction that results in near vision.
Glands			
Sweat	α	Stimulates local secretion.	Stimulates generalized secretion.
Lacrimal (tear)	—	No known functional innervation.	Secretion.
Salivary	α	Vasoconstriction, which decreases secretion.	Stimulates secretion and vasodilation.
Gastric	—	Vasoconstriction, which inhibits secretion.	Stimulates secretion.
Intestinal	—	Vasoconstriction, which inhibits secretion.	Stimulates secretion.
Adrenal medulla	—	No known functional innervation.	Promotes epinephrine and norepinephrine secretion.
Fat cells	β	Promotes lipolysis.	No known functional innervation.
Lungs (bronchial muscle)	β	Dilation.	Constriction.
Heart	β	Increases rate and strength of contraction; constricts coronary vessels that supply blood to heart muscle cells.	Decreases rate and strength of contraction; dilates coronary vessels.
Arterioles			
Skin and mucosa	α	Constriction.	No known functional innervation for most.
Skeletal muscle	α β	Constriction or dilation.	No known functional innervation.
Abdominal viscera	α β	Constriction.	No known functional innervation for most.
Cerebral	α	Slight constriction.	No known functional innervation.
Systemic veins	α β	Constriction and dilation.	No known functional innervation.
Liver	β	Promotes glycogenolysis and gluconeogenesis; decreases bile secretion.	Promotes glycogenesis and gluconeogenesis; increases bile secretion.
Gallbladder and ducts	—	Relaxation.	Contraction.
Stomach	α β	Decreases motility and tone; contracts sphincters.	Increases motility and tone; relaxes sphincters.
Intestines	α β	Decreases motility and tone; contracts sphincters.	Increases motility and tone; relaxes sphincters.
Kidney	β	Constriction of blood vessels that results in decreased urine volume; secretion of renin.	No known functional innervation.
Ureter	α	Increases motility.	Decreases motility.
Pancreas	α β	Inhibits secretion of enzymes and insulin; promotes secretion of glucagon.	Promotes secretion of enzymes and insulin.

EXHIBIT 16-3 *(Continued)*

VISCERAL EFFECTOR	RECEPTOR	EFFECT OF SYMPATHETIC STIMULATION	EFFECT OF PARASYMPATHETIC STIMULATION
Spleen	—	Contraction and discharge of stored blood into general circulation.	No known functional innervation.
Urinary bladder	α β	Relaxation of muscular wall; contraction of internal sphincter.	Contraction of muscular wall; relaxation of internal sphincter.
Arrector pili of hair follicles	α	Contraction that results in erection of hairs.	No known functional innervation.
Uterus	α β	Inhibits contraction if nonpregnant; stimulates contraction if pregnant.	Minimal effect.
Sex organs	α	In male, vasoconstriction of ductus (vas) deferens, seminal vesicle, prostate; results in ejaculation. In female, reverse uterine peristalsis.	Vasodilation and erection in both sexes; secretion in females.

whereas sympathetic impulses inhibit them. The actions of the two systems are carefully integrated to help maintain homeostasis. A summary of the activities of the autonomic nervous system is presented in Exhibit 16-3.

The parasympathetic division is primarily concerned with activities that restore and conserve body energy. It is a **rest-repose system.** Under normal body conditions, for instance, parasympathetic impulses to the digestive glands and the smooth muscle of the digestive system dominate over sympathetic impulses. Thus energy-supplying food can be digested and absorbed by the body.

The sympathetic division, in contrast, is primarily concerned with processes involving the expenditure of energy. When the body is in homeostasis, the main function of the sympathetic division is to counteract the parasympathetic effects just enough to carry out normal processes requiring energy. During extreme stress, however, the sympathetic dominates the parasympathetic. When people are confronted with a stress condition, for example, their bodies become alert and they sometimes perform feats of unusual strength. Fear stimulates the sympathetic division.

Activation of the sympathetic division sets into operation a series of physiological responses collectively called the **fight-or-flight response.** It produces the following effects.

1. The pupils of the eyes dilate.
2. The heart rate increases.
3. The blood vessels of the skin and viscera constrict.
4. The remainder of the blood vessels dilate. This causes a faster flow of blood into the dilated blood vessels of skeletal muscles, cardiac muscle, lungs, and brain—organs involved in fighting off danger.
5. Rapid breathing occurs as the bronchioles dilate to allow faster movement of air in and out of the lungs.
6. Blood sugar level rises as liver glycogen is converted to glucose to supply the body's additional energy needs.
7. The medulla of the adrenal gland is stimulated to produce

epinephrine and norepinephrine, hormones that intensify and prolong the sympathetic effects noted previously.

8. Processes that are not essential for meeting the stress situation are inhibited. For example, muscular movements of the gastrointestinal tract and digestive secretions are slowed down or even stopped.

CLINICAL APPLICATION

If the sympathetic trunk is cut on one side, the sympathetic supply to that side of the head is removed, resulting in **Horner's syndrome,** in which the patient exhibits (on the affected side): ptosis (drooping of the upper eyelid), slight elevation of the lower eyelid; narrowing of the palpebral fissure, enophthalmos (the eye appears sunken), miosis (constricted pupil), anhydrosis (lack of sweating), and flushing of the skin.

VISCERAL AUTONOMIC REFLEXES

A **visceral autonomic reflex** adjusts the activity of a visceral effector. In other words, it results in the contraction of smooth or cardiac muscle or secretion by a gland. Such reflexes assume a key role in activities such as regulating heart action, blood pressure, respiration, digestion, defecation, and urinary bladder functions.

A visceral autonomic reflex arc consists of the following components.

1. **Receptor.** The receptor is the distal end of an afferent neuron in an exteroceptor or enteroceptor.
2. **Afferent neuron.** This neuron, either a somatic afferent or visceral afferent neuron, conducts the sensory impulse to the spinal cord or brain.
3. **Association neurons.** These neurons are found in the central nervous system.
4. **Visceral efferent preganglionic neuron.** In the thoracic and abdominal regions, this neuron is in the lateral gray horn

of the spinal cord. The axon passes through the ventral root of the spinal nerve, the spinal nerve itself, and the white ramus communicans. It then enters a sympathetic trunk or prevertebral ganglion, where it synapses with a postganglionic neuron. In the cranial and sacral regions, the visceral efferent preganglionic axon leaves the central nervous system and passes to a terminal ganglion, where it synapses with a postganglionic neuron. The role of the visceral efferent preganglionic neuron is to convey a motor impulse from the brain or spinal cord to an autonomic ganglion.

5. **Visceral efferent postganglionic neuron.** This neuron conducts a motor impulse from a visceral efferent preganglionic neuron to the visceral effector.

6. **Visceral effector.** A visceral effector is smooth muscle, cardiac muscle, or a gland.

The basic difference between a somatic reflex arc and a visceral autonomic reflex arc is that in a somatic reflex arc only one efferent neuron is involved. In a visceral autonomic reflex arc, two efferent neurons are involved.

Visceral sensations do not always reach the cerebral cortex. Most remain at subconscious levels. Under normal conditions, you are not aware of muscular contractions of the digestive organs, heartbeat, changes in the diameter of blood vessels, and pupil dilation and constriction. Your body adjusts such visceral activities by visceral reflex arcs whose centers are in the spinal cord or lower regions of the brain. Among such centers are the cardiac, respiratory, vasomotor, swallowing, and vomiting centers in the medulla and the temperature control center in the hypothalamus. Stimuli delivered by somatic or visceral afferent neurons synapse in these centers, and the returning motor impulses conducted by visceral efferent neurons bring about an adjustment in the visceral effector without conscious recognition. The impulses are interpreted and acted on subconsciously. Some visceral sensations do give rise to conscious recognition: hunger, nausea, and fullness of the urinary bladder and rectum.

CONTROL BY HIGHER CENTERS

The autonomic nervous system is not a separate nervous system. Although little is known about the specific centers in the brain that regulate specific autonomic functions, it is known that axons from many parts of the central nervous system are connected to both the sympathetic and the parasympathetic divisions of the autonomic nervous system and thus exert considerable control over it. Autonomic centers in the cerebral cortex are connected to autonomic centers of the thalamus, for example. These, in turn, are connected to the hypothalamus. In this hierarchy of command, the thalamus sorts incoming impulses before they reach the cerebral cortex. The cerebral cortex then turns over control and integration of visceral activities to the hypothalamus. It is at the level of the hypothalamus that the major control and integration of the autonomic nervous system is exerted.

The hypothalamus receives input from areas of the nervous system concerned with emotions, visceral functions, olfaction, gustation, as well as changes in temperature, osmolarity, and levels of various substances in blood. Anatomically, the hypothalamus is connected to both the sympathetic and the parasympathetic divisions of the au-

tonomic nervous system by axons of neurons whose dendrites and cell bodies are in various hypothalamic nuclei. The axons form tracts from the hypothalamus to sympathetic and parasympathetic nuclei in the brain stem and spinal cord through relays in the reticular formation. The posterior and lateral portions of the hypothalamus appear to control the sympathetic division. When these areas are stimulated, there is an increase in visceral activities— an increase in heart rate, a rise in blood pressure due to vasoconstriction of blood vessels, an increase in the rate and depth of respiration, dilation of the pupils, and inhibition of the digestive tract. On the other hand, the anterior and medial portions of the hypothalamus seem to control the parasympathetic division. Stimulation of these areas results in a decrease in heart rate, lowering of blood pressure, constriction of the pupils, and increased motility of the digestive tract.

Control of the autonomic nervous system by the cerebral cortex occurs primarily during emotional stress. In extreme anxiety, which can result from either conscious or subconscious stimulation in the cerebral cortex, the cortex can stimulate the hypothalamus. This, in turn, stimulates the cardiac and vasomotor centers of the medulla, which increases heart rate and blood pressure. If the cortex is stimulated by an extremely unpleasant sight, the stimulation causes vasodilation of blood vessels, a lowering of blood pressure, and fainting.

Evidence of even more direct control of visceral responses is provided by data gathered from studies of biofeedback and meditation.

BIOFEEDBACK

In the simplest terms, **biofeedback** is a process in which people get constant signals, or feedback, about visceral body functions such as blood pressure, heart rate, and muscle tension. By using special monitoring devices, they can control these visceral functions consciously.

In a study conducted at the Menninger Foundation,* subjects suffering from migraine headaches received instructions in the use of a monitor that registers the skin temperature of the right index finger. Subjects were also given a typewritten sheet containing two sets of phrases. The first set was designed to help them relax the entire body. The second set was designed to bring about an increased flow of blood in the hands. The subjects practiced raising their skin temperature at home for 5 to 15 minutes a day. When skin temperature increased, the monitor emitted a high-pitched sound. In time, the monitor was abandoned.

Once the subjects learned how to vasodilate their blood vessels, the migraine headaches lessened. Since migraine headaches are believed to involve a distension of blood vessels in the head, the shunting of blood from head to hands relieved the distension and thus the pain.

Other experiments have shown that biofeedback can be applied to childbirth. Women were given monitors

---

* Much of the following discussion of the use of biofeedback for the treatment of migraine headaches is based on information provided by Dr. Joseph D. Sargent of the Menninger Foundation, Topeka, Kansas.

hooked up to their fingers and arms to measure electrical conductivity of the skin and skeletal muscle tension. Both conductivity and tension increased with nervousness and made labor difficult. Muscle tension was recorded as a sirenlike sound that became louder with nervousness. Skin conductivity was recorded as a crackling noise that also increased with nervousness. The monitors kept the women informed of their nervousness. This was the biofeedback. Having pleasant thoughts reduced the sound levels. The reward was less nervousness. The results of the study indicate that the women in a state of reduced nervousness needed less medication during labor, and labor time itself was shortened.

There is no way to determine where biofeedback will lead. Perhaps the outstanding contribution of biofeedback research has been to demonstrate that the autonomic nervous system is not autonomous. Visceral responses can be controlled. Current therapeutic applications of biofeedback include treatment of asthma, Raynaud's disease, hypertension, gastrointestinal disorders, fecal incontinence, anxiety, pain, and neuromuscular rehabilitation.

MEDITATION

Yoga, which literally means union, is defined as a higher state of consciousness achieved through a fully rested and relaxed body and a fully awake and relaxed mind. One widely practiced technique for achieving higher consciousness is called **transcendental meditation (TM).** One sits in a comfortable position with the eyes closed and concentrates on a suitable sound or thought.

Research indicates that transcendental meditation can alter physiological responses. Oxygen consumption decreases drastically along with carbon dioxide elimination. Subjects have experienced a reduction in metabolic rate and blood pressure. Researchers have also observed a decrease in heart rate, an increase in the intensity of alpha brain waves, a sharp decrease in the amount of lactic acid in the blood, and an increase in the skin's electrical resistance. These last four responses are characteristic of a highly relaxed state of mind. Alpha waves are found in the EEGs of almost all individuals in a resting, but awake, state; they disappear during sleep.

These responses have been called an **integrated response**—essentially, a hypometabolic state due to inactivation of the sympathetic division of the autonomic nervous system. The response is the exact opposite of the fight-or-flight response, which is a hyperactive state of the sympathetic division. The existence of the integrated response suggests that the central nervous system does exert some control over the autonomic nervous system.

STUDY OUTLINE

Somatic Efferent and Autonomic Nervous System (p. 357)
1. The somatic efferent nervous system produces conscious movement in skeletal muscles.
2. The autonomic nervous system, or visceral efferent nervous system, regulates visceral activities, that is, activities of smooth muscle, cardiac muscle, and glands, and it usually operates without conscious control.
3. It is regulated by centers in the brain, in particular by the cerebral cortex, the hypothalamus, and the medulla oblongata.
4. A single somatic efferent neuron synapses on skeletal muscles; in the autonomic nervous system, there are two efferent neurons—one from the CNS to a ganglion and one from a ganglion to a visceral effector.
5. Somatic efferent neurons release acetylcholine (ACh) and autonomic efferent neurons release either acetylcholine (ACh) or norepinephrine (NE).

Structure of the Autonomic Nervous System (p. 358)
1. The autonomic nervous system consists of visceral efferent neurons organized into nerves, ganglia, and plexuses.
2. It is entirely motor. All autonomic axons are efferent fibers.
3. Efferent neurons are preganglionic (with myelinated axons) and postganglionic (with unmyelinated axons).
4. The autonomic system consists of two principal divisions: sympathetic (thoracolumbar) and parasympathetic (craniosacral).
5. Autonomic ganglia are classified as sympathetic trunk ganglia (on sides of spinal column), prevertebral ganglia (anterior to spinal column), and terminal ganglia (near or inside visceral effectors).

Physiology of the Autonomic Nervous System (p. 363)
1. Autonomic fibers release neurotransmitters at synapses. On the basis of the neurotransmitter produced, these fibers may be classified as cholinergic or adrenergic.
2. Cholinergic fibers release acetylcholine (ACh). Adrenergic fibers produce norepinephrine (NE).
4. Acetylcholine (ACh) interacts with nicotinic receptors on postganglionic neurons and muscarinic receptors on certain visceral effectors.
5. Norepinephrine (NE) generally interacts with alpha receptors on visceral effectors, and epinephrine generally interacts with alpha and beta receptors on visceral effectors.
6. Sympathetic responses are widespread and, in general, concerned with energy expenditure. Parasympathetic responses are restricted and are typically concerned with energy restoration and conservation.

Visceral Autonomic Reflexes (p. 366)
1. A visceral autonomic reflex adjusts the activity of a visceral effector.
2. A visceral autonomic reflex arc consists of a receptor, afferent neuron, association neuron, visceral efferent preganglionic neuron, visceral efferent postganglionic neuron, and visceral effector.

Control by Higher Centers (p. 367)
1. The hypothalamus controls and integrates the autonomic nervous system. It is connected to both the sympathetic and the parasympathetic divisions.
2. Biofeedback is a process in which people learn to monitor visceral functions and to control them consciously. It has

been used to control heart rate, alleviate migraine headaches, and make childbirth easier.

3. Yoga is a higher consciousness achieved through a fully rested and relaxed body and a fully awake and relaxed mind.

4. Transcendental meditation (TM) produces the following physiological responses: decreased oxygen consumption and carbon dioxide elimination, reduced metabolic rate, decrease in heart rate, increase in the intensity of alpha brain waves, a sharp decrease in the amount of lactic acid in the blood, and an increase in the skin's electrical resistance.

REVIEW QUESTIONS

1. What are the principal components of the autonomic nervous system? What is its general function? Why is it called involuntary?

2. What are the principal differences between the voluntary nervous system and the autonomic nervous system?

3. Relate the role of visceral efferent fibers and visceral effectors to the autonomic nervous system.

4. Distinguish between preganglionic neurons and postganglionic neurons with respect to location and function.

5. What is an autonomic ganglion? Describe the location and function of the three types of autonomic ganglia. Define white and gray rami communicantes.

6. On what basis are the sympathetic and parasympathetic divisions of the autonomic nervous system differentiated anatomically and functionally?

7. Discuss the distinction between cholinergic and adrenergic fibers of the autonomic nervous system.

8. How is acetylcholine (ACh) related to nicotinic and muscarinic receptors?

9. How are alpha and beta receptors related to norepinephrine (NE) and epinephrine?

10. Give examples of the antagonistic effects of the sympathetic and parasympathetic divisions of the autonomic nervous system.

11. Summarize the principal functional differences between the voluntary nervous system and the autonomic nervous system.

12. Give the *sympathetic response* in a fear situation for each of the following body parts: hair follicles, iris of eye, lungs, spleen, adrenal medulla, kidneys, urinary bladder, stomach, intestines, gallbladder, liver, heart, arterioles of the abdominal viscera, skeletal muscles, and skin and mucosa.

13. Define a visceral autonomic reflex and give three examples.

14. Describe a complete visceral autonomic reflex in proper sequence.

15. Describe how the hypothalamus controls and integrates the autonomic nervous system.

16. Define biofeedback. Explain how it could be useful.

17. What is transcendental meditation (TM)? How is the integrated response related to the autonomic nervous system?

The Special Senses

17

- Locate the receptors for olfaction and describe the neural pathway for smell.
- Identify the gustatory receptors and describe the neural pathway for taste.
- Explain the structure and physiology of the accessory visual structures.
- List and describe the structural divisions of the eye.
- Discuss retinal image formation by describing refraction, accommodation, constriction of the pupil, convergence, and inverted image formation.
- Define emmetropia, myopia, hypermetropia, and astigmatism.
- Diagram and discuss the rhodopsin cycle responsible for light sensitivity of rods.
- Identify the afferent pathway of light impulses to the brain.
- Describe the anatomical subdivisions of the ear.
- List the principal events in the physiology of hearing.
- Identify the receptor organs for static and dynamic equilibrium.
- Contrast the causes and symptoms of cataracts, glaucoma, conjunctivitis, trachoma, deafness, labyrinthine disease, Ménière's syndrome, vertigo, otitis media, and motion sickness.
- Define medical terminology associated with the sense organs.

T he special senses—smell, taste, sight, hearing, and equilibrium—have receptor organs that are structurally more complex than receptors for general sensations. The sense of smell is the least specialized, as opposed to the sense of sight, which is the most specialized. Like the general senses, however, the special senses allow us to detect changes in our environment.

OLFACTORY SENSATIONS

STRUCTURE OF RECEPTORS

The receptors for the **olfactory** (ol-FAK-tō-rē) **sense** are located in the nasal epithelium in the superior portion of the nasal cavity on either side of the nasal septum (Figure 17-1). The nasal epithelium consists of three principal kinds of cells: supporting, olfactory, and basal. The **supporting (sustentacular) cells** are columnar epithelial cells of the mucous membrane lining the nose. The **olfactory cells** are bipolar neurons whose cell bodies lie between the supporting cells. The distal (free) end of each olfactory cell contains a dendrite that terminates in a swelling **(olfactory vesicle)** from which six to eight cilia, called **olfactory hairs,** radiate. The hairs are believed to react to odors in the air and then to stimulate the olfactory cells, thus initiating the olfactory response. The proximal (basal) part of each olfactory cell contains a single process that represents its axon. **Basal cells** lie between the bases of the supporting cells and are believed to produce new supporting cells. Within the connective tissue beneath the olfactory epithelium are **olfactory (Bowman's) glands** that produce mucus, which is carried to the surface of the epithelium by ducts. The secretion moistens the surface of the olfactory epithelium and serves as a solvent for odoriferous substances. The continuous secretion of mucus also serves to freshen the surface film of fluid and prevents continuous stimulation of olfactory hairs by the same odor.

STIMULATION OF RECEPTORS

In order for a substance to be smelled it must be volatile, that is, capable of entering into a gaseous state so that the gaseous particles can enter the nostrils. You have probably noticed that smell occurs in cycles each time inspiration occurs and that sensitivity to smell is greatly increased by sniffing. The substance to be smelled must be water-soluble so that it can dissolve in the mucus to make contact with olfactory cells. The substance must also be lipid-soluble. Since the plasma membranes of olfactory hairs are largely lipid, the substance to be smelled must be dissolve in the lipid covering to make contact with olfactory hairs in order to initiate an impulse.

Many attempts have been made to distinguish and classify the primary sensations of smell. One classification includes 7 classes of primary sensations: camphoraceous, musky, floral, pepperminty, ethereal, pungent, and putrid. Recent data suggest there may be as many as 50 or more primary sensations of smell.

It is believed that olfactory cells react to olfactory stimuli in the same way that most sensory receptors react to their specific stimuli: first a generator potential is developed, followed by initiation of a nerve impulse. The **chemical theory** assumes that there are different receptor chemicals in the membranes of olfactory hairs, each capable of reacting with a particular olfactory substance (stimulus). The interaction between chemical receptor and substance alters the permeability of the plasma membrane so that a generator potential is developed, followed by initiation of a nerve impulse.

ADAPTATION AND ODOR THRESHOLDS

The sensation of smell happens quickly. Olfactory cells respond in milliseconds. Adaptation to odors also occurs rapidly and appears to involve the central nervous system. Olfactory receptors themselves adapt about 50 percent in the first second or so after stimulation, but adapt very slowly thereafter. Yet we know from common experience that complete adaptation to odors occurs in about a minute after exposure to a strong atmosphere, a rate far faster than can be explained on the basis of receptor adaptation alone. It has been suggested that most adaptation to odor involves a psychological component in the central nervous system. The mechanism and site of psychological adaptation are unknown.

One of the major characteristics of smell is its low threshold. Only a minute quantity of a substance need be present in air for it to be smelled (minimal identifiable odor or MIO). A good example is the substance methyl mercaptan, which can be detected in concentrations as low as 1/25,000,000,000 mg per milliliter of air.

OLFACTORY PATHWAY

The unmyelinated axons of the olfactory cells unite to form the **olfactory (I) nerves,** which pass through foramina in the cribriform plate of the ethmoid bone (Figure 17-1). The olfactory (I) nerves terminate in paired masses of gray matter called the **olfactory bulbs.** The olfactory bulbs lie beneath the frontal lobes of the cerebrum on either side of the crista galli of the ethmoid bone. The first synapse of the olfactory neural pathway occurs in the olfactory bulbs between the axons of the olfactory (I) nerves and the dendrites of neurons inside the olfactory bulbs. Axons of these neurons run posteriorly to form the **olfactory tract.** From here, impulses are conveyed to the primary olfactory area of the cerebral cortex. In the cerebral cortex, the impulses are interpreted as odor and give rise to the sensation of smell. Impulses related to olfaction do not pass through the thalamus.

Both the supporting cells of the nasal epithelium and tear glands are innervated by branches of the trigeminal (V) nerve. The nerve receives stimuli of pain, cold, heat, tickling, and pressure. Olfactory stimuli such as pepper, ammonia, and chloroform are irritating and may cause tearing because they stimulate the lacrimal and nasal mucosal receptors of the trigeminal (V) nerve as well as the olfactory neurons.

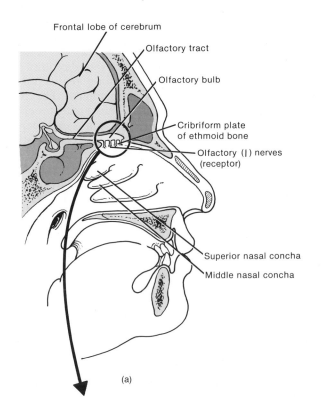

Frontal lobe of cerebrum
Olfactory tract
Olfactory bulb
Cribriform plate of ethmoid bone
Olfactory (I) nerves (receptor)
Superior nasal concha
Middle nasal concha

(a)

GUSTATORY SENSATIONS

STRUCTURE OF RECEPTORS

The receptors for **gustatory** (GUS-ta-tō'-rē) **sensations,** or sensations of taste, are located in the taste buds (Figure 17-2). The nearly 2,000 taste buds are most numerous on the tongue, but they are also found on the soft palate and in the throat. The **taste buds** are oval bodies consisting of three kinds of cells: supporting, gustatory, and basal. The **supporting (sustentacular) cells** are a specialized epithelium that forms a capsule, inside each of which are 4 to 20 **gustatory cells.** Each gustatory cell contains a hairline process **(gustatory hair)** that projects to the external surface through an opening in the taste bud called the **taste pore.** Gustatory cells make contact with taste stimuli through the taste pore. **Basal cells** are found at the periphery of the taste bud near the basal lamina. These cells produce supporting and gustatory cells whose average life span is about 10 days.

Taste buds are found in some connective tissue elevations on the tongue called **papillae** (pa-PIL-ē). The papil-

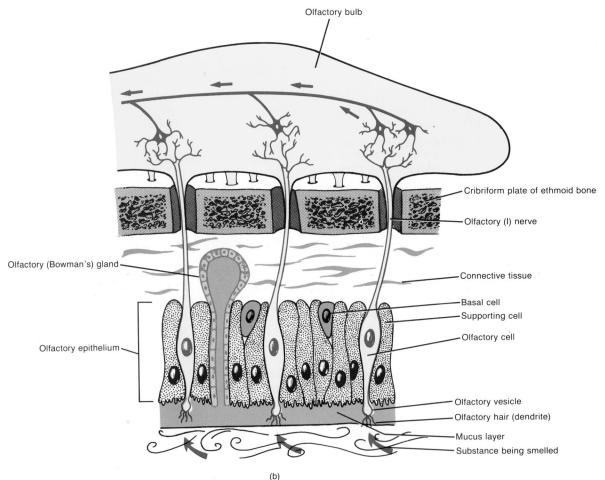

Olfactory bulb

Cribriform plate of ethmoid bone

Olfactory (I) nerve

Connective tissue

Basal cell
Supporting cell

Olfactory cell

Olfactory (Bowman's) gland

Olfactory epithelium

Olfactory vesicle
Olfactory hair (dendrite)
Mucus layer
Substance being smelled

(b)

FIGURE 17–1 Olfactory receptors. (a) Location of receptors in nasal cavity. (b) Enlarged aspect of olfactory receptors.

Connective tissue

Olfactory epithelium

Olfactory (Bowman's) gland

Nucleus of olfactory cell

Nucleus of supporting cell

Olfactory hairs (dendrites)

(c)

FIGURE 17-1 (Continued) (c) Photomicrograph of the olfactory mucosa at a magnification of 300×. (Copyright © 1985 by Michael H. Ross. Used by permission).

lae give the upper surface of the tongue its rough appearance. **Circumvallate** (ser-kum-VAL-āt) or **vallate papillae,** the largest type, are circular and form an inverted V-shaped row at the posterior portion of the tongue. **Fungiform** (FUN-ji-form; meaning mushroom-shaped) **papillae** are knoblike elevations found primarily on the tip and sides of the tongue. All circumvallate and most fungiform papillae contain taste buds. **Filiform** (FIL-i-form) **papillae** are pointed threadlike structures that cover the anterior two-thirds of the tongue. They rarely contain taste buds.

STIMULATION OF RECEPTORS

For gustatory cells to be stimulated, the substances we taste must be in solution in saliva so they can enter taste pores. Once the taste substance makes contact with plasma membranes of the gustatory hairs, the generator potential is developed. This is presumed to occur as a result of a membrane receptor–taste substance interaction. Then the generator potential initates a nerve impulse.

Despite the many substances we seem to taste, there are basically only four primary taste sensations: sour, salt, bitter, and sweet. All other "tastes," such as chocolate, pepper, and coffee, are combinations of these four modified by accompanying olfactory sensations.

Persons with colds or allergies sometimes complain that they cannot taste their food. Although their taste sensations may be operating normally, their olfactory sensations are not. This illustrates that much of what we think of as taste is actually smell. Odors from foods pass upward into the nasopharynx and stimulate the olfactory system. In fact, a given concentration of a substance will stimulate the olfactory system thousands of times more than it stimulates the gustatory system.

Each of the four primary tastes is caused by a different response to different chemicals. Certain regions of the tongue react more strongly than others to certain taste sensations. Although the tip of the tongue reacts to all four primary taste sensations, it is highly sensitive to sweet and salty substances. The posterior portion of the tongue is highly sensitive to bitter substances. The lateral edges of the tongue are more sensitive to sour substances (Figure 17-2c).

ADAPTATION AND TASTE THRESHOLDS

Adaptation to taste occurs rapidly. Complete adaptation can occur in 1 to 5 minutes of continuous stimulation. As with odors, receptor adaptation alone cannot account for the speed of complete adaptation. Although taste receptors have a period of rapid adaptation during the first 2 to 3 seconds after contact with food, adaptation is slow thereafter. Adaptation of receptors to smell contributes to taste adaptation, but still does not account for its speed. As with adaptation to odors, adaptation to taste involves a psychological adaptation in the central nervous system.

The threshold for taste varies for each of the primary tastes. The threshold for bitter substances, as measured by quinine, is lowest. This may have a protective function. The threshold for sour substances, as measured by hydrochloric acid, is somewhat higher. The thresholds for salt substances, as measured by sodium chloride, and sweet substances, as measured by sucrose, are about the same, and are higher than both bitter and sour substances.

GUSTATORY PATHWAY

The cranial nerves that supply afferent fibers to taste buds are the facial (VII), which supplies the anterior two-thirds of the tongue; the glossopharyngeal (IX), which supplies the posterior one-third of the tongue; and the vagus (X), which supplies the throat and epiglottis. Taste impulses are conveyed from the gustatory cells in taste buds along the nerves to the medulla and then to the thalamus. They terminate in the primary gustatory area in the parietal lobe of the cerebral cortex.

VISUAL SENSATIONS

The structures related to vision are the eyeball, the optic (II) nerve, the brain, and a number of accessory structures. The study of the structure, function, and diseases of the eye is known as **ophthalmology** (of'-thal-MOL-ō-jē; *ophthalmo* = eye; *logos* = study of). A physician who specializes in the diagnosis and treatment of eye disorders with drugs, surgery, and corrective lenses is known as an **ophthalmologist,** whereas an **optometrist**

Circumvallate papilla

Taste buds

Glands

Glands

(a)

Circumvallate papilla

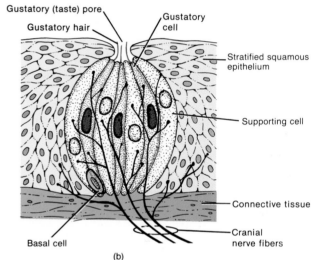

Gustatory (taste) pore

Gustatory hair

Gustatory cell

Stratified squamous epithelium

Supporting cell

Connective tissue

Cranial nerve fibers

Basal cell

(b)

FIGURE 17-2 Gustatory receptors. (a) Photomicrograph of a circumvallate papilla containing taste buds at a magnification of 50×. (Copyright © 1983 by Michael H. Ross. Used by permission). (b) Diagram of the structure of a taste bud. (c) Locations of papillae and four taste zones.

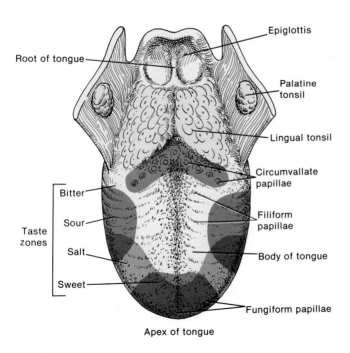

Epiglottis

Root of tongue

Palatine tonsil

Lingual tonsil

Circumvallate papillae

Filiform papillae

Body of tongue

Fungiform papillae

Apex of tongue

Taste zones

Bitter

Sour

Salt

Sweet

(c)

is a specialist with a doctorate in optometry who is licensed to examine and test the eyes and treat visual defects by prescribing corrective lenses. An **optician** is a technician who fits, adjusts, and dispenses corrective lenses on prescription of an ophthalmologist or optometrist.

ACCESSORY STRUCTURES OF EYE

Among the **accessory structures** are the eyebrows, eyelids, eyelashes and the lacrimal apparatus (Figure 17-3). The **eyebrows** form a transverse arch at the junction of the upper eyelid and forehead. Structurally, they resemble the hairy scalp. The skin of the eyebrows is richly supplied with sebaceous glands. The hairs are generally coarse and directed laterally. Deep to the skin of the eyebrows are the fibers of the orbicularis oculi muscles. The eyebrows help to protect the eyeballs from foreign objects, perspiration, and the direct rays of the sun.

(a)

(b)

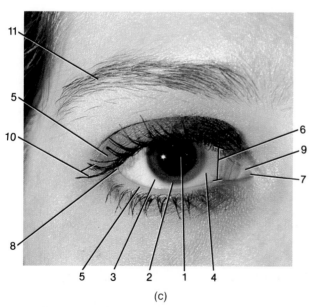

(c)

FIGURE 17-3 Accessory structures of the eye. (a) Anterior view. (b) Sagittal section of the eyelids and anterior portion of the eyeball. (c) Surface anatomy of the eye. (Copyright © 1982 by Gerard J. Tortora. Courtesy of Lynne Tortora.)

1. **Pupil.** Opening of center of iris of eyeball for light transmission.
2. **Iris.** Circular, pigmented muscular membrane behind cornea.
3. **Sclera.** "White" of eye, a coat of fibrous tissue that covers entire eyeball except for cornea.
4. **Conjunctiva.** Membrane that covers exposed surface of eyeball and lines eyelids.
5. **Palpebrae (eyelids).** Folds of skin and muscle lined by conjunctiva.
6. **Palpebral fissure.** Space between eyelids when they are open.
7. **Medial commissure.** Site of union of upper and lower eyelids near nose.
8. **Lateral commissure.** Site of union of upper and lower eyelids away from nose.
9. **Lacrimal caruncle.** Fleshy, yellowish projection of medial commissure that contains modified sweat and sebaceous glands.
10. **Eyelashes.** Hairs on margins of eyelids, usually arranged in two or three rows.
11. **Eyebrows.** Several rows of hairs superior to upper eyelids.

The upper and lower **eyelids,** or **palpebrae** (PAL-pe-brē), have several important roles. They shade the eyes during sleep, protect the eyes from excessive light and foreign objects, and spread lubricating secretions over the eyeballs. The upper eyelid is more movable than the lower and contains in its superior region a special levator muscle known as the **levator palpebrae superioris.** The space between the upper and lower eyelids that exposes the eyeball is called the **palpebral fissure.** Its angles are known as the **lateral commissure** (KOM-i-shūr), which is narrower and closer to the temporal bone, and the **medial commissure,** which is broader and nearer the nasal bone. In the medial commissure, there is a small reddish elevation, the **lacrimal caruncle** (KAR-ung-kul), contain-

ing sebaceous (oil) and sudoriferous (sweat) glands. A whitish material secreted by the caruncle collects in the medial commissure.

From superficial to deep, each eyelid consists of epidermis, dermis, subcutaneous areolar connective tissue, fibers of the orbicularis oculi muscle, a tarsal plate, tarsal glands, and a conjunctiva. The **tarsal plate** is a thick fold of connective tissue that forms much of the inner wall of each eyelid and gives form and support to the eyelids. Embedded in grooves on the deep surface of each tarsal plate is a row of elongated tarsal glands known as **tarsal** or **Meibomian** (mī-BŌ-mē-an) **glands.** These are modified sebaceous glands, and their oily secretion helps keep the eyelids from adhering to each other. Infection of the tarsal glands produces a tumor or cyst on the eyelid called a **chalazion** (ka-LĀ-zē-on). The **conjunctiva** (kon'-junk-TĪ-va) is a thin mucous membrane. It is called the **palpebral conjunctiva** when it lines the inner aspect of the eyelids. It is called the **bulbar (ocular) conjunctiva** when it is reflected from the eyelids onto the anterior surface of the eyeball to the periphery of the cornea. When the blood vessels of the bulbar conjunctiva are dilated and congested due to local irritation or infection, the person has bloodshot eyes.

Projecting from the border of each eyelid, anterior to the tarsal glands, is a row of short, thick hairs, the **eyelashes.** In the upper lid, they are long and turn upward; in the lower lid, they are short and turn downward. Sebaceous glands at the base of the hair follicles of the eyelashes called **sebaceous ciliary glands** or **glands of Zeis (ZĪS)** pour a lubricating fluid into the follicles. Infection of these glands is called a **sty.**

The **lacrimal apparatus** is a term used for a group of structures that manufactures and drains away tears. These structures are the lacrimal glands, the excretory lacrimal ducts, the lacrimal canals, the lacrimal sacs, and the nasolacrimal ducts. A **lacrimal gland** is a compound tubuloacinar gland located at the superior lateral portion of each orbit. Each is about the size and shape of an almond. Leading from the lacrimal glands are 6 to 12 **excretory lacrimal ducts** that empty lacrimal fluid, or tears, onto the surface of the conjunctiva of the upper lid. From here the lacrimal fluid passes medially and enters two small openings called **puncta lacrimalia** that appear as two small pores, one in each papilla of the eyelid, at the medial commissure of the eye. The lacrimal secretion then passes into two ducts, the **lacrimal canals,** and is next conveyed into the lacrimal sac. The lacrimal canals are located in the lacrimal grooves of the lacrimal bones. The **lacrimal sac** is the superior expanded portion of the **nasolacrimal duct,** a canal that transports the lacrimal secretion into the inferior meatus of the nose.

The **lacrimal secretion** is a watery solution containing salts, some mucus, and a bactericidal enzyme called **lysozyme.** It cleans, lubricates, and moistens the eyeball. After being secreted by the lacrimal glands, it is spread over the surface of the eyeball by the blinking of the eyelids. Usually, 1 ml per day is produced.

CLINICAL APPLICATION

Normally, the lacrimal secretion is carried away by evaporation or by passing into the lacrimal canals and then into the nasal cavities as fast as it is produced. If, however, an irritating substance makes contact with the conjunctiva, the lacrimal glands are stimulated to oversecrete. Tears then accumulate more rapidly than they can be carried away. This is a protective mechanism, since the tears dilute and wash away the irritating substance. "Watery" eyes also occur when an inflammation of the nasal mucosa, such as a cold, obstructs the nasolacrimal ducts so that drainage of tears is blocked.

STRUCTURE OF EYEBALL

The adult **eyeball** measures about 2.5 cm (1 inch) in diameter. Of its total surface area, only the anterior one-sixth is exposed. The remainder is recessed and protected by the orbit into which it fits. Anatomically, the eyeball can be divided into three layers: fibrous tunic, vascular tunic, and retina or nervous tunic (Figure 17-4).

Fibrous Tunic

The **fibrous tunic** is the outer coat of the eyeball. It can be divided into two regions: the posterior portion is the sclera and the anterior portion is the cornea. The **sclera** (SKLE-ra), the "white of the eye," is a white coat of dense fibrous tissue that covers all the eyeball except the anterior colored portion. The sclera gives shape to the eyeball and protects its inner parts. Its posterior surface is pierced by the optic (II) nerve. The **cornea** (KOR-nē-a) is a nonvascular, transparent fibrous coat that covers the iris, the colored part of the eye. The cornea's outer surface is covered by an epithelial layer continuous with the epithelium of the bulbar conjunctiva. At the junction of the sclera and cornea is a venous sinus known as the **scleral venous sinus** or **canal of Schlemm.** (Surface features of the eye are shown in Figure 17-3c.)

CLINICAL APPLICATION

Each year, about 10,000 **corneal transplants (keratoplasty)** are performed in the United States. Corneal transplants are considered to be the most successful type of transplantation. The surgical procedure is performed under an operating microscope at magnifications of between 6× and 40×. A general or local anesthesia is used. The defective cornea, usually 7 to 8 mm (about 0.3 inch) in diameter, is excised with microcorneal scissors and replaced with a donor cornea of similar diameter. The transplanted cornea is sewn into place with nylon sutures. Patients usually remain hospitalized for 3 to 5 days and are later fitted with hard or soft contact lenses or eyeglasses.

In the process of normal vision, light rays pass through the cornea, as well as other structures, to form a clear image on the retina (nervous tunic). In myopia (nearsightedness), the cornea is too curved and images of distant objects fall short of the retina, causing blurred vision. Many people with myopia may be helped by a surgical procedure called **radial keratotomy.** In the procedure, several microscopic incisions, like the spikes of a wheel, are made in the cornea. This flattens the cornea and improves vision without using corrective lenses.

Vascular Tunic

The **vascular tunic** is the middle layer of the eyeball and is composed of three portions: the choroid, the ciliary body, and the iris. Collectively, these three structures are called the **uvea** (YOO-vē-a). The **choroid** (KŌ-royd),

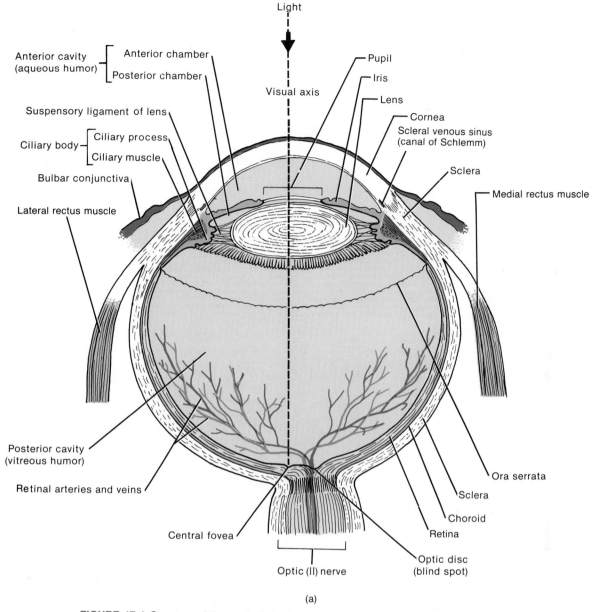

Light

Anterior cavity (aqueous humor) — Anterior chamber — Posterior chamber

Visual axis

Pupil
Iris
Lens
Cornea
Scleral venous sinus (canal of Schlemm)
Sclera
Medial rectus muscle

Suspensory ligament of lens

Ciliary body — Ciliary process — Ciliary muscle

Bulbar conjunctiva

Lateral rectus muscle

Posterior cavity (vitreous humor)

Retinal arteries and veins

Central fovea

Optic (II) nerve

Ora serrata
Sclera
Choroid
Retina
Optic disc (blind spot)

(a)

FIGURE 17-4 Structure of the eyeball. (a) Diagram of gross structure in transverse section.

the posterior portion of the vascular tunic, is a thin, dark brown membrane that lines most of the internal surface of the sclera. It contains numerous blood vessels and a large amount of pigment. The choroid absorbs light rays so they are not reflected within the eyeball. Through its blood supply, it nourishes the retina. Like the sclera, it is pierced by the optic (II) nerve at the back of the eyeball.

In the anterior portion of the vascular tunic, the choroid becomes the **ciliary** (SIL-ē-ar'-ē) **body.** It is the thickest portion of the vascular tunic. It extends from the **ora serrata** (Ō-ra ser-RĀ-ta) of the retina (inner tunic) to a point just behind the sclerocorneal junction. The ora serrata is simply the jagged margin of the retina. The ciliary body consists of the ciliary processes and ciliary muscle. The **ciliary processes** consist of protrusions or folds on the internal surface of the ciliary body that

secrete aqueous humor. The **ciliary muscle** is a smooth muscle that alters the shape of the lens for near or far vision.

The **iris** is the third portion of the vascular tunic. It consists of circular and radial smooth muscle fibers arranged to form a doughnut-shaped structure. The black hole in the center of the iris is the **pupil,** the area through which light enters the eyeball. The iris is suspended between the cornea and the lens and is attached at its outer margin to the ciliary process. A principal function of the iris is to regulate the amount of light entering the eyeball. When the eye is stimulated by bright light, the circular muscles of the iris contract and decrease the size of the pupil. The pupil constricts. When the eye must adjust to dim light, the radial muscles of the iris contract and increase the pupil's size. The pupil dilates.

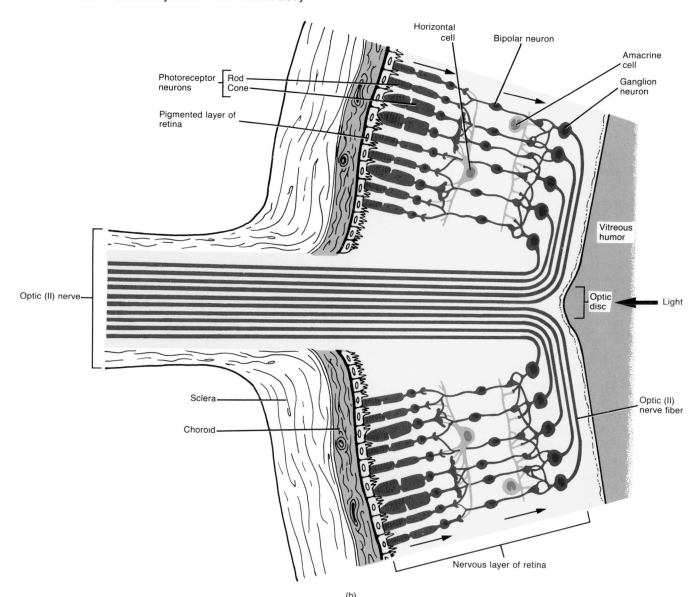

Photoreceptor neurons {Rod / Cone}

Pigmented layer of retina

Horizontal cell

Bipolar neuron

Amacrine cell

Ganglion neuron

Vitreous humor

Optic (II) nerve

Optic disc

Light

Sclera

Choroid

Optic (II) nerve fiber

Nervous layer of retina

(b)

LIGHT

Ganglion neurons

Inner plexiform layer

Inner nuclear layer

Outer plexiform layer

Outer nuclear layer

Rods and cones

Pigmented epithelium

Choroid

(c)

Figure 17-4 (*Continued*) (b) Diagram of the microscopic structure of the retina exaggerated for emphasis. (c) Photomicrograph of a portion of the retina at a magnification of 300×. The outer nuclear layer contains nuclei of photoreceptor neurons; the outer plexiform layer is where photoreceptor and bipolar neurons synapse; the inner nuclear layer contains the nuclei of bipolar neurons; and the inner plexiform layer is where bipolar and ganglion neurons synapse. (Copyright © 1983 by Michael H. Ross. Used by permission.)

Retina (Nervous Tunic)

The third and inner coat of the eye, the **retina (nervous tunic),** lies only in the posterior portion of the eye. Its primary function is image formation. The retina is the only place in the body where blood vessels can be seen directly. This is done by use of a special reflecting light called an ophthalmoscope. It is possible for an ophthalmologist to examine the retina and detect vascular changes associated with hypertension, atherosclerosis, and diabetes. The retina consists of an inner nervous tissue layer and an outer pigmented layer. The retina covers the choroid. At the edge of the ciliary body, it ends in a scalloped border called the ora serrata. This is where the nervous layer or visual portion of the retina ends. The pigmented layer extends anteriorly over the back of the ciliary body and the iris as the nonvisual portion of the retina.

CLINICAL APPLICATION

Detachment of the retina may occur in trauma, such as a blow to the head. The tear in the retina causes blindness in the corresponding field of vision. The actual detachment occurs between the sensory part of the retina and the underlying pigmented layer. Fluid accumulates between these layers, forcing the thin, pliable retina to billow out toward the vitreous humor. The retina may be reattached by photocoagulation by laser beam, cryosurgery, or scleral resection. (A laser is an instrument for amplifying light into an intense, highly concentrated beam that is strong enough to burn tissue.)

The nervous layer of the retina contains three zones of neurons. These three zones, named in the order in which they conduct impulses, are the **photoreceptor neurons,** the **bipolar neurons,** and the **ganglion neurons.** The dendrites of the photoreceptor neurons are called rods and cones because of their shapes. They are visual receptors highly specialized for stimulation by light rays. Functionally, rods and cones develop generator potentials. **Rods** are specialized for vision in dim light. They also allow us to discriminate between different shades of dark and light, and permit us to see shapes and movement. **Cones** are specialized for color vision and sharpness of vision (*visual acuity*). They are stimulated only by bright light. This is why we cannot see color by moonlight. It is estimated that there are 7 million cones and somewhere between 10 and 20 times as many rods. Cones are most densely concentrated in the **central fovea,** a small depression in the center of the macula lutea. The **macula lutea** (MAK-yoo-la LOO-tē-a), or yellow spot, is in the exact center of the posterior portion of the retina, corresponding to the visual axis of the eye. The fovea is the area of sharpest vision because of the high concentration of cones. Rods are absent from the fovea and macula and increase in density toward the periphery of the retina.

CLINICAL APPLICATION

In a disease called **senile macular degeneration (SMD),** new blood vessels grow over the macula lutea. The effect ranges from distorted vision to blindness. SMD accounts for nearly all new cases of blindness in people over 65. Its cause is unknown. Laser beam treatment has been used effectively in arresting blood vessel proliferation and restoring normal vision in some cases. In order to determine if SMD exists, a simple test can be performed without any assistance or special instruments. Stare, one eye at a time (covering the other with your hand), at any long straight line, such as a door frame. If either eye perceives the line as bent or twisted, or if a black spot appears, a physician should be informed immediately.

When information has passed through the photoreceptor neurons, it is conducted across synapses to the bipolar neurons in the intermediate zone of the nervous layer of the retina. From here it is passed to the ganglion neurons. These cells transmit their signals through optic (II) nerve fibers to the brain in the form of nerve impulses. Many rods connect with one bipolar neuron, and many of these bipolar neurons transmit impulses to one ganglion cell. This greatly lowers visual acuity, but it permits summation effects to occur so that low levels of light can stimulate a ganglion cell that would not respond had it been connected directly to a cone. The synaptic connections thus contribute much to the difference in visual acuity and light sensitivity.

The axons of the ganglion neurons extend posteriorly to a small area of the retina called the **optic disc (blind spot).** This region contains openings through which the axons of the ganglion neurons exit as the optic (II) nerve. Since it contains no rods or cones, an image striking it cannot be perceived. Thus it is called the blind spot.

Lens

In addition to the fibrous tunic, vascular tunic, and retina, the eyeball itself contains the lens, just behind the pupil and iris. The **lens** is constructed of numerous layers of protein fibers arranged like the layers of an onion. Normally, the lens is perfectly transparent. It is enclosed by a clear connective tissue capsule and held in position by **suspensory ligaments.** A loss of transparency of the lens is known as a **cataract.**

Interior

The interior of the eyeball is a large cavity divided into two smaller ones by the lens. The **anterior cavity,** the division anterior to the lens, is further divided into the **anterior chamber,** which lies behind the cornea and in front of the iris, and the **posterior chamber,** which lies behind the iris and in front of the suspensory ligaments and lens. The anterior cavity is filled with a watery fluid, similar to cerebrospinal fluid, called the **aqueous humor.** The fluid is believed to be secreted into the posterior chamber by choroid plexuses of the ciliary processes of the ciliary bodies behind the iris. Once the fluid is formed, it permeates the posterior chamber and then passes forward between the iris and the lens, through the pupil, into the anterior chamber. From the anterior chamber, the aqueous humor, which is continually produced, is drained off into the scleral venous sinus (canal of Schlemm) and then into the blood. The anterior chamber thus serves a function similar to the subarachnoid space around the brain and spinal cord. The scleral venous sinus is analogous to a venous sinus of the dura mater. The pressure in the eye, called **intraocular pressure,** is produced mainly by the aqueous humor. The intraocular pressure, along with the vitreous humor, maintains the shape of the eyeball and keeps the retina smoothly applied to the choroid so the retina will form clear images. Normal intraocular pressure (about 16 mm Hg) is maintained by drainage of the aqueous humor through the scleral venous sinus. Excessive intraocular pressure, called **glaucoma** (glaw-KŌ-ma), results in degeneration of the retina and blindness. Besides maintaining intraocular pressure, the aqueous humor is also the principal link between the cardiovascular system and the lens and cornea. Neither the lens nor the cornea has blood vessels.

The second, and larger, cavity of the eyeball is the **posterior cavity.** It lies between the lens and the retina and contains a jellylike substance called the **vitreous humor.** This substance contributes to intraocular pressure, helps to prevent the eyeball from collapsing, and holds the retina flush against the internal portions of the eyeball. The vitreous humor, unlike the aqueous humor, does not undergo constant replacement. It is formed during embryonic life and is not replaced thereafter.

A summary of structures associated with the eyeball is presented in Exhibit 17-1.

PHYSIOLOGY OF VISION

Before light can reach the rods and cones of the retina, it must pass through the cornea, aqueous humor, pupil, lens, and vitreous humor. For vision to occur, light reaching the rods and cones must form an image on the retina. The resulting nerve impulses must then be conducted to the visual area of the cerebral cortex. In discussing the physiology of vision, let us first consider retinal image formation.

EXHIBIT 17-1   SUMMARY OF STRUCTURES ASSOCIATED WITH THE EYEBALL

STRUCTURE	FUNCTION
Fibrous tunic	
Sclera	Provides shape and protects inner parts.
Cornea	Admits and refracts light.
Vascular tunic	
Choroid	Provides blood supply and absorbs light.
Ciliary body	Secretes aqueous humor and alters shape of lens for near or far vision (accommodation).
Iris	Regulates amount of light that enters eyeball.
Retina (nervous tunic)	Receives light, converts light into impulses, and transmits impulses to the optic (II) nerve.
Lens	Refracts light.
Anterior cavity	Contains aqueous humor that helps maintain shape of eyeball, keeps retina applied to choroid, and refracts light.
Posterior cavity	Contains vitreous humor that helps maintain shape of eyeball, keeps retina applied to choroid, and refracts light.

Retinal Image Formation

The formation of an image on the retina requires four basic processes, all concerned with focusing light rays: (1) refraction of light rays, (2) accommodation of the lens, (3) constriction of the pupil, and (4) convergence of the eyes. Accommodation and pupil size are functions of the smooth muscle cells of the ciliary muscle and the dilator and sphincter muscles of the iris. They are termed *intrinsic eye muscles* since they are inside the eyeball. Convergence is a function of the voluntary muscles attached to the outside of the eyeball called the *extrinsic eye muscles* (see Figure 11-6).

• *Refraction of Light Rays*   When light rays traveling through a transparent medium (such as air) pass into a second transparent medium with a different density (such as water), they bend at the surface of the two media. This is **refraction** (Figure 17-5a). The eye has four such media of refraction: cornea, aqueous humor, lens, and vitreous humor. Light rays entering the eye from the air are refracted at the following points: (1) the anterior surface of the cornea as they pass from the lighter air into the denser cornea, (2) the posterior surface of the cornea as they pass into the less dense aqueous humor, (3) the anterior surface of the lens as they pass from the aqueous humor into the denser lens, and (4) the posterior surface of the lens as they pass from the lens into the less dense vitreous humor.

The degree of refraction that takes place at each surface in the eye is very precise. When an object is 6 m (20 ft) or more away from the viewer, the light rays reflected from the object are nearly parallel to one another. The parallel rays must be bent sufficiently to fall exactly on the central fovea, where vision is sharpest. Light rays that are reflected from near objects are divergent rather than parallel. As a result, they must be refracted toward each other to a greater extent. This change in refraction is brought about by the lens of the eye (Figures 17-5b and 17-6).

• *Accommodation of the Lens*   If the surface of a lens curves outward, as in a convex lens, the lens will refract incoming rays toward each other so they eventually intersect. The greater the curve, the more acutely it bends the rays toward each other. Conversely, when the surface of a lens curves inward, as in a concave lens, the rays bend away from each other. The lens of the eye is biconvex. Furthermore, it has the unique ability to change the focusing power of the eye by becoming moderately curved at one moment and greatly curved the next. When the eye is focusing on a close object, the lens curves greatly in order to bend the rays toward the central fovea. This increase in the curvature of the lens is called **accommodation** (Figure 17-6). In near vision, the ciliary muscle contracts, pulling the ciliary process and choroid forward toward the lens. This action releases the tension on the lens and suspensory ligament. Because of its elasticity, the lens shortens, thickens, and bulges. In far vision, the ciliary muscle is relaxed and the lens is flatter. With aging, the lens loses elasticity and, therefore, its ability to accommodate.

Light ray (before refraction)

Air

Water

Light ray (after refraction)

(a) Refraction of light rays

Vitreous humor

Central fovea

Cornea

Aqueous humor

Lens

(b) Normal (emmetropic) eye

Lengthened eyeball

Normal eyeball

— eyeball too long

Nearsighted (myopic) eye, uncorrected

(c)

Concave lens

Nearsighted (myopic) eye, corrected

Normal eyeball

Shortened eyeball

— eyeball too short

Farsighted (hypermetropic) eye, uncorrected

(d)

Convex lens

Farsighted (hypermetropic) eye, corrected

Irregular cornea

Astigmatism from an irregular cornea

(e)

Irregular lens

Astigmatism from an irregular lens

FIGURE 17-5 Normal and abnormal refraction in the eyeball. (a) Refraction of light rays passing from air into water. (b) In the normal or emmetropic eye, light rays from an object are bent sufficiently by the four refracting media and converged on the central fovea. A clear image is formed. (c) In the nearsighted or myopic eye, the image is focused in front of the retina. The condition may result from an elongated eyeball or thickened lens. Correction is by use of a concave lens that diverges entering light rays so that they have to travel further through the eyeball and are focused directly on the retina. (d) In the farsighted or hypermetropic eye, the image is focused behind the retina. The condition results from a shortened eyeball or a thin lens. Correction is by a convex lens that converges entering light rays so that they focus directly on the retina. (e) An astigmatism is an irregular curvature of the cornea (shown on left) or lens (shown on right). As a result, horizontal and vertical rays are focused at two different points on the retina, and the image is not focused on the area of sharpest vision of the retina. This results in blurred or distorted vision. Suitable glasses correct the refraction of an astigmatic eye.

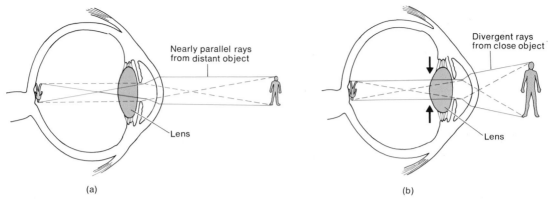

Nearly parallel rays
from distant object

Lens

(a)

Divergent rays
from close object

Lens

(b)

FIGURE 17-6 Accommodation. (a) For objects 6 m (20 ft) or more away. (b) For objects closer than 6 m.

CLINICAL APPLICATION

The normal eye, known as an **emmetropic** (em'-e-TROP-ik) **eye,** can sufficiently refract light rays from an object 6 m (20 ft) away to focus a clear image on the retina. Many individuals, however, do not have this ability because of abnormalities related to improper refraction. Among these abnormalities are **myopia** (mī-Ō-pē-a) or nearsightedness, **hypermetropia** (hī'-per-mē-TRO-pē-a) or farsightedness, and **astigmatism** (a-STIG-ma-tizm) or irregularities in the surface of the lens or cornea. The conditions are illustrated and explained in Figure 17-5c–e. The inability to focus on nearby objects due to loss of elasticity of the lens with age is called **presbyopia** (prez-bē-Ō-pē-a).

Almost 100 million Americans have various kinds of errors of vision. Although most people still wear eyeglasses, contact lenses are becoming increasingly popular. A **contact lens** is a lens that is worn in contact with the cornea, beneath the eyelids in front of the pupil. The outer surface of the lens is spherical and is designed to correct the visual defect, while the inner surface is precisely ground to fit the curvature of the cornea. The lens adheres to the cornea by capillary action on the normal layer of tear fluid.

• *Constriction of Pupil*   The circular muscle fibers of the iris also assume a function in the formation of clear retinal images. Part of the accommodation mechanism consists of the contraction of the dilator and sphincter muscles of the iris to constrict the pupil. **Constriction of the pupil** means narrowing the diameter of the hole through which light enters the eye. This action occurs simultaneously with accommodation of the lens and prevents light rays from entering the eye through the periphery of the lens. Light rays entering at the periphery would not be brought to focus on the retina and would result in blurred vision. The pupil, as noted earlier, also constricts in bright light to protect the retina from sudden or intense stimulation.

• *Convergence*   Because of the alignment of the orbits in the skull, many animals see a set of objects off to

the left through one eye and an entirely different set off to the right through the other. This characteristic doubles their field of vision and allows them to detect predators behind them. In humans, both eyes focus on only one set of objects—a characteristic called **single binocular vision.**

Single binocular vision occurs when light rays from an object are directed toward corresponding points on the two retinas. When we stare straight ahead at a distant object, the incoming light rays are aimed directly at both pupils and are refracted to identical spots on the retinas of both eyes. But as we move closer to the object, our eyes must rotate medially for the light rays from the object to hit the same points on both retinas. The term **convergence** refers to this medial movement of the two eyeballs so they are both directed toward the object being viewed. The nearer the object, the greater the degree of convergence necessary to maintain single binocular vision. Convergence is brought about by the coordinated action of the extrinsic eye muscles.

• *Inverted Image*   Images are focused upside down on the retina. They also undergo mirror reversal, that is, light reflected from the right side of an object hits the left side of the retina and vice versa. Note in Figure 17-5b that reflected light from the top of the object crosses light from the bottom of the object and strikes the retina below the central fovea. Reflected light from the bottom of the object crosses light from the top of the object and strikes the retina above the central fovea. The reason we do not see an inverted world is that the brain learns early in life to coordinate visual images with the exact locations of objects. The brain stores memories of reaching and touching objects and automatically turns visual images right-side-up and right-side-around.

Stimulation of Photoreceptors

• *Excitation of Rods*   After an image is formed on the retina by refraction, accommodation, constriction of the pupil, and convergence, light impulses must be converted into nerve impulses. The initial step is the development of generator potentials by rods and cones. To see how this occurs, we will first examine the role of photopigments.

A **photopigment** is a substance that can absorb light and undergo structural changes that can lead to the development of a generator potential. The photopigment in rods is called **rhodopsin (visual purple),** which is sensitive to low levels of illumination. Rhodopsin is composed of **scotopsin** (a protein) and **retinal** (a derivative of vitamin A) (Figure 17-7). A very important characteristic of retinal is that it can exist in two forms, depending on whether or not light is present. In the absence of light, retinal occurs as *cis*-retinal, in which form it structurally fits the scotopsin portion of rhodopsin. In the presence of light, *cis*-retinal undergoes an immediate change in structure from a curved molecule to a straight molecule called all-*trans*-retinal. Since all-*trans*-retinal has an altered physical structure, it no longer fits the scotopsin portion of rhodopsin, and it actually pulls away from scotopsin, forming a very unstable substance called **bathorhodopsin (prelumirhodopsin)** that rapidly decomposes sequentially to **lumirhodopsin, metarhodopsin I, metarhodopsin II,** and **pararhodopsin.** All of these substances are loose combinations of scotopsin and all-*trans*-retinal. Ultimately, paraphodopsin decomposes completely into scotopsin and all-*trans*-retinal. At the point where metarhodopsin I is con-

verted to metarhodopsin II, the generator potential develops in rods.

Following decomposition of rhodopsin into scotopsin and all-*trans*-retinal in the presence of light, rhodopsin is once again reformed. In this process, which occurs in the absence of light, the all-*trans*-retinal is converted back to *cis*-retinal, in the presence of an enzyme. Then *cis*-retinal combines with scotopsin to re-form rhodopsin. This compound is stable until its decomposition is again triggered by light energy.

The generator potential that develops from rhodopsin breakdown in rods is different from generator potentials that occur in other types of sensory receptors because stimulation results in hyperpolarization rather than depolarization. In rods, there is a decrease in the leakage of Na$^+$ ions to the interior, even though Na$^+$ ions are continually pumped out. The net loss of Na$^+$ ions from rods results in increased negativity inside the membrane of the rod, resulting in hyperpolarization.

Rhodopsin is highly light sensitive—even the light rays from the moon or a candle will break down some of it and thereby allow us to see. The rods, then, are specialized for night vision. However, they are of only limited help

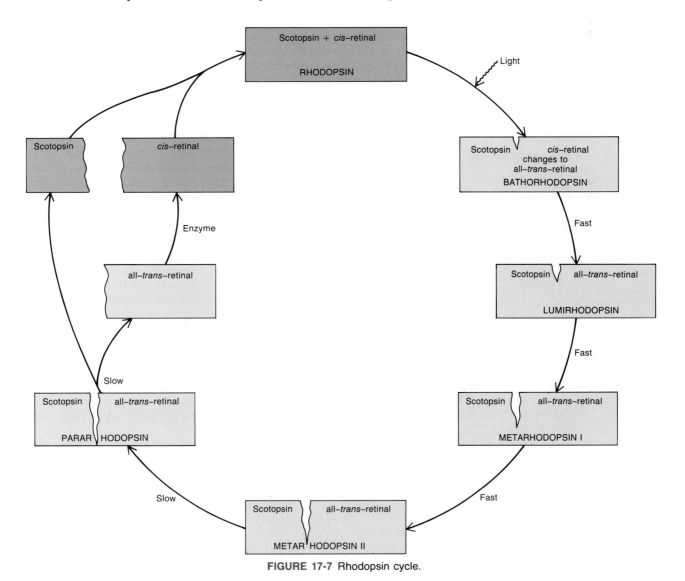

FIGURE 17-7 Rhodopsin cycle.

for daylight vision. In bright light, the rhodopsin is broken down faster than it can be manufactured. In dim light, production is able to keep pace with a slower rate of breakdown. These characteristics of rhodopsin are responsible for the experience of having to adjust to a dark room after walking in from the sunshine. The normal period of adjustment is the time it takes for the completely dissociated rhodopsin to re-form. **Night blindness,** which is also referred to as **nyctalopia** (nik'-ta-LŌ-pē-a), is the lack of normal night vision following the adjustment period. It is most often caused by vitamin A deficiency.

• *Excitation of Cones* Cones are the receptors for bright light and color. As in rods, photopigment decomposition produces generator potentials as a result of hyperpolarization. The photosensitive pigments in cones are almost the same as those in rods. Both contain retinal, but the protein portions in cones are called **photopsins,** which are slightly different from the scotopsin in rods. Unlike rhodopsin, the photopigments of the cones require bright light for their breakdown and they re-form quickly. It is believed that there are three types of cones, each containing a different combination of retinal and photopsin. Each combination has a different maximum absorption of light of a different wavelength, and thus each responds best to light of a given color. One type of cone responds best to red light, the second to green light, and the third to blue light. Just as an artist can obtain almost any color by mixing colors, it is believed that cones can perceive any color by differential stimulation. Stimulation of a cone by two or more colors may produce any combination of colors.

CLINICAL APPLICATION

If a single group of color receptive cones is missing from the retina, an individual cannot distinguish some colors from others and is said to be **color-blind.** The most common type is **red-green color blindness,** in which the cones best receptive to red light and to green light are missing. As a result, the person cannot distinguish between red and green. Color blindness is an inherited condition that affects males far more frequently than females. The inheritance of the condition is discussed in Chapter 29 and illustrated in Figure 29-18.

• *Light and Dark Adaptation* If an individual is exposed to bright light for an extended period of time, a considerable amount of the photosensitive pigment in both rods and cones is broken down into retinal and opsin (scotopsin or photopsin). In addition, most of the retinal in both rods and cones is converted to vitamin A. As a result, the concentration of the photosensitive pigment in both is reduced considerably, and the sensitivity of the eye to light is greatly reduced. This is called **light adaptation.**

Conversely, if an individual is exposed to darkness for an extended period of time, a considerable amount of retinal and opsins in the rods and cones is converted into the photosensitive pigments. Also, large amounts of vitamin A are converted into retinal, which is then converted into photosensitive pigments upon combination with opsins. As a result, the concentration of photosensitive pigments in rods and cones is increased considerably, and the sensitivity of the eye to light is greatly increased. This is called **dark adaptation.**

Visual Pathway

Examination of Figure 17-4b reveals that the retina (nervous tunic) is composed of three principal zones of neurons: photoreceptor (rods and cones), bipolar, and ganglion. For light to stimulate rods and cones to produce generator potentials, light must pass through ganglion and bipolar cells before reaching rods and cones. Note also that there are two special types of cells present in the retina (nervous tunic) called **horizontal cells** and **amacrine cells.**

Once generator potentials are developed by rods and cones, the potentials are transmitted unchanged through the bodies of the rods and cones. These generator potentials induce signals in both bipolar neurons and horizontal cells, probably by means of neurotransmitters. In response to the neurotransmitters, bipolar neurons become excited and horizontal cells become inhibited. Functionally, bipolar neurons transmit the excitatory visual signal from rods and cones to ganglion cells (see Figure 17-4b). Horizontal cells transmit inhibitory signals to bipolar neurons in the areas lateral to excited rods and cones. This lateral inhibition enhances contrasts in the visual scene between areas of the retina (nervous tunic) that are strongly stimulated and adjacent areas that are weakly stimulated. Horizontal cells also assist in the differentiation of various colors. Amacrine cells, which are also excited by bipolar neurons, synapse with ganglion cells and transmit information to them that signals a change in the level of illumination of the retina (nervous tunic).

When bipolar neurons transmit excitatory visual signals to ganglion cells, the ganglion cells become depolarized and initiate nerve impulses. The cell bodies of the ganglion cells lie in the retina, and their axons leave the eyeball via the **optic (II) nerve** (Figure 17-8). The axons pass through the **optic chiasma** (kī-AZ-ma), a crossing point of the optic (II) nerves. Some fibers cross to the opposite side. Others remain uncrossed. On passing through the optic chiasma, the fibers, now part of the **optic tract,** enter the brain and terminate in the lateral geniculate nucleus of the thalamus. Here the fibers synapse with third-order neurons whose axons pass to the visual areas located in the occipital lobes of the cerebral cortex.

Analysis of the afferent pathway to the brain reveals that the visual field of each eye is divided into two regions: the **nasal (medial) half** and the **temporal (lateral) half.** For each eye, light rays from an object in the nasal half of the visual field fall on the temporal half of the retina. Light rays from an object in the temporal half of the visual field fall on the nasal half of the retina (Figure 17-8). Also, light rays from objects at the top of the

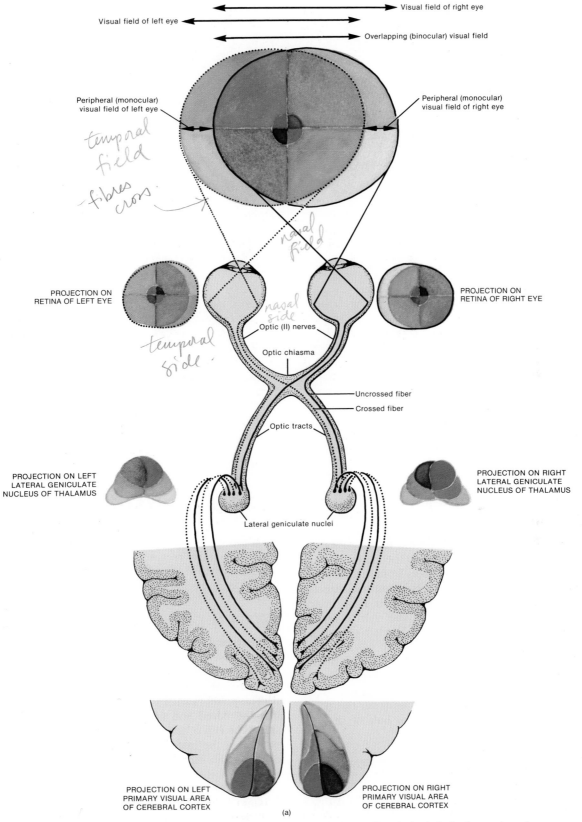

Visual field of right eye

Visual field of left eye

Overlapping (binocular) visual field

Peripheral (monocular) visual field of left eye

Peripheral (monocular) visual field of right eye

temporal field - fibres cross

nasal field

nasal side

temporal side

PROJECTION ON RETINA OF LEFT EYE

PROJECTION ON RETINA OF RIGHT EYE

Optic (II) nerves

Optic chiasma

Uncrossed fiber

Crossed fiber

Optic tracts

PROJECTION ON LEFT LATERAL GENICULATE NUCLEUS OF THALAMUS

PROJECTION ON RIGHT LATERAL GENICULATE NUCLEUS OF THALAMUS

Lateral geniculate nuclei

PROJECTION ON LEFT PRIMARY VISUAL AREA OF CEREBRAL CORTEX

PROJECTION ON RIGHT PRIMARY VISUAL AREA OF CEREBRAL CORTEX

(a)

FIGURE 17-8 Afferent pathway for visual impulses. (a) Diagram. The dark circle in the center of the visual fields is the macula lutea. The center of the macula lutea is the central fovea, the area of sharpest vision.

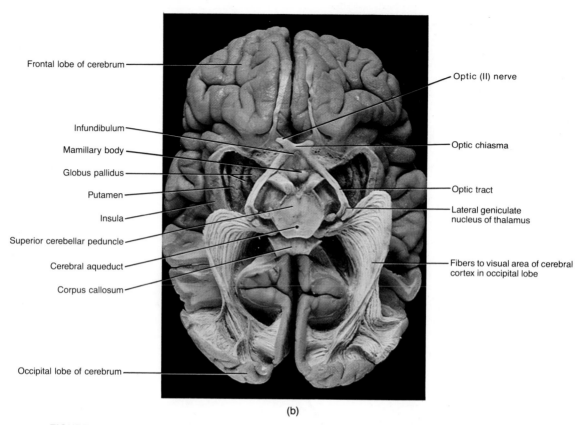

Frontal lobe of cerebrum

Infundibulum

Mamillary body

Globus pallidus

Putamen

Insula

Superior cerebellar peduncle

Cerebral aqueduct

Corpus callosum

Occipital lobe of cerebrum

Optic (II) nerve

Optic chiasma

Optic tract

Lateral geniculate nucleus of thalamus

Fibers to visual area of cerebral cortex in occipital lobe

(b)

FIGURE 17-8 (*Continued*) (b) Photograph as seen from the ventral aspect of the brain. (Courtesy of N. Gluhbegovic and T. H. Williams, *The Human Brain: A Photographic Guide,* Harper & Row, Publishers, Inc., Hagerstown, MD, 1980.)

visual field of each eye fall on the inferior portion of the retina, and light rays from objects at the bottom of the visual field fall on the superior portion of the retina. In the optic chiasma, nerve fibers from the nasal halves of the retinas cross and continue on to the lateral geniculate nuclei of the thalamus; nerve fibers from the temporal halves of the retinas do not cross, but continue directly on to the lateral geniculate nuclei. As a result, the primary visual area of the cerebral cortex of the right occipital lobe interprets visual sensations from the left side of an object via impulses from the temporal half of the retina of the right eye and the nasal half of the retina of the left eye. The primary visual area of the cerebral cortex of the left occipital lobe interprets visual sensations from the right side of an object via impulses from the nasal half of the right eye and the temporal half of the left eye.

CLINICAL APPLICATION

A **scotoma** (skō-TŌ-ma) or **blind spot** in the field of vision, other than the normal blind spot (optic disc), may indicate a brain tumor along one of the afferent pathways. For instance, a symptom of a tumor in the right optic tract might be an inability to see the left side of a normal field of vision without moving the eyeball.

AUDITORY SENSATIONS AND EQUILIBRIUM

In addition to containing receptors for sound waves, the ear also contains receptors for equilibrium. Anatomically, the ear is divided into three principal regions: the external (outer) ear, the middle ear, and the internal (inner) ear.

EXTERNAL OR OUTER EAR

The **external (outer) ear** is structurally designed to collect sound waves and direct them inward (Figure 17-9a,b). It consists of the pinna, the external auditory canal, and the tympanic membrane, also called the eardrum.

The **pinna (auricle)** is a flap of elastic cartilage shaped like the flared end of a trumpet and covered by thick skin. The rim of the pinna is called the **helix;** the inferior portion is the **lobule.** The pinna is attached to the head by ligaments and muscles. Additional surface anatomy features of the external ear are shown in Figure 17-9a.

The **external auditory canal (meatus)** is a tube about 2.5 cm (1 inch) in length that lies in the external auditory meatus of the temporal bone. It leads from the pinna to the eardrum. The wall of the canal consists of bone lined with cartilage that is continuous with the cartilage of the pinna. The cartilage in the external auditory canal is covered with thin, highly sensitive skin. Near the exterior opening, the canal contains a few hairs and specialized sebaceous glands called **ceruminous** (se-ROO-mi-nus)

(a)

1. **Pinna.** Portion of external ear not contained in head, also called auricle or trumpet.
2. **Tragus.** Cartilaginous projection anterior to external opening to ear.
3. **Antitragus.** Cartilaginous projection opposite tragus.
4. **Concha.** Hollow of auricle.
5. **Helix.** Superior and posterior free margin of auricle.
6. **Antihelix.** Semicircular ridge posterior and superior to concha.
7. **Triangular fossa.** Depression in superior portion of antihelix.
8. **Lobule.** Inferior portion of pinna devoid of cartilage.
9. **External auditory canal (meatus).** Canal extending from external ear to eardrum.

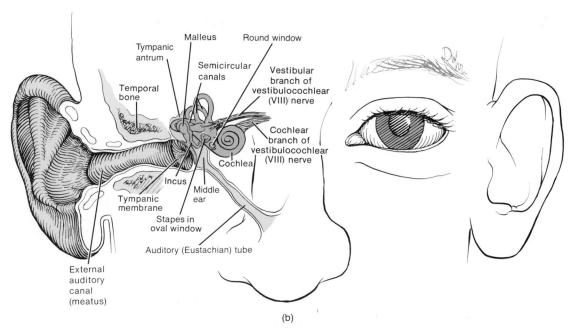

(b)

FIGURE 17-9 Structure of the auditory apparatus. (a) Surface anatomy of the right ear seen in lateral view. (Copyright © 1982 by Gerard J. Tortora. Courtesy of James Borghesi.) (b) Divisions of the right ear into external, middle, and internal portions seen in a frontal section through the right side of the skull.

glands, that secrete **cerumen** (earwax). The combination of hairs and cerumen (se-ROO-min) helps to prevent foreign objects from entering the ear.

The **tympanic** (tim-PAN-ik) **membrane (eardrum)** is a thin, semitransparent partition of fibrous connective tissue between the external auditory canal and the middle ear. Its external surface is concave and covered with skin. Its internal surface is convex and covered with a mucous membrane.

CLINICAL APPLICATION

Rupture of the tympanic membrane may result from foreign bodies, pressure, or infection. Severe bleeding and escape of cerebrospinal fluid through a ruptured tympanic membrane may occur following a severe blow to the head and indicate skull fracture.

(c)

FIGURE 17-9 (Continued) (c) Details of the middle ear and bony labyrinth of the internal ear. (d) Photograph of auditory ossicles of the middle ear. (Courtesy of C. Yokochi and J. W. Rohen, *Photographic Anatomy of the Human Body,* 2nd ed., 1978, IGAKU-SHOIN, Ltd., Tokyo, New York.)

MIDDLE EAR

The **middle ear,** also called the **tympanic cavity,** is a small, epithelial-lined, air-filled cavity hollowed out of the temporal bone (Figure 17-9b–d). The cavity is separated from the external ear by the eardrum and from the internal ear by a thin bony partition that contains two small openings: the oval window and the round window.

The posterior wall of the cavity communicates with the mastoid air cells of the temporal bone through a chamber called the **tympanic antrum.** This anatomical fact explains why a middle ear infection may spread to the temporal bone, causing mastoiditis, or even to the brain.

The anterior wall of the cavity contains an opening that leads directly into the **auditory (Eustachian) tube.** The auditory tube connects the middle ear with the nasopharynx of the throat. Through this passageway, infections may travel from the throat and nose to the ear. The function of the tube is to equalize air pressure on both sides of the tympanic membrane. Abrupt changes in external or internal air pressure might otherwise cause the eardrum to rupture. During swallowing and yawning, the tube opens to allow atmospheric air to enter or leave the middle ear until the internal pressure equals the external pressure. If the pressure is not relieved, intense pain, hearing impairment, tinnitus, and vertigo could develop. Any sudden pressure changes against the eardrum may be equalized by deliberately swallowing or pinching the nose closed, closing the mouth, and gently forcing air from the lungs into the nasopharynx.

Extending across the middle ear are three exceedingly small bones called **auditory ossicles** (OS-si-kuls). The bones, named for their shape, are the malleus, incus, and stapes, commonly called the hammer, anvil, and stirrup,

respectively. They are connected by synovial joints. The "handle" of the **malleus** is attached to the internal surface of the tympanic membrane. Its head articulates with the body of the incus. The **incus** is the intermediate bone in the series and articulates with the head of the stapes. The base or footplate of the **stapes** fits into a small opening in the thin bony partition between the middle and inner ear. The opening is called the **fenestra vestibuli** (fe-NES-tra ves-TIB-yoo-lī), or **oval window.** Directly below the oval window is another opening, the **fenestra cochlea** (fe-NES-tra KŌK-lē-a), or **round window.** This opening is enclosed by a membrane called the **secondary tympanic membrane.** auditory ossicles are attached to the tympanic cavity by means of ligaments.

In addition to the ligaments, there are also two muscles attached to the ossicles. The *tensor tympani muscle* draws the malleus medially, thus increasing tension on the tympanic membrane and reducing the amplitude of oscillations. This action tends to prevent damage to the inner ear when exposed to loud sounds.

The *stapedius muscle* is the smallest of all skeletal muscles. Its action is to draw the stapes posteriorly, thus reducing the oscillatory range of the stapes (preventing excessive movement). Like the tensor tympani muscle, the stapedius muscle has a protective function in that it dampens (checks) large vibrations that result from loud noises. It is for this reason that paralysis of the stapedius muscle is associated with *hyperacusia* (acuteness of hearing).

INTERNAL OR INNER EAR

The **internal (inner) ear** is also called the **labyrinth** (LAB-i-rinth) because of its complicated series of canals (Figure 17-10). Structurally, it consists of two main divisions: a bony labyrinth and a membranous labyrinth that fits in the bony labyrinth. The **bony labyrinth** is a series of cavities in the petrous portion of the temporal bone. It can be divided into three areas named on the basis of shape: the vestibule, cochlea, and semicircular canals. The bony labyrinth is lined with periosteum and contains a fluid called **perilymph.** This fluid surrounds the **membranous labyrinth,** a series of sacs and tubes lying inside and having the same general form as the bony labyrinth. The membranous labyrinth is lined with epithelium and contains a fluid called **endolymph.**

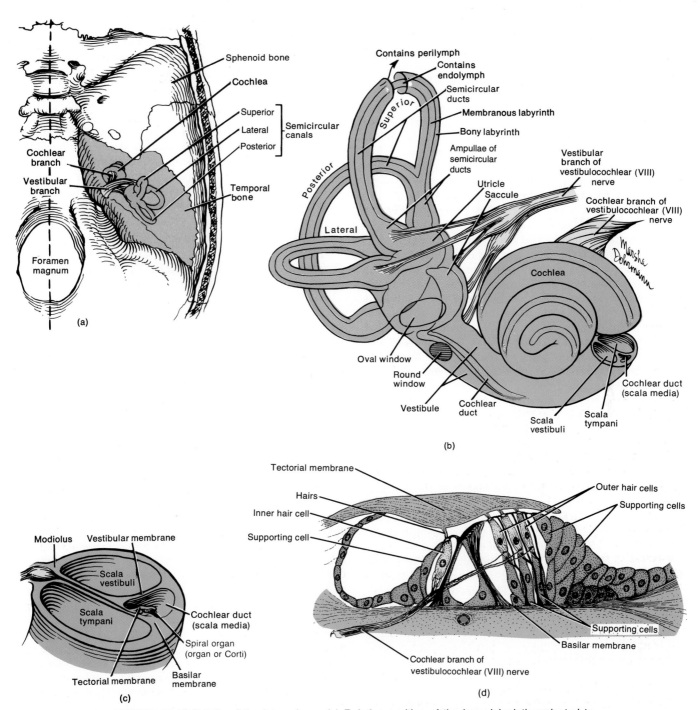

FIGURE 17-10 Details of the internal ear. (a) Relative position of the bony labyrinth projected to the inner surface of the floor of the skull. (b) The outer, blue area belongs to the bony labyrinth. The inner, rust-colored area belongs .to the membranous labyrinth. (c) Cross section through the cochlea. (d) Enlargement of the spiral organ (organ of Corti).

The **vestibule** constitutes the oval central portion of the bony labyrinth. The membranous labyrinth in the vestibule consists of two sacs called the **utricle** and **saccule.** These sacs are connected to each other by a small duct.

Projecting upward and posteriorly from the vestibule are the three bony **semicircular canals.** Each is arranged at approximately right angles to the other two. On the basis of their positions, they are called the superior, posterior, and lateral canals. One end of each canal enlarges into a swelling called the **ampulla** (am-POOL-la). The portions of the membranous labyrinth that lie inside the bony semicircular canals are called the **semicircular ducts (membranous semicircular canals).** These structures are almost identical in shape to the semicircular canals and communicate with the utricle of the vestibule.

Lying in front of the vestibule is the **cochlea** (KŌK-lē-a), so designated because of its resemblance to a snail's shell. The cochlea consists of a bony spiral canal that makes about 2¾ turns around a central bony core called the **modiolus.** A cross section through the cochlea shows the canal is divided into three separate channels by partitions that together have the shape of the letter Y. The stem of the Y is a bony shelf that protrudes into the canal; the wings of the Y are composed mainly of membranous labyrinth. The channel above the bony partition is the **scala vestibuli;** the channel below is the **scala tympani.** The cochlea adjoins the wall of the vestibule, into which the scala vestibuli opens. The scala tympani terminates at the round window. The perilymph of the vestibule is continuous with that of the scala vestibuli. The third channel (between the wings of the Y) is the membranous labyrinth: the **cochlear duct (scala media).** The cochlear duct is separated from the scala vestibuli by the **vestibular membrane.** It is separated from the scala tympani by the **basilar membrane.**

Resting on the basilar membrane is the **spiral organ (organ of Corti),** the organ of hearing. The spiral organ is a series of epithelial cells on the inner surface of the basilar membrane. It consists of a number of supporting cells and hair cells, which are the receptors for auditory sensations. The inner hair cells are medially placed in a single row and extend the entire length of the cochlea. The outer hair cells are arranged in several rows throughout the cochlea. The hair cells have long hairlike processes at their free ends that extend into the endolymph of the cochlear duct. The basal ends of the hair cells are in contact with fibers of the cochlear branch of the vestibulocochlear (VIII) nerve. Projecting over and in contact with the hair cells of the spiral organ is the **tectorial membrane,** a delicate and flexible gelatinous membrane.

CLINICAL APPLICATION

Hair cells of the spiral organ are easily damaged by **exposure to high-intensity noises** such as those produced by engines of jet planes, revved-up motorcycles, and loud music. As a result of the noises, hair cells become arranged in disorganized patterns or they and their supporting cells may degenerate.

SOUND WAVES

Sound waves result from the alternate compression and decompression of air molecules. They originate from a vibrating object, much the same way that waves travel over the surface of water. The sounds heard most acutely by human ears are those from sources that vibrate at frequencies between 1,000 and 4,000 cycles per second (Hz). The entire range extends from 20 to 20,000 Hz.

The frequency of vibration of the source of the sound in directly related to pitch. The greater the vibration, the higher the pitch. Also, the greater the force of the vibration, the louder the sound. Relative sound intensities (loudness) are usually measured in units called **decibels (dB).** In the sound intensity range for normal communication, the human ear can detect approximately a 1-dB change in sound intensity. The hearing threshold, that is, the point at which an average young adult can just detect sound from silence is 0 dB. Between 115 and 120 dB is the threshold of pain. Just to give you some idea of how decibels relate to common sounds in the range between 0 and 120 dB, rustling leaves have a decibel rating of 15, normal conversation 45, crowd noise 60, a vacuum cleaner 75, and a pneumatic drill 90.

PHYSIOLOGY OF HEARING

The events involved in the physiology of hearing sound waves are as follows (Figure 17-11):

1. Sound waves that reach the ear are directed by the pinna into the external auditory canal.

2. When the waves strike the tympanic membrane, the alternate compression and decompression of the air causes the membrane to vibrate. The distance the membrane moves is always very small and is relative to the force and velocity with which air molecules strike it. It vibrates slowly in response to low-frequency sounds and rapidly in response to high-frequency sounds.

3. The central area of the tympanic membrane is connected to the malleus, which also starts to vibrate. The vibration is then picked up by the incus, which transmits the vibration to the stapes.

4. As the stapes moves back and forth, it pushes the oval window in and out.

If sound waves passed directly to the oval window without passing through the tympanic membrane and auditory bones, hearing would be inadequate. A minimal amount of sound energy is required to transmit sound waves through the perilymph of the cochlea. Since the tympanic membrane has a surface area about 22 times larger than that of the oval window, it can collect about 22 times more sound energy. This energy is sufficient to transmit sound waves through the perilymph.

5. The movement of the oval window sets up waves in the perilymph.

6. As the window bulges inward, it pushes the perilymph of the scala vestibuli, and waves are propagated through the scala vestibuli into the fluid of the scala tympani.

7. This pressure pushes the vestibular membrane inward and increases the pressure of the endolymph inside the cochlear duct.

8. The basilar membrane gives under the pressure and bulges out into the scala tympani.

9. The sudden pressure in the scala tympani pushes the perilymph toward the round window, causing it to bulge back into the middle ear. Conversely, as the sound waves subside,

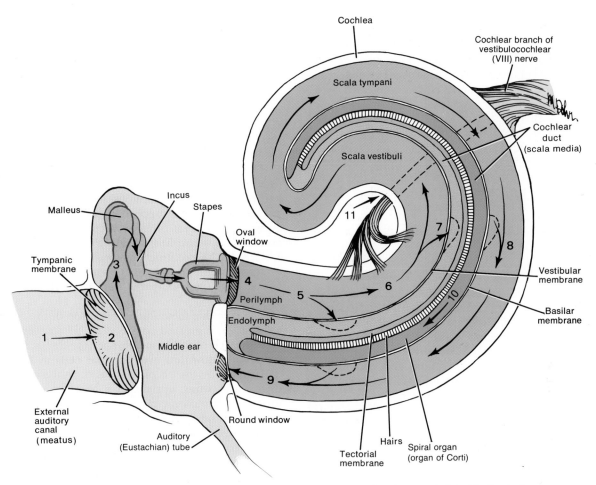

FIGURE 17-11 Physiology of hearing. The numbers correspond to the events listed in the text.

the stapes moves backward and the procedure is reversed. That is, the fluid moves in the opposite direction along the same pathway, and the basilar membrane bulges into the cochlear duct.

10. When the basilar membrane vibrates, the hair cells of the spiral organ are moved against the tectorial membrane. The movement of the hairs develops generator potentials that ultimately lead to the generation of nerve impulses. Differences in pitch are related to sound waves of various frequencies that cause specific regions of the basilar membrane to vibrate more intensely than others. The membrane is less flexible at the base of the cochlea (portion closer to the oval window) and more flexible near the apex of the cochlea. High-frequency or high-pitched sounds cause the basilar membrane to vibrate near the base of the cochlea, and low-frequency or low-pitched sounds cause the basilar membrane to vibrate near the apex of the cochlea. In addition, loudness is determined by the intensity of sound waves. High-intensity sound waves cause greater vibration of the basilar membrane.

11. The impulses are then passed on to the cochlear branch of the vestibulocochlear (VIII) nerve (see Figure 17-10b) and cochlear nuclei in the medulla. Here most impulses cross to the opposite side and then travel to the inferior colliculus of the midbrain, medial geniculate body of the thalamus, and finally to the auditory area of the temporal lobe of the cerebral cortex.

The function of hair cells is to convert a mechanical force (stimulus) into an electrical signal (nerve impulse). It is believed to occur as follows. When the hairs at the top of the cell are moved, channels in the hair cell membranes open, permitting

a rapid influx of potassium (K^+) ions. This depolarizes the hair cell membrane, producing the generator potential. Depolarization spreads through the cell and causes calcium channels in the base of the hair cell to open, resulting in an influx of calcium (Ca^{2+}) ions. At the base of the hair cell are vesicles that store neurotransmitter. The Ca^{2+} ions cause the vesicles to fuse with the basal portion of the hair cell's membrane, thus releasing neurotransmitter from the hair cell, which excites a sensory nerve fiber at the base of the hair cell. From here, the nerve impulse is carried to the brain via the cochlear branch of the vestibulocochlear (VIII) nerve. Although the exact nature of the neurotransmitter is not known, there is some evidence that it might be glutamate or gamma aminobutyric acid (GABA).

CLINICAL APPLICATION

Artificial ears are cochlear implants that translate sounds into electronic signals that can be interpreted by the brain. They take the place of hair cells of the spiral organ, which normally convert sound waves into electrical signals carried to the brain. The implants are used for individuals with sensorineural deafness, that is, deafness due to disease or injury that has destroyed hair cells of the spiral organ. Such people make up the majority of those who have no hearing in either ear.

The device consists of electrodes implanted in the cochlea that are connected to a plug attached to the outside of the head. A microprocessor pack, about the size of a small transistor radio, is worn on the person's belt or in a shirt pocket and connected by a cord to the plug. Sound waves enter a tiny microphone in the ear, similar to that of a usual hearing aid, and travel down the cord into the pack where they are converted into electrical signals. The signals then travel to the electrodes in the cochlea where they stimulate nerve endings and transmit the signals to the brain via the cochlear branch of the vestibulocochlear (VIII) nerve. The sounds heard are crude compared to normal hearing.

PHYSIOLOGY OF EQUILIBRIUM

There are two kinds of **equilibrium.** One, called **static equilibrium,** refers to the orientation of the body (mainly the head) relative to the ground (gravity). The second kind, **dynamic equilibrium,** is the maintenance of body position (mainly the head) in response to sudden movements such as rotation, acceleration, and deceleration. The receptor organs for equilibrium are the maculae of the saccule and the utricle and cristae in the semicircular ducts.

Static Equilibrium

The walls of both the utricle and saccule contain a small, flat, plaquelike region called a **macula** (Figure 17-12). The maculae are the receptors that are concerned with static equilibrium. They provide sensory information regarding the orientation of the head in space and are essential for maintaining posture.

Microscopically, the two maculae resemble the spiral organ. They consist of differentiated neuroepithelial cells that are innervated by the vestibular branch of the vestibulocochlear (VIII) nerve. The maculae are anatomically located in planes perpendicular to one another and possess

two kinds of cells: **hair (receptor) cells** and **supporting cells.** Two shapes of hair cells have been identified: one is more flask-shaped and the other is more cylindrical. Both show long extensions of the cell membrane consisting of many *stereocilia* (they are actually microvilli) and one *kinocilium* (a conventional cilium) anchored firmly to its basal body and extending beyond the longest microvilli. Some of the microvilli reach lengths of over 100 μm, and some receptor cells have over 80 such projections.

The columnar supporting cells of the maculae are scattered between the hair cells. Floating directly over the hair cells is a thick, gelatinous, glycoprotein layer, probably secreted by the supporting cells, called the **otolithic membrane.** A layer of calcium carbonate crystals, called **otoliths** (*oto* = ear; *lithos* = stone), extends over the entire surface of the otolithic membrane. The specific gravity of these otoliths is about 3, which makes them much denser than the endolymph fluid that fills the rest of the utricle.

The otolithic membrane sits on top of the macula like a discus on a greased cookie sheet. If you tilt your head forward, the otolithic membrane (the discus in our analogy) slides downhill over the hair cells in the direction determined by the tilt of your head. In the same manner, if you are sitting upright in a car that suddenly jerks forward, the otolithic membrane, due to its inertia, slides backward and stimulates the hair cells. As the otoliths move, they pull on the gelatinous layer which pulls on the stereocilia and makes them bend. The movement of the stereocilia initiates a nerve impulse that is then transmitted to the vestibular branch of the vestibulocochlear (VIII) nerve (see Figure 17-10b).

Most of the vestibular branch fibers enter the brain stem and terminate in the vestibular nuclear complex in the medulla. The remaining fibers enter the flocculonodular lobe of the cerebellum through the inferior cerebellar peduncle. Bidirectional pathways connect the vestibular nuclei and cerebellum. Fibers from all the vestibular nuclei form the medial longitudinal fasciculus which extends from the brain stem into the cervical portion of the spinal cord. The fasciculus sends impulses to the nuclei of cranial nerves that control eye movements [oculomotor (III), trochlear (IV), and abducens (VI)] and to the accessory (XI) nerve nucleus that helps control movements of the head and neck. In addition, fibers from the lateral vestibular nucleus form the vestibulospinal tract, which conveys impulses to skeletal muscles that regulate body tone in response to head movements. Various pathways between the vestibular nuclei, cerebellum, and cerebrum enable the cerebellum to assume a key role in helping the body maintain static equilibrium. The cerebellum continuously receives updated sensory information from the utricle and saccule concerning static equilibrium. Using this information, the cerebellum monitors and makes corrective adjustments in the motor activities that originate in the cerebral cortex. Essentially, the cerebellum sends continuous impulses to the motor areas of the cerebrum, in response to input from the utricle and saccule, causing the motor system to increase or decrease its impulses to specific skeletal muscles in order to maintain static equilibrium.

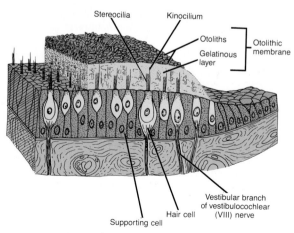

Stereocilia

Kinocilium

Otoliths

Gelatinous layer

Otolithic membrane

Vestibular branch of vestibulocochlear (VIII) nerve

Hair cell

Supporting cell

FIGURE 17-12 Structure of the macula.

Dynamic Equilibrium

The cristae in the three semicircular ducts maintain dynamic equilibrium (Figure 17-13). The ducts are positioned at right angles to one another in three planes: frontal (the superior duct), sagittal (the posterior duct), and lateral (the lateral duct). This positioning permits detection of an imbalance in three planes. In the ampulla, the dilated portion of each duct, there is a small elevation called the **crista.** Each crista is composed of a group of **hair (receptor) cells** and **supporting cells** covered by a mass of gelatinous material called the **cupula.** When the head moves, the endolymph in the semicircular ducts flows over the hairs and bends them. The movement of the hairs stimulates sensory neurons, and the impulses pass over the vestibular branch of the vestibulocochlear (VIII) nerve. The impulses follow the same pathways as those involved in static equilibrium and are eventually sent to the muscles that must contract to maintain body balance in the new position.

A summary of the structures of the ear related to hearing and equilibrium is presented in Exhibit 17-2.

EXHIBIT 17-2 SUMMARY OF STRUCTURES OF THE EAR RELATED TO HEARING AND EQUILIBRIUM

STRUCTURE	FUNCTION
External (outer) ear	
Pinna (auricle)	Collects sound waves.
External auditory canal (meatus)	Directs sound waves to tympanic membrane.
Tympanic membrane	Sound waves cause it to vibrate which, in turn, causes the malleus to vibrate.
Middle ear	
Auditory (Eustachian) tube	Equalizes pressure on both sides of the tympanic membrane.
Auditory ossicles	Transmit sound waves from tympanic membrane to oval window.
Internal (inner) ear	
Utricle	Contains macula, receptor for static equilibrium.
Saccule	Contains macula, receptor for static equilibrium.
Semicircular ducts	Contain cristae, receptors for dynamic equilibrium.
Cochlea	Contains a series of fluids, channels, and membranes that transmit sound waves to the spiral organ (organ of Corti), the organ of hearing; the spiral organ generates nerve impulses and transmits them to the cochlear branch of the vestibulocochlear (VIII) nerve.

FIGURE 17-13 Semicircular ducts and dynamic equilibrium. (a) Position of the cristae relative to the membranous ampullae. (b) Enlarged aspect of a crista.

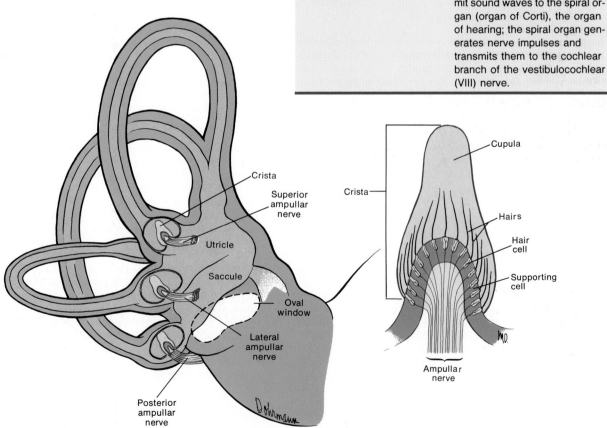

(a)   (b)

DISORDERS: HOMEOSTATIC IMBALANCES

The special sense organs can be altered or damaged by numerous disorders. The causes of disorder range from congenital origins to the effects of old age. Here we discuss a few common disorders of the eyes and ears.

Cataract

The most prevalent disorder resulting in blindness is cataract formation. A **cataract** is a clouding of the lens or its capsule so that it becomes opaque or milk white. Two basic processes involved in cataract formation are the breakdown of the normal lens protein and an influx of water into the lens. As a result, the light from an object, which normally passes directly through the lens to produce a sharp image, produces only a degraded image. If the cataract is severe enough, no image at all is produced. The ultraviolet (UV) light in sunlight is also an important factor in cataract development. UV light causes the deposition of minute brown opaque granules in the lens that result in clouding of the lens.

Glaucoma

The second most common cause of blindness, especially in the elderly, is **glaucoma.** This disorder is characterized by an abnormally high pressure due to a build up of aqueous humor inside the eyeball. The aqueous humor does not return into the bloodstream through the scleral venous sinus (canal of Schlemm) as quickly as it is formed. The fluid accumulates and, by compressing the lens into the vitreous humor, puts pressure on the neurons of the retina. If the pressure continues over a long period of time, glaucoma can progress from mild visual impairment to a point where neurons of the retina are destroyed, resulting in degeneration of the optic disc, visual field defects, and blindness. Treatment is through drugs or laser surgery.

Conjunctivitis (Pinkeye)

Many different eye inflammations exist, but the most common type is **conjunctivitis (pinkeye),** an inflammation of the conjunctiva, the membrane that lines the insides of the eyelids and covers the cornea. Conjunctivitis can be caused by microorganisms, most often the pneumococci or staphylococci bacteria. In such cases, the inflammation is very contagious. This epidemic type, common in children, is rarely serious. Conjunctivitis may also be caused by a number of irritants, in which case the inflammation is not contagious. Irritants include dust, smoke, wind, pollutants in the air, and excessive glare. The condition may be acute or chronic.

Trachoma

A serious form of chronic contagious conjunctivitis is known as **trachoma** (tra-KŌ-ma). It is caused by an organism, called the TRIC agent, that has characteristics of both viruses and bacteria. Trachoma is characterized by many granulations or fleshy projections on the eyelids. If untreated, these projections can irritate and inflame the cornea and reduce vision. The disease produces an excessive growth of subconjunctival tissue and the invasion of blood vessels into the upper half of the front of the cornea. The disease progresses until it covers the entire cornea, bringing about a loss of vision because of corneal opacity.

Antibiotics, such as tetracycline and the sulfa drugs, kill the organisms that cause trachoma and have reduced the seriousness of this infection.

Deafness

Deafness is the lack of the sense of hearing or significant hearing loss. It is generally divided into two principal types: sensori-neural deafness and conduction deafness. *Sensorineural deafness* is caused by impairment of the cochlea or cochlear branch of the vestibulocochlear (VIII) nerve. *Conduction deafness* is caused by impairment of the external and middle ear mechanisms for transmitting sounds into the cochlea. Among the factors that contribute to deafness are atherosclerosis, which reduces blood supply to the ears; repeated exposure to loud noise, which destroys hair cells of the spiral organ (organ of Corti); certain drugs such as streptomycin; disease; impacted cerumen; injury to the tympanic membrane; and aging, which results in thickening of the tympanic membrane, stiffening of the auditory ossicles, and decreased numbers of hair cells due to diminished cell division.

Labyrinthine Disease

Labyrinthine (lab'-i-RIN-thēn) **disease** refers to a malfunction of the inner ear that is characterized by deafness, tinnitus (ringing in the ears), vertigo (hallucination of movement), nausea, and vomiting. There may also be a blurring of vision, nystagmus (rapid, involuntary movement of the eyeballs), and a tendency to fall in a certain direction.

Among the causes of labyrinthine disease are: (1) infection of the middle ear, (2) trauma from brain concussion producing hemorrhage or splitting of the labyrinth, (3) cardiovascular diseases such as atherosclerosis and blood vessel disturbances, (4) congenital malformation of the labyrinth, (5) excessive formation of endolymph, (6) allergy, (7) blood abnormalities, and (8) aging.

Ménière's Syndrome

Ménière's (men-YAIRZ) **syndrome** is one type of labyrinthine disorder that is characterized by fluctuating hearing loss, attacks of vertigo, and roaring tinnitus. Etiology of Ménière's syndrome is unknown. It is now thought that there is either an overproduction or underabsorption of endolymph in the cochlear duct. The hearing loss is caused by distortions in the basilar membrane of the cochlea. The classic type of Ménière's syndrome involves both the semicircular canals and the cochlea. The natural course of Ménière's syndrome may stretch out over a period of years, with the end result being almost total destruction of hearing.

Vertigo

Vertigo (*vertex* = whorl) is a sensation of motion in which the world is revolving or the person is revolving in space. Depending on its cause, vertigo may be classified as: (1) *peripheral,* which originates in the ear, has a sudden onset, and lasts from minutes to hours; (2) *central,* which is caused by an abnormality of the central nervous system (CNS) and may persist for more than 3 weeks; and (3) *psychogenic,* in which the cause is of psychological origin.

Otitis Media

Otitis media is an acute infection of the middle ear cavity with a reddening and outward bulging of the eardrum, which may rupture. Abnormal function of the auditory (Eustachian) tube appears to be the most important mechanism in the pathogenesis of middle ear disease. This allows bacteria from the nasopharynx, which are a primary cause of middle ear infection, to enter the middle ear. Hearing loss is by far the most prevalent complication of otitis media.

Motion Sickness

Motion sickness is a functional disorder brought on by repetitive angular, linear, or vertical motion and characterized by various symptoms, primarily nausea and vomiting. Warning symptoms include yawning, salivation, pallor, hyperventilation, profuse cold sweating, and prolonged drowsiness. Specific kinds of motion sickness are seasickness, airsickness, carsickness, and space-, train-, or swing sickness. Etiology is excessive stimulation of the vestibular apparatus by motion.

MEDICAL TERMINOLOGY

Achromatopsia (*a* = without; *chrom* = color) Complete color blindness.

Ametropia (*ametro* = disproportionate; *ops* = eye) Refractive defect of the eye resulting in an inability to focus images properly on the retina.

Anopsia (*opsia* = vision) A defect of vision.

Astereognosis (*stereos* = solid; *gnosis* = knowledge) Loss of ability to recognize objects or to appreciate their form by touching them.

Audiometer (*audire* = to hear; *metron* = to measure) An instrument used to measure hearing by producing acoustic stimuli of known frequency and intensity.

Blepharitis (*blepharo* = eyelid; *itis* = inflammation of) An inflammation of the eyelid.

Dyskinesia (*dys* = difficult; *kinesis* = movement) Abnormality of motor function characterized by involuntary, purposeless movements.

Eustachitis An inflammation or infection of the auditory (Eustachian) tube.

Keratitis (*kerato* = cornea) An inflammation or infection of the cornea.

Kinesthesis (*aisthesis* = sensation) The sense of perception of movement.

Labyrinthitis An inflammation of the labyrinth (inner ear).

Myringitis (*myringa* = eardrum) An inflammation of the eardrum; also called tympanitis.

Nystagmus A constant, rapid involuntary movement of the eyeballs, possibly caused by a disease of the central nervous system.

Otalgia (*oto* = ear; *algia* = pain) Earache.

Otosclerosis (*sclerosis* = hardening) Pathological process in which new bone is deposited around the oval window. The result may be immobilization of the stapes, leading to deafness.

Presbyopia (*presby* = old) Inability to focus on nearby objects due to loss of elasticity of the crystalline lens. The loss is usually caused by aging.

Ptosis (*ptosis* = fall) Falling or drooping of the eyelid. (This term is also used for the slipping of any organ below its normal position.)

Retinoblastoma (*blast* = bud; *oma* = tumor) A common tumor arising from immature retinal cells and accounting for 2 percent of childhood malignancies.

Strabismus An eye muscle disorder, commonly called "crossed eyes," in which the eyeballs do not move in unison. It may be caused by lack of coordination of the extrinsic eye muscles.

Tinnitus A ringing in the ears.

STUDY OUTLINE

Offactory Sensations (p. 371)

1. The receptors for olfaction, the olfactory cells, are in the nasal epithelium.
2. Substances to be smelled must be volatile, water-soluble, and lipid-soluble.
3. Adaptation to odors occurs quickly, and the threshold of smell is low.
4. Olfactory cells convey impulses to olfactory (I) nerves, olfactory bulbs, olfactory tracts, and cerebral cortex.

Gustatory Sensations (p. 372)

1. The receptors for gustation, the gustatory cells, are located in taste buds.
2. Substances to be tasted must be in solution in saliva.
3. The four primary tastes are salt, sweet, sour, and bitter.
4. Adaptation to taste occurs quickly, and the threshold varies with the taste involved.
5. Gustatory cells convey impulses to cranial nerves V, VII, IX, and X, medulla, thalamus, and cerebral cortex.

Visual Sensations (p. 373)

1. Accessory structures of the eyes include the eyebrows, eyelids, eyelashes, and the lacrimal apparatus.
2. The eye is constructed of three coats: (a) fibrous tunic (sclera and cornea), (b) vascular tunic (choroid, ciliary body, and iris), and (c) retina (nervous tunic), which contains rods and cones.

3. The anterior cavity contains aqueous humor; the posterior cavity contains vitreous humor.
4. The refractive media of the eye are the cornea, aqueous humor, lens, and vitreous humor.
5. Retinal image formation involves refraction of light, accommodation of the lens, constriction of the pupil, convergence, and inverted image formation.
6. Improper refraction may result from myopia (nearsightedness), hypermetropia (farsightedness), and astigmatism (corneal or lens abnormalities).
7. Rods and cones develop generator potentials and ganglion cells initiate nerve impulses.
8. Impulses from ganglion cells are conveyed through the retina to the optic (II) nerve, the optic chiasma, the optic tract, the thalamus, and the cortex.

Auditory Sensations and Equilibrium (p. 386)

1. The ear consists of three anatomical subdivisions: (a) the external or outer ear (pinna, external auditory canal, and tympanic membrane), (b) the middle ear (auditory or Eustachian tube, ossicles, oval window, and round window), and (c) the internal or inner ear (bony labyrinth and membranous labyrinth). The internal ear contains the spiral organ (organ of Corti), the organ of hearing.
2. Sound waves enter the external auditory canal, strike the tympanic membrane, pass through the ossicles, strike the oval window, set up waves in the perilymph, strike the ves-

tibular membrane and scala tympani, increase pressure in the endolymph, strike the basilar membrane, and stimulate hairs on the spiral organ. A sound impulse is then initiated.

3. Static equilibrium is the orientation of the body relative to the pull of gravity. The maculae of the utricle and saccule are the sense organs of static equilibrium.

4. Dynamic equilibrium is the maintenance of body position in response to movement. The cristae in the semicircular ducts are the sense organs of dynamic equilibrium.

Disorders: Homeostatic Imbalances (p. 394)

1. Cataract is the loss of transparency of the lens or capsule.
2. Glaucoma is abnormally high intraocular pressure, which destroys neurons of the retina.

3. Conjunctivitis (pinkeye) is an inflammation of the conjunctiva.
4. Trachoma is a chronic, contagious inflammation of the conjunctiva.
5. Deafness is the lack of the sense of hearing or significant hearing loss. It is classified as sensorineural and conduction.
6. Labyrinthine disease is basically a malfunction of the inner ear that has a variety of causes.
7. Ménière's syndrome is the malfunction of the inner ear that may cause deafness and loss of equilibrium.
8. Vertigo is a sensation of motion that is classified as peripheral, central, or psychogenic.
9. Otitis media is an acute infection of the middle ear cavity.
10. Motion sickness is a functional disorder precipitated by repetitive angular, linear, or vertical motion.

REVIEW QUESTIONS

1. What are the necessary conditions for substances to be smelled?
2. Outline the two principal theories of olfaction.
3. Discuss the origin and path of an impulse that results in smelling.
4. How are papillae related to taste buds? Describe the structure and location of the papillae.
5. How are gustatory receptors stimulated?
6. Describe the thresholds for the four primary tastes.
7. Discuss how an impulse for taste travels from a taste bud to the brain.
8. Describe the structure and importance of the following accessory structures of the eye: eyelids, eyelashes, and eyebrows.
9. What is the function of the lacrimal apparatus? Explain how it operates.
10. By means of a labeled diagram, indicate the principal anatomical structures of the eye.
11. How do extrinsic and intrinsic eye muscles differ?
12. Describe the location and contents of the chambers of the eye. What is intraocular pressure? How is the scleral venous sinus (canal of Schlemm) related to this pressure?
13. Explain how each of the following events is related to the physiology of vision: (a) refraction of light, (b) accommodation of the lens, (c) constriction of the pupil, (d) convergence, and (e) inverted image formation.
14. Distinguish emmetropia, myopia, hypermetropia, and astigmatism by means of a diagram.

15. How are rods and cones excited? Relate your discussion to the rhodopsin cycle by means of a diagram.
16. What is night blindness? What causes it?
17. Distinguish between light adaptation and dark adaptation.
18. Describe the path of a visual impulse from the optic (II) nerve to the brain.
19. Define visual field. Relate the visual field to image formation on the retina.
20. Diagram the principal parts of the outer, middle, and inner ear. Describe the function of each part labeled.
21. What are sound waves? How are sound intensities measured?
22. Explain the events involved in the transmission of sound from the pinna to the spiral organ (organ of Corti).
23. What is the afferent pathway for sound impulses from the cochlear branch of the vestibulocochlear (VIII) nerve to the brain?
24. Compare the function of the maculae in the saccule and utricle in maintaining static equilibrium with the role of the cristae in the semicircular ducts in maintaining dynamic equilibrium.
25. Describe the path of an impulse that results in static and dynamic equilibrium.
26. Define each of the following: cataract, glaucoma, conjunctivitis, trachoma, deafness, labyrinthine disease, Ménière's syndrome, vertigo, otitis media, and motion sickness.
27. Refer to the medical terminology associated with the sense organs. Be sure that you can define each term listed.

The Endocrine System

18

■ **STUDENT OBJECTIVES**

- Discuss the functions of the endocrine system in maintaining homeostasis.
- Define an endocrine gland and an exocrine gland, and list the endocrine glands of the body.
- Describe how hormones are classified according to their chemistry.
- Explain the mechanisms of hormonal action by interaction with plasma membrane receptors and intracellular receptors.
- Identify the role of prostaglandins in hormonal action.
- Describe the control of hormonal secretions via feedback cycles and explain several examples.
- Describe the structural and functional division of the pituitary gland into the adenohypophysis and the neurohypophysis.
- Discuss how the pituitary gland and hypothalamus are structurally and functionally related.
- List the hormones of the adenohypophysis, their principal actions, and their associated hypothalamic regulating factors.
- Describe the release of hormones stored in the neurohypophysis and their principal actions.
- Discuss the symptoms of pituitary dwarfism, giantism, acromegaly, and diabetes insipidus as pituitary gland disorders.
- Explain the synthesis, storage, and release of thyroid hormones.
- Discuss the principal actions and control of thyroid gland hormones.
- Discuss the symptoms of cretinism, myxedema, exophthalmic goiter, and simple goiter as thyroid gland disorders.
- Explain the principal actions and control of the parathyroid hormone.
- Discuss the symptoms of tetany and osteitis fibrosa cystica as parathyroid gland disorders.
- Describe the subdivision of the adrenal (suprarenal) glands into cortical and medullary portions.
- Distinguish the effects of adrenal cortical mineralocorticoids, glucocorticoids, and gonadocorticoids on physiological activities, and explain how the hormones are controlled.
- Discuss the symptoms of aldosteronism, Addison's disease, Cushing's syndrome, and adrenogenital syndrome as adrenal cortical disorders.
- Identify the function of the adrenal medullary secretions as supplements of sympathetic responses.
- Discuss the symptoms of pheochromocytomas as an adrenal medullary disorder.
- Compare the principal actions of the pancreatic hormones and describe how they are controlled.
- Discuss the symptoms of diabetes mellitus and hyperinsulinism as endocrine disorders of the pancreas.
- Describe the possible functions of hormones of the pineal gland.
- Describe the functions of the hormones of the thymus gland in immunity.
- Describe the development of the endocrine system.
- Describe the effects of aging on the endocrine system.
- Define the general adaptation syndrome and compare homeostatic responses and stress responses.
- Identify the body reactions during the alarm, resistance, and exhaustion stages of stress.
- Define medical terminology associated with the endocrine system.

397

Two regulatory systems are involved in transmitting messages and correlating various body functions: the nervous system and endocrine system. The nervous system controls homeostasis through electrical impulses delivered over neurons. The endocrine system affects bodily activities by releasing chemical messengers, called hormones, into the bloodstream. Whereas the nervous system sends messages to a specific set of target cells (muscle cells, gland cells, or other neurons), the endocrine system sends messages to target cells in virtually any part of the body. The nervous system causes muscles to contract and glands to secrete; the endocrine system brings about changes in the metabolic activities of body tissues. Neurons tend to act over short distances within a few milliseconds; hormones can take up to several hours or more to bring about their responses. Also, the effects of nervous system stimulation are generally brief compared to the effects of endocrine stimulation.

Obviously, the body could not function if the two great control systems were to pull in opposite directions. The nervous and endocrine systems coordinate their activities like an interlocking supersystem. Certain parts of the nervous system stimulate or inhibit the release of hormones, and hormones, in turn, are quite capable of stimulating or inhibiting the flow of nerve impulses.

Although the effects of hormones are many and varied, their actions can be categorized into four broad areas:

1. They help to control the internal environment by regulating its chemical composition and volume.

2. They respond to marked changes in environmental conditions to help the body cope with emergency demands such as infection, trauma, emotional stress, dehydration, starvation, hemorrhage, and temperature extremes.

3. They assume a role in the smooth, sequential integration of growth and development.

4. They contribute to the basic processes of reproduction, including gamete production, fertilization, nourishment of the embryo and fetus, delivery, and nourishment of the newborn.

The science concerned with the structure and functions of the endocrine glands and the diagnosis and treatment of disorders of the endocrine system is called **endocrinology** (*endo* = within; *krinein* = to separate; *logos* = study of).

The developmental anatomy of the endocrine system is considered later in the chapter.

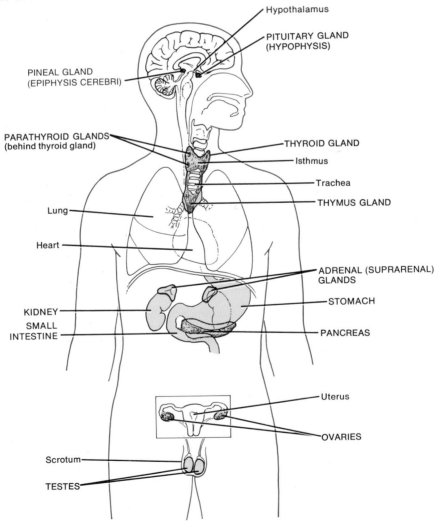

FIGURE 18-1 Location of many endocrine glands, organs containing endocrine tissue, and associated structures.

ENDOCRINE GLANDS

The endocrine glands make up the **endocrine system.** The body contains two kinds of glands: exocrine and endocrine. **Exocrine glands** secrete their products onto a free surface or into ducts. The ducts carry the secretions into body cavities, into the lumens of various organs, or to the body's surface. Exocrine glands include sweat, sebaceous, mucous, and digestive glands. **Endocrine glands** by contrast, secrete their products (hormones) into the extracellular space around the secretory cells, rather than into ducts. The secretion then passes into capillaries to be transported in the blood. The endocrine glands of the body include the pituitary (hypophysis), thyroid, parathyroids, adrenals (suprarenals), pineal (epiphysis cerebri), and thymus gland. In addition, there are several organs of the body that contain endocrine tissue. These include the pancreas, ovaries, testes, kidneys, stomach, small intestine, and placenta. The location of many organs of the endocrine system and endocrine-containing organs is illustrated in Figure 18-1.

CHEMISTRY OF HORMONES

The secretions of endocrine glands are called **hormones** (*hormone* = set in motion). Although hormones are chemically diverse, they can be grouped into three principal classes: amines, proteins and peptides, and steroids.

1. Amines. Structurally, these are the simplest hormone molecules. They are modified from the amino acid tyrosine. Examples of amines are thyroid hormones (triiodothyronine and tetraiodothyronine) secreted by the thyroid gland and epinephrine and norepinephrine (NE), together called catecholamines, secreted by the medulla of the adrenal gland (Figure 18-2a).

2. Proteins and **peptides.** These hormones consist of chains of amino acids, which may vary in size from relatively simple, as in the case of oxytocin (OT) secreted by the hypothalamus, to fairly complex, as in the case of insulin secreted by the pancreas (Figure 18-2b). Protein and peptide hormones are synthesized in the manner described for protein synthesis in Chapter

Other endocrine glands that synthesize protein or peptide hormones include the anterior pituitary, thyroid (a hormone called calcitonin), and parathyroids. Protein and peptide hormones are water-soluble.

3. Steroids. These hormones are derived from cholesterol. Examples include aldosterone, cortisol, and androgen secreted by the cortex of the adrenal gland, testosterone secreted by the testes, and estrogens and progesterone secreted by the ovaries (Figure 18-2c). Steroid hormones are lipid-soluble and are produced in both mitochondria and agranular endoplasmic reticulum.

As you will learn in Chapter 29, all tissues and organs of the body are derived from three embryonic tissues called the endoderm, mesoderm, and ectoderm. Endocrine glands that originate from the endoderm include the parathyroids and pancreas. Endocrine glands of ectodermal origin include the adrenal medulla and pituitary gland. Endocrine glands of mesodermal origin include the adrenal cortex, ovaries, and testes.

The one thing all hormones have in common is the function of maintaining homeostasis by changing the physiological activities of cells.

MECHANISM OF HORMONAL ACTION

OVERVIEW

The amount of hormone released by an endocrine gland or tissue is determined by the body's *need* for the hormone at any given time (Figure 18-3). This is the basis on which the endocrine system operates. Hormone-producing cells have available to them information from sensing and signaling systems that permit the hormone-producing cells to regulate the amount and duration of hormone release. Secretion is normally regulated so that there is no overproduction or underproduction of a particular hormone.

Once a hormone is released by a secretory cell, it is carried by the blood to **target cells,** cells that respond to the hormone. Whereas catecholamines, protein, and peptide hormones are generally carried in free form in blood, steroid and thyroid hormones are carried by one or more plasma proteins synthesized in the liver. Target cells contain receptors that bind the hormone so that it can produce the effect. Once the desired response of target cells occurs, the desired response must be recognized by the secretory cell by some type of feedback signal. Finally, hormones that have accomplished their goals are degraded by target cells or removed by the liver or kidneys.

Triiodothyronine (T$_3$)

Tetraiodothyronine or Thyroxine (T$_4$)

Epinephrine

Norepinephrine (NE)

(a) AMINES

FIGURE 18-2 Representative hormones based on chemical structure.

Oxytocin (OT)

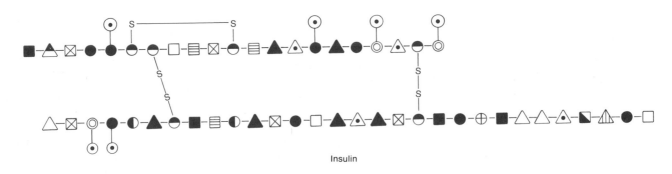

Insulin

(b) PROTEINS AND PEPTIDES

FIGURE 18-2 (*Continued*) Representative hormones based on chemical structure.

RECEPTORS

Cells that produce hormones represent only a limited number of cell types. By contrast, practically all cells of the body are target cells. As a rule, most of the over 50 or so hormones affect a wide range of target cells. Since all body cells are exposed to equal concentrations of hormones, why is it that some target cells respond and others do not? The answer is **receptors,** large protein molecules found in the plasma membrane (integral proteins), cytoplasm, and nucleus of target cells. Receptors are quite specific in that they recognize only certain hormones. Because of the specific complementary structure of hormone and receptor, only certain hormones bind to certain receptors (see Figure 18-4). It is for this very reason that a hormone influences some target cells but not others. In addition, different target cells can possess the same receptors, but the responses of the cells are

Cholesterol

Aldosterone

Testosterone

Progesterone

(c) STEROIDS

FIGURE 18-2 (*Continued*) Representative hormones based on chemical structure.

quite different from each other. For example, the hormone insulin acts on fat cells to stimulate glucose transport and lipid synthesis; it acts on liver cells to stimulate amino acid transport and glycogen synthesis; and it acts on pancreatic cells to inhibit their secretion of a hormone called glucagon.

Once a hormone is bound to a specific receptor, the combination activates a chain of events within the target cell in which the physiological effects of the hormone are expressed. Receptors, like other cell proteins, are constantly synthesized and degraded and change in concentration and affinity in response to changes within the body. There are generally 2,000 to 100,000 receptors per target cell.

The binding of many hormones to specific receptors can occur in two quite different ways. One involves interaction of hormones with plasma membrane receptors; the other involves interaction of hormones with intracellular receptors.

INTERACTION WITH PLASMA MEMBRANE RECEPTORS

One mechanism of hormone action involves an interaction at the cell surface between the hormone and a receptor site on the plasma membrane. A hormone released from an endocrine gland circulates in the blood, reaches a target cell, and brings a specific message to that cell. The hormone is called the **first messenger.** To give the cell its message, the hormone must attach to a specific receptor site (integral protein) on the plasma membrane.

Once a hormone attaches to a specific receptor, there is an increase in the synthesis of cyclic adenosine-3', 5'-monophosphate, or simply cyclic AMP (see Figure 2-15). Cyclic AMP is synthesized from ATP, the main energy-storing chemical in cells, in a process that requires the presence of an enzyme, *adenylate cyclase,* on the inner surface of the plasma membrane. When the first messenger attaches to its receptor, there is an activation of adenylate cyclase in the plasma membrane. Then adenylate cyclase converts ATP into cyclic AMP in the cytoplasm of the cell (Figure 18-4a). Cyclic AMP then acts as a **second messenger,** altering cell function according to the message indicated by the hormone. Cyclic AMP has also

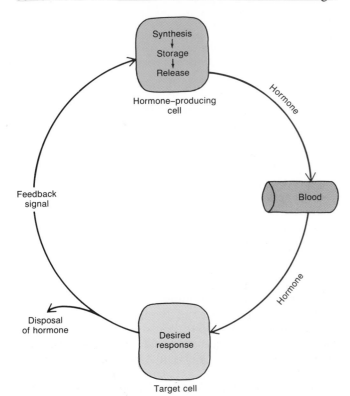

FIGURE 18-3 Overview of physiology of the endocrine system.

FIGURE 18-4 Proposed mechanisms of hormonal action. (a) Interaction with plasma membrane receptors in which there is an increase in the synthesis of cyclic AMP. (b) Activation of genes by a steroid hormone.

been shown to serve as a second messenger for some neurotransmitters and other substances, as well as for hormones.

Cyclic AMP does not directly produce a given physiological response, as indicated by the hormone. Instead, cyclic AMP activates one or more enzymes collectively known as **protein kinases,** which may be free in the cytoplasm or membrane-bound. Protein kinases have the ability to add a phosphate group (phosphorylation) from ATP to a protein. As a result, the protein, usually an enzyme itself, is altered, thus catalyzing a physiological response indicated by the hormone. Such responses include activating enzymes, inducing secretion, activating protein synthesis, and altering plasma membrane permeability. The reason that hormones are so effective in low concentrations is that the original activation of adenylate cyclase initiates a cascade effect in which there is an amplification of products. That is, each hormone molecule activates the formation of several protein kinases. These, in turn, induce the formation of many molecules of another enzyme, and so on. The net effect is that a small amount of hormone produces a response of great magnitude.

A unique property of protein kinases is their ability to catalyze a wide variety of responses. Each of the different protein kinases has a different protein as a substrate, and each exists within different target cells and within different organelles of the same target cell. As a result, one protein kinase might be involved with glycogen synthesis, another with lipid breakdown, another with protein synthesis, and so on. Moreover, protein kinases can inhibit as well as activate enzymes.

High levels of cyclic AMP persist only briefly because they are rapidly degraded by *phosphodiesterase.* At least some physiological responses of many hormones exert their effects through the increased synthesis of cyclic AMP. These include antidiuretic hormone (ADH), oxytocin (OT), follicle-stimulating hormone (FSH), luteinizing hormone (LH), thyroid-stimulating hormone (TSH), adrenocorticotropic hormone (ACTH), calcitonin (CT), parathyroid hormone (PTH), glucagon, epinephrine, and norepinephrine (NE).

Along with cyclic AMP, calcium ions (Ca^{2+}) may sometimes be involved as second messengers. The binding of some hormones to their receptors causes the entry of calcium ions into cells via open calcium channels in the plasma membrane. Also, cyclic AMP causes the release of calcium ions from mitochondria. Once the level of calcium ions is increased, they bind to an intracellular protein called **calmodulin.** Once activated, calmodulin can, in turn, activate or inhibit certain enzymes, many of which are protein kinases. In addition to calcium ions, cyclic 3', 5'-guanosine monophosphate (cGMP) can also serve as a second messenger, eliciting physiological responses that are usually the opposite of those of cyclic AMP.

INTERACTION WITH INTRACELLULAR RECEPTORS

Steroid hormones and thyroid hormones both alter cell function by activation of genes, but by somewhat different mechanisms. Since steroid hormones are lipid-soluble, they easily pass through the plasma membrane of the target cell. Upon entering the cell, a steroid hormone binds to an intracellular protein receptor site in the cytoplasm, activating the receptor so that the hormonal-receptor complex moves into the nucleus of the cell (Figure 18-4b). The complex interacts with specific genes of the nuclear DNA and activates them to form the proteins, usually enzymes, necessary to produce the effect that is characteristic of the hormone.

Thyroid hormones also enter target cells and interact with genes to alter cell function. However, thyroid hormones enter the nucleus in their free state and then combine with their nuclear receptors.

PROSTAGLANDINS (PGs) AND HORMONES

Prostaglandins (pros'-ta-GLAN-dins) or **PGs** are membrane-associated, biologically active lipids that are secreted into blood in minute quantities and are potent in their action. Prostaglandins are also called *local* or *tissue hormones* because their site of action is the immediate area in which they are produced. This differentiates them from *circulating hormones,* which act on distant targets. In addition, prostaglandins are synthesized not by specialized endocrine tissues, as are circulating hormones, but by nearly every mammalian cell and tissue. Chemical and mechanical stimuli, as well as anaphylaxis, lead to prostaglandin release.

Chemically, prostaglandins are composed of 20-carbon fatty acids containing 5 carbon atoms joined to form a cyclopentane ring. They are classified into several groups designated by letters A through I. Thus, prostaglandins are designated as PGA through PGI, respectively. In addition, within each group there are further subdivisions based on the number of double bonds in the fatty acids. Thus PGE_2 has two double bonds and PGE_1 has one. Prostaglandins are believed to be regulators or modulators of cell metabolism. Varying with the tissue and species, prostaglandins either increase or decrease cyclic AMP formation. In this way, prostaglandins can alter the responses of cells to a hormone whose action involves cyclic AMP. In this respect, prostaglandins might be modulators of cyclic AMP-induced responses. Prostaglandins are rapidly inactivated, especially in the lungs, liver, and kidneys.

The broad range of prostaglandin biological activity in relation to smooth muscle, secretion, blood flow, reproduction, platelet function, respiration, nerve impulse transmission, fat metabolism, immune response, and other life processes, as well as their involvement in inflammation, neoplasia, and other conditions, indicates their importance in both normal physiology and pathology. Certain drugs such as aspirin and acetaminophen (Tylenol) inhibit prostaglandin synthesis and thus reduce fever and decrease pain.

What has intrigued investigators even more than the physiological role of prostaglandins has been their pharmacological effects and their implications for potential therapy. These effects include lowering or raising blood pressure; reducing gastric secretion; bronchodilation or bronchoconstriction; stimulating or inhibiting platelet aggregation; contracting or relaxing intestinal and uterine smooth muscle; shrinking nasal passages; blocking or augmenting norepinephrine release; mediating inflammation; increasing intraocular pressure; causing sedation, stupor, and fever; inducing labor in pregnancy; stimulating steroid production; promoting natriuresis (excretion of sodium in the urine), promoting diuresis, potentiating the pain-producing effect of kinins; and many others.

CONTROL OF HORMONAL SECRETIONS: FEEDBACK CONTROL

As indicated earlier, the amount of hormone released by an endocrine gland or tissue is determined by the body's need for the hormone at any given time. Most hormones are released in short bursts, with little or no release between bursts. When properly stimulated, an endocrine gland will release hormone in more frequent bursts, and thus blood levels of the hormone increase. Conversely, in the absence of stimulation, bursts are minimal or inhibited, and thus blood levels of the hormone decrease. Secretion is normally regulated so that there is no overproduction or underproduction of a particular hormone. This regulation is one of the very important ways that the body attempts to maintain homeostasis. Unfortunately, there are times when the regulating mechanism does not operate properly and hormonal levels are excessive or deficient. When this happens, disorders result.

The typical way in which hormonal secretions are regulated is by **negative feedback control** (Chapter 1). As applied to hormones, information regarding the hormone level or its effect is fed back to the gland, which then responds accordingly. Here we will describe three negative feedback systems in which the input that stimulates or inhibits hormonal secretion varies in each situation.

In one type of negative feedback system, the regulation of hormonal secretion does not involve direct participation by the nervous system. For example, blood calcium level is controlled in part by parathyroid hormone (PTH), produced by the parathyroid glands. If, for some reason, blood calcium level is low, this serves as a stimulus for the parathyroids to release more PTH (see Figure 18-14). PTH then exerts its effects in various parts of the body until the blood calcium level is raised to normal. A high blood calcium level serves as a stimulus for the parathyroids to cease their production of PTH. In the absence of the hormone, other mechanisms take over until blood calcium level is lowered to normal. Note that in negative feedback control the body's response (increased or decreased calcium level) is opposite (negative) to the stimulus (low or high calcium level). Other hormones that are regulated without direct involvement of the nervous system include calcitonin (CT), produced by the thyroid gland; insulin by the pancreas; and aldosterone by the adrenal cortex.

In other negative feedback systems, the hormone is released as a direct result of nerve impulses that stimulate the endocrine gland. Epinephrine and norepinephrine (NE) are released from the adrenal medulla in response to sympathetic nerve impulses. Antidiuretic hormone (ADH) is released from the posterior pituitary in response to nerve impulses from the hypothalamus (see Figure 26-10).

There are also negative feedback systems that involve the nervous system through chemical secretions from the hypothalamus, called **regulating factors** (see Figure 18-7). Some regulating factors, called **releasing factors,** stimulate the release of the hormone into the blood, so that it can exert its influence. Other regulating factors, called **inhibiting factors,** prevent the release of the hormone.

One of the few exceptions to the rule of negative feedback control is oxytocin (OT). The regulating system for the release of OT from the pituitary in response to nerve impulses from the hypothalamus is a positive feedback cycle, that is, the output intensifies the input (see Figure 18-9). Another is the luteinizing hormone (LH) surge that results in ovulation (Chapter 28).

As we discuss the effects of various hormones in this chapter, we will also describe how the secretions of the hormones are controlled. At that time you will be able to see which type of negative feedback system is operating.

PITUITARY (HYPOPHYSIS)

The hormones of the **pituitary gland,** also called the **hypophysis** (hī-POF-i-sis), regulate so many body activities that the pituitary has been nicknamed the "master gland." It is a round structure and surprisingly small, measuring about 1.3 cm (0.5 inch) in diameter. The pituitary gland lies in the sella turcica of the sphenoid bone and is attached to the hypothalamus of the brain via a stalklike structure, the **infundibulum** (see Figure 18-5).

The pituitary gland is divided structurally and functionally into an anterior lobe and a posterior lobe. Both are connected to the hypothalamus. The **anterior lobe** constitutes about 75 percent of the total weight of the gland. It is derived from the ectoderm from an embryological invagination of pharyngeal epithelium (see Figure 18-24b). Accordingly, the anterior lobe contains many glandular epithelial cells and forms the glandular part of the pituitary. A system of blood vessels connects the anterior lobe with the hypothalamus.

The **posterior lobe** is also derived from the ectoderm, but from an outgrowth of the hypothalamus of the brain (see Figure 18-24b). Accordingly, the posterior lobe contains axonic ends of neurons whose cell bodies are located in the hypothalamus. The nerve fibers that terminate in the posterior lobe are supported by cells called pituicytes. Other nerve fibers connect the posterior lobe directly with the hypothalamus.

Between the lobes is a small, relatively avascular zone, the **pars intermedia.** Although much larger and more clearly defined in structure and function in some lower animals, its role in humans is obscure.

ADENOHYPOPHYSIS

The anterior lobe of the pituitary is also called the **adenohypophysis** (ad'-e-nō-hī-POF-i-sis). It releases hormones that regulate a whole range of bodily activities from growth to reproduction. The release of these hormones is either stimulated or inhibited by chemical secretions from the hypothalamus called **regulating factors.** Regulating factors, which will be considered along with each of the anterior pituitary hormones, constitute an important link between the nervous and endocrine systems.

The hypothalamic regulating factors are delivered to the adenohypophysis in the following way. The blood supply to the adenohypophysis and infundibulum is derived principally from several *superior hypophyseal* (hī'-po-FIZ-ē-al) *arteries.* These arteries are branches of the internal carotid and posterior communicating arteries (Figure 18-5). The superior hypophyseal arteries form a network or plexus of capillaries, the *primary plexus,* in the infundibulum near the inferior portion of the hypothalamus. Regulating factors from the hypothalamus diffuse into this plexus. This plexus drains into the *hypophyseal portal veins* that pass down the infundibulum. At the inferior portion of the infundibulum, the veins form a *secondary plexus* in the adenohypophysis. From this plexus, hormones of the adenohypophysis pass into the anterior hypophyseal veins for distribution to tissue cells. Such a delivery system permits regulating factors to act quickly on the adenohypophysis without first circulating through the heart. The short route prevents dilution or destruction of the regulating factors.

When the adenohypophysis receives proper stimulation from the hypothalamus via regulating factors, its glandular cells secrete any one of seven hormones. Recently, special staining techniques have established the division of glandular cells into five principal types (Figure 18-6):

1. **Somatotroph cells** produce **growth hormone (GH),** which controls general body growth.

2. **Lactotroph cells** synthesize **prolactin (PRL),** which initiates milk production by the mammary glands.

3. **Corticolipotroph cells** synthesize **adrenocorticotropic hormone (ACTH),** which stimulates the adrenal cortex to secrete its hormones, and **melanocyte-stimulating hormone (MSH),** which is related to skin pigmentation.

4. **Thyrotroph cells** manufacture **thyroid-stimulating hormone (TSH),** which controls the thyroid gland.

5. **Gonadotroph cells** produce **follicle-stimulating hormone (FSH),** which stimulates the production of eggs and sperm in the ovaries and testes, respectively, and **luteinizing hormone (LH),** which stimulates other sexual and reproductive activities.

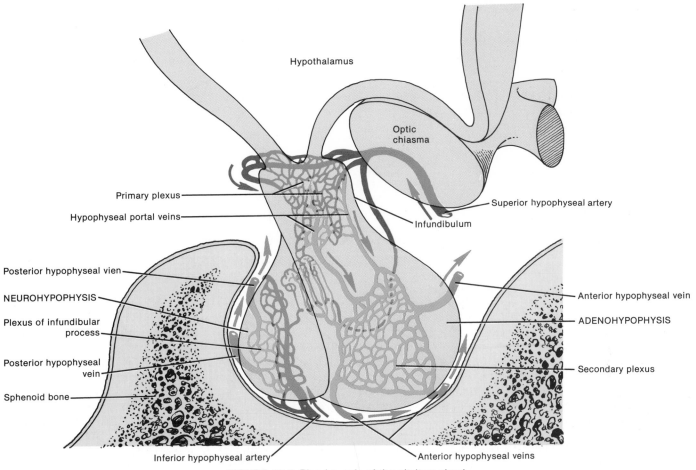

FIGURE 18-5 Blood supply of the pituitary gland.

Except for the growth hormone (GH), melanocyte-stimulating hormone (MSH), and prolactin (PRL), all the secretions are referred to as **tropic hormones** (*trop* = turn on), which means that they stimulate other endocrine glands. Follicle-stimulating hormone (FSH) and luteinizing hormone (LH) are also called **gonadotropic** (gō-nad-ō-TRŌ-pik) **hormones** because they regulate the functions of the gonads. The gonads (ovaries and testes) are the endocrine glands that produce sex steroid hormones.

Growth Hormone (GH)

Growth hormone (GH), also known as **somatotropin** (sō'-ma-tō-TRŌ-pin) and **somatotropic hormone (STH),** turns on body cells to grow. Its principal function is to act on the skeleton and skeletal muscles, in particular, to increase their rate of growth and maintain their size once growth is attained. GH causes cells to grow and multiply by directly increasing the rate at which amino acids enter cells and are built up into proteins. GH is considered to be a hormone of protein anabolism, since it increases the rate of protein synthesis. GH also promotes fat catabolism, that is, it causes cells to switch from burning carbohydrates to burning fats for energy. For example, it stimulates adipose tissue to release fat, and it stimulates other cells to break down the released fat molecules. At the

same time, GH accelerates the rate at which glycogen stored in the liver is converted into glucose and released into the blood. Since the cells are using fats for energy, however, they do not consume as much glucose. The result is an increase in blood sugar level, a condition called **hyperglycemia** (hī'-per-glī-SĒ-mē-a). This process is called the **diabetogenic** (dī'-a-bet'-ō-JEN-ik) **effect** because it mimics the elevated blood glucose level of diabetes mellitus.

GH seems to produce many of its effects by converting other factors into growth-promoting substances called **somatomedins** and **insulinlike growth factors (IGF).** These are small peptides made in the liver under the influence of GH. Both mediate most of the effects of GH and are structurally and functionally similar to insulin. However, their growth-promoting effects are much more potent than those of insulin.

The control of GH secretion is not yet clearly understood. Its release from the anterior pituitary is apparently controlled by at least two regulating factors from the hypothalamus: **growth hormone releasing factor (GHRF)** and **growth hormone inhibiting factor (GHIF),** or **somatostatin.** When GHRF is released by the hypothalamus into the bloodstream, it circulates to the anterior pituitary and stimulates the release of GH. On the other hand, GHIF inhibits the release of GH.

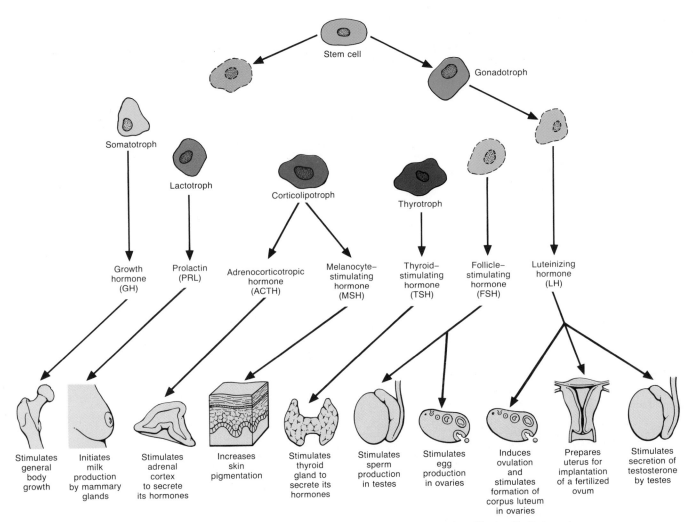

FIGURE 18-6 Cells of the adenohypophysis as revealed by special strains. Most cells that produce growth hormone (somatotrophs) and prolactin (lactotrophs) are separate cells. However, some normal and tumor cells are single cells that produce both growth hormone and prolactin. The corticolipotroph cell produces adrenocorticotropic hormone and melanocyte-stimulating hormone. Thyrotroph cells synthesize thyroid-stimulating hormone. Most gonadotroph cells produce both follicle-stimulating hormone and luteinizing hormone. However, a few separate cells may exist, some producing follicle-stimulating hormone and some producing luteinizing hormone. (Adapted from a slide provided by Calvin Ezrin, M.D., Clinical Professor of Medicine, U.C.L.A., and Adjunct Professor of Pathology, University of Toronto.)

Among the stimuli that promote GH secretion is **hypoglycemia,** that is, low blood sugar level. Other promoting stimuli are listed in Exhibit 18-1. When blood sugar level is low, the hypothalamus is stimulated to secrete GHRF (Figure 18-7a). Upon reaching the anterior pituitary, GHRF causes the anterior pituitary to release GH. Somatomedins, under the influence of GH, raise blood sugar level by converting glycogen into glucose and releasing it into the blood. As soon as blood sugar level returns to normal, GHRF secretion shuts off.

One stimulus that inhibits GH secretion is hyperglycemia, high blood sugar level (Figure 18-7a). Other inhibiting stimuli are listed in Exhibit 18-1. An abnormally high blood sugar level stimulates the hypothalamus to secrete the regulating factor GHIF (somatostatin). GHIF inhibits the release of GHRF and thus the secretion of GH. As a result, blood sugar level decreases. You will see later

that certain endocrine cells of the pancreas, called delta cells, also secrete GHIF (somatostatin), and it can inhibit the secretion of pancreatic hormones (insulin and glucagon) just as it inhibits secretion of GH.

The regulation of secretion of GH illustrates two phenomena that are typical of secretions of the adenohypophysis. First, each hormone is believed to be controlled by its own regulating factor from the hypothalamus. In some cases the regulating factor stimulates hormone secretion; in other cases, it inhibits secretion. Second, secretion is generally regulated through negative feedback systems. Since hormones are chemical regulators of homeostasis, these feedback systems are hardly surprising. Continued heavy secretion of a hormone would overshoot the goal and send the body out of balance in the opposite direction.

FIGURE 18-7 Growth hormone (GH). (a) Regulation of the secretion of growth hormone. Like other hormones of the adenohypophysis, the secretion of GH is controlled by regulating factors. Like most hormones of the body, GH secretion and inhibition involve negative feedback systems. (b) Photograph of an individual with acromegaly. (Courtesy of Lester V. Bergman & Associates, Inc.)

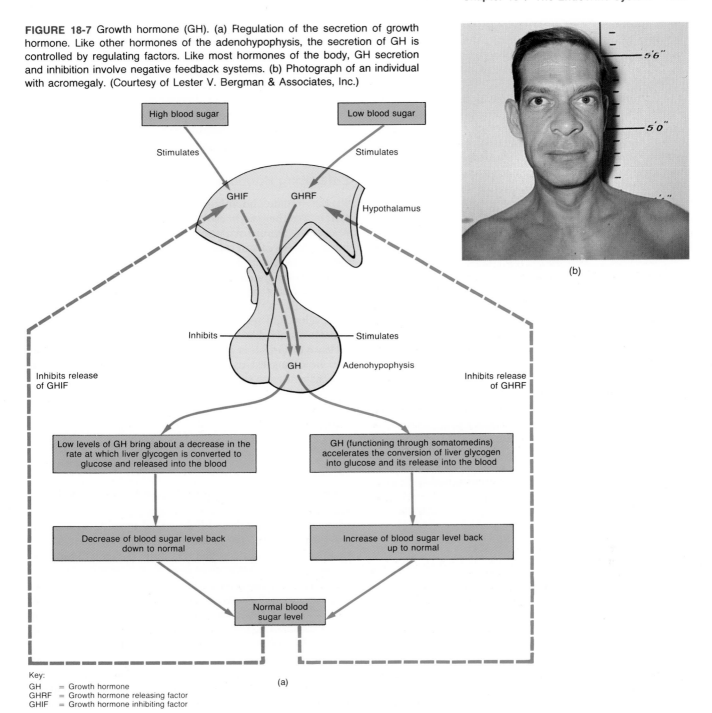

(b)

High blood sugar

Stimulates

Low blood sugar

Stimulates

GHIF   GHRF

Hypothalamus

Inhibits ——————— Stimulates

GH

Adenohypophysis

Inhibits release of GHIF

Inhibits release of GHRF

Low levels of GH bring about a decrease in the rate at which liver glycogen is converted to glucose and released into the blood

GH (functioning through somatomedins) accelerates the conversion of liver glycogen into glucose and its release into the blood

Decrease of blood sugar level back down to normal

Increase of blood sugar level back up to normal

Normal blood sugar level

(a)

Key:
GH   = Growth hormone
GHRF = Growth hormone releasing factor
GHIF = Growth hormone inhibiting factor

CLINICAL APPLICATION

Disorders of the endocrine system, in general, involve **hyposecretion** (underproduction) of hormones or **hypersecretion** (overproduction).

Among the clinically interesting disorders related to the adenohypophysis are those involving GH. If GH is hyposecreted during the growth years, bone growth is slow and the epiphyseal plates close before normal height is reached. This condition is called **pituitary dwarfism.** Other organs of the body also fail to grow, and the pituitary dwarf is childlike in many physical respects. Treatment requires administration of GH during childhood before the epiphyseal plates close. Dwarfism is also caused by other conditions, in which administration of GH is not corrective.

Hypersecretion of GH during childhood results in **giantism (gigantism),** an abnormal increase in the length of long bones. As a result, the person grows to be very large, but body proportions are about normal. Hypersecretion during adulthood is called **acromegaly** (ak'-rō-MEG-a-lē), which is shown in Figure 18-7b. Acromegaly cannot produce further lengthening of the long bones because the epiphyseal plates are already closed. Instead, the bones of the hands, feet, cheeks, and jaws thicken. Other tissues also grow. The eyelids, lips, tongue, and nose enlarge and the skin thickens and furrows, especially on the forehead and soles of the feet.

Thyroid-Stimulating Hormone (TSH)

Thyroid-stimulating hormone (TSH), also called **thyrotropin** (thī-rō-TRO-pin), stimulates the synthesis and secretion of the hormones produced by the thyroid gland. Secretion is controlled by a regulating factor produced by the hypothalamus called **thyrotropin releasing factor (TRF).** Release of TRF depends on blood levels of thyroxine and the body's metabolic rate, among other factors, and operates according to a negative feedback system.

Adrenocorticotropic Hormone (ACTH)

Adrenocorticotropic hormone (ACTH) is also called **adrenocorticotropin** (ad-rē-nō-kor'-ti-kō-TRO-pin). Its tropic function is to control the production and secretion of certain adrenal cortex hormones. Secretion of ACTH is governed by a regulating factor produced by the hypothalamus called **corticotropin releasing factor (CRF).** Release of CRF depends on a number of stimuli and hormones, and operates as a negative feedback system.

Follicle-Stimulating Hormone (FSH)

In the female, **follicle-stimulating hormone (FSH)** is transported from the adenohypophysis by the blood to the ovaries, where it initiates the development of ova each month. FSH also stimulates cells in the ovaries to secrete estrogens, or female sex hormones. In the male, FSH stimulates the testes to initiate sperm production. Secretion of FSH is subject to a regulating factor produced by the hypothalamus called **gonadotropin releasing factor (GnRF).** GnRF is released in response to estrogens and possibly progesterone in the female and to testosterone in the male, and involves a negative feedback system.

Luteinizing Hormone (LH)

In the female, **luteinizing** (LOO-tē-in'-īz-ing) **hormone (LH),** together with estrogens, stimulates the ovary to release an ovum (ovulation) and prepares the uterus for implantation of a fertilized ovum. It also stimulates formation of the corpus luteum in the ovary, which secretes progesterone (another female sex hormone) and readies the mammary glands for milk secretion. In the male, LH stimulates the interstitial endocrinocytes in the testes to develop and secrete large amounts of testosterone. Secretion of LH, like that of FSH, is controlled by GnRF. Release of GnRF is governed by a negative feedback system involving estrogens, progesterone, and testosterone. Recent evidence suggests that the placenta, formed during pregnancy, may be an extrahypothalamic source of GnRF. The GnRF it secretes has been designated as **placental luteotropic releasing factor (pLRF).** Its role in pregnancy will be discussed in Chapter 29.

Prolactin (PRL)

Prolactin (PRL) or the **lactogenic hormone,** together with other hormones, initiates and maintains milk secretion by the mammary glands. The actual ejection of milk by the mammary glands is controlled by a hormone stored in the posterior lobe called oxytocin. Together, milk secretion and ejection are referred to as lactation. PRL acts directly on tissues. By itself, it has little effect; it requires preparation by estrogens, progesterone, corticosteroids, growth hormone, thyroxine, and insulin. When the mammary glands have been primed by these hormones, PRL brings about milk secretion.

PRL has both an inhibitory and an excitatory negative control system. During menstrual cycles, **prolactin inhibiting factor (PIF),** a regulating factor from the hypothalamus, inhibits the release of PRL from the anterior pituitary. As the levels of estrogens and progesterone fall during the late secretory phase of the menstrual cycle, the secretion of PIF diminishes and the blood level of PRL rises. However, its rising level does not last long enough to have much effect on the breasts, which may be tender because of the presence of PRL just before menstruation. As the menstrual cycle starts again and the level of estrogens rises again, PIF is again secreted and the PRL level drops.

PRL levels rise during pregnancy. Apparently a regulating factor from the hypothalamus, called **prolactin releasing factor (PRF),** stimulates PRL secretion after long periods of inhibition. PRL levels fall after delivery and rise again during breast feeding. Sucking reduces hypo-

---

EXHIBIT 18-1 REPRESENTATIVE STIMULI THAT INFLUENCE GROWTH HORMONE (GH) SECRETION	
PROMOTES	**INHIBITS**
Hypoglycemia	Hyperglycemia
Decreased fatty acids	Increased fatty acids
Increased amino acids	Decreased amino acids
Low levels of GH	High levels of GH
Somatomedins	Growth hormone inhibiting factor (GHIF)
Stages 3 and 4 of NREM sleep	REM sleep
Stress	Emotional deprivation
Vigorous physical exercise	Obesity
Estrogens	Hypothyroidism
Glucagon	
Insulin	
Glucocorticoids	
Dopamine (DA)	
Acetylcholine (ACh)	

thalamic secretion of PIF. Mechanical stimulation of the nonlactating female breast brings about increased secretion of PRL; stimulation of the male breast does not.

Melanocyte-Stimulating Hormone (MSH)

The **melanocyte-stimulating hormone (MSH)** increases skin pigmentation by stimulating the dispersion of melanin granules in melanocytes. In the absence of the hormone, the skin may be pallid. An excess of MSH may cause darkening of the skin. Secretion of MSH is stimulated by a hypothalamic regulating factor called **melanocyte-stimulating hormone releasing factor (MRF)**. It is inhibited by a **melanocyte-stimulating hormone inhibiting factor (MIF)**.

Before leaving our discussion of the adenohypophysis, it should be noted that several hormones, as well as neuropeptides, are produced by cells of the adenohypophysis

from a large protein precursor molecule called **pro-opiomelanocortin.** When this molecule is split, it can give rise to adrenocorticotropic hormone (ACTH), melanocyte-stimulating hormone (MSH), enkephalin, endorphin, and beta-lipotropin (β-LPH). This last substance is believed to have a role in stimulating aldosterone secretion by the adrenal gland and lipid breakdown in cells.

A summary of anterior pituitary hormones, their principal actions, associated hypothalamic regulating factors, and related disorders is presented in Exhibit 18-2.

NEUROHYPOPHYSIS

In a strict sense, the posterior lobe, or **neurohypophysis,** is not an endocrine gland, since it does not synthesize hormones. The posterior lobe consists of cells called **pituicytes** (pi-TOO-i-sītz), which are similar in appearance to the neuroglia of the nervous system. It also contains

---

EXHIBIT 18-2   ANTERIOR PITUITARY HORMONES, PRINCIPAL ACTIONS, ASSOCIATED HYPOTHALAMIC REGULATING FACTORS, AND DISORDERS

HORMONE	PRINCIPAL ACTIONS	ASSOCIATED HYPOTHALAMIC REGULATING FACTORS	DISORDERS
Growth hormone (GH)	Growth of body cells; protein anabolism.	Growth hormone releasing factor (GHRF); Growth hormone inhibiting factor (GHIF).	Hyposecretion of GH during the growth years results in pituitary dwarfism; hypersecretion of GH during the growth years results in giantism; hypersecretion of GH during adulthood results in acromegaly.
Thyroid-stimulating hormone (TSH)	Controls secretion of hormones by thyroid gland.	Thyrotropin releasing factor (TRF).	
Adrenocorticotropic hormone (ACTH)	Controls secretion of some hormones by adrenal cortex.	Corticotropin releasing factor (CRF).	
Follicle-stimulating hormone (FSH)	In female, initiates development of ova and induces ovarian secretion of estrogens. In male, stimulates testes to produce sperm.	Gonadotropin releasing factor (GnRF).	
Luteinizing hormone (LH)	In female, together with estrogens, stimulates ovulation and formation of progesterone-producing corpus luteum, prepares uterus for implantation, and readies mammary glands to secrete milk. In male, stimulates interstitial cells in testes to develop and produce testosterone.	Gonadotropin releasing factor (GnRF).	
Prolactin (PRL)	Together with other hormones, initiates and maintains milk secretion by the mammary glands.	Prolactin inhibiting factor (PIF); Prolactin releasing factor (PRF).	
Melanocyte-stimulating hormone (MSH)	Stimulates dispersion of melanin granules in melanocytes.	Melanocyte-stimulating hormone releasing factor (MRF); Melanocyte-stimulating hormone inhibiting factor (MIF).	

axon terminals of secretory neurons of the hypothalamus (Figure 18-8). Such neurons are called **neurosecretory cells.** The cell bodies of the neurons originate in nuclei in the hypothalamus. The fibers project from the hypothalamus, form the **hypothalamic-hypophyseal tract,** and terminate on blood capillaries in the neurohypophysis. The cell bodies of the neurosecretory cells produce two hormones: **oxytocin (OT)** and **antidiuretic hormone (ADH).** OT is produced primarily in the paraventricular nucleus and ADH is synthesized primarily in the supraoptic nucleus.

Following their production, the hormones are transported through the neuron fibers to the neurohypophysis and are stored in the axon terminals where they become bound to small proteins called *neurophysins.* The transport from the cell bodies to the axon terminations takes about 10 hours. Neurophysins aid in storing the hormones and are important in the release mechanism. The bound hormone is released from the nerve endings in response to action potentials reaching the axon terminals. Thus these fibers perform two tasks. First, they act as conduits for the transport of the hormone molecules from their site of production in the hypothalamic nucleus to their site of secretion in the neurohypophysis. Second, the fibers carry the releasing nerve impulses down to the axon terminals where the hormone is stored. The mechanism of hormone release is a familiar one. The action potential depolarizes the axon terminal membrane and Ca^{2+} enters

the nerve endings. When released into the blood, the hormone-neurophysin bonds are broken, probably due to a pH change.

This long-distance arrangement of synthesis, transport, and secretion may seem excessive, but it has a very important advantage: the neurohypophysis can store much more hormone than the hypothalamus. For example, the neurohypophysis can store sufficient ADH to maintain a maximum antidiuretic condition for a whole week. The supraoptic nucleus in the hypothalamus contains enough ADH for only about 8 to 9 hours.

The blood supply to the neurohypophysis is from the *inferior hypophyseal arteries,* derived from the internal carotid arteries. In the neurohypophysis, the inferior hypophyseal arteries form a plexus of capillaries called the *plexus of the infundibular process.* From this plexus hormones stored in the neurohypophysis pass into the *posterior hypophyseal veins* for distribution to tissue cells.

Oxytocin (OT)

Oxytocin (ok'-sē-TŌ-sin) or **OT** stimulates the contraction of the smooth muscle cells in the pregnant uterus and the contractile cells around the ducts of the mammary glands. It is released in large quantities just prior to giving birth (Figure 18-9). When labor begins, the cervix of the uterus is distended. This distension initiates afferent impulses to the neurosecretory cells of the paraventricular nucleus in the hypothalamus that stimulate the synthesis of OT. The OT is transported by neurophysins to the neurohypophysis. The impulses also cause the neurohypophysis to release OT into the blood. It is then carried by the blood to the uterus to reinforce uterine contractions. As the contractions become more forceful, the resulting afferent impulses stimulate the synthesis of more OT. In addition, as the baby's head passes through the uterine cervix, further distension of the cervix triggers the release of more OT. Thus, a positive feedback cycle is established. The cycle is broken by the birth of the infant. Note also that the afferent part of the cycle is neural, whereas the efferent part is hormonal.

OT is used clinically to induce labor (the trademark for such preparations is Pitocin). It is also used in the immediate postpartum period to increase uterine tone and control hemorrhage.

OT affects milk ejection. Milk formed by the glandular cells of the breasts is stored until the baby begins active sucking. For about 30 seconds to 1 minute after nursing begins, the baby receives no milk. During this latent period, afferent impulses from the nipple are transmitted to the hypothalamus. According to a mechanism similar to that involved in forming and releasing OT for uterine muscle contractions, OT flows from the neurohypophysis via the blood to the mammary glands, where it stimulates smooth muscle cells to contract and eject milk.

OT is inhibited by progesterone, but works together with estrogens. Estrogens and progesterone inhibit, via PIF, the release of PRL. PRL accumulates in the pituitary gland during gestation. The sucking stimulation that produces the release of OT also inhibits the release of PIF.

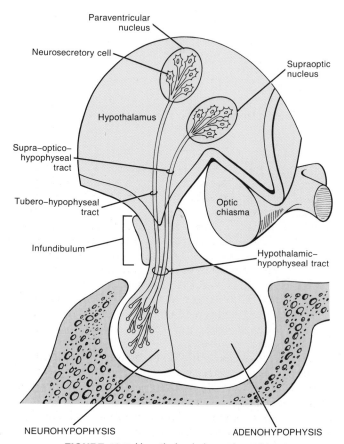

Paraventricular nucleus

Neurosecretory cell

Supraoptic nucleus

Hypothalamus

Supra–optico–hypophyseal tract

Tubero–hypophyseal tract

Optic chiasma

Infundibulum

Hypothalamic–hypophyseal tract

NEUROHYPOPHYSIS   ADENOHYPOPHYSIS

FIGURE 18-8 Hypothalamic-hypophyseal tract.

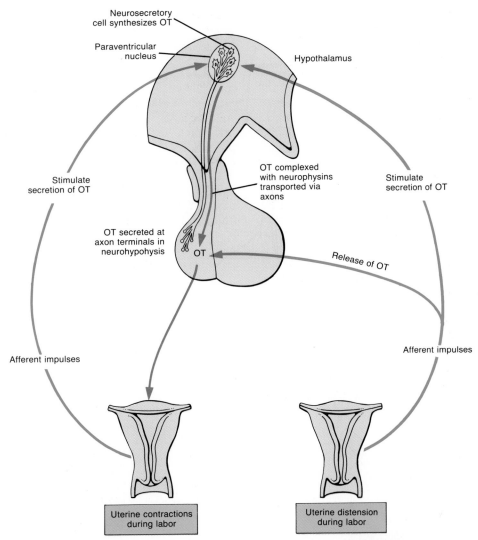

FIGURE 18-9 Regulation of the secretion of oxytocin (OT) during labor.

Antidiuretic Hormone (ADH)

An **antidiuretic** is any chemical substance that prevents excessive urine production. The principal physiological activity of **antidiuretic hormone (ADH)** is its effect on urine volume. ADH causes the kidneys to remove water from newly forming urine and return it to the bloodstream, thus decreasing urine volume. This involves an increase in the permeability of plasma membranes of kidney water-reabsorbing cells so that more water passes from urine into kidney cells. In the absence of ADH, urine output may be increased 10-fold.

ADH can also raise blood pressure by bringing about constriction of arterioles. For this reason, ADH is also referred to as **vasopressin.** This effect is noted if there is a severe loss of blood volume due to hemorrhage.

The amount of ADH normally secreted varies with the body's needs (Figure 18-10). When the body is dehydrated, the concentration of water in the blood falls below normal limits as the salt-to-water ratio changes. Receptors in the hypothalamus called *osmoreceptors* detect the low water concentration in the plasma and stimulate the neurosecretory cells of the supraoptic nucleus in the hypothalamus to synthesize ADH. The hormone is transported by neurophysins to the neurohypophysis. It is then released into the bloodstream and transported to the kidneys. The kidneys respond by decreasing urine output, and water is conserved. ADH also decreases the rate at which perspiration is produced during dehydration. By contrast, if the blood contains a higher-than-normal water concentration, the receptors detect the increase and hormone secretion is stopped. The kidneys can then release large quantities of urine, and the volume of body fluid is brought down to normal.

Secretion of ADH can also be altered by a number of other conditions. Pain, stress, trauma, anxiety, acetylcholine (ACh), nicotine, and drugs, such as morphine, tranquilizers, and some anesthetics, stimulate secretion of the hormone. Alcohol inhibits secretion and thereby increases urine output. This may be why thirst is one symptom of a hangover.

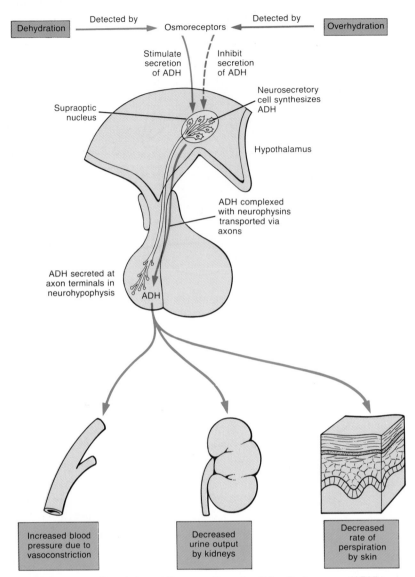

FIGURE 18-10 Regulation of the secretion of antidiuretic hormone (ADH).

The principal abnormality associated with dysfunction of the neurohypophysis is **diabetes insipidus** (in-SIP-i-dus). Diabetes means overflow and insipidus means tasteless. This disorder should not be confused with diabetes mellitus (*meli* = honey), a disorder of the pancreas characterized by sugar in the urine. Diabetes insipidus is the result of a hyposecretion of ADH, usually caused by damage to the neurohypophysis or the supraoptic nucleus in the hypothalamus. Symptoms include excretion of large amounts of urine and subsequent thirst. Diabetes insipidus is treated by administering ADH.

A summary of posterior pituitary hormones, their principal actions, control of secretion, and a related disorder is presented in Exhibit 18-3.

THYROID

The **thyroid gland** is located just below the larynx. The right and left **lateral lobes** lie one on either side of the trachea. The lobes are connected by a mass of tissue called an **isthmus** (IS-mus) that lies in front of the trachea, just below the cricoid cartilage (Figure 18-11). The **pyramidal lobe,** when present, extends upward from the isthmus. The gland has a rich blood supply, receiving about 80 to 120 ml of blood per minute.

Histologically, the thyroid gland is composed of spherical sacs called **thyroid follicles** (Figure 18-12). The wall

EXHIBIT 18-3 SUMMARY OF POSTERIOR PITUITARY HORMONES, PRINCIPAL ACTIONS, CONTROL OF SECRETION, AND DISORDER

HORMONE	PRINCIPAL ACTIONS	CONTROL OF SECRETION	DISORDER
Oxytocin (OT)	Stimulates contraction of smooth muscle cells of pregnant uterus during labor and stimulates contraction of contractile cells of mammary glands for milk ejection.	Neurosecretory cells of hypothalamus secrete OT in response to uterine distension and stimulation of nipples.	
Antidiuretic hormone (ADH)	Principal effect is to decrease urine volume; also raises blood pressure by constricting arterioles during severe hemorrhage.	Neurosecretory cells of hypothalamus secrete ADH in response to low water concentration of the blood, pain, stress, trauma, anxiety, acetylcholine (ACh), nicotine, morphine, and tranquilizers; alcohol inhibits secretion.	Hyposecretion of ADH results in diabetes insipidus.

of each follicle consists of two types of cells. Those that reach the surface of the lumen of the follicle are called **follicular cells,** and those that do not reach the lumen are called **parafollicular** or **C cells.** When the cells are inactive, they tend to be cuboidal, but when actively secreting hormones, they become more columnar. The follic-

ular cells manufacture **thyroxine** (thī-ROK-sēn) or **T₄,** since it contains four atoms of iodine, and **triiodothyronine** (trī-ī′-od-ō-THĪ-rō-nēn) or **T₃,** since it contains three atoms of iodine. Together these hormones are referred to as the **thyroid hormones.** Thyroxine is normally secreted in greater quantity than triiodothyronine, but tri-

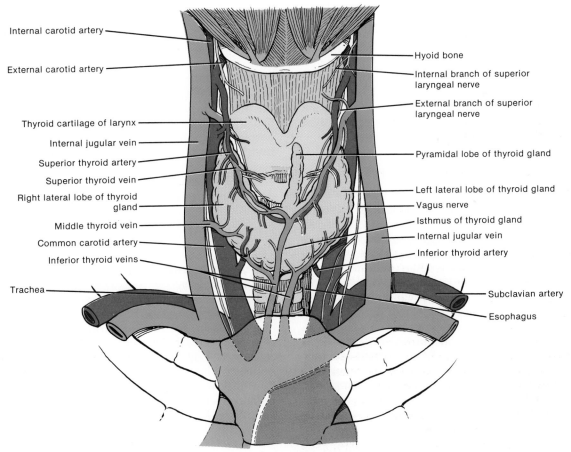

FIGURE 18-11 Location and blood supply of the thyroid gland in anterior view. The pyramidal lobe of the thyroid, when present, may be attached to the hyoid bone by muscle.

Epithelium
of follicle

Thyroid follicle

Thyroglobulin (TGB)

(a)

Parafollicular (C) cell

Follicular cell

Epithelium
of follicle

Blood capillary

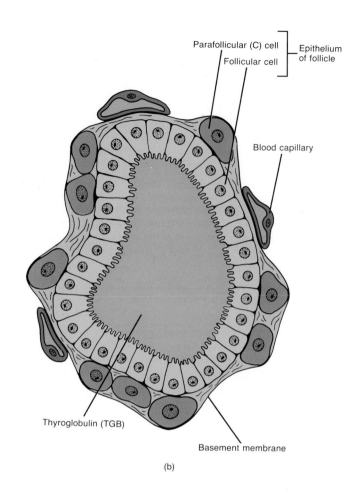

Thyroglobulin (TGB)

Basement membrane

(b)

FIGURE 18-12 Histology of the thyroid gland. (a) Photomicrograph at a magnification of 230×. (Copyright © 1983 by Michael H. Ross. Used by permission.) (b) Diagram showing details of a single thyroid follicle.

iodothyronine is 3 to 4 times more potent. Moreover, in peripheral tissues, especially the liver and lungs, about one-third of triiodothyronine is converted to thyroxine. Both hormones are functionally similar. The parafollicular cells produce **calcitonin** (kal-si-TŌ-nin) or **CT.**

FORMATION, STORAGE, AND RELEASE OF THYROID HORMONES

One of the thyroid gland's unique features is its ability to store hormones and release them in a steady flow over a long period of time. The first stage in the synthesis of thyroid hormones is the active transport of iodine ions from the blood into the follicular cells. Iodide (I⁻) is the ionized form of iodine as it appears in the blood. Under normal conditions, the iodide concentration in the cells is about 40 times that of the blood; during periods of maximal activity, the concentration can increase to over 300 times that of the blood. In the follicular cells, the iodide is oxidized to iodine. Through a series of enzymatically controlled reactions, the iodine combines with the amino acid tyrosine to form the thyroid hormones. This combination occurs within a large glycoprotein molecule, called **thyroglobulin (TGB),** which is secreted by the follicular cells into the follicle.

The thyroid hormones are an integral part of TGB and, as such, are stored (sometimes for months) until needed. The entire complex in the follicle, consisting of

TGB and the stored hormones, constitutes the **thyroid colloid.** Before the thyroid hormones can be released into the blood, droplets of colloid are absorbed into the follicular cells by endocytosis (pinocytosis) and the thyroid hormones are split from TGB. Following their diffusion into the blood, most of the thyroid hormones combine with plasma proteins, mainly **thyroxine-binding globulin (TBG).** Thyroid hormones combined with plasma proteins are referred to as **protein-bound iodine (PBI).**

FUNCTION AND CONTROL OF THYROID HORMONES

The thyroid hormones have three principal effects on the body: (1) regulation of metabolism, (2) regulation of growth and development, and (3) regulation of the activity of the nervous system. With respect to the regulation of metabolism, the thyroid hormones stimulate virtually all aspects of carbohydrate and lipid catabolism in most cells of the body. They also increase the rate of protein synthesis. Since their overall effect is to increase catabolism, they increase the basal metabolic rate. The energy produced raises body temperature as heat is given off, and this phenomenon is called the **calorigenic effect.**

The thyroid hormones help to regulate tissue growth and development, especially in children. They work with GH to accelerate body growth, particularly the growth of nervous tissue. Deficiency of the hormones during fetal development can result in fewer and smaller neurons, defective myelination of axons, and mental retardation. During the early years of life, deficiency of the hormones results in small stature and poor development of certain organs.

Finally, the thyroid hormones increase the reactivity of the nervous system. This results in increased blood flow, increased and more forceful heartbeats, increased blood pressure, increased motility of the gastrointestinal tract, and increased nervousness.

The secretion of thyroid hormones is stimulated by several factors (Figure 18-13a). If thyroid hormone levels in the blood fall below normal or the metabolic rate decreases, chemical sensors in the hypothalamus detect the change in blood chemistry and stimulate the hypothalamus to secrete a regulating factor called thyrotropin releasing factor (TRF). TRF stimulates the adenohypophysis to secrete thyroid-stimulating hormone (TSH). Then TSH stimulates the thyroid to release thyroid hormones

until the metabolic rate returns to normal. Conditions that increase the body's need for energy—a cold environment, high altitude, pregnancy—also trigger this feedback system and increase the secretion of thyroid hormones.

Thyroid activity can be inhibited by a number of other factors. When large amounts of certain sex hormones (estrogens and androgens) are circulating in the blood, for example, TSH secretion diminishes. Aging slows down the activities of most glands, and thyroid production may decrease.

CLINICAL APPLICATION

Hyposecretion of thyroid hormones during the growth years results in **cretinism** (KRĒ-tin-izm), which is shown in Figure 8-13b. Two outstanding clinical symptoms of the cretin are dwarfism and mental retardation.

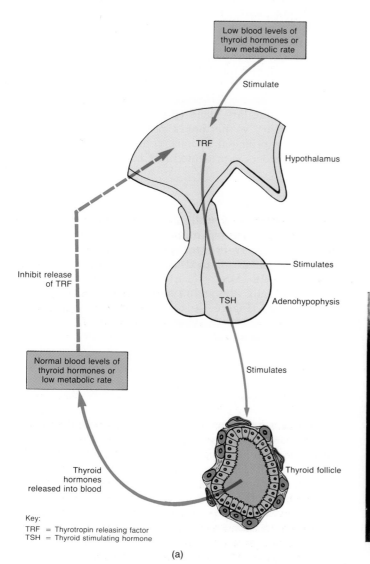

Key:
TRF = Thyrotropin releasing factor
TSH = Thyroid stimulating hormone

(a)

FIGURE 18-13 Thyroid hormones. (a) Regulation of the s
of an individual with cretinism. (c) Photograph of exop
V. Bergman & Associates, Inc.)

The first is caused by failure of the skeleton to grow and mature; the second is caused by failure of the brain to develop fully. Recall that one function of thyroid hormones is to control tissue growth and development. Cretins also exhibit retarded sexual development and a yellowish skin color. Flat pads of fat develop, giving the cretin a characteristic round face and thick nose; a large, thick, protruding tongue; and protruding abdomen. Because the energy-producing metabolic reactions are slow, the cretin has a low body temperature and suffers from general lethargy. Carbohydrates are stored rather than utilized, and heart rate is also slow. If the condition is diagnosed early, the symptoms can be eliminated by administering thyroid hormones.

Hypothyroidism during the adult years produces **myxedema** (mix-e-DĒ-ma). A hallmark of this disorder is an edema that causes the facial tissues to swell and look puffy. Like the cretin, the person with myxedema suffers from slow heart rate, low body temperature, muscular weakness, general lethargy, and a tendency to gain weight easily. The long-term effect of a slow heart rate may overwork the heart muscle, causing the heart to enlarge. Because the brain has already reached maturity, the person with myxedema does not experience mental retardation. However, in moderately severe cases, nerve reactivity may be dulled so that the person lacks mental alertness. Myxedema occurs eight times more frequently in females than in males. Its symptoms are alleviated by the administration of thyroid hormones.

Hypersecretion of thyroid hormones gives rise to **exophthalmic** (ek'-sof-THAL-mik) **goiter** (GOY-ter). This disease, like myxedema, is also more frequent in females. One of its primary symptoms is an enlarged thyroid, called a **goiter,** which may be two to three times its original size. Two other symptoms are an edema behind the eye, which causes the eye to protrude **(exophthalmos),** which is shown in Figure 8-13c, and an abnormally high metabolic rate. The high metabolic rate produces a range of effects that are generally opposite to those of myxedema—increased pulse, high body temperature, and moist, flushed skin. The person loses weight and is usually full of "nervous" energy. The thyroid hormones also increase the responsiveness of the nervous system, causing the person to become irritable and exhibit tremors of the extended fingers. Hyperthyroidism is usually treated by administering drugs that suppress thyroid hormone synthesis, or by surgically removing part of the gland.

Goiter is a symptom of many thyroid disorders. It may also occur if the gland does not receive enough iodine to produce sufficient thyroxine for the body's needs. The follicular cells then enlarge in a futile attempt to produce more thyroid hormones, and they secrete large quantities of thyroglobulin (TGB). This condition is called **simple goiter.** Simple goiter is most often caused by a lower-than-average amount of iodine in the diet. It may also develop if iodine intake is not increased during certain conditions that put a high demand on the body for thyroxine, such as frequent exposure to cold and high-fat and protein diets.

CALCITONIN (CT)

The hormone produced by the parafollicular cells of the thyroid gland in **calcitonin** (kal-si-TŌ-nin) or **CT.** It is involved in the homeostasis of blood calcium level. CT lowers the amount of calcium and phosphate in the blood by inhibiting bone breakdown and accelerating the absorption of calcium by the bones. It appears to exert its effect in lowering calcium and phosphate blood levels by inhibiting osteoclastic action and the parathyroid hormone (PTH). If CT is administered to a person with a normal level of blood calcium, it causes **hypocalcemia** (low blood calcium level). Hypocalcemia is also a complication of magnesium deficiency. If CT is given to a person with **hypercalcemia** (high blood calcium level), the level returns to normal. CT is used to treat postmenopausal osteoporosis, usually in conjunction with adequate calcium and vitamin dietary intake. It is suspected that the blood calcium level directly controls the secretion of CT according to a negative feedback system that does not involve the pituitary gland (Figure 18-14).

A summary of hormones produced by the thyroid gland, their principal actions, control of secretion, and related disorders is presented in Exhibit 18-4.

PARATHYROIDS

Typically embedded on the posterior surfaces of the lateral lobes of the thyroid are small, round masses of tissue called the **parathyroid glands.** Usually, two parathyroids, superior and inferior, are attached to each lateral thyroid lobe (Figure 18-15).

Histologically, the parathyroids contain two kinds of epithelial cells (Figure 18-16). The more numerous cells, called **principal (chief) cells,** are believed to be the major synthesizer of **parathyroid hormone (PTH).** Some researchers believe that the other kind of cell, called an **oxyphil cell,** synthesizes a reserve capacity of hormone.

PARATHYROID HORMONE (PTH)

Parathyroid hormone (PTH) or **parathormone** controls the homeostasis of ions in the blood, especially calcium and phosphate ions. If adequate amounts of vitamin D are present, PTH increases the rate of calcium, some magnesium, and phosphate absorption from the gastrointestinal tract into the blood. It also leads to the activation of vitamin D. Moreover, PTH increases the number and activity of osteoclasts, or bone-destroying cells. As a result, bone tissue is broken down and calcium and phosphate are released into the blood. PTH produces two changes in the kidneys. It increases the rate at which the kidneys remove calcium and magnesium from urine

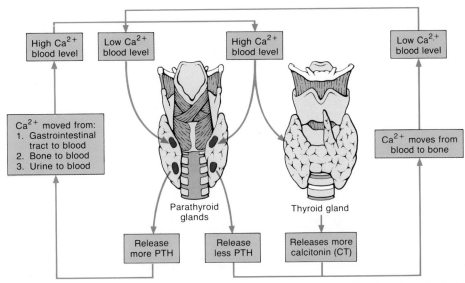

FIGURE 18-14 Regulation of the secretion of the parathyroid hormone (PTH) and calcitonin (CT).

that is being formed and returns them to the blood. It accelerates the transportation of phosphate from the blood into the urine for elimination. More phosphate is lost through the urine than is gained from the bones.

The overall effect of PTH with respect to ions, then, is to decrease blood phosphate level and increase blood calcium level. As far as blood calcium level is concerned, PTH and CT are antagonists.

PTH secretion is not controlled by the pituitary gland. When the calcium level of the blood falls, more PTH is released (see Figure 18-14). Conversely, when the calcium level of the blood rises, less PTH (and more CT) is secreted. This is another example of a negative feedback control system that does not involve the pituitary gland.

CLINICAL APPLICATION

A normal amount of calcium in the extracellular fluid is necessary to maintain the resting state of neurons. A deficiency of calcium caused by *hypoparathyroidism* causes neurons to depolarize without the usual stimulus. As a result, nervous impulses increase and result in muscle twitches, spasms, and convulsions. This condition is called **tetany.** The effects of hypocalcemic tetany are observed in the *Trousseau* (troo-SŌ) and *Chvostek* (VOS-tek) *signs.* Trousseau sign is observed when the binding of a blood pressure cuff around the

EXHIBIT 18-4 SUMMARY OF THYROID GLAND HORMONES, PRINCIPAL ACTIONS, CONTROL OF SECRETION, AND DISORDERS

HORMONE	PRINCIPAL ACTIONS	CONTROL OF SECRETION	DISORDERS
Thyroid hormones			
Thyroxine (T_4)	Regulates metabolism, growth and development, and activity of nervous system.	Thyrotropin releasing factor (TRF) is released from hypothalamus in response to low thyroid hormone levels, low metabolic rate, cold, pregnancy, and high altitudes; TRF secretion is inhibited in response to high thyroid hormone levels, high metabolic rate, high levels of estrogens and androgens, and aging.	Hyposecretion of thyroid hormones during the growth years results in cretinism; hypothyroidism during adult years results in myxedema; hypersecretion of thyroid hormones results in exophthalmic goiter; iodine deficiency produces simple goiter.
Triiodothyronine (T_3)	Same as above.	Same as above.	
Calcitonin (CT)	Lowers blood levels of calcium by accelerating calcium absorption by bones.	High blood calcium levels stimulate secretion; low levels inhibit secretion.	

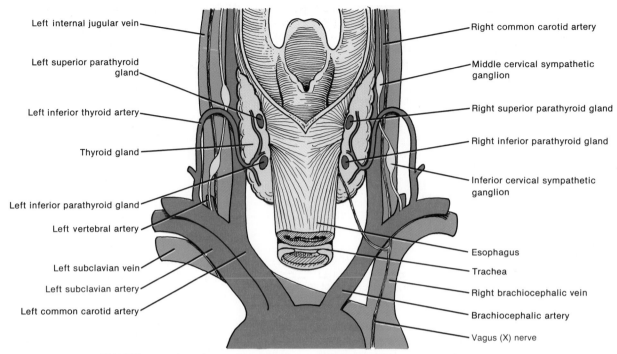

FIGURE 18-15 Location and blood supply of the parathyroid glands in posterior view.

arm produces contraction of the fingers and inability to open the hand. The Chvostek sign is a contracture of the facial muscles elicited by tapping the facial nerves at the angle of the jaw. Hypoparathyroidism results from surgical removal of the parathyroids or from damage caused by parathyroid disease, infection, hemorrhage, or mechanical injury.

Hyperparathyroidism causes demineralization of bone. This condition is called **osteitis fibrosa cystica,** because the areas of destroyed bone tissue are replaced

FIGURE 18-16 Histology of the parathyroid glands. (a) Photomicrograph at a magnification of 180×. (Copyright © 1983 by Michael H. Ross. Used by permission.) (b) Diagram showing details of a portion of a parathyroid gland in relation to a portion of the thyroid gland.

EXHIBIT 18-5   SUMMARY OF PARATHYROID GLAND HORMONE, PRINCIPAL ACTIONS, CONTROL OF SECRETION, AND DISORDERS

HORMONE	PRINCIPAL ACTIONS	CONTROL OF SECRETION	DISORDERS
Parathyroid hormone (PTH)	Increases blood calcium level and decreases blood phosphate level by increasing rate of calcium absorption from gastrointestinal (GI) tract into blood; increases number and activity of osteoclasts; increases calcium absorption by kidneys; increases phosphate excretion by kidneys; and activates vitamin D.	Low blood calcium levels stimulate secretion; high levels inhibit secretion.	Hypoparathyroidism results in tetany. Hyperparathyroidism produces osteitis fibrosa cystica.

by cavities that fill with fibrous tissue. The bones thus become deformed and are highly susceptible to fracture. Hyperparathyroidism is usually caused by a tumor in the parathyroids.

A summary of the principal actions, control of secretion, and disorders related to parathyroid hormone (PTH) is presented in Exhibit 18-5.

ADRENALS (SUPRARENALS)

The body has two **adrenal (suprarenal) glands,** one of which is located superior to each kidney (Figure 18-17). Each adrenal gland is structurally and functionally differentiated into two sections: the outer **adrenal cortex,** which makes up the bulk of the gland, and the inner **adrenal**

medulla (Figure 18-18). Whereas the adrenal cortex is derived from mesoderm, the adrenal medulla is derived from the ectoderm. Covering the gland are an inner, thick layer of fatty connective tissue and an outer, thin fibrous capsule. The adrenals, like the thyroid, are among the most vascular organs of the body.

ADRENAL CORTEX

Histologically, the cortex is subdivided into three zones (Figure 18-18). Each zone has a different cellular arrangement and secretes different groups of steroid hormones. The outer zone, directly underneath the connective tissue capsule, is referred to as the **zona glomerulosa.** It comprises about 15 percent of the total cortical volume. Its cells are arranged in arched loops or round balls. Its primary secretions are a group of hormones called mineralocorticoids (min'-er-al-ō-KOR-ti-koyds).

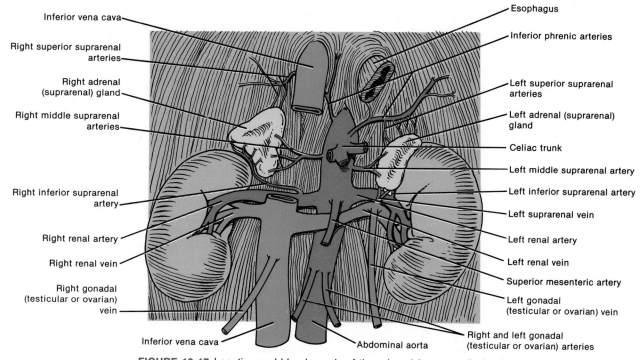

FIGURE 18-17 Location and blood supply of the adrenal (suprarenal) glands.

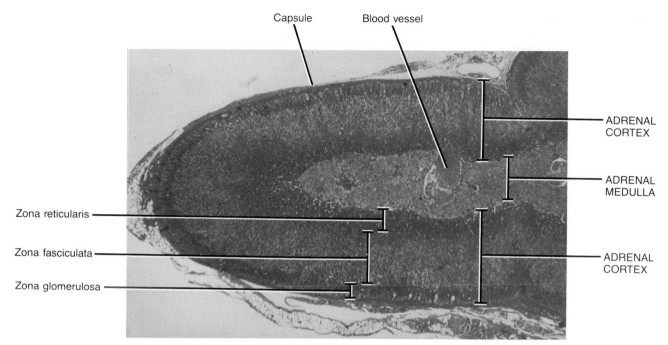

Capsule   Blood vessel

ADRENAL CORTEX

ADRENAL MEDULLA

ADRENAL CORTEX

Zona reticularis

Zona fasciculata

Zona glomerulosa

FIGURE 18-18 Histology of the adrenal (suprarenal) glands. Photomicrograph of a section of the adrenal gland showing its subdivisions and zones at a magnification of 25×. (Copyright © 1983 by Michael H. Ross. Used by permission.)

The middle zone, or **zona fasciculata,** is the widest of the three zones and consists of cells arranged in long, straight cords. The zona fasciculata secretes mainly glucocorticoids (gloo'-kō-KOR-ti-koyds).

The inner zone, the **zona reticularis,** contains cords of cells that branch freely. This zone synthesizes minute amounts of hormones, mainly the sex hormones called gonadocorticoids (gō-na-dō-KOR-ti-koyds), and of these primarily male hormones called androgens.

Mineralocorticoids

Mineralocorticoids help control water and electrolyte homeostasis, particularly the concentrations of sodium and potassium. Although the adrenal cortex secretes at least three different hormones classified as mineralocorticoids, one of these hormones is responsible for about 95 percent of the mineralocorticoid activity—**aldosterone** (al-do-STĒR-ōn). Aldosterone acts on the tubule cells in the kidneys and causes them to increase their reabsorption of sodium. As a result, sodium ions are removed from the urine and returned to the blood. In this manner, aldosterone prevents rapid depletion of sodium from the body. On the other hand, aldosterone decreases reabsorption of potassium, so large amounts of potassium are lost in the urine.

These two basic functions—conservation of sodium and elimination of potassium—cause a number of secondary effects. For example, a large proportion of the sodium reabsorption occurs through an exchange reaction whereby positive hydrogen ions pass into the urine to replace the positive sodium ions. Since this mechanism removes hydrogen ions, it makes the blood less acidic

and prevents acidosis. The movement of sodium ions also sets up a positively charged field in the blood vessels around the kidney tubules. As a result, negatively charged chloride and bicarbonate ions are drawn out of the urine and back into the blood. Finally, the increase in sodium ion concentration in the blood vessels causes water to move by osmosis from the urine into the blood, when ADH is present. In summary, aldosterone causes potassium excretion and sodium reabsorption. The sodium reabsorption leads to the elimination of H$^+$ ions; the retention of Na$^+$, Cl$^-$, and HCO$_3^-$ ions; and the retention of water.

The control of aldosterone secretion is complex. Apparently, several mechanisms operate. One of these is the **renin-angiotensin** (an'-jē-ō-TEN-sin) **pathway** (Figure 18-19). A decrease in blood volume from dehydration, Na$^+$ deficiency, or hemorrhage brings about a drop in blood pressure. The low blood pressure stimulates certain kidney cells, called juxtaglomerular cells, to secrete into the blood an enzyme called *renin* (RĒ-nin) (see Figure 26-9). In this pathway, renin converts *angiotensinogen,* a plasma protein produced by the liver, into *angiotensin I,* which is then converted into *angiotensin II* by a plasma enzyme in the lungs. Angiotensin II stimulates the adrenal cortex to produce more aldosterone. Aldosterone brings about increased Na$^+$ reabsorption and water follows. This reabsorption leads to an increase in extracellular fluid volume and a restoration of blood pressure to normal. Since angiotensin II is a powerful vasoconstrictor, this action also helps to elevate blood pressure.

A second mechanism for the control of aldosterone involves potassium ion concentration. An increased K$^+$ concentration in extracellular fluid directly stimulates al-

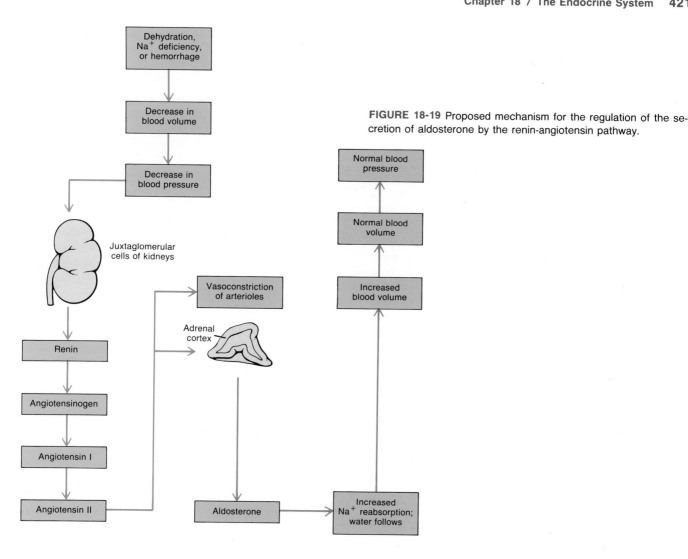

FIGURE 18-19 Proposed mechanism for the regulation of the secretion of aldosterone by the renin-angiotensin pathway.

dosterone secretion by the adrenal cortex and causes the elimination of excess K^+ by the kidneys. A decreased K^+ concentration in extracellular fluid decreases aldosterone production, and thus less K^+ than usual is eliminated by the kidneys.

Adrenocorticotropic hormone (ACTH) from the anterior pituitary has a minor effect on the secretion of aldosterone. Only in the complete absence of ACTH is there a mild to moderate deficiency of aldosterone. The major controlling effect of ACTH is on glucocorticoids, as will be discussed shortly.

CLINICAL APPLICATION

Hypersecretion of the mineralocorticoid aldosterone results in **aldosteronism,** characterized by a decrease in the body's potassium concentration. If potassium depletion is great, neurons cannot depolarize and muscular paralysis results. Hypersecretion also brings about excessive retention of sodium and water. The water increases the volume of the blood and causes high blood pressure. It also increases the volume of the interstitial fluid, producing edema.

Glucocorticoids

The **glucocorticoids** are a group of hormones concerned with normal metabolism and resistance to stress. Three glucocorticoids are **cortisol (hydrocortisone), corticosterone,** and **cortisone.** Of the three, cortisol is the most abundant and is responsible for about 95 percent of glucocorticoid activity. The glucocorticoids have the following effects on the body.

1. Glucocorticoids work with other hormones in promoting normal metabolism. Their role is to make sure enough energy is provided. They increase the rate at which proteins are catabolized and amino acids are removed from cells, primarily muscle cells, and transported to the liver. The amino acids may be synthesized into new proteins, such as the enzymes needed for metabolic reactions. If the body's reserves of glycogen and fat are low, the liver may convert the amino acids to glucose. This conversion of a substance other than carbohydrate into glucose is called **gluconeogenesis** (gloo'-kō-nē'-ō-JEN-e-sis). Glucocorticoids also release fatty acids from adipose tissue for conversion to glucose.

2. Glucocorticoids work in many ways to provide resistance to stress. A sudden increase in available glucose by way of gluconeogenesis from amino acids makes the body more alert. Additional glucose gives the body energy for combating a range

of stresses: fright, temperature extremes, high altitude, bleeding, infection, surgery, trauma, and almost any debilitating disease. Glucocorticoids also make the blood vessels more sensitive to vessel-constricting chemicals. They thereby raise blood pressure. This effect is advantageous if the stress happens to be blood loss, which causes a drop in blood pressure.

3. Glucocorticoids are antiinflammatory compounds. They reduce the number of mast cells, stabilize lysosomal membranes to inhibit the release of histamines, decrease blood capillary permeability, and depress phagocytosis by monocytes. Unfortunately, they also retard connective tissue regeneration and are thereby responsible for slow wound healing. High doses of glucocorticoids cause atrophy of the thymus gland, spleen, and lymph nodes, thus depressing immune responses. However, high doses may be useful in the treatment of rheumatism.

The control of glucocorticoid secretion is a typical negative feedback mechanism (Figure 18-20a). The two principal stimuli are stress and low blood level of glucocorticoids. Either condition stimulates the hypothalamus to secrete a regulating factor called **corticotropin releasing factor (CRF).** This secretion initiates the release of ACTH from the anterior lobe of the pituitary. ACTH is carried through the blood to the adrenal cortex, where it then stimulates glucocorticoid secretion.

CLINICAL APPLICATION

Hyposecretion of glucocorticoids results in the condition called **Addison's disease (primary adrenal insufficiency).** Clinical symptoms include mental lethargy, weight loss, and hypoglycemia, which leads to muscular weakness. Increased potassium and decreased sodium lead to low blood pressure and dehydration.

Cushing's syndrome is a hypersecretion of glucocorticoids, especially cortisol and cortisone (Figure 8-20b). The condition is characterized by the redistribution of fat. The result is spindly legs accompanied by a

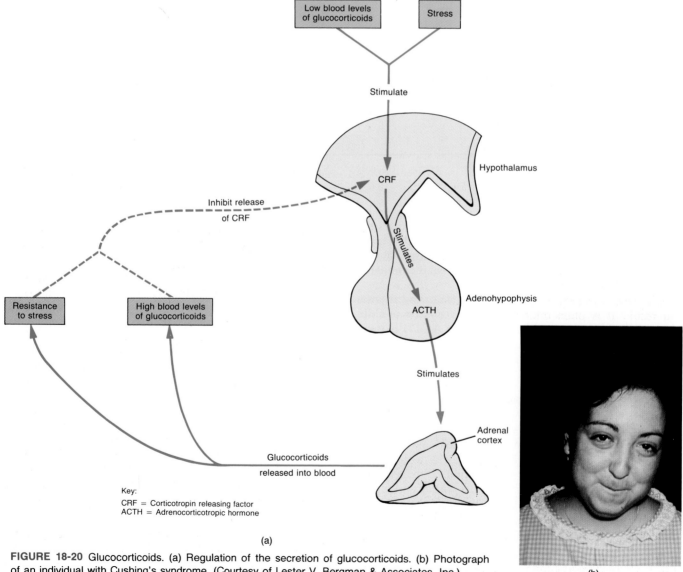

Key:
CRF = Corticotropin releasing factor
ACTH = Adrenocorticotropic hormone

(a)

(b)

FIGURE 18-20 Glucocorticoids. (a) Regulation of the secretion of glucocorticoids. (b) Photograph of an individual with Cushing's syndrome. (Courtesy of Lester V. Bergman & Associates, Inc.)

"moon face," "buffalo hump" on the back, and pendulous abdomen. Facial skin is flushed, and the skin covering the abdomen develops stretch marks. The individual also bruises easily, and wound healing is poor.

Gonadocorticoids

The adrenal cortex secretes both male and female **gonadocorticoids (sex hormones).** These are estrogens and androgens. Estrogens are several closely related female sex hormones that are also produced by the ovaries and placenta. Androgens are male sex hormones. An important androgen, called testosterone, is produced by the testes. The concentration of sex hormones secreted by normal adult adrenals is usually so low that their effects are insignificant.

CLINICAL APPLICATION

The **adrenogenital syndrome** usually refers to a group of enzyme deficiencies that block the synthesis of glucocorticoids. In an attempt to compensate, the anterior pituitary secretes more ACTH. As a result, excess androgenic (male) hormones are produced, causing *virilism,* or masculinization. For instance, the female develops extremely virile characteristics such as growth of a beard, development of a much deeper voice, occasionally the development of baldness, development of a masculine distribution of hair on the body and on the pubis, growth of the clitoris that resembles a penis, and deposition of proteins in the skin and muscles producing typical masculine characteristics. Such virilism may also result from tumors of the adrenal gland called *virilizing adenomas* (*aden* = gland; *oma* = tumor).

In the prepubertal male, the syndrome causes the same characteristics as in the female, plus rapid development of the male sexual organs and creation of male sexual desires. In the adult male, the virilizing characteristics of the adrenogenital syndrome are usually completely obscured by the normal virilizing characteristics of the testosterone secreted by the testes. As a result, it is often difficult to make a diagnosis of adrenogenital syndrome in the male adult. However, an occasional adrenal tumor secretes sufficient quantities of feminizing hormones that the male patient develops **gynecomastia** (*gyneca* = woman, *mast* = breast), which means excessive growth of the male mammary glands. Such a tumor is called a *feminizing adenoma.*

ADRENAL MEDULLA

The adrenal medulla consists of hormone-producing cells, called **chromaffin** (krō-MAF-in) **cells,** which surround large blood-containing sinuses. Chromaffin cells develop from the same source as the postganglionic cells of the sympathetic division of the nervous system. They are directly innervated by preganglionic cells of the sympathetic division of the autonomic nervous system and may be regarded as postganglionic cells that are specialized to

secrete. In all other visceral effectors, preganglionic sympathetic fibers first synapse with postganglionic neurons before innervating the effector. In the adrenal medulla, however, the preganglionic fibers pass directly into the chromaffin cells of the gland. The secretion of hormones from the chromaffin cells is directly controlled by the autonomic nervous system, and innervation by the preganglionic fibers allows the gland to respond rapidly to a stimulus.

Epinephrine and Norepinephrine (NE)

The two principal hormones synthesized by the adrenal medulla are **epinephrine** and **norepinephrine (NE),** also called adrenaline and noradrenaline, respectively. Epinephrine constitutes about 80 percent of the total secretion of the gland and is more potent in its action than norepinephrine. Both hormones are **sympathomimetic** (sim'-pa-thō-mi-MET-ik), that is, they produce effects that mimic those brought about by the sympathetic division of the autonomic nervous system. To a large extent, they are responsible for the fight-or-flight response. Like the glucocorticoids of the adrenal cortices, these hormones help the body resist stress. However, unlike the cortical hormones, the medullary hormones are not essential for life.

Under stress, impulses received by the hypothalamus are conveyed to sympathetic preganglionic neurons, which cause the chromaffin cells to increase their output of epinephrine and norepinephrine. Epinephrine increases blood pressure by increasing heart rate and constricting the blood vessels. It accelerates the rate of respiration, dilates respiratory passageways, decreases the rate of digestion, increases the efficiency of muscular contractions, increases blood sugar level, and stimulates cellular metabolism. Hypoglycemia may also stimulate medullary secretion of epinephrine and norepinephrine.

CLINICAL APPLICATION

Tumors of the chromaffin cells of the adrenal medulla, called **pheochromocytomas** (fē-ō-krō'-mō-sī-TŌ-mas), cause hypersecretion of the medullary hormones. Such tumors are usually benign. The oversecretion causes rapid heart rate, headache, high blood pressure, high levels of sugar in the blood and urine, an elevated basal metabolic rate (BMR), flushing of the face, nervousness, sweating, decreased gastrointestinal (GI) motility, and vertigo. Since the medullary hormones create the same effects as sympathetic nervous stimulation, hypersecretion puts the individual into a prolonged version of the fight-or-flight response. This condition ultimately wears out the body, and the individual eventually suffers from general weakness. The only definitive treatment of pheochromocytomas is surgical removal of the tumor(s).

A summary of the hormones produced by the adrenal glands, their principal actions, control of secretion, and related disorders is presented in Exhibit 18-6.

EXHIBIT 18-6   SUMMARY OF HORMONES PRODUCED BY THE ADRENAL GLANDS, PRINCIPAL ACTIONS, CONTROL OF SECRETION, AND DISORDERS

HORMONE	PRINCIPAL ACTIONS	CONTROL OF SECRETION	DISORDERS
Adrenal Cortical Hormones **Mineralocorticoids (mainly aldosterone)**	Increase blood levels of sodium and water and decrease blood levels of potassium.	Decreased blood volume or sodium levels initiate renin-angiotensin pathway to stimulate aldosterone secretion; increased blood levels of potassium stimulate aldosterone secretion; ACTH has only a minor effect in promoting aldosterone secretion.	Hypersecretion of aldosterone results in aldosteronism.
Glucocorticoids (mainly cortisol)	Help promote normal metabolism, resistance to stress, and counter inflammatory response.	ACTH is released in response to stress and low blood levels of glucocorticoids.	Hyposecretion of glucocorticoids produces Addison's disease. Hypersecretion results in Cushing's syndrome.
Gonadocorticoids	Concentrations secreted by adults are so low that their effects are usually insignificant.	Discussed in detail in Chapter 28.	Androgenital syndrome inhibits synthesis of glucocorticoids that results in excess production of ACTH and androgens, causing virilism. The release of sufficient feminizing hormones in males causes gynecomastia.
Adrenal Medullary Hormones **Epinephrine**	Sympathomimetic, that is, produces effects that mimic those of ANS during stress.		Hypersecretion of medullary hormones results in a prolonged fight-or-flight response.
Norepinephrine (NE)	Same as above.		

PANCREAS

The **pancreas** can be classified as both an endocrine and an exocrine gland. We shall treat its endocrine functions at this point; its exocrine functions are discussed in the chapter on the digestive system (Chapter 24). The pancreas is a flattened organ located posterior and slightly inferior to the stomach (Figure 18-21). The adult pancreas consists of a head, body, and tail.

The endocrine portion of the pancreas consists of clusters of cells called **pancreatic islets** or **islets of Langerhans** (LAHNG-er-hanz) (Figure 18-22). Three kinds of cells are found in these clusters: (1) **alpha cells,** which secrete the hormone glucagon; (2) **beta cells,** which secrete the hormone insulin, and (3) **delta cells,** which secrete growth hormone inhibiting factor (GHIF) or somatostatin. The islets are infiltrated by blood capillaries and by cells (acini) that form the exocrine part of the gland. Glucagon and insulin are the endocrine secretions of the pancreas and are concerned with regulation of blood sugar level.

GLUCAGON

The product of the alpha cells is **glucagon** (GLOO-ka-gon), a hormone whose principal physiological activity

is to increase the blood glucose level (Figure 18-23). Glucagon does this by accelerating the conversion of glycogen in the liver into glucose (glycogenolysis) and the conversion in the liver of other nutrients, such as amino acids, glycerol, and lactic acid, into glucose (gluconeogenesis). The liver then releases the glucose into the blood, and the blood sugar level rises. Secretion of glucagon is directly controlled by the level of blood sugar via a negative feedback system. When the blood sugar level falls below normal, chemical sensors in the alpha cells of the islets stimulate the cells to secrete glucagon. When blood sugar rises, the cells are no longer stimulated and production slackens. If for some reason the self-regulating device fails and the alpha cells secrete glucagon continuously, hyperglycemia may result. Exercise and largely or entirely protein meals that raise the amino acid level of the blood also cause an increase in glucagon secretion. Glucagon secretion is inhibited by GHIF (somatostatin).

INSULIN

The beta cells of the islets produce the hormone **insulin.** Insulin also increases the buildup of proteins in cells. Its chief physiological action is opposite that of glucagon.

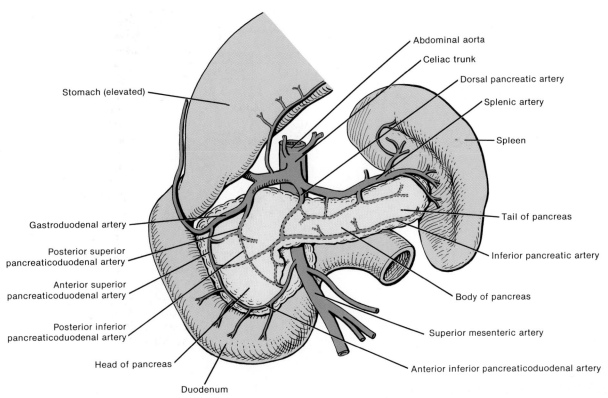

FIGURE 18-21 Location and blood supply of the pancreas.

Insulin decreases blood sugar level in several ways (Figure 18-23). It accelerates the transport of glucose from the blood into cells, especially skeletal muscle cells. Glucose entry into cells depends on the presence of insulin receptors on the surfaces of target cells. It also accelerates the conversion of glucose into glycogen (glycogenesis). Insulin also decreases glycogenolysis and gluconeogenesis, stimulates the conversion of glucose or other nutrients into fatty acids (lipogenesis), and helps stimulate protein synthesis.

(a)

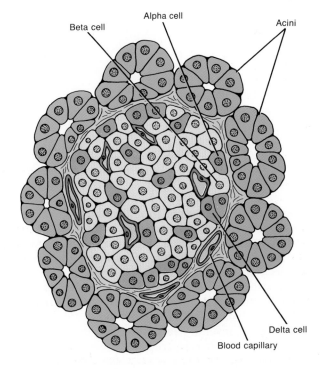

(b)

FIGURE 18-22 Histology of the pancreas. (a) Photomicrograph at a magnification of 600×. (Copyright © 1983 by Michael H. Ross. used by permission.) (b) Diagram.

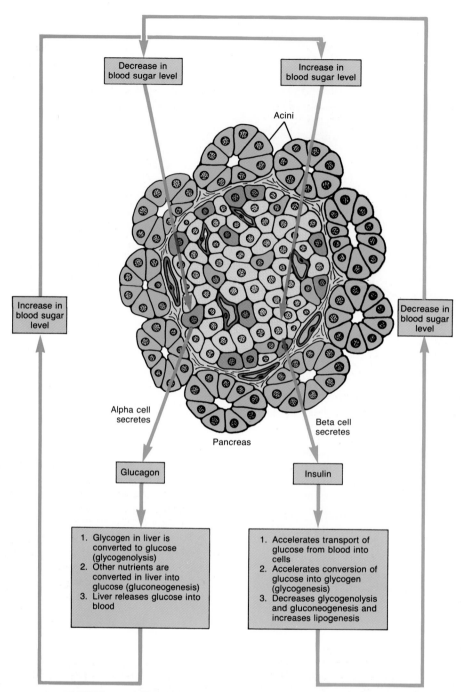

Decrease in blood sugar level

Increase in blood sugar level

Acini

Increase in blood sugar level

Decrease in blood sugar level

Alpha cell secretes

Beta cell secretes

Pancreas

Glucagon

Insulin

1. Glycogen in liver is converted to glucose (glycogenolysis)
2. Other nutrients are converted in liver into glucose (gluconeogenesis)
3. Liver releases glucose into blood

1. Accelerates transport of glucose from blood into cells
2. Accelerates conversion of glucose into glycogen (glycogenesis)
3. Decreases glycogenolysis and gluconeogenesis and increases lipogenesis

FIGURE 18-23 Regulation of the secretion of glucagon and insulin.

The regulation of insulin secretion, like that of glucagon secretion, is directly determined by the level of sugar in the blood and is based on a negative feedback system. Other hormones can indirectly affect insulin production, however. For instance, **GH** raises blood glucose level, and the rise in glucose level triggers insulin secretion. **ACTH**, by stimulating the secretion of glucocorticoids, brings about hyperglycemia and also indirectly stimulates the release of insulin. Gastrointestinal hormones like stomach and intestinal gastrin, secretin, cholecystokinin (CCK), and gastric inhibitory peptide (GIP) also stimulate insulin secretion. **GHIF** (somatostatin) inhibits the secretion of insulin.

CLINICAL APPLICATION

Diabetes mellitus (MEL-i-tus) is not a single hereditary disease, but a heterogeneous group of diseases, all of which ultimately lead to an elevation of glucose in the blood (hyperglycemia) and excretion of glucose in the urine as hyperglycemia increases. Diabetes mellitus is also characterized by the three "polys": an inability to reabsorb water, resulting in increased urine production (*polyuria*); excessive thirst (*polydipsia*); and excessive eating (*polyphagia*).

Two major types of diabetes mellitus have been distinguished: type I and type II. **Type I diabetes,** which occurs abruptly, is characterized by an absolute deficiency of insulin due to a marked decline in the number of insulin-producing beta cells (perhaps caused by the autoimmune destruction of beta cells), even though target cells contain insulin receptors. Type I diabetes is called *insulin-dependent diabetes* because periodic administration of insulin is required to treat it. It is also known as *juvenile-onset diabetes* because it most commonly develops in people younger than age 20, though it persists throughout life. Although people who develop type I diabetes appear to have certain genes that make them more susceptible, some triggering factor is required. Viral infection seems to be such a factor. The deficiency of insulin accelerates the breakdown of the body's reserve of fat resulting in the production of organic acids called ketones. This causes a form of acidosis called *ketosis,* which lowers the pH of the blood and can result in death. The catabolism of stored fats and proteins also causes weight loss. As lipids are transported by the blood from storage depots to hungry cells, lipid particles are deposited on the walls of blood vessels. The deposition leads to atherosclerosis and a multitude of cardiovascular problems including cerebrovascular insufficiency, ischemic heart disease, peripheral vascular disease, and gangrene. One of the major complications of diabetes is loss of vision due to cataracts (excessive blood sugar chemically attaches to lens proteins, causing cloudiness) or damage to blood vessels of the retina. Severe kidney problems also may result from damage to renal blood vessels.

Type II diabetes is much more common than type I, representing more than 90 percent of all cases. Type II diabetes most often occurs in people who are over 40 and overweight. Since type II diabetes usually occurs later in life, it is called *maturity-onset diabetes.* Clinical symptoms are mild, and the high glucose levels in the blood can usually be controlled by diet alone or with antidiabetic drugs such as *glyburide* (DiaBeta). Many type II diabetics have a sufficiency or even a surplus of insulin in the blood. For these individuals, diabetes arises not from a shortage of insulin but probably from defects in the molecular machinery that mediates the action of insulin on its target cells. Cells in many parts of the body, especially skeletal muscles and the liver, become less sensitive to insulin because they have fewer insulin receptors. Type II diabetes is therefore called *non-insulin-dependent diabetes.*

Hyperinsulinism is much rarer than hyposecretion and is generally the result of a malignant tumor in an islet. The principal symptom is a decreased blood glucose level, which stimulates the secretion of epinephrine, glucagon, and GH. As a consequence, anxiety, sweating, tremor, increased heart rate, and weakness occur. Moreover, brain cells do not have enough glucose to function efficiently. This condition leads to mental disorientation, convulsions, unconsciousness, shock, and eventually to death, as the vital centers in the medulla are affected.

A summary of the hormones produced by the pancreas, their principal actions, control of secretion, and related disorders is presented in Exhibit 18-7.

OVARIES AND TESTES

The female gonads, called the **ovaries,** are paired oval bodies located in the pelvic cavity. The ovaries produce female sex hormones called **estrogens** and **progesterone.** These hormones are responsible for the development and maintenance of the female sexual characteristics. Along with the gonadotropic hormones of the pituitary, the sex hormones also regulate the menstrual cycle, maintain pregnancy, and prepare the mammary glands for lactation. The ovaries (and placenta) also produce a hormone called **relaxin,** which relaxes the symphysis pubis, helps

EXHIBIT 18-7 SUMMARY OF HORMONES PRODUCED BY THE PANCREAS, PRINCIPAL ACTIONS, CONTROL OF SECRETION, AND DISORDERS

HORMONE	PRINCIPAL ACTIONS	CONTROL OF SECRETION	DISORDERS
Glucagon	Raises blood sugar level by accelerating breakdown of glycogen into glucose in liver (glycogenolysis) and conversion of other nutrients into glucose in liver (gluconeogenesis) and releasing glucose into blood.	Decreased blood level of glucose, exercise, and largely protein meals stimulate glucagon secretion; GHIF (somatostatin) inhibits glucagon secretion.	
Insulin	Lowers blood sugar level by accelerating transport of glucose into cells, converting glucose into glycogen (glycogenesis), and decreasing glycogenolysis and gluconeogenesis; also increases lipogenesis and stimulates protein synthesis.	Increased blood level of glucose, GH, ACTH, and gastrointestinal hormones stimulate insulin secretion; GHIF (somatostatin) inhibits insulin secretion.	An absolute deficiency of insulin or a defect in the molecular machinery that mediates the action of insulin on target cells produces diabetes mellitus. Hypersecretion of insulin results in hyperinsulinism.

dilate the uterine cervix toward the end of pregnancy, and plays a role in increasing sperm motility.

The male has two oval glands, called **testes,** that lie in the scrotum. The testes produce **testosterone,** the primary male sex hormone, that stimulates the development and maintenance of the male sexual characteristics. The testes also produce the hormone **inhibin** that inhibits secretion of FSH. The detailed structure of the ovaries and testes and the specific roles of gonadotropic hormones and sex hormones will be discussed in Chapter 28.

A summary of hormones produced by the ovaries and testes and their principal actions is presented in Exhibit 18-8.

PINEAL (EPIPHYSIS CEREBRI)

The endocrine gland attached to the roof of the third ventricle is known as the **pineal gland** (because of its resemblance to a pine cone) or **epiphysis cerebri** (see Figure 18-1). The gland is covered by a capsule formed by the pia mater and consists of masses of **neuroglial cells** and parenchymal secretory cells called **pinealocytes.** Around the cells are scattered postganglionic sympathetic fibers. The pineal gland starts to calcify at about the time of puberty. Such calcium deposits are referred to as **brain sand.** Contrary to a once widely held belief, there is no evidence that the pineal atrophies with age and that the presence of brain sand is an indication of atrophy. In fact, the presence of brain sand may indicate increased secretory activity.

Although many anatomical facts concerning the pineal gland have been known for years, its physiology is still somewhat obscure. One hormone secreted by the pineal gland is **melatonin,** which appears to inhibit reproductive activities by inhibiting gonadotropic hormones. Melatonin is produced during darkness; its formation is interrupted when light reaches the eyes. Light entering the eyes stimulates retinal neurons that transmit impulses to the pineal gland that inhibit melatonin secretion. The release of melatonin is based on a circadian rhythm, that is, a cycle of active and nonactive periods determined by internal mechanisms and reoccurring about every 24 hours. Some evidence also exists that the pineal secretes a second hormone called **adrenoglomerulotropin** (a-drē'-nō-glō-mer'-yoo-lō-TRŌ-pin). This hormone may stimulate the adrenal cortex to secrete aldosterone. Other substances found in the pineal gland include norepinephrine (NE), serotonin, histamine, GnRF, and GABA.

A summary of hormones produced by the pineal gland and their principal actions is presented in Exhibit 18-9.

THYMUS

Because of its role in immunity, the details of the structure and functions of the **thymus gland** are discussed in Chapter 22, which deals with the lymphatic system and immunity. At this point, only its hormonal role in immunity will be discussed.

One of the groups of white blood cells is referred to as lymphocytes. These, in turn, are divided into two types, called T cells and B cells, on the basis of their specific roles in immunity. Hormones produced by the thymus gland, called **thymosin, thymic humoral factor (THF), thymic factor (TF),** and **thymopoietin,** promote the maturation of T cells.

A summary of hormones produced by the thymus gland and their principal actions is presented in Exhibit 18-10.

DEVELOPMENTAL ANATOMY OF THE ENDOCRINE SYSTEM

The development of the endocrine system is not as localized as the development of other systems. The endocrine organs develop in widely separated parts of the embryo.

The *pituitary gland* (*hypophysis*) has a double origin from different regions of the **ectoderm.** The *neurohypophysis* (posterior lobe) is derived from an outgrowth of ectoderm called the **neurohypophyseal bud,** located on the floor of the hypothalamus (Figure 18-24a,b). The *infundibulum,* also an outgrowth of the neurohypophyseal bud,

EXHIBIT 18-8 SUMMARY OF HORMONES OF THE OVARIES AND TESTES AND THEIR PRINCIPAL ACTIONS

HORMONE	PRINCIPAL ACTIONS
Ovarian hormones **Estrogens and progesterone**	Development and maintenance of female sexual characteristics. Together with gonadotropic hormones of the adenohypophysis, they also regulate the menstrual cycle, maintain pregnancy, and prepare the mammary glands for lactation.
Relaxin	Relaxes symphysis pubis, helps dilate uterine cervix near the end of pregnancy, and plays a role in increasing sperm motility.
Testicular hormones **Testosterone**	Development and maintenance of male sexual characteristics.
Inhibin	Inhibits secretion of FSH to control sperm production.

EXHIBIT 18-9 SUMMARY OF HORMONES OF THE PINEAL GLAND AND THEIR PRINCIPAL ACTIONS

HORMONE	PRINCIPAL ACTIONS
Melatonin	May inhibit reproductive activities by inhibiting gonadotropic hormones.
Adrenoglomerulotropin	May stimulate the adrenal cortex to secrete aldosterone.

EXHIBIT 18-10 SUMMARY OF HORMONES PRODUCED BY THE THYMUS GLAND AND THEIR PRINCIPAL ACTIONS

HORMONES	PRINCIPAL ACTIONS
Thymosin, thymic humoral factor (THF), thymic factor (TF), and thymopoietin	Promote maturation of T cells.

connects the neurohypophysis to the hypothalamus. The *adenohypophysis* (anterior lobe) is derived from an outgrowth of **ectoderm** from the roof of the stomodeum (mouth) called the **hypophyseal (Rathke's) pouch.** The pouch grows toward the neurohypophyseal bud, and the pouch loses its connection with the roof of the mouth.

The *thyroid gland* develops as a midventral outgrowth of **endoderm,** called the **thyroid diverticulum,** from the floor of the pharynx at the level of the second pair of pharyngeal pouches (Figure 18-24b). The outgrowth projects inferiorly and differentiates into the right and left lateral lobes and the isthmus of the gland.

The *parathyroid glands* develop as outgrowths from the third and fourth **pharyngeal pouches** (Figure 18-24a).

The adrenal cortex and adrenal medulla have completely different embryological origins. The *adrenal cortex* is derived from intermediate **mesoderm** from the same region that produces the gonads (see Figure 28-25). The *adrenal medulla* is **ectodermal** in origin and is derived from the **neural crest,** which also produces sympathetic ganglia and other structures of the nervous system (see Figure 14-13c).

The *pancreas* develops from **endoderm** from dorsal and ventral outgrowths of the part of the **foregut** that later becomes the duodenum (see Figure 24-23). The two outgrowths eventually fuse to form the pancreas. The

FIGURE 18-24 Development of the endocrine system. (a) Position of the neurohypophyseal bud, hypophyseal (Rathke's) pouch, and thyroid diverticulum. (b) Development of the pituitary gland.

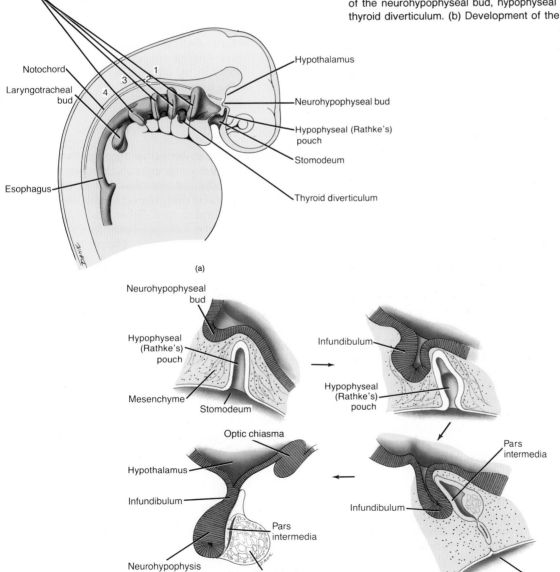

origin of the ovaries and testes is discussed in the section on the reproductive system.

The *pineal gland* arises from **ectoderm** of the **diencephalon** (see Figure 14-14b), as an outgrowth between the thalamus and colliculi.

The *thymus gland* arises from **endoderm** from the third **pharyngeal pouches** (Figure 18-24b).

OTHER ENDOCRINE TISSUES

Before leaving our discussion of hormones, it should be noted that body tissues other than endocrine glands also secrete hormones. The gastrointestinal tract synthesizes several hormones that regulate digestion in the stomach and small intestine. Among these hormones are **stomach gastrin, enteric gastrin, secretin, cholecystokinin (CCK), enterocrinin,** and **gastric inhibitory peptide (GIP).** The actions of these hormones and their control of secretion are summarized in Exhibit 24-4.

The placenta produces **human chorionic gonadotropin (HCG), estrogens, progesterone, relaxin,** and **human chorionic somatomammotropin (HCS),** all of which are related to pregnancy (Chapter 29).

When the kidneys (and liver to a lesser extent) become hypoxic, it is believed that they release an enzyme called **renal erythropoietic factor.** It is secreted into the blood where it acts on a plasma protein to bring about the production of a hormone called **erythropoietin** (ē-rith'-rō-POY-ē-tin), which stimulates red blood cell production (Chapter 19). The kidneys also help bring about the activation of the hormone **vitamin D.**

The skin produces vitamin D in the presence of sunlight (Chapter 5).

Recent research indicates that cardiac muscle fibers of the atria (upper chambers) of the heart produce a peptide hormone called **atrial natriuretic factor (ANF).** Cardiac muscle fibers secrete ANF when they are stretched, as might occur in response to increased blood volume, a factor that increases blood pressure. The general effect of ANF is to modify the renin-angiotensin pathway by somehow inhibiting the secretion of renin by the juxtaglomerular cells of the kidneys and thus the secretion of aldosterone by the adrenal cortex. Because of this inhibition, there is an increase in sodium and water excretion and blood vessel dilation; the result is a decrease in blood pressure. ANF also acts directly on the kidneys to bring about increases in sodium and water excretion and on the hypothalamus to inhibit the secretion of ADH. Again, these conditions lead to a decrease in blood pressure. The overall function of ANF is therefore to lower blood pressure.

AGING AND THE ENDOCRINE SYSTEM

The endocrine system exhibits a variety of changes, and many researchers look to this system with the hope of finding the key to the aging process. Disorders of the endocrine system are not frequent, but when they do occur, most often they are related to pathologic changes rather than age. Diabetes mellitus and thyroid disorders are the most important endocrine problems that have a significant effect on health and function.

STRESS AND THE GENERAL ADAPTATION SYNDROME

Homeostatic mechanisms are geared toward counteracting the everyday stresses of living. If they are successful, the internal environment maintains a uniform chemistry, temperature, and pressure. If a stress is extreme, unusual, or long-lasting, however, the normal mechanisms may not be sufficient. In this case, the stress triggers a wide-ranging set of bodily changes called the **general adaptation syndrome.** Hans Selye, who is the world's acknowledged authority on stress, was responsible for the concept of the general adaptation syndrome. Unlike the homeostatic mechanisms, the general adaptation syndrome does not maintain a normal internal environment. In fact, it does just the opposite. For instance, blood pressure and blood sugar level are raised above normal. The purpose of these changes in the internal environment is to gear up the body to meet an emergency.

It should be pointed out that it is impossible to remove all stress from our everyday lives. In fact, some stress prepares us to meet certain challenges. Thus, some stress is productive, whereas other stress is harmful. In this regard, the term **eustress** is used to distinguish productive stress, and the term **distress** is used to refer to harmful stress. One recent finding suggests that distress lowers resistance to infection by temporarily inhibiting certain components of the immune system.

STRESSORS

The hypothalamus can be called the body's watchdog. It has sensors that detect changes in the chemistry, temperature, and pressure of the blood. It is informed of emotions through tracts that connect it with the emotional centers of the cerebral cortex. When the hypothalamus senses stress, it initiates a chain of reactions that produce the general adaptation syndrome. The stresses that produce the syndrome are called **stressors.** A stressor may be almost any disturbance—heat or cold, environmental poisons, poisons given off by bacteria during a raging infection, heavy bleeding from a wound or surgery, or a strong emotional reaction. Stressors differ among different people and even in the same person at different times.

When a stressor appears, it stimulates the hypothalamus to initiate the syndrome through two pathways. The first pathway is stimulation of the sympathetic nervous system and adrenal medulla. This stimulation produces an immediate set of responses called the alarm reaction. The second pathway, called the resistance reaction, involves the anterior pituitary gland and adrenal cortex. The resistance reaction is slower to start, but its effects last longer.

ALARM REACTION

The **alarm reaction,** or **fight-or-flight response,** is the body's initial reaction to a stressor (Figure 18-25a). It is actually a complex of reactions initiated by the hypothalamic stimulation of the sympathetic nervous system and the adrenal medulla. The responses of the visceral

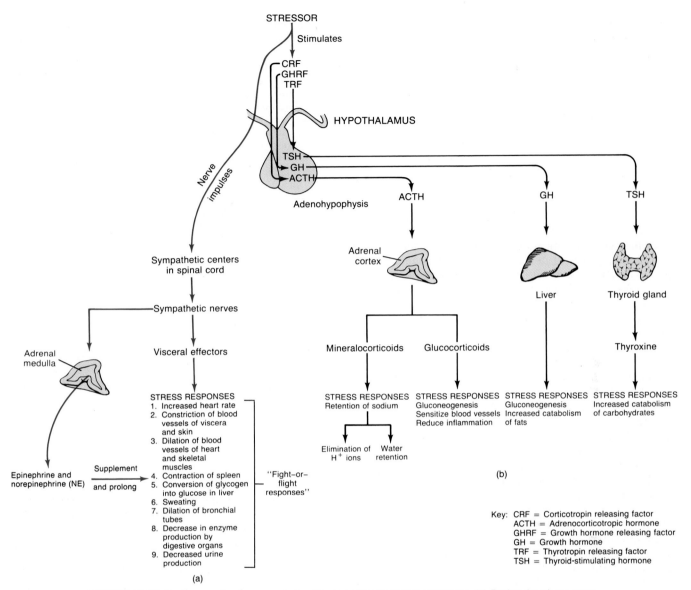

FIGURE 18-25 Responses to stressors during the general adaptation syndrome. (a) During the alarm stage. (b) During the resistance stage. Colored arrows indicate immediate reactions. Black arrows indicate long-term reactions.

effectors are immediate and short-lived. They are designed to counteract a danger by mobilizing the body's resources for immediate physical activity. In essence, the alarm reaction brings tremendous amounts of glucose and oxygen to the organs that are most active in warding off danger. These are the brain, which must become highly alert; the skeletal muscles, which may have to fight off an attacker; and the heart, which must work furiously to pump enough materials to the brain and muscles. The hyperglycemia associated with sympathetic activity is produced by liver glycogenolysis, stimulated by epinephrine and norepinephrine (NE) from the adrenal medulla and liver gluconeogenesis, stimulated by fat mobilization and glucose-sparing by GH and protein mobilization by glucocorticoids.

Among the stress responses that characterize the alarm stage are the following.

1. The heart rate and the strength of cardiac muscle contraction increase. This response circulates substances in the blood very quickly to areas where they are needed to combat the stress.

2. Blood vessels supplying the skin and viscera, except the heart and lungs, undergo constriction. At the same time, blood vessels supplying the skeletal muscles and brain undergo dilation. These responses route more blood to organs active in the stress responses while decreasing blood supply to organs that do not assume an immediate, active role.

3. The spleen contracts and discharges stored blood into the general circulation to provide additional blood. Moreover, red blood cell production is accelerated and the ability of the blood to clot is increased. These preparations are made to combat bleeding.

4. The liver transforms large amounts of stored glycogen into glucose and releases it into the bloodstream. The glucose is broken down by the active cells to provide the energy needed to meet the stressor.

5. Sweat production increases. This response helps lower body temperature, which is elevated as circulation increases and body catabolism increases. Profuse sweating also helps to eliminate wastes produced as a result of accelerated catabolism.

6. The rate of breathing increases, and the respiratory passageways widen to accommodate more air. This response enables the body to acquire more oxygen, which is needed in the decomposition reactions of catabolism. It also allows the body to eliminate more carbon dioxide, which is produced as a side product during catabolism.

7. Production of saliva, stomach enzymes, and intestinal enzymes decreases. This reaction takes place since digestive activity is not essential for counteracting the stress.

8. Sympathetic impulses to the adrenal medulla increase its secretion of epinephrine and norepinephrine. These hormones supplement and prolong many sympathetic responses—increasing heart rate and strength, constricting blood vessels, accelerating the rate of breathing, widening respiratory passageways, increasing the rate of catabolism, decreasing the rate of digestion, and increasing blood sugar level.

If you group the stress responses of the alarm stage by function, you will note that they are designed to increase circulation rapidly, promote catabolism for energy production, and decrease nonessential activities. If the stress is great enough, the body mechanisms may not be able to cope and death can result. Note that during the alarm reaction digestive, urinary, and reproductive activities are inhibited.

RESISTANCE REACTION

The second stage in the stress responses is the **resistance reaction** (Figure 18-25b). Unlike the short-lived alarm reaction that is initiated by nervous impulses from the hypothalamus, the resistance reaction is initiated by regulating factors secreted by the hypothalamus and is a long-term reaction. The regulating factors are CRF, GHRF, and TRF.

CRF stimulates the adenohypophysis to increase its secretion of ACTH. ACTH stimulates the adrenal cortex to secrete more of its hormones. The adrenal cortex is also indirectly stimulated by the alarm reaction. During the alarm reaction, kidney activity is cut back because of decreased blood circulation to the kidneys, since it is not essential for meeting sudden danger. The resultant decrease in urine production stimulates the secretion of the mineralocorticoids.

The mineralocorticoids secreted by the adrenal cortex bring about the conservation of sodium ions by the body. A secondary effect of sodium conservation is the elimination of hydrogen ions. The hydrogen ions build up in high concentrations as a result of increased catabolism and tend to make the blood more acidic. Thus, during stress, a lowering of body pH is prevented. Sodium retention also leads to water retention, thus maintaining the high blood pressure that is typical of the alarm reaction. It also helps make up for fluid lost through severe bleeding.

The glucocorticoids, which are produced in high concentrations during stress, bring about the following reactions.

1. The glucocorticoids accelerate protein catabolism and the conversion of amino acids into glucose so that the body has a large supply of energy long after the immediate stores of glucose have been used up. The glucocorticoids also stimulate the removal of proteins from cell structures and stimulate the liver to break them down into amino acids. The amino acids can then be rebuilt into enzymes that are needed to catalyze the increased chemical activities of the cells or be converted to glucose.

2. The glucocorticoids make blood vessels more sensitive to stimuli that bring about their constriction. This response counteracts a drop in blood pressure caused by bleeding.

3. The glucocorticoids also inhibit the production of fibroblasts, which develop into connective tissue cells. Injured fibroblasts release chemicals that play a role in stimulating the inflammatory response. Thus the glucocorticoids reduce inflammation and prevent it from becoming disruptive rather than protective. Unfortunately, through their effect on fibroblasts, the glucocorticoids also discourage connective tissue formation. Wound healing is therefore slow during a prolonged resistance stage.

Two other regulating factors are secreted by the hypothalamus in response to a stressor: TRF and GHRF. TRF causes the adenohypophysis to secrete TSH; GHRF causes it to secrete GH. TSH stimulates the thyroid to secrete thyroxine, which increases the catabolism of carbohydrates. HGH stimulates the catabolism of fats and the conversion of glycogen to glucose. The combined actions of TSH and GH increase catabolism and thereby supply additional energy for the body.

The resistance stage of the general adaptation syndrome allows the body to continue fighting a stressor long after the effects of the alarm reaction have dissipated. It increases the rate at which life processes occur. It also provides the energy, functional proteins, and circulatory changes required for meeting emotional crises, performing strenuous tasks, fighting infection, or resisting the threat of bleeding to death. During the resistance stage, blood chemistry returns to nearly normal. The cells use glucose at the same rate it enters the bloodstream. Thus blood sugar level returns to normal. Blood pH is brought under control by the kidneys as they excrete more hydrogen ions. However, blood pressure remains abnormally high because the retention of water increases the volume of blood.

All of us are confronted by stressors from time to time, and we have all experienced the resistance stage. Generally, this stage is successful in seeing us through a stressful situation, and our bodies then return to normal. Occasionally, the resistance stage fails to combat the stressor, however, and the body "gives up." In this case, the general adaptation syndrome moves into the stage of exhaustion.

EXHAUSTION

A major cause of exhaustion is loss of potassium ions. When the mineralocorticoids stimulate the kidney to retain sodium ions, potassium and hydrogen ions are traded off for sodium ions and secreted in the urine. As the chief positive ion in cells, potassium is partly responsible for controlling the water concentration of the cytoplasm. As the cells lose more and more potassium, they function less and less effectively. Finally they start to die. This condition is called the **stage of exhaustion.** Unless it is rapidly reversed, vital organs cease functioning and the person dies. Another cause of exhaustion is depletion of the adrenal glucocorticoids. In this case, blood glucose level suddenly falls and the cells do not receive enough nutrients. A final cause of exhaustion is weakening organs.

A long-term or strong resistance reaction puts heavy demands on the body, particularly on the heart, blood vessels, and adrenal cortex. They may not be up to handling the demands, or they may suddenly fail under the strain. In this respect, ability to handle stressors is determined to a large degree by general health.

STRESS AND DISEASE

Scientists have long known about the effects of environmental stress on laboratory animals. Although the exact role of stress in human diseases is not known, it is becoming quite clear that stress can lead to certain diseases. Among the conditions that are related to stress are gastritis, ulcerative colitis, irritable bowel syndrome, peptic ulcers, hypertension, asthma, rheumatoid arthritis, migraine headaches, anxiety, and depression. And, as noted earlier, stress is believed to increase susceptibility to infection by temporarily inhibiting certain components of the immune system. It has also been shown that individuals under stress are at a greater risk of developing chronic disease or dying prematurely.

MEDICAL TERMINOLOGY

Hyperplasia (*hyper* = over, *plas* = grow) Increase in the number of cells due to an increase in the frequency of cell division.
Hypoplasia (*hypo* = under) Defective development of tissue.
Neuroblastoma (*neuro* = nerve) Malignant tumor arising from the adrenal medulla associated with metastases to bones.
Thyroid storm An aggravation of all symptoms of hyperthyroidism characterized by unregulated hypermetabolism with fever and rapid heart rate; results from trauma, surgery, and unusual emotional stress or labor.

STUDY OUTLINE

Endocrine Glands (p. 399)
1. Both the endocrine and nervous systems assume a role in maintaining homeostasis.
2. Hormones help regulate the internal environment, respond to stress, help regulate growth and development, and contribute to reproductive processes.
3. Exocrine glands (sweat, sebaceous, digestive) secrete their products through ducts into body cavities or onto body surfaces.
4. Endocrine glands secrete hormones into the blood.

Chemistry of Hormones (p. 399)
1. Chemically, hormones are classified as amines, proteins and peptides, and steroids.
2. Amines are modified from the amino acid tyrosine (e.g. thyroid hormones); protein and peptide hormones consist of chains of amino acids (e.g. oxytocin); and steroid hormones are derived from cholesterol (e.g. aldosterone).

Mechanism of Hormonal Action (p. 399)
1. The amount of hormone released is determined by the body's need for the hormone.
2. Cells that respond to the effects of hormones are called target cells.
3. Receptors are found in the plasma membrane, cytoplasm, and nucleus of target cells.
4. The combination of hormone and receptor activates a chain of events in a target cell in which the physiological effects of the hormone are expressed.
5. One mechanism of hormonal action involves interaction of hormone with plasma membrane receptors; the second messenger may be cyclic AMP, calcium ions, or cyclic GMP.
6. Another mechanism of hormonal action involves interaction of a hormone with intracellular receptors.
7. Prostaglandins (PG) are membrane-associated lipids that can increase or decrease cyclic AMP formation and thus modulate hormone responses that use cyclic AMP.

Control of Hormonal Secretions: Feedback Control (p. 403)
1. A negative feedback control mechanism prevents overproduction or underproduction of a hormone.
2. Hormone secretions are controlled by levels of circulating hormone itself, nerve impulses, and regulating factors.

Pituitary (Hypophysis) (p. 404)
1. The pituitary is located in the sella turcica and is differentiated into the adenohypophysis (the anterior lobe and glandular portion), the neurohypophysis (the posterior lobe and nervous portion), and pars intermedia (avascular zone between lobes).
2. Hormones of the adenohypophysis are released or inhibited by regulating factors produced by the hypothalamus.
3. The blood supply to the adenohypophysis is from the superior hypophyseal arteries. It transports hypothalamic regulating factors.
4. Histologically, the adenohypophysis consists of somatotroph cells that produce growth hormone (GH); lactotroph cells that produce prolactin (PRL); thyrotroph cells that secrete thyroid-stimulating hormone (TSH); gonadotroph cells that synthesize follicle-stimulating hormone (FSH) and luteinizing hormone (LH); and corticolipotroph cells that secrete adrenocorticotropin hormone (ACTH) and melanocyte-stimulating hormone (MSH).
5. GH stimulates body growth through somatomedins and insulinlike growth factors (IGF) and is controlled by GHIF (growth hormone inhibiting factor or somatostatin) and GHRF (growth hormone releasing factor).
6. Disorders associated with improper levels of GH are pituitary dwarfism, giantism, and acromegaly.
7. TSH regulates thyroid gland activities and is controlled by TRF (thyrotropin releasing factor).
8. ACTH regulates the activities of the adrenal cortex and is controlled by CRF (corticotropin releasing factor).
9. FSH regulates the activities of the ovaries and testes and is controlled by GnRF (gonadotropin releasing factor).

10. LH regulates female and male reproductive activities and is controlled by GnRF.
11. PRL helps initiate milk secretion and is controlled by PIF (prolactin inhibiting factor) and PRF (prolactin releasing factor).
12. MSH increases skin pigmentation and is controlled by MRF (melanocyte-stimulating hormone releasing factor) and MIF (melanocyte-stimulating hormone inhibiting factor).
13. The neural connection between the hypothalamus and neurohypophysis is via the hypothalamic-hypophyseal tract.
14. Hormones made by the hypothalamus and stored in the neurohypophysis are oxytocin or OT (stimulates contraction of uterus and ejection of milk) and antidiuretic hormone or ADH (stimulates water reabsorption by the kidneys and arteriole constriction).
15. OT secretion is controlled by uterine distension and sucking during nursing; ADH is controlled primarily by water concentration.
16. A disorder associated with dysfunction of the neurohypophysis is diabetes insipidus.

Thyroid (p. 412)
1. The thyroid gland is located below the larynx.
2. Histologically, the thyroid consists of thyroid follicles composed of follicular cells, which secrete the thyroid hormones thyroxine (T_4) and triiodothyronine (T_3), and parafollicular cells, which secrete calcitonin (CT).
3. Thyroid hormones are synthesized from iodine and tyrosine within thyroglobulin (TGB) and carried in the blood with plasma proteins, mostly thyroxine-binding globulin (TBG).
4. Thyroid hormones regulate the rate of metabolism, growth and development, and the reactivity of the nervous system. Secretion is controlled by TRF.
5. Cretinism, myxedema, exophthalmic goiter, and simple goiter are disorders associated with the thyroid gland.
6. Calcitonin (CT) lowers the blood level of calcium. Secretion is controlled by its own level in blood.

Parathyroids (p. 416)
1. The parathyroids are embedded on the posterior surfaces of the lateral lobes of the thyroid.
2. The parathyroids consist of principal and oxyphil cells.
3. Parathyroid hormone (PTH) regulates the homeostasis of calcium and phosphate by increasing blood calcium level and decreasing blood phosphate level. Secretion is controlled by its own level in blood.
4. Tetany and osteitis fibrosa cystica are disorders associated with the parathyroid glands.

Adrenals (Suprarenals) (p. 419)
1. The adrenal glands are located superior to the kidneys. They consist of an outer cortex and inner medulla.
2. Histologically, the cortex is divided into a zona glomerulosa, zona fasciculata, and zona reticularis; the medulla consists of chromaffin cells.
3. Cortical secretions are mineralocorticoids, glucocorticoids, and gonadocorticoids.
4. Mineralocorticoids (e.g., aldosterone) increase sodium and water reabsorption and decrease potassium reabsorption. Secretion is controlled by the renin-angiotensin pathway and blood level of potassium.
5. A dysfunction related to aldosterone secretion is aldosteronism.
6. Glucocorticoids (e.g., cortisol) promote normal metabolism, help resist stress, and serve as antiinflammatories. Secretion is controlled by CRF.
7. Disorders associated with glucocorticoid secretion are Addison's disease and Cushing's syndrome.

8. Gonadocorticoids secreted by the adrenal cortex have minimal effects. Excessive production results in adrenogenital syndrome.
9. Medullary secretions are epinephrine and norepinephrine (NE), which produce effects similar to sympathetic responses. They are released under stress.
10. Tumors of medullary chromaffin cells are called pheochromocytomas.

Pancreas (p. 424)
1. The pancreas is posterior and slightly inferior to the stomach.
2. Histologically, it consists of pancreatic islets or islets of Langerhans (endocrine cells) and acini (enzyme-producing cells). Three types of cells in the endocrine portion are alpha cells, beta cells, and delta cells.
3. Alpha cells secrete glucagon, beta cells secrete insulin, and delta cells secrete growth hormone inhibiting factor (GHIF) or somatostatin.
4. Glucagon increases blood sugar level. Secretion is controlled by its own level in the blood.
5. Insulin decreases blood sugar level. Secretion is controlled by its own level in the blood.
6. Disorders associated with insulin production are diabetes mellitus and hyperinsulinism.

Ovaries and Testes (p. 427)
1. Ovaries are located in the pelvic cavity and produce sex hormones related to development and maintenance of female sexual characteristics, menstrual cycle, pregnancy, and lactation.
2. Testes lie inside the scrotum and produce sex hormones related to the development and maintenance of male sexual characteristics.

Pineal (Epiphysis Cerebri) (p. 428)
1. The pineal is attached to the roof of the third ventricle.
2. Histologically, it consists of secretory parenchymal cells called pinealocytes, neuroglial cells, and scattered postganglionic sympathetic fibers. Calcified deposits are referred to as brain sand.
3. It secretes melatonin (possibly regulates reproductive activities by inhibiting gonadotropic hormones) and adrenoglomerulotropin (may stimulate adrenal cortex to secrete aldosterone).

Thymus (p. 428)
1. The thymus gland secretes several hormones related to immunity.
2. Thymosin, thymic humoral factor (THF), thymic factor (TF), and thymopoietin promote the maturation of T cells.

Developmental Anatomy of the Endocrine System (p. 428)
1. The development of the endocrine system is not as localized as other systems.
2. The pituitary, adrenal medulla, and pineal gland develop from ectoderm; the adrenal cortex develops from mesoderm; and the thyroid, pancreas, and thymus develop from endoderm.

Other Endocrine Tissues (p. 430)
1. The gastrointestinal (GI) tract synthesizes stomach and intestinal gastrin, secretin, cholecystokinin (CCK), enterocrinin, and gastric inhibitory peptide (GIP).
2. The placenta produces human chorionic gonadotropin (HCG), estrogens, progesterone, relaxin, and human chorionic somatomammotropin (HCS).
3. The kidneys release an enzyme that produces erythropoietin.
4. The skin synthesizes vitamin D.

Aging and the Endocrine System (p. 430)

1. Most endocrine disorders are related to pathologies rather than age.
2. Diabetes mellitus and thyroid disorders are among the more important endocrine disorders.

Stress and the General Adaptation Syndrome (p. 430)

1. If the stress is extreme or unusual, it triggers a wide-ranging set of bodily changes called the general adaptation syndrome.
2. Unlike the homeostatic mechanisms, this syndrome does not maintain a constant internal environment.

Stressors (p. 430)

1. The stresses that produce the general adaptation syndrome are called stressors.
2. Stressors include surgical operations, poisons, infections, fever, and strong emotional responses.

Alarm Reaction (p. 430)

1. The alarm reaction is initiated by nerve impulses from the hypothalamus to the sympathetic division of the autonomic nervous system and adrenal medulla. Responses are the immediate and short-lived fight-or-flight responses that increase circulation, promote catabolism for energy production, and decrease nonessential activities.

Resistance Reaction (p. 432)

1. The resistance reaction is initiated by regulating factors secreted by the hypothalamus.
2. The regulating factors are CRF, GHRF, and TRF.
3. CRF stimulates the adenohypophysis to increase its secretion of ACTH, which in turn stimulates the adrenal cortex to secrete hormones.
4. Resistance reactions are long-term and accelerate catabolism to provide energy to counteract stress.
5. Glucocorticoids are produced in high concentrations during stress. They create many distinct physiological effects.

Exhaustion (p. 432)

1. The stage of exhaustion results from dramatic changes during alarm and resistance reactions.
2. Exhaustion is caused mainly by loss of potassium, depletion of adrenal glucocorticoids, and weakened organs, and if stress is too great, it may lead to death.

Stress and Disease (p. 433)

1. It appears that stress can lead to certain diseases.
2. Among stress-related conditions are gastritis, peptic ulcers, hypertension, and migraine headaches.

REVIEW QUESTIONS

1. Contrast endocrine and nervous system control of homeostasis.
2. Distinguish between an endocrine gland and an exocrine gland. What are the four principal actions of hormones?
3. What is a hormone? Distinguish between tropic and gonadotropic hormones.
4. Describe the chemical classification of hormones. Give an example of each.
5. Explain how receptors are related to hormones.
6. Describe the mechanism of hormonal action involving (a) interaction with plasma membrane receptors and (b) interaction with intracellular receptors.
7. Define a prostaglandin. Give several functions of prostaglandins. How are prostaglandins related to hormonal action?
8. How are negative feedback systems related to hormonal control? Discuss three models of operation.
9. In what respect is the pituitary gland actually two glands?
10. Describe the histology of the adenohypophysis. Why does the anterior lobe of the gland have such an abundant blood supply?
11. What hormones are produced by the adenohypophysis? What are their functions? How are they controlled?
12. Relate the importance of regulating factors to secretions of the adenohypophysis.
13. Describe the clinical symptoms of pituitary dwarfism, giantism, and acromegaly.
14. Discuss the histology of the neurohypophysis and the function and regulation of its hormones.
15. What are the clinical symptoms of diabetes insipidus?
16. Describe the location and histology of the thyroid gland.
17. How are the thyroid hormones made, stored, and secreted?
18. Discuss the physiological effects of the thyroid hormones. How are these hormones regulated?
19. Discuss the clinical symptoms of cretinism, myxedema, exophthalmic goiter, and simple goiter.
20. Describe the function and control of calcitonin (CT).
21. Where are the parathyroids located? What is their histology?
22. What are the functions of parathyroid hormone (PTH)?
23. Discuss the clinical symptoms of tetany and osteitis fibrosa cystica.
24. Compare the adrenal cortex and adrenal medulla with regard to location and histology.
25. Describe the hormones produced by the adrenal cortex in terms of type, normal function, and control.
26. Describe the clinical symptoms of aldosteronism, Addison's disease, Cushing's syndrome, and adrenogenital syndrome.
27. What relationship does the adrenal medulla have to the autonomic nervous system? What is the action of adrenal medullary hormones?
28. What is a pheochromocytoma?
29. Describe the location of the pancreas and the histology of the pancreatic islets (islets of Langerhans).
30. What are the actions of glucagon and insulin? How are the hormones controlled?
31. Describe the clinical symptoms of diabetes mellitus and hyperinsulinism. Distinguish the two types of diabetes mellitus.
32. Why are the ovaries and testes considered to be endocrine glands?
33. Where is the pineal gland located? What are its assumed functions?
34. How are hormones of the thymus gland related to immunity?
35. Describe the development of the endocrine system.
36. List the hormones secreted by the gastrointestinal tract, placenta, kidneys, and skin.
37. Describe the effects of aging on the endocrine system.
38. Define the general adaptation syndrome. What is a stressor?
39. How do homeostatic responses differ from stress responses?
40. Outline the reactions of the body during the alarm stage, resistance stage, and stage of exhaustion when placed under stress. What is the central role of the hypothalamus during stress?
41. Refer to the glossary of medical terminology associated with the endocrine system. Be sure that you can define each term.

U N I T   IV

Maintenance of the Human Body

This unit explains how the body maintains homeostasis on a day-to-day basis. In these chapters you will be studying the interrelations among the cardiovascular, lymphatic, respiratory, digestive, and urinary systems. You will also learn about metabolism, fluid and electrolyte balance, and acid–base dynamics.

19 / The Cardiovascular System: The Blood

20 / The Cardiovascular System: The Heart

21 / The Cardiovascular System: Vessels and Routes

22 / The Lymphatic System and Immunity

23 / The Respiratory System

24 / The Digestive System

25 / Metabolism

26 / The Urinary System

27 / Fluid, Electrolyte, and Acid–Base Dynamics

The Cardiovascular System: The Blood

19

STUDENT OBJECTIVES

- Contrast the general roles of blood, lymph, and interstitial fluid in maintaining homeostasis.
- Define the principal physical characteristics of blood and its functions in the body.
- Compare the origins of the formed elements in blood.
- Discuss the structure of erythrocytes and their function in the transport of oxygen and carbon dioxide.
- Explain the importance of a reticulocyte count and hematocrit.
- List the structural features and types of leucocytes.
- Explain the significance of a differential count.
- Discuss the role of leucocytes in phagocytosis and antibody production.
- Discuss the structure of thrombocytes and explain their role in blood clotting.
- List the components of plasma and explain their importance.
- Explain how the body attempts to prevent blood loss.
- Identify the stages involved in blood clotting.
- Explain the various factors that promote and inhibit blood clotting.
- Define clotting time, bleeding time, and prothrombin time.
- Explain ABO and Rh blood grouping.
- Define the antigen-antibody reaction as the basis for ABO blood grouping.
- Define the antigen-antibody reaction of the Rh blood grouping system.
- Define erythroblastosis fetalis as a harmful antigen-antibody reaction.
- Compare the location, composition, and function of interstitial fluid and lymph.
- Contrast the causes of nutritional, pernicious, hemorrhagic, hemolytic, aplastic, and sickle-cell anemia.
- Define polycythemia and describe the importance of hematocrit in its diagnosis.
- Identify the clinical symptoms of infectious mononucleosis (IM) and leukemia.
- Define medical terminology associated with blood.

The more specialized cells become, the less capable they are of carrying on an independent existence. For instance, specialized cells are less able to protect themselves from extreme temperatures, toxic chemicals, and changes in pH. They cannot go looking for food or devour whole bits of food. And, if they are firmly implanted in a tissue, they cannot move away from their own wastes. The substance that bathes cells and carries out these vital functions for them is called **interstitial fluid** (also known as **intercellular** or **tissue fluid**).

The interstitial fluid, in turn, must be serviced by blood and lymph. The blood picks up oxygen from the lungs, nutrients from the digestive tract, hormones from the endocrine glands, and enzymes from still other parts of the body. It transports these substances to all the tissues where they diffuse from the capillaries into the interstitial fluid. In the interstitial fluid, the substances are passed on to the cells and exchanged for wastes.

Since the blood services all the tissues of the body, it can be an important medium for the transport of disease-causing organisms. To protect itself from the spread of disease, the body has a lymphatic system—a collection of vessels containing a fluid called lymph. The lymph picks up materials, including wastes, from the interstitial fluid, cleanses them of bacteria, and returns them to the blood. The blood then carries the wastes to the lungs, kidneys, and sweat glands, where they are eliminated from the body. The blood also takes wastes to the liver, where they are detoxified.

Blood inside blood vessels, interstitial fluid around body cells, and lymph inside lymph vessels constitute the **internal environment** of the human organism. Because the body cells are too specialized to adjust to more than very limited changes in their environment, the internal environment must be kept relatively constant. This condition we have called homeostasis. In preceding chapters, we have discussed how the internal environment is kept in homeostasis. Now we will look at that environment itself.

The blood, heart, and blood vessels constitute the **cardiovascular system.** The lymph, lymph vessels, and structures and organs containing lymphatic tissue (large numbers of white blood cells called lymphocytes) make up the **lymphatic system.** We will first take a look at the substance known as blood. The branch of science concerned with the study of blood, blood-forming tissues, and the disorders associated with them is called **hematology** (*hem* = blood; *logos* = study of).

The developmental anatomy of blood and blood vessels is considered in Chapter 21.

PHYSICAL CHARACTERISTICS

The red body fluid that flows through all the vessels except the lymph vessels is called **blood.** Blood is a viscous fluid—it is thicker and more adhesive than water. Water is considered to have a viscosity of 1.0. The viscosity of blood, by comparison, ranges from 4.5 to 5.5. It flows more slowly than water, at least in part because of its viscosity.

The adhesive quality of blood, or its stickiness, may be felt by touching it. Blood is also slightly heavier than water.

Other physical characteristics of blood include a temperature of about 38°C (100.4°F), a pH range of 7.35 to 7.45 (slightly alkaline), and a salt (NaCl) concentration of 0.85 to 0.90 percent.

Blood constitutes about 8 percent of the total body weight. The blood volume of an average-sized male is 5 to 6 liters (5 to 6 qt). An average-sized female has 4 to 5 liters.

The physical characteristics of blood are summarized in Exhibit 19-1.

FUNCTIONS

Despite its simple appearance, blood is a complex liquid that performs a number of critical functions.

   1. It transports: oxygen from the lungs to the cells of the body; carbon dioxide from the cells to the lungs; nutrients from the digestive organs to the cells; waste products from the cells to the kidneys, lungs, and sweat glands; hormones from endocrine glands to the cells; enzymes to various cells.

   2. It regulates: pH through buffers; normal body temperature through the heat-absorbing and coolant properties of its water content; the water content of cells, principally through dissolved sodium ions.

   3. It protects against: blood loss through the clotting mechanism; toxins and foreign microbes through special combat-unit cells.

COMPONENTS

Blood is composed of two portions: formed elements (cells and cell-like structures) and plasma (liquid containing dissolved substances). The formed elements compose about 45 percent of the volume of blood; plasma constitutes about 55 percent (Figure 19-1).

FORMED ELEMENTS

In clinical practice, the most common classification of the **formed elements** of the blood is the following (Figure 19-2).

Erythrocytes (red blood cells)
Leucocytes (white blood cells)

EXHIBIT 19-1 SUMMARY OF PHYSICAL CHARACTERISTICS OF BLOOD	
Viscosity	4.5–5.5
Temperature	38°C (100.4°F)
pH	7.35–7.45
Salinity	0.85–0.90 percent
Total body weight	8 percent
Volume	5–6 liters for males; 4–5 liters for females

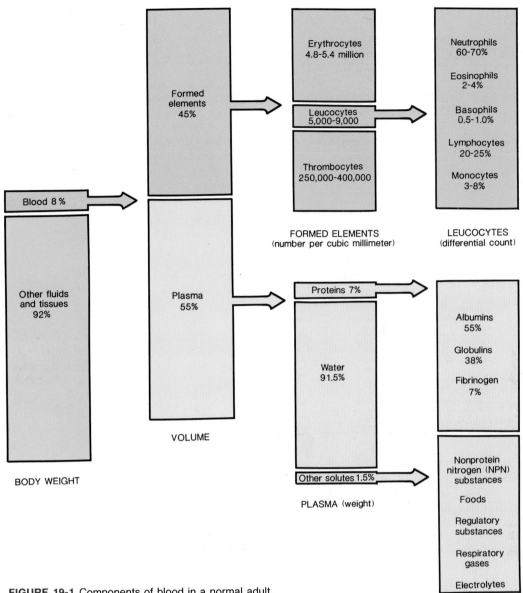

FORMED ELEMENTS
(number per cubic millimeter)

LEUCOCYTES
(differential count)

VOLUME

PLASMA (weight)

BODY WEIGHT

FIGURE 19-1 Components of blood in a normal adult.

Granular leucocytes (granulocytes)
    Neutrophils
    Eosinophils
    Basophils
Agranular leucocytes (agranulocytes)
    Lymphocytes
    Monocytes
Thrombocytes (platelets)

Origin

The process by which blood cells are formed is called **hemopoiesis** (hē-mō-poy-Ē-sis) or **hematopoiesis.** During embryonic and fetal life, there are no clear-cut centers for blood cell production. The yolk sac, liver, spleen, thymus gland, lymph nodes, and bone marrow all participate at various times in producing the formed elements. In the adult, however, we can pinpoint the production process to the red bone marrow in the proximal epiphyses of the humerus and femur, sternum, ribs, vertebrae, and pelvis, and to lymphoid tissue. Red blood cells, granular leucocytes, and platelets are produced in red bone marrow (myeloid tissue). Agranular leucocytes arise from both myeloid tissue and from lymphoid tissue—spleen, tonsils, lymph nodes. Undifferentiated mesenchymal cells in red bone marrow are transformed into **hemocytoblasts** (hē'-mō-SĪ-tō-blasts), immature cells that are eventually capable of developing into mature blood cells (Figure 19-2a). The hemocytoblasts undergo differentiation into five types of cells from which the major types of blood cells develop.

   1. Rubriblasts (proerythroblasts) form mature red blood cells.

   2. Myeloblasts form mature neutrophils, eosinophils, and basophils.

   3. Megakaryoblasts form mature thrombocytes (platelets).

   4. Lymphoblasts form lymphocytes.

   5. Monoblasts form monocytes.

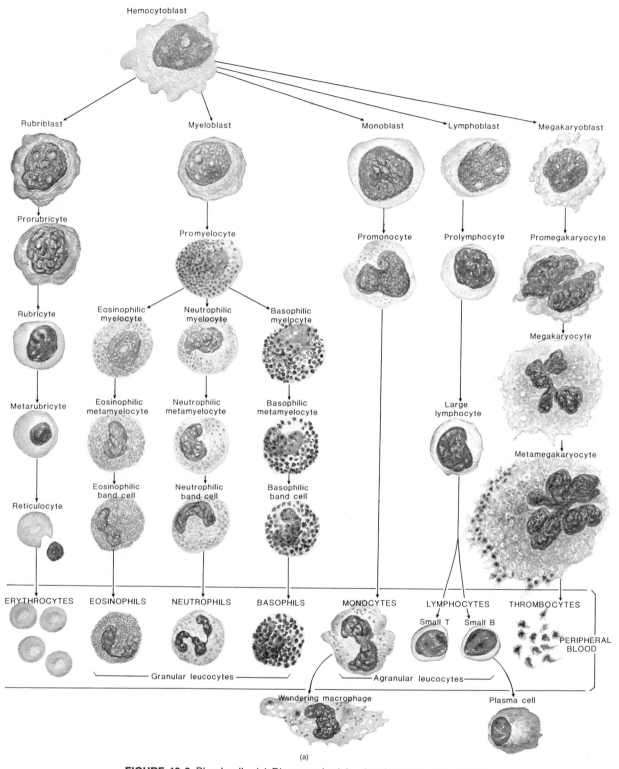

FIGURE 19-2 Blood cells. (a) Diagram of origin, development, and structure.

Erythrocytes

• *Structure* Microscopically, **red blood cells (rbc's)** or **erythrocytes** (e-RITH-rō-sīts) appear as biconcave discs averaging about 8 μm in diameter (Figure 19-2b). Mature red blood cells are quite simple in structure. They lack a nucleus and can neither reproduce nor carry on extensive metabolic activities. The plasma membrane is selectively permeable and consists of protein (stromatin) and lipids (lecithin and cholesterol). The membrane encloses cytoplasm and a red pigment called **hemoglobin.** Hemoglobin, which constitutes about 33 percent of the cell

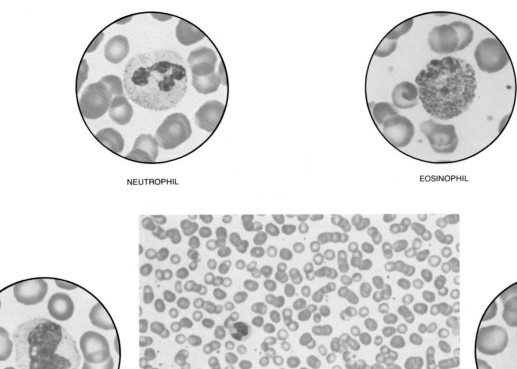

NEUTROPHIL

EOSINOPHIL

MONOCYTE

BASOPHIL

SMALL LYMPHOCYTE

LARGE LYMPHOCYTE

(b)

FIGURE 19-2 (*Continued*) (b) Photomicrograph of blood cells. Shown in the center is a blood smear at a magnification of 400×. The three larger cells with darker-staining nuclei are white blood cells; the more numerous cells are red blood cells, which lack nuclei; and the specklike objects are platelets. The principal types of white blood cells are shown in the large circles at a magnification of 1,800×. These white blood cells are surrounded by red blood cells. Note the platelets, indicated by the arrows. (Courtesy of Michael H. Ross and Edward J. Reith, *Histology: A Text and Atlas,* Copyright © 1985, by Michael H. Ross and Edward J. Reith, Harper & Row, Publishers, Inc., New York.)

weight, is responsible for the red color of blood. Normal values for hemoglobin are 14 to 20 gm/100 ml of blood in infants, 12 to 15 gm/100 ml in adult females, and 14 to 16.5 gm/100 ml in adult males. As you will see later, certain proteins (antigens) on the surfaces of red blood cells are responsible for the various blood groups. The ABO and Rh groups are examples.

• *Functions* The hemoglobin in erythrocytes combines with oxygen to form oxyhemoglobin, and with carbon dioxide to form carbaminohemoglobin, and then transports them through the blood vessels. The hemoglobin molecule consists of a protein called globin and a pigment called heme, which contains iron. As the erythrocytes pass through the lungs, each of the four iron atoms in the hemoglobin molecules combines with a molecule of oxygen. The oxygen is transported in this state to other tissues of the body. In the tissues, the iron-oxygen reaction reverses, and the oxygen is released to diffuse into the interstitial fluid. On the return trip, the globin portion combines with a molecule of carbon dioxide from the interstitial fluid to form carbaminohemoglobin. This complex is transported to the lungs, where the carbon dioxide is released and then exhaled. Although about 23 percent of carbon dioxide is transported by hemoglobin in this manner, the greater portion, about 70 percent, is transported in blood plasma as the bicarbonate ion, HCO_3^- (Chapter 23).

Red blood cells are highly specialized for their transport function. They contain a large number of hemoglobin molecules, which increases their oxygen-carrying capacity. One estimate is that each erythrocyte can carry 280 million molecules of hemoglobin. The biconcave shape of a red blood cell has a much greater surface area than, say, a sphere or a cube. The erythrocyte thus presents the maximum surface area for the diffusion of gas molecules that pass through the membrane to combine with hemoglobin.

CLINICAL APPLICATION

A **blood substitute** called *Fluosol-DA* is a slippery, white liquid that has a very high capacity for oxygen. Since its only function is to transport oxygen, it is actually a hemoglobin substitute. It may eventually turn out to be useful in providing oxygen to tissues which are oxygen-deficient as a result of carbon monoxide poisoning, sickle-cell anemia, strokes, heart attacks, and burns. Its use is limited by the fact that it must be kept frozen until just before use and recipients must receive oxygen by mask. Nevertheless, patients would be free of the risks of transfusion reactions and post-transfusion hepatitis.

• *Life Span and Number* The plasma membrane of a red blood cell becomes fragile and the cell is nonfunctional in about 120 days. The plasma membranes of worn-out red blood cells are removed from circulation by macrophages in the spleen, liver, and bone marrow. The hemoglobin is broken down into hemosiderin, an iron-containing pigment; bilirubin, a non-iron-containing pigment; and globin, a protein. The hemosiderin is stored or used in bone marrow to produce new hemoglobin for new red blood cells. Bilirubin is secreted by the liver into bile, and globin is metabolized by the liver.

A healthy male has about 5.4 million red blood cells per cubic millimeter (mm^3) of blood, and a healthy female has about 4.8 million. The higher value in the male is because of his higher rate of metabolism. To maintain normal quantities of erythrocytes, the body must produce new mature cells at the astonishing rate of 2 million per second. In the adult, production takes place in the red bone narrow in the spongy bone of the cranium, ribs, sternum, bodies of vertebrae, and proximal epiphyses of the humerus and femur.

• *Production* The process by which erythrocytes are formed is called **erythropoiesis** (e-rith'-rō-poy-Ē-sis). Erythropoiesis starts in red bone marrow with the transformation of a hemocytoblast into a rubriblast (see Figure 19-2a). The *rubriblast* (proerythroblast) gives rise to a *prorubricyte* (early erythroblast), which then develops into a *rubricyte* (intermediate erythroblast), the first cell in the sequence that begins to synthesize hemoglobin. The rubricyte next develops into a *metarubricyte* (late erythroblast), in which hemoglobin synthesis is at a maximum. In the next stage, the metarubricyte develops into a *reticulocyte*, a cell in the developmental sequence that contains about 34 percent hemoglobin, retains some endoplasmic reticulum, and loses its nucleus by extrusion. Reticulocytes pass from bone marrow into the bloodstream by squeezing between the endothelial cells of blood capillaries. If a reticulocyte attempts to cross the endothelial barrier before it has lost its nucleus, the nucleus is forced out since it is too rigid to pass through. Reticulocytes generally become *erythrocytes*, or mature red blood cells, within 1 to 2 days after their release from bone marrow. The normal proportion of reticulocytes in blood is between 0.5 and 1.5 percent. Aged erythrocytes are destroyed by fixed phagocytic macrophages in the liver and spleen. The hemoglobin molecules are split apart, the iron is reused, and the rest of the molecule is converted into other substances for reuse or elimination.

Normally erythropoiesis and red cell destruction proceed at the same pace. But if the body suddenly needs more erythrocytes or if erythropoiesis is not keeping up with red blood cell destruction, a homeostatic mechanism steps up erythrocyte production. The mechanism is triggered by the reduced supply of oxygen for body cells that results from a reduced number of erythrocytes. If certain kidney cells become oxygen-deficient, they release an enzyme called **renal erythropoietic factor** that converts a plasma protein into the hormone **erythropoietin**. This hormone circulates through the blood to the red bone marrow, where it stimulates more hemocytoblasts to develop into red blood cells. Erythropoietin is also produced by other body tissues, particularly the liver.

Cellular oxygen deficiency, called **hypoxia** (hī-POKS-ē-a), may occur if you do not breathe in enough oxygen. This deficiency commonly occurs at high altitudes, where the air contains less oxygen. Oxygen deficiency may also occur because of anemia. The term **anemia** means that the number of functional red blood cells or their hemoglobin content is below normal. Consequently, the erythrocytes are unable to transport enough oxygen from the lungs to the cells. Anemia has many causes; lack of iron, lack of certain amino acids, and lack of vitamin B_{12} are but a few. Iron is needed for the oxygen-carrying part of the hemoglobin molecule. The amino acids are needed for the protein, or globin, part. Vitamin B_{12} helps the red bone marrow to produce erythrocytes. This vitamin is obtained from meat, especially liver, but it cannot be absorbed by the lining of the small intestine without the help of another substance—**intrinsic factor** produced by the mucosal cells of the stomach. Intrinsic factor facilitates the absorption of vitamin B_{12}. Once absorbed, it is stored in the liver. (Several types of anemia are described later in the chapter.)

CLINICAL APPLICATION

The rate of erythropoiesis is measured by a procedure called a **reticulocyte** (re-TIK-yoo-lō-sīt) **count.** Some reticulocytes are normally released into the bloodstream before they become mature red blood cells. If the number of reticulocytes in a sample of blood is less than 0.5 percent of the number of mature red blood cells in the sample, erythropoiesis is occurring too slowly. A low reticulocyte count might confirm a diagnosis of anemia, or it might indicate a kidney disease that prevents the kidney cells from producing erythropoietin. If the reticulocytes number more than 1.5 percent of the mature red blood cells, erythropoiesis is abnormally rapid. Any number of problems may be responsible for a high reticulocyte count. These include oxygen deficiency and uncontrolled red blood cell production caused by a cancer in the bone marrow.

Another test with important clinical applications is the hematocrit. **Hematocrit** is the percentage of blood that is made up of red blood cells. It is determined by centrifuging blood and noting the ratio of red blood cells to whole blood. The average hematocrit for males is 40 to 54 percent. The average hematocrit for females is 38 to 47 percent. Anemic blood may have a hematocrit of 15 percent; polycythemic blood (an abnormal increase in the number of functional erythrocytes) may have a hematocrit of 65 percent. Athletes, however, may have higher-than-normal hematocrits, reflecting constant physical activity rather than a pathological condition.

Leucocytes

• *Structure and Types*   Unlike red blood cells, **leucocytes** (LOO-kō-sīts), or **white blood cells (wbc's)** have nuclei and do not contain hemoglobin (see Figure 19-2). Leucocytes fall into two major groups. The first group is the **granular leucocytes.** They develop from red bone marrow, have granules in the cytoplasm, and possess lobed nuclei. The three kinds of granular leucocytes are **neutrophils** or **polymorphs** (10 to 12 μm in diameter), **eosinophils** (10 to 12 μm in diameter), and **basophils** (8 to 10 μm in diameter).

The second principal group of leucocytes is the **agranular leucocytes.** They develop from lymphoid and myeloid tissue, and no cytoplasmic granules can be seen under a light microscope. The two kinds of agranular leucocytes are **lymphocytes** (7 to 15 μm in diameter) and **monocytes** (14 to 19 μm in diameter).

Just as red blood cells have surface proteins, so do white blood cells and all other nucleated cells in the body. These proteins, called **HLA (human leucocyte associated) antigens,** are unique for each person (except for identical twins) and can be used to identify a tissue. If an incompatible tissue is transplanted, it is rejected by the recipient as foreign due, in part, to differences in donor and recipient HLA antigens. The HLA antigens are used to type tissues to help prevent rejection.

• *Functions*   The skin and mucous membranes of the body are continuously exposed to microbes and their toxins. Some of these microbes are capable of invading deeper tissues to cause disease, and once they enter the body, the general function of leucocytes is to combat them by phagocytosis or antibody production. Neutrophils and monocytes are actively **phagocytotic**—they can ingest bacteria and dispose of dead matter (see Figure 3-6a,b). Neutrophils (NOO-trō-fils) are the most active leucocytes in responding to tissue destruction by bacteria. In addition to carrying on phagocytosis, they release the enzyme lysozyme, which destroys certain bacteria. Apparently monocytes (MON-ō-sīts) take longer to reach the site of infection than do neutrophils, but once they arrive, they do so in larger numbers and destroy more microbes. Monocytes that have migrated to infected tissues are called **wandering macrophages.** They clean up cellular debris following an infection.

A number of different chemicals in inflamed tissue cause phagocytes to migrate toward the tissue. This phenomenon is called **chemotaxis.** Among the substances that provide stimuli for chemotaxis are toxins produced by microbes and degenerative products of damaged tissues.

Most leucocytes possess, to some degree, the ability to crawl through the minute spaces between the cells that form the walls of capillaries and through connective and epithelial tissue. This movement, like that of amoebas, is called **diapedesis** (dī'-a-pe-DĒ-sis). First, part of the cell membrane projects outward. Then the cytoplasm and nucleus flow into the projection. Finally, the rest of the membrane snaps up into place. Another projection is made, and so on, until the cell has crawled to its destination (see Figure 22-9b).

Neutrophils also contain **defensins,** amino acids that exhibit a broad range of antibiotic activity against bacteria, fungi, and viruses.

Eosinophils (ē'-ō-SIN-ō-fils) are believed to release substances that combat the effects of histamine and other mediators of inflammation in allergic reactions. Eosinophils leave the capillaries, enter the tissue fluid, and phagocytize antigen-antibody complexes. Eosinophils are also effective against certain parasitic worms. Thus, a high eosinophil count frequently indicates an allergic condition or a parasitic infection.

Basophils (BĀ-sō-fils) are also believed to be involved in allergic reactions. Basophils leave the capillaries, enter the tissues, and become the mast cells of the tissues, liberating heparin, histamine, and serotonin. These substances intensify the overall inflammatory reaction and are involved in hypersensitivity (allergic) reactions (Chapter 22).

Lymphocytes (LIM-fō-sīts) are involved in the production of antibodies. **Antibodies** (AN-ti-bod'-ēz) are special proteins that inactivate antigens. An **antigen** (AN-ti-jen) is any substance that will stimulate the production of antibodies and is capable of reacting specifically with the antibody. Most antigens are proteins, and most are not synthesized by the body. Many of the proteins that make up the cell structures and enzymes of bacteria are antigens. The toxins released by bacteria are also antigens. When antigens enter the body, they react chemically with substances in the lymphocytes and stimulate some lymphocytes, called B cells, to become **plasma cells** (Figure 19-3). The plasma cells then produce antibodies, globulin-type proteins that attach to antigens much as enzymes attach to substrates. Like enzymes, a specific antibody will generally attach only to a certain antigen. However, unlike enzymes, which enhance the reactivity of the substrate, antibodies "cover" their antigens so the antigens cannot come in contact with other chemicals in the body.

In this way, bacterial poisons can be sealed up and rendered harmless. The bacteria themselves are destroyed by the antibodies. This process is called the **antigen-antibody response.** Phagocytes in tissues destroy the antigen-antibody complexes. The antigen-antibody response is discussed in more detail in Chapter 22.

CLINICAL APPLICATION

Multiple myeloma is a malignant disorder of plasma cells in bone marrow. The symptoms are caused by the growing tumor cell mass itself and antibodies produced by the malignant cells. Among the effects are pain and osteoporosis due to multiple osteolytic (bone-dissolving) lesions in bones such as the sternum, ribs, backbone, clavicles, skull, pelvis, and proximal extremities. Other effects include hypercalcemia (excessive blood calcium), anemia, leucopenia (abnormally low level of white blood cells), thrombocytopenia (abnormally low level of thrombocytes), spinal cord compression, enlarged liver, kidney damage, and increased susceptibility to infection.

Other lymphocytes are called **T cells.** One group of T cells, the **killer T cells,** are activated by certain antigens and react by destroying them directly or indirectly by recruiting other lymphocytes and macrophages. T cells

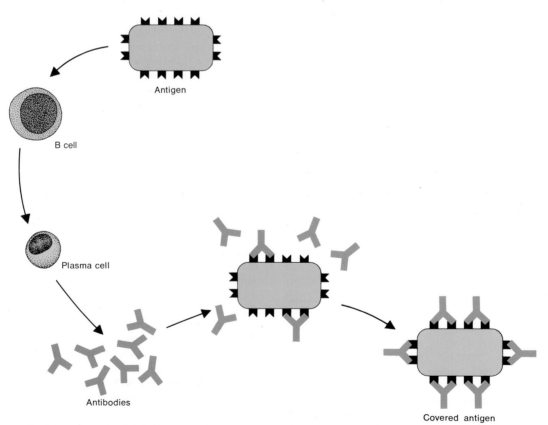

FIGURE 19-3 Antigen-antibody response. An antigen entering the body stimulates a B cell to develop into an antibody-producing plasma cell. The antibodies attach to the antigen, cover it, and render it harmless.

are especially effective against bacteria, viruses, fungi, transplanted cells, and cancer cells. Other types of T cells and their functions are discussed in Chapter 22.

The antigen-antibody response helps us combat infection and gives us immunity to some diseases. It is also responsible for blood types, allergies, and the body's rejection of organs transplanted from an individual with a different genetic makeup.

An increase in the number of white cells present in the blood typically indicates a state of inflammation or infection. Because each type of white cell plays a different role, determining the percentage of each type in the blood assists in diagnosing the condition. A **differential count** is the number of each kind of white cell in 100 white blood cells. A normal differential count falls within the following percentages:

Neutrophils	60–70%
Eosinophils	2–4%
Basophils	0.5–1%
Lymphocytes	20–25%
Monocytes	3–8%
	100%

Particular attention is paid to the neutrophils in a differential count. More often than not, a high neutrophil count indicates damage by invading bacteria. An increase in the number of monocytes generally indicates a chronic infection. Eosinophils and basophils are elevated during allergic reactions. High lymphocyte counts indicate antigen-antibody reactions.

• *Life Span and Number*   Bacteria exist everywhere in the environment and have continuous access to the body through the mouth, nose, and pores of the skin. Furthermore, many cells, especially those of epithelial tissue, age and die, and their remains must be disposed of daily. Even when the body is healthy, the leucocytes actively ingest bacteria and debris. However, a leucocyte can phagocytose only a certain number of substances before they interfere with the leucocyte's normal metabolic activities and bring on its death. Consequently, the life span of most leucocytes is very short. In a healthy body, some white blood cells can live up to several months, but most live only a few days. During a period of infection they may live only a few hours.

Leucocytes are far less numerous than red blood cells, averaging from 5,000 to 9,000 cells per cubic millimeter of blood. Red blood cells, therefore, outnumber white blood cells about 700 to 1. The term **leucocytosis** (loo'-kō-sī-TŌ-sis) refers to an increase in the number of white blood cells. If the increase exceeds 10,000, a pathological condition is usually indicated. An abnormally low level of white blood cells (below 5,000/mm³) is termed **leucopenia** (loo-kō-PĒ-nē-a).

• *Production*   Granular leucocytes are produced in red bone marrow (myeloid tissue); agranular leucocytes are produced in both myeloid and lymphoid tissue. The developmental sequences for the five types of leucocytes are shown in Figure 19-2a.

CLINICAL APPLICATION

Bone marrow, as indicated earlier, contains hemocytoblasts that differentiate into erythrocytes, leucocytes, and platelets. Among the leucocytes are cells that combat infections and cause tissue rejection. Until recently, the use of **bone marrow transplantation,** the transfer of bone marrow from a donor to a recipient, required that the marrow of the donor be obtained from a relative and closely matched to that of the recipient. In cases where the match is not identical, T cells in donor marrow, which are part of the immune system, recognize the recipient's tissues as foreign and attack them. The principal tissues rejected in this way are in the skin, liver, and gastrointestinal (GI) tract. Several recent advances now permit bone marrow transplantation in which the tissue match between donor and recipient is not even close.

Donor marrow is aspirated from the hip bones, mixed with heparin (an anticoagulant), and then passed through screens. The suspension of bone marrow cells is treated to remove T cells and given to the recipient just like a blood transfusion. The cells in the suspension pass through the lungs, enter general circulation, and reseed and grow in the marrow cavities of the recipient's bones.

Bone marrow transplantation has been used to treat aplastic anemia, certain types of leukemia, and severe combined immunodeficiency disease (SCID), an inherited deficiency of infection-fighting blood cells. The technique is also being expanded to treat other kinds of leukemia, non-Hodgkin's lymphoma, thalassemia, multiple myeloma, sickle-cell anemia, and hemolytic anemia.

Thrombocytes

• *Structure*   In addition to the immature cell types that develop into erythrocytes and leucocytes, hemocytoblasts differentiate into still another kind of cell, called a megakaryoblast (see Figure 19-2a). Megakaryoblasts are transformed into megakaryocytes, large cells that shed fragments of cytoplasm. Each fragment becomes enclosed by a piece of the cell membrane and is called a **thrombocyte** (THROM-bō-sīt) or **platelet.** Platelets are round or oval discs without a nucleus. They average from 2 to 4 μm in diameter.

• *Function*   Platelets prevent fluid loss by initiating a chain of reactions that results in blood clotting. This mechanism is described shortly.

• *Life Span and Number*   Like the other formed elements of the blood, platelets have a short life, probably only 5 to 9 days. Between 250,000 and 400,000 platelets appear in each cubic millimeter of blood.

• *Production*   Platelets are produced in red bone marrow according to the developmental sequence shown in Figure 19-2a.

A summary of the formed elements in blood is presented in Exhibit 19-2.

CLINICAL APPLICATION

The most commonly ordered hematology test is the **complete blood count (CBC).** It generally includes determination of hemoglobin, hematocrit, red blood cell count, white blood cell count, differential count, and comments about red blood cell, white blood cell, and platelet morphology.

PLASMA

When the formed elements are removed from blood, a straw-colored liquid called **plasma** is left. Exhibit 19-3 outlines the chemical composition of plasma. Some of the proteins in plasma are also found elsewhere in the body, but in blood they are called **plasma proteins. Albumins,** which constitute 55 percent of plasma proteins, are largely responsible for blood's viscosity. The concentration of albumins is about four times higher in plasma than in interstitial fluid. Along with the electrolytes, albumins also help regulate blood volume by preventing the water in the blood from diffusing into the interstitial fluid. Recall that water moves by osmosis from an area of high water (low solute) concentration to an area of low water (high solute) concentration. **Globulins,** which comprise 38 percent of plasma proteins, are antibody proteins released by plasma cells. Gamma globulin is especially well known because it is able to form an antigen-antibody

complex with the proteins of the hepatitis and measles viruses and the tetanus bacterium, among others. **Fibrinogen** makes up about 7 percent of plasma proteins and takes part in the blood-clotting mechanism along with the platelets.

CLINICAL APPLICATION

Apheresis (a-FER-e-sis; *aphairesis* = removal) refers to a procedure in which blood is withdrawn from the body, its components are selectively separated, the undesirable component is removed, and the remainder is returned to the body. Blood is withdrawn by a needle or catheter, mixed with an anticoagulant, and pumped through a separator where red blood cells, white blood cells, platelets, and plasma are separated by centrifugal force. The process usually takes 3 to 5 hours and no more than 15 percent of the patient's total blood volume is allowed outside the body at any one time.

Apheresis may be used in the treatment of certain diseases such as aplastic anemia, hemolytic anemia, multiple myeloma, thrombocytemia, posttransfusion Rh incompatibility, life-threatening sickle-cell crisis, myasthenia gravis, kidney diseases, systemic lupus-erythematosus (SLE), rheumatoid arthritis (RA), multiple sclerosis (MS), and certain drug overdoses. Because of some potentially serious complications of apheresis (hypovolemia, shock, congestive heart failure, pulmonary edema, muscle twitching, thrombosis), the procedure is used discriminately when more conservative therapy is unsuccessful.

EXHIBIT 19-2 SUMMARY OF THE FORMED ELEMENTS IN BLOOD

FORMED ELEMENT	NUMBER	DIAMETER (in μm)	LIFE SPAN	FUNCTION
Erythrocyte (red blood cell)	4.8 million/mm³ in females; 5.4 million/mm³ in males.	8	120 days.	Transports oxygen and carbon dioxide.
Leucocyte (white blood cell)	5,000–9,000/mm³.		Few hours to a few days.	
Granular				
Neutrophil	60–70% of total.	10–12		Phagocytosis.
Eosinophil	2–4% of total.	10–12		Combats the effects of histamine in allergic reactions, phagocytizes antigen-antibody complexes, and destroys certain parasitic worms.
Basophil	0.5–1% of total.	8–10		Liberates heparin, histamine, and serotonin in allergic reactions that intensify the overall inflammatory response.
Agranular				
Lymphocyte	20–25% of total.	7–15		Immunity (antigen-antibody reactions).
Monocyte	3–8% of total.	14–19		Phagocytosis.
Thrombocyte (platelet)	250,000–400,000/ mm³.	2–4	5–9 days.	Blood clotting.

EXHIBIT 19-3 CHEMICAL COMPOSITION AND DESCRIPTION OF SUBSTANCES IN PLASMA

CONSTITUTENT	DESCRIPTION
WATER	Liquid portion of blood; constitutes about 91.5 percent of plasma. Ninety percent of water derived from absorption from gastrointestinal (GI) tract; 10 percent from cellular respiration. Acts as solvent and suspending medium for solid components of blood and absorbs heat.
SOLUTES	Constitute about 8.5 percent of plasma.
Proteins	
Albumins	Smallest plasma proteins. Produced by liver and provide blood with viscosity, a factor related to maintenance and regulation of blood pressure. Also exert considerable osmotic pressure to maintain water balance between blood and tissues and regulate blood volume.
Globulins	Protein group to which antibodies belong. Gamma globulins attack measles, hepatitis, and polio viruses, and tetanus bacterium.
Fibrinogen	Produced by liver. Plays essential role in clotting.
Nonprotein nitrogen (NPN) substances	Contain nitrogen but are not proteins. Include urea, uric acid, creatine, creatinine, and ammonium salts. Represent breakdown products of protein metabolism and are carried by blood to organs of excretion.
Food substances	Products of digestion passed into blood for distribution to all body cells. Include amino acids (from proteins), glucose (from carbohydrates), and fatty acids and glycerol (from fats).
Regulatory substances	Enzymes, produced by body cells, to catalyze chemical reactions. Hormones, produced by endocrine glands, to regulate growth and development in body.
Respiratory gases	Oxygen and carbon dioxide. Whereas O_2 is more closely associated with hemoglobin or red blood cells, CO_2 is more closely associated with plasma.
Electrolytes	Inorganic salts of plasma. Cations include Na^+, K^+, Ca^{2+}, Mg^{2+}; anions include Cl^-, PO_4^{3-}, SO_4^{2-}, HCO_3^-. Help maintain osmotic pressure, normal pH, physiological balance between tissues and blood.

HEMOSTASIS

Hemostasis refers to the stopping of bleeding. When blood vessels are damaged or ruptured, three basic mechanisms operate to prevent blood loss: (1) vascular spasm, (2) platelet plug formation, and (3) blood coagulation (clotting). Although these mechanisms are useful for preventing hemorrhage in smaller blood vessels, extensive hemorrhage usually requires intervention of some type.

VASCULAR SPASM

When a blood vessel is damaged, the smooth muscle in its wall contracts immediately. Such a **vascular spasm** reduces blood loss for up to 30 minutes, during which time the other hemostatic mechanisms can go into operation. The spasm is probably caused by damage to the vessel wall and from reflexes initiated by pain receptors.

PLATELET PLUG FORMATION

When platelets come into contact with parts of a damaged blood vessel, such as collagen or endothelium, their characteristics change drastically. They begin to enlarge and their shapes become even more irregular. They also become sticky and begin to adhere to the collagen fibers. Their synthesis of ADP and enzymes causes the formation of substances that activate more platelets, causing them to stick to the original platelets. The accumulation and attachment of large numbers of platelets form a **platelet plug.** The plug is very effective in preventing blood loss in a small vessel. Although the platelet plug is initially loose, it becomes quite tight when reinforced by fibrin threads formed during coagulation.

COAGULATION

Normally, blood maintains its liquid state as long as it remains in the vessels. If it is drawn from the body, however, it thickens and forms a gel. Eventually, the gel separates from the liquid. The straw-colored liquid, called **serum,** is simply plasma minus its clotting proteins. The gel is called a **clot** and consists of a network of insoluble fibers in which the cellular components of blood are trapped.

The process of clotting is called **coagulation.** If the blood clots too easily, the result can be thrombosis—clotting in an unbroken blood vessel. If the blood takes too long to clot, a hemorrhage can result.

Clotting involves various chemicals known as **coagulation factors.** In plasma, these factors are called *plasma coagulation factors.* A few *platelet coagulation factors* are released by platelets. One coagulation factor is released by damaged body tissues. The various plasma coagulation factors and their synonyms are listed in Exhibit 19-4.

Clotting is a complex process in which the activated form of one coagulation factor catalyzes the activation of the next factor in the clotting sequence. Once the process is initiated, there is a cascade of events that results in the formation of large quantities of product. For purposes of discussion, we will describe the clotting mechanism as a sequence of three basic stages:

Stage 1. Formation of **prothrombin activator.**
Stage 2. Conversion of prothrombin (a plasma protein formed by the liver) into the enzyme **thrombin** by prothrombin activator.

EXHIBIT 19-4 PLASMA COAGULATION FACTORS AND THEIR SYNONYMS

COAGULATION FACTOR	SYNONYM
I	Fibrinogen
II	Prothrombin
III	Thromboplastin
IV	Calcium ions
V	Proaccelerin, labile factor, or accelerator globulin
VII	Serum prothrombin conversion accelerator (SPCA), stable factor, or proconvertin
VIII	Antihemophilic factor (AHF), antihemophilic factor A, or antihemophilic globulin (AHG)
IX	Christmas factor, plasma thromboplastin component (PTC), or antihemophilic factor B
X	Stuart factor or Stuart-Prower factor
XI	Plasma thromboplastin antecedent (PTA) or antihemophilic factor C
XII	Hageman factor or glass factor
XIII	Fibrin stabilizing factor (FSF)

Stage 3. Conversion of fibrinogen (another plasma protein formed by the liver) **into insoluble fibrin** by thrombin. Fibrin forms the threads of the clot. (Cigarette smoke contains at least two substances that interfere with fibrin formation.)

The formation of prothrombin activator is initiated by the interplay of two mechanisms: the extrinsic and intrinsic pathways of blood clotting.

Extrinsic Pathway

The **extrinsic pathway** of blood clotting has fewer steps than the intrinsic pathway and occurs rapidly, within a matter of seconds if trauma is severe. It is so named because the formation of prothrombin activator is initiated by substances released by damaged blood vessels, or surrounding tissues, *outside* the blood itself. Damaged blood vessels or tissues release a complex of substances called **tissue thromboplastin** (Figure 19-4a). Among the components of tissue thromboplastin are an enzyme and membrane phospholipids. Together, tissue thromboplastin, coagulation factor VIII, and Ca^{2+} ions activate factor X. Once factor X is activated, it reacts with the membrane phospholipids, factor V, and Ca^{2+} ions to form prothrombin activator. This completes the extrinsic pathway and stage 1 of clotting. Next, the prothrombin activator, to-

gether with Ca^{2+} ions, brings about the conversion of prothrombin into thrombin in stage 2. In stage 3, thrombin, in the presence of Ca^{2+} ions, converts fibrinogen, which is soluble, to fibrin, which is insoluble. Thrombin also activates coagulation factor XIII, which strengthens and stabilizes the fibrin clot. Factor XIII occurs in plasma and it is also released by platelets trapped in the clot.

Thrombin has a positive feedback effect, through factor V, in accelerating the formation of prothrombin activator. Once clotting begins and thrombin is formed, a positive feedback cycle is set up in which thrombin accelerates production of prothrombin activator which, in turn, accelerates the production of more thrombin, and so on. If unchecked, a clot would continue to get larger and larger, as a result of the positive feedback cycle. One condition that prevents this is blood flow. The movement of blood within vessels is usually sufficient to carry away factors necessary for clotting, so that their concentrations will not rise enough to promote further clotting. In addition, as you will see shortly, there are certain substances in blood that inhibit or destroy coagulating factors.

Intrinsic Pathway

The **intrinsic pathway** of blood clotting is more complex than the extrinsic pathway and it occurs more slowly, usually requiring several minutes. The intrinsic pathway is so named because the formation of prothrombin activator is initiated by a substance contained *within* the blood itself. The intrinsic pathway is triggered when blood comes into contact with the underlying collagenous fibers of damaged blood vessels (Figure 19-4b). This causes the activation of coagulation factor XII and damage to platelets, resulting in the release of phospholipids by the platelets. Activated factor XII activates factor XI which, in turn, activates factor IX. Once factor IX is activated, it acts together with factor VIII, Ca^{2+} ions, and platelet phospholipids to activate factor X. Once factor X is activated, it reacts with platelet phospholipids, factor V, and Ca^{2+} ions to form prothrombin activator. This completes the intrinsic pathway and stage 1 of clotting. From this point on, the reactions for stages 2 and 3 are similar to those of the extrinsic pathway. Once thrombin is formed, it causes more platelets to adhere to each other, resulting in the release of more platelet phospholipids. This is another example of a positive feedback cycle.

Once the clot is formed, it plugs the ruptured area of the blood vessel and thus prevents hemorrhage (bleeding). Permanent repair of the blood vessel can then take place. In time, fibroblasts form connective tissue in the ruptured area and new endothelial cells repair the lining.

CLINICAL APPLICATION

Hemophilia (*hemo* = blood; *philein* = to love) refers to several different hereditary deficiencies of coagulation in which bleeding may occur spontaneously or after only minor trauma. The effects of all the forms of the disorder are so similar that they are hardly dis-

FIGURE 19-4 Blood clotting. (a) Extrinsic pathway. (b) Intrinsic pathway. (c) Scanning electron micrograph of fibrin threads and red blood cells at a magnification of about 1,500 to 2,000×. (Courtesy of Fisher Scientific Company and S.T.E.M. Laboratories, Inc., Copyright 1975.)

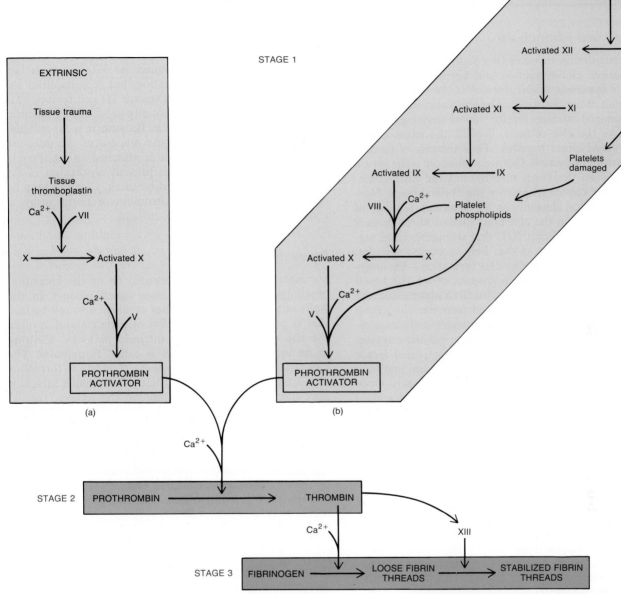

STAGE 1

EXTRINSIC

Tissue trauma

Tissue thromboplastin

Ca^{2+} VII

X → Activated X

Ca^{2+} V

PROTHROMBIN ACTIVATOR

(a)

INTRINSIC

Contact with collagen of damaged blood vessels

Activated XII ← XII

Activated XI ← XI

Activated IX ← IX

VIII Ca^{2+} Platelet phospholipids

Platelets damaged

Activated X ← X

V Ca^{2+}

PHROTHROMBIN ACTIVATOR

(b)

Ca^{2+}

STAGE 2 PROTHROMBIN → THROMBIN

Ca^{2+} XIII

STAGE 3 FIBRINOGEN → LOOSE FIBRIN THREADS → STABILIZED FIBRIN THREADS

tinguishable from one another, but each is a deficiency of a different blood clotting factor. For example, persons with the most common type of hemophilia, hemophilia A (classic hemophilia), lack factor VIII, the antihemophilic factor. Persons with hemophilia B lack factor IX. Those with hemophilia C lack factor XI. Hemophilia A and B occur primarily among males; hemophilia C is a mild form affecting both males and females.

Hemophilia is characterized by spontaneous or traumatic subcutaneous and intramuscular hemorrhaging, nosebleeds, blood in the urine, and joint pain and damage due to joint hemorrhaging. Treatment involves the application of pressure to accessible bleeding sites and transfusions of fresh plasma or the appropriate deficient clotting factor to relieve the bleeding tendency

(c)

for several days. The inheritance of hemophilia is discussed in Chapter 29.

Retraction and Fibrinolysis

Normal coagulation involves two additional events after clot formation: clot retraction and fibrinolysis. **Clot retraction** or **syneresis** (si-NER-e-sis) is the consolidation or tightening of the fibrin clot. The fibrin threads attached to the damaged surfaces of the blood vessel gradually contract. As the clot retracts, it pulls the edges of the damaged vessel closer together. Thus the risk of hemorrhage is further decreased. During retraction, some serum escapes between the fibrin threads, but the formed elements in blood remain trapped in the fibrin threads. Normal clot retraction depends on an adequate number of platelets. Platelets in the clot bind various fibrin threads together and release factor XIII that strengthens and stabilizes the clot and factors that help compress the clot.

The second event following clot formation—**fibrinolysis** (fī-brin-OL-i-sis)—involves dissolution of the blood clot. When a clot is formed, an inactive plasma enzyme called *plasminogen* is incorporated into the clot. Both body tissues and blood contain substances that can activate plasminogen to *plasmin*, an active plasma enzyme. Among these substances are thrombin, activated factor XII, lysosomal enzymes from damaged tissues, and substances from the endothelial lining of blood vessels. Once plasmin is formed, it can dissolve the clot by digesting fibrin threads and inactivating substances such as fibrinogen, prothrombin, and factors V, VIII, and XII. In addition to dissolving large clots in tissues, plasmin is an important enzyme in removing very small clots in small blood vessels.

CLINICAL APPLICATION

A recently developed drug used by physicians to **dissolve blood clots** is a normal enzyme referred to as *tissue plasminogen activator* (*t-PA*). This substance is a genetically engineered, artificial version of an enzyme normally found in very small amounts in the body. Under normal body conditions, t-PA dissolves small blood clots before they cause problems, and usually before an individual is aware of the problem. Physiologically, t-PA initiates a process that converts plasminogen into plasmin, which degrades the blood clot's fibrin network. After the clot dissolves, plasmin, t-PA, and the degraded fibrin fibers are carried away by the blood. The t-PA is broken down by the liver and excreted.

Other enzymes used to dissolve blood clots are streptokinase and urokinase. However, they are less desirable than t-PA because they must be introduced directly into the clot via catheterization; t-PA can be administered intravenously. Also, t-PA acts only on blood clots, whereas the other enzymes act throughout the bloodstream and can cause excess bleeding. Moreover, since t-PA is a naturally occurring protein it is not likely to cause antibody production or an allergic reaction.

Clot formation is a vital mechanism that prevents excessive loss of blood from the body. To form clots, the body needs calcium and vitamin K. Vitamin K is not involved in actual clot formation, but it is required for the synthesis of prothrombin (factor II) and factors VII, IX, and X. The vitamin is normally produced by bacteria that live in the large intestine. Because it is fat-soluble, it can be absorbed through the mucosa of the intestine and into the blood only if it is attached to fat. People suffering from disorders that prevent absorption of fat often experience uncontrolled bleeding. Clotting may be encouraged by applying a thrombin or fibrin spray, a rough surface such as gauze, or heat.

Prevention

Unwanted clotting may be brought on by the formation of cholesterol-containing masses called *plaques* in the walls of the blood vessels. They result in a rough surface that is perfect for the adhesion of platelets and is often the site of clotting via the intrinsic pathway. Clotting in an unbroken blood vessel is called **thrombosis.** The clot itself is a **thrombus** (*thrombo* = clot). A thrombus may dissolve spontaneously, but if it remains intact, it may damage tissues by cutting off the oxygen supply. Equally serious is the possibility that the thrombus will become dislodged and be carried with the blood to a smaller vessel. In a smaller vessel, the clot may block the circulation to a vital organ. A blood clot, bubble of air, fat from broken bones, or a piece of debris transported by the bloodstream is called an **embolus** (*em* = in; *bolus* = a mass). When an embolus becomes lodged in a vessel and cuts off circulation, the condition is called an **embolism.**

No matter how healthy the body is, occasional rough spots appear on uncut vessel walls. In fact, it is believed that blood clotting is a continuous process inside blood vessels that is continually combated by clot-preventing and clot-dissolving mechanisms. Blood contains antithrombic substances—substances that prevent thrombin formation.

In general, any chemical substance that prevents clotting is an **anticoagulant.** *Heparin* is a quick-acting anticoagulant that blocks the clotting mechanism by inhibiting the conversion of prothrombin to thrombin and prevents most thrombus formation. It is used in open-heart surgery to prevent clotting. It is extracted from donated bovine lung tissue and from bovine and porcine intestinal mucosa. The pharmaceutical preparation *warfarin* (*Coumadin*) may be given to patients who are thrombosis-prone. It acts as an antagonist to vitamin K and thus lowers the level of prothrombin. Warfarin is slower acting than heparin and is used primarily as a preventative. *CPD* (*citrate phosphate dextrose*), *ACD* (*acid citrate dextrose*), and *EDTA* (*ethylenediamine tetracetic acid*) are used by laboratories and blood banks to prevent blood samples from clotting. These substances react with Ca^{2+}

to form insoluble compounds. In this way, the blood Ca^{2+} is tied up and is no longer free to catalyze the conversion of prothrombin to thrombin.

Tests

• *Clotting Time*   The time required for blood to coagulate is known as **clotting time.** This time is used as an index of a person's blood-clotting properties. One method for determining clotting time involves drawing a 5 ml sample of blood from a vein into a silicone-coated syringe and placing 1 ml into each of two uncoated glass tubes and each of two silicone-coated tubes. The tubes are then submerged in a water bath at 37°C (98.6°F). The first uncoated glass tube is tilted through a 45° angle every minute until it can be inverted with no flow of blood down the side of the tube. The second uncoated glass tube and the silicone-coated tubes are tilted in a similar manner every 3 minutes. Clotting is initiated when the platelets break up upon coming into contact with the glass. When the clot adheres to the tube, the end point is reached and the time is recorded. The normal range of clotting time for the uncoated glass tube is 5 to 10 minutes and 25 to 45 minutes for the silicone-coated tubes. Prolonged clotting time may be due to a defect in one or more of the stages of clotting, a deficiency of a specific clotting factor (usually factor VIII or IX), or the presence of anticoagulants. Blood taken from individuals with hemophilia clots very slowly or not at all.

• *Bleeding Time*   **Bleeding time** is the time required for the cessation of bleeding from a small skin puncture. In one technique, a blood pressure cuff is placed around the arm and inflated to 40 mm Hg to produce a constant blood pressure in capillaries. The skin of the forearm is then punctured. The edge of a piece of gauze is used to collect blood by touching the drop every 30 seconds. Normal bleeding time by this technique is 4 to 8 minutes.

Unlike clotting time, which involves only the breakdown of platelets, bleeding time also involves constriction of injured blood vessels and all the stages of clot formation.

• *Prothrombin Time*   **Prothrombin time** is a test used to determine the amount of prothrombin in the blood. A blood sample is treated with CPD or a similar compound to tie up the Ca^{2+} and make the blood incoagulable. Then Ca^{2+}, thromboplastin, and plasma coagulation factors V and VII are all mixed with the blood sample. The length of time required for the blood to clot is the prothrombin time and depends on the amount of prothrombin in the sample. Normal prothrombin time is about 11 to 15 seconds.

GROUPING (TYPING)

The surfaces of erythrocytes contain genetically determined antigens called **agglutinogens** (ag'-loo-TIN-ō-jens) or **isoantigens.** There are at least 300 blood group systems that can be detected on the surface of red blood cells. The two major blood group classifications—ABO and Rh⁻—are the ones we will discuss in detail. Among the others are the Lewis, Kell, Kidd, and Duffy systems.

ABO

The **ABO blood grouping** is based on two agglutinogens symbolized as *A* and *B* (Figure 19-5). Individuals whose erythrocytes manufacture only agglutinogen *A* are said to have blood type A. Those who manufacture only agglutinogen *B* are type B. Individuals who manufacture both *A* and *B* are type AB. Those who manufacture neither are type O.

These four blood types are not equally distributed. The incidence in the white population in the United States is: type A, 41 percent; type B, 10 percent; type AB, 4 percent; type O, 45 percent. Among blacks, the frequencies

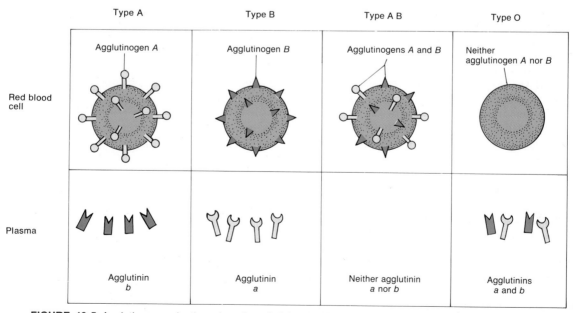

FIGURE 19-5 Agglutinogens (antigens) and agglutinins (antibodies) involved in the ABO blood grouping.

are: type A, 27 percent; type B, 20 percent; type AB, 7 percent; and type O, 46 percent.

The blood plasma of many people contains genetically determined antibodies referred to as **agglutinins** (a-GLOO-ti-nins) or **isoantibodies.** These are agglutinin *a* (anti-A), which attacks agglutinogen A, and agglutinin *b* (anti-B), which attacks B. The agglutinins formed by each of the four blood types ae shown in Figure 19-5. You do not have agglutinins that attack the agglutinogens of your own erythrocytes, but you do have an agglutinin against any agglutinogen you do not synthesize. In an incompatible blood transfusion, the donated erythrocytes are attacked by the recipient agglutinins, causing the blood cells to **agglutinate** (clump). Agglutinated cells become lodged in small capillaries throughout the body and, over a period of hours, the cells swell, rupture, and release hemoglobin into the blood. Such a reaction as it relates to erythrocytes is called **hemolysis** (*lysis* = dissolve). The degree of agglutination depends on the titer (strength or amount) of agglutinin in the blood. This reaction is another example of an antigen-antibody response (Figure 19-6).

Blood transfusions are most frequently given when blood volume is low, for example, during circulatory shock. Other indications for blood transfusions include anemia, hemophilia, and hemolytic disease of the newborn (erythroblastosis fetalis). When blood is transfused, care must be taken to avoid incompatibilities that can result in agglutination, for agglutinated cells can block blood vessels and may lead to kidney or brain damage and death, and the liberated hemoglobin may also cause kidney damage.

As an example of an incompatible blood transfusion, consider what happens if an individual with type A blood receives a transfusion of type B blood. The recipient's blood (type A) contains *A* agglutinogens and *b* agglutinins. The donor's blood (type B) contains *B* agglutinogens and *a* agglutinins. Given this situation, two things can happen. First, the *b* agglutinins in the recipient's plasma will attack the *B* agglutinogens on the donor's erythrocytes, causing hemolysis of the red blood cells. Second, the *a* agglutinins in the donor's plasma will attack the *A* agglutinogens on the recipient's erythrocytes, causing the hemolysis. However, the latter reaction is usually not serious because the donor's *a* agglutinins become so diluted in the recipient's plasma that they do not cause any significant hemolysis of the recipient's erythrocytes. Thus, a person with type A blood may not receive type B or AB blood. But, a person with A blood may receive type A or type O blood. The interactions of the four

(a)   (b)

FIGURE 19-6 Compatibility of blood types as viewed through a microscope. (a) Since the blood type shown has no agglutinins to attack the agglutinogens, no agglutination (clumping) occurs. (b) But if a blood type has incompatible agglutinins, they will attack the agglutinogens and agglutination (clumping) occurs. (Photomicrographs Copyright © 1987 by Michael H. Ross. Used by permission.)

blood types of the ABO system are summarized in Exhibit 19-5.

Individuals with type AB blood do not have any agglutinins in their plasma, and such people are sometimes called universal recipients because they can theoretically receive blood from donors of all four blood types; there are no agglutinins to attack donated erythrocytes. Individuals with type O blood have no agglutinogens on their erythrocytes and are sometimes referred to as universal donors because they can theoretically donate blood to recipients of all four blood types. Type O persons requiring blood may receive only type O blood. In practice, use of the terms *universal recipient* and *universal donor* is misleading and dangerous because there are other agglutinogens and agglutinins in blood, besides those associated with the ABO system, that can cause transfusion problems. Thus, blood should be carefully matched before transfusion, except in extreme emergency.

Knowledge of blood types is also used in disproving paternity, linking suspects to crimes, and as part of anthropology studies to establish a relationship among races.

In about 80% of the population (called **secretors**), soluble antigens of the ABO type appear in saliva and other bodily fluids. In criminal investigations it has been possible to type such fluids from saliva residues on a cigarette, or from semen in cases of rape.

CLINICAL APPLICATION

Scientists recently converted type B blood into type O blood by using an enzyme called alpha galactosidase.

EXHIBIT 19-5   SUMMARY OF ABO SYSTEM INTERACTIONS

BLOOD TYPE	AGGLUTINOGEN (ANTIGEN)	AGGLUTININ (ANTIBODY)	COMPATIBLE DONOR BLOOD TYPES	INCOMPATIBLE DONOR BLOOD TYPES
A	*A*	*b*	A, O	B, AB
B	*B*	*a*	B, O	A, AB
AB	*A, B*	Neither *a* nor *b*	A, B, AB, O	—
O	Neither *A* nor *B*	*a, b*	O	A, B, AB

The enzyme is isolated from green coffee beans and has the ability to remove a sugar from the surface of red blood cells, thus **changing type B into type O blood.** Preliminary tests on animals and a few human volunteers showed no adverse reactions. It is hoped that further testing will confirm the safety of the converted blood and that other conversions can be made so that blood banks could juggle their supplies as needed.

Rh

The **Rh system** of blood classification is so named because it was first worked out in the blood of the *Rhesus* monkey. Like the ABO grouping, the Rh system is based on agglutinogens that lie on the surfaces of erythrocytes. Individuals whose erythrocytes have the Rh agglutinogens are designated *Rh$^+$*. Those who lack Rh agglutinogens are designated *Rh$^-$*. It is estimated that 85 percent of whites and 88 percent of blacks in the United States are Rh$^+$, whereas 15 percent of whites and 12 percent of blacks are Rh$^-$.

Under normal circumstances, human plasma does not contain anti-Rh agglutinins. However, if an Rh$^-$ person receives Rh$^+$ blood, the body starts to make anti-Rh agglutinins that will remain in the blood. If a second transfu-

sion of Rh$^+$ blood is given later, the previously formed anti-Rh agglutinins will react against the donated blood and a severe reaction may occur.

CLINICAL APPLICATION

One of the most common problems with Rh incompatibility arises from pregnancy (Figure 19-7). During pregnancy, a small amount of the fetus's blood may leak from the placenta (afterbirth) into the mother's bloodstream, with the greatest transfer occurring at delivery. If the fetus is Rh$^+$ and the woman is Rh$^-$, she, upon exposure to the Rh$^+$ fetal cells, will make anti-Rh agglutinins. If the woman becomes pregnant again, her anti-Rh agglutinins will cross the placenta and make their way into the bloodstream of the baby. If the fetus is Rh$^-$, no problem will occur, since Rh$^-$ blood does not have the Rh agglutinogen. If the fetus is Rh$^+$, hemolysis may occur in the fetal blood. The hemolysis brought on by fetal-maternal incompatibility is called **hemolytic disease of newborn (erythroblastosis fetalis).** When a baby is born with this condition, all the blood is slowly removed and replaced with Rh$^-$ blood. It is even possible to transfuse blood into the

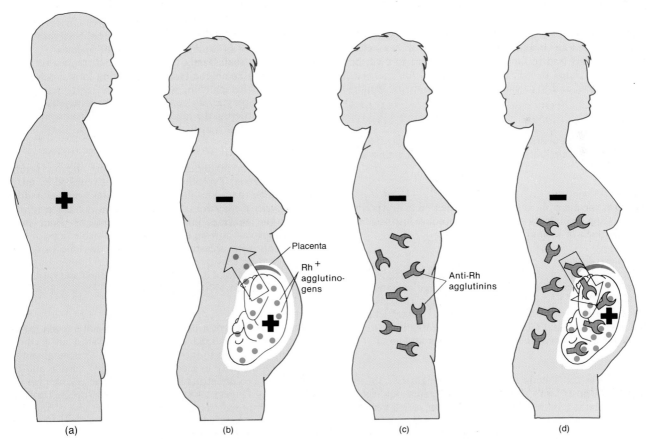

FIGURE 19-7 Development of hemolytic disease of newborn (erythroblastosis fetalis). (a) Rh$^+$ father. (b) Rh$^-$ mother and Rh$^+$ fetus. If an Rh$^-$ female is impregnated by an Rh$^+$ male and the fetus is Rh$^+$, fetal Rh$^+$ agglutinogens may enter the maternal blood via the placenta during delivery. (c) Upon exposure to the fetal Rh$^+$ agglutinogens, the mother will make anti-Rh agglutinins. (d) If the female becomes pregnant again, her anti-Rh agglutinins will cross the placenta into the fetal blood. If the fetus is Rh$^+$, hemolytic disease of newborn (erythroblastosis fetalis) will result.

unborn child if the disease is suspected. More important, though, is the fact that the disorder can be prevented with an injection of an anti-Rh gamma$_2$-globulin agglutinin preparation, administered to Rh$^-$ mothers right after delivery, miscarriage, or abortion. These agglutinins tie up the fetal agglutinogens so the mother cannot respond to the foreign agglutinogens by producing agglutinins. Thus the fetus of the next pregnancy is protected. In the case of an Rh$^+$ mother and an Rh$^-$ child, there are no complications, since the fetus cannot make agglutinins. In cases where Rh incompatibility is not implicated in hemolytic disease of the newborn, other blood groups, such as the Kell, Kidd, and Duffy groups are frequently implicated.

INTERSTITIAL FLUID AND LYMPH

Whole blood does not flow into the tissue spaces; it remains in closed vessels. However, certain constituents of the plasma do move through the capillary walls, and once they move out of the blood, they are called interstitial fluid. Interstitial fluid and lymph are basically the same.

The major difference between the two is location. When the fluid bathes the cells, it is called **interstitial fluid, intercellular fluid,** or **tissue fluid.** When it flows through the lymphatic vessels, it is called **lymph.**

Both fluids are similar in composition to plasma. The principal chemical difference is that they contain less protein than plasma because the larger protein molecules are not easily filtered through the cells that form the walls of the capillaries. The transfer of materials between blood and interstitial fluid occurs by osmosis, diffusion, and filtration across the cells that make up the capillary walls.

Interstitial fluid and lymph also differ from plasma in that they contain variable numbers of leucocytes. Leucocytes can enter the tissue fluid by diapedesis, and the lymphoid tissue itself is one of the sites of agranular leucocyte production. Like plasma, interstitial fluid and lymph lack erythrocytes and platelets.

Other substances, especially organic molecules, in interstitial fluid and lymph vary in relation to the location of the sample analyzed. The lymph vessels that drain the organs of the digestive tract, for example, contain a great deal of lipid absorbed from food.

DISORDERS: HOMEOSTATIC IMBALANCES

Anemia
Anemia is a sign, not a diagnosis. Many kinds of anemia exist, all characterized by insufficient erythrocytes or hemoglobin. These conditions lead to fatigue and intolerance to cold, both of which are related to lack of oxygen needed for energy and heat production, and to paleness, which is due to low hemoglobin content.

Nutritional Anemia
Nutritional anemia arises from an inadequate diet, one that provides insufficient amounts of iron, the necessary amino acids, or vitamin B$_{12}$.

Pernicious Anemia
Pernicious anemia is the insufficient production of erythrocytes resulting from an inability of the body to produce intrinsic factor because of lack of vitamin B$_{12}$.

Hemorrhagic Anemia
An excessive loss of erythrocytes through bleeding is called **hemorrhagic anemia.** Common causes are large wounds, stomach ulcers, and heavy menstrual bleeding. If bleeding is extraordinarily heavy, the anemia is termed acute. Excessive blood loss can be fatal. Slow, prolonged bleeding is apt to produce a chronic anemia; the chief symptom is fatigue.

Hemolotyic Anemia
If erythrocyte cell membranes rupture prematurely, the cells remain as "ghosts" and their hemoglobin pours out into the plasma. A characteristic sign of this condition, called **hemolytic anemia,** is distortion in the shape of erythrocytes that are progressing toward hemolysis. There may also be a sharp increase in the number of reticulocytes, since the destruction of red blood cells stimulates erythropoiesis.

The premature destruction of red blood cells may result from inherent defects, such as hemoglobin defects, abnormal red blood cell enzymes, or defects of the red blood cell membrane. Agents that may cause hemolytic anemia are parasites, toxins,

and antibodies from incompatible blood (Rh$^-$ mother and Rh$^+$ fetus, for instance). Hemolytic disease of newborn (erythroblastosis fetalis) is an example of a hemolytic anemia.

The term **thalassemia** (thal'-a-SĒ-mē-a) represents a group of hereditary hemolytic anemias, resulting from a defect in the synthesis of hemoglobin, which produces extremely thin and fragile erythrocytes. It occurs primarily in populations from countries bordering the Mediterranean Sea. Treatment generally consists of blood transfusions.

Aplastic Anemia
Destruction or inhibition of the red bone marrow results in **aplastic anemia.** Typically, the marrow is replaced by fatty tissue, fibrous tissue, or tumor cells. Toxins, gamma radiation, and certain medications are causes. Many of the medications inhibit the enzymes involved in hemopoiesis. Bone marrow transplants can now be done with a reasonable hope of success in patients with aplastic anemia. They are done early, before the victim is sensitized by transfusions. Immunosuppressive drugs are given for 4 days before the transplant and with decreasing frequency for 100 days afterward.

Sickle-Cell Anemia
The erythrocytes of a person with **sickle-cell anemia** manufacture an abnormal kind of hemoglobin. When an erythrocyte gives up its oxygen to the interstitial fluid, its hemoglobin tends to lose its integrity in places of low oxygen tension and forms long, stiff, rodlike structures that bend the erythrocyte into a sickle shape (Figure 19-8). The sickled cells rupture easily. Even though erythropoiesis is stimulated by the loss of the cells, it cannot keep pace with the hemolysis. The individual consequently suffers from a hemolytic anemia that reduces the amount of oxygen that can be supplied to the tissues. Prolonged oxygen reduction may eventually cause extensive tissue damage. Furthermore, because of the shape of the sickled cells, they tend to get stuck in blood vessels and can cut off blood supply to an organ altogether.

Sickle-cell anemia is characterized by several symptoms. In

Sickled
erythrocyte

Normal
erythrocyte

FIGURE 19-8 Scanning electron micrograph of erythrocytes in sickle-cell anemia at a magnification of about 2,000×. (Courtesy of Fisher Scientific Company and S.T.E.M. Laboratories, Inc., Copyright © 1975.)

young children, hand-foot syndrome is present, in which there is swelling and pain in the wrists and feet. Older patients experience pain in the back and extremities without swelling and abdominal pain. Complications include neurological disorders (meningitis, seizures, stroke), impaired pulmonary function, orthopedic abnormalities (femoral head necrosis, osteomyelitis), genitourinary tract disorders (involuntary urination, blood in urine, kidney failure), ocular disturbances (hemorrhage, detached retina, blindness), and obstetric complications (convulsions, coma, infection).

Sickle-cell anemia is inherited. The gene responsible for the tendency of the erythrocytes to sickle during hypoxia also seems to prevent erythrocytes from rupturing during a malarial crisis. The gene also alters the permeability of the plasma membranes of sickled cells, causing potassium to leak out. Low-levels of potassium kill the malarial parasites that infect sickled cells. Sickle-cell genes are found primarily among populations, or descendants of populations, that live in the malaria belt around the world, including parts of Mediterranean Europe and subtropical Africa and Asia. A person with only one of the sickling genes is said to have sickle-cell trait. Such an individual has a high resistance to malaria—a factor that may have tremendous survival value—but does not develop the anemia. Only people who inherit a sickling gene from both parents get sickle-cell anemia.

Treatment consists of administration of analgesics to relieve pain, antibiotics to counter infections, and transfusion therapy.

Polycythemia

The term **polycythemia** (pol'-ē-sī-THĒ-mē-a) refers to an abnormal increase in the number of red blood cells. The hematocrit is an important procedure in diagnosing the condition. When the hematocrit increases, especially above 55, it is very suggestive of polycythemia. There is an increased blood viscosity associated with the elevated hematocrit. The increased viscosity causes a rise in blood pressure, and contributes to thrombosis and hemorrhage. The thrombosis results from too many red blood cells piling up as they try to enter smaller vessels. The hemorrhage is due to widespread hyperemia (unusually large amount of blood in an organ part).

Infectious Mononucleosis (IM)

Infectious mononucleosis (IM) is a contagious disease primarily affecting lymphoid tissue throughout the body. It is caused by the *Epstein-Barr virus* (*EBV*), the same agent that has been linked to Burkitt's lymphoma, nasopharyngeal carcinoma, and Hodgkin's disease. It occurs mainly in children and young adults, with the peak incidence at 15 to 20 years of age. The virus most commonly enters the body through intimate oral contact, multiplies in lymphoid tissues, and spreads into the blood where it infects and multiplies in B lymphocytes, the primary host cells. As a result of this infection, the B lymphocytes become enlarged and abnormal in appearance and resemble monocytes, the primary reason for which the disease receives its name mononucleosis. Infectious mononucleosis is characterized by an elevated white blood cell count with an abnormally high percentage of lymphocytes. Symptoms include sore throat, lymphadenopathy (enlarged and tender lymph nodes), fever, brilliant red throat and soft palate, stiff neck, cough, and malaise. The spleen may also enlarge. Secondary complications involving the liver, eye, heart, kidneys, and nervous system may develop. There is no cure for infectious mononucleosis, and treatment consists of watching for and treating complications. Usually the disease runs its course in a few weeks, and the individual generally suffers no permanent ill effects.

Leukemia

Clinically, **leukemia** is classified on the basis of the duration and character of the disease, that is, acute or chronic. In its simplest terms, acute leukemia refers to a malignant disease of blood-forming tissues characterized by uncontrolled production and accumulation of immature leucocytes and many of the cells fail to reach maturity. In chronic leukemia, there is an accumulation of mature leucocytes in the bloodstream because they do not die at the end of their normal life span. The *human T-cell leukemia-lymphoma virus-1* (*HTLV-1*) is strongly associated with leukemia. Leukemia is also classified according to the identity and site of origins of the predominant cell involved, such as myelocytic (myelogenous, myeloblastic, granulocytic), lymphocytic (lymphogenous, lymphatic), and monocytic.

In acute leukemia, the anemia and bleeding problems commonly seen result from the crowding out of normal bone marrow cells by the over production of immature cells, preventing normal production of red blood cells and platelets. One cause of death from acute leukemia is internal hemorrhaging, especially cerebral hemorrhage that destroys the vital centers in the brain. Perhaps the most frequent cause of death is uncontrolled infection due to lack of mature or normal white blood cells. The abnormal accumulation of immature leucocytes may be reduced by using x rays and antileukemic drugs. Partial or complete remissions may be induced, with some lasting as long as 15 years.

MEDICAL TERMINOLOGY

Autologous (*auto* = self) **transfusion** Donating one's own blood before elective surgery to ensure an abundant supply and reduce transfusion complications.

Citrated whole blood Whole blood protected from coagulation by CPD (citrate phosphate dextrose) or a similar compound.

Direct (immediate) transfusion (*trans* = through) Transfer of blood directly from one person to another without exposing the blood to air.

Exchange transfusion Removing blood from the recipient while simultaneously replacing it with donor blood. This method is used for erythroblastosis fetalis and poisoning.

Gamma globulin (immune serum globulin) Solution of globulins from nonhuman blood consisting of antibodies that react with specific pathogens, such as measles, epidemic hepatitis, tetanus, and possibly poliomyelitis viruses. It is prepared by injecting the specific virus into animals, removing blood from the animals after antibodies have accumulated, isolating antibodies, and injecting them into a human for short-term immunity.

Hemochromatosis (*heme* = iron; *chroma* = color) Disorder of iron metabolism characterized by excess deposits of iron in tissues, especially the liver and pancreas that result in bronze coloration of the skin, cirrhosis, diabetes mellitus, and bone and joint abnormalities.

Hemorrhage (*rrhage* = bursting forth) Bleeding, either internal (from blood vessels into tissues) or external (from blood vessels directly to the surface of the body).

Indirect (mediate) transfusion Transfer of blood from a donor to a container and then to the recipient, permitting blood to be stored for an emergency. The blood may be separated into its components so that a patient will receive only a needed part.

Platelet concentrates A preparation of platelets obtained from freshly drawn whole blood and used for transfusions in platelet-deficiency disorders such as hemophilia.

Reciprocal transfusion Transfer of blood from a person who has recovered from a contagious infection into the vessels of a patient with the same infection. An equal amount of blood is returned from the patient to the well person. This method allows the patient to receive antibody-bearing lymphocytes from the recovered person.

Septicemia (*sep* = decay; *emia* = condition of blood) Toxins or disease-causing bacteria in the blood. Also called "blood poisoning."

Thrombocytopenia (*thrombo* = clot; *penia* = poverty) Very low platelet count that results in a tendency to bleed from capillaries.

Transfusion Transfer of whole blood, blood components (red blood cells only or plasma only), or bone marrow directly into the bloodstream.

Venesection (*veno* = vein) Opening of vein for withdrawal of blood. Although **phlebotomy** (*phlebo* = vein; *tome* = to cut) is a synonym for venesection, in clinical practice, phlebotomy refers to therapeutic bloodletting, as to remove a pint of blood to lower the viscosity of blood of a patient with polycythemia.

Whole blood Blood containing all formed elements, plasma, and plasma solutes in natural concentration.

STUDY OUTLINE

Physical Characteristics (p. 440)

1. The cardiovascular system consists of blood, the heart, and blood vessels. The lymphatic system consists of lymph, lymph vessels, and lymph glands.
2. Physical characteristics of blood include viscosity, 4.5 to 5.5; temperature, 38°C (100.4°F); pH, 7.35 to 7.45; and a salt (NaCl) concentration of 0.85–0.90 percent. Blood constitutes about 8 percent of body weight.

Functions (p. 440)

1. Blood transports oxygen, carbon dioxide, nutrients, wastes, hormones, and enzymes.
2. It helps to regulate pH, body temperature, and water content of cells.
3. It prevents blood loss through clotting and combats toxins and microbes through special combat-unit cells.

Components (p. 440)

1. The formed elements in blood include erythrocytes (red blood cells), leucocytes (white blood cells), and thrombocytes (platelets).
2. Blood cells are formed by a process called hemopoiesis.
3. Red bone marrow (myeloid tissue) is responsible for producing red blood cells, granular leucocytes, and platelets; lymphoid tissue and myeloid tissue produce agranular leucocytes.

Erythrocytes (p. 442)

1. Erythrocytes are biconcave discs without nuclei and containing hemoglobin.
2. The function of red blood cells is to transport oxygen and carbon dioxide.
3. Red blood cells live about 120 days. A healthy male has about 5.4 million/mm³ of blood; a healthy female, about 4.8 million/mm³.
4. Erythrocyte formation, called erythropoiesis, occurs in adult red marrow of certain bones.
5. A reticulocyte count is a diagnostic test that indicates the rate of erythropoiesis.

6. A hematocrit measures the percentage of red blood cells in whole blood.

Leucocytes (p. 445)

1. Leucocytes are nucleated cells. Two principal types are granular (neutrophils, eosinophils, basophils) and agranular (lymphocytes and monocytes).
2. The general function of leucocytes is to combat inflammation and infection. Neutrophils and monocytes (wandering macrophages) do so through phagocytosis.
3. Eosinophils combat the effects of histamine in allergic reactions, phagocytize antigen-antibody complexes, and combat parasitic worms; basophils liberate heparin, histamine, and serotonin in allergic reactions that intensify the inflammatory response.
4. Lymphocytes, in response to the presence of foreign substances called antigens, differentiate into tissue plasma cells that produce antibodies. Antibodies attach to the antigens and render them harmless. This antigen-antibody response combats infection and provides immunity.
5. A differential count is a diagnostic test in which white blood cells are enumerated.
6. White blood cells usually live for only a few hours or a few days. Normal blood contains 5,000 to 9,000/mm³.

Thrombocytes (p. 447)

1. Thrombocytes are disc-shaped structures without nuclei.
2. They are formed from megakaryocytes and are involved in clotting.
3. Normal blood contains 250,000 to 400,000/mm³.

Plasma (p. 448)

1. The liquid portion of blood, called plasma, consists of 91.5 percent water and 8.5 percent solutes.
2. Principal solutes include proteins (albumins, globulins, fibrinogen), nonprotein nitrogen (NPN) substances, foods, enzymes and hormones, respiratory gases, and electrolytes.

Hemostasis (p. 449)

1. Hemostasis refers to the prevention of blood loss.
2. It involves vascular spasm, platelet plug formation, and blood coagulation.
3. In vascular spasm, the smooth muscle of a blood vessel wall contracts to stop bleeding.
4. Platelet plug formation involves the clumping of platelets to stop bleeding.
5. A clot is a network of insoluble protein (fibrin) in which formed elements of blood are trapped.
6. The chemicals involved in clotting are known as coagulation factors. There are two kinds: plasma and platelet coagulation factors.
7. Blood clotting involves two pathways: the intrinsic and the extrinsic.
8. Normal coagulation also involves clot retraction (tightening of the clot) and fibrinolysis (dissolution of the clot).
9. Clotting in an unbroken blood vessel is called thrombosis. A thrombus that moves from its site of origin is called an embolus.
10. Anticoagulatants (e.g., heparin) prevent clotting.
11. Clinically important clotting tests are clotting time (time required for blood to coagulate), bleeding time (time required for the cessation of bleeding from a small skin puncture), and prothrombin time (time required for the blood to coagulate, which depends on the amount of prothrombin in the blood sample).

Grouping (Typing) (p. 453)

1. ABO and Rh systems are based on antigen-antibody responses.
2. In the ABO system, agglutinogens (antigens) A and B determine blood type. Plasma contains agglutinins (antibodies) that clump agglutinogens that are foreign to the individual.
3. In the Rh system, individuals whose erythrocytes have Rh agglutinogens are classified as Rh^+. Those who lack the antigen are Rh^-.
4. A disorder due to Rh incompatibility between mother and fetus is called hemolytic disease of newborn (erythroblastosis fetalis).

Interstitial Fluid and Lymph (p. 456)

1. Interstitial fluid bathes body cells, whereas lymph is found in lymphatic vessels.
2. These fluids are similar in chemical composition. They differ chemically from plasma in that both contain less protein and a variable number of leucocytes. Like plasma, they contain no platelets or erythrocytes.

Disorders: Homeostatic Imbalances (p. 456)

1. Anemia is a decreased erythrocyte count or hemoglobin deficiency. Kinds of anemia include nutritional, pernicious, hemorrhagic, hemolytic, aplastic, and sickle-cell anemia.
2. Polycythemia is an abnormal increase in the number of erythrocytes.
3. Infectious mononucleosis (IM) is a contagious disease that primarily affects lymphoid tissue. It is characterized by an elevated white blood cell count, with an abnormally high percentage of lymphocytes. The cause is the Epstein-Barr virus (EBV).
4. Leukemia is a malignant disease of blood-forming tissues characterized by the uncontrolled production of white blood cells that interferes with normal clotting and vital body activities.

REVIEW QUESTIONS

1. How are blood, interstitial fluid, and lymph related to the maintenance of homeostasis?
2. Distinguish between the cardiovascular system and lymphatic system.
3. List the principal physical characteristics of blood.
4. List the functions of blood and their relationship to other systems of the body.
5. Distinguish between plasma and formed elements.
6. Describe the origin of blood cells.
7. Describe the microscopic appearance of erythrocytes. What is the essential function of erythrocytes?
8. Define erythropoiesis. Relate erythropoiesis to red blood cell count. What factors accelerate and decelerate erythropoiesis?
9. What is a reticulocyte count? What is its diagnostic significance?
10. Define hematocrit. Compare the hematocrits of anemic and polycythemic blood.
11. Describe the classification of leucocytes. What are their functions?
12. What is the importance of diapedesis, chemotaxis, and phagocytosis in fighting bacterial invasion?
13. What is multiple myeloma? Describe bone marrow transplantation and its use in the treatment of diseases such as multiple myeloma.
14. What is a differential count? What is its significance? Distinguish between leucocytosis and leucopenia.
15. Describe the antigen-antibody response. How is it protective?
16. Describe the structure and function of thrombocytes.
17. Compare erythrocytes, leucocytes, and thrombocytes with respect to size, number per mm^3, and life span.
18. What are the major constituents in plasma? What do they do? What is the difference between plasma and serum?
19. What is apheresis? What are its clinical applications?
20. Define hemostasis. Explain the mechanism involved in vascular spasm and platelet plug formation.
21. Briefly describe the process of clot formation. What is fibrinolysis? Why does blood usually not remain clotted in vessels?
22. How do the extrinsic and intrinsic pathways of blood clotting differ?
23. What are the pathways involved in blood clotting?
24. Define the following: thrombus, embolus, anticoagulant, clotting time, bleeding time, prothrombin time.
25. What is the basis for ABO blood grouping? What are agglutinogens and agglutinins?
26. What is the basis for the Rh system? How does hemolytic disease of newborn (erythroblastosis fetalis) occur? How may it be prevented?
27. Compare interstitial fluid and lymph with regard to location, chemical composition, and function.
28. Define anemia. Contrast the causes of nutritional, pernicious, hemorrhagic, hemolytic, aplastic, and sickle-cell anemias.
29. What is infectious mononucleosis (IM)?
30. What is leukemia, and what are the causes of some of its symptoms?
31. Refer to the glossary of medical terminology associated with blood. Be sure that you can define each term.

The Cardiovascular System: The Heart

20

STUDENT OBJECTIVES

- Describe the location of the heart and identify its borders.
- Describe the structure of the pericardium.
- Contrast the structure and location of the epicardium, myocardium, and endocardium of the heart wall.
- Identify and describe the chambers, great vessels, and valves of the heart.
- Describe the surface anatomy features of the heart.
- Discuss the route of blood in coronary (cardiac) circulation.
- Explain the structural and functional features of the conduction system of the heart.
- Describe an electrocardiogram (ECG) and explain its significance.
- Explain the pressure changes associated with blood flow through the heart.
- Describe the principal events of a cardiac cycle.
- Contrast the sounds of the heart and their clinical significance.
- Define cardiac output (CO) and explain what determines it.
- Define Starling's law of the heart.
- Contrast the effects of sympathetic and parasympathetic stimulation of the heart.
- Define the role of baroreceptors in reflex pathways in controlling heart rate.
- Explain how chemicals, temperature, emotions, sex, and age affect heart rate.
- Define circulatory shock and explain the homeostatic mechanisms that compensate for it.
- Describe the structure and function of an artificial heart.
- Explain how heart disorders are diagnosed via cardiac catheterization and echocardiography.
- Describe the operation of a heart-lung machine.
- List the risk factors involved in heart disease.
- Describe the development of the heart.
- Describe how atherosclerosis and coronary artery spasm contribute to coronary artery disease (CAD).
- Contrast coarctation of the aorta, patent ductus arteriosus, septal defects, valvular stenosis, and tetralogy of Fallot as congenital heart defects.
- Define atrioventricular (AV) block, atrial flutter, atrial fibrillation, and ventricular fibrillation as abnormalities of the conduction system of the heart (arrhythmias).
- Define congestive heart failure (CHF) and cor pulmonale (CP).
- Define medical terminology associated with the heart.

The **heart** is the center of the cardiovascular system. Whereas the term *cardio* refers to the heart, the term *vascular* refers to blood vessels (or an abundant blood supply). The heart is a hollow, muscular organ that weighs about 342 grams (11 oz) and beats over 100,000 times a day to pump 3,784 liters (1,000 gallons) of blood per day through over 60,000 miles of blood vessels. The blood vessels form a network of tubes that carry blood from the heart to the tissues of the body and then return it to the heart. The specific aspects of the heart that we will consider are its location, covering, wall and chambers, great vessels, valves, surface anatomy, conduction system, electrocardiogram (ECG), cardiac cycle (heartbeat), cardiac output (CO), and autonomic control. Several disorders related to the heart will also be considered.

The study of the normal heart and disease associated with it is known as **cardiology.**

The developmental anatomy of the heart is considered at the end of the chapter.

LOCATION

The heart is situated obliquely between the lungs and is a component of the mediastinum, the mass of tissue between the pleurae of the lungs that extends from the sternum to the vertebral column. About two-thirds of the mass of the heart lies to the left of the body's midline (Figure 20-1). The heart is shaped like a blunt cone about the size of your closed fist—12 cm (5 inches) long, 9 cm (3½ inches) wide at its broadest point, and 6 cm (2½ inches) thick.

Its pointed end, the *apex,* is formed by the tip of the left ventricle, and projects inferiorly, anteriorly, and to the left. The *left border* is formed almost entirely by the left ventricle, although the left atrium forms part of the upper end of the border. The *superior border,* where the great vessels enter and leave the heart, is formed by both atria. The *base* of the heart projects superiorly, posteriorly, and to the right. It is formed by the atria, mostly the left atrium. The *right border* is formed by the right

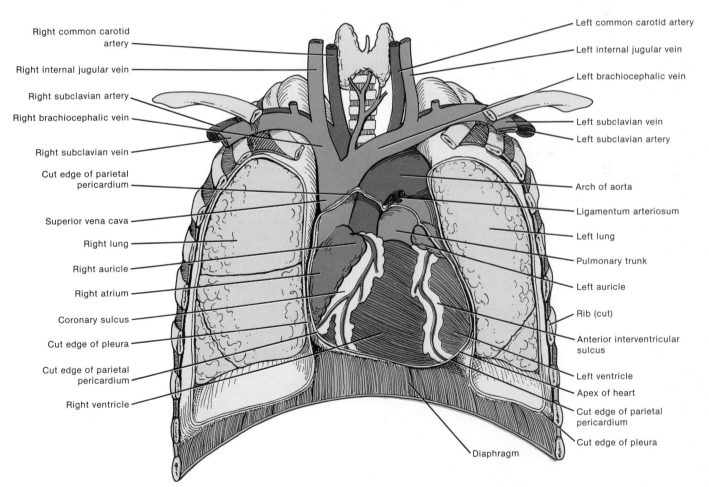

FIGURE 20-1 Position of the heart and associated blood vessels in the thoracic cavity. In this and subsequent illustrations, vessels that carry oxygenated blood are shown in red; vessels that carry deoxygenated blood are shown in blue.

atrium. The *inferior border* is formed by the right ventricle and slightly by the left ventricle. The *sternocostal (anterior) surface* is formed mainly by the right ventricle and right atrium, whereas the *diaphragmatic (inferior) surface* is formed by the left and right ventricles, mostly the left.

PERICARDIUM

The heart is enclosed and held in place by the **pericardium,** an ingenious structure designed to confine the heart to its position in the mediastinum, yet allow it sufficient freedom of movement so that it can contract vigorously and rapidly when the need arises.

The pericardium consists of two portions referred to as the fibrous pericardium and the serous pericardium (Figure 20-2a). The outer *fibrous pericardium* consists of very heavy fibrous connective tissue. The fibrous pericardium resembles a bag that rests on the diaphragm with its open end fused to the connective tissues of the great vessels entering and leaving the heart. The fibrous pericardium prevents overdistension of the heart, provides a tough protective membrane around the heart, and anchors the heart in the mediastinum. The inner *serous pericardium* is a thinner, more delicate membrane that forms a double layer around the heart (Figure 20-2b). The *parietal layer* of the serous pericardium is directly beneath the fibrous pericardium. It constitutes the outer layer of the serous pericardium. The *visceral layer* of the serous pericardium, also called the *epicardium,* is beneath the parietal layer, attached to the myocardium (muscle) of the heart. Between the parietal and visceral layers of the serous pericardium is a thin film of serous fluid that holds the two parts together much like a thin film of water does to two microscope slides. The serous fluid, known as *pericardial fluid,* prevents friction between the membranes as the heart moves. The space occupied by the pericardial fluid is a potential space, called the *pericardial cavity.*

An inflammation of the pericardium is known as **pericarditis.** Pericarditis with a buildup of pericardial fluid or extensive bleeding into the pericardium, if untreated, is a life-threatening condition. Since the pericardium cannot stretch to accommodate the excessive fluid or blood buildup, the heart is subjected to compression. This compression is known as **cardiac tamponade** (tam'-pon-ĀD) and can result in cardiac failure.

HEART WALL

The wall of the heart (Figure 20-2a) is divided into three layers: the epicardium (external layer), myocardium (middle layer), and endocardium (inner layer). The **epicardium** (visceral layer of serous pericardium) is the thin, transparent outer layer of the wall. It is composed of serous tissue and mesothelium.

The **myocardium,** which is cardiac muscle tissue, constitutes the bulk of the heart. Cardiac muscle fibers are involuntary, striated, and branched, and the tissue is arranged in interlacing bundles of fibers. The myocardium is responsible for the contraction of the heart.

The **endocardium** is a thin layer of endothelium overlying a thin layer of connective tissue. It lines the inside of the myocardium and covers the valves of the heart and the tendons that hold them open. It is continuous with the endothelial lining of the large blood vessels of the heart.

Inflammation of the epicardium, myocardium, and endocardium is referred to as **epicarditis, myocarditis,** and **endocarditis,** respectively.

CHAMBERS OF THE HEART

The interior of the heart is divided into four **chambers,** or cavities, which receive the circulating blood (Figure 20-3). The two upper chambers are called the right and left **atria.** Each atrium has an appendage called an *auricle* (OR-i-kul; *auris* = ear), so named because its shape resembles a dog's ear. The auricle increases the atrium's surface area. The lining of the atria is smooth, except for their anterior walls and the lining of the auricles, which contain projecting muscle bundles that are parallel to one another and resemble the teeth of a comb: the *musculi pectinati* (MUS-kyoo-lī pek-ti-NA-tē). These bundles give the lining of the auricles a ridged appearance.

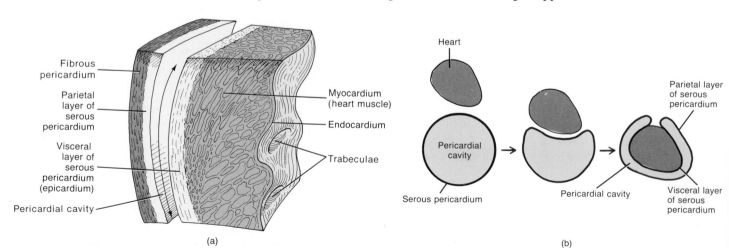

(a)

(b)

FIGURE 20-2 Pericardium and heart wall. (a) Structure of the pericardium and heart wall. (b) Relationship of the serous pericardium to the heart.

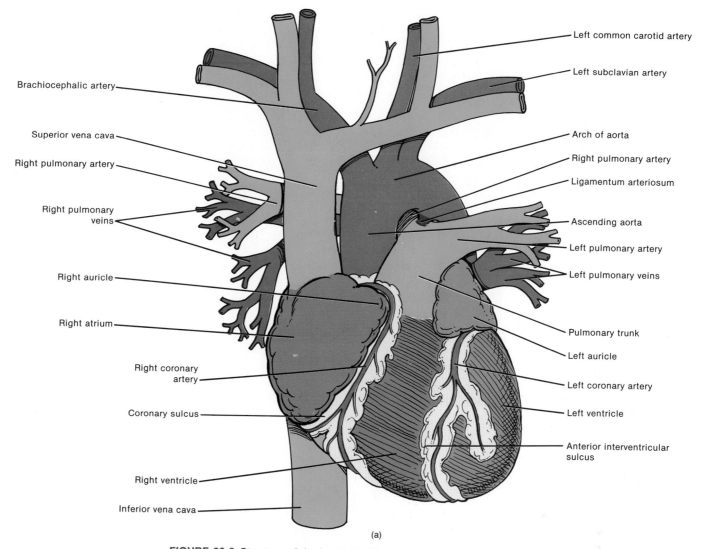

Left common carotid artery

Left subclavian artery

Arch of aorta

Right pulmonary artery

Ligamentum arteriosum

Ascending aorta

Left pulmonary artery

Left pulmonary veins

Pulmonary trunk

Left auricle

Left coronary artery

Left ventricle

Anterior interventricular sulcus

Brachiocephalic artery

Superior vena cava

Right pulmonary artery

Right pulmonary veins

Right auricle

Right atrium

Right coronary artery

Coronary sulcus

Right ventricle

Inferior vena cava

(a)

FIGURE 20-3 Structure of the heart. (a) Diagram of anterior external view.

The atria are separated by a partition called the *interatrial septum.* A prominent feature of this septum is an oval depression, the *fossa ovalis,* which corresponds to the site of the foramen ovale, an opening in the interatrial septum of the fetal heart. The fossa ovalis faces the opening of the inferior vena cava and is located in the septal wall of the right atrium.

The two lower chambers are the right and left **ventricles.** They are separated by an *interventricular septum.*

The muscle tissue of the atria and ventricles is separated by connective tissue that also forms the valves. This "cardiac skeleton" effectively divides the myocardium into two separate muscle masses. Externally, a groove known as the *coronary sulcus* (SUL-kus) separates the atria from the ventricles. It encircles the heart and houses the cornary sinus and circumflex branch of the left coronary artery. The *anterior interventricular sulcus* and *posterior interventricular sulcus* separate the right and left ventricles externally. The sulci contain coronary blood vessels and a variable amount of fat (Figure 20-3a, c).

GREAT VESSELS OF THE HEART

The right atrium receives blood from all parts of the body except the lungs. It receives the blood through three veins. In general, the **superior vena cava** brings blood from parts of the body superior to the heart; in general, the **inferior vena cava** brings blood from parts of the body inferior to the heart; and the **coronary sinus** drains blood from most of the vessels supplying the wall of the heart (Figure 20-3c). The right atrium then delivers the blood into the **right ventricle,** which pumps it into the **pulmonary trunk.** The pulmonary trunk divides into a **right** and **left pulmonary artery,** each of which carries blood to the lungs. In the lungs, the blood releases its carbon dioxide and takes on oxygen. Blood returns to the heart via four **pulmonary veins** that empty into the left atrium. The blood then passes into the left ventricle, which pumps the blood into the **ascending aorta.** From here the blood is passed into the **coronary arteries, arch of the aorta, thoracic aorta,** and **abdominal aorta.** These

Brachiocephalic artery

Left common carotid artery

Left subclavian artery

Arch of aorta

Ligamentum arteriosum

Superior vena cava

Right atrium

Right ventricle

Pulmonary trunk

Left auricle

Anterior interventricular sulcus

Left ventricle

(b)

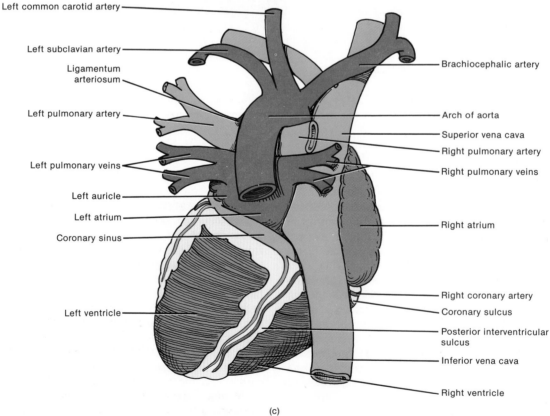

Left common carotid artery

Left subclavian artery

Ligamentum arteriosum

Left pulmonary artery

Left pulmonary veins

Left auricle

Left atrium

Coronary sinus

Left ventricle

Brachiocephalic artery

Arch of aorta

Superior vena cava

Right pulmonary artery

Right pulmonary veins

Right atrium

Right coronary artery

Coronary sulcus

Posterior interventricular sulcus

Inferior vena cava

Right ventricle

(c)

FIGURE 20-3 (*Continued*) (b) Photograph of anterior external view. (Courtesy of C. Yokochi and J. W. Rohen, *Photographic Anatomy of the Human Body,* 2nd ed., 1978, IGAKU-SHOIN, Ltd., Tokyo, New York.) (c) Diagram of posterior external view.

Brachiocephalic artery

Right pulmonary artery

Ascending aorta

Superior vena cava

Right pulmonary vein

Pulmonary trunk

Pulmonary semilunar valve

Right pulmonary vein

Fossa ovalis

Right atrium

Valve of coronary sinus

Valve of inferior vena cava

Tricuspid valve

Chordae tendineae

Right ventricle

Trabeculae carneae

Inferior vena cava

Left common carotid artery

Left subclavian artery

Arch of aorta

Ligamentum arteriosum

Left pulmonary artery

Left pulmonary veins

Left atrium

Aortic semilunar valve

Bicuspid valve

Left ventricle

Interventricular septum

Papillary muscle

Descending thoracic aorta

(d)

FIGURE 20-3 (*Continued*) (d) Diagram of anterior internal view.

blood vessels and their branches transport the blood to all body parts.

During fetal life, there is a temporary blood vessel, called the ductus arteriosus, that connects the pulmonary trunk with the aorta. Its purpose is to redirect blood so that it does not enter the nonfunctioning fetal lungs. The ductus arteriosus normally closes shortly after birth, leaving a remnant known as the **ligamentum arteriosum.**

The thickness of the four chambers varies according to the pressure generated within them (Figure 20-3d). The atria are thin-walled because they need only enough cardiac muscle tissue to deliver the blood into the ventricles with the aid of gravity and a reduced pressure created by the expanding ventricles. The right ventricle has a thicker layer of myocardium than the atria, since it must send blood to the lungs and back around to the left atrium. The left ventricle has the thickest wall, since it must pump blood at high pressure through literally thousands of miles of vessels in the head, trunk, and extremities.

VALVES OF THE HEART

As each chamber of the heart contracts, it pushes a portion of blood into a ventricle or out of the heart through an artery. In order to keep the blood from flowing backward, the heart has structures composed of collagen called **valves.**

ATRIOVENTRICULAR (AV) VALVES

Atrioventricular (AV) valves lie between the atria and ventricles (Figure 20-3d). The right atrioventricular valve between the right atrium and right ventricle is also called the **tricuspid valve** because it consists of three flaps, or cusps. These flaps are fibrous tissues that grow out of the walls of the heart and are covered with endocardium. The pointed ends of the cusps project into the ventricle. Cords called *chordae tendineae* (KOR-dē TEN-di-nē) connect the pointed ends to small conical projections— the *papillary muscles* (muscular columns)—located on the inner surface of the ventricles. The irregular surface of ridges and folds of the myocardium in the ventricles is known as the *trabeculae carneae* (tra-BEK-yoo-lē KAR-nē). The chordae tendineae and their papillary muscles keep the flaps pointing in the direction of the blood flow. The artrioventricular valve between the left atrium and left ventricle is called the **bicuspid (mitral) valve.** It

(e)

Figure 20-3 (*Continued*) (e) Path of blood through the heart.

is also known as the left atrioventricular (AV) valve. It has two cusps that work in the same way as the cusps of the tricuspid valve. Its cusps are also attached by way of the chordae tendineae to papillary muscles.

In order for blood to pass from an atrium to a ventricle, the atrioventricular (AV) valve opens down as the atrium contracts, the papillary muscles relax, and the chordae tendineae slacken (Figure 20-4a). When the ventricle pumps blood out of the heart into an artery, any blood driven back toward the atrium is pushed between the flaps and the ventricle wall (Figure 20-4b). This action drives the cusps upward until their edges meet and close the opening. At the same time, contraction of the papillary muscles and tightening of the chordae tendineae help prevent the valve from swinging upward into the atrium.

SEMILUNAR VALVES

Both arteries that leave the heart have a valve that prevents blood from flowing back into the heart. These are the **semilunar valves** (see Figure 20-3d). The **pulmonary semilunar valve** lies in the opening where the pulmonary trunk leaves the right ventricle. The **aortic semilunar valve** is situated at the opening between the left ventricle and the aorta.

Both valves consist of three semilunar (half-moon or crescent-shaped) cusps. Each cusp is attached by its convex margin to the artery wall. The free borders of the cusps curve outward and project into the opening inside the blood vessel. Like the atrioventricular valves, the semilunar valves permit blood to flow in only one direction—in this case, from the ventricles into the arteries.

SURFACE PROJECTION

The valves of the heart may be identified by surface projection (Figure 20-5). The pulmonary and aortic semilunar valves are represented on the surface by a line about 2.5 cm (1 inch) in length. The pulmonary semilunar valve lies horizontally behind the inner end of the left third costal cartilage and the adjoining part of the sternum. The aortic semilunar valve is placed obliquely behind the left side of the sternum, at the level of the third intercostal space. The tricuspid valve lies behind the sternum, extending from the midline at the level of the fourth costal cartilage down toward the right sixth chondrosternal junction. The bicuspid valve lies behind the left side of the sternum obliquely at the level of the fourth costal cartilage. It is represented by a line about 3 cm in length.

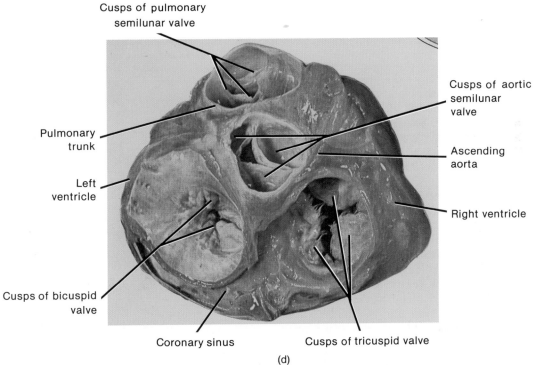

FIGURE 20-4 Function of atrioventricular (AV) valves. (a) Bicuspid valve open. (b) Bicuspid valve closed. The tricuspid valve operates in a similar manner. (c) Photograph of the tricuspid valve. (d) Photograph of the valves of the heart in which the atria have been removed. (Photographs courtesy of C. Yokochi and J. W. Rohen, *Photographic Anatomy of the Human Body,* 2nd ed., 1978, IGAKU-SHOIN, Ltd., Tokyo, New York.)

FIGURE 20-5 Surface projection of the heart. The red dots indicate where heart sounds caused by the respective valves are best heard.

Although heart sounds are produced in part by the closure of valves, they are not necessarily heard best over these valves. Each sound tends to be clearest in a slightly different location closest to the surface of the body (Figure 20-5).

BLOOD SUPPLY

The wall of the heart, like any other tissue, including large blood vessels, has its own blood vessels. Nutrients could not possibly diffuse through all the layers of cells that make up the heart tissue. The flow of blood through the numerous vessels that pierce the myocardium is called the **coronary (cardiac) circulation** (Figure 20-6). The term coronary refers to the fact that the blood vessels of the heart resemble a crown.

The vessels that serve the myocardium include the *left coronary artery,* which originates as a branch of the ascending aorta. This artery runs under the left atrium and divides into the anterior interventricular and circumflex branches. The *anterior interventricular branch* follows the anterior interventricular sulcus and supplies oxygenated blood to the walls of both ventricles. The *circumflex branch* distributes oxygenated blood to the walls of the left ventricle and left atrium.

The *right coronary artery* also originates as a branch of the ascending aorta. It runs under the right atrium and divides into the posterior interventricular and marginal branches. The *posterior interventricular branch* follows the posterior interventricular sulcus and supplies the walls of the two ventricles with oxygenated blood. The *marginal branch* transports oxygenated blood to the myocardium of the right ventricle. The left ventricle receives the most abundant blood supply because of the enormous work it must do.

As blood passes through the coronary system of the heart, it delivers oxygen and nutrients and collects carbon

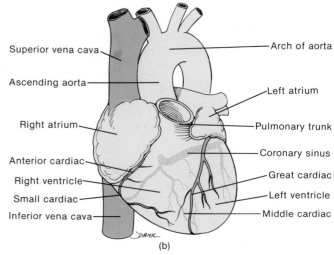

FIGURE 20-6 Coronary (cardiac) circulation. (a) Anterior view of arterial distribution. (b) Anterior view of venous drainage.

dioxide and wastes. Most of the deoxygenated blood, which carries the carbon dioxide and wastes, is collected by a large vein, the *coronary sinus,* which empties into the right atrium. A vascular sinus is a vein with a thin wall that has no smooth muscle to alter its diameter. The principal tributaries of the coronary sinus are the *great cardiac vein,* which drains the anterior aspect of the heart, and the *middle cardiac vein,* which drains the posterior aspect of the heart.

Most parts of the body receive branches from more than one artery, and where two or more arteries supply the same region, they usually connect with each other. This is referred to as *anastomosis.* Anastomoses between arteries provide collateral circulation (alternate routes) for blood to reach a particular organ or tissue. The myocardium contains numerous anastomoses either connecting branches of one coronary artery or between branches of different coronary arteries. When a major coronary blood vessel is about 90 percent obstructed, blood will flow through the collateral vessels. Although most collaterals in the heart are quite small, heart muscle can remain alive as long as it receives as little as 10 to 15 percent of its normal supply.

CLINICAL APPLICATION

Most heart problems result from faulty coronary circulation. If a reduced oxygen supply weakens the cells but does not actually kill them, the condition is called **ischemia** (is-KĒ-mē-a). **Angina pectoris** (an-JĪ-na, or AN-ji-na, PEK-tō-ris), meaning "chest pain," is ischemia of the myocardium. (Pain impulses originating from most visceral muscles are referred to an area on the surface of the body.) Common causes of angina include stress, strenuous exertion after a heavy meal, atherosclerosis, coronary artery spasm, hypertension, anemia, thyroid disease, and aortic stenosis. The symptoms of angina pectoris include chest pain, accompanied by constriction or tightness, labored breathing, and a sensation of foreboding. Sometimes weakness, dizziness, and perspiration occur.

A much more serious problem is **myocardial infarction** (in-FARK-shun), commonly called a heart attack. **Infarction** means the death of an area of tissue because of an interrupted blood supply. Myocardial infarction may result from a thrombus or embolus in one of the coronary arteries. The tissue distal to the obstruction dies and is replaced by noncontractile scar tissue. Thus, the heart muscle loses at least some of its strength. The aftereffects depend partly on the size and location of the infarcted, or dead, area.

It has recently been learned that under some conditions most tissue damage occurs when blood flow is restored (reprofusion) to an area, not when it is interrupted. When blood supply is interrupted, the affected tissue produces an enzyme called *xanthinoxidase.* When blood is restored, the enzyme reacts with oxygen in the blood to produce *superoxide free radicals,* which cause the tissue damage. Attempts are now underway to restore oxygen-starved tissue by preventing the formation of superoxide free radicals using drugs such as superoxide dismutase and allopurinol (also used to treat gout).

Diagnostic tests have been developed that enable physicians to diagnose heart attacks within hours. During a heart attack, certain enzymes are released into the bloodstream and are the strongest indicators that a heart attack has occurred. Among the enzymes are creatine kinase MB isoenzyme (CK-MB), lactic dehydrogenase-1 (LDH-1), serum glutamic oxaloacetic transaminase (SGOT), and creatine phosphokinase (CPK).

CONDUCTION SYSTEM

The heart is innervated by the autonomic nervous system (Chapter 16), but the autonomic neurons only increase or decrease the time it takes to complete a cardiac cycle (heartbeat), that is, they do not initiate contraction. The chamber walls can go on contracting and relaxing, contracting and relaxing, without any direct stimulus from the nervous system. This action is possible because the heart has an intrinsic regulating system called the **conduction system.** The conduction system is composed of specialized muscle tissue that generates and distributes the electrical impulses that stimulate the cardiac muscle fibers to contract. These tissues are the sinoatrial (sinuatrial) node, the atrioventricular (AV) node, the atrioventricular (AV) bundle (of His), the bundle branches, and the conduction myofibers (Purkinje fibers). The cells of the conduction system develop during embryological life from certain cardiac muscle cells.

Before discussing the spread of nerve impulses through the conduction system that results in contraction of the heart, we will first examine a characteristic of nodal cells called **self-excitability,** their ability to spontaneously and rhythmically generate action potentials (nerve impulses). Most smooth muscle fibers and neurons of the central nervous system are also self-excitable. The membranes of cells of the sinoatrial node (described shortly) are very permeable to sodium (Na^+) ions, even in the resting state. As a result, Na^+ ions diffuse through sodium channels into the cell, causing the membrane potential to move toward a more positive value. Once the membrane potential reaches its threshold level, an action potential is generated. After the action potential begins, the membrane becomes less permeable to Na^+ ions, but usually permeable to potassium (K^+) ions. As K^+ ions diffuse out of the cell through potassium channels, the inside of the cell becomes more negative. This reversal of charges stops the action potential. Through the operation of the sodium and potassium pumps, Na^+ ions are actively transported out of the cells and K^+ ions are actively transported into the cells. Then, the influx of Na^+ ions initiates another action potential and the process of self-excitation repeats itself over and over. In humans, the normal resting rate of self-excitation of the sinoatrial node is about 75 times/minute.

The **sinoatrial (sinuatrial) node,** also known as the **SA node,** or **pacemaker,** is a compact mass of cells located in the right atrial wall inferior to the opening of the superior vena cava (Figure 20-7a). The SA node initiates each cardiac cycle and thereby sets the basic pace for the heart rate. The SA node spontaneously depolarizes and generates action potentials faster than other components of the conduction system and myocardium. As a result, nerve impulses from the SA node spread to other areas of the conduction system and myocardium, and stimulate them so frequently that they are not able to generate action potentials at their own inherent rates. Thus, the faster rate of discharge of the SA node sets the rhythm for the rest of the heart, hence its common name pacemaker. The rate set by the SA node may be altered by nerve impulses from the autonomic nervous system or by certain blood-borne chemicals such as thyroid hormones and epinephrine.

Once an action potential is initiated by the SA node, the impulse spreads out over both atria, causing them to contract, and at the same time depolarizing the **atrioventricular (AV) node.** Because of its location near the inferior portion of the interatrial septum, the AV node is one of the last portions of the atria to be depolarized.

From the AV node, a tract of conducting fibers called the **atrioventricular (AV) bundle (bundle of His)** runs through the cardiac skeleton to the top of the interventicular septum. It then continues down both sides of the septum as the **right** and **left bundle branches.** The atrioventricular bundle distributes the action potential over the medial surfaces of the ventricles. Actual contraction of the ventricles is stimulated by **conduction myofibers (Purkinje fibers)** that emerge from the bundle branches and pass into the cells of the myocardium.

ELECTROCARDIOGRAM

Impulse transmission through the conduction system generates electrical currents that can be detected on the body's surface. A recording of the electrical changes that accompany the cardiac cycle is called an **electrocardiogram (ECG** or **EKG).** The instrument used to record the changes is an *electrocardiograph.*

Each portion of the cardiac cycle produces a different electrical impulse. These impulses are transmitted from the electrodes to a recording pen that graphs the impulses as a series of up-and-down waves called *deflection waves.* In a typical record (Figure 20-7b), three clearly recognizable waves accompany each cardiac cycle. The first, called the **P wave,** is a small upward wave. It indicates atrial depolarization—the spread of an impulse from the SA node through the muscle of the two atria. A fraction of a second after the P wave begins, the atria contract. The second wave, called the **QRS wave (complex),** begins as a downward deflection, continues as a large, upright, triangular wave, and ends as a downward wave at its base. This deflection represents ventricular depolarization, that is, the spread of the electrical impulse through the ventricles. The third recognizable deflection is a dome-shaped **T wave.** This wave indicates ventricular repolarization. There is no deflection to show atrial repolarization because the stronger QRS wave masks this event.

In reading an electrocardiogram, it is important to note the size of the deflection waves at certain time intervals. Enlargement of the P wave, for example, indicates enlargement of the atrium, as in mitral stenosis. In this condition, the mitral valve narrows, blood backs up into the atrium, and there is expansion of the atrial wall.

The *P-R interval* is measured from the beginning of the P wave to the beginning of the R wave. It represents the conduction time from the beginning of atrial excitation to the beginning of ventricular excitation. The P-R interval is the time required for an impulse to travel through the atria and atrioventricular node to the remaining conducting tissues. The lengthening of this interval, as in atherosclerotic heart disease and rheumatic fever, occurs because the heart tissue covered by the P-R interval, namely the atria and atrioventricular node, is scarred or inflamed. Thus the impulse must travel at a slower rate and the interval is lengthened. The normal P-R interval covers no more than 0.2 sec.

An enlarged Q wave may indicate a myocardial infarction (heart attack). An enlarged R wave generally indicates enlarged ventricles.

The *S-T segment* begins at the end of the S wave and terminates at the beginning of the T wave. It represents the time between the end of the spread of the impulse through the ventricles and repolarization of the ventricles. The S-T segment is elevated in acute myocardial infarction and depressed when the heart muscle receives insufficient oxygen.

The T wave represents ventricular repolarization. It is flat when the heart muscle is receiving insufficient oxygen, as in atherosclerotic heart disease. It may be elevated when the body's potassium level is increased.

The ECG is useful in diagnosing abnormal cardiac rhythms and conduction patterns and following the course of recovery from a heart attack. It can also detect the presence of fetal life or determine the presence of several fetuses. For some people, ECG monitoring may be done by using a *Holter monitor.* This is an ambulatory system in which the monitoring hardware is worn by an individual while going about everyday routines. The Holter monitor is used especially to detect rhythm disorders in the conduction system. It is also used to correlate rhythm disorders and symptoms and to follow the effectiveness of drugs.

CLINICAL APPLICATION

When major elements of the conduction system are disrupted, an irregular heart rhythm may occur. In one type of rhythm disturbance, the ventricles fail to receive atrial impulses, causing the ventricles and atria to beat independently of each other. In patients with such a condition, normal heart rhythm can be restored and maintained with an **artificial pacemaker,** a device that sends out small electrical charges that stimulate the heart. It consists of three basic parts: a *pulse generator,* which contains the battery cells and produces the impulse; a *lead,* which is a flexible wire connected to the pulse generator that delivers the impulse to the

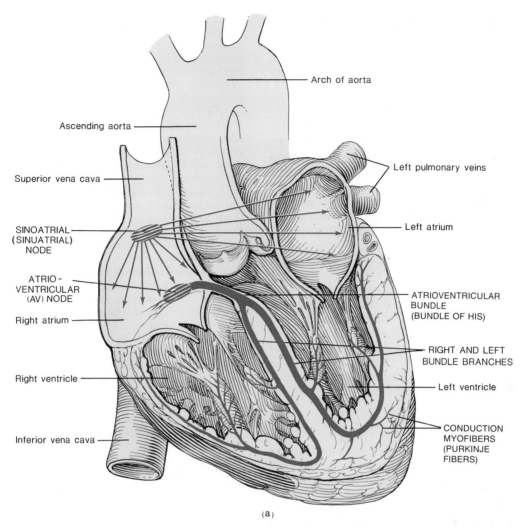

- Arch of aorta
- Ascending aorta
- Superior vena cava
- SINOATRIAL (SINUATRIAL) NODE
- ATRIO-VENTRICULAR (AV) NODE
- Right atrium
- Right ventricle
- Inferior vena cava
- Left pulmonary veins
- Left atrium
- ATRIOVENTRICULAR BUNDLE (BUNDLE OF HIS)
- RIGHT AND LEFT BUNDLE BRANCHES
- Left ventricle
- CONDUCTION MYOFIBERS (PURKINJE FIBERS)

(a)

FIGURE 20-7 Conduction system of the heart. (a) Location of the nodes and bundles of the conduction system. (b) Normal electrocardiogram of a single heartbeat, enlarged for emphasis.

Seconds

(b)

electrode; and an *electrode,* which makes contact with a portion of the heart and delivers the charge to the heart.

BLOOD FLOW THROUGH THE HEART

Two phenomena control the movement of blood through the heart: the opening and closing of the valves and the contraction and relaxation of the myocardium. Both these activities occur without direct stimulation from the nervous system. The valves are controlled by pressure changes in each heart chamber. The contraction of the cardiac muscle is stimulated by its conduction system.

Blood flows from an area of higher pressure to an area of lower pressure. The pressure developed in a heart chamber is related primarily to the chamber's size. For example, if the chamber size decreases, the pressure increases. The pressure in the atria is called **atrial pressure,** that in the ventricles is called **ventricular pressure,** and pressure in the aorta and pulmonary trunk is referred to as **arterial pressure.** The relationship of these pressures to a heartbeat in the left side of the heart is shown in Figure 20-8b. Although the pressures in the right side of the heart are somewhat lower, the same pattern exists.

CARDIAC CYCLE

In a normal heartbeat, the two atria contract while the two ventricles relax. Then, when the two ventricles contract, the two atria relax. The term **systole** (SIS-tō-lē) refers to the phase of contraction; **diastole** (dī-AS-tō-lē) is the phase of relaxation. A **cardiac cycle,** or complete heartbeat, consists of a systole and diastole of both atria plus the systole and diastole of both ventricles.

For purposes of our discussion, we will divide the cardiac cycle into the following phases: As you read the description, refer to Figure 20-8.

1. **Atrial Systole (Contraction)** Under normal conditions, blood flows continuously from the superior vena cava, inferior vena cava, and coronary sinus into the right atrium and from the pulmonary veins into the left atrium. The bulk of blood, about 70 percent, flows passively from the atria into the ventricles, even before atrial contraction occurs. When the SA node fires, the atria depolarize. This is followed by atrial contraction. Atrial depolarization produces the P wave on the ECG. Contraction of the atria forces its remaining blood into the ventricles. This final push accounts for only about 30 percent of the blood passed into the ventricles. Thus, atrial contraction is not really necessary for ventricular filling at normal heart rates. During atrial systole, deoxygenated blood passes through the open tricuspid valve from the right atrium into the right ventricle, and oxygenated blood passes through the open bicuspid valve from the left atrium to the left ventricle.

If you look at the atrial pressure curve in Figure 20-8b, you will notice that there are several pressure elevations, the a, c, and v waves. The *a wave* is produced by atrial contraction. The *c wave* is caused by ventricular contraction, which makes the atrioventricular values bulge into the atria. The *v wave* is due to atrial filling while the atrioventricular valves are closed during ventricular contraction.

2. **Ventricular Filling.** When the ventricles are contracting, the atrioventricular valves are closed and atrial pressure increases as blood fills the atria. During atrial diastole (relaxation), deoxygenated blood from various parts of the body enters the right atrium and oxygenated blood from the lungs enters the left atrium. But, once ventricular contraction is over, ventricular pressure falls. The higher atrial pressure pushes the atrioventricular valves open and blood fills the ventricles.

The major part of **ventricular filling** occurs immediately upon opening of the atrioventricular valves. The first third of ventricular filling is the *period of rapid filling.* During the middle third, called *diastasis,* only a small amount of blood flows into the ventricles. This is blood that continuously empties into the right atrium from the superior vena cava, inferior vena cava, and coronary sinus, and into the left atrium from the pulmonary veins, and passes through the atria directly into the ventricles. During the last third of ventricular filling, the atria contract. As noted earlier, this accounts for about 30 percent of the blood that fills the ventricles.

3. **Ventricular Systole (Contraction).** Near the end of atrial systole, the action potential from the SA node passes to the AV node and through the ventricles, causing them to depolarize and contract. This is represented as the QRS complex in the ECG. The beginning of ventricular contraction coincides with the first heart sound. At the onset of ventricular contraction, there is an abrupt rise in ventricular pressure that causes the atrioventricular valves to close. The first 0.05 seconds of ventricular systole is known as **isovolumetric contraction** because ventricular volume is constant (Figure 20-8c). It is the interval of time between the start of ventricular systole and the opening of the semilunar valves. During this time, there is contraction of the ventricles but no emptying; there is a rapid rise in ventricular pressure.

Once ventricular pressure exceeds arterial pressure, the semilunar valves open and blood is forced from the ventricles into their respective arteries. This is called the **ejection period** and lasts for about 0.25 seconds, until the semilunar valves close. The blood pumped by each ventricle during ejection is about half the content of the ventricle and is known as stroke volume (discussed shortly).

4. **Ventricular Diastole (Relaxation).** At the end of ventricular contraction, the ventricles suddenly begin their relaxation. The period of time, about 0.05 seconds between the opening of the atrioventricular valves and the closing of the semilunar valves is termed **isovolumetric relaxation.** It is characterized by a drastic decrease in ventricular pressure without a change in ventricular volume (Figure 20-8c). The higher arterial pressure causes blood to flow back toward the ventricles. This results in closure of the semilunar valves. Closure of the aortic semilunar valve produces a brief rise in arterial (aortic) pressure (dicrotic notch), shown in Figure 20-8b, and the second heart sound.

Firing of the SA node results in atrial depolarization, followed by atrial contraction, and the beginning of another cardiac cycle.

TIMING

If we assume that the average heart beats 75 times per minute, then each cardiac cycle requires about 0.8 sec. During the first 0.1 sec, the atria contract and the ventricles relax. The atrioventricular valves are open, and the semilunar valves are closed. For the next 0.3 sec, the atria are relaxing and the ventricles are contracting. During the first part of this period, all valves are closed; during the second part, the semilunar valves are open.

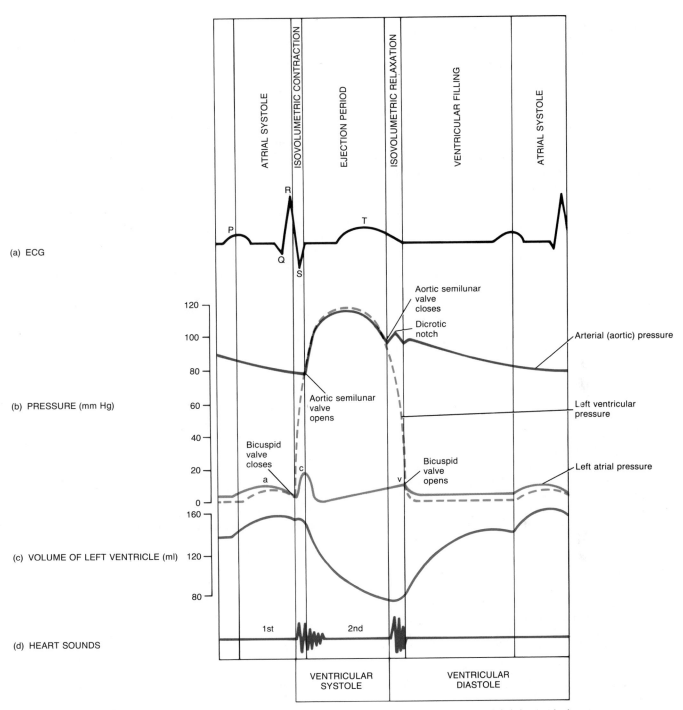

FIGURE 20-8 Cardiac cycle. (a) ECG related to the cardiac cycle. (b) Left atrial, left ventricular, and arterial (aortic) pressure changes along with the opening and closing of valves during the cardiac cycle. The letters a, c, and v represent the elevation waves of the left atrial pressure curve. (c) Left ventricular volume during the cardiac cycle. (d) Heart sounds related to the cardiac cycle.

The last 0.4 sec of the cycle is the *relaxation (quiescent) period* and all chambers are in diastole. In a complete cycle, then, the atria are in systole 0.1 sec and in diastole 0.7 sec; the ventricles are in systole 0.3 sec and in diastole 0.5 sec. For the first part of the relaxation period, all valves are closed; during the latter part, the atrioventricular valves open and blood starts draining into the ventri-cles. When the heart beats faster than normal, the relaxation period is shortened accordingly.

SOUNDS

The act of listening to sounds within the body is called **auscultation** (*auscultare* = to listen; aws-kul-TĀ-shun) and it is usually done with a stethoscope. The sound of the heartbeat comes primarily from turbulence in blood flow created by the closure of the valves, not from the contraction of the heart muscle (Figure 20-8b,d). The first sound, which can be described as a **lubb** ($\overline{oo}$) sound, is a long, booming sound. The lubb is the sound created by the closure of the atrioventricular valves soon after ventricular systole begins. The second sound, which is heard as a short, sharp sound, can be described as a **dupp** ($\breve{u}$) sound. Dupp is the sound created as the semilunar valves close toward the end of ventricular systole. A pause between the second sound and the first sound of the next cycle is about two times longer than the pause between the first and second sound of each cycle. Thus the cardiac cycle can be heard as a lubb, dupp, pause; lubb, dupp, pause; lubb, dupp, pause.

CLINICAL APPLICATION

Heart sounds provide valuable information about the valves. Unusual sounds are called **murmurs.** Some murmurs are caused by the noise made by a little blood bubbling back up into an atrium because of improper closure of an atrioventricular valve.

Frequently a portion of a normal mitral valve is pushed back too far (prolapsed) during contraction due to expansion of the cusps and elongation of the chordae tendineae. Such a condition is called **mitral valve prolapse (MVP),** an inherited disorder resulting from defective collagen. Or, a murmur might be caused by a diseased mitral valve due to rheumatic fever (**valvular heart disease**). In the absence of any complications, many murmurs have no clinical significance.

CARDIAC OUTPUT (CO)

Although the heart possesses properties that enable it to beat independently, its operation is regulated by events occurring in the rest of the body. All body cells must receive a certain amount of oxygenated blood each minute to maintain health and life. When cells are very active, as during exercise, they need even more blood. During rest periods, cellular need is reduced, and the heart cuts back on its output.

The amount of blood ejected from the left ventricle into the aorta per minute is called the **cardiac output (CO)** or **cardiac minute output.** Cardiac output is determined by (1) the amount of blood pumped by the left ventricle during each beat and (2) the number of heartbeats per minute. The amount of blood ejected by a ventricle during each systole is called the **stroke volume (SV).** In a resting adult, stroke volume averages 70 ml and heart rate is about 75 beats per minute. The average cardiac output, then, in a resting adult is

$$\text{Cardiac output} = \text{stroke volume} \times \text{beats per minute}$$
$$= 70 \text{ ml} \times 75/\text{min}$$
$$= 5250 \text{ ml/min} \quad \text{or} \quad 5.25 \text{ liters/min}$$

Factors that increase stroke volume or heart rate tend to increase cardiac output. Factors that decrease stroke volume or heart rate tend to decrease cardiac output.

CLINICAL APPLICATION

Syncope (SIN-kō-pē) or faint refers to an impairment or temporary loss of consciousness, most commonly due to cerebral ischemia (lack of sufficient blood). Clinically, it may be preceded by uneasiness, malaise, lightheadedness, nausea, vertigo, confusion, disturbances in vision, weakness, sweating, or tinnitus. Among the factors that contribute to syncope are altered cardiac output, altered blood oxygen levels, and failure of compensatory cerebral blood flow mechanisms. A sudden change in cardiac output results in an alteration of blood pressure and, if the change is too abrupt for compensation by autoregulatory mechanisms, syncope results. The depth of syncope varies from confusion to temporary loss of consciousness. Although jerking of limbs may occur just after syncope, it is only fleeting and sphincter control is not lost.

STROKE VOLUME

The actual amount of blood ejected by a ventricle with each heartbeat depends on how much blood enters the ventricle during diastole and how much blood is left in the ventricle following its systole.

End-Diastolic Volume (EDV)

The term **end-diastolic volume (EDV)** refers to the volume of blood that enters a ventricle during diastole. As noted earlier, about 70 percent of the blood flows from the atria into the ventricles before atrial systole; atrial contraction causes the additional 30 percent to enter the ventricles. During diastole, the ventricles normally increase their volume to 120 to 130 ml, the end-diastolic volume. This volume is determined principally by the length of ventricular diastole and venous pressure. When heart rate increases, the duration of diastole is shorter, the ventricles may contract before they are adequately filled, and the EDV is reduced. When venous pressure increases, a greater volume of blood is forced into the ventricles, and the EDV is increased. As you will see shortly, this results in a more forceful ventricular contraction.

End-Systolic Volume (ESV)

End-systolic volume (ESV) refers to the volume of blood still left in a ventricle following its systole. Since end-diastolic volume is about 120 to 130 ml and stroke volume is about 70 ml, end-systolic volume is about 50 to 60 ml (end-diastolic volume minus stroke volume).

End-systolic volume is determined principally by arterial pressure and the force of ventricular contraction. Arterial pressure in the aorta and pulmonary trunk just prior to ventricular systole is the pressure that must be overcome for the semilunar valves to open. If arterial pressure is significantly elevated, the resistance prevents the ventricles from pumping out as much as they should and stroke volume decreases.

One factor that determines the force of ventricular contraction is the length of cardiac muscle fibers. Within limits, the greater the length of stretched fibers, the stronger the contraction. This relationship is referred to as **Starling's law of the heart.** The situation is somewhat like stretching a rubber band; the more you stretch it, the harder it contracts. This is similar to the relationship of skeletal muscle fiber length and force of contraction discussed in Chapter 10. When cardiac muscle fibers are moderately stretched, there are numerous cross-bridge connections between thin and thick myofilaments, and the contraction is fairly strong. However, when the fibers are overstretched, there are fewer cross-bridge connections, and the force of contraction is weaker. The operation of Starling's law of the heart is important in maintaining equal blood output from both ventricles. For example, if for some reason the right ventricle begins to pump more blood than the left, the increased blood volume to the left ventricle stretches it, resulting in an increased cardiac output equal to that of the right ventricle. If the right ventricle pumps more blood than the left, without compensation by the left, blood would accumulate in the lungs. The operation of Starling's law of the heart is also important in individuals who have constricted arteries or improperly functioning heart valves. Arterial constriction creates a resistance to blood flow in the aorta (or pulmonary trunk). This prevents normal emptying of the ventricle, causing it to overfill with blood and stretch beyond normal. As a result, the ventricle contracts more forcefully, thus maintaining normal cardiac output. This is what happens to individuals who have hypertension (high blood pressure). A similar situation occurs with improperly functioning heart valves. If blood backflows into an atrium or ventricle, the chamber overfills with blood and contracts harder. Since the increased blood leaks backward during diastole (relaxation), net cardiac output remains normal.

The force of ventricular contraction is also determined by factors such as nervous control, chemicals, temperature, emotions, sex, and age. These factors are considered as part of the discussion dealing with the regulation of heart rate.

CLINICAL APPLICATION

The maximum percentage that the cardiac output can increase above normal is referred to as **cardiac reserve.** For example, during strenuous exercise the cardiac output of a normal adult may increase to about four times normal. Since this is an increase of 400 percent above normal, we say that the cardiac reserve is 400 percent. Well-trained athletes may have a cardiac reserve as high as 600 percent, an increase in cardiac output of about six times normal. Although cardiac reserve depends on many factors, it is markedly influenced by heart conditions such as ischemic heart disease, valvular disorders, and myocardial damage.

HEART RATE

Cardiac output depends on heart rate as well as stroke volume. In fact, changing heart rate is the body's principal mechanism of short-term control over cardiac output and blood pressure. The pacemaker initiates contraction, and left to itself would set an unvarying heart rate. However, the body's need for blood supply varies under different conditions. There are several regulatory mechanisms stimulated by factors such as chemicals present in the body, temperature, emotional state, and age.

During certain pathological conditions, stroke volume may fall dangerously low. If the ventricular myocardium is weak or damaged by an infarction, it cannot contract strongly. Or blood volume may be reduced by excessive bleeding. Stroke volume then falls because the cardiac fibers are not sufficiently stretched. In these cases, the body attempts to maintain a safe cardiac output by increasing the rate and strength of contraction.

The heart rate is regulated by several factors. The most important control of heart rate and strength of contraction is the autonomic nervous system.

Autonomic Control

Within the medulla of the brain is a group of neurons called the **cardioacceleratory center (CAC).** Arising from this center are sympathetic fibers that travel down a tract in the spinal cord and then pass outward in the *cardiac (accelerator) nerves* and innervate the SA node, AV node, and portions of the myocardium (Figure 20-9). When the cardioacceleratory center is stimulated, nerve impulses travel along the sympathetic fibers. This causes them to release norepinephrine, which increases the rate of heartbeat and the strength of contraction.

The medulla also contains a group of neurons that form the **cardioinhibitory center (CIC).** Arising from this center are parasympathetic fibers that reach the heart via the *vagus (X) nerve.* These fibers innervate the SA node and AV node. When this center is stimulated, nerve impulses transmitted along the parasympathetic fibers cause the release of acetylcholine. This decreases the rate of heartbeat.

The autonomic control of the heart is therefore the result of opposing sympathetic (stimulatory) and parasympathetic (inhibitory) influences. Sensory impulses from receptors in different parts of the cardiovascular system act upon the centers so that a balance between stimulation and inhibition is maintained. Nerve cells capable of responding to changes in blood pressure are called **baroreceptors** or **pressure receptors.** These receptors affect the rate of heartbeat and are involved in three reflex pathways: the carotid sinus reflex, the aortic reflex, and the right heart (atrial) reflex.

• *Carotid Sinus Reflex*   The **carotid sinus reflex** is concerned with maintaining normal blood pressure in the brain. The *carotid sinus* is a small widening of the internal carotid artery just above the point where it branches off from the common carotid artery (Figure 20-9). In the wall of the carotid sinus or attached to it are barorecep-

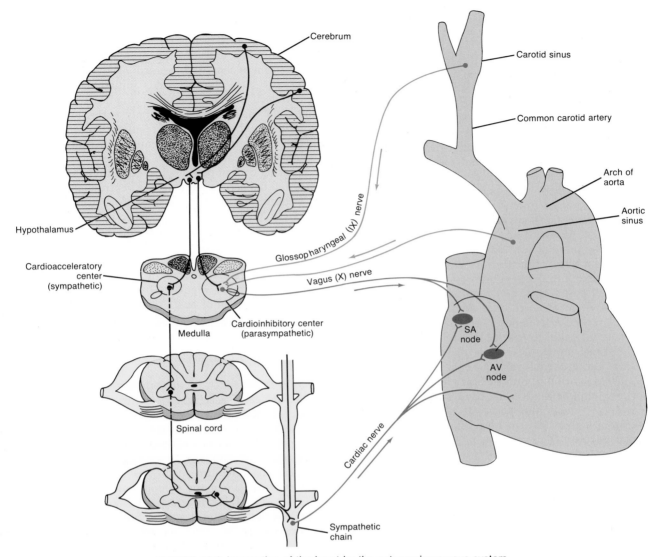

FIGURE 20-9 Innervation of the heart by the autonomic nervous system.

tors. Any increase in blood pressure stretches the wall of the sinus, and the stretching stimulates the baroreceptors. The impulses then travel from the baroreceptors over sensory neurons in the glossopharyngeal (IX) nerves. Within the medulla, the impulses stimulate the cardioinhibitory center and inhibit the cardioacceleratory center. Consequently, more parasympathetic impulses pass from the cardioinhibitory center via the vagus (X) nerves to the heart and fewer sympathetic impulses pass from the cardioacceleratory center via cardiac (accelerator) nerves to the heart. The result is a decrease in heart rate and force of contraction. There is a subsequent decrease in cardiac output and a decrease in arterial blood pressure to normal.

If blood pressure falls, reflex acceleration of the heart takes place. The baroreceptors in the carotid sinus do not stimulate the cardioinhibitory center, and the cardioacceleratory center is free to dominate. The heart then beats faster and more forcefully to restore normal blood pressure. This inverse relationship between blood pressure

and heart rate is referred to as **Marey's law of the heart** (Figure 20-10).

The ability of the carotid sinus reflex (and the aortic reflex to be described shortly) to maintain a relatively constant blood pressure is very important when a person sits or stands from a lying position. Immediately upon moving from a prone to erect position, blood pressure in the head and upper part of the body falls. The decrease in pressure, however, is counteracted by the reflexes. If the pressure was to fall markedly, unconsciousness could occur.

• *Aortic Reflex* The **aortic reflex** is concerned with general systemic blood pressure. It is initiated by baroreceptors in the wall of the arch of the aorta or attached to it (see Figure 20-9), and it operates like the carotid sinus reflex.

• *Right Heart (Atrial) Reflex* The **right heart (atrial) reflex** responds to venous blood pressure. It is initiated

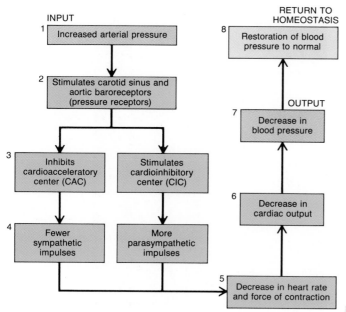

FIGURE 20-10 Control of the heart rate and force of contraction by baroreceptors in the carotid sinus reflex and aortic reflex.

by baroreceptors in the superior and inferior venae cavae and in the right atrium. When venous pressure increases, the baroreceptors send impulses that stimulate the cardioaccelerator center. This action causes the heart rate to increase. This mechanism is called the **Bainbridge reflex.**

Chemicals

Certain chemicals present in the body also have an effect on heart rate. For example, epinephrine, produced by the adrenal medulla in response to sympathetic stimulation, increases the excitability of the SA node. This, in turn, increases the rate and strength of contraction. Elevated levels of potassium or sodium decrease the heart rate and strength of contraction. It appears that excess potassium interferes with the generation of nerve impulses, and excess sodium interferes with calcium participation in muscular contraction. An excess of calcium increases heart rate and strength of contraction.

Temperature

Increased body temperature, such as occurs during fever or strenuous exercise, causes the AV node to discharge impulses faster and thereby increases heart rate. Decreased body temperature resulting from exposure to cold or deliberately cooling the body prior to surgery decreases heart rate and strength of contraction.

Emotions

Strong emotions such as fear, anger, and anxiety, along with a multitude of physiological stressors, increase heart rate through the general adaptation syndrome. Mental states such as depression and grief tend to stimulate the cardioinhibitory center and decrease heart rate.

Sex and Age

Sex is another factor—the heartbeat is somewhat faster in females. Age is yet another factor—the heartbeat is fastest at birth, moderately fast in youth, average in adulthood, and below average in old age.

A summary of the factors that influence cardiac output is presented in Figure 20-11.

CIRCULATORY SHOCK AND HOMEOSTASIS

When cardiac output or blood volume is reduced to the point where body tissues do not receive an adequate blood supply, **circulatory shock** results. Circulatory shock is caused by loss of blood volume through hemorrhage or through the release of histamine due to damage to body tissues (trauma). The characteristic symptoms of circulatory shock are a pale, clammy skin; cyanosis of the ears and fingers; a feeble though rapid pulse; shallow and rapid breathing; lowered body temperature; and mental confusion or unconsciousness.

If the shock is mild, certain homeostatic mechanisms of the cardiovascular system compensate so that no serious damage results. Lowered blood pressure is compensated for by constriction of blood vessels and water retention. Renin is secreted by the kidneys, aldosterone by the

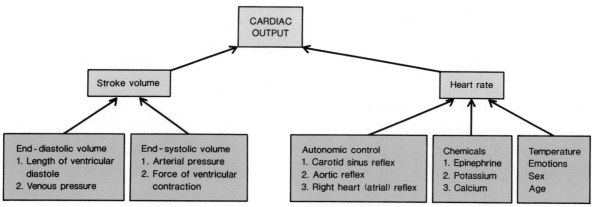

FIGURE 20-11 Summary of factors that influence cardiac output.

adrenal cortex, epinephrine by the adrenal medulla, and ADH by the posterior pituitary. Even though some blood is lost from circulation, blood return to the heart is normal and cardiac output remains essentially unchanged. Veins and many arterioles are constricted during compensation, but there is no constriction of arterioles supplying the heart and brain. As a consequence, blood flow to the heart and brain is normal or nearly so. Compensation is an effective homeostatic mechanism until about 900 ml of blood is lost.

If the shock is severe, death may occur. For instance, if the return of venous blood is greatly diminished by excessive blood loss, the compensatory mechanisms are insufficient. When the cardiac output decreases, the heart fails to pump enough blood to supply its own coronary vessels and the heart muscle weakens. In addition, prolonged vasoconstriction ultimately leads to tissue hypoxia, and vital organs such as the kidneys and liver are damaged. Essentially, the initial shock promotes more shock and a **circulatory shock cycle** is established (Figure 20-12). Once the shock reaches a certain level of severity, damage to the circulatory organs is so extensive that death ensues.

ARTIFICIAL HEART

On December 2, 1982, Dr. Barney B. Clark, a 61-year-old retired dentist, made medical history by becoming the first human to receive an **artificial heart.** The 7½-hour operation was performed by Dr. William DeVries, then a surgeon at the University of Utah Medical Center.

The artificial heart, known as Jarvik-7 after its inventor, Dr. Robert Jarvik, consists of an aluminum base and a pair of rigid plastic chambers that serve as ventricles (Figure 20-13). Each chamber contains a flexible diaphragm driven by compressed air that forces blood past mechanical valves into the major blood vessels. The surgery essentially consisted of removing the ventricles from Dr. Clark's heart but leaving the atria intact. Then Dacron connectors were sutured onto the atria, aorta, and

(a)

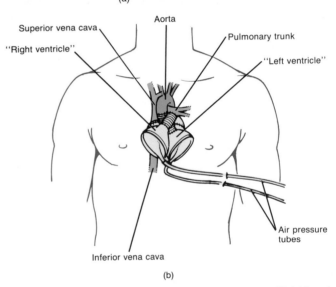

(b)

FIGURE 20-13 Artificial heart. (a) Photograph of an artificial heart prior to implantation. (Courtesy of Bradley Nelson, University of Utah.) (b) Diagram of an implanted heart.

pulmonary trunk. The artificial heart was then snapped into position via the connectors. Essentially, the artificial heart is designed to pump sufficient blood to sustain only moderate activity.

Recently, the artificial heart has come under some criticism because of its possible role in causing several disabling or fatal strokes in its recipients. Research is continuing and clinical trials are being undertaken in the hopes of developing a safer, more efficient and practical artificial heart.

DIAGNOSING HEART DISORDERS

The diagnosis of heart disorders may be made by invasive or noninvasive techniques. Invasive techniques involve

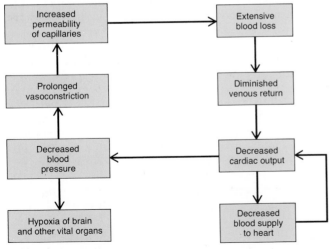

FIGURE 20-12 Circulatory shock cycle. Note how the cycle perpetuates itself until death results.

puncture or incision of the skin or insertion of an instrument or foreign material into the body. None of these is necessary with a noninvasive technique.

An invasive measure available for the study of the cardiovascular system is called **cardiac catheterization.** In the procedure, the tip of a long plastic *catheter,* or tube, is introduced into a vein in the arm or leg. The catheter is radiopaque so that it can be seen with a fluoroscope. With the help of the fluoroscope, it is then threaded through the vena cava and into the right atrium, right ventricle, or pulmonary trunk. The catheter can also be inserted into an artery of the arm or leg and worked up through the aorta to the left atrium and ventricle.

Pressure in the heart and blood vessels is most commonly studied during cardiac catheterization. The procedure is also used to assess left ventricular function, cardiac output (CO), and diastolic properties of the left ventricle. Also, cardiac catherization allows physicians to measure the flow of blood through the heart and blood vessels, the oxygen content of blood, the status of heart valves and the conduction system, and the exact anatomic location of various cardiovascular lesions. Still, other physicians use cardiac catheterization to inject clot-dissolving enzymes into coronary arteries immediately following a heart attack.

The development of cardiac catheterization equipment and techniques has now made it possible to place catheters into the vascular system of almost every organ.

The major noninvasive techniques used to evaluate cardiovascular disorders are physical examination, electrocardiography (ECG), x rays, and a newer technique called **echocardiography** (*echo* = reflected sound). This procedure is a special application of diagnostic ultrasound in which high-frequency sound waves, inaudible to the human ear, are employed to take pictures of the heart. Sound waves aimed at the heart are reflected to a crystal that converts the waves into an image that is displayed on a television-like screen. The image can be photographed, videotaped, or recorded as a graph on paper. Among the clinical uses of echocardiography are evaluating congenital heart disorders, tumors, and valvular disorders; detecting fluid around the heart; and assessing the effects of coronary artery disease on the heart.

HEART-LUNG MACHINE

Some surgical procedures, such as those that involve heart transplantation, open-heart surgery, and coronary artery bypass surgery, make it necessary to use a **heart-lung machine.** This device performs two complex roles simultaneously: it pumps blood (functioning as a heart), and removes carbon dioxide (CO_2) from blood and adds oxygen (O_2) to it (functioning as the lungs). The principle of the heart-lung machine is fairly simple. Deoxygenated blood from the venae cavae is removed from the body and passed through an oxygenator, where carbon dioxide is removed and oxygen is added. Then, the oxygenated blood is passed through a temperature controller, where it may be rewarmed to body temperature or cooled. Next, the blood passes through a filter to remove emboli and

is finally returned to the body into the arterial system at the proper pressure. If necessary, drugs, anesthetics, and transfusions may be added to the circuit.

To surgically repair certain heart abnormalities, it is necessary to slow down a patient's heart. One such method is called **hypothermia** (hī-pō-THER-mē-a), which refers to a low body temperature. In various surgical procedures, it refers to a deliberate cooling of the body to slow metabolism and reduce the oxygen needs of the tissues. Thus, the heart and brain can withstand short periods of interrupted or reduced blood flow. Lost blood is then replaced by transfusion during and after the operation.

RISK FACTORS IN HEART DISEASE

It is estimated that one in every five persons who reaches age 60 will have a myocardial infarction (heart attack). One in every four persons between 30 and 60 has the potential to be stricken. Heart disease is epidemic in this country, despite the fact that some of the causes can be foreseen and prevented. The results of research indicate that people who develop combinations of certain risk factors eventually have heart attacks. Risk factors are characteristics, symptoms, or signs present in a person free of disease that are statistically associated with an excessive rate of development of a disease. Among the risk factors in heart disease are:

1. High blood cholesterol level.
2. High blood pressure.
3. Cigarette smoking.
4. Obesity.
5. Lack of regular exercise.
6. Diabetes mellitus.
7. Genetic predisposition (family history of heart disease at an early age).
8. Sex.
9. Age.

The first five risk factors all contribute to increasing the heart's work load. High blood cholesterol is discussed shortly, and hypertension is discussed in the next chapter. Cigarette smoking, through the effects of nicotine, stimulates the adrenal gland to oversecrete aldosterone, epinephrine, and norepinephrine—powerful vasoconstrictors. Overweight people develop miles of extra capillaries to nourish fat tissue. The heart has to work harder to pump the blood through more vessels. Without exercise, venous return gets less help from contracting skeletal muscles. In addition, regular exercise strengthens the smooth muscle of blood vessels and enables them to assist general circulation. Exercise also increases cardiac efficiency and output. In diabetes mellitus, fat metabolism dominates glucose metabolism. As a result, cholesterol levels get progressively higher and result in plaque formation, a situation that may lead to high blood pressure. High blood pressure drives fat into the vessel wall, encouraging atherosclerosis. Up to age 50, there is a 10 to 15 year lag in the extent of heart disease in females compared to males; after that, the rates of disease in both sexes are similar. Regardless of sex, the incidence of heart disease increases with age.

DEVELOPMENTAL ANATOMY OF THE HEART

The *heart,* a derivative of **mesoderm,** begins to develop before the end of the third week. It begins its development in the ventral region of the embryo beneath the foregut (see Figure 24-23). The first step is the formation of a pair of tubes, the **endothelial (endocardial) tubes,** from mesodermal cells (Figure 20-14). These tubes then unite to form a common tube, the **primitive heart tube.** Next, the primitive heart tube develops into five regions: **ventricle, bulbus cordis, atrium, sinus venosus,** and **truncus arteriosus.** Since the bulbus cordis and ventricle subsequently grow more rapidly than the others, and the heart grows more rapidly than its superior and inferior attachments, the heart assumes a U-shape and later an S-Shape.

The flexures of the heart reorient the regions so that the atrium and sinus venous eventually come to lie superior to the bulbus cordis, ventricle, and truncus anteriosus.

At about the seventh week a partition, the **interatrial septum,** forms in the atrial region, dividing it into a *right* and *left atrium.* The opening in the partition is the **foramen ovale,** which normally closes at birth and later forms a depression called the *fossa ovalis.* An **interventricular septum** also develops and partitions the ventricular region into a *right* and *left ventricle.* The bulbus cordis and truncus arteriosus divide into two vessels, the *aorta* (arising from the left ventricle) and the *pulmonary trunk* (arising from the right ventricle). Recall that the ductus arteriosus is a temporary vessel between the aorta and pulmonary trunk until birth. The great veins of the heart, *superior vena cava* and *inferior vena cava,* develop from the venous end of the primitive heart tube.

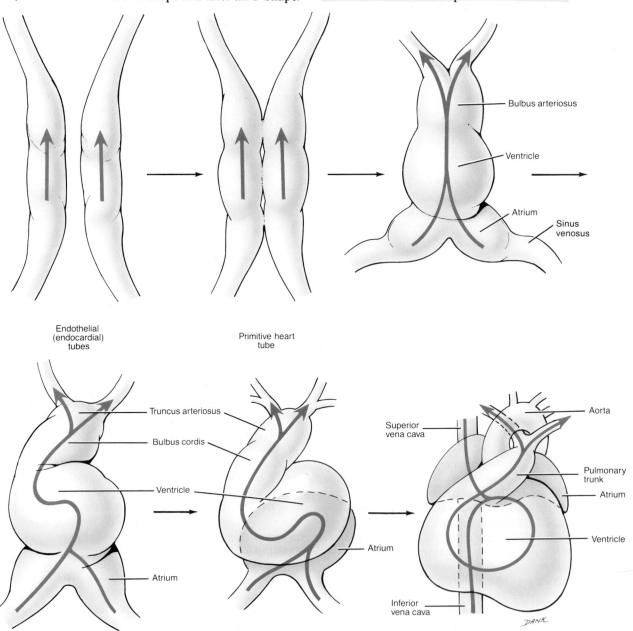

FIGURE 20-14 Development of the heart. The red arrows show the direction of the flow of blood from the venous to the arterial end.

DISORDERS: HOMEOSTATIC IMBALANCES

Coronary Artery Disease (CAD)

Coronary artery disease (CAD) is a condition in which the heart muscle receives an inadequate amount of blood because of an interruption of its blood supply. It is presently the leading cause of death in the United States. Depending on the degree of interruption, symptoms can range from a mild chest pain to a full-scale heart attack. Generally, symptoms manifest themselves when there is about a 75 percent narrowing of coronary artery lumina. The underlying causes of CAD are many and varied. Two of the principal ones are atherosclerosis and coronary artery spasm. Another is a thrombus or an embolus in a coronary artery.

Atherosclerosis

Atherosclerosis (ath'-er-ō-skle-RŌ-sis) is a process in which fatty substances, especially cholesterol and triglycerides (ingested fats), are deposited in the walls of medium-sized and large arteries in response to certain stimuli. It is believed that the first event in atherosclerosis is damage to the endothelial lining of the artery. Contributing factors include high blood pressure (hypertension), carbon monoxide in cigarettes, diabetes mellitus, and a high cholesterol level. Following endothelial damage, monocytes begin to stick to the smooth inner endothelial lining of the tunica interna (inner coat) of the artery. Then, they slip between endothelial cells, penetrate the endothelium, and develop into macrophages. Here, the macrophages begin to take up cholesterol from low-density lipoproteins (described shortly). At the same time, smooth muscle cells in the tunica media (middle coat) of the artery also ingest cholesterol. The more cholesterol there is in the blood, the faster the process occurs. The accumulated cholesterol and cells form a lesion called an **atherosclerotic plaque** (Figure 20-15). The plaque looks like a pearly gray or yellow mound of tissue. As it grows, it may obstruct blood flow in the affected artery and damage the tissues the artery supplies. An additional danger is that the plaque provides a roughened surface that causes blood plate-lets to release *platelet-derived growth factor* (**PDGF**), a hormone that promotes the division of smooth muscle cells. In fact, macrophages and endothelial cells also produce PDGF. This further complicates the process since PDGF causes the lesion to grow larger. Platelets in the area of the atherosclerotic plaque also release clot-forming chemicals. Thus a thrombus may form. If the clot breaks off and forms an emoblus, it may obstruct small arteries and capillaries quite a distance from the site of formation.

Cholesterol is a major building block of all plasma membranes and a key compound for the synthesis of steroid hormones and bile salts. Some cholesterol is present in foods (eggs, dairy products, organ meats, beef, pork, and processed luncheon meats); most is synthesized by the liver. As it turns out, foods containing cholesterol are not the main source of cholesterol in the blood. It is the type of fat eaten. When saturated fats are broken down in the body, the liver uses some of the breakdown products to produce cholesterol. Saturated fats (see Chapter 2) can increase blood cholesterol level as much as 25 percent.

Cholesterol and triglycerides cannot dissolve in water, and thus cannot travel in the blood in their unaltered forms. They are made water-soluble by combining with proteins produced by the liver and intestine, and the complexes thus formed are called **lipoproteins.** The two major classes are called **low-density lipoproteins (LDL)** and **high-density lipoproteins (HDL).** There is an important difference between the two: LDL seems to pick up cholesterol and deposit it in body cells including, under abnormal conditions, smooth muscle cells in arteries. HDL seems to gather cholesterol from body cells and transports it to the liver for elimination. Most tissues of the body contain LDL receptors. The liver, suprarenal (adrenal) glands, and ovaries, because of their large requirement for cholesterol, have the most LDL receptors. Once LDL attaches to its receptor, it is taken into the cell by receptor-mediated endocytosis (Chapter 3). Within the cell, the LDL is broken down and the cholesterol is released to serve the cell's needs. In this way, LDL is

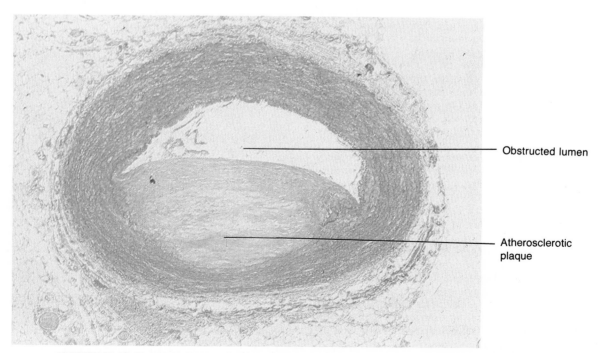

Obstructed lumen

Atherosclerotic plaque

FIGURE 20-15 Photomicrograph of a cross section of an artery partially obstructed by an atherosclerotic plaque. (Courtesy of Lester V. Bergman & Associates, Inc.)

removed from the blood. Once a cell has sufficient cholesterol for its activities, a negative feedback system inhibits the cell from synthesizing new LDL receptors. It is believed that some people have too few LDL receptors, due to various environmental and genetic reasons, and thus are more susceptible to atherosclerosis. A third type of lipoprotein, known as a **very low-density lipoprotein (VLDL),** contains some cholesterol and a large amount of triglyceride (fat). It also plays a role in the development of atherosclerotic plaques, but its role is not clear.

The amount of total cholesterol (HDL, VDL, and VLDL) in blood is one commonly used indicator of risk for CAD. In general, as the total cholesterol level increases above 150 mg/dl (a deciliter is $\frac{1}{10}$ of a liter), the risk of CAD slowly begins to rise. Above 200 mg/dl, the risk rapidly increases. Physicians frequently take no measures to lower a person's cholesterol level until it exceeds 250 mg/dl. Even more important than total cholesterol is the balance between LDL and HDL. Supposedly, the higher the ratio of HDL to total cholesterol (and to LDL), the lower the risk of CAD. Several ways to improve the HDL-LDL ratio are regular sustained exercise, diet (reducing animal fats, especially), and giving up cigarettes.

Unsaturated oils from fish, seal, and whale contain fatty acids referred to as omega-3 fatty acids. These fatty acids appear to have a role in protecting against heart disease. Specifically, they are believed to decrease serum cholesterol and low-density lipoprotein (LDL) levels; decrease the tendency of blood platelets to clump together, thus decreasing the likelihood of blood clot formation; suppress production of clotting factors; and enhance the production of an anticlotting agent. Oily fish, such as salmon, bluefish, herring, sardines, anchovies, and mackeral are good sources of omega-3 fatty acids. Shellfish also contain omega-3 fatty acids.

Diagnosis of atherosclerosis is based upon evaluation of blood pressure, blood tests to determine lipoprotein and glucose levels, a stress ECG, angiography, and nuclear cardiology. The *stress ECG* is done while a person is exercising on a treadmill or bicycle, in order to determine the presence and extent of CAD that might be missed on a resting ECG. *Angiography* (an-jē-OG-ra-fē) involves injecting a dye directly into the blood and then taking x rays of the arteries. The technique is also called *arteriography,* and the film is called an *arteriogram.* Angiography of the blood vessels of the heart is called *coronary angiography. Nuclear cardiology* involves injection of a radioactive substance into a vein followed by analysis of its distribution in the heart. An example of such a test is a *thallium scan.* As the radioisotope thallium 201 circulates through the coronary arteries, a camera placed over the chest picks up the signals from the radioisotope, relays them to a computer for translation, and a picture of the heart is displayed. An area of poor circulation of thallium in the heart indicates blockage of a coronary artery.

The treatment of CAD varies with the nature and urgency of the symptoms. Among the possible treatments are drug therapy, coronary artery bypass surgery, and percutaneous transluminal coronary angioplasty (PTCA). The two most commonly used drugs are *nitroglycerin* and *beta blockers.* Antithrombotic drugs that suppress the activity of platelets are being tested on a trial basis.

Coronary artery bypass surgery is the most direct way of increasing the blood supply to the heart. In a typical operation, a portion of a blood vessel is removed from another part of the body. At one time, the saphenous vein from the leg was frequently used. Now, however, the internal mammary artery from the chest is used. Research has shown that obstructions reform in grafted veins more readily than arteries. Once the heart is exposed, circulation is maintained by a heart-lung machine. Next, a clamp is placed across the aorta above the openings of the coronary arteries. This stops the blood supply to the heart. A cold saline solution containing potassium is then flushed into the heart through a small catheter inserted into the aorta. The saline solution ensures that the myocardium will maintain its natural state, and the potassium and low temperature of the solution discourage any tendency for the heart to start beating prematurely. A segment of the grafted blood vessel is then sutured between the aorta and the unblocked portion of the coronary artery, distal to the obstruction (Figure 20-16). If more than one artery is clogged, additional bypasses may be made. Once the repair is completed, the aortic clamp is removed and, as blood at normal temperature flows into the coronary arteries, the cold saline solution and potassium leave. This normally causes the heart to start beating again. If it does not, a mild shock from a defibrillator is applied from a pair of paddlelike electrodes.

Although coronary bypass surgery increases blood flow and decreases angina, it tends not to increase survival rates. Whether treated with surgery or medication, patients live about equally as long. The unfavorable conditions that initially led to the surgery will in time frequently clog the new blood vessels, or other poor aspects of the circulatory system will take their toll in increased blood pressure, heart enlargement, or stroke.

CAD may also be treated by a procedure referred to as *percutaneous transluminal coronary angioplasty (PTCA).* Like coronary artery bypass surgery, it is an attempt to increase the blood supply to the heart muscle. In this procedure, a balloon catheter is inserted into an artery of an arm or leg and gently guided through the arterial system under x-ray observation. Once in place in an obstructed coronary artery, the balloon is inflated for a few seconds, thus compressing the obstructing material against the vessel wall and permitting increased blood flow. Unfortunately, calcified plaques are not crushed by the balloon, and the balloon can damage the arterial wall. Despite these drawbacks, it is hoped that the procedure can be refined so that it will come into general use in the treatment of atherosclerosis.

Experiments are now underway to vaporize atherosclerotic plaques by using lasers, a procedure called *laser angioplasty.*

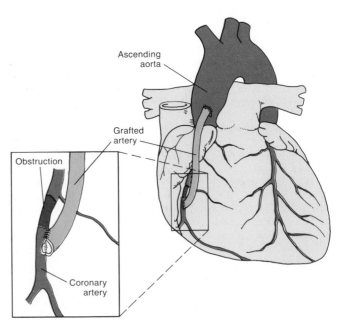

FIGURE 20-16 Coronary bypass surgery. The internal mammary artery is removed from the patient's chest. At a point distal to the obstruction, the grafted artery is sutured to the coronary artery.

Coronary Artery Spasm

Atherosclerosis results in a fixed obstruction to blood flow. Obstruction can also be caused by **coronary artery spasm,** a condition in which the smooth muscle of a coronary artery undergoes a sudden contraction, resulting in vasoconstriction. Coronary artery spasm typically occurs in individuals with atherosclerosis and may result in chest pain during rest (variant angina), chest pain during exertion (typical angina), heart attacks, and sudden death. Although the causes of coronary artery spasm are unknown, several factors are receiving attention. These include smoking, stress, and a vasoconstrictor chemical released by platelets. There is considerable interest in aspirin and other drugs that might inhibit the vasoconstrictor chemical. Coronary artery spasm is treated with long-acting nitrates (nitroglycerin) and calcium-blocking agents.

Congenital Defects

One baby in every 100 is born with a malformed heart. Such an inborn defect is called a **congenital defect.** Many defects are not serious and may go unnoticed for a lifetime. Others heal themselves. But some are life-threatening and must be mended by surgical techniques ranging from simple suturing to replacement of malfunctioning parts with synthetic materials.

One congenital defect that can occur is **coarctation** (kō'-ark-TĀ-shun) **of the aorta.** In this condition, a segment of the aorta is too narrow. As a result, the flow of oxygenated blood to the body is reduced, the left ventricle is forced to pump harder, and high blood pressure develops. The condition may be corrected by insertion of a synthetic graft.

Another common congenital defect is **patent ductus arteriosus.** The ductus arteriosus between the aorta and the pulmonary trunk normally closes shortly after birth in response to increased oxygenation of blood flowing through the ductus arteriosus. Only a few hours after birth, the muscular wall of the ductus arteriosus constricts and, within a week, constriction is sufficient to stop all blood flow. During the second month, the ductus arteriosus becomes occluded by the formation of fibrous tissue in its lumen. In some babies, the ductus arteriosus remains open. As a result, aortic blood flows into the lower-pressure pulmonary trunk, thus increasing the pulmonary trunk blood pressure and overworking both ventricles and the heart. Treatment consists of a surgical procedure in which the ductus arteriosus is ligated or cut so that the cut ends are tied off.

A **septal defect** is an opening in the septum that separates the interior of the heart into a left and right side. **Interatrial septal defect** is failure of the fetal foramen ovale between the two atria to close after birth. Because pressure in the right atrium is low, interatrial septal defect generally allows a good deal of blood to flow from the left atrium to the right without going through systemic circulation. This defect, an example of left-to-right shunt, overloads the pulmonary circulation, produces fatigue, and increases respiratory infections. If it occurs early in life, it inhibits growth because the systemic circulation may be deprived of a considerable portion of the blood destined for the organs and tissues of the body. **Interventricular septal defect** is caused by an incomplete closure of the interventricular septum. Due to the higher pressure in the left ventricle, blood is shunted directly from the left ventricle into the right ventricle. This is another example of a left-to-right shunt. Septal openings can now be sewn shut or covered with synthetic patches.

Valvular stenosis is a narrowing, or *stenosis,* of one of the valves regulating blood flow in the heart. Stenosis may occur in the valve itself, most commonly in the mitral valve from rheumatic heart disease or in the aortic valve from sclerosis or rheumatic fever. Or it may occur near a valve. All stenoses are serious because they place a severe work load on the heart by making it work harder to push the blood through the abnor-mally narrow valve openings. As a result of mitral stenosis, blood pressure is increased. Angina pectoris and heart failure may accompany this disorder. Most stenosed valves are totally replaced with artificial valves.

Tetralogy of Fallot (tet-RAL-ō-jē of fal-Ō) is a combination of four defects: an interventricular septal defect, an aorta that emerges from both ventricles instead of from the left ventricle only, a stenosed pulmonary semilunar valve, and an enlarged right ventricle. The condition is an example of a right-to-left shunt, in which blood is shunted from the right ventricle to the left ventricle without going through pulmonary circulation. Because there is stenosis of the pulmonary semilunar valve, the increased right ventricular pressure forces deoxygenated blood from the right ventricle to enter the left ventricle through the interventricular septum. As a result, deoxygenated blood gets mixed with the oxygenated blood that is pumped into systemic circulation. Also, because the aorta emerges from the right ventricle and the pulmonary artery is stenosed, very little blood gets to the lungs, and pulmonary circulation is bypassed almost completely. The insufficient amount of oxygenated blood in systemic circulation results in *cyanosis* (sī-a-NŌ-sis), a blue or dark purple discoloration of skin. For this reason tetralogy of Fallot is one of the conditions that causes a "blue baby." Lung disorders and suffocation also result in cyanosis.

It is possible to correct cases of tetralogy of Fallot when the patient is of proper age and condition. Open-heart operations are performed in which the narrowed pulmonary valve is cut open and the interventricular septal defect is sealed with a Dacron patch.

Arrhythmias

Arrhythmia (a-RITH-mē-a) is a general term that refers to an abnormality or irregularity in the heart rhythm. More physicians are using the term **dysrhythmia** since it implies an abnormal rhythm, whereas arrhythmia implies no rhythm. An arrhythmia results when there is a disturbance in the conduction system of the heart, either due to faulty production of electrical impulses or faulty conduction of impulses as they pass through the system.

There are many different types of arrhythmias that can occur, some normal and some quite serious. Arrhythmias may be caused by factors such as caffeine, nicotine, alcohol, anxiety, certain drugs, hyperthyroidism, potassium deficiency, and certain heart diseases. Serious arrhythmias can result in cardiac arrest if the heart cannot supply its own oxygen demands, as well as those of the rest of the body. They can be controlled, and the normal heart rhythm can be reestablished, if they are detected and treated early enough.

Heart Block

One serious arrhythmia is called a **heart block.** Perhaps the most common blockage is in the atrioventricular (AV) node, which conducts impulses from the atria to the ventricles. This disturbance is called *atrioventricular (AV) block.* It usually indicates a myocardial infarction, atherosclerosis, rheumatic heart disease, diphtheria, or syphilis. With complete AV block, patients may have vertigo, unconsciousness, or convulsions. These symptoms result from a decreased cardiac output with diminished cerebral blood flow and cerebral hypoxia or lack of sufficient oxygen.

Among the causes of AV block are excessive stimulation by the vagus (X) nerves (which depresses conductivity of the junctional fibers), destruction of the AV bundle as a result of coronary infarct, atherosclerosis, myocarditis, or depression caused by various drugs.

Flutter and Fibrillation

Two other abnormal rhythms are **flutter** and **fibrillation.** In

atrial flutter the artial rhythm averages between 240 and 360 beats per minute. The condition is essentially rapid atrial contractions accompanied by a second-degree AV block. It generally indicates severe damage to heart muscle. Atrial flutter usually becomes fibrillation after a few hours, days, or weeks. **Atrial fibrillation** is asynchronous contraction of the atrial muscles that causes the atria to contract irregularly and still faster. Atrial flutter and fibrillation occur in myocardial infarction, acute and chronic rheumatic heart disease, and hyperthyroidism. Atrial fibrillation results in complete uncoordination of atrial contraction so that atrial pumping ceases altogether. When the muscle fibrillates, the muscle fibers of the atrium quiver individually instead of contracting together. The quivering cancels out the pumping of the atrium. In a strong heart, atrial fibrillation reduces the pumping effectiveness of the heart by only 25 to 30 percent.

Ventricular fibrillation is another abnormality that almost always indicates imminent cardiac arrest and death unless corrected quickly. It is characterized by asynchronous, haphazard, ventricular muscle contractions. The rate may be rapid or slow. The impulse travels to the different parts of the ventricles at different rates. Thus, part of the ventricle may be contracting while other parts are still unstimulated. Ventricular contraction becomes ineffective and circulatory failure and death occur. Ventricular fibrillation may be caused by coronary occlusion. It sometimes occurs during surgical procedures on the heart or pericardium. It may be the cause of death in electrocution.

Ventricular Premature Contraction (VPC)

Another form of arrhythmia arises when a small region of the heart outside the pacemaker becomes more excitable than normal, causing an occasional abnormal impulse to be generated between normal impulses. The region from which the abnormal impulse is generated is called an *ectopic focus*. As a wave of depolarization spreads outward from the ectopic focus, it causes a **ventricular premature contraction (VPC)**, also called a **premature ventricular contraction (PVC)**. The contraction occurs early in diastole before the SA node is normally scheduled to discharge its impulses. A person with such a ventricular premature contraction might feel a thump in the chest. The contrac-

tions may be relatively benign and may be caused by emotional stress, excessive intake of stimulants such as caffeine or nicotine, and lack of sleep. In other cases, the contractions may indicate an underlying pathology.

Congestive Heart Failure (CHF)

Congestive heart failure (CHF) may be defined as a chronic or acute state that results when the heart is not capable of supplying the oxygen demands of the body. Symptoms and signs of CHF include fatigue, peripheral and pulmonary edema, and visceral congestion. These symptoms are produced by diminished blood flow to the various tissues of the body and by accumulation of excess blood in the various organs because the heart is unable to pump out the blood returned to it by the great veins. Both diminished blood flow and accumulation of excess blood in the organs occur together, but certain symptoms result from congestion, while others are produced by poor tissue nutrition. CHF is usually caused by coronary artery disease in which atherosclerosis leads to myocardial ischemia.

Cor Pulmonale (CP)

Cor pulmonale (kor pul-mōn-ALE; *cor* = heart; *pulmon* = lung), or **CP**, refers to right ventricular hypertrophy from disorders that bring about hypertension (high blood pressure) in the pulmonary circulation. Certain disorders such as those involving the respiratory center in the brain, lung diseases, diseases involving airways in the lungs, deformities of the thoracic cage, and neuromuscular disorders can result in hypoxia in the lungs. As a result, vasoconstriction of pulmonary arteries and arterioles occurs and blood flow to the lungs is decreased. Also, any conditions such as tumors, aneurysms, and diseased blood vessels in the lungs can result in decreased blood flow. The reduction in blood flow and increased resistance in the lungs brings about pulmonary hypertension. Ultimately, this causes increased pressure in the pulmonary arteries and pulmonary trunk, resulting in right ventricular hypertrophy.

Individuals with CP may exhibit fatigue, weakness, dyspnea, right upper quadrant (RUQ) pain due to an enlarged liver, warm and cyanotic extremities, cyanosis of the face, distended neck veins, chest pain, and edema.

MEDICAL TERMINOLOGY

Aortic insufficiency An improper closure of the aortic semilunar valve that permits a backflow of blood.

Cardiac arrest Complete stoppage of the heartbeat.

Cardiomegaly (*mega* = large) Heart enlargement. Long-distance runners and weight lifters frequently develop heart enlargement as a natural adaptation to increased work load produced by regular exercise.

Commissurotomy An operation that is performed to widen the opening in a heart valve that has become narrowed by scar tissue.

Compensation A change in the circulatory system made to compensate for some abnormality; an adjustment of size of heart or rate of heartbeat made to counterbalance a defect in structure or function; often used specifically to describe the maintenance of adequate circulation in spite of the presence of heart disease.

Compliance The passive or diastolic stiffness properties of the left ventricle. A hypertrophied or fibrosed heart with a stiff wall, for example, has decreased compliance. Also,

the stiffness properties of the lung.

Constrictive pericarditis A shrinking and thickening of the pericardium that prevents the heart muscle from expanding and contracting normally.

Incompetent valve Any valve that does not close properly, thus permitting a backflow of blood; also called **valvular insufficiency**.

Palpitation A fluttering of the heart or abnormal rate or rhythm of the heart.

Pancarditis (*pan* = all) Inflammation of the whole heart, including the inner layer (endocardium), heart muscle (myocardium), and outer sac (pericardium).

Paroxysmal tachycardia A period of rapid heartbeats that begins and ends suddenly.

Percussion Tapping a part of the body as an aid in diagnosing the condition of parts of the body by the sound obtained.

Stokes-Adams syndrome Sudden attacks of unconsciousness, sometimes with convulsions, which may accompany heart block.

STUDY OUTLINE

Location (p. 461)
1. The heart is situated obliquely between the lungs in the mediastinum.
2. About two-thirds of its mass is to the left of the midline.

Pericardium (p. 462)
1. The pericardium consists of an outer fibrous layer and an inner serous pericardium.
2. The serous pericardium is composed of a parietal and visceral layer.
3. Between the parietal and visceral layers of the serous pericardium is the pericardial cavity, a space filled with pericardial fluid that prevents friction between the two membranes.

Wall; Chambers; Vessels; and Valves (p. 462)
1. The wall of the heart has three layers: epicardium, myocardium, and endocardium.
2. The chambers include two upper atria and two lower ventricles.
3. The blood flows through the heart from the superior and inferior venae cavae and the coronary sinus to the right atrium, through the tricuspid valve to the right ventricle, through the pulmonary trunk to the lungs, through the pulmonary veins into the left atrium, through the bicuspid valve to the left ventricle, and out through the aorta.
4. Valves prevent backflow of blood in the heart.
5. Atrioventricular (AV) valves, between the atria and their ventricles, are the tricuspid valve on the right side of the heart and the bicuspid (mitral) valve on the left.
6. The chordae tendineae and their muscles keep the flaps of the valves pointing in the direction of blood flow.
7. The two arteries that leave the heart both have a semilunar valve.
8. The heart valves may be identified by surface projection.

Blood Supply (p. 468)
1. The coronary (cardiac) circulation takes oxygenated blood through the arterial system of the myocardium.
2. Deoxygenated blood returns to the right atrium via the coronary sinus.
3. Complications of this system are angina pectoris and myocardial infarction.

Conduction System (p. 469)
1. The conduction system consists of tissue specialized for impulse conduction.
2. Components of this system are the sinoatrial (SA) node (pacemaker), atrioventricular (AV) node, atrioventricular (AV) bundle (bundle of His), bundle branches, and conduction myofibers (Purkinje fibers).

Electrocardiogram (p. 470)
1. A cardiac cycle consists of the systole (contraction) and diastole (relaxation) of both atria plus the systole and diastole of both ventricles followed by a short pause.
2. The record of electrical changes during each cardiac cycle is referred to as an electrocardiogram (ECG).
3. A normal ECG consists of a P wave (spread of impulse from SA node over atria), QRS wave (spread of impulse through ventricles), and T wave (ventricular repolarization).
4. The P-R interval represents the conduction time from the beginning of atrial excitation to the beginning of ventricular excitation. The S-T segment represents the time between the end of the spread of the impulse through the ventricles and repolarization of the ventricles.
5. The ECG is invaluable in diagnosing abnormal cardiac rhythms and conduction patterns, detecting the presence of fetal life, determining the presence of several fetuses, and following the course of recovery from a heart attack.
6. An artificial pacemaker may be used to restore an abnormal cardiac rhythm.

Blood Flow Through the Heart (p. 472)
1. Blood flows through the heart from an area of higher to lower pressure.
2. The pressure developed is related to the size and volume of a chamber.
3. The movement of blood through the heart is controlled by the opening and closing of the valves and the contraction and relaxation of the myocardium.

Cardiac Cycle (p. 472)
1. A cardiac cycle consists of the systole (contraction) and diastole (relaxation) of both atria, plus the systole and diastole of both ventricles followed by a short pause.
2. The phases of the cardiac cycle are: (1) atrial systole, (2) ventricular filling, (3) ventricular systole, and (4) ventricular diastole.
3. With an average heartbeat of 75/min, a complete cardiac cycle requires 0.8 sec.
4. The first heart sound (lubb) represents the closing of the atrioventricular valves. The second sound (dupp) represents the closing of semilunar valves.
5. A peculiar sound is called a murmur.

Cardiac Output (CO) (p. 474)
1. Cardiac output (CO) is the amount of blood ejected by the left ventricle into the aorta per minute. It is calculated as follows: CO = stroke volume × beats per minute.
2. Stroke volume (SV) is the amount of blood ejected by a ventricle during each systole.
3. Stroke volume (SV) depends on how much blood enters a ventricle during diastole (end-diastolic volume) and how much blood is left in a ventricle following its systole (end-systolic volume).
4. According to Starling's law of the heart, the greater the length of stretched cardiac fibers, within limits, the stronger the contraction.
5. The maximum percentage that cardiac output can be increased above normal is cardiac reserve.
6. Heart rate and strength of contraction may be increased by sympathetic stimulation from the cardioacceleratory center (CAC) in the medulla and decreased by parasympathetic stimulation from the cardioinhibitory center (CIC) in the medulla.
7. Baroreceptors (pressure receptors) are nerve cells that respond to changes in blood pressure. They act on the cardiac centers in the medulla through three reflex pathways: carotid sinus reflex, aortic reflex, and right heart (atrial) reflex.
8. Other influences on heart rate include chemicals (epinephrine, sodium, potassium), temperature, emotion, sex, and age.

Circulatory Shock and Homeostasis (p. 477)
1. Shock results when cardiac output is reduced or blood volume decreases to the point where body tissues become hypoxic.

2. Mild shock is compensated by vasoconstriction and water retention.
3. In severe shock, venous return is diminished and cardiac output decreases. The heart becomes hypoxic, prolonged vasoconstriction leads to hypoxia of other organs, and the shock cycle is intensified.

Artificial Heart (p. 478)

1. The artificial heart consists of an aluminum base and rigid plastic chambers that serve as ventricles.
2. When it is implanted, the recipient's atria are left intact.

Diagnosing Heart Disorders (p. 478)

1. Diagnosis of heart disorders may be invasive or noninvasive.
2. An invasive technique is cardiac catheterization, which is used to measure pressures and oxygen, ventricular function, cardiac output (CO); locate lesions; inject clot-dissolving enzymes; and dilate coronary blood vessels.
3. A new noninvasive technique is echocardiography, in which sound waves are converted to images that evaluate congenital heart disorders, tumors, and valvular disorders; detect fluid around the heart; and assess the effects of coronary artery disease.

Heart-Lung Machine (p. 479)

1. Hypothermia (deliberate body cooling) and the heart–lung bypass permit open-heart surgery.
2. The heart-lung machine pumps blood (functioning as a heart) and oxygenates blood and removes carbon dioxide (functioning as lungs).

Risk Factors in Heart Disease (p 479)

1. Risk factors in heart disease include high blood cholesterol, high blood pressure, cigarette smoking, obesity, lack of regular exercise, diabetes mellitus, genetic disposition, sex, and age.

Developmental Anatomy of the Heart (p. 480)

1. The heart develops from mesoderm.
2. The endothelial tubes develop into the four-chambered heart and great vessels of the heart.

Disorders: Homeostatic Imbalances (p. 481)

1. Coronary artery disease (CAD) refers to a condition in which the myocardium receives inadequate blood due to atherosclerosis, coronary artery spasm, or thrombi or emboli; it may be treated with drugs or surgically.
2. Congenital heart defects include coarctation of the aorta, patent ductus arteriosus, septal defects, valvular stenosis, and tetralogy of Fallot.
3. Arrhythmias result in heart block, flutter, and fibrillation.
4. Congestive heart failure (CHF) results when the heart cannot supply the oxygen demands of the body.
5. Cor pulmonale (CP) refers to right ventricular hypertrophy secondary to pulmonary hypertension.

REVIEW QUESTIONS

1. Describe the location of the heart in the mediastinum and identify the various borders and surfaces of the heart.
2. Distinguish the subdivisions of the pericardium. What is the purpose of this structure?
3. Compare the three layers of the heart wall according to composition, location, and function.
4. Define atria and ventricles. What vessels enter or exit the atria and ventricles?
5. Describe the principal valves in the heart and how they operate.
6. Describe how the valves of the heart may be identified by surface projection.
7. Describe the route of blood in coronary (cardiac) circulation. Distinguish between angina pectoris and myocardial infarction. How is angina treated?
8. Describe the structure and function of the heart's conducting system.
9. Define and label the deflection waves of a normal electrocardiogram (ECG). Explain why the ECG is an important diagnostic tool.
10. What is a Holter monitor? Why is it used?
11. Describe the operation of an artificial pacemaker.
12. Discuss the pressure changes associated with the flow of blood through the heart.
13. Define cardiac cycle.
14. List the principal events of atrial systole, ventricular filling, ventricular systole, and ventricular diastole.
15. By means of a labeled diagram, relate the events of the cardiac cycle to time. What is the quiescent period?
16. Describe the significance of the heart sounds. What is a heart murmur?
17. What is cardiac output (CO)? How is it calculated?
18. Define stroke volume (SV). Explain how end-diastolic volume (EDV) and end-systolic volume (ESV) are related to stroke volume.
19. What is Starling's law of the heart? What is its significance?
20. Define cardiac reserve. Why is it important?
21. Distinguish between the cardioaccelatory center (CAC) and cardioinhibitory center (CIC) with respect to the regulation of heart rate.
22. What is a baroreceptor (pressure receptor)? Outline the operation of the carotid sinus reflex, aortic reflex, and right heart (atrial) reflex.
23. Explain how each of the following affects heart rate: chemicals, temperature, emotions, sex, and age.
24. Define circulatory shock. What are its symptoms? How is homeostasis restored after mild shock?
25. Describe the effects of severe circulatory shock by drawing a shock cycle.
26. Explain how an artificial heart is implanted and how it operates.
27. How are heart disorders diagnosed? Explain the principle and clinical applications of cardiac catheterization and echocardiography.
28. Describe the operation of the heart-lung machine.
29. Describe the risk factors involved in heart disease.
30. Describe how the heart develops.
31. What is coronary artery disease (CAD)? What factors contribute to it? Explain.
32. Define each of the following congenital heart defects: coarctation of the aorta, patent ductus arteriosus, septal defects, valvular stenosis, and tetralogy of Fallot.
33. Define an arrhythmia. Give several examples.
34. Describe the symptoms of congestive heart failure (CHF) and cor pulmonale (CP).
35. Refer to the glossary of medical terminology associated with the heart. Be sure that you can define each term.

The Cardiovascular System: Vessels and Routes

21

STUDENT OBJECTIVES

- Contrast the structure and function of arteries, arterioles, capillaries, venules, and veins.
- Define a blood reservoir and explain its importance.
- Relate the importance of cardiac output (CO), blood volume, and peripheral resistance to blood pressure.
- Explain the role of the vasomotor center in controlling blood pressure.
- Contrast the roles of baroreceptors and chemoreceptors in regulating blood pressure.
- Describe the effects of epinephrine, norepinephrine (NE), antidiuretic hormone (ADH), angiotensin II, histamine, and kinins on blood pressure.
- Define autoregulation and explain its importance.
- Discuss the various pressures involved in the movement of fluids between capillaries and interstitial spaces.
- Explain how skeletal muscle contractions, valves in veins, and breathing assist in the return of venous blood to the heart.
- Define pulse and identify the arteries where pulse may be felt.
- Define blood pressure (BP) and explain one clinical method for recording systolic and diastolic pressure.
- Contrast the clinical significance of systolic, diastolic, and pulse pressures.
- Identify the principal arteries and veins of systemic circulation.
- Identify the major blood vessels of pulmonary circulation.
- Trace the route of blood involved in hepatic portal circulation and explain its importance.
- Contrast fetal and adult circulation.
- Explain the fate of fetal circulation structures once postnatal circulation is established.
- Explain the effects of exercise on the cardiovascular system.
- Describe the effects of aging on the cardiovascular system.
- Describe the development of blood vessels and blood.
- List the causes and symptoms of aneurysms, coronary artery disease (CAD), hypertension, and deep-venous thrombosis (DVT).
- Define medical terminology associated with the cardiovascular system.

Blood vessels form a network of tubes that carry blood away from the heart, transport it to the tissues of the body, and then return it to the heart. **Arteries** are the vessels that carry blood from the heart to the tissues. Large, elastic arteries leave the heart and divide into medium-sized, muscular vessels that branch out into the various regions of the body. These medium-sized arteries then divide into small arteries, which, in turn, divide into still smaller arteries called **arterioles.** As the arterioles enter a tissue, they branch into countless microscopic vessels called **capillaries.** Through the walls of the capillaries, substances are exchanged between the blood and body tissues. Before leaving the tissue, groups of capillaries reunite to form small veins called **venules.** These, in turn, merge to form progressively larger tubes called veins. **Veins,** then, are blood vessels that convey blood from the tissues back to the heart. Since blood vessels require oxygen and nutrients just like other tissues of the body, they also have blood vessels in their own walls called **vasa vasorum.**

The developmental anatomy of blood vessels and blood is considered later in the chapter.

ARTERIES

Arteries have walls constructed of three coats or tunics and a hollow core, called a *lumen,* through which the blood flows (Figures 21-1 and 21-4). The inner coat of an arterial wall, the **tunica interna (intima),** is composed of a lining of endothelium (simple squamous epithelium) that is in contact with the blood and a layer of elastic tissue called the internal elastic membrane. The middle coat, or **tunica media,** is usually the thickest layer. It consists of elastic fibers and smooth muscle. The outer coat, the **tunica externa (adventitia),** is composed principally of elastic and collagenous fibers. An external elastic membrane may separate the tunica media from the tunica externa.

As a result of the structure of the middle coat especially, arteries have two major properties: elasticity and contractility. When the ventricles of the heart contract and eject blood into the large arteries, the arteries expand to accommodate the extra blood. Then, as the ventricles relax, the elastic recoil of the arteries forces the blood onward. The contractility of an artery comes from its

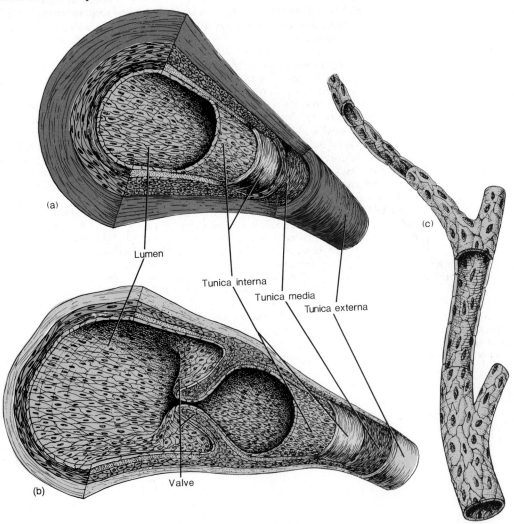

Lumen

Tunica interna

Tunica media

Tunica externa

Valve

(a)

(b)

(c)

FIGURE 21-1 Comparative structure of (a) an artery, (b) a vein, and (c) a capillary. The relative size of the capillary is enlarged for emphasis.

smooth muscle, which is arranged longitudinally and in rings around the lumen somewhat like a doughnut and is innervated by sympathetic branches of the autonomic nervous system. When there is sympathetic stimulation, the smooth muscle contracts, squeezes the wall around the lumen, and narrows the vessel. Such a decrease in the size of the lumen is called **vasoconstriction.** Conversely, when sympathetic stimulation is removed, the smooth muscle fibers relax and the size of the arterial lumen increases. This increase is called **vasodilation** and is usually due to the inhibition of vasoconstriction.

The contractility of arteries also serves a function in stopping bleeding. This is called vascular spasm, one of the three mechanisms involved in hemostasis (Chapter 19). The blood flowing through an artery is under a great deal of pressure. Thus great quantities of blood can be quickly lost from a broken artery. When an artery is cut, its wall constricts so that blood does not escape quite so rapidly. However, there is a limit to how much vasoconstriction can help.

ELASTIC (CONDUCTING) ARTERIES

Large arteries are referred to as **elastic (conducting) arteries.** They include the aorta and brachiocephalic, common carotid, subclavian, vertebral, and common iliac arteries. The wall of elastic arteries is relatively thin in proportion to the diameter of the arteries, and their tunica media contains more elastic fibers and less smooth muscle. As the heart alternately contracts and relaxes, the rate of blood flow tends to be intermittent. When the heart contracts and forces blood into the aorta, the wall of the elastic arteries stretches to accommodate the surge of blood and store the pressure energy. During relaxation of the heart, the wall of the elastic arteries recoils, moving the blood forward in a more continuous flow. Elastic arteries are called conducting arteries because they conduct blood from the heart to the distribution system of arteries.

MUSCULAR (DISTRIBUTING) ARTERIES

Medium-sized arteries are called **muscular (distributing) arteries.** They include the axillary, brachial, radial, intercostal, splenic, mesenteric, femoral, popliteal, and tibial arteries. Their tunica media contains more smooth muscle than elastic fibers, and they are capable of greater vasoconstriction and vasodilation to adjust the volume of blood to suit the needs of the structure supplied. The wall of muscular arteries is relatively thick, mainly due to the large amounts of smooth muscle. Muscular arteries are called distributing arteries because they distribute blood to various parts of the body.

ANASTOMOSES

Most parts of the body receive branches from more than one artery. In such areas the distal ends of the vessels unite. The junction of two or more vessels supplying the same body region is called an **anastomosis** (a-nas-tō-MŌ-sis). Anastomoses may also occur between the origins of veins and between arterioles and venules. Anastomoses between arteries provide alternate routes by which blood can reach a tissue or organ. Thus if a vessel is occluded by disease, injury, or surgery, circulation to a part of the body is not necessarily stopped. The alternate route of blood to a body part through an anastomosis is known as **collateral circulation.** An alternate blood route may also be from nonanastomosing vessels that supply the same region of the body.

Arteries that do not anastomose are known as **end arteries.** Occlusion of an end artery interrupts the blood supply to a whole segment of an organ, producing necrosis (death) of that segment.

CLINICAL APPLICATION

A procedure that provides physicians with a much clearer look at diseased arteries is called **digital subtraction angiography (DSA).** The procedure employs a computer technique that compares an x-ray image of the same region of the body before and after a contrast substance has been introduced intravenously. Any tissue or blood vessels that show up in the first image can be subtracted (erased) from the second image, leaving an unobstructed view of an artery. DSA figuratively lifts an artery out of the body so that it can be studied in isolation. DSA helps in diagnosing lesions in the carotid arteries leading to the brain, a potential cause of strokes, and in evaluating patients before surgery and after coronary bypass and certain transplant operations.

ARTERIOLES

An **arteriole** is a small artery that delivers blood to capillaries. Arterioles closer to the arteries from which they branch have a tunica interna like that of arteries, a tunica media composed of smooth muscle and very few elastic fibers, and a tunica externa composed mostly of elastic and collagenous fibers. As arterioles get smaller in size, the tunics change character so that arterioles closest to capillaries consist of little more than a layer of endothelium surrounded by a few scattered smooth muscle cells (Figure 21-2).

Arterioles play a key role in regulating blood flow from arteries into capillaries. The smooth muscle of arte-

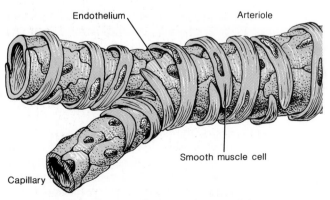

FIGURE 21-2 Structure of an arteriole.

rioles, like that of arteries, is subject to vasoconstriction and vasodilation. During vasoconstriction, blood flow into capillaries is restricted; during vasodilation, the flow is significantly increased. The relationship of arterioles to blood flow will be considered in detail later in the chapter.

CAPILLARIES

Capillaries are microscopic vessels that usually connect arterioles and venules (see Figure 21-1). They are found near almost every cell in the body. The distribution of capillaries in the body varies with the activity of the tissue. For example, in places where activity is higher, such as muscles, liver, kidneys, lungs, and nervous system, there are rich capillary supplies. In areas where activity is lower, such as tendons and ligaments, the capillary supply is not as extensive. The epidermis, cornea of the eye, and cartilage are devoid of capillaries.

The primary function of capillaries is to permit the exchange of nutrients and wastes between the blood and tissue cells. The structure of the capillaries is admirably suited to this purpose. Capillary walls are composed of only a single layer of cells (endothelium) and a basement membrane (see Figure 21-4b). They have no tunica media or tunica externa. Thus a substance in the blood must pass through the plasma membrane of just one cell to reach tissue cells. This vital exchange of materials occurs only through capillary walls—the thick walls of arteries and veins present too great a barrier.

Although capillaries pass directly from arterioles to venules in some places in the body, in other places they form extensive branching networks. These networks increase the surface area for diffusion and thereby allow a rapid exchange of large quantities of materials. In most tissues, blood normally flows through only a small portion of the capillary network when metabolic needs are low. But, when a tissue becomes active, the entire capillary network fills with blood.

The flow of blood through capillaries is regulated by vessels with smooth muscle in their walls. A **metarteriole** (*met* = beyond) is a vessel that emerges from an arteriole, traverses the capillary network, and empties into a venule (Figure 21-3). The proximal portions of metarterioles are surrounded by scattered smooth muscle cells whose contraction and relaxation help to regulate the amount and force of blood. The distal portion of a metarteriole has no smooth muscle cells and is called a *thoroughfare channel*. It serves as a low-resistance channel that increases blood flow. **True capillaries** emerge from arterioles or metarterioles and are not on the direct flow route from arteriole to venule. At their sites of origin, there is a ring of smooth muscle called a **precapillary sphincter** that controls the flow of blood entering a true capillary. The various factors that regulate the contraction of smooth muscle cells and precapillary sphincters are discussed later.

Some capillaries of the body, such as those found in muscle tissue and other locations, are referred to as **continuous capillaries,** so named because the cytoplasm of the endothelial cells is continuous when viewed in cross section through a microscope; the cytoplasm appears as

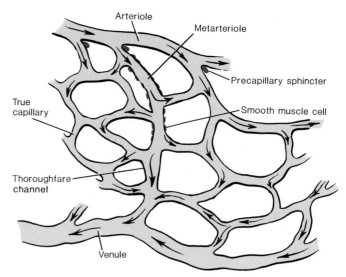

FIGURE 21-3 Details of a capillary network.

an uninterrupted ring, except for the endothelial junction. Other capillaries of the body are referred to as **fenestrated capillaries.** They differ from continuous capillaries in that their endothelial cells have numerous fenestrae (pores) where the cytoplasm is very thin or absent. The fenestrae range from 70 to 100 nm in diameter and are closed by a thin diaphragm, except in the capillaries in the kidneys where they are assumed to be open. Fenestrated capillaries are also found in the villi of the small intestine, choroid plexuses of the ventricles in the brain, ciliary processes of the eyes, and endocrine glands.

Microscopic blood vessels in certain parts of the body, such as the liver, are termed **sinusoids.** They are wider than capillaries and more tortuous. Also, instead of the usual endothelial lining, sinusoids are lined largely by phagocytic cells. In the liver, such cells are called *stellate reticuloendothelial (Kupffer's) cells.* Like capillaries, sinusoids convey blood from arterioles to venules. Other regions containing sinusoids include the spleen, adenohypophysis, parathyroid glands, and adrenal cortex.

VENULES

When several capillaries unite, they form small veins called **venules.** Venules collect blood from capillaries and drain it into veins. The venules closest to the capillaries consist of a tunica interna of endothelium and a tunica externa of connective tissue. As the venules approach the veins, they also contain the tunica media characteristic of veins.

VEINS

Veins are composed of essentially the same three coats as arteries, but they have considerably less elastic tissue and smooth muscle and contain more white fibrous tissue (Figures 21-1 and 21-4). However, they are still distensible enough to adapt to variations in the volume and pressure of blood passing through them.

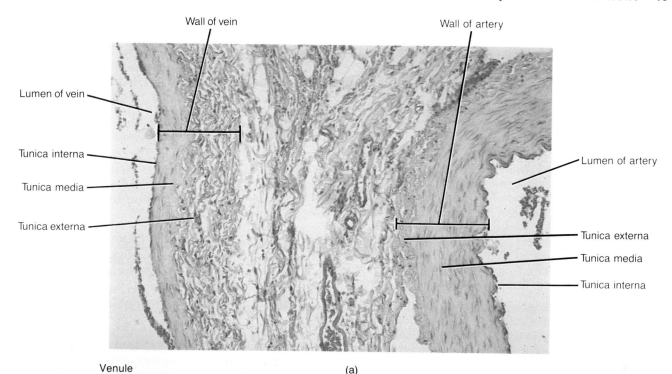

Wall of vein

Lumen of vein

Tunica interna

Tunica media

Tunica externa

Wall of artery

Lumen of artery

Tunica externa

Tunica media

Tunica interna

(a)

Venule

Capillaries

(b)

FIGURE 21-4 Histology of blood vessels. (a) Photomicrograph of a portion of the wall of an artery and its accompanying vein at a magnification of 250×. (Copyright © 1983 by Michael H. Ross. Used by permission.) (b) Photomicrograph of several capillaries uniting to form a venule. Note the red blood cells in single file in the capillary. (Courtesy of Michael H. Ross and Edward J. Reith, *Histology: A Text and Atlas,* Copyright © 1985, by Michael H. Ross and Edward J. Reith, Harper & Row, Publishers, Inc., New York.)

flow rather than in the rapid spurts characteristic of arteries. Most of the structural differences between arteries and veins reflect this pressure difference. For example, veins do not need walls as strong as those of arteries. The low pressure in veins, however, has its disadvantages. When you stand, the pressure pushing blood up the veins in your lower extremities is barely enough to balance the force of gravity pushing it back down. For this reason, many veins, especially those in the limbs, contain valves that prevent backflow (see Figure 21-9).

CLINICAL APPLICATION

In people with weak venous valves, gravity forces large quantities of blood back down into distal parts of the vein. This pressure overloads the vein and pushes its wall outward. After repeated overloading, the walls lose their elasticity and become stretched and flabby. Such dilated and tortuous veins caused by incompetent valves are called **varicose veins.** They may be due to heredity, mechanical factors (prolonged standing and pregnancy), or aging. Because a varicosed wall is not able to exert a firm resistance against the blood, blood tends to accumulate in the pouched-out area of the vein, causing it to swell and forcing fluid into the surrounding tissue. Veins close to the surface of the legs

By the time the blood leaves the capillaries and moves into the veins, it has lost a great deal of pressure. The difference in pressure can be observed in the blood flow from a cut vessel; blood leaves a cut vein in an even are highly susceptible to varicosities. Veins that lie deeper are not as vulnerable because surrounding skeletal muscles prevent their walls from overstretching.

Varicose veins may be treated by several methods, depending on the severity of the condition. These include: (1) frequent periods of rest with elevation of the lower extremities; (2) external pressure with elastic stockings or bandages; (3) a procedure known as sclerotherapy, an intravenous injection of sclerosing chemicals that collapses the veins and prevents blood from flowing into them, thus eliminating the purple-blue discoloration; and (4) surgery, in which the saphenous veins with their incompetent valves are ligated and removed ("stripping").

A mild form of varicose veins in the legs is referred to as **spider burst veins,** a condition that is most commonly treated by sclerotherapy.

A **vascular (venous) sinus** is a vein with a thin endothelial wall that has no smooth muscle to alter its diameter. Surrounding tissue replaces the tunica media and tunica externa to provide support. Intracranial vascular sinuses, which are supported by the dura mater, return cerebrospinal fluid and deoxygenated blood from the brain to the heart. Another example of a vascular sinus is the coronary sinus of the heart.

BLOOD RESERVOIRS

The volume of blood in various parts of the cardiovascular system varies considerably. Veins, venules, and venous sinuses contain about 59 percent of the blood in the system, arteries about 13 percent, pulmonary vessels about 12 percent, the heart about 9 percent, and arterioles and capillaries about 7 percent. Since systemic veins contain so much of the blood, they are referred to as **blood reservoirs.** They serve as storage depots for blood, which can be moved quickly to other parts of the body if the need arises. When there is increased muscular activity, the vasomotor center sends increasing sympathetic impulses to veins that serve as blood reservoirs. The result is vasoconstriction, which permits the distribution of blood from venous reservoirs to skeletal muscles, where it is needed most. A similar mechanism operates in cases of hemorrhage, when blood volume and pressure decrease. Vasoconstriction of veins in venous reservoirs helps to compensate for the blood loss. Among the principal blood reservoirs are the veins of the abdominal organs (especially the liver and spleen) and the veins of the skin.

PHYSIOLOGY OF CIRCULATION

BLOOD FLOW AND BLOOD PRESSURE

Blood flows through its system of closed vessels because of different pressures in various parts of the system. It always flows from regions of higher pressure to regions of lower pressure. The mean (average) pressure in the aorta is about 100 mm Hg (mercury). This pressure continually decreases rapidly through the arterial system and more slowly through the venous system (Figure 21-5). Because of the continuous drop in pressure, blood flows from the aorta (100 mm Hg) to the arteries (100–40 mm Hg) to arterioles (40–25 mm Hg) to capillaries (25–12 mm Hg) to venules (12–8 mm Hg) to veins (10–5 mm Hg) to the venae cavae (2 mm Hg). The pressure in the right atrium is 0 mm Hg.

Other mechanisms aid the flow of blood also. When blood leaves the capillaries, it enters the venules and veins, which are larger in diameter and thereby offer less resistance to flow. Contraction of skeletal muscles around the veins also helps drive blood toward the heart.

Factors That Affect Arterial Blood Pressure

We have defined blood pressure as the pressure exerted by blood on the wall of any blood vessel. In clinical use, however, the term refers to the pressure in arteries. Here we will identify several factors that influence arterial blood pressure.

• *Cardiac Output* (CO) Cardiac output (CO), the amount of blood ejected by the left ventricle into the aorta each minute, is the principal determinant of blood pressure. As noted in Chapter 20, CO is calculated by multiplying stroke volume by heart rate. In a normal, resting adult, it is about 5.25 liters/min (70 ml × 75 beats/min). Blood pressure varies directly with CO. If CO is increased by any increase in stroke volume or heart rate, then blood pressure increases. A decrease in CO causes a decrease in blood pressure.

• *Blood Volume* Blood pressure is directly proportional to the volume of blood in the cardiovascular system. The normal volume of blood in a human body is about 5 liters (5 qt). Any decrease in this volume, as from hemorrhage, decreases the amount of blood that is circulated through the arteries each minute. As a result, blood pressure drops. Conversely, anything that increases blood volume, such as high salt intake and therefore water retention, increases blood pressure.

• *Peripheral Resistance* **Peripheral resistance** refers to the resistance (impedance) to blood flow by the force of friction between blood and the walls of blood vessels. It is related to the viscosity of blood and blood vessel diameter.

The viscosity of blood is a function of the ratio of red blood cells and solutes to fluid. Any condition that increases the viscosity of blood, such as dehydration or an unusually high number of red blood cells, increases blood pressure. A depletion of plasma proteins or red blood cells, as a result of anemia or hemorrhage, decreases blood viscosity and blood pressure.

The smaller the diameter of a vessel, the more resistance it offers the blood. A major function of arterioles is to control peripheral resistance and, therefore, blood pressure and flow, by changing their diameters. The center for this regulation is the vasomotor center in the medulla, which will be described shortly.

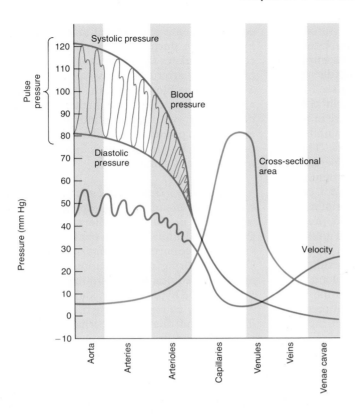

FIGURE 21-5 Relationship among blood pressure, velocity of blood flow, and total cross-sectional area in various blood vessels of the cardiovascular system. The velocity curve (green) and cross-sectional area curve (blue) are not related to the pressures shown in millimeters of mercury (mm Hg).

The factors that affect arterial blood pressure are summarized in Figure 21-6.

Control of Blood Pressure

In Chapter 20 we considered the increase and decrease of heart rate and force of contraction by the cardioacceleratory center (CAC) and the cardioinhibitory center (CIC). We also took a look at how certain chemicals (epinephrine, potassium, sodium, and calcium), temperature, emo-

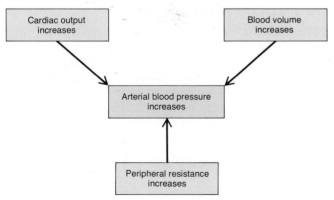

FIGURE 21-6 Summary of factors that affect arterial blood pressure.

tions, sex, and age affect heart rate. By their effect on heart rate and force of contraction, these factors also control blood pressure. Any increase in heart rate and force of contraction increases blood pressure. Conversely, any decrease will lower blood pressure.

At this point we will examine those factors that help to regulate blood pressure by acting on the blood vessels themselves.

• *Vasomotor Center* Within the medulla is a cluster of neurons referred to as the **vasomotor** (*vas* = vessel, *motor* = movement) **center.** The function of this center is to control the diameter of blood vessels, especially arterioles. It continually sends impulses to the smooth muscle in arteriole walls that result in a moderate state of vasoconstriction at all times. This state of tonic contraction, called *vasomotor tone,* assumes a role in maintaining peripheral resistance and blood pressure. The vasomotor center brings about vasoconstriction by increasing the number of sympathetic impulses above normal. It causes vasodilation by decreasing the number of sympathetic impulses below normal. In other words, in this case, the sympathetic division of the ANS can bring about both vasoconstriction and vasodilation.

The vasomotor center is affected by several factors, all of which influence blood pressure.

• *Baroreceptors* It will be recalled that stimulated baroreceptors (pressure receptors) in the wall of the carotid sinus and aorta or attached to them send impulses to the cardiac center that result in increased or decreased cardiac output to help regulate blood pressure (see Figure 20-9). This reflex acts not only on the heart but also on the arterioles. For example, if there is an increase in blood pressure, the baroreceptors stimulate the cardioinhibitory center and inhibit the cardioacceleratory center. The result is a decrease in cardiac output and a decrease in blood pressure. The baroreceptors also send impulses to the vasomotor center. In response, the vasomotor center decreases sympathetic stimulation to the arterioles. The result is vasodilation and a decrease in blood pressure (Figure 21-7). If there is a decrease in blood pressure, the baroreceptors inhibit the cardioinhibitory center and stimulate the cardioacceleratory center. The result is an increase in cardiac output and an increase in blood pressure. Also, the baroreceptors send impulses to the vasomotor center that increase sympathetic stimulation to arterioles. The result is vasoconstriction and an increase in blood pressure.

• *Chemoreceptors* Receptors sensitive to chemicals in the blood are called **chemoreceptors.** Chemoreceptors near or on the carotid sinus and aorta are called *carotid* and *aortic bodies,* respectively. They are sensitive to arterial blood levels of oxygen, carbon dioxide, and hydrogen ions. Given a deficiency of oxygen (hypoxia), an increase in hydrogen ion concentration, or an excess of carbon dioxide (hypercapnia), the chemoreceptors are stimulated and send impulses to the vasomotor center. In response, the vasomotor center increases sympathetic stimulation to arterioles. This brings about vasoconstriction and an increase in blood pressure. As you will see in Chapter 23, chemoreceptors also stimulate the medullary rhythmicity area in response to these chemical stimuli to adjust the rate of respiration.

• *Control by Higher Brain Centers* In response to strong emotions, **higher brain centers,** such as the cerebral cortex, can have a significant influence on blood pressure. For example, during periods of intense anger, the cerebral cortex stimulates the vasomotor center to fire sympathetic impulses to arterioles. This causes vasoconstriction and an increase in blood pressure. When a person is emotionally upset, impulses from higher brain centers cause a decrease in sympathetic stimulation by the vasomotor center. This produces vasodilation and a decrease in blood pressure. A frequent result is fainting because blood flow to the brain is diminished.

• *Chemicals* Several **chemicals** affect blood pressure by causing vasoconstriction. Epinephrine and norepinephrine (NE), produced by the adrenal medulla, increase the rate and force of heart contractions and bring about vasoconstriction of abdominal and cutaneous arterioles. They also bring about dilation of cardiac and skeletal muscle arterioles. Antidiuretic hormone (ADH), produced by the hypothalamus and released from the neurohypophysis, causes vasoconstriction if there is a severe loss of blood due to hemorrhage. Angiotensin II helps to raise blood pressure by stimulating secretion of aldosterone (increases sodium ion concentration and water reabsorption) and causing vasoconstriction due to release of renin. Histamine produced by mast cells and kinins found in plasma are vasodilators that assume key functions during the inflammatory response.

• *Autoregulation* (*Local Control*) **Autoregulation** refers to a local, automatic adjustment of blood flow in a given region of the body in response to the particular needs of the tissue. In most body tissues, oxygen is the principal, though not direct, stimulus for autoregulation. A suggested mechanism for autoregulation is as follows. In response to low oxygen supplies, the cells in the immediate area produce and release *vasodilator substances.* Such substances are thought to include potassium ions, hydrogen ions, carbon dioxide, lactic acid, and adenosine. Once released, the vasodilator substances produce a local dilation of arterioles and relaxation of precapillary sphincters. The result is an increased flow of blood into the tissue, which restores oxygen levels to normal. The autoregulation mechanism is important in meeting the nutritional demands of active tissues, such as muscle tissue, where the demand might increase as much as 10-fold.

Capillary Exchange

For reasons to be discussed shortly, you will see that, although blood pressure decreases consistently from the aorta to the venae cavae, the velocity of blood decreases as it flows through the aorta, arterioles, and capillaries, and then increases as it passes into venules and veins. The velocity of blood flow in capillaries is the slowest in the cardiovascular system, and this is important to allow for the exchange of materials between blood and body tissues.

Blood usually does not flow in a continuous manner through capillary networks. Rather, it flows intermittently because of contraction and relaxation of the smooth muscle cells of metarterioles and the precapillary sphincters of true capillaries. The intermittent contraction and

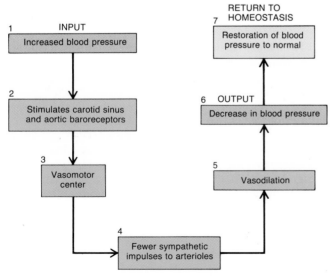

FIGURE 21-7 Vasomotor center control of blood pressure by stimulation of baroreceptors.

relaxation, which may occur 5 to 10 times per minute, is called **vasomotion.** The most important factor in controlling vasomotion is the oxygen concentration in tissues, a mechanism similar to autoregulation in arterioles.

Diffusion is the principal mechanism by which substances are exchanged between capillary blood and body cells. Most substances are exchanged by this process through endothelial junctions of continuous capillaries and fenestrae of fenestrated capillaries. Some larger molecules are exchanged by pinocytosis.

The movement of water and dissolved substances, except proteins, through capillary walls is dependent on several opposing forces, or pressures. Some forces push fluid out of capillaries into the surrounding interstitial (tissue) spaces, resulting in filtration of fluid. To prevent fluid from moving in one direction only and accumulating in interstitial spaces, opposing forces push fluid from interstitial spaces into blood capillaries, resulting in reabsorption of fluid. Let us now see how these forces operate.

First we will consider the hydrostatic pressures involved. These pressures are due to the weight of water in the fluids. Blood pressure in capillaries, called **blood hydrostatic pressure (BHP),** tends to move fluids out of capillaries into interstitial fluid. BHP averages about 30 mm Hg at the arterial end of a capillary and about 15 mm Hg at the venous end, although the pressures may vary considerably, depending on the activity of the arterioles and venules (Figure 21-8). Opposing BHP is the pressure of the interstitial fluid, called **interstitial fluid hydrostatic pressure (IFHP),** which tends to move fluid out of interstitial spaces into capillaries. Because of difficulties in measuring IFHP, its reported values are variable, ranging from small positive to small negative values. For purposes of our discussion, we will assume, however, that the value of IFHP is 0 mm Hg at both ends of capillaries. Regardless of its exact value, the basic principles of fluid movement still apply.

Now let us consider the osmotic pressures involved in fluid movement. Osmotic pressures are due to the presence of nondiffusible proteins in blood and interstitial fluid. **Blood osmotic pressure (BOP),** which tends to move fluid from interstitial spaces into capillaries, averages about 28 mm Hg at both ends of capillaries. Opposing BOP is **interstitial fluid osmotic pressure (IFOP),** which tends to move fluid out of capillaries into interstitial fluid and averages about 6 mm Hg at both ends of capillaries.

Whether fluids leave or enter capillaries depends on how the pressures relate to each other. If the forces that tend to move fluid out of capillaries are greater than the forces that tend to move fluid into capillaries, the fluid will move from capillaries into interstitial spaces (filtration). If, on the other hand, the forces that tend to move fluid out of interstitial spaces into capillaries are greater than the forces that tend to move fluid out of capillaries, then fluid will move from interstitial spaces into capillaries (reabsorption). The term **effective filtration pressure (Peff)** is used to show the direction of fluid movement. It is calculated as follows:

$$P\text{eff} = (BHP + IFOP) - (IFHP + BOP)$$

Substituting values at the arterial end of a capillary,

$$P\text{eff} = (30 + 6) - (0 + 28)$$
$$= (36) - (28)$$
$$= 8 \text{ mm Hg}$$

Substituting measured values at the venous end of a capillary,

$$P\text{eff} = (15 + 6) - (0 + 28)$$
$$= (21) - (28)$$
$$= -7 \text{ mm Hg}$$

FIGURE 21-8 Hydrostatic and osmotic pressure forces involved in moving fluid out of capillaries (filtration) and into capillaries (reabsorption).

Thus, at the arterial end of a capillary, there is a *net outward force* (8 mm Hg), and fluid moves out of the capillary (filtered) into interstitial spaces. At the venous end of a capillary, the negative value (−7 mm Hg) represents a *net inward force,* and fluid moves into the capillary (reabsorbed) from tissue spaces.

Not all the fluid filtered at one end of the capillary is reabsorbed at the other end. When we discuss fluid dynamics in more detail in Chapter 27, you will see that some of the filtered fluid and any proteins that escape from blood into interstitial fluid are returned by the lymphatic system to the cardiovascular system. Under normal conditions there is a state of near equilibrium at the arterial and venous ends of a capillary in which filtered fluid and absorbed fluid, in addition to that returned to the lymphatic system, are nearly equal. This near equilibrium is known as **Starling's law of the capillaries.**

Factors That Aid Venous Return

The establishment of a pressure gradient is the primary reason why blood flows. But a number of other factors help the blood to return through the veins: contractions of the skeletal muscles, valves, and breathing.

• *Velocity of Blood Flow* The velocity of blood flow is inversely related to the cross-sectional area of the blood vessels. Blood flows most rapidly where the cross-sectional area is least (see Figure 21-5). Each time an artery branches, the total cross-sectional area of all the branches is greater than that of the original vessel. When branches combine, the resulting cross-sectional area is less than that of the original branches. The cross-sectional area of the aorta is 2.5 cm², and the velocity of the blood there is 40 cm/sec. Capillaries have a cross-sectional area of 2,500 cm², and the velocity is less than 0.1 cm/sec. In the venae cavae, the cross-sectional area is 8 cm², and the velocity is 5 to 20 cm/sec. The area and velocity for the different types of blood vessels are listed in Exhibit 21-1.

Thus the velocity of blood decreases as it flows from the aorta to arteries to arterioles to capillaries. The velocity of blood in the capillaries is the slowest in the cardiovascular system. Thus there is adequate exchange (diffusion) time between the capillaries and adjacent tissues. As blood vessels leave capillaries and approach the heart,

their cross-sectional area decreases. Therefore, the velocity of blood increases as it flows from capillaries to venules to veins to the heart.

• *Skeletal Muscle Contractions and Valves* The combination of skeletal muscle contractions and valves in veins is important in returning venous blood to the heart. Many veins, especially those in the extremities, contain valves. When skeletal muscles contract, they tighten around the veins running through them, and the valves open. This pressure drives the blood toward the heart—the action is called **milking** (Figure 21-9). When the muscles relax, the valves close to prevent the backflow of blood away from the heart. Individuals who are immobilized through injury or disease cannot take advantage of these contractions. As a result, the return of venous blood to the heart is slower, and the heart has to work harder. For this reason, periodic massage is helpful.

• *Breathing* Another factor that is important in maintaining venous circulation is breathing. During inspiration, the diaphragm moves downward. This causes a decrease in pressure in the chest cavity and an increase in pressure in the abdominal cavity. Once the pressure difference is established, blood moves from the abdominal veins into the thoracic veins. When the pressures reverse during expiration, blood in the veins is prevented from backflowing by the valves.

CHECKING CIRCULATION

PULSE

The alternate expansion and elastic recoil of an artery with each systole of the left ventricle is called the **pulse.** Pulse is strongest in the arteries closest to the heart. It becomes weaker as it passes over the arterial system, and it disappears altogether in the capillaries. The pulse may be felt in any artery that lies near the surface of the body and over a bone or other firm tissue. The radial artery at the wrist is most commonly used. Other arteries that may be used for determining pulse are the:

1. Temporal artery, above and toward the outside of the eye.
2. Facial artery, at the lower jawbone on a line with the corners of the mouth.
3. Common carotid artery, on the side of the neck.
4. Brachial artery, along the inner side of the biceps brachii muscle.
5. Femoral artery, near the pelvic bone.
6. Popliteal artery, behind the knee.

EXHIBIT 21-1 RELATIONSHIP BETWEEN BLOOD VELOCITY AND CROSS-SECTIONAL AREA

VESSEL	CROSS-SECTIONAL AREA (cm²)	VELOCITY (cm/sec)
Aorta	2.5	40
Arteries	20	10–40
Arterioles	40	0.1
Capillaries	2,500	less than 0.1
Venules	250	0.3
Veins	80	0.3–5.0
Venae cavae	8	5–20

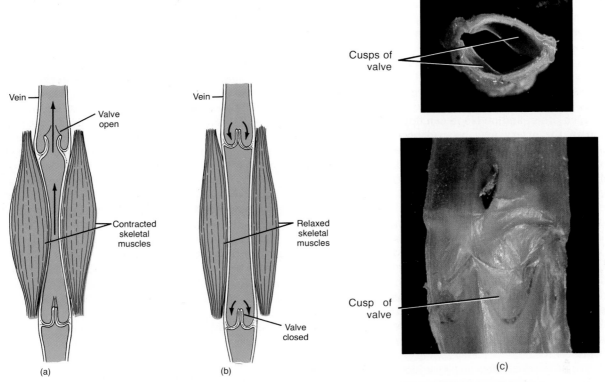

FIGURE 21-9 Role of skeletal muscle contractions and venous valves in returning blood to the heart. (a) When skeletal muscles contract, the valves open, and blood is forced toward the heart. (b) When skeletal muscles relax, the valves close to prevent the backflowing of blood from the heart. (c) Photograph of a one-way valve in a vein (cross section above, longitudinal section below). (Courtesy of J. A. Gosling et al., *Atlas of Human Anatomy with Integrated Text,* Copyright © 1985 by Gower Medical Publishing Ltd.)

7. Posterior tibial artery, behind the medial malleolus of the tibia.

8. Dorsalis pedis artery, over the instep of the foot.

The pulse rate is the same as the heart rate and averages between 70 and 90 beats per minute in the resting state. The term **tachycardia** (tak'-i-KAR-dē-a; *tacky* = fast) is applied to a rapid heart or pulse rate (over 100/min). The term **bradycardia** (brad'-i-KAR-dē-a; *brady* = slow) indicates a slow heart or pulse rate (under 50/min).

Other characteristics of the pulse may give additional information about circulation. For example, the intervals between beats should be equal in length. If a pulse is missed at intervals, the pulse is said to be irregular. Also, each pulse beat should be of equal strength. Irregularities in strength may indicate a lack of muscle tone in the heart or arteries.

MEASUREMENT OF BLOOD PRESSURE (BP)

In clinical use, the term **blood pressure (BP)** refers to the pressure in arteries exerted by the left ventricle when it undergoes systole and the pressure remaining in the arteries when the ventricle is in diastole. Blood pressure is usually taken in the left brachial artery, and it is measured by a *sphygmomanometer* (sfig'-mō-ma-NOM-e-ter; *sphygmo* = pulse). A commonly used sphygmomanometer consists of a rubber cuff attached by a rubber tube to a compressible hand pump or bulb (Figure 21-10).

FIGURE 21-10 Measurement of blood pressure using a sphygmomanometer and a stethoscope. (Copyright © 1982 by Gerard J. Tortora. Courtesy of Geraldine C. Tortora.)

Another tube attaches to the cuff and to a column of mercury or pressure dial marked off in millimeters. This column measures the pressure. The cuff is wrapped around the arm over the brachial artery and inflated by squeezing the bulb. The inflation creates a pressure on the artery. The bulb is squeezed until the pressure in the cuff exceeds the pressure in the artery. At this point, the walls of the brachial artery are compressed tightly against each other, and no blood can flow through. Compression of the artery may be evidenced in two ways. First, if a stethoscope is placed over the artery below the cuff, no sound can be heard. Second, no pulse can be felt by placing the fingers over the radial artery at the wrist.

Next the cuff is deflated gradually until the pressure in the cuff is slightly less than the maximal pressure in the brachial artery. At this point, the artery opens, a spurt of blood passes through, and a sound may be heard through the stethoscope. As cuff pressure is further reduced, the sound suddenly becomes faint. Finally, the sound disappears altogether. When the first sound is heard, a reading on the mercury column is made. This sound corresponds to **systolic blood pressure**—the force with which blood is pushing against arterial walls during ventricular contraction. The pressure recorded on the mercury column when the sounds suddenly become faint is called **diastolic blood pressure.** It measures the force of blood in arteries during ventricular relaxation. Whereas systolic pressure indicates the force of the left ventricular contraction, disastolic pressure provides information about the resistance of blood vessels. The various sounds that are heard while taking blood pressure are called *Korotkoff* (kō-ROT-kof) *sounds.*

CLINICAL APPLICATION

The average blood pressure of a young adult male is about 120 mm Hg systolic and 80 mm Hg diastolic, expressed as 120/80. In young adult females, the pressures are 8 to 10 mm Hg less. The difference between systolic and diastolic pressure is called **pulse pressure.** This pressure, which averages 40 mm Hg, provides information about the condition of the arteries. For example, conditions such as atherosclerosis and patent ductus arteriosus greatly increase pulse pressure. The normal ratio of systolic pressure to diastolic pressure to pulse pressure is about $3:2:1$.

CIRCULATORY ROUTES

The arteries, arterioles, capillaries, venules, and veins are organized into definite routes to circulate the blood throughout the body. We can now look at the basic routes the blood takes as it is transported through its vessels.

Figure 21-11 shows a number of basic **circulatory routes** through which the blood travels. **Systemic circulation** includes all the oxygenated blood that leaves the left ventricle through the aorta and the deoxygenated blood that returns to the right atrium after traveling to all the organs including the nutrient arteries to the lungs. Two of the many subdivisions of the systemic circulation are the **coronary (cardiac) circulation** (see Figure 20-6), which supplies the myocardium of the heart, and the **hepatic portal circulation,** which runs from the digestive tract to the liver. Blood leaving the aorta and traveling through the systemic arteries is a bright red color. As it moves through the capillaries, it loses its oxygen and takes on carbon dioxide, so that the blood in the systemic veins is a dark red color.

When blood returns to the heart from the systemic route, it goes out of the right ventricle through the **pulmonary circulation** to the lungs. In the lungs, it loses its carbon dioxide and takes on oxygen. It is now bright red again. It returns to the left atrium of the heart and reenters the systemic circulation.

Another major route—**fetal circulation**—exists only in the fetus and contains special structures that allow the developing fetus to exchange materials with its mother (see Figure 21-24).

Cerebral circulation (cerebral arterial circle, or circle of Willis) is discussed in Exhibit 21-4.

SYSTEMIC CIRCULATION

The flow of blood from the left ventricle to all parts of the body and back to the right atrium is called the **systemic circulation.** The purpose of systemic circulation is to carry oxygen and nutrients to body tissues and to remove carbon dioxide and other wastes from the tissues. All systemic arteries branch from the *aorta,* which arises from the left ventricle of the heart.

As the aorta emerges from the left ventricle, it passes upward and deep to the pulmonary trunk. At this point, it is called the *ascending aorta.* The ascending aorta gives off two coronary branches to the heart muscle. Then it turns to the left, forming the *arch of the aorta,* before descending to the level of the fourth thoracic vertebra as the *descending aorta.* The descending aorta lies close to the vertebral bodies, passes through the diaphragm, and at the level of the fourth lumbar vertebra, divides into two *common iliac arteries,* which carry blood to the lower extremities. The section of the descending aorta between the arch of aorta and the diaphragm is referred to as the *thoracic aorta.* The section between the diaphragm and the common iliac arteries is termed the *abdominal aorta.* Each section of the aorta gives off arteries that continue to branch into distributing arteries leading to organs, and finally into the arterioles and capillaries that pierce the tissues.

Blood is returned to the heart through the systemic veins. All the veins of the systemic circulation flow into either the *superior* or *inferior venae cavae* or the *coronary sinus.* They in turn empty into the right atrium. The principal arteries and veins of systemic circulation are described and illustrated in Exhibits 21-2 to 21-13 and Figures 21-12 to 21-21.

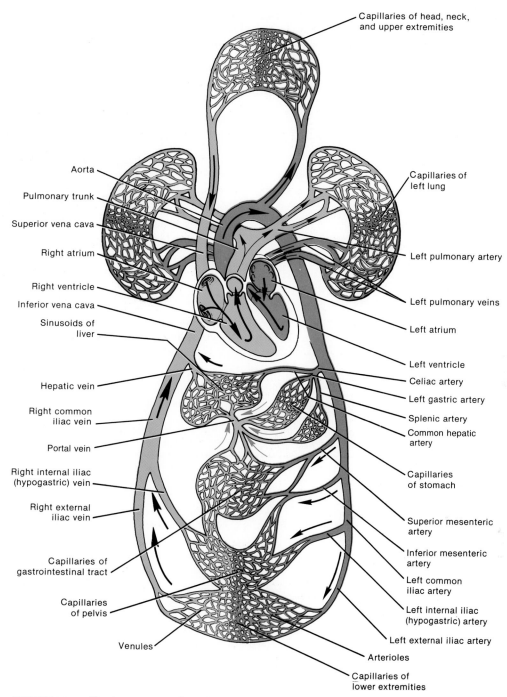

FIGURE 21-11 Circulatory routes. Systemic circulation is indicated by heavy black arrows, pulmonary circulation by thin black arrows in the pulmonary blood vessels, and hepatic portal circulation by thin colored arrows. Refer to Figure 20-6 for the details of coronary (cardiac) circulation and to figure 21-24 for the details of fetal circulation.

EXHIBIT 21-2 AORTA AND ITS BRANCHES (Figure 21-12)

DIVISION OF AORTA	ARTERIAL BRANCH	REGION SUPPLIED
Ascending aorta	Right and left coronary	Heart
Arch of aorta	Brachiocephalic ⟨ Right common carotid / Right subclavian	Right side of head and neck / Right upper extremity
	Left common carotid	Left side of head and neck
	Left subclavian	Left upper extremity
Thoracic aorta	Intercostals	Intercostal and chest muscles and pleurae
	Superior phrenics	Posterior and superior surfaces of diaphragm
	Bronchials	Bronchi of lungs
	Esophageals	Esophagus
Abdominal aorta	Inferior phrenics	Inferior surface of diaphragm
	Celiac → Common hepatic	Liver
	Celiac → Left gastric	Stomach and esophagus
	Celiac → Splenic	Spleen, pancreas, and stomach
	Superior mesenteric	Small intestine, cecum, and ascending and transverse colons
	Suprarenals	Adrenal (suprarenal) glands
	Renals	Kidneys
	Gonadals → Testicular	Testes
	Gonadals → Ovarian	Ovaries
	Inferior mesenteric	Transverse, descending, and sigmoid colons and rectum
	Common iliacs → External iliacs	Lower extremities
	Common iliacs → Internal iliacs (hypogastrics)	Uterus, prostate, muscles of buttocks, and urinary bladder

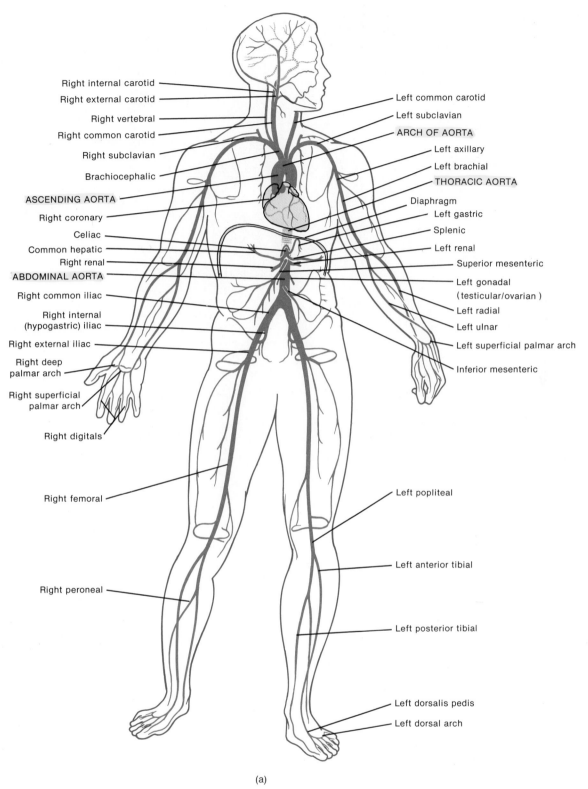

Right internal carotid

Right external carotid

Right vertebral

Right common carotid

Right subclavian

Brachiocephalic

ASCENDING AORTA

Right coronary

Celiac

Common hepatic

Right renal

ABDOMINAL AORTA

Right common iliac

Right internal
(hypogastric) iliac

Right external iliac

Right deep
palmar arch

Right superficial
palmar arch

Right digitals

Right femoral

Right peroneal

Left common carotid

Left subclavian

ARCH OF AORTA

Left axillary

Left brachial

THORACIC AORTA

Diaphragm

Left gastric

Splenic

Left renal

Superior mesenteric

Left gonadal
(testicular/ovarian)

Left radial

Left ulnar

Left superficial palmar arch

Inferior mesenteric

Left popliteal

Left anterior tibial

Left posterior tibial

Left dorsalis pedis

Left dorsal arch

(a)

FIGURE 21-12 Aorta and its principal branches in anterior view. (a) Diagram.

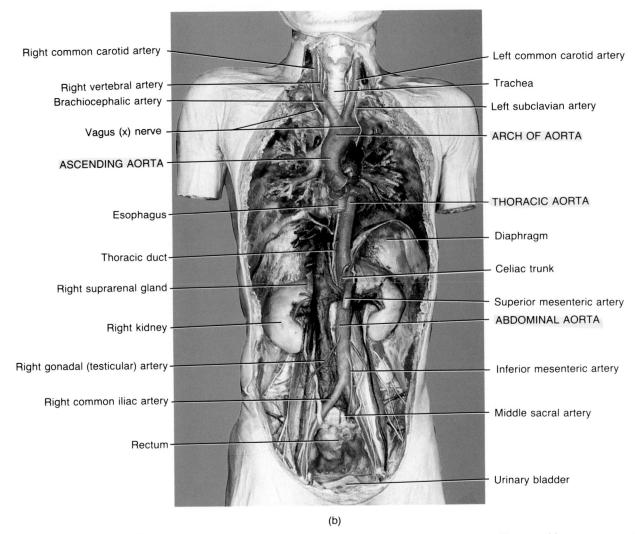

Right common carotid artery

Right vertebral artery

Brachiocephalic artery

Vagus (x) nerve

ASCENDING AORTA

Esophagus

Thoracic duct

Right suprarenal gland

Right kidney

Right gonadal (testicular) artery

Right common iliac artery

Rectum

Left common carotid artery

Trachea

Left subclavian artery

ARCH OF AORTA

THORACIC AORTA

Diaphragm

Celiac trunk

Superior mesenteric artery

ABDOMINAL AORTA

Inferior mesenteric artery

Middle sacral artery

Urinary bladder

(b)

FIGURE 21-12 (*Continued*) (b) Photograph. (Courtesy of C. Yokochi and J. W. Rohen, *Photographic Anatomy of the Human Body,* 2nd ed., 1978, IGAKU-SHOIN, Ltd., Tokyo, New York.)

EXHIBIT 21-3   ASCENDING AORTA (Figure 21-13)

BRANCH	DESCRIPTION AND REGION SUPPLIED
Coronary arteries	Right and left branches arise from ascending aorta just superior to aortic semilunar valve. They form crown around heart, giving off branches to atrial and ventricular myocardium.

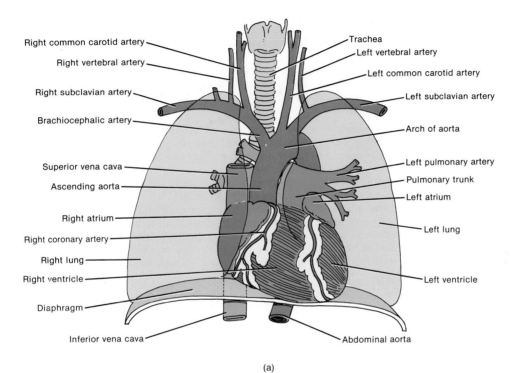

(a)

FIGURE 21-13 Ascending aorta and its branches in anterior view. (a) Diagram. (b) Scheme of distribution.

(b)

EXHIBIT 21-4   ARCH OF AORTA (Figure 21-14)

BRANCH	DESCRIPTION AND REGION SUPPLIED
Brachiocephalic	**Brachiocephalic artery** is first branch off arch of aorta. It divides to form right subclavian artery and right common carotid artery. **Right subclavian artery** extends from brachiocephalic to first rib and then passes into armpit (axilla) and supplies arm, forearm, and hand. Continuation of right subclavian into axilla is called **axillary artery.*** From here, it continues into arm as **brachial artery.** At bend of elbow, brachial artery divides into medial **ulnar** and lateral **radial arteries.** These vessels pass down to palm, one on each side of forearm. In palm, branches of two arteries anastomose to form two palmar arches— **superficial palmar arch** and **deep palmar arch.** From these arches arise **digital arteries,** which supply fingers and thumb. Before passing into axilla, right subclavian gives off major branch to brain called **vertebral artery.** Right vertebral artery passes through foramina of transverse processes of cervical vertebrae and enters skull through foramen magnum to reach undersurface of brain. Here it unites with left vertebral artery to form **basilar artery.** **Right common carotid artery** passes upward in neck. At upper level of larynx, it divides into **right external** and **right internal carotid arteries.** External carotid supplies right side of thyroid gland, tongue, throat, face, ear, scalp, and dura mater. Internal carotid supplies brain, right eye, and right sides of forehead and nose. Anastomoses of left and right internal carotids along with basilar artery form arterial circle at base of brain called **cerebral arterial circle (circle of Willis).** From this anastomosis arise arteries supplying brain. Essentially, circle of Willis is formed by union of **anterior cerebral arteries** (branches of internal carotids) and **posterior cerebral arteries** (branches of basilar artery). Posterior cerebral arteries are connected with internal carotids by **posterior communicating arteries.** Anterior cerebral arteries are connected by **anterior communicating arteries.** Circle of Willis equalizes blood pressure to brain and provides alternate routes for blood to brain, should arteries become damaged.
Left common carotid	**Left common carotid** is second branch off arch of arota (see Figure 21-13). Corresponding to right common carotid, it divides into basically same branches with same names, except that arteries are now labeled ''left'' instead of ''right.''
Left subclavian	**Left subclavian artery** is third branch off arch of aorta (see Figure 21-13). It distributes blood to left vertebral artery and vessels of left upper extremity. Arteries branching from left subclavian are named like those of right subclavian.

* The right subclavian artery is a good example of the practice of giving the same vessel different names as it passes through different regions.

FIGURE 21-14 Arch of the aorta and its branches. (a) Anterior view of the arteries of the right upper extremity. (b) Right lateral view of the arteries of the neck and head. (c) Arteries of the base of the brain. Note the arteries that comprise the cerebral arterial circle (circle of Willis).

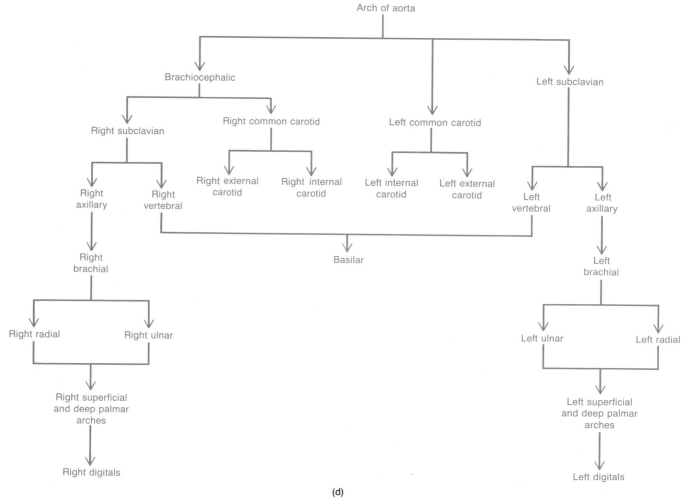

(d)

FIGURE 21-14 (*Continued*) (d) Scheme of distribution.

EXHIBIT 21-5 THORACIC AORTA (Figure 21-15)	
BRANCH	**DESCRIPTION AND REGION SUPPLIED**
	Thoracic aorta runs from fourth to twelfth thoracic vertebra. Along its course, it sends off numerous small arteries to viscera and skeletal muscles of the chest. Branches of an artery that supply viscera are called **visceral branches**. Those that supply body wall structures are **parietal branches**.
VISCERAL Pericardial	Several minute **pericardial arteries** supply blood to dorsal aspect of pericardium.
Bronchial	One right and two left **bronchial arteries** supply the bronchial tubes, areolar tissue of the lungs, bronchial lymph nodes, and esophagus.
Esophageal	Four or five **esophageal arteries** supply the esophagus.
Mediastinal	Numerous small **mediastinal arteries** supply blood to structures in the posterior mediastinum.
PARIETAL Posterior intercostal	Nine pairs of **posterior intercostal arteries** supply the intercostal, pectoral, and abdominal muscles; overlying subcutaneous tissue and skin; mammary glands; and vertebral canal and its contents.
Subcostal	The left and right **subcostal arteries** have a distribution similar to that of the posterior intercostals.
Superior phrenic	Small **superior phrenic arteries** supply the posterior surface of the diaphragm.

Left common carotid — *off aortic arch*
Left vertebral

Right common carotid

Right vertebral

subcl.

Brachiocephalic

Right subclavian

Left subclavian

Arch of aorta

Thoracic aorta

Right axillary

Right intercostal

Right brachial

Left gastric

Diaphragm

Left suprarenal

Splenic

Superior mesenteric

Celiac

Common hepatic

Right suprarenals

Right renal

Left renal

Left gonadal (testicular or ovarian)

Abdominal aorta

Right gonadal (testicular or ovarian)

Right lumbars

Inferior mesenteric

Middle sacral

Right common iliac

Right internal iliac (hypogastric)

Inguinal ligament

Right external iliac

Right femoral

FIGURE 21-15 Thoracic and abdominal aorta and their principal branches in anterior view.

EXHIBIT 21-6 ABDOMINAL AORTA (Figure 21-15)

BRANCH	DESCRIPTION AND REGION SUPPLIED
VISCERAL	
Celiac	**Celiac artery (trunk)** is first visceral aortic branch below diaphragm. It has three branches: (1) **common hepatic artery,** which supplies tissues of liver; (2) **left gastric artery,** which supplies stomach; and (3) **splenic artery,** which supplies spleen, pancreas, and stomach.
Superior mesenteric	**Superior mesenteric artery** distributes blood to small intestine and part of large intestine.
Suprarenals	Right and left **suprarenal arteries** supply blood to adrenal (suprarenal) glands.
Renals	Right and left **renal arteries** carry blood to kidneys.
Gonadals (testicular or ovarian)	Right and left **testicular arteries** extend into scrotum and terminate in testes; right and left **ovarian arteries** are distributed to ovaries.
Inferior mesenteric	**Inferior mesenteric artery** supplies major part of large intestine and rectum.
PARIETAL	
Inferior phrenics	**Inferior phrenic arteries** are distributed to undersurface of diaphragm.
Lumbars	**Lumbar arteries** supply spinal cord and its meninges and muscles and skin of lumbar region of back.
Middle sacral	**Middle sacral artery** supplies sacrum, coccyx, gluteus maximus muscles, and rectum.

EXHIBIT 21-7 ARTERIES OF PELVIS AND LOWER EXTREMITIES (Figure 21-16)

BRANCH	DESCRIPTION AND REGION SUPPLIED
Common iliacs	At about level of fourth lumbar vertebra, abdominal aorta divides into right and left **common iliac arteries.** Each passes downward about 5 cm (2 inches) and gives rise to two branches: internal iliac and external iliac.
Internal iliacs	**Internal iliac (hypogastric) arteries** form branches that supply psoas major, quadratus lumborum, and medial side of each thigh, urinary bladder, rectum, prostate gland, ductus (vas) deferens, uterus, and vagina.
External iliacs	**External iliac arteries** diverge through pelvis, enter thighs, and here become right and left **femoral arteries.** Both femorals send branches back up to genitals and wall of abdomen. Other branches run to muscles of thigh. Femoral continues down medial and posterior side of thigh at back of knee joint, where it becomes **popliteal artery.** Between knee and ankle, popliteal runs down back of leg and is called **posterior tibial artery.** Below knee, **peroneal artery** branches off posterior tibial to supply structures on medial side of fibula and calcaneus. In calf, **anterior tibial artery** branches off popliteal and runs along front of leg. At ankle, it becomes **dorsalis pedis artery.** At ankle, posterior tibial divides into **medial** and **lateral plantar arteries.** These arteries anastomose with dorsalis pedis and supply blood to foot.

FIGURE 21-16 Arteries of the pelvis and right lower extremity. (a) Anterior view. (b) Posterior view.

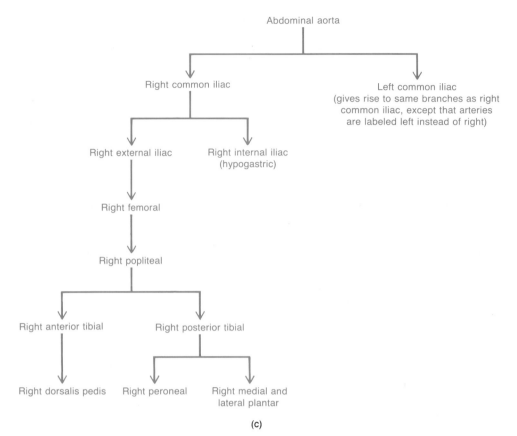

FIGURE 21-16 (*Continued*) (c) Scheme of distribution.

EXHIBIT 21-8 VEINS OF SYSTEMIC CIRCULATION (Figure 21-17)

VEIN	DESCRIPTION AND REGION DRAINED
	All systemic and coronary (cardiac) veins return blood to the right atrium of the heart through one of three large vessels. Return flow in coronary circulation is taken up by **cardiac veins,** which empty into the large vein of the heart, the **coronary sinus.** From here, the blood empties into the right atrium of the heart (see Figure 20-6). Return flow in systemic circulation empties into the superior vena cava or inferior vena cava.
Superior vena cava	Veins that empty into the **superior vena cava** are veins of the head and neck, upper extremities, and some from the thorax.
Inferior vena cava	Veins that empty into the **inferior vena cava** are some from the thorax and veins of the abdomen, pelvis, and lower extremities.

CLINICAL APPLICATION

The inferior vena cava is commonly compressed during the later stages of pregnancy due to the enlargement of the uterus. This produces edema of the ankles and feet and temporary varicose veins.

FIGURE 21-17 Principal veins in anterior view.

EXHIBIT 21-9   VEINS OF HEAD AND NECK (Figure 21-18)

VEIN	DESCRIPTION AND REGION DRAINED
Internal jugulars	Right and left **internal jugular veins** receive blood from face and neck. They arise as continuation of **sigmoid sinuses** at base of skull. Intracranial vascular sinuses are located between layers of dura mater and receive blood from brain. Other sinuses that drain into internal jugular include **superior sagittal sinus, inferior sagittal sinus, straight sinus,** and **transverse (lateral) sinuses.** Internal jugulars descend on either side of neck and pass behind clavicles, where they join with right and left subclavian veins. Unions of internal jugulars and subclavians form right and left brachiocephalic veins. From here blood flows into superior vena cava.
External jugulars	Right and left **external jugular veins** run down neck along outside of internal jugulars. They drain blood from parotid (salivary) glands, facial muscles, scalp, and other superficial structures into subclavian veins.

CLINICAL APPLICATION

In cases of heart failure, the venous pressure in the right atrium may rise. In such patients the pressure in the column of blood in the external jugular vein rises so that, even with the patient at rest and sitting in a chair, the external jugular vein will be visibly distended. Temporary distention of the vein is often seen in healthy adults when the intrathoracic pressure is raised in coughing and physical exertion.

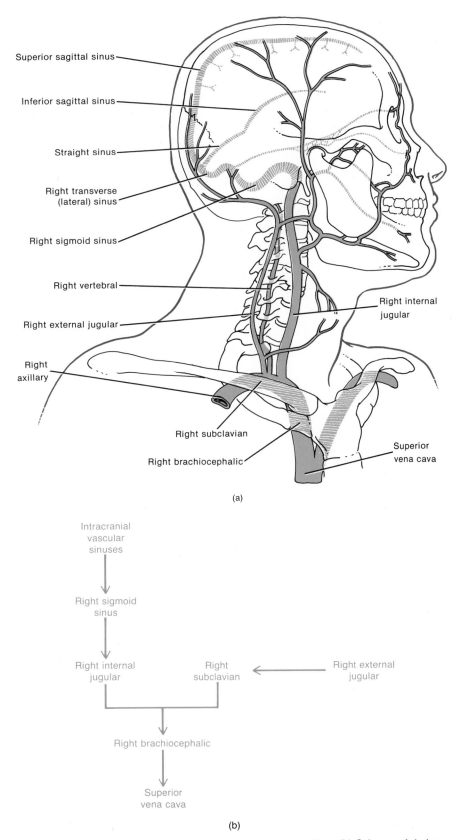

FIGURE 21-18 Veins of the head and neck. (a) Right lateral view. (b) Scheme of drainage.

Right subclavian

Right axillary

Right brachial

Right cephalic

Right accessory
cephalic

Right cephalic

Right digital

Right external jugular

Right internal jugular

Right brachiocephalic

Superior vena cava

Right basilic

Right median cubital

Right basilic

Right median antebrachial

Right palmar
venous arch

(a)

FIGURE 21-19 Veins of the right upper extremity in anterior view. (a) Diagram.

EXHIBIT 21-10	VEINS OF UPPER EXTREMITIES (Figure 21-19)
VEIN	**DESCRIPTION AND REGION DRAINED**
	Blood from each upper extremity is returned to the heart by deep and superficial veins. Both sets of veins contain valves. **Deep veins** are located deep in the body. They usually accompany arteries, and many have the same names as corresponding arteries. **Superficial veins** are located just below the skin and are often visible. They anastomose extensively with each other and deep veins.
SUPERFICIAL Cephalics	**Cephalic vein** of each upper extremity begins in the medial part of **dorsal venous arch** and winds upward around radial border of forearm. In front of elbow, it is connected to basilic vein by the **median cubital vein.** Just below elbow, cephalic vein unites with **accessory cephalic vein** to form cephalic vein of upper extremity. Ultimately, cephalic vein empties into axillary vein.
Basilics	**Basilic vein** of each upper extremity originates in the ulnar part of **dorsal venous arch.** It extends along posterior surface of ulna to point below elbow where it receives **median cubital vein.** If a vein must be punctured for an injection, transfusion, or removal of a blood sample, median cubitals are preferred. The median cubital vein joins the basilic vein to form the axillary vein.
Median antebrachials	**Median antebrachial veins** drain **palmar venous arch,** ascend on ulnar side of anterior forearm, and end in median cubital veins.
DEEP Radials	**Radial veins** receive **dorsal metacarpal veins.**
Ulnars	**Ulnar veins** receive tributaries from **palmar venous arch.** Radial and ulnar veins unite in bend of elbow to form brachial vein.
Brachials	Located on either side of brachial artery, **brachial veins** join into axillary veins.
Axillaries	**Axillary veins** are a continuation of brachials and basilics. Axillaries end at first rib, where they become subclavians.
Subclavians	Right and left **subclavian veins** unite with internal jugulars to form brachiocephalic veins. Thoracic duct of lymphatic system delivers lymph into left subclavian vein at junction with internal jugular. Right lymphatic duct delivers lymph into right subclavian vein at corresponding junction.

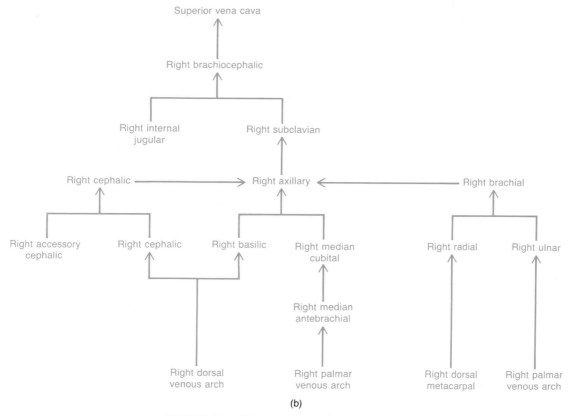

(b)

FIGURE 21-19 (*Continued*) (b) Scheme of drainage.

EXHIBIT 21-11 VEINS OF THORAX (Figure 21-20)

VEIN	DESCRIPTION AND REGION DRAINED
Brachiocephalic	Right and left **brachiocephalic veins,** formed by union of subclavians and internal jugulars, drain blood from head, neck, upper extremities, mammary glands, and upper thorax. Brachiocephalics unite to form superior vena cava.
Azygos veins	**Azygos veins,** besides collecting blood from thorax, may serve as bypass for inferior vena cava that drains blood from lower body. Several small veins directly link azygos veins with inferior vena cava. And large veins that drain lower extremities and abdomen dump blood into azygos. If inferior vena cava or hepatic portal vein becomes obstructed, azygos veins can return blood from lower body to superior vena cava.
Azygos	**Azygos vein** lies in front of vertebral column, slightly right of midline. It begins as continuation of right ascending lumbar vein. It connects with inferior vena cava, right common iliac, and lumbar veins. Azygos receives blood from **right intercostal veins** that drain chest muscles; from hemiazygos and accessory hemiazygos veins; from several **esophageal, mediastinal,** and **pericardial veins;** and from right **bronchial vein.** Vein ascends to fourth thoracic vertebra, arches over right lung, and empties into superior vena cava.
Hemiazygos	**Hemiazygos vein** is in front of vertebral column and slightly left of midline. It begins as continuation of left ascending lumbar vein. It receives blood from lower four or five **intercostal veins** and some **esophageal** and **mediastinal veins.** At level of ninth thoracic vertebra, it joins azygos vein.
Accessory hemiazygos	**Accessory hemiazygos vein** is also in front and to left of vertebral column. It receives blood from three or four **intercostal veins** and left **bronchial vein.** It joins azygos at level of eighth thoracic vertebra.

EXHIBIT 21-12 VEINS OF ABDOMEN AND PELVIS (Figure 21-20)

VEIN	DESCRIPTION AND REGION DRAINED
Inferior vena cava	**Inferior vena cava** is the largest vein of the body. It is formed by union of two common iliac veins that drain lower extremities and abdomen. Inferior vena cava extends upward through abdomen and thorax to right atrium. Numerous small veins enter the inferior vena cava. Most carry return flow from branches of abdominal aorta, and names correspond to names of arteries.
Common iliacs	**Common iliac veins** are formed by union of internal (hypogastric) and external iliac veins and represent distal continuation of inferior vena cava at its bifurcation.
Internal iliacs	Tributaries of **internal iliac (hypogastric) veins** basically correspond to branches of internal iliac arteries. Internal iliacs drain gluteal muscles, medial side of thigh, urinary bladder, rectum, prostate gland, ductus (vas) deferens, uterus, and vagina.
External iliacs	**External iliac veins** are continuation of femoral veins and receive blood from lower extremities and inferior part of anterior abdominal wall.
Renals	**Renal veins** drain kidneys.
Gonadals (testicular or ovarian)	**Testicular veins** drain testes (left testicular vein empties into left renal vein); **ovarian veins** drain ovaries (left ovarian vein empties into left renal vein).
Suprarenals	**Suprarenal veins** drain adrenal (suprarenal) glands (left suprarenal vein empties into left renal vein).
Inferior phrenics	**Inferior phrenic veins** drain diaphragm (left inferior phrenic vein sends tributary to left renal vein).
Hepatics	**Hepatic veins** drain liver.
Lumbars	A series of parallel **lumbar veins** drain blood from both sides of posterior abdominal wall. Lumbars connect at right angles with right and left **ascending lumbar veins,** which form origin of corresponding azygos or hemiazygos vein. Lumbars drain blood into ascending lumbars and then run to inferior vena cava, where they release remainder of flow.

FIGURE 21-20 Veins of the thorax, abdomen, and pelvis in anterior view.

Inferior vena cava

L4

Left common iliac

Right common iliac

Right internal iliac (hypogastric)

Right external iliac

Right femoral

Right great saphenous

Right popliteal

Right small saphenous

Right anterior tibial

Right peroneal

Right great saphenous

Right posterior tibial

Right dorsalis pedis

Right medial plantar

Right dorsal venous arch

Right lateral plantar

Right plantar arch

(a)

(b)

DANK

FIGURE 21-21 Veins of the pelvis and right lower extremity. (a) Anterior view. (b) Posterior view.

EXHIBIT 21-13   VEINS OF LOWER EXTREMITIES (Figure 21-21)

VEIN	DESCRIPTION AND REGION DRAINED
	Blood from each lower extremity is returned by superficial set and deep set of veins.
SUPERFICIAL VEINS **Great saphenous**	**Great saphenous vein,** longest vein in body, begins at medial end of **dorsal venous arch** of foot. It passes in front of medial malleolus and then upward along medial aspect of leg and thigh. It receives tributaries from superficial tissues and connects with deep veins as well. It empties into femoral vein in groin.

CLINICAL APPLICATION

The great saphenous vein is very constant in its position anterior to the medial malleolus. It is frequently used for prolonged administration of intravenous fluids. This is particularly important in very young babies and in patients of any age who are in shock and whose veins are collapsed. It and the small saphenous vein are subject to varicosity.

Small saphenous	**Small saphenous vein** begins at lateral end of dorsal venous arch of foot. It passes behind lateral malleolus and ascends under skin of back of leg. It receives blood from foot and posterior portion of leg. It empties into popliteal vein behind knee.
DEEP VEINS **Posterior tibial**	**Posterior tibial vein** is formed by union of **medial** and **lateral plantar veins** behind medial malleolus. It ascends deep in muscle at back of leg, receives blood from **peroneal vein,** and unites with anterior tibial vein just below knee.
Anterior tibial	**Anterior tibial vein** is upward continuation of **dorsalis pedis veins** in foot. It runs between tibia and fibula and unites with posterior tibial to form popliteal vein.
Popliteal	**Popliteal vein,** just behind knee, receives blood from anterior and posterior tibials and small saphenous vein.
Femoral	**Femoral vein** is upward continuation of popliteal just above knee. Femorals run up posterior of thighs and drain deep structures of thighs. After receiving great saphenous veins in groin, they continue as right and left external iliac veins.

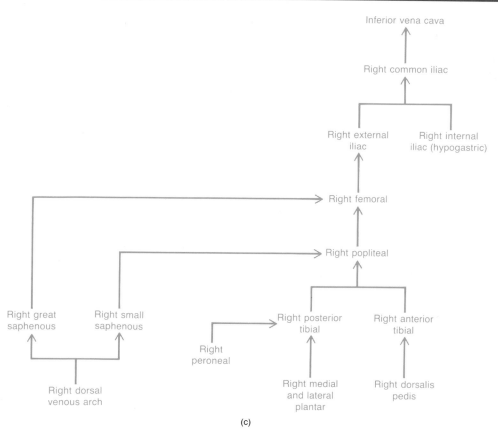

(c)

FIGURE 21-21 (Continued) (c) Scheme of drainage.

PULMONARY CIRCULATION

The flow of deoxygenated blood from the right ventricle to the lungs and the return of oxygenated blood from the lungs to the left atrium is called **pulmonary circulation** (Figure 21-22). The *pulmonary trunk* emerges from the right ventricle and passes upward, backward, and to the left. It then divides into two branches: the *right pulmonary artery* runs to the right lung; the *left pulmonary artery* goes to the left lung. On entering the lungs, the branches divide and subdivide until ultimately they form capillaries around the alveoli in the lungs. Carbon dioxide is passed from the blood into the alveoli to be breathed out of the lungs. Oxygen breathed in by the lungs is passed from the alveoli into the blood. The capillaries unite, venules and veins are formed, and eventually two *pulmonary veins* exit from each lung and transport the oxygenated blood to the left atrium. The pulmonary veins are the only postnatal veins that carry oxygenated blood. Contractions of the left ventricle then send the blood into the systemic circulation.

HEPATIC PORTAL CIRCULATION

Blood enters the liver from two sources. The hepatic artery delivers oxygenated blood from the systemic circulation; the hepatic portal vein delivers deoxygenated blood from the digestive organs. The term **hepatic portal circulation** refers to this flow of venous blood from the digestive organs to the liver before returning to the heart (Figure 21-23). Hepatic portal blood is rich with substances absorbed from the digestive tract. The liver monitors these substances before they pass into the general circulation. For example, the liver stores nutrients such as glucose. It also modifies other digested substances so they may be used by cells, detoxifies harmful substances that have been absorbed by the digestive tract, and destroys bacteria by phagocytosis.

The hepatic portal system includes veins that drain blood from the pancreas, spleen, stomach, intestines, and gallbladder and transport it to the portal vein of the liver. The *hepatic portal vein* is formed by the union of the superior mesenteric and splenic veins. The *superior mesenteric vein* drains blood from the small intestine and portions of the large intestine and stomach. The *splenic vein* drains the spleen and receives tributaries from the stomach, pancreas, and portions of the colon. The tributaries from the stomach are the *gastric, pyloric,* and *gastroepiploic veins.* The *pancreatic veins* come from the pancreas, and the *inferior mesenteric veins* come from the portions of the colon. Before the hepatic portal vein enters the liver, it receives the *cystic vein* from the gallbladder and other veins. Ultimately, blood leaves the liver through the *hepatic veins,* which enter the inferior vena cava.

FIGURE 21-22 Pulmonary circulation. (a) Diagram. (b) Scheme of circulation.

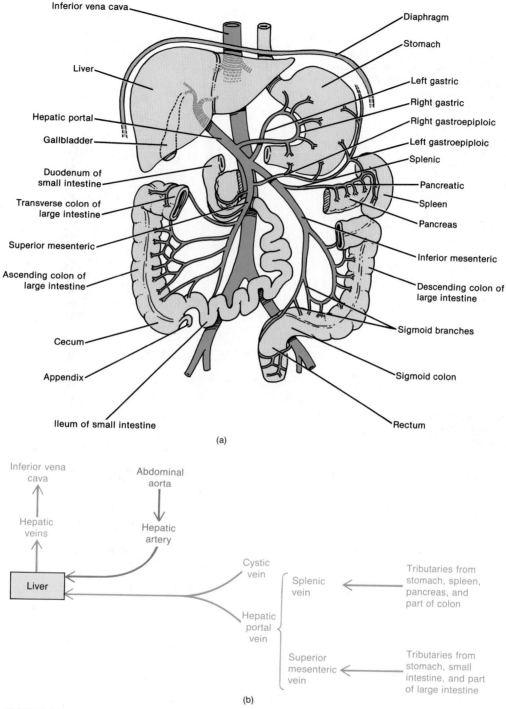

FIGURE 21-23 Hepatic portal circulation. (a) Diagram. (b) Scheme of blood flow through the liver, including arterial circulation. Deoxygenated blood is indicated in blue; oxygenated blood in red.

FETAL CIRCULATION

The circulatory system of a fetus, called **fetal circulation,** differs from an adult's because the lungs, kidneys, and digestive tract of a fetus are nonfunctional. The fetus derives its oxygen and nutrients from the maternal blood, and eliminates its carbon dioxide and wastes into the maternal blood (Figure 21-24).

The exchange of materials between fetal and maternal circulation occurs through a structure called the *placenta* (pla-SEN-ta). It is attached to the umbilicus of the fetus by the umbilical (um-BIL-i-kal) cord, and it communicates with the mother through countless small blood vessels that emerge from the uterine wall. The umbilical cord contains blood vessels that branch into capillaries in the placenta. Wastes from the fetal blood diffuse out

Right common carotid artery

Brachiocephalic artery

Right subclavian artery

Superior vena cava

Ascending aorta

Pulmonary trunk

Right atrium

Foramen ovale

Right ventricle

Liver

Ductus venosus

To mother

From mother

To mother

Placenta

Umbilical cord

Umbilical arteries

Left common carotid artery

Left vertebral artery

Left subclavian artery

Ductus arteriosus

Left atrium

Left ventricle

Abdominal aorta

Superior mesenteric artery

Portal vein

Umbilical vein

Inferior vena cava

Left common iliac artery

Left external iliac artery

Left internal iliac (hypogastric) artery

(a)

FIGURE 21-24 Fetal circulation. (a) Diagram. The various colors represent different degrees of oxygenation of blood ranging from greatest (red) to intermediate (two shades of purple) to least (blue).

of the capillaries, into spaces containing maternal blood (intervillous spaces) in the placenta, and finally into the mother's uterine blood vessels. Nutrients travel the opposite route—from the maternal blood vessels to the intervillous spaces to the fetal capillaries. Normally there is no mixing of maternal and fetal blood since all exchanges occur through capillaries.

Blood passes from the fetus to the placenta via two *umbilical arteries.* These branches of the internal iliac (hypogastric) arteries are included in the umbilical cord. At the placenta, the blood picks up oxygen and nutrients and eliminates carbon dioxide and wastes. The oxygenated blood returns from the placenta via a single *umbilical vein.* This vein ascends to the liver of the fetus, where

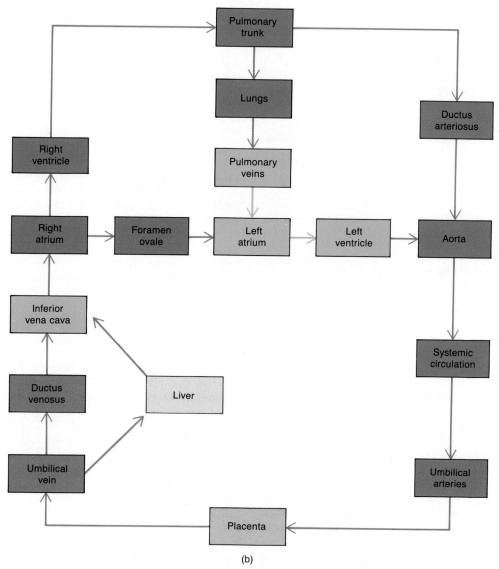

FIGURE 21-24 (*Continued*) (b) Scheme of circulation.

it divides into two branches. Some blood flows through the branch that joins the hepatic portal vein and enters the liver. Although the fetal liver manufactures red blood cells, it does not function in digestion. Therefore, most of the blood flows into the second branch, the *ductus venosus* (DUK-tus ve-NŌ-sus). The ductus venosus eventually passes its blood to the inferior vena cava, bypassing the liver.

In general, circulation through other portions of the fetus is not unlike postnatal circulation. Deoxygenated blood returning from the lower regions is mingled with oxygenated blood from the ductus venosus in the inferior vena cava. This mixed blood then enters the right atrium. The circulation of blood through the upper portion of the fetus is also similar to postnatal flow. Deoxygenated blood returning from the upper regions of the fetus is collected by the superior vena cava, and it also passes into the right atrium.

Most of the blood does not pass through the right ventricle to the lungs, as it does in postnatal circulation, since the fetal lungs do not operate. In the fetus, an opening called the *foramen ovale* (fō-RĀ-men ō-VAL-ē) exists in the septum between the right and left atria. A valve in the inferior vena cava directs about a third of the blood through the foramen ovale so that it may be sent directly into the systemic circulation. The blood that does descend into the right ventricle is pumped into the pulmonary trunk, but little of this blood actually reaches the lungs. Most blood in the pulmonary trunk is sent through the *ductus arteriosus* (ar-tē-rē-Ō-sus). This small vessel connecting the pulmonary trunk with the aorta enables blood in excess of nutrient requirements to bypass the fetal lungs. The blood in the aorta is carried to all parts of the fetus through its systemic branches. When the common iliac arteries branch into the external and internal iliacs (hypogastric), part of the blood flows into the internal iliacs (hypogastrics). It then goes to the umbilical arteries and back to the placenta for another exchange of materials. The only vessel that carries fully oxygenated blood is the umbilical vein.

At birth, when lung, renal, digestive, and liver functions are established, the special structures of fetal circulation are no longer needed and the following changes occur.

1. The umbilical arteries atrophy to become the *lateral umbilical ligaments.*
2. The umbilical vein becomes the *round ligament* of the liver.
3. The placenta is delivered by the mother as the *"afterbirth."*
4. The ductus venosus becomes the *ligamentum venosum,* a fibrous cord in the liver.
5. The foramen ovale normally closes shortly after birth to become the *fossa ovalis,* a depression in the interatrial septum.
6. The ductus arteriosus closes, atrophies, and becomes the *ligamentum arteriosum.*

Anatomical defects resulting from failure of these changes to occur are described in Chapter 20.

EXERCISE AND THE CARDIOVASCULAR SYSTEM

No matter what a person's level of fitness, it can be improved at any age with regular participation in exercise. Of the various types of exercise, some are more effective than others for improving the health of the cardiovascular system because they involve movements of large body muscles. One example is aerobics, in which the sustained movements of large muscles increase the flow of blood to the heart. Brisk walking, running, bicycling, cross-country skiing, and swimming are examples of aerobic exercises.

Sustained exercise increases the oxygen demand of the muscles, and whether the demand is met depends primarily on the adequacy of cardiac output. After several weeks of training, the healthy individual increases cardiac output and thereby increases the rate of oxygen delivery to the tissues.

Physical conditioning also causes an interesting effect upon systemic blood pressure. After a conditioning period, hypertensive individuals show a reduction in systolic pressure amounting to an average of 13 mm Hg. Blood pressure response to a given work load also is less after training, while lower pressure itself helps reduce myocardial oxygen requirements. So conditioning also may be useful in the management of systemic hypertension.

Additional benefits to be gained from physical conditioning are an increase in high-density lipoprotein (HDL), a substance that seems to counter the impact of cholesterol in heart disease, and a decrease in triglyceride levels. Exercise also helps to control weight and increases the body's ability to dissolve blood clots by increasing fibrinolytic activity. Intense exercise increases levels of endorphins, the body's natural painkillers. This may explain the psychological "high" that runners experience with strenuous training and the "low" they feel when they miss regular workouts. Exercise also helps make bones stronger, and

thus may be a factor in inhibiting and treating osteoporosis.

AGING AND THE CARDIOVASCULAR SYSTEM

General changes associated with aging and the cardiovascular system include loss of extensibility of the aorta, reduction in cardiac muscle cell size, progressive loss of cardiac muscular strength, a reduced output of blood by the heart, and an increase in blood pressure. There is an increase in the incidence of coronary artery disease (CAD), the major cause of heart disease and death in older Americans. Congestive heart failure, a set of symptoms associated with impaired pumping performance of the heart, also occurs. Changes in blood vessels such as hardening of the arteries and cholesterol deposits in arteries that serve brain tissue reduce nourishment to the brain and result in the malfunction or death of brain cells.

DEVELOPMENTAL ANATOMY OF BLOOD AND BLOOD VESSELS

Since the human egg and yolk sac have little yolk to nourish the developing embryo, blood and blood vessel formation starts as early as 15 to 16 days. The development begins in the **mesoderm** of the yolk sac, chorion, and body stalk.

Blood vessels develop from isolated masses and cords of mesenchyme in the mesoderm called **blood islands** (Figure 21-25). Spaces soon appear in the islands and become

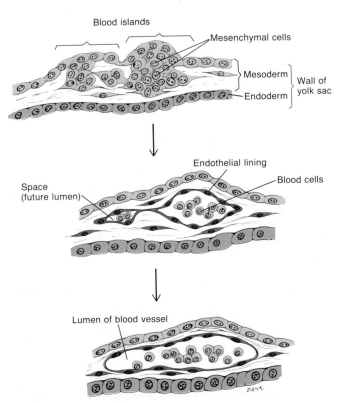

FIGURE 21-25 Development of blood vessels and blood cells from blood islands.

the lumens of the blood vessels. Some of the mesenchymal cells immediately around the spaces give rise to the *endothelial lining of the blood vessels.* Mesenchyme around the endothelium forms the *tunics* (intima, media, externa) of the larger blood vessels. Growth and fusion of blood islands form an extensive network of blood vessels throughout the embryo.

Blood plasma and *blood cells* are produced by the endothelial cells and appear in the blood vessels of the yolk sac and allantois quite early. Blood formation in the embryo itself begins at about the second month in the liver, spleen, bone marrow, and lymph nodes.

DISORDERS: HOMEOSTATIC IMBALANCES

Aneurysm

An **aneurysm** (AN-yoo-rizm) is a thin, weakened section of the wall of an artery or a vein that bulges outward forming a balloonlike sac of the blood vessel. Common causes of aneurysms include atherosclerosis, syphilis, congenital blood vessel defects, and trauma. If an aneurysm goes untreated, it grows larger and larger until the blood vessel wall becomes so thin that it may burst, causing massive hemorrhage with shock, severe pain, stroke, or death, depending on which vessel is involved. Even an unruptured aneurysm can lead to damage by interrupting blood flow or putting pressure on adjacent blood vessels, organs, or bones.

Surgical repair of an aneurysm consists of temporarily clamping the damaged artery above and below the aneurysm and then opening the aneurysm. A graft, usually of Dacron, is then sutured to healthy segments of the artery to reestablish normal blood flow.

Coronary Artery Disease (CAD)

In Chapter 20 it was indicated that the common causes of heart disease are related to inadequate coronary blood supply, anatomical disorders, and arrhythmias. **Coronary artery disease (CAD)** is a condition in which the heart muscle receives inadequate blood because of an interruption of blood supply. Depending on the degree of interruption, symptoms can range from a mild chest pain to a full-scale heart attack. The underlying causes of CAD are many and varied. Two of the principal ones are atherosclerosis and coronary artery spasm, both of which were discussed in detail in Chapter 20.

Hypertension

Hypertension, or high blood pressure, is the most common disease affecting the heart and blood vessels. Statistics indicate that hypertension afflicts one out of every five American adults. Although there is some disagreement as to what defines hypertension, a strong consensus has emerged suggesting that a blood pressure of 120/80 is normal and desirable in a healthy adult. A reading of 140/90 is generally regarded as the threshold of hypertension, while higher values, especially those over 160/95, are classified as dangerously hypertensive.

Primary hypertension (essential hypertension) is a persistently elevated blood pressure that cannot be attributed to any particular organic cause. Approximately 85 percent of all hypertension cases fit this definition. The other 15 percent are *secondary hypertension.* Secondary hypertension has an identifiable underlying cause such as atherosclerosis, kidney disease, and adrenal hypersecretion. Atherosclerosis increases blood pressure by reducing the elasticity of the arterial wall and narrowing the lumen through which the blood can flow. Kidney disease and obstruction of blood flow may cause the kidneys to release renin into the blood. This enzyme catalyzes the formation of angiotensin II from a plasma protein. Angiotensin II is a powerful bloodvessel constrictor—and the most potent agent known for raising blood pressure. It also stimulates aldosterone release. Aldosteronism, the hypersecretion of aldosterone, may also cause an increase in blood pressure. Aldosterone is the adrenal cortex hormone that promotes the retention of salt and water by the kidneys. It thus tends to increase plasma volume. *Pheochromocytoma* (fē-ō-krō'-mō-sī-TŌ-ma) is a tumor of the adrenal medulla. It produces and releases into the blood large quantities of norepinephrine and epinephrine. These hormones also raise blood pressure. Epinephrine causes an increase in heart rate and norepinephrine causes vasoconstriction.

High blood pressure is of considerable concern because of the harm it can do to the heart, brain, and kidneys if it remains uncontrolled. The heart is most commonly affected by high blood pressure. When pressure is high, the heart uses more energy in pumping. Because of the increased effort, the heart muscle thickens and the heart becomes enlarged. The heart also needs more oxygen. If it cannot meet the demands put on it, angina pectoris or even myocardial infarction may develop. Hypertension is also a factor in the development of atherosclerosis. Continued high blood pressure may produce a cerebral vascular accident (CVA), or stroke. In this case, severe strain has been imposed on the cerebral arteries that supply the brain. These arteries are usually less protected by surrounding tissues than are the major arteries in other parts of the body. These weakened cerebral arteries may finally rupture, and a brain hemorrhage follows.

The kidneys are also prime targets of hypertension. The principal site of damage is in the arterioles that supply them. The continual high blood pressure pushing against the walls of the arterioles causes them to thicken, thus narrowing the lumen. The blood supply to the kidneys is thereby gradually reduced. In response, the kidneys may secrete renin, which raises the blood pressure even higher and complicates the problem. The reduced blood flow to the kidney cells may eventually lead to the death of the cells.

Medical science cannot cure hypertension. However, almost all cases of hypertension, whether mild or severe, can be controlled in a number of ways. The overweight person with hypertension will usually be placed on a reducing diet, because blood pressure often falls with weight loss. Treatment often involves the restriction of sodium intake. Sodium restriction curbs fluid retention by the body and also tends to reduce blood volume. Recent evidence suggests that less fat and more potassium and calcium may also lead to a reduction in blood pressure, whereas magnesium deficiency may cause hypertension. Since nicotine is a vasoconstrictor, it elevates blood pressure. Stopping smoking may help to decrease blood pressure. As indicated earlier, exercise can help to reduce hypertension. In recent years, some physicians have advocated relaxation techniques (yoga, meditation, and biofeedback) to treat hypertension. For those individuals who require medication, a number of drugs are available. Many people can be treated with diuretics, which eliminate large amounts of water and sodium, thus decreasing blood volume and reducing blood pressure. Vasodilators are often used in combination with diuretics. They relax the smooth muscle in arterial walls, causing vasodilation and thus lowering blood pressure. Beta blockers are also used to lower blood pressure, often in combination with diuretics.

Deep-Venous Thrombosis (DVT)

Venous thrombosis, the presence of a thrombus in a vein, typically occurs in deep veins of the lower extremities. In such cases, the condition is referred to as **deep-venous thrombosis (DVT)**. The two most serious complications of DVT are pulmonary embolism, in which the thrombus dislodges and finds its way into the pulmonary arterial blood flow, and postphlebitic syndrome, which consists of edema, pain, and skin changes due to destruction of venous valves. Treatment consists of intravenous heparin therapy and elevation of the extremity, fibrinolytic therapy (streptokinase or urokinase), and, in rare cases, thrombectomy.

MEDICAL TERMINOLOGY

Angiocardiography (*angio* = vessel; *cardio* = heart; *graph* = writing) X-ray examination of the heart and great blood vessels after injection of a radiopaque dye into the blood stream.

Aortography X-ray examination of the aorta and its main branches after injection of radiopaque dye.

Arteritis (*itis* = inflammation of) Inflammation of an artery, probably due to an autoimmune response.

Claudication Pain and lameness or limping caused by defective circulation of the blood in the vessels of the limbs.

Coronary endarterectomy The removal of the obstructing area within the lumen of the vessel.

Cyanosis (*cyano* = blue) Slightly bluish, dark purple skin coloration due to oxygen deficiency in systemic blood.

Hypercholesteremia (*hyper* = over; *heme* = blood) An excess of cholesterol in the blood.

Hypotension (*hypo* = below; *tension* = pressure) Low blood pressure; most commonly used to describe an acute drop in blood pressure, as occurs in circulatory shock.

Normotensive Characterized by normal blood pressure.

Occlusion The closure or obstruction of the lumen of a structure such as blood vessel.

Phlebitis (*phleb* = vein) Inflammation of a vein often in a leg.

Postural hypotension Lowering of systemic blood pressure with the assumption of an erect or semierect posture; it is usually a sign of a disease. May be caused by cardiovascular or neurogenic factors.

Raynaud's disease A vascular disorder, primarily of females, characterized by bilateral attacks of ischemia, usually of the fingers and toes, in which the skin becomes pale and exhibits burning and pain; it is brought on by cold or emotional stimuli. If the condition is secondary to another disorder, it is called **Raynaud's phenomenon.**

Shunt A passage between two blood vessels or between the two sides of the heart.

Thrombectomy (*thrombo* = clot) An operation to remove a blood clot from a blood vessel.

Thrombophlebitis Inflammation of a vein with clot formation.

STUDY OUTLINE

Arteries (p. 488)

1. Arteries carry blood away from the heart. The wall of an artery consists of a tunica interna, tunica media (which maintains elasticity and contractility), and tunica externa.
2. Large arteries are referred to as elastic (conducting) arteries, and medium-sized arteries are called muscular (distributing) arteries.
3. Many arteries anastomose—the distal ends of two or more vessels unite. An alternate blood route from an anastomosis is called collateral circulation. Arteries that do not anastomose are called end arteries.

Arterioles (p. 489)

1. Arterioles are small arteries that deliver blood to capillaries.
2. Through constriction and dilation they assume a key role in regulating blood flow from arteries into capillaries.

Capillaries (p. 490)

1. Capillaries are microscopic blood vessels through which materials are exchanged between blood and tissue cells; some capillaries are continuous, others are fenestrated.
2. Capillaries branch to form an extensive capillary network throughout the tissue. This network increases the surface area, allowing a rapid exchange of large quantities of materials.
3. Precapillary sphincters regulate blood flow through capillaries.
4. Microscopic blood vessels in the liver are called sinusoids.

Venules (p. 490)

1. Venules are small vessels that continue from capillaries and merge to form veins.
2. They drain blood from capillaries into veins.

Veins (p. 490)

1. Veins consist of the same three tunics as arteries, but have less elastic tissue and smooth muscle.
2. They contain valves to prevent backflow of blood.
3. Weak valves can lead to varicose veins or hemorrhoids.
4. Vascular (venous) sinuses are veins with very thin walls.

Blood Reservoirs (p. 492)

1. Systemic veins are collectively called blood reservoirs.
2. They store blood, which through vasoconstriction can move to other parts of the body if the need arises.
3. The principal reservoirs are the veins of the abdominal organs (liver and spleen) and skin.

Physiology of Circulation (p. 492)

Blood Flow and Blood Pressure (p. 492)

1. Blood flows from regions of higher to lower pressure. The established pressure gradient is from aorta (100 mm Hg) to arteries (100–40 mm Hg) to arterioles (40–25 mm Hg) to capillaries (25–12 mm Hg) to venules (12–8 mm Hg) to veins (10–5 mm Hg) to venae cavae (2 mm Hg) to right atrium (0 mm Hg).

2. Any factor that increases cardiac output (CO) increases blood pressure.
3. As blood volume increases, blood pressure increases.
4. Peripheral resistance is determined by blood viscosity and blood vessel diameter. Increased viscosity and vasoconstriction increase peripheral resistance and thus increase blood pressure.
5. Factors that determine heart rate and force of contraction, and therefore blood pressure, are the autonomic nervous system through the cardiac center, chemicals, temperature, emotions, sex, and age.
6. Factors that regulate blood pressure by acting on blood vessels include the vasomotor center in the medulla together with baroreceptors, chemoreceptors, and higher brain centers; chemicals; and autoregulation.
7. The movement of water and dissolved substances (except proteins) through capillaries is dependent on hydrostatic and osmotic pressures.
8. The near equilibrium at the arterial and venous ends of a capillary by which fluids exit and enter is called Starling's law of the capillaries.
9. Blood return to the heart is maintained by several factors including skeletal muscular contractions, valves in veins (especially in the extremities), and breathing.

Checking Circulation (p. 496)

Pulse (p. 496)

1. Pulse is the alternate expansion and elastic recoil of an artery with each heartbeat. It may be felt in any artery that lies near the surface or over a hard tissue.
2. A normal rate is between 70 and 80 beats per minute.

Measurement of Blood Pressure (p. 497)

1. Blood pressure is the pressure exerted by blood on the wall of an artery when the left ventricle undergoes systole and then diastole. It is measured by the use of a sphygmomanometer.
2. Systolic blood pressure is the force of blood recorded during ventricular contraction. Diastolic blood pressure is the force of blood recorded during ventricular relaxation. The average blood pressure is 120/80 mm Hg.
3. Pulse pressure is the difference between systolic and diastolic pressure. It averages 40 mm Hg and provides information about the condition of arteries.

Circulatory Routes (p. 498)

1. The largest circulatory route is the systemic circulation.
2. Two of the several subdivisions of the systemic circulation are the coronary (cardiac) circulation and the hepatic portal circulation.
3. Other routes include the cerebral, pulmonary, and fetal circulation.

Systemic Circulation (p. 498)

1. The systemic circulation takes oxygenated blood from the left ventricle through the aorta to all parts of the body, including lung tissue.
2. The aorta is divided into the ascending aorta, the arch of the aorta, and the descending aorta. Each section gives off arteries that branch to supply the whole body.
3. Blood is returned to the heart through the systemic veins. All the veins of the systemic circulation flow into either the superior or inferior venae cavae or the coronary sinus. They in turn empty into the right atrium.

Pulmonary Circulation (p. 520)

1. The pulmonary circulation takes deoxygenated blood from the right ventricle to the lungs and returns oxygenated blood from the lungs to the left atrium.
2. It allows blood to be oxygenated for systemic circulation.

Hepatic Portal Circulation (p. 520)

1. The hepatic portal circulation collects blood from the veins of the pancreas, spleen, stomach, intestines, and gallbladder and directs it into the hepatic portal vein of the liver.
2. This circulation enables the liver to utilize nutrients and detoxify harmful substances in the blood.

Fetal Circulation (p. 521)

1. The fetal circulation involves the exchange of materials between fetus and mother.
2. The fetus derives its oxygen and nutrients and eliminates its carbon dioxide and wastes through the maternal blood supply by means of a structure called the placenta.
3. At birth, when lung, digestive, and liver functions are established, the special structures of fetal circulation are no longer needed.

Exercise and the Cardiovascular System (p. 524)

1. Aerobic exercises provide useful benefits for the cardiovascular system.
2. Among the benefits are increased cardiac output, increased delivery of oxygen to tissues, reduction in systolic blood pressure, increase in high-density lipoprotein (HDL), weight control, and increased ability to dissolve blood clots.

Aging and the Cardiovascular System (p. 524)

1. General changes include loss of elasticity of blood vessels, reduction in cardiac muscle size, and reduced cardiac output.
2. The incidence of coronary artery disease (CAD), congestive heart failure, and atherosclerosis increases with age.

Developmental Anatomy of Blood Vessels and Blood (p. 524)

1. Blood vessels develop from isolated masses of mesenchyme in mesoderm called blood islands.
2. Blood is produced by the endothelium of blood vessels.

Disorders: Homeostatic Imbalances (p. 525)

1. An aneurysm is a thin, weakened section of the wall of an artery or vein that bulges outward forming a balloonlike sac.
2. Hypertension, or high blood pressure, is classified as primary and secondary.
3. Deep-venous thrombosis (DVT) refers to a blood clot in a deep vein, especially in the lower extremities.

REVIEW QUESTIONS

1. Describe the structural and functional differences among arteries, arterioles, capillaries, venules, and veins.
2. Discuss the importance of the elasticity and contractility of arteries.
3. Distinguish between elastic (conducting) and muscular (distributing) arteries in terms of location, histology, and function. What is an anastomosis? What is collateral circulation?
4. Describe how capillaries are structurally adapted for exchanging materials with body cells.
5. Define varicose veins and describe how they are treated.

6. What are blood reservoirs? Why are they important?
7. Why does blood flow faster in arteries and veins than in capillaries?
8. Describe how each of the following affects blood pressure: cardiac output (CO), blood volume, and peripheral resistance.
9. What is the vasomotor center? What are its functions?
10. Discuss how the vasomotor center operates with baroreceptors, chemoreceptors, and higher brain centers to control blood pressures.
11. Explain the effects of different chemicals, such as epinephrine, norepinephrine (NE), antidiuretic hormone (ADH), angiotensin II, histamine, and kinins on blood pressure.
12. What is meant by autoregulation? Describe its probable mechanism.
13. Describe how hydrostatic and osmotic pressures determine fluid movement through capillaries. Set up the equation for effective filtration pressure (P_{eff}) to substantiate your response.
14. What is Starling's law of the capillaries?
15. Describe the factors that assist the return of venous blood to the heart.
16. Define pulse. Where may pulse be felt?
17. Contrast tachycardia, bradycardia, and irregular pulse.
18. What is blood pressure (BP)? Describe how systolic and diastolic blood pressure are recorded by means of a sphygmomanometer.
19. Compare the clinical significance of systolic and diastolic pressure. How are these pressures written?
20. Define pulse pressure. What does this pressure indicate?
21. What is meant by a circulatory route? Define systemic circulation.
22. Diagram the major divisions of the aorta, their principal arterial branches, and the regions supplied.
23. Trace a drop of blood from the arch of the aorta through its systemic circulatory route to the tip of the big toe on your left foot and back to the heart again. Remember that the major branches of the arch are the brachiocephalic artery, left common carotid artery, and left subclavian artery. Be sure to indicate which veins return the blood to the heart.
24. What is the cerebral arterial circle (circle of Willis)? Why is it important?
25. What are visceral branches of an artery? Parietal branches?
26. What major organs are supplied by branches of the thoracic aorta? How is blood returned from these organs to the heart?
27. What organs are supplied by the celiac, superior mesenteric, renal, inferior mesenteric, inferior phrenic, and middle sacral arteries? How is blood returned to the heart?
28. Trace a drop of blood from the brachiocephalic artery into the digits of the right upper extremity and back again to the right atrium.
29. What are the three major groups of systemic veins?
30. Define pulmonary circulation. Prepare a diagram to indicate the route. What is the purpose of the route?
31. What is hepatic portal circulation? Describe the route by means of a diagram. Why is this route significant?
32. Discuss in detail the anatomy and physiology of fetal circulation. Be sure to indicate the function of the umbilical arteries, umbilical vein, ductus venosus, foramen ovale, and ductus arteriosus.
33. What are the effects of exercise on the cardiovascular system?
34. Describe the effects of aging on the cardiovascular system.
35. Describe the development of blood vessels and blood.
36. What is an aneurysm? Why is an aneurysm a serious problem?
37. Compare the causes of primary and secondary hypertension. How does hypertension affect the body? How is hypertension treated?
38. What is deep-venous thrombosis (DVT)?
39. Refer to the glossary of medical terminology associated with blood vessels. Be sure that you can define each term.

The Lymphatic System and Immunity

22

- Describe the components of the lymphatic system and list its functions.
- Describe the structure and origin of lymphatics and contrast them with veins.
- Describe the histological aspects of lymph nodes, tonsils, spleen, and thymus gland, and explain their functions.
- Trace the general plan of lymph circulation from lymphatics into the thoracic duct or right lymphatic duct.
- Discuss how edema develops.
- Explain the difference between nonspecific and specific resistance.
- Discuss the roles of the skin and mucous membranes, antimicrobial substances, phagocytosis, inflammation, and fever as components of nonspecific resistance.
- Define immunity and explain the relationship between an antigen and an antibody.
- Contrast the main characteristics of antigens and antibodies.
- Contrast the role of T cells in cellular immunity and the role of B cells in humoral immunity.
- Explain the role of the skin in immunity.
- Define a monoclonal antibody and explain its importance.
- Discuss the relationship of immunology to cancer.
- Describe the development of the lymphatic system.
- Describe the clinical symptoms of the following disorders: hypersensitivity (allergy), tissue rejection, autoimmune diseases, acquired immune deficiency syndrome (AIDS), severe combined immunodeficiency (SCID), and Hodgkin's disease.
- Define medical terminology associated with the lymphatic system.

The **lymphatic** (lim-FAT-ik) **system** consists of a fluid called lymph, vessels that convey lymph called lymphatics, and a number of structures and organs, all of which contain lymphatic (lymphoid) tissue (see Figure 22-2b). Essentially, lymphatic tissue is a specialized form of reticular connective tissue that contains large numbers of lymphocytes. The stroma (framework) of lymphatic tissue is a meshwork of reticular fibers and reticular cells (fibroblasts and fixed macrophages). One exception to this is the thymus gland, which has a stroma of epithelioreticular tissue (discussed later in the chapter).

Lymphatic tissue is organized in various ways. Accumulations of lymphatic tissue not enclosed by a capsule are referred to as **diffuse lymphatic tissue.** This is the simplest form of lymphatic tissue and is found in the lamina propria (connective tissue) of mucous membranes of the gastrointestinal (GI) tract, respiratory passageways, urinary tract, and reproductive tract. It is also normally found in small amounts in the stroma of almost every organ of the body. **Lymphatic nodules** are unencapsulated oval-shaped concentrations of lymphatic tissue that usually consist of a central, lighter-staining region consisting of large lymphocytes (*germinal center*) and a peripheral, darker-staining region of small lymphocytes (cortex). Most lymphatic nodules are solitary, small, and discrete. Such nodules are found randomly in the lamina propria of mucous membranes of the gastrointestinal (GI) tract, respiratory passageways, urinary tract, and reproductive tract. Some lymphatic nodules occur in multiple, large aggregations in specific parts of the body. Among these are the tonsils in the pharyngeal region and aggregated lymphatic follicles (Peyer's patches) in the ileum of the small intestine (Chapter 24). Aggregations of lymphatic nodules also occur in the appendix. The **lymphatic organs** of the body—the lymph nodes, spleen, and thymus gland—all contain lymphatic tissue enclosed by a connective tissue capsule. Since bone marrow produces lymphocytes, it is also a component of the lymphatic system.

The lymphatic system has several functions. Lymphatics drain protein-containing fluid from tissue spaces that escape from blood capillaries. The proteins, which cannot be directly reabsorbed by blood vessels, are returned to the cardiovascular system by lymphatics. Lymphatics also transport fats from the gastrointestinal (GI) tract to the blood. Lymphatic tissue also functions in surveillance and defense, that is, lymphocytes, with the aid of macrophages, protect the body from foreign cells, microbes, and cancer cells. Lymphocytes recognize foreign cells and substances, microbes, and cancer cells, and respond to them in two general ways. Some lymphocytes (T cells) destroy them directly or indirectly by releasing various substances. Other lymphocytes (B cells) differentiate into plasma cells that secrete antibodies against foreign substances to help eliminate them. Overall, the lymphatic system concentrates foreign substances in certain lymphatic organs, circulates lymphocytes through the organs to make contact with the foreign substances, and destroys the foreign substances and eliminates them from the body.

The developmental anatomy of the lymphatic system is considered later in the chapter.

LYMPHATIC VESSELS

Lymphatic vessels originate as microscopic vessels in spaces between cells called **lymph capillaries** (Figure 22-1a). Lymph capillaries may occur singly or in extensive plexuses. They originate throughout the body, but not in avascular tissue, the central nervous system, splenic pulp, and bone marrow. They are slightly larger and more permeable than blood capillaries.

Lymph capillaries also differ from blood capillaries in that they end blindly; blood capillaries have an arterial and a venous end. In addition, lymph capillaries are structurally adapted to ensure the return of proteins to the circulation when they leak out of blood capillaries. Close examination of lymph capillaries reveals that the endothelial cells making up the capillary wall overlap each other, forming pores (Figure 22-1b). This overlapping arrangement permits fluid to flow easily into the capillary but prevents the flow of fluid out of the capillary, much like a one-way valve would operate. Note also that the outer surfaces of the endothelial cells of the capillary wall are attached to the surrounding tissue by structures called *anchoring filaments.* During edema, there is an excessive accumulation of fluid in the tissue, causing tissue swelling. This swelling produces a pull on the anchoring filaments, opening the pores even more so that more fluid can flow into the lymph capillary.

Just as blood capillaries converge to form venules and veins, lymph capillaries unite to form larger and larger lymph vessels called **lymphatics** (Figure 22-2). Lymphatics resemble veins in structure, but have thinner walls and more valves, and contain lymph nodes at various intervals. Lymphatics of the skin travel in loose subcutaneous tissue and generally follow veins. Lymphatics of

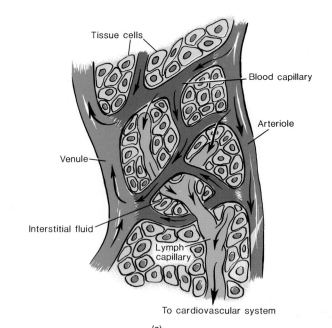

FIGURE 22-1 Lymph capillaries. (a) Relationship of lymph capillaries to tissue cells and blood capillaries.

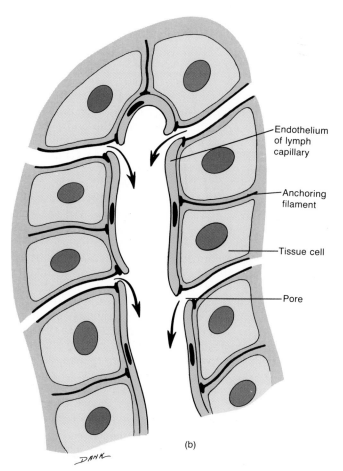

FIGURE 22-1 (*Continued*) (b) Details of a lymph capillary.

the viscera generally follow arteries, forming plexuses around them. Ultimately, lymphatics converge into two main channels—the thoracic duct and the right lymphatic duct. These will be described shortly.

CLINICAL APPLICATION

Lymphangiography (lim-fan'jē-OG-ra-fē) is the x-ray examination of lymphatic vessels and lymph organs after they are filled with a radiopaque substance. Such an x ray is called a **lymphangiogram** (lim-FAN-jē-ō-gram). Lymphagiograms are useful in detecting edema and carcinomas, and in locating lymph nodes for surgical or radiotherapeutic treatment.

LYMPHATIC TISSUE

LYMPH NODES

The oval or bean-shaped structures located along the length of lymphatics are called **lymph nodes.** They are scattered throughout the body, usually in groups, and range from 1 to 25 mm (0.04 to 1 inch) in length. A lymph node contains a slight depression on one side called a *hilus* (HĪ-lus) or *hilum,* where blood vessels and efferent lymphatic vessels leave the node (Figure 22-3). Each node is covered by a *capsule* of dense connective tissue that extends into the node. The capsular extensions are called *trabeculae* (tra-BEK-yoo-lē). Internal to the capsule is a supporting network of reticular fibers and reticular cells (fibroblasts and macrophages). The capsule, trabeculae,

FIGURE 22-2 Lymphatic system. (a) Schematic representation of the relationship of the lymphatic system to the cardiovascular system.

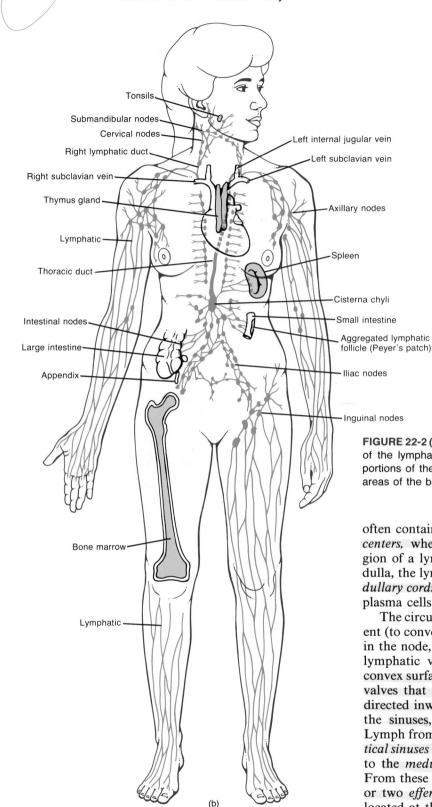

Tonsils
Submandibular nodes
Cervical nodes
Right lymphatic duct
Right subclavian vein
Thymus gland
Lymphatic
Thoracic duct
Intestinal nodes
Large intestine
Appendix
Bone marrow
Lymphatic

Left internal jugular vein
Left subclavian vein
Axillary nodes
Spleen
Cisterna chyli
Small intestine
Aggregated lymphatic follicle (Peyer's patch)
Iliac nodes
Inguinal nodes

(b)

(c)

FIGURE 22-2 (*Continued*) (b) Location of the principal components of the lymphatic system. (c) The light gold area indicates those portions of the body drained by the right lymphatic duct. All other areas of the body are drained by the thoracic duct.

often contain lighter-staining central areas, the *germinal centers*, where lymphocytes are produced. The inner region of a lymph node is called the *medulla*. In the medulla, the lymphocytes are arranged in strands called *medullary cords*. These cords also contain macrophages and plasma cells.

The circulation of lymph through a node involves afferent (to convey toward a center) lymphatic vessels, sinuses in the node, and efferent (to convey away from a center) lymphatic vessels. *Afferent lymphatic vessels* enter the convex surface of the node at several points. They contain valves that open toward the node so that the lymph is directed inward. Once inside the node, the lymph enters the sinuses, which are a series of irregular channels. Lymph from the afferent lymphatic vessels enters the *cortical sinuses* just inside the capsule. From here it circulates to the *medullary sinuses* between the medullary cords. From these sinuses the lymph usually circulates into one or two *efferent lymphatic vessels.* The efferent vessel is located at the hilus of the lymph node. It is wider than the afferent vessels and contains valves that open away from the node to convey the lymph out of the node.

Lymph nodes are scattered throughout the body, usually in groups (see Figure 22-2b). Typically, these groups are arranged in two sets: *superficial* and *deep*.

Lymph passing from tissue spaces through lymphatics on its way back to the cardiovascular system is filtered

and reticular fibers and cells constitute the stroma (framework) of a lymph node. The parenchyma of a lymph node is specialized into two regions: cortex and medulla. The outer *cortex* contains densely packed lymphocytes arranged in masses called *lymphatic nodules*. The nodules

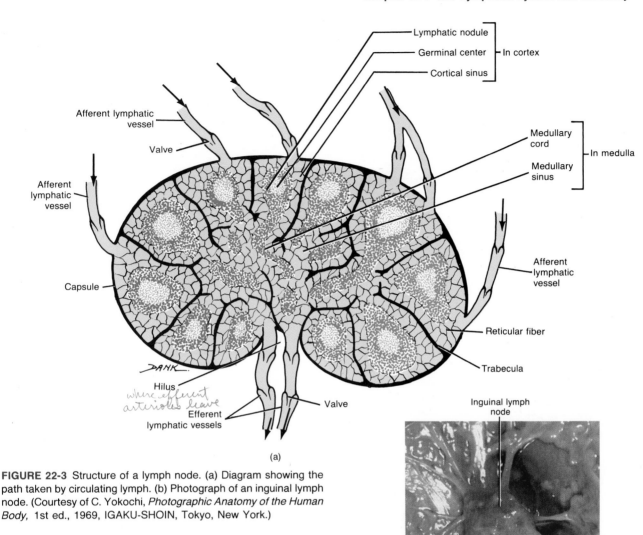

FIGURE 22-3 Structure of a lymph node. (a) Diagram showing the path taken by circulating lymph. (b) Photograph of an inguinal lymph node. (Courtesy of C. Yokochi, *Photographic Anatomy of the Human Body,* 1st ed., 1969, IGAKU-SHOIN, Tokyo, New York.)

through lymph nodes. As lymph passes through the nodes, it is filtered of foreign substances. These substances are trapped by the reticular fibers within the node. Then, macrophages destroy the foreign substances by phagocytosis, T cells may destroy them by releasing various products, and/or B cells may produce antibodies that destroy them. Lymph nodes also produce lymphocytes, some of which can circulate to other parts of the body.

Histological features of lymphatics and lymph nodes are shown in Figure 22-4.

CLINICAL APPLICATION

Knowledge of the location of the lymph nodes and the direction of lymph flow is important in the diagnosis and prognosis of the spread of cancer by **metastasis.** Cancer cells usually spread by way of the lymphatic system and produce aggregates of tumor cells where they lodge. Such secondary tumor sites are predictable by the direction of lymph flow from the organ primarily involved.

TONSILS

Tonsils are multiple aggregations of large lymphatic nodules embedded in a mucous membrane. The tonsils are arranged in a ring at the junction of the oral cavity and pharynx. The single **pharyngeal** (fa-RIN-jē-al) **tonsil** or **adenoid** is embedded in the posterior wall of the nasopharynx (see Figure 23-2b). The paired **palatine** (PAL-a-tīn) **tonsils** are situated in the tonsillar fossae between the pharyngopalatine and glossopalatine arches (see Figure 17-2c). These are the ones commonly removed by a tonsillectomy. The paired **lingual** (LIN-gwal) **tonsils** are located at the base of the tongue and may also have to be removed by a tonsillectomy (see Figure 17-2c).

The tonsils are situated strategically to protect against invasion of foreign substances. Functionally, the tonsils produce lymphocytes and antibodies.

FIGURE 22-4 Histology of lymphatics and lymph nodes. (a) Photomicrograph of a lymphatic at a magnification of 180×. (b) Photomicrograph of a lymph node at a magnification of 25×. (Photomicrographs Copyright © 1983 by Michael H. Ross. Used by permission.)

SPLEEN

The oval **spleen** is the largest mass of lymphatic tissue in the body, measuring about 12 cm (5 inches) in length. It is situated in the left hypochondriac region between the fundus of the stomach and diaphragm (see Figure 1-8d). Its *visceral surface* (Figure 22-5a) contains the contours of the organs adjacent to it—the gastric impression (stomach), renal impression (left kidney), and colic impression (left flexure of colon). The *diaphragmatic* (dī-a-fra-MAT-ik) *surface* is smooth and convex and conforms to the concave surface of the diaphragm to which it is adjacent.

The spleen is surrounded by a capsule of dense connective tissue and scattered smooth muscle fibers (Figure 22-5b). The capsule, in turn, is covered by a serous membrane, the peritoneum. Like lymph nodes, the spleen contains a hilus, trabeculae, and reticular fibers and cells. The capsule, trabeculae, and reticular fibers and cells constitute the stroma of the spleen.

The parenchyma of the spleen consists of two different kinds of tissue called white pulp and red pulp (Figure 22-5b,c). *White pulp* is essentially lymphatic tissue, mostly lymphocytes, arranged around arteries called central arteries. In various areas, the lymphocytes are thickened

(a)

(b)

(c)

FIGURE 22-5 Structure of the spleen. (a) Diagram of visceral surface. (b) Photograph of a section through the spleen showing white and red pulp. (Courtesy of C. Yokochi and J. W. Rohen, *Photographic Anatomy of the Human Body,* 2nd ed., 1978, IGAKU-SHOIN, Ltd., Tokyo, New York.) (c) Photomicrograph of a portion of the spleen at a magnification of 60×. (Copyright © 1983 by Michael H. Ross. Used by permission.)

into lymphatic nodules referred to as *splenic nodules* (*Malpighian corpuscles*). The *red pulp* consists of *venous sinuses* filled with blood and cords of splenic tissue called *splenic (Billroth's) cords.* Veins are closely associated with the red pulp. Splenic cords consist of erythrocytes, macrophages, lymphocytes, plasma cells, and granulocytes.

The splenic artery and vein and the efferent lymphatics pass through the hilus. Since the spleen has no afferent lymphatic vessels or lymph sinuses, it does not filter lymph. One key splenic function related to immunity is the production of B lymphocytes, which develop into antibody-producing plasma cells. The spleen also phagocy-

tizes bacteria and worn-out and damaged red blood cells and platelets. In addition, the spleen stores and releases blood in case of demand, such as during hemorrhage. The release seems to be purely sympathetic. The sympathetic impulses cause the smooth muscle of the capsule of the spleen to contract. During early fetal development, the spleen participates in blood cell formation.

About 10 percent of the population has *accessory spleens.* They are most commonly found near the hilus of the primary spleen or embedded in the tail of the pancreas. In general, accessory spleens are about 1 cm (0.5 inch) in diameter.

THYMUS GLAND

Usually a bilobed lymphatic organ, the **thymus gland** is located in the superior mediastinum, posterior to the sternum and between the lungs (Figure 22-6a). The two **thymic lobes** are held in close proximity by an enveloping layer of connective tissue. Each lobe is enclosed by a connective tissue **capsule.** The capsule gives off extensions into the lobes called **trabeculae,** which divide the lobes into **lobules.** Each lobule consists of a deeply staining peripheral **cortex** and a lighter-staining central **medulla.** The cortex is composed almost entirely of small, medium, and large tightly packed lymphocytes held in place by reticular tissue fibers (Figure 22-6b). Since the reticular tissue differs in origin and structure from that usually found in other lymphatic organs, it is referred to as **epithelioreticular** supporting tissue. The medulla consists mostly of epithelial cells and more widely scattered lymphocytes, and its reticulum is more cellular than fibrous. In addition, the medulla contains characteristic **thymic (Hassall's) corpuscles,** concentric layers of epithelial cells. Their significance is unknown.

The thymus gland is conspicuous in the infant, and it reaches its maximum size of about 40 grams during puberty. After puberty, most of the thymic tissue is replaced by fat and connective tissue. By the time the person reaches maturity, the gland has atrophied substantially.

Its role in immunity is to help produce T cells that destroy invading microbes directly or indirectly by producing various substances. This will be discussed in detail shortly.

LYMPH CIRCULATION

ROUTE

When plasma is filtered by blood capillaries, it passes into the interstitial spaces; it is then known as interstitial fluid. Recall from Chapter 21 that fluid movement between blood capillaries and body cells depends on hydrostatic and osmotic pressures. When this fluid passes from interstitial spaces into lymph capillaries, it is called lymph. Lymph from lymph capillaries is then passed to lymphatics that run toward lymph nodes. At the nodes, afferent vessels penetrate the capsules at numerous points, and the lymph passes through the sinuses of the nodes. Efferent vessels from the nodes either run with afferent vessels into another node of the same group or pass on to another group of nodes. From the most proximal group of each chain of nodes, the efferent vessels unite to form **lymph trunks.** The principal trunks are the *lumbar, intestinal, bronchomediastinal, subclavian,* and *jugular trunks* (Figure 22-7).

Thoracic (Left Lymphatic) Duct

The principal trunks pass their lymph into two main channels, the thoracic duct and the right lymphatic duct. The **thoracic (left lymphatic) duct** is about 38 to 45 cm (15

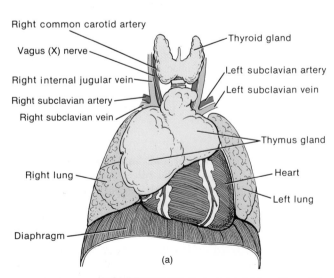

Right common carotid artery
Vagus (X) nerve
Right internal jugular vein
Right subclavian artery
Right subclavian vein
Right lung
Diaphragm
Thyroid gland
Left subclavian artery
Left subclavian vein
Thymus gland
Heart
Left lung

(a)

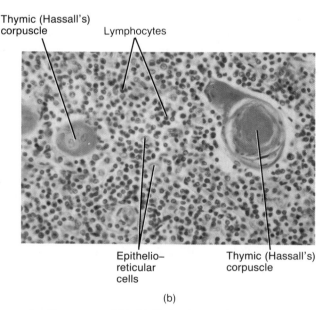

Thymic (Hassall's) corpuscle
Lymphocytes
Epithelio-reticular cells
Thymic (Hassall's) corpuscle

(b)

FIGURE 22-6 Thymus gland. (a) Location of the thymus gland in a young child. (b) Photomicrograph of an enlarged aspect of a thymic corpuscle at a magnification of 600×. (Copyright © 1983 by Michael H. Ross. Used by permission.)

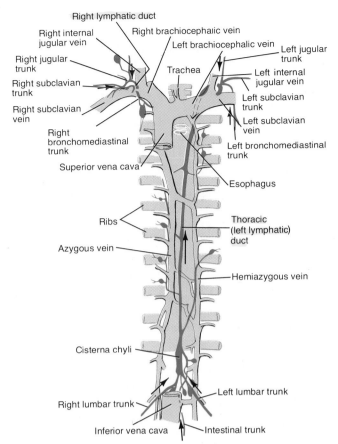

FIGURE 22-7 Relationship of lymph trunks to the thoracic duct and right lymphatic duct. Also see Figure 22-2.

*Labels in figure:*
Right lymphatic duct
Right internal jugular vein
Right brachiocephalic vein
Left brachiocephalic vein
Left jugular trunk
Right jugular trunk
Trachea
Left internal jugular vein
Right subclavian trunk
Right subclavian vein
Left subclavian trunk
Left subclavian vein
Right bronchomediastinal trunk
Left bronchomediastinal trunk
Superior vena cava
Esophagus
Ribs
Thoracic (left lymphatic) duct
Azygous vein
Hemiazygous vein
Cisterna chyli
Right lumbar trunk
Left lumbar trunk
Inferior vena cava
Intestinal trunk

to 18 inches) in length and begins as a dilation in front of the second lumbar vertebra called the *cisterna chyli* (sis-TER-na KĪ-lē). (See Figure 22-2b also.) The thoracic duct is the main collecting duct of the lymphatic system and receives lymph from the left side of the head, neck, and chest, the left upper extremity, and the entire body below the ribs.

The cisterna chyli receives lymph from the right and left lumbar trunks and from the intestinal trunk. The lumbar trunks drain lymph from the lower extremities, wall and viscera of the pelvis, kidneys, suprarenals, and the deep lymphatics from most of the abdominal wall. The intestinal trunk drains lymph from the stomach, intestines, pancreas, spleen, and visceral surface of the liver. In the neck, the thoracic duct also receives lymph from the left jugular, left subclavian, and left bronchomediastinal trunks. The left jugular trunk drains lymph from the left side of the head and neck; the left subclavian trunk drains lymph from the upper left extremity; and the left bronchomediastinal trunk drains lymph from the left side of the head and neck; the left subclavian trunk drains lymph from the upper left extremity; and the left bronchomediastinal trunk drains lymph from the left side of the deeper parts of the anterior thoracic wall, upper part of the anterior abdominal wall, anterior part of the diaphragm, left lung, and left side of the heart.

Right Lymphatic Duct

The **right lymphatic duct** is about 1.25 cm (0.5 inch) long and drains lymph from the upper right side of the body (see Figure 22-2b). The right lymphatic duct collects lymph from its trunks as follows (Figure 22-7). It receives lymph from the right jugular trunk, which drains the right side of the head and neck, from the right subclavian trunk, which drains the right upper extremity, and from the right bronchomediastinal trunk, which drains the right side of the thorax, right lung, right side of the heart, and part of the convex surface of the liver.

Ultimately, the thoracic duct empties all of its lymph into the junction of the left internal jugular vein and left subclavian vein, and the right lymphatic duct empties all of its lymph into the junction of the right internal jugular vein and right subclavian vein. Thus, lymph is drained back into the blood and the cycle repeats itself continuously.

MAINTENANCE

The flow of lymph from tissue spaces to the large lymphatic ducts to the subclavian veins is maintained primarily by the milking action of muscle tissue. Skeletal muscle contractions compress lymph vessels and force lymph toward the subclavian veins. Lymph vessels, like veins, contain valves, and the valves ensure the movement of lymph toward the subclavian veins (see Figure 22-4a).

Another factor that maintains lymph flow is respiratory movements. These movements create a pressure gradient between the two ends of the lymphatic system. Lymph flows from the abdominal region, where the pressure is higher, toward the thoracic region, where it is lower.

CLINICAL APPLICATION

Edema, an excessive accumulation of interstitial fluid in tissue spaces, may be caused by an obstruction, such as an infected node or a blockage of vessels, in the pathway between the lymphatic capillaries and the subclavian veins. Another cause is excessive lymph formation and increased permeability of blood capillary walls. A rise in capillary blood pressure, in which interstitial fluid is formed faster than it is passed into lymphatics, also may result in edema.

NONSPECIFIC RESISTANCE TO DISEASE

The human body continually attempts to maintain homeostasis by counteracting harmful stimuli in the environment. Frequently, these stimuli are disease-producing organisms, called *pathogens* (PATH-ō-jens), or their toxins. The ability to ward off disease is called **resistance.** Lack of resistance is called **susceptibility.** Defenses against disease may be grouped into two broad areas: nonspecific resistance and specific resistance. Nonspecific resistance is inherited and represents a wide variety of body reactions against a wide range of pathogens. Specific resistance or **immunity** involves the production of a specific antibody against a specific pathogen or its toxin. Immunity is developed during a person's life; it is not inherited. We will first consider mechanisms of nonspecific resistance.

SKIN AND MUCOUS MEMBRANES

The skin and mucous membranes of the body are commonly regarded as the first line of defense against disease-causing microorganisms. They possess certain mechanical and chemical factors that are involved in combating the initial attempt of a microbe to cause disease.

Mechanical Factors

The **intact skin,** as noted in Chapter 5, consists of two distinct portions called the dermis and epidermis. If you consider the closely packed cells of the epidermis, its continuous layering, and the presence of keratin, you can see that the intact skin provides a formidable physical barrier to the entrance of microbes. The intact surface of healthy epidermis seems to be rarely penetrated by bacteria. But when the epithelial surface is broken, a subcutaneous infection often develops. The bacteria most likely to cause such an infection are staphylococci, which normally inhabit the hair follicles and sweat glands of the skin. Moreover, when the skin is moist, as in hot, humid climates, dermal infections, especially fungus infections, such as athlete's foot, are quite common.

Mucous membranes, like the skin, also consist of an epithelial layer and an underlying connective tissue layer. But unlike the skin, a mucous membrane lines a body cavity that opens to the exterior. Examples include the membranes lining the entire digestive, respiratory, urinary, and reproductive tracts. The epithelial layer of a mucous membrane secretes a fluid called *mucus,* which prevents the cavities from drying out. Since mucus is slightly viscous, it traps many microbes that enter the respiratory and digestive tracts. The mucous membrane of the nose has mucus-coated **hairs** that trap and filter air containing microbes, dust, and pollutants. The mucous membrane of the upper respiratory tract contains **cilia,** microscopic hairlike projections of the epithelial cells (see Figure 23-4a). These cilia move in such a manner that they pass inhaled dust and microbes that have become trapped in mucus toward the throat. This so-called "ciliary escalator" keeps the "mucus blanket" moving toward the throat at a rate of 1 to 3 cm per hour. Coughing and sneezing speed up the "escalator."

Since some pathogens can thrive on the moist secretions of a mucous membrane, they are able to penetrate the membrane if present in sufficient numbers. This penetration may be related to toxic products produced by the microbes, prior injury by viral infections, or mucosal irritations. Although mucous membranes do inhibit the entrance of many microbes, they are less effective than the skin.

There are several other mechanical factors that help to protect epithelial surfaces of the skin and mucous membranes. One such mechanism that protects the eyes is the **lacrimal** (LAK-ri-mal) **apparatus** (see Figure 17-3a).

It consists of a group of structures that manufactures and drains away tears, which are spread over the surface of the eyeball by blinking. Normally, the tears are carried away by evaporation or pass into the nose as fast as they are produced. The continual washing action of tears helps to keep microbes from settling on the surface of the eye. If an irritating substance or large number of microbes make contact with the eye, the lacrimal glands start to oversecrete. Tears then accumulate more rapidly than they can be carried away. This is also a protective mechanism to dilute and wash away the irritating substance or microbes.

Saliva, produced by the salivary glands, washes microbes from the surfaces of the teeth and the mucous membrane of the mouth, much as tears wash the eyes. This helps to prevent colonization by microbes.

Microbes are additionally prevented from entering the lower respiratory tract by a small lid of cartilage called the **epiglottis** that covers the voice box during swallowing (see Figure 23-2b). The cleansing of the urethra by the **flow of urine** represents another mechanical factor that prevents microbial colonization in the urinary system.

Chemical Factors

Mechanical factors alone do not account for the high degree of resistance of skin and mucous membranes to microbial invasion. Certain chemical factors play an important role. **Sebum,** produced by the sebaceous (oil) glands of the skin, forms a protective film over the surface of the skin and inhibits the growth of certain bacteria. **Perspiration** flushes some microbes from the skin. **Gastric juice** is a collection of hydrochloric acid, enzymes, and mucus produced by the glands of the stomach. The very high acidity of gastric juice (pH 1.2–3.0) is sufficient to preserve the usual sterility of the stomach. The acidity of gastric juice destroys bacteria and almost all important bacterial toxins.

The **acidic pH** of the skin, between 3 and 5, is due in part to the acid products of bacterial metabolism. This acidity probably discourages the growth of many microbes that contact the skin. One of the components of sebum is **unsaturated fatty acids.** These fatty acids kill certain pathogenic bacteria, such as *Streptococcus pyogenes,* the causative agent of scarlet fever and streptococcal sore throat. **Lysozyme** is an enzyme capable of breaking down cell walls of various bacteria under certain conditions. Lysozyme is normally found in perspiration, tears, saliva, nasal secretions, and tissue fluids. (The role of the skin in immunity is considered later in the chapter.)

ANTIMICROBIAL SUBSTANCES

In addition to the mechanical and chemical barriers of the skin and mucous membranes, the body also produces certain antimicrobial substances. Among these are interferon, complement, and properdin.

Interferon (IFN)

Host cells infected with viruses produce a protein called **interferon** (in'-ter-FĒR-on) or **IFN.** Actually, there are three principal types of interferon called alpha, beta, and gamma. In humans, IFN has been shown to be produced by lymphocytes and other leucocytes and fibroblasts. Once released from virus-infected cells, IFN diffuses to uninfected neighboring cells and binds to surface receptors. This somehow induces uninfected cells to synthesize another antiviral protein that inhibits intracellular viral replication. Interferon appears to be the body's first line of defense against infection by many different viruses. It appears to decrease the virulence (disease-producing power) of viruses associated with chicken-pox, genital herpes, rabies, rubella, chronic hepatitis, shingles, eye infections, encephalitis, and at least one type of common cold.

The suggested relationship between viruses and cancer has prompted scientists to study the effects of interferon on cancer patients. Data from a number of clinical trials indicate that large doses of pure interferon produced by genetic-engineering techniques have only limited effects against some types of tumors and no effect against others. Among the human cancers shown to respond in some way to interferon therapy are osteogenic carcinoma, multiple myeloma, and certain types of leukemias and lymphomas. Two recombinant preparations of interferon appear somewhat useful in treating two malignant conditions, melanoma and renal cell cancer, that do not respond well to other therapies.

Interferon use against cancer in clinical trials has produced side effects ranging from mild to life-threatening. For example, patients may experience fatigue, loss of appetite, malaise, fever, chills, acute confusional states, motor slowing, disorientation, seizures, or cardiac complications.

A form of alpha interferon, called alpha-2-interferon, has been incorporated into a nasal spray that may prevent the spread of colds among family members.

Complement

Another antimicrobial substance that is very important to nonspecific resistance (and immunity) is complement. **Complement** is actually a group of eleven proteins found in normal blood serum. The system is called complement because it "complements" certain immune and allergic reactions involving antibodies. These will be discussed shortly. The function of antibody is to recognize the microbe as a foreign organism, form an antigen-antibody complex, and activate complement for attack. The antigen-antibody complex also fixes (attaches) the complement to the surface of the invading microbe. Once complement is activated, it destroys microbes as follows:

1. Some complement proteins initiate a series of reactions leading to **cell lysis** caused by holes in the plasma membrane of the microbe.
2. Some complement proteins interact with receptors on phagocytes, promoting phagocytosis. This is called **opsonization** (op-sō-ni-ZA-shun) or **immune adherence.**
3. Some complement proteins contribute to the development of **inflammation** by causing the release of **histamine** from mast cells, leucocytes, and platelets. Histamine increases the permeability of blood capillaries, a process that enables leucocytes to penetrate tissues in order to combat infection or allergy.

4. Some complement proteins serve as **chemotactic** (kē-mō-TAK-tic) **agents,** attracting large numbers of leucocytes to the area.

Properdin

Properdin (prō-PER-din; *pro* = in behalf of, *perdere* = to destroy), like complement, is also a protein found in serum. It is a complex consisting of three proteins. Properdin, acting together with complement, leads to the destruction of several types of bacteria, the enhancement of phagocytosis, and triggering of inflammatory responses.

PHAGOCYTOSIS

When microbes penetrate the skin and mucous membranes, or bypass the antimicrobial substances in blood, there is another nonspecific resistance of the body called phagocytosis. Very simply, **phagocytosis** (*phagein* = to eat; *cyto* = cell) means the ingestion and destruction of microbes or any foreign particulate matter by certain cells called phagocytes.

Kinds of Phagocytes

The kinds of phagocytes that participate in the process fall into two broad categories: microphages and macrophages. The granulocytes of blood are **microphages** (MĪ-krō-fā-jez). However, not all granulocytes exhibit the same phagocytic capabilities. Neutrophils have the most prominent phagocytic activity. Eosinophils are believed to have some phagocytic capability, and the role of basophils in phagocytosis is debatable.

When an infection occurs, both microphages (especially neutrophils) and monocytes migrate to the infected area. During this migration, the monocytes enlarge and develop into actively phagocytic cells called **macrophages** (MAK-rō-fā-jez). Since these cells leave the blood and migrate to infected areas, they are called **wandering macrophages.** Some of the macrophages, called **fixed macrophages** or **histiocytes** (HIS-tē-ō-sīts) enter certain tissues and organs of the body and remain there. Fixed macrophages are found in the liver (stellate reticuloendothelial cells), lungs (alveolar macrophages), brain (microglia), spleen, lymph nodes, and bone marrow.

Mechanism

For convenience of study, phagocytosis will be divided into two phases: adherence and ingestion. **Adherence (attachment)** is the formation of firm contact between the plasma membrane of the phagocyte and the microbe (or foreign material). In some instances, adherence occurs easily, and the microbe is readily phagocytized (see Figure 3-6a,b). In other cases, adherence is more difficult, but the particle can be phagocytized if the phagocyte traps the particle to be ingested against a rough surface, like a blood vessel, blood clot, or connective tissue fibers, where it cannot slide away. This is sometimes called **nonimmune (surface) phagocytosis.** Or bacteria can be phagocytized if they are first coated with complement or antibody to promote the attachment of the microbe to the phagocyte. This is **opsonization (immune adherence)**. A final factor that helps adherence is chemotaxis, the attraction of phagocytes to microbes by certain chemicals, such as microbial products, components of white blood cells and tissue cells, and chemicals derived from complement.

Following adherence, **ingestion** occurs. In the process of ingestion, projections of the cell membrane of the phagocyte, called pseudopodia, engulf the microbe (Figure 22-8). Once the microbe is surrounded, the membrane folds inward, forming a sac around the microbe called a **phagocytic vesicle.** The vesicle pinches off from the membrane and enters the cytoplasm. Within the cytoplasm, it collides with lysosomes that contain digestive enzymes and bactericidal substances. Upon contact, the vesicular and lysosomal membranes fuse together to form a single, larger structure called a **phagolysosome (digestive vesicle).** Within the phagolysosome, most bacteria are usually killed within 10 to 30 minutes. It is assumed that destruction occurs as a result of the contents of the lysosomes: lactic acid (which lowers the pH in the phagolysosome), the production of hydrogen peroxide, lysozyme, and the destructive capabilities of enzymes that degrade carbohydrates, proteins, lipids, and nucleic acids. Any indigestible materials that cannot be degraded further are found in structures called **residual bodies.** The cell disposes of materials in residual bodies by exocytosis, in which the residual body migrates to the plasma membrane, fuses to it, ruptures, and releases its contents.

Some microbes, such as toxin-producing staphylococci, may be ingested, but are not necessarily killed. In fact, their toxins can actually kill the phagocytes. Other microbes, such as the tubercle bacillus, may multiply within the phagolysosome and eventually destroy the phagocyte. Still other microbes, such as the causative agents of tularemia and brucellosis, may remain dormant in phagocytes for months or years at a time.

INFLAMMATION

When cells are damaged by microbes, physical agents, or chemical agents, the injury sets off an **inflammation (inflammatory response).** The injury may be viewed as a form of stress.

Symptoms

Inflammation is usually characterized by four fundamental symptoms: **redness, pain, heat,** and **swelling.** A fifth symptom can be the **loss of function** of the injured area. Whether loss of function occurs depends on the site and extent of the injury. The inflammatory response serves a protective and defensive role. It is an attempt to neutralize and destroy toxic agents at the site of injury and to prevent their spread to other organs. Thus the inflammatory response is an attempt to restore tissue homeostasis.

The immediate inflammatory response to tissue injury consists of a complicated sequence of physiological and anatomical adjustments. Various body components are involved in the initial response: blood vessels; intercellular fluid mixed with parts of injured cells, called the *exudate* (EKS-yoo-dāt); the cellular components of blood; and the surrounding epithelial and connective tissues. Other factors that affect the inflammatory response are the individual's age and general state of health. Healing processes of all types exert a great demand on the body's store of nutrients. Thus nutrition plays an essential role in healing.

Stages

The inflammatory response is one of the body's internal systems of defense. The response of a tissue to a rusty nail wound is similar to other inflammatory responses, such as the sore throat that results from bacterial or viral invasion. These are the basic stages of response.

1. Vasodilation and increased permeability of blood vessels. Immediately following tissue damage, there is vasodilation and increased permeability of blood vessels in the area of the injury. **Vasodilation** is an increase in diameter in the blood vessels. **Increased permeability** means that substances normally retained in blood are permitted to pass from the blood vessels. Vasodilation allows more blood to go to the damaged area, and increased permeability permits defensive substances in the blood to enter the injured area. Such defensive substances include white blood cells and clot-forming chemicals. The increased blood supply also removes toxic products and dead cells, preventing them from complicating the injury. These toxic substances include waste products released by invading microorganisms. As the blood clots around the site of activity, the microbe and its toxins are prevented from spreading to other parts of the body.

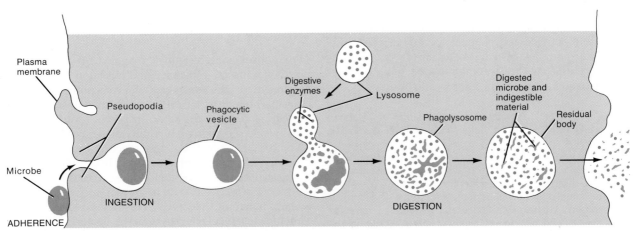

FIGURE 22-8 Phases of phagocytosis.

Vasodilation and increased permeability are caused by the release of certain chemicals by damaged cells in response to injury. One such substance is *histamine* (Figure 22-9a). It is present in many tissues of the body, especially in mast cells in connective tissue, circulating basophils (one type of white blood cell), and blood platelets. Histamine is released in direct response to any injured cells containing it. Neutrophils, another type of white blood cell attracted to the site of injury, can also produce chemicals that cause the release of histamine. Other substances that play a role in vasodilation and increased permeability of blood vessels are *kinins.* These chemicals also attract neutrophils to the injured area. Inflammation also results in an increased synthesis of *prostaglandins (PG)*, especially of the E series. Prostaglandins are potent vasodilators by themselves and they also intensify the effect of histamine and kinins. Prostaglandins also bring about increased permeability of blood vessels.

In addition to the responses in the area of injury, the body may also respond by increasing the metabolic rate and quickening the heartbeat so that more blood circulates to the injured area per minute. Within minutes after the injury, the quickened metabolism and circulation and, especially, the dilation and increased permeability of capillaries produce heat, redness, and swelling. The heat results from the large amount of warm blood that accumulates in the area and, to a certain extent, from

the heat energy produced by the metabolic reactions. The large amounts of blood in the area are also responsible for the redness.

Pain, whether immediate or delayed, is a cardinal symptom of inflammation. It can result from an injury to nerve fibers or from an irritation caused by the release of toxic chemicals from microorganisms. Pain may also be due to increased pressure from edema. Prostaglandins intensify and prolong the pain associated with inflammation. However, kinins affect some nerve endings, causing much of the pain associated with inflammation.

2. Phagocyte migration. Generally, within an hour after the inflammatory process is initiated, phagocytes (microphages and macrophages) appear on the scene. As the flow of blood decreases, microphages (neutrophils) begin to stick to the inner surface of the endothelium (lining) of blood vessels. This is called **margination.** Then, the neutrophils begin to squeeze through the wall of the blood vessel to reach the damaged area. This migration, which resembles amoeboid movement, is called **diapedesis** (di'-a-pe-DĒ-sis). It can take as little as 2 minutes (Figure 22-9b). The movement of neutrophils depends on **chemotaxis** (kē'-mō-TAK-sis; *chemo* = chemicals, *taxis* = arrangement), the attraction of neutrophils by certain chemicals. Neutrophils are attracted by microbes, kinins, and other neutrophils. A steady stream of neutrophils is ensured by the production and release of additional cells from bone marrow. This is brought

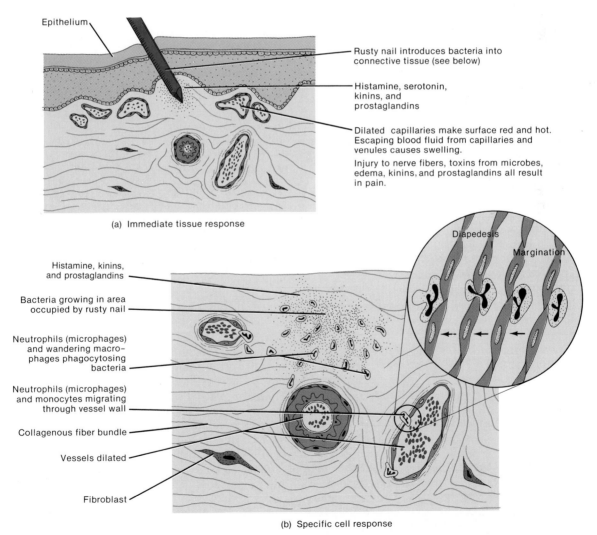

(a) Immediate tissue response

Epithelium

Rusty nail introduces bacteria into connective tissue (see below)

Histamine, serotonin, kinins, and prostaglandins

Dilated capillaries make surface red and hot. Escaping blood fluid from capillaries and venules causes swelling.

Injury to nerve fibers, toxins from microbes, edema, kinins, and prostaglandins all result in pain.

Histamine, kinins, and prostaglandins

Bacteria growing in area occupied by rusty nail

Neutrophils (microphages) and wandering macrophages phagocytosing bacteria

Neutrophils (microphages) and monocytes migrating through vessel wall

Collagenous fiber bundle

Vessels dilated

Fibroblast

Diapedesis

Margination

(b) Specific cell response

FIGURE 22-9 Tissue and cell response to an injury. (a) Immediate tissue response. (b) Specific cell response.

about by a substance called **leucocytosis-promoting factor,** which is released from inflamed tissues. Neutrophils attempt to destroy the invading microbes by phagocytosis. Neutrophils also contain chemicals with antibiotic activity called **defensins,** so named because of their apparent role in preventing and overcoming infections. Defensins are active against bacteria, fungi, and viruses, unlike other antibiotics that are directed against specific microbes.

As the inflammatory response continues, monocytes follow the neutrophils into the infected area. Once in the tissue, monocytes become transformed into wandering macrophages that augment the phagocytic activity of fixed macrophages. In addition, fixed macrophages mobilize under the stimulus of inflammation and, as wandering macrophages, also migrate to the infected area. The neutrophils predominate in the early stages of infection, but tend to die off rapidly. Macrophages enter the picture during a later stage of the infection. They are several times more phagocytic than neutrophils and large enough to engulf tissue that has been destroyed, neutrophils that have been destroyed, and invading microbes.

3. **Release of nutrients.** Nutrients stored in the body are used to support the defensive cells. They are also used in the increased metabolic reactions of the cells under attack.

4. **Fibrin formation.** The blood contains a soluble protein called **fibrinogen** (fī-BRIN-ō-jen). The increased permeability of capillaries causes leakage of fibrinogen to tissues. Fibrinogen is then converted to an insoluble, thick network called **fibrin** (FĪ-brin), which localizes and traps the invading organisms, preventing their spread. This network eventually forms a fibrin clot that prevents hemorrhage and isolates the infected area.

5. **Pus formation.** In all but very mild inflammations, **pyogenesis** (*pyo* = pus; *genesis* = to produce) occurs. **Pus** is a thick fluid that contains living, as well as nonliving, white blood cells and debris from other dead tissue.

CLINICAL APPLICATION

If the pus cannot drain out of the body, an abscess develops. An **abscess** is simply an excessive accumulation of pus in a confined space. Common examples are pimples and boils. When inflamed tissue is shed many times, it produces an open sore, called an **ulcer,** on the surface of an organ or tissue. Ulcers may result from a prolonged inflammatory response to a continuously injured tissue. For instance, overproduction of digestive acids in the stomach may cause a steady erosion of the epithelial tissue lining the stomach. People with poor circulation are susceptible to ulcers in the tissues of their legs. The ulcers develop when the tissues are continuously damaged by a shortage of oxygen and nutrients.

FEVER

The most frequent cause of **fever,** an abnormally high body temperature, is infection from bacteria (and their toxins) and viruses. The high body temperature inhibits some microbial growth and speeds up body reactions that aid repair. (Fever is discussed in more detail in Chapter 25.)

A summary of some of the various components of nonspecific resistance is presented in Exhibit 22-1.

IMMUNITY (SPECIFIC RESISTANCE TO DISEASE)

Despite the variety of mechanisms of nonspecific resistance, they all have one thing in common. They are designed to protect the body from any kind of pathogen. They are not specifically directed against a particular microbe. Specific resistance to disease, called **immunity,** involves the production of a specific type of cell or specific molecule (antibody) to destroy a particular antigen. If antigen 1 invades the body, antibody 1 is produced against it. If antigen 2 invades the body, antibody 2 is produced against it, and so on. The branch of science that deals with the responses of the body when challenged by antigens is referred to as **immunology** (*immunis* = free; im'-yoo-NOL-ō-jē). We will examine the components of specific resistance by first examining the nature of antigens and antibodies.

ANTIGENS

Definition

An **antigen (immunogen)** is any chemical substance that, when introduced into the body, causes the body to produce specific antibodies, which can react with the antigen. Antigens thus have two important characteristics. The first is **immunogenicity** (im'-yoo-nō-jen-IS-it-ē), the ability to stimulate the formation of specific antibodies. The second is **reactivity,** the ability of the antigen to react specifically with the produced antibodies. An antigen with both of these characteristics is called a **complete antigen.**

Characteristics

Chemically, the vast majority of antigens are proteins, nucleoproteins (nucleic acid + protein), lipoproteins (lipid + protein), glycoproteins (carbohydrate + protein), and certain large polysaccharides. In general, they have molecular weights of 10,000 or greater.

The entire microbe, such as a bacterium or virus, or components of microbes may act as an antigen. For example, bacterial structures such as flagella, capsules, and cell walls are also antigenic; bacterial toxins are highly antigenic. Nonmicrobial examples of antigens include pollen, egg white, incompatible blood cells, and transplanted tissues and organs. The myriad of antigens in the environment provides many opportunities for the production of antibodies by the body.

Antibodies do not form against the whole antigen. At specific regions on the surface of the antigen, called **antigenic determinant sites** (Figure 22-10), specific chemical groups of the antigen combine with the antibody. This combination depends upon the size and shape of the determinant site and the manner in which it corresponds to the chemical structure of the antibody. The combination is very much like the lock and key analogy used to describe the combination of enzyme and substrate molecule.

The number of antigenic determinant sites on the surface of an antigen is called *valence.* Most antigens are *multivalent,* that is, they have several antigenic determinant sites. It is believed that in order to induce antibody

EXHIBIT 22-1   SUMMARY OF NONSPECIFIC RESISTANCE

COMPONENT	FUNCTIONS	COMPONENT	FUNCTIONS
Skin and mucous membranes		**Acid pH of skin**	Discourages growth of many microbes.
Mechanical factors		**Unsaturated fatty acids**	Antibacterial substance in sebum.
Intact skin	Forms a physical barrier to the entrance of microbes.	**Lysozyme**	Antimicrobial substance in perspiration, tears, saliva, nasal secretions, and tissue fluids.
Mucous membranes	Inhibit the entrance of many microbes, but not as effective as intact skin.	**Antimicrobial substances**	
Mucus	Traps microbes in respiratory and digestive tracts.	**Interferon (IFN)**	Protects uninfected host cells from viral infection.
Hairs	Filter microbes and dust in nose.	**Complement**	Causes lysis of microbes, promotes phagocytosis, contributes to inflammation, serves as chemotactic agent.
Cilia	Together with mucus, trap and remove microbes and dust from upper respiratory tract.	**Properdin**	Works with complement to bring about same responses as complement.
Lacrimal apparatus	Tears dilute and wash away irritating substances and microbes.	**Phagocytosis**	Ingestion and destruction of foreign particulate matter by microphages and macrophages.
Saliva	Washes microbes from surfaces of teeth and mucous membranes of mouth.	**Inflammation**	Confines and destroys microbes and repairs tissues.
Epiglottis	Prevents microbes and dust from entering trachea.	**Fever**	Inhibits microbial growth and speeds up body reactions that aid repair.
Urine	Washes microbes from urethra.		
Chemical Factors			
Gastric juice	Destroys bacteria and most toxins in stomach.		

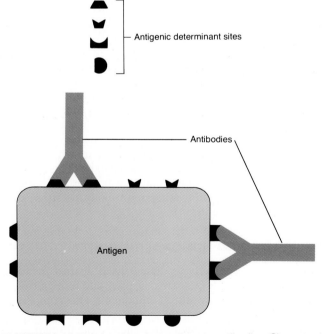

FIGURE 22-10 Relationship of an antigen to antibodies. Since most antigens contain more than one antigenic determinant site, they are referred to as multivalent. Most human antibodies are bivalent; they have two reaction sites that are complementary to the antigenic determinant sites for which they are specific. Each antibody has reaction sites for specific antigenic determinant sites only.

formation, an antigen must have at least two determinant sites (*bivalent*). If an antigen is chemically broken down, it is possible to separate the determinant site. Its molecular weight, compared to that of the antigen as a whole, is small. It might be only 200 to 1,000. An isolated determinant site still has the ability to react with an antibody in response to the original antigen (reactivity), but does not have the ability to stimulate the production of antibodies (immunogenicity) when injected into an animal. Its lower molecular weight prevents it from serving as a complete antigen. A determinant site that has reactivity but not immunogenicity is called a **partial antigen** or **hapten** (*haptein* = to grasp). A hapten can stimulate an immune response if it is attached to a larger carrier molecule so that the combined molecule has two determinant sites. Some drugs of low molecular weight, like penicillin, may combine with proteins of high molecular weight in the body. This combination forms two determinant sites, and the complex becomes antigenic. Antibodies formed against the complex are responsible for the allergic reactions to drugs and other chemicals.

As a rule, antigens are foreign substances. They are not usually part of the chemistry of the body. The body's own substances, recognized as "self," do not act as antigens; substances identified as "nonself," however, stimulate antibody production. However, there are certain conditions in which the distinction between self and nonself breaks down, and antibodies that attack the body are produced. These conditions (to be described later) result in what are called autoimmune disease.

Now let us look at some of the characteristics of antibodies.

ANTIBODIES

Definition

An **antibody** is a protein produced by the body in response to the presence of an antigen and is capable of combining specifically with the antigen. This is essentially the complementary definition of an antigen.

The specific fit of the antibody with antigen depends not only on the size and shape of the antigenic determinant site but also on the corresponding antibody site, much like the lock and key analogy (Figure 22-10). An antibody, like an antigen, also has a valence. Whereas most antigens are multivalent, antibodies are bivalent or multivalent. The majority of human antibodies are bivalent.

Antibodies belong to a group of proteins called globulins, and for this reason they are also known as **immunoglobulins** (im'-yoo-nō-GLOB-yoo-lins) or **Ig.** Five different classes of immunoglobulins are known to exist in humans. These are designated as IgG, IgA, IgM, IgD, and IgE. Each has a distinct chemical structure and a specific biological role. IgG antibodies enhance phagocytosis, neutralize toxins, and protect the fetus and newborn. IgA antibodies provide localized protection on mucosal surfaces; IgM antibodies are especially effective against microbes by causing agglutination and lysis. IgD antibodies may be involved in stimulating antibody-producing cells to manufacture antibodies. IgE antibodies are involved in allergic reactions. It is interesting to note that during stress IgA levels decrease. Depending on an individual's personality, this decrease can lower resistance to infection.

Structure

Since antibodies are proteins, they consist of polypeptide chains. Most antibodies contain two pairs of polypeptide chains (Figure 22-11). Two of the chains are identical to each other and are called *heavy (H) chains.* Each consists of more than 400 amino acids. The other two chains, also identical to each other, are called *light (L) chains,* and each consists of 200 amino acids. The antibody consists of identical halves held together by disulfide bonds (S—S). Each half consists of a heavy and a light chain, also held together by a disulfide bond (S—S). Overall, the antibody molecule sometimes assumes the shape of a letter Y. At other times, it resembles a letter T. The reasons for these differences in shape will be explained shortly.

Within each H and L chain, there are two distinct regions. The tops of the H and L chains, called the *variable portions,* contain the antigen-binding site. The variable portion is different for each kind of antibody, and it is

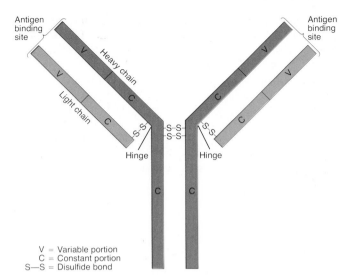

V = Variable portion
C = Constant portion
S—S = Disulfide bond

FIGURE 22-11 Diagrammatic representation of the portions of an antibody molecule.

this portion that allows the antibody to recognize and attach specifically to a particular type of antigen. Since most antibodies have two variable portions for attachment of antigens, they are bivalent. The remainder of each polypeptide chain is called the *constant portion.* The constant portion is the same in all antibodies of the same class and is responsible for the type of antigen-antibody reaction that occurs. However, the constant portion differs from one class of antibody to another, and its structure serves as a basis for distinguishing the classes.

In performing its role in immunity, the antibody is believed to behave as a switch. When antibodies are viewed in the electron microscope before combination with antigen, they resemble the letter T. But after combination with antigen, they appear to be smaller and resemble the letter Y. It is possible that the binding of antigen causes a rearrangement of the structure of antibody. If you examine Figure 22-11 again, you will note that there is a hinge area along the H chains; this hinge provides the antibody molecule with flexibility. When antigen and antibody are combined, the rearrangement of antibody might consist of a pivoting movement in which the H chain moves from a T shape into a Y shape. This pivoting could cause exposure of the constant portion, which then functions in a particular antigen-antibody reaction, such as fixing complement.

CELLULAR AND HUMORAL IMMUNITY

The ability of the body to defend itself against invading agents such as bacteria, toxins, viruses, and foreign tissues consists of two closely allied components. One component consists of the formation of specially sensitized lymphocytes that have the capacity to attach to the foreign agent and destroy it. This is called **cellular (cell-mediated) immunity** and is particularly effective against fungi, parasites, intracellular viral infections, cancer cells, and foreign tissue transplants. In the other component, the body produces circulating antibodies that are capable of attacking an invading agent. This is called **humoral (antibody-mediated) immunity** and is particularly effective against bacterial and viral infections.

Cellular immunity and humoral immunity are the product of the body's lymphoid tissue. The bulk of lymphoid tissue is located in the lymph nodes, but it is also found in the spleen and gastrointestinal (GI) tract and, to a lesser extent, in bone marrow. The placement of lymphoid tissue is strategically designed to intercept an invading agent before it can spread too extensively into the general circulation.

Formation of T Cells and B Cells

Lymphoid tissue consists primarily of lymphocytes that may be distinguished into two kinds. **T cells** are responsible for cellular immunity. **B cells** develop into specialized cells (plasma cells) that produce antibodies and provide humoral immunity.

Both types of lymphocytes are derived originally in the embryo from lymphocytic stem cells in bone marrow (Figure 22-12). Before migrating to their positions in lymphoid tissues, the descendants of the stem cells follow two distinct pathways. About half of them first migrate to the thymus gland, where they are processed to become T cells. The name T cell is derived from the processing that occurs in the thymus gland. The thymus gland in some way confers what is called *immunologic competence* on the T cells. This means that they develop the ability to differentiate into cells that perform specific immune reactions. They then leave the thymus gland and become embedded in lymphoid tissue. Immunologic competence is conferred by the thymus gland shortly before birth

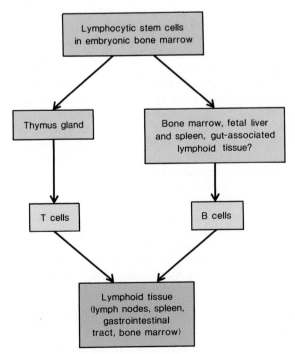

FIGURE 22-12 Origin and differentiation of T cells and B cells.

and for a few months after birth. Removal of the thymus gland from an animal before it has processed the T cells results in a failure to develop cellular immunity. Even after T cells leave the thymus gland to take up residence in other tissues, thymic hormones continue to stimulate them and their descendants.

The remaining stem cells are processed in some unknown area of the body, possibly bone marrow, fetal liver and spleen, or gut-associated lymphoid tissue. They become the B cells. They are given this name because in birds they are processed in the bursa of Fabricius, a small pouch of lymphoid tissue attached to the intestine. Once processed, B cells migrate to lymphoid tissue and take up their positions there. Although both T cells and B cells occupy the same lymphoid tissues, they localize in separate areas of the tissues.

T Cells, B Cells, and Macrophages

Unlike the various nonspecific resistance factors previously discussed, the responses of T cells and B cells very much depend on the ability of the cells to recognize specific antigens. This specific recognition is based on the specific fit between antigenic determinant sites of antigens and variable portions of antibodies. However, for T and B cells to perform their vital functions in immunity, another cell must participate. This cell, which has already been discussed in previous chapters, is called a macrophage. Macrophages process and present antigens to T cells and B cells. The processing probably occurs when the macrophages phagocytize the antigens; the partially digested antigen is then displayed on the surface of the macrophage and presented to the T or B cell for recognition. During the processing and presentation of antigen, macrophages secrete *interleukin 1,* a powerful protein that induces proliferation of T cells and B cells. Once these cells become active, they, in turn, induce macrophages to participate in the destruction of antigens.

T Cells and Cellular Immunity

There are literally thousands of different types of T cells, and each is capable of responding to a specific antigen or set of antigens. This forms the basis for cellular immunity. At any given time, most T cells are inactive. When an antigen enters the body, only the particular T cell specifically programmed to react with the antigen becomes activated. Such an activated T cell is said to be *sensitized.* Activation occurs when macrophages phagocytize the antigen and present it to the T cell. Sensitized T cells increase in size, differentiate, and divide, each differentiated cell giving rise to a *clone,* or population of cells identical to itself (Figure 22-13). Several subpopulations of cells within the clone can be recognized: killer T cells, helper T cells, suppressor T cells, delayed hypersensitivity T cells, amplifier T cells, and memory T cells.

Killer T cells (cytotoxic T cells) leave the lymphoid tissue and migrate to the site of invasion, where several things happen. They attach to the invading cell and secrete *lymphotoxins* that destroy the antigen directly. Next, the lymphotoxins produce holes in the plasma membranes

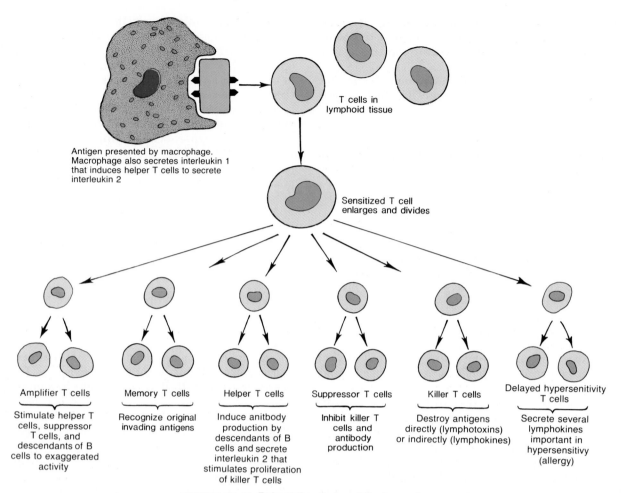

Antigen presented by macrophage. Macrophage also secretes interleukin 1 that induces helper T cells to secrete interleukin 2

T cells in lymphoid tissue

Sensitized T cell enlarges and divides

Amplifier T cells
Stimulate helper T cells, suppressor T cells, and descendants of B cells to exaggerated activity

Memory T cells
Recognize original invading antigens

Helper T cells
Induce anitbody production by descendants of B cells and secrete interleukin 2 that stimulates proliferation of killer T cells

Suppressor T cells
Inhibit killer T cells and antibody production

Killer T cells
Destroy antigens directly (lymphotoxins) or indirectly (lymphokines)

Delayed hypersenitivity T cells
Secrete several lymphokines important in hypersensitivy (allergy)

FIGURE 22-13 Role of T cells in cellular immunity.

of invading cells, resulting in their lysis. Most of the protection provided by T cells is the result of secretion of very powerful proteins called **lymphokines.** For example, the cells can release a lymphokine called *transfer factor,* a substance that reacts with nonsensitized lymphocytes at the site of invasion, causing them to take on the same characteristics as the sensitized killer T cells. In this way, additional lymphocytes are recruited, and the effect of the sensitized killer T cells is intensified. Killer T cells also secrete a lymphokine called *macrophage chemotactic factor,* a substance that attracts macrophages to the site of invasion where they destroy the antigens by phagocytosis. Another lymphokine released by killer T cells, called *macrophage activating factor (MAF),* greatly increases the phagocytic activity of macrophages. A lymphokine called *migration inhibition factor (MIF)* prevents macrophages from migrating away from the site of infection, and a lymphokine called *mitogenic factor* causes uncommitted or nonsensitized lymphocytes to divide more rapidly. In summary, then, killer T cells can destroy antigens directly by releasing lymphotoxins and indirectly by secreting lymphokines that recruit additional lymphocytes, attract macrophages, intensify phagocytosis by macrophages, prevent macrophages from migrating away from the site of infection, and induce rapid division of uncommitted or nonsensitized lymphocytes. Killer T cells also secrete interferon, which exerts inhibition of viral replica-

tion and enhances the killing action of killer T cells themselves. Killer T cells are especially effective against slowly developing bacterial diseases, such as tuberculosis and brucellosis, some viruses, fungi, transplanted cells, and cancer cells.

Helper T cells cooperate with B cells to help amplify antibody production. Although helper T cells do not themselves secrete antibodies, antigens first interact with them before inducing antibody production by the descendants of B cells. Helper T cells also secrete *interleukin 2,* a lymphokine that stimulates proliferation of killer T cells. The secretion of interleukin 2 is stimulated by interleukin 1 that is released by macrophages. Helper T cells also secrete proteins that amplify the inflammatory response and killing by macrophages.

Suppressor T cells dampen parts of the immune response by shutting down certain activities several weeks after infection activates them. They can inhibit secretion of injurious substances by killer T cells and inhibit production of antibodies by plasma cells. The normal ratio of helper T cells to suppressor T cells is 2:1.

Delayed hypersensitivity T cells are the source of several lymphokines, such as MIF and MAF, that are key factors in the hypersensitivity (allergy) response.

Amplifier T cells somehow stimulate helper T cells, suppressor T cells, and descendants of B cells to exaggerated levels of activity.

Memory T cells are programmed to recognize the original invading antigen. Should the pathogen invade the body at a later date, the memory cells initiate a far swifter reaction than during the first invasion. In fact, the second response is so swift that the pathogens are usually destroyed before any signs or symptoms of the disease occur.

Natural Killer (NK) Cells

There is another population of lymphocytes in the body that have certain characteristics similar to killer T cells. These lymphocytes, referred to as **natural killer (NK) cells,** have the ability to kill certain cells spontaneously, without interacting with lymphocytes or antigens. Like killer T cells, NK cells lyse target cells. Although NK cells may be a separate line of T cells, they are not killer T cells. NK cells provide an immediate first line of defense against viruses, cancer cells, and perhaps other microbes. NK cells can produce interferon that inhibits viral replication and enhances their own killing action by lysis. It has been observed that the number of NK cells is decreased in cancer patients and that the decrease is related to the severity of the disease.

B Cells and Humoral Immunity

The body contains not only thousands of different T cells but also thousands of different B cells, each capable of responding to a specific antigen. Whereas killer T cells leave their reservoirs of lymphoid tissue to meet a foreign antigen, B cells respond differently. They differentiate into cells that produce specific antibodies that then circulate in the lymph and blood to reach the site of invasion. When a foreign antigen has been prepared and presented by macrophages to B cells in lymph nodes, the spleen, or lymphoid tissue in the gastrointestinal (GI) tract, B cells specific for the antigen are activated. Some of them enlarge and divide and differentiate into a clone of **plasma cells** under the influence of thymic hormones (Figure 22-14). Plasma cells secrete the antibody. B cells proliferation and differentiation into plasma cells is also influenced by interleukin 1, secreted by macrophages. The phenomenal rate of antibody secretion is about 2,000 molecules per second for each cell, and it occurs for several days until the plasma cell dies (in 4 to 5 days). The activated B cells that do not differentiate into plasma cells remain as **memory B cells,** ready to respond more rapidly and forcefully should the same antigen appear at a future time.

Different antigens stimulate different B cells to develop into plasma cells and their accompanying memory B cells because the B cells of a particular clone are capable of secreting only one kind of antibody. Each specific antigen activates only those B cells already predetermined to secrete antibody specific to that antigen. The specific antigen selects a specific B cell because that particular B cell has on its surface the antibody molecules that it is capable of producing. The surface antibodies serve as receptor sites with which the antigen can combine.

Once the antigen-antibody complex is formed, the antibody activates complement enzymes for attack and fixes the complement to the surface of the antigen.

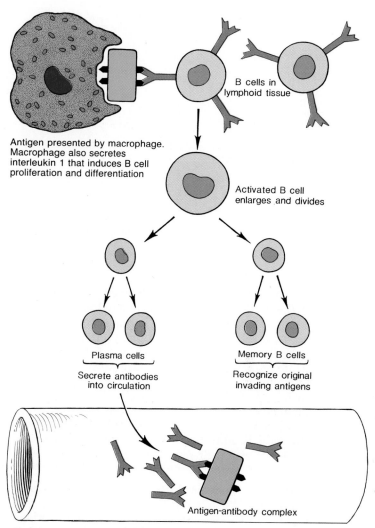

Antigen presented by macrophage. Macrophage also secretes interleukin 1 that induces B cell proliferation and differentiation

B cells in lymphoid tissue

Activated B cell enlarges and divides

Plasma cells
Secrete antibodies into circulation

Memory B cells
Recognize original invading antigens

Antigen-antibody complex

FIGURE 22-14 Role of B cells in humoral immunity. Although not illustrated, helper T cells assist B cells in antibody production.

In most cases, only the interaction of macrophage and B cell is required for the sequence of responses that results in antibody secretion. However, in some cases, interaction of B cells with helper T cells and suppressor T cells is involved. Recall that helper T cells can amplify antibody production, whereas suppressor T cells can inhibit it.

CLINICAL APPLICATION

The immune response of the body, whether cellular or humoral, is much more intense after a second or subsequent exposure to an antigen than after the initial exposure. This can be demonstrated by measuring the amount of antibody in serum, called the *antibody titer* (TĪ-ter). After an initial contact with an antigen, there is a period of several days during which no antibody is present. Then there is a slow rise in the antibody titer, followed by a gradual decline. Such a response of the body to the first contact is called the **primary response** (Figure 22-15). During the primary response, in which the body is said to be primed or sensitized, there is proliferation of immunocompetent lympho-

FIGURE 22-15 Primary and anamnestic (secondary) responses to the injection of an antigen.

cytes. When the antigen is contacted again, whether for the second time or the two hundredth, there is an immediate proliferation of immunocompetent lymphocytes, and the antibody titer is far greater than that for the primary response. This accelerated, more intense response is called the **anamnestic** (an'-am-NES-tik; *anamne* = recall) or **secondary response.** The reason for the anamnestic response is that some of the immunocompetent lymphocytes formed during the primary response remain as memory cells. These memory cells not only add to the pool of cells that can respond to the antigen—their response is more intense as well.

Primary and anamnestic responses occur during microbial infection. When you recover from an infection without taking antibiotics, it is usually because of the primary response. If, at a later time, you contact the same microbe, the anamnestic response could be so swift that the microbes are quickly destroyed and you do not exhibit any signs or symptoms.

You have probably guessed by now that the anamnestic response provides the basis for immunization against certain diseases. When you receive the initial immunization, your body is sensitized. Should you encounter the pathogen again as an infecting microbe or "booster dose," your body experiences the anamnestic response. The booster dose literally boosts the antibody titer to a higher level. Since their effects are not permanent, boosters must be given periodically, and immunization schedules are designed to maintain high antibody titers.

A summary of the various cells involved in immune responses is presented in Exhibit 22-2.

THE SKIN AND IMMUNITY

The skin not only provides nonspecific resistance to disease because of its mechanical and chemical factors but it also is an active component of the immune system. Recall from Chapter 5 that the epidermis of the skin contains four principal kinds of cells: melanocytes, keratinocytes, Langerhans cells, and Granstein cells. The last three types of cells assume a role in immunity, presumably according to the following mechanism.

EXHIBIT 22-2 SUMMARY OF CELLS THAT ARE IMPORTANT IN IMMUNE RESPONSES

CELL	FUNCTION
Macrophage	Phagocytosis; processing and presentation of foreign antigens; secretion of interleukin 1 that stimulates secretion of interleukin 2 by helper T cells (stimulates proliferation of killer T cells) and induces proliferation of B cells.
Killer T cell	Lysis of foreign cells by lymphotoxins and release of various lymphokines that recruit and intensify killer T cell actions (transfer factor), attract macrophages (macrophage chemotactic factor), increase phagocytic activity of macrophages (macrophage activating factor), prevent macrophage migration from site of action (migration inhibition factor), and proliferation of uncommitted or nonsensitized lymphocytes (mitogenic factor). Killer T cells also secrete interferon.
Helper T cell	Cooperates with B cells to amplify antibody production by plasma cells and secrete interleukin 2, which stimulates proliferation of killer T cells.
Suppressor T cell	Inhibits secretion of injurious substances by killer T cells and inhibits antibody production by plasma cells.
Delayed hypersensitivity T cell	Secretes macrophage activating factor and migration inhibition factor, key substances related to hypersensitivity (allergy).
Amplifier T cell	Stimulates helper T cells, suppressor T cells, and B cells to exaggerated levels of activity.
Memory T cell	Remains in lymphoid tissue and recognizes original invading antigens, even years after infection.
Natural killer (NK) cell	Lymphocyte that destroys foreign cells by lysis and produces interferon.
B cell	Differentiates into antibody-producing plasma cell.
Plasma cell	Descendant of B cell that produces antibodies.
Memory B cell	Ready to respond more rapidly and forcefully than initially should the same antigen challenge the body in the future.

If an antigen penetrates the upper layers of keratinized cells of the epidermis, it binds to either Langerhans or Granstein cells. Langerhans cells present the antigen to helper T cells, which are also present in epidermis, thus activating the T cells. In response to interleukin 1 secreted by keratinocytes, T cells are stimulated to produce interleukin 2. This substance causes T cells to proliferate so that the number of T cells responding to antigen is greatly increased. The T cells then enter the lymphatic system and spread throughout the body.

Just as Langerhans cells interact with helper T cells, Granstein cells interact with suppressor T cells in the epidermis. Normally, however, the helper T cell response predominates to destroy the antigens. In those instances where Langerhans cells might be destroyed by ultraviolet radiation or bypassed by antigens that react directly with suppressor T cells, the suppressor T cell response predominates.

MONOCLONAL ANTIBODIES

Scientists have known for many years how to induce laboratory animals (or humans) to produce antibodies. By injecting a particular antigen, antibodies are produced against the antigen by plasma cells (descendants of B cells). However, since the antibodies are produced by many different plasma cells, they vary physically and chemically; they are not pure. What scientists have learned to do is isolate a single B cell and fuse it with a tumor cell that is capable of proliferating endlessly. The resulting hybrid cell is called a **hybridoma cell.** These cells are a long-term source of substantial quantities of pure antibodies called **monoclonal antibodies.** Such antibodies will go after one and only one antigen.

One clinical use for monoclonal antibodies is to measure levels of a drug in a patient's blood. Other uses include the diagnosis of pregnancy, allergies, and diseases such as hepatitis, rabies, and certain sexually transmitted diseases. Compared with conventional diagnostic tests, monoclonal antibodies offer greater sensitivity, specificity, and speed. They have also been used to detect cancer at an early stage and to determine the extent of metastasis. One of the most exciting applications is in the area of treating diseases. Monoclonal antibodies, either alone or in combination with radioactivity, toxins, or drugs, are being used in trial studies to treat cancer. This approach has the advantage of destroying diseased tissues only, while sparing healthy tissues, thus overcoming some of the major adverse effects of chemotherapy and radiation treatment. Monoclonal antibodies may also be useful in preparing vaccines to counteract the rejection associated with transplants and to treat autoimmune diseases. One such disease that has been treated with some success is severe combined immunodeficiency (SCID), the most serious of all immune deficiency diseases. SCID victims are born lacking all immune defenses for fighting disease. This is the rare disease that affected the boy who was internationally known as the "boy in a bubble."

IMMUNOLOGY AND CANCER

When a normal cell becomes transformed into a cancer cell, the tumor cell assumes cell surface components called **tumor-specific antigens.** It is believed that the immune system usually recognizes tumor-specific antigens as "nonself" and destroys the cancer cells carrying them. Such an immune response is called **immunologic surveillance.** Most investigators believe that cellular immunity is the basic mechanism involved in tumor destruction.

In this regard, it is presumed that sensitized killer T cells react with tumor-specific antigens, initiating lysis of tumor cells. Sensitized macrophages may also be involved.

Despite the mechanism of immunologic surveillance, some cancer cells escape destruction, a phenomenon called **immunologic escape.** One possible explanation is that tumor cells may shed their tumor-specific antigens, thus evading recognition. Another theory is that antibodies produced by plasma cells bind to tumor-specific antigens, preventing recognition by killer T cells. It is reasonable to assume that immunologic techniques using monoclonal antibodies and toxins or radioactive compounds may be used in the future to prevent cancer. Hopefully, the antibody would selectively locate the cancer cell, and the attached toxic agent or radioactive substance would destroy the cell but cause little or no damage to healthy tissue. There is great expectation that this approach will become analogous to the use of antibiotics in treating infectious diseases.

DEVELOPMENTAL ANATOMY OF THE LYMPHATIC SYSTEM

The lymphatic system begins its development by the end of the fifth week. *Lymphatic vessels* develop from **lymph sacs** that arise from developing veins. Thus, the lymphatic system is also derived from **mesoderm.**

The first lymph sacs to appear are the paired **jugular lymph sacs** at the junction of the internal jugular and subclavian veins (Figure 22-16). From the jugular lymph sacs, capillary plexuses spread to the *thorax, upper extremities, neck,* and *head.* Some of the plexuses enlarge and form lymphatics in their respective regions. Each jugular lymph sac retains at least one connection with

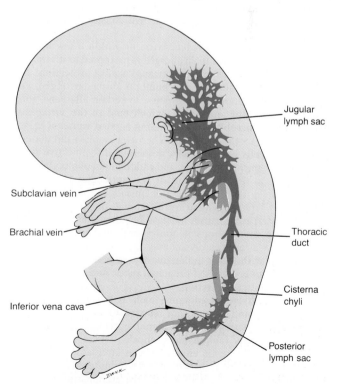

FIGURE 22-16 Development of the lymphatic system.

its jugular vein, the left one developing into the superior portion of the thoracic duct (left lymphatic duct).

The next lymph sac to appear is the unpaired **retroperitoneal lymph sac** at the root of the mesentery of the intestine. It develops from the primitive vena cava and mesonephric (primitive kidney) veins. Capillary plexuses and lymphatics spread from the retroperitoneal lymph sac to the *abdominal viscera* and *diaphragm*. The sac establishes connections with the cisterna chyli but loses its connections with neighboring veins.

At about the time the retroperitoneal lymph sac is developing, another lymph sac, the **cisterna chyli,** develops below the diaphragm on the posterior abdominal wall. It gives rise to the inferior portion of the *thoracic duct*

and the *cisterna chyli* of the thoracic duct. Like the retroperitoneal lymph sac, the cisterna chyli also loses its connections with surrounding veins.

The last of the lymph sacs, the paired **posterior lymph sacs,** develop from the iliac veins at their union with the posterior cardinal veins. The posterior lymph sacs produce capillary plexuses and lymphatics of the *abdominal wall, pelvic region,* and *lower extremity.* The posterior lymph sacs join the cisterna chyli and lose their connections with adjacent veins.

With the exception of the anterior part of the sac from which the cisterna chyli develops, all lymph sacs become invaded by **mesenchymal cells** and are converted into groups of *lymph nodes.*

DISORDERS: HOMEOSTATIC IMBALANCES

The antigen-antibody response is essential to survival and typically provides a state of immunity. Under certain circumstances, however, it may create problems. Three such problems are hypersensitivity (allergy), tissue rejection, and autoimmunity.

Hypersensitivity (Allergy)

A person who is overly reactive to an antigen is said to be **hypersensitive** or **allergic.** Whenever an allergic reaction occurs, there is tissue injury. The antigens that induce an allergic reaction are called **allergens.** Almost any substance can be an allergen for some individual. Common allergens include certain foods (milk, eggs), antibiotics such as penicillin, cosmetics, chemicals in plants such as poison ivy, pollens, dust, molds, and even microbes.

There are four basic types of hypersensitivity reactions: type I (anaphylaxis), type II (cytotoxic), type III (immune complex), and type IV (cell-mediated). The first three involve antibodies; the last involves T cells. Here we will consider only types I and IV.

Type I (anaphylaxis) reactions occur within a few minutes after a person sensitized to an allergen is reexposed to it. *Anaphylaxis* (an'-a-fi-LAK-sis) literally means "against protection" and results from the interaction of humoral antibodies (IgE) with mast cells and basophils. In response to certain allergens, some people produce IgE antibodies, which bind to the surfaces of mast cells and basophils. This binding is what causes a person to be allergic to the allergen. Whereas basophils circulate in blood, mast cells are especially numerous in connective tissue of the skin, respiratory system, and endothelium of blood vessels. In response to the attachment of IgE antibodies to basophils and mast cells, the cells release chemicals called mediators of anaphylaxis, among which are histamine and prostaglandins. Collectively, the mediators increase blood capillary permeability, increase smooth muscle contraction, and increase mucus secretion. As a result, a person may experience edema and redness, along with other inflammatory responses; difficulty in breathing from constricted bronchial tubes; and a "runny" nose from excess mucus secretion.

Some anaphylactic reactions, such as hay fever, bronchial asthma, hives, and eczema, are referred to as localized. Other anaphylactic reactions are considered systemic. An example is acute anaphylaxis (anaphylactic shock), which may produce life-threatening systemic effects such as circulatory shock and asphyxia and can be fatal within a few minutes. The effects of histamine may be counteracted by the administration of epinephrine.

Type IV (cell-mediated) reactions involve T cells and often are not apparent for a day or more. Type IV reactions occur when allergens that bind to tissue cells are phagocytized by macrophages and presented to receptors on the surfaces of T cells. This results in the proliferation of T cells, which respond by destroying the allergens. An example of a type IV reaction that involves the skin is the familiar skin test for tuberculosis.

Tissue Rejection

Transplantation involves the replacement of an injured or diseased tissue or organ. Usually, the body recognizes the proteins in the transplanted tissue or organ as foreign and produces antibodies against them. This phenomenon is known as **tissue rejection.** Rejection can be somewhat reduced by matching donor and recipient HLA antigens and by administering drugs that inhibit the body's ability to form antibodies. Recall from Chapter 19 that white blood cells and other nucleated cells have surface antigens called HLA antigens. These are unique for each person, except identical twins. The more closely matched the HLA antigens between donor and recipient, the less the likelihood of tissue rejection.

Types of Transplants

The most successful transplants are **autografts,** transplants in which one's own tissue is grafted to another part of the body (such as skin grafts for burn treatment or plastic surgery), and **isografts,** transplants in which the donor and recipient have identical genetic backgrounds.

An **allograft (homograft)** is a transplant between individuals of the same species, but with different genetic backgrounds. The success of this type of transplant has been moderate. Frequently, it is used as a temporary measure until the damaged or diseased tissue is able to repair itself. Skin transplants from other individuals and blood transfusions might properly be considered allografts. The one organ allograft that has been quite successful is the thymus. Children born without a thymus can now receive the gland from an aborted fetus. The thymus-deficient child cannot produce antibodies and thus cannot reject the transplant. Rejection later on indicates that the child is manufacturing antibodies and no longer needs the organ.

A **xenograft** is a transplant between animals of different species. Although this type of transplantation is used primarily as a physiological dressing over severe burns, a xenograph performed in October 1984 focused attention on the procedure. Baby Fae, an infant born prematurely with an invariably fatal form of congenital heart disease, was the recipient of a baboon's heart. After 20 days she died from heart and kidney problems related to rejection.

Immunosuppressive Therapy

Until recently, **immunosuppressive drugs** suppressed not only the recipient's immune rejection of the donor organ but also the immune response to all antigens as well. This causes patients to become very susceptible to infectious diseases. A drug called cyclosporine, derived from a fungus, has largely overcome this problem with regard to kidney, heart, and liver transplants. A selective immunosuppressive drug, it inhibits T cells that are responsible for tissue rejection, but has only a minimal effect on B cells. Thus, rejection is avoided and resistance against disease is still maintained.

Autoimmune Diseases

Under normal conditions, the body's immune mechanism is able to recognize its own tissues and chemicals. It normally does not produce T cells or B cells against its own substances. Such recognition of self is called **immunologic tolerance.** Although the mechanism of tolerance is not completely understood, it is believed that suppressor T cells may inhibit the differentiation of B cells into antibody-producing plasma cells, or inhibit helper T cells that cooperate with B cells to amplify antibody production.

At times, however, immunologic tolerance breaks down and the body has difficulty in discriminating between its own antigens and foreign antigens. This loss of immunologic tolerance leads to an **autoimmune disease (autoimmunity).** Such diseases are immunologic responses mediated by antibodies against a person's own tissue antigens. Among human autoimmune diseases are rheumatoid arthritis (RA), systemic lupus erythematosus (SLE), thyroiditis, rheumatic fever, glomerulonephritis, encephalomyelitis, hemolytic and pernicious anemia, Addison's disease, Grave's disease, possibly some forms of diabetes, myasthenia gravis, and multiple sclerosis (MS).

Severe Combined Immunodeficiency (SCID)

Severe combined immunodeficiency (SCID) is a rare immunodeficiency disease in which both B cells and T cells are missing or inactive in providing immunity. Perhaps the most famous patient with SCID was David, the "bubble boy," who lived in a sterile plastic chamber for all but 15 days of his life; he died on February 22, 1984, at age 12. He was the oldest untreated survivor of SCID.

David was placed in the sterile chamber shortly after birth to protect him from microbes that his body could not fight. In an effort to correct his disorder, David underwent a bone marrow transplant from his older sister, who was the most closely but still not perfectly matched donor available. Eighty days after the transplant and still in a germ-free environment, David developed some of the symptoms of infectious mononucleosis (IM), a condition caused by the Epstein-Barr virus (EBV). David was brought out of isolation for easier treatment, with the hope that the transplant would provide the same protection as his sterile plastic chamber. Unfortunately, the transplant was not successful, and David died about 4 months after it was performed. It was discovered that what killed David was not the immediate failure of the transplant but cancer. His B cells had proliferated as a result of the EBV. This is a very clear demonstration of a virus causing a cancer in humans.

Hodgkin's Disease

Hodgkin's disease is a malignant disorder, usually arising in lymph nodes, the cause of which is unknown. The histological diagnosis of the disease is made by the presence of large, malignant, multinucleate cells in the affected lymph node called Reed-Sternberg cells. The disease is initially characterized by a painless, nontender lymph node, most commonly in the neck, but occasionally in the axilla, inguinal, or femoral region. About one-quarter to one-third of patients also have an unexplained and persistent fever and/or night sweats. Fatigue and weight loss are also associated complaints, as is pruritus. Treatment consists of radiation therapy, chemotherapy, and combinations of the two. Hodgkin's disease is considered to be a curable malignancy.

Acquired Immune Deficiency Syndrome (AIDS)

Never before has science been confronted with an epidemic in which the primary disease only lowers the victim's immunity, and then a second unrelated disease produces the symptoms that may result in death. This primary disease, first recognized in June 1981, is called **acquired immune deficiency syndrome (AIDS).**

The cause of AIDS is a virus called human T-lymphotropic virus, type III (HTLV-III). A variant of this virus, known as HTLV-I, causes a rare type of leukemia in humans and attacks and transforms T cells. The reason the AIDS virus lowers the body's immune system is that the virus primarily attacks helper T cells. As a consequence, several key roles of helper T cells in the immune response are inhibited. For example, helper T cells cooperate with B cells to amplify antibody production. The reduction of the number of helper T cells inhibits antibody production by descendants of B cells against the AIDS virus and other microbes. Also, helper T cells secrete interleukin 2 that stimulates proliferation of killer T cells. With the loss of helper T cells, fewer killer T cells are available to destroy various antigens. In AIDS, the ratio of helper T cells to suppressor T cells, which is normally 2:1, is reversed.

Symptoms of AIDS may develop for months or years and include malaise; a low-grade fever or night sweats; coughing; shortness of breath; sore throat; extreme fatigue; muscle aches; unexplained weight loss; enlarged lymph nodes in the neck, axilla, and groin; and blue-violet or brownish spots on the skin, usually on the lower extremities.

The deterioration of host defenses allows for the development of cancer and opportunistic infections of various kinds. The two diseases that most often kill AIDS victims are Kaposi's sarcoma and *Pneumocystis carinii* pneumonia. Kaposi's sarcoma is a deadly form of skin cancer prevalent in equatorial Africa but previously almost unknown in the United States. It arises from endothelial cells of blood vessels and produces painless, purple or brownish lesions that resemble bruises on the skin or on the inside of the mouth, nose, or rectum. *Pneumocystis carinii* pneumonia is a rare form of pneumonia caused by the protozoan *Pneumocystis carinii.* It results in shortness of breath, persistent dry cough, sharp chest pains, and difficulty in breathing. AIDS victims are also subject to a form of herpes that attacks the central nervous system and a bacterial infection that usually causes tuberculosis in chickens and pigs.

Among the high-risk groups who are susceptible to AIDS are homosexual and bisexual males, intravenous drug abusers, sexual partners of individuals in AIDS risk groups, and infants born to mothers at risk. Other risk groups include hemophiliacs under treatment with blood plasma products and Haitian immigrants. It is also found among infants and other patients who have received blood transfusions. (Current testing of blood has virtually eliminated future contamination by blood transfusion or therapy.)

The pattern of AIDS transmission closely resembles the occurrence of hepatitis B (serum hepatitis), a liver disease that commonly strikes homosexual drug addicts using contaminated needles, and sometimes patients getting blood transfusions. Like hepatitis B, AIDS appears to be transmitted by intimate direct contact involving mucosal surfaces or by parenteral (injection) routes. Two fluids considered most highly contagious are semen and blood. Airborne transmission and spread via food, water, insects, and casual contact seem unlikely.

Treatment of AIDS is complicated by the lifestyle of the virus that causes the disease. When it enters the cell, the virus, which consists of a single strand of RNA, manufactures an enzyme that allows it to be transcribed into the host cell's DNA. Once this infection occurs, it is very likely to be for life. Scientists are now at work to develop a vaccine against the virus that causes AIDS.

Some people who carry the AIDS virus will develop the **AIDS-related complex (ARC).** This condition may include fever, diarrhea, malaise, and a generalized swelling of lymph nodes. ARC may follow several courses. It may resolve, persist for a long period, or progress to full-blown AIDS.

MEDICAL TERMINOLOGY

Adenitis (*adeno* = gland; *itis* = inflammation of)   Enlarged, tender, and inflamed lymph nodes resulting from an infection.
Elephantiasis   Great enlargement of a limb (especially lower limbs) and scrotum resulting from obstruction of lymph glands or vessels by a parasitic worm.
Hypersplenism (*hyper* = over)   Abnormal splenic activity involving highly increased blood cell destruction.
Lymphadenectomy (*ectomy* = removal)   Removal of a lymph node.

Lymphadenopathy (*patho* = disease)   Enlarged, sometimes tender lymph glands.
Lymphangioma (*angio* = vessel; *oma* = tumor).   A benign tumor of the lymph vessels.
Lymphangitis   Inflammation of the lymphatic vessels.
Lymphedema (*edema* = swelling)   Accumulation of lymph fluid producing subcutaneous tissue swelling.
Lymphoma   Any tumor composed of lymph tissue.
Lymphostasis (*stasis* = halt)   A lymph flow stoppage.
Splenomegaly (*mega* = large)   Enlarged spleen.

STUDY OUTLINE

Lymphatic Vessels (p. 530)
1. The lymphatic system consists of lymph, lymphatic vessels, and structures and organs that contain lymphatic tissue (specialized reticular tissue containing large numbers of lymphocytes).
2. Among the lymphatic tissue-containing components of the lymphatic system are diffuse lymphatic tissue, lymphatic nodules, and lymphatic organs (lymph nodes, spleen, and thymus gland).
3. Lymphatic vessels begin as blind-ended lymph capillaries in tissue spaces between cells.
4. Lymph capillaries merge to form larger vessels, called lymphatics, which ultimately converge into the thoracic duct or right lymphatic duct.
5. Lymphatics have thinner walls and more valves than veins.

Lymphatic Tissue (p. 531)
1. Lymph nodes are oval structures located along lymphatics.
2. Lymph enters nodes through afferent lymphatic vessels and exits through efferent lymphatic vessels.
3. Lymph passing through the nodes is filtered. Lymph nodes also produce lymphocytes.
4. Tonsils are multiple aggregations of large lymphatic nodules embedded in mucous membranes. They include the pharyngeal, palatine, and lingual tonsils.
5. The spleen is the largest mass of lymphatic tissue in the body and functions in production of lymphocytes and antibodies, phagocytosis of bacteria and worn-out red blood cells, and storage of blood.
6. The thymus gland functions in immunity by producing T cells.

Lymph Circulation (p. 536)
1. The passage of lymph is from interstitial fluid, to lymph capillaries, to lymphatics, to lymph trunks, to the thoracic duct or right lymphatic duct, to the subclavian veins.
2. Lymph flows as a result of skeletal muscle contractions and respiratory movements. It is also aided by valves in the lymphatics.

Nonspecific Resistance to Disease (p. 537)
1. The ability to ward off disease using a number of defenses is called resistance. Lack of resistance is called susceptibility.
2. Nonspecific resistance refers to a wide variety of body responses against a wide range of pathogens.
3. Nonspecific resistance includes mechanical factors (skin, mucous membranes, lacrimal apparatus, saliva, mucus, cilia, epiglottis, and flow of urine), chemical factors (gastric juice, acid pH of skin, unsaturated fatty acids, and lysozyme), antimicrobial substances (interferon, complement, and properdin), phagocytosis, inflammation, and fever.

Immunity (Specific Resistance to Disease) (p. 542)
1. Specific resistance to disease involves the production of a specific lymphocyte or antibody against a specific antigen and is called immunity.
2. Antigens are chemical substances that, when introduced into the body, stimulate the production of antibodies that react with the antigen.
3. Examples of antigens are microbes, microbial structures, pollen, incompatible blood cells, and transplants.
4. Antigens are characterized by immunogenicity, reactivity, and multivalence.
5. Antibodies are proteins produced in response to antigens.
6. Based on chemistry and structure, antibodies are distinguished into five principal classes, each with specific biological roles (IgG, IgA, IgM, IgD, and IgE).
7. Antibodies consist of heavy and light chains and variable and constant portions.
8. Cellular immunity refers to destruction of antigens by T cells, and humoral immunity refers to destruction of antigens by antibodies.
9. T cells are processed in the thymus gland; B cells may be processed in bone marrow, fetal liver and spleen, or gut-associated lymphoid tissue.
10. Macrophages process and present antigens to T cells and

B cells, and secrete interleukin 1 that induces proliferation of T cells and B cells.

11. T cells consist of several subpopulations; killer T cells secrete lymphotoxins that destroy (lyse) antigens directly and lymphokines that destroy antigens indirectly; helper T cells cooperate with B cells to help amplify antibody production and secrete interleukin 2 that stimulates proliferation of killer T cells; suppressor T cells inhibit secretion of injurious substances by killer T cells and antibody production by plasma cells; delayed hypersensitivity T cells produce lymphokines and are important in hypersensitivity responses; and amplifier T cells stimulate helper and suppressor T cells and plasma cells to greater levels of activity.

12. Natural killer (NK) cells are lymphocytes with certain characteristics similar to those of killer T cells.

13. B cells develop into antibody-producing plasma cells under the influence of thymic hormones and interleukin 1; memory B cells recognize the original, invading antigen.

14. The anamnestic response provides the basis for immunization against certain diseases.

15. The immunologic properties of the skin are due to keratinocytes, Langerhans cells, and Granstein cells.

16. Monoclonal antibodies are pure antibodies produced by fusing a B cell with a tumor cell; they are important in diagnosis, detection of disease, treatment, preparing vaccines, and countering rejection by transplants and autoimmune diseases.

17. Cancer cells contain tumor-specific antigens and are frequently destroyed by the body's immune system (immunologic surveillance); some cancer cells escape detection and destruction, a phenomenon called immunologic escape.

Developmental Anatomy of the Lymphatic System (p. 549)

1. Lymphatic vessels develop from lymph sacs, which develop from veins. Thus they are derived from mesoderm.

2. Lymph nodes develop from lymph sacs that become invaded by mesenchymal cells.

Disorders: Homeostatic Imbalances (p. 550)

1. Hypersensitivity is overreactivity to an antigen. Localized anaphylactic reactions include hay fever, asthma, eczema, and hives; acute anaphylaxis is a severe reaction with systemic effects.

2. Tissue rejection of a transplanted tissue or organ involves antibody production against the proteins (antigens) in the transplant. It may be overcome with immunosuppressive drugs.

3. Autoimmune diseases result when the body does not recognize "self" antigens and produces antibodies against them. Several human autoimmune diseases are rheumatoid arthritis (RA), systemic lupus erythematosus (SLE), rheumatic fever, hemolytic and pernicious anemias, myasthenia gravis, and multiple sclerosis (MS).

4. Severe combined immunodeficiency (SCID) is an immunodeficiency disease in which both B cells and T cells are missing or inactive in providing immunity.

5. Hodgkin's disease is a malignant disorder, usually arising in lymph nodes.

6. Acquired immune deficiency syndrome (AIDS) lowers the body's immunity by decreasing the number of helper T cells and reversing the ratio of helper T cells to suppressor T cells. AIDS victims frequently develop Kaposi's sarcoma and *Pneumocystis carinii* pneumonia.

REVIEW QUESTIONS

1. Identify the components and functions of the lymphatic system.
2. How do lymphatic vessels originate? Compare veins and lymphatics with regard to structure.
3. What is a lymphangiogram? What is its diagnostic value?
4. Describe the structure of a lymph node. What functions do lymph nodes serve?
5. Identify the tonsils by location.
6. Describe the location, gross anatomy, histology, and functions of the spleen. What is a splenectomy?
7. Describe the role of the thymus gland in immunity.
8. Construct a diagram to indicate the route of lymph circulation.
9. List and explain the various factors involved in the maintenance of lymph circulation.
10. Define edema. What are some of its causes?
11. Describe the operation of the various mechanical and chemical factors involved in nonspecific resistance.
12. Outline the role of the following antimicrobial substances: interferon (IFN), complement, and properdin.
13. What is phagocytosis? Describe the steps involved in adherence and ingestion.
14. Compare the roles of microphages and macrophages in phagocytosis.
15. Define an inflammation. Describe the principal symptoms associated with inflammation.
16. How are vasodilation and increased permeability of blood vessels initiated during inflammation? Why are they important? What causes edema?
17. Define immunity. Distinguish between an antigen and an antibody.
18. List the various characteristics of antigens.

19. By means of a labeled diagram, describe the structure of an antibody. What are the characteristics of antibodies?
20. Where are T cells and B cells processed? How do they differ in function?
21. What is the function of macrophages in immunity?
22. Describe the role of T cells in cellular immunity. Be sure to cite the roles of each type of T cell.
23. What are natural killer (NK) cells? How do they differ from killer T cells?
24. Describe the role of B cells in humoral immunity.
25. Discuss the importance of the secondary (anamnestic) response of the body to an antigen.
26. Describe the role of the skin in immunity.
27. What are monoclonal antibodies? How are they produced? Why are they clinically important?
28. Explain how immunology is related to cancer.
29. Describe how the lymphatic system develops.
30. Define hypersensitivity (allergy) and distinguish between type I (anaphylaxis) and type IV (cell-mediated) reactions.
31. Why does tissue rejection occur? How is this problem overcome?
32. How are transplants classified?
33. Define an autoimmune disease. Give several examples.
34. What is severe combined immunodeficiency (SCID)?
35. Define Hodgkin's disease.
36. Describe the cause and symptoms of acquired immune deficiency syndrome (AIDS). What are the complications of AIDS?
37. Refer to the glossary of medical terminology associated with the lymphatic system. Be sure that you can define each term.

The Respiratory System

23

■ STUDENT OBJECTIVES

- Identify the organs of the respiratory system.
- Compare the structure and function of the external and internal nose.
- Differentiate the three regions of the pharynx and describe their roles in respiration.
- Describe the structure of the larynx and explain its function in respiration and voice production.
- Explain the structure and function of the trachea.
- Contrast tracheostomy and intubation as alternative methods for clearing obstructed air passageways.
- Describe the location and structure of the tubes that form the bronchial tree.
- Identify the coverings of the lungs and the division of the lungs into lobes.
- Describe the composition of a lobule of the lung.
- Explain the structure of the alveolar-capillary (respiratory) membrane and its function in the diffusion of respiratory gases.
- List the events involved in inspiration and expiration.
- Explain how compliance and airway resistance relate to breathing.
- Define coughing, sneezing, sighing, yawning, sobbing, crying, laughing, and hiccuping as modified respiratory movements.
- Compare the volumes and capacities of air exchanged during respiration.
- Define Boyle's law, Dalton's law, and Henry's law.
- Explain how external and internal respiration occur.
- Describe how the oxygen-carrying capacity of the blood is affected by PO_2, PCO_2, temperature, and DPG.
- Explain how the respiratory gases are transported by the blood.
- Distinguish the various types of hypoxia.
- Explain how the respiratory center functions in establishing the basic rhythm of respiration.
- Describe how various neural and chemical factors may modify the rate of respiration.
- Describe the effects of aging on the respiratory system.
- Describe the development of the respiratory system.
- Define bronchogenic carcinoma (lung cancer), bronchial asthma, bronchitis, emphysema, pneumonia, tuberculosis, respiratory distress syndrome (RDS) of the newborn, respiratory failure, sudden infant death syndrome (SIDS), coryza (common cold), influenza (flu), pulmonary embolism (PE), and pulmonary edema as disorders of the respiratory system.
- Define medical terminology associated with the respiratory system.

Cells need a continuous supply of oxygen to carry out the activities that are vital to their survival. Many of these activities release quantities of carbon dioxide. Since an excessive amount of carbon dioxide produces acid conditions that are poisonous to cells, the gas must be eliminated quickly and efficiently. The two systems that supply oxygen and eliminate carbon dioxide are the cardiovascular system and the respiratory system. The **respiratory system** consists of organs that exchange gases between the atmosphere and blood. These organs are the nose, pharynx, larynx, trachea, bronchi, and lungs (Figure 23-1). The cardiovascular system transports the gases in the blood between the lungs and the cells. The term *upper respiratory system* refers to the nose, throat, and associated structures. The *lower respiratory system* refers to the remainder of the system.

The overall exchange of gases between the atmosphere, the blood, and the cells is **respiration.** Three basic processes are involved. The first process, **pulmonary ventilation,** or breathing, is the inspiration (inflow) and expiration (outflow) of air between the atmosphere and the lungs. The second and third processes involve the exchange of gases within the body. **External respiration** is the exchange of gases between the lungs and blood. **Internal respiration** is the exchange of gases between the blood and the cells.

The respiratory and cardiovascular systems participate equally in respiration. Failure of either system has the same effect on the body: disruption of homeostasis and rapid death of cells from oxygen starvation.

The developmental anatomy of the respiratory system is considered later in the chapter.

(a)

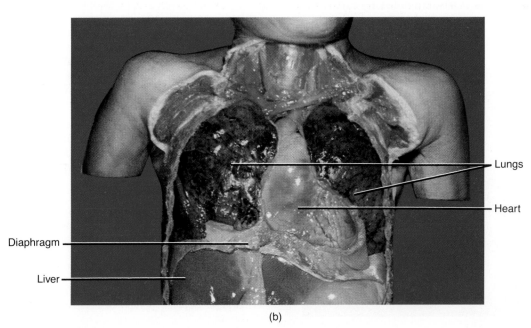

(b)

FIGURE 23-1 Organs of the respiratory system in relation to surrounding structures. (a) Diagram. (b) Photograph. (Courtesy of C. Yokochi and J. W. Rohen, *Photographic Anatomy of the Human Body,* 2nd ed., 1978, IGAKU-SHOIN, Ltd., Tokyo, New York.)

ORGANS

NOSE

Anatomy

The **nose** has an external portion and an internal portion inside the skull (Figure 23-2). The external portion consists of a supporting framework of bone and cartilage covered with skin and lined with mucous membrane. The bridge of the nose is formed by the nasal bones, which hold it in a fixed position. Because it has a framework of pliable cartilage, the rest of the external nose is somewhat flexible. On the undersurface of the external nose are two openings called the **nostrils** or **external nares** (NA-rēz; *sing.*, naris). The surface anatomy of the nose is shown in Figure 23-2a.

The internal portion of the nose is a large cavity in the skull that lies inferior to the cranium and superior to the mouth. Anteriorly, the internal nose merges with the external nose, and posteriorly it communicates with the throat (pharynx) through two openings called the **internal nares (choanae).** Four paranasal sinuses (frontal, sphenoidal, maxillary, and ethmoidal) and the nasolacrimal ducts also open into the internal nose. The lateral walls of the internal nose are formed by the ethmoid, maxillae, and inferior nasal conchae bones. The ethmoid also forms the roof. The floor is formed by the palatine bones and the palatine process of the maxilla, which together comprise the hard palate.

The inside of both the external and internal nose consists of a **nasal cavity,** divided into right and left sides by a vertical partition called the **nasal septum.** The anterior portion of the septum is made primarily of cartilage. The remainder is formed by the vomer and the perpendicular plate of the ethmoid (see Figure 7-7a). The anterior portion of the nasal cavity, just inside the nostrils, is called the **vestibule.** It is surrounded by cartilage. The upper nasal cavity is surrounded by bone.

Physiology

The interior structures of the nose are specialized for three functions: incoming air is warmed, moistened, and filtered; olfactory stimuli are received; and large hollow resonating chambers are provided for speech sounds.

When air enters the nostrils, it passes first through the vestibule. The vestibule is lined by skin containing coarse hairs that filter out large dust particles. The air then passes into the upper nasal cavity. Three shelves formed by projections of the superior, middle, and inferior nasal conchae extend out of the lateral wall of the cavity. The conchae, almost reaching the septum, subdivide each side of the nasal cavity into a series of groovelike passageways—the **superior, middle,** and **inferior meatuses.** Mucous membrane lines the cavity and its shelves. The olfactory receptors lie in the membrane lining the area superior to the superior nasal conchae and is also called the **olfactory region.** Below the olfactory region, the membrane contains capillaries and pseudostratified ciliated columnar cells with many goblet cells. As the air whirls around the conchae and meatuses, it is warmed by the capillaries. Mucus secreted by the goblet cells moistens the air and traps dust particles. Drainage from the lacrimal ducts, and perhaps secretions from the paranasal sinuses, also help moisten the air. The cilia move the mucus-dust packages along the pharynx so they can be eliminated from the body.

PHARYNX

The **pharynx** (FAR-inks), or throat, is a somewhat funnel-shaped tube about 13 cm (5 inches) long that starts at the internal nares and extends partway down the neck (Figure 23-2b). It lies just posterior to the nasal cavity and oral cavity and just anterior to the cervical vertebrae. Its wall is composed of skeletal muscles and lined with mucous membrane. The functions of the pharynx are to serve as a passageway for air and food and to provide a resonating chamber for speech sounds.

The branch of medicine that deals with the diagnosis and treatment of diseases of the ears, nose, and throat is called **otorhinolaryngology** (ō'-tō-rī'-nō-lar'-in-GOL-ō-jē; *otic* = ear; *rhino* = nose).

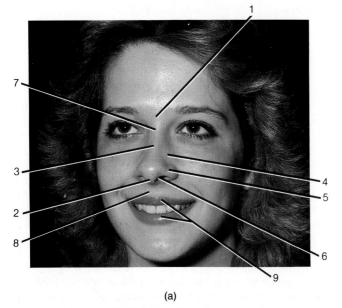

1. **Root.** Superior attachment of nose at forehead located between eyes.
2. **Apex.** Tip of nose.
3. **Dorsum nasi.** Rounded anterior border connecting root and apex; in profile, may be straight, convex, concave, or wavy.
4. **Nasofacial angle.** Point at which side of nose blends with tissues of face.
5. **Ala.** Convex flared portion of inferior lateral surface; unites with upper lip.
6. **External naris.** External opening into nose.
7. **Bridge.** Superior portion of dorsum nasi, superficial to nasal bones.
8. **Philtrum.** Vertical groove in medial portion of upper lip.
9. **Lips.** Upper and lower fleshy borders of oral cavity.

(a)

FIGURE 23-2 Respiratory organs in the head and neck. (a) Surface anatomy of the nose in anterior view. (Copyright © 1982 by Gerard J. Tortora. Courtesy of Lynne Tortora.)

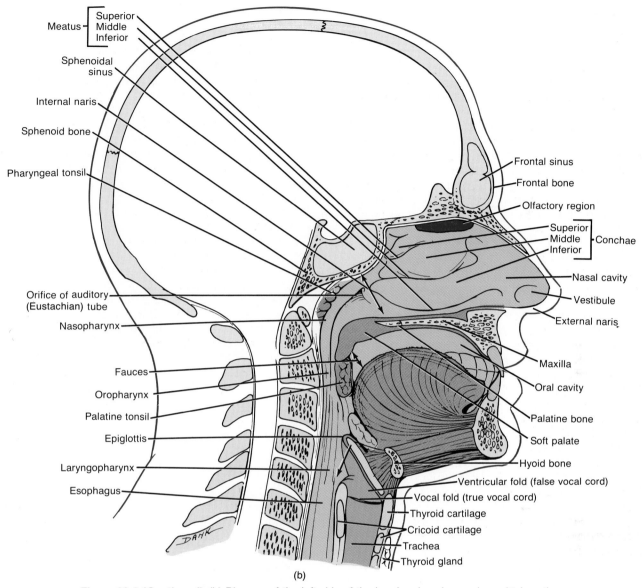

Meatus — Superior
Middle
Inferior

Sphenoidal
sinus

Internal naris

Sphenoid bone

Pharyngeal tonsil

Orifice of auditory
(Eustachian) tube

Nasopharynx

Fauces

Oropharynx

Palatine tonsil

Epiglottis

Laryngopharynx

Esophagus

Frontal sinus
Frontal bone
Olfactory region
Superior
Middle — Conchae
Inferior
Nasal cavity
Vestibule
External naris
Maxilla
Oral cavity
Palatine bone
Soft palate
Hyoid bone
Ventricular fold (false vocal cord)
Vocal fold (true vocal cord)
Thyroid cartilage
Cricoid cartilage
Trachea
Thyroid gland

(b)

Figure 23-2 (*Continued*) (b) Diagram of the left side of the head and neck seen in sagittal section with the nasal septum removed.

The uppermost portion of the pharynx, called the **nasopharynx,** lies posterior to the internal nasal cavity and extends to the plane of the soft palate. There are four openings in its wall: two internal nares and two openings that lead into the auditory (Eustachian) tubes. The posterior wall also contains the pharyngeal tonsil, or adenoid. Through the internal nares the nasopharynx exchanges air with the nasal cavities and receives the packages of dust-laden mucus. It is lined with pseudostratified ciliated epithelium, and the cilia move the mucus down toward the mouth. The nasopharynx also exchanges small amounts of air with the auditory tubes so that the air pressure inside the middle ear equals the pressure of the atmospheric air flowing through the nose and pharynx.

The middle portion of the pharynx, the **oropharynx,** lies posterior to the oral cavity and extends from the soft palate inferior to the level of the hyoid bone. It has only one opening, the *fauces* (FAW-sēz), or opening from the mouth. It is lined by stratified squamous epithelium. This portion of the pharynx is both respiratory and digestive in function, since it is a common passageway for both air and food. Two pairs of tonsils, the palatine and lingual tonsils, are found in the oropharynx. The lingual tonsil lies at the base of the tongue (see also Figure 17-2c).

The lowest portion of the pharynx, the **laryngopharynx** (la-rin'-gō-FAR-inks), extends downward from the hyoid bone and becomes continuous with the esophagus (food tube) posteriorly and the larynx (voice box) anteriorly. Like the oropharynx, the laryngopharynx is a respiratory and a digestive pathway and is lined by stratified squamous epithelium.

Sphenoidal sinus

Sphenoidal bone

Orifice of auditory (Eustachian) tube

Internal naris

Soft palate

Pharynx

Frontal sinus

Frontal bone

Nasal bone

Superior

Middle — Conchae

Inferior

Nasal cavity

Vestibule

External naris

Superior

Middle — Meatuses

Inferior

Tongue   Hard palate

(c)

FIGURE 23-2 (Continued) (c) Photograph of the left side of the nasal cavity seen in sagittal section with the nasal septum removed. (Courtesy of C. Yokochi and J. W. Rohen, *Photographic Anatomy of the Human Body,* 2nd ed., 1978, IGAKU-SHOIN, Ltd., Tokyo, New York.)

LARYNX

The **larynx,** or voice box, is a short passageway that connects the pharynx with the trachea. It lies in the midline of the neck anterior to the fourth through sixth cervical vertebrae.

Anatomy

The wall of the larynx is supported by nine pieces of cartilage (Figure 23-3). Three are single and three are paired. The three single pieces are the thyroid cartilage, epiglottic cartilage (epiglottis), and cricoid cartilage. Of the paired cartilages, the arytenoid cartilages are the most important. The paired corniculate and cuneiform cartilages are of lesser significance.

The **thyroid cartilage (Adam's apple)** consists of two fused plates that form the anterior wall of the larynx and give it its triangular shape. It is larger in males than in females.

The **epiglottis** is a large, leaf-shaped piece of cartilage lying on top of the larynx (see also Figure 23-2b). The "stem" of the epiglottis is attached to the thyroid cartilage, but the "leaf" portion is unattached and free to move up and down like a trap door. During swallowing, there is elevation of the larynx. This causes, the free edge of the epiglottis to form a lid over the glottis, closing it off. The **glottis** is the space between the vocal folds (true

vocal cords) in the larynx. In this way, the larynx is closed off and liquids and foods are routed into the esophagus and kept out of the larynx and air passageways below it. When anything but air passes into the larynx, a cough reflex attempts to expel the material.

The **cricoid** (KRĪ-koyd) **cartilage** is a ring of cartilage forming the inferior wall of the larynx. It is attached to the first ring of cartilage of the trachea.

The paired **arytenoid** (ar'-i-TĒ-noyd) **cartilages** are pyramidal in shape and located at the superior border of the cricoid cartilage. They attach to the vocal folds and pharyngeal muscles, and by their action can move the vocal folds.

The paired **corniculate** (kor-NIK-yoo-lāt) **cartilages** are cone-shaped. One is located at the apex of each arytenoid cartilage. The paired **cuneiform** (kyoo-NĒ-i-form) **cartilages** are rod-shaped cartilages that connect the epiglottis to the arytenoid cartilages.

The epithelium lining the larynx below the vocal folds is pseudostratified. It consists of ciliated columnar cells, goblet cells, and basal cells, and it helps to trap dust not removed in the upper passages.

Voice Production

The mucous membrane of the larynx is arranged into two pairs of folds—an upper pair called the **ventricular folds (false vocal cords)** and a lower pair called simply

(c)

FIGURE 23-3 Larynx. (a) Diagram of anterior view. (b) Diagram of posterior view. (c) Photograph of the true vocal cords, false vocal cords, and open glottis from above. (Courtesy of C. Yokochi and J. W. Rohen, *Photographic Anatomy of the Human Body,* 2nd ed., 1978, IGAKU-SHOIN, Ltd., Tokyo, New York.)

than in females, and therefore they vibrate more slowly. Thus men generally have a lower range of pitch than women.

Sound originates from the vibration of the vocal folds, but other structures are necessary for converting the sound into recognizable speech. The pharynx, mouth, nasal cavity, and parnasal sinuses all act as resonating chambers that give the voice its human and individual quality. By constricting and relaxing the muscles in the wall of the pharynx, we produce the vowel sounds. Muscles of the face, tongue, and lips help us to enunciate words.

CLINICAL APPLICATION

Laryngitis is an inflammation of the larynx that is most often caused by a respiratory infection or irritants such as cigarette smoke. Inflammation of the vocal folds themselves causes hoarseness or loss of voice by interfering with the contraction of the folds or by causing them to swell to the point where they cannot vibrate freely. Many long-term smokers acquire a permanent hoarseness from the damage done by chronic inflammation.

Cancer of the larynx is found almost exclusively in individuals who smoke. The condition is characterized by hoarseness, pain on swallowing, or pain radiating to an ear. Treatment is by radiation therapy and/or surgery.

---

the **vocal folds (true vocal cords)** (Figure 23-3c). When the ventricular folds are brought together, they function in holding the breath against pressure in the thoracic cavity, such as might occur when a person exerts a strain while lifting a heavy weight. The mucous membrane of the vocal folds is lined by nonkeratinized stratified squamous epithelium. Under the membrane lie bands of elastic ligaments stretched between pieces of rigid cartilage like the strings on a guitar. Skeletal muscles of the larynx, called intrinsic muscles, are attached internally to the pieces of rigid cartilage and to the vocal folds themselves (see Figure 11-10). When the muscles contract, they pull the strings of elastic ligaments tight and stretch the vocal folds out into the air passageways so that the glottis is narrowed. If air is directed against the vocal folds, they vibrate and set up sound waves in the column of air in the pharynx, nose, and mouth. The greater the pressure of air, the louder the sound.

Pitch is controlled by the tension on the vocal folds. If they are pulled taut by the muscles, they vibrate more rapidly and a higher pitch results. Lower sounds are produced by decreasing the muscular tension on the vocal folds. Vocal folds are usually thicker and longer in males

TRACHEA

The **trachea** (TRĀ-kē-a), or windpipe, is a tubular passageway for air about 12 cm (4.5 inches) in length and 2.5 cm (1 inch) in diameter. It is located anterior to the esophagus and extends from the larynx to the fifth thoracic vertebra, where it divides into right and left primary bronchi (see Figure 23-5a).

The wall of the trachea consists of a mucosa, submucosa, cartilaginous layer, and adventitia. The tracheal epithelium of the mucosa is pseudostratified. It consists of ciliated columnar cells that reach the luminal surface, goblet cells, and basal cells that do not reach the luminal surface (Figure 23-4a). The epithelium provides the same protection against dust as the membrane lining the larynx. Seromucous glands and their ducts are present in the submucosa. The cartilaginous layer consists of 16 to 20 horizontal incomplete rings of hyaline cartilage that look like a series of letter C's stacked one on top of another. The open parts of the C's face the esophagus and permit it to expand into the trachea during swallowing (Figure 23-4b). Transverse smooth muscle fibers, called the *trachealis muscle,* attach the open ends of the cartilage rings. The open ends of the rings of cartilage are also attached by elastic connective tissue. The solid parts of the C's provide a rigid support so the tracheal wall does not collapse inward and obstruct the air passageway.

At the point where the trachea bifurcates into right and left primary bronchi, there is an internal ridge called the **carina** (ka-RĪ-na). It is formed by a posterior and somewhat inferior projection of the last tracheal cartilage. The mucous membrane of the carina is one of the most sensitive areas of the respiratory system and is associated with the cough reflex. Widening and distortion of the carina, which can be seen in an examination by bronchoscopy, is a serious prognostic sign, since it usually indicates a carcinoma of the lymph nodes around the bifurcation of the trachea. **Bronchoscopy** is the visual examination of the bronchi through a **bronchoscope,** an illuminated, tubular instrument that can be passed through the trachea into the bronchi.

CLINICAL APPLICATION

Occasionally the respiratory passageways are unable to protect themselves from obstruction. The rings of cartilage may be accidentally crushed; the mucous membrane may become inflamed and swell so much that it closes off the air passageway; inflamed membranes secrete a great deal of mucus that may clog the lower respiratory passageways; a large object may be breathed in (aspirated) while the glottis is open; or an aspirated foreign object may cause spasm of the laryngeal muscles. The passageways must be cleared quickly. If the obstruction is above the level of the chest, a **tracheostomy** (trā-kē-OS-tō-mē) may be performed. A midline skin incision is made in the neck from just above the cricoid cartilage to the jugular notch of the sternum. Next, an incision is made in

the trachea below the obstructed area. The patient breathes through a metal or plastic tracheal tube inserted through the incision. Another method is **intubation.** A tube is inserted into the mouth or nose and passed down through the larynx and trachea. The firm wall of the tube pushes back any flexible obstruction, and the inside of the tube provides a passageway for air. If mucus is clogging the trachea, it can be suctioned out through the tube.

BRONCHI

The trachea terminates in the chest by dividing at the sternal angle into a **right primary bronchus** (BRON-kus), which goes to the right lung, and a **left primary bronchus,** which goes to the left lung (Figure 23-5). The right primary bronchus is more vertical, shorter, and wider than the left. As a result, foreign objects in the air passageways are more likely to enter it than the left and frequently lodge in it. Like the trachea, the primary bronchi (BRONG-kē) contain incomplete rings of cartilage and are lined by pseudostratified ciliated epithelium.

On entering the lungs, the primary bronchi divide to form smaller bronchi—the **secondary (lobar) bronchi,** one for each lobe of the lung (the right lung has three lobes; the left lung has two). The secondary bronchi continue to branch, forming still smaller bronchi, called **tertiary (segmental) bronchi** that divide into **bronchioles.** Bronchioles, in turn, branch into even smaller tubes called **terminal bronchioles.** This continuous branching from the trachea resembles a tree trunk with its branches, and is commonly referred to as the **bronchial tree.**

As the branching becomes more extensive in the bronchial tree, several structural changes may be noted. First, rings of cartilage are replaced by plates of cartilage that finally disappear in the bronchioles. Second, as the cartilage decreases, the amount of smooth muscle increases. Third, the epithelium changes from pseudostratified ciliated to simple cuboidal in the terminal bronchioles.

CLINICAL APPLICATION

The fact that the walls of the bronchioles contain a great deal of smooth muscle but no cartilage is clinically significant. During an **asthma attack** the muscles go into spasm. Because there is no supporting cartilage, the spasms can close off the air passageways.

Bronchography (brong-KOG-ra-fē) is a technique for examining the bronchial tree. An intratracheal catheter is passed into the mouth or nose, through the glottis, and into the trachea. Then, an opaque contrast medium, usually containing iodine, is introduced by means of gravity into the trachea and distributed through the bronchial branches. Roentgenograms of the chest in various positions are taken and the developed film, a **bronchogram** (BRONG-kō-gram), provides a picture of the tree (Figure 23-5b).

Ciliated columnar cells   Cilia   Goblet cell

Pseudostratified epithelium

Basement membrane   (a)   Basal cell   Connective tissue (lamina propia)

Thyroid gland

Tracheal cartilage

Esophagus

Trachealis muscle

Tracheal cartilage   Tracheal epithelium   Thyroid gland

(b)

FIGURE 23-4 Trachea. (a) Photomicrograph of an enlarged aspect of the tracheal epithelium at a magnification of 600×. (Copyright © 1983 by Michael H. Ross. Used by permission.) (b) Photomicrograph showing the relationship of the trachea to the esophagus in cross section at a magnification of 3×. (Copyright © 1987 by Michael H. Ross. Used by permission.)

LUNGS

The **lungs** are paired, cone-shaped organs lying in the thoracic cavity. They are separated from each other by the heart and other structures in the mediastinum (see Figure 20-1). Two layers of serous membrane, collectively called the **pleural membrane,** enclose and protect each lung. The outer layer is attached to the wall of the thoracic cavity and is called the **parietal pleura.** The inner layer, the **visceral pleura,** covers the lungs themselves. Between the visceral and parietal pleura is a small potential space, the **pleural cavity,** which contains a lubricating fluid se-

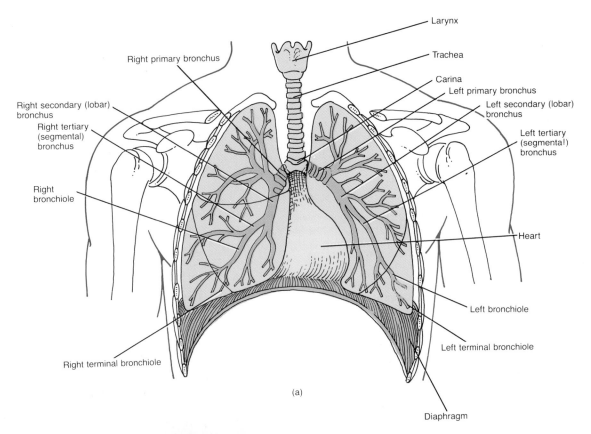

Right primary bronchus

Right secondary (lobar)
bronchus

Right tertiary
(segmental)
bronchus

Right
bronchiole

Right terminal bronchiole

Larynx

Trachea

Carina

Left primary bronchus

Left secondary (lobar)
bronchus

Left tertiary
(segmental)
bronchus

Heart

Left bronchiole

Left terminal bronchiole

Diaphragm

(a)

(b)

FIGURE 23-5 Bronchial tree. (a) Diagram. (b) Anteroposterior bronchogram. (Courtesy of John H. Juhl. From Lester W. Paul and John H. Juhl, *The Essentials of Roentgen Interpretation,* 4th ed., J. B. Lippincott, Philadelphia, 1987.)

to collapse. Fluid can be drained from the pleural cavity by inserting a needle, usually posteriorly through the seventh intercostal space. The needle is passed along the superior border of the lower rib to avoid damage to the intercostal nerves and blood vessels. Below the seventh intercostal space there is danger of penetrating the diaphragm.

Inflammation of the pleural membrane, or **pleurisy,** causes friction during breathing that can be quite painful when the swollen membranes rub against each other.

creted by the membranes (see Figure 1-7). This fluid prevents friction between the membranes and allows them to move easily on one another during breathing.

CLINICAL APPLICATION

In certain conditions, the pleural cavity may fill with air **(pneumothorax),** blood **(hemothorax),** or pus. Air in the pleural cavity, most commonly introduced in a surgical opening of the chest, may cause the lung

Gross Anatomy

The lungs extend from the diaphragm to a point about 1.5 to 2.5 cm (0.75 to 1 inch) superior to the clavicles and lie against the ribs anteriorly and posteriorly. The broad inferior portion of the lung, the **base,** is concave and fits over the convex area of the diaphragm (Figure 23-6). The narrow superior portion of the lung is termed the **apex (cupula).** The surface of the lung lying against the ribs, the **costal surface,** is rounded to match the curvature of the ribs. The **mediastinal (medial) surface** of each lung contains a vertical slit, the **hilus,** through which bronchi, pulmonary vessels, lymphatics, and nerves enter and exit. These structures are held together by the pleura

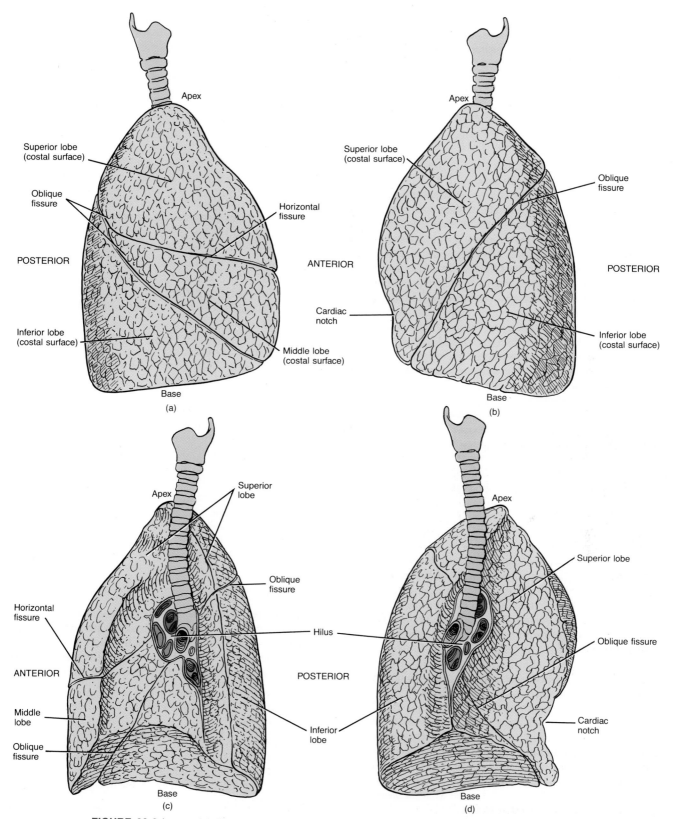

FIGURE 23-6 Lungs. (a) Right lung, lateral view. (b) Left lung, lateral view. (c) Right lung, medial view. (d) Left lung, medial view.

and connective tissue, and they constitute the **root** of the lung. Medially, the left lung also contains a concavity, the **cardiac notch,** in which the heart lies.

The right lung is thicker and broader than the left. It is also somewhat shorter than the left because the diaphragm is higher on the right side to accommodate the liver that lies below it.

Lobes and Fissures

Each lung is divided into lobes by one or more fissures. Both lungs have an **oblique fissure,** which extends downward and forward. The right lung also has a **horizontal fissure.** The oblique fissure in the left lung separates the **superior lobe** from the **inferior lobe.** The upper part of the oblique fissure of the right lung separates the superior lobe from the inferior lobe, whereas the lower part of the oblique fissure separates the inferior lobe from the **middle lobe.** The horizontal fissure of the right lung subdivides the superior lobe, thus forming a middle lobe.

Each lobe receives its own secondary (lobar) bronchus. Thus the right primary bronchus gives rise to three secondary (lobar) bronchi called the **superior, middle,** and **inferior secondary (lobar) bronchi.** The left primary bronchus gives rise to a **superior** and an **inferior secondary (lobar) bronchus.** Within the substance of the lung, the secondary bronchi give rise to the **tertiary (segmental) bronchi,** which are constant in both origin and distribution. The segment of lung tissue that each supplies is called a **bronchopulmonary segment.** Bronchial and pulmonary disorders, such as tumors or abscesses, may be localized in a bronchopulmonary segment and may be surgically removed without seriously disrupting surrounding lung tissue.

Lobules

Each bronchopulmonary segment of the lungs is broken up into many small compartments called **lobules** (Figure 23-7a). Each lobule is wrapped in elastic connective tissue and contains a lymphatic vessel, an arteriole, a venule, and a branch from a terminal bronchiole. Terminal bronchioles subdivide into microscopic branches called **respiratory bronchioles.** As the respiratory bronchioles become more distal, the epithelial lining changes from cuboidal to squamous. Respiratory bronchioles, in turn, subdivide into several (2 to 11) **alveolar ducts (atria).**

Around the circumference of the alveolar ducts are numerous alveoli and alveolar sacs. An **alveolus** (al-VĒ-ō-lus) is a cup-shaped outpouching lined by epithelium and supported by a thin elastic basement membrane. **Alveolar sacs** are two or more alveoli that share a common opening (Figure 23-7a,b). The alveolar walls consist of two principal types of epithelial cells: *squamous pulmonary epithelial cells* and *septal cells* (Figure 23-8a). The squamous pulmonary epithelial cells are the larger of the two types of cells and form a continuous lining of the alveolar wall, except for occasional septal cells. Septal cells are much smaller, somewhat cuboidal in shape, and are dispersed among the squamous pulmonary epithelial cells. Septal cells produce a phospholipid substance called

surfactant (sur-FAK-tant), which lowers surface tension (described shortly). Also found within the alveolar wall are free *alveolar macrophages (dust cells),* highly phagocytic cells that serve to remove dust particles or other debris from alveolar spaces; monocytes, white blood cells that become transformed into alveolar macrophages; and fibroblasts. Also present between lining cells of the alveolar wall are reticular and elastic fibers. Deep to the layer of squamous pulmonary epithelial cells is an elastic basement membrane. Over the alveoli, the arteriole and venule disperse into a capillary network. The blood capillaries consist of a single layer of endothelial cells and basement membrane.

CLINICAL APPLICATION

Many respiratory disorders are treated by means of **nebulization** (neb-yoo-li-ZĀ-shun). This procedure is the administering of medication in the form of droplets that are suspended in air to select areas of the respiratory tract. The patient inhales the medication as a fine mist. The number of droplets suspended in the mist and the area of the respiratory tract that the medication will reach both depend on droplet size. Smaller droplets (approximately 2 μm in diameter) can be suspended in greater numbers than can large droplets and will reach the alveolar ducts and sacs. Larger droplets (approximately 7 to 16 μm in diameter) will be deposited mostly in the bronchi and bronchioles. Droplets of 40 μm and larger will be deposited in the upper respiratory tract—the mouth, pharynx, trachea, and main bronchi. Nebulization therapy can be used with many different types of drugs, such as chemicals that relax the smooth muscle of the respiratory passageways, chemicals that reduce the thickness of mucus, and antibiotics.

Alveolar-Capillary (Respiratory) Membrane

The exchange of respiratory gases between the lungs and blood takes place by diffusion across the alveolar and capillary walls. This membrane, through which the respiratory gases move, is collectively known as the **alveolar-capillary (respiratory) membrane** (Figure 23-8b). It consists of:

1. A layer of squamous pulmonary epithelial cells with septal cells and free alveolar macrophages that constitute the alveolar (epithelial) wall.
2. An epithelial basement membrane underneath the alveolar wall.
3. A capillary basement membrane that is often fused to the epithelial basement membrane.
4. The endothelial cells of the capillary.

Despite the large number of layers, the alveolar-capillary membrane averages only 0.5 μm in thickness. This is of considerable importance to the efficient diffusion of respiratory gases. Moreover, it has been estimated that the lungs contain 300 million alveoli, providing an immense surface area of 70 m² (753 ft²) for the exchange of gases.

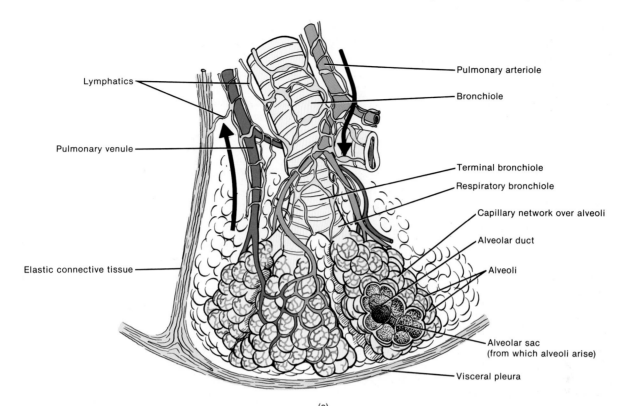

Pulmonary arteriole

Bronchiole

Terminal bronchiole

Respiratory bronchiole

Capillary network over alveoli

Alveolar duct

Alveoli

Alveolar sac
(from which alveoli arise)

Visceral pleura

Lymphatics

Pulmonary venule

Elastic connective tissue

(a)

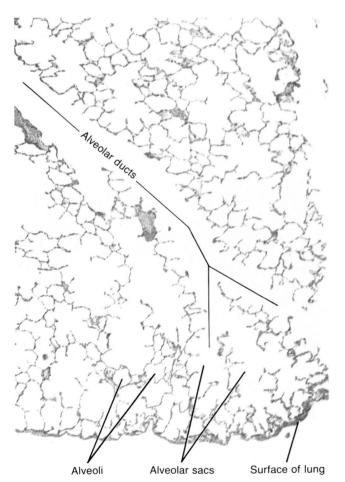

Alveolar ducts

Alveoli   Alveolar sacs   Surface of lung

(b)

FIGURE 23-7 Histology of the lungs. (a) Diagram of a lobule of the lung. (b) Photomicrograph of alveolar ducts, alveolar sacs, and alveoli at a magnification of 55×. (Courtesy of Michael H. Ross and Edward J. Reith, *Histology: A Text and Atlas,* Copyright © 1985, by Michael H. Ross and Edward J. Reith, Harper & Row, Publishers, Inc., New York.)

Blood Supply

The arterial supply of the lungs is derived from the pulmonary trunk. It divides into a left pulmonary artery that enters the left lung and a right pulmonary artery that enters the right lung. The venous return of the oxygenated blood is by way of the pulmonary veins, typically two in number on each side—the right and left superior and inferior pulmonary veins. All four veins drain into the left atrium (see Figure 21-22).

RESPIRATION

The principal purpose of **respiration** is to supply the cells of the body with oxygen and remove the carbon dioxide produced by cellular activities. The three basic processes of respiration are pulmonary ventilation, external respiration, and internal respiration.

PULMONARY VENTILATION

Pulmonary ventilation (breathing) is the process by which gases are exchanged between the atmosphere and lung alveoli. Air flows between the atmosphere and lungs for the same reason that blood flows through the body—a pressure gradient exists. We breathe in when the pressure inside the lungs is less than the air pressure in the atmo-

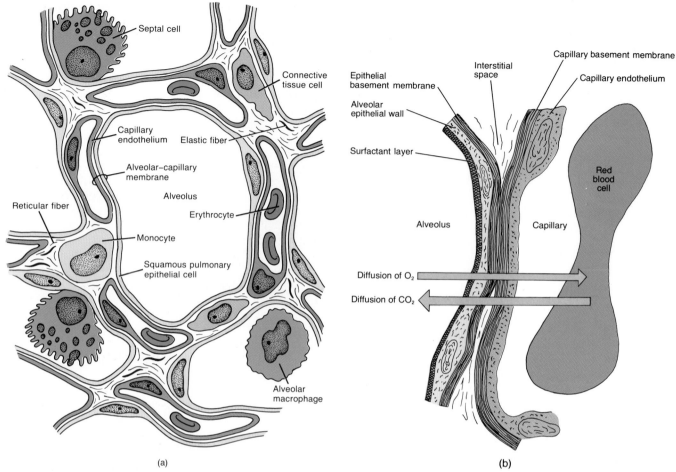

FIGURE 23-8 Structure of an alveolus. (a) Details of an alveolus. (b) Details of alveolar-capillary membrane.

sphere. We breathe out when the pressure inside the lungs is greater than the pressure in the atmosphere. Let us examine the mechanics of pulmonary ventilation by first looking at inspiration.

Inspiration

Breathing in is called **inspiration (inhalation).** Just before each inspiration, the air pressure inside the lungs equals the pressure of the atmosphere, which is about 760 mm Hg, or 1 atmosphere (atm), at sea level. For air to flow into the lungs, the pressure inside the lungs must become lower than the pressure in the atmosphere. This condition is achieved by increasing the volume of the lungs.

The pressure of a gas in a closed container is inversely proportional to the volume of the container. If the size of a closed container is increased, the pressure of the air inside the container decreases. If the size of the container is decreased, then the pressure inside it increases. This is referred to as **Boyle's law** and may be demonstrated as follows. Suppose we place a gas in a cylinder that has a movable piston and a pressure gauge, and the initial pressure is 1 atm (Figure 23-9). This pressure is created by the gas molecules striking the wall of the container. If the piston is pushed down, the gas is concen-

trated in a smaller volume. This means that the same number of gas molecules are striking less wall space. The gauge shows that the pressure doubles as the gas is compressed to half its volume. In other words, the same number of molecules in half the space produces twice the

FIGURE 23-9 Boyle's law. The volume of gas varies inversely with the pressure. If the volume is decreased to one-fourth, what happens to the pressure?

pressure. Conversely, if the piston is raised to increase the volume, the pressure decreases. Thus, the volume of a gas varies inversely with pressure (assuming that temperature is constant). Boyle's law applies to the operation of a bicycle pump and the blowing up of a balloon. Differences in pressure force air into our lungs when we inhale and force the air out when we exhale.

In order for inspiration to occur, the lungs must be expanded. This increases lung volume and thus decreases the pressure in the lungs. The first step toward increasing lung volume involves contraction of the principal inspiratory muscles—the diaphragm and external intercostals (Figure 23-10; see also Figure 11-12). The diaphragm is a sheet of skeletal muscle that forms the floor of the thoracic cavity. Contraction of the diaphragm causes it

to flatten, lowering its dome. This increases the vertical diameter of the thoracic cavity and accounts for the movement of more than two-thirds of the air that enters the lungs during inspiration. At the same time the diaphragm contracts, the external intercostals contract. As a result, the ribs are pulled upward and the sternum is pushed forward. This increases the anterior-posterior diameter of the thoracic cavity. Advanced pregnancy, excessive obesity, or confining abdominal clothing can prevent a complete descent of the diaphragm.

The term applied to normal quiet breathing is **eupnea** (yoop-NĒ-a; *eu* = normal). Eupnea involves shallow, deep, or combined shallow and deep breathing. Shallow (chest) breathing is called **costal breathing.** It consists of an upward and outward movement of the chest as a result of contraction of the external intercostal muscles. Deep (abdominal) breathing is called **diaphragmatic breathing.** It consists of the outward movement of the abdomen as a result of the contraction and descent of the diaphragm. During *forced inspiration,* accessory muscles of inspiration also participate in increasing the size of the thoracic cavity. Contraction of the sternocleidomastoids elevates the sternum, and contraction of the scalenes elevates the superior ribs. Inspiration is referred to as an active process because it is initiated by muscle contraction.

FIGURE 23-10 Pulmonary ventilation: muscles of inspiration and expiration. (a) Inspiratory muscles and their actions (left), and expiratory muscles and their actions (right). (b) Changes in size of thoracic cavity during inspiration (blue) and expiration (black).

During normal breathing, the pressure between the two pleural layers, called **intrapleural (intrathoracic) pressure,** is always subatmospheric. (It may become temporarily positive only during one modified respiratory movements such as coughing or straining during defecation.) Just before inspiration, it is about 756 mm Hg (Figure 23-11). The overall increase in the size of the thoracic cavity causes intrapleural pressure to fall to about 754 mm Hg. Consequently, the walls of the lungs are sucked outward by the partial vacuum. Expansion of the lungs is further aided by movement of the pleura. The parietal and visceral pleurae are normally strongly attached to each other due to surface tension created by their moist adjoining surfaces. As the thoracic cavity expands, the

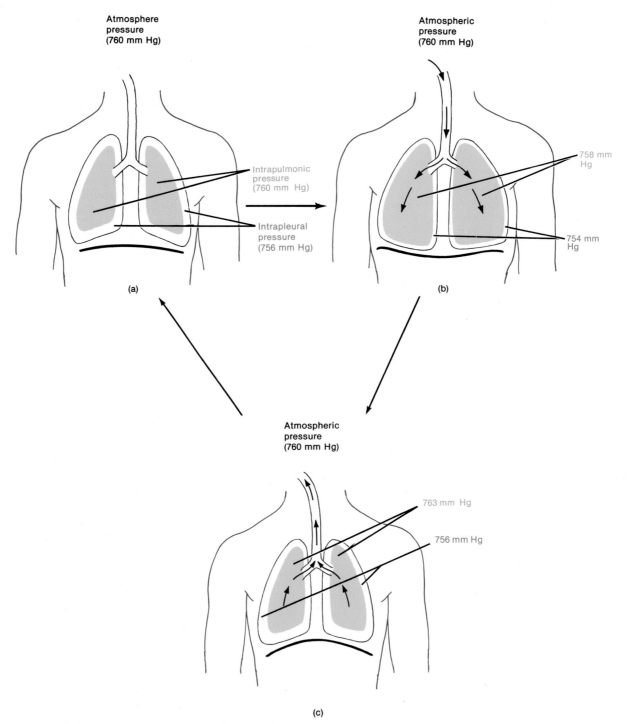

FIGURE 23-11 Pulmonary ventilation: pressure changes. (a) Lungs and pleural cavity just before inspiration. (b) Chest expanded and intrapleural pressure decreased; lungs pulled outward and intrapulmonic pressure decreased. (c) Chest relaxes, intrapleural pressure rises, and lungs snap inward. Intrapulmonic pressure raised, forcing air out until intrapulmonic pressure equals atmospheric pressure (a).

parietal pleura lining the cavity is pulled in all directions, and the visceral pleura is pulled along with it.

When the volume of the lungs increases, the pressure inside the lungs, called the **intrapulmonic (intraalveolar) pressure,** drops from 760 to 758 mm Hg. A pressure gradient is thus established between the atmosphere and the alveoli. Air rushes from the atmosphere into the lungs, and an inspiration takes place. Air continues to move into the lungs until intrapulmonic pressure equals atmospheric pressure.

Inspiration may be summarized as follows:

Diaphragm and external
intercostal muscles
contract
↓
Thoracic cavity increases
in size and lungs expand
↓
Intrapulmonic pressure
decreases to 758 mm Hg
↓
Inspiration

Expiration

Breathing out, called **expiration (exhalation),** is also achieved by a pressure gradient, but in this case the gradient is reversed so that the pressure in the lungs is greater than the pressure of the atmosphere. Normal expiration, unlike inspiration, is a passive process since no muscular contractions are involved. Expiration starts when the inspiratory muscles relax. As the external intercostals relax, the ribs move downward, and as the diaphragm relaxes, its dome moves upward. These movements decrease the vertical and anterior-posterior diameters of the thoracic cavity, and it returns to its resting size (see Figure 23-10).

Expiration becomes active during higher levels of ventilation and when air movement out of the lungs is impeded. During these times, contraction of the internal intercostals moves the ribs downward, and contraction of the abdominal muscles moves the inferior ribs downward and compresses the abdominal viscera, thus forcing the diaphragm upward.

As intrapleural pressure returns to its preinspiration level (756 mm Hg), the walls of the lungs are no longer sucked out. The elastic basement membranes of the alveoli and elastic fibers in bronchioles and alveolar ducts recoil into their relaxed shape, and lung volume decreases. Intrapulmonic pressure increases to 763 mm Hg, and air moves from the area of higher pressure in the alveoli to the area of lower pressure in the atmosphere (see Figure 23-11).

Expiration may be summarized as follows:

Diaphragm and external
intercostal muscles relax
↓
Thoracic cavity decreases
in size and lungs recoil
↓

Intrapulmonic pressure
increases to 763 mm Hg
↓
Expiration

Atelectasis (Collapsed Lung)

It was noted earlier that intrapleural pressure is normally subatmospheric. The pleural cavities are sealed off from the outside environment and cannot equalize their pressure with that of the atmosphere. Nor can the diaphragm and rib cage move inward enough to bring the intrapleural pressure up to atmospheric pressure. Maintenance of a low intrapleural pressure is vital to the functioning of the lungs. The alveoli are so elastic that at the end of an expiration they attempt to recoil inward and collapse on themselves like the walls of a deflated balloon. A collapsed lung or portion of a lung is called **atelectasis** (at'-ē-LEK-ta-sis; *ateles* = incomplete, *ektasis* = dilation). Such a collapse, which would obstruct the movement of air, is prevented by the slightly lower pressure in the pleural cavities that keeps the alveoli slightly inflated.

Another factor preventing the collapse of alveoli is the presence of **surfactant,** the phospholipid produced by the septal cells of the alveolar walls. Surfactant decreases surface tension in the lungs. That is, it forms a thin lining on the alveoli and prevents them from sticking together following expiration. Thus, as alveoli become smaller, for example following expiration, the tendency of alveoli to collapse is minimized because the surface tension does not increase. As you will see later, a deficiency of surfactant in infants results in a disorder called respiratory distress syndrome (RDS) of the newborn.

Compliance

Compliance refers to the ease with which the lungs and thoracic wall can be expanded. High compliance means that the lungs and thoracic wall expand easily; low compliance means that they resist expansion. Compliance is related to two principal factors: elasticity and surface tension. The presence of elastic fibers in lung tissue results in high compliance. If surface tension within lung tissue were high, the tissues would resist expansion, but surfactant lowers surface tension and thus increases compliance. Any condition that destroys lung tissue, causes it to become fibrotic or edematous, causes a deficiency in surfactant, or in any way impedes lung expansion or contraction decreases compliance.

Airway Resistance

The walls of the respiratory passageways, especially the bronchi and bronchioles, offer some resistance to the normal flow of air into the lungs. The muscular contraction of normal inspiration not only expands the thoracic cavity but also helps to overcome resistance to airflow. Any condition that obstructs the air passageways increases resistance, and more pressure to force air through is required. During a forced expiration, as in coughing, straining, or playing a wind instrument, intrapleural pressure may increase from its normally subatmospheric (negative) value to a positive one. This greatly increases airway resistance because it results in compression of the airways. This is an important consideration for people with chronic obstructive pulmonary disease in which airway resistance is high even during rest.

CLINICAL APPLICATION

Sleep apnea (ap-NĒ-a; *a* = without; *pnoia* = breath) is a term used to describe frequent and prolonged periods during which breathing stops during sleep. Even in normal individuals, there are alterations in the pattern of breathing and short periods of apnea. However, when apnea occurs frequently and for long periods of time, sleep is disturbed. Since sleep apnea often occurs during REM sleep, the person feels fatigued.

MODIFIED RESPIRATORY MOVEMENTS

Respirations also provide humans with methods for expressing emotions such as laughing, yawning, sighing, and sobbing. Moreover, respiratory air can be used to expel foreign matter from the upper air passages through actions such as sneezing and coughing. Some of the modified respiratory movements that express emotion or clear the air passageways are listed in Exhibit 23-1. All these movements are reflexes, but some of them also can be initiated voluntarily.

PULMONARY AIR VOLUMES AND CAPACITIES

In clinical practice, the word **respiration** means one inspiration plus one expiration. The healthy adult averages about 12 respirations a minute while at rest. During each respiration, the lungs exchange various amounts of air with the atmosphere. A lower-than-normal amount of exchange is usually a sign of pulmonary malfunction.

Spirometry

The apparatus commonly used to measure the amount of air exchanged during breathing and rate of ventilation is a **spirometer** or **respirometer** (Figure 23-12). A spirometer consists of a weighted drum inverted over a chamber of water. The drum usually contains oxygen or air. A tube connects the air-filled chamber with the subject's mouth. During inspiration, air is removed from the chamber, the drum sinks, and an upward deflection is recorded by a stylus on graph paper on the rotating drum. During expiration, air is added, the drum rises, and a downward deflection is recorded. The record is called a **spirogram** (Figure 23-13).

Pulmonary Volumes

During the process of normal quiet breathing, about 500 ml of air moves into the respiratory passageways with

EXHIBIT 23-1 MODIFIED RESPIRATORY MOVEMENTS

MOVEMENT	COMMENT
Coughing	A long-drawn and deep inspiration followed by a complete closure of the glottis, which results in a strong expiration that suddenly pushes the glottis open and sends a blast of air through the upper respiratory passages. Stimulus for this reflex act may be a foreign body lodged in the larynx, trachea, or epiglottis.
Sneezing	Spasmodic contraction of muscles of expiration which forcefully expels air through the nose and mouth. Stimulus may be an irritation of the nasal mucosa.
Sighing	A long-drawn and deep inspiration immediately followed by a shorter but forceful expiration.
Yawning	A deep inspiration through the widely opened mouth producing an exaggerated depression of the lower jaw. It may be stimulated by drowsiness or fatigue, but precise stimulus-receptor cause is unknown.
Sobbing	A series of convulsive inspirations followed by a single prolonged expiration. The glottis closes earlier than normal after each inspiration so only a little air enters the lungs with each inspiration.
Crying	An inspiration followed by many short convulsive expirations, during which the glottis remains open and the vocal cords vibrate; accompanied by characteristic facial expressions.
Laughing	The same basic movements as crying, but the rhythm of the movements and the facial expressions usually differ from those of crying. Laughing and crying are sometimes indistinguishable.
Hiccuping	Spasmodic contraction of the diaphragm followed by a spasmodic closure of the glottis to produce a sharp inspiratory sound. Stimulus is usually irritation of the sensory nerve endings of the digestive tract.

each inspiration. The same amount moves out with each expiration. This volume of air inspired (or expired) is called **tidal volume** (Figure 23-13). Only about 350 ml of the tidal volume actually reaches the alveoli. The other 150 ml remains in air spaces of the nose, pharynx, larynx, trachea, and bronchi and is known as **dead air volume.** The total air taken in during 1 minute is called the **minute volume of respiration.** It is calculated by multiplying the tidal volume by the normal breathing rate per minute. An average volume would be 500 ml times 12 respirations per minute, or 6,000 ml/min.

By taking a very deep breath, we can inspire a good deal more than 500 ml. This excess inhaled air, called

FIGURE 23-12 Spirometers. (a) Collins spirometer. This type of spirometer is the one commonly used in college biology laboratories. (Photographed in a biology laboratory at Bergen Community College by Gerard J. Tortora.) (b) Ohio 842 spirometer. This instrument is a highly sophisticated spirometer that utilizes a computerized mechanism for recording results. (Courtesy of Lenny Patti.)

the **inspiratory reserve volume,** averages 3,100 ml above the 500 ml of tidal volume. Thus the respiratory system can pull in 3,600 ml of air. In fact, even more air can be pulled in if inspiration follows forced expiration.

If we inhale normally and then exhale as forcibly as possible, we should be able to push out 1,200 ml of air in addition to the 500 ml tidal volume. This extra 1,200 ml is called the **expiratory reserve volume.**

Even after the expiratory reserve volume is expelled, a good deal of air remains in the lungs because the lower intrapleural pressure keeps the alveoli slightly inflated, and some air also remains in the noncollapsible air passageways. This air, the **residual volume,** amounts to about 1,200 ml.

Opening the thoracic cavity allows the intrapleural pressure to equal the atmospheric pressure, forcing out

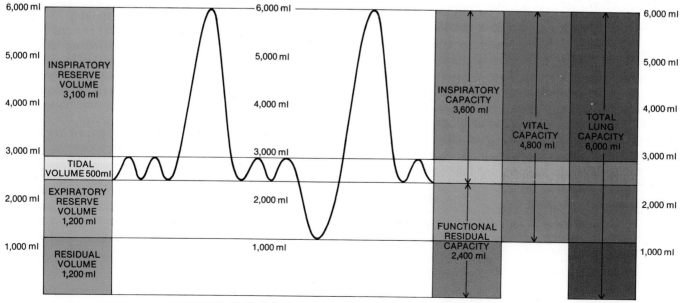

FIGURE 23-13 Spirogram of pulmonary volumes and capacities.

some of the residual volume. The air remaining is called the **minimal volume.** Minimal volume provides a medical and legal tool for determining whether a baby was born dead or died after birth. The presence of minimal volume can be demonstrated by placing a piece of lung in water and watching it float. Fetal lungs contain no air, and so the lung of a stillborn will not float in water.

Pulmonary Capacities

Lung capacity can be calculated by combining various lung volumes. **Inspiratory capacity,** the total inspiratory ability of the lungs, is the sum of tidal volume plus inspiratory reserve volume (3,600 ml). **Functional residual capacity** is the sum of residual volume plus expiratory reserve volume (2,400 ml). **Vital capacity** is the sum of inspiratory reserve volume, tidal volume, and expiratory reserve volume (4,800 ml). Finally, **total lung capacity** is the sum of all volumes (6,000 ml).

CLINICAL APPLICATION

The **measurement of respiratory volumes and capacities** is an essential tool for determining how well the lungs are functioning. Spirometry is indicated for an individual exhibiting labored breathing and is used in the diagnosis of respiratory disorders such as bronchial asthma and emphysema. For instance, during the early stages of emphysema, many of the alveoli lose their elasticity. During expiration they fail to snap inward, and consequently they fail to force out a normal amount of air. Thus the residual volume is increased at the expense of the expiratory reserve volume. Pulmonary infections can cause inflammation and an accumulation of fluid in the air spaces of the lungs (pulmonary edema). The fluid reduces the amount of space available for air and consequently decreases the vital capacity.

EXCHANGE OF RESPIRATORY GASES

As soon as the lungs fill with air, oxygen diffuses from the alveoli into the blood, through the interstitial fluid, and finally into the cells. Carbon dioxide diffuses in the opposite direction—from the cells, through interstitial fluid to the blood, and to the alveoli. To understand how respiratory gases are exchanged in the body, you need to know a few gas laws.

Dalton's Law

According to **Dalton's law,** each gas in a mixture of gases exerts its own pressure as if all the other gases were not present. This *partial pressure* is denoted as P. The total pressure of the mixture is calculated by simply adding all the partial pressures. Atmospheric air is a mixture of several gases—oxygen, carbon dioxide, nitrogen, water vapor, and a number of other gases that appear in such small quantities that we will ignore them. Atmospheric pressure is the sum of the pressures of all these gases:

Atmospheric pressure = $PO_2 + PCO_2 + PN_2 + PH_2O$ (760 mm Hg)

We can determine the partial pressure exerted by each component in the mixture by multiplying the percentage of the mixture the particular gas constitutes by the total pressure of the mixture. For example, to find the partial pressure of oxygen in the atmosphere, multiply the percentage of atmospheric air composed of oxygen (21 percent) by the total atmospheric pressure (760 mm Hg):

Atmospheric $PO_2 = 21\% \times 760$ mm Hg
$= 159.60$ or 160 mm Hg

Since the percentage of CO_2 in the atmosphere is 0.04,

Atmospheric $PCO_2 = 0.04\% \times 760$ mm Hg
$= 0.3040$ or 0.3 mm Hg

The partial pressures of the respiratory gases and nitrogen in the atmosphere, alveoli, blood, and tissue cells are shown in Exhibit 23-2. These partial pressures are important in determining the movement of oxygen and carbon dioxide between the atmosphere and lungs, the lungs and blood, and the blood and body cells. When a mixture of gases diffuses across a selectively permeable membrane, each gas diffuses from the area where its partial pressure is greater to the area where its partial pressure is less. Every gas is on its own and behaves as if the other gases in the mixture did not exist.

The amounts of respiratory gases vary in inspired (atmospheric), alveolar, and expired air (Exhibit 23-3). Inspired (atmospheric) air contains about 21 percent oxygen and 0.04 percent carbon dioxide. Expired air contains less oxygen (about 16 percent) and more carbon dioxide (about 4.5 percent) than inspired air. Compared to alveolar air, expired air contains more oxygen (about 16 percent versus 14 percent) and less carbon dioxide (about 4.5 percent versus 5.5 percent) because some of the expired air is dead air volume that has not participated in gaseous

---

EXHIBIT 23-2 PARTIAL PRESSURES (mm Hg) OF RESPIRATORY GASES AND NITROGEN IN ATMOSPHERIC AIR, ALVEOLAR AIR, BLOOD, AND TISSUE CELLS

	ATMOSPHERIC AIR (SEA LEVEL)	ALVEOLAR AIR	DEOXYGENATED BLOOD	OXYGENATED BLOOD	TISSUE CELLS
PO_2	160	105	40	105	40
PCO_2	0.3	40	45	40	45
PN_2	597	569	569	569	569

EXHIBIT 23-3 APPROXIMATE PERCENTAGE OF OXYGEN AND CARBON DIOXIDE IN INSPIRED AIR, ALVEOLAR AIR, AND EXPIRED AIR

	INSPIRED AIR	ALVEOLAR AIR	EXPIRED AIR
Oxygen	21	15	16
Carbon dioxide	0.04	5.50	3.6

exchange. Expired air is actually a mixture of inspired and alveolar air.

Henry's Law

You have probably noticed that a bottle of soda makes a hissing sound when the top is removed, and bubbles rise to the surface for some time afterward. The gas dissolved in carbonated beverages is carbon dioxide. The ability of a gas to stay in solution depends on its partial pressure and solubility coefficient, that is, its physical or chemical attraction for water. The solubility coefficient of carbon dioxide is high (0.57), that of oxygen is lower (0.024), and that of nitrogen is still lower (0.012). The higher the partial pressure of a gas over a liquid and the higher the solubility coefficient, the more gas will stay in solution. Since the soda is bottled under pressure and capped, the CO_2 remains dissolved as long as the bottle is unopened. Once you remove the cap, the pressure is released and the gas begins to bubble out. This phenomenon is explained by **Henry's law:** The quantity of a gas that will dissolve in a liquid is proportional to the partial pressure of the gas and its solubility coefficient, when the temperature remains constant.

Henry's law explains two conditions resulting from changes in the solubility of nitrogen in body fluids. Even though the air we breathe contains about 79 percent nitrogen, this gas has no known effect on bodily functions since very little of it dissolves in blood plasma because of its low solubility coefficient at sea level pressure. But, when a deep-sea diver, scuba (**self-contained underwater breathing apparatus**) diver, or caisson worker (person who builds tunnels under rivers) breathes air under high pressure, the nitrogen in the mixture can affect the body. Partial pressure is a function of total pressure, and therefore the partial pressure of all the components of a mixture increases as the total increases. Since the partial pressure of nitrogen is higher in a mixture of compressed air than in air at sea level pressure, a considerable amount of nitrogen goes into solution in plasma and interstitial fluid. Excessive amounts of dissolved nitrogen may produce giddiness and other symptoms similar to alcohol intoxication. The condition is called **nitrogen narcosis** or "rapture of the depths." The greater the depth, the more severe the condition.

If a diver is brought to the surface slowly, the dissolved nitrogen can be eliminated through the lungs. However, if a diver ascends too rapidly, the nitrogen comes out of solution too quickly to be eliminated by respiration.

Instead it forms gas bubbles in the tissues that result in **decompression sickness (caisson disease** or **bends).** The effects of decompression sickness typically result from bubbles in nervous tissue and can be mild or severe, depending on the amount of bubbles formed. Symptoms include joint pain, especially in the arms and legs, dizziness, shortness of breath, extreme fatigue, paralysis, and unconsciousness. Decompression sickness can be prevented by a slow ascent or by the use of a special decompression tank within 5 minutes after arriving at the surface. The use of helium-oxygen mixtures instead of air containing nitrogen may reduce the dangers of decompression sickness since helium is only about 40 percent as soluble as nitrogen in blood.

CLINICAL APPLICATION

A major clinical application of Henry's law is **hyperbaric** (*hyper* = over; *baros* = pressure) **oxygenation (HBO).** Using pressure to cause more oxygen to dissolve in the blood is an effective technique in treating patients infected by anaerobic bacteria, such as those that cause tetanus and gangrene. (Anaerobic bacteria cannot live in the presence of free oxygen.) A person undergoing hyperbaric oxygenation is placed in a hyperbaric chamber, which contains oxygen at a pressure of 3 to 4 atm (2,280 to 3,040 mm Hg). The body tissues pick up the oxygen, and the bacteria are killed. Hyperbaric chambers may also be used for treating certain heart disorders, carbon monoxide poisoning, gas embolisms, crush injuries, cerebral edema, certain hard to treat bone infections, smoke inhalation, drowning, asphyxia, vascular insufficiencies, and burns.

EXTERNAL RESPIRATION

External respiration is the exchange of oxygen and carbon dioxide between the alveoli of the lungs and pulmonary blood capillaries (Figure 23-14a). It results in the conversion of *deoxygenated blood* (more CO_2 than O_2) coming from the heart to *oxygenated blood* (more O_2 than CO_2) returning to the heart. During inspiration, atmospheric air containing oxygen enters the alveoli. Deoxygenated blood is pumped from the right ventricle through the pulmonary arteries into the pulmonary capillaries overlying the alveoli. The PO_2 of alveolar air is 105 mm Hg. The PO_2 of the deoxygenated blood entering the pulmonary capillaries is only 40 mm Hg. As a result of this difference in PO_2, oxygen diffuses from the alveoli into the deoxygenated blood until equilibrium is reached, and the PO_2 of the now oxygenated blood is 105 mm Hg. While oxygen diffuses from the alveoli into deoxygenated blood, carbon dioxide diffuses in the opposite direction. On arriving at the lungs, the PCO_2 of pulmonary deoxygenated blood is 45 mm Hg, while that of the alveoli is 40 mm Hg. Because of this difference in PCO_2, carbon dioxide diffuses from pulmonary deoxygenated blood into the alveoli until the PCO_2 of the blood decreases to 40 mm Hg, the PCO_2 of pulmonary oxygenated blood. Thus the PO_2 and PCO_2 of oxygenated blood leaving the lungs

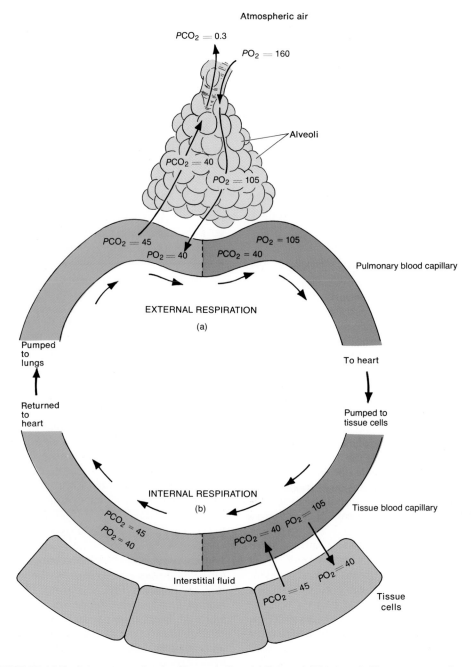

FIGURE 23-14 Partial pressures involved in respiration. (a) External. (b) Internal. All pressures are in mm Hg.

are the same as in alveolar air. The carbon dioxide that diffuses into the alveoli is eliminated from the lungs during expiration.

External respiration is aided by several anatomic adaptations. The total thickness of the alveolar-capillary (respiratory) membranes is only 0.5 μm. Thicker membranes would inhibit diffusion. The surface area over which diffusion may occur is large. The total surface area of the alveoli is about 70 m² (753 ft²), many more times the total surface area of the skin. Lying over the alveoli are countless capillaries—so many that 900 ml of blood is able to participate in gas exchange at any time. Finally,

the capillaries are so narrow that the red blood cells must flow through them in single file. This feature gives each red blood cell maximum exposure to the available oxygen.

CLINICAL APPLICATION

The efficiency of external respiration depends on several factors. One of the most important is altitude. As long as alveolar PO_2 is higher than venous blood PO_2, oxygen diffuses from the alveoli into the blood.

As a person ascends in altitude, the atmospheric PO_2 decreases, the alveolar PO_2 decreases correspondingly, and less oxygen diffuses into the blood. For example, at sea level, PO_2 is 160 mm Hg. At 10,000 ft, it decreases to 110 mm Hg, at 20,000 ft to 73 mm Hg, and at 50,000 ft to 18 mm Hg. The common symptoms of **high altitude sickness (acute mountain sickness)**—shortness of breath, nausea, dizziness—are attributable to the low concentrations of oxygen in the blood.

Another factor that affects external respiration is the total surface area available for O_2–CO_2 exchange. Any **pulmonary disorder** that decreases the functional surface area formed by the alveolar-capillary membranes decreases the efficiency of external respiration.

A third factor that influences external respiration is the **minute volume of respiration.** Certain drugs, such as morphine, slow down the respiration rate, thereby decreasing the amount of oxygen and carbon dioxide that can be exchanged between the alveoli and the blood.

INTERNAL RESPIRATION

As soon as external respiration is completed, oxygenated blood leaves the lungs through the pulmonary veins and returns to the heart. From here it is pumped from the left ventricle into the aorta and through the systemic arteries to tissue cells. The exchange of oxygen and carbon dioxide between tissue blood capillaries and tissue cells is called **internal respiration** (Figure 23-14b). It results in the conversion of oxygenated blood into deoxygenated blood. Oxygenated blood in tissue capillaries has a PO_2 of 105 mm Hg, whereas tissue cells have a PO_2 of 40 mm Hg. Because of this difference in PO_2, oxygen diffuses from the oxygenated blood through interstitial fluid and into tissue cells until the PO_2 in blood decreases to 40 mm Hg, the PO_2 of tissue capillary deoxygenated blood. At rest, only about 25 percent of the available oxygen in oxygenated blood actually enters tissue cells. This amount is sufficient to support the needs of resting cells. During exercise, more oxygen is released. While oxygen diffuses from the tissue blood capillaries into tissue cells, carbon dioxide diffuses in the opposite direction. The PCO_2 of tissue cells is 45 mm Hg, while that of tissue capillary oxygenated blood is 40 mm Hg. As a result, carbon dioxide diffuses from tissue cells through interstitial fluid into the oxygenated blood until the PCO_2 in the blood increases to 45 mm Hg, the PCO_2 of tissue capillary deoxygenated blood. The deoxygenated blood now returns to the heart. From here it is pumped to the lungs for another cycle of external respiration.

TRANSPORT OF RESPIRATORY GASES

The transportation of respiratory gases between the lungs and body tissues is a function of the blood. When oxygen and carbon dioxide enter the blood, certain physical and chemical changes occur that aid in gas transport and exchange.

Oxygen

Under normal resting conditions, each 100 ml of oxygenated blood contains 20 ml of oxygen. Oxygen does not dissolve easily in water and, therefore, very little oxygen is carried in the dissolved state in the water in blood plasma. In fact, 100 ml of oxygenated blood contains only about 3 percent oxygen dissolved in plasma. The remainder of the oxygen, about 97 percent, is carried in chemical combination with hemoglobin in red blood cells (Figure 23-15).

Hemoglobin consists of a protein portion called globin and a pigment portion called heme. The heme portion contains four atoms of iron, each capable of combining with a molecule of oxygen. Oxygen and hemoglobin combine in an easily reversible reaction to form **oxyhemoglobin.**

$$\underset{\substack{\text{Reduced (deoxygenated)}\\\text{hemoglobin}\\\text{(uncombined}\\\text{hemoglobin)}}}{\text{HB}} + \underset{\text{Oxygen}}{O_2} \rightleftharpoons \underset{\substack{\text{Oxyhemoglobin}\\\text{(combined}\\\text{hemoglobin)}}}{HbO_2}$$

• *Hemoglobin and PO_2* The most important factor that determines how much oxygen combines with hemoglobin is PO_2. When Hb (reduced or deoxygenated hemoglobin) is completely converted to HbO_2, it is referred to as **fully saturated.** When hemoglobin consists of a mixture of Hb and HbO_2, it is **partially saturated.** The **percent saturation of hemoglobin** is the percent of HbO_2 in total hemoglobin. The degree of saturation of hemoglobin with oxygen is illustrated in Figure 23-16, the oxygen-hemoglobin dissociation curve. Note that when the PO_2 is high, hemoglobin binds with large amounts of oxygen and is almost fully saturated. When PO_2 is low, hemoglobin is only partially saturated and oxygen is released from hemoglobin. In other words, the greater the PO_2, the more oxygen will combine with hemoglobin, until the available hemoglobin molecules are saturated. Therefore, in pulmonary capillaries, where PO_2 is high, a lot of oxygen binds with hemoglobin, but in tissue capillaries, where the PO_2 is lower, hemoglobin does not hold as much oxygen, and the oxygen is released for diffusion into the tissue cells. Recall that only about 25 percent of the available oxygen splits from hemoglobin under resting conditions.

Note in the oxygen-hemoglobin dissociation curve that when the PO_2 is between 60 and 100 mm Hg, hemoglobin is 90 percent or more saturated with oxygen. Thus, blood picks up a nearly full load of oxygen from the lungs even when the PO_2 there is as low as 60 mm Hg. This explains why people can still perform well at high altitudes or when they have certain cardiac and pulmonary diseases, even though PO_2 may drop to as low as 60 mm Hg. Hemoglobin is 100 percent saturated with oxygen at a PO_2 of about 250 mm Hg. Note also in the curve that at a considerably lower PO_2 of 40 mm Hg hemoglobin is 75 percent saturated with oxygen, but drops to only 13 percent saturated at 10 mm Hg. This means that between 10 and 40 mm Hg large amounts of oxygen are released in response to only small changes in PO_2. In

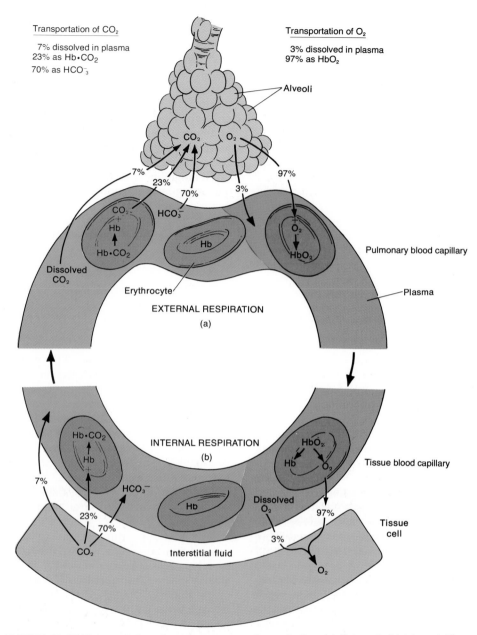

Transportation of CO₂

7% dissolved in plasma
23% as Hb·CO₂
70% as HCO₃⁻

Transportation of O₂

3% dissolved in plasma
97% as HbO₂

FIGURE 23-15 Transportation of respiratory gases in respiration. (a) External. (b) Internal. The role of red blood cells in the transportation of carbon dioxide is detailed in Figure 23-19.

active tissues such as contracting muscles, PO_2 may decrease to well below 40 mm Hg. At such pressures, nearly all of the oxygen is released from hemoglobin, oxygen required for contracting muscles.

• *Hemoglobin and pH* The amount of oxygen released from hemoglobin is determined by several factors in addition to PO_2. For example, in an acid environment, oxygen splits more readily from hemoglobin (Figure 23-17). This is referred to as the **Bohr effect,** and is based on the belief that when hydrogen ions bind to hemoglobin they alter the structure of hemoglobin and thereby decrease its oxygen-carrying capacity.

Low blood pH (acidic condition) results from the presence of lactic acid, a reaction product of muscle contrac-

tion, and high PCO_2. As carbon dioxide is taken up by the blood, much of it is temporarily converted to carbonic acid. This conversion is catalyzed by an enzyme in red blood cells called *carbonic anhydrase.*

$$\underset{\substack{\text{Carbon}\\\text{dioxide}}}{CO_2} + \underset{\text{Water}}{H_2O} \underset{\substack{\text{carbonic}\\\text{anhydrase}}}{\rightleftharpoons} \underset{\substack{\text{Carbonic}\\\text{acid}}}{H_2CO_3} \rightleftharpoons \underset{\substack{\text{Hydrogen}\\\text{ion}}}{H^+} + \underset{\substack{\text{Bicarbonate}\\\text{ion}}}{HCO_3^-}$$

The carbonic acid thus formed in red blood cells dissociates into hydrogen ions and bicarbonate ions. As the hydrogen ion concentration increases, pH decreases. Thus, an increased PCO_2 produces a more acid environment that helps to split oxygen from hemoglobin.

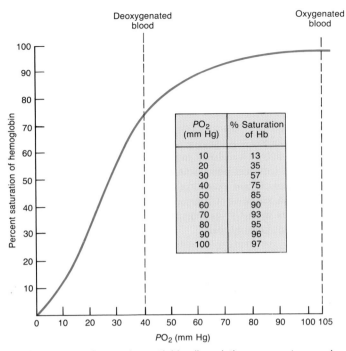

FIGURE 23-16 Oxygen-hemoglobin dissociation curve at normal body temperature showing the relationship between hemoglobin saturation and PO_2. As PO_2 increases, more oxygen combines with hemoglobin. The percentages of saturation of Hb shown in the table are approximate.

FIGURE 23-17 Oxygen-hemoglobin dissociation curve at normal body temperature showing the relationship between hemoglobin saturation and pH (PCO_2). As pH increases (PCO_2 decreases), more oxygen combines with hemoglobin and less is available to tissues. Or, stated another way, as pH decreases (PCO_2 increases), less oxygen combines with hemoglobin and more is available to tissues. These relationships are emphasized by the broken lines.

• *Hemoglobin and Temperature*   Within limits, as temperature increases, so does the amount of oxygen released from hemoglobin (Figure 23-18). Heat energy is a by-product of the metabolic reactions of all cells, and contracting muscle cells release an especially large amount of heat. Splitting the oxyhemoglobin molecule is another example of how homeostatic mechanisms adjust body activities to cellular needs. Active cells require more oxygen, and active cells liberate more acid and heat. The acid and heat, in turn, stimulate the oxyhemoglobin to release its oxygen.

• *Hemoglobin and DPG*   A final factor that helps to release oxygen from hemoglobin is a substance found in red blood cells called **2,3-diphosphoglycerate,** or simply **DPG.** It is an intermediate compound formed in red blood cells during glycolysis (Chapter 25). It has the ability to combine reversibly with hemoglobin and thus alter its structure to release oxygen. The greater the level of DPG, the more oxygen is released from hemoglobin. The production of DPG is greatest when there is decreased oxygen delivery to tissues. Thus, DPG enhances oxygen delivery to tissues and helps to maintain the release of oxygen by hemoglobin. Cells that are actively metabolizing will have increased levels of CO_2 (decreased pH), increased temperature, and high levels of DPG. Such cells can then get more oxygen more quickly.

• *Fetal Hemoglobin*   Fetal hemoglobin differs from adult hemoglobin in structure and in its affinity for oxygen. Fetal hemoglobin, which has a higher affinity for

oxygen, can carry as much as 20 to 30 percent more oxygen than maternal hemoglobin. Thus, as the maternal blood enters the placenta, oxygen is readily transferred to fetal blood.

FIGURE 23-18 Oxygen-hemoglobin dissociation curve showing the relationship between hemoglobin saturation and temperature. As temperature increases, less oxygen combines with hemoglobin.

CLINICAL APPLICATION

Carbon monoxide (CO) is a colorless and odorless gas found in exhaust fumes from automobiles and in tobacco smoke. It is a by-product of burning carbon-containing materials such as coal and wood. One of its interesting features is that it combines with hemoglobin very much as oxygen does, except that the combination of carbon monoxide and hemoglobin is over 200 times as tenacious as the combination of oxygen and hemoglobin. In addition, in concentrations as small as 0.1 percent ($PCO = 0.5$ mm Hg), carbon monoxide will combine with half the hemoglobin molecules. Thus, the oxygen-carrying capacity of the blood is reduced by one-half. Increased levels of carbon monoxide lead to hypoxia, and the result is **carbon monoxide poisoning.** The condition may be treated by administering pure oxygen ($PO_2 = 600$ mm Hg), which slowly replaces the carbon monoxide combined with the hemoglobin.

• *Oxygen Toxicity* Astronauts live in a cabin environment that is 100 percent oxygen maintained at a pressure of one-third of an atmosphere. This results in an inspired PO_2 of about 260 mm Hg, which is over 100 mm Hg higher than the normal level of 160 mm Hg. Breathing pure oxygen can be hazardous. Animal experiments have shown that when guinea pigs breathe 100 percent oxygen at atmospheric pressure, they develop pulmonary edema. In patients breathing 100 percent oxygen, there has been evidence of impaired gas exchange. So far, none of the astronauts has shown evidence of oxygen toxicity.

One reported hazard of breathing 100 percent oxygen has been observed in premature infants. When baby incubators were provided with 100 percent oxygen, blindness occurred in some of the infants. The formation of fibrous tissue behind the lens is triggered by a local vasoconstriction caused by the high PO_2. The hazard is removed simply by reducing the percent of oxygen in the incubator to below 40 percent.

Hypoxia

Hypoxia refers to a general low level of oxygen availability; physiologically, it refers to a deficiency of oxygen at the tissue level. Based on the cause, we can classify hypoxias as follows:

1. **Hypoxic hypoxia.** This is caused by a low PO_2 in arterial blood. The condition may be the result of high altitude, obstructions in air passageways, or fluid in the lungs.
2. **Anemic hypoxia.** In this case, there is too little functioning hemoglobin in the blood. Among the causes are hemorrhage, anemia, or failure of hemoglobin to carry its normal complement of oxygen, as in carbon monoxide (CO) poisoning.
3. **Stagnant hypoxia.** This condition results from the inability of blood to carry oxygen to tissues fast enough to sustain their needs. It may be caused by heart failure or circulatory shock, both of which inhibit adequate delivery of oxygen to tissues.

4. **Histotoxic hypoxia.** In this condition, the blood delivers adequate oxygen to tissues, but the tissues are unable to use it properly. A commonly cited cause is cyanide poisoning, in which cyanide blocks the metabolic machinery of cells related to oxygen utilization.

Carbon Dioxide

Under normal resting conditions, each 100 ml of deoxygenated blood contains 4 ml of carbon dioxide. The CO_2 is carried by the blood in several forms (see Figure 23-15). The smallest percentage, about 7 percent, is dissolved in plasma. Upon reaching the lungs, it diffuses into the alveoli. A somewhat higher percentage, about 23 percent, combines with the globin portion of hemoglobin to form **carbaminohemoglobin.**

$$\underset{\text{Hemoglobin}}{\text{Hb}} + \underset{\text{Carbon dioxide}}{CO_2} \rightleftharpoons \underset{\text{Carbaminohemoglobin}}{\text{Hb} \cdot CO_2}$$

The formation of carbaminohemoglobin is greatly influenced by PCO_2. For example, in tissue capillaries PCO_2 is relatively high, and this encourages the formation of carbaminohemoglobin. But in pulmonary capillaries, PCO_2 is relatively low, and the CO_2 readily splits apart from globin and enters the alveoli by diffusion.

The greatest percentage of CO_2, about 70 percent, is transported in plasma as bicarbonate ions. The reaction that brings about this method of transportation of CO_2 is the same one noted earlier.

$$\underset{\substack{\text{Carbon}\\\text{dioxide}}}{CO_2} + \underset{\text{Water}}{H_2O} \underset{\text{carbonic anhydrase}}{\rightleftharpoons} \underset{\substack{\text{Carbonic}\\\text{acid}}}{H_2CO_3} \rightleftharpoons \underset{\substack{\text{Hydrogen}\\\text{ion}}}{H^+} + \underset{\substack{\text{Bicarbonate}\\\text{ion}}}{HCO_3^-}$$

As CO_2 diffuses into tissue capillaries and enters the red blood cells, it reacts with water, in the presence of carbonic anhydrase, to form carbonic acid. The carbonic acid dissociates into H^+ ions and HCO_3^- ions. The H^+ ions combine mainly with hemoglobin. The HCO_3^- ions leave the red blood cells and enter the plasma. In exchange, chloride ions (Cl^-) diffuse from plasma into the red blood cells. This exchange of negative ions maintains the ionic balance between plasma and red blood cells and is known as the **chloride shift** (Figure 23-19a). The Cl^- ions that enter red blood cells combine with potassium ions (K^+) to form the salt potassium chloride (KCl). The HCO_3^- ions that enter plasma from the red blood cells combine with sodium (Na^+), the principal positive ion is extracellular fluid, to form sodium bicarbonate ($NaHCO_3$). The net effect of all of these reactions is that CO_2 is carried from tissue cells as bicarbonate ions in plasma.

Deoxygenated blood returning to the lungs contains CO_2 dissolved in plasma, CO_2 combined with globin as carbaminohemoglobin, and CO_2 incorporated in bicarbonate ions. In the pulmonary capillaries, the events are reversed. The CO_2 dissolved in plasma diffuses into the alveoli. The CO_2 combined with globin splits from the globin and diffuses into the alveoli. The CO_2 carried as bicarbonate is released as follows (Figure 23-19b). As the hemoglobin in pulmonary blood picks up oxygen,

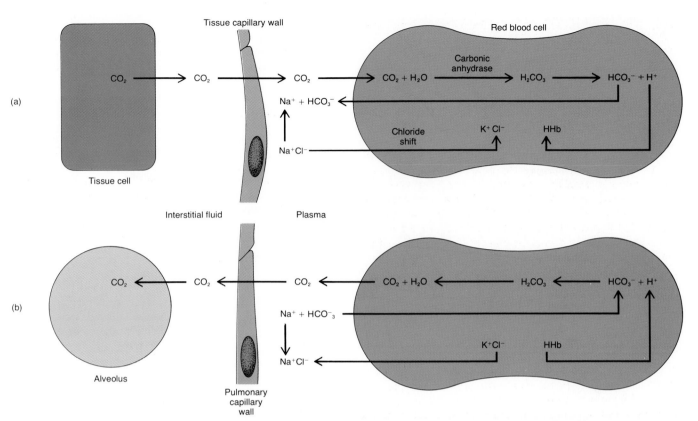

FIGURE 23-19 Transportation of carbon dioxide. (a) Between tissue cells and tissue capillaries. (b) Between pulmonary capillaries and alveoli.

the H^+ ions are released from hemoglobin. The Cl^- ions simultaneously split from K^+ ions, and HCO_3^- ions reenter the red blood cells after splitting from Na^+ ions. The H^+ and HCO_3^- ions recombine to form H_2CO_3, which splits into CO_2 and H_2O. The CO_2 leaves the red blood cells and diffuses into the alveoli. The direction of the carbonic acid reaction depends mostly on $P\mathrm{CO}_2$. In tissue capillaries, where $P\mathrm{CO}_2$, is high, bicarbonate is formed. In pulmonary capillaries, where $P\mathrm{CO}_2$ is low, CO_2 and H_2O are formed.

In our discussion of hemoglobin and $P\mathrm{O}_2$, it was indicated that the greater the $P\mathrm{O}_2$ the greater the amount of oxygen that will combine with hemoglobin, until the available hemoglobin molecules are saturated (see Figure 23-16). The same relationship, however, does not hold true for CO_2. That is, as the $P\mathrm{CO}_2$ increases, the amount of bound CO_2 also increases. It is for this reason that saturation of hemoglobin with CO_2 does not occur.

CONTROL OF RESPIRATION

The basic rhythm of respiration is controlled by portions of the nervous system in the medulla and pons. In response to the demands of the body, this rhythm can be modified. We will first examine the principal mechanisms involved in the nervous control of the rhythm of respiration.

NERVOUS CONTROL

The size of the thorax is affected by the action of the respiratory muscles. These muscles contract and relax as a result of nerve impulses transmitted to them from centers in the brain. The area from which nerve impulses are sent to respiratory muscles is located bilaterally in the reticular formation of the brain stem; it is referred to as the **respiratory center.** The respiratory center consists of a widely dispersed group of neurons that is functionally divided into three areas: (1) the medullary rhythmicity area, in the medulla; (2) the pneumotaxic (noo-mō-TAK-sik) area, in the pons; and (3) the apneustic (ap-NOO-stik) area, in the pons (Figure 23-20).

Medullary Rhythmicity Area

The function of the **medullary rhythmicity area** is to control the basic rhythm of respiration. In the normal resting state, inspiration usually lasts for about 2 seconds and expiration for about 3 seconds. This is the basic rhythm of respiration. Within the medullary rhythmicity area are found both inspiratory and expiratory neurons that comprise inspiratory and expiratory areas, respectively. Let us first consider the proposed role of the inspiratory neurons in respiration.

The basic rhythm of respiration is determined by nerve impulses generated in the inspiratory area (Figure 23-

FIGURE 23-20 Approximate location of areas of the respiratory center.

21a). At the beginning of expiration, the inspiratory area is inactive, but after 3 seconds, it suddenly and automatically becomes active. This activity seems to result from an intrinsic excitability of the inspiratory neurons. In fact, when all incoming nerve connections to the inspiratory area are cut or blocked, the area still rhythmically discharges impulses that result in inspiration. Nerve impulses from the active inspiratory area last for about 2 seconds and travel to the muscles of inspiration. The impulses reach the diaphragm by the phrenic nerves, and the external intercostal muscles by the intercostal nerves. When the impulses reach the inspiratory muscles, the muscles contract and inspiration occurs. At the end of 2 seconds, the inspiratory muscles become inactive again, and the cycle repeats itself over and over.

It is believed that the expiratory neurons remain inactive during most normal quiet respiration. During quiet respiration, inspiration is accomplished by active contraction of the inspiratory muscles, and expiration results from passive recoil of the lungs and thoracic wall as the inspiratory muscles relax. However, during high levels of ventilation, it is believed that impulses from the inspiratory area activate the expiratory area (Figure 23-21b). Impulses discharged from the expiratory area cause con-

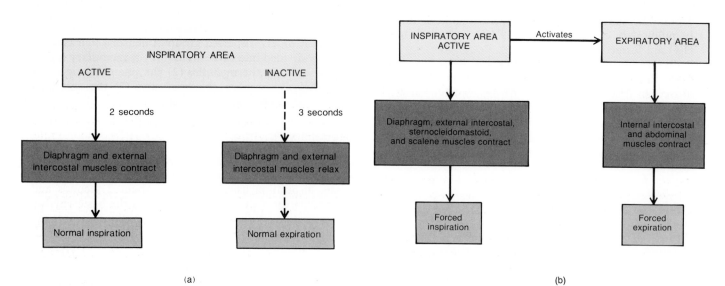

(a)

(b)

FIGURE 23-21 Proposed role of the medullary rhythmicity area in controlling the basic rhythm of respiration. (a) During normal quiet breathing. (b) During high levels of respiration.

traction of the internal intercostals and abdominal muscles that decrease the size of the thoracic cavity.

Pneumotaxic Area

Although the medullary rhythmicity area controls the basic rhythm of respiration, other parts of the nervous system help coordinate the transition between inspiration and expiration. One of these is the **pneumotaxic area** in the upper pons (see Figure 23-20). It continuously transmits inhibitory impulses to the inspiratory area. The major effect of these impulses is to help turn off the inspiratory area before the lungs become too full of air. In other words, the impulses limit inspiration and thus facilitate expiration.

Apneustic Area

Another part of the nervous system that coordinates the transition between inspiration and expiration is the **apneustic area** in the lower pons (see Figure 23-20). It sends stimulatory impulses to the inspiratory area that activate it and prolong inspiration, thus inhibiting expiration. This occurs when the pneumotaxic area is inactive. When the penumotaxic area is active, it overrides the apneustic area.

REGULATION OF RESPIRATORY CENTER ACTIVITY

Although the basic rhythm of respiration is set and coordinated by the respiratory center, the rhythm can be modified in response to the demands of the body by neural input to the center.

Cortical Influences

The respiratory center has connections with the cerebral cortex, which means we can voluntarily alter our pattern of breathing. We can even refuse to breathe at all for a short time. Voluntary control is protective because it enables us to prevent water or irritating gases from entering the lungs. The ability to stop breathing is limited by the buildup of CO_2 in the blood, however. When the PCO_2 increases to a certain level, the inspiratory area is stimulated, impulses are sent to inspiratory muscles, and breathing resumes whether or not the person wishes. It is impossible for people to kill themselves by holding their breath.

Inflation Reflex

Located in the walls of bronchi and bronchioles throughout the lungs are **stretch receptors.** When the receptors become overstretched, impulses are sent along the vagus (X) nerves to the inspiratory area and apneustic area. In response, the inspiratory area is inhibited and the apneustic area is inhibited from activating the inspiratory area. The result is that expiration follows. As air leaves the lungs during expiration, the lungs deflate and the stretch receptors are no longer stimulated. Thus, the inspiratory and apneustic areas are no longer inhibited and

a new inspiration begins. This reflex is referred to as the **inflation (Hering-Breuer) reflex.** Some evidence suggests that the reflex is mainly a protective mechanism for preventing overinflation of the lungs rather than a key component in the regulation of respiration.

Chemical Stimuli

Certain chemical stimuli determine how fast we breathe. The ultimate goal of the respiratory system is to maintain proper levels of carbon dioxide and oxygen, and the system is highly responsive to changes in the blood levels of either. Although it is convenient to speak of carbon dioxide as the most important chemical stimulus for regulating the rate of respiration, it is actually hydrogen ions that assume this role. For example, carbon dioxide (CO_2) in the blood combines with water (H_2O) to form carbonic acid (H_2CO_3). But the carbonic acid quickly breaks down into H^+ ions and bicarbonate (HCO_3^-) ions. Any increase in CO_2 will cause an increase in H^+ ions, and any decrease in CO_2 will cause a decrease in H^+ ions. In effect, it is primarily the H^+ ions that alter the rate of respiration rather than the CO_2 molecules. Although the following discussion refers to the levels of CO_2 and their effect on respiration, keep in mind that it is really the H^+ ions that cause the effects.

Within the medulla is an area, called the **chemosensitive area,** that is highly sensitive to blood concentrations of CO_2. Outside the central nervous system are **chemoreceptors,** which are sensitive to changes in CO_2 and O_2 levels in the blood. They are located within the carotid bodies near the bifurcation of the common carotid arteries and in the aortic bodies near the arch of the aorta. The *carotid bodies* are small, oval nodules about 4 to 5 mm long, located in the space between the internal and external carotid arteries. The *aortic bodies* are clustered in the region between the arch of the aorta and the dorsal surface of the pulmonary artery. The afferent nerve fibers from the carotid bodies join with those from the carotid sinus to form the carotid sinus nerve, which joins the glossopharyngeal (IX) nerve. The afferent fibers from aortic bodies run into the vagus (X) nerve.

Under normal circumstances, arterial blood PCO_2 is 40 mm Hg. If there is even a slight increase in PCO_2—a condition called **hypercapnia**—the chemosensitive area in the medulla and chemoreceptors in the carotid and aortic sinuses are stimulated (Figure 23-22). Stimulation of the chemosensitive area and chemoreceptors causes the inspiratory area to become highly active, and the rate of respiration increases. This increased rate, **hyperventilation,** allows the body to expel more CO_2 until the PCO_2 is lowered to normal. If arterial PCO_2 is lower than 40 mm Hg, the chemosensitive area and chemoreceptors are not stimulated and stimulatory impulses are not sent to the inspiratory area. Consequently, the area sets its own moderate pace until CO_2 accumulates and the PCO_2 rises to 40 mm Hg. A slow rate of respiration is called **hypoventilation.**

The oxygen chemoreceptors are sensitive only to large decreases in the PO_2 because hemoglobin remains about 85 percent or more saturated at PO_2 values all the way

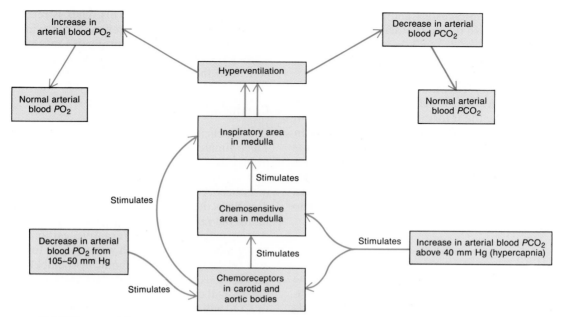

FIGURE 23-22 Effects of increased blood levels of PCO_2 (blue arrows) and decreased blood levels of PO_2 down to 50 mm Hg (red arrows).

down to 50 mm Hg (see Figure 23-16). If arterial PO_2 falls from a normal of 105 mm Hg to about 50 mm Hg, the oxygen chemoreceptors become stimulated and send impulses to the inspiratory area and respiration increases (Figure 23-22). But if the PO_2 falls much below 50 mm Hg, the cells of the inspiratory area suffer oxygen starvation and do not respond well to any chemical receptors. They send fewer impulses to the inspiratory muscles, and the respiration rate decreases or breathing ceases altogether.

Other Influences

The carotid and aortic sinuses also contain baroreceptors (pressure receptors) that are stimulated by a rise in blood pressure. Although these baroreceptors are concerned mainly with the control of circulation (see Figure 20-10), they help control respiration. For example, a sudden rise in blood pressure decreases the rate of respiration, and a drop in blood pressure increases the respiratory rate.

Other factors that control respiration are:

1. An increase in body temperature, as during a fever or vigorous muscular exercise, increases the rate of respiration. A decrease in body temperature decreases respiratory rate. A sudden cold stimulus such as plunging into cold water causes **apnea,** a temporary cessation of breathing.

2. A sudden, severe pain brings about apnea, but a prolonged pain triggers the general adaptation syndrome and increases respiratory rate.

3. Stretching the anal sphincter muscle increases the respiratory rate. This technique is sometimes employed to stimulate respiration during emergencies.

4. Irritation of the pharynx or larynx by touch or chemicals brings about an immediate cessation of breathing followed by coughing.

A summary of stimuli that affect the rate of respiration is presented in Exhibit 23-4.

AGING AND THE RESPIRATORY SYSTEM

The cardiovascular and respiratory systems operate as a unit, and damage or disease in one of these organ systems is often secondarily reflected in the other. In pulmonary heart disease, the right side of the heart enlarges in response to certain lung diseases. This condition is known as right ventricular hypertrophy, or cor pulmonale. With advancing age, the airways and tissues of the

EXHIBIT 23-4 SUMMARY OF STIMULI THAT AFFECT THE RATE OF RESPIRATION

STIMULI THAT INCREASE RATE OF RESPIRATION	STIMULI THAT DECREASE (OR INHIBIT) RATE OF RESPIRATION
Increase in arterial blood PCO_2 above 40 mm Hg	Decrease in arterial blood PO_2 below 40 mm Hg
Decrease in arterial blood PO_2 from 105 mm Hg to 50 mm Hg	Decrease in arterial blood PO_2 below 50 mm Hg
Decrease in blood pressure Increase in body temperature	Increase in blood pressure Decrease in body temperature decreases rate of respiration, and sudden cold stimulus causes apnea
Prolonged pain	Severe pain causes apnea
Stretching anal sphincter	Irritation of pharynx or larynx by touch or chemicals causes apnea

respiratory tract, including the air sacs, become less elastic and more rigid. In addition to the lungs becoming less elastic, the chest wall also becomes more rigid. As a result, there is a decrease in pulmonary lung capacity. In fact, vital capacity (the amount of air moved by maximal inspiration followed by maximal expiration) can decrease as much as 35 percent by age 70. Also, there is a decrease in blood levels of oxygen, decreased activity of alveolar macrophages, and diminished ciliary action of the epithelium lining the respiratory tract. Due to all of these age-related factors, elderly people are more susceptible to pneumonia, bronchitis, emphysema, and other pulmonary disorders.

DEVELOPMENTAL ANATOMY OF THE RESPIRATORY SYSTEM

The development of the mouth and pharynx is considered in Chapter 24 on the digestive system. Here we will consider the remainder of the respiratory system.

At about 4 weeks, the respiratory system begins as an outgrowth of the **endoderm** of the foregut (precursor of some digestive organs) just behind the pharynx. This outgrowth is called the **laryngotracheal bud** (see Figure 18-24). As the bud grows, it elongates and differentiates into the future *larynx* and other structures as well. Its proximal end maintains a slit-like opening into the pharynx called the *glottis*. The middle portion of the bud gives rise to the *trachea*. The distal portion divides into two **lung buds,** which grow into the *bronchi* and *lungs* (Figure 23-23).

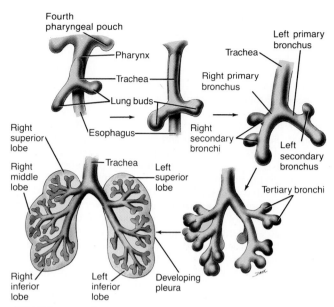

FIGURE 23-23 Development of the bronchial tubes and lungs.

As the lung buds develop, they branch and rebranch and give rise to all the *bronchial tubes.* After the sixth month, the closed terminal portions of the tubes dilate and become the *alveoli* of the lungs. The smooth muscle, cartilage, and connective tissues of the bronchial tubes and the pleural sacs of the lungs are contributed by **mesenchymal (mesodermal) cells.**

DISORDERS: HOMEOSTATIC IMBALANCES

Bronchogenic Carcinoma (Lung Cancer)

As part of ordinary breathing, many irritating substances are inhaled. Almost all pollutants, including inhaled smoke, have an irritating effect on the bronchial tubes and lungs and may be regarded as stresses or irritating stimuli. The effects of irritation on the bronchial epithelium are important clinically because a common lung cancer, **bronchogenic carcinoma,** starts in the walls of the bronchi.

Microscopic examination of the pseudostratified epithelium of a bronchial tube reveals three kinds of cells, all of which are in contact with the basement membrane. The columnar cells contain cilia and extend upward from the basement membrane and reach the lumenal surface. At intervals between the ciliated columnar cells are the mucus-secreting goblet cells. The basal cells do not reach the lumenal surface. The basal cells divide continuously, replacing the columnar cells as they wear down and are sloughed off.

The constant irritation by inhaled smoke and pollutants causes an enlargement of the goblet cells of the bronchial epithelium. They respond by secreting excessive mucus. The basal cells also respond to the stress by undergoing cell division so fast that they push into the area occupied by the goblet and columnar cells. As many as 20 rows of basal cells may be produced. Many researchers believe that if the stress is removed at this point, the epithelium can return to normal.

If the stress persists, more and more mucus is secreted and the cilia become less effective. As a result, mucus is not carried toward the throat but remains trapped in the bronchial tubes.

The individual then develops a "smoker's cough." Moreover, the constant irritation from the pollutant slowly destroys the alveoli, which are replaced with thick, inelastic connective tissue. Mucus that has accumulated becomes trapped in the air sacs. Millions of sacs rupture, reducing the diffusion surface for the exchange of oxygen and carbon dioxide. The individual has now developed emphysema. If the stress is removed at this point, there is little chance for improvement. Alveolar tissue that has been destroyed cannot be repaired. But removal of the stress can stop further destruction of lung tissue.

If the stress still continues, the emphysema gets progressively worse, and the basal cells of the bronchial tubes continue to divide and break through the basement membrane. At this point the stage is set for bronchogenic carcinoma. Columnar and goblet cells disappear and may be replaced with squamous cancer cells. If this happens, the malignant growth spreads throughout the lung and may block a bronchial tube. If the obstruction occurs in a large bronchial tube, very little oxygen enters the lung, and disease-producing bacteria thrive on the mucoid secretions. In the end, the patient may develop emphysema, carcinoma, and a host of infectious diseases. Treatment involves surgical removal of the diseased lung. However, metastasis (spreading) of the growth through the lymphatics or blood vessels may result in new growths in other parts of the body such as the brain and liver.

Other factors may be associated with lung cancer. For instance, breast, stomach, and prostate malignancies can metastasize to the lungs. People who apparently have not been exposed to pollutants do occasionally develop bronchogenic carcinoma. However, the occurrence of bronchogenic carcinoma is probably over 20 times higher in heavy cigarette smokers than it is in nonsmokers.

Bronchial Asthma

Bronchial asthma is a reaction, usually allergic, characterized by attacks of wheezing and difficult breathing. Attacks are brought on by spasms of the smooth muscles that lie in the walls of the smaller bronchi and bronchioles, causing the passageways to close partially. The patient has trouble exhaling, and the alveoli may remain inflated during expiration. Usually the mucous membranes that line the respiratory passageways become irritated and secrete excessive amounts of mucus that may clog the bronchi and bronchioles and worsen the attack. About three out of four asthma victims are allergic to edible or airborne substances as common as wheat or dust. Others are sensitive to the proteins of harmless bacteria that inhabit the paranasal sinuses, nose, and throat. Asthma might also have a psychosomatic origin. (The anaphylactic mechanism involved in bronchial asthma may be reviewed in Chapter 22.)

Bronchitis

Bronchitis is inflammation of the bronchi characterized by hypertrophy and hyperplasia of seromucous glands and goblet cells lining the bronchial airways. The typical symptom is a productive cough in which a thick greenish-yellow sputum is raised. This secretion signifies the presence of the underlying infection that is causing excessive secretion of mucus. Cigarette smoking remains the most important cause of chronic bronchitis, that is, bronchitis that lasts for at least 3 months of the year for 2 successive years. Other factors that influence development of the disease are family history of bronchitis or other lung disease, air pollution, carbon monoxide, respiratory infections, and deficient antibodies, especially IgA antibodies.

Emphysema

In **emphysema** (em'-fi-SĒ-ma), the alveolar walls lose their elasticity and remain filled with air during expiration. The name means "blown up," or "full of air." Reduced forced expiratory volume is the first symptom. Later, alveoli in other areas of the lungs are damaged. Many alveoli may merge to form larger air sacs with a reduced overall volume. The lungs become permanently inflated because they have lost elasticity. To adjust to the increased lung size, the size of the chest cage increases, resulting in a "barrel chest." The patient has to work voluntarily to exhale. Oxygen diffusion does not occur as easily across the damaged alveolar-capillary membrane, blood O_2 is somewhat lowered, and any mild exercise that raises the oxygen requirements of the cells leaves the patient breathless. As the disease progresses, the alveoli are replaced with thick fibrous connective tissue. Even carbon dioxide does not diffuse easily through this fibrous tissue. If the blood cannot buffer all the hydrogen ions that accumulate, the blood pH drops or unusually high amounts of carbon dioxide may dissolve in the plasma. High carbon dioxide levels produce acid conditions that are toxic to brain cells. Consequently, the inspiratory area becomes less active and the respiration rate slows down, further aggravating the problem. The compressed and damaged capillaries around the deteriorating alveoli may no longer be able to receive blood. As a result, resistance to blood flow increases in the pulmonary trunk and the right ventricle overworks as it attempts to force blood through the remaining capillaries.

Emphysema is generally caused by a long-term irritation. Air pollution, occupational exposure to industrial dust, and cigarette smoke are the most common irritants. Cigarette smoke not only deactivates a protein that is seemingly crucial in preventing emphysema but also prevents the repair of affected lung tissue. Chronic bronchial asthma also may produce alveolar damage. Cases of emphysema are becoming more and more frequent in the United States. The irony is that the disease can be prevented and the progressive deterioration can be stopped by eliminating the harmful stimuli.

Diseases such as bronchial asthma, bronchitis, and emphysema have in common some degree of obstruction of the air passageways. The term **chronic obstructive pulmonary disease (COPD)** is used to refer to these disorders.

Pneumonia

The term **pneumonia** means an acute infection or inflammation of the alveoli. In this disease, the alveolar sacs fill up with fluid and dead white blood cells, reducing the amount of air space in the lungs. (Remember that one of the cardinal signs of inflammation is edema.) Oxygen has difficulty diffusing through the inflamed alveoli, and the blood O_2 may be drastically reduced. Blood CO_2 usually remains normal because carbon dioxide diffuses through the alveoli more easily than oxygen does.

The most common cause of pneumonia is the pneumococcus bacterium (*Streptococcus pneumoniae*), but other bacteria or fungi, protozoans, or viruses may be the source of trouble.

Tuberculosis

The bacterium, *Mycobacterium tuberculosis,* produces an inflammation called **tuberculosis.** This disease still ranks as the number-one killer in the communicable disease category. Tuberculosis most often affects the lungs and the pleurae. The bacteria destroy parts of the lung tissue, and the tissue is replaced by fibrous connective tissue. Because the connective tissue is inelastic and thick, the affected areas of the lungs do not snap back during expiration and larger amounts of air are retained. Gases no longer diffuse easily through the fibrous tissue.

Tuberculosis bacteria are spread by inhalation. Although they can withstand exposure to many disinfectants, they die quickly in sunlight. Thus, tuberculosis is sometimes associated with crowded, poorly lit housing. Many drugs are successful in treating tuberculosis. Rest, sunlight, and good diet are vital parts of treatment.

Respiratory Distress Syndrome (RDS) of the Newborn

Respiratory distress syndrome (RDS) of the newborn, also called **glassy-lung disease** or **hyaline membrane disease (HMD)**, is responsible for approximately 20,000 newborn infant deaths per year. Before birth, the respiratory passages are filled with fluid. Part of this fluid is amniotic fluid inhaled during respiratory movements in utero. The remainder is produced by the submucosal glands and the goblet cells of the respiratory epithelium.

At birth, this fluid-filled passageway must become an air-filled passageway, and the collapsed primitive alveoli (terminal sacs) must expand and function in gaseous exchange. The success of this transition depends largely on surfactant, the phospholipid produced by alveolar cells that lowers surface tension in the fluid layer lining the primitive alveoli once air enters the lungs. Surfactant is present in the fetus's lungs as early as 23 weeks, and by 28 to 32 weeks the amount present is sufficient to prevent alveolar collapse during breathing. Surfactant is produced continuously by septal alveolar cells. The presence of surfactant can be detected by amniocentesis.

Although in a normal, full-term infant the second and subsequent breaths require less respiratory effort than the first, breathing is not completely normal until about 40 minutes after birth. The entire lung is not inflated fully with the first one or two breaths. In fact, for the first 7 to 10 days, small areas of the lungs may remain uninflated.

In the newborn whose lungs are deficient in surfactant, the

effort required for the first breath is essentially the same as that required in normal newborns. However, the surface tension of the alveolar fluid is 7 to 14 times higher than the surface tension of alveolar fluid with a monomolecular layer of surfactant. Consequently, during expiration after the first inspiration, the surface tension of the alveoli increases as the alveoli deflate. The alveoli collapse almost to their original uninflated state.

Idiopathic RDS of the newborn usually appears within a few hours after birth. (An idiopathic condition is one that occurs spontaneously in an individual from an unknown or obscure cause.) Affected infants show difficult and labored breathing with withdrawal of the intercostal and subcostal spaces. Death may occur soon after onset of respiratory difficulty or may be delayed for a few days, although many infants survive. At autopsy, the lungs are underinflated and areas of atelectasis (collapse of lung) are prominent. If the infant survives for at least a few hours after developing respiratory distress, the alveoli are often filled with a fluid of high-protein content that resembles a hyaline (or glassy) membrane. RDS of the newborn occurs frequently in premature infants and also in infants of diabetic mothers, particularly if the diabetes is untreated or poorly controlled.

A treatment called PEEP (positive end expiratory pressure) is used to treat RDS of the newborn. It consists of passing a tube through the air passages to the top of the lungs to provide needed oxygen-rich air at continuous pressures of up to 14 mm Hg. Continuous pressure keeps the baby's alveoli open and available for gas exchange. The addition of human surfactant, obtained from the amniotic fluid of infants born by caesarian section, to the oxygen-rich air counters the effects of RDS of the newborn more effectively and lessens lung damage.

Respiratory Failure

Respiratory failure refers to a condition in which the respiratory system cannot supply sufficient oxygen to maintain metabolism or cannot eliminate enough carbon dioxide to prevent respiratory acidosis. Specifically, respiratory failure occurs when the PO_2 of arterial blood is less than 50 mm Hg or when the PCO_2 of arterial blood is greater than 50 mm Hg. Respiratory failure always causes dysfunction in other organs as well.

Among the causes of respiratory failure are lung disorders (chronic obstructive pulmonary disease, pneumonia, and adult respiratory distress syndrome); mechanical disorders that disturb the chest wall or neuromusculature (anesthetic blocking agents, cervical spinal cord injuries, myasthenia gravis, and amyotrophic lateral sclerosis); depression of the respiratory center by drugs, strokes, or trauma; and carbon monoxide poisoning.

Symptoms of respiratory failure include disorientation, malaise, headache, muscular weakness, difficulty with sleep, palpitations, breathlessness, coughing, tachycardia, dysrhythmia, cyanosis, hypertension, edema, stupor, and coma. Treatment consists of providing adequate oxygenation and reversing the respiratory acidosis.

Sudden Infant Death Syndrome (SIDS)

Sudden infant death syndrome (SIDS), also called crib death or cot death, kills approximately 10,000 American babies every year. It kills more infants between the ages of 1 week and 12 months than any other disease. SIDS occurs without warning. Although about half of its victims had an upper respiratory infection within 2 weeks of death, the babies tended to be remarkably healthy. Two other findings are important: the apparently *silent* nature of death and the *disarray* frequently found at the scene.

Given this death scene, it is not surprising that for centuries "crib death" was mistakenly thought to be due to suffocation. Most medical people, however, believe that a healthy child *can-*

not smother in its bedclothes. The diaper of a SIDS victim is usually full of stool and urine, and the urinary bladder is almost always empty. External examination reveals only a blood-tinged froth that often exudes from the nostrils. These signs all point to a sudden lethal episode associated with tonic motor activity, not to a gradually deepening coma leading to death.

The similarity of findings in SIDS cases strengthens the theory that there is one underlying cause. The seasonal distribution of SIDS incidences, being lowest during the summer months and highest in the late fall and winter, strongly suggests an infectious agent, viruses being the most likely possibility. Most cases occur at a time in life when gamma globulin levels are low and the infant is at a critical period of susceptibility.

Many hypotheses have been proposed as explanations for SIDS. Although the precise pathogenic mechanisms have yet to be clarified, several valuable observations have emerged as a result of investigations during the past decade. Most, if not all, such cases appear to share a common mechanism ("final common pathway"). Thus, SIDS is probably a single disease entity. According to one study, a likely cause of death is laryngospasm, possibly triggered by a previous mild, nonspecific virus infection of the upper respiratory tract. Another possible factor includes periods of prolonged apnea as a result of malfunction of the respiratory center. Hypoxia may also be involved. Other suspected causes are an allergic reaction, bacterial poisoning, excess dopamine (DA), and overheating.

Coryza (Common Cold) and Influenza (Flu)

Several viruses are responsible for **coryza (common cold).** A group of viruses, called rhinoviruses, is responsible for about 40 percent of all colds in adults. Typical symptoms include sneezing, excessive nasal secretion, and congestion. The uncomplicated common cold is not usually accompanied by a fever. Complications include sinusitis, ear infections, and laryngitis. Since common colds are caused by viruses, antibiotics are of no use in treatment. Although over-the-counter drugs may lessen the severity of certain symptoms of the common cold, they do not lessen recovery time. As we noted in Chapter 22, an experimental antiviral nasal spray containing alpha-2-interferon may prevent the spread of colds among family members. The interferon is effective against rhinoviruses.

Influenza (flu) is also caused by a virus. Its symptoms include chills, fever (usually higher than 101°F), headache, and muscular aches. Coldlike symptoms appear as the fever subsides. Diarrhea is not a symptom of influenza.

Pulmonary Embolism (PE)

Pulmonary embolism (PE) refers to the presence of a blood clot or other foreign substance in a pulmonary arterial vessel that obstructs circulation to lung tissue. An embolus may arise in a distant site, such as a leg vein, and pass through the right side of the heart to a lung via a pulmonary artery. The immediate effect of a PE is complete or partial obstruction of the pulmonary arterial blood flow to the lung resulting in dysfunction of the affected lung tissue. A large embolus can produce death in a few minutes. PE rarely develops in patients without one or more risk factors. Among these are immobilization in bed for more than 3 days; heart disease (especially with congestive heart failure); trauma, such as fractures; malignant disease; obesity; pregnancy; polycythemia; and the use of oral contraceptives.

Among the clinical findings associated with PE are sudden and unexplained dyspnea (labored breathing), chest pain, apprehension, rapid breathing, rales, cough, deep-venous thrombosis, pulmonary infarction as evidenced by hemoptysis (expectoration of blood or blood-stained sputum), and syncope.

Treatment of PE consists of intravenous (IV) injections of heparin (an anticoagulant); administration of oral anticoagulants, such as warfarin; and administration of clot-dissolving enzymes, such as streptokinase and urokinase. In some cases surgery is indicated to reverse the lethal effect of massive PE by embolectomy.

Pulmonary Edema

Pulmonary edema refers to an abnormal accumulation of interstitial fluid in the interstitial spaces and alveoli of the lungs. The edema may arise from increased pulmonary capillary permeability (pulmonary origin) or increased pulmonary capillary pressure (cardiac origin). This latter cause may coincide with congestive heart failure. The most common symptom is dyspnea. Others include wheezing, tachypnea (rapid respirations), restlessness, feeling of suffocation, cyanosis, pallor (paleness), and diaphoresis (excessive perspiration). Treatment consists of administering oxygen, drugs that dilate the bronchioles and lower blood pressure, and drugs that correct acid-base imblance; suctioning of airways; and mechanical ventilation.

MEDICAL TERMINOLOGY

Abdominal thrust (Heimlich) maneuver   First aid procedure designated to clear the air passageways of obstructing objects. It is performed by applying a quick upward thrust that causes sudden elevation of the diaphragm and forceful, rapid expulsion of air in the lungs; this action forces air out the trachea to eject the obstructing object.

Asphyxia (*sphyxis* = pulse)   Oxygen starvation due to low atmospheric oxygen or interference with ventilation, external respiration, or internal respiration.

Aspiration (*spirare* = breathe)   The act of breathing, especially breathing in; drawing of a substance in or out by suction.

Bronchiectasis (*ektasis* = dilation)   A chronic dilation of the bronchi or bronchioles.

Cardiopulmonary resuscitation (CPR)   The artificial establishment of normal or near normal respiration and circulation. The **A, B, C's** of cardiopulmonary resuscitation are **Airway, Breathing,** and **Circulation,** meaning the rescuer must establish an airway, provide artificial ventilation if breathing has stopped, and reestablish circulation if there is inadequate cardiac action.

Cheyne-Stokes respiration   A repeated cycle of irregular breathing beginning with shallow breaths that increase in depth and rapidity, then decrease and cease altogether for 15 to 20 seconds. Cheyne-Stokes is normal in infants. It is also often seen just before death from pulmonary, cerebral, cardiac, and kidney disease, and is referred to as the "death rattle."

Diphtheria (*diphthera* = membrane)   An acute bacterial infection that causes the mucous membranes of the oropharynx, nasopharynx, and larynx to enlarge and become leathery. Enlarged membranes may obstruct airways and cause death from asphyxiation.

Dyspnea (*dys* = painful, difficult; *pnoia* = breath)   Labored breathing (short-windedness).

Hypoxia (*hypo* = below, under)   Reduction in oxygen supply to cells.

Orthopnea (*ortho* = straight)   Inability to breathe in a horizontal position.

Pneumonectomy (*pneumo* = lung; *tome* = cutting)   Surgical removal of a lung.

Rales   Sounds sometimes heard in the lungs that resemble bubbling or rattling. May be caused by air or an abnormal secretion in the lungs.

Respirator   An apparatus that qualifies air breathed in through it or a device for providing artificial respiration or to assist in pulmonary ventilation.

Rhinitis (*rhino* = nose)   Chronic or acute inflammation of the mucous membrane of the nose.

STUDY OUTLINE

Organs (p. 556)

1. Respiratory organs include the nose, pharynx, larynx, trachea, bronchi, and lungs.
2. They act with the cardiovascular system to supply oxygen and remove carbon dioxide from the blood.

Nose (p. 556)

1. The external portion is made of cartilage and skin and is lined with mucous membrane. Openings to the exterior are the external nares.
2. The internal portion communicates with the nasopharynx through the internal nares and the paranasal sinuses.
3. The nasal cavity is divided by a septum. The anterior portion of the cavity is called the vestibule.
4. The nose is adapted for warming, moistening, and filtering air, for olfaction, and for speech.

Pharynx (p. 556)

1. The pharynx (throat) is a muscular tube lined by a mucous membrane.

2. The anatomic regions are nasopharynx, oropharynx, and laryngopharynx.
3. The nasopharynx functions in respiration. The oropharynx and laryngopharynx function both in digestion and in respiration.

Larynx (p. 558)

1. The larynx (voice box) is a passageway that connects the pharynx with the trachea.
2. It contains the thyroid cartilage (Adam's apple); the epiglottis, which prevents food from entering the larynx; the cricoid cartilage, which connects the larynx and trachea; and the paired arytenoid, corniculate, and cuneiform cartilages.
3. The larynx contains vocal folds, which produce sound. Taut folds produce high pitches, and relaxed ones produce low pitches.

Trachea (p. 560)

1. The trachea (windpipe) extends from the larynx to the primary bronchi.

2. It is composed of smooth muscle and C-shaped rings of cartilage and is lined with pseudostratified epithelium.
3. Two methods of removing obstructions from the respiratory passageways are tracheostomy and intubation.

Bronchi (p. 560)

1. The bronchial tree consists of the trachea, primary bronchi, secondary bronchi, tertiary bronchi, bronchioles, and terminal bronchioles. Walls of bronchi contain rings of cartilage; walls of bronchioles do not.
2. A bronchogram is a roentgenogram of the tree after introduction of an opaque contrast medium usually containing iodine.

Lungs (p. 561)

1. Lungs are paired organs in the thoracic cavity. They are enclosed by the pleural membrane. The parietal pleura is the outer layer; the visceral pleura is the inner layer.
2. The right lung has three lobes separated by two fissures; the left lung has two lobes separated by one fissure and a depression, the cardiac notch.
3. The secondary bronchi give rise to branches called segmental bronchi, which supply segments of lung tissue called bronchopulmonary segments.
4. Each bronchopulmonary segment consists of lobules, which contain lymphatics, arterioles, venules, terminal bronchioles, respiratory bronchioles, alveolar ducts, alveolar sacs, and alveoli.
5. Gas exchange occurs across the alveolar-capillary (respiratory) membranes.

Respiration (p. 565)

Pulmonary Ventilation (p. 565)

1. Pulmonary ventilation or breathing consists of inspiration and expiration.
2. The movement of air into and out of the lungs depends on pressure changes governed in part by Boyle's law, which states that the volume of a gas varies inversely with pressure, assuming that temperature is constant.
3. Inspiration occurs when intrapulmonic pressure falls below atmospheric pressure. Contraction of the diaphragm and external intercostal muscles increases the size of the thorax, thus decreasing the intrapleural pressure so that the lungs expand. Expansion of the lungs decreases intrapulmonic pressure, so that air moves along the pressure gradient from the atmosphere into the lungs.
4. Expiration occurs when intrapulmonic pressure is higher than atmospheric pressure. Relaxation of the diaphragm and external intercostal muscles increases intrapleural pressure, lung volume decreases, and intrapulmonic pressure increases so that air moves from the lungs to the atmosphere.
5. During forced inspiration, accessory muscles of inspiration (sternocleidomastoids and scalenes) are also used.
6. Forced expiration employs contraction of the internal intercostals and abdominal muscles.
7. Compliance is the ease with which the lungs and thoracic wall expand.
8. The walls of the respiratory passageways offer some resistance to breathing.

Modified Respiratory Movements (p. 570)

1. Modified respiratory movements are used to express emotions and to clear air passageways.
2. Coughing, sneezing, sighing, yawning, sobbing, crying, laughing, and hiccuping are types of modified respiratory movements.

Pulmonary Air Volumes and Capacities (p. 570)

1. Air volumes exchanged during breathing and rate of respiration are measured with a spirometer.
2. Among the pulmonary air volumes exchanged in ventilation are tidal volume, inspiratory reserve, expiratory reserve, residual volume, and minimal volumes.
3. Pulmonary lung capacities, the sum of two or more volumes, include inspiratory, functional residual, vital, and total.
4. The minute volume of respiration is the total air taken in during 1 minute (tidal volume times 12 respirations per minute).

Exchange of Respiratory Gases (p. 572)

1. The partial pressure of a gas is the pressure exerted by that gas in a mixture of gases. It is symbolized by P.
2. According to Dalton's law, each gas in a mixture of gases exerts its own pressure as if all the other gases were not present.
3. Henry's law states that the quantity of a gas that will dissolve in a liquid is proportional to the partial pressure of the gas and its solubility coefficient, when the temperature remains constant.
4. A major clinical application of Henry's law is hyperbaric oxygenation (HBO).

External Respiration; Internal Respiration (p. 573; p. 575)

1. In internal and external expiration O_2 and CO_2 move from areas of their higher partial pressure to areas of their lower partial pressures.
2. External respiration is the exchange of gases between alveoli and pulmonary blood capillaries. It is aided by a thin alveolar-capillary (respiratory) membrane, a large alveolar surface area, and a rich blood supply.
3. Internal respiration is the exchange of gases between tissue blood capillaries and tissue cells.

Transport of Respiratory Gases (p. 575)

1. In each 100 ml of oxygenated blood, 3 percent of the O_2 is dissolved in plasma and 97 percent is carried with hemoglobin as oxyhemoglobin (HbO_2).
2. The association of oxygen and hemoglobin is affected by PO_2, PCO_2, temperature, and DPG.
3. Hypoxia refers to oxygen deficiency at the tissue level and is classified as hypoxic, anemic, stagnant, and histotoxic.
4. In each 100 ml of deoxygenated blood, 7 percent of CO_2 is dissolved in plasma, 23 percent combines with hemoglobin as carbaminohemoglobin ($Hb \cdot CO_2$), and 70 percent is converted to the bicarbonate ion (HCO_3^-).
5. Carbon monoxide poisoning occurs when CO combines with hemoglobin instead of oxygen. The result is hypoxia.

Control of Respiration (p. 579)

Nervous Control (p. 579)

1. The respiratory center consists of a medullary rhythmicity area (inspiratory and expiratory area), pneumotaxic area, and apneustic area.
2. The inspiratory area has an intrinsic excitability that sets the basic rhythm of respiration.
3. The pneumotaxic and apneustic areas coordinate the transition between inspiration and expiration.

Regulation of Respiratory Center Activity (p. 581)

1. Respirations may be modified by a number of factors, both in the brain and outside.
2. Among the modifying factors are cortical influences, the inflation reflex, chemical stimuli (O_2 and CO_2 levels), blood

pressure, temperature, pain, and irritation to the respiratory mucosa.

Aging and the Respiratory System (p. 582)

1. Aging results in decreased vital capacity, decreased blood level of O_2, and diminished alveolar macrophage activity.
2. Elderly people are more susceptible to pneumonia, emphysema, bronchitis, and other pulmonary disorders.

Developmental Anatomy of the Respiratory System (p. 583)

1. The respiratory system begins as an outgrowth of endoderm called the laryngotracheal bud.
2. Smooth muscle, cartilage, and connective tissue of the bronchial tubes and pleural sacs develop from mesoderm.

Disorders: Homeostatic Imblances (p. 583)

1. In bronchogenic carcinoma, bronchial epithelial cells are replaced by cancer cells after constant irritation has disrupted the normal growth, division, and function of the epithelial cells.
2. Bronchial asthma is characterized by spasms of smooth muscle in bronchial tubes that result in partial closure of air passageways; inflammation; inflated alveoli; and excess mucus production.
3. Bronchitis is an inflammation of the bronchial tubes.

4. Emphysema is characterized by deterioration of alveoli leading to loss of their elasticity. Symptoms are reduced expiratory volume, inflated lungs at the end of expiration, and enlarged chest.
5. Pneumonia is an acute inflammation or infection of alveoli.
6. Tuberculosis is an inflammation of pleura and lungs produced by the organism *Mycobacterium tuberculosis*.
7. Respiratory distress syndrome (RDS) of the newborn is an infant disorder in which surfactant is lacking and alveolar ducts and alveoli have a glassy appearance.
8. Respiratory failure means the respiratory system cannot supply sufficient oxygen or eliminate sufficient carbon dioxide.
9. Sudden infant death syndrome (SIDS) has recently been linked to laryngospasm, possibly triggered by a viral infection of the upper respiratory tract.
10. Coryza (common cold) is caused by viruses and is usually not accompanied by a fever, whereas influenza (flu) is usually accompanied by a fever greater than 101°F.
11. Pulmonary embolism (PE) refers to a blood clot or other foreign substance in a pulmonary arterial vessel that obstructs blood flow to lung tissue.
12. Pulmonary edema is an accumulation of interstitial fluid in interstitial spaces and alveoli of the lungs.

REVIEW QUESTIONS

1. What organs make up the respiratory system? What function do the respiratory and cardiovascular systems have in common?
2. Describe the structure of the external and internal nose, and describe their functions in filtering, warming, and moistening air.
3. What is the pharynx? Differentiate the three regions of the pharynx and indicate their roles in respiration.
4. Describe the structure of the larynx and explain how it functions in respiration and voice production.
5. Describe the location and structure of the trachea. What is tracheostomy? Intubation?
6. What is the bronchial tree? Describe its structure. What is a bronchogram?
7. Where are the lungs located? Distinguish the parietal pleura from the visceral pleura. What is pleurisy?
8. Define each of the following parts of a lung: base, apex, costal surface, medial surface, hilus, root, cardiac notch, and lobe.
9. What is a lobule of the lung? Describe its composition and function in respiration.
10. What is a bronchopulmonary segment?
11. Describe the histology of the alveolar-capillary (respiratory) membrane.
12. Define nebulization and describe its clinical application.
13. What are the basic differences among pulmonary ventilation, external respiration, and internal respiration?
14. Discuss the basic steps involved in inspiration and expiration. Be sure to include values for all pressures involved.
15. Distinguish between quiet and forced inspiration and expiration.
16. Describe how compliance and airway resistance relate to pulmonary ventilation.
17. Define the various kinds of modified respiratory movements.
18. What is a spirometer? Define the various lung volumes and

capacities. How is the minute volume of respiration calculated?
19. Define the partial pressure of a gas. How is it calculated?
20. Define Boyle's law, Dalton's law, and Henry's law.
21. How does decompression sickness occur? What is hyperbaric oxygenation (HBO)?
22. Construct a diagram to illustrate how and why the respiratory gases move during external and internal respiration.
23. What factors affect external respiration?
24. Describe the relationship between hemoglobin and PO_2, PCO_2, temperature, and DPG.
25. Define hypoxia and distinguish the principal types.
26. Explain how CO_2 is picked up by tissue capillary blood and then discharged into the alveoli.
27. How does the medullary rhythmicity area function in controlling respiration? How are the apneustic area and pneumotaxic area related to the control of respiration?
28. Explain how each of the following modifies respiration: cerebral cortex, inflation reflex, CO_2, O_2, blood pressure, temperature, pain, and irritations of the respiratory mucosa.
29. How does the control of respiration demonstrate the principle of homeostasis?
30. Describe the effects of aging on the respiratory system.
31. Describe the development of the respiratory system.
32. For each of the following disorders, list the principal clinical symptoms: bronchogenic carcinoma, bronchial asthma, bronchitis, emphysema, pneumonia, tuberculosis, respiratory distress syndrome (RDS) of the newborn, respiratory failure, sudden infant death syndrome (SIDS), coryza (common cold), influenza (flu), pulmonary embolism (PE), pulmonary edema, and carbon monoxide poisoning.
33. Refer to the glossary of medical terminology associated with the respiratory system. Be sure that you can define each term.

The Digestive System

24

- Describe the mechanism that regulates food intake.
- Define digestion and distinguish between the chemical and mechanical phases.
- Identify the organs of the gastrointestinal (GI) tract and the accessory organs of digestion.
- Discuss the structure of the wall of the gastrointestinal (GI) tract.
- Explain the structure of the mouth and its role in mechanical digestion.
- Describe the location, histology, and functions of the salivary glands.
- Identify the mechanisms that regulate the secretion of saliva.
- Identify the parts of a typical tooth and compare deciduous and permanent dentitions.
- Discuss the stages of swallowing.
- Describe the role of the esophagus in digestion.
- Describe the anatomy and histology of the stomach, and explain the relationship between its structural features and digestion.
- Discuss the factors that control the secretion of gastric juice and gastric emptying.
- Describe the anatomy and histology of the pancreas and its role in digestion.
- Explain how pancreatic secretion is regulated.
- Describe the anatomy and histology of the liver and its role in digestion.
- Explain how bile secretion is regulated.
- Discuss the role of the gallbladder in digestion.
- Discuss the structural features of the small intestine that adapt it for digestion and absorption.
- Explain how small intestinal secretions are controlled.
- Define absorption and explain how the end products of digestion are absorbed.
- Describe the anatomy and histology of the large intestine.
- Define the processes involved in the formation of feces and defecation.
- Describe the effects of aging on the digestive system.
- Describe the development of the digestive system.
- Describe the clinical symptoms of the following disorders: dental caries, periodontal disease, peritonitis, peptic ulcers, appendicitis, gastrointestinal (GI) tumors, diverticulitis, cirrhosis, hepatitis, gallstones, anorexia nervosa, and bulimia.
- Define medical terminology associated with the digestive system.

Food is vital for life because it is the source of energy that drives the chemical reactions occurring in every cell. Energy is needed for muscle contraction, the conduction of nerve impulses, and the secretory and absorptive activities of many cells. Food as it is consumed, however, is not in a state suitable for use as an energy source by any cell. The food must be broken down into molecule-sized pieces so that it can be transported through the cell membranes. The breaking down of food molecules for use by body cells is called **digestion,** and the organs that collectively perform this function comprise the **digestive system.**

The medical specialty that deals with the structure, function, diagnosis, and treatment of diseases of the stomach and intestines is called **gastroenterology** (gas'-trō-en'-ter-OL-ō-jē; *gastro* = stomach; *enteron* = intestines).

The developmental anatomy of the digestive system is considered later in the chapter.

REGULATION OF FOOD INTAKE

Within the hypothalamus are two centers related to food intake. One is a cluster of nerve cells in the lateral nuclei called the **feeding (hunger) center.** When this area is stimulated in animals, they begin to eat heartily, even if they are already full. The second center is a cluster of neurons in the ventromedial nuclei of the hypothalamus referred to as the **satiety center.** When this center is stimulated in animals, it causes them to stop eating, even if they have been starved for days. Apparently the feeding center is constantly active, but it is inhibited by the satiety center. Other parts of the brain that assume a function in feeding and satiety are the brain stem, amygdala, and limbic system.

One substance that is related to food intake is glucose. According to the *glucostatic theory,* when blood glucose levels are low, feeding increases. It is believed that in response to low levels of blood glucose the rate of glucose utilization by neurons of the satiety center becomes low. As a result, the activity of the neurons decreases to the point where the satiety center can no longer inhibit the feeding center, and the individual eats. Conversely, when blood glucose levels are high, the activity of the neurons of the satiety center is sufficiently high to inhibit the feeding center, and feeding is depressed. Low levels of amino acids in the blood also enhance feeding, whereas high levels depress eating. This mechanism, however, is not as powerful as that for glucose.

Lipids are also believed to be related to the regulation of food intake. It has been observed that as the amount of adipose tissue increases in the body, the rate of feeding decreases. It is theorized that some substance or substances, perhaps fatty acids, are released from stored fats in proportion to the total fat content of the body. The released substances then activate neurons of the satiety center, which, in turn, inhibit the feeding center.

Another factor that affects food intake is body temperature. Whereas a cold environment enhances eating, a warm environment depresses it. Food intake is also regulated by distension of the gastrointestinal tract, particularly the stomach and duodenum. When these organs are stretched, a reflex is initiated that activates the satiety center and depresses the feeding center. It has also been shown that the hormone cholecystokinin (CCK), secreted when fat enters the small intestine, inhibits eating.

DIGESTIVE PROCESSES

The digestive system prepares food for consumption by the cells through five basic activities.

1. **Ingestion,** taking food into the body (eating).
2. **Movement** of food along the digestive tract.
3. **Digestion,** the breakdown of food by both chemical and mechanical processes.
4. **Absorption,** the passage of digested food from the digestive tract into the cardiovascular and lymphatic systems for distribution to cells.
5. **Defecation,** the elimination of indigestible substances from the body.

Chemical digestion is a series of catabolic reactions that break down the large carbohydrate, lipid, and protein molecules that we eat into molecules usable by body cells. These products of digestion are small enough to pass through the walls of the digestive organs, into the blood and lymph capillaries, and eventually into the body's cells. **Mechanical digestion** consists of various movements that aid chemical digestion. Food is prepared by the teeth before it can be swallowed. Then the smooth muscles of the stomach and small intestine churn the food so it is thoroughly mixed with the enzymes that catalyze the reactions.

ORGANIZATION

The organs of digestion are traditionally divided into two main groups. First is the **gastrointestinal (GI) tract** or **alimentary canal,** a continuous tube running through the ventral body cavity and extending from the mouth to the anus (Figure 24-1). The relationship of the digestive organs to the nine regions of the abdominopelvic cavity may be reviewed in Figure 1-8b. The length of a tract taken from a cadaver is about 9 m (30 ft). In a living person it is somewhat shorter because the muscles in its wall are in a state of tone. Organs composing the gastrointestinal tract include the mouth, pharynx, esophagus, stomach, small intestine, and large intestine. The GI tract contains the food from the time it is eaten until it is digested and prepared for elimination. Muscular contractions in the wall of the GI tract break down the food physically by churning it. Secretions produced by cells along the tract break down the food chemically.

The second group of organs composing the digestive system consists of the **accessory structures**—the teeth, tongue, salivary glands, liver, gallbladder, and pancreas. Teeth protrude into the GI tract and aid in the physical

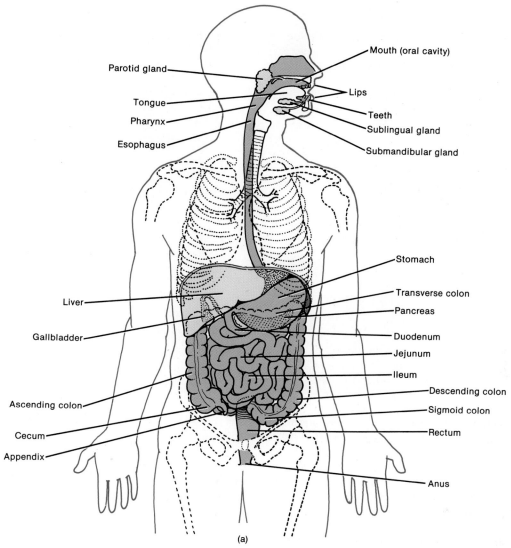

(a)

FIGURE 24-1 Organs of the digestive system and related structures. (a) Diagram.

breakdown of food. The other accessory structures, except for the tongue, lie totally outside the tract and produce or store secretions that aid in the chemical breakdown of food. These secretions are released into the tract through ducts.

GENERAL HISTOLOGY

The wall of the GI tract, especially from the esophagus to the anal canal, has the same basic arrangment of tissues. The four coats or tunics of the tract from the inside out are the mucosa, submucosa, muscularis, and serosa or adventitia (Figure 24-2).

Mucosa

The **mucosa,** or inner lining of the tract, is a mucous membrane attached to a thin layer of visceral muscle. Two layers compose the membrane: a *lining epithelium,* which is in direct contact with the contents of the GI tract, and an underlying layer of loose connective tissue

called the *lamina propria.* Under the lamina propria is smooth muscle called the *muscularis mucosae.*

The epithelial layer is composed of nonkeratinized cells that are stratified in the mouth and esophagus, but are simple throughout the rest of the tract. The functions of the stratified epithelium are protection and secretion. The functions of the simple epithelium are secretion and absorption.

The lamina propria is made of loose connective tissue containing many blood and lymph vessels and scattered lymphatic nodules, masses of lymphatic tissue that are not encapsulated. This layer supports the epithelium, binds it to the muscularis mucosae, and provides it with a blood and lymph supply. The blood and lymph vessels are the avenues by which nutrients in the tract reach the other tissues of the body. The lymphatic tissue also protects against disease. Remember that the GI tract is in contact with the outside environment and contains food that often carries harmful bacteria. Unlike the skin, the mucous membrane of the tract is not protected from bacterial entry by keratin.

Right colic (hepatic) flexure

Taenia coli

Ascending colon

Haustra

Cecum

Vermiform appendix

Rectum

Esophagus

Stomach

Left colic (splenic) flexure

Transverse colon

Greater omentum

Jejunum

Descending colon

Ileum

Sigmoid colon

(b)

FIGURE 24-1 (Continued) (b) Photograph. (Courtesy of C. Yokochi and J. W. Rohen, *Photographic Anatomy of the Human Body*, 2nd ed., 1978, IGAKU-SHOIN, Ltd., Tokyo, New York.)

The muscularis mucosae contains smooth muscle fibers that throw the mucous membrane of the intestine into small folds that increase the digestive and absorptive area. With one exception, which will be described later, the other three coats of the intestine contain no glandular epithelium.

Submucosa

The **submucosa** consists of loose connective tissue that binds the mucosa to the third tunic, the muscularis. It is highly vascular and contains a portion of the *submucous plexus (plexus of Meissner)*, which is part of the autonomic nerve supply to the muscularis mucosae. This plexus is also important in controlling secretions by the GI tract.

Muscularis

The **muscularis** of the mouth, pharynx, and esophagus consists in part of skeletal muscle that produces voluntary swallowing. Throughout the rest of the tract, the muscularis consists of smooth muscle that is generally found in two sheets: an inner ring of circular fibers and an outer sheet of longitudinal fibers. Contractions of the smooth muscles help to break down food physically, mix it with digestive secretions, and propel it through the tract. The muscularis also contains the major nerve supply to the alimentary tract—the *myenteric plexus (plexus of Auer-*

bach), which consists of fibers from both autonomic divisions. This plexus mostly controls GI motility.

Serosa

The **serosa** is the outermost layer of most portions of the canal. It is a serous membrane composed of connective tissue and epithelium. This layer is also called the **visceral peritoneum** and forms a portion of the peritoneum, which we will describe in detail.

PERITONEUM

The **peritoneum** (per'-i-tō-NĒ-um) is the largest serous membrane of the body. Serous membranes are also associated with the heart (pericardium) and lungs (pleurae). Serous membranes consist of a layer of simple squamous epithelium (called mesothelium) and an underlying supporting layer of connective tissue. The **parietal peritoneum** lines the wall of the abdominal cavity. The **visceral peritoneum** covers some of the organs and constitutes their serosa. The potential space between the parietal and visceral portions of the peritoneum is called the **peritoneal cavity** and contains serous fluid. In certain diseases, the peritoneal cavity may become distended by several liters of fluid so that it forms an actual space. Such an accumulation of serous fluid is called **ascites** (a-SĪ-tēz). As you will see later, some organs lie on the posterior abdominal wall and are covered by peritoneum on their anterior

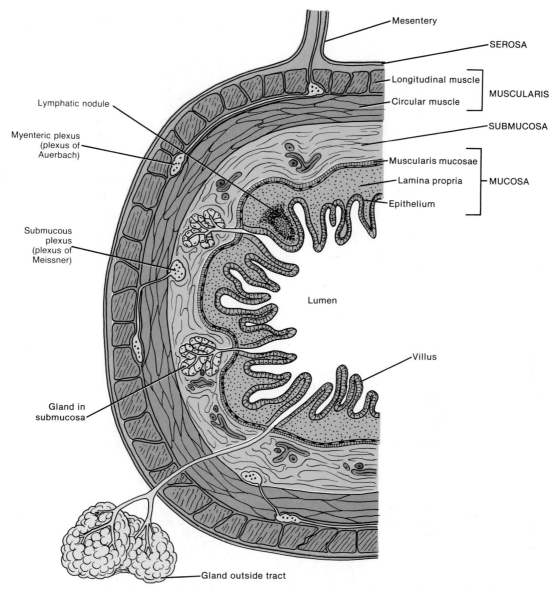

Mesentery

SEROSA

Longitudinal muscle ⎱
Circular muscle   ⎰ MUSCULARIS

SUBMUCOSA

Muscularis mucosae ⎱
Lamina propria   ⎰ MUCOSA
Epithelium

Lymphatic nodule

Myenteric plexus
(plexus of
Auerbach)

Submucous
plexus
(plexus of
Meissner)

Lumen

Gland in
submucosa

Villus

Gland outside tract

FIGURE 24-2 Portion of the gastrointestinal tract seen in cross section showing the various layers and related structures.

surfaces only. Such organs, including the kidneys and pancreas, are said to be **retroperitoneal.**

Unlike the pericardium and pleurae, the peritoneum contains large folds that weave in between the viscera. The folds bind the organs to each other and to the walls of the cavity and contain the blood and lymph vessels and the nerves that supply the abdominal organs. One extension of the peritoneum is called the **mesentery** (MEZ-en-ter'-ē). It is an outward fold of the serous coat of the small intestine (Figure 24-3). The tip of the fold is attached to the posterior abdominal wall. The mesentery binds the small intestine to the wall. A similar fold of parietal peritoneum, called the **mesocolon** (mez'-ō-KŌ-lon), binds the large intestine to the posterior body wall. It also carries blood vessels and lymphatics to the intestines.

Other important peritoneal folds are the falciform ligament, the lesser omentum, and the greater omentum. The **falciform** (FAL-si-form) **ligament** attaches the liver to the anterior abdominal wall and diaphragm. The **lesser omentum** (ō-MENT-um) arises as two folds in the serosa of the stomach and duodenum suspending the stomach and duodenum from the liver. The **greater omentum** is a four-layered fold in the serosa of the stomach that hangs down like an apron over the front of the intestines. It then passes up to part of the large intestine (the transverse colon), wraps itself around it, and finally attaches to the parietal peritoneum of the posterior wall of the abdominal cavity. Because the greater omentum contains large quantities of adipose tissue, it commonly is called the "fatty apron." The greater omentum contains numerous lymph nodes. If an infection occurs in the intestine, plasma cells

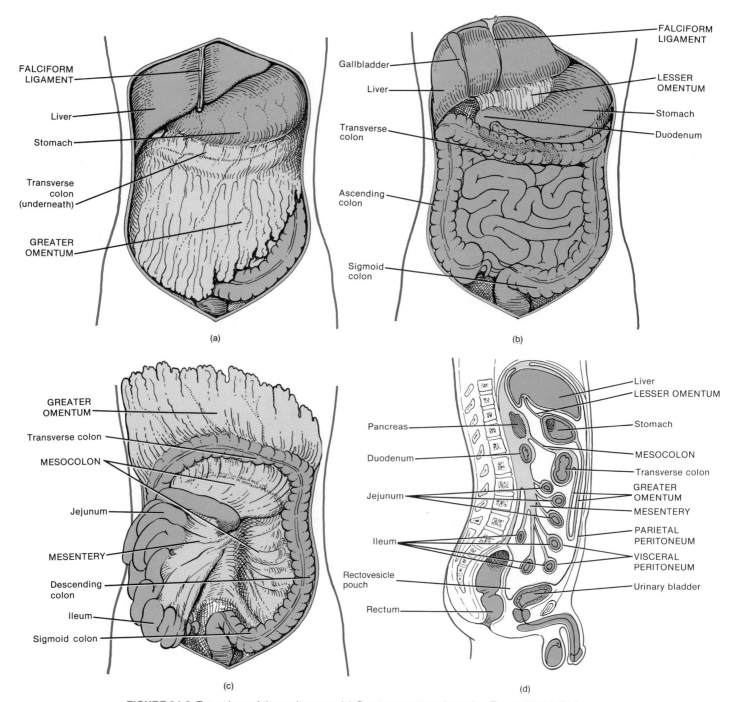

FIGURE 24-3 Extensions of the peritoneum. (a) Greater omentum (see also Figure 24-1b). (b) Lesser omentum. The liver and gallbladder have been lifted. (c) Mesentery. The greater omentum has been lifted. (d) Sagittal section through the abdomen and pelvis indicating the relationship of the peritoneal extensions to each other.

formed in the lymph nodes combat the infection and help prevent it from spreading to the peritoneum.

MOUTH (ORAL CAVITY)

The **mouth,** also referred to as the **oral,** or **buccal** (BUK-al), **cavity,** is formed by the cheeks, hard and soft palates, and tongue (Figure 24-4). Forming the lateral walls of the oral cavity are the **cheeks**—muscular structures covered on the outside by skin and lined by nonkeratinized stratified squamous epithelium. The anterior portions of the cheeks terminate in the superior and inferior lips.

The **lips (labia)** are fleshy folds surrounding the orifice of the mouth. They are covered on the outside by skin and on the inside by a mucous membrane. The transition zone where the two kinds of covering tissue meet is called the **vermilion** (ver-MIL-yon). This portion of the lips is nonkeratinized, and the color of the blood in the underlying blood vessels is visible through the transparent surface layer of the vermilion. The inner surface of each lip is attached to its corresponding gum by a midline fold of mucous membrane called the **labial frenulum** (LĀ-bē-al FREN-yoo-lum).

The orbicularis oris muscle and connective tissue lie between the external integumentary covering and the internal mucosal lining. During chewing, the cheeks and lips help to keep food between the upper and lower teeth. They also assist in speech.

The **vestibule** of the oral cavity is bounded externally by the cheeks and lips and internally by the gums and teeth. The **oral cavity proper** extends from the vestibule to the **fauces** (FAW-sēs), the opening between the oral cavity and the pharynx or throat.

The **hard palate,** the anterior portion of the roof of the mouth, is formed by the maxillae and palatine bones, is covered by mucous membrane, and forms a bony partition between the oral and nasal cavities. The **soft palate** forms the posterior portion of the roof of the mouth. It is an arch-shaped muscular partition between the oropharynx and nasopharynx and is lined by mucous membrane.

Hanging from the free border of the soft palate is a conical muscular process called the **uvula** (OO-vyoo-la).

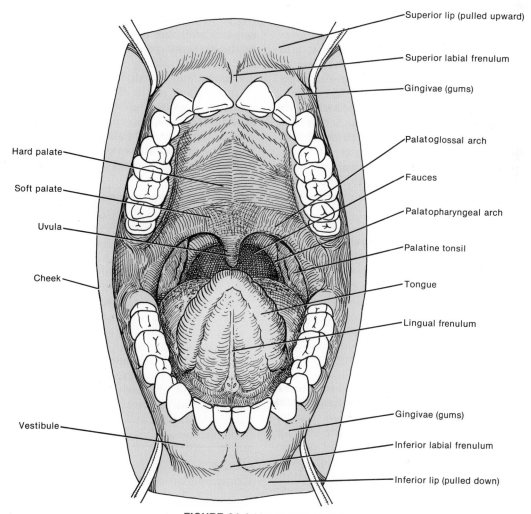

FIGURE 24-4 Mouth (oral cavity).

Labels (clockwise from top right):
- Superior lip (pulled upward)
- Superior labial frenulum
- Gingivae (gums)
- Palatoglossal arch
- Fauces
- Palatopharyngeal arch
- Palatine tonsil
- Tongue
- Lingual frenulum
- Gingivae (gums)
- Inferior labial frenulum
- Inferior lip (pulled down)
- Vestibule
- Cheek
- Uvula
- Soft palate
- Hard palate

On either side of the base of the uvula are two muscular folds that run down the lateral side of the soft palate. Anteriorly, the **palatoglossal arch (anterior pillar)** extends inferiorly, laterally, and anteriorly to the side of the base of the tongue. Posteriorly, the **palatopharyngeal (PAL-a-tō-fa-rin'-jē-al) arch (posterior pillar)** projects inferiorly, laterally, and posteriorly to the side of the pharynx. The palatine tonsils are situated between the arches, and the lingual tonsil is situated at the base of the tongue. At the posterior border of the soft palate, the mouth opens into the oropharynx through the fauces.

TONGUE

The **tongue,** together with its associated muscles, forms the floor of the oral cavity. It is an accessory structure of the digestive system composed of skeletal muscle covered with mucous membrane (see Figure 17-2). The tongue is divided into symmetrical lateral halves by a median septum that extends throughout its entire length and is attached inferiorly to the hyoid bone. Each half of the tongue consists of an identical complement of extrinsic and intrinsic muscles.

The **extrinsic muscles** of the tongue originate outside the tongue and insert into it. They include the hyoglossus, chondroglossus, genioglossus, styloglossus, and palatoglossus (see Figure 11-7). The extrinsic muscles move the tongue from side to side and in and out. These movements maneuver food for chewing, shape the food into a rounded mass, called a **bolus,** and force the food to the back of the mouth for swallowing. They also form the floor of the mouth and hold the tongue in position. The **intrinsic muscles** originate and insert within the tongue and alter the shape and size of the tongue for speech and swallowing. The intrinsic muscles include the longitudinalis superior, longitudinalis inferior, transversus linguae, and verticalis linguae. The **lingual frenulum,** a fold of mucous membrane in the midline of the undersurface of the tongue, aids in limiting the movement of the tongue posteriorly.

CLINICAL APPLICATION

If the lingual frenulum is too short, tongue movements are restricted, speech is faulty, and the person is said to be "tongue-tied." This congenital problem is referred to as **ankyloglossia** (ang'-ki-lō-GLOSS-ē-a). It can be corrected by cutting the lingual frenulum.

The upper surface and sides of the tongue are covered with **papillae** (pa-PIL-ē), projections of the lamina propria covered with epithelium. **Filiform papillae** are conical projections distributed in parallel rows over the anterior two-thirds of the tongue. They are whitish and contain no taste buds. **Fungiform papillae** are mushroomlike elevations distributed among the filiform papillae, and are more numerous near the tip of the tongue. They appear as red dots on the surface of the tongue, and most of them contain taste buds. **Circumvallate papillae,** 10 to 12 in number, are arranged in the form of an inverted V on the posterior surface of the tongue, and all of them contain taste buds. Note the taste zones of the tongue in Figure 17-2c.

SALIVARY GLANDS

Saliva is a fluid that is continuously secreted by glands in or near the mouth. Ordinarily, just enough saliva is secreted to keep the mucous membranes of the mouth moist, but when food enters the mouth, secretion increases so the saliva can lubricate, dissolve, and begin the chemical breakdown of the food. The mucous membrane lining the mouth contains many small glands, the **buccal glands,** that secrete small amounts of saliva. However, the major portion of saliva is secreted by the **salivary glands,** accessory structures that lie outside the mouth and pour their contents into ducts that empty into the oral cavity. There are three pairs of salivary glands: parotid, submandibular (submaxillary), and sublingual glands (Figure 24-5a).

The **parotid glands** are located under and in front of the ears between the skin and the masseter muscle. They are compound tubuloacinar glands. Each secretes into the oral cavity vestibule via a duct, called the parotid (Stensen's) duct, that pierces the buccinator muscle to open into the vestibule opposite the upper second molar tooth. The **submandibular glands,** which are compound acinar glands, are found beneath the base of the tongue in the posterior part of the floor of the mouth (Figure 24-5b). Their ducts, the submandibular (Wharton's) ducts, run superficially under the mucosa on either side of the midline of the floor of the mouth and enter the oral cavity proper just behind the central incisors. The **sublingual glands,** also compound acinar glands, are anterior to the submandibular glands, and their ducts, the lesser sublingual (Rivinus's) ducts, open into the floor of the mouth in the oral cavity proper.

CLINICAL APPLICATION

Although any of the salivary glands may become infected as a result of a nasopharyngeal infection, the parotids are typically the target of the mumps virus (myxovirus). **Mumps** is an inflammation and enlargement of the parotid glands accompanied by moderate fever, malaise, and extreme pain in the throat, especially when swallowing sour foods or acid juices. Swelling occurs on one or both sides of the face, just anterior to the ramus of the mandible. In about 20 to 35 percent of males past puberty, the testes may also become inflamed, and although it rarely occurs, sterility is a possible consequence. In some people, aseptic meningitis, pancreatitis, and hearing loss may also occur as complications.

Composition of Saliva

The fluids secreted by the buccal glands and the three pairs of salivary glands constitute **saliva.** Amounts of sa-

(a)

(b)

FIGURE 24-5 Salivary glands. (a) Location of the salivary glands. (b) Photomicrograph of the subman-dibular gland showing mostly serous acini and a few mucous acini at a magnification of 120×. The parotids consist of all serous acini, and the sublinguals consist of mostly mucous acini and a few serous acini. (Copyright © 1983 by Michael H. Ross. Used by permission.)

liva secreted daily vary considerably but range from 1,000 to 1,500 ml. Chemically, saliva is 99.5 percent water and 0.5 percent solutes. Among the solutes are salts—chlorides, bicarbonates, and phosphates of sodium and potassium. Some dissolved gases and various organic substances including urea and uric acid, serum albumin and globulin, mucin, the bacteriolytic enzyme lysozyme, and the digestive enzyme salivary amylase are also present.

Each saliva-producing gland supplies different proportions of ingredients to saliva. The parotids contain cells that secrete a watery serous liquid containing the enzyme salivary amylase. The submandibular glands contain cells similar to those found in the parotids and some mucous cells. Therefore, they secrete a fluid that is thickened with mucus, but still contains quite a bit of enzyme. The sublingual glands contain mostly mucous cells, so they secrete a much thicker fluid that contributes only a small amount of enzyme to the saliva.

The water in saliva provides a medium for dissolving foods so they can be tasted and for initiating digestive reactions. The chlorides in the saliva activate the salivary amylase. The bicarbonates and phosphates buffer chemicals that enter the mouth and keep the saliva at a slightly acidic pH of 6.35 to 6.85. Urea and uric acid are found in saliva because the saliva-producing glands (like the sweat glands of the skin) help the body to get rid of wastes. Mucin is a protein that forms mucus when dissolved in water. Mucus lubricates the food so it can be easily turned in the mouth, formed into a ball, and swallowed. The enzyme lysozyme destroys bacteria, thereby protecting the mucous membrane from infection and thus the teeth from decay.

Secretion of Saliva

Salivation is entirely under nervous control. Normally, moderate amounts of saliva are continuously secreted in response to parasympathetic stimulation to keep the mucous membranes moist and to lubricate the movements of the tongue and lips during speech. The saliva is then swallowed and reabsorbed to prevent fluid loss. Dehydration causes the salivary and buccal glands to cease secreting saliva to conserve water. The subsequent feeling of dryness in the mouth promotes sensations of thirst. This phenomenon is also noted during fear or anxiety, when sympathetic stimulation dominates.

Food stimulates the glands to secrete heavily. When food is taken into the mouth, chemicals in the food stimulate receptors in taste buds on the tongue. Rolling a dry, indigestible object over the tongue produces friction, which also may stimulate the receptors. Impulses are conveyed from the receptors to two salivary nuclei in the brain stem called the **superior** and **inferior salivatory nuclei.** The nuclei are located at about the junction of the medulla and pons. Returning parasympathetic autonomic impulses from the nuclei activate the secretion of saliva.

The smell, sight, touch, or sound of food preparation also stimulates increased saliva secretion. These stimuli constitute psychological activation and involve learned behavior. Memories stored in the cerebral cortex that associate the stimuli with food are stimulated. The cortex

sends impulses to the nuclei in the brain stem via extrapyramidal pathways, and the salivary glands are activated. Psychological activation of the glands has some benefit to the body because it allows the mouth to start chemical digestion as soon as the food is ingested.

Salivation also occurs in response to swallowing irritating foods or during nausea. Reflexes originating in the stomach and upper small intestine stimulate salivation. This mechanism presumably helps to dilute or neutralize the irritating substance.

Saliva continues to be secreted heavily some time after food is swallowed. This flow of saliva washes out the mouth and dilutes and buffers the chemical remnants of irritating substances.

TEETH

The **teeth (dentes)** are accessory structures of the digestive system located in sockets of the alveolar processes of the mandible and maxillae. The alveolar processes are covered by the **gingivae** (jin-JI-vē) or gums, which extend slightly into each socket forming the gingival sulcus (Figure 24-6). The sockets are lined by the **periodontal ligament,** which consists of dense fibrous connective tissue and is attached to the socket walls and the cemental surface of the roots. Thus it anchors the teeth in position and also acts as a shock absorber to dissipate the forces of chewing.

A typical tooth consists of three principal portions. The **crown** is the portion above the level of the gums. The **root** consists of one to three projections embedded in the socket. The **neck** is the constricted junction line of the crown and the root.

Teeth are composed primarily of **dentin,** a bonelike substance that gives the tooth its basic shape and rigidity. The dentin encloses a cavity. The enlarged part of the cavity, the **pulp cavity,** lies in the crown and is filled with **pulp,** a connective tissue containing blood vessels, nerves, and lymphatics. Narrow extensions of the pulp cavity run through the root of the tooth and are called **root canals.** Each root canal has an opening at its base, the **apical foramen.** Through the foramen enter blood vessels bearing nourishment, lymphatics affording protection, and nerves providing sensation. The dentin of the crown is covered by **enamel** that consists primarily of calcium phosphate and calcium carbonate. Enamel is the hardest substance in the body and protects the tooth from the wear of chewing. It is also a barrier against acids that easily dissolve the dentin. The dentin of the root is covered by **cementum,** another bonelike substance, which attaches the root to the periodontal ligament.

The branch of dentistry that is concerned with the prevention, diagnosis, and treatment of diseases that affect the pulp, root, periodontal ligament, and alveolar bone is known as **endodontics** (en'-dō-DON-tiks; *endo* = within; *odous* = tooth).

Dentitions

Everyone has two **dentitions,** or sets of teeth. The first of these—the **deciduous teeth, milk teeth,** or **baby teeth**—

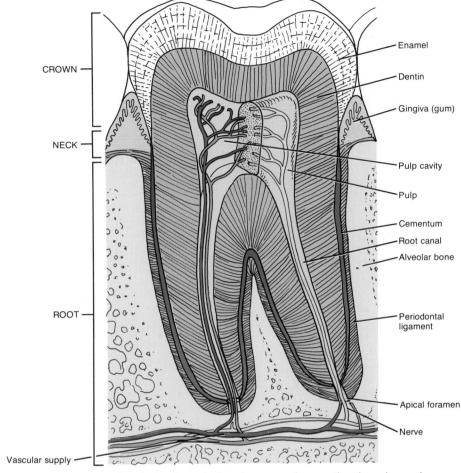

CROWN

NECK

ROOT

Vascular supply

Enamel

Dentin

Gingiva (gum)

Pulp cavity

Pulp

Cementum

Root canal

Alveolar bone

Periodontal ligament

Apical foramen

Nerve

FIGURE 24-6 Parts of a typical tooth as seen in a section through a molar.

begin to erupt at about 6 months of age, and one pair appears at about each month thereafter until all 20 are present. Figure 24-7a illustrates the deciduous teeth. The incisors, which are closest to the midline, are chisel-shaped and adapted for cutting into food. They are referred to as **central** or **lateral incisors** on the basis of their position. Next to the incisors, moving posteriorly, are the **cuspids (canines),** which have a pointed surface called a cusp. Cuspids are used to tear and shred food. The incisors and cuspids have only one root apiece. Behind them lie the **first** and **second molars,** which have four cusps. Upper molars have three roots; lower molars have two roots. The molars crush and grind food.

All the deciduous teeth are lost—generally between 6 and 12 years of age—and are replaced by the **permanent dentition** (Figure 24-7b). The permanent dentition contains 32 teeth that appear between the age of 6 and adulthood. It resembles the deciduous dentition with the following exceptions. The deciduous molars are replaced with the **first** and **second premolars (bicuspids),** which have two cusps and one root (upper first bicuspids have two roots) and are used for crushing and grinding. The permanent molars erupt into the mouth behind the bicuspids. They do not replace any deciduous teeth and erupt as the jaw grows to accommodate them—the **first molars**

at age 6, the **second molars** at age 12, the **third molars (wisdom teeth)** after age 18. The human jaw has become smaller through time and often does not afford enough room behind the second molars for the eruption of the third molars. In this case, the third molars remain embedded in the alveolar bone and are said to be "impacted." Most often they cause pressure and pain and must be surgically removed. In some individuals, third molars may be dwarfed in size or may not develop at all.

DIGESTION IN THE MOUTH

Mechanical

Through chewing, or **mastication,** the tongue manipulates food, the teeth grind it, and the food is mixed with saliva. As a result, the food is reduced to a soft, flexible bolus that is easily swallowed.

Chemical

The enzyme **salivary amylase** (AM-i-lās), formerly known as ptyalin, initiates the breakdown of starch. This is the only chemical digestion that occurs in the mouth. Carbohydrates are either monosaccharide and disaccharide sug-

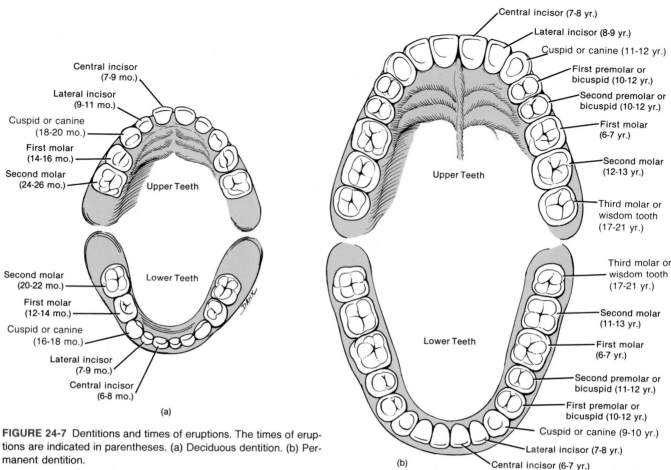

FIGURE 24-7 Dentitions and times of eruptions. The times of eruptions are indicated in parentheses. (a) Deciduous dentition. (b) Permanent dentition.

ars or polysaccharide starches (see Chapter 2). Most of the carbohydrates we eat are polysaccharides. Since only monosaccharides can be absorbed into the bloodstream, ingested disaccharides and polysaccharides must be broken down. The function of salivary amylase is to break the chemical bonds between some of the monosaccharides in the starches to reduce the long-chain polysaccharides to the disaccharide maltose. Food usually is swallowed too quickly for all of the starches to be reduced to disaccharides in the mouth. However, salivary amylase in the swallowed food continues to act on starches for another 15 to 30 minutes in the stomach before the stomach acids eventually inactivate it.

Exhibit 24-1 summarizes digestion in the mouth.

DEGLUTITION

Swallowing, or **deglutition** (dē-gloo-TISH-un), is a mechanism that moves food from the mouth to the stomach. It is facilitated by saliva and mucus and involves the mouth, pharynx, and esophagus. Swallowing is conveniently divided into three stages: (1) the voluntary stage, in which the bolus is moved into the oropharynx; (2) the pharyngeal stage, the involuntary passage of the bolus through the pharynx into the esophagus; and (3) the esophageal stage, the involuntary passage of the bolus through the esophagus into the stomach (described in conjunction with the esophagus).

Swallowing starts when the bolus is forced to the back of the mouth cavity and into the oropharynx by the movement of the tongue upward and backward against the palate. This represents the **voluntary stage** of swallowing.

With the passage of the bolus into the oropharynx, the involuntary **pharyngeal stage** of swallowing begins (Figure 24-8b). The respiratory passageways close and breathing is temporarily interrupted. The bolus stimulates receptors in the oropharynx, which send impulses to the **deglutition center** in the medulla and lower pons of the brain stem. The returning impulses cause the soft palate and uvula to move upward to close off the nasopharynx, and the larynx is pulled forward and upward under the tongue. As the larynx rises, it meets the epiglottis, which seals off the glottis. The movement of the larynx also pulls the vocal cords together, further sealing off the respiratory tract, and widens the opening between the laryngopharynx and esophagus. The bolus passes through the laryngopharynx and enters the esophagus in 1 to 2 seconds. The respiratory passageways then reopen and breathing resumes.

ESOPHAGUS

The **esophagus** (e-SOF-a-gus), the third principal organ involved in deglutition, is a muscular, collapsible tube that lies behind the trachea. It is about 23 to 25 cm (10 inches) long and begins at the end of the laryngophar-

EXHIBIT 24-1   DIGESTION IN THE MOUTH

STRUCTURE	ACTIVITY	RESULT
Cheeks	Keep food between teeth during mastication.	Foods uniformly chewed.
Lips	Keep food between teeth during mastication.	Foods uniformly chewed.
Tongue		
Extrinsic muscles	Move tongue from side to side and in and out.	Food maneuvered for mastication, shaped into bolus, and maneuvered for deglutition.
Intrinsic muscles	Alter shape of tongue.	Deglutition and speech.
Taste buds	Serve as receptors for food stimulus.	Secretion of saliva stimulated by nerve impulses from taste buds to salivatory nuclei in brain stem to salivary glands.
Buccal glands	Secrete saliva.	Lining of mouth and pharynx moistened and lubricated.
Salivary glands	Secrete saliva.	Lining of mouth and pharynx moistened and lubricated. Saliva softens, moistens, and dissolves food, coats food with mucin, cleanses mouth and teeth. Salivary amylase reduces polysaccharides to the disaccharide maltose.
Teeth	Cut, tear, and pulverize food.	Solid foods reduced to smaller particles for swallowing.

ynx, passes through the mediastinum anterior to the vertebral column, pierces the diaphragm through an opening called the *esophageal hiatus,* and terminates in the superior portion of the stomach.

HISTOLOGY

The *mucosa* of the esophagus consists of nonkeratinized stratified squamous epithelium, lamina propria, and a muscularis mucosae (Figure 24-9a, b). The *submucosa* contains connective tissue and blood vessels. The *muscu-*

laris of the upper third is striated, the middle third is striated and smooth, and the lower third is smooth. The outer layer is known as the *adventitia* (ad-ven-TISH-ya) rather than the serosa because the loose connective tissue of the layer is not covered by epithelium (mesothelium) and because the connective tissue merges with the connective tissue of surrounding structures.

ACTIVITIES

The esophagus does not produce digestive enzymes and does not carry on absorption. It secretes mucus and transports food to the stomach. The passage of food from the laryngopharynx into the esophagus is regulated by a sphincter (thick circle of muscle around an opening) at the entrance to the esophagus called the **upper esophageal** (e-sof'-a-JĒ-al) **sphincter.** It consists of the cricopharyngeus muscle attached to the cricoid cartilage. The elevation of the larynx during the pharyngeal stage of swallowing causes the sphincter to relax and the bolus enters the esophagus. The sphincter also relaxes during expiration.

During the **esophageal stage** of swallowing, food is pushed through the esophagus by involuntary muscular movements called **peristalsis** (per'-i-STAL-sis) (see Figure 24-8c). Peristalsis is a function of the muscularis and is controlled by the medulla. In the section of the esophagus lying just above and around the top of the bolus, the circular muscle fibers contract. The contraction constricts the esophageal wall and squeezes the bolus downward. Meanwhile, longitudinal fibers lying around the bottom of and just below the bolus also contract. Contraction of the longitudinal fibers shortens this lower section, pushing its walls outward so it can receive the bolus. The contractions are repeated in a wave that moves down the esophagus, pushing the food toward the stomach. Passage of the bolus is further facilitated by glands secreting mucus. The passage of solid or semisolid food from the mouth to the stomach takes 4 to 8 seconds. Very soft foods and liquids pass through in about 1 second.

Just above the level of the diaphragm, the esophagus is slightly narrowed. This narrowing has been attributed to a physiological sphincter in the inferior part of the esophagus known as the **lower esophageal (gastroesophageal) sphincter.** The lower esophageal sphincter relaxes during swallowing and thus aids the passage of the bolus from the esophagus into the stomach.

CLINICAL APPLICATION

If the lower esophageal sphincter fails to relax normally as food approaches, the condition is called **achalasia** (ak'-a-LĀ-zē-a; *a* = without; *chalasis* = relaxation). As a result, food passage from the esophagus into the stomach is greatly impeded. A whole meal may become lodged in the esophagus, entering the stomach very slowly. Distension of the esophagus results in chest pain that is often confused with pain originating from the heart. The condition is caused by malfunction of the myenteric plexus.

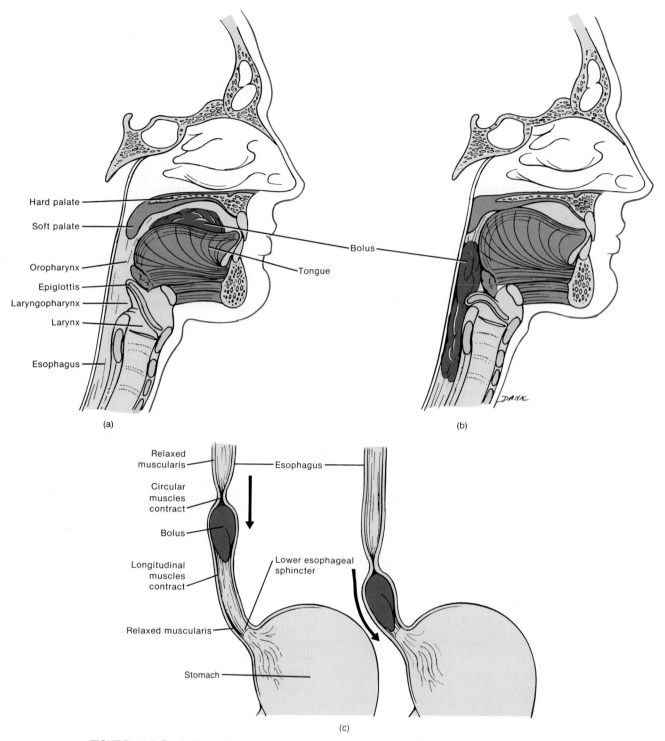

FIGURE 24-8 Deglutition. (a) Position of structures prior to deglutition. (b) During the pharyngeal stage of deglutition, the tongue rises against the palate, the nose is closed off, the larynx rises, the epiglottis seals off the larynx, and the bolus is passed into the esophagus. (c) Esophageal stage of deglutition.

If, on the other hand, the lower esophageal sphincter fails to close adequately after food has entered the stomach, the stomach contents can enter the lower esophagus. Hydrochloric acid from the stomach contents can irritate the esophageal wall, resulting in a burning sensation. The sensation is known as **heartburn** because it is experienced in the region over the heart, although it is not related to any cardiac problem. Heartburn can be treated by taking antacids (Tums, Gelusil, Rolaids, Maalox) that neutralize the hydrochloric acid and lessen the severity of the burning sensation and discomfort.

(a)

(b)

FIGURE 24-9 Histology of the esophagus. (a) Photomicrograph of a portion of the wall of the esophagus at a magnification of 60×. (b) Photomicrograph of an enlarged aspect of the mucosa of the esophagus at a magnification of 180×. (Photomicrographs Copyright © 1983 by Michael H. Ross. Used by permission.)

Exhibit 24-2 summarizes the digestion-related activities of the pharynx and esophagus.

EXHIBIT 24-2 DIGESTIVE ACTIVITIES OF THE PHARYNX AND ESOPHAGUS		
STRUCTURE	**ACTIVITY**	**RESULT**
Pharynx	Pharyngeal stage of deglutition.	Moves bolus from oropharynx to laryngopharynx and into esophagus; closes air passageways.
	Relaxation of upper esophageal sphincter.	Moves bolus from laryngopharynx into esophagus.
Esophagus	Esophageal stage of deglutition (peristalsis).	Forces bolus down esophagus.
	Relaxation of lower esophageal sphincter. Secretion of mucus.	Moves bolus into stomach. Lubricates esophagus for smooth passage of bolus.

STOMACH

The **stomach** is a J-shaped enlargement of the GI tract directly under the diaphragm in the epigastric, umbilical, and left hypochondriac regions of the abdomen (see Figure 1-8b). The superior portion of the stomach is a continuation of the esophagus. The inferior portion empties into the duodenum, the first part of the small intestine. Within each individual, the position and size of the stomach vary continually. For instance, the diaphragm pushes the stomach downward with each inspiration and pulls it upward with each expiration. Empty, it is about the size of a large sausage, but it can stretch to accommodate large amounts of food.

ANATOMY

The stomach is divided into four areas: cardia, fundus, body, and pylorus (Figure 24-10). The **cardia** surrounds the lower esophageal sphincter. The rounded portion above and to the left of the cardia is the **fundus.** Below the fundus is the large central portion of the stomach, called the **body.** The narrow, inferior region is the **pylorus.** The concave medial border of the stomach is called the **lesser curvature,** and the convex lateral border is the

(a)

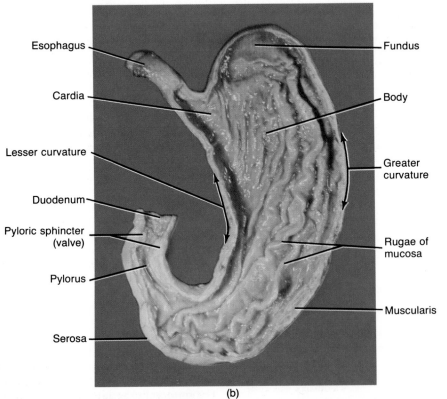

(b)

FIGURE 24-10 External and internal anatomy of the stomach. (a) Diagram in anterior view. (b) Photograph of the internal surface in anterior view showing rugae. (Courtesy of C. Yokochi and J. W. Rohen, *Photographic Anatomy of the Human Body,* 2nd ed., 1978, IGAKU-SHOIN, Ltd., Tokyo, New York.)

greater curvature. The pylorus communicates with the duodenum of the small intestine via a sphincter called the **pyloric sphincter (valve).**

CLINICAL APPLICATION

Two abnormalities of the pyloric spincter can occur in infants. **Pylorospasm** is characterized by failure of the muscle fibers encircling the opening to relax normally. It can be caused by hypertrophy or continuous spasm of the sphincter and usually occurs between the second and twelfth weeks of life. Ingested food does not pass easily from the stomach to the small intestine, the stomach becomes overly full, and the infant vomits frequently to relieve the pressure. Pylorospasm is treated by adrenergic drugs that relax the muscle fibers of the sphincter. **Pyloric stenosis** is a narrowing of the pyloric sphincter caused by a tumor-like mass that apparently is formed by enlargement of the circular muscle fibers. It must be surgically corrected.

HISTOLOGY

The stomach wall is composed of the same four basic layers as the rest of the alimentary canal, with certain modifications. When the stomach is empty, the *mucosa* lies in large folds, called **rugae** (ROO-jē), that can be seen with the naked eye. Microscopic inspection of the mucosa reveals a layer of simple columnar epithelium containing many narrow openings that extend down into the lamina propria (Figure 24-11b, c). These pits—**gastric glands**—are lined with several kinds of secreting cells: zymogenic, parietal, mucous, and enteroendocrine. The **zymogenic (peptic) cells** secrete the principal gastric enzyme precursor, pepsinogen. Hydrochloric acid, involved in the conversion of pepsinogen to the active enzyme pepsin, and intrinsic factor, involved in the absorption of vitamin B_{12} for red blood cell production, are produced by the **parietal (oxyntic) cells.** You may recall from Chapter 19 that inability to produce intrinsic factor can result in pernicious enemia. The **mucous cells** secrete mucus. Secretions of the zymogenic, parietal, and mucous cells are collectively called **gastric juice.** The **enteroendocrine cells** secrete stomach gastrin, a hormone that stimulates secretion of hydrochloric acid and pepsinogen, contracts the lower esophageal sphincter, mildly increases motility of the GI tract, and relaxes the pyloric sphincter.

CLINICAL APPLICATION

The general term **endoscopy** refers to the visual inspection of any cavity of the body using an endoscope, an illuminated tube with lenses. Endoscopes can be used to visualize the entire gastrointestinal tract. In fact, endoscopes can be fitted with special devices that also remove foreign objects from the esophagus and stomach, dilate strictures in the esophagus, remove small gallstones, temporarily stop bleeding, biopsy lesions, and remove polyps from the colon.

Endoscopic examination of the stomach is called **gastroscopy.** During the procedure, the patient is anesthesized and the endoscope is passed into the stomach, which is then inflated with air. The gastric mucosa may be examined for any abnormalities and, if necessary, a gastric mucosal biopsy may be performed.

The *submucosa* of the stomach is composed of loose areolar connective tissue, which connects the mucosa to the muscularis.

The *muscularis,* unlike that in other areas of the alimentary canal, has three layers of smooth muscle: an outer longitudinal layer, a middle circular layer, and an inner oblique layer. This arrangement of fibers allows the stomach to contract in a variety of ways to churn food, break it into small particles, mix it with gastric juice, and pass it to the duodenum.

The *serosa* covering the stomach is part of the visceral peritoneum. At the lesser curvature, the two layers of the visceral peritoneum come together and extend upward to the liver as the lesser omentum. At the greater curvature, the visceral peritoneum continues downward as the greater omentum hanging over the intestines.

DIGESTION IN THE STOMACH

Mechanical

Several minutes after food enters the stomach, gentle, rippling, peristaltic movements called **mixing waves** pass over the stomach every 15 to 25 seconds. These waves macerate food, mix it with the secretions of the gastric glands, and reduce it to a thin liquid called **chyme** (kīm). Few mixing waves are observed in the fundus, which is primarily a storage area. Foods may remain in the fundus for an hour or more without becoming mixed with gastric juice. During this time, salivary digestion continues.

As digestion proceeds in the stomach, more vigorous mixing waves begin at the body of the stomach and intensify as they reach the pylorus. The pyloric sphincter normally remains almost, but not completely, closed. As food reaches the pylorus, each mixing wave forces a small amount of the gastric contents into the duodenum through the pyloric sphincter. Most of the food is forced back into the body of the stomach, where it is subjected to further mixing. The next wave pushes it forward again and forces a little more into the duodenum. The forward and backward movement of the gastric contents are responsible for almost all of the mixing in the stomach.

Chemical

The principal chemical activity of the stomach is to begin the digestion of proteins. In the adult, digestion is achieved primarily through the enzyme **pepsin.** Pepsin breaks certain peptide bonds between the amino acids making up proteins. Thus a protein chain of many amino acids is broken down into smaller fragments called *pep-*

Ruga

Mucosa

Submucosa

Muscularis

(a)

tides. Pepsin is most effective in the very acidic environment of the stomach (pH 2). It becomes inactive in an alkaline environment.

What keeps pepsin from digesting the protein in stomach cells along with the food? First, pepsin is secreted in an inactive form called **pepsinogen,** so it cannot digest the proteins in the zymogenic cells that produce it. It is not converted into active pepsin until it comes in contact with the hydrochloric acid secreted by the parietal cells. Second, the stomach cells are protected by mucus, especially after pepsin has been activated. The mucus coats the mucosa to form a barrier between it and the gastric juices.

Another enzyme of the stomach is **gastric lipase.** Gastric lipase splits the butterfat molecules found in milk. This enzyme operates best at a pH of 5 to 6 and has a limited role in the adult stomach. Adults rely almost exclusively on an enzyme secreted by the pancreas into the small intestine to digest fats.

FIGURE 24-11 Histology of the stomach. (a) Photomicrograph of a portion of the fundic wall of the stomach at a magnification of 18×. (Copyright © 1983 by Michael H. Ross. Used by permission.) (b) Diagram of gastric glands from the fundic wall of the stomach. (c) Photomicrograph of an enlarged aspect of the mucosa of the fundic wall of the stomach at a magnification of 180×. (Copyright © 1983 by Michael H. Ross. Used by permission.)

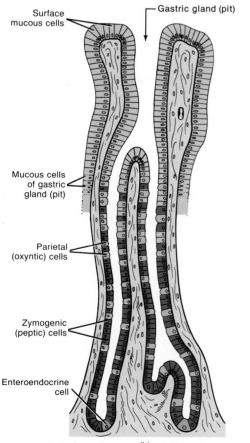

Surface mucous cells

Gastric gland (pit)

Mucous cells of gastric gland (pit)

Parietal (oxyntic) cells

Zymogenic (peptic) cells

Enteroendocrine cell

(b)

Mucous cells of gastric gland (pit)

Parietal (oxyntic) cells

Zymogenic (peptic) cells

(c)

The infant stomach also secretes **rennin,** which is important in the digestion of milk. Rennin and calcium act on the casein of milk to produce a curd. The coagulation prevents too rapid a passage of milk from the stomach. Rennin is absent in the gastric secretions of adults.

The principal activities of gastric digestion are summarized in Exhibit 24-3.

REGULATION OF GASTRIC SECRETION

Stimulation

The secretion of gastric juice is related by both nervous and hormonal mechanisms (Figure 24-12). Parasympathetic impulses from nuclei in the medulla are transmitted via the vagus (X) nerves and stimulate the gastric glands to secrete pepsinogen, hydrochloric acid, mucus, and stomach gastrin. Stomach gastrin is also secreted by gastric glands in response to certain foods that enter the stomach.

• *Cephalic (Reflex) Phase*   The **cephalic (reflex) phase** of gastric secretions occurs before food enters the stomach and prepares the stomach for digestion. The sight, smell, taste, or thought of food initiates this reflex. Nerve impulses from the cerebral cortex or feeding center in the hypothalamus send impulses to the medulla. The medulla relays impulses over the parasympathetic fibers in the vagus (X) nerve to stimulate the gastric glands to secrete.

• *Gastric Phase*   Once the food reaches the stomach, both nervous and hormonal mechanisms ensure that gastric secretion continues. This is the **gastric phase** of secretion. Food of any kind causes distension and stimulates receptors in the wall of the stomach. These receptors send impulses to the medulla and back to the gastric glands, and they may send messages to the glands as well. The impulses stimulate the flow of gastric juice. Emotions such as anger, fear, and anxiety may slow down

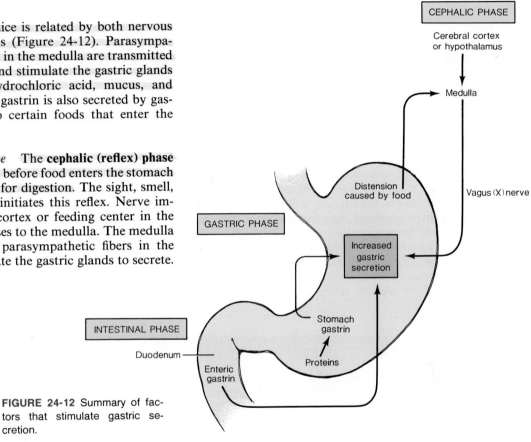

FIGURE 24-12 Summary of factors that stimulate gastric secretion.

EXHIBIT 24-3 SUMMARY OF GASTRIC DIGESTION		
STRUCTURE	**ACTIVITY**	**RESULT**
Mucosa		
Zymogenic (peptic) cells	Secrete pepsinogen.	Precursor of pepsin is produced.
Parietal (oxyntic) cells	Secrete hydrochloric acid.	Converts pepsinogen into pepsin, which digests proteins into peptides.
	Secrete intrinsic factor.	Required for absorption of vitamin B_{12} and erythrocyte formation.
Mucous cells	Secrete mucus.	Prevents digestion of stomach wall.
Enteroendocrine cells	Secrete stomach gastrin.	Stimulates gastric secretion, contracts lower esophageal sphincter, increases motility of the stomach, and relaxes pyloric sphincter.
Muscularis	Mixing waves.	Macerate food, mix it with gastric juice, reduce food to chyme, and force chyme through pyloric sphincter.
Pyloric sphincter (valve)	Opens to permit passage of chyme into duodenum.	Prevents backflow of food from duodenum to stomach.

digestion in the stomach because they stimulate the sympathetic nervous system, which inhibits the impulses of the parasympathetic fibers.

Protein foods and alcohol stimulate the pyloric mucosa to secrete the hormone **stomach gastrin.** It is absorbed into the bloodstream, circulated thorough the body, and finally reaches its target cells, the gastric glands, where it stimulates secretion of large amounts of gastric juice. It also contracts the lower esophageal sphincter, increases motility of the GI tract, and relaxes the pyloric sphincter and ileocecal sphincter.

• *Intestinal Phase* Some investigators believe that when partially digested proteins leave the stomach and enter the duodenum, they stimulate the duodenal mucosa to release **enteric gastrin,** a hormone that stimulates the gastric glands to continue their secretion. This constitutes the **intestinal phase** of secretion. However, this mechanism produces relatively small amounts of gastric juice.

Inhibition

Even though chyme stimulates gastric secretion during the gastric phase, it can inhibit secretion during the intestinal phase. For example, the presence of food in the small intestine during the intestinal phase initiates an **enterogastric reflex** in which nerve impulses carried to the medulla from the duodenum return to the stomach and inhibit gastric secretion. These impulses ultimately inhibit parasympathetic stimulation and stimulate sympathetic activity. Stimuli that initiate this reflex are distension of the duodenum, the presence of acid or partially digested proteins in food in the duodenum, or irritation of the duodenal mucosa.

Several intestinal hormones also inhibit gastric secretion. In the presence of acid, partially digested proteins, fats, hypertonic or hypotonic fluids, or irritating substances in chyme, the intestinal mucosa releases **secretin, cholecystokinin (CCK),** and **gastric inhibiting peptide (GIP).** All three hormones inhibit gastric secretion and decrease motility of the GI tract (Exhibit 24-4). Secretin and cholecystokinin are also important in the control of pancreatic and intestinal secretion, and cholecystokinin also helps regulate secretion of bile from the gallbladder (see Exhibit 24-5).

REGULATION OF GASTRIC EMPTYING

Gastric emptying is stimulated by two principal factors: nerve impulses in response to distension, and stomach gastrin released in the presence of certain types of foods. As noted earlier, during the gastric phase of secretion, distension and the presence of partially digested proteins and alcohol stimulate secretion of gastric juice and stomach gastrin. In the presence of stomach gastrin, the lower esophageal sphincter contracts, the motility of the stomach increases, and the pyloric sphincter relaxes. The net effect of these actions is stomach emptying (Figure 24-13a).

The stomach empties all its contents into the duodenum within 2 to 6 hours after ingestion. Foods rich in carbohydrate spend the least time in the stomach. Protein

---

EXHIBIT 24-4 HORMONAL CONTROL OF GASTRIC SECRETION, PANCREATIC SECRETION, AND SECRETION AND RELEASE OF BILE

HORMONE	WHERE PRODUCED	STIMULANT	ACTION
Stomach gastrin	Pyloric mucosa.	Partially digested proteins in stomach.	Stimulates secretion of gastric juice, constricts lower esophageal sphincter, increases motility of GI tract, and relaxes pyloric sphincter and ileocecal sphincter.
Enteric gastrin	Intestinal mucosa.	Partially digested proteins in chyme in small intestine.	Same as above.
Secretin	Intestinal mucosa.	Acid, partially digested proteins, fats, hypertonic or hypotonic fluids, or irritants in chyme in small intestine.	Inhibits secretion of gastric juice, decreases motility of the GI tract, stimulates secretion of pancreatic juice rich in sodium bicarbonate ions, stimulates secretion of bile by hepatic cells of liver, and stimulates secretion of intestinal juice.
Cholecystokinin (CCK)	Intestinal mucosa.	Same as for secretin.	Inhibits secretion of gastric juice, decreases motility of the GI tract, stimulates the secretion of pancreatic juice rich in digestive enzymes, causes ejection of bile from the gallbladder and opening of the sphincter of the hepatopancreatic ampulla (sphincter of Oddi), and stimulates secretion of intestinal juice.
Gastric inhibiting peptide (GIP)	Intestinal mucosa.	Same as for secretin.	Inhibits secretion of gastric juice and decreases motility of the GI tract.

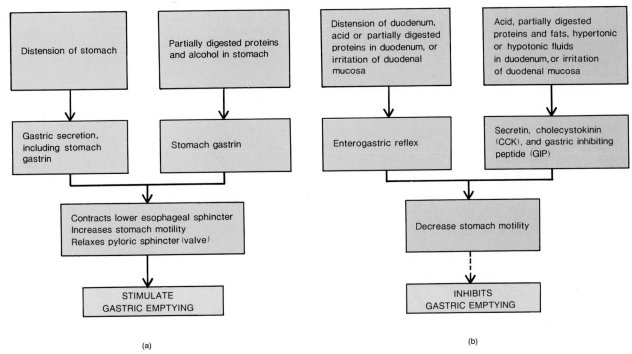

FIGURE 24-13 Factors that (a) stimulate gastric emptying and (b) inhibit gastric emptying.

foods are somewhat slower, and emptying is slowest after a meal containing large amounts of fat.

Stomach emptying is inhibited by the enterogastric reflex and hormones released in response to certain constituents in chyme. The enterogastric reflex not only inhibits gastric secretion, it also inhibits gastric motility. The hormones secretin, cholecystokinin (CCK), and gastric inhibiting peptide (GIP) also inhibit gastric secretion and inhibit gastric motility (Figure 24-13b). The rate of stomach emptying is limited to the amount of chyme that the small intestine can process.

CLINICAL APPLICATION

Excessive gastric emptying in the wrong direction sometimes occurs. **Vomiting** is the forcible expulsion of the contents of the upper GI tract (stomach and sometimes duodenum) through the mouth. The strongest stimuli for vomiting are irritation and distension of the stomach. Other stimuli include unpleasant sights and dizziness. Impulses are transmitted to the vomiting center in the medulla, and returning impulses to the upper GI tract organs, diaphragm, and abdominal muscles bring about the vomiting act. Basically, vomiting involves squeezing the stomach between the diaphragm and abdominal muscles and expelling of the contents through open esophageal sphincters. Prolonged vomiting, especially in infants and elderly people, can be serious because the loss of gastric juice and fluids can lead to disturbances in fluid and acid–base balance.

ABSORPTION

The stomach wall is permeable to the passage of most materials into the blood, so most substances are not absorbed until they reach the small intestine. However, the stomach does participate in the absorption of some water, electrolytes, certain drugs (especially aspirin), and alcohol.

PANCREAS

The next organ of the GI tract involved in the breakdown of food is the small intestine. Chemical digestion in the small intestine depends not only on its own secretions but also on activities of three accessory structures of digestion outside the alimentary canal: the pancreas, liver, and gallbladder. We will first consider the activities of the accessory structures and then examine their contributions to digestion in the small intestine.

ANATOMY

The **pancreas** is a soft, oblong tubuloacinar gland about 12.5 cm (6 inches) long and 2.5 cm (1 inch) thick. It lies posterior to the greater curvature of the stomach and is connected by a duct (usually two) to the duodenum (Figure 24-14). The pancreas is divided into a head, body, and tail. The **head** is the expanded portion near the C-shaped curve of the duodenum. Moving superiorly and to the left of the head are the centrally located **body** and the terminal tapering **tail.**

The pancreas is linked to the small intestine usually by two ducts. Pancreatic secretions pass from the secret-

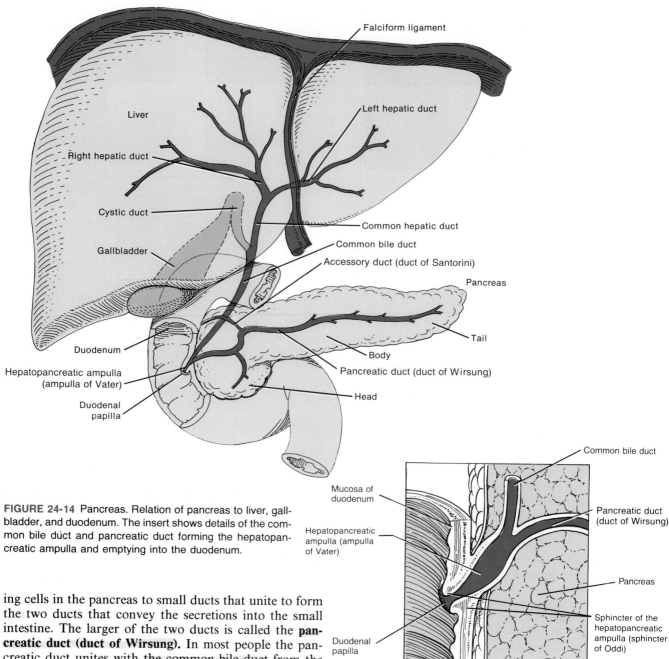

FIGURE 24-14 Pancreas. Relation of pancreas to liver, gallbladder, and duodenum. The insert shows details of the common bile duct and pancreatic duct forming the hepatopancreatic ampulla and emptying into the duodenum.

ing cells in the pancreas to small ducts that unite to form the two ducts that convey the secretions into the small intestine. The larger of the two ducts is called the **pancreatic duct (duct of Wirsung).** In most people the pancreatic duct unites with the common bile duct from the liver and gallbladder and enters the duodenum in a common duct, called the **hepatopancreatic ampulla (ampulla of Vater).** The ampulla opens on an elevation of the duodenal mucosa known as the **duodenal papilla,** about 10 cm (4 inches) below the pylorus of the stomach. The smaller of the two ducts is the **accessory duct (duct of Santorini),** which leads from the pancreas and empties into the duodenum about 2.5 cm (1 inch) above the ampulla.

HISTOLOGY

The pancreas is made up of small clusters of glandular epithelial cells. About 1 percent of the cells, the **pancreatic islets (islets of Langerhans),** form the endocrine portion of the pancreas and consist of alpha, beta, and delta cells

that secrete hormones (glucagon, insulin, and somatostatin, respectively). The functions of these hormones may be reviewed in Chapter 18. The remaining 99 percent of the cells, called **acini** (AS-i-nē), are the exocrine portions of the organ (see Figure 18-22). Secreting cells of the acini release a mixture of digestive enzymes called pancreatic juice.

PANCREATIC JUICE

Each day the pancreas produces 1,200 to 1,500 ml (about 1.2 to 1.5 qt) of **pancreatic juice,** a clear, colorless liquid. It consists mostly of water, some salts, sodium bicarbonate, and enzymes. The sodium bicarbonate gives pancreatic juice a slightly alkaline pH (7.1 to 8.2) that stops the action of pepsin from the stomach and creates the proper environment for the enzymes in the small intestine. The enzymes in pancreatic juice include a carbohydrate-digesting enzyme called *pancreatic amylase;* several protein-digesting enzymes called *trypsin* (TRIP-sin), *chymotrypsin* (kī'-mō-TRIP-sin) and *carboxypolypeptidase* (kar-bok'-sē-polē'-PEP-ti-dās); the principal fat-digesting enzyme in the adult body called *pancreatic lipase;* and nucleic acid–digesting enzymes called *ribonuclease* and *deoxyribonuclease.*

Just as pepsin is produced in the stomach in an inactive form (pepsinogen), so too are the protein-digesting enzymes of the pancreas. This prevents the enzymes from digesting cells of the pancreas. The active enzyme trypsin is secreted in an inactive form called *trypsinogen* (trip-SIN-ō-jen). Its activation to trypsin is accomplished in the small intestine by an enzyme secreted by the intestinal mucosa when chyme comes in contact with the mucosa. The activating enzyme is called *enterokinase* (en'-ter-ō-KĪ-nās). Chymotrypsin is activated in the small intestine by trypsin from its inactive form, *chymotrypsinogen.* Carboxypolypeptidase is also activated in the small intestine by trypsin. Its inactive form is called procarboxypolypeptidase.

REGULATION OF PANCREATIC SECRETIONS

Pancreatic secretion, like gastric secretion, is regulated by both nervous and hormonal mechanisms (Figure 24-15). When the cephalic and gastric phases of gastric secretion occur, parasympathetic impulses are simultaneously transmitted along the vagus (X) nerves to the pancreas that result in the secretion of pancreatic enzymes.

In response to chyme in the small intestine, especially chyme that contains partially digested proteins, fats, hypertonic or hypotonic fluids, or irritating substances, the

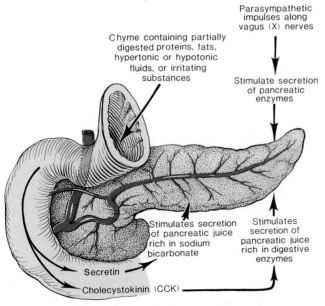

Chyme containing partially digested proteins, fats, hypertonic or hypotonic fluids, or irritating substances

Parasympathetic impulses along vagus (X) nerves

Stimulate secretion of pancreatic enzymes

Stimulates secretion of pancreatic juice rich in sodium bicarbonate

Stimulates secretion of pancreatic juice rich in digestive enzymes

Secretin

Cholecystokinin (CCK)

FIGURE 24-15 Control of pancreatic secretion.

small intestinal mucosa secretes secretin and cholecystokinin (CCK), two hormones that affect pancreatic secretion. Secretin stimulates the pancreas to secrete pancreatic juice that is rich in sodium bicarbonate ions. Cholecystokinin (CCK) stimulates a pancreatic secretion rich in digestive enzymes (see Exhibit 24-4).

LIVER

The **liver** weighs about 1.4 kg (about 3 lb) in the average adult. It is located under the diaphragm and occupies most of the right hypochondrium and part of the epigastrium of the abdomen (see Figure 1-8c).

ANATOMY

The liver is almost completely covered by peritoneum and completely covered by a dense connective tissue layer that lies beneath the peritoneum. It is divided into two principal lobes—the **right lobe** and the **left lobe**—separated by the **falciform ligament** (Figure 24-16). Associated with the right lobe are the inferior **quadrate lobe** and posterior **caudate lobe.** The falciform ligament is a reflection of the parietal peritoneum, which extends from the undersurface of the diaphragm to the superior surface of the liver, between the two principal lobes of the liver. In the free border of the falciform ligament is the **ligamentum teres (round ligament).** It extends from the liver to the umbilicus. The ligamentum teres is a fibrous cord derived from the umbilical vein of the fetus.

Bile, one of the liver's products, enters **bile capillaries** or **canaliculi** (kan'-a-LIK-yoo-lī) that empty into small ducts. These small ducts eventually merge to form the larger **right** and **left hepatic ducts,** which unite to leave the liver as the **common hepatic duct** (Figures 24-14 and 24-16b). Further on, the common hepatic duct joins the **cystic duct** from the gallbladder. The two tubes become the **common bile duct.** The common bile duct and pancreatic duct enter the duodenum in a common duct called the **hepatopancreatic ampulla (ampulla of Vater).**

HISTOLOGY

The lobes of the liver are made up of numerous functional units called **lobules,** which may be seen under a microscope (Figure 24-17). A lobule consists of cords of **hepatic (liver) cells** arranged in a radial pattern around a **central vein.** Between the cords are endothelial-lined spaces called **sinusoids,** through which blood passes. The sinusoids are also partly lined with phagocytic cells, termed **stellate reticuloendothelial (Kupffer's) cells,** that destroy worn-out white and red blood cells and bacteria. The liver contains sinusoids instead of typical capillaries.

BLOOD SUPPLY

The liver receives a double supply of blood. From the hepatic artery it obtains oxygenated blood, and from the hepatic portal vein it receives deoxygenated blood containing newly absorbed nutrients (see Figures 21-23 and 24-17). Branches of both the hepatic artery and the he-

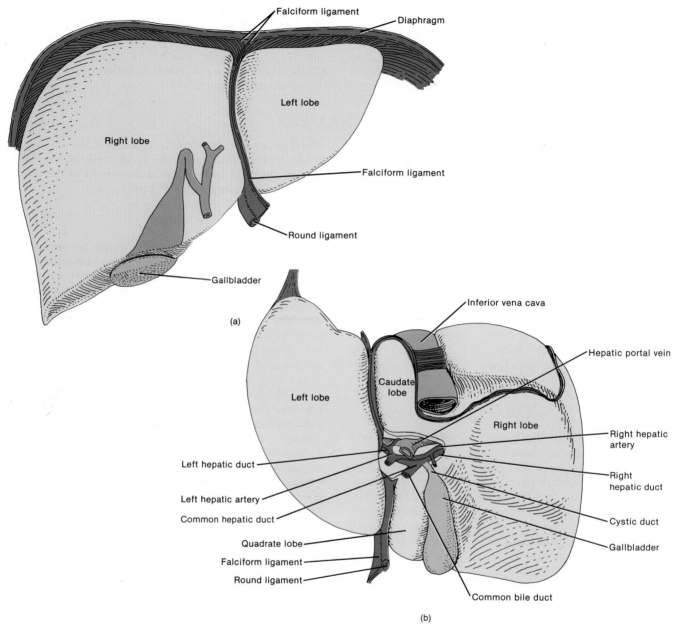

FIGURE 24-16 External anatomy of the liver. (a) Anterior view. (b) Posteroinferior view.

patic portal vein carry the blood into the sinusoids of the lobules, where oxygen, most of the nutrients, and certain poisons are extracted by the hepatic cells. Nutrients are stored or used to make new materials. The poisons are stored or detoxified. Products manufactured by the hepatic cells and nutrients needed by other cells are secreted back into the blood. The blood then drains into the central vein and eventually passes into a hepatic vein. Unlike the other products of the liver, bile normally is not secreted into the bloodstream.

BILE

Each day the hepatic cells secrete 800 to 1,000 ml (about 1 qt) of **bile,** a yellow, brownish, or olive-green liquid.

It has a pH of 7.6 to 8.6. Bile consists mostly of water and bile salts, cholesterol, a phospholipid called lecithin, bile pigments, and several ions.

Bile is partially an excretory product and partially a digestive secretion. Bile salts assume a role in **emulsification,** the breakdown of fat globules into a suspension of fat droplets about 1 μm in diameter, and absorption of fats following their digestion. Cholesterol is made soluble in bile by bile salts and lecithin. The principal bile pigment is **bilirubin.** When red blood cells are broken down, iron, globin, and bilirubin are released. The iron and globin are recycled, but some of the bilirubin is excreted into the bile ducts. Bilirubin eventually is broken down in the intestine, and one of its breakdown products (urobilinogen) gives feces their color.

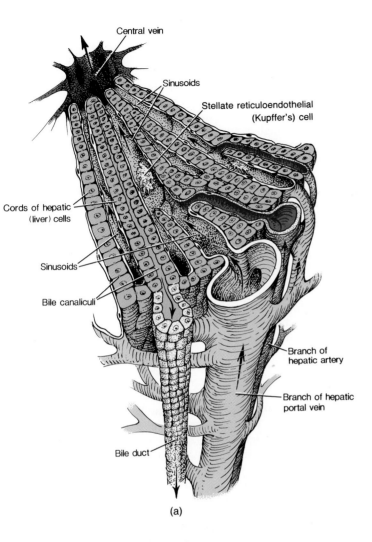

Central vein

Sinusoids

Stellate reticuloendothelial (Kupffer's) cell

Cords of hepatic (liver) cells

Sinusoids

Bile canaliculi

Branch of hepatic artery

Branch of hepatic portal vein

Bile duct

(a)

CLINICAL APPLICATION

If insufficient bile salts or lecithin are present in bile, or if there is excessive cholesterol, the cholesterol precipitates out of solution and crystalizes to form **gallstones (biliary calculi)**. The problems associated with gallstone formation and the diagnosis and treatment of gallstones are discussed in detail at the end of the chapter.

If the liver is unable to remove bilirubin from the blood because of increased destruction of red blood cells or obstruction of bile ducts, large amounts of bilirubin circulate through the bloodstream and collect in other tissues, giving the skin and eyes a yellow color. This condition is called **jaundice**. If the jaundice is due to damaged red blood cells, it is called **hemolytic jaundice;** if it is due to obstruction in the biliary system, it is known as **obstructive jaundice.** Since the liver of a newborn functions poorly for the first week or so, large amounts of bilirubin are excreted into blood instead of being incorporated into bile in the liver. The result is a type of jaundice called **neonatal (physiological) jaundice.**

REGULATION OF BILE SECRETION

The rate at which bile is secreted is determined by several factors. Vagal stimulation can increase the production of bile to more than twice the normal rate. Secretin, the hormone that stimulates the synthesis of pancreatic juice rich in sodium bicarbonate, also stimulates the secretion of bile (see Exhibit 24-4). Within limits, as blood flow through the liver increases, so does the secretion of bile.

Approximate boundary of lobule

Sinusoid

Central vein of liver lobule

Cord of hepatic (liver) cells

(b)

FIGURE 24-17 Histology of the liver. (a) Diagram of the microscopic appearance of a portion of a liver lobule. (b) Photomicrograph of a liver lobule at a magnification of 65×. (Copyright © 1983 by Michael H. Ross. Used by permission.)

Finally, the presence of large amounts of bile salts in the blood also increases the rate of bile production.

FUNCTIONS OF THE LIVER

The liver performs many vital functions. Among these are the following:

1. The liver manufactures bile salts, which are used in the small intestine for the emulsification and absorption of fats, cholesterol, phospholipids, and lipoproteins.

2. The liver, together with mast cells, manufactures the anticoagulant heparin and most of the other plasma proteins, such as prothrombin, fibrinogen, and albumin.

3. The stellate reticuloendothelial (Kupffer's) cells of the liver phagocytize worn-out red and white blood cells and some bacteria.

4. Liver cells contain enzymes that either break down poisons or transform them into less harmful compounds. When amino acids are burned for energy, for example, they leave behind toxic nitrogenous wastes (such as ammonia) that are converted to urea by the liver cells. Moderate amounts of urea are harmless to the body and are easily excreted by the kidneys and sweat glands.

5. Newly absorbed nutrients are collected in the liver. Depending on the body's needs, it can change any excess monosaccharides into glycogen or fat, both of which can be stored, or it can transform glycogen, fat, and protein into glucose.

6. The liver stores glycogen, copper, iron, and vitamins A, B_{12}, D, E, and K. It also stores some poisons that cannot be broken down and excreted. (High levels of DDT are found in the livers of animals, including humans, who eat sprayed fruits and vegetables.)

7. The liver and kidneys participate in the activation of vitamin D.

GALLBLADDER

The **gallbladder** is a pear-shaped sac about 7 to 10 cm (3 to 4 inches) long. It is located in a fossa of the visceral surface of the liver (see Figures 24-14a and 24-16).

HISTOLOGY

The inner wall of the gallbladder consists of a mucous membrane arranged in rugae resembling those of the stomach. The middle, muscular coat of the wall consists of smooth muscle fibers. Contraction of these fibers by hormonal stimulation ejects the contents of the gallbladder into the cystic duct. The outer coat is the visceral peritoneum.

FUNCTION

The function of the gallbladder is to store and concentrate bile (up to 10-fold) until it is needed in the small intestine. In the concentration process, water and many ions are absorbed by the gallbladder mucosa. Bile from the liver enters the small intestine through the common bile duct. When the small intestine is empty, a valve around the hepatopancreatic ampulla (ampulla of Vater) called the **sphincter of the hepatopancreatic ampulla (sphincter of Oddi),** closes, and the backed-up bile overflows into the cystic duct to the gallbladder for storage.

EMPTYING OF THE GALLBLADDER

In order for the gallbladder to eject bile into the small intestine to participate in the digestive process, the muscularis must contract to force bile into the common bile duct and the sphincter of the hepatopancreatic ampulla must relax. Chyme entering the duodenum that contains particularly high concentrations of fats or partially digested proteins stimulates the intestinal mucosa to secrete cholecystokinin (CCK). This hormone brings about contraction of the muscularis coupled with relaxation of the sphincter of the hepatopancreatic ampulla, resulting in emptying of the gallbladder (see Exhibit 24-4).

SMALL INTESTINE

The major portions of digestion and absorption occur in a long tube called the **small intestine.** The small intestine begins at the pyloric sphincter of the stomach, coils through the central and lower part of the abdominal cavity, and eventually opens into the large intestine. It averages 2.5 cm (1 inch) in diameter and about 6.35 m (21 ft) in length.

ANATOMY

The small intestine is divided into three segments (see Figure 24-1). The **duodenum** (doo'-ō-DĒ-num), the shortest part, originates at the pyloric sphincter of the stomach and extends about 25 cm (10 inches) until it merges with the jejunum. The **jejunum** (jē-JOO-num) is about 2.5 m (8 ft) long and extends to the ileum. The final portion of the small intestine, the **ileum** (IL-ē-um), measures about 3.6 m (12 ft) and joins the large intestine at the **ileocecal** (il'-ē-ō-SĒ-kal) **sphincter (valve).**

HISTOLOGY

The wall of the small intestine is composed of the same four tunics that make up most of the GI tract. However, both the mucosa and the submucosa are modified to allow the small intestine to complete the processes of digestion and absorption (Figure 24-18).

Plicae circulares

(a)

FIGURE 24-18 Small intestine. Shown are various structures that adapt the small intestine for digestion and absorption. (a) Photograph of a section of the jejunum cut open to expose the plicae circulares. (Courtesy of C. Yokochi, *Photographic Anatomy of the Human Body,* 1st ed., 1969, IGAKU-SHOIN, Ltd., Tokyo, New York.)

Villi
Plica circularis
Mucosa
Submucosa
Muscularis
Serosa

(b)

FIGURE 24-18 (*Continued*) (b) Villi in relation to the layers. (c) Enlarged aspect of several villi. (d) Scanning electron micrograph of a cross section of the small intestine at a magnification of 30×. (From *Tissues and Organs: A Text-Atlas of Scanning Electron Microscopy* by Richard G. Kessel and Randy H. Kardon. W. H. Freeman and Company. Copyright © 1979.)

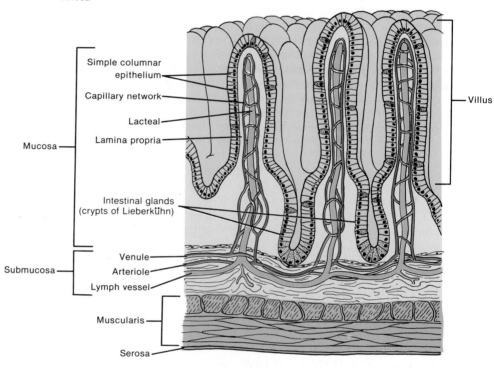

Mucosa

Simple columnar epithelium
Capillary network
Lacteal
Lamina propria

Intestinal glands (crypts of Lieberkühn)

Villus

Submucosa

Venule
Arteriole
Lymph vessel

Muscularis

Serosa

(c)

Serosa
Muscularis
Submucosa
Villi of mucosa
Lumen

(d)

The mucosa contains many pits lined with glandular epithelium. These pits—the **intestinal glands (crypts of Lieberkühn)**—secrete intestinal juice. The submucosa of the duodenum contains **duodenal (Brunner's) glands,** which secrete an alkaline mucus to protect the wall of the small intestine from the action of the enzymes and to aid in neutralizing acid in the chyme. Some of the epithelial cells in the mucosa and submucosa have been transformed to goblet cells, which secrete additional mucus.

Since almost all the absorption of nutrients occurs in the small intestine, its structure is specially adapted for this function. Its length alone provides a large surface area for absorption and that area is further increased by modifications in the structure of its wall. The epithelium covering and lining the mucosa consists of simple columnar epithelium. These epithelial cells, except those transformed into goblet cells, contain **microvilli,** fingerlike projections of the plasma membrane. Larger amounts of digested nutrients diffuse into the intestinal wall because the microvilli increase the surface area of the plasma membrane. They also increase the surface area for digestion.

The mucosa lies in a series of **villi,** projections 0.5 to 1 mm high, giving the intestinal mucosa its velvety appearance. The enormous number of villi (10–40 per square millimeter) vastly increases the surface area of the epithelium available for absorption and digestion. Each villus has a core of lamina propria, the connective tissue layer of the mucosa. Embedded in this connective tissue are an arteriole, a venule, a capillary network, and a **lacteal** (LAK-tē-al) or lymphatic vessel. Nutrients that diffuse through the epithelial cells that cover the villus are able to pass through the capillary walls and the lacteal and enter the cardiovascular and lymphatic systems.

In addition to the microvilli and villi, a third set of projections called **plicae circulares** (PLĪ-kē SER-kyoo-lar-es), or **circular folds,** further increases the surface area

for absorption and digestion. The plicae are permanent ridges, about 10 mm (0.4 inch) high, in the mucosa. Some of the folds extend all the way around the intestine, and others extend only partway around. The folds begin near the proximal portion of the duodenum and terminate at about the midportion of the ileum. The plicae circulares enhance absorption by causing the chyme to spiral, rather than to move in a straight line, as it passes through the small intestine. Since the plicae circulares and villi decrease in size in the distal ileum, most absorption occurs in the duodenum and jejunum.

The muscularis of the small intestine consists of two layers of smooth muscle. The outer, thinner layer contains longitudinally arranged fibers. The inner, thicker layer contains circularly arranged fibers. Except for a major portion of the duodenum, the serosa (or visceral peritoneum) completely covers the small intestine. Additional histological aspects of the small intestine are shown in Figure 24-19.

There is an abundance of lymphatic tissue in the form of lymphatic nodules, masses of lymphatic tissue not covered by a capsule wall. **Solitary lymphatic nodules** are most numerous in the lower part of the ileum. Groups of lymphatic nodules, referred to as **aggregated lymphatic follicles (Peyer's patches),** are numerous in the ileum.

INTESTINAL JUICE

Intestinal juice is a clear yellow fluid secreted in amounts of about 2 to 3 liters (about 2 to 3 qt) a day. It has a pH of 7.6, which is slightly alkaline, and contains water and mucus. The juice is rapidly reabsorbed by the villi and provides a vehicle for the absorption of substances from chyme as they come in contact with the villi. The intestinal enzymes are formed in the epithelial cells that line the villi, and most digestion by enzymes of the small intestine occurs within the cells on the surfaces of their microvilli. In addition, as small intestinal cells containing

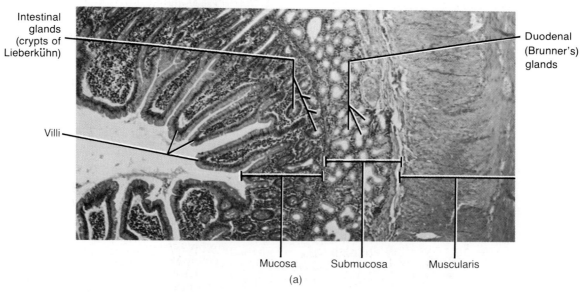

Intestinal glands (crypts of Lieberkühn)

Villi

Duodenal (Brunner's) glands

Mucosa Submucosa Muscularis

(a)

FIGURE 24-19 Histology of the small intestine. (a) Photomicrograph of a portion of the wall of the duodenum at a magnification of 40×.

Simple columnar epithelium

Lamina propria

Goblet cells

Intestinal glands (crypts of Lieberkühn)

Muscularis mucosae

(b)

FIGURE 24-19 (*Continued*) (b) Photomicrograph of an enlarged aspect of two villi from the ileum at a magnification of 120×. (Photomicrographs copyright © 1983 by Michael H. Ross. Used by permission.)

the tract. Segmentation starts with the contractions of circular muscle fibers in a portion of the small intestine, an action that constricts the intestine into segments. Next, muscle fibers that encircle the middle of each segment also contract, dividing each segment again. Finally, the fibers that contracted first relax, and each small segment unites with an adjoining small segment so that large areas are reformed. This sequence of events is repeated 12 to 16 times a minute, sloshing the chyme back and forth (Figure 24-20). Segmentation depends mainly on intestinal distension, which initiates impulses to the central nervous system. Returning parasympathetic impulses increase motility. Sympathetic impulses decrease intestinal motility as distension decreases.

Peristalsis propels the chyme onward through the intestinal tract. Peristaltic contractions in the small intestine are normally very weak compared to those in the esophagus or stomach. Chyme moves through the small intestine at a rate of about 1 cm/min. Thus, chyme remains in the small intestine for 3 to 5 hours. Peristalsis, like segmentation, is initiated by distension and controlled by the autonomic nervous system.

Chemical

In the mouth, salivary amylase converts starch (polysaccharide) to maltose (disaccharide). In the stomach, pepsin converts proteins to peptides (small proteins). Thus, chyme entering the small intestine contains partially digested carbohydrates, partially digested proteins, and

enzymes slough off into the lumen of the intestine, they break apart and release small quantities of enzymes that digest some food in the chyme. Thus, most digestion by enzymes of the small intestine occurs in or on the epithelial cells that line the villi, rather than in the lumen, as in other parts of the gastrointestinal tract. Among the enzymes produced by small intestinal cells are three carbohydrate-digesting enzymes called *maltase, sucrase,* and *lactase;* several protein-digesting enzymes called *peptidases;* and two nucleic acid–digesting enzymes, *ribonuclease* and *deoxyribonuclease.*

DIGESTION IN THE SMALL INTESTINE

Mechanical

The movements of the small intestine are arbitrarily divided into two types: segmentation and peristalsis. **Segmentation** is the major movement of the small intestine. It is strictly a localized contraction in areas containing food. It mixes chyme with the digestive juices and brings the particles of food into contact with the mucosa for absorption. It does not push the intestinal contents along

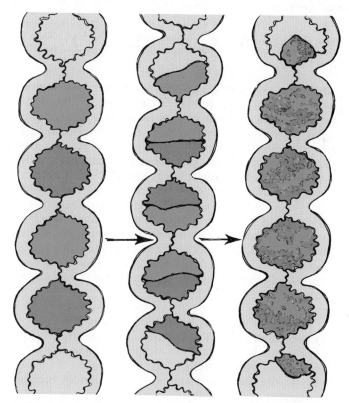

FIGURE 24-20 Segmentation. Localized contractions thoroughly mix the contents of the small intestine.

essentially undigested lipids. The completion of the digestion of carbohydrates, proteins, and lipids is a collective effort of pancreatic juice, bile, and intestinal juice in the small intestine.

• *Carbohydrates* Even though the action of **salivary amylase** may continue in the stomach for some time, its activity is blocked by the acid pH of the stomach. Thus, few starches are reduced to maltose by the time chyme leaves the stomach. Any starches not already broken down into the disaccharide are converted by **pancreatic amylase,** an enzyme in pancreatic juice that acts in the small intestine.

Sucrose and lactose, two disaccharides, are ingested as such and are not acted upon until they reach the small intestine. Three enzymes in the intestinal juice digest the disaccharides into monosaccharides. **Maltase** splits maltose into two molecules of glucose. **Sucrase** breaks sucrose into a molecule of glucose and a molecule of fructose. **Lactase** digests lactose into a molecule of glucose and a molecule of galactose. This completes the digestion of carbohydrates.

CLINICAL APPLICATION

In some individuals the mucosal cells of the small intestine fail to produce lactase, which is essential for the digestion of lactose. This condition is called **lactose intolerance.** Its symptoms include diarrhea, gas, bloating, and abdominal cramps following consumption of milk and other dairy products. The deficiency occurs in 6 to 8 percent of Caucasians, 70 to 75 percent of blacks, and 90 to 95 percent of Asians. It sometimes occurs following gastric surgery.

• *Proteins* Protein digestion starts in the stomach, where proteins are fragmented by the action of **pepsin** into peptides. Enzymes found in pancreatic juice continue the digestion. **Trypsin** and **chymotrypsin** continue to break down proteins into peptides. Although pepsin, trypsin, and chymotrypsin all convert whole proteins into peptides, their actions differ somewhat since each splits peptide bonds between different amino acids. **Carboxypolypeptidase** acts on peptides and breaks the peptide bond that attaches the terminal amino acid to the carboxyl (acid) end of the peptide. Protein digestion is completed by the **peptidases. Aminopeptidase** acts on peptides and breaks the peptide bonds that attach amino acids to the amino end of the peptide. **Dipeptidase** splits dipeptides (two amino acids joined by a peptide bond) into amino acids that can be absorbed.

• *Lipids* In an adult, almost all lipid digestion occurs in the small intestine. The first step in the process involves the preparation of neutral fats (triglycerides) by bile salts. Neutral fats, or just simply fats, are the most abundant lipids in the diet. They are called triglycerides because they consist of a molecule of glycerol and three molecules of fatty acid (see Figure 2-11). Bile salts break the globules of fat into droplets about 1 μm in diameter. This process

is called **emulsification.** It is necessary so that the fat-splitting enzyme can get at the lipid molecules. In the second step, **pancreatic lipase,** an enzyme found in pancreatic juice, hydrolyzes each fat molecule into fatty acids and monoglycerides, end products of fat digestion. Lipase removes two of the three fatty acids from glycerol; the third remains attached to the glycerol, thus forming monoglycerides.

• *Nucleic Acids* Both intestinal juice and pancreatic juice contain **nucleases** that digest nucleotides into their constituent pentoses and nitrogen bases. **Ribonuclease** acts on ribonucleic acid nucleotides, and **deoxyribonuclease** acts on deoxyribonucleic acid nucleotides.

A summary of digestive enzymes in terms of source, substrate acted on, and product is presented in Exhibit 24-5.

REGULATION OF INTESTINAL SECRETION

The most important means for regulating small intestinal secretion is local reflexes in response to the presence of chyme. Also, secretin and cholecystokinin (CCK) stimulate the production of intestinal juice.

ABSORPTION

All the chemical and mechanical phases of digestion from the mouth down through the small intestine are directed toward changing food into forms that can pass through the epithelial cells lining the mucosa into the underlying blood and lymph vessels. These forms are monosaccharides (glucose, fructose, and galactose), amino acids, fatty acids, glycerol, and glycerides. Passage of these digested nutrients from the alimentary canal into the blood or lymph is called **absorption.**

About 90 percent of all absorption of nutrients takes place throughout the length of the small intestine. The other 10 percent occurs in the stomach and large intestine. Any undigested or unabsorbed material left in the small intestine is passed on to the large intestine. Absorption of materials in the small intestine occurs specifically through the villi and depends on diffusion, facilitated diffusion, osmosis, and active transport.

Carbohydrates

Essentially all carbohydrates are absorbed as monosaccharides. Glucose and galactose are transported into epithelial cells of the villi by an active process that is coupled with the active transport of sodium. It appears that the carrier protein has receptor sites for both glucose and sodium, and unless both sites are simultaneously filled, neither substance is transported. Fructose is transported by facilitated diffusion. Transported monosaccharides then move out of the epithelial cells by diffusion and enter the capillaries of the villi. From here they are transported in the bloodstream to the liver via the hepatic portal system. After their passage through the liver, they move through the heart and then enter general circulation (Figure 24-21).

EXHIBIT 24-5   SUMMARY OF DIGESTIVE ENZYMES

ENZYME	SOURCE	SUBSTRATE	PRODUCT
Salivary amylase	Salivary glands.	Starches (polysaccharides).	Maltose (disaccharide).
Pepsin (activated from pepsinogen by hydrochloric acid)	Stomach (zymogenic cells).	Proteins.	Peptides.
Pancreatic amylase	Pancreas.	Starches (polysaccharides).	Maltose (disaccharide).
Trypsin (activated from trypsinogen by enterokinase)	Pancreas.	Proteins.	Peptides.
Chymotrypsin (activated from chymotrypsinogen by trypsin)	Pancreas.	Proteins.	Peptides.
Carboxypolypeptidase (activated from procarboxypolypeptidase by trypsin)	Pancreas.	Terminal amino acid at carboxyl (acid) end of peptides.	Peptides and amino acids.
Pancreatic lipase	Pancreas.	Neutral fats (triglycerides) that have been emulsified by bile salts.	Fatty acids and monoglycerides.
Maltase	Small intestine.	Maltose.	Glucose.
Sucrase	Small intestine.	Sucrose.	Glucose and fructose.
Lactase	Small intestine.	Lactose.	Glucose and galactose.
Peptidases			
Aminopeptidase	Small intestine.	Terminal amino acids at amino end of peptides.	Amino acids.
Dipeptidase	Small intestine.	Dipeptides.	Amino acids.
Nucleases			
Ribonuclease	Pancreas and small intestine.	Ribonucleic acid nucleotides.	Pentoses and nitrogen bases.
Deoxyribonuclease	Pancreas and small intestine.	Deoxyribonucleic acid nucleotides.	Pentoses and nitrogen bases.

Proteins

Most proteins are absorbed as amino acids, and the process occurs mostly in the duodenum and jejunum. Amino acid transport into epithelial cells of the villi is an active transport process also coupled with active sodium transport. There appear to be several (perhaps four) different transport systems for amino acids, one each for the various chemical groups of amino acids. Amino acids move out of the epithelial cells by diffusion to enter the bloodstream. They follow the same route as that taken by monosaccharides. At times, dipeptides and tripeptides can be taken in by epithelial cells by active transport. Most of them are hydrolyzed to amino acids in the cells and then passed into the capillaries of the villi (Figure 24-21).

Lipids

As a result of emulsification and fat digestion, neutral fats (triglycerides) are broken down into monoglycerides and fatty acids. Recall that lipase removes two of the three fatty acids from glycerol during fat digestion; the other fatty acid remains attached to glycerol, thus forming monoglycerides. Short-chain fatty acids (those with less than 10 to 12 carbon atoms) pass into the epithelial cells by diffusion and follow the same route taken by monosaccharides and amino acids (Figure 24-21).

Most fatty acids are long-chain fatty acids. They and the monoglycerides are transported differently. Bile salts form spherical aggregates called **micelles** (mī-SELZ). They are about 2.5 nm in diameter and consist of 20 to 50 molecules of bile salt. Despite their large size, micelles have the ability to dissolve in water in the intestinal fluid. During fat digestion, fatty acids and monoglycerides dissolve in the center of the micelles, and it is in this form that they reach the epithelial cells of the villi. On coming into contact with the surfaces of the epithelial cells, fatty acids and monoglycerides diffuse into the cells, leaving the micelles behind in chyme. The micelles continually repeat this ferrying function. The majority of bile salts in the small intestine are ultimately reabsorbed in the ileum to be returned by the blood to the liver for resecretion. This cycle is called **enterohepatic circulation.** Insufficient bile salts, due to obstruction of the biliary ducts or removal of the gallbladder, can result in the loss of up to 40 percent of lipids in feces due to improper lipid

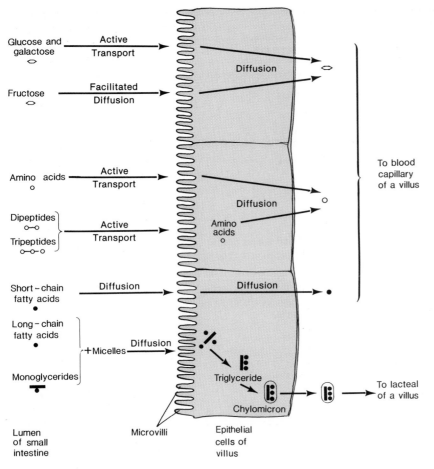

Glucose and galactose

Active Transport

Fructose

Facilitated Diffusion

Diffusion

Amino acids

Active Transport

Dipeptides

Tripeptides

Active Transport

Amino acids

Diffusion

Short–chain fatty acids

Diffusion

Diffusion

Long–chain fatty acids

Monoglycerides

+ Micelles

Diffusion

Triglyceride

Chylomicron

To lacteal of a villus

To blood capillary of a villus

Lumen of small intestine

Microvilli

Epithelial cells of villus

(a)

FIGURE 24-21 Absorption. (a) Movement of digested nutrients through mucosal cells of the villi. For simplicity, all digested foods are shown in the lumen of the small intestine, even though some nutrients are digested within or on the surfaces of small intestinal epithelial cells. (b) Movement of digested nutrients into the cardiovascular and lymphatic systems.

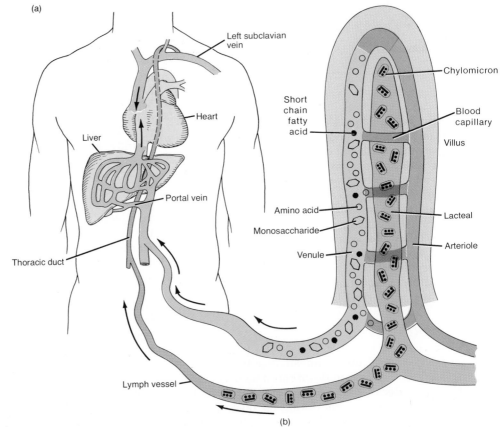

Left subclavian vein

Heart

Liver

Portal vein

Thoracic duct

Lymph vessel

Short chain fatty acid

Amino acid

Monosaccharide

Venule

Chylomicron

Blood capillary

Villus

Lacteal

Arteriole

(b)

absorption. Also, when lipids are not absorbed properly, the fat-soluble vitamins (A, D, E, K) are not adequately absorbed.

Within the epithelial cells, many monoglycerides are further digested by lipase in the cells to glycerol and fatty acids. Then, the fatty acids and glycerol are recombined to form triglycerides in the smooth endoplasmic reticulum of the epithelial cell. The triglycerides, still in the endoplasmic reticulum, aggregate into globules along with phospholipids and cholesterol and become coated with proteins. These masses are called **chylomicrons.** The protein coat keeps the chylomicrons suspended and from sticking to each other. The chylomicrons leave the epithelial cell and enter the lacteal of a villus. From here, they are transported by way of lymphatic vessels to the thoracic duct and enter the cardiovascular system at the left subclavian vein. Finally, they arrive at the liver through the hepatic artery (Figure 24-21).

Water

The total volume of fluid that enters the small intestine each day is about 9 liters (about 9 qt). This fluid is derived from ingestion of liquids (about 1.5 liters) and from various gastrointestinal secretions (about 7.5 liters). Roughly 8 to 8.5 liters of the fluid in the small intestine is absorbed; the remainder, about 0.5 to 1.0 liter, passes into the large intestine. There most of it is also absorbed.

The absorption of water by the small intestine occurs by osmosis from the lumen of the small intestine through epithelial cells and into the blood capillaries in the villi. The normal rate of absorption is about 200 to 400 ml/

hour. Water can move across the intestinal mucosa in both directions. The absorption of water from the small intestine is associated with the absorption of electrolytes and digested foods in order to maintain an osmotic balance with the blood.

Electrolytes

The electrolytes absorbed by the small intestine are mostly constituents of gastrointestinal secretions. Some are also components of ingested foods and liquids. Sodium is able to move in and out of epithelial cells by diffusion. It can also move into mucosal cells by active transport for removal from the small intestine. Chloride, iodide, and nitrate ions can passively follow sodium ions or be actively transported. Calcium ions are also actively transported, and their movement depends on parathyroid hormone (PTH) and vitamin D. Other electrolytes such as iron, potassium, magnesium, and phosphate also move by active transport.

Vitamins

Fat-soluble vitamins, such as A, D, E, and K, are absorbed along with ingested dietary fats in micelles. In fact, they cannot be absorbed unless they are ingested with some fat. Most water-soluble vitamins, such as the B vitamins and C, are absorbed by diffusion. Vitamin B_{12}, you may recall, requires combination with intrinsic factor produced by the stomach for its absorption.

A summary of the digestive and absorptive activities of the small intestine is presented in Exhibit 24-6.

EXHIBIT 24-6 SUMMARY OF DIGESTION AND ABSORPTION IN THE SMALL INTESTINE	
STRUCTURE	**ACTIVITY**
Pancreas	Delivers pancreatic juice into the duodenum via the pancreatic duct (see Exhibit 24-5 for pancreatic enzymes and their functions).
Liver	Produces bile which is necessary for emulsification of fats.
Gallbladder	Stores, concentrates, and delivers bile into the duodenum via the common bile duct.
Small Intestine	
Mucosa and submucosa	
Intestinal glands	Secrete intestinal juice (see Exhibit 24-5 for intestinal enzymes and their functions).
Duodenal (Brunner's) glands	Secrete mucus for protection and lubrication.
Microvilli	Fingerlike projections of epithelial cells that increase surface area for absorption and digestion.
Villi	Projections of mucosa that are the sites of absorption of digested food and also increase the surface area for absorption and digestion.
Plicae circulares	Circular folds of mucosa and submucosa that increase surface area for absorption and digestion.
Muscularis	
Segmentation	Consists of alternating contractions of circular fibers that produce segmentation and resegmentation of portions of the small intestine; mixes chyme with digestive juices and brings food into contact with the mucosa for absorption.
Peristalsis	Consists of mild waves of contraction and relaxation of circular and longitudinal muscle passing the length of the small intestine; moves chyme toward ileocecal valve.

LARGE INTESTINE

The overall functions of the large intestine are the completion of absorption, the manufacture of certain vitamins, the formation of feces, and the expulsion of feces from the body.

ANATOMY

The **large intestine** is about 1.5 m (5 ft) in length and averages 6.5 cm (2.5 inches) in diameter. It extends from the ileum to the anus and is attached to the posterior abdominal wall by its **mesocolon** of visceral peritoneum. Structurally, the large intestine is divided into four principal regions: cecum, colon, rectum, and anal canal (Figure 24-22a).

The opening from the ileum into the large intestine is guarded by a fold of mucous membrane called the **ileocecal sphincter (valve).** This structure allows materials from the small intestine to pass into the large intestine. Hanging below the ileocecal valve is the **cecum,** a blind pouch about 6 cm (2.5 inches) long. Attached to the cecum is a twisted, coiled tube, measuring about 8 cm (3 inches) in length, called the **vermiform appendix** (*vermis* = worm). The visceral peritoneum of the appendix, called the **mesoappendix,** attaches the appendix to the inferior part of the ileum and adjacent part of the posterior abdominal wall.

The open end of the cecum merges with a long tube called the **colon.** The colon is divided into ascending, transverse, descending, and sigmoid portions. The **ascending colon** ascends on the right side of the abdomen, reaches the undersurface of the liver, and turns abruptly to the left. Here it forms the **right colic (hepatic) flexure.** The colon continues across the abdomen to the left side as the **transverse colon.** It curves beneath the lower end of the spleen on the left side as the **left colic (splenic) flexure** and passes downward to the level of the iliac crest as the **descending colon.** The **sigmoid colon** begins at the left iliac crest, projects inward to the midline, and terminates as the rectum at about the level of the third sacral vertebra.

The **rectum,** the last 20 cm (8 inches) of the GI tract, lies anterior to the sacrum and coccyx. The terminal 2 to 3 cm (1 inch) of the rectum is called the **anal canal** (Figure 24-22b). The mucous membrane of the anal canal is arranged in longitudinal folds called **anal columns** that contain a network of arteries and veins. The opening of the anal canal to the exterior is called the **anus.** It is guarded by an internal sphincter of smooth muscle (involuntary) and an external sphincter of skeletal muscle (voluntary). Normally the anus is closed except during the elimination of the wastes of digestion.

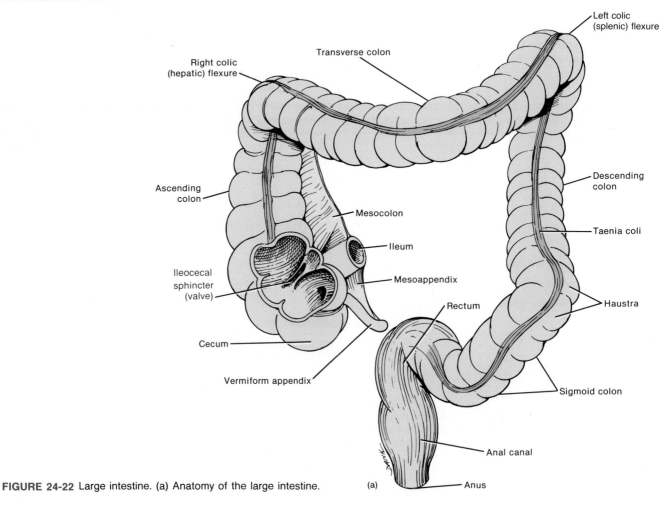

FIGURE 24-22 Large intestine. (a) Anatomy of the large intestine.

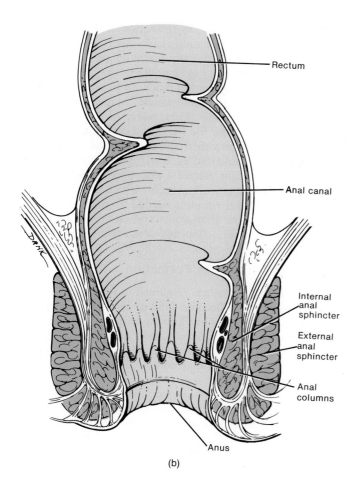

Rectum

Anal canal

Internal anal sphincter

External anal sphincter

Anal columns

Anus

(b)

The medical specialty that deals with the diagnosis and treatment of disorders of the rectum and anus is called **proctology** (*proct* = rectum; *logos* = study of).

CLINICAL APPLICATION

As noted in Chapter 21, varicosities in any veins involve inflammation and enlargement. Varacosities of the rectal veins are known as **hemorrhoids (piles).** Initially contained within the anus (first degree), they gradually enlarge until they prolapse or extend outward on defecation (second degree) and finally remain prolapsed through the anal orifice (third degree). Hemorrhoids may be caused by constipation. Repeated straining during defecation forces blood down into the superior hemorrhoidal plexus, increasing pressure in these veins. Constipation is related to low-fiber diets, especially in North America. One hypothesis links increased intraabdominal pressure, caused by straining during evacuation of firm feces, directly to hemorrhoids or varicose veins. The most widely used treatment for hemorrhoids that bleed consistently and cause severe pain and discomfort is *rubber band ligation,* in which a rubber band is tied around the hemorrhoid, cutting off its blood supply. In a few days, the hemorrhoid dries up and falls off. An even newer treatment is called *infrared photocoagulation,* in which a very high-energy light beam coagulates the hemorrhoid.

FIGURE 24-22 (*Continued*) (b) Anal canal seen in longitudinal section. (c) Histology of the large intestine. Shown is a photomicrograph of an enlarged aspect of the mucosa of the large intestine at a magnification of 140×. (Copyright © 1983 by Michael H. Ross. Used by permission.)

HISTOLOGY

The wall of the large intestine differs from that of the small intestine in several respects. No villi or permanent circular folds are found in the mucosa, which does, however, contain simple columnar epithelium with numerous goblet cells (Figure 24-22c). The columnar cells function primarily in water absorption. The goblet cells secrete

Simple columnar epithelium

Goblet cells

Muscularis mucosae

Opening of intestinal gland

Lamina propria

Intestinal glands

(c)

mucus that lubricates the colonic contents as they pass through the colon. Both columnar and mucous cells are located in long, straight, tubular intestinal glands that extend the full thickness of the mucosa. Solitary lymphatic nodules also are found in the mucosa. The submucosa of the large intestine is similar to that found in the rest of the alimentary canal. The muscularis consists of an external layer of longitudinal muscles and an internal layer of circular muscles. Unlike other parts of the digestive tract, portions of the longitudinal muscles are thickened, forming three conspicuous longitudinal bands referred to as **taeniae coli** (TĒ-nē-a KŌ-lī). Each band runs the length of most of the large intestine (Figure 24-22a). Tonic contractions of the bands gather the colon into a series of pouches called **haustra** (HAWS-tra), which give the colon its puckered appearance. The serosa of the large intestine is part of the visceral peritoneum. Small pouches of visceral peritoneum filled with fat are attached to taeniae coli and are called **epiploic appendages.**

DIGESTION IN THE LARGE INTESTINE

Mechanical

The passage of chyme from the ileum into the cecum is regulated by the action of the ileocecal valve. The valve normally remains mildly contracted so that the passage of chyme into the cecum is usually a slow process. Immediately following a meal, there is a **gastroileal reflex** in which ileal peristalsis is intensified and any chyme in the ileum is forced into the cecum. The hormone stomach gastrin also relaxes the valve. Whenever the cecum is distended, the degree of contraction of the ileocecal valve is intensified.

Movements of the colon begin when substances enter through the ileocecal valve. Since chyme moves through the small intestine at a fairly constant rate, the time required for a meal to pass into the colon is determined by gastric evacuation time. As food passes through the ileocecal valve, it fills the cecum and accumulates in the ascending colon.

One movement characteristic of the large intestine is **haustral churning.** In this process, the haustra remain relaxed and distended while they fill up. When the distension reaches a certain point, the walls contract and squeeze the contents into the next haustrum. **Peristalsis** also occurs, although at a slower rate than in other portions of the tract (3 to 12 contractions per minute). A final type of movement is **mass peristalsis,** a strong peristaltic wave that begins in about the middle of the transverse colon and drives the colonic contents into the rectum. Food in the stomach initiates this reflex action in the colon. Thus mass peristalsis usually takes place three or four times a day, during a meal or immediately after.

Chemical

The last stage of digestion occurs through bacterial, not enzymatic, action. Mucus is secreted by the glands of the large intestine, but no enzymes are secreted. Chyme is prepared for elimination by the action of bacteria. These bacteria ferment any remaining carbohydrates and release hydrogen, carbon dioxide, and methane gas. These gases contribute to flatus (gas) in the colon. They also convert remaining proteins to amino acids and break down the amino acids into simpler substances: indole, skatole, hydrogen sulfide, and fatty acids. Some of the indole and skatole is carried off in the feces and contributes to their odor. The rest are absorbed and transported to the liver, where they are converted to less toxic compounds and excreted in the urine. Bacteria also decompose bilirubin to simpler pigments (urobilinogen), which give feces their brown color. Several vitamins needed for normal metabolism, including some B vitamins and vitamin K, are synthesized by bacterial action and absorbed.

Absorption and Feces Formation

By the time the chyme has remained in the large intestine 3 to 10 hours, it has become solid or semisolid as a result of absorption and is now known as **feces.** Chemically, feces consist of water, inorganic salts, sloughed off epithelial cells from the mucosa of the alimentary canal, bacteria, products of bacterial decomposition, and undigested parts of food.

Although most water absorption occurs in the small intestine, the large intestine absorbs enough to make it an important organ in maintaining the body's water balance. Of the 0.5 to 1.0 liter that enters the large intestine, all but about 100 ml is absorbed. The absorption is greatest in the cecum and ascending colon. The large intestine also absorbs electrolytes, including sodium and chloride.

Defecation

Mass peristaltic movements push fecal material from the sigmoid colon into the rectum. The resulting distension of the rectal wall stimulates pressure-sensitive receptors, initiating a reflex for **defecation,** the emptying of the rectum. The defecation reflex occurs as follows. In response to distension of the rectal wall, the receptors send impulses to the sacral spinal cord. Motor impulses from the cord travel along parasympathetic nerves back to the descending colon, sigmoid colon, rectum, and anus. Contraction of the longitudinal rectal muscles shortens the rectum, thereby increasing the pressure inside it. The pressure along with voluntary contractions of the diaphragm and abdominal muscles forces the internal sphincter open, and the feces are expelled through the anus. The external sphincter is voluntarily controlled. If it is voluntarily relaxed, defecation occurs; if it is voluntarily constricted, defecation can be postponed. Voluntary contractions of the diaphragm and abdominal muscles aid defecation by increasing the pressure inside the abdomen, which pushes the walls of the sigmoid colon and rectum inward. If defecation does not occur, the feces back into the sigmoid colon until the next wave of mass peristalsis again stimulates the pressure-sensitive receptors, creating the desire to defecate.

In infants, the defecation reflex causes automatic emptying of the rectum without the voluntary control of the external anal sphincter. In certain instances of spinal cord injury, the reflex is abolished and defecation requires supportive measures, such as cathartics (laxatives).

CLINICAL APPLICATION

Diarrhea refers to frequent defecation of liquid feces caused by increased motility of the intestines. Since chyme passes too quickly through the small intestine and feces pass too quickly through the large intestine, there is not enough time for absorption. Like vomiting, diarrhea can result in dehydration and electrolyte imbalances. Diarrhea may be caused by stress and microbes that irritate the gastrointestinal mucosa.

Constipation refers to infrequent or difficult defecation. It is caused by decreased motility of the intestines in which feces remain in the colon for prolonged periods of time. As it does so, there is considerable water absorption and feces become dry and hard. Constipation may be caused by improper bowel habits, spasms of the colon, insufficient bulk in the diet, lack of exercise, and emotions. Usual treatment for constipation is a mild cathartic (laxative) that induces defecation.

Activities of the large intestine are summarized in Exhibit 24-7.

AGING AND THE DIGESTIVE SYSTEM

Overall general changes associated with aging of the digestive system include decreasing secretory mechanisms, decreasing motility (muscular movement) of the digestive organs, loss of strength and tone of the muscular tissue and its supporting structures, changes in neurosensory feedback on enzyme and hormone release, and the diminished response to pain and internal sensations. Specific changes include reduced sensitivity to mouth irritations and sores, loss of taste, pyorrhea, difficulty in swallowing, hiatal hernia, cancer of the esophagus, gastritis, peptic ulcer, and gastric cancer. Changes in the small intestine include duodenal ulcers, appendicitis, malabsorption, and maldigestion. Other pathologies that increase in incidence are gallbladder problems, jaundice, cirrhosis, and acute pancreatitis. Large-intestinal changes such as constipation, cancer of the colon or rectum, hemorrhoids, and diverticular disease of the colon also occur.

DEVELOPMENTAL ANATOMY OF THE DIGESTIVE SYSTEM

About the fourteenth day after fertilization, the cells of the endoderm form a cavity referred to as the **primitive gut** (Figure 24-23). Soon after the mesoderm forms and splits into two layers (somatic and splanchnic), the splanchnic mesoderm associates with the endoderm of the primitive gut. Thus, the primitive gut has a double-layered wall. The **endodermal layer** gives rise to the *epithelial lining* and *glands* of most of the gastrointestinal tract, and the **mesodermal layer** produces its *smooth muscle* and *connective tissue*.

The primitive gut elongates, and about the latter part of the third week, it differentiates into an anterior **foregut**, a central **midgut**, and a posterior **hindgut**. Until the fifth week of development, the midgut opens into the yolk sac. After that time, the yolk sac constricts, detaches from the midgut, and the midgut seals. In the region of the foregut, a depression consisting of **ectoderm**, the **stomodeum**, appears. This develops into the *oral cavity*. The **oral membrane** that separates the foregut from the stomodeum ruptures during the fourth week of development, so that the foregut is continuous with the outside of the embryo through the oral cavity. Another depression consisting of **ectoderm**, the **proctodeum**, forms in the hindgut and goes on to develop into the *anus*. The **cloacal membrane**, which separates the hindgut from the proctodeum, ruptures, so that the hindgut is continuous with the outside of the embryo through the anus. Thus, the digestive tract forms a continuous tube from the mouth to the anus.

The foregut develops into the *pharynx, esophagus, stomach,* and a *portion of the duodenum*. The midgut

---

EXHIBIT 24-7   DIGESTIVE ACTIVITIES OF LARGE INTESTINE

STRUCTURE	ACTION	FUNCTION
Mucosa	Secretes mucus.	Lubricates colon and protects mucosa.
	Absorbs water and other soluble compounds.	Maintains water balance; solidifies feces. Vitamins and electrolytes absorbed and toxic substances sent to liver to be detoxified.
	Bacterial activity.	Breaks down undigested carbohydrates, proteins, and amino acids into products that can be expelled in feces or absorbed and detoxified by liver. Certain B vitamins and vitamin K synthesized.
Muscularis	Haustral churning.	Contents moved from haustrum to haustrum by muscular contractions.
	Peristalsis.	Contents moved along length of colon by contractions of circular and longitudinal muscles.
	Mass peristalsis.	Contents forced into sigmoid colon and rectum by strong peristaltic wave.
	Defecation.	Feces eliminated by contractions in sigmoid colon and rectum.

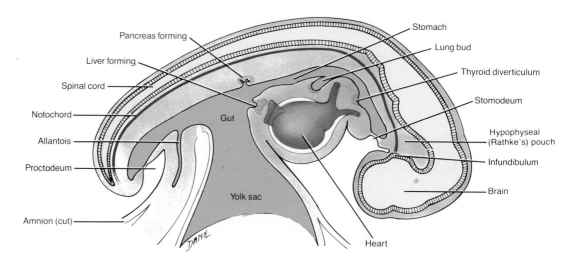

FIGURE 24-23 Development of the digestive system.

is transformed into the *remainder of the duodenum,* the *jejunum,* the *ileum,* and *portions of the large intestine* (cecum, appendix, ascending colon, and most of the transverse colon). The hindgut develops into the *remainder of the large intestine,* except for a portion of the anal canal that is derived from the proctodeum.

As development progresses, the endoderm at various places along the foregut develops into hollow buds that grow into the mesoderm. These buds will develop into the *salivary glands, liver, gallbladder,* and *pancreas.* Each of the glands retains a connection with the gastrointestinal tract via a duct.

DISORDERS: HOMEOSTATIC IMBALANCES

Dental Caries

Dental caries, or tooth decay, involve a gradual demineralization (softening) of the enamel and dentin. If untreated, various microorganisms may invade the pulp, causing inflammation and infection with subsequent death (necrosis) of the pulp and abscess of the alveolar bone surrounding the root's apex. Such teeth are treated by root canal therapy.

The process of dental caries is initiated when bacteria act on sugars, giving off acids that demineralize the enamel. Microbes that digest sugar into lactic acid are common in the mouth cavity. One that seems to be cariogenic (caries causing) is the bacterium *Streptococcus mutans.* **Dextran,** a sticky polysaccharide produced from sucrose, forms a capsule around the bacteria causing them to stick to the teeth. Masses of bacterial cells, dextran, and other debris adhering to teeth are collectively called **dental plaque.** Saliva cannot reach the tooth surface to buffer the acid because the plaque covers the teeth. Brushing the teeth immediately after eating removes the plaque from flat surfaces before the bacteria have a chance to go to work. Dentists also suggest that the plaque between the teeth be removed every 24 hours with dental floss or by flushing with a water irrigation device.

Preventive measures other than brushing, flossing, and irrigation include prenatal diet supplements (chiefly vitamin D, calcium, and phosphorus), fluoride treatments to protect against acids during the period when teeth are being calcified, and dental sealing. In this last procedure, pits and fissures that serve as reservoirs for dental plaque are sealed by the application of a permanent, durable plastic sealant. The sealant is applied to the prepared biting surfaces of the molar teeth. Dental sealants are used primarily for children and adolescents, although some adults could benefit from them.

Periodontal Disease

Periodontal disease is a collective term for a variety of conditions characterized by inflammation and degeneration of the gingivae, alveolar bone, periodontal ligament, and cementum. One such condition is called **pyorrhea.** The initial symptoms are enlargement and inflammation of the soft tissue and bleeding gums. Without treatment, the soft tissue may deteriorate and the alveolar bone may be resorbed, causing loosening of the teeth and recession of the gums.

Periodontal diseases are frequently caused by poor oral hygiene; local irritants, such as bacteria, impacted food, and cigarette smoke; or by a poor "bite." The latter may put a strain on the tissues supporting the teeth. Periodontal diseases may also be caused by allergies, vitamin deficiencies (especially vitamin C), and a number of systemic disorders, especially those that affect bone, connective tissue, or circulation.

Peritonitis

Peritonitis is an acute inflammation of the serous membrane lining the abdominal cavity and covering the abdominal viscera. One possible cause is contamination of the peritoneum by pathogenic bacteria from the external environment. This contamination could result from accidental or surgical wounds in the abdominal wall or from perforation or rupture of organs with consequent exposure to the outside environment. Another possible cause is perforation of the walls of organs that contain bacteria or chemicals beneficial to the organ but toxic to the peritoneum. For example, the large intestine contains colonies of bacteria that live on undigested nutrients and break them down so they can be eliminated. But if the bacteria enter the peritoneal cavity, they attack the cells of the peritoneum for food and produce acute infection. Moreover, the peritoneum has no natural barriers that keep it from being irritated or digested by chemical substances such as bile and digestive enzymes.

Although it contains a great deal of lymphatic tissue and can combat infection fairly well, the peritoneum is in contact with most of the abdominal organs. If infection gets out of hand, it may destroy vital organs and bring on death. For these reasons, perforation of the alimentary canal from an ulcer is considered serious. A surgeon planning to do extensive surgery on the colon may give the patient high doses of antibiotics for several days prior to surgery to kill intestinal bacteria and reduce the risk of peritoneal contamination.

Peptic Ulcers

An **ulcer** is a craterlike lesion in a membrane. Ulcers that develop in areas of the alimentary canal exposed to acid gastric juice are called **peptic ulcers.** Peptic ulcers occasionally develop in the lower end of the esophagus, but most occur on the lesser curvature of the stomach, where they are called **gastric ulcers,** or in the first part of the duodenum, where they are called **duodenal ulcers.** Most peptic ulcers are duodenal.

Hypersecretion of acid gastric juice seems to be the immediate cause of duodenal ulcers. In gastric ulcer patients, because the stomach wall is highly adapted to resist gastric juice through the secretion of mucus, the cause may be hyposecretion of mucus. Hypersecretion of pepsin also may contribute to ulcer formation.

Among the factors believed to stimulate an increase in acid secretion are emotions, certain foods or medications (alcohol, coffee, aspirin), and overstimulation of the vagus (X) nerve. Normally, the mucous membrane lining the stomach and duodenal walls resists the secretions of hydrochloric acid and pepsin. In some people, however, this resistance breaks down and an ulcer develops.

The most common complication of peptic ulcers is bleeding. Another is perforation, erosion of the ulcer all the way through the wall of the stomach or duodenum. Perforation allows bacteria and partially digested food to pass into the peritoneal cavity, producing peritonitis. A third complication is obstruction.

Appendicitis

Appendicitis is an inflammation of the vermiform appendix. It is preceded by obstruction of the lumen of the apppendix by fecal material, inflammation, a foreign body, carcinoma of the cecum, stenosis, or kinking of the organ. The infection that follows may result in edema, ischemia, gangrene, and perforation. Rupture of the appendix develops into peritonitis. Loops of the intestines, the omentum, and the parietal peritoneum may become adherent and form an abscess, either at the site of the appendix or elsewhere in the abdominal cavity.

Typically, appendicitis begins with referred pain in the umbilical region of the abdomen followed by anorexia (lack or loss of appetite for food), nausea, and vomiting. After several hours, the pain localizes in the right lower quadrant (RLQ) and is continuous, dull or severe, and intensified by coughing, sneezing, or body movements.

Early appendectomy (removal of the appendix) is recommended in all suspected cases because it is safer to operate than to risk gangrene, rupture, and peritonitis. Appendectomy may be performed through a muscle-splitting incision in the right lower quadrant (RLQ) in which the cecum is brought into the incision. The base of the appendix is tied, the appendix is excised, and the stump is usually cauterized and then invaginated into the cecum.

Tumors

Both benign and malignant **tumors** can occur in all parts of the gastrointestinal tract. The benign growths are much more common, but malignant tumors are responsible for 30 percent of all deaths from cancer in the United States. Carcinoma of the colon and rectum is one of the most common malignant diseases, ranking second to that of the lungs in males and lungs and breasts in females. Over 50 percent of colorectal cancers occur in the rectum and sigmoid colon.

Malignant tumors of the rectum may be detected by *digital rectal examination.* The easiest method for detecting cancer of the colon is *fecal occult blood testing,* in which a sample of stool is tested for the presence of occult (hidden) blood, a sign that a malignant tumor might be present. Direct visualization of the rectum and sigmoid colon for the detection of carcinomas is possible by using a *flexible fiberoptic sigmoidoscope.* By using a *flexible fiberoptic endoscope* a physician can visualize the entire gastrointestinal tract. The endoscopic examination of the colon is known as *colonoscopy.* Both the sigmoidoscope and endoscope can also be used to magnify, photograph, biopsy, and remove polyps.

Another test in a routine examination for intestinal disorders is the filling of the gastrointestinal tract with barium, which is either swallowed or given in an enema. Barium, a mineral, shows up on x rays the same way that calcium appears in bones. Tumors as well as ulcers can be diagnosed this way. The only definitive treatment of gastrointestinal carcinomas, if they cannot be removed using the endoscope, is surgery.

Diverticulitis

Diverticula are saclike outpouchings of the wall of the colon in places where the muscularis has become weak. The development of diverticula is called **diverticulosis.** Many people who develop diverticulosis are asymptomatic and experience no complications. About 15 percent of people with diverticulosis will eventually develop an inflammation within diverticula, a condition known as **diverticulitis.**

Research indicates that diverticula form because of lack of sufficient bulk in the colon during segmentation. The powerful contractions, working against insufficient bulk, create a pressure so high that it causes the colonic walls to blow out.

The increase in diverticular disease has been attributed to a shift to low-fiber diets. Patients treated for diverticular disease with high-fiber diets show marked relief of symptoms.

Treatment consists of bed rest, cleansing enemas, and drugs to reduce infection. In severe cases, portions of the affected colon may require surgical removal and temporary colostomy.

Cirrhosis

Cirrhosis refers to a distorted or scarred liver as a result of chronic inflammation. The parenchymal (functional) liver cells are replaced by fibrous or adipose connective tissue, a process called stromal repair. The liver has a high capacity for parenchymal regeneration, and stromal repair occurs whenever a parenchymal cell is killed or cells are damaged continuously for a long time. The symptoms of cirrhosis include jaundice, edema in the legs, uncontrolled bleeding, and increased sensitivity to drugs. Cirrhosis may be caused by hepatitis (inflammation of the liver), certain chemicals that destroy liver cells, parasites that infect the liver, and alcoholism.

Hepatitis

Hepatitis refers to inflammation of the liver and can be caused by viruses, drugs, and chemicals, including alcohol. Clinically several types are recognized.

Hepatitis A (infectious hepatitis) is caused by hepatitis A virus and is spread by fecal contamination of food, clothing, toys, eating utensils, and so forth (fecal-oral route). It is generally a mild disease of children and young adults characterized by anorexia, malaise, nausea, diarrhea, fever, and chills. Eventually jaundice appears. It does not cause lasting liver damage. Most people recover in 4 to 6 weeks.

Hepatitis B (serum hepatitis) is caused by hepatitis B virus and is spread primarily by contaminated syringes and transfusion equipment. It can also be spread by any secretion of fluid by the body (tears, saliva, semen). Hepatitis B can produce chronic liver inflammation and can persist for years or even a lifetime. Persons who harbor the active hepatitis B virus are at risk for cirrhosis and also become carriers.

Non-A, non-B (NANB) hepatitis is a form of hepatitis that cannot be traced to either hepatitis A or hepatitis B viruses. It is clinically similar to hepatitis B and is often spread by blood transfusions. It is believed to account for considerably more posttransfusion hepatitis than that related to hepatitis B.

Gallstones

The cholesterol in bile may crystallize at any point between bile canaliculi, where it is first apparent, and the hepatopancreatic ampulla, where the bile enters the duodenum. The fusion of single crystals is the beginning of 95 percent of all **gallstones (biliary calculi).** Following their formation, gallstones gradually grow in size and number and may cause minimal, intermittent, or complete obstruction to the flow of bile from the gallbladder into the duct system. If obstruction of the outlet occurs and the gallbladder cannot empty as it normally does after eating, the pressure within it increases and the individual may have intense pain or discomfort **(biliary colic).** Complete obstruction of the flow of bile into the duodenum may result in death.

Treatment of gallstones consists of using gallstone-dissolving drugs or surgery. One such drug is methyl tert-butyl ether (MTBE), a cholesterol solvent which, under local anesthesia, is pumped in and out of the gallbladder through a catheter to dissolve the gallstones. For individuals in which the procedure is not indicated, removal of the gallbladder and its contents (cholecystectomy) is necessary.

Anorexia Nervosa

Anorexia nervosa is a disorder characterized by loss of appetite and bizarre patterns of eating. The subconsciously self-imposed starvation appears to be a response to emotional conflicts about self-identification and acceptance of a normal adult sex role. The disorder is found predominantly in young, single females. The physical consequence of the disorder is severe and progressive starvation. Amenorrhea (absence of menstruation) and a lowered basal metabolic rate reflect the depressant effects of the starvation. Individuals may become emaciated and may ultimately die of starvation or one of its complications. Also associated with the disorder are osteoporosis, depression, brain abnormalities coupled with impaired mental performance, and a tendency to frequently void large amounts of urine. Treatment consists of psychotherapy and dietary regulation.

Bulimia

A disorder that typically affects single, middle-class, young, white females is known as **bulimia** (*bous* = ox; *limos* = hunger), or **binge-purge syndrome.** It is characterized by uncontrollable overeating followed by forced vomiting or overdoses of laxatives. This binge-purge cycle occurs in response to fears of being overweight, stress, depression, and physiological disorders such as hypothalamic tumors.

Bulimia can upset the body's electrolyte balance and increase susceptibility to flu, salivary gland infections, dry skin, acne, muscle spasms, loss of hair, kidney and liver diseases, tooth decay, ulcers, hernias, constipation, and hormone imbalances. Treatment of bulimia combines nutrition counseling, psychotherapy, and medical treatment.

Dietary Fiber (Roughage) and GI Disorders

A deficiency in our diet that has received recent attention is lack of dietary fiber or roughage. Dietary fiber consists of indigestible substances, such as cellulose, lignin, and pectin, found in fruits, vegetables, grains, and beans. The bulk of our intake consists of extensively purified carbohydrates—mainly starches and sugars—as well as substantial amounts of fats and oils. People who choose a fiber-rich, unrefined diet will greatly reduce their chances of developing disease of overnutrition (such as obesity, diabetes, gallstones, and coronary heart disease), diseases of the underworked mouth (caries and periodontal disease), and diseases of an underfed large bowel (constipation, varicose veins, hemorrhoids, colon spasm, diverticulitis, appendicitis, and large intestinal cancer). Each of these conditions is directly related to the digestion and metabolism of food and the operation of the digestive system. The daily addition of a small amount of bran or other foods high in fiber would overcome the roughage deficiency.

MEDICAL TERMINOLOGY

Botulism (*botulus* = sausage) A type of food poisoning caused by a toxin produced by *Clostridium botulinum.* The bacterium is ingested in improperly cooked or preserved foods. The toxin inhibits nerve impulse conduction at synapses by inhibiting the release of acetylcholine. Symptoms include paralysis, nausea, vomiting, blurred or double vision, difficulty in speech, difficulty in swallowing, dryness of the mouth, and general weakness.

Cholecystitis (*chole* = bile; *kystis* = bladder; *itis* = inflammation of) Inflammation of the gallbladder that often leads to infection. Some cases are caused by obstruction of the cystic duct with bile stones. Stagnating bile salts irritate the mucosa. Dead mucosal cells provide a medium for the growth of bacteria.

Cholelithiasis (*lithos* = stone) The presence of gallstones.

Colitis Inflammation of the colon and rectum. Inflammation of the mucosa reduces absorption of water and salts, producing watery, bloody feces, and, in severe cases, dehydration and salt depletion. Irritated muscularis spasms produce cramps.

Colostomy (*stomoun* = provide an opening) An incision of the colon to create an artificial opening or "stoma" to the exterior. This opening serves as a substitute anus through which feces are eliminated. A temporary colostomy may be done to allow a badly inflamed colon to rest and heal. If the rectum is removed for malignancy, the colostomy provides a permanent outlet for feces.

Dysphagia (*dys* = abnormal; *phagein* = to eat) Difficulty in swallowing that may be caused by inflammation, paralysis, obstruction, or trauma.

Enteritis (*enteron* = intestine) An inflammation of the intestine, particularly the small intestine.

Flatus Excessive amounts of air (gas) in the stomach or intestine, usually expelled through the anus. If the gas is expelled through the mouth, it is called **eructation** or **belching** (burping). Flatus may result from gas released during the breakdown of foods in the stomach or from swallowing air or gas-containing substances such as carbonated drinks.

Gastrectomy (*gastro* = stomach; *tome* = excision) Removal of a portion of or the entire stomach.

Hernia Protrusion of an organ or part of an organ through a membrane or cavity wall, usually the abdominal cavity. *Diaphragmatic (hiatal) hernia* is the protrusion of the lower esophagus, stomach, or intestine into the thoracic cavity through the opening in the diaphragm (esophageal hiatus) that allows passage of the esophagus. *Umbilical hernia* is the protrusion of abdominal organs through the navel area of the abdominal wall. *Inguinal hernia* is the protrusion of the hernial sac containing the intestine into the inguinal opening. It may extend into the scrotal compartment, causing strangulation of the herniated part.

Inflammatory bowel disease (IBD) Disorder that exists in two forms: (1) Crohn's disease (inflamed intestine) and (2) ulcerative colitis (inflammation of the colon consisting of ulcerations and usually accompanied by rectal bleeding).

Irritable bowel syndrome Disease of the entire gastrointestinal tract characterized by abnormal muscular contractions, especially a spastic colon, excessive mucus in stools, and alternating diarrhea and constipation.

Nausea (*nausia* = seasickness) Discomfort preceding vomiting. Possibly, it is caused by distension or irritation of the gastrointestinal tract, most commonly the stomach.

Pancreatitis Inflammation of the pancreas. The pancreas secretes active trypsin instead of trypsinogen, and the trypsin digests the pancreatic cells and blood vessels.

STUDY OUTLINE

Regulation of Food Intake (p. 590)

1. Two centers in the hypothalamus related to regulation of food intake are the feeding center and satiety center; the feeding center is constantly active, but may be inhibited by the satiety center.

2. Among the stimuli that affect the feeding and satiety centers are glucose, amino acids, lipids, body temperature, distension, and cholecystokinin (CCK).

Digestive Processes (p. 590)
1. Food is prepared for use by cells by five basic activities: ingestion, movement, mechanical and chemical digestion, absorption, and defecation.
2. Chemical digestion is a series of catabolic reactions that break down the large carbohydrate, lipid, and protein molecules of food into molecules that are usable by body cells.
3. Mechanical digestion consists of movements that aid chemical digestion.
4. Absorption is the passage of end products of digestion from the digestive tract into blood or lymph for distribution to cells.
5. Defecation is emptying of the rectum.

Organization (p. 590)
1. The organs of digestion are usually divided into two main groups: those composing the gastrointestinal (GI) tract, or alimentary canal, and accessory structures.
2. The GI tract is a continuous tube running through the ventral body cavity from the mouth to the anus.
3. The accessory structures include the teeth, tongue, salivary glands, liver, gallbladder, and pancreas.
4. The basic arrangement of tissues in the alimentary canal from the inside outward is the mucosa, submucosa, muscularis, and serosa (peritoneum).
5. Extensions of the peritoneum include the mesentery, mesocolon, falciform ligament, lesser omentum, and greater omentum.

Mouth (Oral Cavity) (p. 595)
1. The mouth is formed by the cheeks, palates, lips, and tongue, which aid mechanical digestion.
2. The vestibule is the space between the cheeks and lips and teeth and gums.
3. The oral cavity proper extends from the vestibule to the fauces.

Tongue (p. 596)
1. The tongue, together with its associated muscles, forms the floor of the oral cavity. It is composed of skeletal muscle covered with mucous membrane.
2. The upper surface and sides of the tongue are covered with papillae. Some papillae contain taste buds.

Salivary Glands (p. 596)
1. The major portion of saliva is secreted by the salivary glands, which lie outside the mouth and pour their contents into ducts that empty into the oral cavity.
2. There are three pairs of salivary glands: the parotid, submandibular (submaxillary), and sublingual glands.
3. Saliva lubricates food and starts the chemical digestion of carbohydrates.
4. Salivation is entirely under nervous control.

Teeth (p. 598)
1. The teeth, or dentes, project into the mouth and are adapted for mechanical digestion.
2. A typical tooth consists of three principal portions: crown, root, and cervix.
3. Teeth are composed primarily of dentin covered by enamel, the hardest substance in the body.
4. There are two dentitions—deciduous and permanent.

Digestion in the Mouth (p. 599)
1. Through mastication food is mixed with saliva and shaped into a bolus.
2. Salivary amylase converts polysaccharides (starches) to disaccharides (maltose).

Deglutition (p. 600)
1. Deglutition or swallowing moves a bolus from the mouth to the stomach.
2. It consists of a voluntary stage, pharyngeal stage (involuntary) and esophageal stage (involuntary).

Esophagus (p. 600)
1. The esophagus is a collapsible, muscular tube that connects the pharynx to the stomach.
2. It passes a bolus into the stomach by peristalsis.
3. It contains an upper and lower esophageal sphincter.

Stomach (p. 603)

Anatomy; Histology (p. 603; p. 605)
1. The stomach begins at the bottom of the esophagus and ends at the pyloric sphincter.
2. The anatomic subdivisions of the stomach include the cardia, fundus, body, and pylorus.
3. Adaptations of the stomach for digestion include rugae; glands that produce mucus, hydrochloric acid, a protein-digesting enzyme, intrinsic factor, and stomach gastrin; and a three-layered muscularis for efficient mechanical movement.

Digestion in the Stomach (p. 605)
1. Mechanical digestion consists of mixing waves.
2. Chemical digestion consists of the conversion of proteins into peptides by pepsin.

Regulation of Gastric Secretion (p. 607)
1. Gastric secretion is regulated by nervous and hormonal mechanisms.
2. Stimulation occurs in three phases: cephalic (reflex), gastric, and intestinal.

Regulation of Gastric Emptying (p. 608)
1. Gastric emptying is stimulated in response to distension, and stomach gastrin released in response to the presence of certain types of foods.
2. Gastric emptying is inhibited by the enterogastric reflex and hormones (secretin, CCK, and GIP).

Absorption (p. 609)
1. The stomach wall is impermeable to most substances.
2. Among the substances absorbed are some water, certain electrolytes and drugs, and alcohol.

Pancreas (p. 609)
1. The pancreas is divisible into a head, body, and tail, and is connected to the duodenum via the pancreatic duct (duct of Wirsung) and accessory duct (duct of Santorini).
2. Pancreatic islets (islets of Langerhans) secrete hormones, and acini secrete pancreatic juice.
3. Pancreatic juice contains enzymes that digest starch (pancreatic amylase), proteins (trypsin, chymotrypsin, and carboxypeptidase), fats (pancreatic lipase), and nucleotides (nucleases).
4. Pancreatic secretion is regulated by nervous and hormonal mechanisms.

Liver (p. 611)

1. The liver is divisible into left and right lobes; associated with the right lobe are the caudate and quadrate lobes.
2. The lobes of the liver are made up of lobules that contain hepatocytes (liver cells), sinusoids, stellate reticuloendothelial (Kupffer's) cells, and a central vein.
3. Hepatocytes of the liver produce bile that is transported by a duct system to the gallbladder for storage.
4. Bile's contribution to digestion is the emulsification of neutral fats.
5. The liver also functions in producing heparin and plasma proteins, phagocytosis, detoxification, conversion of nutrients, and storage of minerals, vitamins, and glycogen.
6. Bile secretion is regulated by nervous and hormonal mechanisms.

Gallbladder (p. 614)

1. The gallbladder is a sac located in a fossa on the visceral surface of the liver.
2. The gallbladder stores and concentrates bile.
3. Bile is ejected into the common bile duct under the influence of cholecystokinin (CCK).

Small Intestine (p. 614)

Anatomy; Histology (p. 614)

1. The small intestine extends from the pyloric sphincter to the ileocecal sphincter.
2. It is divided into duodenum, jejunum, and ileum.
3. It is highly adapted for digestion and absorption. Its glands produce enzymes and mucus, and the microvilli, villi, and plicae circulares of its wall provide a large surface area for digestion and absorption.
4. Intestinal enzymes break down foods inside epithelial cells of the mucosa.

Intestinal Juice; Digestion (p. 616; p. 617)

1. Intestinal enzymes break down maltose to glucose (maltase), sucrose to glucose and fructose (sucrase), lactose to glucose and galactose (lactase), terminal amino acids at the amino ends of peptides (aminopeptidase), dipeptides to amino acids (dipeptidase), and nucleotides to pentoses and nitrogen bases (nucleases).
2. Mechanical digestion in the small intestine involves segmentation and peristalsis.

Regulation of Intestinal Secretion (p. 618)

1. The most important mechanism is local reflexes.
2. Hormones also assume a role.

Absorption (p. 618)

1. Absorption is the passage of the end products of digestion from the alimentary canal into the blood or lymph.
2. Monosaccharides, amino acids, and short-chain fatty acids pass into the blood capillaries.
3. Long-chain fatty acids and monoglycerides are absorbed as part of micelles, resynthesized to triglycerides, and transported as chylomicrons.
4. Chylomicrons are taken up by the lacteal of a villus.
5. The small intestine also absorbs water, electrolytes, and vitamins.

Large Intestine (p. 622)

Anatomy; Histology (p. (p. 622; p. 623)

1. The large intestine extends from the ileocecal sphincter to the anus.

2. Its subdivisions include the cecum, colon, rectum, and anal canal.
3. The mucosa contains numerous goblet cells, and the muscularis consists of taeniae coli.

Digestion in the Large Intestine (p. 624)

1. Mechanical movements of the large intestine include haustral churning, peristalsis, and mass peristalsis.
2. The last stages of chemical digestion occur in the large intestine through bacterial, rather than enzymatic, action. Substances are further broken down and some vitamins are synthesized.

Absorption and Feces Formation (p. 624)

1. The large intestine absorbs water, electrolytes, and vitamins.
2. Feces consists of water, inorganic salts, epithelial cells, bacteria, and undigested foods.

Defecation (p. 624)

1. The elimination of feces from the large intestine is called defecation.
2. Defecation is a reflex action aided by voluntary contractions of the diaphragm and abdominal muscles.

Aging and the Digestive System (p. 625)

1. General changes include decreased secretory mechanisms, decreased motility, and loss of tone.
2. Specific changes include loss of taste, pyorrhea, hernias, ulcers, constipation, hemorrhoids, and diverticular diseases.

Developmental Anatomy of the Digestive System (p. 625)

1. The endoderm of the primitive gut forms the epithelium and glands of most of the gastrointestinal tract.
2. The mesoderm of the primitive gut forms the smooth muscle and connective tissue of the gastrointestinal tract.

Disorders: Homeostatic Imbalances (p. 627)

1. Dental caries are started by acid-producing bacteria that reside in dental plaque.
2. Periodontal diseases are characterized by inflammation and degeneration of gingivae, alveolar bone, periodontal membrane, and cementum.
3. Peritonitis is inflammation of the peritoneum.
4. Peptic ulcers are craterlike lesions that develop in the mucous membrane of the alimentary canal in areas exposed to gastric juice.
5. Appendicitis is an inflammation of the vermiform appendix resulting from obstruction of the lumen of the appendix by inflammation, a foreign body, carcinoma of the cecum, stenosis, or kinking of the organ.
6. Tumors of the gastrointestinal (GI) tract may be detected by sigmoidoscopy, colonoscopy, and barium x ray.
7. Diverticulitis is inflammation of diverticula in the colon.
8. Cirrhosis is a condition in which parenchymal cells of the liver damaged by chronic inflammation are replaced by fibrous or adipose connective tissue.
9. Hepatitis is an inflammation of the liver. Types include hepatitis A; hepatitis B; non-A, non-B (NANB) hepatitis.
10. The fusion of individual crystals of cholesterol is the beginning of 95 percent of all gallstones. Gallstones can cause obstruction to the outflow of bile in any portion of the duct system.
11. Anorexia nervosa is a disorder characterized by loss of appetite.
12. Bulimia is a binge-purge syndrome of behavior in which uncontrollable overeating is followed by forced vomiting or overdoses of laxatives.

REVIEW QUESTIONS

1. Describe how food intake is regulated.
2. Define digestion. Distinguish between chemical and mechanical digestion.
3. Identify the organs of the gastrointestinal (GI) tract in sequence. How does the gastrointestinal (GI) tract differ from the accessory structures of digestion?
4. Describe the structure of each of the four coats of the gastrointestinal (GI) tract.
5. What is the peritoneum? Describe the location and function of the mesentery, mesocolon, falciform ligament, lesser omentum, and greater omentum.
6. What structures form the oral cavity?
7. Make a simple diagram of the tongue. Indicate the location of the papillae and the four taste zones. What is ankyloglossia?
8. Describe the location of the salivary glands and their ducts. What are buccal glands?
9. How are salivary glands distinguished histologically?
10. Describe the composition of saliva and the role of each of its components in digestion. What is the pH of saliva?
11. How is salivary secretion regulated?
12. What are the principal portions of a typical tooth? What are the functions of each part?
13. Compare deciduous and permanent dentitions with regard to number of teeth and time of eruption.
14. Contrast the functions of incisors, cuspids, premolars, and molars.
15. What is a bolus? How is it formed?
16. Define deglutition. List the sequence of events involved in passing a bolus from the mouth to the stomach. Be sure to discuss the voluntary, pharyngeal, and esophageal stages of swallowing.
17. Describe the location and histology of the esophagus. What is its role in digestion?
18. Explain the operation of the upper and lower esophageal sphincters. How is achalasia related to heartburn?
19. Describe the location of the stomach. List and briefly explain the anatomical features of the stomach.
20. Distinguish between pyloric stenosis and pylorospasm.
21. What is the importance of rugae, zymogenic cells, parietal cells, mucous cells, and enteroendocrine cells in the stomach?
22. Describe mechanical digestion in the stomach.
23. What is the role of pepsin? Why is it secreted in an inactive form?
24. Outline the factors that stimulate and inhibit gastric secretion. Be sure to discuss the cephalic, gastric, and intestinal phases.
25. How is gastric emptying stimulated and inhibited?
26. Explain how vomiting occurs.
27. Describe the role of the stomach in absorption.
28. Where is the pancreas located? Describe the duct system connecting the pancreas to the duodenum.
29. What are pancreatic acini? Contrast their functions with those of the pancreatic islets (islets of Langerhans).
30. Describe the composition of pancreatic juice and the digestive functions of each component.
31. How is pancreatic juice secretion regulated?
32. Where is the liver located? What are its principal functions?
33. Describe the anatomy of the liver. Draw a labeled diagram of a liver lobule.
34. How is blood carried to and from the liver?
35. Once bile has been formed by the liver, how is it collected and transported to the gallbladder for storage?
36. What is the function of bile?
37. How is bile secretion regulated? Define jaundice and distinguish two principal types.
38. Where is the gallbladder located? How is it connected to the duodenum?
39. Describe the function of the gallbladder. How is emptying of the gallbladder regulated?
40. What are the subdivisions of the small intestine? How are the mucosa and submucosa of the small intestine adapted for digestion and absorption?
41. Describe the movements in the small intestine.
42. Explain the function of each enzyme in intestinal juice.
43. How is small intestinal secretion regulated?
44. Define absorption. How are the end products of carbohydrate and protein digestion absorbed? How are the end products of fat digestion absorbed?
45. What routes are taken by absorbed nutrients to reach the liver?
46. Describe the absorption of water, electrolytes, and vitamins by the small intestine.
47. What are the principal subdivisions of the large intestine? How does the muscularis of the large intestine differ from that of the rest of the gastrointestinal tract? What are haustra?
48. Describe the mechanical movements that occur in the large intestine.
49. How is the ileocecal sphincter regulated?
50. Explain the activities of the large intestine that change its contents into feces.
51. Define defecation. How does it occur?
52. Distinguish between diarrhea and constipation.
53. Describe the effects of aging on the digestive system.
54. Describe the development of the digestive system.
55. Describe the causes (where known) and clinical symptoms for: dental caries, periodontal disease, peritonitis, peptic ulcers, appendicitis, tumors, diverticulitis, cirrhosis, hepatitis, gallstones, anorexia nervosa, and bulimia.
56. Refer to the glossary of medical terminology associated with the digestive system. Be sure that you can define each term.

Metabolism

■ **STUDENT OBJECTIVES**

- Define a nutrient and list the functions of the six principal classes of nutrients.
- Define metabolism and contrast the roles of anabolism and catabolism.
- Define oxidation and reduction and their importance in metabolism.
- Describe the characteristics and importance of enzymes in metabolism.
- Explain the fate of absorbed carbohydrates.
- Describe the complete oxidation of glucose via glycolysis, the Krebs cycle, and the electron transport chain.
- Define glycogenesis, glycogenolysis, and gluconeogenesis.
- Explain the fate of absorbed lipids.
- Discuss fat storage in adipose tissue.
- Explain how glycerol may be converted to glucose.
- Explain the catabolism of fatty acids via beta oxidation.
- Define ketosis and list its effects on the body.
- Explain the fate of absorbed proteins.
- Discuss the catabolism of amino acids.
- Explain the mechanism involved in protein synthesis.
- Distinguish between the absorptive and postabsorptive (fasting) states.
- Explain how various hormones regulate metabolism.
- Compare the sources, functions, and importance of minerals in metabolism.
- Define a vitamin and differentiate between fat-soluble and water-soluble vitamins.
- Compare the sources, functions, and deficiency symptoms and disorders of the principal vitamins.
- Describe abnormalities associated with megadoses of several minerals and vitamins.
- Explain how the caloric value of foods is determined.
- Describe the various mechanisms that produce body heat.
- Define basal metabolic rate (BMR) and explain several factors that affect it.
- Describe the various ways that body heat is lost.
- Explain how normal body temperature is maintained.
- Describe fever, heat cramp, heatstroke, and heat exhaustion as abnormalities of temperature regulation.
- Define obesity, starvation, phenylketonuria (PKU), cystic fibrosis, celiac disease, and kwashiorkor.

Nutrients are chemical substances in food that provide energy, form new body components, or assist in the functioning of various body processes. There are six principal classes of nutrients: carbohydrates, lipids, proteins, minerals, vitamins, and water. Carbohydrates, proteins, and lipids are digested by enzymes in the gastrointestinal tract. The end products of digestion that ultimately reach body cells are monosaccharides, amino acids, fatty acids, glycerol, and monoglycerides. Some are used to synthesize new structural molecules in cells or to synthesize new regulatory molecules, such as hormones and enzymes. Most are used to produce energy to sustain life processes. This energy is used for processes such as active transport, DNA replication, synthesis of proteins and other molecules, muscle contraction, and nerve impulse conduction. Until the energy is needed, it is stored in ATP (adenosine triphosphate).

Some minerals and many vitamins are part of enzyme systems that catalyze the reactions undergone by carbohydrates, proteins, and lipids.

Water has five major functions. It is an excellent solvent and suspending medium, it participates in hydrolysis reactions, it acts as a coolant, it lubricates, and it helps to maintain a constant body temperature due to its ability to release and absorb heat slowly.

METABOLISM

Metabolism (me-TAB-ō-lizm) refers to all the chemical reactions of the body. The body's metabolism may be thought of as an energy-balancing act between anabolic (synthetic) and catabolic (degradative) reactions.

ANABOLISM

In living cells, those chemical reactions that combine simple substances into more complex molecules are collectively known as **anabolism** (a-NAB-ō-lizm). Overall, an anabolic process requires energy, which is supplied by catabolic reactions (Figure 25-1). One example of an anabolic process is the formation of peptide bonds between amino acids, thereby building up the amino acids into the protein portions of cytoplasm, enzymes, hormones, and antibodies. Fats also participate in the body's anabolism. For instance, fats can be built into the phospholipids that form the plasma membrane. They are also part of the steroid hormones. Through anabolism, simple sugars are also built up into polysaccharides.

CATABOLISM

The chemical reactions that break down complex organic compounds into simple ones are collectively known as **catabolism** (ka-TAB-ō-lizm). Catabolic reactions release the available chemical energy in organic molecules. This energy is then stored in ATP. Catabolic reactions release energy that can be used to drive anabolic reactions. The role of ATP in the relationship between catabolic and anabolic processes is shown in Figure 25-1.

Chemical digestion is a catabolic process in which the breaking of bonds of food molecules releases energy. Another example is a process called oxidation (cellular respiration). It involves the breakdown of absorbed nutrients resulting in a release of energy.

OXIDATION-REDUCTION REACTIONS

Oxidation is the removal of electrons and hydrogen ions (hydrogen atoms) from a molecule or, less commonly, the addition of oxygen to a molecule. In many cellular oxidations, two electrons and hydrogen ions are removed at the same time. Most oxidations are actually dehydrogenations, that is, they involve the loss of hydrogen atoms. An example of an oxidation is the conversion of lactic acid into pyruvic acid:

$$\underset{\text{Lactic acid}}{\overset{\displaystyle \text{COOH}}{\underset{\displaystyle \text{CH}_3}{|\;\text{CHOH}\;|}}} \xrightarrow{\;-2\text{H (oxidation)}\;} \underset{\text{Pyruvic acid}}{\overset{\displaystyle \text{COOH}}{\underset{\displaystyle \text{CH}_3}{|\;\text{C}{=}\text{O}\;|}}}$$

As you will see later, when a substance is oxidized, the freed hydrogen atoms are transferred to another compound by substances that work with enzymes called coenzymes.

Reduction is the addition of electrons and hydrogen ions (hydrogen atoms) to a molecule or, less commonly, the removal of oxygen from a molecule. Reduction is the opposite of oxidation. An example of reduction is the conversion of pyruvic acid into lactic acid:

$$\underset{\text{Pyruvic acid}}{\begin{matrix}\text{COOH}\\\text{C}{=}\text{O}\\\text{CH}_3\end{matrix}} \xrightarrow{\;+2\text{H (reduction)}\;} \underset{\text{Lactic acid}}{\begin{matrix}\text{COOH}\\\text{CHOH}\\\text{CH}_3\end{matrix}}$$

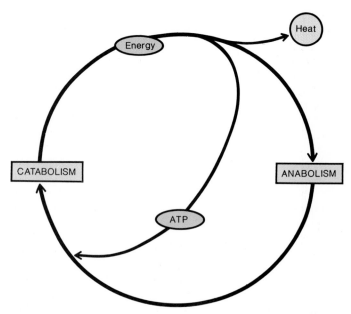

Figure 25-1 Relationship of ATP to catabolism and anabolism.

Within a cell, oxidation and reduction reactions are always coupled, that is, whenever a substance is oxidized, another is almost simultaneously reduced. This coupling of reactions is simply referred to as oxidation-reduction.

One point to remember about oxidation-reduction reactions is that oxidation is usually an energy-producing reaction. Cells take nutrients (energy sources) and degrade them from highly reduced compounds (with many hydrogen atoms) to highly oxidized compounds (with many oxygen atoms or multiple bonds). For example, when a cell oxidizes a molecule of glucose ($C_6H_{12}O_6$), the energy in the chemical bonds of the glucose molecule is removed in a stepwise manner and ultimately trapped by ATP. ATP is then an energy source for energy-requiring reactions. Compounds such as glucose that have many hydrogen atoms are highly reduced compounds and thus contain more potential energy than oxidized compounds.

METABOLISM AND ENZYMES

In Chapter 2 we noted that chemical reactions occur when chemical bonds are made or broken, and in order for reactions to occur, reactants must collide. The effectiveness of collision depends on the velocity of the reactants, the amount of energy required for reaction to occur (activation energy), and the specific configuration of the colliding particles. In body cells, enzymes serve as catalysts that speed up chemical reactions by increasing the frequency of collision, lowering activation energy, and properly orienting colliding molecules. A *catalyst* is a chemical substance that alters the rate of a chemical reaction without becoming part of the products of the reaction or being used up. Enzymes speed up chemical reactions without an increase in temperature or pressure, changes that could seriously disrupt or even kill cells.

Characteristics of Enzymes

As noted in Chapter 2, all enzymes are proteins, ranging in molecular weight from 10,000 to somewhere in the millions. Each of the thousand or more known enzymes has a characteristic three-dimensional shape with a specific surface structure due to its primary, secondary, and tertiary protein organization.

Enzymes are extremely efficient as catalysts of biochemical reactions. They are capable of catalyzing reactions at rates that are 10^8 to 10^{10} times (up to 10 billion times) more rapid than those of comparable reactions occurring without enzymes. The *turnover number*, that is, the number of substrate molecules metabolized per enzyme molecule per second, is generally between 1 and 10,000, and can be as high as 500,000.

Enzymes are very specific in the reactions they catalyze and the molecules with which they react. Such molecules are called **substrates.** From the large number of diverse molecules in the cell, an enzyme must "find" the correct substrate.

Enzymes are also subject to various controls by the cellular machinery. For example, their rate of synthesis at any given time is under the control of the cell's genes and is influenced by various other molecules in the cell. Many enzymes exist in both active and inactive forms in the cell. The rate at which the inactive form becomes active or the active form becomes inactive is determined by the cellular environment at any given time.

The names of enzymes usually end in the suffix *-ase.* All enzymes can be grouped according to the types of chemical reactions they catalyze. For example, *oxidoreductases* are involved with oxidation-reduction reactions. (Enzymes that remove hydrogen are called *dehydrogenases;* those that add oxygen are called *oxidases.*) *Transferases* transfer groups of atoms, *hydrolases* add water, *isomerases* rearrange atoms within a molecule, and *ligases* join two molecules in a reaction in which ATP is broken down.

Parts of an Enzyme

Some enzymes consist only of proteins. Most enzymes, however, contain a protein called an **apoenzyme** that is inactive without a nonprotein component called the **cofactor.** Together, the apoenzyme and cofactor are an activated **holoenzyme,** or whole enzyme. If the cofactor is removed, the apoenzyme will not function. The cofactor can be a metal ion or a complex organic molecule called a **coenzyme.**

Examples of metal ions that serve as cofactors are iron, magnesium, zinc, and calcium. Their function is to serve as a bridge between enzyme and substrate so that the enzyme can bring about transformation of the substrate. Coenzymes assist in the enzyme's activity by accepting atoms removed from the substrate or donating atoms needed by the substrate. Many coenzymes are derived from vitamins. Three coenzymes commonly used by living cells to carry hydrogen atoms are nicotinamide adenine dinucleotide (NAD) and nicotinamide adenine dinucleotide phosphate (NADP), both of which are derivatives of the B vitamin niacin, and flavin adenine dinucleotide (FAD), a derivative of vitamin B_2 (riboflavin).

Mechanism of Enzymatic Action

Exactly how enzymes lower activation energy is not completely understood. However, the following general sequence is believed to occur (Figure 25-2). The surface of the substrate makes contact with a specific region on the surface of the enzyme molecule known as the **active site.** A temporary intermediate compound called an **enzyme-substrate complex** forms. The substrate molecule is transformed (by rearrangement of existing atoms, breakdown of the substrate molecule, or combination of several substrate molecules) and the transformed substrate molecules, now called the products of the reaction, move away from the enzyme molecule. After the reaction is completed, the products of the reaction move away from the enzyme and the enzyme is free to attach to another substrate molecule.

We will now consider the metabolism of carbohydrates, lipids, and proteins in body cells.

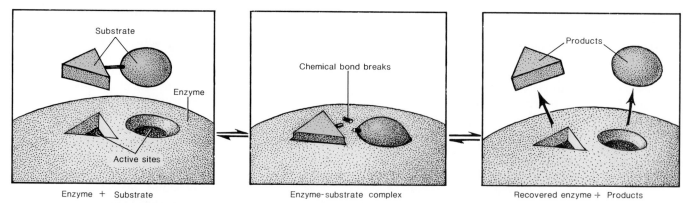

FIGURE 25-2 Enzyme action in a decomposition reaction. The enzyme and substrate molecules combine to form an enzyme-substrate complex. During combination, the substrate is changed into products. Once the products are formed, the enzyme is recovered and may be used again to catalyze a similar reaction.

CARBOHYDRATE METABOLISM

During the process of digestion, polysaccharides and di-saccharides are hydrolyzed to become the monosaccha-rides—glucose, fructose, and galactose—which are absorbed into the capillaries of the villi of the small intestine. They are then carried through the hepatic portal vein to the liver, where fructose and galactose are converted to glucose. The liver is the only organ that has the necessary enzymes to make this conversion. Thus the story of carbohydrate metabolism is really the story of glucose metabolism.

FATE OF CARBOHYDRATES

Since glucose is the body's preferred source of energy, the fate of absorbed glucose depends on the body cells' energy needs. If the cells require immediate energy, the glucose is oxidized by the cells. Each gram of carbohydrate produces about 4.0 kilocalories (kcal). (The kilocalorie content of a food is a measure of the heat it releases upon oxidation. Determination of caloric value is described later in the chapter.) The glucose not needed for immediate use is handled in several ways. First, the liver can convert excess glucose to glycogen and then store it. Skeletal muscle cells can also store glycogen. Second, if the glycogen storage areas are filled up, the liver cells can transform the glucose to fat that can be stored in adipose tissue. Later, when the cells need more energy, the glycogen and fat can be converted back to glucose, which is released into the bloodstream so it can be transported to cells for oxidation. Third, excess glucose can be excreted in the urine. Normally, this happens only when a meal containing mostly carbohydrates and no fats are eaten. Without the inhibiting effect of fats, the stomach empties its contents quickly and the carbohydrates are all digested at the same time. As a result, large numbers of monosaccharides suddenly flood into the bloodstream. Since the liver is unable to process all of them simultaneously, the blood glucose level rises and the condition of hyperglycemia may result in glucose in the urine (glycosuria).

GLUCOSE MOVEMENT INTO CELLS

Before glucose can be used by body cells, it must first pass through the plasma membrane and then enter the cytoplasm. The process by which this occurs is facilitated diffusion (Chapter 3), and the rate of glucose transport is greatly increased by insulin (Chapter 18). Immediately upon entry into cells, glucose combines with a phosphate group, produced by the breakdown of ATP. This addition of a phosphate group to a molecule is called *phosphorylation.* The product of this enzymatically catalyzed reaction is glucose-6-phosphate (see Figure 25-8). In most cells of the body, phosphorylation serves to capture glucose in the cell so that it cannot move back out. Liver cells, kidney tubule cells, and intestinal epithelial cells have the necessary enzyme (phosphatase) to remove the phosphate group and transport glucose out of the cell (see Figure 25-8).

GLUCOSE CATABOLISM

The **oxidation** of glucose is also known as **cellular respiration.** It occurs in every cell in the body and provides the cell's chief source of energy. The complete oxidation of glucose to carbon dioxide and water produces large amounts of energy. It occurs in three successive stages: glycolysis, the Krebs cycle, and the electron transport chain (see Figure 25-7).

Glycolysis

The term **glycolysis** (glī-KOL-i-sis; *glyco* = sugar, *lysis* = breakdown) refers to a series of chemical reactions in the cytoplasm of a cell that convert a six-carbon molecule of glucose into two three-carbon molecules of pyruvic acid. The simplified overall reaction for glycolysis is:

$$C—C—C—C—C—C \longrightarrow 2\ C—C—C + 2\ ATP$$

Glucose — Pyruvic acid

The details of glycolysis are shown in Figure 25-3a. The essential features of the process are:

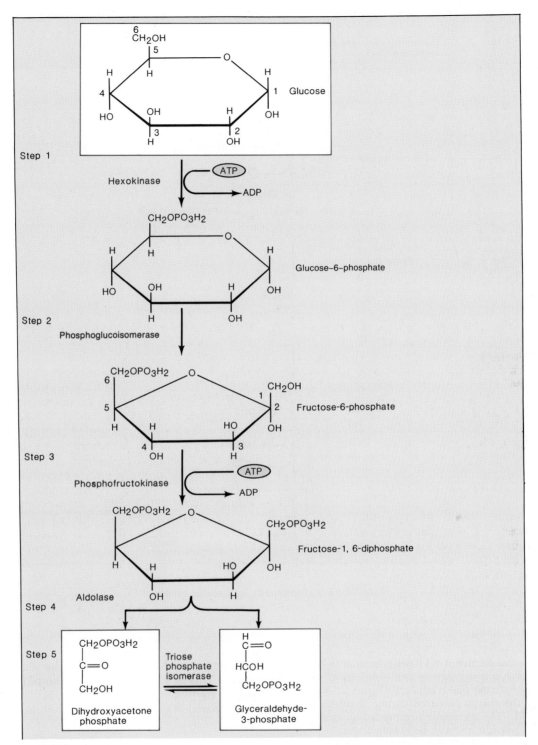

FIGURE 25-3 Glycolysis. As a result of glycolysis, there is a net gain of two molecules of ATP. (a) Detailed version.

1. Each of the 10 reactions is catalyzed by a specific enzyme. (The names of the enzymes are indicated in the rust color.)

2. All of the reactions are reversible.

3. The first three reactions involve the addition of a phosphate group (phosphorylation) to glucose, its conversion to fructose, and the addition of another phosphate group to fructose to prime the pathway. This involves an input of energy in which two molecules of ATP are converted to ADP.

4. The double phosphorylated molecule of fructose splits into two three-carbon compounds, glyceraldehyde-3-phosphate and dihydroxyacetone phosphate. These compounds are interconvertible.

5. Each of the two three-carbon compounds is degraded to form a molecule of pyruvic acid. In the process, each three-carbon compound generates two molecules of ATP.

6. Since two molecules of ATP are needed to get glycolysis

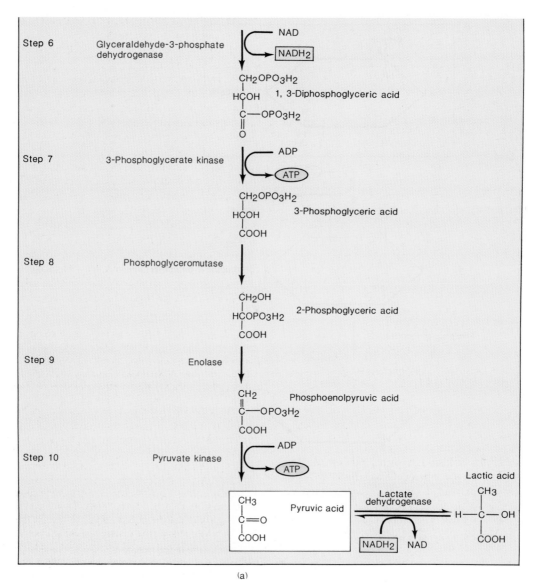

(a)

FIGURE 25-3 (*Continued*) (a) Detailed version.

started, and four molecules of ATP are generated to complete the process, there is a *net gain of two molecules of ATP for each molecule of glucose that is oxidized* (Figure 25-3b).

7. Most of the energy produced during glycolysis is used to generate ATP. The remainder is expended as heat energy, some of which helps to maintain body temperature.

The fate of pyruvic acid depends on the availability of oxygen. If anaerobic conditions exist in cells (as during strenuous exercise), pyruvic acid is reduced by the addition of two hydrogen atoms to form lactic acid (Figure 25-3a). The lactic acid may be transported to the liver, where it is converted back to pyruvic acid. Or, it may remain in the cells until aerobic conditions are restored (recall the mechanisms for the repayment of the oxygen debt described in Chapter 10), and it is then converted to pyruvic acid in the cells.

Under aerobic conditions, the process of the complete oxidation of glucose continues, and pyruvic acid is oxidized to form carbon dioxide and water in two sets of reactions: the Krebs cycle and the electron transport chain. This is known as **aerobic respiration.** However, before pyruvic acid can enter the Krebs cycle, it must be prepared. This process occurs in mitochondria.

Formation of Acetyl Coenzyme A

Each step in the oxidation of glucose requires a different enzyme and often a coenzyme as well. We are interested in only one coenzyme at this point: a substance called **coenzyme A (CoA),** which contains a derivative of pantothenic acid, a B vitamin.

During the transitional step between glycolysis and the Krebs cycle, pyruvic acid is prepared for entrance into the cycle. Essentially, pyruvic acid is converted to a two-carbon compound by the loss of carbon dioxide. The loss of a molecule of CO_2 by a substance is called

(b)

FIGURE 25-3 (Continued) (b) Simplified version.

decarboxylation (dē-kar-bok'-si-LĀ-shun). (It is indicated in the diagram by the arrow labeled CO_2.) This two-carbon fragment, called an **acetyl group,** attaches itself to coenzyme A, and the whole complex is called **acetyl coenzyme A** (Figure 25-4). This substance is now prepared to enter the Krebs cycle.

Krebs Cycle

The **Krebs cycle** is also called the **citric acid cycle,** or the **tricarboxylic acid (TCA) cycle.** It is a series of reactions that occur in the matrix of mitochondria of cells (Figure 25-5). Coenzyme A carries an acetyl group to a mitochondrion, detaches itself, and goes back into the cytoplasm to pick up another fragment. The acetyl group combines with oxaloacetic acid to form citric acid. From this point the Krebs cycle consists mainly of a series of

decarboxylation and oxidation-reduction reactions, each controlled by a different enzyme.

Let us look at the decarboxylation reactions first. As we have seen (Figure 25-4), in preparation for entering the cycle, pyruvic acid is decarboxylated to form an acetyl group. In the cycle, isocitric acid loses a molecule of CO_2 to form α-ketoglutaric acid. In the next step, α-ketoglutaric acid is decarboxylated (and picks up a molecule of CoA) to form succinyl CoA. Thus each time pyruvic acid enters the Krebs cycle three molecules of CO_2 are liberated by decarboxylation. The molecules of CO_2 leave the mitochondria, diffuse through the cytoplasm of the cell to the plasma membrane, and then diffuse into the blood by internal respiration. Eventually the CO_2 is transported by the blood to the lungs by external respiration. Once it reaches the lungs, it is exhaled.

Now let us look at the oxidation-reduction reactions. Remember, when a molecule is oxidized, it loses hydrogen atoms (electrons and hydrogen ions). When a molecule is reduced, it gains hydrogen atoms. Every oxidation is coupled with a reduction.

1. In the conversion of pyruvic acid to acetyl CoA (Figure 25-4), pyruvic acid loses two hydrogen atoms, that is, it is oxidized. The hydrogen atoms are picked up by a coenzyme called **nicotinamide adenine dinucleotide (NAD).** NAD contains two adenine nucleotides and is derived from a B vitamin called niacin. Since NAD picks up the hydrogen atoms, it is reduced and represented as $NADH_2$ (indicated in the diagram by the curved arrow entering and then leaving the reaction).

2. Isocitric acid is oxidized to form α-ketoglutaric acid. NAD is reduced to $NADH_2$.

3. α-Ketoglutaric acid is oxidized in the conversion to succinyl CoA. Again, NAD is reduced to $NADH_2$.

4. Succinic acid is oxidized to fumaric acid. In this case, the coenzyme that picks up the two hydrogen atoms is **flavin adenine dinucleotide (FAD).** FAD, like NAD, contains two adenine nucleotides, but it is derived from vitamin B_2 (riboflavin). It belongs to a class of compounds called **flavoproteins.** FAD is reduced to $FADH_2$.

5. Malic acid is oxidized to oxalocetic acid. NAD is reduced to $NADH_2$.

Thus, each time a molecule of pyruvic acid enters the Krebs cycle, four molecules of $NADH_2$ and one molecule of $FADH_2$ are produced by the various oxidation-reduction reactions. These reduced coenzymes are very important because they now contain the stored energy originally in glucose and then in pyruvic acid. In the next stage in the complete oxidation of glucose, the electron trans-

FIGURE 25-4 Formation of acetyl coenzyme A from pyruvic acid for entrance into the Krebs cycle.

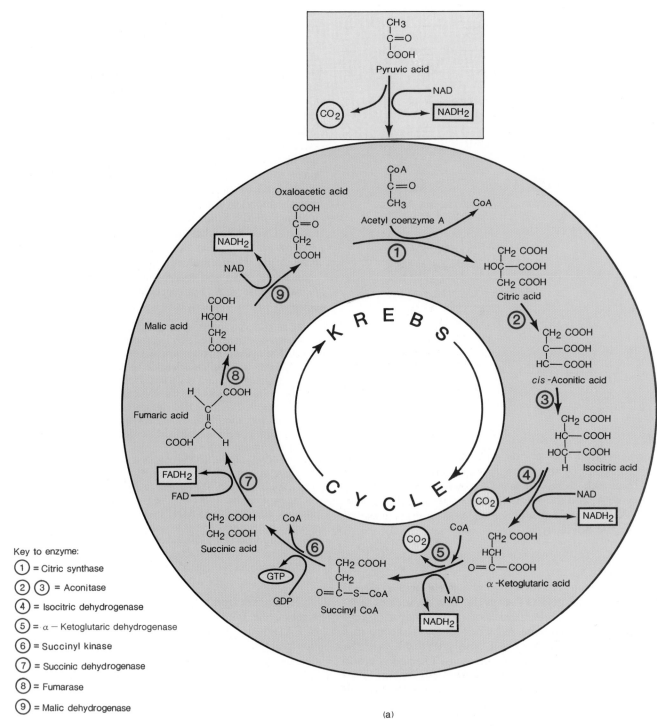

Key to enzyme:
① = Citric synthase
② ③ = Aconitase
④ = Isocitric dehydrogenase
⑤ = α – Ketoglutaric dehydrogenase
⑥ = Succinyl kinase
⑦ = Succinic dehydrogenase
⑧ = Fumarase
⑨ = Malic dehydrogenase

(a)

FIGURE 25-5 Krebs cycle. The net results of the Krebs cycle are the production of reduced coenzymes ($NADH_2$ and $FADH_2$), which contain stored energy; the generation of GTP, a high-energy compound that is used to produce ATP; and the formation of CO_2, which is transported to the lungs for expiration. (a) Detailed version.

port chain, the energy in the coenzymes is transferred to **ATP** for storage.

Electron Transport Chain

The **electron transport chain** is a series of oxidation-reduction reactions that occur on mitochondrial cristae in

which the energy in $NADH_2$ and $FADH_2$ is liberated and transferred to ATP for storage. In the electron transport chain, three types of carrier molecules are alternately oxidized and reduced, and participate in the generation of ATP: (1) the coenzyme **FAD,** (2) a coenzyme called **coenzyme Q (ubiquinone),** and (3) **cytochromes** (SĪ-tō-krōms), red protein pigments that have an iron-containing

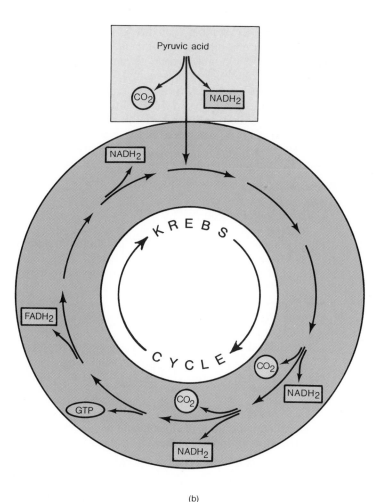

(b)

FIGURE 25-5 (*Continued*) (b) Simplified version.

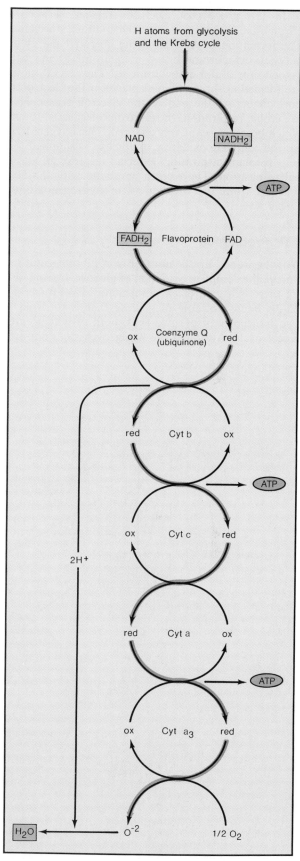

group capable of alternating between a reduced form
(Fe^{2+}) and an oxidized form (Fe^{3+}).

The first step in the electron transport chain involves
the transfer of hydrogen atoms from $NADH_2$ (mostly
from the Krebs cycle) to FAD (Figure 25-6). $NADH_2$
is oxidized to NAD and FAD is reduced to $FADH_2$.
The importance of the hydrogen transfer from $NADH_2$
to FAD is that it releases energy. This energy is used
to produce ATP from ADP. The hydrogen atoms are
then used to reduce coenzyme Q. After that step, the
hydrogen atoms do not stay intact. They ionize into hydrogen ions and electrons.

$$H \longrightarrow H^+ + e^-$$
Hydrogen atom   Hydrogen ion   Electron

In the next step of the electron transport system, the
electrons (e^-) from the hydrogen atoms are passed succes-
sively from one cytochrome to another—from cytochrome b to cytochrome c to cytochrome a and finally
to cytochrome a_3 (cytochrome oxidase). Each cytochrome
in the electron transport system is alternately reduced,
as it picks up electrons, and oxidized, as it gives up electrons. Because of the involvement of cytochromes in the
electron transport chain, it is also known as the **cytochrome system.**

FIGURE 25-6 Electron transport chain. The successive transfer of
hydrogen atoms and electrons results in the formation of ATP at
three different places.

In the process of electron transfer between cytochromes, more energy is liberated and then stored in ATP. The formation of ATP from the liberated energy occurs between cytochrome b and cytochrome c, and between cytochrome a and cytochrome a₃. Thus, ATP molecules are formed at three different places in the electron transport chain. At the end of the electron transport chain in aerobic respiration, the electrons are passed to oxygen, which becomes negatively charged. This oxygen, required for the complete oxidation of glucose, is supplied to body cells through inspiration, external respiration, and internal respiration. The oxygen combines with the H^+ ions released from the breakdown of hydrogen atoms and forms water.

ATP Production

Now let us add up the number of ATP molecules formed during the complete oxidation of glucose. During glycolysis there is a gain of two molecules of ATP (see Figure 25-3). Each molecule of glucose that is oxidized yields two molecules of pyruvic acid, both of which enter the Krebs cycle. Although the Krebs cycle does not produce

any ATP directly, it does generate two molecules of a high-energy compound called **guanosine triphosphate (GTP)**. GTP is produced from guanosine diphosphate (GDP), using the energy of succinyl coenzyme A (see Figure 25-5). Ultimately, the energy incorporated in GTP is used to produce ATP from ADP. Thus, two molecules of ATP are formed indirectly in the Krebs cycle. The bulk of the ATP is generated in the electron transport chain, where a total of 34 ATP molecules are produced. This ATP accounting can be summarized as follows:

Process	*Net Gain of ATP*	
Glycolysis		
(1) oxidation of glucose to pyruvic acid	2 ATP	
(2) production of 2 NADH₂	6 ATP	(in electron transport chain)
Formation of acetyl CoA produces 2 NADH₂	6 ATP	(in electron transport chain)
Krebs Cycle		
(1) oxidation of succinyl CoA to succinic acid	2 GTP	(equivalent of ATP)
(2) production of 6 NADH₂	18 ATP	(in electron transport chain)
(3) production of 2 FADH₂	4 ATP	(in electron transport chain)
TOTAL 38 ATP		

The complete oxidation of a molecule of glucose can be summarized as follows (Figure 25-7):

$$C_6H_{12}O_6 \;+\; 6O_2 \;\longrightarrow\; 38ATP \;+\; 6CO_2 + 6H_2O$$

| Glucose | Oxygen | Adenosine triphosphate | Carbon dioxide | Water |

FIGURE 25-7 Summary of the complete oxidation of glucose.

About 43 percent of the energy originally in glucose is captured by ATP, the remainder is given off as heat. Compared to man-made machines, this energy yield is quite efficient.

Glycolysis, the Krebs cycle, and especially the electron transport chain provide all the ATP for cellular activities. And because the Krebs cycle and electron transport chain are aerobic processes, the cells cannot carry on their activities for long without sufficient oxygen.

GLUCOSE ANABOLISM

Most of the glucose in the body is catabolized to supply energy. However, some glucose participates in a number of anabolic reactions. One is the synthesis of glycogen from many glucose molecules. Another is the manufacture of glucose from the breakdown products of proteins and lipids.

Glucose Storage: Glycogenesis

If glucose is not needed immediately for energy, it is combined with many other molecules of glucose to form a long-chain molecule called glycogen. This process is called **glycogenesis** (*glyco* = sugar; *genesis* = origin). The body can store about 500 g (about 1.1 lb) of glycogen in the liver and skeletal muscle cells. Roughly 80 percent of the glycogen is stored in the skeletal muscles.

In the process of glycogenesis (Figure 25-8), glucose that enters cells is first phosphorylated to glucose-6-phosphate. This is then converted to glucose-1-phosphate, then to uridine diphosphate glucose, and finally to glycogen. Glycogenesis is stimulated by insulin from the pancreas.

Glucose Release: Glycogenolysis

When the body needs energy, the glycogen stored in the liver is broken down into glucose and released into the

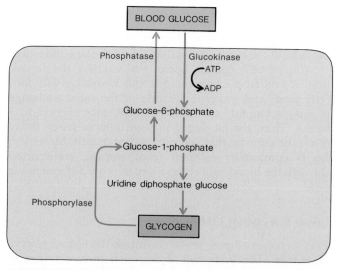

FIGURE 25-8 Glycogenesis and glycogenolysis. The glycogenesis pathway, the conversion of glucose into glycogen, is indicated by red arrows. The glycogenolysis pathway, the conversion of glycogen into glucose, is indicated by blue arrows.

bloodstream to be transported to cells, where it will be catabolized. The process of converting glycogen back to glucose is called **glycogenolysis** (*lysis* = breakdown). Glycogenolysis usually occurs between meals.

Glycogenolysis does not occur exactly by the reverse processes involved in glycogenesis (Figure 25-8). The first step in glycogenolysis involves the splitting off of glucose molecules from the branched glycogen molecule by phosphorylation to form glucose-1-phosphate. Phosphorylase, the enzyme that catalyzes this reaction, is activated by the hormones glucagon from the pancreas and epinephrine from the adrenal medulla. Glucose-1-phosphate is then converted to glucose-6-phosphate and finally to glucose. Phosphatase, the enzyme that converts glucose-6-phosphate into glucose and releases glucose into the blood, is present in liver cells, but absent in skeletal muscle cells. In skeletal muscle cells, glycogen is broken down into glucose-1-phosphate, which is then catabolized for energy via glycolysis and the Krebs cycle. Moreover, pyruvic acid and lactic acid produced by glycolysis in muscle cells can be converted to glucose in the liver. In this way, muscle glycogen is an indirect source of blood glucose.

Formation of Glucose from Proteins and Fats: Gluconeogenesis

When your liver runs low on glycogen, it is time to eat. If you do not eat, your body starts catabolizing fats and proteins. Actually, the body normally catabolizes some of its fats and a few of its proteins. But large-scale fat and protein catabolism does not happen unless you are starving, eating meals that contain very few carbohydrates, or suffering from an endocrine disorder.

Both fat molecules and protein molecules may be converted in the liver to glucose. The process by which glucose is formed from noncarbohydrate sources is called **gluconeogenesis** (gloo'-kō-nē'-ō-JEN-e-sis).

In the process of gluconeogenesis, moderate quantities of glucose can be formed from amino acids and the glycerol portion of fat molecules (Figure 25-9). About 60 percent of the amino acids in the body can undergo this conversion. Amino acids such as alanine, cysteine, glycine, serine, and threonine are converted to pyruvic acid. The pyruvic acid may be resynthesized into glucose or enter the Krebs cycle. Glycerol may be converted into glyceraldehyde-3-phosphate, which may also be resynthesized into glucose or form pyruvic acid. Figure 25-9 also shows how gluconeogenesis is related to other metabolic reactions.

Gluconeogenesis is stimulated by cortisol, one of the glucocorticoid hormones of the adrenal cortex, and thyroxine from the thyroid gland. Cortisol mobilizes proteins from body cells, making them available in the form of amino acids, thus supplying a pool of amino acids for gluconeogenesis. Thyroxine also mobilizes proteins and may mobilize fats from fat depots by making glycerol available for gluconeogenesis. Gluconeogenesis is also stimulated by epinephrine, glucagon, and growth hormone (GH).

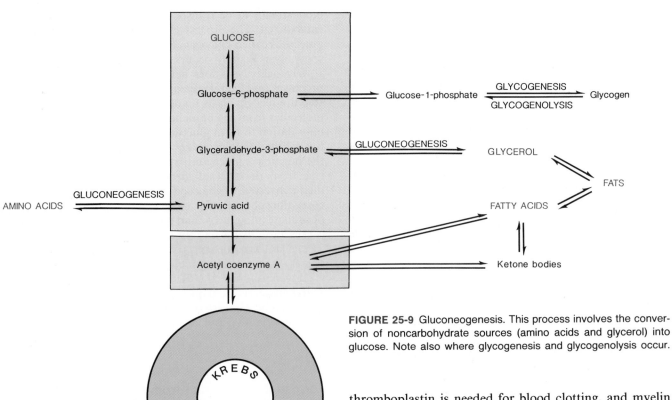

FIGURE 25-9 Gluconeogenesis. This process involves the conversion of noncarbohydrate sources (amino acids and glycerol) into glucose. Note also where glycogenesis and glycogenolysis occur.

thromboplastin is needed for blood clotting, and myelin sheaths speed up nerve impulse conduction. Cholesterol, another lipid, is used in the synthesis of bile salts and steroid hormones (adrenocortical hormones and sex hormones). The various functions of lipids in the body may be reviewed in Exhibit 2-4.

LIPID METABOLISM

Lipids are second to carbohydrates as a source of energy. More frequently they are used as building blocks to form essential structures. When neutral fats (triglycerides) are eaten, they are ultimately digested into fatty acids and monoglycerides. Short-chain fatty acids enter the blood capillaries in villi, whereas long-chain fatty acids and monoglycerides are carried in micelles to epithelial cells of the villi for entrance. Once inside, they are further digested to glycerol and fatty acids, recombined to form triglycerides, and transported as chylomicrons through the lacteals of villi into the thoracic duct, subclavian vein, and to the liver.

FATE OF LIPIDS

Lipids, such as fatty acids and glycerol, like carbohydrates, may be oxidized to produce ATP. Each gram of fat produces about 9.0 kilocalories. If the body has no immediate need to utilize fats this way, they are stored in adipose tissue (fat depots) throughout the body and in the liver. Other lipids are used as structural molecules or to synthesize other essential substances. For example, phospholipids are constituents of plasma membranes and substrates for the synthesis of prostaglandins, lipoproteins are used to transport cholesterol throughout the body,

FAT STORAGE

The major function of adipose tissue is to store fats until they are needed for energy in other parts of the body. It also insulates and protects. About 50 percent of stored fat is deposited in subcutaneous tissue—approximately 12 percent around the kidneys, 10 to 15 percent in the omenta, 20 percent in genital areas, and 5 to 8 percent between muscles. Fat is also stored behind the eyes, in the sulci of the heart, and in the folds of the large intestine.

Adipose cells contain lipases that catalyze the deposition of fats from chylomicrons and hydrolyze fats into fatty acids and glycerol. Because of the rapid exchange of fatty acids, fats are renewed about once every 2 to 3 weeks. Thus, the fat stored in your adipose tissue today is not the same fat that was present last month. Moreover, fat is continually released from storage, transported through the blood, and redeposited in other adipose tissue cells.

LIPID CATABOLISM

Fats stored in adipose tissue constitute the largest reserve of energy. The body can store much more fat than it can glycogen. Moreover, the energy yield of fats is more than twice that of carbohydrates. Nevertheless, fats are only the body's second-favorite source of energy because they are more difficult to catabolize than carbohydrates.

Glycerol

Before fat molecules can be metabolized as an energy source, they must be released from fat depots and split into glycerol and fatty acids. Both these processes are stimulated by growth hormone (GH). The glycerol and fatty acids are then catabolized separately (Figure 25-10).

Glycerol is converted easily by many cells of the body to glyceraldehyde-3-phosphate, one of the compounds also formed during the catabolism of glucose. The cells then transform glyceraldehyde-3-phosphate into glucose or continue the catabolic sequence to pyruvic acid. This is one example of gluconeogenesis. Since glyceraldehyde-3-phosphate is an intermediate product in the conversion of glycerol to glucose, glycerol thus enters the glycolytic pathway to be used for energy.

Fatty Acids

Fatty acids are catabolized differently, and the process occurs in the matrix of mitochondria. The first step in fatty-acid catabolism involves a series of reactions called **beta oxidation.** Through a series of complex reactions involving dehydration, hydration, and cleavage, enzymes remove pairs of carbon atoms at a time from the long chain of carbon atoms comprising a fatty acid. The result of beta oxidation is a series of two-carbon fragments, **acetyl coenzyme A (CoA).** Since the majority of fatty acids have even numbers of carbon atoms, the number

of acetyl CoA molecules produced is easily calculated by dividing the number of carbon atoms in the fatty acid by two. Thus, palmitic acid, a fatty acid with 16 carbon atoms, will produce acetyl CoA molecules upon beta oxidation.

In the second step of fatty acid catabolism, the acetyl CoA formed as a result of beta oxidation enters the Krebs cycle (Figure 25-10). In terms of energy yield, a 16-carbon fatty acid, such as palmitic acid, can yield a net of 192 ATPs upon its complete oxidation via the Krebs cycle and electron transport chain.

As part of normal fatty acid catabolism, the liver can take acetyl CoA molecules, two at a time, and condense them to form a substance called **acetoacetic acid** which is converted mostly into B-hydroxybutyric acid and partially into **acetone.** These substances are collectively known as **ketone bodies,** and their formation is called **ketogenesis** (Figure 25-10). They then leave the liver, enter the bloodstream, and diffuse into other body cells where they are broken down into acetyl CoA, which enters the Krebs cycle for oxidation. During periods of excessive beta oxidation, large amounts of acetyl CoA are produced. This might occur following a meal rich in fats or during fasting or starvation because essentially no carbohydrates are available for catabolism. It may also occur in diabetes mellitus because insulin is not available to stimulate glycogenesis and glucose transport into cells. When a diabetic becomes seriously insulin deficient, one of the telltale signs is a sweet smell of acetone on the breath.

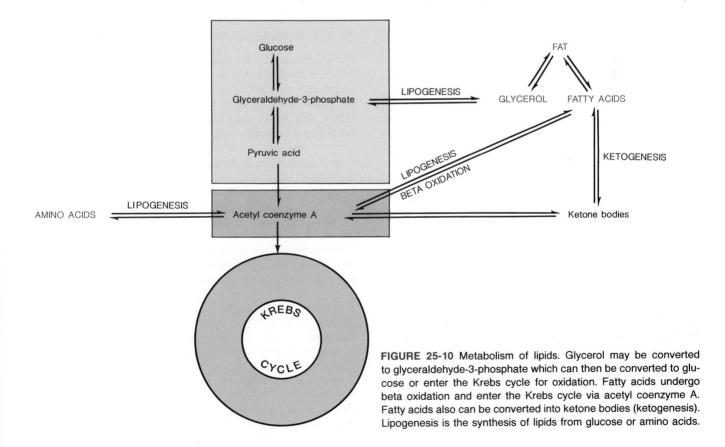

FIGURE 25-10 Metabolism of lipids. Glycerol may be converted to glyceraldehyde-3-phosphate which can then be converted to glucose or enter the Krebs cycle for oxidation. Fatty acids undergo beta oxidation and enter the Krebs cycle via acetyl coenzyme A. Fatty acids also can be converted into ketone bodies (ketogenesis). Lipogenesis is the synthesis of lipids from glucose or amino acids.

Since the body prefers glucose as a source of energy, ketone bodies are generally produced in very small quantities (0.3 to 2.0 mg/100 ml blood). When the number of ketone bodies in the blood rises above normal—a condition called **ketosis**—the ketone bodies, most of which are acids, must be buffered by the body. If too many accumulate, they use up the body's buffers and the blood pH falls. Thus extreme or prolonged ketosis can lead to **acidosis**, or abnormally low blood pH.

LIPID ANABOLISM: LIPOGENESIS

Liver cells can synthesize lipids from glucose or amino acids through a process called **lipogenesis** (Figure 25-10). Lipogenesis occurs when a greater quantity of carbohydrate enters the body than can be used for energy or stored as glycogen. The excess carbohydrate is synthesized into fats. The steps in the conversion of glucose to lipids are complex and involve the formation of glyceraldehyde-3-phosphate, which can be converted to glycerol, and acetyl CoA, which can be converted to fatty acids (see Figure 25-9). The process is enhanced by insulin. The resulting glycerol and fatty acids can undergo anabolic reactions to become fat that can be stored, or go through a series of anabolic reactions that produce other lipids such as lipoproteins, phospholipids, and cholesterol.

Many amino acids can be converted into acetyl CoA, which can be converted into fats (Figure 25-10). When people have more proteins in their diets than can be utilized as such, much of the excess protein is converted to and stored as fats.

PROTEIN METABOLISM

During the process of digestion, proteins are broken down into their constituent amino acids. The amino acids are then absorbed by the blood capillaries in villi and transported to the liver via the hepatic portal vein.

FATE OF PROTEINS

Amino acids enter body cells by active transport. This process is stimulated by growth hormone (GH) and insulin. Almost immediately after entrance, they are synthesized into proteins. Generally, the body uses very little protein for energy, as long as it ingests or stores sufficient amounts of carbohydrates and fats. Each gram of protein produces about 4.0 kcal. Many proteins function as enzymes. Other proteins are involved in transportation (hemoglobin); others serve as antibodies, clotting chemicals (fibrinogen), hormones (insulin), and contractile elements in muscle cells (actin and myosin). Several proteins serve as structural components of the body (collagen, elastin, keratin, and nucleoproteins). The various functions of proteins in the body may be reviewed in Exhibit 2-5.

PROTEIN CATABOLISM

A certain amount of protein catabolism occurs in the body each day, although much of this is only partial catabolism. Proteins are extracted from worn-out cells, such as red blood cells, and broken down into free amino acids. Some amino acids are converted into other amino acids, peptide bonds are reformed, and new proteins are made.

If other energy sources are used up or if other sources are adequate and protein intake is high, the liver can convert protein to fat or glucose, or oxidize it to carbon dioxide and water. However, before amino acids can be catabolized, they must first be converted to various substances that can enter the Krebs cycle. One such conversion consists of removing the amino group (NH_2) from the amino acid, a process called **deamination**. The liver cells then convert the NH_2 to ammonia (NH_3) and finally to urea, which is excreted in the urine. Other conversions are decarboxylation and dehydrogenation. The fate of the remaining part of the amino acid depends on what kind of amino acid it is. Figure 25-11 shows that amino acids enter the Krebs cycle at different points. They may be converted to pyruvic acid, acetyl CoA, α-ketoglutaric acid, succinyl CoA, fumaric acid, oxaloacetic acid, or acetoacetyl CoA. The point is that amino acids can be altered in various ways to enter the Krebs cycle at various locations.

The gluconeogenesis of amino acids into glucose may be reviewed in Figure 25-9. The conversion of amino acids into fatty acids or ketone bodies is shown in Figure 25-10.

PROTEIN ANABOLISM

Protein anabolism involves the formation of peptide bonds between amino acids to produce new proteins. Protein anabolism, or synthesis, is carried out on the ribosomes of almost every cell in the body, directed by the cells' DNA and RNA (see Figures 3-14 and 3-15). Protein synthesis is stimulated by growth hormone (GH), thyroxine, and insulin. The synthesized proteins are the primary constituents of enzymes, antibodies, clotting chemicals, hormones, structural components of cells, and so forth. Because proteins are a primary ingredient of most cell structures, high-protein diets are essential during the growth years, during pregnancy, and when tissue has been damaged by disease or injury.

Of the naturally occurring amino acids, 10 are referred to as **essential amino acids**. These amino acids cannot be synthesized by the human body from molecules present in the body. They are synthesized by plants or bacteria, and so foods containing these amino acids are "essential" for human growth and must be part of the diet. **Nonessential amino acids** can be synthesized by body cells by a process called **transamination**, the transfer of an amino group from an amino acid to a substance such as pyruvic acid or an acid of the Krebs cycle. Once the appropriate essential and nonessential amino acids are present in cells, protein synthesis occurs rapidly.

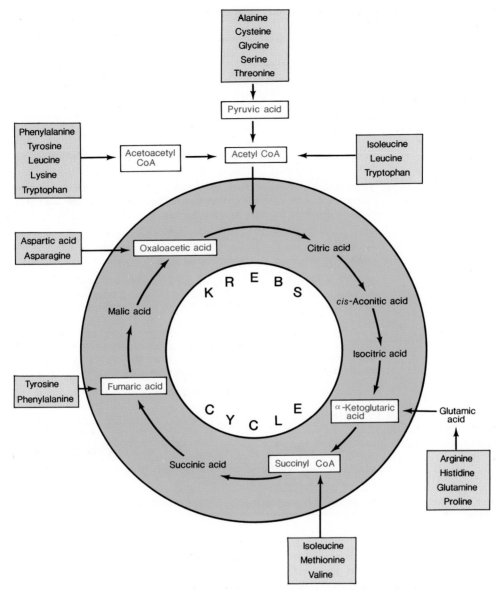

FIGURE 25-11 Various points at which amino acids, shown in gold boxes, enter the Krebs cycle for oxidation.

A summary of carbohydrate, lipid, and protein metabolism is presented in Exhibit 25-1.

ABSORPTIVE AND POSTABSORPTIVE STATES

The body actually alternates between two metabolic states, referred to as absorptive and postabsorptive. During the **absorptive state,** ingested nutrients are entering the cardiovascular and lymphatic systems from the gastrointestinal tract. During the **postabsorptive (fasting) state,** absorption is complete and the energy needs of the body must be satisfied by nutrients already in the body. By analyzing the major events of both states we can also gain a better understanding of how the metabolic pathways are interrelated.

ABSORPTIVE STATE

An average meal usually requires about 4 hours for complete absorption and, given 3 meals a day, the body spends about 12 hours each day in the absorptive state. The other 12 hours, during late morning, late afternoon, and most of the evening are spent in the postabsorptive state.

During the absorptive state, glucose transported to the liver is mostly converted to fats or glycogen; little is oxidized for energy in the liver. Some fat synthesized in the liver remains there, but most of it is released into the blood for storage in adipose tissue. Adipose tissue cells also take up glucose not picked up by the liver and convert it into fat for storage. Some blood glucose is stored as glycogen in skeletal muscles. The majority of blood glucose is used by body cells for oxidation to carbon dioxide, water, and ATP (Figure 25-12).

EXHIBIT 25-1   SUMMARY OF METABOLISM

PROCESS	COMMENT
CARBOHYDRATE METABOLISM	
Glucose catabolism	Complete oxidation of glucose, also referred to as cellular respiration, is chief source of energy in cells. The process requires glycolysis, Krebs cycle, and electron transport chain. The complete oxidation of 1 molecule of glucose results in the net production of 38 molecules of ATP.
Glycolysis	Conversion of glucose into pyruvic acid results in the production of some ATP. Reactions do not require oxygen.
Krebs cycle	Cycle includes series of oxidation-reduction reactions in which coenzymes (NAD and FAD) pick up hydrogen atoms from oxidized organic acids and some ATP is produced. CO_2 and H_2O are by-products. Reactions are aerobic.
Electron transport chain	Third set of reactions in glucose catabolism is another series of oxidation-reduction reactions, in which electrons are passed between FAD, coenzyme Q, and cytochromes, and most of the ATP is produced. Reactions are aerobic.
Glucose anabolism	Some glucose is converted into glycogen (glycogenesis) for storage if not needed immediately for energy. Glycogen can be converted to glucose (glycogenolysis) if needed for energy. The conversion of fats and proteins into glucose is called gluconeogenesis.
LIPID METABOLISM	
Catabolism	Glycerol may be converted into glucose (gluconeogenesis) or catabolized via glycolysis. Fatty acids are catabolized via beta oxidation into acetyl CoA that is catabolized in the Krebs cycle. Acetyl CoA can also be converted into ketone bodies.
Anabolism	The synthesis of lipids from glucose and amino acids is lipogenesis. Fats are stored in adipose tissue.
PROTEIN METABOLISM	
Catabolism	Amino acids are oxidized via the Krebs cycle after conversion by processes such as deamination. Ammonia resulting from the conversions is converted into urea in the liver and excreted in urine. Amino acids may be converted into glucose (gluconeogenesis) and fatty acids or ketone bodies.
Anabolism	Protein synthesis is directed by DNA and utilizes the cells' RNA and ribosomes.

During the absorptive state, most fat is stored in adipose tissue; only a small portion is used for synthesis. Adipose cells derive the fat from chylomicrons, from synthetic activities of the liver, and from their own synthetic reactions (Figure 25-12).

Many absorbed amino acids that enter liver cells are converted to carbohydrates called keto acids. These are catabolized to carbon dioxide and water to provide energy. Some keto acids are also synthesized into fatty acids in liver cells. These fatty acids may be synthesized into fats and stored in adipose tissue. Some amino acids that enter liver cells are used to synthesize proteins, such as plasma proteins. Amino acids not taken up by liver cells enter other cells of the body, such as muscle cells, for synthesis into proteins.

POSTABSORPTIVE (FASTING) STATE

The principal concern of the body during the postabsorptive state is to maintain normal blood glucose level (70 to 110 mg/100 ml of blood), even though no absorption is occurring. The maintenance of this level is especially important to the functioning of the nervous system, which cannot use any other nutrient for energy.

In general, blood glucose concentrations are maintained by utilization of various glucose sources and by glucose-sparing reactions such as the utilization of fat or possibly protein for energy. A major source of blood glucose during fasting is liver glycogen, which can provide only about a 4-hour supply of glucose. Liver glycogen is continually being formed and broken down as needed in response to the concentration of glucose in the blood (Figure 25-13). Another source of blood glucose is glycerol, produced by the hydrolysis of triglycerides, primarily in adipose tissue. However, only the glycerol produced from fat breakdown can form glucose; the fatty acids cannot be utilized because acetyl CoA cannot be readily converted to pyruvic acid. But, the fatty acids are oxidized directly by entering the Krebs cycle as acetyl CoA, thereby sparing the use of glucose for energy. A third source of glucose may be muscle glycogen during periods of vigorous exercise when significant amounts of lactic acid are produced by anaerobic glycolysis in muscle. Muscle lactic acid is released into the blood and carried to the liver, where it is converted back to glucose and released into the blood. Finally, tissue protein, primarily from skeletal muscle, can contribute to blood glucose. During fasting, large amounts of amino acids from protein

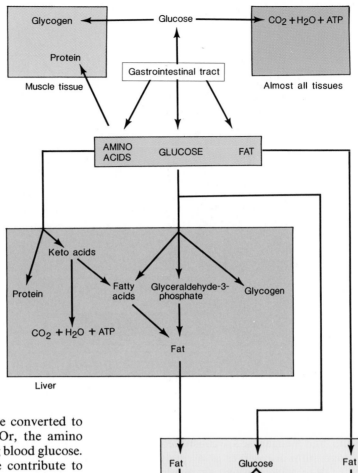

FIGURE 25-12 Principal pathways during the absorptive state.

breakdown are released from muscle to be converted to glucose in the liver by gluconeogenesis. Or, the amino acids may be oxidized directly, thus sparing blood glucose. During fasting, amino acids from muscle contribute to blood glucose after liver glycogen and fat stores are depleted.

Despite all of these mechanisms for supplying blood glucose, they cannot maintain blood glucose level for very long. Thus, a major body adjustment must be made during the postabsorptive state. Although the nervous system continues to utilize blood glucose normally, all other body tissues reduce their oxidation of glucose and switch over to fat as their energy source. Accordingly, fats in adipose tissue are broken down and fatty acids are released into the blood. The fatty acids are picked up by body cells, except nervous tissue, and oxidized to carbon dioxide, water, and ATP. The liver converts fatty acids to ketone bodies, which enter most body cells and are then oxidized to carbon dioxide, water, and ATP. As a result of this glucose sparing and fat utilization, an individual can fast for several weeks, provided water is consumed.

REGULATION OF METABOLISM

Absorbed nutrients have several alternatives, based upon the needs of the body. They may be oxidized for energy, stored, or converted. The pathway taken by a particular nutrient is enzymatically controlled and is regulated by hormones. In the discussion of the metabolism of carbohydrates, lipids, and proteins, reference has been made to the hormonal regulation involved. The actions of the principal hormones related to metabolism are summarized in Exhibit 25-2.

Hormones are the primary regulators of metabolism. However, hormonal control is ineffective without the proper minerals and vitamins. Some minerals and many vitamins are components of the enzyme systems that catalyze the metabolic reactions.

MINERALS

Minerals are inorganic substances. They may appear in combination with each other or in combination with organic compounds. Minerals constitute about 4 percent of the total body weight, and they are concentrated most heavily in the skeleton. Minerals known to perform functions essential to life include calcium, phosphorus, sodium, manganese, cobalt, copper, zinc, selenium, and chromium. Other minerals—aluminum, silicon, arsenic, nickel—are present in the body, but their functions have not yet been determined.

Calcium and phosphorus form part of the structure of bone. But since minerals do not form long-chain com-

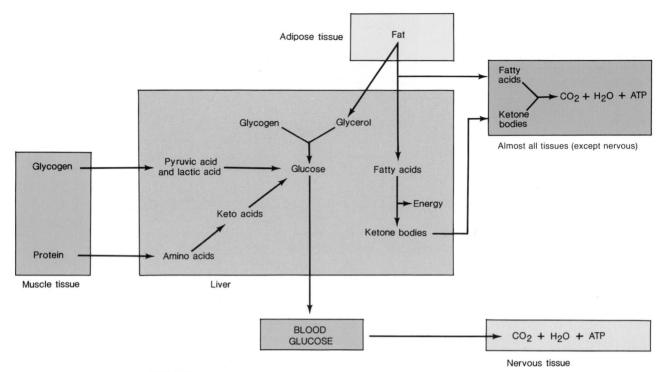

FIGURE 25-13 Principal pathways during the postabsorptive (fasting) state.

pounds, they are otherwise poor building materials. Their chief role is to help regulate body processes. Calcium, iron, magnesium, and manganese are constituents of some coenzymes. Magnesium also serves as a catalyst for the conversion of ADP to ATP. Without these minerals, metabolism would stop and the body would die. Minerals such as sodium and phosphorus work in buffer systems. Sodium helps regulate the osmosis of water and, along with other ions, is involved in the generation of nerve impulses. Exhibit 25-3 describes the functions of some

EXHIBIT 25-2   SUMMARY OF HORMONAL REGULATION OF THE METABOLISM OF CARBOHYDRATES, LIPIDS, AND PROTEINS

HORMONE	ACTIONS
Insulin	Increases glucose uptake by cells, especially muscle and adipose cells. Stimulates conversion of glucose into glycogen (glycogenesis). Stimulates fat synthesis and inhibits fat breakdown. Stimulates active transport of amino acids into cells, especially muscle cells. Stimulates protein synthesis.
Glucagon	Stimulates conversion of glycogen into glucose (glycogenolysis). Stimulates conversion of noncarbohydrates into glucose (gluconeogensis). Stimulates breakdown of fats (fat mobilization).
Epinephrine	Stimulates conversion of glycogen into glucose (glycogenolysis). Stimulates conversion of noncarbohydrates into glucose (gluconeogenesis). Stimulates breakdown of fats (fat mobilization).
Growth hormone (GH)	Stimulates active transport of amino acids into body cells, especially muscle cells. Stimulates protein synthesis. Stimulates conversion of noncarbohydrates into glucose (gluconeogenesis). Stimulates breakdown of fats (fat mobilization).
Thyroxine	Stimulates protein synthesis. Stimulates conversion of noncarbohydrates into glucose (gluconeogenesis). Stimulates breakdown of fats (fat mobilization).
Cortisol	Stimulates conversion of noncarbohydrates into glucose (gluconeogenesis).
Testosterone	Increases protein deposition in body cells, especially muscle cells.

EXHIBIT 25-3   MINERALS VITAL TO THE BODY

MINERAL	COMMENTS	IMPORTANCE
Calcium	Most abundant cation in body. Appears in combination with phosphorus in ratio of 2:1.5. About 99 percent is stored in bone and teeth. Remainder stored in muscle, other soft tissues, and blood plasma. Blood calcium level controlled by calcitonin (CT) and parathyroid hormone (PTH). Absorption occurs only in the presence of vitamin D. Most is excreted in feces and small amount in urine. Sources are milk, egg yolk, shellfish, green leafy vegetables.	Formation of bones and teeth, blood clotting, normal muscle and nerve activity, endocytosis and exocytosis, cellular motility, chromosome movement prior to cell division, glycogen metabolism, and synthesis and release of neurotransmitters.
Phosphorus	About 80 percent found in bones and teeth. Remainder distributed in muscle, brain cells, blood. More functions than any other mineral. Blood phosphorous level controlled by calcitonin (CT) and parathyroid hormone (PTH). Most excreted in urine, small amount eliminated in feces. Sources are dairy products, meat, fish, poultry, nuts.	Formation of bones and teeth. Constitutes a major buffer system of blood. Plays important role in muscle contraction and nerve activity. Component of many enzymes. Involved in transfer and storage of energy (ATP). Component of DNA and RNA.
Iron	About 66 percent found in hemoglobin of blood. Remainder distributed in skeletal muscles, liver, spleen, enzymes. Normal losses of iron occur by shedding of hair, epithelial cells, and mucosal cells, and in sweat, urine, feces, and bile. Sources are meat, liver, shellfish, egg yolk, beans, legumes, dried fruits, nuts, cereals.	As component of hemoglobin, carries O_2 to body cells. Component of cytochromes involved in formation of ATP from catabolism.
Iodine	Essential component of thyroid hormones. Excreted in urine. Sources are seafood, cod-liver oil, and vegetables grown in iodine-rich soils or iodized salt.	Required by thyroid gland to synthesize thyroid hormones, hormones that regulate metabolic rate.
Copper	Some stored in liver and spleen. Most excreted in feces. Sources include eggs, whole-wheat flour, beans, beets, liver, fish, spinach, asparagus.	Required with iron for synthesis of hemoglobin. Component of enzyme necessary for melanin pigment formation.
Sodium	Most found in extracellular fluids, some in bones. Excreted in urine and perspiration. Normal intake of NaCl (table salt) supplies required amounts.	As most abundant cation in extracellular fluid, strongly affects distribution of water through osmosis. Part of bicarbonate buffer system. Functions in nerve impulses conduction.
Potassium	Principal cation in intracellular fluid. Most is excreted in urine. Normal food intake supplies required amounts.	Functions in transmission of nerve impulses and muscle contraction.
Chlorine	Found in extracellular and intracellular fluids. Principal anion of extracellular fluid. Most excreted in urine. Normal intake of NaCl supplies required amounts.	Assumes role in acid–base balance of blood, water balance, and formation of HCl in stomach.
Magnesium	Component of soft tissues and bone. Excreted in urine and feces. Widespread in various foods.	Required for normal functioning of muscle and nervous tissue. Participates in bone formation. Constituent of many coenzymes.
Sulfur	Constituent of many proteins (such as insulin) and some vitamins (thiamine and biotin). Excreted in urine. Sources include beef, liver, lamb, fish, poultry, eggs, cheese, beans.	As component of hormones and vitamins, regulates various body activities.
Zinc	Important component of certain enzymes. Widespread in many foods, especially meats.	As a component of carbonic anhydrase, important in carbon dioxide metabolism. Necessary for normal growth and wound healing, proper functioning of prostate gland, normal taste sensations and appetite, and normal sperm counts in males. As a component of peptidases, it is involved in protein digestion.
Fluorine	Component of bones, teeth, other tissues.	Appears to improve tooth structure and inhibit cariogenesis.
Manganese	Some stored in liver and spleen. Most excreted in feces.	Activates several enzymes. Needed for hemoglobin synthesis, urea formation, growth, reproduction, and lactation.
Cobalt	Constituent of vitamin B_{12}.	As part of B_{12}, required for maturation of erythropoeisis.

EXHIBIT 25-3 *(Continued)*

MINERAL	COMMENTS	IMPORTANCE
Chromium	Found in high concentrations in brewer's yeast. Also found in wine and some brands of beer.	Necessary for the proper utilization of dietary sugars and other carbohydrates by optimizing the production and effects of insulin. Helps increase blood levels of HDL, while decreasing levels of LDL.
Selenium	Found in seafood, meat, chicken, grain cereals, egg yolk, milk, mushrooms, and garlic.	An antioxidant. Prevents chromosome breakage and may assume a role in preventing certain birth defects.

minerals vital to the body. Note that the body generally uses the ions of the minerals rather than the nonionized form. Some minerals, such as chlorine, are toxic or even fatal to the body if ingested in the nonionized form.

VITAMINS

Organic nutrients required in minute amounts to maintain growth and normal metabolism are called vitamins. Unlike carbohydrates, fats, or proteins, vitamins do not provide energy or serve as building materials. The essential function of vitamins is the regulation of physiological processes. Of the vitamins whose functions are known, most serve as coenzymes.

Most vitamins cannot be synthesized by the body. They may be ingested in foods or pills. Other vitamins, such as vitamin K, are produced by bacteria in the gastrointestinal tract. The body can assemble some vitamins if the raw materials called *provitamins* are provided. Vitamin A is produced by the body from the provitamin carotene, a chemical present in spinach, carrots, liver, and milk.

No single food contains all the required vitamins—one of the best reasons for eating a balanced diet.

The term **avitaminosis** refers to a deficiency of any vitamin in the diet. **Hypervitaminosis** refers to an excess of one or more vitamins. Conditions related to megadoses of certain vitamins and minerals are discussed shortly.

On the basis of solubility, vitamins are divided into two principal groups: fat-soluble and water-soluble. **Fat-soluble** vitamins are absorbed along with ingested dietary fats by the small intestine as micelles. In fact, they cannot be absorbed unless they are ingested with some fat. Fat-soluble vitamins are generally stored in cells, particularly liver cells, so reserves can be built up. Examples of fat-soluble vitamins are vitamins A, D, E, and K. **Water-soluble** vitamins, by contrast, are absorbed along with water in the gastrointestinal tract and dissolve in the body fluids. Excess quantities of these vitamins are excreted in the urine. Thus the body does not store water-soluble vitamins well. Examples of water-soluble vitamins are the B vitamins and vitamin C.

Exhibit 25-4 lists the principal vitamins, their sources, functions, and related disorders.

EXHIBIT 25-4   THE PRINCIPAL VITAMINS

VITAMIN	COMMENT AND SOURCE	FUNCTION	DEFICIENCY SYMPTOMS AND DISORDERS
FAT-SOLUBLE			
A	Formed from provitamin carotene (and other provitamins) in intestinal tract. Requires bile salts and fat for absorption. Stored in liver. Sources of carotene and other provitamins include yellow and green vegetables; sources of vitamin A include fish-liver oils, milk, butter.	Maintains general health and vigor of epithelial cells.	Deficiency results in atrophy and keratinization of epithelium, leading to dry skin and hair, increased incidence of ear, sinus, respiratory, urinary, and digestive infections, inability to gain weight, drying of cornea with ulceration (**xerophthalmia**), nervous disorders, and skin sores.
		Essential for formation of rhodopsin, light-sensitive chemical in rods of retina.	**Night blindness** or decreased ability for dark adaptation.
		Growth of bones and teeth by apparently helping to regulate activity of osteoblasts and osteoclasts.	Slow and faulty development of bones and teeth.

EXHIBIT 25-4 (Continued)

VITAMIN	COMMENT AND SOURCE	FUNCTION	DEFICIENCY SYMPTOMS AND DISORDERS
D	In presence of sunlight, 7-dehy-drocholesterol in the skin is converted to cholecalciferol (vitamin D_3). In the liver, cholecalciferol is converted to 25-hydroxycholecalciferol. In the kidneys, 25-hydroxycholecalciferol is converted to 1,25-dihydroxycalciferol, the active form of vitamin D. Dietary vitamin D requires moderate amounts of bile salts and fat for absorption. Stored in tissues to slight extent. Most excreted via bile. Sources include fish-liver oils, egg yolk, fortified milk.	Essential for absorption and utilization of calcium and phosphorus from gastrointestinal tract. May work with parathyroid hormone (PTH) that controls calcium metabolism.	Defective utilization of calcium by bones leads to **rickets** in children and **osteomalacia** in adults. Possible loss of muscle tone.
E (tocopherols)	Stored in liver, adipose tissue, and muscles. Requires bile salts and fat for absorption. Sources include fresh nuts and wheat germ, seed oils, green leafy vegetables.	Believed to inhibit catabolism of certain fatty acids that help form cell structures, especially membranes. Involved in formation of DNA, RNA, and red blood cells. May promote wound healing, contribute to the normal structure and functioning of the nervous system, prevent scarring and reduce the severity of visual loss associated with retrolental fibroplasia (an eye disease in premature infants caused by too much oxygen in incubators) by functioning as an antioxidant. Believed to help protect liver from toxic chemicals like carbon tetrachloride.	May cause the oxidation of unsaturated fats resulting in abnormal structure and function of mitochondria, lysosomes, and plasma membranes, a possible consequence being hemolytic anemia. Deficiency also causes muscular dystrophy in monkeys and sterility in rats.
K	Produced in considerable quantities by intestinal bacteria. Requires bile salts and fat for absorption. Stored in liver and spleen. Other sources include spinach, cauliflower, cabbage, liver.	Coenzyme believed essential for synthesis of prothrombin by liver and several clotting factors. Also known as antihemorrhagic vitamin.	Delayed clotting time results in excessive bleeding.
WATER-SOLUBLE			
B_1 (thiamine)	Rapidly destroyed by heat. Not stored in body. Excessive intake eliminated in urine. Sources include whole-grain products, eggs, pork, nuts, liver, yeast.	Acts as coenzyme for many different enzymes involved in carbohydrate metabolism of pyruvic acid to CO_2 and H_2O. Essential for synthesis of acetylcholine.	Improper carbohydrate metabolism leads to buildup of pyruvic and lactic acids and insufficient energy for muscle and nerve cells. Deficiency leads to two syndromes: (1) **beri-beri**—partial paralysis of smooth muscle of GI tract causing digestive disturbances, skeletal muscle paralysis, atrophy of limbs; (2) **polyneuritis**—due to degeneration of myelin sheaths reflexes related to kinesthesia are impaired, impairment of sense of touch, decreased intestinal motility, stunted growth in children, and poor appetite.

EXHIBIT 25-4 *(Continued)*

VITAMIN	COMMENT AND SOURCE	FUNCTION	DEFICIENCY SYMPTOMS AND DISORDERS
B_2 (riboflavin)	Not stored in large amounts in tissues. Most is excreted in urine. Small amounts supplied by bacteria of GI tract. Other sources include yeast, liver, beef, veal, lamb, eggs, whole-grain products, asparagus, peas, beets, peanuts.	Component of certain coenzymes (e.g., FAD) concerned with carbohydrate and protein metabolism, especially in cells of eye, integument, mucosa of intestine, blood.	Deficiency may lead to improper utilization of oxygen resulting in blurred vision, cataracts, and corneal ulcerations. Also dermatitis and cracking of skin, lesions of intestinal mucosa, and development of one type of anemia.
Niacin (nicotinamide)	Derived from amino acid tryptophan. Sources include yeast, meats, liver, fish, whole-grain products, peas, beans, nuts.	Essential component of coenzyme (NAD) concerned with energy-releasing reactions. In lipid metabolism, inhibits production of cholesterol and assists in fat breakdown.	Principal deficiency is **pellagra,** characterized by dermatitis, diarrhea, and psychological disturbances.
B_6 (pyridoxine)	Formed by bacteria of GI tract. Stored in liver, muscle, brain. Other sources include salmon, yeast, tomatoes, yellow corn, spinach, whole-grain products, liver, yogurt.	May function as coenzyme in fat metabolism. Essential coenzyme for normal amino acid metabolism. Assists production of circulating antibodies.	Most common deficiency symptom is dermatitis of eyes, nose, and mouth. Other symptoms are retarded growth and nausea.
B_{12} (cyanocobalamin)	Only B vitamin not found in vegetables; only vitamin containing cobalt. Absorption from GI tract dependent on HCl and intrinsic factor secreted by gastric mucosa. Sources include liver, kidney, milk, eggs, cheese, meat.	Coenzyme necessary for red blood cell formation, formation of amino acid methionine, entrance of some amino acids into Krebs cycle, and manufacture of choline (chemical similar in function to acetylcholine).	Pernicious anemia and malfunction of nervous system due to degeneration of axons of spinal cord.
Pantothenic acid	Stored primarily in liver and kidneys. Some produced by bacteria of GI tract. Other sources include kidney, liver, yeast, green vegetables, cereal.	Constituent of coenzyme A essential for transfer of pyruvic acid into Krebs cycle, conversion of lipids and amino acids into glucose, and synthesis of cholesterol and steroid hormones.	Experimental deficiency tests indicate fatigue, muscle spasms, neuromuscular degeneration, insufficient production of adrenal steroid hormones.
Folic acid	Synthesized by bacteria of GI tract. Other sources include green leafy vegetables and liver.	Component of enzyme systems synthesizing purines and pyrimidines built into DNA and RNA. Essential for normal production of red and white blood cells.	Production of abnormally large red blood cells (macrocytic anemia).
Biotin	Synthesized by bacteria of GI tract. Other sources include yeast, liver, egg yolk, kidneys.	Essential coenzyme for conversion of pyruvic acid to oxaloacetic acid and synthesis of fatty acids and purines.	Mental depression, muscular pain, dermatitis, fatigue, nausea.
C (ascorbic acid)	Rapidly destroyed by heat. Some stored in glandular tissue and plasma. Sources include cirtrus fruits, tomatoes, green vegetables.	Exact role not understood. Promotes many metabolic reactions, particularly protein metabolism, including laying down of collagen in formation of connective tissue. As coenzyme may combine with poisons, rendering them harmless until excreted. Works with antibodies. Promotes wound healing.	Scurvy; anemia; many symptoms related to poor connective tissue growth and repair including tender swollen gums, loosening of teeth (alveolar processes also deteriorate), poor wound healing, bleeding (vessel walls fragile because of connective tissue degeneration), and retardation of growth.

Abnormal conditions related to several minerals and vitamins are presented in Exhibit 25-5.

METABOLISM AND BODY HEAT

We will now consider the relationship of foods to body heat, mechanisms of heat gain and loss, and the regulation of body temperature.

MEASURING HEAT

Heat is a form of energy that can be measured as **temperature** and expressed in units called calories. A **calorie (cal)** is the amount of heat energy required to raise the temperature of 1 g of water 1°C, from 14° to 15°C. Since the calorie is a small unit and large amounts of energy are stored in foods, the **kilocalorie (kcal)** is used instead. A kilocalorie is equal to 1,000 cal and is defined as the amount of heat required to raise the temperature of 1,000 g of water 1°C. The kilocalorie is the unit we use to express the heating value of foods and to measure the body's metabolic rate.

The apparatus used to determine the caloric value of foods is called a **calorimeter.** A weighed sample of a dehydrated food is burned completely in an insulated metal container. The energy released by the burning food is absorbed by the container and transferred to a known volume of water that surrounds the container. The change in the water's temperature is directly related to the number of kilocalories released by the food. Knowing the caloric value of foods is important; if we know the amount of energy the body uses for various activities, we can adjust our food intake. In this way we can control body weight by taking in only enough kilocalories to sustain our activities.

PRODUCTION OF BODY HEAT

Most of the heat produced by the body comes from oxidation of the food we eat. The rate at which this heat is produced—the **metabolic rate**—is also measured in kilocalories. Among the factors that affect metabolic rate are the following:

1. **Exercise.** During strenuous exercise, the metabolic rate may increase to as much as 15 times the normal rate. In well-trained athletes the rate may increase up to 20 times.

2. **Nervous system.** In a stress situation, the sympathetic nervous system is stimulated and the nerves release norepinephrine (NE), which increases the metabolic rate of body cells.

3. **Hormones.** In addition to norepinephrine (NE), several other hormones affect metabolic rate. Epinephrine is secreted in stress situations. Increased secretions of thyroid hormones increase the metabolic rate. Testosterone and growth hormone (GH) also increase metabolic rate.

4. **Body temperature.** The higher the body temperature, the higher the metabolic rate. Each 1°C rise in temperature increases the rate of biochemical reactions by about 10 percent. The metabolic rate may be substantially increased during fever.

5. **Ingestion of food.** The ingestion of food can raise metabolic rate by as much as 10 to 20 percent. This effect is called **specific dynamic action (SDA)** and is greatest with proteins and less with carbohydrates and lipids.

6. **Age.** The metabolic rate of a child, in relation to its size, is about double that of an elderly person, as the high rates of reactions related to growth metabolic rate decrease with age.

7. **Others.** Other factors that affect metabolic rate are sex

---

EXHIBIT 25-5   ABNORMAL CONDITIONS ASSOCIATED WITH MEGADOSES OF SELECTED MINERALS AND VITAMINS*

SUBSTANCE	ABNORMALITY
MINERALS	
Calcium	Depresses nerve function, causes drowsiness, extreme lethargy, calcium deposits, kidney stones.
Iron	Damage to liver, heart, and pancreas.
Zinc	Masklike fixed expression, difficulty in walking, slurred speech, hand tremor, involuntary laughter.
Cobalt	Goiter, polycythemia, and heart damage.
Selenium	Nausea, vomiting, fatigue, irritability, and loss of fingernails and toenails.
VITAMINS	
A	Blurred vision, dizziness, ringing in the ears, headache, skin rash, nausea, vomiting, diarrhea, hair loss, menstrual irregularities, fatigue, liver damage, abnormal bone growth, and damage to nervous system.
D	Calcium deposits, deafness, nausea, loss of appetite, kidney stones, weak bones, hypertension, high cholesterol.
E	Thrombophlebitis, pulmonary embolism, hypertension, severe fatigue, breast tenderness, and slow wound healing.
Niacin	Flushing, peptic ulcers, liver dysfunction, gout, arrhythmias, and hyperglycemia.
B_6	Impaired sense of position and vibration, diminished tendon reflexes, numbness and loss of sensations in hands and feet, difficulty in walking.
C	Dependence on megadoses may lead to scurvy when withdrawn, kidney stones, diarrhea, hemolysis, and blood clotting.

* A **megadose** is defined as taking 10 or more times the adult recommended daily allowances (RDA's) set by the National Research Council.

(lower in females, except during pregnancy and lactation), climate (higher in tropical regions), sleep (lower), and malnutrition (lower).

BASAL METABOLIC RATE

Since many factors affect metabolic rate, it is measured under standard conditions designed to reduce or eliminate those factors as much as possible. These conditions of the body are called the **basal state,** and the measurement obtained is the **basal metabolic rate (BMR).** The person should not exercise for 30 to 60 minutes before the measurement is taken. The individual must be completely at rest, but awake. Air temperature should be comfortable. The person should fast for at least 12 hours. Body temperature should be normal. Basal metabolic rate is a measure of the rate at which the body breaks down foods (and therefore releases heat). BMR is also a measure of how much thyroxine the thyroid gland is producing, since thyroxine regulates the rate of food breakdown and is not a controllable factor under basal conditions.

The basal metabolic rate is most often measured indirectly by measuring oxygen consumption using a respirometer. If a given amount of food releases a given amount of heat energy when it is oxidized, it must combine with a given amount of oxygen. Thus, by measuring the amount of oxygen needed for the metabolism of foods, we can determine how many kilocalories are produced. The amount of heat energy released when 1 liter of oxygen combines with carbohydrates is 5.05 kcal; with fats, the heat released is 4.70 kcal; with proteins, the heat released is 4.60 kcal. The average of the three values is 4.825 kcal. Therefore, every time a liter of oxygen is consumed, 4.825 kcal are produced.

Basal metabolic rate is usually expressed in kilocalories per square meter of body surface area per hour (kcal/m^2/hr). Suppose you use 1.8 liters of oxygen in 6 minutes as recorded on a respirometer. This means that your oxygen consumption in an hour would be 18 liters (1.8 × 10). Your basal metabolic rate would be 18 × 4.825 kcal or 86.85 kcal/hr. To express the kilocalories per square meter of body surface, a standardized chart is used. Such a chart shows square meters of body surface relative to height in centimeters and weight in kilograms. If you weigh 75 kg and are 190 cm tall, your body surface area is 2 m^2. Your basal metabolic rate is equal to 86.85 kcal divided by 2, or about 43.43 kcal/m^2/hr.

The normal basal metabolic rates for various age groups by sex are also listed in standardized charts. Suppose a 27-year-old male has a recorded basal metabolic rate of 45.7 kcal/m^2/hr. The chart would show that his basal metabolic rate should be about 40.3. What does this mean? The recorded basal metabolic rate is 5.4 kcal/m^2/hr, or about 14 percent, above normal. Since a basal metabolic rate between +15 and −15 percent is considered normal, this male is judged to be normal. Values above or below 15 percent may indicate an excess or deficiency of thyroid hormones. When the thyroid is secreting extreme quantities of thyroid hormones, the basal metabolic rate can go as high as +100 percent. If, on the other hand, the thyroid is secreting very little of its thyroid hormones, the basal metabolic rate may be as low as −50 percent.

LOSS OF BODY HEAT

Body heat is produced by the oxidation of foods we eat. This heat must be removed continuously or body temperature would rise steadily. The principal routes of heat loss include radiation, conduction, convection, and evaporation.

Radiation

Radiation is the transfer of heat as infrared heat rays from one object to another without physical contact. Your body loses heat by the radiation of heat waves to cooler objects nearby such as ceilings, floors, and walls. If these objects are at a higher temperature, you absorb heat by radiation. Incidentally, the air temperature has no relationship to the radiation of heat to and from objects. Skiers can remove their shirts in bright sunshine, even though the air temperature is very low, because the radiant heat from the sun is adequate to warm them. In a room at 21°C (70°F), about 60 percent of heat loss is by radiation.

Conduction

Another method of heat transfer is **conduction.** In this process, body heat is transferred to a substance or object in contact with the body, such as chairs, clothing, or jewelry. About 3 percent of body heat is lost via conduction.

Convection

Convection is the transfer of heat by the movement of a liquid or gas between areas of different temperature. When cool air makes contact with the body, it becomes warmed and is carried away by convection currents. Then more cool air makes contact with the body and is carried away. The faster the air moves, the faster the rate of convection. About 15 percent of body heat is lost to the air by convection.

Evaporation

Evaporation is the conversion of a liquid to a vapor. Water has a high heat of evaporation. The *latent heat of evaporation* is the amount of heat necessary to evaporate 1 g of water at 30°C (86°F). Because of water's high latent heat of evaporation, every gram of water evaporating from the skin takes with it a great deal of heat—about 0.58 kcal per gram of water. Under normal conditions, about 22 percent of heat loss occurs through evaporation. The evaporation of only 150 ml of water per hour is enough to remove all the heat produced by the body under basal conditions. Under extreme conditions, about 4 liters (1 gal) of perspiration is produced each hour and this volume can remove 2,000 kcal of heat from the body. This is approximately 32 times the basal level of heat production. The rate of evaporation is inversely related to relative humidity, the ratio of the actual amount of moisture in the air to the greatest amount it can hold at a given temperature. The higher the relative humidity, the lower the rate of evaporation.

A summary of heat-producing and heat-losing mechanisms is presented in Exhibit 25-6.

BODY TEMPERATURE REGULATION

If the amount of heat production equals the amount of heat loss, you maintain a constant body temperature near 37°C (98.6°F). If your heat-producing mechanisms generate more heat than is lost by your heat-losing mechanisms, your body temperature rises. If your heat-losing mechanisms give off more heat than is generated by heat-producing mechanisms, your temperature falls.

Hypothalamic Thermostat

Body temperature is regulated by mechanisms that attempt to keep heat production and heat loss in balance. A center of control of these mechanisms is found in the hypothalamus in a group of neurons in the anterior portion referred to as the **preoptic area.** If blood temperature increases, the neurons of the preoptic area fire impulses more rapidly. If something causes the blood's temperature to decrease, these neurons fire impulses more slowly. The preoptic area is adjusted to maintain normal body temperature, and thus serves as your thermostat.

EXHIBIT 25-6 SUMMARY OF HEAT-PRODUCING AND HEAT-LOSING MECHANISMS	
MECHANISM	**COMMENT**
Heat production	Most body heat is produced by the oxidation of foods, and the rate at which it is produced is called the metabolic rate. The rate is affected by exercise, strong sympathetic stimulation, hormones, and body temperature.
Heat loss	
Radiation	Transfer of heat from the body to an object without physical contact. Example is losing heat to a cool object such as a floor.
Conduction	Transfer of heat from the body to any object in physical contact with the body such as clothing.
Convection	When cool air makes contact with the body, it is warmed and carried away by convection currents.
Evaporation	The conversion of a liquid to a vapor in which the evaporating substance (e.g., perspiration) removes heat from the body.

Impulses from the preoptic area are sent to other portions of the hypothalamus known as the heat-losing center and the heat-promoting center. The **heat-losing center,** when stimulated by the preoptic area, sets into operation a series of responses that lower body temperature. The **heat-promoting center,** when stimulated by the preoptic area, sets into operation a series of responses that raise body temperature. The heat-losing center is mainly parasympathetic in function; the heat-promoting center is primarily sympathetic.

Mechanisms of Heat Production

Suppose the environmental temperature is low or blood temperature falls below normal. Both stresses stimulate the preoptic area. The preoptic area, in turn, activates the heat-promoting center. In response, the heat-promoting center discharges impulses that automatically set into operation a number of responses designed to increase body heat and bring body temperature back up to normal.

- *Vasoconstriction*   Impulses from the heat-promoting center stimulate sympathetic nerves that cause blood vessels of the skin to constrict. The net effect of vasoconstriction is to decrease the flow of warm blood from the internal organs to the skin, thus decreasing the transfer of heat from the internal organs to the skin. This reduction in heat loss helps raise the internal body temperature.

- *Sympathetic Stimulation*   Another response triggered by the heat-promoting center is the sympathetic stimulation of metabolism. The heat-promoting center stimulates sympathetic nerves leading to the adrenal medulla. This stimulation causes the medulla to secrete epinephrine and norepinephrine (NE) into the blood. The hormones, in turn, bring about an increase in cellular metabolism, a reaction that also increases heat production. This effect is called **chemical thermogenesis.**

- *Skeletal Muscles*   Heat production is also increased by responses of skeletal muscles. For example, stimulation of the heat-promoting center causes stimulation of parts of the brain that increase muscle tone and hence heat production. As the muscle tone increases, the stretching of the agonist muscle initiates the stretch reflex and the muscle contracts. This contraction causes the antagonist muscle to stretch, and it too develops a stretch reflex. The repetitive cycle—called **shivering**—increases the rate of heat production. During maximal shivering, body heat production can rise up to about four times normal.

- *Thyroxine*   Another body response that increases heat production is increased production of thyroxine. A cold environmental temperature causes the secretion of the regulating factor TRF produced by the preoptic area of the hypothalamus. The TRF in turn stimulates the anterior pituitary to secrete TSH, which causes the thyroid to release thyroxine into the blood. Since increased levels of thyroxine increase the metabolic rate, body temperature is increased.

Mechanisms of Heat Loss

Now suppose some stress raises body temperature above normal. The stress or higher temperature of the blood stimulates the preoptic area, which in turn stimulates the heat-losing center and inhibits the heat-promoting center. Instead of blood vessels in the skin constricting, they dilate. The skin becomes warm, and the excess heat is lost to the environment. At the same time, the metabolic rate and muscle tone are decreased. All these responses reverse the heat-promoting effects and bring body temperature down to normal.

When the body is subjected to high environmental temperatures or strenuous exercise, the high temperature of the blood signals the hypothalamus, which activates the heat-losing center. In response, impulses are sent out to the sweat glands of the skin, and they produce more perspiration. As the perspiration evaporates from the surface of the skin, the skin is cooled.

BODY TEMPERATURE ABNORMALITIES

A **fever** is an abnormally high body temperature. The most frequent cause of fever is infection from bacteria (and their toxins) and viruses. Other causes are heart attacks, tumors, tissue destruction by x rays or trauma, and reactions to vaccines. Any substances capable of producing fever are called **exogenous** (*exo* = outside; *gennan* = to produce) **pyrogens** (*pyro* = fire). The mechanism of fever production is believed to occur as follows. Exogenous pyrogens cause certain phagocytes, namely monocytes and macrophages, to synthesize small proteins called **endogenous** (*endo* = within) **pyrogens.** Endogenous pyrogens circulate to the anterior hypothalamus and induce neurons of the preoptic area to secrete prostaglandins, particularly of the E series. Prostaglandins reset the hypothalamic thermostat at a higher temperature. Both aspirin and acetaminophen (Tylenol) reduce fever by inhibiting synthesis of prostaglandins.

Suppose that as a result of exogenous pyrogens the thermostat is set at 39.4°C (103°F). Now the heat-promoting mechanisms (vasoconstriction, increased metabolism, shivering) are operating at full force. Thus, even though body temperature is climbing higher than normal, say 38.3°C (101°F), the skin remains cold and shivering occurs. This condition, called a **chill,** is a definite sign that body temperature is rising. After several hours, body temperature reaches the setting of the thermostat and the chills disappear. But the body will continue to regulate temperature at 39.4°C (103°F) until the stress is removed. When the stress is removed, the thermostat is reset at normal—37°C (98.6°F). Since body temperature remains high in the beginning, the heat-losing mechanisms (vasodilation and sweating) go into operation to decrease body temperature. The skin becomes warm and the person begins to sweat. This phase of the fever is called the **crisis,** and indicates that body temperature is falling.

Up to a point, fever is beneficial. The high body temperature is believed to inhibit the growth of some bacteria and viruses. Fever also increases heart rate so that white blood cells are delivered to sites of infection more rapidly. Moreover, heat speeds up the rate of chemical reactions. This increase may help body cells to repair themselves more quickly during a disease. Among the complications of fever are dehydration, acidosis, and permanent brain damage. As a rule, death results if body temperature rises to 44.4 to 45.5°C (112 to 114°F). On the other end of the scale, death usually results when body temperature falls to 21.2 to 23.9°C (70 to 75°F).

In addition to fever, several other conditions are associated with abnormal body temperature:

1. **Heat cramps.** These occur as a result of profuse sweating that removes water and salt (NaCl) from the body. The salt loss causes painful contractions of muscles called heat cramps. The cramps tend to occur in muscles used while working, but do not appear until the person relaxes after work. Salted liquids usually lead to rapid improvement.

2. **Heatstroke (sunstroke).** This condition is brought about when the temperature and relative humidity are high, making it difficult for the body to lose heat by radiation or evaporation. As a result, there is a decreased flow of blood to the skin, perspiration is greatly reduced, and body temperature rises sharply. The skin is thus dry and hot—the temperature may reach 110°F. Brain cells are quickly affected and may be destroyed permanently. Treatment must be undertaken immediately and consists of cooling the body by immersing the victim in cool water and by administering fluids and electrolytes.

3. **Heat exhaustion (heat prostration).** The body temperature is generally normal, or a little below, and the skin is cool and clammy (moist) due to profuse perspiration. Unlike heatstroke, in heat exhaustion the heart is generally at fault. Heat exhaustion is normally characterized by fluid and electrolyte loss, especially salt. The salt loss results in muscle cramps, dizziness, vomiting, and fainting. Fluid loss may cause low blood pressure. Complete rest and salt tablets are recommended.

DISORDERS: HOMEOSTATIC IMBALANCES

Obesity

In affluent technological nations such as ours, the demand for physical work is steadily decreasing. This same environment provides a variety of appetizing, rich foods that are easily obtained. One consequence of these conditions is **obesity,** defined as a body weight 10 to 20 percent above a desirable standard as the result of an excessive accumulation of fat. There is little doubt that even moderate obesity is hazardous to health. It is implicated as a risk factor in cardiovascular disease, hypertension, pulmonary disease, diabetes mellitus (type II), arthritis, varicose veins, and gallbladder disease.

Classification

Clinically, obesity is classified as hypertrophic and hyperplastic. In **hypertrophic (adult-onset) obesity,** there is an increase in the amount of fat in adipocytes, but no increase in the number of fat cells. Such individuals tend to be thin or average in weight until about age 20 to 40, at which time weight gain begins. The gain may be associated with an imbalance between caloric intake and utilization. People with hypertrophic obesity tend to have fat distributed in central locations ("middle-age spread"), and their problem is easier to treat. The second category of obesity is **hyperplastic (lifelong) obesity,** in which there is an increase in the number of adipocytes, as well as an increase in the amount of fat within them. These individuals tend to be obese as children and have a large spurt in weight gain during adolesence. After adolesence, the number of adipocytes remains about the same throughout life. Such people are generally grossly obese, and the distribution of fat is both central and peripheral. Treatment is considerably more difficult.

Causes

In a relatively few cases, obesity may result from trauma or tumors in the food-regulating centers in the hypothalamus. In most cases of obesity, no specific cause can be identified. Contributing factors include eating habits taught early in life, overeating to relieve tension, and social customs. Recently, obesity has been strongly linked to genetic factors.

Treatment

Reduction of body weight involves keeping caloric intake well below energy expenditure. The goals during weight decrease are:

1. Loss of body fat with a minimal accompanying breakdown of lean tissue.
2. Maintenance of physical and emotional fitness during the reducing period.
3. Establishment of eating and exercise habits to maintain weight at the recommended level.

If dietary restrictions and drugs still fail to control the patient's obesity, the only alternative may be surgery. Three surgical procedures for overweight are the jejunoileal (intestinal) bypass, gastroplasty (gastric stapling), and gastric banding. The suitable patients for these procedures are individuals whose weight is more than 100 pounds over the ideal body weight for height or 100 percent over the ideal.

Starvation

Starvation may arise from inadequate intake of nutrients or the inability to digest, absorb, or metabolize ingested nutrients. Starvation may be voluntary, as in fasting or anorexia nervosa, or involuntary, as in deprivation or disease (e.g. diabetes mellitus, cancer). Regardless of the cause, starvation is characterized by the loss of energy stores in the form of glycogen, fats, and proteins.

In the first stage of starvation, carbohydrate stores in the body are depleted first. During approximately the first day of fasting, low glucose levels stimulate glucagon secretion by the pancreas. As a result, glycogen is converted to glucose (glycogenolysis) and released from the liver. This restores blood glucose levels to normal and makes the glucose available for use by body cells, including brain cells.

Once glycogen stores are depleted, the next stage of starvation occurs. During this time, the primary energy source for most body cells is fatty acids from lipid stores. As the liver metabolizes in the fatty acids, ketone bodies are produced in large quantities and transported to body cells. Even brain cells use ketone bodies as a source of energy. However, since body cells are limited in the amount of ketone bodies they can metabolize, excess ketone bodies appear in the blood, resulting in a condition called ketosis. This, in turn, leads to metabolic acidosis, a decrease in the pH of blood below normal, since the body cannot neutralize the excess ketone bodies. As you will see in Chapter 27, metabolic acidosis results in depression of the central nervous system that may lead to coma. Ketosis and metabolic acidosis are frequently associated with crash dieting, low-carbohydrate diets, high-protein diets, and other fad diets. During the early stages of starvation, large quantities of muscle

protein not essential to cellular functioning are broken down to amino acids. These, in turn, are converted by the liver into glucose (gluconeogenesis). Although this glucose is used to maintain a fairly normal blood sugar level, muscle and other tissues use ketone bodies as an energy source. The length of the second stage of starvation is primarily determined by the amount of stored fat in the body.

When fat reserves are depleted, the third stage of starvation occurs. During this time, even the proteins needed to maintain cellular functions are broken down as a source of energy. It is estimated that once protein stores are depleted to about one-half of their normal level, death results. The third stage of starvation occurs very rapidly once it has begun, frequently leading to death in 24 hours.

Phenylketonuria (PKU)

By definition, **phenylketonuria** (fen'-il-kē'-tō-NOO-rē-a) or **PKU** is a genetic error of metabolism characterized by an elevation of the amino acid phenylalanine in the blood. It is frequently associated with mental retardation. The DNA of children with phenylketonuria lacks the gene that normally programs the manufacture of the enzyme phenylalanine hydroxylase. This enzyme is necessary for converting phenylalanine into the amino acid tryosine, an amino acid that enters the Krebs cycle. As a result, phenylalanine cannot be metabolized, and what is not used in protein synthesis builds up in the blood. High levels of phenylalanine are toxic to the brain during the early years of life when the brain is developing. Mental retardation can be prevented, when the condition is detected early, by restricting the child to a diet that supplies only the amount of phenylalanine necessary for growth.

Cystic Fibrosis

Cystic fibrosis is an inherited disease of the exocrine glands that affects the pancreas, respiratory passageways, and salivary and sweat glands. It is the most common lethal genetic disease of Caucasians—5 percent of the population are thought to be genetic carriers. The cause of cystic fibrosis has recently been linked to an inability of chloride (Cl^-) ions to cross epithelial cells in affected regions of the body.

Among the most common signs and symptoms are pancreatic insufficiency, pulmonary involvement, and cirrhosis of the liver. It is characterized by the production of thick exocrine secretions that do not drain easily from the respiratory passageways. The buildup of the secretions leads to inflammation and replacement of injured cells with connective tissue that blocks these passageways. One of the prominent features is blockage of the pancreatic ducts so that the digestive enzymes cannot reach the intestine. Since pancreatic juice contains the only fat-digestion enzyme, the person fails to absorb fats or fat-soluble vitamins and thus suffers from vitamin A, D, and K deficiency diseases. Calcium needs fat to be absorbed, so tetany also may result.

A child suffering from cystic fibrosis is given pancreatic extract and large doses of vitamins A, D, and K. The therapeutic diet is low, but not lacking, in fats and high in carbohydrates and proteins that can be used for energy and can also be converted by the liver into the lipids essential for life processes.

Celiac Disease

Celiac disease results in malabsorption by the intestinal mucosa due to the ingestion of gluten. *Gluten* is the water-insoluble protein fraction of wheat, rye, barley, and oats. In susceptible persons, ingestion of gluten induces destruction of villi and inhibition of enzyme secretion accompanied by a variable amount of malabsorption. The condition is easily remedied by administering a diet that excludes all cereal grains except rice and corn.

Kwashiorkor

Some dietary proteins are referred to as complete proteins, that is, they contain adequate amounts of essential amino acids. Sources of complete proteins are primarily animal products such as milk, meat, fish, poultry, and eggs. Incomplete proteins lack essential amino acids. An example is zein, the protein in corn, which lacks the essential amino acids tryptophan and lysine. The diet of many African natives consists largely of corn meal. As a result, many African children especially develop a protein deficiency disorder called **kwashiorkor** (kwash-ē-OR-kor). It is characterized by hypoprotein edema of the abdomen, lethargy, failure to grow, and sometimes mental retardation.

STUDY OUTLINE

Metabolism (p. 634)

1. Nutrients are chemical substances in food that provide energy, act as building blocks in forming new body components, or assist in the functioning of various body processes.
2. There are six major classes of nutrients: carbohydrates, lipids, proteins, minerals, vitamins, and water.
3. Metabolism refers to all chemical reactions of the body and has two phases: catabolism and anabolism.
4. Anabolism consists of a series of synthetic reactions whereby small molecules are built up into larger ones that form the body's structural and functional components. Anabolic reactions use energy.
5. Catabolism is the term for decomposition reactions that provide energy.
6. Oxidation is the removal of hydrogen atoms from a substance; reduction is the addition of hydrogen atoms to a substance; oxidation-reduction reactions are coupled.
7. Metabolic reactions are catalyzed by enzymes, proteins that are very efficient, specific for their substrates, and subject to cellular controls. Names of enzymes typically end in *-ase*.
8. A complete enzyme, consisting of an apoenzyme and cofactor, is known as a holoenzyme.
9. The essential function of an enzyme is to catalyze chemical reactions.

Carbohydrate Metabolism (p. 636)

1. During digestion, polysaccharides and disaccharides are converted to monosaccharides, which are absorbed through capillaries in villi and transported to the liver via the hepatic portal vein.
2. Carbohydrate metabolism is primarily concerned with glucose metabolism.

Fate of Carbohydrates (p. 636)

1. Some glucose is oxidized by cells to provide energy; it moves into cells by faciliated diffusion and becomes phosphorylated to glucose-6-phosphate; insulin stimulates glucose movement into cells.
2. Excess glucose can be stored by the liver and skeletal muscles as glycogen or converted to fat.
3. Glucose excreted in the urine can produce glycosuria.

Glucose Catabolism (p. 636)

1. Glucose oxidation is also called cellular respiration.
2. The complete oxidation of glucose to CO_2 and H_2O involves glycolysis, the Krebs cycle, and the electron transport chain.

Glycolysis (p. 636)

1. Glycolysis refers to the breakdown of glucose into two molecules of pyruvic acid.
2. When oxygen is in short supply, pyruvic acid is converted to lactic acid; under aerobic conditions, pyruvic acid enters the Krebs cycle.
3. As a result of glycolysis, there is a net production of 2 molecules of ATP.

Krebs Cycle (p. 639)

1. Pyruvic acid is prepared for entrance into the Krebs cycle by conversion to a two-carbon compound (acetyl group) followed by the addition of coenzyme A to form acetyl coenzyme A.
2. The Krebs cycle involves decarboxylations and oxidations and reductions of various organic acids.
3. Each molecule of pyruvic acid that enters the Krebs cycle produces 3 molecules of CO_2, 4 molecules of $NADH_2$, 1 molecule of $FADH_2$, and 1 molecule of GTP.
4. The energy originally in glucose and then pyruvic acid is primarily in the reduced coenzymes $NADH_2$ and $FADH_2$.

Electron Transport Chain (p. 640)

1. The electron transport chain is a series of oxidation-reduction reactions in which the energy in $NADH_2$ and $FADH_2$ is liberated and transferred to ATP for storage.
2. The carrier molecules involved include FAD, coenzyme Q, and cytochromes.
3. The electron transport chain yields 34 molecules of ATP and H_2O.
4. The complete oxidation of glucose can be represented as follows:

$$C_6H_{12}O_6 + 6O_2 \rightarrow 38ATP + 6CO_2 + 6H_2O$$

Glucose Anabolism (p. 643)

1. The conversion of glucose to glycogen for storage in the liver and skeletal muscle is called glycogenesis. The process occurs in the liver and is stimulated by insulin.
2. The body can store about 500 g of glycogen.
3. The conversion of glycogen back to glucose is called glycogenolysis.
4. It occurs between meals and is stimulated by glucagon and epinephrine.
5. Gluconeogenesis is the conversion of fat and protein molecules into glucose. It is stimulated by cortisol, thyroxine, epinephrine, glucagon, and growth hormone (GH).
6. Glycerol may be converted to glyceraldehyde-3-phosphate, and some amino acids may be converted to pyruvic acid.

Lipid Metabolism (p. 644)

1. During digestion, fats are ultimately broken down into fatty acids and monoglycerides.
2. Long-chain fatty acids and monoglycerides are carried in micelles for entrance into villi, digested to glycerol and fatty acids in epithelial cells, recombined to form triglycerides, and transported by chylomicrons through the lacteals of villi into the thoracic duct.

Fate of Lipids (p. 644)

1. Some fats may be oxidized to produce ATP.
2. Some fats are stored in adipose tissue.
3. Other lipids are used as structural molecules or to synthesize essential molecules. Examples include phospholipids of plasma membranes, lipoproteins that transport cholesterol,

thromboplastin for blood clotting, and cholesterol used to synthesize bile salts and steroid hormones.

Fat Storage (p. 644)
1. Fats are stored in adipose tissue, mostly in the subcutaneous layer.
2. Adipose cells contain lipases that catalyze the deposition of fats from chylomicrons and hydrolyze fats into fatty acids and glycerol.

Lipid Catabolism (p. 644)
1. Fat is released from depots and split into fatty acids and glycerol under the influence of growth hormone (GH).
2. Glycerol can be converted into glucose by conversion into glyceraldehyde-3-phosphate.
3. In beta oxidation, carbon atoms are removed in pairs from fatty acid chains; the resulting molecules of acetyl coenzyme A enter the Krebs cycle.
4. The formation of ketone bodies by the liver is a normal phase of fatty acid catabolism, but an excess of ketone bodies, called ketosis, may cause acidosis.

Lipid Anabolism: Lipogenesis (p. 646)
1. The conversion of glucose or amino acids into lipids is called lipogenesis. The process is stimulated by insulin.
2. The intermediary links in lipogenesis are glyceraldehyde-3-phospate and acetyl coenzyme A.

Protein Metabolism (p. 646)
1. During digestion, proteins are hydrolyzed into amino acids.
2. Amino acids are absorbed by the capillaries of villi and enter the liver via the hepatic portal vein.

Fate of Proteins (p. 646)
1. Amino acids, under the influence of growth hormone (GH) and insulin, enter body cells by active transport.
2. Inside cells, amino acids are synthesized into proteins that functions as enzymes, hormones, structural elements, and so forth. Very little protein is used as a source of energy.

Protein Catabolism (p. 646)
1. Before amino acids can be catabolized, they must be converted to substances that can enter the Krebs cycle; these conversions involve deamination, decarboxylation, and hydrogenation.
2. Amino acids may also be converted into glucose, fatty acids, and ketone bodies.

Protein Anabolism (p. 646)
1. Protein synthesis is stimulated by growth hormone (GH), thyroxine, and insulin.
2. The process is directed by DNA and RNA and carried out on the ribosomes of cells.

Absorptive and Postabsorptive States (p. 647)
1. During the absorptive state, ingested nutrients enter the blood and lymph from the GI tract.
2. During the absorptive state, most blood glucose is used by body cells for oxidation. Glucose transported to the liver is converted to glycogen or fat. Most fat is stored in adipose tissue. Amino acids in liver cells are converted to carbohydrates, fats, and proteins.
3. During the postabsorptive (fasting) state, absorption is complete and the energy needs of the body are satisfied by nutrients already present in the body.
4. The major concern of the body during the postabsorptive state is to maintain normal blood glucose level. This involves conversion of liver and skeletal muscle glycogen into glucose,

conversion of glycerol into glucose, and conversion of amino acids into glucose. The body also switches from glucose oxidation to fatty acid oxidation.

Regulation of Metabolism (p. 649)
1. Absorbed nutrients may be oxidized, stored, or converted, based on the needs of the body.
2. The pathway taken by a particular nutrient is enzymatically controlled and is regulated by hormones (see Exhibit 25-2).

Minerals (p. 649)
1. Minerals are inorganic substances that help regulate body processes.
2. Minerals known to perform essential functions are calcium, phosphorus, sodium, chlorine, potassium, magnesium, iron, sulfur, iodine, manganese, cobalt, copper, zinc, selenium, and chromium. Their functions are summarized in Exhibit 25-3.

Vitamins (p. 652)
1. Vitamins are organic nutrients that maintain growth and normal metabolism. Many function in enzyme systems.
2. Fat-soluble vitamins are absorbed with fats and include A, D, E, and K.
3. Water-soluble vitamins are absorbed with water and include the B vitamins and vitamin C.
4. The functions and deficiency disorders of the principal vitamins are summarized in Exhibit 25-4.
5. Abnormalities associated with megadoses of selected minerals and vitamins are summarized in Exhibit 25-5.

Metabolism and Body Heat (p. 655)
1. A kilocalorie (kcal) is the amount of energy required to raise the temperature of 1,000 g of water 1°C from 14° to 15°C.
2. The kilocalorie is the unit of heat used to express the caloric value of foods and to measure the body's metabolic rate.
3. The apparatus used to determine the caloric value of foods is called a calorimeter.

Production of Body Heat (p. 655)
1. Most body heat is a result of oxidation of the food we eat. The rate at which this heat is produced is known as the metabolic rate.
2. Metabolic rate is affected by exercise, the nervous system, hormones, body temperature, ingestion of food, age, sex, climate, sleep, and malnutrition.
3. Measurement of the metabolic rate under basal conditions is called the basal metabolic rate (BMR).
4. BMR is expressed in kilocalories per square meter of body area per hour ($kcal/m^2/hr$).

Loss of Body Heat (p. 656)
1. Radiation is the transfer of heat as infrared heat rays from one object to another without physical contact.
2. Conduction is the transfer of body heat to a substance or object in contact with the body.
3. Convection is the transfer of body heat by the movement of air that has been warmed by the body.
4. Evaporation is the conversion of a liquid to a vapor.

Body Temperature Regulation (p. 657)
1. A normal body temperature is maintained by a delicate balance between heat-producing and heat-losing mechanisms.
2. The hypothalamic thermostat is the preoptic area.
3. Mechanisms that produce heat are vasoconstriction, sympathetic stimulation, skeletal muscle contraction, and thyroxine production.
4. Mechanisms of heat loss include vasodilation, decreased met-

abolic rate, decreased skeletal muscle contraction, and perspiration.

Body Temperature Abnormalities (p. 658)

1. Fever is an abnormally high body temperature caused by prostaglandins, with exogenous and endogenous pyrogens and mediators.
2. Heat cramps are painful skeletal muscle contractions due to loss of salt and water.
3. Heatstroke results in decreased blood flow to skin, reduced perspiration, and high body temperature. Fluid therapy and body cooling are indicated.
4. Heat exhaustion results in a normal or below normal body temperature, profuse perspiration, nausea, cramps, and dizziness. Rest and salt tablets are indicated.

Disorders: Homeostatic Imbalances (p. 658)

1. Obesity is defined as a body weight 10 to 20 percent above desirable standard as the result of excessive accumulation of fat. It is classified as hypertrophic and hyperplastic.
2. Starvation progresses through three stages that involve depletion of carbohydrates, proteins, and lipids.
3. Phenylketonuria (PKU) is a genetic error of metabolism characterized by an elevation of phenylalanine in the blood.
4. Cystic fibrosis is a metabolic disease of the exocrine glands in which absorption of vitamins A, D, and K and calcium is inadequate.
5. Celiac disease is a condition in which the ingestion of gluten causes morphological changes in the small intestinal mucosa resulting in malabsorption.
6. Kwashiorkor is a protein deficiency disorder characterized by hypoprotein edema, lethargy, failure to grow, and sometimes mental retardation.

REVIEW QUESTIONS

1. Define a nutrient. List the six classes of nutrients and indicate the function of each.
2. What is metabolism? Distinguish between anabolism and catabolism, and give examples of each.
3. Define oxidation and reduction, and give an example of each.
4. What are the functions of enzymes as catalysts in biochemical reactions.
5. List and explain the characteristics of enzymes. Describe the parts of an enzyme.
6. Briefly describe the mechanism of enzyme action.
7. How are carbohydrates absorbed and what are their fates in the body? How does glucose move into body cells?
8. Define glycolysis. Describe its principal events and outcome.
9. Describe how acetyl coenzyme A is formed.
10. Outline the principal events and outcomes of the Krebs cycle.
11. Explain what happens in the electron transport chain.
12. Summarize the outcomes of the complete oxidation of a molecule of glucose.
13. Define glycogenesis and glycogenolysis. Under what circumstances does each occur?
14. Why is gluconeogenesis important? Give specific examples to substantiate your answer.
15. How are fats absorbed and what are their fates in the body? Where are fats stored in the body?
16. Explain the principal events of the catabolism of glycerol and fatty acids.
17. What are ketone bodies? What is ketosis?
18. Define lipogenesis and explain its importance.
19. How are proteins absorbed and what are their fates in the body?
20. Relate deamination to amino acid catabolism.
21. Summarize the major steps involved in protein synthesis.
22. Distinguish between essential and nonessential amino acids.
23. What is the absorptive state? Outline its principal events.
24. What is the postabsorptive (fasting) state? Outline its principal events.
25. Indicate the roles of the following hormones in the regulation of metabolism: insulin, glucagon, epinephrine, growth hormone (GH), thyroxine, cortisol, and testosterone.
26. What is a mineral? Briefly describe the functions of the following minerals: calcium, phosphorus, iron, iodine, copper, sodium, potassium, chlorine, magnesium, sulfur, zinc, fluorine, manganese, cobalt, chromium, and selenium.
27. Define a vitamin. Explain how we obtain vitamins. Distinguish between a fat-soluble and a water-soluble vitamin.
28. Compare avitaminosis and hypervitaminosis.
29. For each of the following vitamins, indicate its principal function and effect of deficiency: A, D, E, K, B_1, B_2, niacin, B_6, B_{12}, pantothenic acid, folic acid, biotin, and C.
30. Describe abnormalities associated with megadoses of several minerals and vitamins.
31. Define a kilocalorie. How is the unit used?
32. How is the caloric value of foods determined?
33. What is metabolic rate? What factors affect it?
34. Define basal metabolic rate (BMR). How is it measured?
35. Define each of the following mechanisms of heat loss: radiation, conduction, convection, and evaporation.
36. Explain how body temperature is regulated by describing the mechanisms of heat production and heat loss.
37. Contrast fever, heat cramps, heatstroke, and heat exhaustion as body temperature abnormalities.
38. What causes fever? What are its benefits?
39. What is obesity? How is it classified? How is it treated?
40. Describe the stages of starvation.
41. Define phenylketonuria (PKU), cystic fibrosis, celiac disease, and kwashiorkor.

The Urinary System

26

STUDENT OBJECTIVES

- Identify the external and internal gross anatomical features of the kidneys.
- Define the structural adaptations of a nephron for urine formation.
- Describe the blood and nerve supply to the kidneys.
- Describe the structure and function of the juxtaglomerular apparatus.
- Discuss the process of urine formation through glomerular filtration, tubular reabsorption, and tubular secretion.
- Define the forces that support and oppose the filtration of blood in the kidneys.
- Explain the mechanism and importance of tubular reabsorption.
- Compare the obligatory and facultative reabsorption of water.
- Explain the mechanism and importance of tubular secretion.
- Explain how the kidneys produce dilute and concentrated urine.
- Discuss the operational principle of hemodialysis.
- Compare the effects of blood pressure, blood concentration, temperature, diuretics, and emotions on urine volume.
- List and describe the physical characteristics of urine.
- List the normal chemical constituents of urine.
- Define albuminuria, glycosuria, hematuria, pyuria, ketosis, bilirubinuria, urobilinogenuria, casts, and renal calculi.
- Discuss the structure and physiology of the ureters.
- Describe the structure and physiology of the urinary bladder.
- Explain the physiology of the micturition reflex.
- Compare the causes of incontinence and retention.
- Explain the structure and physiology of the urethra.
- Describe the effects of aging on the urinary system.
- Describe the development of the urinary system.
- Discuss the causes of gout, glomerulonephritis, pyelitis, pyelonephritis, cystitis, nephrosis, polycystic disease, renal failure, and urinary tract infections (UTIs).
- Define medical terminology associated with the urinary system.

The metabolism of nutrients results in the production of wastes by body cells, including carbon dioxide and excess water and heat. Protein catabolism produces toxic nitrogenous wastes such as ammonia and urea. In addition, many of the essential ions such as sodium, chloride, sulfate, phosphate, and hydrogen tend to accumulate in excess of the body's needs. All the toxic materials and the excess essential materials must be eliminated.

The primary function of the **urinary system** is to help keep the body in homeostasis by controlling the composition and volume of blood. It does so by removing and restoring selected amounts of water and solutes. Two kidneys, two ureters, one urinary bladder, and a single urethra make up the system (Figure 26-1). The kidneys regulate the composition and volume of the blood and remove wastes from the blood in the form of urine. They excrete selected amounts of various wastes, assume a role in erythropoiesis by forming renal erythropoietic factor, help control blood pH, help regulate blood pressure by secreting renin (which activates the renin-angiotensin pathway), and participate in the activation of vitamin D. Urine is excreted from each kidney through its ureter and is stored in the urinary bladder until it is expelled from the body through the urethra. Other systems that aid in waste elimination are the respiratory, integumentary, and digestive systems (see Exhibit 26-3).

The specialized branch of medicine that deals with structure, function, and diseases of the male and female urinary systems and the male reproductive system is known as **urology** (yoo-ROL-ō-jē; *uro* = urine or urinary tract; *logos* = study of).

The developmental anatomy of the urinary system is considered at the end of the chapter.

KIDNEYS

The paired **kidneys** are reddish organs that resemble kidney beans in shape. They are found just above the waist between the parietal peritoneum and the posterior wall of the abdomen. Since they are external to the peritoneal lining of the abdominal cavity, their placement is described as *retroperitoneal* (re'-trō-per-i-tō-NĒ-al). Other retroperitoneal structures include the ureters and adrenal (suprarenal) glands. Relative to the vertebral column, the kidneys are located between the levels of the last thoracic and third lumbar vertebrae and are partially protected by the eleventh and twelfth pairs of ribs. The right kidney is slightly lower than the left because of the large area occupied by the liver.

EXTERNAL ANATOMY

The average adult kidney measures about 10 to 12 cm (4 to 5 inches) long, 5.0 to 7.5 cm (2 to 3 inches) wide, and 2.5 cm (1 inch) thick. Its concave medial border faces the vertebral column. Near the center of the concave border is a notch called the **hilus,** through which the ureter leaves the kidney. Blood and lymph vessels and nerves also enter and exit the kidney through the hilus (Figure 26-2). The hilus is the entrance to a cavity in the kidney called the **renal sinus.**

Three layers of tissue surround each kidney. The innermost layer, the **renal capsule,** is a smooth, transparent, fibrous membrane that can easily be stripped off the kidney and is continuous with the outer coat of the ureter at the hilus. It serves as a barrier against trauma and the spread of infection to the kidney. The second layer,

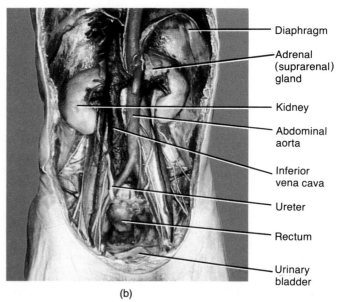

(a) (b)

FIGURE 26-1 Organs of the male urinary system in relation to surrounding structures. (a) Diagram. (b) Photograph. (Courtesy of C. Yokochi and J. W. Rohen, *Photographic Anatomy of the Human Body,* 2nd ed., 1978, IGAKU-SHOIN, Ltd., Tokyo, New York.)

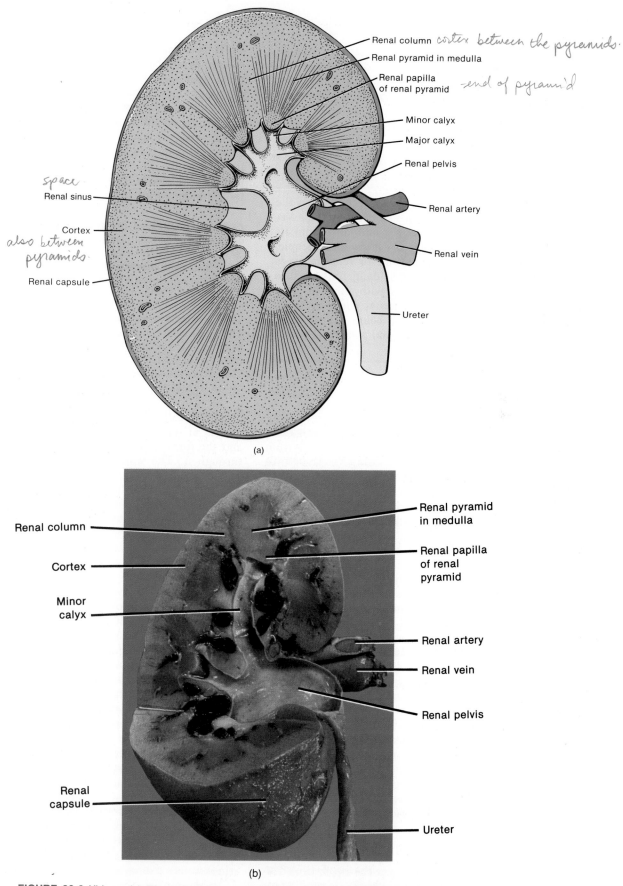

Renal column *cortex between the pyramids.*
Renal pyramid in medulla
Renal papilla of renal pyramid *end of pyramid*
Minor calyx
Major calyx
Renal pelvis
Renal artery
Renal vein

space
Renal sinus
Cortex
also between pyramids.
Renal capsule
Ureter

(a)

Renal column
Cortex
Minor calyx
Renal capsule

Renal pyramid in medulla
Renal papilla of renal pyramid
Renal artery
Renal vein
Renal pelvis
Ureter

(b)

FIGURE 26-2 Kidney. (a) Diagram of a coronal section of the right kidney illustrating the internal anatomy. (b) Photograph of a coronal and a cross section of the right kidney illustrating the internal anatomy. (Courtesy of C. Yokochi and J. W. Rohen, *Photographic Anatomy of the Human Body*, 2nd ed., 1978, IGAKU-SHOIN, Ltd., Tokyo, New York.)

the **adipose capsule,** is a mass of fatty tissue surrounding the renal capsule. It also protects the kidney from trauma and holds it firmly in place within the abdominal cavity. The outermost layer, the **renal fascia,** is a thin layer of fibrous connective tissue that anchors the kidney to its surrounding structures and to the abdominal wall.

CLINICAL APPLICATION

Floating kidney, or **ptosis,** (TŌ-sis) occurs when the kidney is no longer held in place securely by the adjacent organs or its covering of fat and slips from its normal position. Individuals, especially thin people, in whom either the adipose capsule or renal fascia is deficient, may develop ptosis. It is dangerous because it may cause kinking of the ureter with reflux of urine and retrograde pressure. Pain occurs if the ureter is twisted. Also, if the kidneys drop below the rib cage, they become susceptible to blows and penetrating injuries.

INTERNAL ANATOMY

A coronal (frontal) section through a kidney reveals an outer, reddish area called the **cortex** and an inner, reddish-brown region called the **medulla** (Figure 26-2). Within the medulla are 5 to 14 striated, triangular structures termed **renal (medullary) pyramids.** The striated appearance is due to the presence of straight tubules and blood vessels. The bases of the pyramids face the cortical area, and their apices, called **renal papillae,** are directed toward the center of the kidney. The cortex is the smooth-textured area extending from the renal capsule to the bases of the pyramids and into the spaces between them. It is divided into an outer cortical zone and an inner juxtamedullary zone. The cortical substance between the renal pyramids forms the **renal columns.**

Together the cortex and renal pyramids constitute the parenchyma of the kidney. Structurally, the parenchyma of each kidney consists of approximately 1 million microscopic units called nephrons, collecting ducts, and their associated vascular supply. Nephrons are the functional units of the kidney. They help regulate blood composition and form urine.

In the renal sinus of the kidney is a large cavity called the **renal pelvis.** The edge of the pelvis contains cuplike extensions called **major** and **minor calyces** (KĀ-li-sēz). There are 2 or 3 major calyces and 8 to 18 minor calyces. Each minor calyx collects urine from collecting ducts of the pyramids. From the major calyces, the urine drains into the pelvis and out through the ureter.

NEPHRON

The functional unit of the kidney is the **nephron** (NEF-ron) (Figure 26-3). Essentially, a nephron is a **renal tubule** and its vascular component. It begins as a double-walled cup, called the **glomerular (Bowman's) capsule,** lying in the cortex of the kidney. The outer wall, or *parietal layer,* is composed of simple squamous epithelium (Figure 26-

4). It is separated from the inner wall, known as the *visceral layer,* by the *capsular space.* The visceral layer consists of epithelial cells called podocytes. It surrounds a capillary network called the **glomerulus** (glō-MER-yoo-lus). Collectively, the glomerular capsule and its enclosed glomerulus constitute a **renal corpuscle** (KŌR-pus-sul).

The visceral layer of the glomerular capsule and the endothelium of the glomerulus form an **endothelial-capsular membrane.** This membrane consists of the following parts, listed here in the order in which substances filtered by the kidney must pass through them.

1. **Endothelium of the glomerulus.** This single layer of endothelial cells has completely open pores (fenestrated) averaging 50 to 100 nm in diameter.
2. **Basement membrane of the glomerulus.** This extracellular membrane lies beneath the endothelium and contains no pores. It consists of fibrils in a glycoprotein matrix. It serves as the dialyzing membrane.
3. **Epithelium of the visceral layer of the glomerular capsule.** These epithelial cells, because of their peculiar shape, are called **podocytes.** The podocytes contain footlike structures called **pedicels** (PED-i-sels). The pedicels are arranged parallel to the circumference of the glomerulus and cover the basement membrane except for spaces between them called **filtration slits (slit pores).**

The endothelial-capsular membrane filters water and solutes in the blood. Large molecules, such as proteins, and the formed elements in blood do not normally pass through it. The water and solutes that are filtered out of the blood pass into the capsular space between the visceral and parietal layers of the glomerular capsule and then into the renal tubule.

The glomerular capsule opens into the first section of the renal tubule, called the **proximal convoluted tubule,** which also lies in the cortex. Convoluted means the tubule is coiled rather than straight; proximal signifies that the glomerular capsule is the origin of the tubule. The wall of the proximal convoluted tubule consists of cuboidal epithelium with microvilli. These cytoplasmic extensions, like those of the small intestine, increase the surface area for reabsorption and secretion.

Nephrons are frequently classified into two kinds. A **cortical nephron** usually has its glomerulus in the outer cortical zone, and the remainder of the nephron rarely penetrates the medulla. A **juxtamedullary nephron** usually has its glomerulus close to the corticomedullary junction, and other parts of the nephron penetrate deeply into the medulla (see Figure 26-3).

In the juxtamedullary nephron, the renal tubule straightens, becomes thinner, and dips into the medulla, where it is called the **descending (thin) limb of the loop of the nephron.** This section consists of squamous epithelium. The tubule then increases in diameter as it bends into a U-shaped structure called the **loop of the nephron (loop of Henle).** It then ascends toward the cortex as the **ascending (thick) limb of the loop of the nephron,** which consists of cuboidal and low columnar epithelium.

In the cortex, the tubule again becomes convoluted. Because of its distance from the point of origin at the glomerular capsule, this section is referred to as the **distal convoluted tubule.** The cells of the distal tubule, like those of the proximal tubule, are cuboidal. Unlike the cells of

FIGURE 26-3 Nephrons. (a) Juxtamedullary nephron. (b) Cortical nephron. The nephrons are colored pink.

the proximal tubule, however, the cells of the distal tubule have few microvilli. In a cortical nephron, the proximal section runs into the distal tubule without the intervening limbs and loop of the nephron. In both types of nephron, the distal tubule terminates by merging with a straight **collecting duct.**

In the medulla, the collecting ducts receive the distal tubules of several nephrons, pass through the renal pyramids, and open at the renal papillae into the minor calyces through a number of large **papillary ducts.** On the average, there are 30 papillary ducts per renal papilla. Cells of the collecting ducts are cuboidal; those of the papillary ducts are columnar.

The histology of a nephron and glomerulus is shown in Figure 26-5.

BLOOD AND NERVE SUPPLY

The nephrons are largely responsible for removing wastes from the blood and regulating its fluid and electrolyte content. Thus they are abundantly supplied with blood vessels. The right and left **renal arteries** transport about one-fourth the total cardiac output to the kidneys (Figure 26-6). Approximately 1,200 ml passes through the kidneys every minute.

Before or immediately after entering through the hilus, the renal artery divides into several branches that enter the parenchyma and pass as the **interlobar arteries** between the renal pyramids in the renal columns. At the bases of the pyramids, the interlobar arteries arch between the medulla and cortex; here they are known as the **arcu-**

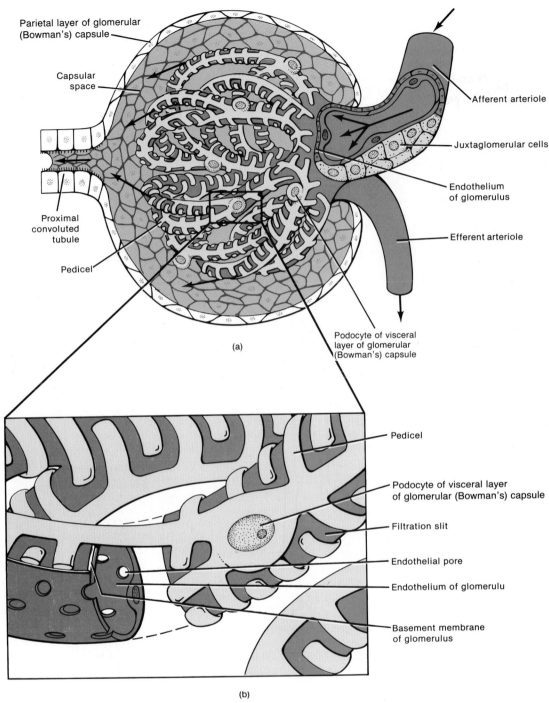

Parietal layer of glomerular
(Bowman's) capsule

Capsular
space

Proximal
convoluted
tubule

Pedicel

Afferent arteriole

Juxtaglomerular cells

Endothelium
of glomerulus

Efferent arteriole

Podocyte of visceral
layer of glomerular
(Bowman's) capsule

(a)

Pedicel

Podocyte of visceral layer
of glomerular (Bowman's) capsule

Filtration slit

Endothelial pore

Endothelium of glomerulu

Basement membrane
of glomerulus

(b)

FIGURE 26-4 Endothelial-capsular membrane. (a) Parts of a renal corpuscle. (b) Enlarged aspect of a portion of the endothelial-capsular membrane. The size of the filtration slits has been exaggerated for emphasis.

ate arteries. Divisions of the arcuate arteries produce a series of **interlobular arteries,** which enter the cortex and divide into **afferent arterioles** (see Figure 26-3).

One afferent arteriole is distributed to each glomerular capsule, where the arteriole divides into the tangled capillary network called the **glomerulus.** The glomerular capillaries then reunite to form an **efferent arteriole,** which leads away from the capsule and is smaller in diameter than the afferent arteriole. This variation in diameter helps

raise the glomerular pressure. The afferent-efferent arteriole situation is unique because blood usually flows out of capillaries into venules and not into other arterioles.

Each efferent arteriole of a cortical nephron divides to form a network of capillaries, called the **peritubular capillaries,** around the convoluted tubules. The efferent arteriole of a juxtamedullary nephron also forms peritubular capillaries. In addition, it forms long loops of thin-walled vessels called **vasa recta** that dip down alongside

Podocyte of visceral
layer of glomerular
(Bowman's) capsule

Pedicels

(c)

Filtration slits

FIGURE 26-4 (Continued) (c) Scanning electron micrograph of a podocyte at a magnification of 7,800×. (From *Tissues and Organs: A Text-Atlas of Scanning Electron Microscopy* by Richard G. Kessel and Randy H. Kardon. W. H. Freeman and Company. Copyright © 1979.)

Macula
densa

Renal corpuscle

Capsular
space

Distal
convoluted
tubules

Glomerulus

Glomerular
(Bowman's)
capsule

(a)

the loop of the nephron into the medullary region of the papilla.

The peritubular capillaries eventually reunite to form **interlobular veins.** The blood then drains through the **arcuate veins** to the **interlobar veins** running between the pyramids and leaves the kidney through a single **renal vein** that exits at the hilus. The vasa recta pass blood into the interlobular veins. From here, it goes to the arcuate veins, the interlobar veins, and then into the renal vein.

The nerve supply to the kidneys is derived from the **renal plexus** of the sympathetic division of the autonomic nervous system. Nerves from the plexus accompany the renal arteries and their branches and are distributed to the vessels. Because the nerves are vasomotor, they regulate the circulation of blood in the kidney by regulating the diameters of the arterioles.

JUXTAGLOMERULAR APPARATUS

The smooth muscle cells of the tunica media adjacent to the afferent arteriole (and sometimes efferent arteriole) are modified in several ways. Their nuclei become rounded (instead of elongated), and their cytoplasm contains granules (instead of myofibrils). Such modified cells are called **juxtaglomerular cells.** The cells of the distal convoluted tubule adjacent to the afferent and efferent arteriole become considerably narrower and taller. Collectively, these

Distal
convoluted
tubules

Proximal
convoluted
tubules

(b)

FIGURE 26-5 Histology of a nephron. (a) Photomicrograph of the cortex of the kidney showing a renal corpuscle and surrounding renal tubules at a magnification of 400×. (b) Photomicrograph of the cortex of the kidney showing renal tubules at a magnification of 400×. (Photomicrographs copyright © 1983 by Michael H. Ross. Used by permission.)

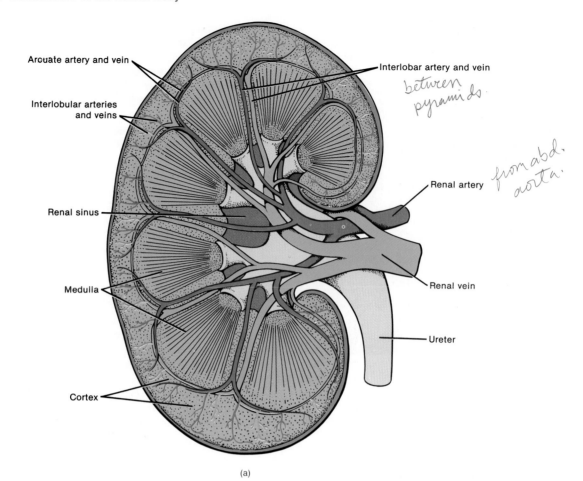

Arcuate artery and vein

Interlobular arteries and veins

Renal sinus

Medulla

Cortex

Interlobar artery and vein

between pyramids

Renal artery

from abd. aorta.

Renal vein

Ureter

(a)

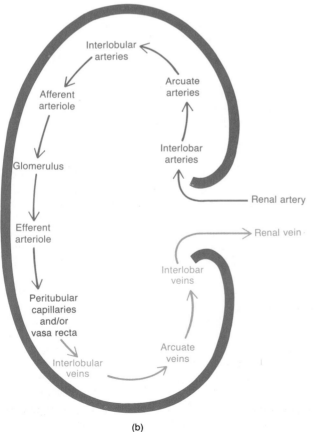

(b)

FIGURE 26-6 Blood supply of the right kidney. (a) Diagram in coronal section. (b) Scheme of circulation. This view is designed to show the *sequence* of blood flow, not the anatomical location of blood vessels, which is shown in (a).

cells are known as the **macula densa.** Together with the modified cells of the afferent arteriole they constitute the **juxtaglomerular apparatus** (Figure 26-7), which helps regulate renal blood pressure. This will be described shortly.

PHYSIOLOGY

The major work of the urinary system is done by the nephrons. The other parts of the system are primarily passageways and storage areas. Nephrons carry out three important functions: (1) they control blood concentration and volume by removing selected amounts of water and solutes; (2) they help regulate blood pH; and (3) they also remove toxic wastes from the blood. As the nephrons go about these activities, they remove many materials from the blood, return the ones that the body requires, and eliminate the remainder. The eliminated materials are collectively called **urine.** The entire volume of blood in the body is filtered by the kidneys approximately 60 times a day. Urine formation requires three principal processes: glomerular filtration, tubular reabsorption, and tubular secretion.

FIGURE 26-7 Juxtaglomerular apparatus. (a) External view. (b) Cells of the juxtaglomerular apparatus seen in cross section. The macula densa adjacent to the efferent arteriole is not illustrated.

Glomerular Filtration

The first step in the production of urine is called **glomerular filtration.** Filtration—the forcing of fluids and dissolved substances through a membrane by pressure—occurs in the renal corpuscle of the kidneys across the endothelial-capsular membrane. When blood enters the glomerulus, the blood pressure forces water and dissolved blood components (plasma) through the endothelial pores of the capillaries, basement membrane, and on through the filtration slits of the adjoining visceral wall of the glomerular capsule (see Figure 26-4). The resulting fluid is the **filtrate.** In a healthy person, the filtrate consists of all the materials present in the blood except for the formed elements and most proteins, which are too large to pass through the endothelial-capsular barrier. Exhibit 26-1 compares the constituents of plasma, glomerular filtrate, and urine during a 24-hour period. Although the

values shown are typical, they vary considerably according to diet. The chemicals listed in plasma are those present in glomerular blood plasma before filtration. The chemicals listed in the filtrate immediately after the glomerular capsule are those that pass from the glomerular blood plasma through the endothelial-capsular membrane before reabsorption. The chemicals in the filtrate are the ones that have been filtered.

Renal corpuscles are especially structured for filtering blood. First, each capsule contains a tremendous length of highly coiled glomerular capillaries presenting a vast surface area for filtration. Second, the endothelial-capsular membrane is structurally adapted for filtration. Although the endothelial pores generally do not restrict the passage of substances, the basement membrane permits the passage of smaller molecules. Thus water, glucose, vitamins, amino acids, small proteins, nitrogenous wastes, and ions pass into the glomerular capsule. Large

EXHIBIT 26-1 CHEMICALS IN PLASMA, FILTRATE, AND URINE DURING 24-HOUR PERIOD*

CHEMICAL	PLASMA	FILTRATE IMMEDIATELY AFTER PASSING INTO GLOMERULAR CAPSULE	REABSORBED FROM FILTRATE	URINE
Water	180,000 ml	180,000 ml	178,000 ml	1,000 ml
Proteins	7,000–9,000	10–20	10–20	0†
Chloride (Cl⁻)	630	630	625	5
Sodium (Na⁺)	540	540	537	3
Bicarbonate (HCO₃⁻)	300	300	299.7	0.3
Glucose	180	180	180	0
Urea	53	53	28	25
Potassium (K⁺)	28	28	24	4
Uric acid	8.5	8.5	7.7	0.8
Creatinine	1.5	1.5	0	1.5

* All values, except for water, are expressed in grams. The chemicals are arranged in sequence from highest to lowest concentration in plasma.

† Although trace amounts of protein (170 to 250 mg) normally appear in urine, we will assume for purposes of discussion that all of it is reabsorbed from filtrate.

proteins and the formed elements in blood do not normally pass through the basement membrane. The filtration slits permit only the occasional passage of very small plasma proteins such as albumins. Third, the efferent arteriole is smaller in diameter than the afferent arteriole, so there is resistance to the outflow of blood from the glomerulus. Consequently, blood pressure is higher in the glomerular capillaries than in other capillaries. Glomerular blood pressure averages about 60 mm Hg, whereas the blood pressure of other capillaries averages only 30 mm Hg. Fourth, the endothelial-capsular membrane separating the blood from the space in the glomerular capsule is very thin (0.1 μm).

The filtering of the blood depends on a number of opposing pressures. The chief one is the **glomerular blood hydrostatic pressure.** *Hydrostatic (hydro = water) pressure* is the force that a fluid under pressure exerts against the walls of its container. Glomerular blood hydrostatic pressure means the blood pressure in the glomerulus (Figure 26-8). This pressure tends to move fluid out of the glomeruli at a force averaging about 60 mm Hg.

However, glomerular blood hydrostatic pressure is opposed by two other forces. The first of these, **capsular hydrostatic pressure,** develops in the following way. When the filtrate is forced into the space between the walls of the glomerular capsule, it meets with two forms of resistance: the walls of the capsule and the fluid that has already filled the renal tubule. As a result, some filtrate is pushed back into the capillary. The amount of "push" is the capsular hydrostatic pressure. It usually measures about 20 mm Hg.

The second force opposing filtration into the glomerular capsule is the **blood colloidal osmotic pressure.** *Os-*

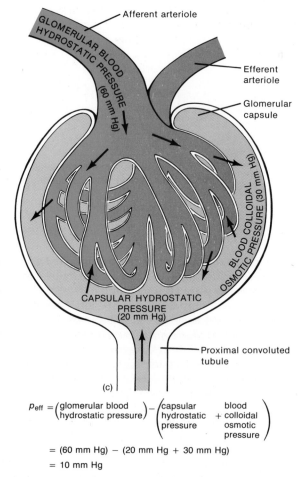

$$p_{eff} = \binom{\text{glomerular blood}}{\text{hydrostatic pressure}} - \binom{\text{capsular} \quad \text{blood}}{\text{hydrostatic} + \text{colloidal}}_{\text{pressure} \quad \text{osmotic} \atop \text{pressure}}$$

$$= (60 \text{ mm Hg}) - (20 \text{ mm Hg} + 30 \text{ mm Hg})$$

$$= 10 \text{ mm Hg}$$

FIGURE 26-8 Forces involved in effective filtration pressure (Peff).

motic pressure is the pressure required to prevent the movement of pure water into a solution containing solutes when the solutions are separated by a selectively permeable membrane. The greater the solute concentration of the solution, the greater its osmotic pressure. Hydrostatic pressure develops because of a force outside a solution. Osmotic pressure develops because of the concentration of the solution itself. Since the blood contains a much higher concentration of proteins than the filtrate does, water moves out of the filtrate and back into the blood vessel. This blood osmotic pressure is normally about 30 mm Hg.

To determine how much filtration finally occurs, we have to subtract the forces that oppose filtration from the glomerular blood hydrostatic pressure. The net result is called the **effective filtration pressure**, which is abbreviated *P*eff.

$$P\text{eff} = \begin{pmatrix} \text{glomerular} \\ \text{blood} \\ \text{hydrostatic} \\ \text{pressure} \end{pmatrix} - \begin{pmatrix} \text{capsular} \\ \text{hydrostatic} + \\ \text{pressure} \end{pmatrix} \begin{matrix} \text{blood} \\ \text{colloidal} \\ \text{osmotic} \\ \text{pressure} \end{matrix}$$

By substituting the values just discussed, a normal *P*eff may be calculated as follows.

$$P\text{eff} = (60 \text{ mm Hg}) - (20 \text{ mm Hg} + 30 \text{ mm Hg})$$
$$= (60 \text{ mm Hg}) - (50 \text{ mm Hg})$$
$$= 10 \text{ mm Hg}$$

This means that a pressure of about 10 mm Hg causes a normal amount of plasma to filter from the glomerulus into the glomerular capsule. This produces about 125 ml of filtrate per minute in both kidneys. The term *filtration fraction* refers to the percentage of plasma entering the nephrons that actually becomes glomerular filtrate. Although filtration fraction averages between 16 and 20 percent, the value does vary considerably in both health and disease.

Certain conditions may alter these pressures and thus the *P*eff. In some forms of kidney disease, such as glomerulonephritis, glomerular capillaries become so permeable that the plasma proteins are able to pass from the blood into the filtrate. As a result, the capsular filtrate exerts an osmotic pressure that draws water out of the blood. Thus, if a capsular osmotic pressure develops, the *P*eff will increase. At the same time, blood colloid osmotic pressure decreases, further increasing the *P*eff.

The *P*eff is also affected by changes in the general arterial blood pressure. Severe hemorrhaging produces a drop in general blood pressure, which also decreases the glomerular blood hydrostatic pressure. If the blood pressure falls to the point where the hydrostatic pressure in the glomeruli reaches 50 mm Hg, no filtration occurs, because the glomerular blood hydrostatic pressure equals the opposing forces. Such a condition is called **anuria** (a-NOO-rē-a), a daily urine output of less than 50 ml. It may be caused by insufficient pressure to accomplish filtration or by inflammation of the glomeruli so that plasma is prevented from entering the glomerulus.

A final factor that may affect the *P*eff is the regulation of the size of the afferent and efferent arterioles. In this case, glomerular blood hydrostatic pressure is regulated

separately from the general blood pressure. Sympathetic impulses and small doses of epinephrine cause constriction of both afferent and efferent arterioles. However, intense sympathetic impulses and large doses of epinephrine cause greater constriction of afferent than efferent arterioles. This intense stimulation results in a decrease in glomerular hydrostatic pressure even though blood pressure in other parts of the body may be normal or even higher than normal. Intense sympathetic stimulation is most likely to occur during the alarm reaction of the general adaptation syndrome. Blood may also be shunted away from the kidneys during hemorrhage.

Tubular Reabsorption

The amount of filtrate that flows out of all the renal corpuscles of both kidneys every minute is called the **glomerular filtration rate (GFR)**. In the normal adult, this rate is about 125 ml/min—about 180 liters (48 gal) a day. But as the filtrate passes through the renal tubules, about 99 percent of it is reabsorbed into the blood. Thus only about 1 percent of the filtrate actually leaves the body (about 1 liter a day). The movement of the filtrate back into the blood of the peritubular capillaries or vasa recta is called **tubular reabsorption.**

Tubular reabsorption, carried out by epithelial cells throughout the renal tubule, is a very discriminating process. Only specific amounts of certain substances are reabsorbed, depending on the body's needs at the time. The maximum amount of a substance that can be reabsorbed under any condition is called the substance's **tubular maximum (Tm)**. Materials that are reabsorbed include water, glucose, amino acids, and ions such as Na^+, K^+, Ca^{2+}, Cl^-, HCO_3^-, and HPO_4^{2-}. Tubular reabsorption allows the body to retain most of its nutrients. Wastes such as urea are only partially reabsorbed. Exhibit 26-1 compares the values for the chemicals in the filtrate immediately after passing into the glomerular (Bowman's) capsule with those reabsorbed from the filtrate. It will give you an idea of how much of the various substances the kidneys reabsorb.

Reabsorption is carried out through both passive and active transport mechanisms. It is believed that glucose is reabsorbed by an active process involving a carrier system. The carrier, probably an enzyme, exists in the membranes of the tubular epithelial cells in a fixed and limited amount. Normally, all the glucose filtered by the glomeruli (125 mg/100 ml of filtrate/min) is reabsorbed by the tubules.

CLINICAL APPLICATION

The capacity of the glucose carrier system is limited. If the plasma concentration of glucose is above normal (180–200 mg/100 ml), the glucose transport mechanism cannot reabsorb it all and the excess remains in the urine. If there is a malfunction in the tubular carrier mechanism, glucose appears in the urine even though the blood sugar level is normal. This condition is called **glycosuria**. As you will see later, glycosuria is a symptom of untreated diabetes mellitus.

Most sodium ions are actively transported (reabsorbed) from the proximal convoluted tubules and limbs of the loop of the nephron. The reabsorption of the remaining Na⁺ occurs in the distal convoluted tubules and collecting ducts. Na⁺ reabsorption varies with its concentration in extracellular fluid. When the Na⁺ concentration of the blood is low, there is a drop in blood pressure, and the **renin-angiotensin pathway** goes into operation. The juxtaglomerular cells of the kidneys secrete an enzyme called renin, which converts angiotensinogen (synthesized by the liver) into angiotensin I. As it passes through the lungs, angiotensin I is converted into angiotension II. Angiotensin II stimulates the zona glomerulosa of the adrenal cortex to produce aldosterone. Aldosterone brings about increased Na⁺ and water reabsorption by the distal convoluted tubules and collecting ducts (Figure 26-9). Extracellular fluid volume increases and blood pressure is restored to normal. In the absence of aldosterone, the Na⁺ in the distal convoluted tubules and collecting ducts is not reabsorbed. It thus passes into urine for excretion.

When Na⁺ ions move out of the tubules into the peritubular blood, the blood momentarily becomes more electropositive than the filtrate. Chloride, a negatively charged ion, follows the positive sodium ion out of the tubule by electrostatic attraction. Thus the movement of Na⁺ ions influences the movement of Cl⁻ ions and other anions into the blood.

Cl⁻ ions, and not Na⁺ ions, are actively transported from the ascending limb of the loop of the nephron. In this portion of the nephron, then, Cl⁻ ions move actively and Na⁺ ions follow passively. Whether Na⁺ ions move actively and Cl⁻ ions follow passively, or Cl⁻ ions move actively and Na⁺ ions follow passively, the end result is the same.

Water reabsorption is controlled by sodium transport. As Na⁺ ions are transported from the proximal convoluted tubules into the blood, the osmotic pressure of the blood becomes higher than that of the filtrate. Water follows the Na⁺ ions into the blood in order to reestablish the osmotic equilibrium. About 80 percent of the water is reabsorbed by this method from the proximal convoluted tubule. This portion of reabsorbed water occurs as a function of osmosis, so it is referred to as **obligatory reabsorption.** The proximal convoluted tubule has no control over osmosis because it is always permeable to water (Figure 26-10).

Passage of most of the remaining water in the filtrate can be regulated. The permeability of the cells of the distal and collecting tubules is controlled by the antidiuretic hormone (ADH), produced by the hypothalamus. When the blood-water concentration is low, the posterior pituitary releases ADH, a hormone that increases the permeability of the plasma membranes of the distal tubule and collecting duct cells. As the membranes become more permeable, more water molecules pass into the cells and then into blood. This type of absorption is called *facultative* (FAK-ul-tā'-tiv), meaning that it occurs under some conditions but not others. **Facultative reabsorption** is responsible for about 20 percent of the water in the filtrate.

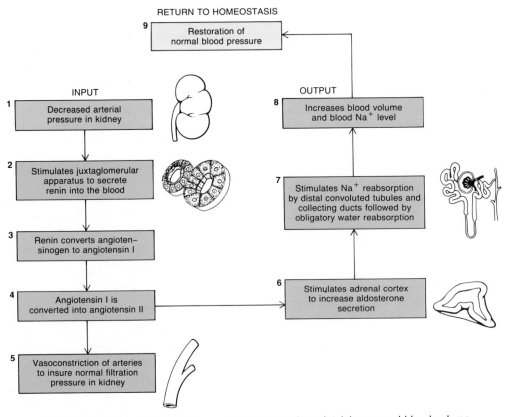

FIGURE 26-9 Role of the juxtaglomerular apparatus in maintaining normal blood volume.

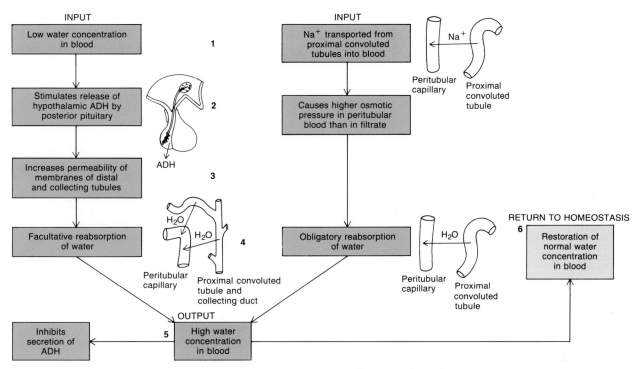

FIGURE 26-10 Factors that control water reabsorption.

It is a major mechanism for controlling water content of the blood (Figure 26-10).

Within limits, glomerular filtration rate (GFR) remains fairly constant because of the operation of two negative feedback systems that originate in the juxtaglomerular apparatus. One involves vasodilation of afferent arterioles; the other involves vasoconstriction of efferent arterioles. When there is a low rate of glomerular filtration, there is an excess reabsorption of chloride (Cl^-) ions from the filtrate. In response to diminished levels of Cl^- ions, cells of the macula densa cause vasodilation of afferent arterioles. This increases the rate of blood flow into the glomerulus, thus raising glomerular filtration rate back to normal. Diminished levels of Cl^- ions also cause the juxtaglomerular cells to release renin. This initiates a response (renin-angiotensin pathway) in which vasoconstriction of the efferent arterioles causes glomerular pressure to rise, thus also raising glomerular filtration rate back to normal.

Tubular Secretion

The third process involved in urine formation is **tubular secretion.** Whereas tubular reabsorption removes substances from the filtrate into the blood, tubular secretion adds materials to the filtrate from the blood. These secreted substances include potassium and hydrogen ions, ammonia, creatinine, and the drugs penicillin and paraaminohippuric acid. Tubular secretion has two principal effects. It rids the body of certain materials, and it controls the blood pH.

The body has to maintain normal blood pH (7.35 to 7.45) despite the fact that a normal diet provides more acid-producing foods than alkali-producing foods. To raise blood pH, the renal tubules secrete hydrogen and ammonium ions into the filtrate. Both these substances make the urine acidic.

Secretion of the hydrogen ions occurs through the formation of carbonic acid (Figure 26-11a). A certain amount of carbon dioxide normally diffuses from the peritubular blood into the cells of the distal tubules and collecting ducts. Once inside the epithelial cells, CO_2 combines with water to form carbonic acid (H_2CO_3), which then dissociates into H^+ ions and bicarbonate ions (HCO_3^-). A low blood pH stimulates the cells to secrete the H^+ ion into the urine. As the H^+ enters the urine, it displaces another positive ion, usually Na^+, and forms a weak acid or a salt of the acid that is eliminated in urine. The displaced Na^+ or other positive ion diffuses from the urine into the tubule cell, where it combines with the bicarbonate ion to form sodium bicarbonate ($NaHCO_3$), which then is absorbed into the blood. Not only is H^+ eliminated by the body and Na^+ conserved, but Na^+ is conserved in the form of $NaHCO_3$, which can buffer other H^+ ions in the blood.

A second mechanism for raising blood pH is secretion of the ammonium ion (Figure 26-11b). Ammonia, in certain concentrations, is a poisonous waste product derived from the deamination of amino acids. The liver converts much of the ammonia to a less toxic compound called urea. Urea and ammonia both become part of the glomerular filtrate and are subsequently expelled from the body. Any ammonia produced by the deamination of amino acids in the tubule is secreted into the urine. When ammonia (NH_3) forms in the distal and collecting tubule cells, it combines with H^+ to form the ammonium ion (NH_4^+).

FIGURE 26-11 Role of the kidneys in maintaining blood pH. (a) Acidification of urine and conservation of sodium bicarbonate ($NaHCO_3$) by the elimination of hydrogen ions (H^+). (b) Acidification of urine and conservation of sodium bicarbonate ($NaHCO_3$) by the elimination of ammonium ions (NH_4^+).

(The H^+ may come from the dissociation of H_2CO_3.) The cells secrete NH_4^+ into the filtrate, where it takes the place of a positive ion, usually Na^+, in a salt and is eliminated. The displaced Na^+ diffuses into the renal cells and combines with HCO_3^- to form $NaHCO_3$.

As a result of hydrogen and ammonium ion secretion, urine normally has an acidic pH of 6. The relationship of renal tubule ion excretion to blood pH level is summarized in Figure 26-12. Exhibit 26-2 summarizes filtration, reabsorption, and secretion in the nephrons.

MECHANISM OF URINE DILUTION

The rate at which water is lost from the body depends on antidiuretic hormone (ADH). As noted earlier, ADH controls the permeability of collecting ducts to water.

FIGURE 26-12 Summary of kidney mechanisms that maintain the homeostasis of blood pH.

In the absence of ADH, the ducts are virtually impermeable to water, and water is expelled into urine. However, in the presence of ADH, the ducts become quite permeable to water, and water is reabsorbed back into blood. Thus, less water is expelled into urine.

For the kidneys to produce a dilute urine, they must form a urine that contains more water than normal for the amount of solutes. This is accomplished when the renal tubules absorb more solutes than water. The kidneys produce a dilute urine as follows. The normal concentration of glomerular filtrate as it enters the proximal convoluted tubule in the cortex of the kidney is about 300 milliosmols (mOsm) per liter (Figure 26-13a).* A milliosmol represents the concentration of substances in the filtrate, mainly NaCl. It is a measure of the osmotic properties of a solution. The concentration of glomerular filtrate is isotonic to plasma. (See Chapter 3 for a discussion of isotonic solutions.) The ascending limb of the loop of the nephron (loop of Henle) is quite impermeable to water, but actively reabsorbs chloride (Cl^-) ions. The ions pass into interstitial fluid (fluid between tubules, ducts, and capillaries) and then into peritubular capillaries. This causes peritubular blood to become more electronegative than the filtrate, passively drawing out positive ions, mostly sodium (Na^+) ions. As a result, the concentration of these ions in the filtrate is reduced from its original concentration of 300 mOsm to about 100 mOsm, but the water remains. The filtrate leaving the ascending limb is thus diluted. As the filtrate continues through the distal convoluted tubule and collecting ducts, some additional ions are reabsorbed, producing an even more dilute urine. In these regions, Na^+ ions are actively reabsorbed and negative ions, mainly Cl^-, are passively reabsorbed. In

---

* A **milliosmol** is a unit of osmotic pressure equal to one-thousandth gram molecular weight of a substance divided by the number of particles (or ions) into which a substance dissociates in one liter of solution.

EXHIBIT 26-2 FILTRATION, REABSORPTION, AND SECRETION

REGION OF NEPHRON	ACTIVITY
Renal corpuscle (endothelial-capsular membrane)	Filtration of glomerular blood under hydrostatic pressure results in formation of filtrate, free of plasma proteins and cellular elements of blood.
Proximal convoluted tubule and descending and ascending limbs of the loop of the nephron	Reabsorption of physiologically important solutes such as Na^+, K^+, Cl^-, HCO_3^-, and glucose. Obligatory reabsorption of water by osmosis (proximal convoluted tubule).
Distal convoluted tubule	Reabsorption of Na^+. Facultative reabsorption of water under control of ADH. Secretion of H^+, NH_3, K^+, creatinine, and certain drugs. Conservation of $NaHCO_3$.
Collecting duct	Facultative reabsorption of water under control of ADH.

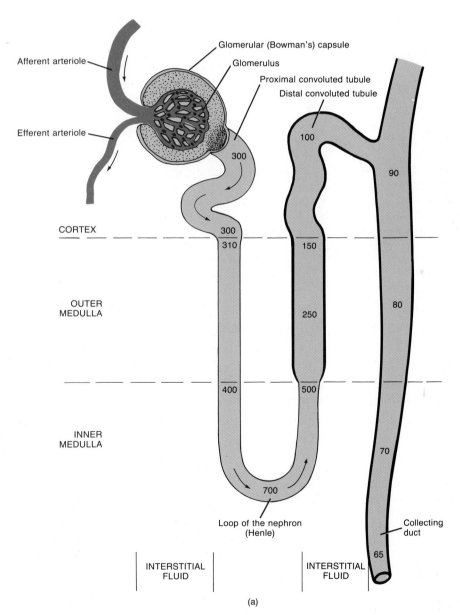

(a)

FIGURE 26-13 Mechanism of urine dilution and urine concentration: countercurrent multiplier. All concentrations are in milliosmols (mOsm). (a) Urine dilution. The portions of the renal tubules that are indicated by heavy lines represent areas that are virtually impermeable to water in the absence of ADH and where various ions are reabsorbed.

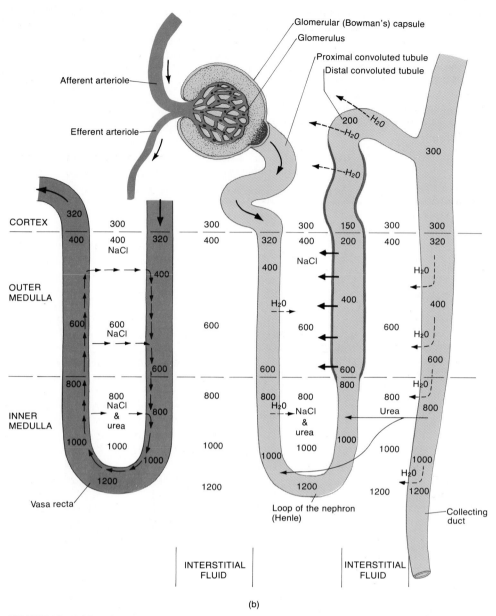

Afferent arteriole

Efferent arteriole

Glomerular (Bowman's) capsule

Glomerulus

Proximal convoluted tubule

Distal convoluted tubule

CORTEX

OUTER MEDULLA

INNER MEDULLA

Vasa recta

Loop of the nephron (Henle)

Collecting duct

INTERSTITIAL FLUID

INTERSTITIAL FLUID

(b)

FIGURE 26-13 (*Continued*) (b) Urine concentration. The illustration on the left represents the cardiovascular component (vasa recta), while the illustration on the right represents the nephron component.

the absence of ADH, urine may be only one-fourth as concentrated as blood plasma and glomerular filtrate. By the time the dilute filtrate passes into the terminal collecting ducts, its concentration can be as low as 65 to 70 milliosmols. The excretion of dilute urine, therefore, is the result of reabsorption of solutes, but retention of water, in the absence of ADH. Such a dilute urine is said to be *hyposmotic* (*hypotonic*) to blood plasma.

MECHANISM OF URINE CONCENTRATION

During times of low water intake, the kidneys still must eliminate wastes and excess ions while conserving water. Basically, this is accomplished by increasing the volume

of water that is reabsorbed into blood and thus producing the excretion of a more concentrated urine. Such urine is said to be *hyperosmotic* (*hypertonic*) to blood plasma. Whereas diluting urine simply involves the reabsorption of more solutes than water, concentrating urine is somewhat more complex. The ability of the kidneys to produce hyperosmotic urine depends in part on the countercurrent mechanism.

The excretion of concentrated urine begins with a high concentration of solutes in the interstitial fluid in the medulla of the kidney. If you examine Figure 26-13b, you will note that the solute concentration of the interstitial fluid in the kidney increases from about 300 mOsm in the cortex to about 1,200 mOsm in the inner medulla.

The high medullary concentration is maintained by two principal factors. The first is solute reabsorption from various parts of the renal tubules:

1. In the ascending limb of the loop of the nephron, Cl^- ions are actively transported from the filtrate into the interstitial fluid of the outer medulla. Na^+ ions, as well as other cations, follow passively. All the ions become concentrated in the fluid and are carried downward into the inner medulla by the blood flowing in the vasa recta (Figure 26-13b).

2. In the collecting duct, Na^+ ions are actively transported from the filtrate into the interstitial fluid of the inner medulla. Cl^- ions follow passively.

3. In the presence of high levels of ADH, the collecting duct becomes very permeable to water. Water quickly moves out of the ducts by osmosis into the interstitial fluid of the inner medulla, greatly increasing the concentration of urea in the collecting duct. As a result, urea moves passively by diffusion into the interstitial fluid of the medulla. This movement of urea increases the concentration of urea in the interstitial fluid above that in the filtrate in the loop of the nephron. Now, because of this difference in concentration, some urea diffuses into the filtrate in the loop of the nephron, thus raising its concentration of urea (Figure 26-13b). As the filtrate moves into the ascending limb and distal convoluted tubule, both of which are virtually impermeable to urea, and then into the collecting duct, more water moves out of the collecting duct by osmosis in the presence of ADH. This *further* increases the concentration of urea in the collecting duct and *more* urea diffuses into the interstitial fluid of the inner medulla and the cycle repeats itself. Thus, in the presence of ADH, water movement from the collecting duct can concentrate urea to very high levels as the cycle repeats itself.

The second factor that maintains the high medullary concentration of solutes (sodium, chloride, and urea) is the **countercurrent mechanism,** which is based on the anatomical arrangement of the juxtamedullary nephrons and vasa recta.

If you examine Figure 26-13b, you will note that the descending limb of the loop of the nephron carries filtrate downward from the cortex deep into the medulla. The ascending limb of the loop of the nephron carries filtrate upward from deep in the medulla toward the cortex. Thus, we have a situation in which fluid flowing in one tube runs parallel and counter (opposite) to fluid flowing in another tube. This is a countercurrent flow. The descending limb is relatively permeable to water and relatively impermeable to solutes. Since the interstitial fluid outside the descending limb is more concentrated than the filtrate within it, water moves out of the descending limb by osmosis. This water movement causes the concentration of the filtrate to increase. As the filtrate continues down the descending limb, more water leaves it by osmosis and the concentration of the filtrate increases even more. In fact, as the filtrate passes through the loop of the nephron (loop of Henle), its concentration is 1,200 mOsm per liter. As noted earlier, the ascending limb is relatively inpermeable to water, but Na^+ and Cl^- ions move from the limb into the interstitial fluid of the medulla. As the filtrate moves upward through the ascending limb and the ions move out, the concentration of filtrate in the ascending limb progressively decreases. Near the cortex, the concentration has decreased to 200 mOsm per liter,

a concentration below that of plasma. The overall effect of the countercurrent flow is that filtrate is progressively concentrated as it flows down the descending limb, progressively diluted as it moves up the ascending limb, and NaCl is highly concentrated in the medulla. Remember that urea is also concentrated there as well.

If you again examine Figure 26-13b, you can see that the vasa recta also consists of a descending portion and an ascending portion that are parallel to each other. The various solutes remain concentrated within the medulla because of the arrangement of the ascending and descending portions of the vasa recta. Just as filtrate flows in opposite directions in the limbs of the nephron, blood flows in opposite directions in the ascending and descending portions of the vasa recta. This again is a countercurrent flow. Blood entering the vasa recta has a solute concentration of about 300 mOsm per liter. As it flows down the descending portion into the medulla, where the interstitial fluid becomes more and more concentrated, NaCl and urea diffuse into blood. As the blood becomes more concentrated, it flows into the ascending portion of the vasa recta, where the interstitial fluid becomes less and less concentrated. As a result, NaCl and urea diffuse from the blood into the interstitial fluid so that blood leaving the vasa recta has a concentration only slightly higher than blood entering. The essential feature of the countercurrent flow in the vasa recta is that high medullary concentrations of solutes are maintained, since blood passing through the vasa recta removes very few solutes. Moreover, the flow of blood in the vasa recta is very slow, and this sluggish flow also helps prevent the rapid removal of solutes from the medulla.

The actual excretion of concentrated urine depends on the presence of ADH. When ADH is present, there is a rapid movement of water from the collecting ducts into interstitial fluid. This causes a progressive increase in the solute concentration in the collecting ducts. Thus, a concentrated urine is excreted, that is, one with a high solute concentration (up to 1,200 mOsm) and little water. Urine can be four times more concentrated than blood plasma and glomerular filtrate.

HEMODIALYSIS THERAPY

If the kidneys are so impaired by disease or injury that they are unable to excrete nitrogenous wastes and regulate pH and electrolyte concentration of the plasma, the blood must be filtered by an artificial device. Such filtering of the blood is called **hemodialysis.** *Dialysis* means using a selectively permeable membrane to separate large nondiffusible particles from smaller diffusible ones. One of the best-known devices for accomplishing dialysis is the kidney machine (Figure 26-14). A tube connects it with the patient's radial artery. The blood is pumped from the artery through the tubes to one side of a selectively permeable dialyzing membrane made of cellulose acetate. The other side of the membrane is continually washed with an artificial solution called the dialysate. The blood that passes through the artificial kidney is treated with an anticoagulant. Only about 500 ml of the patient's blood is in the machine at a time. This volume is easily compen-

FIGURE 26-14 Operation of an artificial kidney. The blood route is indicated in red and blue. The route of the dialysate is indicated in gold.

sated for by vasoconstriction and increased cardiac output.

All substances (including wastes) in the blood except protein molecules and blood cells can diffuse back and forth across the selectively permeable membrane. The electrolyte level of the plasma is controlled by keeping the dialysate electrolytes at the same concentration found in normal plasma. Any excess plasma electrolytes move down the concentration gradient and into the dialysate. If the plasma electrolyte level is normal, it is in equilibrium with the dialysate and no electrolytes are gained or lost. Since the dialysate contains no wastes, substances such as urea move down the concentration gradient and into the dialysate. Thus wastes are removed and normal electrolyte balance is maintained.

A great advantage of the kidney machine is that nutrition can be bolstered by placing large quantities of glucose in the dialysate. While the blood gives up its wastes, the glucose diffuses into the blood. Thus the kidney machine beautifully accomplishes the principal function of the fundamental unit of the kidney—the nephron.

There are obvious drawbacks to the artificial kidney, however. Anticoagulants must be added to the blood during dialysis. A large amount of the patient's blood must flow through this apparatus to make the treatment effective, and so the slow rate at which the blood can be processed makes the treatment time-consuming. To date, no artificial kidney has been implanted permanently.

A recent development, called **continuous ambulatory peritoneal dialysis (CAPD),** promises to make hemodialysis more convenient and less time-consuming for many patients. CAPD uses the peritoneum instead of cellulose acetate as the dialyzing membrane. Since the peritoneum is a selectively permeable membrane, it permits rapid bidirectional transfer of substances. A catheter is placed in the patient's peritoneal cavity and connected to a supply of dialysate. Gravity feeds the solution into the abdominal cavity from its plastic container. When the process is complete, the dialysate is returned from the abdominal cavity to the plastic container and then discarded.

HOMEOSTASIS

Excretion is one of the primary ways in which the volume, pH, and chemistry of the body fluids are kept in homeostasis. The kidneys assume a good deal of the burden for excretion, but they share this responsibility with several other organ systems.

OTHER EXCRETORY ORGANS

The lungs, integument, and gastrointestinal tract all perform special excretory functions (Exhibit 26-3). A major responsibility for regulating body temperature through the excretion of water is assumed by sudoriferous (sweat) glands of the skin. The lungs maintain blood-gas homeostasis through the elimination of carbon dioxide. One way in which the kidneys maintain homeostasis is by coordinating their activities with other excretory organs. When the integument increases its excretion of water, the renal tubules increase their reabsorption of water, and blood volume is maintained. When the lungs fail to eliminate enough carbon dioxide, the kidneys attempt to compensate. They change some of the carbon dioxide into sodium bicarbonate, which becomes part of the blood buffer systems.

URINE

The by-product of the kidneys' activities is **urine.** It gets its name from one of its constituents called uric acid. In a healthy person its volume, pH, and solute concentration vary with the needs of the internal environment. During certain pathological conditions, the characteristics of urine may change drastically. An analysis of the volume and physical and chemical properties of urine tells us much about the state of the body.

Volume

The volume of urine eliminated per day in the normal adult varies between 1,000 and 2,000 ml (1 to 2 qt). Urine volume is influenced by a number of factors: blood pres-

EXHIBIT 26-3 EXCRETORY ORGANS AND PRODUCTS ELIMINATED

EXCRETORY ORGANS	PRODUCTS ELIMINATED	
	PRIMARY	SECONDARY
Kidneys	Water, nitrogenous wastes from protein catabolism, and inorganic salts.	Heat and carbon dioxide.
Lungs	Carbon dioxide.	Heat and water.
Skin (sudoriferous glands)	Heat.	Carbon dioxide, water, salts, and urea.
Gastrointestinal (GI) tract	Solid wastes and secretions.	Carbon dioxide, water, salts, and heat.

sure, blood concentration, diet, temperature, diuretics, mental state, and general health.

• *Blood Pressure*   The cells of the juxtaglomerular apparatus are particularly sensitive to changes in blood pressure. When renal blood pressure falls below normal, the juxtaglomerular apparatus secretes renin and the renin-angiotensin pathway is activated (see Figure 26-9). As a result, facultative reabsorption of water increases, blood volume increases, and urine volume decreases. By raising the blood pressure, the juxtaglomerular apparatus ensures that the kidney cells receive enough oxygen and that the glomerular hydrostatic pressure is high enough to maintain a normal *P*eff. The juxtaglomerular apparatus also regulates blood pressure throughout the body.

• *Blood Concentration*   The concentrations of water and solutes in the blood also affect urine volume. If you have gone without water all day and the water concentration of your blood becomes low, osmotic receptors in the hypothalamus stimulate the posterior pituitary to release ADH. The hormone stimulates the cells of the distal convoluted tubule and collecting duct to allow water out of the filtrate and into the blood by facultative reabsorption. Thus urine volume decreases and water is conserved.

If you have just drunk an excessive amount of liquid, urine volume may be increased through two mechanisms. First, the blood-water concentration increases above normal. This means that the osmoreceptors in the hypothalamus are no longer stimulated to secrete ADH, and facultative water reabsorption stops. Second, the excess water causes the blood pressure to rise. In response, the renal vessels dilate, more blood is brought to the glomeruli, and the filtration rate increases.

The concentration of sodium ions in the blood also influences urine volume. Sodium concentration affects aldosterone secretion, which in turn affects both sodium reabsorption and the obligatory reabsorption of water.

• *Temperature*   When the internal or external temperature rises above normal, the rate of perspiration increases and cutaneous vessels dilate and fluid diffuses from the capillaries to the surface of the skin. As water volume decreases, ADH is secreted and facultative reabsorption increases. In addition, the increase in temperature stimulates the abdominal vessels to constrict, so the blood flow in the glomeruli and filtration decrease. Both mechanisms reduce the volume of urine.

If the body is exposed to low temperatures, the cutaneous vessels constrict and the abdominal vessels dilate. More blood is shunted to the glomeruli, glomerular blood hydrostatic pressure increases, and urine volume increases.

• *Diuretics*   Certain chemicals increase urine volume by inhibiting the facultative reabsorption of water. Such chemicals are called **diuretics,** and the abnormal increase in urine flow is called *diuresis.* Some diuretics act directly on the tubular epithelium as they are carried through the kidneys. Others act indirectly by inhibiting the secretion of ADH as they circulate through the brain. Coffee, tea, and alcoholic beverages are diuretics.

• *Emotions*   Some emotional states can affect urine volume. Nervousness, for example, can cause an enormous discharge of urine because impulses from the brain cause an increase in blood pressure, resulting in an increased glomerular filtration rate.

Physical Characteristics

• *Color*   Normal urine is usually a yellow or amber-colored, transparent liquid with a characteristic odor. The color is caused by urochrome, a pigment derived from the metabolism of bile. The color of urine varies considerably with the ratio of solutes to water in the urine. The less water there is, the darker the urine. Fever decreases urine volume in the same way that high environmental temperatures do, sometimes making the urine quite concentrated. It is not uncommon for a feverish person to have dark yellow or brown urine. The color of urine may also be affected by diet, such as a reddish color from beets, and by the presence of abnormal constituents, such as certain drugs. A red or brown to black color may indicate the presence of red blood cells or hemoglobin from bleeding in the urinary system.

• *Turbidity*   Fresh urine is usually transparent. Turbid (cloudy) urine does not necessarily indicate a pathological condition, since turbidity may result from mucin secreted by the lining of the urinary tract. The presence of mucin above a critical level usually denotes an abnormality, however.

• *Odor*   The odor of urine may vary. For example, some individuals have inherited the ability to form a substance called methyl mercaptan from the digestion of asparagus. This substance gives urine a characteristic odor. In cases of diabetes, urine has a "sweetish" odor because of the presence of acetone. Stale urine develops the odor of ammonia due to ammonium carbonate formation as a result of urea decomposition.

• *pH*   Normal urine is slightly acid. It ranges between pH 4.6 and 8.0. Variations in urine pH are closely related to diet. These variations are due to differences in the end products of metabolism. Whereas a high-protein diet increases acidity, a diet composed largely of vegetables increases alkalinity. High altitude, fasting, and exercise also cause variations in urinary pH. Ammonium carbonate forms in standing urine. Since it can dissociate into ammonium ions and form a strong base, the presence of ammonium carbonate tends to make urine more alkaline.

• *Specific Gravity*   *Specific gravity* is the ratio of the weight of a volume of a substance to the weight of an equal volume of distilled water. Water has a specific gravity of 1.000. The specific gravity of urine depends on the amount of solid materials in solution and ranges from 1.001 to 1.035 in normal urine. The greater the concentration of solutes, the higher the specific gravity. Above normal specific gravity readings may be due to the presence of blood cells, casts, or bacteria in urine (described later).

The physical characteristics of urine are summarized in Exhibit 26-4.

EXHIBIT 26-4 PHYSICAL CHARACTERISTICS OF NORMAL URINE

CHARACTERISTIC	DESCRIPTION
Volume	1–2 liters in 24 hours, but varies considerably.
Color	Yellow or amber, but varies with concentration and diet.
Turbidity	Transparent when freshly voided, but becomes turbid upon standing.
Odor	Aromatic, but becomes ammonialike upon standing.
pH	4.6–8.0, average 6.0; varies considerably with diet.
Specific gravity	1.001–1.035.

Chemical Composition

Water accounts for about 95 percent of the total volume of urine. The remaining 5 percent consists of solutes derived from cellular metabolism and outside sources such as drugs. The solutes are described in Exhibit 26-5.

CLINICAL APPLICATION

Several screening **tests for renal function** are available. The most commonly ordered test is the *blood urea nitrogen (BUN)*, which measures blood levels of nitrogen in urea. Urea is an end product of protein metabolism produced in the liver and excreted in the urine, Although the BUN is relatively insensitive to mild degrees of renal dysfunction, it is a good clinical indicator of significant renal dysfunction.

The *serum creatinine test* measures the level of creatinine in blood. Creatinine is derived from the conversion of creatine phosphate in skeletal muscle tissue. Like the BUN, serum creatinine levels are elevated only when there is significant renal damage.

Renal clearance tests measure the volume of plasma that is cleared of a substance in a given period of time. The tests are used to assess glomerular filtration rates and renal blood flow. Clearance studies have been devised using injected substances such as inulin and naturally occurring substances such as creatinine and urea.

Abnormal Constituents

If the body's chemical processes are not operating efficiently, traces of substances not normally present may appear in the urine, or normal constituents may appear in abnormal amounts. Analyzing the physical, chemical, and microscopic properties of urine often provides information that aids diagnosis. Such an analysis is called a **urinalysis (UA).**

- *Albumin* Albumin is a normal constituent of plasma, but it usually appears in only very small amounts because the particles are too large to pass through the pores in the capillary walls. The presence of excessive albumin in the urine—**albuminuria**—indicates an increase in the permeability of the glomerular membrane. Conditions that lead to albuminuria include injury to the glomerular membrane as a result of disease, increased blood pressure, and irritation of kidney cells by substances such as bacterial toxins, ether, or heavy metals. Other proteins, such as globulin and fibrinogen, may also appear in the urine under certain conditions.

- *Glucose* The presence of sugar in the urine is termed **glycosuria.** Normal urine contains such small amounts of glucose that clinically it may be considered absent. The most common cause of glycosuria is a high blood sugar level. Remember that glucose is filtered into the glomerular (Bowman's) capsule. Later, in the proximal convoluted tubules, the tubule cells actively transport the glucose back into the blood. However, the number of glucose carrier molecules is limited. If more carbohydrates are ingested than can be used or stored as glycogen or fat, more sugar is filtered into the glomerular capsule than can be removed by the carriers. This condition, called **temporary (alimentary) glycosuria,** is not considered pathological. Another nonpathological cause is emotional stress. Stress can cause excessive amounts of epinephrine to be secreted. Epinephrine stimulates the breakdown of glycogen and the liberation of glucose from the liver. A **pathological glycosuria** results from diabetes mellitus. In this case, there is a frequent or continuous elimination of glucose because the pancreas fails to produce sufficient insulin. When glycosuria occurs with a normal blood sugar level, the problem lies in failure of the kidney tubular cells to reabsorb glucose.

- *Erythrocytes* The appearance of red blood cells in the urine is called **hematuria.** Hematuria generally indicates a pathological condition. One cause is acute inflammation of the urinary organs as a result of disease or irritation from kidney stones. Other causes include tumors, trauma, and kidney disease. Whenever blood is found in the urine, additional tests are performed to ascertain the part of the urinary tract that is bleeding. One should also make sure the sample was not contaminated with menstrual blood from the vagina. Some long distance runners may also develop "red urine" due to damaged blood vessels in the bottom of the feet. As a result of the damage, blood appears in the urine.

- *Leucocytes* The presence of leucocytes and other components of pus in the urine, referred to as **pyuria,** indicates infection in the kidney or other urinary organs. Again the source of the pus must be located, and care should be taken that the urine is not contaminated.

- *Ketone Bodies* Ketone (acetone) bodies appear in normal urine in small amounts. Their appearance in high quantities, a condition called **ketosis (acetonuria),** may indicate abnormalities. It may be caused by diabetes melli-

EXHIBIT 26-5   PRINCIPAL SOLUTES IN URINE OF ADULT MALE ON MIXED DIET

CONSTITUENT	AMOUNT (g)*	COMMENTS
ORGANIC		
Urea	25.0	Comprises 60–90 percent of all nitrogenous material. Derived primarily from deamination of proteins (ammonia combines with CO_2 to form urea).
Creatinine	1.5	Normal alkaline constituent of blood. Derived primarily from creatine (nitrogenous substance in muscle tissue).
Uric acid	0.8	Product of catabolism of nucleic acids derived from food or cellular destruction. Because of insolubility, tends to crystallize and is common component of kidney stones.
Hippuric acid	0.7	Form in which benzoic acid (toxic substance in fruits and vegetables) is believed to be eliminated from body. High-vegetable diets increase quantity of hippuric acid excreted.
Indican	0.01	Potassium salt of indole. Indole results from putrefaction of protein in large intestine and is carried by blood to liver, where it is probably changed to indican (less poisonous substance).
Ketone bodies	0.04	Also called acetone bodies. Normally found in small amounts. In cases of diabetes and acute starvation, ketone bodies appear in high concentrations.
Other substances	2.9	May be present in minute quantities, depending on diet and general health. Include carbohydrates, pigments, fatty acids, mucin, enzymes, and hormones.
INORGANIC		
NaCl	15.0	Principal inorganic salt. Amount excreted varies with intake.
K^+	3.3	Occurs as chloride, sulfate, and phosphate salts.
SO_4^{2-}	2.5	Derived from amino acids.
PO_4^{3-}	2.5	Occurs as sodium compounds (monosodium and disodium phosphate) that serve as buffers in blood.
NH_4^+	0.7	Occurs as ammonium salts. Derived from protein catabolism and from glutamine in kidneys. Amount produced by kidney may vary with need of body for conserving Na^+ ions to offset acidity of blood and tissue fluids.
Mg^{2+}	0.1	Occurs as chloride, sulfate, and phosphate salts.
Ca^{2+}	0.3	Occurs as chloride, sulfate, and phosphate salts.

* Values are for a urine sample collected over 24 hours.

tus, starvation, or simply too little carbohydrate in the diet. Whatever the cause, excessive quantities of fatty acids are oxidized in the liver, and the ketone bodies are filtered from the plasma into the glomerular capsule.

- *Bilirubin*   As noted in Chapter 24, bilirubin is a product of the hemolysis of red blood cells and the breakdown of hemoglobin by reticuloendothelial cells. The pigment gives bile its major pigmentation. Bilirubin is carried to liver cells by an albumin molecule and is referred to as **unconjugated (free) bilirubin.** When it reaches the liver, the albumin is released. The bilirubin is then combined with certain substances in liver cells (glucuronic acid or sulfate) and is referred to as **conjugated bilirubin.** In this form it is secreted by the liver into the biliary system and then into the small intestine. Intestinal bacteria convert conjugated bilirubin into urobilogen, which gives feces its characteristic color. Above normal levels of bilirubin in urine are referred to as **bilirubinuria.** A routine urinalysis measures total bilirubin; it does not distinguish between unconjugated and conjugated. A normal level of total bilirubin rules out any significant liver dysfunction or excessive hemolysis of red blood cells. If the level is elevated, tests that differentiate between unconjugated and conjugated bilirubin may be performed. An increase in unconjugated bilirubin is more frequently associated with excessive hemolysis of red blood cells. An increase in conjugated bilirubin is more likely a result of liver dysfunction or obstruction of the biliary system.

- *Urobilinogen*   Part of the urobilinogen formed in the intestine is excreted with feces. Another portion is absorbed and returned to the liver where it is metabolized and excreted in bile. A small percentage of the urobilinogen that is absorbed escapes the liver and enters general circulation. The kidneys handle this amount of urobilinogen as though it were a foreign substance, that is, the tubular cells do not actively reabsorb the filtered urobilinogen. The presence of urobilinogen in urine is called **urobilinogenuria.** Thus, traces of urobilinogen in urine are normal. When urobilinogenuria is above normal, it indicates an increase in the production of bilirubin and inability of the liver to remove reabsorbed urobilinogen from the blood. Conditions that contribute to increased urobilinogen are hemolytic and pernicious anemia, infectious hepatitis, biliary obstruction, jaundice, cirrhosis, congestive heart failure, and infectious mononucleosis.

• *Casts* Microscopic examination of urine may reveal **casts**—tiny masses of material that have hardened and assumed the shape of the lumens of the tubules and were then flushed out of the tubules by a buildup of filtrate behind them. Casts are named after the substances that compose them or for their appearance. There are white-blood-cell casts, red-blood-cell casts, epithelial casts that contain cells from the walls of the tubes, granular casts that contain decomposed cells which form granules, and fatty casts from cells that have become fatty.

• *Renal Calculi* Occasionally, the crystals of salts found in urine may solidify into insoluble stones called **renal calculi (kidney stones)** (Figure 26-15). They may be formed in any portion of the urinary tract from the kidney tubules to the external opening. Conditions leading to stone formation include the ingestion of excessive mineral salts, a decrease in the amount of water, abnormally alkaline or acidic urine, and overactivity of the parathyroid glands. Common constituents of stones are calcium oxalate, uric acid, and calcium phosphate crystals. Calcium oxalate crystals are the most common type. A protein isolated from urine called glycoprotein crystal-growth inhibitor (GCI) inhibits formation of calcium oxalate stones. Individuals who do not synthesize GCI form the calcium oxalate stones. Kidney stones usually form in the pelvis of the kidney, where they cause pain, hematuria, and pyuria.

For kidney stones that become painful or obstructive, surgical removal is the typical alternative. However, there are now two new methods of treatment that do not involve conventional surgery. One procedure, called *extracorporeal shock wave lithotripsy* (*ESWL*), involves the use of ultrasound waves generated by an instrument called a lithotripter. In this procedure, a patient is placed in a water bath and is subjected to ultrasound waves. Once the stones are shattered, the fragments are eliminated via the urine. The other procedure is called *percutaneous ultrasonic lithotripsy* (*PUL*). A tube resembling a cystoscope is inserted into the kidney through a small opening in the back. The surgeon can then pick out small stones whole, or shatter large stones with ultrasound waves, and remove the fragments by suction.

• *Microbes* In a properly collected and processed specimen the finding of bacteria may be of considerable importance. If bacteria are seen in a centrifuged specimen, but not in the unspun sample, it suggests a bacterial count of less than 10,000/ml. The presence of bacteria in an unspun sample suggests a count greater than 100,000/ml. The various bacteria in urine are identified by different microbiological tests. The number and type of bacteria vary with specific infections in the urinary tract. The most common fungus to appear in urine is *Candida albicans,* a common cause of vaginitis. The most frequent protozoan seen in urine is *Trichomonas vaginalis,* a cause of vaginitis in females and urethritis in males.

URETERS

Once urine is formed by the nephrons and collecting ducts, it drains through papillary ducts into the calyces surrounding the renal papillae. The minor calyces join to become the major calyces that unite to become the renal pelvis. From the pelvis, the urine drains into the ureters and is carried by peristalsis to the urinary bladder. From the urinary bladder, the urine is discharged from the body through the single urethra.

STRUCTURE

The body has two **ureters** (YOO-re-ters)—one for each kidney. Each ureter is an extension of the pelvis of the kidney and extends 25 to 30 cm (10 to 12 inches) to the urinary bladder (see Figure 26-1). As the ureters descend, their thick walls increase in diameter, but at their widest point they measure less than 1.7 cm (0.5 inch) in diameter. Like the kidneys, the ureters are retroperitoneal in placement. The ureters enter the urinary bladder at the superior lateral angle of its base.

Although there are no anatomical valves at the openings of the ureters into the urinary bladder, there is a functional one that is quite effective. Since the ureters pass under the urinary bladder for several centimeters, pressure in the urinary bladder compresses the ureters and prevents backflow of urine when pressure builds up in the urinary bladder during urination. When this physiological valve is not operating, it is possible for cystitis (urinary bladder inflammation) to develop into kidney infection.

HISTOLOGY

Three coats of tissue form the wall of the ureters. The inner coat, or mucosa, is a mucous membrane with transitional epithelium. The solute concentration and pH of urine differ drastically from the internal environment of cells that form the wall of the ureters. Mucus secreted by the mucosa prevents the cells from coming in contact with urine. Throughout most of the length of the ureters, the second or middle coat, the muscularis, is composed

FIGURE 26-15 Photograph of a renal calculus adjacent to a dime for size comparison. (Courtesy of Matt Iacobino.)

of inner longitudinal and outer circular layers of smooth muscle. The muscularis of the lower third of the ureters also contains a layer of outer longitudinal muscle. Peristalsis is the major function of the muscularis. The third, or external, coat of the ureters is a fibrous coat. Extensions of the fibrous coat anchor the ureters in place.

PHYSIOLOGY

The principal function of the ureters is to transport urine from the renal pelvis into the urinary bladder. Urine is carried through the ureters primarily by peristaltic contractions of the muscular walls of the ureters, but hydrostatic pressure and gravity also contribute. Peristaltic waves pass from the kidney to the urinary bladder, varying in rate from 1 to 5/min, depending on the amount of urine formation.

URINARY BLADDER

The **urinary bladder** is a hollow muscular organ situated in the pelvic cavity posterior to the symphysis pubis. In the male, it is directly anterior to the rectum. In the female, it is anterior to the vagina and inferior to the uterus. It is a freely movable organ held in position by folds of the peritoneum. The shape of the urinary bladder depends on how much urine it contains. When empty, it looks like a deflated balloon. It becomes spherical when slightly distended. As urine volume increases, it becomes pear-shaped and rises into the abdominal cavity.

STRUCTURE

At the base of the urinary bladder is a small triangular area, the **trigone** (TRĪ-gōn), that points anteriorly (Figure 26-16). The opening to the urethra is found in the apex of this triangle. At the two points of the base, the ureters drain into the bladder. It is easily identified because the mucosa is firmly bound to the muscularis so that the trigone is typically smooth.

HISTOLOGY

Four coats make up the wall of the urinary bladder (Figure 26-17). The mucosa, the innermost coat, is a mucous membrane containing transitional epithelium. Transitional epithelium is able to stretch—a marked advantage for an organ that must continually inflate and deflate. Rugae (folds in the mucosa) are also present. The second coat, the submucosa, is a layer of connective tissue that connects the mucosa and muscular coats. The third coat—a muscular one called the **detrusor** (de-TROO-ser) **muscle**—consists of three layers: inner longitudinal, middle circular, and outer longitudinal muscles. In the area around the opening to the urethra, the circular fibers form an **internal sphincter** muscle. Below the internal sphincter is the **external sphincter,** which is composed of skeletal muscle. The outermost coat, the serous coat, is formed by the peritoneum and covers only the superior surface of the organ.

FIGURE 26-16 Urinary bladder and female urethra.

CLINICAL APPLICATION

Cystoscopy (sis-TOS-kō-pē), direct visual examination of the urinary tract, is a commonly employed procedure to evaluate various disorders of the urinary bladder. The instrument used for the examination is called a **cystoscope** (*cyst* = urinary bladder; *skopein* + to examine), a narrow, hollow, metallic tube with a tele-

FIGURE 26-17 Histology of the urinary bladder. Photomicrograph of a portion of the wall of the urinary bladder at a magnification of 50×. The details of the mucosa of the urinary bladder are shown in Exhibit 4-1, Stratified transitional. (Copyright © 1983 by Michael H. Ross. Used by permission.)

scopic lens and fiberoptic lighting. Cystoscopy is generally performed under local anesthesia in which the cystoscope is gently introduced into the urethral orifice and slowly advanced into the urinary bladder. Cystoscopy is a valuable procedure in evaluating strictures (narrowed areas) in the urethra, proper emptying of the urinary bladder, the presence of kidney stones, tumors, and the potential need for surgery.

PHYSIOLOGY

Urine is expelled from the urinary bladder by an act called **micturition** (mik'-too-RISH-un), commonly known as urination or voiding. This response is brought about by a combination of involuntary and voluntary nervous impulses. The average capacity of the urinary bladder is 700 to 800 ml. When the amount of urine in the bladder exceeds 200 to 400 ml, stretch receptors in the urinary bladder wall transmit impulses to the lower portion of the spinal cord. These impulses initiate a conscious desire to expel urine and a subconscious reflex referred to as the **micturition reflex.** Parasympathetic impulses transmitted from the sacral area of the spinal cord reach the urinary bladder wall and internal urethral sphincter, bringing about contraction of the detrusor muscle of the urinary bladder and relaxation of the internal sphincter. Then the conscious portion of the brain sends impulses to the external sphincter, the sphincter relaxes, and urination takes place. Although emptying of the urinary bladder is controlled by reflex, it may be initiated voluntarily and stopped at will because of cerebral control of the external sphincter.

CLINICAL APPLICATION

A lack of voluntary control over micturition is referred to as **incontinence.** In infants about 2 years old and under, incontinence is normal because neurons to the external sphincter muscle are not completely developed. Infants void whenever the urinary bladder is sufficiently distended to arouse a reflex stimulus. Proper training overcomes incontinence if the latter is not caused by emotional stress or irritation of the urinary bladder.

Involuntary micturition in the adult may occur as a result of unconsciousness, injury to the spinal nerves controlling the urinary bladder, irritation due to abnormal constituents in urine, disease of the urinary bladder, and inability of the detrusor muscle to relax due to emotional stress.

Retention, a failure to void urine, may be due to an obstruction in the urethra or neck of the urinary bladder, nervous contraction of the urethra, or lack of sensation to urinate.

URETHRA

The **urethra** is a small tube leading from the floor of the urinary bladder to the exterior of the body (see Figure 26-16). In females, it lies directly posterior to the symphysis pubis and is embedded in the anterior wall of the vagina. Its undilated diameter is about 6 mm (0.25 inch), and its length is approximately 3.8 cm (1.5 inches). The female urethra is directed obliquely inferiorly and anteriorly. The opening of the urethra to the exterior, the **urethral orifice** is located between the clitoris and vaginal opening.

In males, the urethra is about 20 cm (8 inches) long. Immediately below the urinary bladder it passes vertically through the prostate gland, then pierces the urogenital diaphragm, and finally pierces the penis and takes a curved course through its body (see Figures 28-1 and 28-10).

HISTOLOGY

The wall of the female urethra consists of three coats: an inner mucous coat that is continuous externally with that of the vulva, an intermediate thin layer of spongy tissue containing a plexus of veins, and an outer muscular coat that is continuous with that of the urinary bladder and consists of circularly arranged fibers of smooth muscle.

The male urethra is composed of two coats. An inner mucous membrane is continuous with the mucous membrane of the urinary bladder. An outer submucous tissue connects the urethra with the structures through which it passes.

PHYSIOLOGY

The urethra is the terminal portion of the urinary system. It serves as the passageway for discharging urine from the body. The male urethra also serves as the duct through which reproductive fluid (semen) is discharged from the body.

AGING AND THE URINARY SYSTEM

The effectiveness of kidney function decreases with aging, and by age 70 the filtering mechanism is only about one-half what it was at age 40. Urinary incontinence and urinary tract infections are two major problems associated with aging of the urinary system. Other pathologies include polyuria (excessive urine production), nocturia (excessive urination at night), increased frequency of urination, dysuria (painful urination), retention of urine (failure to void urine), and hematuria (blood in the urine). Changes and diseases in the kidney include acute and chronic kidney inflammations and renal calculi (kidney stones). The prostate gland is often implicated in various disorders of the urinary tract, and cancer of the prostate is the most frequent malignancy in elderly males.

DEVELOPMENTAL ANATOMY OF THE URINARY SYSTEM

As early as the third week of development, a portion of the mesoderm along the posterior half of the dorsal side of the embryo, the **intermediate mesoderm,** differentiates into the kidneys. Three pairs of kidneys differentiate within the intermediate mesoderm in successive time periods: pronephros, mesonephros, and metanephros (Figure 26-18). Only the last remains as the functional kidneys of the adult.

The first kidney to form, the **pronephros,** is the superior of the three. Associated with its formation is a tube, the **pronephric duct.** This duct empties into the **cloaca,** which is the dilated caudal end of the gut derived from **endoderm.** The pronephros begins to degenerate during the fourth week and is completely gone by the sixth week. The pronephric ducts, however, remain.

The pronephros is replaced by the second kidney, the **mesonephros.** With its appearance, the retained portion of the pronephric duct, which connects to the mesonephros, becomes known as the **mesonephric duct.** The mesonephros begins to degenerate by the sixth week and is just about completely gone by the eighth week.

At about the fifth week, an outgrowth, called a **ureteric bud,** develops from the distal end of the mesonephric duct near the cloaca. This bud is the developing **metanephros.** As it grows toward the head of the embryo, its end widens to form the *pelvis* of the kidney with its *calyces* and associated *collecting tubules.* The unexpanded portion

FIGURE 26-18 Development of the urinary system.

of the bud, the **metanephric duct,** becomes the *ureter.* The *nephrons,* the functional units of the kidney, arise from the intermediate mesoderm around each ureteric bud.

During development, the cloaca divides into a **urogenital sinus,** into which urinary and genital ducts empty, and a *rectum* that discharges into the anal canal. The *urinary bladder* develops from the urogenital sinus. In the female, the *urethra* develops from lengthening of the short duct that extends from the urinary bladder to the urogenital sinus. The *vestibule,* into which the urinary and genital ducts empty, is also derived from the urogenital sinus. In the male, the urethra is considerably longer and more complicated, but is also derived from the urogenital sinus.

DISORDERS: HOMEOSTATIC IMBALANCES

Gout
Gout is a hereditary condition associated with an excessively high level of uric acid in the blood. When nucleic acids are catabolized, a certain amount of uric acid is produced as a waste. Some people seem to produce excessive amounts of uric acid, and others seem to have trouble excreting normal amounts. In either case, uric acid accumulates in the body and tends to solidify into crystals that are deposited in the joints and kidney tissue. When the crystals are deposited in the joints, the condition is called gouty arthritis. Gout is aggravated by excessive use of diuretics, dehydration, and starvation.

Glomerulonephritis (Bright's Disease)
Glomerulonephritis (Bright's disease) is an inflammation of the kidney that involves the glomeruli. One of the most common causes of glomerulonephritis is an allergic reaction to the toxins given off by streptococci bacteria that have recently infected another part of the body, especially the throat. The glomeruli become so inflamed, swollen, and engorged with blood that the glomerular membranes become highly permeable and allow blood cells and proteins to enter the filtrate. Thus the urine contains many erythrocytes and much protein. The glomeruli may be permanently changed, leading to chronic renal disease and renal failure.

Pyelitis and Pyelonephritis
Pyelitis is an inflammation of the kidney pelvis and its calyces. **Pyelonephritis,** an inflammation of one or both kidneys, involves the nephrons and the renal pelvis. The disease is generally a complication of infection elsewhere in the body. In females, it is often a complication of lower urinary tract infections. The causative agent in about 75 percent of the cases is the bacterium *Escherichia coli.* Should pyelonephritis become chronic, scar tissue forms in the kidneys and severely impairs their function.

Cystitis
Cystitis is an inflammation of the urinary bladder involving principally the mucosa and submucosa. It may be caused by bacterial infection, chemicals, or mechanical injury.

Nephrosis
Nephrosis is a condition in which the glomular membrane leaks, allowing large amounts of protein to escape from the blood into the urine. Water and sodium then accumulate in the body creating edema especially around the ankles and feet, abdomen, and eyes. Nephrosis is more common in children than adults, but occurs in all ages. Although it cannot always be cured, certain synthetic steroid hormones such as cortisone and prednisone, which are similar to the natural hormones, secreted by the adrenal glands, can suppress some forms of it.

Polycystic Disease
Polycystic disease may be caused by a defect in the renal tubular system that deforms nephrons and results in cystlike dilations along their course. It is the most common inherited disorder of the kidneys. The kidney tissue is riddled with cysts, small holes, and fluid-filled bubbles ranging in size from a pinhead to the diameter of an egg. These cysts gradually increase until they squeeze out the normal tissue, interfering with kidney function and causing uremia. The chief symptom is weight gain. The kidneys themselves may enlarge from their normal 0.25 kg (0.5 lb) to as much as 14 kg (30 lb). Many people lead normal lives without ever knowing they have the disease, however, and the condition is sometimes discovered only after death. Although the disease is progressive, its advance can be slowed by diet, drugs, and fluid intake. Kidney failure as a result of polycystic disease seldom occurs before the mid-forties and frequently can be postponed until the sixties.

Renal Failure
Renal failure, a decrease or cessation of glomerular filtration, is classified into two types: acute and chronic. **Acute renal failure (ARF)** is a clinical syndrome in which the kidneys abruptly stop working entirely or almost entirely. The hallmark of ARF is suppression of urine flow, usually categorized by *oliguria* (olig = scanty), daily urine output less than 500 ml, or *anuria,* daily urine output less than 50 ml. One cause of ARF is low blood volume or decreased cardiac output. Acute tubular necrosis (damage to renal tubules by ischemia or toxins) and a kidney stone can also cause ARF.

Chronic renal failure (CRF) refers to the progressive and generally irreversible decline in glomerular filtration rate (GFR) that may result from chronic glomerulonephritis, pyelonephritis, congenital polycystic disease, and traumatic loss of kidney tissue, among others. CRF develops in three stages. In the first stage, diminished renal reserve, nephrons are destroyed until about 75 percent of the functioning nephrons are lost. At this stage, the person may show no symptoms since the remaining nephrons enlarge and take over the function of those that have been lost. When more nephrons are lost, the balance between glomerular filtration and tubular reabsorption is destroyed, and any change in diet or fluid intake brings on the symptoms. Once 75 percent of the nephrons are lost, the person enters the second stage, called renal insufficiency. In this stage, there is a decrease in GFR and increased blood levels of nitrogenous wastes and creatine. Also, the kidneys cannot effectively concentrate or dilute urine. The final stage, called end stage renal failure (uremia), occurs when about 90 percent of the nephrons have been lost. At this stage, GFR diminishes to about 10 percent normal, and blood levels of nitrogenous wastes and creatine increase further. The low GFR results in oliguria. Individuals with CRF are candidates for hemodialysis therapy and kidney transplantation.

Among the effects of renal failure are edema from salt and water retention; acidosis due to inability of the kidneys to excrete acidic substances; increased levels of nonprotein nitrogen (NPN) substances, especially urea, due to impaired renal excretion of metabolic waste products; elevated potassium levels that can lead to cardiac arrest; anemia, since the kidneys no longer produce renal erythropoietic factor required for red blood cell production; and osteomalacia, since the kidneys are no longer able to convert vitamin D to its active form for calcium absorption from the small intestine.

Urinary Tract Infections (UTIs)

The term **urinary tract infection (UTI)** is used to describe either an infection of a part of the urinary system or the presence of large numbers of microbes in urine. Included are *significant bacteriuria* (the presence of bacteria in urine in significant numbers to indicate active infection), *asymptomatic bacteriuria* (the multiplication of large numbers of bacteria in urine without producing symptoms), *urethritis* (inflammation of the urethra), *cystitis* (inflammation of the urinary bladder), and *pyelonephritis* (inflammation of the kidneys).

Acute UTI is much more common in females than males and is most often caused by bacteria (*Escherichia coli*). Among the individuals at increased risk are pregnant women and those with renal disease, hypertension, and diabetes. Symptoms associated with UTI include burning on urination or painful urination, urinary urgency and frequency, pubic and back pain, passage of cloudy or blood-tinged urine, chills, fever, nausea, vomiting, and urethral discharge, usually in males.

MEDICAL TERMINOLOGY

Azotemia (*azo* = nitrogen-containing; *emia* = condition of blood) Presence of urea or other nitrogenous elements in the blood.

Cystocele (*cyst* = bladder; *cele* = cyst) Hernia of the urinary bladder.

Dysuria (*dys* = painful; *uria* = urine) Painful urination.

Enuresis (*enourein* = to void urine) Bed-wetting; may be due to faulty toilet training, to some psychological or emotional disturbance, or rarely to some physical disorder.

Intravenous pyelogram or **IVP** (*intra* = within; *veno* = vein; *pyelo* = pelvis of kidney; *gram* = written or recorded) X-ray film of the kidneys after injection of a dye.

Nephroblastoma (*neph* = kidney; *blastos* = germ or forming; *oma* = tumor) Embryonal carcinosarcoma; a malignant tumor arising from epithelial and connective tissue.

Nephrocele Hernia of the kidney.

Polyuria (*poly* = much) Excessive urine.

Stricture Narrowing of the lumen of a canal or hollow organ, as the ureter or urethra.

Uremia (*emia* = condition of blood) Toxic levels of urea in the blood resulting from severe malfunction of the kidneys.

Urethritis Inflammation of the urethra, caused by highly acid urine, the presence of bacteria, or constriction of the urethral passage.

STUDY OUTLINE

1. The primary function of the urinary system is to regulate the concentration and volume of blood by removing and restoring selected amounts of water and solutes. It also excretes wastes.
2. The organs of the urinary system are the kidneys, ureters, urinary bladder, and urethra.

Kidneys (p. 664)

External Anatomy; Internal Anatomy (p. 664)

1. The kidneys are retroperitoneal organs attached to the posterior abdominal wall.
2. Three layers of tissue surround the kidneys: renal capsule, adipose capsule, and renal fascia.
3. Internally, the kidneys consist of a cortex, medulla, pyramids, papillae, columns, calyces, and a pelvis.
4. The nephron is the functional unit of the kidneys.
5. Each juxtamedullary nephron consists of a glomerular (Bowman's) capsule, glomerulus, proximal convoluted tubule, descending limb of the loop of the nephron, loop of the nephron (loop of Henle), ascending limb of the loop of the nephron, distal convoluted tubule, and collecting duct. The limbs and loop of the nephron are absent in a cortical nephron.
6. The filtering unit of a nephron is the endothelial-capsular membrane. It consists of the glomerular endothelium, glomerular basement membrane, and epithelium (podocytes) of the visceral layer of the glomerular capsule.
7. The extensive flow of blood through the kidney begins in the renal artery and terminates in the renal vein.
8. The nerve supply to the kidney is derived from the renal plexus.

9. The juxtaglomerular apparatus consists of the juxtaglomerular cells and the macula densa of the distal convoluted tubule.

Physiology (p. 670)

1. Nephrons are the functional units of the kidneys. They help form urine and regulate blood composition.
2. The nephrons form urine by glomerular filtration, tubular reabsorption, and tubular secretion.
3. The primary force behind glomerular filtration is hydrostatic pressure.
4. Filtration of blood depends on the force of glomerular blood hydrostatic pressure in relation to two opposing forces: capsular hydrostatic pressure and blood colloidal osmotic pressure. This relationship is called effective filtration pressure (Peff).
5. If glomerular blood hydrostatic pressure falls to 50 mm Hg, renal suppression occurs because the glomerular blood hydrostatic pressure exactly equals the opposing pressures.
6. Most substances in plasma are filtered by the glomerular capsule. Normally, blood cells and most proteins are not filtered.
7. Tubular reabsorption retains substances needed by the body, including water, glucose, amino acids, and ions. The maximum of a substance that can be absorbed is called tubular maximum.
8. About 80 percent of the reabsorbed water is returned by obligatory reabsorption, the rest by facultative reabsorption.
9. When glomerular filtration rate is low, vasodilation of afferent arterioles and vasoconstriction of efferent arterioles brings the rate back to normal.
10. Chemicals not needed by the body are discharged into the

urine by tubular secretion. Included are ions, nitrogenous wastes, and certain drugs.

11. The kidneys help maintain blood pH by excreting H^+ and NH_4^+ ions. In exchange, the kidneys conserve sodium bicarbonate.

12. The kidneys produce dilute urine in the absence of ADH; renal tubules absorb more solutes than water.

13. The kidneys secrete concentrated urine in the presence of ADH; large amounts of water are reabsorbed from the filtrate into interstitial fluid, increasing solute concentration. The countercurrent mechanism also contributes to the secretion of concentrated urine.

Hemodialysis Therapy (p. 679)

1. Filtering blood through an artificial device is called hemodialysis.

2. The kidney machine filters the blood of wastes and adds nutrients; a recent variation is called continuous ambulatory peritoneal dialysis (CAPD).

Homeostasis (p. 680)

1. Besides the kidneys, the lungs, integument, and alimentary canal assume excretory functions.

2. Urine volume is influenced by blood pressure, blood concentration, temperature, diuretics, and emotions.

3. The physical characteristics of urine evaluated in a urinalysis (UA) are color, odor, turbidity, pH, and specific gravity.

4. Chemically, normal urine contains about 95 percent water and 5 percent solutes. The solutes include urea, creatinine, uric acid, hippuric acid, indican, ketone bodies, salts, and ions.

5. Abnormal constituents diagnosed through urinalysis include albumin, glucose, erythrocytes, leucocytes, ketone bodies, bilirubin, urobilinogen, casts, renal calculi, and microbes.

Ureters (p. 684)

1. The ureters are retroperitoneal and consist of a mucosa, muscularis, and fibrous coat.

2. The ureters transport urine from the renal pelvis to the urinary bladder, primarily by peristalsis.

Urinary Bladder (p. 685)

1. The urinary bladder is posterior to the symphysis pubis. Its function is to store urine prior to micturition.

2. Histologically, the urinary bladder consists of a mucosa (with rugae), a muscularis (detrusor muscle), and a serous coat.

3. A lack of control over micturition is called incontinence; failure to void urine is referred to as retention.

Urethra (p. 686)

1. The urethra is a tube leading from the floor of the urinary bladder to the exterior.

2. Its function is to discharge urine from the body.

Aging and the Urinary System (p. 686)

1. With age, kidney function decreases.

2. Common problems related to aging include incontinence, urinary tract infections, prostate disorders, and renal calculi.

Developmental Anatomy of the Urinary System (p. 686)

1. The kidneys develop from intermediate mesoderm.

2. They develop in the following sequence: pronephros, mesonephros, metanephros.

Disorders: Homeostatic Imbalances (p. 688)

1. Gout is a high level of uric acid in the blood.

2. Glomerulonephritis is an inflammation of the glomeruli of the kidney.

3. Pyelitis is an inflammation of the kidney pelvis and calyces; pyelonephritis is an interstitial inflammation of one or both kidneys.

4. Cystitis is an inflammation of the urinary bladder.

5. Nephrosis leads to protein in the urine due to glomerular membrane permeability.

6. Polycystic disease is an inherited kidney disease in which nephrons are deformed.

7. Renal failure is classified as acute and chronic.

8. The term urinary tract infection (UTI) refers to either an infection of a part of the urinary system or the presence of large numbers of microbes in urine.

REVIEW QUESTIONS

1. What are the functions of the urinary system? What organs compose the system?

2. Describe the location of the kidneys. Why are they said to be retroperitoneal?

3. Prepare a labeled diagram that illustrates the principal external and internal features of the kidney?

4. What is a nephron? List and describe the parts of a nephron from the glomerular (Bowman's) capsule to the collecting duct.

5. Describe the details of the endothelial-capsular membrane.

6. Distinguish between cortical and juxtaglomerular nephrons. How are nephrons supplied with blood?

7. Describe the structure and importance of the juxtaglomerular apparatus.

8. What is glomerular filtration? Define filtrate.

9. Set up an equation to indicate how effective filtration pressure is calculated.

10. What are the major chemical differences among plasma, filtrate, and urine?

11. Define tubular reabsorption. Why is the process physiologically important? What is glomerular filtration rate (GFR)?

12. What chemical substances are normally reabsorbed by the kidneys? What is tubular maximum (Tm)?

13. Describe how glucose and sodium are reabsorbed by the kidneys. Where does the process occur?

14. How is chloride reabsorption related to sodium reabsorption?

15. Distinguish obligatory from facultative reabsorption of water. How is facultative reabsorption controlled?

16. Define tubular secretion. Why is it important? List some substances that are secreted.

17. Explain the mechanisms by which the kidneys help to control body pH.

18. Describe how the kidneys produce a dilute or concentrated urine.

19. Define the countercurrent multiplier mechanism. Why is it important?

20. What is hemodialysis? Briefly describe the operation of an artificial kidney. What is continuous ambulatory peritoneal dialysis (CAPD)?

21. Contrast the functions of the lungs, integument, and gastrointestinal tract as excretory organs.

22. What is urine? Describe the effects of blood pressure, blood concentration, temperature, diuretics, and emotions on the volume of urine formed.

23. Describe the following physical characteristics of normal urine: color, turbidity, odor, pH, and specific gravity.

24. Describe the chemical composition of normal urine.

25. Define each of the following: albuminuria, glycosuria, hematuria, pyuria, ketosis, bilirubinuria, urobilinogenuria, casts, and renal calculi.

26. Describe the structure, histology, and function of the ureters.

27. How is the urinary bladder adapted to its storage function?

28. What is micturition? What is cystoscopy?

29. Contrast the causes of incontinence and retention.

30. Compare the location of the urethra in the male and female.

31. Describe the effects of aging on the urinary system.

32. Describe the development of the urinary system.

33. Define each of the following: gout, glomerulonephritis, pyelitis, pyelonephritis, cystitis, nephrosis, polycystic disease, renal failure, and urinary tract infections (UTIs).

34. Refer to the glossary of medical terminology associated with the urinary system. Be sure that you can define each term.

Fluid, Electrolyte, and Acid–Base Dynamics

27

The term **body fluid** refers to the body water and its dissolved substances. Fluid composes 45 to 75 percent of the body weight.

FLUID COMPARTMENTS AND FLUID BALANCE

About two-thirds of the fluid is located in cells and is termed **intracellular fluid (ICF).** The other third, called **extracellular fluid (ECF),** includes all the rest of the body fluids—interstitial fluid; plasma and lymph; cerebrospinal fluid; GI tract fluids; synovial fluid; the fluids of the eyes and ears; pleural, pericardial, and peritoneal fluids; and the glomerular filtrate.

Body fluids are separated into distinct compartments whose walls are the selectively permeable membranes provided by the plasma membranes of cells. A compartment may be as small as the interior of a single cell or as large as the combined interiors of the heart and vessels. Fluids exist in compartments, but keep in mind that they are in constant motion from one compartment to another. In the healthy individual, the volume of fluid in each compartment remains relatively stable—another example of homeostasis.

Water constitutes the bulk of all the body fluids. When we say that the body is in *fluid balance,* we mean it contains the required amount of water distributed to the various compartments according to their needs.

Osmosis is the primary way in which water moves in and out of body compartments. The concentration of solutes in the fluids is therefore a major determinant of fluid balance. Most solutes in body fluids are electrolytes—compounds that dissociate into ions. Fluid balance, then, means water balance, but it also implies electrolyte balance. The two are inseparable.

WATER

Water is by far the largest single constituent of the body, making up from 45 to 75 percent of the total body weight. The percentage varies from person to person and depends primarily on the amount of fat present and age. Since fat is basically water-free, lean people have a greater proportion of water to total body weight than fat people. Water proportion also decreases with age. An infant has the highest amount of water per body weight. In a normal adult male, water averages about 65 percent of the body weight. Females have more subcutaneous fat than males, and in a normal female water averages about 55 percent.

FLUID INTAKE AND OUTPUT

The primary source of body fluid is water derived from ingested liquids (1,600 ml) and foods (700 ml) that has been absorbed from the alimentary canal. This water, called *preformed water,* amounts to about 2,300 ml/day. Another source of fluid is *metabolic water,* the water produced through catabolism. This amounts to about 200 ml/day. Thus total fluid input averages about 2,500 ml/day.

There are several avenues of fluid output. The kidneys on the average lose about 1,500 ml/day, the skin about 500 ml/day, the lungs about 300 ml/day, and the GI tract about 200 ml/day. Fluid output thus totals 2,500 ml/day. Under normal circumstances, fluid intake equals fluid output, so the body maintains a constant volume (Exhibit 27-1).

REGULATION OF INTAKE

Fluid intake is regulated by thirst. According to one theory, when water loss is greater than water intake, the resulting *dehydration* stimulates thirst through both local and general responses. Locally, it leads to a decrease in the flow of saliva that produces a dryness of the mucosa of the mouth and pharynx (Figure 27-1). Dryness is interpreted by the brain as a sensation of thirst. In addition, dehydration raises blood osmotic pressure. Receptors in the thirst center of the hypothalamus are stimulated by the increase in osmotic pressure and initiate impulses that are also interpreted as a sensation of thirst. The response, a desire to drink fluids, thus balances the fluid loss.

The initial quenching of thirst results from wetting the mucosa of the mouth and pharynx, but the major inhibition of thirst is believed to occur as a result of distension of the stomach or intestine and a decrease in osmotic pressure in fluids of the hypothalamus. Apparently, stimulated stretch receptors in the wall of the intestine send impulses that inhibit the thirst center in the hypothalamus.

REGULATION OF OUTPUT

Under normal circumstances, fluid output is adjusted by antidiuretic hormone (ADH) and aldosterone, both of which regulate urine production (Chapter 26). Under abnormal conditions, other factors may influence output heavily. If the body is dehydrated, blood pressure falls, glomerular filtration rate decreases accordingly, and water is conserved. Conversely, excessive blood fluid results in an increase of blood pressure, glomerular filtration rate, and fluid output. Hypertension produces the same effect. Hyperventilation leads to increased fluid output through the loss of water vapor by the lungs. Vomiting and diarrhea result in fluid loss from the gastrointestinal tract. Finally, fever and destruction of extensive areas of the skin from burns bring about excessive water loss through the skin.

EXHIBIT 27-1 SUMMARY OF FLUID INTAKE AND OUTPUT PER DAY

TAKE		OUTPUT	
Ingested liquids	1,600 ml	Kidneys	1,500 ml
Ingested foods	700 ml	Skin	500 ml
Metabolic water	200 ml	Lungs	300 ml
Total	2,500 ml	GI tract	200 ml
		Total	2,500 ml

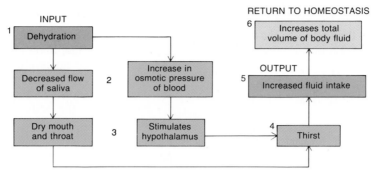

FIGURE 27-1 Regulation of fluid volume by the adjustment of intake to output.

ELECTROLYTES

The body fluids contain a variety of dissolved chemicals. Some of these chemicals are compounds with covalent bonds, that is, the atoms that compose the molecule share electrons and do not form ions. Such compounds are called **nonelectrolytes.** Nonelectrolytes include most organic compounds, such as glucose, urea, and creatine. Other compounds, called **electrolytes,** have at least one ionic bond. When they dissolve in a body fluid, they dissociate into positive and negative ions. The positive ions are called cations; the negative ions are anions. Acids, bases, and salts are elecrolytes. Most electrolytes are inorganic compounds, but a few are organic. For example, some proteins form ionic bonds. When the protein is put in solution, the ion detaches and the rest of the protein molecule carries the opposite charge.

Electrolytes serve three general functions in the body. First, many are essential minerals. Second, they control the osmosis of water between body compartments. Third, they help maintain the acid–base balance required for normal cellular activites.

CONCENTRATION

During osmosis, water moves to the area with the greater number of particles in solution. A particle may be a whole molecule or an ion. An electrolyte exerts a far greater effect on osmosis than a nonelectrolyte because an electrolyte molecule dissociates into at least two particles, both of them charged. Suppose the nonelectrolyte glucose and two different electrolytes are placed in solution:

$$C_6H_{12}O_6 \xrightarrow{H_2O} C_6H_{12}O_6$$
Glucose

$$NaCl \xrightarrow{H_2O} Na^+ + Cl^-$$
Sodium chloride

$$CaCl_2 \xrightarrow{H_2O} Ca^{2+} + Cl^- + Cl^-$$
Calcium chloride

Because glucose does not break apart when dissolved in water, a molecule of glucose contributes only one particle to the solution. Sodium chloride, on the other hand, contributes two ions, or particles, and calcium chloride contributes three. Thus calcium chloride has three times as great an effect on solute concentration as glucose.

Just as important, once the electrolyte dissociates, its ions can attract other ions of the opposite charge. If equal amounts of Ca^{2+} and Na^+ are placed in solution, the calcium ions will attract about twice as many chloride ions to their area as the sodium ions.

To determine how much effect an electrolyte has on concentration, we must look at the concentrations of its individual ions. The concentration of an ion is commonly expressed in **milliequivalents per liter (meq/liter)**—the number of electrical charges in each liter of solution. The meq/liter equals the number of ions in solution times the number of charges the ions carries. Ion concentration can be calculated as shown in Exhibit 27-2.

DISTRIBUTION

Figure 27-2 compares the principal chemical constituents of plasma, interstitial fluid, and intracellular fluid. The chief difference between plasma and interstitial fluid is that plasma contains quite a few protein anions, whereas interstitial fluid has hardly any. Since normal capillary membranes are practically impermeable to protein, the protein stays in the plasma and does not move out of the blood into the interstitial fluid. Plasma also contains more Na^+ ions but fewer Cl^- ions than the interstitial fluid. In most other respects the two fluids are similar.

Intracellular fluid varies considerably from extracellular fluid, however. In extracellular fluid, the most abundant cation is Na^+, and the most abundant anion is Cl^-. In intracellular fluid, the most abundant cation is K^+, and the most abundant anion is HPO_4^{2-} (phosphate). Also, there are more protein anions in intracellular fluid than in extracellular fluid.

FUNCTIONS AND REGULATION

Sodium

Sodium (Na^+), the most abundant extracellular ion, represents about 90 percent of extracellular cations. Sodium is necessary for the transmission of impulses in nervous and muscle tissue. Its movement also plays a significant role in fluid and electrolyte balance.

EXHIBIT 27-2   CALCULATING CONCENTRATION OF IONS IN SOLUTION

The number of milliequivalents of an ion in each liter of solution is expressed by the following equation:

$$\text{meq/liter} = \frac{\text{milligrams of ion per liter of solution}}{\text{atomic weight}} \times \text{number of charges on one ion}$$

The atomic weight of an element is the number of protons and neutrons in an atom. Dividing the total weight of a solute by its atomic weight tells us how many ions there are in solution. The atomic weight of calcium is 40, whereas that of sodium is 23. Calcium is therefore a heavier element, and 100 g of calcium contains fewer atoms than does 100 g of sodium. The atomic weights of the elements can be found in a periodic table.

Using the preceding formula, we can calculate the milliequivalents per liter for calcium. In 1 liter of plasma there are normally 100 mg of calcium. Thus, by substituting this value in the formula, we arrive at:

$$\text{meq/liter} = \frac{100}{\text{atomic weight}} \times \text{number of charges}$$

The atomic weight of calcium is 40, and its number of charges is 2. By substituting these values we arrive at:

$$\text{meq/liter} = \frac{100}{40} \times 2 = 5$$

Let us now find the milliequivalents per liter of plasma for sodium.

Milligrams of ion per liter = 3,300

Number of charges = 1

Atomic weight = 23

$$\text{meq/liter} = \frac{3,300}{23} \times 1$$

$$= 143.0$$

Even though calcium has a greater number of charges than sodium, the body retains many more sodium ions than calcium ions. Therefore, the milliequivalent for sodium in plasma is higher.

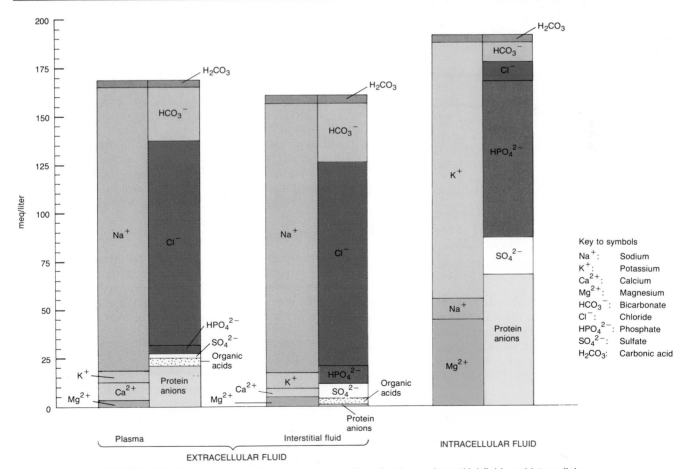

FIGURE 27-2 Comparison of electrolyte concentrations in plasma, interstitial fluid, and intracellular fluid. The height of each column represents the total electrolyte concentration.

Sodium loss from the body may occur through excessive perspiration, certain diuretics, and burns. Such a loss can result in **hyponatremia** (hī'-pō-na-TRĒ-mē-a; *natrium* = sodium), a lower-than-normal blood sodium level. Hyponatremia is characterized by muscular weakness, headache, hypotension, tachycardia, and circulatory shock. Severe sodium loss can result in mental confusion, stupor, and coma.

The sodium level in the blood is controlled primarily by the hormone aldosterone from the adrenal cortex. Aldosterone acts on the distal convoluted tubules and collecting ducts of the kidneys and causes them to increase their reabsorption of sodium. The sodium thus moves from the filtrate back into the blood and establishes an osmotic gradient. Aldosterone is secreted in response to reduced blood volume or cardiac output, decreased extracellular sodium, increased extracellular potassium, and physical stress.

Chloride

Chloride (Cl⁻) is mainly an extracellular anion. However, it can easily diffuse between the extracellular and intracellular compartments. This movement makes chloride important in regulating osmotic pressure differences between compartments. In the gastric mucosal glands, chloride combines with hydrogen to form hydrochloric acid.

An abnormally low level of chloride in the blood, called **hypochloremia** (hī'-pō-klō-RĒ-mē-a) may be caused by excessive vomiting, dehydration, and certain diuretics. Symptoms include muscle spasms, alkalosis, depressed respirations, and even coma.

Part of the regulation of chloride is indirectly under the control of aldosterone. Aldosterone regulates sodium reabsorption and chloride follows sodium passively.

Potassium

Potassium (K⁺) is the most abundant cation in intracellular fluid. It helps maintain the fluid volume in cells and control pH. Potassium also assumes a key role in the functioning of nervous and muscle tissue. When potassium ions move out of the cell, they are replaced by sodium ions and hydrogen ions. This shift of hydrogen ions helps regulate pH.

A lower-than-normal level of potassium, called **hypokalemia** (hī'-pō-ka-LĒ-mē-a; *kalium* = potassium),

may result from vomiting, diarrhea, high sodium intake, kidney disease, and use of some diuretics. Symptoms include cramps and fatigue, flaccid paralysis, mental confusion, increased urine output, shallow respirations, and changes in the electrocardiogram, including a lengthening of the Q-T interval and flattening of the T-wave.

The blood level of potassium is under the control of mineralocorticoids, mainly aldosterone. The mechanism is exactly opposite that of sodium. When sodium concentration is low, aldosterone secretion increases and more sodium is reabsorbed. But when potassium concentration is high, more aldosterone is secreted and more potassium is excreted. This process occurs in the distal tubules and collecting ducts of the kidneys.

Calcium and Phosphate

Calcium (Ca²⁺) and phosphate (HPO₄²⁻) are stored in bones and teeth and released when needed. Calcium is principally an extracellular electrolyte; phosphate is principally an intracellular electrolyte. Calcium is a structural component of bones and teeth. It is also required for blood clotting, neurotransmitter release, muscle contraction, and normal heartbeat. Phosphate is an important structural component of bones and teeth. In addition, it is necessary for the formation of nucleic acids (DNA and RNA), the synthesis of high-energy compounds (ATP and creatine phosphate), and buffering reactions.

Calcium and phosphate blood levels are regulated by several hormones. Parathyroid hormone (PTH) is released when the calcium blood level is low. PTH stimulates osteoclasts to release calcium and phosphate into the blood, increases the absorption of calcium from the gastrointestinal tract, and causes renal tubular cells to excrete phosphate. Calcitonin (CT), from the thyroid gland, decreases the blood level of calcium by stimulating osteoblasts and inhibiting osteoclasts. In the presence of calcitonin, osteoblasts remove calcium and phosphate from the blood and deposit them in bone.

Magnesium

Magnesium (Mg²⁺) is primarily an intracellular cation. It is important for the sodium-potassium pump. It activates enzyme systems needed to produce cellular energy by the breakdown of ATP into ADP and enzyme systems involved in essential reactions in the liver and bone tissue.

Symptoms of **hypomagnesemia (magnesium deficiency)** include increased neuromuscular and central nervous system irritability leading to tremor, tetany, and possible convulsions. Diuretic-induced hypomagnesemia can result in cardiac arrhythmias. **Hypermagnesemia (magnesium excess)** may cause central nervous system depression, coma, and hypotension.

Magnesium level is regulated by aldosterone. When magnesium concentration is low, increased aldosterone secretion acts on the kidneys so that more magnesium is reabsorbed.

MOVEMENT OF BODY FLUIDS

BETWEEN PLASMA AND INTERSTITIAL COMPARTMENTS

The movement of fluid between plasma and interstitial compartments occurs across capillary membranes. This movement was discussed in detail in Chapter 21 (see Figure 21-8), but we will review it here. Basically, the fluid movement is dependent on four principal pressures: (1) blood hydrostatic pressure (BHP), (2) interstitial fluid hydrostatic pressure (IFHP), (3) blood osmotic pressure (BOP), and (4) interstitial fluid osmotic pressure (IFOP).

The difference between the two forces that move fluid out of plasma and the two forces that push it into plasma is the **effective filtration pressure (Peff).** The Peff at the arterial end of a capillary is 8 mm Hg, while that at the venous end is −7 mm Hg. Thus, at the arterial end of a capillary, fluid moves out (filtered) from plasma into the interstitial compartment at a pressure of 8 mm Hg; at the venous end of a capillary, fluid moves in (reabsorbed) from the interstitial to plasma compartment at a pressure of −7 mm Hg. Thus, not all of the fluid filtered at one end of the capillary is reabsorbed at the other. The fluid not reabsorbed and any proteins that escape from capillaries pass into lymph capillaries. From here, the fluid (lymph) moves through lymphatics to the thoracic duct or right lymphatic duct for entrance into the cardiovascular system via the subclavian veins. Under normal conditions, there is a state of near equilibrium at the arterial and venous ends of a capillary in which filtered fluid and absorbed fluid, as well as that picked up by the lymphatic system, are nearly equal. This near equilibrium is **Starling's law of the capillaries.**

CLINICAL APPLICATION

Occasionally the balance between interstitial fluid and plasma is disrupted. **Edema,** the abnormal increase in interstitial fluid resulting in tissue swelling, is an example of fluid imbalance (Figure 27-3). Hypertension may cause edema by raising the blood hydrostatic pressure Another cause is inflammation. As part of the inflammatory response, capillaries become more permeable and allow proteins to leave the plasma and enter the interstitial fluid. Consequently, the blood osmotic pressure falls and the interstitial fluid osmotic pressure rises.

BETWEEN INTERSTITIAL AND INTRACELLULAR COMPARTMENTS

Intracellular fluid has a higher osmotic pressure than interstitial fluid. In addition, the principal cation inside the cell is K^+, whereas the principal cation outside is Na^+

FIGURE 27-3 Conditions that produce edema.

(see Figure 27-2). Normally, the higher intracellular osmotic pressure is balanced by forces that move water out of the cell, so the amount of water inside the cell does not change. When a fluid imbalance between these two compartments occurs, it is usually caused by a change in the Na^+ or K^+ concentration.

Sodium balance in the body normally is controlled by aldosterone and ADH. ADH regulates extracellular fluid electrolyte concentration by adjusting the amount of water reabsorbed into the blood by the distal convoluted tubules and collecting ducts of the kidneys. Aldosterone regulates extracellular fluid volume by adjusting the amount of sodium reabsorbed by the blood from the kidneys. Certain conditions, however, may result in an eventual decrease in the sodium concentration in interstitial fluid. For instance, during sweating the skin excretes sodium as well as water. Sodium also may be lost through vomiting and diarrhea. Coupled with low sodium intake, these conditions can quickly produce a sodium deficit (Figure 27-4). The decrease in sodium concentration in the interstitial fluid lowers the interstitial fluid osmotic pressure and establishes an effective filtration pressure gradient between the interstitial fluid and the intracellular fluid. Water moves from the interstitial fluid into the cells, producing two results that can be quite serious.

The first result, an increase in intracellular water concentration, called **overhydration,** is particularly disruptive to nerve cell function. In fact, severe overhydration, or **water intoxication,** produces neurological symptoms ranging from disoriented behavior to convulsions, coma, and even death. The second result of the fluid shift is a loss of interstitial fluid volume that leads to a decrease in the interstitial fluid hydrostatic pressure. As the interstitial hydrostatic pressure drops, water moves out of the plasma, resulting in a loss of blood volume that may lead to circulatory shock.

ACID–BASE BALANCE

Before reading this section, you might want to review the discussion of acids, bases, and pH in Chapter 2. In

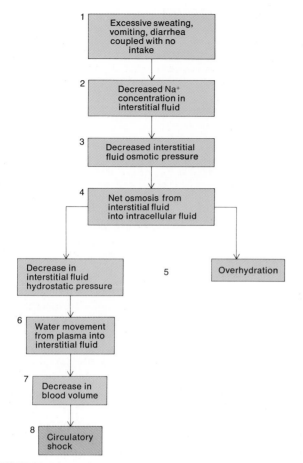

1. Excessive sweating, vomiting, diarrhea coupled with no intake

2. Decreased Na⁺ concentration in interstitial fluid

3. Decreased interstitial fluid osmotic pressure

4. Net osmosis from interstitial fluid into intracellular fluid

5. Overhydration

Decrease in interstitial fluid hydrostatic pressure

6. Water movement from plasma into interstitial fluid

7. Decrease in blood volume

8. Circulatory shock

FIGURE 27-4 Interrelations between fluid imbalance and electrolyte imbalance.

Carbonic Acid–Bicarbonate

The **carbonic acid–bicarbonate buffer system** is an important regulator of blood pH. This system is based on the weak acid, carbonic acid, and a weak base, primarily sodium bicarbonate. The following equations illustrate the mechanism.

$$\underset{\substack{\text{Hydrochloric} \\ \text{acid} \\ \text{(strong acid)}}}{\text{HCl}} + \underset{\substack{\text{Sodium} \\ \text{bicarbonate} \\ \text{(weak base)}}}{\text{NaHCO}_3} \rightleftharpoons \underset{\substack{\text{Sodium} \\ \text{chloride} \\ \text{(salt)}}}{\text{NaCl}} + \underset{\substack{\text{Carbonic acid} \\ \text{(weak acid)}}}{\text{H}_2\text{CO}_3}$$

$$\underset{\substack{\text{Sodium} \\ \text{hydroxide} \\ \text{(strong base)}}}{\text{NaOH}} + \underset{\substack{\text{Carbonic acid} \\ \text{(weak acid)}}}{\text{H}_2\text{CO}_3} \rightleftharpoons \underset{\text{Water}}{\text{H}_2\text{O}} + \underset{\substack{\text{Sodium} \\ \text{bicarbonate} \\ \text{(weak base)}}}{\text{NaHCO}_3}$$

Normal body processes tend to acidify the blood rather than make it more alkaline, and the body needs more bicarbonate salt than it needs carbonic acid. In fact, when extracellular pH is normal (7.4), bicarbonate molecules outnumber carbonic acid by 20 to 1.

Phosphate

The **phosphate buffer system** acts in essentially the same manner as the bicarbonate buffer system. Its two components are sodium dihydrogen phosphate and sodium monohydrogen phosphate. The dihydrogen phosphate ion acts as the weak acid and is capable of buffering strong bases.

$$\underset{\substack{\text{Sodium} \\ \text{hydroxide} \\ \text{(strong base)}}}{\text{NaOH}} + \underset{\substack{\text{Sodium dihydrogen} \\ \text{phosphate} \\ \text{(weak acid)}}}{\text{NaH}_2\text{PO}_4} \rightleftharpoons \underset{\text{Water}}{\text{H}_2\text{O}} + \underset{\substack{\text{Sodium} \\ \text{monohydrogen} \\ \text{phosphate} \\ \text{(weak base)}}}{\text{Na}_2\text{HPO}_4}$$

The monohydrogen phosphate ion acts as the weak base and is capable of buffering strong acids.

$$\underset{\substack{\text{Hydrochloric} \\ \text{acid} \\ \text{(strong} \\ \text{acid)}}}{\text{HCl}} + \underset{\substack{\text{Sodium} \\ \text{monohydrogen} \\ \text{phosphate} \\ \text{(weak base)}}}{\text{Na}_2\text{HPO}_4} \rightleftharpoons \underset{\substack{\text{Sodium} \\ \text{chloride} \\ \text{(salt)}}}{\text{NaCl}} + \underset{\substack{\text{Sodium dihydrogen} \\ \text{phosphate} \\ \text{(weak acid)}}}{\text{NaH}_2\text{PO}_4}$$

The phosphate buffer system is an important regulator of pH both in red blood cells and in the kidney tubular fluids. NaH_2PO_4 is formed when excess H^+ ions in the kidney tubules combine with Na_2HPO_4. In this reaction, the sodium released from Na_2HPO_4 forms sodium bicarbonate (NaHCO_3) and is passed into the blood. The H^+ ion that replaces sodium becomes part of the NaH_2PO_4 that is passed into the urine. This reaction is one of the mechanisms by which the kidneys help maintain pH by the acidification of urine.

Hemoglobin-Oxyhemoglobin

The **hemoglobin-oxyhemoglobin buffer system** is an effective method for buffering carbonic acid in the blood. When blood moves from the arterial end of a capillary to the venous end, the carbon dioxide given up by body cells enters the erythrocytes and combines with water to form

addition to controlling water movement, electrolytes also help regulate the body's acid–base balance. The overall acid–base balance is maintained by controlling the hydrogen ion concentration of body fluids, particularly extracellular fluid. In a healthy person, the pH of the extracellular fluid is stabilized between 7.35 and 7.45. Homeostasis of this narrow range is essential to survival and depends on three major mechanisms: buffer systems, respirations, and kidney excretion.

BUFFER SYSTEMS

Most **buffer systems** of the body consist of a weak acid and a weak base, and they function to prevent drastic changes in the pH of a body fluid by changing strong acids and bases into weak acids and bases. Buffers work within fractions of a second. Recall that a strong acid dissociates into H^+ ions more easily than does a weak acid. Strong acids, therefore, lower pH more than weak ones because strong acids contribute more H^+ ions. Similarly, strong bases raise pH more than weak ones because strong bases dissociate more easily into OH^- ions. The principal buffer systems of the body fluids are the carbonic acid–bicarbonate system, the phosphate system, the hemoglobin-oxyhemoglobin system, and the protein system.

carbonic acid (Figure 27-5). Simultaneously, oxyhemoglo-bin gives up its oxygen to the body cells, and some be-comes reduced hemoglobin, and carries a negative charge. The hemoglobin anion attracts the hydrogen ion from the carbonic acid and becomes an acid that is even weaker than carbonic acid. When the hemoglobin-oxyhemoglobin system is active, the exchange reaction that occurs shows why the erythrocyte tends to give up its oxygen when PCO_2 is high.

Protein

The **protein buffer system** is the most abundant buffer in body cells and plasma. Proteins are composed of amino acids. An amino acid is an organic compound that con-tains at least one carboxyl group (COOH) and at least one amine group (NH_2). The carboxyl group acts like an acid and can dissociate in this way.

$$NH_2—\overset{\overset{\displaystyle R}{|}}{\underset{\underset{\displaystyle H}{|}}{C}}—COO^-H^+$$

The hydrogen ion is then able to react with any excess hydroxide ion in the solution to form water.

The amine group has a tendency to act as a base.

$$COOH—\overset{\overset{\displaystyle R}{|}}{\underset{\underset{\displaystyle H}{|}}{C}}—NH_3^+OH^-$$

The hydroxide ion can dissociate, react with excess hydro-gen ions, and also form water. Thus proteins act as both acidic and basic buffers.

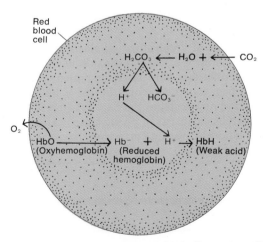

FIGURE 27-5 Hemoglobin-oxyhemoglobin buffer system. Oxyhem-oglobin gives up its oxygen in an acid medium and buffers the acid. The bicarbonate (HCO_3^-) may remain in the cell and combine with potassium (K^+), or it may move out of the cell and combine with sodium (Na^+). In this way, much of the carbon dioxide is carried back to the lungs in the form of potassium bicarbonate ($KHCO_3$) or sodium bicarbonate ($NaHCO_3$).

RESPIRATIONS

Respirations also assume a role in maintaining the pH of the body. An increase in the carbon dioxide concentra-tion in body fluids as a result of cellular respiration lowers the pH. This is illustrated by the following equation:

$$CO_2 + H_2O \rightleftharpoons H_2CO_3 \rightleftharpoons H^+ + HCO_3^-$$

Conversely, a decrease in the carbon dioxide concentra-tion of body fluids raises the pH.

The pH of body fluids may be adjusted by a change in the rate of breathing, an adjustment that usually takes from 1 to 3 minutes. If the rate of breathing is increased, more carbon dioxide is exhaled and the blood pH rises. Slowing down the respiration rate means less carbon diox-ide is exhaled, and the blood pH falls. Doubling the breathing rate increases the pH by about 0.23. Thus it can be increased from 7.4 to 7.63. Reducing the breathing rate to one-quarter its normal rate lowers the pH by 0.4. Thus it can be decreased from 7.4 to 7.0. If you consider that breathing rate can be altered up to eight times the normal rate, it should become obvious that alterations in the pH of body fluids may be greatly influenced by respiration.

The pH of body fluids, in turn, affects the rate of breathing (Figure 27-6). If, for example, the blood be-comes more acidic, the increase in hydrogen ions stimu-lates the respiratory center in the medulla and respirations increase. The same effect is achieved if the blood concen-tration of carbon dioxide increases. On the other hand, if the pH of the blood increases, the respiratory center is inhibited and respirations decrease. A decrease in the carbon dioxide concentration of blood has the same effect. The respiratory mechanism normally can eliminate more acid or base than can all the buffers combined.

KIDNEY EXCRETION

Since the role of the kidneys in maintaining pH has al-ready been discussed in Chapter 26, we shall simply refer you to that section. Review Figure 26-11 very carefully.

A summary of the mechanisms that maintain body pH is presented in Exhibit 27-3.

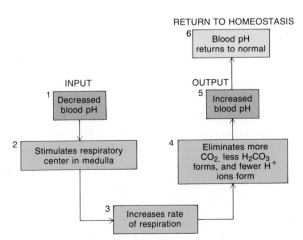

FIGURE 27-6 Relationship between pH and respirations.

EXHIBIT 27-3 MECHANISMS THAT MAINTAIN BODY pH

MECHANISM	COMMENTS
Buffer systems	Consist of weak acid and weak base.
	Prevent drastic changes in body fluid pH.
Carbonic acid–bicarbonate	Important regulator of blood pH.
Phosphate	Important buffer in red blood cells and kidney tubule cells.
Hemoglobin-oxyhemoglobin	Buffers carbonic acid in blood.
Protein	Most abundant buffer in body cells and plasma.
Respirations	Regulate CO_2 level of body fluids.
Increased	Raises pH.
Decreased	Lowers pH.
Kidneys	Excrete H^+ and NH_4^+ and conserve bicarbonate.

ACID–BASE IMBALANCES

The normal blood pH range is 7.35 to 7.45. Any considerable deviation from this value falls under the category of acidosis or alkalosis. **Acidosis** is a condition in which blood pH ranges from 7.35 to 6.80. **Alkalosis** is a pH range of 7.45 to 8.00.

Respiratory acidosis is caused by hypoventilation. It occurs as a result of any condition that decreases the movement of carbon dioxide from the blood to the alveoli of the lungs and therefore causes a buildup of carbon dioxide, carbonic acid, and hydrogen ions. Such conditions include emphysema, pulmonary edema, injury to the respiratory center of the medulla, or disorders of the muscles involved in breathing. In uncompensated respiratory acidosis, the normal bicarbonate–carbonic acid ratio shifts from 20:1 to 10:1 or 8:1, and the blood pH decreases.

Respiratory alkalosis is caused by hyperventilation. It occurs as a result of any condition that stimulates the respiratory center. Such conditions include oxygen deficiency due to high altitude, severe anxiety, and aspirin overdose. In uncompensated respiratory alkalosis, the normal bicarbonate–carbonic acid ratio shifts from 20:1 to 20:0.5, and the blood pH increases.

Metabolic acidosis results from an abnormal increase in acidic metabolic products, other than carbon dioxide, and from the loss of bicarbonate ions from the body. Ketosis is a good example of metabolic acidosis brought on by an increase in the production of acidic metabolic products. Acidosis due to loss of bicarbonate may occur with diarrhea and renal tubular dysfunction. In uncompensated metabolic acidosis, the ratio of bicarbonate to carbonic acid is 12.5:1.

Metabolic alkalosis is caused by a nonrespiratory loss of acid by the body or excessive intake or alkaline drugs. Excessive vomiting of gastric contents results in a substantial loss of hydrochloric acid and is probably the most frequent cause of metabolic alkalosis. In uncompensated metabolic alkalosis, the ratio of bicarbonate to carbonic acid is 31.6:1.

The principal physiological effect of acidosis is depression of the central nervous system through depression of synaptic transmission. If the blood pH falls below 7, depression of the nervous system is so acute that the individual becomes disoriented and comatose. In fact, patients with severe acidosis usually die in a state of coma. On the other hand, the major physiological effect of alkalosis is overexcitability of the nervous system through facilitation of synaptic transmission. The overexcitability occurs both in the central nervous system and in peripheral nerves. Because of the overexcitability, nerves conduct impulses repetitively, even when not stimulated by normal stimuli, resulting in nervousness, muscle spasms, and even convulsions.

CLINICAL APPLICATION

The primary **treatment for respiratory acidosis** is aimed at increasing the exhalation of carbon dioxide. Excessive secretions may be suctioned out of the respiratory tract, and artificial respiration may be given. **Treatment of metabolic acidosis** consists of intravenous solutions of sodium bicarbonate and correcting the cause of acidosis.

Treatment of respiratory alkalosis is aimed at increasing the level of carbon dioxide in the body. One corrective measure is to have the patient breathe into a paper bag and then rebreathe the exhaled mixture of CO_2 and oxygen from the bag. **Treatment for metabolic alkalosis** consists of giving a medication containing the chloride ion and correcting the cause of alkalosis.

STUDY OUTLINE

Fluid Compartments and Fluid Balance (p. 693)
1. Body fluid is water and its dissolved substances.
2. About two-thirds of the body's fluid is located in cells and is called intracellular fluid (ICF).
3. The other third is called extracellular fluid (ECF). It includes interstitial fluid; plasma and lymph; cerebrospinal fluid; GI tract fluids; synovial fluid; fluids of the eyes and ears; pleural, pericardial, and peritoneal fluids; and the glomerular filtrate.

4. Fluid balance means that the various body compartments contain the required amount of water.
5. Fluid balance and electrolyte balance are inseparable.

Water (p. 693)
1. Water is the largest single constituent in the body, varying from 45 to 75 percent of body weight, depending on the amount of fat present and age.

2. Primary sources of fluid intake are ingested liquids and foods and water produced by catabolism.
3. Avenues of fluid output are the kidneys, skin, lungs, and GI tract.
4. The stimulus for fluid intake is dehydration resulting in thirst sensations. Under normal conditions, fluid output is adjusted by aldosterone and ADH.

Electrolytes (p. 694)

1. Electrolytes are chemicals that dissolve in body fluids and dissociate into either cations (positive ions) or anions (negative ions).
2. Electrolyte concentration is expressed in milliequivalents per liter (meq/liter).
3. Electrolytes have a greater effect on osmosis than nonelectrolytes.
4. Plasma, interstitial fluid, and intracellular fluid contain varying kinds and amounts of electrolytes.
5. Electrolytes are needed for normal metabolism, proper fluid movement between compartments, and regulation of pH.
6. Sodium is the most abundant extracellular ion. It is involved in nerve impulse transmission, muscle contraction, and fluid and electrolyte balance. Its level is controlled by aldosterone.
7. Chloride is mainly an extracellular anion. It assumes a role in regulating osmotic pressure and forming HCl. Its level is controlled indirectly by aldosterone.
8. Potassium is the most abundant cation in intracellular fluid. It is involved in maintaining fluid volume, nerve impulse conduction, muscle contraction, and regulating pH. Its level is controlled by aldosterone.
9. Calcium is principally an extracellular ion that is a structural component of bones and teeth. It also functions in blood clotting, neurotransmitter release, muscle contraction, and heartbeat. Its level is controlled by parathyroid hormone (PTH) and calcitonin (CT).
10. Phosphate is principally an intracellular ion that is a structural component of bones and teeth. It is also required for the synthesis of nucleic acids and ATP, and for buffer reactions. Its level is controlled by PTH and CT.
11. Magnesium is primarily an intracellular electrolyte that activates several enzyme systems. Its level is controlled by aldosterone.

Movement of Body Fluids (p. 697)

1. At the arterial end of a capillary, fluid moves from plasma into interstitial fluid. At the venous end, fluid moves in the opposite direction.
2. The state of near equilibrium at the arterial and venous ends of a capillary between filtered fluid and absorbed fluid, as well as that picked up by the lymphatic system, is referred to as Starling's law of the capillaries.
3. Fluid movement between interstitial and intracellular compartments depends on the movement of sodium and potassium and the secretion of aldosterone and ADH.
4. Fluid imbalance may lead to edema and overhydration (water intoxication).

Acid–Base Balance (p. 697)

1. The overall acid–base balance of the body is maintained by controlling the H^+ concentration of body fluids, especially extracellular fluid.
2. The normal pH of extracellular fluid is 7.35 to 7.45.
3. Homeostasis of pH is maintained by buffers, respirations, and kidney excretion.
4. The important buffer systems include: carbonic acid–bicarbonate, phosphate, hemoglobin-oxyhemoglobin, and protein.
5. An increase in rate of respirations, increases pH; a decrease in rate, decreases pH.

Acid–Base Imbalances (p. 700)

1. Acidosis is a blood pH between 7.35 and 6.80. Its principal effect is depression of the CNS.
2. Alkalosis is a blood pH between 7.45 and 8.00. Its principal effect is overexcitability of the CNS.
3. Respiratory acidosis is caused by hypoventilation; metabolic acidosis results from an abnormal increase in acid metabolic products (other than CO_2) and loss of bicarbonate.
4. Respiratory alkalosis is caused by hyperventilation; metabolic alkalosis results from nonrespiratory loss of acid or excess intake of alkaline drugs.

REVIEW QUESTIONS

1. Define body fluid. List several compartments and describe how they are separated.
2. What is meant by fluid balance? How are fluid balance and electrolyte balance related?
3. Describe the avenues of fluid intake and fluid output. Be sure to indicate volumes in each case.
4. Discuss the role of thirst in regulating fluid intake.
5. Explain how aldosterone and ADH adjust normal fluid output. What are some abnormal routes of fluid output?
6. Define a nonelectrolyte and an electrolyte. Give specific examples of each.
7. Distinguish between a cation and an anion. Give several examples of each.
8. How is the ionic concentration of a fluid expressed? Calculate the ionic concentration of sodium in a body fluid.
9. Describe the functions of electrolytes in the body.
10. Describe some of the major differences in the electrolytic concentrations of the three major fluid compartments in the body.
11. Name three important extracellular electrolytes and three important intracellular electrolytes.
12. Indicate the function and regulation of each of the following electrolytes: sodium, chloride, potassium, calcium, phosphate, and magnesium.
13. Describe the physiological effects of hyponatremia, hypochloremia, hypokalemia, hypomagnesemia, and hypermagnesemia.
14. Explain the forces involved in moving fluid between plasma and interstitial fluid. Summarize these forces by setting up an equation to express effective filtration pressure (P eff).
15. What is Starling's law of the capillaries?
16. Explain the factors involved in fluid movement between the interstitial fluid and the intracellular fluid.
17. Define edema. What are some of its causes?
18. Explain how the following buffer systems help to maintain the pH of body fluids: carbonic acid–bicarbonate, phosphate, hemoglobin-oxyhemoglobin, and protein.
19. Describe how respirations are related to the maintenance of pH.
20. Briefly discuss the role of the kidneys in maintaining pH.
21. Define acidosis and alkalosis. Distinguish between respiratory and metabolic acidosis and alkalosis.
22. What are the principal physiological effects of acidosis and alkalosis?
23. How are acidosis and alkalosis treated?

UNIT V
Continuity

This unit is designed to show you how the human organism is adapted for reproduction. It also traces the developmental sequence involved in pregnancy. Principles of inheritance and birth control conclude the unit.

28 / The Reproductive Systems

29 / Development and Inheritance

The Reproductive Systems

28

■ **STUDENT OBJECTIVES**

- Define reproduction and classify the organs of reproduction by function.
- Explain the structure, histology, and functions of the testes.
- Define meiosis and explain the principal events of spermatogenesis.
- Describe the physiological effects of testosterone and inhibin.
- Describe the seminiferous tubules, straight tubules, and rete testis as components of the duct system of the testes.
- Describe the location, structure, histology, and functions of the ductus epididymis, ductus (vas) deferens, and ejaculatory duct.
- Describe the three anatomical subdivisions of the male urethra.
- Explain the location and functions of the seminal vesicles, prostate gland, and bulbourethral (Cowper's) glands, the accessory reproductive glands.
- Discuss the chemical composition of semen (seminal fluid).
- Explain the structure and functions of the penis.
- Describe the location, histology, and functions of the ovaries.
- Describe the principal events of oogenesis.
- Explain the location, structure, histology, and functions of the uterine (Fallopian) tubes.
- Explain the histology and blood supply of the uterus.
- Compare the principal events of the menstrual and ovarian cycles.
- Discuss the physiological effects of estrogens, progesterone, and relaxin.
- Describe the location, structure, and functions of the vagina.
- Describe the components of the vulva and explain their functions.
- Explain the structure, development, and histology of the mammary glands.
- Explain the roles of the male and female in sexual intercourse.
- Contrast the various kinds of birth control and their effectiveness.
- Describe the effects of aging on the reproductive systems.
- Describe the development of the reproductive systems.
- Explain the symptoms and causes of sexually transmitted diseases (STDs) such as gonorrhea, syphilis, genital herpes, trichomoniasis, and nongonococcal urethritis (NGU).
- Describe the symptoms and causes of male disorders (prostate dysfunctions, impotence, and infertility) and female disorders (amenorrhea, dysmenorrhea, premenstrual syndrome [PMS], toxic shock syndrome [TSS], ovarian cysts, endometriosis, infertility, breast tumors, cervical cancer, and pelvic inflammatory disease [PID]).
- Define medical terminology associated with the reproductive systems.

Reproduction is the mechanism by which the thread of life is sustained. In one sense, reproduction is the process by which a single cell duplicates its genetic material, allowing an organism to grow and repair itself; thus reproduction maintains the life of the individual. But reproduction is also the process by which genetic material is passed from generation to generation. In this regard, reproduction maintains the continuation of the species.

The organs of the male and female reproductive systems may be grouped by function. The testes and ovaries, also called **gonads,** function in the production of gametes—sperm cells and ova, respectively. The gonads also secrete hormones. The production of gametes and their discharge into ducts classifies the gonads as exocrine glands, whereas their production of hormones classifies them as endocrine glands. The **ducts** transport, receive, and store gametes. Still other reproductive organs, called **accessory glands,** produce materials that support gametes.

The developmental anatomy of the reproductive systems is considered later in the chapter.

MALE REPRODUCTIVE SYSTEM

The organs of the male reproductive system are the testes, or male gonads, which produce sperm; a number of ducts that either store or transport sperm to the exterior; accessory glands that add secretions constituting the semen; and several supporting structures, including the penis (Figure 28-1).

SCROTUM

The **scrotum** is a cutaneous outpouching of the abdomen consisting of loose skin and superficial fascia (Figure 28-1). It is the supporting structure for the testes. Externally, it looks like a single pouch of skin separated into lateral portions by a medium ridge called the **raphe** (RĀ-fē). Internally, it is divided by a septum into two sacs, each containing a single testis. The septum consists of superficial fascia and contractile tissue called the **dartos** (DAR-tōs), which consists of bundles of smooth muscle fibers. The dartos is also found in the subcutaneous tissue of

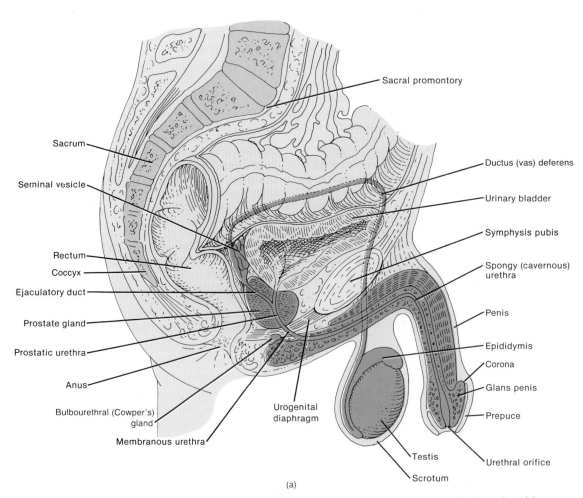

Sacral promontory

Sacrum

Seminal vesicle

Rectum

Coccyx

Ejaculatory duct

Prostate gland

Prostatic urethra

Anus

Bulbourethral (Cowper's) gland

Membranous urethra

Urogenital diaphragm

Ductus (vas) deferens

Urinary bladder

Symphysis pubis

Spongy (cavernous) urethra

Penis

Epididymis

Corona

Glans penis

Prepuce

Urethral orifice

Testis

Scrotum

(a)

FIGURE 28-1 Male organs of reproduction and surrounding structures seen in sagittal section. (a) Diagram.

Sacral promontory

Sacrum

Ductus (vas) deferens

Rectum

Coccyx

Seminal vesicle

Ejaculatory duct

External anal sphincter

Anus

Scrotum

Ureter

Rectus abdominis muscle

Urinary bladder

Prostate gland

Symphysis pubis

Spongy (cavernous) urethra

Penis

Corona

Glans penis

Urethral orifice

(b)

FIGURE 28-1 (Continued) (b) Photograph. (Courtesy of C. Yokochi and J. W. Rohen, *Photographic Anatomy of the Human Body,* 2nd ed., 1978, IGAKU-SHOIN, Ltd., Tokyo, New York.)

the scrotum and is directly continuous with the subcutaneous tissue of the abdominal wall. The dartos causes wrinkling of the skin of the scrotum.

The location of the scrotum and the contraction of its muscle fibers regulate the temperature of the testes. The production and survival of sperm require a temperature that is lower than body temperature. Because the scrotum is outside the body cavities, it supplies an environment about 3°C below body temperature. The **cremaster (krē-MAS-ter) muscle** (see Figure 28-8), a small, circular band of skeletal muscle, elevates the testes during sexual arousal and on exposure to cold, moving them closer to the pelvic cavity where they can absorb body heat. Exposure to warmth reverses the process.

TESTES

The **testes,** or **testicles,** are paired oval glands measuring about 5 cm (2 inches) in length and 2.5 cm (1 inch) in diameter (Figure 28-2). Each weighs between about 10 and 15 g. The testes develop high on the embryo's posterior abdominal wall, and usually begin their descent into the scrotum through the inguinal canals during the latter half of the seventh month (see Figure 28-8).

CLINICAL APPLICATION

When the testes do not descend, the condition is referred to as **cryptorchidism** (krip-TOR-ki-dizm). The condition occurs in about 3 percent of full-term infants and about 30 percent of premature infants. Cryptorchidism results in sterility because the cells involved in the initial development of sperm cells are destroyed by the higher body temperature of the pelvic cavity. Also, the probability of cancer of the testes is 30 to 50 times greater in cryptorchid testes. Undescended testes can be placed in the scrotum by administering hormones or by surgical means prior to puberty without ill effects. Sometimes, undescended testes undergo spontaneous descent.

The testes are covered by a dense layer of white fibrous tissue, the **tunica albuginea** (al'-byoo-JIN-ē-a), that extends inward and divides each testis into a series of internal compartments called **lobules.** Each of the 200 to 300 lobules contains one to three tightly coiled tubules, the convoluted **seminiferous tubules,** that produce sperm by a process called **spermatogenesis.** This process is considered shortly.

(a)

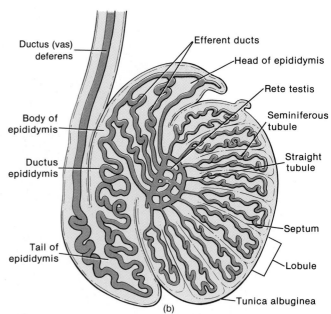

(b)

FIGURE 28-2 Anatomy of the testes. (a) Photograph of external view in the scrotum. (Courtesy of C. Yokochi and J. W. Rohen, *Photographic Anatomy of the Human Body,* 2nd ed., 1978, IGAKU-SHOIN, Ltd., Tokyo, New York.) (b) Diagram of a sagittal section illustrating the internal anatomy.

A cross section through a seminiferous tubule reveals that it is lined with cells in various stages of development (Figure 28-3). The most immature cells, the **spermatogonia,** are located against the basement membrane. These cells collectively constitute the germinal epithelium. Toward the lumen of the tube, one can see layers of progressively more mature cells. In order of advancing maturity, these are primary spermatocytes, secondary spermatocytes, and spermatids. By the time a **sperm cell,** or **spermatozoon** (sper'-ma-tō-ZŌ-on), has nearly reached maturity, it is in the lumen of the tubule and begins to be moved through a series of ducts. Embedded between the developing sperm cells in the tubules are **sustentacular** (sus'-ten-TAK-yoo-lar), or **Sertoli, cells.** Just internal to the basement membrane, each sustentacular cell is joined to another by junctional points that form a **blood-testis barrier.** The barrier is important because spermatozoa and developing cells develop surface antigens that are recognized as foreign by the immune system, and the barrier prevents an immune response against the antigens by isolating the cells from the blood. Such an immune response is seen following vasectomy (described shortly) in which sperm-specific antibodies, produced in cells of the immune system, are exposed to spermatozoa that no longer remain isolated in the reproductive tract. Sustentacular cells support and protect developing spermatogenic cells; nourish spermatocytes, spermatids, and spermatozoa; phagocytize degenerating spermatogenic cells; control movements of spermatogenic cells and the release of spermatozoa into the lumen of the seminiferous tubule; and secrete the hormone inhibin and

androgen-binding protein, a hormone required for sperm production that concentrates testosterone in the seminiferous tubule. Between the seminiferous tubules are clusters of **interstitial endocrinocytes (interstitial cells of Leydig).** These cells secrete the male hormone testosterone, the most important androgen.

Spermatogenesis

The process by which the testes produce spermatozoa by meiosis is called **spermatogenesis** (sper'-ma-tō-JEN-e-sis). Before reading the following discussion of spermatogenesis, you should review the details of meiosis in Chapter 3 (see Figure 3-17). At this point, a few key concepts should be kept in mind.

1. In sexual reproduction, a new organism is produced by the union and fusion of sex cells called **gametes.** Male gametes, produced in the testes, are called sperm cells, and female gametes, produced in the ovaries, are called ova.

2. The cell resulting from the union and fusion of gametes is called a **zygote,** which contains a mixture of chromosomes (DNA) from the two parents. Through repeated mitotic cell divisions, a zygote develops into a new organism.

3. Gametes differ from all other body cells (somatic cells) in that they contain the **haploid** (one-half) **chromosome number,** symbolized as *n.* In humans, this is 23 chromosomes, a single set of chromosomes. Uninucleated somatic cells contain the **diploid chromosome number,** symbolized 2*n.* In humans, this is 46 chromosomes, two sets of chromosomes.

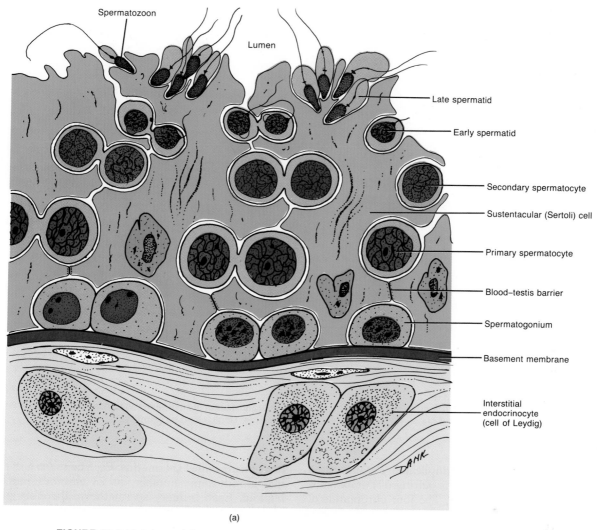

FIGURE 28-3 Histology of the testes. (a) Diagram of a cross section of a portion of a seminiferous tubule showing the stages of spermatogenesis.

4. In a diploid cell, two chromosomes that belong to a pair are called **homologous chromosomes (homologues).**

5. If gametes were diploid (2*n*), like somatic cells, the zygote would contain twice the diploid number (4*n*), and with every succeeding generation the chromosome number would continue to double.

6. This continual doubling of the chromosome number does not occur because of meiosis, a process of cell division by which gametes produced in the testes and ovaries receive the haploid chromosome number. Thus, when gametes fuse, the zygote contains the diploid chromosome number (2*n*).

In humans, spermatogenesis takes about 2 to 3 weeks. The seminiferous tubules are lined with immature cells called **spermatogonia** (sper'-ma-tō-GŌ-nē-a), or sperm mother cells (Figures 28-3a and 28-4). Spermatogonia contain the diploid chromosome number and represent a heterogenous group of cells in which three subtypes can be distinguished. These are referred to as pale type A, dark type A, and type B, and are distinguished by the appearance of their nuclear chromatin. Pale type A spermatogonia remain relatively undifferentiated and ca-

pable of extensive mitotic division. Following division, some of the daughter cells remain undifferentiated and serve as a reservoir of precursor cells to prevent depletion of the stem cell population. Such cells remain near the basement membrane. The remainder of the daughter cells differentiate into type B spermatogonia. These cells lose contact with the basement membrane of the seminiferous tubule, undergo certain developmental changes, and become known as **primary spermatocytes** (SPER-ma-tō-sītz'). Primary spermatocytes, like spermatogonia, are diploid, that is, they have 46 chromosomes. Dark type A spermatogonia are believed to represent reserve stem cells, only becoming activated if pale type A cells become critically depleted.

• *Reduction Division* (*Meiosis I*)   Each primary spermatocyte enlarges before dividing. Then two nuclear divisions take place as part of meiosis. In the first, DNA is replicated and 46 chromosomes (each made up of two chromatids) form and move toward the equatorial plane of the nucleus. There they line up by homologous pairs

Spermatids

Spermatozoa

Blood vessel

Interstitial endocrinocytes (cells of Leydig)

Primary spermatocytes

Lumen of seminiferous tubule

Basement membrane

Spermatogonia

(b)

FIGURE 28-3 (*Continued*) (b) Photomicrograph of an enlarged aspect of several seminiferous tubules at a magnification of 350×. (Copyright © 1983 by Michael H. Ross. Used by permission.)

so that there are 23 pairs of duplicated chromosomes in the center of the nucleus. This pairing of homologous chromosomes is called **synapsis.** The four chromatids of each homologous pair then twist around each other to form a **tetrad.** In a tetrad, portions of one chromatid may be exchanged with portions of another. This process, called **crossing-over,** permits an exchange of genes among chromatids (see Figure 3-18) that results in the recombination of genes. Thus, the spermatozoa eventually produced may be genetically unlike each other and unlike the cell that produced them—hence the great variation among humans. Next the meiotic spindle forms and the chromosomal microtubules produced by the centromeres of the paired chromosomes extend toward the poles of the cell. As the pairs separate, one member of each pair migrates to opposite poles of the dividing nucleus. The cells formed by the first nuclear division (reduction division) are called **secondary spermatocytes.** Each cell has 23 chromosomes—the haploid number. Each chromosome of the secondary spermatocytes, however, is made up of two chromatids. Moreover, the genes of the chromosomes of secondary spermatocytes may be rearranged as a result of crossing-over.

• *Equatorial Division (Meiosis II)*   The second nuclear division of meiosis is equatorial division. There is no replication of DNA. The chromosomes (each composed of two chromatids) line up in single file around the equatorial plane, and the chromatids of each chromosome separate from each other. The cells formed from the equatorial

division are called **spermatids.** Each contains half the original chromosome number, or 23 chromosomes, and is haploid. Each primary spermatocyte therefore produces four spermatids by meiosis (reduction division and equatorial division). Spermatids lie close to the lumen of the seminiferous tubule.

During spermatogenesis, a very interesting and unique process occurs. As the sperm cells proliferate, they fail to complete cytoplasmic separation (cytokinesis) so that all of the daughter cells, except for the least-differentiated spermatogonia, remain continuous via cytoplasmic bridges. These cytoplasmic bridges persist until development of the spermatozoa is complete, at which point they float out individually into the lumen of the seminiferous tubule. Thus, the offspring of an original spermatogonium remain in cytoplasmic communication through their entire development. This pattern of development undoubtedly accounts for the synchronized production of spermatozoa in any given area of a seminiferous tubule. This pattern may have survival value in that half the spermatozoa contain an X chromosome and half a Y chromosome. The X chromosome probably carries many essential genes that are lacking on the Y chromosome and, if it were not for the cytoplasmic bridges between the developing sperm, it may be that the Y-bearing spermatozoon could not survive, with the result that no males could be produced in the next generation.

• *Spermiogenesis*   The final stage of spermatogenesis, called **spermiogenesis** (sper'-mē-ō-JEN-e-sis), involves the

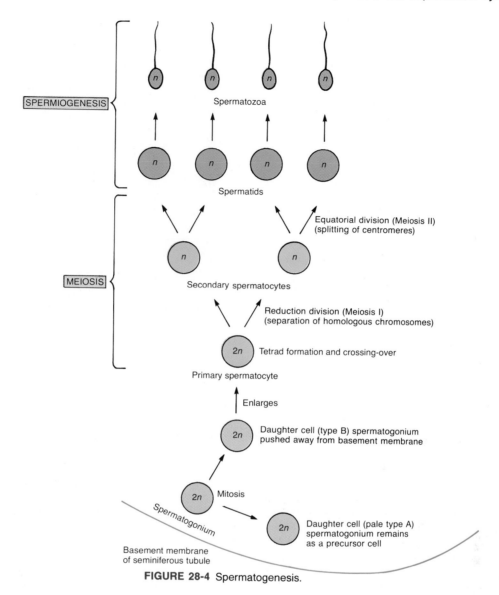

FIGURE 28-4 Spermatogenesis.

maturation of spermatids into spermatozoa. Each spermatid embeds in a sustentacular (Sertoli) cell and develops a head with an acrosome (described shortly) and a flagellum (tail). Sustentacular cells extend from the basement membrane to the interior of the seminiferous tubule where they nourish the developing spermatids. Since there is no cell division in spermiogenesis, each spermatid develops into a single **spermatozoon (sperm cell).**

Spermatozoa enter the lumen of the seminiferous tubule and migrate to the ductus epididymis, where in 18 hours to 10 days they complete their maturation and become capable of fertilizing an ovum. Spermatozoa are also stored in the ductus (vas) deferens. Here, they can retain their fertility for up to several months.

Spermatozoa

Spermatozoa are produced or matured at the rate of about 300 million per day and, once ejaculated, have a life expec-

tancy of about 48 hours within the female reproductive tract. A spermatozoon is highly adapted for reaching and penetrating a female ovum. It is composed of a head, a midpiece, and a tail (Figure 28-5). Within the **head** are the nuclear material and a dense granule called the **acrosome,** which develops from the Golgi complex and contains enzymes (hyaluronidase and proteinases) that facilitate penetration of the sperm cell into the ovum. The acrosome is basically a specialized lysosome. Numerous mitochondria in the **midpiece** carry on the metabolism that provides energy for locomotion. The **tail,** a typical flagellum, propels the sperm along its way.

Testosterone and Inhibin

Secretions of the anterior pituitary gland assume a major role in the developmental changes associated with puberty. At the onset of puberty the anterior pituitary starts to secrete gonadotropic hormones called follicle-stimulat-

FIGURE 28-5 Spermatozoa. (a) Diagram of the parts of a spermatozoon. (b) Scanning electron micrograph of several spermatozoa at a magnification of 2,000×. (Courtesy of Fisher Scientific Company and S.T.E.M. Laboratories, Inc., Copyright © 1975.)

ing hormone (FSH) and luteinizing hormone (LH). Their release is controlled from the hypothalamus by gonadotropin releasing factor (GnRF). Once secreted, the gonadotropic hormones have profound effects on male reproductive organs. FSH acts on the seminiferous tubules to initiate spermatogenesis and stimulate sustentacular cells. LH also assists the seminiferous tubules to develop mature sperm, but its chief function is to stimulate the interstitial endocrinocytes to secrete the hormone testosterone (tes-TOS-te-rōn).

Testosterone is synthesized from cholesterol or acetyl coenzyme A in the testes. It is the principal male hormone (androgen) and has a number of effects on the body. It controls the development, growth, and maintenance of the male sex organs. It also stimulates bone growth, protein anabolism, sexual behavior, final maturation of sperm, and the development of male secondary sex characteristics. These characteristics, which appear at puberty, include muscular and skeletal development resulting in wide shoulders and narrow hips; body hair patterns that include pubic hair, axillary and chest hair (within hereditary limits), facial hair, and temporal hairline recession; and enlargement of the thyroid cartilage of the larynx producing deepening of the voice. Testosterone also stimulates descent of the testes just prior to birth.

The interaction of LH with testosterone illustrates the operation of another negative feedback system (Figure 28-6). LH stimulates the production of testosterone, but once the testosterone concentration in the blood reaches a certain level, it inhibits the release of GnRF by the hypothalamus. This inhibition, in turn, inhibits the release of LH by the anterior pituitary. Thus testosterone production is decreased. However, once the testosterone concentration in the blood decreases to a certain level, GnRF

is released by the hypothalamus. This release of GnRF stimulates the release of LH by the anterior pituitary and stimulates testosterone production. Thus the testosterone-LH cycle is complete.

Inhibin is a protein hormone that has a direct effect on the anterior pituitary by inhibiting the secretion of

FIGURE 28-6 Secretion, physiological effects, and control of testosterone and inhibin.

FSH. FSH brings about spermatogenesis and stimulates sustentacular cells. Once the degree of spermatogenesis required for male reproductive functions has been achieved, sustentacular cells secrete inhibin. The hormone feeds back negatively to the anterior pituitary to inhibit FSH and thus to decrease spermatogenesis (Figure 28-6). If spermatogenesis is proceeding too slowly, lack of inhibin production permits FSH secretion and an increased rate of spermatogenesis.

DUCTS

Ducts of the Testis

Following their production, spermatozoa are moved through the convoluted seminiferous tubules to the **straight tubules** (see Figure 28-2b). The straight tubules lead to a network of ducts in the testis called the **rete** (RĒ-tē) **testis.** Some of the cells lining the rete testis possess cilia that probably help move the sperm along. The sperm are next transported out of the testis.

Epididymis

The sperm are transported out of the testis through a series of coiled **efferent ducts** in the epididymis that empty into a single tube called the ductus epididymis. At this point, the sperm are morphologically mature.

The **epididymis** (ep'-i-DID-i-mis) is a comma-shaped organ that lies along the posterior border of the testis (see Figures 28-1 and 28-2) and consists mostly of a tightly coiled tube, the **ductus epididymis.** The larger, superior portion of the epididymis is known as the **head.** It consists of the efferent ducts that empty into the ductus epididymis. The **body** of the epididymis contains the ductus epididymis. The **tail** is the smaller, inferior portion. Within the tail, the ductus epididymis continues as the ductus deferens.

The ductus epididymis measures about 6 m (20 ft) in length and 1 mm in diameter. It is tightly packed within the epididymis, which measures only about 3.8 cm (1.5 inches). The ductus epididymis is lined with pseudostratified columnar epithelium, and its wall contains smooth muscle. The free surfaces of the columnar cells contain long, branching microvilli called *stereocilia* (Figure 28-7).

Functionally, the ductus epididymis is the site of sperm maturation. They require between 18 hours and 10 days to complete their maturation, that is, to become capable of fertilizing an ovum. The ductus epididymis also stores spermatozoa and propels them toward the urethra during ejaculation by peristaltic contraction of its smooth muscle. Spermatozoa may remain in storage in the ductus epididymis for up to 4 weeks. After that, they are reabsorbed.

Ductus (Vas) Deferens

Within the tail of the epididymis, the ductus epididymis becomes less convoluted, its diameter increases, and at this point it is referred to as the **ductus (vas) deferens** or **seminal duct** (see Figure 28-2). The ductus deferens, about 45 cm long (18 inches), ascends along the posterior

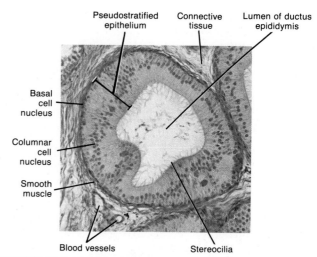

FIGURE 28-7 Histology of the ductus epididymis. Photomicrograph of the ductus epididymis seen in cross section at a magnification of 160×. (Copyright © 1983 by Michael H. Ross. Used by permission.

border of the testis, penetrates the inguinal canal, and enters the pelvic cavity, where it loops over the side and down the posterior surface of the urinary bladder (see Figure 28-1). The dilated terminal portion of the ductus deferens is known as the **ampulla** (am-POOL-la). The ductus deferens is lined with pseudostratified epithelium and contains a heavy coat of three layers of muscle. Functionally, the ductus deferens stores sperm for up to several months and propels them toward the urethra during ejaculation by peristaltic contractions of the muscular coat.

CLINICAL APPLICATION

One method of sterilization of males is called **vasectomy,** a relatively uncomplicated procedure, typically performed under local anesthesia, in which a portion of each ductus deferens is removed. In the procedure, an incision is made in the scrotum, the ducts are located, each is tied in two places, and the portion between the ties is excised. Although sperm production continues in the testes, the sperm cannot reach the exterior because the ducts are cut, and the sperm degenerate. Sperm cells are destroyed by phagocytosis. Vasectomy has no effect on sexual desire and performance, and if performed correctly, it is virtually 100 percent effective.

Traveling with the ductus deferens as it ascends in the scrotum are the testicular artery, autonomic nerves, veins that drain the testes (pampiniform plexus), lymphatics, and the cremaster muscle. These structures constitute the **spermatic cord,** a supporting structure of the male reproductive system (Figure 28-8). The cremaster muscle, which also surrounds the testes, elevates the testes during sexual stimulation and exposure to cold. The spermatic cord and ilioinguinal nerve pass through the **inguinal** (IN-gwin-al) **canal** in the male. The canal is an oblique passageway in the anterior abdominal wall just superior and

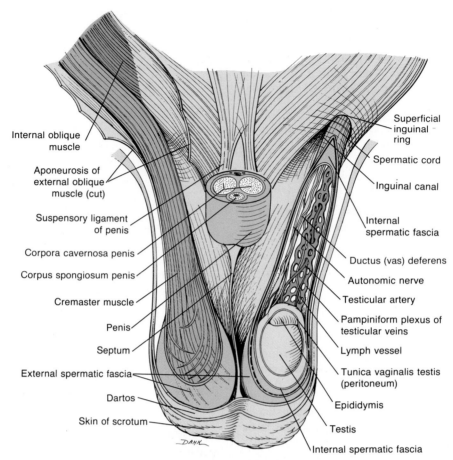

FIGURE 28-8 Spermatic cord and inguinal canal. The left spermatic cord has been opened to expose its contents.

parallel to the medial half of the inguinal ligament. The canal is about 4 to 5 cm (1.6 to 2.0 inches) in length. It originates at the *deep (abdominal) inguinal ring,* a slitlike opening in the aponeurosis of the transversus abdominis muscle. The canal terminates at the *superficial (subcutaneous) inguinal ring,* a triangular opening in the aponeurosis of the external oblique muscle. In the female, the round ligament of the uterus and ilioinguinal nerve pass through the inguinal canal.

CLINICAL APPLICATION

The inguinal canal represents a weak spot in the abdominal wall. It is frequently the site of an **inguinal hernia**—a rupture or separation of a portion of the abdominal wall resulting in the protrusion of a part of an organ. Since the inguinal canal is smaller in females, inguinal hernias occur much less frequently in females.

Ejaculatory Duct

Posterior to the urinary bladder are the **ejaculatory** (e-JAK-yoo-la-tō'-rē) **ducts** (Figure 28-9). Each duct is about 2 cm (1 inch) long and is formed by the union of the duct from the seminal vesicle and ductus deferens.

The ejaculatory ducts eject spermatozoa into the prostatic urethra.

Urethra

The **urethra** is the terminal duct of the system, serving as a passageway for spermatozoa or urine. In the male, the urethra passes through the prostate (PROS-tāt) gland, the urogenital diaphragm, and the penis. It measures about 20 cm (8 inches) in length and is subdivided into three parts (see Figures 28-1 and 28-9). The **prostatic urethra** is 2 to 3 cm (1 inch) long and passes through the prostate gland. It continues inferiorly, and as it passes through the urogenital diaphragm, a muscular partition between the two ischiopubic rami, it is known as the **membranous urethra.** The membranous portion is about 1 cm (0.5 inch) in length. As it passes through the corpus spongiosum of the penis, it is known as the **spongy (cavernous) urethra.** This portion is about 15 cm (6 inches) long. The spongy urethra enters the bulb of the penis and terminates at the external **urethral orifice.**

ACCESSORY SEX GLANDS

Whereas the ducts of the male reproductive system store and transport sperm cells, the **accessory sex glands** secrete the liquid portion of semen. The paired **seminal vesicles**

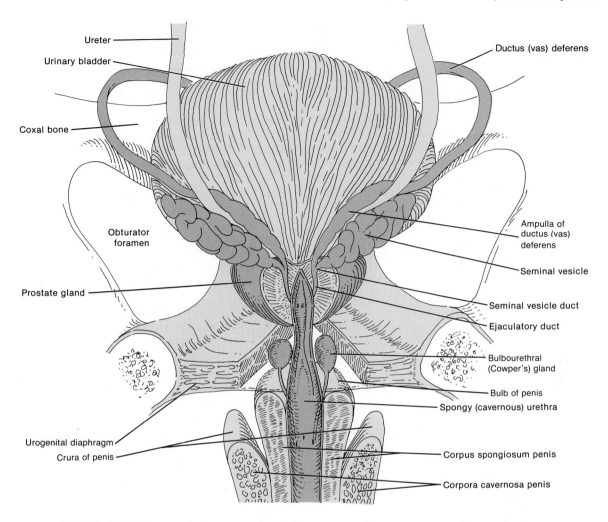

Ureter
Urinary bladder
Coxal bone
Obturator foramen
Prostate gland
Urogenital diaphragm
Crura of penis

Ductus (vas) deferens
Ampulla of ductus (vas) deferens
Seminal vesicle
Seminal vesicle duct
Ejaculatory duct
Bulbourethral (Cowper's) gland
Bulb of penis
Spongy (cavernous) urethra
Corpus spongiosum penis
Corpora cavernosa penis

FIGURE 28-9 Male reproductive organs in relation to surrounding structures seen in posterior view.

(VES-i-kuls) are convoluted pouchlike structures, about 5 cm (2 inches) in length, lying posterior to and at the base of the urinary bladder in front of the rectum (see Figure 28-9). They secrete an alkaline, viscous fluid, rich in the sugar fructose, and pass it into the ejaculatory duct. This secretion contributes to sperm viability. It constitutes about 60 percent of the volume of semen.

The **prostate gland** is a single, doughnut-shaped gland about the size of a chestnut (Figure 28-9). It is inferior to the urinary bladder and surrounds the superior portion of the urethra. The prostate secretes an alkaline fluid into the prostatic urethra through numerous prostatic ducts. The prostatic secretion constitutes 13 to 33 percent of the volume of semen and contributes to sperm motility. Presumably, sperm motility is not optimal in an acid environment. The alkalinity of the prostatic secretion neutralizes the acidity of vaginal and other secretions, and thus promotes maximal motility.

The paired **bulbourethral** (bul'-bō-yoo-RĒ-thral), or **Cowper's glands** are about the size of peas. They are located beneath the prostate on either side of the membranous urethra (Figure 28-9). The bulbourethral glands secrete an alkaline substance that protects sperm by neutralizing the acid environment of the urethra. Their ducts open into the spongy urethra.

SEMEN (SEMINAL FLUID)

Semen (seminal fluid) is a mixture of sperm and the secretions of the seminal vesicles, the prostate gland, and the bulbourethral glands. The average volume of semen for each ejaculation is 2.5 to 5 ml, and the average range of spermatozoa ejaculated is 50 to 100 million/ml. When the number of spermatozoa falls below 20 million/ml, the male is likely to be sterile. The very large number is required because only a small percentage eventually reach the ovum. And, although only a single spermatozoon fertilizes an ovum, fertilization seems to require the combined action at the ovum of a larger number of them. The intercellular material of the cells covering the ovum presents a barrier to the sperm. This barrier is digested by the hyaluronidase and proteinases secreted by the acrosomes of sperm. However, it appears that a single sperm does not produce enough of these enzymes to dissolve the barrier. A passageway through which one may enter can be created only by the action of many.

Semen has a slightly alkaline pH of 7.20 to 7.60. The prostatic secretion gives semen a milky appearance, and fluids from the seminal vesicles and bulbourethral glands give it a mucoid consistency. Semen provides spermatozoa with a transportation medium and nutrients. It neutralizes the acid environment of the male urethra and the female vagina. It also contains enzymes that activate sperm after ejaculation.

Semen has been shown to contain an antibiotic, called *seminalplasmin,* which has the ability to destroy a number of bacteria. Its antimicrobial activity has been described as similar to that currently exerted by penicillin, streptomycin, and tetracyclines. Since both semen and the lower female reproductive tract contain bacteria, seminalplasmin may keep these bacteria under control to help ensure fertilization.

Once ejaculated into the vagina, liquid semen coagulates rapidly because of a clotting enzyme produced by the prostate that acts on a substance produced by the seminal vesicle. This clot liquefies in about 5 to 20 minutes because of another enzyme produced by the prostate gland. Abnormal or delayed liquefaction of coagulated semen may cause complete or partial immobilization of spermatozoa, thus inhibiting their movement through the cervix of the uterus.

CLINICAL APPLICATION

Semen analysis is the most valuable test in evaluating sterility. Among the criteria analyzed are the following:

1. **Volume.** A low volume might suggest an anatomical or functional destruction or inflammation.
2. **Motility.** This refers to the percentage of motile spermatozoa (40 to 60 percent) and quality of movement (forward and progressive).
3. **Count.** Sperm counts below 20 million/ml could indicate sterility.
4. **Liquefaction.** Delayed liquefaction of more than 2 hours suggests inflammation of accessory sex glands or enzyme defects in the glands.
5. **Morphology.** No more than about 35 percent of spermatozoa should have abnormal morphology.
6. **Autoagglutination.** Agglutination does not occur normally.
7. **pH.** A rise in pH could indicate prostatitis.
8. **Fructose.** This sugar is present in a normal ejaculate. Its absence indicates obstruction or congenital absence of the ejaculatory ducts or seminal vesicles.

A normal semen analysis does not guarantee fertility; the absence of spermatozoa or zero motility are the only definitive signs of sterility.

PENIS

The **penis** is used to introduce spermatozoa into the vagina (Figure 28-10). The penis is cylindrical in shape and consists of a body, root, and glans penis. The **body** of the penis is composed of three cylindrical masses of tissue, each bound by fibrous tissue (**tunica albuginea**). The two dorsolateral masses are called the **corpora cavernosa penis.** The smaller midventral mass, the **corpus spongiosum penis,** contains the spongy urethra. All three masses are enclosed by fascia and loose-fitting skin and consist of erectile tissue permeated by blood sinuses. Under the influence of sexual stimulation, the arteries supplying the

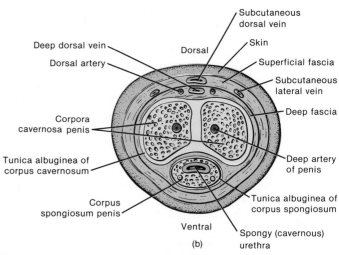

FIGURE 28-10 Internal structure of the penis. (a) Coronal section. (b) Cross section.

penis dilate, and large quantities of blood enter the blood sinuses. Expansion of these spaces compresses the veins draining the penis so most entering blood is retained. These vascular changes result in an **erection,** a parasympathetic reflex. The penis returns to its flaccid state when the arteries constrict and pressure on the veins is relieved. Details of erection are presented later in the chapter. During ejaculation, which is a sympathetic reflex, the smooth muscle sphincter at the base of the urinary bladder is closed because of the higher pressure in the urethra caused by expansion of the corpus spongiosum penis. Thus urine is not expelled during ejaculation, and semen does not enter the urinary bladder.

The **root** of the penis is the attached portion and consists of the **bulb of the penis,** the expanded portion of the base of the corpus spongiosum penis, and the **crura of the penis,** the separated and tapered portion of the corpora cavernosa penis. The bulb of the penis is attached to the inferior surface of the urogenital diaphragm and enclosed by the bulbocavernosus muscle. Each crus (pleural is *crura*) of the penis is attached to the ischial and pubic rami and surrounded by the ischiocavernosus muscle (see Figure 11-14).

The distal end of the corpus spongiosum penis is a slightly enlarged region called the **glans penis,** which means shaped like an acorn. The margin of the glans penis is referred to as the **corona.** Covering the glans is the loosely fitting **prepuce** (PRĒ-pyoos), or **foreskin.**

CLINICAL APPLICATION

Circumcision (*circumcido* = to cut around) is a surgical procedure in which part or all of the prepuce is removed. It is usually performed by the third or fourth day after birth (or on the eighth day as part of a Jewish religious rite).

FEMALE REPRODUCTIVE SYSTEM

The female organs of reproduction include the ovaries, which produce ova (eggs) and progesterone, estrogens, and relaxin; the uterine (Fallopian) tubes, which transport the ova to the uterus (womb); the vagina; and external organs that constitute the vulva, or pudendum (Figure 28-11). The mammary glands also are considered part of the female reproductive system.

The specialized branch of medicine that deals with the diagnosis and treatment of diseases of the female reproductive system is called **gynecology** (gī'-ne-KOL-ō-jē; *gyneco* = woman).

OVARIES

The **ovaries,** or female gonads, are paired glands resembling unshelled almonds in size and shape. They are homolo-

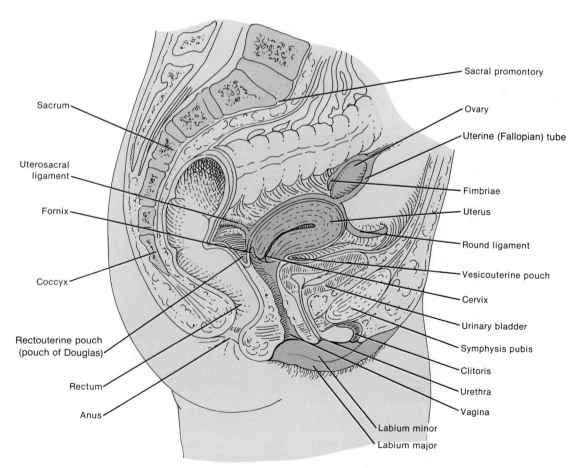

FIGURE 28-11 Female organs of reproduction and surrounding structures seen in sagittal section.

gous to the testes. They descend to the brim of the pelvis during the third month of development. They are positioned in the upper pelvic cavity, one on each side of the uterus. The ovaries are maintained in position by a series of ligaments (Figure 28-12). They are attached to the broad ligament of the uterus, which is itself part of the parietal peritoneum, by a double-layered fold of peritoneum called the **mesovarium,** which surrounds the ovary and ovarian ligament. The ovaries are anchored to the uterus by the **ovarian ligament** and are attached to the pelvic wall by the **suspensory ligament.** Each ovary also contains a **hilus,** the point of entrance for blood vessels and nerves.

The microscope reveals that each ovary consists of the following parts (Figure 28-13).

1. **Germinal epithelium.** A layer of simple cuboidal epithelium that covers the free surface of the ovary.

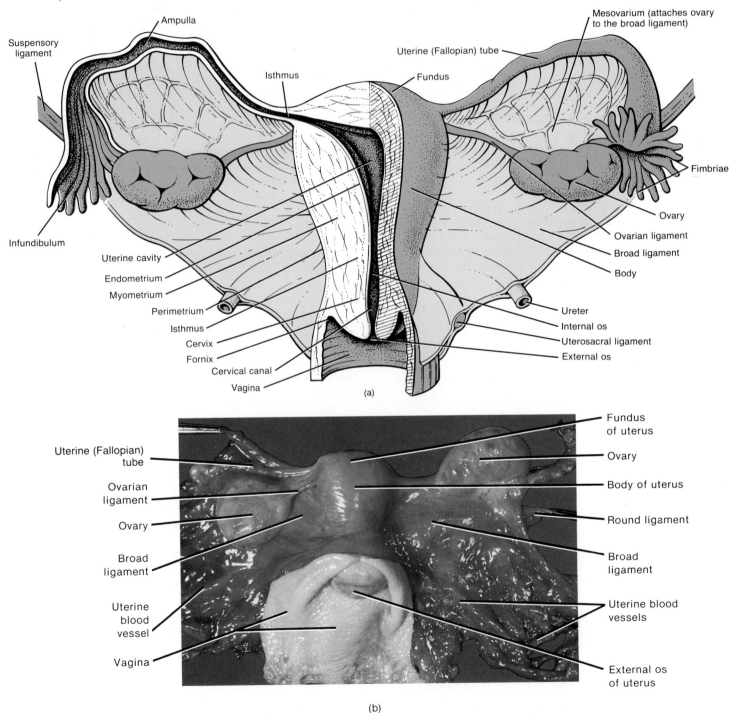

FIGURE 28-12 Uterus and associated structures seen in anterior view. (a) Diagram. The left side of the figure has been sectioned to show internal structures. (b) Photograph. (Courtesy of C. Yokochi, *Photographic Anatomy of the Human Body,* 1st ed., 1969, IGAKU-SHOIN, Ltd., Tokyo, New York.)

(a)

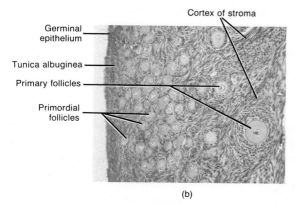

(b)

FIGURE 28-13 Histology of the ovary. (a) Diagram of the parts of an ovary seen in sectional view. The arrows indicate the sequence of developmental stages that occurs as part of the ovarian cycle. (b) Photomicrograph of the cortex of an ovary at a magnification of 60×. (c) Photomicrograph of an enlarged aspect of a secondary follicle at a magnification of 160×. The theca interna and theca externa are connective tissue coverings around the secondary follicle. (Photomicrographs copyright © 1983 by Michael H. Ross. Used by permission.)

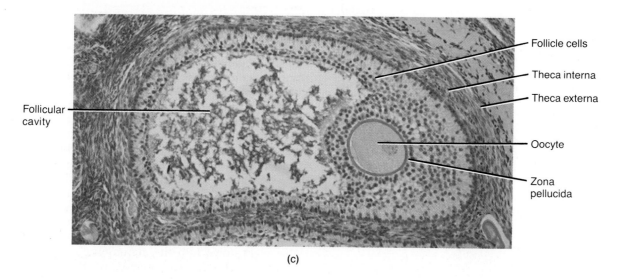

(c)

2. Tunica albuginea. A capsule of collagenous connective tissue immediately deep to the germinal epithelium.

3. Stroma. A region of connective tissue deep to the tunica albuginea and composed of an outer, dense layer called the *cortex* and an inner, loose layer known as the *medulla*. The cortex contains ovarian follicles.

4. Ovarian follicles. Oocytes (immature ova) and their surrounding tissues in various stages of development.

5. **Vesicular ovarian (Graafian) follicle.** A relatively large, fluid-filled follicle containing an immature ovum and its surrounding tissues. The follicle secretes hormones called estrogens.

6. **Corpus luteum.** Glandular body that develops from a vesicular ovarian follicle after extrusion of an ovum (ovulation). The corpus luteum produces the hormones progesterone, estrogens, and relaxin.

Oogenesis

The formation of a haploid ovum by meiosis in the ovary is referred to as **oogenesis** (ō'-ō-JEN-e-sis). With some exceptions, oogenesis occurs in essentially the same manner as spermatogenesis. It involves meiosis and maturation.

• *Reduction Division (Meiosis I)*   The precursor cell in oogenesis is a diploid cell called the **oogonium** (ō'-o-GŌ-nē-um), or egg mother cell (Figure 28-14). Early in the prenatal development of the ovaries, oogonia proliferate by mitosis. But, as development continues, oogonia in the primary follicles lose their ability to carry on mitosis. At about the third month of prenatal development, oogonia develop into larger cells also containing the diploid number of chromosomes and called **primary oocytes** (Ō-o-sītz). They remain in this stage until their follicular cells respond to FSH from the anterior pituitary, which, in turn, has responded to GnRF from the hypothalamus.

Starting with puberty, several follicles respond each month to the rising level of FSH. As the preovulatory phase of the menstrual cycle proceeds and LH is secreted

from the anterior pituitary, one of the follicles reaches a stage in which the diploid primary oocyte undergoes reduction division. Synapsis, tetrad formation, and crossing-over occur, and two cells of unequal size, both with 23 chromosomes of two chromatids each, are produced. The smaller cell, called the **first polar body,** is essentially a packet of discarded nuclear material. The larger cell, known as the **secondary oocyte,** receives most of the cytoplasm.

• *Equatorial Division (Meiosis II)*   At ovulation, the secondary oocyte with its polar body and surrounding supporting cells is discharged. The discharged secondary oocyte enters the uterine (Fallopian) tube and, if spermatozoa are present and fertilization occurs, the second division, the equatorial division, takes place after fertilization.

• *Maturation*   The secondary oocyte produces two cells of unequal size, both of them haploid. The larger cell is called an **ootid** and eventually develops into an **ovum,** or mature egg; the smaller is the **second polar body.**

The first polar body may undergo another division to produce two polar bodies. If it does, meiosis of the primary oocyte results in a single haploid ovum and three polar bodies. In any event, all polar bodies disintegrate. Thus each oogonium produces a single ovum, whereas each spermatocyte produces four spermatozoa.

UTERINE (FALLOPIAN) TUBES

The female body contains two **uterine (Fallopian) tubes,** also called **oviducts,** that extend laterally from the uterus and transport the ova from the ovaries to the uterus (see Figure 28-12). Measuring about 10 cm (4 inches) long, the tubes are positioned between the folds of the broad ligaments of the uterus. The funnel-shaped open distal end of each tube, called the **infundibulum,** lies close to the ovary but is not attached to it and is surrounded by a fringe of fingerlike projections called **fimbriae** (FIM-brē-ē). From the infundibulum, the uterine tube extends medially and inferiorly and attaches to the superior lateral angle of the uterus. The **ampulla** of the uterine tube is the widest, longest portion, making up about two-thirds of its length. The **isthmus** of the uterine tube is the short, narrow, thick-walled portion that joins the uterus.

Histologically, the uterine tubes are composed of three layers. The internal **mucosa** contains ciliated columnar cells and secretory cells, which are believed to aid the movement and nutrition of the ovum. The middle layer, the **muscularis,** is composed of a thick, circular region of smooth muscle and an outer, thin, longitudinal region of smooth muscle. Peristaltic contractions of the muscularis and the ciliary action of the mucosa help move the ovum down into the uterus. The outer layer of the uterine tubes is a serous membrane, the **serosa.**

About once a month an immature ovum ruptures from the surface of the ovary near the infundibulum of the uterine tube, a process called **ovulation.** The ovum is swept into the tube by the ciliary action of the epithelium of the infundibulum. It is then moved along the tube by

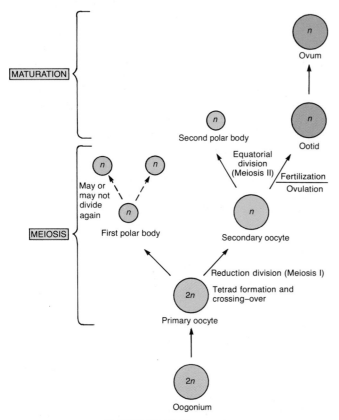

FIGURE 28-14 Oogenesis.

ciliary action supplemented by the peristaltic contractions of the muscularis. If the ovum is fertilized by a sperm cell, it usually occurs in the ampulla of the uterine tube. Fertilization may occur at any time up to about 24 hours following ovulation. The fertilized ovum (zygote), now referred to as a blastocyst, descends into the uterus within 7 days. An unfertilized ovum disintegrates.

CLINICAL APPLICATION

Ectopic (*ektopos* = displaced) **pregnancy (EP)** refers to the development of an embryo or fetus outside the uterine cavity. The majority occur in the uterine tube, usually in the ampullar and infundibular portions. Some occur in the ovaries, abdomen, uterine cervix, and broad ligaments. The basic cause of a tubal pregnancy is impaired passage of the fertilized ovum through the uterine tube related to factors such as pelvic inflammatory disease (PID), previous uterine tube surgery, previous ectopic pregnancy, repeated elective abortions, pelvic tumors, and developmental abnormalities. Ectopic pregnancy may be characterized by one or two missed periods, followed by vaginal bleeding and acute pelvic pain.

UTERUS

The site of menstruation, implantation of a fertilized ovum, development of the fetus during pregnancy, and labor is the **uterus.** Situated between the urinary bladder and the rectum, the uterus is shaped like an inverted pear (see Figures 28-11 and 28-12). Before the first pregnancy, the adult uterus measures approximately 7.5 cm (3 inches) long, 5 cm (2 inches) wide, and 2.5 cm (1 inch) thick.

Anatomical subdivisions of the uterus include the dome-shaped portion above the uterine tubes called the **fundus,** the major tapering central portion called the **body,** and the inferior narrow portion opening into the vagina called the **cervix.** Between the body and the cervix is the **isthmus** (IS-mus), a constricted region about 1 cm (0.5 inch) long. The interior of the body of the uterus is called the **uterine cavity,** and the interior of the narrow cervix is called the **cervical canal.** The junction of the isthmus with the cervical canal is the **internal os.** The **external os** is the place where the cervix opens into the vagina.

CLINICAL APPLICATION

Early diagnosis of **cancer of the uterus** is accomplished by the *Papanicolaou* (pap'-a-NIK-ō-la-oo) *test,* or Pap smear. In this generally painless procedure, a few cells from the part of the vagina surrounding the cervix and the cervix are removed with a swab and examined microscopically. Malignant cells have a characteristic appearance and indicate an early stage of cancer, even before symptoms occur. Estimates indicate that the Pap smear is more than 90 percent reliable in detecting cancer of the cervix.

To rule out invasive carcinoma, a *cone biopsy* of the cervix is performed. A cone biopsy is a hospital procedure in which an inverted cone of tissue is excised. It requires an anesthetic and is usually done only when abnormal cells have been detected. In another procedure, *punch biopsy* is combined with an *endocervical curettage* (ku-re-TAZH), or *ECC;* this combination has a high degree of diagnostic accuracy. In a punch biopsy, a disc or segment of tissue is excised. Curretage is a procedure in which the cervix is dilated and the endometrium (lining) of the uterus is scraped with a spoon-shaped instrument called a curette. This procedure is commonly called a *D and C.* If the carcinoma has spread beyond the lining, treatment may involve complete or partial removal of the uterus, called a *hysterectomy,* or radiation treatment.

Normally the uterus is flexed between the uterine body and the cervix. This is called **anteflexion.** In this position, the body of the uterus projects anteriorly and slightly superiorly over the urinary bladder, and the cervix projects inferiorly and posteriorly and enters the anterior wall of the vagina at nearly a right angle. Several structures that are either extensions of the parietal peritoneum or fibromuscular cords, referred to as ligaments, maintain the position of the uterus. The paired **broad ligaments** are double folds of parietal peritoneum attaching the uterus to either side of the pelvic cavity. Uterine blood vessels and nerves pass through the broad ligaments. The paired **uterosacral ligaments,** also peritoneal extensions, lie on either side of the rectum and connect the uterus to the sacrum. The **cardinal (lateral cervical) ligaments** extend below the bases of the broad ligaments between the pelvic wall and the cervix and vagina. These ligaments contain smooth muscle, uterine blood vessels, and nerves, and are the chief ligaments that maintain the position of the uterus and help to keep it from dropping down into the vagina. The **round ligaments** are bands of fibrous connective tissue between the layers of the broad ligament. They extend from a point on the uterus just below the uterine tubes to a portion of the external genitalia. Although the ligaments normally maintain the anteflexed position of the uterus, they also afford the uterine body some movement. As a result, the uterus may become malpositioned. A posterior tilting of the uterus is called **retroflexion.**

Histologically, the uterus consists of three layers of tissue. The outer layer, the **perimetrium,** or **serosa,** is part of the visceral peritoneum. Laterally, it becomes the broad ligament. Anteriorly, it is reflected over the urinary bladder and forms a shallow pouch, the **vesicouterine** (ves'-i-kō-YOO-ter-in) **pouch** (see Figure 28-11). Posteriorly, it is reflected onto the rectum and forms a deep pouch, the **rectouterine** (rek-tō-YOO-ter-in) **pouch**—the lowest point in the pelvic cavity.

The middle layer of the uterus, the **myometrium,** forms the bulk of the uterine wall (Figure 28-15). This layer consists of three layers of smooth muscle fibers and is

Lumen of uterus

Simple columnar epithelium

Stratum functionalis

Endometrial glands

Stratum basalis

Myometrium

FIGURE 28-15 Histology of the uterus. Photomicrograph of a portion of the uterine wall in which the stratum functionalis is in an early stage of proliferation at a magnification of 25×. (Copyright © 1983 by Michael H. Ross. Used by permission.)

thickest in the fundus and thinnest in the cervix. During childbirth, coordinated contractions of the muscles help to expel the fetus from the body of the uterus.

The inner layer of the uterus, the **endometrium,** is a mucous membrane composed of two principal layers. The *stratum functionalis,* the layer closer to the uterine cavity, is shed during menstruation. The second layer, the *stratum basalis* (bā-SAL-is), is permanent and produces a new functionalis following menstruation. The endometrium contains numerous glands.

CLINICAL APPLICATION

More and more physicians are beginning to use **colposcopy** (kol-POS-ko-pē) in order to evaluate the status of the mucosa of the vagina and cervix. Colposcopy is the direct examination of the vaginal and cervical mucosa with a magnifying device (a colposcope) similar to a low-power binocular microscope. Various instruments that magnify the mucous membrane from about 6 to 40 times its actual size are commercially available. The application of a 3 percent solution of acetic acid removes mucus and enhances the appearance of mucosal columnar epithelium.

Blood is supplied to the uterus by branches of the internal iliac artery called *uterine arteries.* Branches called *arcuate arteries* are arranged in a circular fashion in the myometrium and give off *radial arteries* that penetrate

deeply into the myometrium (Figure 28-16). Just before the branches enter the endometrium, they divide into two kinds of arterioles. The *straight arteriole* terminates in the basalis and supplies it with the materials necessary to regenerate the functionalis. The *spiral arteriole* penetrates the functionalis and changes markedly during the menstrual cycle. The uterus is drained by the *uterine veins.*

ENDOCRINE RELATIONS: MENSTRUAL AND OVARIAN CYCLES

The principal events of the menstrual cycle can be correlated with those of the ovarian cycle and changes in the endometrium. All are hormonally controlled events.

The **menstrual cycle** is a series of changes in the endometrium of a nonpregnant female. Each month the endometrium is prepared to receive an already fertilized ovum that eventually normally develops into an embryo and then into a fetus until delivery. If no fertilization occurs, the stratum functionalis portion of the endometrium is shed. The **ovarian cycle** is a monthly series of events associated with the maturation of an ovum.

Hormonal Control

The menstrual cycle, ovarian cycle, and other changes associated with puberty in the female are controlled by a regulating factor from the hypothalamus called gonadotropin releasing factor (GnRF). Its influence is shown in Figure 28-17. GnRF stimulates the release of follicle-stimulating hormone (FSH) from the anterior pituitary. FSH stimulates the initial development of the ovarian follicles and the secretion of estrogens by the follicles. GnRF also stimulates the release of another anterior pituitary hormone—the luteinizing hormone (LH), which stimulates the further development of ovarian follicles, brings about ovulation, and stimulates the production of estrogens, progesterone, and relaxin by ovarian cells.

At least six different estrogens have been isolated from the plasma of human females. However, only three are present in significant quantities. These are *beta* (β)-*estradiol, estrone,* and *estriol.* Of these, β-estradiol exerts the major effect. It is synthesized from cholesterol or acetyl coenzyme A in the ovaries. As reference is made to estrogens in subsequent discussions, keep in mind that β-estradiol is the principal estrogen.

Estrogens, the hormones of growth, have three main functions. First is the development and maintenance of female reproductive structures, especially the endometrial lining of the uterus, secondary sex characteristics, and the breasts. The secondary sex characteristics include fat distribution to the breasts, abdomen, mons pubis, and hips; voice pitch; broad pelvis; and hair pattern. Second, they control fluid and electrolyte balance. Third, they increase protein anabolism. In this regard, estrogens are synergistic with growth hormone (GH). High levels of estrogens in the blood inhibit the release of GnRF by the hypothalamus, which in turn inhibits the secretion of FSH by the anterior pituitary gland. This inhibition provides the basis for the action of one kind of contraceptive pill.

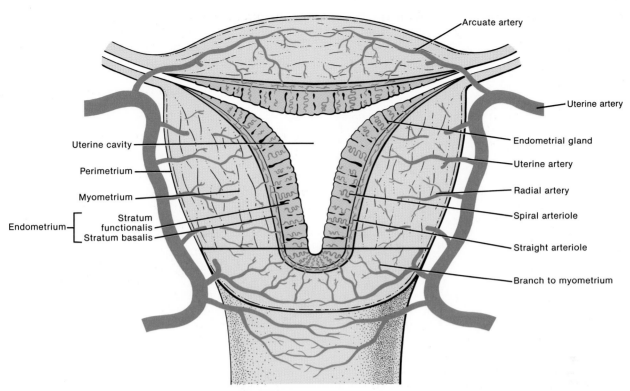

FIGURE 28-16 Blood supply of the uterus.

Progesterone, the hormone of maturation, works with estrogens to prepare the endometrium for implantation of a fertilized ovum and the mammary glands for milk secretion. High levels of progesterone also inhibit GnRF and prolactin (PRL). Progesterone, like estrogens, is synthesized from cholesterol or acetyl coenzyme A in the ovaries.

Relaxin goes into operation near the end of pregnancy. It relaxes the symphysis pubis and helps dilate the uterine cervix to facilitate delivery. The hormone also plays a role in increasing sperm motility.

Menstrual Phase (Menstruation)

The duration of the menstrual cycle ranges from 24 to 35 days. For this discussion, we will assume an average duration of 28 days. Events occurring during the menstrual cycle may be divided into three phases: the men-

strual phase, the preovulatory phase, and the postovulatory phase (Figure 28-18).

The **menstrual phase,** also called **menstruation** or the **menses,** is the periodic discharge of 25 to 65 ml of blood,

FIGURE 28-17 Secretion and physiological effects of estrogens, progesterone, and relaxin.

FIGURE 28-18 Correlation of menstrual and ovarian cycles with the hypothalamic and anterior pituitary gland hormones. In the cycle shown, fertilization and implantation have not occurred.

tissue fluid, mucus, and epithelial cells. It is caused by a sudden reduction in estrogens and progesterone and lasts for approximately the first 5 days of the cycle. The discharge is associated with endometrial changes in which the functionalis layer degenerates and patchy areas of bleeding develop. Small areas of the functionalis detach one at a time (total detachment would result in hemorrhage), the uterine glands discharge their contents and collapse, and tissue fluid is discharged. The menstrual flow passes from the uterine cavity to the cervix and through the vagina to the exterior. Generally the flow terminates by the fifth day of the cycle. At this time the entire functionalis has been shed and the endometrium is very thin because only the basalis remains.

During the menstrual phase, the ovarian cycle is also in operation. Ovarian follicles, called **primary follicles,** begin their development. At birth each ovary contains

about 200,000 such follicles, each consisting of a potential ovum surrounded by a layer of cells. During the early part of each menstrual phase 20 to 25 primary follicles start to produce very low levels of estrogens. A clear membrane, the **zona pellucida,** also develops around the potential ovum. Toward the end of the menstrual phase (days 4 to 5) about 20 of the primary follicles develop into **secondary follicles,** as the cells of the surrounding layer increase in number and differentiate, secreting follicular (fō-LIK-yoo-lar) fluid. This fluid forces an immature ovum to the edge of the secondary follicle and fills the follicular cavity. The production of estrogens by the secondary follicles elevates the level of estrogens in the blood slightly. Ovarian follicle development is the result of GnRF secretion by the hypothalamus, which in turn stimulates FSH production by the anterior pituitary. During this part of the cycle, FSH secretion is relatively high.

Although a number of follicles begin development each cycle, only one attains maturity. The others undergo atresia (death).

Preovulatory Phase

The **preovulatory phase,** the second phase of the menstrual cycle, is the time between menstruation and ovulation. This phase of the menstrual cycle is more variable in length than the other phases. It lasts from days 6 to 13 in a 28-day cycle.

FSH and LH stimulate the ovarian follicles to produce more estrogens, and this increase in estrogens stimulates the repair of the endometrium. Cells of the stratum basalis undergo mitosis and produce a new stratum functionalis. As the endometrium thickens, the short, straight endometrial glands develop and the arterioles coil and lengthen as they penetrate the functionalis. The thickness of the endometrium approximately doubles to about 4 to 6 mm. Because of the proliferation of endometrial cells, the preovulatory phase is also termed the **proliferative phase.** Still another name is the **follicular phase** because of increasing secretion of estrogens by the developing follicle. Functionally, estrogens are the dominant ovarian hormones during this phase of the menstrual cycle (Figure 28-19).

During the preovulatory phase, one of the secondary follicles in the ovary matures into a **vesicular ovarian (Graafian) follicle,** a follicle ready for ovulation. During the maturation process, the follicle increases its estrogen production. Early in the preovulatory phase, FSH is the dominant hormone of the anterior pituitary, but close to the time of ovulation, LH is secreted in increasing quantities (Figure 28-19). Moreover, small amounts of progesterone may be produced by the vesicular ovarian follicle a day or two before ovulation.

Ovulation

Ovulation, the rupture of the vesicular ovarian follicle with release of the immature ovum into the pelvic cavity, usually occurs on day 14 in a 28-day cycle. Just prior to ovulation, the high level of estrogens that developed during the preovulatory phase inhibits GnRF production by the hypothalamus. This, in turn, inhibits FSH secretion by the anterior pituitary via a negative feedback cycle. Concurrently, the high level of estrogens acts in a positive feedback cycle to cause the anterior pituitary to release a surge of LH. Without this surge of LH, ovulation will not occur. (An over-the-counter home test that detects the LH surge associated with ovulation is now available. The test predicts ovulation a day in advance.) Following ovulation, the vesicular ovarian follicle collapses and blood within it forms a clot called the **corpus hemorrhagicum.** The clot is eventually absorbed by the remaining follicular cells. In time, the follicular cells enlarge, change character, and form the **corpus luteum,** or yellow body.

Postovulatory Phase

The **postovulatory phase** of the menstrual cycle is the most constant in duration and lasts from days 15 to 28 in a 28-day cycle. It represents the time between ovulation and the onset of the next menses. Following ovulation, LH secretion stimulates the development of the corpus luteum. The corpus luteum then secretes increasing quantities of estrogens and progesterone. Progesterone is responsible for preparing the endometrium to receive a fertilized ovum. Preparatory activities include secretory activity of the endometrial glands that causes them to appear tortuously coiled, vascularization of the superficial endometrium, thickening of the endometrium, glycogen storage, and an increase in the amount of tissue fluid. These preparatory changes are maximal about 1 week after ovulation, and they correspond to the anticipated arrival of the fertilized ovum. During the postovulatory phase, FSH secretion again gradually increases and LH secretion decreases. The functionally dominant ovarian hormone during this phase is progesterone (see Figure 28-19). The relationship of progesterone to prostaglandins in causing painful menstruation will be considered at the end of the chapter.

If fertilization and implantation do not occur, the rising levels of progesterone and estrogens from the corpus luteum inhibit GnRF and LH secretion. As a result the corpus luteum degenerates and becomes the **corpus albicans.** The decreased secretion of progesterone and estrogens by the degenerating corpus luteum then initiates another menstrual period. In addition, the decreased levels of progesterone and estrogens in the blood bring about a new output of the anterior pituitary hormones—especially FSH in response to an increased output of GnRF by the hypothalamus. Thus a new ovarian cycle is initiated. A summary of these hormonal interactions is presented in Figure 28-20.

If however, fertilization and implantation do occur, the corpus luteum is maintained for 3 to 4 months into the pregnancy, during which time it secretes estrogens and progesterone. The corpus luteum is maintained by **human chorionic** (kō-rē-ON-ik) **gonadotropin (HCG),** a hormone produced by the developing placenta. The placenta itself secretes estrogens to support pregnancy and

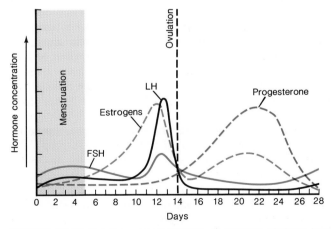

FIGURE 28-19 Relative concentrations of anterior pituitary hormones (FSH and LH) and ovarian hormones (estrogens and progesterone) during a normal menstrual cycle.

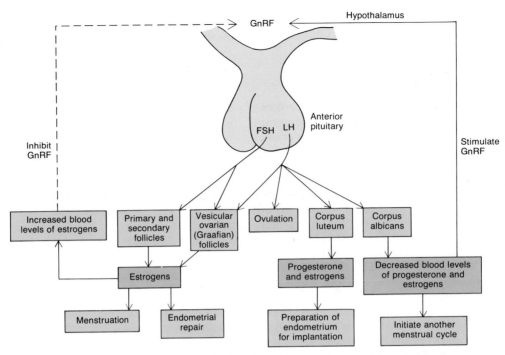

FIGURE 28-20 Summary of hormonal interactions of the menstrual and ovarian cycles.

progesterone to support pregnancy and breast development for lactation. Once the placenta begins its secretion, the role of the corpus luteum becomes minor.

Menarche and Menopause

The menstrual cycle normally occurs once each month from **menarche** (me-NAR-kē), the first menses, to **menopause,** the last menses. The advent of menopause is signaled by the **climacteric** (klī-mak-TER-ik)—menstrual cycles become less frequent. The climacteric, which typically begins between ages 40 and 50, results from the failure of the ovaries to respond to the stimulation of gonadotropic hormones from the anterior pituitary. Some women experience hot flashes, copious sweating, headache, hair loss, muscular pains, and emotional instability. In the postmenopausal woman there will be some atrophy of the ovaries, uterine tubes, uterus, vagina, external genitalia, and breasts.

The cause of menopause is related to a decreasing ability of aging ovaries to respond to FSH and LH. As a result there is a decrease in the production of estrogens by the ovaries. Throughout a woman's sexual life, some of the primary ovarian follicles grow into vesicular ovarian follicles with each sexual cycle, and eventually most of them degenerate. As the number of primary follicles diminishes, the production of estrogens by the ovary decreases.

VAGINA

The **vagina** serves as a passageway for the menstrual flow. It is also the receptacle for the penis during coitus, or sexual intercourse, and the lower portion of the birth canal. It is a muscular, tubular organ lined with mucous

membrane and measures about 10 cm (4 inches) in length, extending from the cervix to the vestibule (see Figures 28-11 and 28-12a). Situated between the urinary bladder and the rectum, it is directed superiorly and posteriorly, where it attaches to the uterus. A recess, called the **fornix,** surrounds the vaginal attachment to the cervix. The dorsal recess, called the posterior fornix, is deeper than the ventral and two lateral fornices. The fornices make possible the use of contraceptive diaphragms.

Histologically, the mucosa of the vagina is continuous with that of the uterus and consists of stratified squamous epithelium and connective tissue that lies in a series of transverse folds, the **rugae.** Most of the lubrication of the vagina during sexual intercourse is a result of a mucoid secretion produced by the vaginal epithelium. The muscularis is composed of longitudinal smooth muscle that can stretch considerably. This distension is important because the vagina receives the penis during sexual intercourse and serves as the lower portion of the birth canal. At the lower end of the vaginal opening, the **vaginal orifice,** is a thin fold of vascularized mucous membrane called the **hymen,** which forms a border around the orifice, partially closing it (see Figure 28-21).

CLINICAL APPLICATION

Sometimes the hymen completely covers the orifice, a condition called **imperforate** (im-PER-fō-rāt) **hymen.** Surgery is required to open the orifice to permit the discharge of the menstrual flow.

The mucosa of the vagina contains large amounts of glycogen, which upon decomposition produces organic

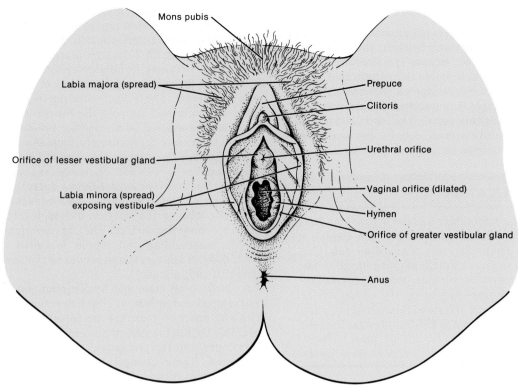

Mons pubis

Labia majora (spread)

Orifice of lesser vestibular gland

Labia minora (spread)
exposing vestibule

Prepuce

Clitoris

Urethral orifice

Vaginal orifice (dilated)

Hymen

Orifice of greater vestibular gland

Anus

FIGURE 28-21 Vulva.

acids. These acids create a low pH environment that retards microbial growth. However, the acidity is also injurious to sperm cells. Semen neutralizes the acidity of the vagina to ensure survival of the sperm.

VULVA

The term **vulva** (VUL-va), or **pudendum** (pyoo-DEN-dum), is a collective designation for the external genitalia of the female (Figure 28-21). Its components are as follows.

The **mons pubis (veneris),** an elevation of adipose tissue covered by coarse pubic hair, is situated over the symphysis pubis. It lies anterior to the vaginal and urethral openings. From the mons pubis, two longitudinal folds of skin, the **labia majora** (LĀ-bē-a ma-JŌ-ra), extend inferiorly and posteriorly. The labia majora, the female homologue of the scrotum, contain an abundance of adipose tissue and sebaceous (oil) and sudoriferous (sweat) glands; they are covered by pubic hair on their upper outer surfaces. Medial to the labia majora are two folds of skin called the **labia minora** (MĪ-nō-ra). Unlike the labia majora, the labia minora are devoid of pubic hair and fat and have few sudoriferous glands. They do, however, contain numerous sebaceous glands.

The **clitoris** (KLI-to-ris) is a small, cylindrical mass of erectile tissue and nerves. It is located at the anterior junction of the labia minora. A layer of skin called the **prepuce** (foreskin) is formed at the point where the labia minora unite and covers the body of the clitoris. The exposed portion of the clitoris is the **glans.** The clitoris is homologous to the glans penis of the male. Homologous

means that two organs correspond in structure, position, and origin. Like the penis, the clitoris is capable of enlargement upon tactile stimulation and assumes a role in sexual excitement of the female.

The cleft between the labia minora is called the **vestibule.** Within the vestibule are the hymen, vaginal orifice, urethral orifice, and the openings of several ducts. The **vaginal orifice** occupies the greater portion of the vestibule and is bordered by the hymen. Anterior to the vaginal orifice and posterior to the clitoris is the **urethral orifice.** On either side of the urethral orifice are the openings of the ducts of the **paraurethral (Skene's) glands.** These glands are embedded in the wall of the urethra and secrete mucus. On either side of the vaginal orifice itself are the **greater vestibular (Bartholin's) glands.** These glands open by ducts into a groove between the hymen and labia minora and produce a mucoid secretion that supplements lubrication during sexual intercourse. A number of **lesser vestibular glands,** whose orifices are microscopic, open into the vestibule. The paraurethral glands are homologous to the male prostate. The greater vestibular glands are homologous to the male bulbourethral glands.

CLINICAL APPLICATION

One important sign in the **diagnosis of pregnancy** is a bluish discoloration of the vulva and vagina due to venous congestion. The discoloration appears at about the eighth to twelfth week and increases in intensity as the pregnancy progresses.

PERINEUM

The **perineum** (per'-i-NĒ-um) is the diamond-shaped area at the inferior end of the trunk between the thighs and buttocks of both males and females. It is bounded anteriorly by the symphysis pubis, laterally by the ischial tuberosities, and posteriorly by the coccyx. A transverse line drawn between the ischial tuberosities divides the perineum into an anterior **urogenital** (yoo'-rō-JEN-i-tal) **triangle** that contains the external genitalia and a posterior **anal triangle** that contains the anus (Figure 28-22).

CLINICAL APPLICATION

In the female, the region between the vagina and anus is known as the **clinical perineum.** If the vagina is too small to accommodate the head of an emerging fetus, the skin, vaginal epithelium, subcutaneous fat, and superficial transverse perineal muscle of the clinical perineum may tear. Moreover, the tissues of the rectum may be damaged. To avoid this, a small incision called an **episiotomy** (e-piz'-ē-OT-ō-mē) is made in the perineal skin and underlying tissues just prior to delivery. After delivery the episiotomy is sutured in layers.

MAMMARY GLANDS

The **mammary glands** are modified sudoriferous (sweat) glands (branched tubuloalveolar) that lie over the pectoralis major and serratus anterior muscles and are attached to them by a layer of connective tissue (Figure 28-23). Internally, each mammary gland consists of 15 to 20 **lobes,** or compartments, separated by adipose tissue. The amount of adipose tissue determines the size of the breasts. However, breast size has nothing to do with the amount of milk produced. In each lobe are several smaller compartments called **lobules,** composed of connective tissue in which milk-secreting cells referred to as **alveoli** are embedded (Figure 28-24). Alveoli are arranged in grape-like clusters. Between the lobules are strands of connective tissue called the **suspensory ligaments of the breast (Cooper's ligaments).** These ligaments run between the skin and deep fascia and support the breast. Alveoli convey the milk into a series of **secondary tubules.** From here the milk passes into the **mammary ducts.** As the mammary ducts approach the nipple, they expand to form sinuses called **ampullae** (am-POOL-ē), where milk may be stored. The ampullae continue as **lactiferous ducts** that terminate in the **nipple.** Each lactiferous duct conveys milk from one of the lobes to the exterior, although some may join before reaching the surface. The circular pigmented area of skin surrounding the nipple is called the **areola** (a-RĒ-ō-la). It appears rough because it contains modified sebaceous (oil) glands.

At birth, both male and female mammary glands are undeveloped and appear as slight elevations on the chest. With the onset of puberty, the female breasts begin to develop—the ductile system matures, extensive fat deposition occurs, and the areola and nipple grow and become pigmented. These changes are correlated with an increased output of estrogens by the ovary. Further mammary development occurs at sexual maturity with the onset of ovulation and the formation of the corpus luteum. During adolescence, increased levels of progesterone cause the alveoli to proliferate, enlarge, and become secretory. Also, fat deposition continues, increasing the size of the glands. Although the changes in mammary gland development are associated with estrogens and progesterone secretion by the ovaries, ovarian secretion is ultimately controlled by FSH, which is secreted in response to GnRF by the hypothalamus.

FIGURE 28-22 Perineum. Borders seen in the female.

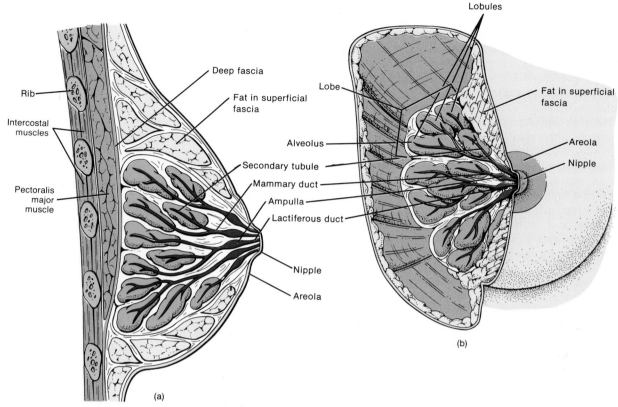

FIGURE 28-23 Mammary glands. (a) Sagittal section. (b) Anterior view, partially sectioned.

FIGURE 28-24 Histology of the mammary glands. Photomicrograph of several alveoli in a nonlactating mammary gland at a magnification of 60×. (Copyright © 1983 by Michael H. Ross. Used by permission.)

The essential function of the mammary glands is milk secretion and ejection, together called **lactation.** The secretion of milk is due largely to the hormone prolactin (PRL), with contributions from progesterone and estrogens. The ejection of milk occurs in the presence of oxytocin (OT). Lactation is considered in detail in Chapter 29.

CLINICAL APPLICATION

Early detection—especially by breast self-examination and mammography—is still the most promising method to increase the survival rate for **breast cancer.** It is estimated that 95 percent of breast cancer is first detected by women themselves. Each month after the menstrual period the breasts should be thoroughly examined for lumps, puckering of the skin, or discharge.

The most effective screening technique for routinely detecting tumors less than 1.27 cm (0.5 inch) in diameter is *mammography.* One reason that mammography is so useful is that it can detect small calcium deposits, called microcalcifications, in breast tissue. Such calcifications frequently indicate the presence of a tumor. The mammographic image, called a mammogram, can be obtained in either of two ways. In one procedure, x rays are beamed onto a specially coated metal plate and the blue-on-white image produced provides the physician with detail of thicker as well as thinner portions of the breast. This procedure is called *xeroradiography.* In the other procedure, called *film-screen mammography,* the x rays are beamed onto a fluorescent screen, which exposes a film in contact with it. It uses a very low dose of radiation and can be adjusted to emphasize different kinds of tissue within the breast.

One of the most recent breast cancer detecting procedures is *ultrasound.* The procedure is performed while the patient lies on her stomach in a specially designed hospital bed with her breasts immersed in a tank of water. Ultrasound produces images using a device that first emits a pulse of high-frequency sound and then records the echo on a monitor. Although ultrasound can neither detect microcalcifications or tumors less than 1 cm in diameter, it can be used to determine whether a lump is a benign cyst or a malignant tumor.

Another technique combines computed tomography with mammography (*CT/M*) and appears to overcome some of the limitations of mammography. The procedure is based upon the fact that breast carcinoma has an abnormal affinity for iodide. CT scans are made before and after the rapid intravenous infusion of an iodide contrast material. Comparison of the initial density of a suspected lesion with the density following infusion of the iodide gives an indication of the status of the tumor. CT/M affords definitive diagnostic help in instances where the mammographic and physical examinations are inconclusive and apppears to be a significantly improved method of breast cancer diagnosis.

Long-term studies are now underway to determine the effectiveness of an experimental home device for breast cancer detection. It is called a **breast cancer screening indicator.** It is composed of two plastic discs that contain heat-sensitive chemicals. The discs are worn inside a female's brassiere for 15 minutes each month so that she can observe color changes that could indicate a breast abnormality. The screening device is said to detect tumors as small as 2 mm in diameter, their size about 2 to 10 years before they become palpable or can be seen by mammogram.

SEXUAL INTERCOURSE

Sexual intercourse, or **copulation** (in humans, called **coitus**), is the process by which spermatozoa are deposited in the vagina.

CLINICAL APPLICATION

Artificial insemination (*in* = into; *seminatus* = sown; *semen* = seed) refers to the deposition of seminal fluid within the vagina or cervix by artificial means. If the husband's seminal fluid is used, the process is known as **homologous insemination.** Such a procedure might be carried out if there is a developmental anomaly that prevents placement of the penis in the vagina or normal ejaculation. A small volume of seminal fluid is also an indication for homologous insemination. If seminal fluid from a donor is used, the process is called **heterologous insemination.** In such cases, an anonymous donor is selected on the basis of race, blood type compatibility, physical appearance, general health, and genetic background.

MALE SEXUAL ACT

Erection

The male role in the sexual act starts with **erection,** the enlargement and stiffening of the penis. An erection may be initiated in the cerebrum by stimuli such as anticipation, memory, and visual sensation, or it may be a reflex brought on by stimulation of the touch receptors in the penis, especially in the glans. In any case, parasympathetic impulses that pass from the sacral portion of the spinal cord to the penis cause dilation of the arteries of the penis, allowing blood to fill the cavernous spaces of the spongy bodies.

Lubrication

Parasympathetic impulses from the sacral cord also cause the bulbourethral (Cowper's) glands to secrete mucus, which affords some **lubrication** for intercourse. The mucus flows through the urethra. The major portion of lubricating fluid is produced by the female. Without satisfactory lubrication, the male sexual act is difficult since unlubricated intercourse causes pain impulses that inhibit rather than promote coitus.

Orgasm

Tactile stimulation of the penis brings about emission and ejaculation. When sexual stimulation becomes intense, rhythmic sympathetic impulses leave the spinal cord at the levels of the first and second lumbar vertebrae and pass to the genital organs. These impulses cause peristaltic contractions of the ducts in the testes, epididymides, and ductus (vas) deferens that propel spermatozoa into the urethra—a process called **emission.** Simultaneously, peristaltic contractions of the seminal vesicles and prostate expel seminal and prostatic fluid along with the spermatozoa. All these mix with the mucus of the bulbourethral glands, resulting in the fluid called semen. Other rhythmic impulses sent from the spinal cord at the levels of the first and second sacral vertebrae reach the skeletal muscles at the base of the penis, and the penis expels the semen from the urethra to the exterior. The propulsion of semen from the urethra to the exterior constitutes an **ejaculation.** A number of sensory and motor activities accompany ejaculation, including a rapid heart rate, an increase in blood pressure, an increase in respiration, and pleasurable sensations. These activities, together with the muscular events involved in ejaculation, are referred to as an **orgasm.**

During sexual intercourse, ejaculation introduces millions of sperm into the vagina. Next, the sperm move into the cervix where muscular contractions aid their movement into the uterus. Inside the uterus, rhythmic contractions of the muscular wall greatly aid the sperm as they swim toward the uterine (Fallopian) tubes. There is some evidence that suggests the hormone oxytocin (OT) may be released from the posterior pituitary during orgasm and/or that prostaglandins in seminal fluid function to cause the uterine contractions. Of the total number of spermatozoa that enter the vagina, less than 1 percent come in proximity to the ovum.

FEMALE SEXUAL ACT

Erection

The female role in the sex act, like that of the male, also involves erection, lubrication, and orgasm. Stimulation of the female, as in the male, depends on both psychic and tactile responses. Under appropriate conditions, stimulation of the female genitalia, especially the clitoris, results in erection and widespread sexual arousal. This response is controlled by parasympathetic impulses sent from the sacral spinal cord to the external genitalia.

Lubrication

Parasympathetic impulses from the sacral spinal cord also cause the bulk of **lubrication** of the vagina. The impulses result in the production of a mucoid fluid from the epithelium of the vaginal mucosa. Some mucus is also produced by the greater vestibular (Bartholin's) glands. As was noted earlier, lack of sufficient lubrication results in pain impulses that inhibit rather than promote coitus.

Orgasm (Climax)

When tactile stimulation of the genitalia reaches maximum intensity, reflexes are initiated that cause the female **orgasm (climax).** Female orgasm is analogous to male ejaculation and may assume a role in fertilization of an ovum. The perineal muscles contract rhythmically from spinal reflexes similar to those that occur in the male ejaculation.

BIRTH CONTROL

Methods of **birth control** include removal of the gonads and uterus, sterilization, and mechanical and chemical contraception.

REMOVAL OF GONADS AND UTERUS

Castration (removal of the testes), **hysterectomy** (removal of the uterus), and **oophorectomy** (ō'-of-ō-REK-tō-mē; removal of the ovaries) are all absolute preventive meth-

ods. Once performed, these operations cannot be reversed, and it is impossible to produce offspring. However, removal of the testes or ovaries has adverse effects because of the importance of these organs in the endocrine system. Generally these operations are performed only if the organs are diseased. Castration before puberty prevents the development of secondary sex characteristics.

STERILIZATION

One means of **sterilization** of males is **vasectomy** (discussed earlier in the chapter). Sterilization in females generally is achieved by performing a **tubual ligation** (lī-GĀ-shun). An incision is made into the abdominal cavity, the uterine (Fallopian) tubes are squeezed, and a small loop called a knuckle is made. A suture is tied tightly at the base of the knuckle and the knuckle is then cut. After 4 or 5 days the suture is digested by body fluids and the two severed ends of the tubes separate. The ovum thus is prevented from passing to the uterus, and the sperm cannot reach the ovum. Sterilization normally does not affect sexual performance or enjoyment.

Another method of sterilizing women is the **laparoscopic** (lap'-a-rō-SKŌ-pik) **technique.** After a woman receives local or general anesthesia, a harmless gas is introduced into her abdomen to create a gas bubble. The bubble expands the abdominal cavity and pushes the intestines away from the pelvic organs, permitting safe easy access to the uterine tubes. The doctor makes a small incision at the lower rim of the umbilicus and inserts a laparoscope to view the inside of the abdominal cavity and the uterine tubes. The tubes can be closed with this instrument or a second incision can be made at the pubic hairline to insert a cautery forceps. Once the uterine tubes are sealed, the instrument is removed, the gas is released, and the incision is covered with a bandage. After a few hours the patient can usually go home.

CONTRACEPTION

Contraception is the prevention of fertilization without destroying fertility by natural, mechanical, or chemical means.

Natural

The **natural methods** include complete or periodic abstinence. An example of periodic abstinence is the rhythm method, which takes advantage of the fact that a fertilizable ovum is available only during a period of 3 to 5 days in each menstrual cycle. During this time the couple refrains from intercourse. Its effectiveness is limited by the fact that few women have absolutely regular cycles. Moreover, some women occasionally ovulate during the "safe" times of the month, such as during menstruation.

Mechanical

Mechanical methods of contraception include the use of the condom by the male and the diaphragm by the female.

The **condom** is a nonporous, elastic (rubber or similar material) covering placed over the penis that prevents deposition of sperm in the female reproductive tract. The **diaphragm** is a dome-shaped structure that fits over the cervix and is generally used in conjunction with a sperm-killing chemical. The diaphragm stops the sperm from passing into the cervix. The chemical kills the sperm cells. Toxic shock syndrome (TSS) and recurrent urinary tract infections are associated with diaphragm use in some females.

Another mechanical method of contraception is an **intrauterine devide (IUD).** The device is a small object made of plastic, copper, or stainless steel that is inserted into the cavity of the uterus. It is not clear how IUDs operate. Some investigators believe they cause changes in the uterine lining that, in turn, produce a substance which destroys either the sperm or the fertilized ovum. Some IUDs release contraceptive agents in minute amounts; one secretes progesterone. The dangers associated with the use of IUDs in some females include pelvic inflammatory disease (PID) and infertility. Because of this, IUDs are rapidly declining in popularity. Lawsuits resulting from damage claims have caused manufacturers of copper IUDs in the United States to stop production and sales.

Chemical

Chemical methods of contraception include spermicidal and hormonal methods. Various foams, creams, jellies, suppositories, and douches make the vagina and cervix unfavorable for sperm survival. A recent spermicidal development is a **contraceptive sponge,** sold under the trade name Today®. It is a nonprescription, polyurethane sponge that contains a spermicide called nonoxynol-9, also used in a variety of contraceptive foams and jellies. The sponge is placed in the vagina where it releases spermicide for up to 24 hours and also acts as a physical barrier to sperm. The sponge is equal to the diaphragm in effectiveness. Some cases of toxic shock syndrome (TSS) have been reported among users of the contraceptive sponge.

Modified human chorionic gonadotropin (HCG), a hormone produced by the placenta, has shown promise in laboratory animals as a chemical method of birth control. In its modified form, HCG prevents implantation of a fertilized egg or terminates an already established pregnancy. Modified versions of gonadotropin releasing factor (GnRF) are also being tested as contraceptives that inhibit ovulation.

The hormonal method, **oral contraceptive** or **OC** (the pill), has found rapid and widespread use. Although several pills are available, the one most commonly used contains a high concentration of progesterone and a low concentration of estrogens. These two hormones act on the anterior pituitary to decrease the secretion of FSH and LH by inhibiting GnRF by the hypothalamus. The low levels of FSH and LH are not adequate to initiate follicle maturation and ovulation. In the absence of a mature ovum, pregnancy cannot occur.

Women for whom all oral contraceptives are contraindicated include those with a history of thromboembolic disorders (predisposition to blood clotting), cerebral blood vessel damage, hypertension, liver malfunction, heart disease, or cancer of the breast or reproductive system. Recent reports link pill users with an increased risk of infertility. About 40 percent of all pill users experience side effects—generally minor problems such as nausea, weight gain, headache, irregular menses, spotting between periods, and amenorrhea. The statistics on the life-threatening conditions associated with the pill such as blood clots, heart attacks, liver tumors, and gallbladder disease are somewhat more reassuring. For all the problems combined, fewer than 3 deaths occurred per 100,000 users under age 30, 4 among those 30 to 35, 10 among those 35 to 39, and 18 among women over 40. The major exception is that women who take the pill and smoke face far higher odds of developing heart attack and stroke than do nonsmoking pill users.

A summary of birth control methods is presented in Exhibit 28-1.

CLINICAL APPLICATION

The quest for an efficient male oral contraceptive has been disappointing until recently. An oral contraceptive, **gossypol,** which is derived from cottonseed oil, has achieved a high efficacy (power to produce effects) rate of 99.9 percent in clinical tests in China. Its action is to inhibit an enzyme required for spermatogenesis. Levels of blood luteinizing hormone (LH) and testosterone remained unchanged and potency was not impaired. The number and morphology of spermatozoa gradually recover and fertility is restored to normal within 3 months after termination of therapy in most individuals. In some cases, fertility is not regained when gossypol is stopped. Gossypol also lowers potassium levels in the blood.

Modified versions of GnRF have also been shown to inhibit testosterone production, with a resultant decrease in sperm count and motility. However, the development of impotence has made this approach unpopular.

AGING AND THE REPRODUCTIVE SYSTEMS

Although there are major age-specific physical changes in structure and function, few age-specific disorders are associated with the reproductive system. In the male, the decreasing production of testosterone produces less muscle strength, fewer viable sperm, and decreased sexual desire. However, abundant spermatozoa may be found even in old age. Most of the pathologies produce uneventful recoveries, except for prostate problems that could become serious and fatal.

The female reproductive system becomes less efficient, possibly as a result of less frequent ovulation and the

EXHIBIT 28-1   BIRTH CONTROL METHODS

METHOD	COMMENTS
Removal of gonads and uterus	Irreversible sterility. Generally performed if organs are diseased rather than as contraceptive method because of importance of hormones produced by gonads.
Sterilization	Procedure involving severing ductus (vas) deferens in males (vasectomy) and uterine (Fallopian) tubes in females (tubal ligation and laparoscopic technique).
Natural contraception	Abstinence from intercourse during time of month woman is fertile. Under ideal circumstances, effectiveness in women with regular menstrual cycles may approach that of mechanical and chemical contraceptives. Extremely difficult to determine fertile period. Effectiveness can be increased by recording body temperature each morning before getting up; a small rise in temperature indicates that ovulation has occurred 1 or 2 days earlier and ovum can no longer be fertilized.
Mechanical contraception **Condom**	Thin, strong sheath of rubber or similar material worn by male to prevent sperm from entering vagina. Failures caused by sheath tearing or slipping off after climax or not putting the sheath on soon enough. If used correctly and consistently, effectiveness similar to that of diaphragm.
Diaphragm	Flexible rubber dome inserted into vagina to cover cervix, providing barrier to sperm. Usually used with spermicidal cream or jelly. Must be left in place at least 6 hours after intercourse and may be left in place as long as 24 hours. Must be fitted by physician or other trained personnel and refitted every 2 years and after each pregnancy. Offers high level of protection if used with spermicide; rate of 2 to 3 pregnancies per 100 women per year is estimated for consistent users. If not used consistently, much higher pregnancy rates must be expected. Occasional failures caused by improper insertion or displacement during sexual intercourse.
Intrauterine device (IUD)	Small object (loop, coil, T, or 7) made of plastic, copper, or stainless steel and inserted into uterus by physician. May be left in place for long periods of time (some must be changed every 2 to 3 years). Does not require continued attention by user. Some women cannot use them because of expulsion, bleeding, or discomfort. Not recommended for women who have not had children because uterus is too small and cervical canal too narrow. Use of IUDs has diminished because of potential problems such as pelvic inflammatory disease (PID) and infertility. No longer manufactured in the United States.
Chemical contraception **Foams, creams, jellies, suppositories, vaginal douches, contraceptive sponge**	Sperm-killing chemicals inserted into vagina to coat vaginal surfaces and cervical opening. Provide protection for about 1 hour. Effective when used alone, but significantly more effective when used with diaphragm or condom. The contraceptive sponge is made of polyurethane and releases a spermicide for up to 24 hours. It rates with the diaphragm in effectiveness. A few cases of toxic shock syndrome (TSS) have been reported among users of the contraceptive sponge.
Oral contraceptives (OC)	Except for total abstinence or surgical sterilization, most effective contraceptive known. Side effects include nausea, occasional light bleeding between periods, breast tenderness or enlargement, fluid retention, and weight gain. Should not be used by women who have cardiovascular conditions (thromboembolic disorders, cerebrovascular disease, heart disease, hypertension), liver malfunction, cancer or neoplasia of breast or reproductive organs, or by women who smoke. Pill users may have an increased risk of infertility.

declining ability of the uterine (Fallopian) tubes and uterus to support the young embryo. There is a decrease in the production of progesterone and estrogens. Menopause is only one of a series of phases that leads to reduced fertility, irregular or absent menstruation, and a variety of physical changes. Uterine cancer peaks at about 65 years of age, but cervical cancer is more common in younger women, and breast cancer is the leading cause of death among women between the ages of 40 and 60. Prolapse (falling down or sinking) of the uterus is possibly the commonest complaint among female geriatric patients.

DEVELOPMENTAL ANATOMY OF THE REPRODUCTIVE SYSTEMS

The *gonads* develop from the **intermediate mesoderm.** By the sixth week, they appear as bulges that protrude into the ventral body cavity (Figure 28-25). The gonads develop near the mesonephric ducts. A second pair of ducts, the **paramesonephric (Müller's) ducts,** develop lateral to the mesonephric ducts. Both sets of ducts empty into the urogenital sinus. At about the eighth week, the gonads are clearly differentiated into ovaries or testes.

FIGURE 28-25 Development of the internal reproductive systems.

In the male embryo, the *testes* connect to the mesonephric duct through a series of tubules. These tubules become the *seminiferous tubules.* Continued development of the mesonephric ducts produces the *efferent ducts, ductus epididymis, ductus (vas) deferens, ejaculatory ducts,* and *seminal vesicle.* The *prostate* and *bulbourethral (Cowper's) glands* are **endodermal** outgrowths of the urethra. Shortly after the gonads differentiate into testes, the paramesonephric ducts degenerate without contributing any functional structures to the male reproductive system.

In the female embryo, the gonads develop into *ovaries.* At about the same time, the distal ends of the paramesonephric ducts fuse to form the *uterus* and *vagina.* The unfused portions become the *uterine (Fallopian) tubes.* The *greater (Bartholin's)* and *lesser vestibular glands* develop from **endodermal** outgrowths of the vestibule. The mesonephric ducts in the female degenerate without contributing any functional structures to the female reproductive system.

The *external genitals* of both male and female embryos also remain undifferentiated until about the eighth week. Before differentiation, all embryos have an elevated region, the **genital tubercle,** a point between the tail (future coccyx) and the umbilical cord, where the mesonephric and paramesonephric ducts open to the exterior (Figure 28-26). The tubercle consists of a **urethral groove** (opening into the urogenital sinus), paired **urethral folds,** and paired **labioscrotal swellings.**

In the male embryo, the genital tubercle elongates and develops into a *penis.* Fusion of the urethral folds forms the *spongy (cavernous) urethra* and leaves an opening only at the distal end of the penis, the *urethral orifice.* The labioscrotal swellings develop into the *testes.* In the female, the genital tubercle gives rise to the *clitoris.* The urethral folds remain open as the *labia minora,* and the labioscrotal swellings become the *labia majora.* The urethral groove becomes the *vestibule.*

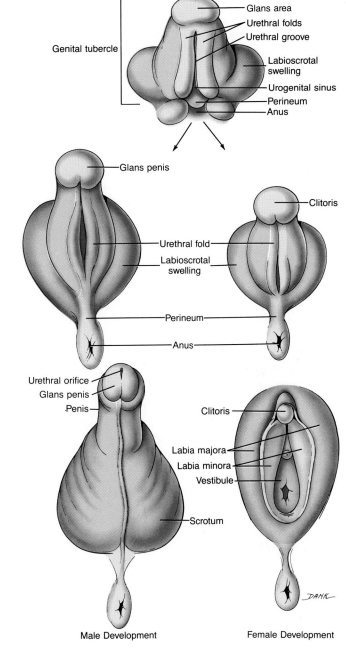

FIGURE 28-26 Development of the external genitals.

Male Development   Female Development

DISORDERS: HOMEOSTATIC IMBALANCES

Sexually Transmitted Diseases (STDs)

The general term **sexually transmitted disease (STD)** is applied to any of the large group of diseases that can be spread by sexual contact. The group includes conditions traditionally specified as **venereal diseases (VD),** from Venus, goddess of love, such as gonorrhea, syphilis, and genital herpes, and several other conditions that are contracted sexually, or may be contracted otherwise, but are then transmitted to a sexual partner.

Gonorrhea

Gonorrhea is an infectious sexually transmitted disease that affects primarily the mucous membrane of the urogenital tract, the rectum, and occasionally the eyes. The disease is caused by the bacterium *Neisseria gonorrhoeae.* Discharges from the involved mucous membranes are the source of infection, and the bacteria are transmitted by direct contact, usually sexual or during passage of a newborn through the birth canal.

Males usually suffer inflammation of the urethra with pus and painful urination. Fibrosis sometimes occurs in an advanced stage, causing stricture of the urethra. There also may be involvement of the epididymis and prostate gland. In females, infection may occur in the urethra, vagina, and cervix, and there may be a discharge of pus. However, infected females often harbor the disease without any symptoms until it has progressed to a more advanced stage. If the uterine (Fallopian) tubes become involved, pelvic inflammation may follow. Peritonitis, or inflammation of the peritoneum, is a very serious disorder. The infection should be treated and controlled immediately because, if neglected, sterility or death may result. Although antibiotics have greatly reduced the mortality rate of acute peritonitis, it is estimated that between 50,000 and 80,000 women are made sterile by gonorrhea every year as a result of scar tissue formation that closes the uterine tubes. If the bacteria are transmitted to the eyes of a newborn in the birth canal, blindness can result.

Administration of a 1 percent silver nitrate solution or penicillin in the infant's eyes prevents infection. Penicillin or tetracycline are the drugs of choice for the treatment of gonorrhea in adults.

Syphilis

Syphilis is a sexually transmitted disease caused by the bacterium *Treponema pallidum.* It is acquired through sexual contact or transmitted through the placenta to a fetus. The disease progresses through several stages: primary, secondary latent, and sometimes tertiary. During the *primary stage,* the chief symptom is an open sore, called a *chancre,* at the point of contact. The chancre heals within 1 to 5 weeks. From 6 to 24 weeks later, symptoms such as a skin rash, fever, and aches in the joints and muscles usher in the *secondary stage.* These symptoms also eventually disappear (in about 4 to 12 weeks), and the disease ceases to be infectious, but a blood test for the presence of the bacteria generally remains positive. During this "symptomless" period, called the *latent stage,* the bacteria may invade body organs. When signs of organ degeneration appear, the disease is said to be in the *tertiary stage.*

If the syphilis bacteria attack the organs of the nervous system, the tertiary stage is called *neurosyphilis.* Neurosyphilis may take different forms, depending on the tissue involved. For instance, about 2 years after the onset of the disease, the bacteria may attack the meninges, producing meningitis. The blood vessels that supply the brain may also become infected. In this case, symptoms depend on the parts of the brain destroyed by oxygen and glucose starvation. Cerebellar damage is manifested by uncoordinated movements in such activities as writing. As the motor areas become extensively damaged victims may be unable to control urine and bowel movements. Eventually, they may become bedridden, unable even to feed themselves. Damage to the cerebral cortex produces memory loss and personality changes that range from irritability to hallucinations.

Infection of the fetus with syphillis can occur after the fifth month. Infection of the mother is not necessarily followed by fetal infection, provided that the placenta remains intact. But once the bacteria gain access to fetal circulation, there is nothing to impede their growth and multiplication. As many as 80 percent of children born to untreated syphilitic mothers will be infected in the uterus if the fetus is exposed at the onset or in the early stages of the disease. About 25 percent of the fetuses will die within the uterus. Most of the survivors will arrive prematurely, but 30 percent will die shortly after birth. Of the infected and untreated children surviving infancy, about 40 percent will develop symptomatic syphilis during their lifetimes.

Syphilis can be treated with antibiotics (penicillin) during the primary, secondary, and latent periods. Certain forms of neurosyphilis may also be successfully treated, but the prognosis for others is very poor. Noticeable symptoms do not always appear during the first two stages of the disease. Syphilis, however, is usually diagnosed through a blood test whether noticeable symptoms appear or not. The importance of these blood tests and follow-up treatments cannot be overemphasized.

Genital Herpes

Another sexually transmitted disease, **genital herpes,** is now common in the United States. The sexual transmission of the herpes simplex virus is well established. Unlike syphilis and gonorrhea, genital herpes is incurable. Type I herpes simplex virus is the virus that causes the majority of infections above the waist such as cold sores. Type II herpes simplex virus causes most infections below the waist such as painful genital blisters on the prepuce, glans penis, and penile shaft in males and on the vulva or sometimes high up in the vagina in females. The blisters disappear and reappear in most patients, but the disease itself remains in the body.

Genital herpes virus infection causes considerable discomfort, and there is an extraordinarily high rate of recurrence of the disease. The infection is usually characterized by fever, lymphadenopathy, and numerous clusters of genital blisters. For pregnant women with genital herpes symptoms at the time of delivery, a cesarean section will usually prevent complications in the child. Complications range from a mild asymptomatic infection to CNS damage to death.

Treatment of the symptoms involves pain medication, saline compresses, sexual abstinence for the duration of the eruption, and use of a drug called acyclovir. This drug interferes with viral DNA replication but not with host cell DNA replication. Acyclovir speeds the healing and sometimes reduces the pain of initial genital herpes infections and shortens the duration of lesions in patients with recurrent genital herpes.

Trichomoniasis

The microorganism *Trichomonas vaginalis,* a flagellated protozoan (one-celled animal), causes **trichomoniasis,** an inflammation of the mucous membrane of the vagina in females and the urethra in males. Symptoms include vaginal discharge and severe vaginal itch in women. Men can have it without symptoms, but can transmit it to women nonetheless. Sexual partners must be treated simultaneously.

Nongonococcal Urethritis (NGU)

The term **nongonococcal urethritis (NGU),** also known as **nonspecific urethritis (NSU),** refers to an inflammation of the urethra not caused by the bacterium *Neisseria gonorrhoeae.* The bacterium *Chlamydia trachomatis* is known to cause some cases of NGU. NGU affects both males and females. The urethra swells and narrows, impeding the flow of urine. Both urination and the urgency to urinate increase. Urination is accompanied by burning pain, and there may be a purulent (pus-containing) discharge.

Although nonmicrobial factors such as trauma (passage of a catheter) or chemical agents (alcohol and certain chemotherapeutic agents) can cause this condition, it is estimated that at least 40 percent of the cases of NGU are acquired sexually. In fact, NGU may be the most common sexually transmitted disease in the United States today. Although it is not a disease which must be reported to state departments of health and exact data are lacking, the Centers for Disease Control estimates that 4 to 9 million Americans have NGU. Since the symptoms are often mild in males, and females are usually asymptomatic, many cases of infection go untreated. Complications are not common, but can be serious. Males may develop inflammation of the epididymis. In females, inflammation may cause sterility by blocking the cervix or uterine (Fallopian) tubes. As in gonorrhea, the bacteria can be passed from mother to infant during birth, infecting the eyes. NGU of chlamydial origin responds to treatment with tetracycline.

Male Disorders

Prostate

The prostate gland is susceptible to infection, enlargement, and benign and malignant tumors. Because the prostate surrounds the urethra, any of these disorders can obstruct the flow of urine. Prolonged obstruction may result in serious changes in the urinary bladder, ureters, and kidneys and may perpetuate urinary tract infections. Therefore, if the obstruction cannot be relieved by other means, surgical removal of part of or the entire gland is indicated. The surgical procedure is called **prostatectomy** (pros'-ta-TEK-tō-mē).

Acute and chronic infections of the prostate gland are common in postpubescent males, often in association with inflammation of the urethra. In **acute prostatitis,** the prostate gland becomes swollen and tender. Appropriate antibiotic therapy, bed rest, and above-normal fluid intake are effective treatment.

Chronic prostatitis is one of the most common chronic infections in men of the middle and later years. On examination, the prostate gland feels enlarged, soft, and extremely tender. The surface outline is irregular and may be hard. This disease frequently produces no symptoms, but the prostate is believed to harbor infectious microorganisms responsible for some allergic conditions, arthritis, and inflammation of nerves (neuritis), muscles (myositis), and the iris (iritis).

An **enlarged prostate** gland, increasing to two to four times larger than normal, occurs in approximately one-third of all males over age 60. The cause is unknown, and the enlarged condition usually can be detected by rectal examination.

Tumors of the male reproductive system usually involve the prostate gland. Carcinoma of the prostate is the second leading cause of death from cancer in men in the United States, and it is responsible for approximately 19,000 deaths annually. Its incidence is related to age, race, occupation, geography, and ethnic origin. Both benign and malignant growths are common in elderly men. Both types of tumors put pressure on the urethra, making urination painful and difficult. At times, the excessive back pressure destroys kidney tissue and gives rise to an increased susceptibility to infection. Therefore, even when the tumor is benign, surgery is indicated.

Sexual Functional Abnormalities

Impotence is the inability of an adult male to attain or hold an erection long enough for normal intercourse. Impotence could be the result of physical abnormalities of the penis, systemic disorders such as syphilis, vascular disturbances, neurological disorders, testosterone deficiency, or psychic factors such as fear of causing pregnancy, fear of sexually transmitted diseases, religious inhibitions, and emotional immaturity.

Infertility (sterility) is an inability to fertilize the ovum. It does not imply impotence. Male fertility requires production of adequate amounts of viable, normal spermatozoa by the testes, unobstructed transportation of sperm through the seminal tract, and satisfactory deposition in the vagina. The tubules of the testes are sensitive to many factors—x rays, infections, toxins, malnutrition—that may cause degenerative changes and produce male sterility. If inadequate spermatozoa production is suspected, a sperm analysis should be performed. At least one type of fertility may be improved by administration of vitamin C.

Female Disorders

Menstrual Abnormalities

Because menstruation reflects not only the health of the uterus but also the health of the endocrine glands that control it, the ovaries and the pituitary gland, disorders of the female reproductive system frequently involve menstrual disorders.

Amenorrhea is the absence of menstruation. If a woman has never menstruated, the condition is called **primary amenorrhea.** Primary amenorrhea can be caused by endocrine disorders, most often in the pituitary gland and hypothalamus, or by a genetically caused abnormal development of the ovaries or uterus. **Secondary amenorrhea,** the skipping of one or more periods, is commonly experienced by women at some time during their lives. Changes in body weight, either gains or losses, often cause amenorrhea. Obesity may disturb ovarian function, and similarly, the extreme weight loss that characterizes anorexia nervosa often leads to a suspension of menstrual flow. When amenorrhea is unrelated to weight, analysis of levels of estrogens often reveals deficiencies of pituitary and ovarian hormones. Amenorrhea may also be caused by continuous involvement in rigorous athletic training.

Dysmenorrhea is painful menstruation caused by forceful contraction of the uterus. It is often accompanied by nausea, vomiting, diarrhea, headache, fatigue, and nervousness. Some cases are caused by pathological conditions such as uterine tumors, ovarian cysts, endometriosis, and pelvic inflammatory disease (PID). However, other cases of dysmenorrhea are not related to any pathologies. Although the cause of these cases is unknown, it appears that they may be triggered by an overproduction of prostaglandins by the uterus. Prostaglandins are known to stimulate uterine contractions, but they cannot do so in the presence of high levels of progesterone. As we have noted earlier, progesterone levels are high during the last half of the menstrual cycle. During this time, prostaglandins are apparently inhibited by progesterone from producing uterine contractions. However, if pregnancy does not occur, progesterone levels drop rapidly and prostaglandin production increases. This causes the uterus to contract and slough off its lining and may result in dysmenorrhea. The other symptoms of dysmenorrhea—nausea, vomiting, diarrhea, and headache—may be due to prostaglandin-stimulated contractions of the smooth muscle of the stomach, intestines, and blood vessels in the brain. Drugs that inhibit prostaglandin synthesis (naproxen and ibuprofen) are used to treat dysmenorrhea.

Abnormal uterine bleeding includes menstruation of excessive duration or excessive amount, too frequent menstruation, intermenstrual bleeding, and postmenopausal bleeding. These abnormalities may be caused by disordered hormonal regulation, emotional factors, fibroid tumors of the uterus, and systemic diseases.

Premenstrual syndrome (PMS) is a term usually reserved for severe physical and emotional distress occurring late in the postovulatory phase of the menstrual cycle and sometimes overlapping with menstruation. Signs and symptoms include edema, breast swelling and tenderness, abdominal distension, backache, joint pain, constipation, skin eruptions, fatigue and lethargy, greater need for sleep, depression or anxiety, irritability, headache, poor coordination and clumsiness, and cravings for sweet or salty foods. The basic cause of PMS is unknown. One postulated mechanism involves excessive levels of estrogens or inadequate levels of progesterone or dopamine (DA). Other proposed causes are vitamin B_6 deficiency and altered glucose metabolism (hypoglycemia).

Toxic Shock Syndrome (TSS)

Toxic shock syndrome (TSS) is primarily a disease of previously healthy, young, menstruating females who use tampons. It is also recognized in males, children, and nonmenstruating females. Clinically, TSS is characterized by high fever up to 40.6°C (105°F), sore throat or very tender mouth, headache, fatigue, irritability, muscle soreness and tenderness, conjunctivitis, diarrhea and vomiting, abdominal pain, vaginal irritation, and erythematous rash. Other symptoms include lethargy, unresponsiveness, memory loss, hypotension, peripheral vasoconstriction, respiratory distress syndrome, intravascular coagulation, decreased platelet count, renal failure, circulatory shock, and liver involvement.

It is now clear that toxin-producing strains of the bacterium *Staphylococcus aureus* are necessary for development of the disease. Actually, it appears that a virus has become incorporated into *S. aureus,* causing the bacterium to produce the toxins. Apparently, certain high-absorbency tampons provide a substrate upon which the bacteria grow and produce toxins. Initial therapy is directed at correcting all homeostatic imbalances as quickly as possible. Antistaphylococcal antibiotics, such as penicillin or clindamycin, are also administered. In severe cases, high doses of corticosteroids are also administered.

Ovarian Cysts
Ovarian cysts are fluid-containing tumors of the ovary. Follicular cysts may occur in the ovaries of elderly women, in ovaries that have inflammatory diseases, and in menstruating females. They have thin walls and contain a serous albuminous material. Cysts may also arise from the corpus luteum or the endometrium.

Endometriosis
Endometriosis is the growth of endometrial tissue outside the uterus. The tissue enters the pelvic cavity via the open uterine (Fallopian) tubes and may be found in any of several sites—on the ovaries, rectouterine pouch, surface of the uterus, sigmoid colon, pelvic and abdominal lymph nodes, cervix, abdominal wall, kidneys, and urinary bladder. One theory for the development of endometriosis is that there is regurgitation of menstrual flow through the uterine tubes. Endometriosis is common in women 30 to 40 years of age. Symptoms include premenstrual pain or unusual menstrual pain. The unusual pain is caused by the displaced tissue sloughing off at the same time the normal uterine endometrium is being shed during menstruation. Infertility can be a consequence. Treatment usually consists of hormone therapy or surgery. Endometriosis disappears at menopause or when the ovaries are removed.

Infertility
Female inferility, or the inability to conceive, occurs in about 10 percent of married females in the United States. Once it is established that ovulation occurs regularly, the reproductive tract is examined for functional and anatomical disorders to determine the possibility of union of the sperm and the ovum in the uterine tube.

Disorders Involving the Breasts
The breasts of females are highly susceptible to cysts and tumors. Men are also susceptible to breast tumors, but certain breast cancers are 100 times more common in women.

In the female, the benign **fibroadenoma** is a common tumor of the breast. It occurs most frequently in young women. Fibroadenomas have a firm rubbery consistency and are easily moved about within the mammary tissue.The usual treatment is excision of the growth. The breast itself is not removed.

Breast cancer has one of the highest fatality rates of all cancers affecting women, but it is rare in men. In the female, breast cancer is rarely seen before age 30, and its occurrence rises rapidly after menopause. Breast cancer is generally not painful until it becomes quite advanced, so often it is not discovered early or, if noted, is ignored. Any lump, no matter how small, should be reported to a doctor at once. Treatment for breast cancer may involve hormone therapy, chemotherapy, lumpectomy (removal of just the tumor and immediate surrounding tissue), a modified or radical mastectomy, or a combination of these. A *radical mastectomy* involves removal of the affected breast along with the underlying pectoral muscles and the axillary lymph nodes. Metastasis of cancerous cells is usually through the lymphatics or blood. Radiation treatment and chemotherapy may follow the surgery to ensure the destruction of any stray cancer cells.

Among the factors that clearly increase the risk of breast cancer development are: (1) a family history of breast cancer, especially in a mother or sister; (2) never having a child or having a first child after age 34; (3) previous cancer in one breast; (4) exposure to ionizing radiation; and (5) an abnormal mammogram.

Cervical Cancer
Another common disorder of the female reproductive tract is **cervical cancer,** carcinoma of the cervix of the uterus. The condition starts with *cervical dysplasia* (dis-PLĀ-sē-a), a change in the shape, growth, and number of the cervical cells. If the condition is minimal, the cells may regress to normal. If it is severe, it may progress to cancer. Cervical cancer may be detected in most cases in its earliest stages by a Pap smear. Depending on the progress of the disease, treatment may consist of excision of lesions, radiotherapy, chemotherapy, and hysterectomy.

Pelvic Inflammatory Disease (PID)
Pelvic inflammatory disease (PID) is a collective term for any extensive bacterial infection of the pelvic organs, especially the uterus, uterine (Fallopian) tubes, or ovaries. A vaginal or uterine infection may spread into the uterine tube (*salpingitis*) or even farther into the abdominal cavity, where it infects the peritoneum (*peritonitis*). PID is most commonly caused by gonorrhea, but any bacterium can trigger infection. Often the early symptoms of PID, which include increased vaginal discharge and pelvic pain, occur just after menstruation. As infection spreads, fever may develop in advanced cases along with painful abscesses of the reproductive organs. Early treatment with antibiotics (tetracycline or penicillin) can stop the spread of PID.

MEDICAL TERMINOLOGY

Leukorrhea (*leuco* = white; *rrhea* = discharge)   A nonbloody vaginal discharge that may occur at any age and affects most women at some time.

Neoplasia (*neo* = new; *plas* = form, grow)   A condition characterized by the presence of new growths (tumors).

Oophorectomy (*oophoro* = ovary; *ectomy* = removal of)   Excision of an ovary. Bilateral oophorectomy refers to the removal of both ovaries.

Pruritus   Itching.

Salpingectomy (*salpingo* = tube)   Excision of a uterine (Fallopian) tube.

Smegma (*smegma* = soap)   The secretion, consisting principally of desquamated epithelial cells, found chiefly about the external genitalia and especially under the foreskin of the male.

Vaginitis   Inflammation of the vagina.

STUDY OUTLINE

Male Reproductive System (p. 706)

1. Reproduction is the process by which genetic material is passed on from one generation to the next.
2. The organs of reproduction are grouped as: gonads (produce gametes), ducts (transport and store gametes), and accessory glands (produce materials that support gametes).
3. The male structures of reproduction include the testes, ductus epididymis, ductus (vas) deferens, ejaculatory duct, urethra, seminal vesicles, prostate gland, bulbourethral (Cowper's) glands, and penis.

Scrotum (p. 706)

1. The scrotum is a cutaneous outpouching of the abdomen that supports the testes.
2. It regulates the temperature of the testes by contraction of the cremaster muscle to elevate them closer to the pelvic cavity.

Testes (p. 707)

1. The testes are oval-shaped glands (gonads) in the scrotum containing seminiferous tubules, in which sperm cells are made; sustentacular (Sertoli) cells, which nourish sperm cells; and interstitial endocrinocytes (cells of Leydig), which produce the male sex hormone testosterone.
2. Failure of the testes to descend is called cryptorchidism.
3. Ova and sperm are collectively called gametes, or sex cells, and are produced in gonads.
4. Uninucleated somatic cells divide by mitosis, the process in which each daughter cell receives the full complement of 23 chromosome pairs (46 chromosomes). Somatic cells are said to be diploid ($2n$).
5. Immature gametes divide by meiosis, in which the pairs of chromosomes are split so that the mature gamete has only 23 chromosomes. It is said to be haploid (n).
6. Spermatogenesis occurs in the testes. It results in the formation of four haploid spermatozoa.
7. The spermatogenesis sequence consists of reduction division, equatorial division, and spermiogenesis.
8. Mature spermatozoa consist of a head, midpiece, and tail. Their function is to fertilize an ovum.
9. At puberty, GnRF stimulates anterior pituitary secretion of FSH and LH. FSH initiates spermatogenesis, and LH assists spermatogenesis and stimulates production of testosterone.
10. Testosterone controls the growth, development, and maintenance of sex organs; stimulates bone growth, protein anabolism, and sperm maturation; and stimulates development of male secondary sex characteristics.
11. Inhibin is produced by sustentacular cells. Its inhibition of FSH helps to regulate the rate of spermatogenesis.

Ducts (p. 713)

1. The duct system of the testes includes the seminiferous tubules, straight tubules, and rete testis.
2. Sperm are transported out of the testes through the efferent ducts.
3. The ductus epididymis is lined by stereocilia and is the site of sperm maturation and storage.
4. The ductus (vas) deferens stores sperm and propels them toward the urethra during ejaculation.
5. Alteration of the ductus (vas) deferens to prevent fertilization is called vasectomy.
6. The ejaculatory ducts are formed by the union of the ducts from the seminal vesicles and ductus (vas) deferens and eject spermatozoa into the prostatic urethra.
7. The male urethra is subdivided into three portions: prostatic, membranous, and spongy (cavernous).

Accessory Sex Glands (p. 714)

1. The seminal vesicles secrete an alkaline, viscous fluid that constitutes about 60 percent of the volume of semen and contributes to sperm viability.
2. The prostate gland secretes an alkaline fluid that constitutes about 13 to 33 percent of the volume of semen and contributes to sperm motility.
3. The bulbourethral (Cowper's) gland secretes mucus for lubrication and a substance that neutralizes acid.
4. Semen (seminal fluid) is a mixture of spermatozoa and accessory gland secretions that provides the fluid in which spermatozoa are transported, provides nutrients, and neutralizes the acidity of the male urethra and female vagina.

Penis (p. 716)

1. The penis is the male organ of copulation that consists of a root, body, and glans penis.
2. Expansion of its blood sinuses under the influence of sexual excitation is called erection.

Female Reproductive System (p. 717)

1. The female organs of reproduction include the ovaries (gonads), uterine tubes, uterus, vagina, and vulva.
2. The mammary glands are considered part of the reproductive system.

Ovaries (p. 717)

1. The ovaries are female gonads located in the upper pelvic cavity, on either side of the uterus.
2. They produce ova, discharge ova (ovulation), and secrete estrogens, progesterone, and relaxin.
3. Oogenesis occurs in the ovaries. It results in the formation of a single haploid ovum.
4. The oogenesis sequence consists of reduction division, equatorial division, and maturation.

Uterine (Fallopian) Tubes (p. 720)

1. The uterine (Fallopian) tubes transport ova from the ovaries to the uterus and are the normal sites of fertilization.
2. Implantation outside the uterus (pelvic or tubular) is called an ectopic pregnancy.

Uterus (p. 721)

1. The uterus is an inverted, pear-shaped organ that functions in menstruation, implantation of a fertilized ovum, development of a fetus during pregnancy, and labor.
2. The uterus is normally held in position by a series of ligaments.
3. Histologically, the uterus consists of an outer perimetrium, middle myometrium, and inner endometrium.

Endocrine Relations: Menstrual and Ovarian Cycles (p. 722)

1. The function of the menstrual cycle is to prepare the endometrium each month for the reception of a fertilized egg. The ovarian cycle is associated with the maturation of an ovum each month.
2. The menstrual and ovarian cycles are controlled by GnRF, which stimulates the release of FSH and LH.

3. FSH stimulates the initial development of ovarian follicles and secretion of estrogens by the ovaries. LH stimulates further development of ovarian follicles, ovulation, and the secretion of estrogens and progesterone by the ovaries.
4. Estrogens stimulate the growth, development, and maintenance of female reproductive structures; stimulate the development of secondary sex characteristics; regulate fluid and electrolyte balance; and stimulate protein anabolism.
5. Progesterone works with estrogens to prepare the endometrium for implantation and the mammary glands for milk secretion.
6. Relaxin relaxes the symphysis pubis and helps dilate the uterine cervix to facilitate delivery, and increases sperm motility.
7. During the menstrual phase, the functionalis layer of the endometrium is shed with a discharge of blood, tissue fluid, mucus, and epithelial cells. Primary follicles develop into secondary follicles.
8. During the preovulatory phase, endometrial repair occurs. A secondary follicle develops into a vesicular (Graafian) follicle. Estrogens are the dominant ovarian hormones.
9. Ovulation is the rupture of a vesicular (Graafian) follicle and the release of an immature ovum into the pelvic cavity brought about by inhibition of FSH and a surge of LH.
10. During the postovulatory phase, the endometrium thickens in anticipation of implantation. Progesterone is the dominant ovarian hormone.
11. If fertilization and implantation do not occur, the corpus luteum degenerates and low levels of estrogens and progesterone initiate another menstrual and ovarian cycle.
12. If fertilization and implantation do occur, the corpus luteum is maintained by placental HCG, and the corpus luteum and placenta secrete estrogens and progesterone to support pregnancy and breast development for lactation.
13. The female climacteric is the time immediately before menopause, the cessation of the sexual cycles.

Vagina (p. 726)
1. The vagina is a passageway for the menstrual flow, the receptacle for the penis during sexual intercourse, and the lower portion of the birth canal.
2. It is capable of considerable distension to accomplish its functions.

Vulva (p. 727)
1. The vulva is a collective term for the external genitals of the female.
2. It consists of the mons pubis, labia majora, labia minora, clitoris, vestibule, vaginal and urethral orifices, and the paraurethral (Skene's), greater vestibular (Bartholin's), and lesser vestibular glands.

Perineum (p. 728)
1. The perineum is a diamond-shaped area at the inferior end of the trunk between the thighs and buttocks.
2. An incision in the perineal skin prior to delivery is called an episiotomy.

Mammary Glands (p. 728)
1. The mammary glands are modified sweat glands (branched tubuloalveolar) over the pectoralis major muscles. Their function is to secrete and eject milk (lactation).
2. Mammary gland development is dependent on estrogens and progesterone.
3. Milk secretion is due mainly to PRL, and milk ejection is stimulated by OT.

Sexual Intercourse (p. 730)
1. The role of the male in the sex act involves erection, lubrication, and orgasm.
2. The female role also involves erection, lubrication, and orgasm (climax).

Birth Control (p. 731)
1. Methods include removal of gonads and uterus, sterilization (vasectomy, tubal ligation, laparoscopic technique), and contraception (natural, mechanical, and chemical).
2. Contraceptive pills of the combination type contain estrogens and progesterone in concentrations that decrease the secretion of FSH and LH, and thereby inhibit ovulation.

Aging and the Reproductive Systems (p. 732)
1. In the male, decreased levels of testosterone decrease muscle strength, sexual desire, and viable sperm; prostate disorders are common.
2. In the female, levels of progesterone and estrogens decrease, resulting in changes in menstruation; uterine and breast cancer increase in incidence.

Developmental Anatomy of the Reproductive Systems (p. 733)
1. The gonads develop from intermediate mesoderm and are differentiated into ovaries or testes by about the eighth week.
2. The external genitals develop from the genital tubercle.

Disorders: Homeostatic Imbalances (p. 735)
1. Sexually transmitted diseases (STDs) are diseases spread by sexual contact and include gonorrhea, syphilis, genital herpes, trichomoniasis, and nongonococcal urethritis (NGU).
2. Conditions that affect the prostate are prostatitis, enlarged prostate, and tumors.
3. Impotence is the inability of the male to attain or hold an erection long enough for intercourse.
4. Infertility is the inability of a male's sperm to fertilize an ovum.
5. Menstrual disorders include amenorrhea, dysmenorrhea, abnormal bleeding, and premenstrual syndrome (PMS).
6. Toxic shock syndrome (TSS) includes widespread homeostatic imbalances and is a reaction to toxins produced by *Staphylococcus aureus.*
7. Ovarian cysts are tumors that contain fluid.
8. Endometriosis refers to the growth of uterine tissue outside the uterus.
9. Female infertility is the inability of the female to conceive.
10. The mammary glands are susceptible to benign fibroadenomas and malignant tumors. The removal of a malignant breast, pectoral muscles, and lymph nodes is called a radical mastectomy.
11. Cervical cancer can be diagnosed by a Pap test.
12. Pelvic inflammatory disease (PID) refers to bacterial infection of pelvic organs.

REVIEW QUESTIONS

1. Define reproduction. Describe how the reproductive organs are classified and list the male and female organs of reproduction.
2. Describe the function of the scrotum in protecting the testes from temperature fluctuations. What is cryptorchidism?
3. Describe the internal structure of a testis. Where are the sperm cells made?
4. Describe the principal events of spermatogenesis. Why is meiosis important? Distinguish between haploid and diploid cells.
5. Identify the principal parts of a spermatozoon. List the functions of each.
6. Explain the effects of FSH and LH on the male reproductive system. How are these hormones controlled by GnRF?
7. Describe the physiological effects of testosterone and inhibin on the male reproductive system. How is testosterone level controlled?
8. Which ducts are involved in transporting sperm within the testes?
9. Describe the location, structure, and functions of the ductus epididymis, ductus (vas) deferens, and ejaculatory duct.
10. What is a vasectomy?
11. What is the spermatic cord? What is an inguinal hernia?
12. Give the location of the three subdivisions of the male urethra.
13. Trace the course of spermatozoa through the male system of ducts from the seminiferous tubules through the urethra.
14. Briefly explain the locations and functions of the seminal vesicles, prostate gland, and bulbourethral (Cowper's) glands.
15. What is semen? What is its function? What criteria are evaluated in a semen analysis?
16. How is the penis structurally adapted as an organ of copulation? What is circumcision?
17. How does an erection occur?
18. How are the ovaries held in position in the pelvic cavity? Describe the microscopic structure of an ovary.
19. What are the functions of the ovaries?
20. Describe the principal events of oogenesis.
21. Where are the uterine (Fallopian) tubes located? What is their function? What is an ectopic pregnancy?
22. Diagram the principal parts of the uterus.
23. Describe the arrangement of ligaments that hold the uterus in its normal position. What is retroflexion?
24. Discuss the blood supply to the uterus. Why is an abundant blood supply important?
25. Describe the histology of the uterus.
26. Define menstrual cycle and ovarian cycle. What is the function of each?
27. What is the function of each of the following in the menstrual and ovarian cycles: GnRF, FSH, LH, estrogens, progesterone, relaxin?
28. Briefly outline the major events of each phase of the menstrual cycle and correlate them with the events of the ovarian cycle.
29. Prepare a labeled diagram of the principal hormonal interactions involved in the menstrual and ovarian cycles.
30. What is menarche? Define female climacteric and menopause.
31. What is the function of the vagina? Describe its histology.
32. List the parts of the vulva and the functions of each part.
33. What is the perineum? Define episiotomy.
34. Describe the structure of the mammary glands. How are they supported?
35. Describe the passage of milk from the alveolar cells of the mammary gland to the nipple.
36. Explain the roles of estrogens and progesterone in the development of the mammary glands.
37. Define lactation. How is it controlled?
38. How is breast cancer detected?
39. Explain the role of the male's erection, lubrication, and orgasm in the sex act. How do the female's erection, lubrication, and orgasm (climax) contribute to the sex act?
40. Briefly describe the various methods of birth control and the effectiveness of each.
41. Explain the effects of aging on the reproductive systems.
42. Describe the development of the reproductive systems.
43. Define a sexually transmitted disease (STD). Describe the cause, clinical symptoms, and treatment of gonorrhea, syphilis, genital herpes, trichomoniasis, and nongonococcal urethritis (NGU).
44. Describe several disorders that affect the prostate gland.
45. What are some of the causes of amenorrhea, dysmenorrhea, and abnormal uterine bleeding?
46. Describe the clinical symptoms of premenstrual syndrome (PMS) and toxic shock syndrome (TSS).
47. What are ovarian cysts? Define endometriosis.
48. How is breast cancer treated?
49. What is cervical cancer? Relate the condition to cervical dysplasia.
50. Define pelvic inflammatory disease (PID).
51. Refer to the glossary of medical terminology at the end of the chapter. Be sure that you can define each term.

Development and Inheritance

29

Developmental anatomy is the study of the sequence of events from the fertilization of an ovum to the formation of an adult organism. As we look at the sequence from fertilization to birth, we will consider fertilization, implantation, embryonic development, fetal growth, gestation, parturition, and labor.

DEVELOPMENT DURING PREGNANCY

Once spermatozoa and ova are developed through meiosis and maturation, and the spermatozoa are deposited in the vagina, pregnancy can occur. **Pregnancy** is a sequence of events that normally includes fertilization, implantation, embryonic growth, and fetal growth that terminates in birth.

FERTILIZATION AND IMPLANTATION

Fertilization

The term **fertilization** is applied to the penetration of an ovum by a spermatozoon and the subsequent union of the sperm nucleus and the nucleus of the ovum. Of the hundreds of millions of sperm cells introduced into the vagina, very few, perhaps only several hundred to several thousand, arrive in the vicinity of the ovum. Fertilization normally occurs in the uterine (Fallopian) tube when the ovum is about one-third of the way down the tube, usually within 24 hours after ovulation. Peristaltic contractions and the action of cilia transport the ovum through the uterine tube. The mechanism by which sperm reach the uterine tube is apparently related to several factors. Sperm probably swim up the female tract by means of whiplike movements of their flagella. In addi-

tion, the acrosome of sperm produces an enzyme called *acrosin,* which stimulates sperm motility and migration within the female reproductive tract. Finally, sperm are probably transported by muscular contractions of the uterus.

In addition to assisting in the transport of sperm, the female reproductive tract also confers on sperm the capacity to fertilize an ovum. Although sperm undergo maturation in the epididymis, they are still not able to fertilize an ovum until they have remained in the female reproductive tract for several hours. The functional changes that sperm undergo in the female reproductive tract that allow them to fertilize an ovum are referred to as **capacitation.** During this process, it is believed that the enzymes hyaluronidase and proteinases are secreted by the acrosomes. The enzymes help dissolve the intercellular materials covering the ovum, the gelatinous covering called the **zona pellucida** (pe-LOO-si-da) and several layers of cells, the innermost of which are follicle cells, known as the **corona radiata** (Figure 29-1a). Once this is accomplished, only one spermatozoon enters and fertilizes an ovum, because once union is achieved, the electrical changes in the surface of the ovum block the entry of other sperm, and enzymes produced by the fertilized ovum alter receptor sites so that sperm already bound are detached and others are prevented from binding. In this way, polyspermy, the fertilization of an ovum by more than one spermatozoon, is prevented.

When a spermatozoon has entered the ovum, the tail is shed and the nucleus in the head develops into a structure called the **male pronucleus.** The nucleus of the ovum develops into a **female pronucleus** (Figure 29-1b). After the pronuclei are formed, they fuse to produce a **segmentation nucleus.** The segmentation nucleus contains 23 chro-

(a)

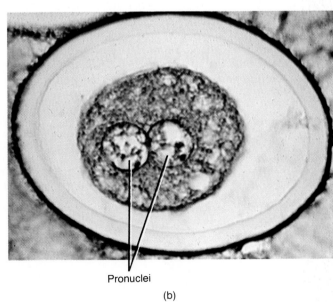

(b)

FIGURE 29-1 Fertilization and implantation. (a) Photomicrograph of a spermatozoon moving through the zona pellucida on its way to reach the nucleus of the ovum. (Courtesy of Popperfoto.) (b) Photomicrograph showing male and female pronuclei. (Courtesy of Carolina Biological Supply Company.)

mosomes from the male pronucleus and 23 chromosomes from the female pronucleus. Thus the fusion of the haploid pronuclei restores the diploid number. The fertilized ovum, consisting of a segmentation nucleus, cytoplasm, and enveloping membrane, is called a **zygote.**

CLINICAL APPLICATION

Dizygotic (fraternal) twins are produced from the independent release of two ova and the subsequent fertilization of each by different spermatozoa. They are the same age and are in the uterus at the same time, but they are genetically as dissimilar as any other siblings. They may or may not be the same sex. **Monozygotic (identical) twins** are derived from a single fertilized ovum that splits at an early stage in development. They contain the same genetic material and are always the same sex.

Formation of the Morula

Immediately after fertilization, rapid cell division of the zygote takes place. This early division of the zygote is called **cleavage.** During this time the dividing cells are contained by the zona pellucida. Although cleavage increases the number of cells, it does not result in an increase in the size of the developing organism.

The first cleavage is completed after about 36 hours, and each succeeding division takes slightly less time (Figure 29-2). By the second day after conception, the second cleavage is completed. By the end of the third day there are 16 cells. The progressively smaller cells produced are called **blastomeres** (BLAS-tō-mērz). A few days after fertilization the successive cleavages have produced a solid mass of cells, the **morula** (MOR-yoo-la) or mulberry, which is about the same size as the original zygote.

Development of the Blastocyst

As the number of cells in the morula increases, it moves from the original site of fertilization down through the ciliated uterine tube toward the uterus and enters the

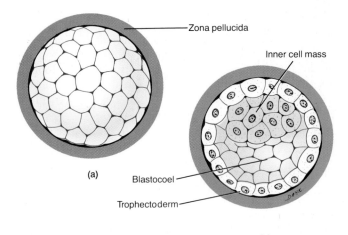

FIGURE 29-3 Blastocyst. (a) External view. (b) Internal view.

uterine cavity. By this time, the dense cluster of cells is altered to form a hollow ball of cells. The mass is now referred to as a **blastocyst** (Figure 29-3).

The blastocyst is differentiated into an outer covering of cells called the **trophectoderm** (trōf-EK-tō-derm), an **inner cell mass,** and an internal fluid-filled cavity called the **blastocoel** (BLAS-tō-sēl). The trophectoderm ultimately forms part of the membranes composing the fetal portion of the placenta; the inner cell mass develops into the embryo.

Implantation

The blastocyst remains free within the cavity of the uterus from 2 to 4 days before it actually attaches to the uterine wall. During this time, nourishment is provided by secretions of the endometrium, sometimes called uterine milk. The attachment of the blastocyst to the endometrium occurs 7 to 8 days after fertilization and is called **implantation** (Figure 29-4). At this time, the endometrium is in its postovulatory phase. During implantation, the cells of the trophectoderm secrete enzymes that enable the blastocyst to penetrate the uterine lining. The release of enzymes by the trophectoderm digests and liquefies

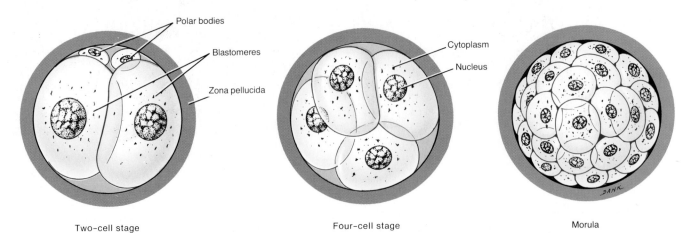

FIGURE 29-2 Formation of the morula.

FIGURE 29-4 Implantation. (a) External view of the blastocyst in relation to the endometrium of the uterus about 5 days after fertilization. (b) Internal view of the blastocyst in relation to the endometrium about 6 days after fertilization. (c) Internal view of the blastocyst at implantation about 7 days after fertilization. (d) Photomicrograph of implantation. (From *From Conception to Birth: The Drama of Life's Beginnings* by Roberts Rugh, Landrum B. Shettles with Richard Einhorn. Copyright © 1971 by Roberts Rugh and Landrum B. Shettles. By permission of Harper & Row, Publishers, Inc.)

Labels in figure:
- Endometrium of uterus
- Orifice of duct of uterine gland
- Blastocyst
- (a)
- Endometrium of uterus (sectioned)
- Trophectoderm
- Inner cell mass
- Blastocoel
- (b)
- Endometrium of uterus (sectioned)
- Blastocoel
- Amniotic cavity
- Ectoderm
- Endoderm
- Embryonic disc
- (c)
- Blastocyst
- Endometrium
- (d)

the endometrial cells. The fluid and nutrients further nourish the burrowing blastocyst for about a week after implantation. Eventually, the blastocyst becomes buried in the endometrium, usually on the posterior wall of the fundus or body of the uterus. The blastocyst becomes oriented so that the inner cell mass is toward the endometrium. Eventually, nutrients are delivered through the placenta for the subsequent growth and development of the embryo and fetus.

A summary of the principal events associated with fertilization and implantation is shown in Figure 29-5.

CLINICAL APPLICATION

In the early months of pregnancy, some females develop **hyperemesis gravidarum (morning sickness)**, characterized by nausea and vomiting. Although the exact cause is unknown, it is possible that the degenerative products of digested portions of the endometrium during implantation may be responsible. Another possible cause is high levels of human chorionic gonadotropin (HCG) secreted by the placenta.

EXTERNAL HUMAN FERTILIZATION

On July 12, 1978, Louise Joy Brown was born near Manchester, England. Her birth was the first recorded case of **external human fertilization (in vitro fertilization)**—fertilization in a glass dish. The procedure developed for external human fertilization is carried out as follows. The female is given follicle-stimulating hormone (FSH) soon after menstruation, so that several ova, rather than the typical single one, will be produced (superovulation). Administration of luteinizing hormone (LH) may also ensure

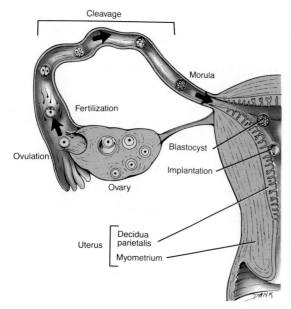

FIGURE 29-5 Summary of events associated with fertilization and implantation.

the maturation of the ova. Next, a small incision is made near the umbilicus, and the ova are aspirated from the follicles and placed in a medium that simulates the fluids in the female reproductive tract. The ova are then transferred to a solution of the male's sperm. Once fertilization has taken place, the fertilized ovum is put in another medium and is observed for cleavage. When the fertilized ovum reaches the 8-cell or 16-cell stage, it is introduced into the uterus for implantation and subsequent growth. The growth and developmental sequences that occur are similar to those in internal fertilization. It is also possible to freeze unused embryos (cryopreservation) to permit parents a successive pregnancy several years later or allow a second attempt at implantation if the first attempt is unsuccessful.

EMBRYO TRANSFER

Embryo transfer is an alternate to external human fertilization. It is a procedure in which a husband's seminal fluid is used to artificially inseminate a fertile ovum donor and, following fertilization, the morula or blastocyst is transferred from the donor to the infertile wife who carries it to term. Embryo transfer is indicated for females who are infertile, who have surgically untreatable blocked uterine (Fallopian) tubes; or who are afraid of passing on their own genes because they are carriers of a serious genetic disorder.

In the procedure, the donor is monitored to ascertain the time of ovulation by checking her blood levels of luteinizing hormone (LH) and by ultrasound. The wife is also monitored to make sure that her ovarian cycle is synchronized with that of the donor. Once ovulation occurs in the donor, she is artificially inseminated with the husband's seminal fluid. Four days later, a morula or blastocyst is flushed from the donor's uterus through a soft plastic catheter and transferred to the uterus of the wife, where it grows and develops until the time of birth. Embryo transfer is an office procedure that requires no anesthesia and may be performed in a few minutes.

GAMETE INTRAFALLOPIAN TRANSFER (GIFT) AND TRANSVAGINAL OOCYTE RETRIEVAL

Recently two alternatives to external human fertilization and embryo transfer have been developed. The first is called **gamete intrafallopian transfer (GIFT)**. The technique is essentially an attempt to mimic the normal process of conception by uniting sperm and ovum in the prospective mother's uterine (Fallopian) tubes. In the procedure, the female is given FSH and LH to stimulate the production of several ova. The ova are aspirated with a laparoscope fitted with a suction device, mixed with a solution of the male's sperm outside the body, and then immediately inserted into the uterine (Fallopian) tubes.

In the second technique, called **transvaginal oocyte retrieval**, the female is given hormones to produce several ova, a needle is placed through the vaginal wall and guided to the ovaries by ultrasound, suction is applied to the needle, and the ova are removed and placed in a solution outside the body. The male's sperm are added to the solution, and the fertilized ova are then implanted in the uterus.

EMBRYONIC DEVELOPMENT

The first 2 months of development are generally considered the **embryonic period.** During this period the developing human is called an **embryo.** The study of development from the fertilized egg through the eighth week in utero is referred to as **embryology.** The months of development after the second month are considered the **fetal period,** and during this time the developing human is called a **fetus.** By the end of the embryonic period the rudiments of all the principal adult organs are present, the embryonic membranes are developed, and the placenta is functioning.

BEGINNINGS OF ORGAN SYSTEMS

Following implantation, the inner cell mass of the blastocyst begins to differentiate into the three **primary germ layers:** ectoderm, endoderm, and mesoderm. They are the embryonic tissues from which all tissues and organs of the body will develop. The various movements of cell groups leading to the establishment of the primary germ layers are referred to as **gastrulation.**

In the human, the germ layers form so quickly that it is difficult to determine the exact sequence of events. Before implantation, a layer of **ectoderm** (the trophectoderm) already has formed around the blastocoel (see Figure 29-4b). The trophectoderm will become part of the chorion, one of the fetal membranes. Within 8 days after fertilization, the top layer of cells of the inner cell mass proliferates and forms the amnion (another fetal membrane) and a space, the **amniotic (amnionic) cavity,** over

the inner cell mass. The bottom layer of the inner cell mass develops into the **endoderm.**

About the twelfth day after fertilization, striking changes appear (Figure 29-6a). The cells below the amniotic cavity are called the **embryonic disc.** They will form the embryo. At this stage, the embryonic disc contains ectodermal and endodermal cells; the mesodermal cells are scattered external to the disc. The cells of the endodermal layer have been dividing rapidly, so that groups of them now extend downward in a circle, forming the yolk sac, another fetal membrane. The cells of the **mesoderm,** which develop between the ectodermal and endodermal layers, also have been dividing, and many have left the area of the embryonic disc and can be seen around the structures that are becoming fetal membranes.

About the fourteenth day, the cells of the embryonic disc differentiate into three distinct layers: the upper ectoderm, the middle mesoderm, and the lower endoderm (Figure 29-6b). The mesoderm in the disc soon splits into two layers, and the space between the layers becomes the **extraembryonic coelom.**

As the embryo develops (Figure 29-6c), the endoderm becomes the epithelial lining of the digestive tract, respiratory tract, and a number of other organs. The mesoderm forms the peritoneum, muscle, bone, and other connective tissue. The ectoderm develops into the skin and nervous system. Exhibit 29-1 provides more details about the fates of these primary germ layers.

Embryonic Membranes

During the embryonic period, the **embryonic membranes** form (Figure 29-7). These membranes lie outside the embryo and protect and nourish the embryo and, later, the

---

EXHIBIT 29-1 STRUCTURES PRODUCED BY THE THREE PRIMARY GERM LAYERS

ENDODERM	MESODERM	ECTODERM
Epithelium of digestive tract (except the oral cavity and anal canal) and the epithelium of its glands.	All skeletal, most smooth, and all cardiac muscle.	All nervous tissue.
		Epidermis of skin.
Epithelium of urinary bladder, gallbladder, and liver.	Cartilage, bone, and other connective tissues.	Hair follicles, arrector pili muscles, nails, and epithelium of skin glands (sebaceous and sudoriferous).
	Blood, bone marrow, and lymphoid tissue.	
Epithelium of pharynx, auditory (Eustachian) tube, tonsils, larynx, trachea, bronchi, and lungs.	Endothelium of blood vessels and lymphatics.	Lens, cornea, and optic (II) nerve of eye and internal eye muscles.
Epithelium of thyroid, parathyroid, pancreas, and thymus glands.	Dermis of skin.	Internal and external ear.
	Fibrous tunic and vascular tunic of eye.	Neuroepithelium of sense organs.
Epithelium of prostate and bulbourethral (Cowper's) glands, vagina, vestibule, urethra, and associated glands such as the greater (Bartholin's) vestibular and lesser vestibular glands.	Middle ear.	Epithelium of oral cavity, nasal cavity, paranasal sinuses, salivary glands, and anal canal.
	Mesothelium of ventral body and joint cavities.	Epithelium of pineal gland, hypophysis, and adrenal medulla.
	Epithelium of kidneys and ureters.	
	Epithelium of adrenal cortex.	
	Epithelium of gonads and genital ducts.	

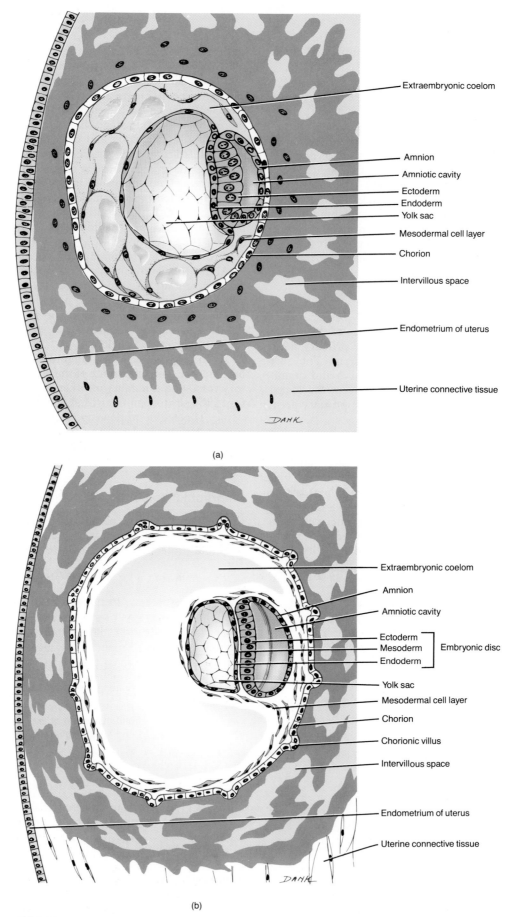

(a)

(b)

FIGURE 29-6 Formation of the primary germ layers and associated structures. (a) Internal view of the developing embryo about 12 days after fertilization. (b) Internal view of the developing embryo about 14 days after fertilization.

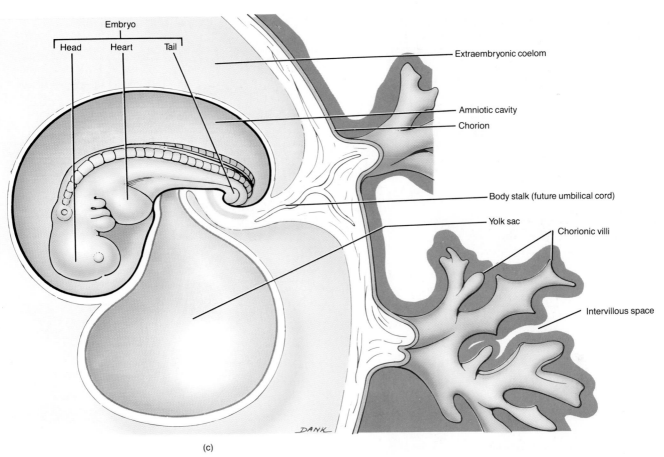

(c)

FIGURE 29-6 (*Continued*) (c) External view of the developing embryo about 25 days after fertilization.

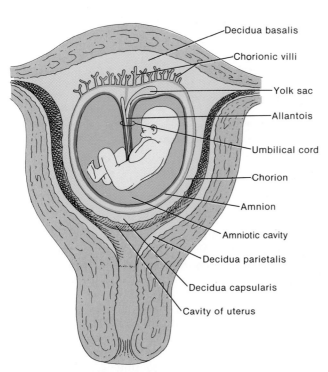

FIGURE 29-7 Embryonic membranes.

fetus. The membranes are the yolk sac, amnion, chorion, and allantois.

The **yolk sac** is an endoderm-lined membrane that, in many species, provides the primary or exclusive nutrient for the embryo (Figure 29-8; see also Figures 29-6c, and 29-7). However, the human embryo receives its nourishment from the endometrium, and the yolk sac remains small. During an early stage of development it becomes a nonfunctional part of the umbilical cord.

The **amnion** is a thin, protective membrane that initially overlies the embryonic disc and is formed by the eighth day following fertilization. As the embryo grows, the amnion entirely surrounds the embryo and becomes filled with **amniotic fluid** (Figure 29-8). Amniotic fluid serves as a shock absorber for the fetus. The amnion usually ruptures just before birth and with its fluid constitutes the "bag of waters."

The **chorion** (KŌ-rē-on) is derived from the trophectoderm of the blastocyst and its associated mesoderm. It surrounds the embryo and, later, the fetus. Eventually the chorion becomes the principal embryonic part of the placenta, the structure through which materials are exchanged between mother and fetus. The amnion also surrounds the fetus and eventually fuses to the inner layer of the chorion.

The **allantois** (a-LAN-tō-is) is a small vascularized membrane. Later its blood vessels serve as connections

FIGURE 29-8 Ten-week fetus in which the amnion, yolk sac, umbilical cord, and placenta are clearly visible. (From *From Conception to Birth: The Drama of Life's Beginnings* by Roberts Rugh, Landrum B. Shettles with Richard Einhorn. Copyright © 1971 by Roberts Rugh and Landrum B. Shettles. By permission of Harper & Row, Publishers, Inc.)

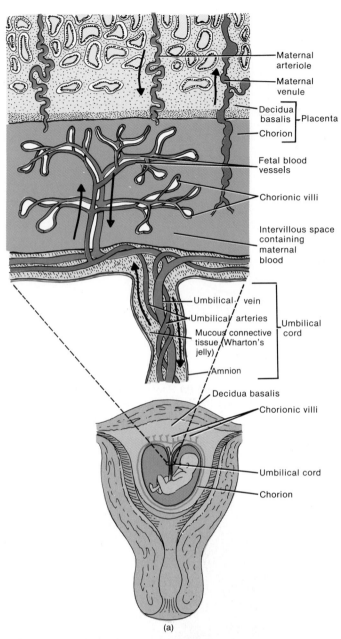

FIGURE 29-9 Placenta and umbilical cord. (a) Diagram of the structure of the placenta and umbilical cord.

in the placenta between mother and fetus. This connection is the umbilical cord.

Placenta and Umbilical Cord

Development of the placenta, the third major event of the embryonic period, is accomplished by the third month of pregnancy. The **placenta** has the shape of a flat cake when fully developed and is formed by the chorion of the embryo and a portion of the endometrium (decidua basalis) of the mother (Figure 29-9). It provides an exchange of nutrients and wastes between fetus and mother and secretes hormones necessary to maintain pregnancy.

If implantation occurs, a portion of the endometrium becomes modified and is known as the **decidua** (dē-SID-yoo-a). The decidua includes all but the deepest layer of the endometrium and is shed when the fetus is delivered. Different regions of the decidua, all areas of the stratum functionalis, are named on the basis of their positions relative to the site of the implanted, fertilized ovum (Figure 29-10). The *decidua parietalis* (pa-rī-e-TAL-is) is the portion of the modified endometrium that lines the entire pregnant uterus, except for the area where the placenta is forming. The *decidua capsularis* is the portion

of the endometrium between the embryo and the uterine cavity. The *decidua basalis* is the portion of the endometrium between the chorion and the stratum basalis of the uterus. The decidua basalis becomes the maternal part of the placenta.

During embryonic life, fingerlike projections of the chorion, called **chorionic villi** (kō'-rē-ON-ik VIL-ī), grow into the decidua basalis of the endometrium (see also Figures 29-6 and 29-7). These will contain fetal blood vessels of the allantois. They continue growing until they are bathed in maternal blood sinuses called **intervillous** (in-ter-VIL-us) **spaces.** Thus, maternal and fetal blood vessels are brought into proximity. It should be noted, however, that maternal and fetal blood do not normally

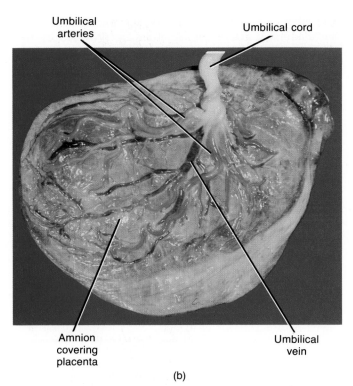

Umbilical
arteries

Umbilical cord

Amnion
covering
placenta

Umbilical
vein

(b)

FIGURE 29-9 (*Continued*) (b) Photograph of the fetal aspect of the placenta. Although the umbilical arteries are red, they carry deoxygenated blood. Similarly, although the umbilical vein is blue, it carries oxygenated blood. (Photograph courtesy of C. Yokochi and J. W. Rohen, *Photographic Anatomy of the Human Body,* 2nd ed., 1978, IGAKU-SHOIN, Ltd., Tokyo, New York.)

mix. Oxygen and nutrients from the mother's blood diffuse into the capillaries of the villi. From the capillaries the nutrients circulate into the umbilical vein. Wastes leave the fetus through the umbilical arteries, pass into the capillaries of the villi, and diffuse into the maternal blood. The **umbilical cord** consists of an outer layer of amnion containing the umbilical arteries and umbilical vein, supported internally by mucous connective tissue (Wharton's jelly) from the allantois.

CLINICAL APPLICATION

At delivery, the placenta detaches from the uterus and is referred to as the **afterbirth.** At this time, the umbilical cord is severed, leaving the baby on its own. The scar that marks the site of the entry of the fetal umbilical cord into the abdomen is the **umbilicus (navel).**

Pharmaceutical houses use human placenta as a source of hormones, drugs, and blood. Portions of placentas are also used for burn coverage. The placental and umbilical cord veins are used in blood vessel grafts.

FETAL GROWTH

During the **fetal period,** organs established by the primary germ layers grow rapidly. The organism takes on a human appearance. A summary of changes associated with the fetal period is presented in Exhibit 29-2.

Placenta

Umbilical cord

Amnion

(c)

FIGURE 29-9 (*Continued*) (c) Twelve-week fetus showing the relationship of the placenta to the umbilical cord. (From *From Conception to Birth: The Drama of Life's Beginnings* by Roberts Rugh, Landrum B. Shettles with Richard Einhorn. Copyright © 1971 by Roberts Rugh and Landrum B. Shettles. By permission of Harper & Row, Publishers, Inc.)

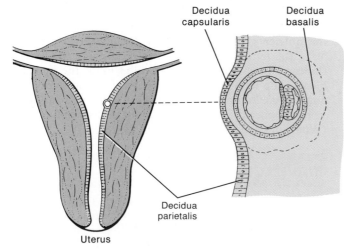

FIGURE 29-10 Regions of the decidua.

HORMONES OF PREGNANCY

The corpus luteum is maintained for at least the first 3 or 4 months of pregnancy, during which time it continues to secrete **estrogens** and **progesterone.** Both these hormones maintain the lining of the uterus during pregnancy and prepare the mammary glands to secrete milk. The amount secreted by the corpus luteum, however, is only slightly more than that produced after ovulation in a normal menstrual cycle. The high levels of estrogens and progesterone needed to maintain pregnancy and develop the mammary glands for lactation are provided by the placenta.

The chorion of the placenta secretes **human chorionic gonadotropin (HCG).** The primary role of HCG seems to be to maintain the activity of the corpus luteum, especially with regard to continuous progesterone secretion—an activity necessary for the continued attachment of the fetus to the lining of the uterus (Figure 29-11).

CLINICAL APPLICATION

Human chorionic gonadotropin is excreted in the urine of pregnant women from about the eighth day of pregnancy, reaching its peak of excretion about the eighth week of pregnancy. The HCG level decreases sharply during the fourth and fifth months and then levels off until childbirth. Excretion of HCG in the urine serves as the basis for most home **pregnancy tests.** HCG can be detected in blood by a laboratory test even before a period is missed.

The placenta begins to secrete estrogens and progesterone no later than the sixtieth day of pregnancy. They are secreted in increasing quantities until the time of birth. By the fourth month, when the placenta is established, the secretion of HCG is greatly reduced because the secretions of the corpus luteum are no longer essential.

EXHIBIT 29-2 CHANGES ASSOCIATED WITH FETAL GROWTH

END OF MONTH	APPROXIMATE SIZE AND WEIGHT	REPRESENTATIVE CHANGES
1	0.6 cm (3/16 inch)	Eyes, nose, and ears not yet visible. Backbone and vertebral canal form. Small buds that will develop into arms and legs form. Heart forms and starts beating. Body systems begin to form.
2	3 cm (1¼ inches) 1 g (1/30 oz)	Eyes far apart, eyelids fused, nose flat. Ossification begins. Limbs become distinct as arms and legs. Digits are well formed. Major blood vessels form. Many internal organs continue to develop.
3	7.5 cm (3 inches) 28 g (1 oz)	Eyes almost fully developed but eyelids still fused, nose develops bridge, and external ears are present. Ossification continues. Appendages are fully formed and nails develop. Heartbeat can be detected. Body systems continue to develop.
4	18 cm (6½–7 inches) 113 g (4 oz)	Head large in proportion to rest of body. Face takes on human features and hair appears on head. Skin bright pink. Many bones ossified, and joints begin to form. Continued development of body systems.
5	25–30 cm (10–12 inches) 227–454 g (½–1 lb)	Head less disproportionate to rest of body. Fine hair (lanugo) covers body. Skin still bright pink. Rapid development of body systems.
6	27–35 cm (11–14 inches) 567–781 g (1¼–1½ lb)	Head becomes even less disproportionate to rest of body. Eyelids separate and eyelashes form. Skin wrinkled and pink.
7	32–42 cm (13–17 inches) 1,135–1,362 g (2½–3 lb)	Head and body become more proportionate. Skin wrinkled and pink. Seven-month fetus (premature baby) is capable of survival.
8	41–45 cm (16½–18 inches) 2,043–2,270 g (4½–5 lb)	Subcutaneous fat deposited. Skin less wrinkled. Testes descend into scrotum. Bones of head are soft. Chances of survival much greater at end of eighth month.
9	50 cm (20 inches) 3,178–3,405 g (7–7½ lb)	Additional subcutaneous fat accumulates. Lanugo shed. Nails extend to tips of fingers and maybe even beyond.

(a)

(b)

FIGURE 29-11 Hormones of pregnancy. (a) Summary of sources and functions. (b) Blood levels of HCG, estrogens, and progesterone.

The placenta supplies the levels of estrogens and progesterone needed to maintain the pregnancy. The placental hormones thus take over management of the mother's body in preparation for parturition (birth) and lactation. Following delivery, estrogens and progesterone in the blood decrease to normal levels.

Recent evidence indicates that the placenta produces a regulating factor called **placental luteotropic releasing factor (pLRF).** It is chemically similar to gonadotropin releasing factor (GnRF) produced by the hypothalamus, which causes the synthesis and release of luteinizing hormone (LH) by the anterior pituitary. The proposed function of pLRF is to stimulate the secretion of HCG by the placenta.

Another hormone produced by the chorion of the placenta is **human chorionic somatomammotropin (HCS).**

Its secretion begins at about the same time as that of HCG, but its pattern of secretion is quite different. The rate of secretion of HCS increases in proportion to placental mass, reaching maximum levels after 32 weeks and remaining relatively constant after that. HCS is believed to stimulate some development of breast tissue for lactation, enhance growth by causing protein deposition in tissues, and regulate certain aspects of metabolism. For example, HCS causes decreased utilization of glucose by the mother, thus making more available for fetal metabolism. Also, HCS promotes the release of fatty acids from fat depots, providing an alternative source of energy for the mother's metabolism.

Relaxin is a hormone produced by the placenta and ovaries. Its physiological role is to relax the symphysis pubis and ligaments of the sacroiliac and sacrococcygeal joints and help dilate the uterine cervix toward the end of pregnancy. Both of these actions assist in delivery. The hormone also helps to increase sperm motility.

GESTATION

The time the zygote, embryo, or fetus is carried in the female reproductive tract is called **gestation** (jes-TĀ-shun). The total human gestation period is about 280 days from the beginning of the last menstrual period. The specialized branch of medicine that deals with pregnancy, labor, and the period of time immediately following delivery is called **obstetrics** (ob-STET-riks; *obstetrix* = midwife).

By about the end of the third month of gestation, the uterus occupies most of the pelvic cavity, and as the fetus continues to grow, the uterus extends higher and higher into the abdominal cavity. In fact, toward the end of a full-term pregnancy, the uterus occupies practically all of the abdominal cavity, reaching above the costal margin nearly to the xiphoid process of the sternum (Figure 29-12), causing displacement of the maternal intestines, liver, and stomach upward, elevation of the diaphragm, and widening of the thoracic cavity. In the pelvic cavity, there is compression of the ureters and urinary bladder. In addition, the breasts enlarge in anticipation of lactation, and the areolae around the nipples become darkly pigmented.

In addition to the anatomical changes associated with pregnancy, there are also certain pregnancy-induced

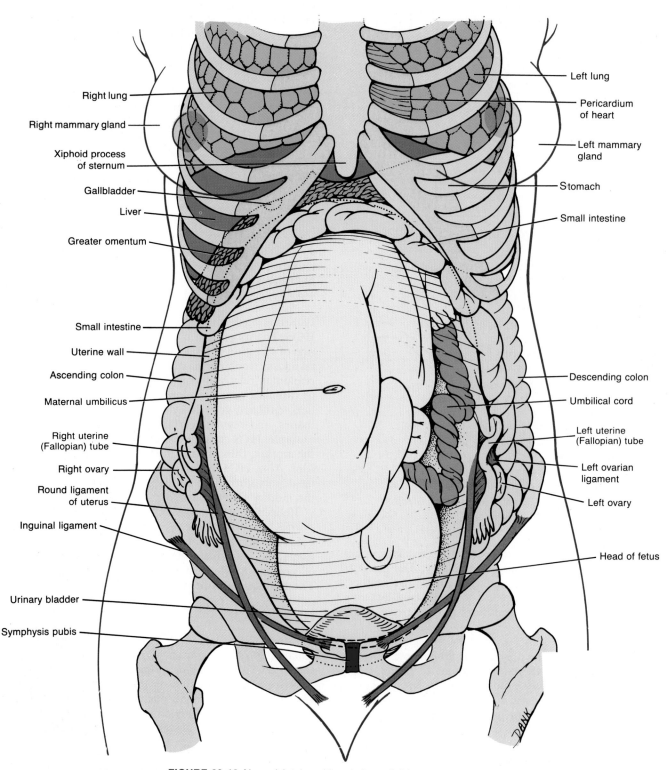

FIGURE 29-12 Normal fetal position during a full-term pregnancy.

physiological changes. For example, there is a rise in cardiac output (CO) by 30 to 40 percent by the twenty-seventh week due to increased maternal blood flow to the placenta and increased metabolism; an increase in pulse rate by about 15 beats per minute; and an increase in blood volume up to 30 to 50 percent, mostly during the latter half of pregnancy. In the supine position, the enlarged uterus may compress the aorta, resulting in diminished blood flow to the uterus. Hormonal changes associated with pregnancy and compression of the inferior vena cava can also produce varicose veins. Pulmonary function is also altered during pregnancy in that functional residual capacity is lowered and tidal volume is increased. This latter condition is due to hypocapnia and compensated respiratory alkalosis. With regard to the gastrointestinal (GI) system, there is a general decrease in motility that can result in constipation and a delay in gastric emptying time. Pressure on the urinary bladder by the enlarging uterus can produce urinary symptoms, such as frequency, urgency, and stress incontinence.

CLINICAL APPLICATION

Physicians can see into the uterus of a pregnant woman without exposing her to the known dangers of x rays and without pain or intrusion. One such technique, called **ultrasound** (also **sonography** or **ultrasonography**), uses high-frequency, inaudible sound waves, which are directed into the abdomen of the mother-to-be and then reflected back to a receiver. The reflected waves give a visual "echo" of what's inside the uterus. This echo is transformed electronically into an image on a screen. The potential risks of ultrasonography, such as decreased immune response, changes in plasma membrane functions, and breakdown of macromolecules, are currently under investigation. Another technique is the **Doppler detector,** a fetal monitor that registers the baby's heart rate. It is used routinely in many hospitals during labor.

In the United States today, most obstetricians prescribe ultrasound examinations only when there is some clinical question about the normal progress of the pregnancy. By far the most common use of diagnostic ultrasound is to determine true fetal age when the date of conception is unknown or mistaken by the mother. It is also used to evaluate fetal growth, determine fetal position, determine the time of ovulation in infertile females to enhance the chances of pregnancy, ascertain the cause of vaginal bleeding, diagnose ectopic and multiple pregnancies, and as an adjunct to special procedures such as amniocentesis. Ultrasound is not used routinely to determine the sex of a fetus; it is only performed for a specific medical indication.

Ultrasound has gained a wide application beyond obstetrics, including the detection of tumors, gallstones, and other abnormal internal masses; heart disorders; and brain hemorrhages.

PRENATAL DIAGNOSTIC TECHNIQUES

AMNIOCENTESIS

Amniocentesis is a technique of withdrawing some of the amniotic fluid that bathes the developing fetus to diagnose genetic disorders or to determine fetal maturity or well-being. Using ultrasound, the position of the fetus and placenta are first determined. About 10 to 20 ml of fluid is removed by hypodermic needle puncture of the uterus, usually 16 to 20 weeks after conception (Figure 29-13). Cells and fluid are subjected to microscopic examination and biochemical testing to determine abnormalities in chromosome number or structure and biochemical defects. Close to 300 chromosomal disorders and over 50 inheritable biochemical defects can be detected through amniocentesis, including hemophilia, certain muscular dystrophies, Tay-Sachs disease, myelocytic leukemia, Klinefelter's and Turner's syndromes, sickle-cell anemia, thalassemia, and cystic fibrosis. When both parents are known or suspected to be genetic carriers of any one of these disorders, or when maternal age approaches 40, amniocentesis is advised.

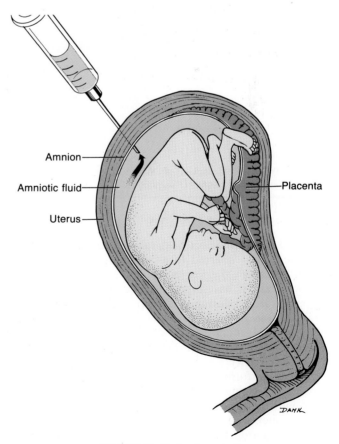

Amnion

Amniotic fluid

Uterus

Placenta

FIGURE 29-13 Amniocentesis.

One chromosome disorder that may be diagnosed through amniocentesis is **Down's syndrome** (Figure 29-14). This disorder is characterized by mental retardation, retarded physical development (short stature and stubby fingers), distinctive facial structures (large tongue, flat profile, broad skull, slanting eyes, and round head), and malformation of the heart, ears, hands, and feet. Sexual maturity is rarely attained. Individuals with the disorder usually have 47 chromosomes instead of the normal 46 (an extra chromosome in the twenty-first pair).

FIGURE 29-14 Photograph of an individual with Down's syndrome. (Courtesy of Lester V. Bergman & Associates, Inc.)

CHORIONIC VILLI SAMPLING (CVS)

A new test used to detect prenatal genetic defects is now available and is referred to as **chorionic villi sampling (CVS).** Although the test picks up the same defects as amniocentesis, it has several advantages. First of all, it can be performed during the first trimester of pregnancy, usually at 8 to 10 weeks gestation. Also, results are available within 24 hours. Moreover, the procedure does not require penetration of the abdominal wall, uterine wall, or amniotic cavity.

CVS is performed as follows. A catheter is placed through the vagina into the uterus and then to the chorionic villi under ultrasound guidance. About 30 mg of tissue is suctioned out and prepared for chromosomal analysis. Chorion cells and fetal cells contain identical genetic information. The safety of the procedure is believed to be comparable to that for amniocentesis.

PARTURITION AND LABOR

The term **parturition** (par'-too-RISH-un) refers to birth. Parturition is accompanied by a sequence of events commonly called **labor.** The onset of labor is apparently related to a complex interaction of many factors. Just prior to birth, the muscles of the uterus contract rhythmically and forcefully. Both placental and ovarian hormones seem to play a role in these contractions. Since progesterone inhibits uterine contractions, labor cannot take place until its effects are diminished. At the end of gestation, the level of estrogens in the mother's blood is sufficient to overcome the inhibiting effects of progesterone and labor commences. It has been suggested that some factor released by the placenta, fetus, or mother rather suddenly overcomes the inhibiting effects of progesterone so that estrogens can exert their effect. Prostaglandins may also play a role in labor. Oxytocin (OT) from the posterior pituitary gland also stimulates uterine contractions (see Figure 18-9), and relaxin assists by relaxing the symphysis pubis and helping to dilate the uterine cervix.

Uterine contractions occur in waves, quite similar to peristaltic waves, that start at the top of the uterus and move downward. These waves expel the fetus. **True labor** begins when pains occur at regular intervals. The pains correspond to uterine contractions. As the interval between contractions shortens, the contractions intensify. Another sign of true labor in some females is localization of pain in the back, which is intensified by walking. A reliable indication of true labor is the "show" and dilation of the cervix. The "show" is a discharge of a blood-containing mucus that accumulates in the cervical canal during labor. In **false labor,** pain is felt in the abdomen at irregular intervals. The pain does not intensify and is not altered significantly by walking. There is no "show" and no cervical dilation.

Labor can be divided into three stages (Figure 29-15).

1. The **stage of dilation** is the time from the onset of labor to the complete dilation of the cervix (Fig. 29-15). During this stage there are regular contractions of the uterus, usually a rupturing of the amniotic sac, and complete dilation (10 cm) of the cervix. If the amniotic sac does not rupture spontaneously, it is done artificially.

2. The **stage of expulsion** is the time from complete cervical dilation to delivery.

3. The **placental stage** is the time after delivery until the placenta or "afterbirth" is expelled by powerful uterine contractions. These contractions also constrict blood vessels that were torn during delivery. In this way, the possibility of hemorrhage is reduced.

Pudendal (pyoo-DEN-dal) **nerve block** is used for procedures such as episiotomy. The primary innervation to the skin and muscles of the perineum is the pudendal

FIGURE 29-15 Parturition. (a) Fetal position prior to birth. (b) Dilation. Protrusion of amnionic sac through partly dilated cervix (left). Amnionic sac ruptured and complete dilation of cervix (right). (c) Stage of expulsion. (d) Placental stage.

nerve. In the transvaginal approach, the needle is passed through the lateral vaginal wall to a point just medial to the ischial spine. The anesthesia results in loss of the anal reflex, relaxation of the muscles of the floor of the pelvis, and loss of sensation to the vulva and lower one-third of the vagina.

Various deformities of the female pelvis may be responsible for **dystocia** (dis-TŌ-sē-a), that is, difficult labor. Pelvic deformities may be congenital or acquired from disease, fractures, or poor posture. Among other conditions associated with difficult labor are malposition of the fetus, malpresentation of the fetus, and premature rupture of the fetal membranes.

(e)

FIGURE 29-15 (*Continued*) (e) Sequential photographs of parturition. Top: Emergence of the infant's head. (Courtesy of Kinne, Photo Researchers, Inc.) Bottom left: Delivery of the infant. (Courtesy of Thomas, Photo Researchers, Inc.) Bottom right: Delivery of the placenta. (Courtesy of McCartney, Photo Researchers, Inc.)

CLINICAL APPLICATION

If either dystocia or prolonged labor are indicated, it may be necessary to deliver the baby via a **cesarean** (*caedere* = to cut) **section.** In this procedure, a low, horizontal incision is made through the abdominal wall and uterus, through which the baby and placenta are removed.

ADJUSTMENTS OF THE INFANT AT BIRTH

During pregnancy, the embryo and later the fetus is totally dependent on the mother for its existence. The mother supplies the fetus with oxygen and nutrients, eliminates its carbon dioxide and other wastes, and protects it against shocks, temperature changes, and certain harmful microbes. However, a delivery of a physiologically immature baby carries certain risks.

A **premature infant** or "preemie" is generally considered to be one who weighs less than 2,500 gm (5 lb, 8 oz) at birth. It appears that poor prenatal care, a history of an earlier premature delivery, and mother's age below 16 or above 35 increase the risk of premature delivery. The problems related to survival of premature infants are due to the fact that they are not yet ready to take over functions the mother's body should be performing, particularly those related to respiration and digestion. The major problem with delivery of an infant under 36 weeks is infant respiratory distress syndrome (RDS), a condition that can be helped by a ventilator that delivers oxygen until the lungs can operate on their own. It is also possible to supply nutrients intravenously until the digestive system matures.

At birth, a physiologically mature baby becomes self-supporting, and the newborn's body systems must make various adjustments. Following are some changes that occur in the respiratory and cardiovascular systems.

RESPIRATORY SYSTEM

The respiratory system is fairly well developed at least 2 months before birth as evidenced by the fact that premature babies delivered at 7 months are able to breathe and cry. The fetus depends entirely on the mother for obtaining oxygen and eliminating carbon dioxide. The fetal lungs are either collapsed or partially filled with amniotic fluid, which is absorbed at birth. After delivery the baby's supply of oxygen from the mother is stopped. Circulation in the baby continues, and as the blood level of carbon dioxide increases, the respiratory center in the medulla is stimulated. This causes the respiratory muscles to contract and the baby draws its first breath. Since the first inspiration is unusually deep because the lungs contain no air, the baby exhales vigorously and naturally cries. A full-term baby may breathe 45 times a minute for the first 2 weeks after exposure to air. The rate is gradually reduced until it approaches a normal rate.

CARDIOVASCULAR SYSTEM

Following the first inspiration by the baby, the cardiovascular system must make several adjustments. The foramen ovale of the fetal heart between atria closes at the moment of birth. This diverts deoxygenated blood to the lungs for the first time. The foramen ovale is closed by two flaps of heart tissue that fold together and permanently fuse. The remnant of the foramen ovale is the fossa ovalis. Once the lungs begin to function, the ductus arteriosus is shut off by contractions of the muscles in its wall. The ductus arteriosus generally does not completely and irreversibly close for about 3 months following birth. Incomplete closing, as you already know, results in patent ductus arteriosus.

The ductus venosus of fetal circulation connects the umbilical vein directly with the inferior vena cava. It forces any remaining umbilical blood directly into the fetal liver and from there to the heart. When the umbilical cord is severed, all visceral blood of the fetus goes directly to the fetal heart via the inferior vena cava. This shunting of blood usually occurs within minutes after birth, but may take a week or two to complete. The ligamentum venosum, the remnant of the ductus venosus, is well established by the eighth postnatal week.

At birth, the infant's pulse may be from 120 to 160 per minute, and may go as high as 180 following excitation. Several days after birth, there is a greater independent need for oxygen, which stimulates an increase in the rate of erythrocyte and hemoglobin production. This increase usually lasts for only a few days. Moreover, the white blood cell count at birth is very high, sometimes as much as 45,000 cells per cubic millimeter, but this decreases rapidly by the seventh day.

Finally, the infant's liver may not be adjusted at birth to control the production of bile pigment. As a result of this and other complicating factors, a temporary jaundice may result in as many as 50 percent of normal newborns by the third or fourth day after birth.

POTENTIAL HAZARDS TO THE DEVELOPING EMBRYO AND FETUS

The developing fetus is susceptible to a number of potential hazards that can be transmitted from the mother.

Such hazards include infectious microbes, chemicals and drugs, alcohol, and cigarette smoking. In addition, certain environmental conditions and pollutants can damage the fetus or even cause fetal death.

CHEMICALS AND DRUGS

Since the placenta is known to be an ineffective barrier between the maternal and fetal circulations, actually any drug or chemical dangerous to the infant may be considered potentially dangerous to the fetus when given to the mother. Many chemicals and drugs have been proven to be toxic and teratogenic to the developing embryo and fetus. A **teratogen** (*terato* = monster) is any agent or influence that causes physical defects in the developing embryo. Examples are pesticides, defoliants, industrial chemicals, some hormones, antibiotics, oral anticoagulants, anticonvulsants, antitumor agents, thyroid drugs, thalidomide, diethylstilbestrol (DES), LSD, and marijuana. In addition, accumulating evidence suggests that many agents assumed to be nonteratogenic can produce long-term adverse effects on reproductive ability and neurobehavioral development.

IRRADIATION

Ionizing radiations are potent teratogens. Treatment of pregnant mothers with large doses of x rays and radium during the embryo's susceptible period of development may cause microcephaly (small size of head in relation to the rest of the body), mental retardation, and skeletal malformations. Caution is advised for diagnostic x rays during the first trimester of pregnancy.

ALCOHOL

Alcohol has been a suspected teratogen for centuries, but only recently has a relationship been recognized between maternal alcohol intake and the characteristic pattern of malformations in the fetus. The term applied to the effects of intrauterine exposure to alcohol is **fetal alcohol syndrome (FAS).** Studies to date have indicated that in the general population the incidence of FAS may exceed 1 per 1,000 live births, and may be by far the number one fetal teratogen. The symptoms shown by children may include slow growth before and after birth, small head, facial irregularities such as narrow eye slits and sunken nasal bridge, defective heart and other organs, malformed arms and legs, genital abnormalities, and mental retardation. There are also behavioral problems, such as hyperactivity, extreme nervousness, and a poor attention span.

CIGARETTE SMOKING

The latest evidence not only indicates a causal relationship between cigarette smoking during pregnancy and low infant birth weight, but also points to a strong probable association between smoking and a higher fetal and infant mortality rate. Infants nursing from smoking mothers have also been found to have an increased incidence of gastrointestinal disturbances. Other pathologies of infants of smoking mothers include an increased incidence of respiratory problems during the first year of life, including bronchitis and pneumonia. Cigarette smoking may be teratogenic and cause cardiac abnormalities and anencephaly (a developmental anomaly with absence of neural tissue in the cranium). Maternal smoking also appears to be a significant etiologic factor in the development of cleft lip and palate, and has been tentatively linked with sudden infant death syndrome (SIDS).

LACTATION

The term **lactation** refers to the secretion and ejection of milk by the mammary glands. The major hormone in promoting lactation is **prolactin (PRL)** from the anterior pituitary gland. It is released in response to the hypothalamic prolactin releasing factor (PRF). Even though PRL levels increase as the pregnancy progresses, there is no milk secretion because estrogens and progesterone cause the hypothalamus to release prolactin-inhibiting factor (PIF). Following delivery, the levels of estrogens and progesterone in the mother's blood decrease and the inhibition is removed.

The principal stimulus in maintaining prolactin secretion during lactation is the sucking action of the infant. Sucking initiates impulses from receptors in the nipples to the hypothalamus. The impulses inhibit PIF production, and PRL is released by the anterior pituitary. The sucking action also initiates impulses to the posterior pituitary via the hypothalamus. These impulses stimulate the release of the hormone **oxytocin (OT)** by the posterior pituitary gland. This hormone induces certain cells surrounding the outer walls of the alveoli to contract, thereby compressing the alveoli. The compression moves milk from the alveoli of the mammary gland into the ducts, where it can be sucked. This process is referred to as **milk letdown.**

During late pregnancy and the first few days after birth, the mammary glands secrete a cloudy fluid called **colostrum.** Although it is not as nutritious as true milk, since it contains less lactose and virtually no fat, it serves adequately until the appearance of true milk on about the fourth day. Colostrum and maternal milk are thought to contain antibodies that protect the infant during the first few months of life.

Following birth of the infant, PRL level starts to return to the nonpregnant level, but each time the mother nurses the infant, nerve impulses from the nipples to the hypothalamus cause the release of PRF and the secretion of a 10-fold increase in PRL by the anterior pituitary that lasts about an hour. PRL acts on the mammary glands to provide milk for the next nursing period. If this surge of PRL is blocked by injury or disease, or if nursing is discontinued, the mammary glands lose their ability to secrete milk in a few days. However, milk secretion can continue for several years if the child continues to suckle. Milk secretion normally decreases considerably within 7 to 9 months.

Lactation often prevents the occurrence of female ovarian cycles for the first few months following delivery, although there is no guarantee. Ovulation will normally precede the resumption of the period so there is always an unknown factor. The contraceptive effect of breast feeding is, therefore, not a very effective birth control measure. However, the suppression of ovulation during lactation is believed to occur as follows. During breast feeding, neural input from the nipple reaches the hypothalamus and causes it to produce beta (β)-endorphin. This, in turn, suppresses the release of gonadotropin releasing factor (GnRF). Consequently, there is a decreased production in LH and ovulation is inhibited.

CLINICAL APPLICATION

Advocates of **breast feeding** feel that it offers the following advantages to the infant.

1. It establishes early and prolonged contact between mother and infant.
2. The infant is more in control of intake.
3. Fats and iron in human milk are better absorbed than those in cow's milk, and the amino acids in human milk are more readily metabolized. Also, the lower sodium content of human milk is more suited to the infant's needs.
4. Breast feeding provides important antibodies that prevent gastroenteritis. Immunity to respiratory infections and meningitis is also greater.
5. Premature infants benefit from breast feeding because the milk produced by mothers of premature infants seems to be specially adapted to the infant's needs by having a higher protein content than the milk of mothers of full-term infants.
6. There is less likelihood of an allergic reaction in the baby to the milk of the mother.
7. The act of sucking on the breast promotes development of the jaw, facial muscles, and teeth.

INHERITANCE

Inheritance is the passage of hereditary traits from one generation to another. It is the process by which you acquired your characteristics from your parents and will transmit your characteristics to your children. The branch of biology that deals with inheritance is called **genetics.**

GENOTYPE AND PHENOTYPE

The nuclei of all human cells except gametes contain 23 pairs of chromosomes—the diploid number. One chromosome from each pair comes from the mother, and the other comes from the father. Homologues, the two chromosomes in a pair, contain genes that control the same traits. If a chromosome contains a gene for height, its homologue will contain a gene for height.

The relationship of genes to heredity is illustrated admirably by the disorder called phenylketonuria or PKU (see Figure 29-16). People with PKU are unable to manufacture the enzyme phenylalanine hydroxylase. It is believed the PKU is brought about by an abnormal gene,

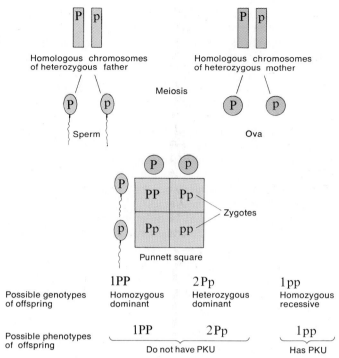

FIGURE 29-16 Inheritance of phenylketonuria (PKU).

Exhibit 29-3 lists some of the variety of simple inherited structural and functional traits in humans. The classification of traits in this manner is rather arbitrary and somewhat artificial, since structure and function are intimately related. Nevertheless, some organizational scheme is needed that gives an indication of the wide spectrum of inherited human characteristics. Many traits were at one time thought to be inherited in a simple Mendelian manner and were classified as "dominant" or "recessive." Then further research revealed that some traits were actually complex and that the genetic basis was correspondingly complicated. No wonder students of human genetics prefer to work with enzymes and antigens, in which the path from a gene to its product is much shorter, more direct, and less influenced by other genes and by environment. (If you would like to find out about the inheritance of some particular traits, one good source is McKusick's *Mendelian Inheritance in Man*, 1975.)

Normal traits do not always dominate over abnormal ones, but genes for severe disorders are more frequently recessive than dominant. People who have severe disorders often do not live long enough to pass the abnormal gene on to the next generation.

CLINICAL APPLICATION

An exception is **Huntington's chorea**—a major disorder caused by a dominant gene and characterized by

which can be symbolized as p. The normal gene will be symbolized as P. The chromosome concerned with directions for phenylalanine hydroxylase production will have either p or P on it. Its homologue will also have p or P. Thus every individual will have one of the following genetic makeups, or **genotypes** (JĒ-nō-tīps): PP, Pp, or pp. Although people with genotypes of Pp have the abnormal gene, only those with genotype pp suffer from the disorder because the normal gene, when present, dominates over and inhibits the abnormal one. A gene that dominates is called the **dominant gene,** and the trait expressed is said to be a dominant trait. The gene that is inhibited is called the **recessive gene,** and the trait it controls is called the recessive trait.

By tradition, we symbolize the dominant gene with a capital letter and the recessive one with a lowercase letter. An individual with the same genes on homologous chromosomes (for example, PP or pp) is said to be **homozygous** for the trait. An individual with different genes on homologous chromosomes (for example, Pp), is said to be **heterozygous** for the trait. **Phenotype** (FĒ-nō-tīp) refers to how the genetic makeup is expressed in the body. A person with Pp has a different genotype from one with PP, but both have the same phenotype—which in this case is normal production of phenylalanine hydroxylase.

To detemine how gametes containing haploid chromosomes unite to form diploid fertilized eggs, special charts called **Punnett squares** are used. Usually, the male gametes (sperm cells) are placed at the side of the chart and the female gametes (ova) at the top (as in Figure 29-16). The four spaces on the chart represent the possible combinations of male and female gametes that could form fertilized eggs.

EXHIBIT 29-3 SELECTED HEREDITARY TRAITS IN HUMANS

DOMINANT	RECESSIVE
Curly hair	Straight hair
Dark brown hair	All other colors
Coarse body hair	Fine body hair
Pattern baldness (dominant in males)	Baldness (recessive in females)
Normal skin pigmentation	Albinism
Brown eyes	Blue or gray eyes
Near or farsightedness	Normal vision
Normal hearing	Deafness
Normal color vision	Color blindness
Broad lips	Thin lips
Large eyes	Small eyes
Polydactylism (extra digits)	Normal digits
Brachydactylism (short digits)	Normal digits
Syndactylism (webbed digits)	Normal digits
Hypertension	Normal blood pressure
Diabetes insipidus	Normal excretion
Huntington's chorea	Normal nervous system
Normal mentality	Schizophrenia
Migraine headaches	Normal
Normal resistance to disease	Susceptibility to disease
Enlarged spleen	Normal spleen
Enlarged colon	Normal colon
A or B blood factor	O blood factor
Rh blood factor	No Rh blood factor

degeneration of nervous tissue, usually leading to mental disturbance and death. The first signs of Huntington's chorea do not occur until adulthood, very often after the person has already produced offspring.

INHERITANCE OF SEX

Microscopic examination of the chromosomes in cells reveals that one pair differs in males and in females (Figure 29-17a). In females, the pair consists of two chromosomes designated as X chromosomes. One X chromosome is present in males, but its mate is a chromosome called a Y chromosome. The XX pair in the female and the XY pair in the male are called the **sex chromosomes.** All other chromosomes are called **autosomes.**

The sex chromosomes are responsible for the sex of the individual (Figure 29-17b). When a spermatocyte undergoes meiosis to reduce its chromosome number, one daughter cell will contain the X chromosome and the other contain the Y chromosome. Oocytes have no Y chromosomes and produce only X-containing ova. If the ovum is subsequently fertilized by an X-bearing sperm,

the offspring normally will be female (XX). Fertilization by a Y sperm normally produces a male (XY). Thus, sex is determined at fertilization. Although X and Y sperm are produced in equal amounts, more males are born than females. It is speculated that Y sperm swim faster than X sperm because X sperm contain more genetic material.

COLOR BLINDNESS AND X-LINKED INHERITANCE

The sex chromosomes also are responsible for the transmission of a number of nonsexual traits. Genes for these traits appear on X chromosomes, but many of these genes are absent from Y chromosomes. This feature produces a pattern of heredity that is different from the pattern described earlier. Let us consider color blindness. The gene for **color blindness** is a recessive one designated c. Normal color vision, designated C, dominates. The C/c genes are located on the X chromosome. The Y chromosome does not contain the segment of DNA that programs this aspect of vision. Thus the ability to see colors depends entirely on the X chromosomes. The genetic possibilities are:

X^cX^c	Normal female
X^cX^c	Normal female carrying the recessive gene
X^cX^c	Color-blind female
X^cY	Normal male
X^cY	Color-blind male

Only females who have two X^c chromosomes are color blind. In X^cX^c females the trait is inhibited by the normal, dominant gene. Males, on the other hand, do not have a second X chromosome that would inhibit the trait. Therefore all males with an X^c chromosome will be color blind. The inheritance of color blindness is illustrated in Figure 29-18.

(a)

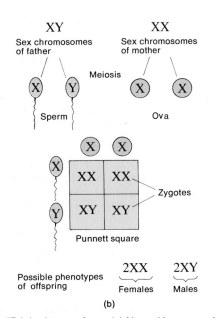

(b)

FIGURE 29-17 Inheritance of sex. (a) Normal human male chromosomes. Sex chromosomes are the twenty-third pair, indicated in the blue colored box. (b) Sex determination.

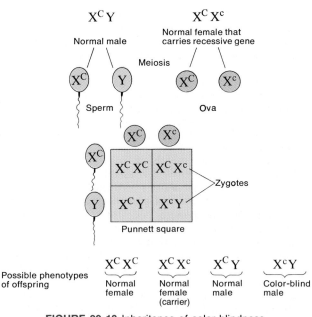

FIGURE 29-18 Inheritance of color blindness.

Traits inherited in the manner just described are called **X-linked traits.** Another X-linked trait is **hemophilia**— a condition in which the blood fails to clot or clots very slowly after an injury (Chapter 19). Like the trait for color blindness, hemophilia is caused by a recessive gene. If H represents normal clotting and h represents abnormal clotting, then X^{h}X^h females will have the disorder. Males with X^{H}Y will be normal; males with X^{h}Y will be hemophiliac. Actually, clotting time varies somewhat among hemophiliacs, so the condition may be affected by other genes as well.

A few other X-linked traits in humans are nonfunctional sweat glands, certain forms of diabetes, some types of deafness, uncontrollable rolling of the eyeballs, absence of central incisors, night blindness, one form of cataract, white forelocks, juvenile glaucoma, and juvenile muscular dystrophy.

MEDICAL TERMINOLOGY

Abortion   Premature expulsion from the uterus of the products of conception—embryo or nonviable fetus.

Cautery   Application of a caustic (burning) substance or instrument for the destruction of tissue.

Colpotomy (*colp* = vagina; *tome* = cutting)   Incision of the vagina.

Culdoscopy (*skopein* = to examine)   A procedure in which a culdoscope (endoscope) is used to view the pelvic cavity. The approach is through the vagina.

Hermaphroditism   Presence of both male and female sex organs in one individual.

Karyotype (*karyon* = nucleus)   The chromosomal elements typical of a cell, drawn in their true proportions, based on the average of measurements determined in a number of cells. Useful in judging whether or not chromosomes are normal in number and structure.

Lethal gene (*lethum* = death)   A gene that, when expressed, results in death either in the embryonic state or shortly after birth.

Lochia   The discharge from the birth canal consisting initially of blood and later of serous fluid occurring after childbirth. The discharge is derived from the former placental site and may last up to about 2 to 4 weeks.

Mutation (*mutare* = change)   A permanent, heritable change in a gene that causes it to have a different effect than it had previously.

Preeclampsia   A syndrome characterized by sudden hypertension, large amounts of protein in urine, and generalized edema; might be related to an autoimmune or allergic reaction due to the presence of a fetus; when the condition is also associated with convulsions and coma, it is referred to as **eclampsia.** This condition is dangerous and, if untreated, it can result in the death of both mother and fetus.

Puerperal (*puer* = child, *parere* = to bring forth) **fever**   Infectious disease of childbirth, also called puerperal sepsis and childbed fever. The disease results from an infection originating in the birth canal and affects the endometrium. It may spread to other pelvic structures and lead to septicemia.

STUDY OUTLINE

Development During Pregnancy (p. 743)
1. Pregnancy is a sequence of events that includes fertilization, implantation, embryonic growth, fetal growth, and birth.
2. Its various events are hormonally controlled.

Fertilization and Implantation (p. 743)
1. Fertilization refers to the penetration of the ovum by a sperm cell and the subsequent union of the sperm and ovum nuclei to form a zygote.
2. Penetration is facilitated by hyaluronidase and proteinases produced by sperm.
3. Normally only one sperm fertilizes an ovum.
4. Early rapid cell division of a zygote is called cleavage, and the cells produced by cleavage are called blastomeres.
5. The solid mass of cells produced by cleavage is a morula.
6. The morula develops into a blastocyst, a hollow ball of cells differentiated into a trophectoderm (future embryonic membranes) and inner cell mass (future embryo).
7. The attachment of a blastocyst to the endometrium is called implantation; it occurs by enzymatic degradation of the endometrium.
8. External human fertilization refers to the fertilization of an ovum outside the body and the subsequent implantation of the zygote.
9. In embryo transfer, a husband's seminal fluid is used to artifically inseminate a fertile ovum donor, and following fertilization, the morula or blastocyst is transferred from the donor to the infertile wife.
10. Alternatives to external human fertilization and embryo transfer include gamete intrafallopian transfer (GIFT) and transvaginal oocyte retrieval.

Embryonic Development (p. 747)
1. During embryonic growth, the primary germ layers and embryonic membranes are formed and the placenta is functioning.
2. The primary germ layers—ectoderm, mesoderm, and endoderm—form all tissues of the developing organism.
3. Embryonic membranes include the yolk sac, amnion, chorion, and allantois.
4. Fetal and maternal materials are exchanged through the placenta.

Fetal Growth (p. 751)
1. During the fetal period, organs established by the primary germ layers grow rapidly.
2. The principal changes associated with fetal growth are summarized in Exhibit 29-2.

Hormones of Pregnancy (p. 752)

1. Pregnancy is maintained by human chorionic gonadotropin (HCG), estrogens, and progesterone.
2. Placental luteotropic releasing factor (pLRF) stimulates secretion of HCG.
3. Human chorionic somatomammotropin (HCS) assumes a role in breast development, protein anabolism, and glucose and fatty acid catabolism.
4. Relaxin relaxes the symphysis pubis and helps dilate the uterine cervix toward the end of pregnancy.

Gestation (p. 753)

1. The time an embryo or fetus is carried in the uterus is called gestation.
2. Human gestation lasts about 280 days from the beginning of the last menstrual period.
3. During gestation, several anatomical and physiological changes occur.

Prenatal Diagnostic Techniques (p. 755)

1. Amniocentesis is the withdrawal of amniotic fluid. It can be used to diagnose inherited biochemical defects and chromosomal disorders, such as hemophilia, Tay-Sachs disease, sickle-cell anemia, and Down's syndrome.
2. Down's syndrome is a chromosomal abnormality characterized by mental retardation and retarded physical development.
3. Chorionic villi sampling (CVS) involves withdrawal of chorionic villi for chromosomal analysis.
4. CVS can be done sooner than amniocentesis, and the results are available sooner.

Parturition and Labor (p. 756)

1. Parturition refers to birth and is accompanied by a sequence of events called labor.

2. The birth of a baby involves dilation of the cervix, expulsion of the fetus, and delivery of the placenta.

Adjustments of the Infant at Birth (p. 758)

1. The fetus depends on the mother for oxygen and nutrients, removal of wastes, and protection.
2. Following birth the respiratory and cardiovascular systems undergo changes in adjusting to self-supporting postnatal life.

Potential Hazards to the Developing Embryo and Fetus (p. 759)

1. The developing embryo and fetus is susceptible to many potential hazards that can be transmitted from the mother.
2. Examples are infections, microbes, chemicals and drugs, alcohol, and smoking.

Lactation (p. 760)

1. Lactation refers to the secretion and ejection of milk by the mammary glands.
2. Secretion is influenced by prolaction (PRL), estrogens, and progesterone.
3. Ejection is influenced by oxytocin (OT).

Inheritance (p. 760)

1. Inheritance is the passage of hereditary traits from one generation to another.
2. The genetic makeup of an organism is called its genotype. The traits expressed are called its phenotype.
3. Dominant genes control a particular trait; expression of recessive genes is inhibited by dominant genes.
3. Sex is determined by the Y chromosome of the male at fertilization.
4. Color blindness and hemophilia primarily affect males because there are no counterbalancing dominant genes on the Y chromosomes.

REVIEW QUESTIONS

1. Define developmental anatomy.
2. Define fertilization. Where does it normally occur? How is a morula formed?
3. Explain how dizygotic (fraternal) and monozygotic (identical) twins are produced.
4. Describe the components of a blastocyst.
5. What is implantation? How does the fertilized ovum implant itself? What causes morning sickness?
6. Describe the procedure for external human fertilization, embryo transfer, gamete intrafallopian transfer (GIFT), and transvaginal oocyte retrieval.
7. Define the embryonic period and the fetal period.
8. List several body structures formed by the endoderm, mesoderm, and ectoderm.
9. What is an embryonic membrane? Describe the functions of the four embryonic membranes.
10. Explain the importance of the placenta and umbilical cord to fetal growth.
11. Outline some of the major developmental changes during fetal growth.
12. List the hormones involved in pregnancy and describe the functions of each.
13. Define gestation and parturition.
14. Describe several anatomical and physiological changes that occur during gestation.
15. What is amniocentesis? What is its value?
16. Describe chorionic villi sampling (CVS). What are its advantages over amniocentesis?
17. Distinguish between false and true labor. Describe what happens during the stage of dilation, the stage of expulsion, and the placental stage of delivery.
18. Discuss the principal respiratory and cardiovascular adjustments made by an infant at birth.
19. Explain in detail some of the potential hazards for the developing embryo and fetus.
20. What is lactation? Name the hormones involved and their functions.
21. Define inheritance. What is genetics?
22. Define the following terms: genotype, phenotype, dominant, recessive, homozygous, heterozygous.
23. What is a Punnett square?
24. List several dominant and recessive traits inherited in humans.
25. Set up Punnett squares to show the inheritance of the following traits: sex, color blindness, hemophilia.
26. What is X-linked inheritance?
27. Refer to the glossary of medical terminology associated with development and inheritance. Be sure that you can define each term.

Appendix A: Measurements

UNITS OF MEASUREMENT

When you measure something, you are comparing it with some standard scale to determine its *magnitude*. How long is it? How much does it weigh? How fast is it going? Some measurements are made directly by comparing the unknown quantity with the known unit of the same kind, for example, weighing a patient on a scale and taking the reading directly in pounds. Other measurements are indirect and are done by calculation, for example, counting a person's blood cells in a certain number of squares on a microscope slide and then calculating the total blood count.

Regardless of how a measurement is taken, it always requires two things: a *number* and a *unit*. When recording the weight of a patient, you would not just say 145. You have to give both the number (145) and the unit (pounds). When you count blood cells, you report the measurement as 10,000 (number) white blood cells per cubic millimeter of blood (unit).

All the units in use can be expressed in terms of one of three special units called **fundamental units.** These fundamental units have been established arbitrarily as length, mass, and time. Mass is perhaps an unfamiliar term to you. **Mass** is the amount of matter an object contains. The mass of this textbook is the same whether it is measured in a laboratory, under the sea, on top of a mountain, or even on the moon. No matter where you take it, it still has the same quantity of matter. **Weight** on the other hand, is determined by the pull of gravity on an object. This textbook will not have the same weight on earth as on the moon because of the differences in gravitation. However, as long as we are dealing only with earthbound objects, weight and mass may be considered synonymous terms because the force of gravity on the surface of the earth is nearly constant. Thus weight remains nearly the same regardless of where the measurements are taken.

All units other than the fundamental ones are **derived units**—they can always be written as some combination of the three fundamental units. For example, units of volume are derived from units of length (the volume of a cube = length × width × height). Units of speed are combinations of distance and time (miles per hour).

Units are grouped into systems of measurement. The two principal systems of measurement commonly used in this country are the U.S. and the metric systems. The apothecary system is used by physicians and pharmacists.

U.S. SYSTEM

The **U.S. system** of measurement is used in everyday household work, industry, and some fields of engineering. The fundamental units in the U.S. system are the foot (length), the pound (mass), and the second (time).

The basic problem with the U.S. system is that there is no *uniform* progression from one unit to another. If you want to convert a measurement of 2½ yd to feet, you have to multiply it by 3 because there are 3 ft in a yard. If you want to convert the same length to inches, you have to multiply by 3 and then by 12 (or by 36) because there are 12 inches in a foot. In other words, to convert one unit of length to another, it is necessary to use *different* numbers each time. Conversions in the metric system are much easier since they are based on progressions of the number 10.

Exhibit A-1 lists U.S. units of measurement.

EXHIBIT A-1 U.S. UNITS OF MEASUREMENT	
FUNDA-MENTAL OR DERIVED UNIT	**UNITS AND U.S. EQUIVALENTS**
Length	1 inch = 0.083 foot 1 foot (ft) = 12 inches = 0.333 yard 1 yard (yd) = 3 ft = 36 inches 1 mile (mi) = 1,760 yd = 5,280 ft
Mass	1 grain (gr) = 0.002285 ounce 1 dram (dr) = 27.34 gr = 0.063 ounce 1 ounce (oz) = 16 dr = 437.5 gr 1 pound (lb) = 16 oz = 7,000 gr 1 ton = 2,000 lb
Time	1 second (sec) = 1/86,400 of a day 1 minute (min) = 60 sec 1 hour (hr) = 60 min = 3,600 sec 1 day = 24 hr = 1,440 min = 86,400 sec
Volume	1 fluidram (fl dr) = 0.125 fluidounce 1 fluidounce (fl oz) = 8 fl dr = 0.0625 quart = 0.008 gallon 1 pint (pt) = 16 fl oz = 128 fl dr 1 quart (qt) = 2 pt = 32 fl oz = 256 fl dr = 0.25 gallon 1 gallon (gal) = 4 qt = 8 pt = 128 fl oz = 1.024 fl dr

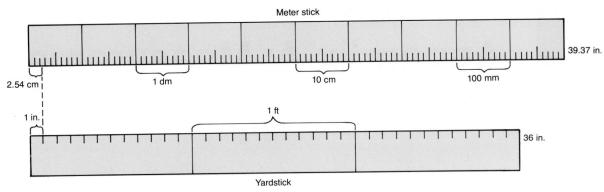

FIGURE A-1 Metric and U.S. units of length.

METRIC SYSTEM

The **metric system,** introduced in France in 1790, is now used by all major countries except the United States. Scientific observations are almost universally expressed in metric units.

LENGTH

The standard of length in the metric system is the *meter* (m). It was originally defined in 1790 as one ten-millionth of the distance from the North Pole to the Equator. In 1889 it was redefined as the distance measured at 0°C between two lines on a bar of platinum-iridium kept at the International Bureau of Weights and Measures in France. Finally, in 1960 the meter was redefined as the length of 1,650,763.73 light waves emitted by atoms of the gas krypton under strictly specified conditions. The meter is equal to 39.37 inches.

A major advantage of the metric system is that units are related to one another by factors of 10. Thus 1 m is 10 decimeters

(dm) or 100 centimeters (cm) or 1,000 millimeters (mm). Conversion from one unit to another is simple. Figure A-1 illustrates the differences between metric and U.S. conversions by comparing the meter stick and the yardstick. Exhibit A-2 lists the metric units of length with U.S. equivalents.

Since numbers with many zeros (very large numbers or very small fractions) are cumbersome to work with, they are expressed in *exponential form,* that is, as powers of 10. The form of exponential notation is

$$M \times 10^n$$

You can determine M and n in two steps. For example, how is 0.0000000001 written in exponential form? First, determine M by moving the decimal point so that only one nonzero digit is to the left of it:

0.0000000001.

The digit to the left of the decimal is 1; therefore $M = 1$. Second, determine n by counting the number of places you moved the decimal point. If you moved the point to the left, make the number positive; if you moved it to the right, it is

EXHIBIT A-2   METRIC UNITS OF LENGTH AND SOME U.S. EQUIVALENTS

METRIC UNIT	MEANING OF PREFIX	METRIC EQUIVALENT	U.S. EQUIVALENT
1 kilometer (km)	kilo = 1,000	1,000 m = 10^3 m	3,280.84 ft = 0.62 mi 1 mi = 1.61 km
1 hectometer (hm)	hecto = 100	100 m = 10^2 m	328 ft
1 dekameter (dam)	deka = 10	10 m = 10^0 m	32.8 ft
1 meter (m)	Standard unit of length		39.37 inches = 3.28 ft = 1.09 yd
1 decimeter (dm)	deci = 1/10	0.1 m = 10^{-1} m	3.94 inches
1 centimeter (cm)	centi = 1/100	0.01 m = 10^{-2} m	0.394 inch 1 inch = 2.54 cm
1 millimeter (mm)	milli = 1/1,000	0.001 m = 1/10 cm = 10^3 m	0.0394 inches
1 micrometer (μm) [formerly micron (μ)]	micro = 1/1,000,000	0.000,001 m = 1/10,000 cm = 10^{-6} m	3.94×10^{-5} inches
1 nanometer (nm) [formerly millimicron (mμ)]	nano = 1/1,000,000,000	0.000,000,001 m = 1/10,000,000 cm = 10^{-9} m	3.94×10^{-8} inches

negative. Since you moved the decimal point 10 places to the right, $n = -10$. Thus

$$0.0000000001 = 1 \times 10^{-10}$$

Now do a problem on your own. The wavelength of yellow light is about 0.000059 cm. Convert the centimeters to exponential form. If your answer is 5.9×10^{-5}, you are ready to continue. If you got the wrong answer, reread the discussion.

When we are working with a very large number, the same rules apply, but our exponential value will be positive rather than negative. Refer to Exhibit A-2. Note that 1 km equals 1,000 m. Even though 1,000 is not a cumbersome number, we can still convert it into exponential form. First, move the decimal point so there is only one nonzero digit to the left of it to determine M:

$$1.\underset{\smile}{000}.$$

Now, because the decimal has been moved three places to the left, n equals +3 or simply 3. Thus

$$1 \text{ km} = 1 \times 10^3 \text{ m}$$

Do another problem on your own. The speed of light is about 30,000,000,000 cm/sec. Convert the centimeters to exponential form. Your answer should be 3×10^{10} cm.

Review Exhibit A-2 and note some common metric and U.S. equivalents. Note also the exponential forms.

MASS

Now let us look at the second fundamental unit of the metric system: mass. The standard unit of mass is the *kilogram* (kg). A kilogram is defined as the mass of a platinum-iridium cylinder kept at the International Bureau of Weights and Measures in France. The standard pound is defined in terms of standard kilogram: 1 lb equals 0.4536 kg.

Exhibit A-3 lists metric units of mass and some U.S. equivalents.

EXHIBIT A-3   METRIC UNITS OF MASS AND SOME U.S. EQUIVALENTS

METRIC UNIT	METRIC EQUIVALENT	U.S. EQUIVALENT
1 kilogram (kg)	1,000 g	2.205 lb 1 ton = 907 kg
1 hectogram (hg)	100 g	
1 dekagram (dag)	10 g	0.0353 oz
1 gram (gr)	1 g	1 lb = 453.6 g 1 oz = 28.35 g
1 decigram (dg)	0.1 g	
1 centigram (cg)	0.01 g	
1 milligram (mg)	0.001 g	0.015 gr
1 microgram (μg)	0.000,001 g	
1 nanogram (ng)	0.000,000,001 g	
1 picogram (pg)	0.000,000,000,001 g	

TIME

The third fundamental unit of both the metric and the U.S. systems is time. The standard of time is the *second*. Formerly, the second was defined as 1/86,400 of a mean solar day. (A mean solar day is the average of the lengths of all days throughout the year.) Currently, the second is defined as the time required for 9,192,631,770 vibrations of cesium atoms when they are vibrating in a specific manner. Units of time are used in measuring pulse and heart rate, metabolic rate, x-ray exposure, and intervals between medications.

Exhibit A-1 lists the units of time.

VOLUME

Units of volume, or capacity, are derived units based on length. *Volume* in the U.S. system may be expressed as cubic feet (ft³), cubic inches (inch³), and cubic yards (yd³), or as a unit of volume such as the quart. Volume in the metric system may be expressed in cubic units of length such as cubic centimeters (cm³) or in terms of the basic unit of volume, the *liter*. A liter is defined as the volume occupied by 1,000 g of pure water at 4°C. Since 1 cm³ of water at this temperature weighs 1 g, then 1,000 g of water occupies a volume of 1,000 cm³. This means that a litter is equal to 1,000 cm³ and 1 milliliter (ml) is equal to 1 cm³. Because of this relationship, many volume-measuring devices, such as hypodermic needles, may be graduated in either milliliters or cubic centimeters.

Exhibit A-4 lists metric units of volume and some U.S. equivalents.

APOTHECARY SYSTEM

In addition to the U.S. and metric systems, there is the **apothecary system**. This system is commonly used by physicians prescribing medications and by pharmacists preparing them. Exhibit A-5 lists the important units and equivalents of the apothecary system. Note that the units of mass have the same names as in the U.S. system, but they are not equivalent (1 oz = 28.35 g; 1 oz ap = 30 g) and they do not have the same relationship to one another (1 oz = 16 dr; 1 oz ap = 8 dr ap). The units of volume are the same in both systems.

EXHIBIT A-4   METRIC UNITS OF VOLUME AND SOME EQUIVALENTS

METRIC UNIT	METRIC EQUIVALENT	U.S. EQUIVALENT
1 liter	1,000 ml	33.81 fl oz = 1.057 qt 946 ml = 1 qt
1 milliliter (ml)	0.001 liter	0.0338 fl oz 30 ml = 1 fl oz 5 ml = 1 teaspoon
1 cubic centimeter (cm³)	0.999972 ml	0.0338 fl oz

EXHIBIT A-5   APOTHECARY SYSTEM OF MASS AND VOLUME WITH METRIC EQUIVALENTS

FUNDAMENTAL UNIT	APOTHECARY UNIT AND CONVERSION	METRIC EQUIVALENT
Mass	1 grain (gr) = 0.002083 ounce	1 g = 15 gr
	1 dram (dr ap) = 60 gr	4 g = 1 dr
	1 ounce (oz ap) = 8 dr ap	30 g = 1 oz
	1 pound (lb ap) = 12 oz ap	1 kg = 32 oz
Volume	1 fluidram (fl dr) = 60 minims (min)	1 ml (or cm^3 = 15 min
	1 fluidounce (fl oz) = 8 fl dr	4 ml (or cm^3) = 1 fl dr
	1 pint (pt) = 16 fl oz	30 ml (or cm^3) = 1 fl oz
		500 ml (or cm^3) = 1 pt
		1,000 ml (or cm^3) = 1 qt

FREQUENTLY USED CONVERSIONS BASED ON MILLIGRAM

1,000 mg (1 g)	= 15 grains
600 mg (0.6 g)	= 10 grains
300 mg (0.3 g)	= 5 grains
60 mg (0.06 g)	= 1 grain
30 mg (0.03 g)	= 0.50 (1/2) grain
20 mg (0.02 g)	= 0.33 (1/3) grain
10 mg (0.01 g)	= 0.166 (1/6) grain
5 mg (0.005 g)	= 0.083 (1/12) grain
4 mg (0.004 g)	= 0.66 (1/15) grain
1 mg (0.001 g)	= 0.016 (1/60) grain
0.5 mg (0.0005 g)	= 0.0083 (1/120) grain
0.1 mg (0.0001 g)	= 0.0016 (1/600) grain

Appendix B: Normal Values for Selected Blood and Urine Tests

KEY TO SYMBOLS

mm³ = cubic millimeter	mEq/1 = milliequivalent per liter
dl = deciliter	mg = milligram
g = gram	ml = milliliter
> = greater than	mm = millimeter
IU = international unit	mm Hg = millimeter of mercury
kg = kilogram	mM = millimole
< = less than	mOsm = milliosmole
1 = liter	% = percent
μg = microgram	U = unit

EXHIBIT B-1   WHOLE BLOOD (WB), SERUM (S), AND PLASMA (P) TESTS

TEST (SPECIMEN)	NORMAL VALUES	CLINICAL IMPLICATION
Albumin (S)	3.5–5.5 g/dl	Values increase in nephritis, fever, trauma, and severe anemia and leukemia; values decrease following severe burns.
Alcohol, ethyl (CH_3CH_2OH) (S, P, or WB)	Subclinical: < 10 mg/dl < 0.005% Toxic Level: > 0.45%	Levels between 50–150 mg/dl exert unmistakable effects on behavior, especially driving.
Aminotransferases (S) Aspartate amino-transferase	3–21 IU/l 5–25 Reitman-Frankel units	Values increase in myocardial infarction, liver disease, trauma to skeletal muscles, and severe burns; values decrease in beriberi and uncontrolled diabetes mellitus with acidosis.
Alanine amino-transferase	5–24 IU/l 5–35 Reitman-Frankel units	Values increase in liver disease.
Ammonia (P)	56–150 μg/dl	Values increase in liver disease, heart failure, emphysema, pneumonia, cor pulmonale, and hemolytic disease of newborn (erythroblastosis fetalis).
Amylase (S)	60–180 Somogyi units/dl	Values increase in acute pancreatitis, mumps, and obstruction of pancreatic duct; values decrease in hepatitis, cirrhosis, burns, and toxemia of pregnancy.
Bilirubin (S)	Direct: 0.1–0.4 mg/dl Indirect: 0.2–1.0 mg/dl Total: 0.3–1.4 mg/dl Children: 0.2–0.8 mg/dl Newborn: 1.0–12.0 mg/dl	Values increase in jaundice due to hepatic, obstructive, or hemolytic causes.
Bleeding time (WB)	4–8 minutes (Simplate)	Values increase in thrombocytopenia, severe liver disease, leukemia, and aplastic anemia.
Blood urea nitrogen (BUN) (S)	8–23 mg/dl	Values increase in kidney disease, shock, dehydration, diabetes, and acute myocardial infarction (MI); values decrease in liver failure, impaired absorption, and overhydration.
Calcium (Ca^{2+}) (S)	Adults: 4.3–5.3 mEq/l Children: 6 mEq/l	Values increase in cancer, hyperparathyroidism, Addison's disease, hyperthyroidism, and Paget's disease; values decrease in hypoparathyroidism, chronic renal failure, osteomalacia, rickets, and diarrhea.
Carbon dioxide (CO_2), content (WB)	Arterial: 19–24 mM Venous: 22–26 mM	Values increase in severe vomiting, emphysema, and aldosteronism; values decrease in severe diarrhea, starvation, and acute renal failure.
Carbon dioxide, partial pressure ($P CO_2$) (WB)	Arterial: 40 mm Hg Venous: 45 mm Hg	Values increase in hypoventilation, obstructive lung disease, and emphysema; values decrease in hyperventilation, hypoxia, and pregnancy.
Carcinoembryonic antigen (CEA) (P)	0–25 mg/ml	Values increase in carcinoma of the liver and pancreas, inflammatory bowel disease (IBD), cirrhosis, and chronic cigarette smoking.
Carotene, beta (S)	40–200 μg/dl (varies with diet)	Values increase in myxedema, diabetes mellitus, and excessive dietary intake; values decrease in fat malabsorption, liver disease, and poor dietary intake.
Chloride (Cl^-) (S)	95–108 mEq/l	Values increase in dehydration, Cushing's syndrome, and anemia; values decrease in severe vomiting, severe burns, diabetic acidosis, and fever.
Cholesterol (S) Total	150–250 mg/dl (varies with diet, sex and age)	Values increase in diabetes mellitus, cardiovascular disease, nephrosis, and hypothyroidism; values decrease in liver disease, hyperthyroidism, fat malabsorption, pernicious anemia, severe infections, and terminal stages of cancer.
HDL cholesterol (P)	29–77 mg/dl	
LDL cholesterol (P)	62–185 mg/dl	
VLDL cholesterol (P)	0–40 mg/dl	
Clotting (coagulation) time (WB)	5–10 minutes (Lee-White)	Values increase in afibrinogenemia and hyperheparanemia.

EXHIBIT B-1   *(Continued)*

TEST (SPECIMEN)	NORMAL VALUES	CLINICAL IMPLICATION
Cortisol (hydro-cortisone) (P)	8 A.M.–10 A.M.: 5–25 μg/dl 4 P.M.–6 P.M.: 3–13 μg/dl	Values increase in hyperthyroidism, stress, obesity, and Cushing's syndrome; values decrease in hypothyroidism, liver disease, and Addison's disease.
Creatine (S or P)	Male: 0.2–0.6 mg/dl Female: 0.6–1.0 mg/dl	Values increase in nephritis, muscular dystrophy, damage to muscle tissue, and pregnancy.
Creatine kinase (CK) (S)	Male: 5–35 U/ml Female: 5–25 U/ml	Values increase in myocardial infarction, progressive muscular dystrophy, myxedema, convulsions, hypothyroidism, and pulmonary edema.
Creatinine (S)	0.6–1.5 mg/dl	Values increase in impaired renal function, giantism, and acromegaly; values decrease in muscular dystrophy.
Erythrocyte Sedimentation Rate (ESR) (WB)	(Westergren) Male: Under 50 years, 15 mm/hr Over 50 years, 20 mm/hr Female: Under 50 years, 20 mm/hr Over 50 years, 30 mm/hr	Values increase in infection, carcinoma, tissue destruction, and nephritis; values decrease in sickle-cell anemia and congestive heart failure (CHF).
Fetal hemoglobin (WB)	Newborns: 60–90% Before age 2: 0–4% Adults: 0–2%	Values increase in thalassemia, sickle-cell anemia, and leakage of fetal blood into maternal bloodstream.
Gamma-glutamyl transferase (GGT) (S)	Male: 4–23 IU/l Female: 3–5–13 IU/l	Values increase in cirrhosis of the liver, metastatic cancer of the liver, cholelithiasis, congestive heart failure (CHF), and alcoholism.
Globulins (S)	2.3–3.5 g/dl	Values increase in chronic infections.
Glucose (S)	70–110 mg/dl	Values increase in diabetes mellitus, acute stress, hyperthyroidism, chronic liver disease, and nephritis; values decrease in Addison's disease, hypothyroidism, and cancer of the pancreas.
Hemoglobin (S or P)	Male: 14–16.5 gm/100 ml Female: 12–15 gm/100 ml Newborn: 14–20 gm/100 ml	Values increase in polycythemia, congestive heart failure (CHF), chronic obstructive pulmonary disease, and at high altitudes; values decrease in anemia, hyperthyroidism, cirrhosis of the liver, and severe hemorrhage.
Hematocrit (WB)	Male: 40–54% Female: 38–47%	Values increase in polycythemia, severe dehydration, and shock; values decrease in anemia, leukemia, cirrhosis, and hyperthyroidism.
Immunoglobins (S) IgG IgA IgM IgD IgE	 800–1,801 mg/dl 113–563 mg/dl 54–222 mg/dl 0.5–3.0 mg/dl 0.01–0.04 mg/dl	IgG values increase in infections of all types, liver disease, and severe malnutrition; IgA values increase in cirrhosis of the liver, chronic infections, and autoimmune disorders and decrease in immunologic deficiency states; IgM values increase in trypanosomiasis and decrease in lymphoid aplasia; IgD values increase in chronic infections and myelomas; IgE values increase in hay fever, asthma, and anaphylactic shock.
Iron, total (S)	60–150 μg/dl	Values increase in liver disease and various anemias; values decrease in iron-deficiency anemia.
Ketone bodies (acetone) (S or P)	Negative: 0.3–2.0 mg/dl Toxic level: 20 mg/dl	Values increase in ketoacidosis, fever, anorexia, fasting, starvation, high-fat diet, and following vomiting.
Lactic acid ($CH_3 \cdot CHOH \cdot COOH$) (WB)	Arterial: 3–7 mg/dl Venous: 5–20 mg/dl	Values increase during muscular activity, congestive heart failure (CHF), shock, and severe hemorrhage.
Lactic dehydrogenase (LDL) (S)	150–450 Wroblewski units 71–207 IU/l	Values increase in myocardial infarction, liver disease, skeletal muscle necrosis, and extensive cancer.
Lipids (S) Total Cholesterol Triglycerides Phospholipids Fatty acids (free)	 400–800 mg/dl 150–250 mg/dl 10–190 mg/dl 150–380 mg/dl 9.0–15.0 mM/l	Values increase in hyperlipidemia, diabetes mellitus, and hypothyroidism; values decrease in fat malabsorption.
Mucoprotein (S)	80–200 mg/dl	Values increase in cancer, infections, and rheumatoid arthritis; values decrease in hepatitis A (infectious) and cirrhosis.

EXHIBIT B-1 (*Continued*)

TEST (SPECIMEN)	NORMAL VALUES	CLINICAL IMPLICATION
Osmolality (S)	275–300 mOsm/kg	Values increase in cirrhosis, congestive heart failure (CHF), and high-protein diets; values decrease in aldosteronism, diabetes insipidus, and hyper-calcemia.
Oxygen (O₂) content (WB)	Arterial: 15–23 ml/dl	Values increase in polycythemia; values decrease in chronic obstructive lung disease.
Oxygen, partial pressure (*PO₂*) (WB)	Arterial: 105 mm Hg	Values increase in polycythemia and hyperventilation; values decrease in anemias, insufficient atmospheric oxygen, and hypoventilation.
pH (WB)	Arterial: 7.35–7.45	Values increase in vomiting, hyperventilation, excessive bicarbonate, and lack of oxygen; values decrease in renal failure, diabetic ketoacidosis, hypoxia, airway obstruction, and shock.
Phosphatase (S)		
acid	4–13 U/dl (King-Armstrong)	Values increase in prostatic cancer, some liver disease, hyperparathyroidism, myocardial infarction (MI), and pulmonary embolism.
alkaline	4–13 U/dl (King-Armstrong)	Values increase in some liver diseases, hyperparathyroidism, and pregnancy.
Phosphorous, inorganic (S)	Adult: 1.8–4.1 mEq/l Children: 2.3–4.1 mEq/l	Value increase in renal disease, hypoparathyroidism, hypocalcemia, bone tumors, Addison's disease, and acromegaly; values decrease in hyper-parathyroidism, rickets, osteomalacia, and diabetic coma.
Platelet count (WB)	250,000–400,000/mm³	Values increase in cancer, trauma, heart disease, and cirrhosis; values decrease in anemias, allergic conditions, and during cancer chemotherapy.
Protein (S) Total Albumin Globulin A/G ratio	6.0–7.8 g/dl 3.2–4.5 g/dl 2.3–3.5 g/dl 1.5:1 to 2.5:1	Total protein values increase in dehydration, shock, systemic lupus erythematosis, (SLE), rheumatoid arthritis (RA), chronic infections, and chronic liver disease; total protein values decrease in insufficient protein intake, hemorrhage, malabsorption, diarrhea, and severe burns.
Protein-bound iodine (PBI) (S)	4.0–8.0 μ/dl	Values increase in hyperthyroidism and acute thyroiditis; values decrease in hypothyroidism, chronic thyroiditis, myxedema, and nephrosis.
Prothrombin time (WB)	11–15 seconds (depends on the concentration of thrombin reagent used)	Values increase in prothrombin and vitamin K deficiency, liver disease, and hypervitaminosis A.
Red blood cell count (WB)	Male: 5.4 million/mm³ Female: 4.8 million/mm³	Values increase in polycythemia, dehydration, and following hemorrhaging; values decrease in systemic lupus erythematosis (SLE), anemias, and Addison's disease.
Reticulocyte count (WB)	0.5–1.5%	Values increase in hemolytic anemia, metastatic carcinoma, and leukemia; values decrease in iron-deficiency and pernicious anemia and radiation therapy.
Sodium (Na⁺) (S)	136–142 mEq/l	Values increase in dehydration, aldosteronism, coma, Cushing's disease, and diabetes insipidus; values decrease in severe burns, vomiting, diarrhea, Addison's disease, nephritis, excessive sweating, and edema.
Thyroxine (T₄) (S)	2.9–6.4 μg/dl	Values increase in hyperthyroidism; values decrease in hypothyroidism.
Thyroxine-binding globulin (TBG) (S)	10–26 μg/dl	Values increase in hypothyroidism; values decrease in hyperthyroidism.
Uric acid (S)	Male: 4.0–8.5 mg/dl Female: 2.7–7.3 mg/dl	Values increase in impaired renal function, gout, metastatic cancer, shock, and starvation; values decrease in persons treated with uricosuric drugs.
White blood cell count, dif-ferential (WB) Neutrophils Eosinophils Basophils Lymphocytes Monocytes	60–70% 2–4% 0.5–1% 20–25% 3–8%	Neutrophils increase in acute infections; eosinophils and basophils increase in allergic reactions; lymphocytes increase during antigen-antibody reactions; monocytes increase in chronic infections.
White blood cell count, total (WB)	5000–9000/mm³	Values increase in acute infections, trauma, malignant diseases, and cardiovascular diseases; values decrease in diabetes mellitus, anemias, and following cancer chemotherapy.

EXHIBIT B-2   URINE TESTS

TEST	SAMPLE	NORMAL VALUES	CLINICAL IMPLICATION
Ammonia nitrogen	24 hour	20–70 mEq/l	Values increase in diabetes mellitus; values decrease in liver diseases.
Amylase	2 hour	35–260 Somogyi units/hr	Values increase in pancreatitis and choledocholethiasis.
Bilirubin	Random	Negative–0.002 mg/dl	Values increase in liver disease and obstructive biliary disease.
Blood, occult	Random	Negative	Values increase in renal disease; extensive burns, transfusion reactions, and hemolytic anemia.
Calcium (Ca^{2+})	Random 24 hour Average diet: Low Ca^{2+} diet: High Ca^{2+} diet:	10 mg/dl 100–240 mg/24 hr 150 mg/24 hr 240–300 mg/24 hr	Values increase in hyperparathyroidism, metastatic malignancies, and primary cancer of breasts and lungs; values decrease in hypoparathyroidism and vitamin D deficiency.
Casts Epithelial	24 hour	Occasional	Values increase in nephrosis and heavy metal poisoning.
Granular	24 hour	Occasional	Values increase in nephritis and pyelonephritis.
Hyaline	24 hour	Occasional	Values increase in glomerular membrane damage and fever.
Red blood cell	24 hour	Occasional	Values increase in pyelonephritis, kidney stones, and cystitis.
White blood cell	24 hour	Occasional	Values increase in kidney infections.
Chloride (Cl^-)	24 hour	110–254 mEq/24 hr	Values increase in Addison's disease, dehydration, and starvation; values decrease in pyloric obstruction, diarrhea, and emphysema.
Color	Random	Yellow, straw, amber	Varies with many disease states, hydration, and diet.
Concentration test	Random, and after restriction	Sp.Gr.: 1.025 Osmolality: 850 mOsm/l	Values decrease in renal disease, hyperparathyroidism, and bone diseases.
Creatinine	24 hour	Male: 1.0–2.0 g/24 hr Female: 0.8–1.8 g/24 hr	Values increase in infections; values decrease in muscular atrophy, anemia, and kidney diseases.
Creatinine clearance	24 hour	115–120 ml/min	Values increase in renal diseases.
Glucose	Random	Negative	Values increase in diabetes mellitus, brain injury, and myocardial infarction.
Hydroxycorticosteroids	24 hour	Male: 5–15 mg/24 hr Female: 2–13 mg/24 hr	Values increase in Cushing's syndrome, burns, and infections; values decrease in Addison's disease.
Ketone bodies (acetone)	Random	Negative	Values increase in diabetic acidosis, fever, anorexia, fasting, and starvation.
17-keto-steroids (KS)	24 hour	Male: 8–25 mg/24 hr Female: 5–15 mg/24 hr	Values decrease in surgery, burns, infections, adrenogenital syndrome, and Cushing's syndrome.
Odor	Random	Aromatic	Becomes acetonelike in diabetic ketosis.
Osmolality	24 hour	500–800 mOsm/kg water	Values increase in cirrhosis, congestive heart failure (CHF), and high protein diets; values decrease in aldosteronism, diabetes, insipidus, and hypokalemia.
pH	Random	4.6–8.0	Values increase in urinary tract infections and severe alkalosis; values decrease in acidosis, emphysema, starvation, and dehydration.
Phenylpyruvic acid	Random	Negative	Values increase in phenylketonuria (PKU).

EXHIBIT B-2 *(Continued)*

TEST	SAMPLE	NORMAL VALUES	CLINICAL IMPLICATION
Potassium (K⁺)	24 hour	25–100 mEq/l	Values increase in chronic renal failure, dehydration, starvation, and Cushing's syndrome; values decrease in diarrhea, malabsorption syndrome, and adrenal cortical insufficiency.
Protein (albumin)	Random	Negative	Values increase in nephritis, fever, severe anemias, trauma, and hyperthyroidism.
Sodium (Na⁺)	24 hour	75–200 mg/24 hr	Values increase in dehydration, starvation, and diabetic acidosis; values decrease in diarrhea, acute renal failure, emphysema, and Cushing's syndrome.
Specific gravity	Random	1.016–1.022 (normal fluid intake) 1.001–1.035 (range)	Values increase in diabetes mellitus and excessive water loss; values decrease in absence of antidiuretic hormone (ADH) and severe renal damage.
Urea	Random	25–35 g/24 hr	Values increase in response to increased protein intake; values decrease in impaired renal function.
Urea clearance	24 hour	64–99 ml/min (Maximum) 41–65 ml/min (Standard) or more than 75% of normal clearance	Values increase in renal diseases.
Uric acid	24 hour	0.4–1.0 g/24 hr	Values increase in gout, leukemia, and liver disease; values decrease in kidney disease.
Urobilinogen	2 hour	0.3–1.0 Ehrlich units	Values increase in anemias, hepatitis A (infectious), biliary disease, and cirrhosis; values decrease in cholelithiasis and renal insufficiency.
Volume, total	24 hour	1000–2000 ml/24 hr	Varies with many factors.

Normal values researched and provided by Professor John Lo Russo, Bergen Community College, Division of Allied Health Sciences. The values in the exhibit are not intended to be definitive since they may vary among different laboratories.

Appendix C: Abbreviations

Many terms, especially medical terms, are commonly expressed in abbreviated form. In order to familiarize you with some of these abbreviations, an alphabetical list has been prepared. Most of the terms listed have been used in the book; some terms not referred to in the book have been included because of their frequent use.

ACS	acute confusional state
AD	Alzheimer's disease
AID	automatic implantable defibrillation; artificial insemination by donor
AIDS	acquired immune deficiency syndrome
AMI	acute myocardial infarction
ALS	amyotrophic lateral sclerosis
ANS	autonomic nervous system
ARF	acute renal failure
ATP	adenosine triphosphate
AV	atrioventricular
BBB	blood–brain barrier
BEAM	brain electrical activity mapping
BMR	basal metabolic rate
BP	blood pressure
BS	blood sugar
BUN	blood urea nitrogen
CABG	coronary artery bypass grafting
CAC	cardioacceleratory center
CAD	coronary artery disease
CAPD	continuous ambulatory peritoneal dialysis
CBC	complete blood count
CCCC	closed-chest cardiac compression
CF	cystic fibrosis
CHD	coronary heart disease
CHF	congestive heart failure
CIC	cardioinhibitory center
CNS	central nervous system
CO	cardiac output; carbon monoxide
COAD	chronic obstructive airways disease
COPD	chronic obstructive pulmonary disease
CP	cor pulmonale
CPR	cardiopulmonary resuscitation
CRF	chronic renal failure
C-section	cesarean section
CSF	cerebrospinal fluid
CT (CAT)	computed tomography (computed axial tomography)
CUSA	cavitron ultrasonic surgical aspirator
CVA	cerebrovascular accident
CVP	central venous pressure
CVS	chronic villi sampling
D & C	dilation and curretage
DBP	diastolic blood pressure
DES	diethylstilbestrol
DH	delayed hypersensitivity
DMD	Duchenne muscular dystrophy
DMSO	dimethyl sulfoxide
DNA	deoxyribonucleic acid
DNR	do not resuscitate
DSR	dynamic spatial reconstructor
DVT	deep-venous thrombosis
Dx	diagnosis

EBV	Epstein-Barr virus
ECF	extracellular fluid
ECG (EKG)	electrocardiogram
ECT	electroconvulsive therapy
EDV	end-diastolic volume
EEG	electroencephalogram
EM	electron micrograph
EMG	electromyogram
EOG	electrooculogram
EP	ectopic pregnancy
EPSP	excitatory postsynaptic potential
ER	endoplasmic reticulum
ESR	erythrocyte sedimentation rate
ESRD	end-stage renal disease
ESV	end-systolic volume
ESWL	extracorporeal shock wave lithotripsy
FAS	fetal alcohol syndrome
FUO	fever of unknown origin
GFR	glomerular filtration rate
GI	gastrointestinal
Hb	hemoglobin
HBO	hyperbaric oxygenation
HBV	hepatitis B virus
HDCV	human diploid cell vaccine
HDL	high-density lipoprotein
HF	heart failure
HR	heart rate
HSV	herpes simplex virus
IBS	irritable bowel syndrome
ICF	intracellular fluid
ICU	intensive care unit
ID	intradermal
IM	intramuscular; infectious mononucleosis
INF	interferon
IPSP	inhibitory postsynaptic potential
IUCD	intrauterine contraceptive device
IUD	intrauterine device
IV	intravenous
LDL	low-density lipoprotein
LLQ	left lower quadrant
LUQ	left upper quadrant
LV	left ventricular
MRI	magnetic resonance imaging
meq/liter	milliequivalents per liter
MG	myasthenia gravis
MI	myocardial infarction
mm^3	cubic millimeter
mm Hg	millimeters of mercury
MS	multiple sclerosis
MVP	mitral valve prolapse
NGU	nongonococcal urethritis
NLMC	nocturnal leg muscle cramping
NPN	nonprotein nitrogen

NREM	nonrapid eye movement		Rx	prescription
OC	oral contraceptive		SA	sinoatrial (sinuatrial)
OTC	over-the-counter		SCA	sickle-cell anemia
PCT	positron computed tomography		SCID	severe combined immunodeficiency syndrome
PE	pulmonary embolism		SIDS	sudden infant death syndrome
P_{eff}	effective filtration pressure		SLE	systemic lupus erythematosus
PEMFs	pulsating electromagnetic fields		SMD	senile macular degeneration
PET	positron emission tomography		SNS	somatic nervous system
PG	prostaglandin		SPF	sun protection factor
PID	pelvic inflammatory disease		STD	sexually transmitted disease
PKU	phenylketonuria		SV	stroke volume
PMNs	polymorphonuclear leucocytes		TIA	transient ischemic attack
PMS	premenstrual syndrome		Tm	tubular maximum
PNS	peripheral nervous system		TM	transcendental meditation
PPNG	penicillinase-producing *Neisseria gonorrhea*		TM joint	temporomandibular joint
PTA	percutaneous transluminal angioplasty		t-PA	tissue plasminogen activator
PUL	percutaneous ultrasonic lithotripsy		TPN	total parenteral nutrition
RA	rheumatoid arthritis		TSS	toxic shock syndrome
RAS	reticular activating system		Tx	treatment
RBC	red blood cell		UA	urinalysis
RDS	respiratory distress syndrome		URI	upper respiratory infection
REM	rapid eye movement		UTI	urinary tract infection
RES	reticuloendothelial system		UV	ultraviolet
Rh	*Rhesus*		VD	venereal disease
RIG	rabies immune globulin		VF	ventricular fibrillation
RLQ	right lower quadrant		VPC	ventricular premature contraction
RM	radical mastectomy		VSD	ventricular septal defect
RNA	ribonucleic acid		VT	ventricular tachycardia
RUQ	right upper quadrant		WBC	white blood cell

Appendix D: Eponyms Used in This Text

Eponymous terms are those named after a person. In general, eponyms should be avoided where possible, since they are totally nondescriptive, often vague, and do not necessarily indicate that the person whose name is used actually contributed anything very original. However, since eponyms are still in frequent use, this glossary has been prepared to indicate which current terms have been used to replace eponyms in this book. In the body of the text eponyms are cited in parentheses, immediately following the current terms where they are used for the first time in a chapter or later in the book. In addition, although eponyms are included in the index, they have been cross referenced to their current terminology.

EPONYM	CURRENT TERMINOLOGY
Adam's apple	thyroid cartilage
ampulla of Vater (VA-ter)	hepatopancreatic ampulla
Bartholin's (BAR-tō-linz) gland	greater vestibular gland
Billroth's (BIL-rōtz) cord	splenic cord
Bowman's (BŌ-manz) capsule	glomerular capsule
Bowman's (BŌ-manz) gland	olfactory gland
Broca's (BRŌ-kaz) area	motor speech area
Brunner's (BRUN-erz) gland	duodenal gland
bundle of His (HISS)	atrioventricular (AV) bundle
canal of Schlemm (SHLEM)	scleral venous sinus
circle of Willis (WIL-is)	cerebral arterial circle
Cooper's (KOO-perz) ligament	suspensory ligament of the breast
Cowper's (KOW-perz) gland	bulbourethral gland
crypt of Lieberkühn (LĒ-ber-kyoon)	intestinal gland
duct of Rivinus (re-VĒ-nus)	lesser sublingual duct
duct of Santorini (san'-tō-RĒ-nē)	accessory duct
duct of Wirsung (VĒR-sung)	pancreatic duct
end organ of Ruffini (roo-FĒ-nē)	type II cutaneous mechanoreceptor
Eustachian (yoo-STĀ-kē-an) tube	auditory tube
Fallopian (fal-LŌ-pē-an) tube	uterine tube
gland of Zeis (ZĪS)	sebaceous ciliary gland
Golgi (GOL-jē) tendon organ	tendon organ
Graafian (GRAF-ē-an) follicle	vesicular ovarian follicle
Hassall's (HAS-alz) corpuscle	thymic corpuscle
Haversian (ha-VĒR-shun) canal	central canal
Haversian (ha-VĒR-shun) system	osteon
interstitial cell of Leydig (LĪ-dig)	interstitial endocrinocyte
islet of Langerhans (LANG-er-hanz)	pancreatic islet
Kupffer's (KOOP-ferz) cells	stellate reticuloendothelial cell
loop of Henle (HEN-lē)	loop of the nephron
Malpighian (mal-PIG-ē-an) corpuscle	splenic nodule
Meibomian (mi-BŌ-mē-an) gland	tarsal gland
Meissner's (MĪS-nerz) corpuscle	corpuscle of touch
Merkel's (MER-kelz) disc	tactile disc
Müller's (MIL-erz) duct	paramesonephric duct
Nissl (NIS-l) bodies	chromatophilic substance
node of Ranvier (ron-VĒ-ā)	neurofibral node
organ of Corti (KOR-tē)	spiral organ
Pacinian (pa-SIN-ē-an) corpuscle	lamellated corpuscle
Peyer's (PĪ-erz) patches	aggregated lymphatic follicles
plexus of Auerbach (OW-er-bak)	myenteric plexus
plexus of Meissner (MĪS-ner)	submucous plexus
pouch of Douglas	rectouterine pouch
Purkinje (pur-KIN-jē) fiber	conduction myofiber

EPONYM	CURRENT TERMINOLOGY
Rathke's (rath-KEZ) pouch	hypophyseal pouch
Schwann (SCHVON) cell	neurolemmocyte
Sertoli (ser-TŌ-lē) cell	sustentacular cell
Skene's (SKĒNZ) gland	paraurethral gland
sphincter of Oddi (OD-dē)	sphincter of the hepatopancreatic ampulla
Stensen's (STEN-senz) duct	parotid duct
Volkmann's (FŌLK-manz) canal	perforating canal
Wharton's (HWAR-tunz) duct	submandibular duct
Wharton's (HWAR-tunz) jelly	mucous connective tissue
Wormian (WER-mē-an) bone	sutural bone

Glossary of Prefixes, Suffixes, and Combining Forms

Many medical terms are "compound" words, that is, they are made up of one or more word roots or combining forms of word roots with prefixes or suffixes. For example, *leucocyte* (white blood cell) is a combination of *leuco,* the combining form for the word root meaning "white," and *cyt,* the word root meaning "cell." Learning the medical meanings of the fundamental word parts will enable you to analyze many long, complicated terms.

The following list includes some of the most commonly used word roots, combining forms, prefixes, and suffixes used in making medical terms and an example for each.

WORD ROOTS AND COMBINING FORMS

Acou-, Acu- hearing Acoustics (a-KOO-stiks), the science of sounds or hearing.

Acr-, Acro- extremity Acromegaly (ak'-rō-MEG-a-lē), hyperplasia of the nose, jaws, fingers, and toes.

Aden-, Adeno- gland Adenoma (ad-en-'Ō-ma), a tumor with a glandlike structure.

Alg-, Algia- pain Neuralgia (nyoo-RAL-ja), pain along the course of a nerve.

Angi- vessel Angiocardiography (an'-jē-ō-kard-ē-OG-ra-fē), roentgenography of the great blood vessels and heart after intravenous injection of radiopaque fluid.

Arthr-, Arthro- joint Arthropathy (ar-THROP-a-thē), disease of a joint.

Aut-, Auto- self Autolysis (aw-TOL-i-sis), destruction of cells of the body by their own enzymes, even after death.

Bio- life, living Biopsy (BĪ-op-sē), examination of tissue removed from a living body.

Blast- germ, bud Blastocyte (BLAS-tō-sīt), an embryonic or undifferentiated cell.

Blephar- eyelid Blepharitis (blef-a-RĪT-is), inflammation of the eyelids.

Brachi- arm Brachialis (brā-kē-AL-is), muscle that flexes the forearm.

Bronch- trachea, windpipe Bronchoscopy (bron-KOS-kō-pē), direct visual examination of the bronchi.

Bucc-, cheek Buccocervical (bū-kō-SER-vi-kal), pertaining to the cheek and neck.

Capit- head Decapitate (dē-KAP-i-tāt), to remove the head.

Carcin- cancer Carcinogenic (kar-sin-ō-JEN-ik), causing cancer.

Cardi-, Cardia-, Cardio- heart Cardiogram (KARD-ē-o-gram), a recording of the force and form of the heart's movements.

Cephal- head Hydrocephalus (hī-drō-SEF-a-lus), enlargement of the head due to an abnormal accumulation of fluid.

Cerebro- brain Cerebrospinal (se-rē'-brō-SPĪN-al) fluid, fluid contained within the cranium and spinal canal.

Cheil- lip Cheilosis (kī-LŌ-sis), dry scaling of the lips.

Chole- bile, gall Cholecystogram (kō-lē-SIS-tō-gram), roentgenogram of the gallbladder.

Chondr-, Chondri-, Chondrio- cartilage Chondrocyte (KON-drō-sīt), a cartilage cell.

Chrom-, Chromat-, Chromato- color Hyperchromic (hī-per-KRŌ-mik), highly colored.

Crani- skull Craniotomy (krā-nē-OT-o-mē), surgical opening of the skull.

Cry-, Cyro- cold Cryosurgery (krī-ō-SERJ-e-rē), surgical procedure using a very cold liquid nitrogen probe.

Cut- skin Subcutaneous (sub-kyoo-TĀ-nē-us), under the skin.

Cysti-, Cysto- sac, bladder Cystoscope (SIS-tō-skōp), instrument for interior examination of the bladder.

Cyt-, Cyto-, Cyte- cell Cytology (sī-TOL-ō-jē), the study of cells.

Dactyl-, Dactylo- digits (usually fingers, but sometimes toes) Polydactylism (pol-ē-DAK-til-ism), above normal number of fingers or toes.

Derma-, Dermato- skin Dermatosis (der-ma-TŌ-sis), any skin disease.

Entero- intestine Enteritis (ent-e-RĪT-is), inflammation of the intestine.

Erythro- red Erythrocyte (e-RITH-rō-sīt), red blood cell.

Galacto- milk Galactose (ga-LAK-tose), a milk sugar.

Gastr- stomach Gastrointestinal (gas'-trō-in-TES-tin-al), pertaining to the stomach and intestine.

Gloss-, Glosso- tongue Hypoglossal (hī'-pō-GLOS-al), located under the tongue.

Glyco- sugar Glycosuria (glī'-kō-SHUR-ē-a), sugar in the urine.

Gyn-, Gyne-, Gynec- female, women Gynecology (gīn'-e-KOL-ō-jē), the medical specialty dealing with disorders of the female.

Hem-, Hemat- blood Hematoma (hē'-ma-TŌ-ma), a tumor or swelling filled with blood.

Hepar-, Hepato- liver Hepatitis (hep-a-TĪT-is), inflammation of the liver.

Hist-, Histio- tissue Histology (his-TOL-ō-jē), the study of tissues.

Hydr- water Hydrocele (HĪ-drō-sēl), accumulation of fluid in a saclike cavity.

Hyster- uterus Hysterectomy (his'-te-REK-tō-mē), surgical removal of the uterus.

Ileo- ileum Ileocecal (il'-ē-ō-SĒ-kal) valve, folds at the opening between ileum and cecum.

Ilio- ilium Iliosacral (il'-ē-ō-SA-kral), pertaining to ilium and sacrum.

Lachry-, Lacri- tears Nasolacrimal (nā-zō-LAK-rim-al), pertaining to the nose and lacrimal apparatus.

Laparo- loin, flank, abdomen Laparoscopy (lap'-a-ROS-kō-pē), examination of the interior of the abdomen by means of a laparoscope.

Leuco-, Leuko- white Leucocyte (LYOO-kō-sīt), white blood cell.

Lip-, Lipo- fat Lipoma (lī-PŌ-ma), a fatty tumor.

Mamm- breast Mammography (ma-MOG-ra-fē), roentgenography of the mammary gland.

Mast- breast Mastitis (ma-STĪT-is), inflammation of the mammary gland.

Meningo- membrane Meningitis (men-in-JĪT-is), inflammation of the membranes of spinal cord and brain.

Metro- uterus Endometrium (en'-dō-MĒ-trē-um), lining of the uterus.

Morpho- form, shape Morphology (mor-FOL-o-jē), the study of form and structure of things.

Myelo- marrow, spinal cord Poliomyelitis (pō'-lē-ō-mī'-a-LĪT-is), inflammation of the gray matter of the spinal cord.

Myo- muscle Myocardium (mī'-ō-KARD-ē-um), heart muscle.

Necro- corpse, dead Necrosis (ne-KRŌ-sis), death of areas of tissue surrounded by healthy tissue.

Nephro- kidney Nephrosis (ne-FRŌ-sis), degeneration of kidney tissue.

Neuro- nerve Neuroblastoma (nyoor'-ō-blas-TŌ-ma), malignant tumor of the nervous system composed of embryonic nerve cells.

Oculo- eye Binocular (bī-NOK-yoo-lar), pertaining to the two eyes.

Odont- tooth Orthodontic (or-thō-DONT-ik), pertaining to the proper positioning and relationship of the teeth.

Oo- egg Oocyte (Ō-ō-sīt), original egg cell.

Ophthalm- eye Ophthalmology (of'-thal-MOL-o-jē), the study of the eye and its diseases.

Oss-, Osseo-, Osteo- bone Osteoma (os-tē-Ō-ma), bone tumor.

Oto- ear Otosclerosis (ō'-tō-skle-RŌ-sis), formation of bone in the labyrinth of the ear.

Patho- disease Pathogenic (path'-ō-JEN-ik), causing disease.

Ped- children Pediatrician (pēd-ē-a-TRISH-an), medical specialist in the treatment of children.

Phag-, Phago- to eat Phagocytosis (fag'-ō-sī-TŌ-sis), the process by which cells ingest particulate matter.

Philic-, Philo- to like, have an affinity for Hydrophilic (hī-drō-FIL-ik), having an affinity for water.

Phleb- vein Phlebitis (fli-BĪT-is), inflammation of the veins.

Pneumo- lung, air Pneumothorax (nyoo-mō-THŌR-aks), air in the thoracic cavity.

Pod- foot Podiatry (po-DĪ-a-trē), the diagnosis and treatment of foot disorders.

Procto- anus, rectum Proctoscopy (prok-TOS-kō-pē), instrumental examination of the rectum.

Psycho- soul, mind Psychiatry (sī-KĪ-a-trē), treatment of mental disorders.

Pyo- pus Pyuria (pī-YOOR-ē-a), pus in the urine.

Rhin- nose Rhinitis (rī-NĪT-is), inflammation of nasal mucosa.

Scler-, Sclero- hard Atherosclerosis (ath'-er-ō-skle-RŌ-sis), hardening of the arteries.

Sep-, Septic- toxic condition due to microorganisms Septicemia (sep'-ti-SĒ-mē-a), presence of bacterial toxins in the blood (blood poisoning).

Soma-, Somato- body Somatotropic (sō-mat-ō-TRŌ-pik), having a stimulating effect on body growth.

Sten- narrow Stenosis (ste-NŌ-sis), narrowing of a duct or canal.

Stasis-, Stat- stand still Homeostasis (hō'-mē-ō-STĀ-sis), achievement of a steady state.

Therm- heat Thermometer (ther-MOM-et-er), instrument used to measure and record heat.

Tox-, Toxic- poison Toxemia (tok-SĒ-mē-a), poisonous substances in the blood.

Trich- hair Trichosis (trik-Ō-sis), disease of the hair.

Viscer- organ Visceral (VIS-e-ral), pertaining to the abdominal organs.

Zoo- animal Zoology (zō-OL-o-jē), the study of animals.

PREFIXES

A-, An- without, lack of, deficient Anesthesia (an-'es-THĒ-zha), without sensation.

Ab- away from, from Abnormal (ab-NOR-mal), away from normal.

Ad- to, near, toward Adduction (a-DUK-shun), movement of an extremity toward the axis of the body.

Ambi- both sides Ambidextrous (am'-bi-DEK-strus), able to use either hand.

Ante- before Antepartum (ant-ē-PAR-tum), before delivery of a baby.

Anti- against Anticoagulant (an-tī-kō-AG-yoo-lant), a substance that prevents coagulation of blood.

Bi- two, double, both Biceps (BĪ-seps), a muscle with two heads of origin.

Brachy- short Brachyesophagus (brā-kē-e-SOF-a-gus), short esophagus.

Brady- slow Bradycardia (brād'-ē-KARD-ē-a), abnormal slowness of the heartbeat.

Cata- down, lower, under, against Catabolism (ka-TAB-a-lizm), metabolic breakdown into simpler substances.

Circum- around Circumrenal (ser-kum-RĒN-al), around the kidney.

Col-, Con-, Com- with, together Congenital (kon-JEN-i-tal), existing at birth.

Contra- against, opposite Contraception (kon-tra-SEP-shun), the prevention of conception.

Crypt- hidden, concealed Cryptorchidism (krip-TOR-ka-dizm'), undescended or hidden testes.

Di-, Diplo- two Diploid (DIP-loyd), having double the haploid number of chromosomes.

Dis- separation, apart, away from Disarticulate (dis'-ar-TIK-yoo-lāt'), to separate at a joint.

Dys- painful, difficult Dyspnea (disp-NĒ-a), difficult breathing.

E-, Ec-, Ex- out from, out of Eccentric (ek-SEN-trik), not located at the center.

Ecto-, Exo- outside Ectopic pregnancy (ek-TOP-ik), gestation outside the uterine cavity.

Em-, En- in, on Empyema (em'-pī-Ē-ma), pus in a body cavity.

End-, Endo- inside Endocardium (en'-dō-KARD-ē-um), membrane lining the inner surface of the heart.

Epi- upon, on, above Epidermis (ep'-i-DER-mis), outermost layer of skin.

Eu- well Eupnea (YOOP-nē-a), normal breathing.

Ex-, Exo- out, away from Exocrine (EK-sō-krin), excreting outwardly or away from.

Extra- outside, beyond, in addition to Extracellular (ek'-stra-SEL-yoo-lar), outside of the cell.

Hemi- half Hemiplegia (hem-ē-PLĒ-jē-a), paralysis of only half of the body.

Heter-, Hetero- other, different Heterogeneous (het'-e-rō-JEN-ē-us), composed of different substances.

Homeo-, Homo- unchanging, the same, steady Homeostasis (hō'-mē-ō-STĀ-sis), achievement of a steady state.

Hyper- beyond, excessive Hyperglycemia (hī-per-glī-SĒ-mē-a), excessive amount of sugar in the blood.

Hypo- under, below, deficient Hypodermic (hī'-pō-DER-mik), below the skin or dermis.

Idio- self, one's own, separate Idiopathic (id'-ē-ō-PATH-ik), a disease without recognizable cause.

Inter- among, between Intercostal (int'-er-KOS-tal), between the ribs.

Intra- within, inside Intracellular (in'-tra-SEL-yoo-lar), inside the cell.

Iso- equal, like Isogenic (ī-sō-JEN-ik), alike in morphological development.

Macro- large, great Macrophage (MAK-rō-fāj), large phagocytic cell.

Mal- bad, abnormal Malnutrition (mal'-noo-TRISH-un), lack of necessary food substances.

Mega-, Megalo- great, large Megakaryocyte (meg'-a-KAR-ē-ō-sīt), giant cell of bone marrow.

Meta- after, beyond Metacarpus (met'-a-KAR-pus), the part of the hand between the wrist and fingers.

Micro- small Microtome (MĪ-krō-tōm), instrument for preparing very thin slices of tissue for microscopic examination.

Neo- new Neonatal (nē-ō-NĀT-al), pertaining to the first 4 weeks after birth.

Oligo- small, deficient Oliguria (ol-ig-YOO-rē-a), abnormally small amount of urine.

Ortho- straight, normal Orthopnea (or-thop-NĒ-a), inability to breathe in any position except when straight or erect.

Para- near, beyond, apart from, beside Paranasal (par-a-NĀ-zal), near the nose.

Per- through Percutaneous (per'-kyoo-TĀ-nē-us), through the skin.

Peri- around Pericardium (per'-i-KARD-ē-um), membrane or sac around the heart.

Poly- much, many Polycythemia (pol'-i-sī-THĒ-mē-a), an excess of red blood cells.

Post- after, beyond Postnatal (pōst-NĀT-al), after birth.

Pre-, Pro- before, in front of Prenatal (prē-NĀT-al), before birth.

Pseud-, Pseudo- false Pseudoangina (soo'-dō-an-JĪ-na), false angina.

Retro- backward, located behind Retroperitoneal (re'-trō-per'-it-on-Ē-al), located behind the peritoneum.

Semi- half Semicircular (sem'-i-SER-kyoo-lar) canals, canals in the shape of a half circle.

Sub- under, beneath, below Submucosa (sub'-myoo-KŌ-sa), tissue layer under a mucous membrane.

Super- above, beyond Superficial (soo-per-FISH-al), confined to the surface.

Supra- above, over Suprarenal (soo-pra-RĒN-al), adrenal gland above the kidney.

Sym-, Syn- with, together, joined Syndrome (SIN-drōm), all the symptoms of a disease considered as a whole.

Tachy- rapid Tachycardia (tak'-i-KARD-ē-a), rapid heart action.

Trans- across, through, beyond Transudation (trans-yoo-DĀ-shun), oozing of a fluid through pores.

Tri- three Trigone (TRĪ-gon), a triangular space, as at the base of the bladder.

SUFFIXES

-able capable of, having ability to Viable (VĪ-a-bal), capable of living.

-ac, -al pertaining to Cardiac (KARD-ē-ak), pertaining to the heart.

-ary connected with Ciliary (SIL-ē-ar-ē), resembling any hairlike structure.

-asis, -asia, -esis, -osis condition or state of Hemostasis (hē-mō-STĀ-sis), stopping of bleeding or circulation.

-cel, -cele swelling, an enlarged space or cavity Meningocele (men-IN-gō-sēl), enlargement of the meninges.

-cid, -cide, -cis cut, kill, destroy Germicide (jer-mi-SĪD), a substance that kills germs.

-ectasia, -ectasis stretching, dilation Bronchiectasis (bron-kē-EK-ta-sis), dilation of a bronchus or bronchi.

-ectomize, ectomy excision of, removal of Thyroidectomy (thī-royd-EK-to-mē), surgical removal of a thyroid gland.

-emia condition of blood Lipemia (lip-Ē-mē-a), abnormally high concentration of fat in the blood.

-ferent carry Efferent (EF-e-rent), carrying away from a center.

-form shape Fusiform (FYOO-zi-form), spindle-shaped.

-gen agent that produces or originates Pathogen (PATH-o-jen), microorganism or substance capable of producing a disease.

-genic produced from, producing Pyogenic (pī-ō-JEN-ik), producing pus.

-gram record, that which is recorded Electrocardiogram (e-lek'-trō-KARD-ē-ō-gram), record of heart action.

-graph instrument for recording Electroencephalograph (e-lek'-trō-en-SEF-a-lō-graf), instrument for recording electrical activity of the brain.

-ia state, condition Hypermetropia (hī'-per-me-TRŌ-pē-a), condition of farsightedness.

-iatrics, iatry medical practice specialties Pediatrics (pēd-ē-A-triks), medical science relating to care of children and treatment of their diseases.

-ism condition, state Rheumatism (ROO-ma-tizm), inflammation, especially of muscle and joints.

-itis inflammation Neuritis (nyoo-RĪT-is), inflammation of a nerve or nerves.

-logy, -ology the study or science of Physiology (fiz-ē-OL-o-jē), the study of function of body parts.

-lyso, -lysis solution, dissolve, loosening Hemolysis (hē-MOL-i-sis), dissolution of red blood cells.

-malacia softening Osteomalacia (os'-tē-ō-ma-LA-shē-a), softening of bone.

-oma tumor Fibroma (fi-BRŌ-ma), tumor composed mostly of fibrous tissue.

-ory pertaining to Sensory (SENS-o-rē), pertaining to sensation.

-ose full of Adipose (AD-i-pōz), characterized by presence of fat.

-pathy disease Neuropathy (nyoo-ROP-a-thē), disease of the peripheral nervous system.

-penia deficiency Thrombocytopenia (throm'-bō-sīt'-o-PĒ-nē-a), deficiency of thrombocytes in the blood.

-phobe, -phobia fear of, aversion to Hydrophobia (hī'-drō-FŌ-bē-a), fear of water.

-plasia, -plasty development, formation Rhinoplasty (RĪ-nō-plas-tē), surgical reconstruction of the nose.

-plegia, -plexy stroke, paralysis Apoplexy (AP-o-plek-sē), sudden loss of consciousness and paralysis.

-pnea to breathe Apnea (AP-nē-a), temporary absence of respiration, following a period of overbreathing.

-poiesis production Hematopoiesis (he-mat'-a-poy-Ē-sis), formation and development of red blood cells.

-rrhea flow, discharge Diarrhea (dī-a-RĒ-a), abnormal frequency of bowel evacuation, the stools with a more or less fluid consistency.

-scope instrument for viewing Bronchoscope (BRON-ko-skōp), instrument used to examine the interior of a bronchus.

-stomy creation of a mouth or artificial opening Tracheostomy (trā-kē-OST-o-mē), creation of an opening in the trachea.

-tomy cutting into, incision into Appendectomy (ap-en-DEK-to-mē), surgical removal of the appendix.

-trophy state relating to nutrition or growth Hypertrophy (hī-PER-trō-fē), excessive growth of an organ or part.

-tropic turning toward, influencing, changing Gonadotropic (gō-nad-a-TRŌ-pic), influencing the gonads.

-uria urine Polyuria (pol-ē-YOOR-ē-a), excessive secretion of urine.

Glossary of Terms

PRONUNCIATION KEY

1. The strongest accented syllable appears in capital letters, for example, bilateral (bī-LAT-er-al) and diagnosis (dī-ag-NŌ-sis).
2. If there is a secondary accent, it is noted by a single quote mark ('), for example, constitution (kon'-sti-TOO-shun) and physiology (fiz'-ē-OL-ō-jē). Any additional secondary accents are also noted by a single quote mark, for example, decarboxylation (dē'-kar-bok'-si-LĀ-shun).
3. Vowels marked with a line above the letter are pronounced with the long sound, as in the following common words:

   ā as in māke
   ē as in bē
   ī as in īvy
   ō as in pōle

4. Vowels not so marked are pronounced with the short sound, as in the following words:

   e as in bet
   i as in sip
   o as in not
   u as in bud

5. Other phonetic symbols are used to indicate the following sounds:

   a as in above
   oo as in sue
   yoo as in cute
   oy as in oil

Abatement (a-BĀT-ment) A decrease in the seriousness of a disorder or in the severity of pain or other symptoms.

Abdomen (ab-DŌ-men) The area between the diaphragm and pelvis.

Abdominal (ab-DŌM-i-nal) **cavity** Superior portion of the abdominopelvic cavity that contains the stomach, spleen, liver, gallbladder, pancreas, small intestine, and most of the large intestine.

Abdominal thrust maneuver A first-aid procedure for choking. Employs a quick, upward thrust against the diaphragm that forces air out of the lungs with sufficient force to eject any lodged material. Also called the **Heimlich** (HĪM-lik) **maneuver.**

Abdominopelvic (ab-dom'-i-nō-PEL-vic) **cavity** Inferior component of the ventral body cavity that is subdivided into an upper abdominal cavity and a lower pelvic cavity.

Abduction (ab-DUK-shun) Movement away from the axis or midline of the body or one of its parts.

Abortion (a-BOR-shun) The premature loss or removal of the embryo or nonviable fetus; any failure in the normal process of developing or maturing.

Abscess (AB-ses) A localized collection of pus and liquefied tissue in a cavity.

Absorption (ab-SORP-shun) The taking up of liquids by solids or of gases by solids or liquids; intake of fluids or other substances by cells of the skin or mucous membranes; the passage of digested foods from the GI tract into blood or lymph.

Absorptive state Metabolic state during which ingested nutrients are being absorbed by the blood or lymph from the GI tract.

Accessory duct A duct of the pancreas that empties into the duodenum about 2.5 cm (1 inch) superior to the ampulla of Vater (hepatopancreatic ampulla). Also called the **duct of Santorini** (san'-tō-RE-nē).

Accommodation (a-kom-ō-DĀ-shun) A change in the curvature of the eye lens to adjust for vision at various distances; focusing.

Acetabulum (as'-e-TAB-yoo-lum) The rounded cavity on the external surface of the coxal bone that receives the head of the femur.

Acetylcholine (as'-ē-til-KŌ-lēn) **(ACh)** A neurotransmitter, liberated at synapses in the central nervous system, that stimulates skeletal muscle contraction.

Achille's tendon *See* **Calcaneal tendon.**

Acid (AS-id) A proton donor, or substance that dissociates into hydrogen ions (H^+) and anions, characterized by an excess of hydrogen ions and a pH less than 7.

Acidosis (as-i-DŌ-sis) A condition in which blood pH is between 7.35 and 6.80.

Acini (AS-i-nē) Masses of cells in the pancreas that secrete digestive enzymes.

Acne (AK-nē) Inflammation of sebaceous glands that usually begins at puberty; the basic acne lesions in order of increasing severity are comedones, papules, pustules, and cysts.

Acoustic (a-KOOS-tik) Pertaining to sound or the sense of hearing.

Acquired immune deficiency syndrome (AIDS) A deficiency of helper T cells and a reversed ratio of helper T cells to suppressor T cells that results in fever or night sweats, coughing, sore throat, fatigue, body aches, weight loss, and enlarged lymph nodes. Caused by a virus called human-T lymphotropic virus, type III (HTLV-III).

Acromegaly (ak'-rō-MEG-a-lē) Condition caused by hypersecretion of growth hormone (GH) during adulthood characterized by thickened bones and enlargement of other tissues.

Acrosome (AK-rō-sōm) A dense granule in the head of a spermatozoon that contains enzymes that facilitate the penetration of a spermatazoon into an ovum.

Actin (AK-tin) The contractile protein that makes up thin myofilaments in muscle fiber.

Action potential A wave of negativity that self-propagates along the outside surface of the membrane of a neuron. Also called a **nerve impulse.**

Active transport The movement of substances, usually ions, across cell membranes, against a concentration gradient, requiring the expenditure of energy.

Acuity (a-KYOO-i-tē) Clearness or sharpness, usually of vision.

Acupuncture (AK-ū-punk'-chur) The insertion of a needle into a tissue for the purpose of drawing fluid or relieving pain. It is also an ancient Chinese practice employed to cure illnesses by inserting needles into specific locations of the skin.

Acute (a-KYOOT) Having rapid onset, severe symptoms, and a short course; not chronic.

Adam's apple *See* **Thyroid cartilage.**

Adaptation (ad'-ap-TĀ-shun) The adjustment of the pupil of the eye to light variations. The property by which a neuron relays a decreased frequency of action potentials from a receptor even though the strength of the stimulus remains constant. The decrease in perception of a sensation over time while the stimulus is still present.

Addison's (AD-i-sonz) **disease** Disorder caused by hyposecretion of glucocorticoids characterized by muscular weakness, mental lethargy, weight loss, low blood pressure, and dehydration.

Adduction (ad-DUK-shun) Movement toward the axis or midline of the body or one of its parts.

Adenohypophysis (ad'-e-nō-hī-POF-i-sis) The anterior portion of the pituitary gland.

Adenoids (AD-e-noyds) The pharyngeal tonsils.

Adenosine triphosphate (a-DEN-ō-sēn trī-FOS-fāt) **(ATP)** The universal energy-carrying molecule manufactured in all living cells as a means of capturing and storing energy. It consists of the purine base *adenine* and the five-carbon sugar *ribose,* to which are added, in linear array, three *phosphate* molecules.

Adenylate cyclase (a-DEN-i-lāt SĪ-klās) An enzyme in the postsynaptic membrane that is activated when certain neurotransmitters bind to their receptors; the enzyme converts ATP into cyclic AMP.

Adherence (ad-HĒR-ens) Firm contact between the plasma membrane of a phagocyte and an antigen.

Adhesion (ad-HĒ-zhun) Abnormal joining of parts to each other.

Adipocyte (AD-i-pō-sīt) Fat cell, derived from a fibroblast.

Adrenal cortex (a-DRĒ-nal KOR-teks) The outer portion of an adrenal gland, divided into three zones, each of which has a different cellular arrangement and secretes different hormones.

Adrenal (a-DRĒ-nal) **glands** Two glands located superior to each kidney. Also called the **suprarenal** (soo'-pra-RĒ-nal) **glands.**

Adrenal medulla (me-DUL-a) The inner portion of an adrenal gland, consisting of cells that secrete epinephrine and norepinephrine (NE) in response to the stimulation of preganglionic sympathetic neurons.

Adrenergic (ad'-ren-ER-jik) **fiber** A nerve fiber that, when stimulated, releases norepinephrine (noradrenaline) at a synapse.

Adrenocorticotropic (ad-rē'-nō-kor-ti-kō-TRŌP-ik) **hormone (ACTH)** A hormone produced by the adenohypophysis (anterior lobe) of the pituitary gland that influences the production and secretion of certain hormones of the adrenal cortex.

Adrenoglomerulotropin (a-drē'-nō-glō-mer'-yoo-lō-TRŌ-pin) A hormone secreted by the pineal gland that may stimulate aldosterone secretion.

Adventitia (ad-ven-TISH-ya) The outermost covering of a structure or organ.

Aerobic (air-Ō-bik) Requiring molecular oxygen.

Afferent arteriole (AF-er-ent ar-TĒ-rē-ōl) A blood vessel of a kidney that breaks up into the capillary network called a glomerulus; there is one afferent arteriole for each glomerulus.

Afferent neuron (NOO-ron) A neuron that carries an impulse toward the central nervous system. Also called a **sensory neuron.**

Afterimage Persistence of a sensation even though the stimulus has been removed.

Agglutination (a-gloo'-ti-NĀ-shun) Clumping of microorganisms or blood corpuscles; typically an antigen-antibody reaction.

Agglutinin (a-GLOO-ti-nin) A specific principle or antibody in blood serum capable of causing the clumping of bacteria, blood corpuscles, or particles. Also called **isoantibody.**

Agglutinogen (ag'-loo-TIN-ō-gen) A genetically determined antigen located on the surface of erythrocytes; basis for the ABO grouping and Rh system of blood classification. Also called an **isoantigen.**

Aggregated lymphatic follicles Aggregated lymph nodules that are most numerous in the ileum. Also called **Peyer's (PĪ-erz) patches.**

Aging Progressive failure of the body's homeostatic adaptive responses.

Agnosia (ag-NŌ-zē-a) A loss of the ability to recognize the meaning of stimuli from the various senses (visual, auditory, touch).

Agraphia (a-GRAF-ē-a) An inability to write.

Albinism (AL-bin-izm) Abnormal, nonpathological, partial or total absence of pigment in skin, hair, and eyes.

Albumin (al-BYOO-min) The most abundant (60 percent) and smallest of the plasma proteins, which functions primarily to regulate osmotic pressure of plasma.

Albuminuria (al-byoo'-min-UR-ēa) Presence of albumin in the urine.

Aldosterone (al-do-STĒR-ōn) A mineralocorticoid produced by the adrenal cortex that brings about sodium and water reabsorption and potassium excretion.

Aldosteronism (al'-do-STER-ōn-izm') Condition caused by hypersecretion of aldosterone characterized by muscular paralysis, high blood pressure, and edema.

Alimentary (al-i-MEN-ta-rē) Pertaining to nutrition.

Alkaline (AL-ka-līn) Containing more hydroxyl ions than hydrogen ions to produce a pH of more than 7.

Alkalosis (al-ka-LŌ-sis) A condition in which blood pH is between 7.45 and 8.00.

Allantois (a-LAN-tō-is) A small, vascularized membrane between the chorion and amnion of the fetus.

Allergen (AL-er-jen) An antigen that evokes a hypersensitivity reaction.

Allergic (a-LER-jik) Pertaining to or sensitive to an allergen.

All-or-none principle In muscle physiology, muscle fibers of a motor unit contract to their fullest extent or not at all. In neuron physiology, if a stimulus is strong enough to initiate an action potential, an impulse is transmitted along the entire neuron at a constant and minimum strength.

Alpha (AL-fa) **cell** A cell in the pancreatic islets (islets of Langerhans) in the pancreas that secretes glucagon.

Alpha receptor Receptor found on visceral effectors innervated by most sympathetic postganglionic axons; in general, stimulation of alpha receptors leads to excitation.

Altitude sickness Disorder caused by decreased levels of alveolar PO_2 as altitude increases and characterized by shortness of breath, nausea, and dizziness.

Alveolar-capillary (al-VĒ-ō-lar) **membrane** Structure in the lungs consisting of the alveolar wall and basement membrane and a capillary endothelium and basement membrane through which the diffusion of respiratory gases occurs. Also called the **respiratory membrane.**

Alveolar duct Branch of a respiratory bronchiole around which alveoli and alveolar sacs are arranged.

Aveolar macrophage (MAK-rō-fāj) Cell found in the alveolar walls of the lungs that is highly phagocytic. Also called a **dust cell.**

Aveolar sac A collection or cluster of alveoli that share a common opening.

Alveolus (al-VĒ-ō-lus) A small hollow or cavity; an air sac in the lungs; milk-secreting portion of a mammary gland.

Alzheimer's (ALTZ-hī-merz) **disease (AD)** Disabling neurological disorder characterized by dysfunction and death of specific cerebral neurons resulting in widespread intellectual impairment, personality changes, and fluctuations in alertness.

Amenorrhea (ā-men-ō-RĒ-a) Absence of menstruation.

Amino acid An organic acid, containing a carboxyl group (COOH) and an amino group (NH₂), that is the basic unit from which proteins are formed.

Amnesia (am-NĒ-zē-a) A lack or loss of memory.

Amniocentesis (am'-nē-ō-sen-TĒ-sis) Removal of amniotic fluid by inserting a needle transabdominally into the amniotic cavity.

Amnion (AM-nē-on) The innermost fetal membrane; a thin transparent sac that holds the fetus suspended in amniotic fluid. Also called the **"bag of waters."**

Amphiarthrosis (am'-fē-ar-THRŌ-sis) Articulation midway between diarthrosis and synarthrosis, in which the articulating bony surfaces are separated by an elastic substance to which both are attached, so that the mobility is slight, but may be exerted in all directions.

Ampulla (am-POOL-la) A saclike dilation of a canal.

Ampulla of Vater *See* **Hepatopancreatic ampulla.**

Amyotrophic (a-mē-ō-TROF-ik) **lateral sclerosis (ALS)** Progressive neuromuscular disease characterized by degeneration of motor cells in the spinal cord that leads to muscular weakness. Also called **Lou Gehrig's disease.**

Anabolism (a-NAB-ō-lizm) Synthetic energy-requiring reactions whereby small molecules are built up into larger ones.

Anaerobic (an-AIR-ō-bik) Not requiring molecular oxygen.

Anal (Ā-nal) **canal** The terminal 2 or 3 cm (1 inch) of the rectum; opens to the exterior of the anus.

Anal column A longitudinal fold in the mucous membrane of the anal canal that contains a network of arteries and veins.

Analgesia (an-al-JĒ-zē-a) Absence of normal sense of pain.

Anal triangle The subdivision of the male or female perineum that contains the anus.

Anamnestic (an'-am-NES-tik) **response** Accelerated, more intense production of antibodies upon a subsequent exposure to an antigen after the initial exposure.

Anaphase (AN-a-fāz) The third stage of mitosis in which the chromatids that have separated at the centromeres move to opposite poles.

Anaphylaxis (an'-a-fi-LAK-sis) A hypersensitivity (allergic) reaction in which IgE antibodies attach to mast cells and basophils, causing them to produce mediators of anaphylaxis (histamine and prostaglandins) that bring about increased blood permeability, increased smooth muscle contraction, and increased mucus production. Examples are hayfever, bronchial asthma, hives, and anaphylactic shock.

Anastomosis (a-nas-tō-MŌ-sis) An end-to-end union or joining together of blood vessels, lymphatics, or nerves.

Anatomical (an'-a-TOM-i-kal) **position** A position of the body universally used in anatomical descriptions in which the body is erect, facing the observer, the upper extremities are at the sides, and palms of the hands are facing forward.

Anatomy (a-NAT-ō-mē) The structure or study of structure of the body and the relationship of its parts to each other.

Androgen (AN-drō-jen) Substance producing or stimulating male characteristics, such as the male hormone testosterone.

Anemia (a-NĒ-mē-a) Condition of the blood in which the number of functional red blood cells or their hemoglobin content is below normal.

Anesthesia (an'-es-THĒ-zē-a) A total or partial loss of feeling or sensation, usually defined with respect to loss of pain sensation.

Aneurysm (AN-yoo-rizm) A saclike enlargement of a blood vessel caused by a weakening of its wall.

Angina pectoris (an-JĪ-na *or* AN-ji-na PEK-tō-ris) A pain in the chest related to reduced coronary circulation that may or may not involve heart or artery disease.

Angiography (an-jē-OG-ra-fē) X-ray examination of blood vessels after injection of a radiopaque substance into the common carotid or vertebral artery; used to demonstrate cerebral blood vessels and may detect brain tumors with specific vascular patterns.

Angiotensin (an-jē-ō-TEN-sin) Either of two forms of a protein associated with regulation of blood pressure. Angiotensin I is produced by the action of renin on angiotensinogen and is converted by the action of a plasma enzyme into angiotensin II, which controls aldosterone secretion by the adrenal cortex.

Anion (AN-ī-on) A negatively charged ion. An example is Cl⁻.

Ankyloglossia (ang'-ki-lō-GLOSS-ē-a) "Tongue-tied"; restriction of tongue movements by a short lingual frenulum.

Anomaly (a-NOM-a-lē) An abnormality that may be a developmental (congenital) defect; a variant from the usual standard.

Anopsia (an-OP-sē-a) A defect of vision.

Anorexia nervosa (an-ō-REK-sē-a ner-VŌ-sa) A disorder characterized by loss of appetite and bizarre patterns of eating.

Anosmia (an-OZ-mē-a) Loss of the sense of smell.

Anoxia (an-OK-sē-a) Deficiency of oxygen.

Antagonist (an-TAG-ō-nist) A muscle that has an action opposite that of the agonist and yields to the movement of the prime mover (agonist).

Antepartum (an-tē-PAR-tum) Before delivery of the child; occurring (to the mother) before childbirth.

Anterior (an-TĒR-ē-or) Nearer to or at the front of the body. Also called **ventral.**

Anterior root The structure composed of axons of motor or efferent fibers that emerges from the anterior aspect of the spinal cord and extends laterally to join a posterior root, forming a spinal nerve. Also called a **ventral root.**

Antibiotic (an'-ti-bī-OT-ik) Literally, "antilife"; a chemical produced by a microorganism that is able to inhibit the growth of or kill other microorganisms.

Antibody (AN-ti-bod'-ē) A substance produced by certain cells in the presence of a specific antigen that combines with that antigen to neutralize, inhibit, or destroy it.

Anticoagulant (an-tī-cō-AG-yoo-lant) A substance that is able to delay, suppress, or prevent the clotting of blood.

Antidiuretic (an'-ti-dī-yoo-RET-ik) Substance that inhibits urine formation.

Antidiuretic hormone (ADH) Hormone produced by neurosecretory cells in the paraventricular nucleus of the hypothalamus that stimulates water reabsorption from kidney cells into the blood and vasoconstriction of arterioles.

Antigen (AN-ti-jen) Any substance that when introduced into the tissues or blood induces the formation of antibodies or reacts with them.

Anulus fibrosus (AN-yoo-lus-fī-BRŌ-sus) A ring of fibrous tissue and fibrocartilage that encircles the pulpy substance (nucleus pulposus) of an intervertebral disc.

Anuria (a-NOO-rē-a) A daily urine output of less than 50 ml.

Anus (Ā-nus) The distal end and outlet of the rectum.

Aorta (ā-OR-ta) The main systemic trunk of the arterial system of the body; emerges from the left ventricle.

Aortic (ā-OR-tik) **body** Receptor on or near the arch of the aorta that responds to alterations in blood levels of oxygen, carbon dioxide, and hydrogen ions.

Aortic reflex A reflex concerned with maintaining normal general systemic blood pressure.

Aphasia (a-FĀ-zē-a) Loss of ability to express oneself properly through speech or loss of verbal comprehension.

Apheresis (a-FER-ē-sis) Procedure in which blood is removed from the body, components are selectively separated, the undesired component is removed, and the remainder is returned to the body.

Apnea (ap-NĒ-a) Temporary cessation of breathing.

Apneustic (ap-NOO-stik) **area** Portion of the respiratory center in the pons that sends stimulatory impulses to the inspiratory area that activate and prolong inspiration and inhibit expiration.

Apocrine (AP-ō-krin) **gland** A type of gland in which the secretory products gather at the free end of the secreting cell and are pinched off, along with some of the cytoplasm, to become the secretion, as in mammary glands.

Aponeurosis (ap'-ō-noo-RŌ-sis) A sheetlike layer of dense, regularly arranged connective tissue joining one muscle with another or with bone.

Appendicitis (a-pen-di-SĪ-tis) Inflammation of the vermiform appendix.

Appositional (ap'-ō-ZISH-o-nal) **growth** Growth due to surface deposition of material, as in the growth in diameter of cartilage and bone. Also called **exogenous** (eks-OJ-e-nus) **growth.**

Aqueous humor (AK-wē-us HYOO-mor) The watery fluid that fills the anterior cavity of the eye.

Arachnoid (a-RAK-noyd) The middle of the three coverings (meninges) of the brain.

Arachnoid villus (VIL-us) Berrylike tuft of arachnoid that protrudes into the superior saggital sinus and through which the cerebrospinal fluid enters the bloodstream.

Arbor vitae (AR-bōr VĒ-tē) The treelike appearance of the white matter tracts of the cerebellum when seen in midsagittal section. A series of branching ridges within the cervix of the uterus.

Arch of the aorta (ā-OR-ta) The most superior portion of the aorta, lying between the ascending and descending segments of the aorta.

Areflexia (a'-rē-FLEK-sē-a) Absence of reflexes.

Areola (a-RĒ-ō-la) Any tiny space in a tissue. The pigmented ring around the nipple of the breast.

Arm The portion of the upper extremity from the shoulder to the elbow.

Arrector pili (a-REK-tor PI-lē) Smooth muscles attached to hairs; contraction pulls the hairs into a more vertical position, resulting in "goose bumps."

Arrythmia (a-RITH-mē-a) Irregular heart rhythm. Also called a **dysrhythmia.**

Arteriogram (ar-TĒR-ē-ō-gram) A roentgenogram of an artery after injection of a radiopaque substance into the blood.

Arteriole (ar-TĒ-rē-ōl) A small arterial branch that delivers blood to a capillary.

Artery (AR-ter-ē) A blood vessel that carries blood away from the heart.

Arthritis (ar-THRĪ-tis) Inflammation of a joint.

Arthrology (ar-THROL-ō-jē) The study or description of joints.

Arthroscopy (ar-THROS-co-pē) Surgical technique in which an arthroscope is inserted into a small incision, usually in the knee, to repair torn cartilage.

Articular (ar-TIK-yoo-lar) **capsule** Sleevelike structure around a synovial joint composed of a fibrous capsule and a synovial membrane.

Articular cartilage (KAR-ti-lij) Hyaline cartilage attached to articular bone surfaces.

Articular disc Fibrocartilage pad between articular surfaces of bones of some synovial joints. Also called a **meniscus** (men-IS-cus).

Articulation (ar-tik'-yoo-LĀ-shun) A joint.

Artificial insemination (in-sem'-i-NĀ-shun) The deposition of seminal fluid within the vagina or cervix by artificial means. It may be homologous (using the husband's semen) or heterologous (using a donor's semen).

Artificial pacemaker A device that generates and delivers electrical signals to the heart to maintain a regular heart rhythm.

Arytenoid (ar'-i-TĒ-noyd) **cartilages** A pair of small cartilages of the larynx that articulates with the cricoid cartilage.

Ascending colon (KŌ-lon) The portion of the large intestine that passes upward from the cecum to the lower edge of the liver where it bends at the right colic (hepatic) flexure to become the transverse colon.

Ascites (as-SĪ-tēz) Serous fluid in the peritoneal cavity.

Aseptic (ā-SEP-tik) Free from any infectious or septic material.

Asphyxia (as-FIX-ē-a) Unconsciousness due to interference with the oxygen supply of the blood.

Aspirate (AS-pir-āt) To remove by suction.

Association area A portion of the cerebral cortex connected by many motor and sensory fibers to other parts of the cortex. The association areas are concerned with motor patterns, memory, concepts of word-hearing and word-seeing, reasoning, will, judgment, and personality traits.

Association neuron (NOO-ron) A nerve cell lying completely within the central nervous system that carries impulses from sensory neurons to motor neurons. Also called an **interneuron** or **connecting neuron.**

Astereognosis (as-ter'-ē-ōg-NŌ-sis) Inability to recognize objects or forms by touch.

Asthenia (as-THĒ-nē-a) Lack or loss of strength; debility.

Astigmatism (a-STIG-ma-tizm) An irregularity of the lens or cornea of the eye causing the image to be out of focus and producing faulty vision.

Astrocyte (AS-trō-sīt) A neuroglial cell having a star shape that supports neurons in the brain and spinal cord and attaches the neurons to blood vessels.

Ataxia (a-TAK-sē-a) A lack of muscular coordination, lack of precision.

Atelectasis (at'-ē-LEK-ta-sis) A collapsed or airless state of all or part of the lung, which may be acute or chronic.

Atherosclerosis (ath'-er-ō-skle-RŌ-sis) A process in which fatty substances (cholesterol and triglycerides) are deposited in the walls of medium and large arteries in response to certain stimuli (hypertension, carbon monoxide, dietary cholesterol). Following endothelial damage, monocytes stick to the tunica interna, develop into macrophages, and take up cholesterol and low-density lipoproteins. Smooth muscle cells in the tunica media ingest cholesterol. This results in the formation of an atherosclerotic plaque that decreases the size of the arterial lumen.

Atom Unit of matter that comprises a chemical element; consists of a nucleus and electrons.

Atomic weight Total number of protons and neutrons in an atom.

Atresia (a-TRĒ-zē-a) Abnormal closure of a passage, or absence of a normal body opening.

Atrial fibrillation (Ā-trē-al fib-ri-LĀ-shun) Asynchronous contraction of the atria that results in the cessation of atrial pumping.

Atrioventricular (AV) (ā'-trē-ō-ven-TRIK-yoo-lar) **bundle** The portion of the conduction system of the heart that begins at the atrioventricular (AV) node, passes through the cardiac skeleton separating the atria and the ventricles, then runs a short distance down the interventricular septum before splitting into right and left bundle branches. Also called the **bundle of His** (HISS).

Atrioventricular (AV) node The portion of the conduction system of the heart made up of a compact mass of conducting cells located near the orifice of the coronary sinus in the right atrial wall.

Atrioventricular (AV) valve A structure made up of membranous flaps or cusps that allows blood to flow in one direction only, from an atrium into a ventricle.

Atrium (Ā-trē-um) A superior chamber of the heart.

Atrophy (AT-rō-fē) Wasting away or decrease in size of a part, due to a failure, abnormality of nutrition, or lack of use.

Auditory ossicle (AW-di-tō-rē OS-si-kul) One of the three small bones of the middle ear called the malleus, incus, and stapes.

Auditory tube The tube that connects the middle ear with the nose and nasopharynx of the throat. Also called the **Eustachian** (yoo-STĀ-kē-an) **tube.**

Auricle (OR-i-kul) The flap or pinna of the ear. An appendage of an atrium of the heart.

Auscultation (aws-kul-TĀ-shun) Examination by listening to sounds in the body.

Autoimmunity An immunologic response against a person's own tissue antigens.

Autolysis (aw-TOL-i-sis) Spontaneous self-destruction of cells by their own digestive enzymes upon death or a pathological process.

Autonomic ganglion (aw'-tō-NOM-ik GANG-lē-on) A cluster of sympathetic or parasympathetic cell bodies located outside the central nervous system.

Autonomic nervous system (ANS) Visceral efferent neurons, both sympathetic and parasympathetic, that transmit impulses from the central nervous system to smooth muscle, cardiac muscle, and glands; so named because this portion of the nervous system was thought to be self-governing or spontaneous.

Autonomic plexus (PLEK-sus) An extensive network of sympathetic and parasympathetic fibers; the cardiac, celiac, and pelvic plexuses are located in the thorax, abdomen, and pelvis, respectively.

Autopsy (AW-top-sē) The examination of the body after death.

Autoregulation (aw-tō-reg-yoo-LĀ-shun) A local, automatic adjustment of blood flow in a given region of the body in response to tissue needs.

Autosome (AW-tō-sōm) Any chromosome other than the pair of sex chromosomes.

Avitaminosis (ā-vī-ta-min-Ō-sis) A deficiency of a vitamin in the diet.

Axilla (ak-SIL-a) The small hollow beneath the arm where it joins the body at the shoulders. Also called the **armpit.**

Axon (AK-son) The process of a nerve cell that carries an impulse away from the cell body.

Axon terminal Terminal branch of an axon and its collateral. Also called a **telodendrium** (tel-ō-DEN-drē-um).

Babinski (ba-BIN-skē) **sign** Extension of the great toe, with or without fanning of the other toes, in response to stimulation of the outer margin of the sole of the foot; normal up to 1½ years of age.

Back The posterior part of the body; the dorsum.

Bainbridge (BĀN-bridge) **reflex** The increased heart rate that follows increased pressure or distension of the right atrium.

Ball-and-socket joint A synovial joint in which the rounded surface of one bone moves within a cup-shaped depression or fossa of another bone, as in the shoulder or hip joint. Also called a **spheroid** (SFĒ-roid) **joint.**

Baroreceptor (bar'-ō-re-SEP-tor) Nerve cell capable of responding to changes in blood pressure. Also called a **pressoreceptor.**

Bartholin's gland *See* **Greater vestibular gland.**

Basal metabolic (BĀ-sal met'-a-BOL-ik) **rate (BMR)** The rate of metabolism measured under standard or basal conditions.

Basal ganglia (GANG-glē-a) Paired clusters of cell bodies that make up the central gray matter in each cerebral hemisphere, including the caudate nucleus, lentiform nucleus, claustrum, and amygdaloid body. Also called **cerebral nuclei** (SER-e-bral NOO-klē-ī).

Base The broadest part of a pyramidal structure. A nonacid or a proton acceptor, characterized by excess of hydroxide ions and a pH greater than 7. A ring-shaped, nitrogen-containing organic molecule that is one of the components of a nucleotide, for example, adenine, guanine, cytosine, thymine, and uracil.

Basement membrane Thin, extracellular layer consisting of basal lamina secreted by epithelial cells and reticular lamina secreted by connective tissue cells.

Basilar (BAS-i-lar) **membrane** A membrane in the cochlea of the inner ear that separates the cochlear duct from the scala tympani and on which the spiral organ (organ of Corti) rests.

Basophil (BĀ-sō-fil) A type of white blood cell characterized by a pale nucleus and large granules that stain readily with basic dyes.

B cell A lymphocyte that develops into a plasma cell that produces antibodies or a memory cell.

Belly The abdomen. The gaster or prominent, fleshy part of a skeletal muscle.

Benign (be-NĪN) Not malignant.

Beta (BĀ-ta) **cell** A cell in the pancreatic islets (islets of Langerhans) in the pancreas that secretes insulin.

Beta receptor Receptor found on visceral effectors innervated by most sympathetic postganglionic axons; in general, stimulation of beta receptors leads to inhibition.

Bicuspid (bī-KUS-pid) **valve** Atrioventricular (AV) valve on the left side of the heart. Also called the **mitral valve.**

Bifurcate (bī-FUR-kāt) Having two branches or divisions; forked.

Bilateral (bī-LAT-er-al) Pertaining to two sides of the body.

Bile (bīl) A secretion of the liver that emulsifies fats.

Biliary calculi (BIL-ē-er-ē CAL-kyoo-lē) Gallstones formed by the crystallization of cholesterol in bile.

Bilirubin (bil-ē-ROO-bin) A red pigment that is one of the end products of hemoglobin breakdown in the liver cells and is excreted as a waste material in the bile.

Bilirubinuria (bil-ē-roo-bi-NOO-rē-a) The presence of above normal levels of bilirubin in urine.

Biliverdin (bil-ē-VER-din) A green pigment that is one of the first products of hemoglobin breakdown in the liver cells and is converted to bilirubin or excreted as a waste material in bile.

Biofeedback Process by which an individual gets constant signals (feedback) about various visceral body functions.

Biopsy (BĪ-op-sē) Removal of tissue or other material from the living body for examination, usually microscopic.

Blastocel (BLAS-tō-sēl) The fluid-filled cavity within the blastocyst.

Blastocyst (BLAS-tō-sist) In the development of an embryo, a hollow ball of cells that consists of a blastocel (the internal cavity), trophectoderm (outer cells), and inner cell mass.

Blastomere (BLAS-tō-mēr) One of the cells resulting from the cleavage of a fertilized ovum.

Blastula (BLAS-tyoo-la)   An early stage in the development of a zygote.

Bleeding time   The time required for the cessation of bleeding from a small skin puncture as a result of platelet disintegration and blood vessel constriction; ranges from 4 to 8 minutes.

Blepharism (BLEF-a-rizm)   Spasm of the eyelids; continuous blinking.

Blind spot   Area in the retina at the end of the optic (II) nerve in which there are no light receptor cells.

Blood   The fluid that circulates through the heart, arteries, capillaries, and veins and that constitutes the chief means of transport within the body.

Blood–brain barrier (BBB)   A special mechanism that prevents the passage of materials from the blood to the cerebrospinal fluid and brain.

Blood island   Isolated mass and cord of mesenchyme in the mesoderm from which blood vessels develop.

Blood pressure (BP)   Force exerted by blood as it presses against and attempts to stretch blood vessels, especially arteries; clinically, a measure of the pressure in arteries during ventricular systole and ventricular diastole.

Blood reservoir (REZ-er-vwar)   Systemic veins that contain large amounts of blood that can be moved quickly to parts of the body requiring the blood.

Blood-testis barrier   A barrier formed by sustentacular (Sertoli) cells that prevents an immune response against antigens produced by spermatozoa and developing cells by isolating the cells from the blood.

Body cavity   A space within the body that contains various internal organs.

Bolus (BŌ-lus)   A soft, rounded mass, usually food, that is swallowed.

Bony labyrinth (LAB-i-rinth)   A series of cavities within the petrous portion of the temporal bone, forming the vestibule, cochlea, and semicircular canals of the inner ear.

Bowman's capsule   See **Glomerular capsule.**

Bradycardia (brad'-i-KAR-dē-a)   A slow heartbeat or pulse rate.

Brachial plexus (BRĀ-kē-al PLEK-sus)   A network of nerve fibers of the anterior rami of spinal nerves C5, C6, C7, C8, and T1. The nerves that emerge from the brachial plexus supply the upper extremity.

Brain   A mass of nervous tissue located in the cranial cavity.

Brain electrical activity mapping (BEAM)   Noninvasive procedure that measures and displays the electrical activity of the brain; used primarily to diagnose epilepsy.

Brain sand   Calcium deposits in the pineal gland that are laid down starting at puberty.

Brain stem   The portion of the brain immediately superior to the spinal cord, made up of the medulla oblongata, pons, and midbrain.

Bright's disease   See **Glomerulonephritis.**

Broca's (BRŌ-kaz) **area**   Motor area of the brain in the frontal lobe that translates thoughts into speech. Also called the **motor speech area.**

Bronchi (BRONG-kē)   Branches of the respiratory passageway including primary bronchi (the two divisions of the trachea), secondary or lobar bronchi (divisions of the primary that are distributed to the lobes of the lung), and tertiary or segmental bronchi (divisions of the secondary that are distributed to bronchopulmonary segments of the lung).

Bronchial asthma (BRONG-kē-al AZ-ma)   Usually allergic reaction characterized by smooth muscle spasms in bronchi resulting in wheezing and difficult breathing.

Bronchial tree   The trachea, bronchi, and their branching structures.

Bronchiectasis (brong'-kē-EK-ta-sis)   A chronic disorder in which there is a loss of the normal tissue and expansion of lung air passages; characterized by difficult breathing, coughing, expectoration of pus, and foul breath.

Bronchiole (BRONG-kē-ol)   Branch of a tertiary bronchus further dividing into terminal bronchioles (distributed to lobules of the lung), which divide into respiratory bronchioles (distributed to aveolar sacs).

Bronchitis (brong-KĪ-tis)   Inflammation of the bronchi characterized by hypertrophy and hyperplasia of seromucous glands and goblet cells that line the bronchi and resulting in a productive cough.

Bronchogenic carcinoma (brong'-kō-JEN-ik kar'-si-NŌ-ma)   Cancer originating in the bronchi.

Bronchogram (BRONG-kō-gram)   A roentgenogram of the lungs and bronchi

Bronchopulmonary (brong'-kō-PUL-mō-ner'-ē) **segment**   One of the smaller divisions of a lobe of a lung supplied by its own branches of a bronchus.

Bronchoscope (BRONG-kō-skōp)   An instrument used to examine the interior of the bronchi of the lungs.

Bronchus (BRONG-kus)   One of the two large branches of the trachea. *Plural,* **bronchi** (BRONG-kē).

Brunner's gland   See **Duodenal gland.**

Buccal (BUK-al)   Pertaining to the cheek or mouth.

Buffer (BUF-er) **system**   A pair of chemicals, one a weak acid and one a weak base, that resists changes in pH.

Bulb of penis   Expanded portion of the base of the corpus spongiosum penis.

Bulbourethral (bul'-bō-yoo-RĒ-thral) **gland**   One of a pair of glands located inferior to the prostate gland on either side of the urethra that secretes an alkaline fluid into the cavernous urethra. Also called a **Cowper's** (KOW-perz) **gland.**

Bulimia (boo-LIM-ē-a)   A disorder characterized by uncontrollable overeating followed by forced vomiting or overdoses of laxatives.

Bullae (BYOOL-ē)   Blisters beneath or within the epidermis.

Bundle branch   One of the two branches of the atrioventricular (AV) bundle made up of specialized muscle fibers that transmit electrical impulses to the ventricles.

Bundle of His   See **Atrioventricular (AV) bundle.**

Bunion (BUN-yun)   Lateral deviation of the great toe that produces inflammation and thickening of the bursa, bone spurs, and calluses.

Burn   An injury caused by heat (fire, steam), chemicals, electricity, or the ultraviolet rays of the sun.

Bursa (BUR-sa)   A sac or pouch of synovial fluid located at friction points, especially about joints.

Bursitis (bur-SĪ-tis)   Inflammation of a bursa.

Buttocks (BUT-oks)   The two fleshy masses on the posterior aspect of the lower trunk, formed by the gluteal muscles.

Cachexia (kah-KEK-sē-ah)   A state of ill health, malnutrition, and wasting.

Calcaneal tendon   The tendon of the soleus and gastrocnemius muscles at the back of the heel. Also called the **Achilles** (a-KIL-ēz) **tendon.**

Calcitonin (kal-si-TŌ-nin) **(CT)**   A hormone produced by the thyroid gland that lowers the calcium and phosphate levels of the blood by inhibiting bone breakdown and accelerating calcium absorption by bones.

Calculus (KAL-kyoo-lus)   A stone, or insoluble mass of crystallized salts or other material, formed within the body, as in the gallbladder, kidney, or urinary bladder.

Callus (KAL-lus)   A growth of new bone tissue in and around a fractured area, ultimately replaced by mature bone. An acquired, localized thickening.

Calorie (KAL-ō-rē)   A unit of heat. A calorie (cal) is the standard unit and is the amount of heat necessary to raise 1 g of water 1°C from 14° to 15°C. The kilocalorie (kcal), used in metabolic and nutrition studies, is the amount of heat necessary to raise 1,000 g of water 1°C; it is equal to 1,000 cal.

Calyx (KĀL-iks) Any cuplike division of the kidney pelvis. *Plural,* **calyces** (KĀ-li-sēz).

Canaliculus (kan'-a-LIK-yoo-lus) A small channel or canal, as in bones, where they connect the lacunae. *Plural,* **canaliculi** (kan'-a-LIK-yoo-lī).

Canal of Schlemm *See* **Scleral venous sinus.**

Cancer (KAN-ser) A malignant tumor of epithelial origin tending to infiltrate and give rise to new growths or metastases. Also called **carcinoma** (kar'-si-NŌ-ma).

Capacitation (ka'-pas-i-TĀ-shun) The functional changes that sperm undergo in the female reproductive tract that allows them to fertilize an ovum.

Capillary (KAP-i-lar'-ē) A microscopic blood vessel located between an arteriole and venule through which materials are exchanged between blood and body cells.

Carbohydrate (kar'-bō-HĪ-drāt) An organic compound containing carbon, hydrogen, and oxygen in a particular amount and arrangement and comprised of sugar subunits; usually has the formula $(CH_2O)n$.

Carbon monoxide poisoning Hypoxia due to increased levels of carbon monoxide as a result of its preferential and tenacious combination with hemoglobin compared to oxygen.

Carcinogen (kar-SIN-ō-jen) Any substance that causes cancer.

Cardiac catheterization (kath'-e-ter-i-ZĀ-shun) An invasive procedure in which a catheter is introduced into a vein or artery and threaded, via fluoroscopy, through blood vessels to reach the heart.

Cardiac (KAR-dē-ak) **cycle** A complete heartbeat consisting of systole and diastole of both atria plus systole and diastole of both ventricles.

Cardiac muscle An organ specialized for contraction, composed of cardiac muscle cells, forming the wall of the heart, and stimulated by an intrinsic conduction system and visceral efferent neurons.

Cardiac notch An angular notch in the anterior border of the left lung.

Cardiac output (CO) The volume of blood pumped from one ventricle of the heart (usually measured from the left ventricle) in 1 min; about 5.2 liters/min under normal resting conditions.

Cardiac reserve The maximum percentage that cardiac output can increase above normal.

Cardiac tamponade (tam'-pon-ĀD) Compression of the heart due to excessive fluid or blood in the pericardial sac that could result in cardiac failure.

Cardinal ligament A ligament of the uterus, extending laterally from the cervix and vagina as a continuation of the broad ligament.

Cardioacceleratory (kar-dē-ō-ak-SEL-er-a-tō-rē) **center (CAC)** A group of neurons in the medulla from which cardiac nerves (sympathetic) arise; impulses along the nerves release epinephrine that increases the rate and force of heartbeat.

Cardioinhibitory (kar-dē-ō-in-HIB-i-tō-rē) **center (CIC)** A group of neurons in the medulla from which parasympathetic fibers that reach the heart via the vagus (X) nerve arise; impulses along the nerves release acetylcholine that decreases the rate and force of heartbeat.

Cardiology (kar-de-OL-ō-jē) The study of the heart and diseases associated with it.

Cardiopulmonary resuscitation (rē-sus-i-TĀ-shun) **(CPR)** A technique employed to restore life or consciousness to a person apparently dead or dying; includes external respiration (exhaled air respiration) and external cardiac massage.

Carotid (ka-ROT-id) **body** Receptor on or near the carotid sinus that responds to alterations in blood levels of oxygen, carbon dioxide, and hydrogen ions.

Carotid sinus A dilated region of the internal carotid artery immediately above the bifurcation of the common carotid artery that contains receptors that monitor blood pressure.

Carotid sinus reflex A reflex concerned with maintaining normal blood pressure in the brain.

Carpus (KAR-pus) A collective term for the eight bones of the wrist.

Cartilage (KAR-ti-lij) A type of connective tissue consisting of chondrocytes in lacunae embedded in a dense network of collagenous and elastic fibers and a matrix of chondroitin sulfate.

Cartilaginous (kar'-ti-LAJ-i-nus) **joint** A joint without a joint cavity where the articulating bones are held tightly together by cartilage, allowing little or no movement.

Cast A small mass of hardened material formed within a cavity in the body and then discharged from the body; can originate in different areas and be composed of various materials.

Castration (kas-TRĀ-shun) The removal of the testes.

Catabolism (ka-TAB-ō-lizm) Chemical reactions that break down complex organic compounds into simple ones with the release of energy.

Cataract (KAT-a-rakt) Loss of transparency of the lens of the eye or its capsule or both.

Catheter (KATH-i-ter) A tube that can be inserted into a body cavity through a canal or into a blood vessel; used to remove fluids, such as urine and blood, and to introduce diagnostic materials or medication.

Cation A positively charged ion. An example is Na^+.

Cauda equina (KAW-da ē-KWĪ-na) A taillike collection of roots of spinal nerves at the inferior end of the spinal canal.

Cecum (SĒ-kum) A blind pouch at the proximal end of the large intestine to which the ileum is attached.

Cell The basic structural and functional unit of all organisms; the smallest structure capable of performing all the activities vital to life.

Cell inclusion A lifeless, often temporary, constituent in the cytoplasm of a cell as opposed to an organelle.

Cellular immunity That component of immunity in which specially sensitized lymphocytes (T cells) attach to antigens to destroy them. Also called **cell-mediated immunity.**

Cementum (se-MEN-tum) Calcified tissue covering the root of a tooth.

Center An area in the brain where a particular function is localized.

Center of ossification (os'-i-fi-KĀ-shun) An area in the cartilage model of a future bone where the cartilage cells hypertrophy and then secrete enzymes that result in the calcification of their matrix, resulting in the death of the cartilage cells, followed by the invasion of the area by osteoblasts that then lay down bone.

Central canal A circular channel running longitudinally in the center of an osteon (Haversian system) of mature compact bone, containing blood and lymph vessels and nerves. Also called a **Haversian** (ha-VER-shun) **canal.** A microscopic tube running the length of the spinal cord in the gray commissure.

Central fovea (FŌ-vē-a) A cuplike depression in the center of the macula lutea of the retina, containing cones only; the area of clearest vision.

Central nervous system (CNS) Brain and spinal cord.

Centrioles (SEN-trē-ōlz) Paired, cylindrical organelles within a centrosome, each consisting of a ring of microtubules and arranged at right angles to each other; function in cell division.

Centromere (SEN-trō-mēr)   The clear, constricted portion of a chromosome where the two chromatids are joined; serves as the point of attachment for the chromosomal microtubules.

Centrosome (SEN-trō-sōm)   A rather dense area of cytoplasm, near the nucleus of a cell, containing a pair of centrioles.

Cephalic (se-FAL-ik)   Pertaining to the head; superior in position.

Cerebellar peduncle (ser-e-BEL-ar pe-DUNG-kul)   A bundle of nerve fibers connecting the cerebellum with the brain stem.

Cerebellum (ser-e-BEL-um)   The portion of the brain lying posterior to the medulla and pons, concerned with coordination of movements.

Cerebral aqueduct (SER-e-bral AK-we-dukt)   A channel through the midbrain connecting the third and fourth ventricles and containing cerebrospinal fluid.

Cerebral cortex   The surface of the cerebral hemispheres, 2 to 4 mm thick, consisting of six layers of nerve cell bodies (gray matter) in most areas.

Cerebral palsy (PAL-zē)   A group of nonprogressive motor disorders caused by damage to motor areas of the brain (cerebral cortex, basal ganglia, and cerebellum) during fetal life, birth, or infancy.

Cerebral peduncle (pe-DUNG-kul)   One of a pair of nerve fiber bundles located on the ventral surface of the midbrain, conducting impulses between the pons and the cerebral hemispheres.

Cerebrospinal (se-rē'-brō-SPĪ-nal) **fluid** (CSF)   A fluid produced in the choroid plexuses of the ventricles of the brain that circulates in the ventricles and the subarachnoid space around the brain and spinal cord.

Cerebrovascular (se-rē'-brō-VAS-kyoo-lar) **accident** (CVA)   Destruction of brain tissue (infarction) resulting from disorders of blood vessels that supply the brain. Also called a **stroke.**

Cerebrum (SER-ē-brum)   The two hemispheres of the forebrain, making up the largest part of the brain.

Ceruminous (se-ROO-mi-nus) **gland**   A modified sudoriferous (sweat) gland in the external auditory meatus that secretes cerumen (ear wax).

Cervical dysplasia (dis-PLĀ-sē-a)   A change in the shape, growth, and number of cervical cells of the uterus that, if severe, may progress to cancer.

Cervical ganglion (SER-vi-kul GANG-glē-on)   A cluster of nerve cell bodies of postganglionic sympathetic neurons located in the neck, near the vertebral column.

Cervical plexus (PLEK-sus)   A network of neuron fibers formed by the anterior rami of the first four cervical nerves.

Cervix (SER-viks)   Neck; any constricted portion of an organ, especially the lower cylindrical part of the uterus.

Ceasarean (se-SA-rē-an) **section**   Procedure in which a low, horizontal incision is made through the abdominal wall and uterus for removal of the baby and placenta.

Chemonucleolysis (kē'-mō-noo'-klē-OL-i-sis)   Dissolution of the nucleus pulposus of an intervertebral disc by injection of a proteolytic enzyme (chymopapin) to relieve the pressure and pain associated with a herniated (slipped) disc.

Chemoreceptor (kē'-mō-re-SEP-tor)   Receptor outside the central nervous system on or near the carotid and aortic bodies that detects the presence of chemicals.

Chemotaxis (kē-mō-TAK-sis)   Attraction of phagocytes to microbes by a chemical stimulus.

Chemotherapy (kē'-mō-THER-a-pē)   The treatment of illness or disease by chemicals.

Chiropractic (kī'-rō-PRAK-tik)   A system of treating disease by using one's hands to manipulate body parts, mostly the vertebral column.

Chloride shift   Diffusion of bicarbonate ions from the red blood cells into plasma and of chloride ions from plasma into red blood cells that maintains ionic balance between red blood cells and plasma.

Cholecystectomy (kō'-lē-sis-TEK-tō-mē)   Surgical removal of the gallbladder.

Cholesterol (kō-LES-te-rol)   Classified as a lipid, the most abundant steroid in animal tissues; located in cell membranes and used for the synthesis of steroid hormones and bile salts.

Cholinergic (kō'-lin-ER-jik) **fiber**   A nerve ending that liberates acetylcholine at a synapse.

Cholinesterase (kō'-lin-ES-ter-ās)   An enzyme that hydrolyzes acetylcholine.

Chondrocyte (KON-drō-sīt)   Cell of mature cartilage.

Chondroitin (kon-DROY-tin) **sulfate**   An amorphous matrix material found outside the cell.

Chordae tendineae (KOR-dē TEN-di-nē)   Cords that connect the heart valves with the papillary muscles.

Chorion (KŌ-rē-on)   The outermost fetal membrane; serves a protective and nutritive function.

Chorionic villi sampling (CVS)   The removal of a sample of chorionic villi tissue by means of a catheter to analyze the tissue for prenatal genetic defects.

Chorionic villus (kō'-rē-ON-ik VIL-lus)   Fingerlike projection of the chorion that grows into the decidua basalis of the endometrium and contains fetal blood vessels.

Choroid (KŌ-royd)   One of the vascular coats of the eyeball.

Choroid plexus (PLEK-sus)   A vascular structure located in the roof of each of the four ventricles of the brain; produces cerebrospinal fluid.

Chromaffin (krō-MAF-in) **cell**   Cell that has an affinity for chrome salts, due in part to the presence of the precursors of the neurotransmitter epinephrine; found, among other places, in the adrenal medulla.

Chromatid (KRŌ-ma-tid)   One of a pair of identical connected nucleoprotein strands that are joined at the centromere and separate during cell division, each becoming a chromosome of one of the two daughter cells.

Chromatin (KRŌ-ma-tin)   The threadlike mass of the genetic material consisting principally of DNA, which is present in the nucleus of a nondividing or interphase cell.

Chromatolysis (krō'-ma-TOL-i-sis)   The breakdown of chromatophilic substance (Nissl bodies) into finely granular masses, a reaction characteristic of the axon reaction of peripheral nerve damage.

Chromatophilic substance   Rough endoplasmic reticulum in the cell bodies of neurons that functions in protein synthesis. Also called **Nissl bodies.**

Chromosomal microtubule (mī-krō-TOOB-yool)   Microtubule formed during prophase of mitosis that originates from centromeres, extends from a centromere to a pole of the cell, and assists in chromosomal movement; constitutes a part of the mitotic spindle.

Chromosome (KRŌ-mō-sōm)   One of the 46 small, dark-staining bodies that appear in the nucleus of a human diploid cell during cell division.

Chronic (KRON-ik)   Long-term or frequently recurring; applied to a disease that is not acute.

Chylomicron (kī'-lō-MĪ-kron)   Aggregate of triglycerides, phospholipids, and cholesterol coated with protein that enters a lacteal of a villus during absorption.

Chyme (kīm)   The semifluid mixture of partly digested food and digestive secretions found in the stomach and small intestine during digestion of a meal.

Cicatrix (SIK-a-triks)   A scar left by a healed wound.

Ciliary (SIL-ē-ar'-ē) **body**   One of the three portions of the vascular tunic of the eyeball, the others being the choroid and the iris; includes the ciliary muscle and the ciliary processes.

Cilium (SIL-ē-um) A hair or hairlike process projecting from a cell that may be used to move the entire cell or to move substances along the surface of the cell.

Circadian (ser-KĀ-dē-an) **rhythm** A cycle of active and nonactive periods in organisms determined by internal mechanisms and repeating about every 24 hours.

Circulation time Time required for blood to pass from the right atrium, through pulmonary circulation, back to the left ventricle, through systemic circulation to the foot, and back again to the right atrium; normally about 1 min.

Circulatory (SER-kyoo-la-to'-rē) **shock** A condition in which body tissues, including critical organs, do not receive an adequate blood supply because of reduced cardiac output or reduced blood volume; signs include lowered blood pressure, feeble and rapid pulse, shallow and rapid breathing, paleness or cyanosis, and mental confusion or unconsciousness.

Circumcision (ser'-kum-SIZH-un) Removal of the foreskin, the fold over the glans penis.

Circumduction (ser'-kum-DUK-shun) A movement at a synovial joint in which the distal end of a bone moves in a circle while the proximal end remains relatively stable.

Circumvallate papilla (ser'-kum-VAL-āt pa-PIL-a) One of the circular projections that is arranged in an inverted V-shaped row at the posterior portion of the tongue; the largest of the elevations on the upper surface of the tongue containing taste buds.

Cirrhosis (si-RŌ-sis) A liver disorder in which the parenchymal cells are destroyed and replaced by connective tissue.

Cisterna chyli (sis-TER-na KĪ-lē) The origin of the thoracic duct.

Cleavage The rapid mitotic divisions following the fertilization of an ovum, resulting in an increased number of progressively smaller cells, called blastomeres, so that the overall size of the zygote remains the same.

Cleft palate Condition in which the palatine processes of the maxillae do not unite before birth; cleft lip, a split in the upper lip, is often associated with cleft palate.

Climacteric (klī-mak-TER-ik) Cessation of the reproductive function in the female or diminution of testicular activity in the male.

Climax The peak period or moments of greatest intensity during sexual excitement.

Clitoris (KLI-to-ris) An erectile organ of the female that is homologous to the male penis.

Clot The end result of a series of biochemical reactions that changes liquid plasma into a gelatinous mass; specifically, the conversion of fibrinogen into a tangle of polymerized fibrin molecules.

Clot retraction (rē-TRAK-shun) The consolidation of a fibrin clot to pull damaged tissue together.

Clotting time The time required for blood to coagulate as a result of the breakup of platelets upon coming into contact with glass; ranges from 5 to 15 minutes in uncoated glass tubes.

Coagulation (cō-ag-yoo-LĀ-shun) Process by which a blood clot is formed.

Coarctation (kō'-ark-TĀ-shun) **of the aorta** Congenital condition in which the aorta is too narrow and results in reduced blood supply, increased ventricular pumping, and high blood pressure.

Coccygeal plexus (PLEK-sus) A network of nerves formed by the anterior rami of the coccygeal and the fourth and fifth sacral nerves. Fibers from the plexus supply the skin in the region of the coccyx.

Coccyx (KOK-six) The fused bones at the end of the vertebral column.

Cochlea (KŌK-lē-a) A winding, cone-shaped tube forming a portion of the inner ear and containing the spiral organ (organ of Corti).

Cochlear duct The membranous cochlea consisting of a spirally arranged tube enclosed in the bony cochlea and lying along its outer wall. Also called the **scala media** (SCA-la MĒ-dē-a).

Coenzyme A nonprotein organic molecule that is associated with and activates an enzyme; many are derived from vitamins. An example is nicotinamide adenine dinucleotide (NAD), derived from the B vitamin niacin.

Coitus (KŌ-i-tus) Sexual intercourse. Also called **copulation** (cop-yoo-LĀ-shun).

Collagen (KOL-a-jen) A protein that is the main organic constituent of connective tissue.

Collateral circulation The alternate route taken by blood through an anastomosis.

Collision theory Theory that explains how chemical reactions occur. It states that all atoms, ions, and molecules are constantly moving and colliding, and the energy transferred during collision might disrupt their electron structures so that chemical bonds are broken or formed.

Colon The division of the large intestine consisting of ascending, transverse, descending, and sigmoid portions.

Color blindness Any deviation in the normal perception of colors, resulting from the lack of one or more of the photopigments of the cones.

Colostomy (kō-LOS-tō-mē) The surgical creation of a new opening from the colon to the body surface.

Colostrum (kō-LOS-trum) A thin, cloudy fluid secreted by the mammary glands a few days prior to or after delivery before true milk is secreted.

Colposcopy (kol-POS-kō-pē) Direct examination of the vaginal and cervical mucosa using a magnifying device.

Coma (KŌ-ma) Profound unconsciousness from which one cannot be aroused, even by strong stimulation.

Commissure (KOM-i-shūr) The angular junction of the eyelids at either corner of the eyes.

Common bile duct A tube formed by the union of the common hepatic duct and the cystic duct that empties bile into the duodenum at the hepatopancreatic ampulla (ampulla of Vater).

Compact (dense) bone Bone tissue with no apparent spaces in which the layers of lamellae are fitted tightly together. Compact bone is found immediately deep to the periosteum and external to spongy bone.

Complement A group of interacting plasma proteins that is activated by certain antigen-antibody reactions and follows a specific sequence of reactions on the surface of a cell on which the antibody is fixed so that the membrane is lysed.

Complete blood count (CBC) Hematology test that usually includes hemoglobin determination, hematocrit, red and white blood cell count, differential count, and comments about blood cell morphology.

Compliance The ease with which the lungs and thoracic wall can be expanded.

Compound A substance that can be broken down into two or more other substances by chemical means.

Computed tomography (tō-MOG-ra-fē) **(CT)** X-ray technique that provides a cross-sectional picture of any area of the body. Also called **computed axial tomography (CAT).**

Concha (KONG-ka) A scroll-like bone found in the skull. *Plural,* **conchae** (KONG-kē).

Concussion (kon-KUSH-un) Traumatic injury to the brain that produces no visible bruising, but may result in abrupt, temporary loss of consciousness.

Conduction myofiber Muscle fiber in the subendocardial tissue of the heart specialized for conducting an impulse to the myocardium; part of the conduction system of the heart.

Also called a **Purkinje** (pur-KIN-jē) **fiber.**

Cone   The light-sensitive receptor in the retina concerned with color vision.

Congenital (kon-JEN-i-tal)   Present at the time of birth.

Congestive heart failure (CHF)   Chronic or acute state that results when the heart is not capable of supplying the oxygen demands of the body.

Conjunctiva (kon'-junk-TĪ-va)   The delicate membrane covering the eyeball and lining the eyes.

Conjunctivitis (kon-junk'-ti-VĪ-tis)   Inflammation of the conjunctiva, the delicate membrane covering the eyeball and lining the eyelids.

Connective tissue   The most abundant of the four tissue types in the body, performing the functions of binding and supporting; consists of relatively few cells in a great deal of intercellular substance.

Constipation (con-sti-PĀ-shun)   Infrequent or difficult defecation caused by decreased motility of the intestines.

Contact inhibition   Phenomenon by which migration of a growing cell is stopped when it makes contact with another cell of its own kind.

Continuous microtubules (mī-krō-TOOB-yoolz)   Microtubules formed during prophase of mitosis that originate from the vicinity of the centrioles, grow toward each other, extend from one pole of the cell to another; and assist in chromosomal movement; constitute a part of the mitotic spindle.

Contraception (kon'-tra-SEP-shun)   The prevention of conception or impregnation without destroying fertility.

Contractility (kon'-trak-TIL-i-tē)   The ability of muscle tissue to shorten.

Contralateral (kon'-tra-LAT-er-al)   On the opposite side; affecting the opposite side of the body.

Conus medullaris (KŌ-nus med-yoo-LAR-is)   The tapered portion of the spinal cord below the lumbar enlargement.

Convergence (con-VER-jens)   An anatomical arrangement in which the synaptic end bulbs of several presynaptic neurons terminate on one postsynaptic neuron. The medial movement of the two eyeballs so that both are directed toward a near object being viewed in order to produce a single image.

Convulsion (con-VUL-shun)   Violent, involuntary, tetanic contractions of an entire group of muscles.

Cornea (KOR-nē-a)   The transparent fibrous coat that covers the iris of the eye.

Coronal (kō-RŌ-nal) **plane**   A plane that runs vertical to the ground and divides the body into anterior and posterior portions. Also called **frontal plane.**

Corona radiata   Innermost layer of follicle cells surrounding an ovum.

Coronary (KOR-ō-na-rē) **artery disease (CAD)**   A condition in which the heart muscle receives inadequate blood due to an interruption of its blood supply.

Coronary artery spasm   A condition in which the smooth muscle of a coronary artery undergoes a sudden contraction, resulting in vasoconstriction.

Coronary circulation   The pathway followed by the blood from the ascending aorta through the blood vessels supplying the heart and returning to the right atrium. Also called **cardiac circulation.**

Coronary sinus (SĪ-nus)   A wide venous channel on the posterior surface of the heart that collects the blood from the coronary circulation and returns it to the right atrium.

Corpora quadrigemina (KOR-por-a kwad-ri-JEM-in-a)   Four small elevations (superior and inferior colliculi) on the dorsal region of the midbrain concerned with visual and auditory functions.

Cor pulmonale (kor pul-mōn-ALE) **(CP)**   Right ventricular hypertrophy from disorders that bring about hypertension in pulmonary circulation.

Corpus albicans (KOR-pus AL-bi-kanz)   A white fibrous patch in the ovary that forms after the corpus luteum regresses.

Corpus callosum (kal-LŌ-sum)   The great commissure of the brain between the cerebral hemispheres.

Corpuscle of touch   The sensory receptor for the sensation of touch; found in the dermal papillae, especially in palms and soles. Also called a **Meissner's** (MĪS-nerz) **corpuscle.**

Corpus luteum (LOO-tē-um)   A yellow endocrine gland in the ovary formed when a follicle has discharged its ovum; secretes estrogens, progesterone, and relaxin.

Corpus striatum (strī-Ā-tum)   An area in the interior of each cerebral hemisphere composed of the caudate and lentiform nuclei of the basal ganglia and white matter of the internal capsule, arranged in a striated manner.

Cortex (KOR-teks)   An outer layer of an organ. The convoluted layer of gray matter covering each cerebral hemisphere.

Costal cartilage (KOS-tal KAR-ti-lij)   Hyaline cartilage that attaches a rib to the sternum.

Countercurrent mechanism   One mechanism involved in the ability of the kidneys to produce a hyperosmotic urine.

Cowper's gland   See **Bulbourethral gland.**

Cramp   A spasmodic, especially a tonic, contraction of one of many muscles, usually painful.

Cranial (KRĀ-nē-al) **cavity**   A subdivision of the dorsal body cavity formed by the cranial bones and containing the brain.

Cranial nerve   One of 12 pairs of nerves that leave the brain, pass through foramina in the skull, and supply the head, neck, and part of the trunk; each is designated by a Roman numeral and a name.

Craniosacral (krā-nē-ō-SĀ-kral) **outflow**   The fibers of parasympathetic preganglionic neurons, which have their cell bodies located in nuceli in the brain stem and in the lateral gray matter of the sacral portion of the spinal cord.

Craniotomy (krā'-nē-OT-ō-mē)   Any operation on the skull, as for surgery on the brain or decompression of the fetal head in difficult labor.

Cranium (KRĀ-nē-um)   The skeleton of the skull that protects the brain and the organs of sight, hearing, and balance; includes the frontal, parietal, temporal, occipital, sphenoid, and ethmoid bones.

Crenation (krē-NĀ-shun)   The shrinkage of red blood cells into knobbed, starry forms when placed in a hypertonic solution.

Cretinism (KRĒ-tin-izm)   Severe congenital thyroid deficiency during childhood leading to physical and mental retardation.

Crista (KRIS-ta)   A crest or ridged structure. A small elevation in the ampulla of each semicircular duct that serves as a receptor for dynamic equilibrium.

Cross-matching   A test performed prior to blood transfusion to check for compatibility of blood types. Red blood cells from donor are mixed with a sample of recipient's serum and then rbc's from recipient are mixed with a sample of donor's serum. If no agglutination occurs, the bloods are deemed compatible.

Crus (krus) **of penis**   Separated, tapered portion of the corpora cavernosa penis. *Plural,* **crura** (KROO-ra).

Cryosurgery (KRĪ-ō-ser-jer-ē)   The destruction of tissue by application of extreme cold.

Crypt of Lieberkühn   See **Intestinal gland.**

Cryptorchidism (krip-TOR-ki-dizm)   The condition of undescended testes.

Cupula (KUP-yoo-la)   A mass of gelatinous material covering the hair cells of a crista, a receptor in the ampulla of a semicircular canal stimulated when the head moves.

Curvature (KUR-va-tūr)   A nonangular deviation of a straight line, as in the greater and lesser curvatures of the stomach. Abnormal curvatures of the vertebral column include kyphosis, lordosis, and scoliosis.

Cushing's syndrome   Condition caused by a hypersecretion of glucocorticoids characterized by spindly legs, "moon face," "buffalo hump," pendulous abdomen, flushed facial skin, and poor wound healing.

Cutaneous (kyoo-TĀ-nē-us)   Pertaining to the skin.

Cyanosis (sī-a-NŌ-sis)   Slightly bluish or dark purple discoloration of the skin and the mucous membrane due to an oxygen deficiency.

Cyclic adenosine-3′,5′-monophosphate (cyclic AMP)   Molecule formed from ATP by the action of the enzyme adenylate cyclase; serves as an intracellular messenger for some hormones.

Cystic (SIS-tik) **duct**   The duct that transports bile from the gallbladder to the common bile duct.

Cystitis (sis-TĪ-tis)   Inflammation of the urinary bladder.

Cystoscope (SIS-tō-skōp)   An instrument used to examine the inside of the urinary bladder.

Cytochrome (SĪ-tō-krōm)   A protein containing an iron portion capable of alternating between a reduced form and an oxidized form.

Cytokinesis (sī-tō-ki-NĒ-sis)   Division of the cytoplasm.

Cytology (sī-TOL-o-jē)   The study of cells.

Cytoplasm (SĪ-tō-plazm)   Substance within a cell's plasma membrane and external to its nucleus. Also called **protoplasm.**

Cytoskeleton   Complex internal structure of cytoplasm consisting of microfilaments, microtubules, and intermediate filaments.

Dartos (DAR-tōs)   The contractile tissue under the skin of the scrotum.

Dead air volume   The volume of air that is inhaled but remains in spaces in the upper respiratory system and does not reach the alveoli to participate in gas exchange; about 150 ml.

Deafness   Lack of the sense of hearing or a significant hearing loss.

Debility (dē-BIL-i-tē)   Weakness of tonicity in functions or organs of the body.

Decibel (DES-i-bel) **(db)**   A unit that measures relative sound intensity (loudness).

Decidua (dē-SID-yoo-a)   That portion of the endometrium of the uterus (all but the deepest layer) that is modified for pregnancy and shed after childbirth.

Deciduous (dē-SID-yoo-us)   Falling off or being shed seasonally or at a particular stage of development. In the body, referring to the first set of teeth.

Decompression sickness   A condition characterized by joint pains and neurologic symptoms; follows from a too-rapid reduction of environmental pressure or decompression, so that nitrogen that dissolved in body fluid under pressure comes out of solution as bubbles that form air emboli and occlude crucial blood vessels. Also called **caisson (KĀ-son) disease** or **bends.**

Decubitus (dē-KYOO-bi-tus) **ulcer**   Tissue destruction due to a constant deficiency of blood to tissues overlying a bony projection that has been subjected to prolonged pressure against an object such as a bed, cast, or splint. Also called **bedsore, pressure sore,** or **trophic ulcer.**

Decussation (dē′-ku-SĀ-shun)   A crossing over; usually refers to the crossing of most of the fibers in the large motor tracts to opposite sides in the medullary pyramids.

Deep   Away from the surface of the body.

Deep fascia (FASH-ē-a)   A sheet of connective tissue wrapped around a muscle to hold it in place.

Deep inguinal (IN-gwi-nal) **ring**   A slitlike opening in the aponeurosis of the transversus abdominis muscle that represents the origin of the inguinal canal.

Defecation (def-e-KĀ-shun)   The discharge of feces from the rectum.

Defibrillation (dē-fib-ri-LĀ-shun)   Delivery of a very strong electrical current to the heart in an attempt to stop ventricular fibrillation.

Deglutition (dē-gloo-TISH-un)   The act of swallowing.

Dehydration (dē-hī-DRĀ-shun)   Excessive loss of water from the body or its parts.

Delta cell   A cell in the pancreatic islets (islets of Langerhans) in the pancreas that secretes somatostatin.

Demineralization (de-min′-er-al-i-ZĀ-shun)   Loss of calcium and phosphorus from bones.

Dendrite (DEN-drīt)   A nerve cell process carrying an impulse toward the cell body.

Dental caries (KA-rēz)   Gradual demineralization of the enamel and dentin of a tooth that may invade the pulp and alveolar bone. Also called **tooth decay.**

Dentin (DEN-tin)   The osseous tissues of a tooth enclosing the pulp cavity.

Dentition (den-TI-shun)   The eruption of teeth. The number, shape, and arrangement of teeth.

Deoxyribonucleic (dē-ok′-sē-ri′-bō-nyoo-KLĒ-ik) **acid (DNA)**   A nucleic acid in the shape of a double helix constructed of nucleotides consisting of one of four nitrogen bases (adenine, cytosine, guanine, or thymine), deoxyribose, and a phosphate group; encoded in the nucleotides is genetic information.

Depolarization (dē-pō-lar-i-ZĀ-shun)   Used in neurophysiology to describe the reduction of voltage across a cell membrane; expressed as a movement toward less negative (more positive) voltages on the interior side of the cell membrane.

Depression (dē-PRESS-shun)   Movement in which a part of the body moves downward.

Dermal papilla (pa-PILL-a)   Fingerlike projection of the papillary region of the dermis that may contain blood capillaries or corpuscles of touch (Meissner's corpuscles).

Dermatology (der-ma-TOL-ō-jē)   The medical specialty dealing with diseases of the skin.

Dermatome (DER-ma-tōm)   An instrument for incising the skin or cutting thin transplants of skin. The cutaneous area developed from one embryonic spinal cord segment and receiving most of its innervation from one spinal nerve.

Dermis (DER-mis)   A layer of dense connective tissue lying deep to the epidermis; the true skin or corium.

Descending colon (KŌ-lon)   The part of the large intestine descending from the left colic (splenic) flexure to the level of the left iliac crest.

Detritus (de-TRI-tus)   Any broken-down or degenerative tissue or carious matter.

Detrusor muscle (de-TROO-ser)   Muscle in the wall of the urinary bladder.

Developmental anatomy   The study of development from the fertilized egg to the adult form. The branch of anatomy called embryology is generally restricted to the study of development from the fertilized egg through the eighth week in utero.

Diabetes insipidus (dī-a-BĒ-tēz in-SIP-i-dus)   Condition caused by hyposecretion of antidiuretic hormone (ADH) and characterized by excretion of large amounts of urine and thirst.

Diabetes mellitus (MEL-i-tus)   Hereditary condition caused by hyposecretion of insulin and characterized by hyperglycemia, increased urine production, excessive thirst, and excessive eating.

Diagnosis (dī-ag-NŌ-sis)   Recognition of disease states from signs and symptoms by inspection, palpation, laboratory tests, and other means.

Dialysis (dī-AL-i-sis)   The process of separating crystalloids (smaller particles) from colloids (larger particles) by the difference in their rates of diffusion through a selectively permeable membrane.

Diapedesis (dī-a-pe-DĒ-sis) The passage of white blood cells through intact blood vessel walls.

Diaphragm (DĪ-a-fram) Any partition that separates one area from another, especially the dome-shaped skeletal muscle between the thoracic and abdominal cavities.

Diaphysis (dī-AF-i-sis) The shaft of a long bone.

Diarrhea (dī-a-RĒ-a) Frequent defecation of liquid feces caused by increased motility of the intestines.

Diarthrosis (dī-'ar-THRŌ-sis) Articulation in which opposing bones move freely, as in a hinge joint.

Diastole (dī-AS-tō-lē) In the cardiac cycle, the phase of relaxation or dilation of the heart muscle, especially of the ventricles.

Diastolic (di-as-TOL-ik) **blood pressure** The force exerted by blood on arterial walls during ventricular relaxation; the lowest blood pressure measured in the large arteries, about 80 mm Hg under normal conditions for a young, adult male.

Diencephalon (dī-'en-SEF-a-lon) A part of the brain consisting primarily of the thalamus and the hypothalamus.

Differential (dif'-fer-EN-shal) **count** Determination of the number of each kind of white blood cell in a sample of 100 cells for diagnostic purposes.

Differentiation (dif'-e-ren'-shē-Ā-shun) Acquisition of specific functions different from those of the original general type.

Diffusion (dif-YOO-zhun) A passive process in which there is a net or greater movement of molecules or ions from a region of high concentration to a region of low concentration until equilibrium is reached.

Digestion (di-JES-chun) The mechanical and chemical breakdown of food to simple molecules that can be absorbed by the body.

Dilate (DĪ-lāte) To expand or swell.

Diploid (DIP-loyd) Having the number of chromosomes characteristically found in the somatic cells of an organism.

Diplopia (di-PLŌ-pē-a) Double vision.

Dislocation (dis-lō-KĀ-shun) Displacement of a bone from a joint with tearing of ligaments, tendons, and articular capsules. Also called **luxation** (luks-Ā-shun).

Dissect (DIS-sekt) To separate tissues and parts of a cadaver (corpse) or an organ for anatomical study.

Dissociation (dis'-sō-sē-Ā-shun) Separation of inorganic acids, bases, and salts into ions when dissolved in water. Also called **ionization** (ī'-on-i-ZĀ-shun).

Distal (DIS-tal) Farther from the attachment of an extremity to the trunk or a structure; farther from the point of origin.

Diuretic (dī-yoo-RET-ik) A chemical that increases urine volume by inhibiting facultative reabsorption of water.

Divergence (di-VER-jens) An anatomical arrangement in which the synaptic end bulbs of one presynaptic neuron terminate on several postsynaptic neurons.

Diverticulitis (dī-ver-tik-yoo-LĪ-tis) Inflammation of diverticula, saclike outpouchings of the colonic wall, when the muscularis becomes weak.

Diverticulum (dī-ver-TIK-yoo-lum) A sac or pouch in the wall of a canal or organ, especially in the colon.

Dominant gene A gene that is able to override the influence of the complementary gene on the homologous chromosome; the gene that is expressed.

Dorsal body cavity Cavity near the dorsal surface of the body that consists of a cranial cavity and vertebral canal.

Dorsiflexion (dor'-si-FLEK-shun) Flexion of the foot at the ankle.

Down's syndrome An inherited defect due to an extra copy of chromosome 21. Symptoms include mental retardation; a small skull, flattened front to back; a short, flat nose; short fingers; and a widened space between first two digits of the hand and foot. Also called **trisomy 21.**

Duct of Santorini *See* **Accessory duct.**

Duct of Wirsung *See* **Pancreatic duct.**

Ductus arteriosus (DUK-tus ar-tē-rē-Ō-sus) A small vessel connecting the pulmonary trunk with the aorta; found only in the fetus.

Ductus deferens (DEF-er-ens) The duct that conducts spermatozoa from the epididymis to the ejaculatory duct. Also called the **vas deferens** or **seminal duct.**

Ductus epididymis (ep'-i-DID-i-mis) A tightly coiled tube inside the epididymis, distinguished into a head, body, and tail, in which spermatozoa undergo maturation.

Ductus venosus (ve-NŌ-sus) A small vessel in the fetus that helps the circulation bypass the liver.

Duodenal gland Gland in the submucosa of the duodenum that secretes an alkaline mucus to protect the lining of the small intestine from the action of enzymes and to help neutralize the acid in chyme. Also called **Brunner's** (BRUN-erz) **gland.**

Duodenal papilla (doo-ō-DĒ-nal pa-PILL-a) An elevation on the duodenal mucosa that receives the hepatopancreatic ampulla (ampulla of Vater).

Duodenum (doo'-ō-DĒ-num) The first portion of the small intestine.

Dura mater (DYOO-ra MĀ-ter) The outer membrane (meninx) covering the brain and spinal cord.

Dynamic equilibrium (ē-kwi-LIB-rē-um) The maintenance of body position, mainly the head, in response to sudden movements such as rotation, acceleration, and deceleration.

Dynamic spatial reconstructor (DSR) An x-ray machine that has the ability to construct moving, three-dimensional, life-size images of all or part of an internal organ from any view desired.

Dysfunction (dis-FUNK-shun) Absence of complete normal function.

Dyslexia (dis-LEK-sē-a) Impairment of ability to comprehend written language.

Dysmenorrhea (dis'-men-ō-RĒ-a) Painful menstruation.

Dystrophia (dis-TRŌ-fē-a) Progressive weakening of a muscle.

Ectoderm The outermost of the three primary germ layers that gives rise to the nervous system and the epidermis of skin and its derivatives.

Ectopic (ek-TOP-ik) Out of the normal location.

Eczema (EK-ze-ma) A skin rash characterized by itching, swelling, blistering, oozing, and scaling of the skin.

Edema (e-DĒ-ma) An abnormal accumulation of fluid in body tissues.

Effector (e-FEK-tor) The organ of the body, either a muscle or a gland, that responds to a motor neuron impulse.

Efferent arteriole (EF-er-ent ar-TĒ-rē-ōl) A vessel of the renal vascular system that transports blood from the glomerulus to the peritubular capillary.

Efferent (EF-er-ent) **ducts** A series of coiled tubes that transport spermatozoa from the rete testis to the epididymis.

Efferent neuron (NOO-ron) A neuron that conveys impulses from the brain and spinal cord to effectors that may be either muscles or glands. Also called a **motor neuron.**

Effusion (e-FYOO-zhun) The escape of fluid from the lymphatics or blood vessels into a cavity or into tissues.

Ejaculation (e-jak-yoo-LĀ-shun) The reflex ejection or expulsion of semen from the penis.

Ejaculatory (e-JAK-yoo-la-tō'-rē) **duct** A tube that transports spermatozoa from the ductus (vas) deferens to the prostatic urethra.

Elasticity (e-las-TIS-i-tē) The ability of tissue to return to its original shape after contraction or extension.

Electrocardiogram (e-lek'-trō-KAR-dē-ō-gram) **(ECG or EKG)** A recording of the electrical changes that accompany the cardiac cycle.

Electroencephalogram (e-lek'-trō-en-SEF-a-lō-gram) **(EEG)** A recording of the electrical impulses of the brain.

Electrolyte (ē-LEK-trō-līt) Any compound that separates into ions when dissolved in water and is able to conduct electricity.

Electron transport chain A series of oxidation-reduction reactions in the catabolism of glucose in which energy is released and transferred for storage to ATP.

Eleidin (el-Ē-i-din) A translucent substance found in the skin.

Elevation (el-e-VĀ-shun) Movement in which a part of the body moves upward.

Ellipsoidal (e-lip-SOY-dal) **joint** A synovial joint structured so that an oval-shaped condyle of one bone fits into an elliptical cavity of another bone, permitting side-to-side and back-and-forth movements, as at the joint at the wrist between the radius and carpals. Also called a **condyloid** (KON-di-loid) **joint.**

Embolism (EM-bō-lizm) Obstruction or closure of a vessel by an embolus.

Embolus (EM-bō-lus) A blood clot, bubble of air, fat from broken bones, mass of bacteria, or other debris or foreign material transported by the blood.

Embryo (EM-brē-ō) The young of any organism in an early stage of development; in humans, the developing organism from fertilization to the end of the eighth week in utero.

Embryology (em'-brē-OL-ō-jē) The study of development from the fertilized egg to the end of the eighth week in utero.

Embryo transfer A procedure in which a husband's semen is used to artificially inseminate a fertile ovum donor and the morula or blastocyst is then transferred from the donor to the infertile wife who carries it to term.

Emesis (EM-e-sis) Vomiting.

Emmetropia (em'-e-TRŌ-pē-a) The ideal optical condition of the eyes.

Emphysema (em'-fi-SĒ-ma) A swelling or inflation of air passages due to loss of elasticity in the alveoli.

Emulsification (ē-mul'-si-fi-KĀ-shun) The dispersion of large fat globules to smaller uniformly distributed particles.

Enamel (e-NAM-el) The hard, white substance covering the crown of a tooth.

End-diastolic (dī-a-STOL-ik) **volume (EDV)** The volume of blood, about 120 to 130 ml, that enters a ventricle during diastole (relaxation).

Endocardium (en-dō-KAR-dē-um) The layer of the heart wall, composed of endothelium and smooth muscle, that lines the inside of the heart and covers the valves and tendons that hold the valves open.

Endochondral ossification (en'-dō-KON-dral os'-i-fi-KĀ-shun) The replacement of cartilage by bone. Also called **intracartilaginous** (in'-tra-kar'-ti-LAJ-i-nus) **ossification.**

Endocrine (EN-dō-krin) **gland** A gland that secretes hormones into the blood; a ductless gland.

Endocrinology (en'-dō-kri-NOL-ō-jē) The science concerned with the structure and functions of endocrine glands and the diagnosis and treatment of disorders of the endocrine system.

Endocytosis (en'-dō-sī-TŌ-sis) The uptake into a cell of large molecules and particles in which a segment of plasma membrane surrounds the substance, encloses it, and brings it in; includes phagocytosis, pinocytosis, and receptor-mediated endocytosis.

Endoderm The innermost of the three primary germ layers of the developing embryo that gives rise to the gastrointestinal tract, urinary bladder and urethra, and respiratory tract.

Endodontics (en'-dō-DON-tiks) The branch of dentistry concerned with the prevention, diagnosis, and treatment of diseases that affect the pulp, root, periodontal ligament, and alveolar bone.

Endogenous (en-DOJ-e-nus) Growing from or beginning within the organism.

Endolymph (EN-dō-lymf') The fluid within the membranous labyrinth of the inner ear.

Endometriosis (en'-dō-MĒ-trē-ō'-sis) The growth of endometrial tissue outside the uterus.

Endometrium (en'-dō-MĒ-trē-um) The mucous membrane lining the uterus.

Endomysium (en'-dō-MĪZ-ē-um) Invagination of the perimysium separating each individual muscle cell.

Endoneurium (en'-dō-NYOO-rē-um) Connective tissue wrapping around individual nerve fibers.

Endoplasmic reticulum (en'-dō-PLAZ-mik re-TIK-yoo-lum) **(ER)** A network of channels running through the cytoplasm of a cell that serves in intracellular transportation, support, storage, synthesis, and packaging of molecules. Portions of ER where ribosomes are attached to the outer surface are called granular or rough reticulum; portions that have no ribosomes are called agranular or smooth reticulum.

End organ of Ruffini *See* **cutaneous mechanoreceptor.**

Endorphin (en-DOR-fin) A neuropeptide in the central nervous system that acts as a painkiller.

Endoscope (EN-dō-skōp') An illuminated tube with lenses used to look inside hollow organs such as the stomach (gastroscope) or urinary bladder (cystoscope).

Endoscopy (en-DOS-kō-pē) The visual examination of any cavity of the body using an endoscope, an illuminated tube with lenses.

Endosteum (en-DOS-tē-um) The membrane that lines the medullary cavity of bones.

Endothelial-capsular (en-dō-THĒ-lē-al) **membrane** A filtration membrane in a nephron consisting of the endothelium and basement membrane of the glomerulus and the epithelium of the visceral layer of the glomerular (Bowman's) capsule.

Endothelial (en'-dō-THĒ-lē-al) **tube** A derivative of mesoderm that develops beneath the foregut of an embryo before the end of the third week and eventually develops into the heart.

Endothelium (en'-dō-THĒ-lē-um) The layer of simple squamous epithelium that lines the cavities of the heart and blood and lymphatic vessels.

End-systolic (sis-TO-lik) **volume (ESV)** The volume of blood, about 50 to 60 ml, remaining in a ventricle following its systole (contraction).

Enkephalin (en-KEF-a-lin) A neuropeptide found in the central nervous system that acts as a painkiller.

Enteroendocrine (en-ter-ō-ENDŌ-krin) **cell** A stomach cell that secretes the hormone stomach gastrin.

Enterogastric (en-te-rō-GAS-trik) **reflex** A reflex that inhibits gastric secretion; initiated by food in the small intestine.

Enuresis (en'-yoo-RĒ-sis) Involuntary discharge of urine, complete or partial, after age 3.

Enzyme (EN-zīm) A substance that affects the speed of chemical changes; an organic catalyst, usually a protein.

Eosinophil (ē'-ō-SIN-ō-fil) A type of white blood cell characterized by granular cytoplasm readily stained by eosin.

Ependyma (e-PEN-de-ma) Neuroglial cells that line ventricles of the brain and probably assist in the circulation of cerebrospinal fluid (CSF). Also called **ependymocytes** (e-PEN-di-mō-sītz).

Epidemiology (ep'-i-dē-mē-OL-ō-jē) Medical science concerned with the occurrence and distribution of disease in human populations.

Epidermis (ep'-i-DERM-is) The outermost layer of skin, composed of stratified squamous epithelium.

Epididymis (ep'-i-DID-i-mis) A comma-shaped organ that lies along the posterior border of the testis and contains the ductus epididymis, in which sperm undergo maturation. *Plural,* **epididymides** (ep'-i-DID-i-mi-dēz).

Epidural space (ep'-i-DOO-ral) A space between the spinal dura mater and the vertebral canal, containing loose connective tissue and a plexus of veins.

Epiglottis (ep'-i-GLOT-is) A large, leaf-shaped piece of cartilage lying on top of the larynx, with its "stem" attached to the thyroid cartilage and its "leaf" portion unattached and free to move up and down to cover the glottis.

Epilepsy (EP-i-lep'-sē) Neurological disorder characterized by short, periodic attacks of motor, sensory, or psychological malfunction.

Epimysium (ep'-i-MĪZ-ē-um) Fibrous connective tissue around muscles.

Epinephrine (ep-ē-NEF-rin) Hormone secreted by the adrenal medulla that produces actions similar to those that result from sympathetic stimulation. Also called **adrenaline** (a-DREN-a-lin).

Epineurium (ep'-i-NYOO-rē-um) The outermost covering around the entire nerve.

Epiphyseal (ep'-i-FIZ-ē-al) **line** The remnant of the epiphyseal plate in a long bone.

Epiphyseal (ep'-i-FIZ-ē-al) **plate** The plate between the epiphysis and diaphysis.

Epiphysis (ē-PIF-i-sis) The end of a long bone, usually larger in diameter than the shaft (the diaphysis).

Epiphysis cerebri (se-RĒ-brē) Pineal gland.

Episiotomy (e-piz'-ē-OT-ō-mē) Incision of the clinical perineum at the end of second stage of labor to avoid tearing.

Epistaxis (ep'-i-STAK-sis) Hemorrhage from the nose; nosebleed.

Epithelial (ep'-i-THĒ-lē-al) **tissue** The tissue that forms glands or the outer part of the skin and lines blood vessels, hollow organs, and passages that lead externally from the body.

Eponychium (ep'-ō-NIK-ē-um) Narrow band of stratum corneum at the proximal border of a nail that extends from the margin of the nail wall. Also called the **cuticle.**

Erection (ē-REK-shun) The enlarged and stiff state of the penis (or clitoris) resulting from the engorgement of the spongy erectile tissue with blood.

Eructation (e-ruk'-TĀ-shun) The forceful expulsion of gas from the stomach. Also called **belching.**

Erythema (er'-e-THĒ-ma) Skin redness usually caused by engorgement of the capillaries in the lower layers of the skin.

Erythematosus (er-i'-them-a-TŌ-sis) Pertaining to redness.

Erythrocyte (e-RITH-rō-sīt) Red blood cell.

Erythropoiesis (e-rith'-rō-poy-Ē-sis) The process by which erythrocytes are formed.

Erythropoietin (ē-rith'-rō-POY-ē-tin) A hormone formed from a plasma protein that stimulates red blood cell production.

Esophagus (e-SOF-a-gus) A hollow muscular tube connecting the pharynx and the stomach.

Essential amino acids Those 10 amino acids that cannot be synthesized by the human body at an adequate rate to meet its needs and therefore must be obtained from the diet.

Estrogens (ES-tro-jens) Female sex hormones produced by the ovaries concerned with the development and maintenance of female reproductive structures and secondary sex characteristics, fluid and electrolyte balance, and protein anabolism. Examples are β-estradiol, estrone, and estriol.

Etiology (ē'-tē-OL-ō-jē) The study of the causes of disease, including theories of origin and the organisms, if any, involved.

Euphoria (yoo-FŌR-ē-a) A subjectively pleasant feeling of well-being marked by confidence and assurance.

Eupnea (yoop-NĒ-a) Normal quiet breathing.

Eustachian tube *See* **Auditory tube.**

Euthanasia (yoo'-tha-NĀ-zē-ā) The practice of ending a life in case of incurable disease.

Eversion (ē-VER-zhun) The movement of the sole outward at the ankle joint.

Exacerbation (eg-zas'-er-BĀ-shun) An increase in the severity of symptoms or of disease.

Excitability (ek-sīt'-a-BIL-i-tē) The ability of muscle tissue to receive and respond to stimuli; the ability of nerve cells to respond to stimuli and convert them into nerve impulses.

Excitatory postsynaptic potential (EPSP) The slight decrease in negative voltage seen on the postsynaptic membrane when it is stimulated by a presynaptic terminal. The EPSP is a localized event that decreases in strength from the point of excitation.

Excrement (EKS-kre-ment) Material cast out from the body as waste, especially fecal matter.

Excretion (eks-KRĒ-shun) The process of eliminating waste products from a cell, tissue, or the entire body; or, the products excreted.

Exocrine (EK-sō-krin) **gland** A gland that secretes substances into ducts that empty at covering or lining epithelium or directly onto a free surface.

Exocytosis (ex'-ō-sī-TŌ-sis) A process of discharging cellular products too big to go through the membrane. Particles for export are enclosed by Golgi membranes when they are synthesized. Vesicles pinch off from the Golgi complex and carry the enclosed particles to the interior surface of the cell membrane, where the vesicle membrane and cell membrane fuse and the contents of the vesicle are discharged.

Exogenous (ex-SOJ-e-nus) Originating outside an organ or part.

Exophthalmic goiter (ek'-sof-THAL-mik GOY-ter) Condition caused by hypersecretion of thyroid hormones characterized by protrusion of the eyeballs and an enlarged thyroid.

Expiration (ek-spi-RĀ-shun) Breathing out; expelling air from the lungs into the atmosphere. Also called **exhalation.**

Expiratory (eks-PĪ-ra-tō-rē) **reserve volume** The volume of air in excess of tidal volume that can be forcibly exhaled; about 1,200 ml.

Extensibility (ek-sten'-si-BIL-i-tē) The ability of tissue to be stretched when pulled.

Extension (ek-STEN-shun) An increase in the anterior angle between two bones, except in extension of the knee and toes, in which the posterior angle is involved; restoring a body part to its anatomical position after flexion.

External auditory (AW-di-tōr-ē) **canal** or **meatus** (mē-Ā-tus) A canal in the temporal bone that leads to the middle ear.

External ear The outer ear, consisting of the pinna, the external auditory canal, and the tympanic membrane or eardrum.

External nares (NA-rēz) The external nostrils, or the openings into the nasal cavity on the exterior of the body.

External respiration The exchange of respiratory gases between the lungs and blood.

Exteroceptor (eks'-ter-ō-SEP-tor) A receptor adapted for the reception of stimuli from outside the body.

Extracellular fluid (ECF) Fluid outside of body cells, such as interstitial fluid and plasma.

Extravasation (eks-trav-a-SĀ-shun) The escape of fluid from a vessel into the tissues, especially blood, lymph, or serum.

Extrinsic (ek-STRIN-sik) Of external origin.

Extrinsic clotting pathway Sequence of reactions leading to blood clotting that is initiated by the release of a substance *outside* blood itself, from damaged blood vessels or surrounding tissues.

Exudate (EKS-yoo-dāt) Escaping fluid or semifluid material that oozes from a space that may contain serum, pus, and cellular debris.

Eyebrow The hairy ridge above the eye.

Face The anterior aspect of the head.

Falciform ligament (FAL-si-form LIG-a-ment) A sheet of parietal peritoneum between the two principal lobes of the liver. The ligamentum teres, or remnant of the umbilical vein, lies within its fold.

Facilitated diffusion (fa-SIL-i-tā-ted dif-YOO-zhun) Diffusion in which a substance not soluble by itself in lipids is transported across a selectively permeable membrane by combining with a carrier substance.

Facilitation (fa-sil-i-TĀ-shun) The process in which a nerve cell membrane is partially depolarized by a subliminal stimulus so that a subsequent subliminal stimulus can further depolarize the membrane to reach the threshold of nerve impulse initiation.

Facultative (FAK-ul-tā-tive) **water reabsorption** The absorption of water from distal convoluted tubules and collecting ducts of nephrons in response to antidiuretic hormone (ADH).

Fallopian tube *See* **Uterine tube.**

Falx cerebelli (falks ser'-e-BEL-lē) A small triangular process of the dura mater attached to the occipital bone in the posterior cranial fossa and projecting inward between the two cerebellar hemispheres.

Falx cerebri (SER-e-brē) A fold of the dura mater extending down into the longitudinal fissure between the two cerebral hemispheres.

Fascia (FASH-ē-a) A fibrous membrane covering, supporting, and separating muscles.

Fascicle (FAS-i-kul) A small bundle or cluster, especially of nerve or muscle fibers. Also called **fasciculus** (fa-SIK-yoo-lus); *plural,* **fasciculi** (fa-SIK-yoo-lī).

Fat A lipid compound formed from one molecule of glycerol and three molecules of fatty acids; the body's most highly concentrated source of energy. Adipose tissue, composed of loose connective tissue and cells specialized for fat storage and present in the form of soft pads between various organs for support, protection, and insulation.

Fauces (FAW-sēz) The opening from the mouth into the pharynx.

Febrile (FĒ-bril) Feverish; pertaining to a fever.

Feces (FĒ-sēz) Material discharged from the rectum and made up of bacteria, excretions, and food residue. Also called **stool.**

Fenestra cochlea (fe-NES-tra KŌK-lē-a) A small opening between the middle and inner ear, directly below the oval window, covered by the second tympanic membrane. Also called the **round window.**

Fenestra vestibuli (ves-TIB-yoo-lī) A small opening between the middle and inner ear into which the footplate of the stapes fits. Also called the **oval window.**

Fertilization (fer'-ti-li-ZĀ-shun) Penetration of an ovum by a spermatozoon and subsequent union of the nuclei of the cells.

Fetal (FĒ-tal) **alcohol syndrome (FAS)** Term applied to the effects of intrauterine exposure to alcohol, such as slow growth, defective organs, and mental retardation.

Fetal circulation The circulatory system of the fetus, including the placenta and special blood vessels involved in the exchange of materials between fetus and mother.

Fetus (FĒ-tus) The latter stages of the developing young of an animal; in humans, the developing organism in utero from the beginning of the third month to birth.

Fever An elevation in body temperature above its normal temperature of 37° C (98.6°F).

Fibrillation (fi-bri-LĀ-shun) Irregular twitching of individual muscle fibers (cells) or small groups of muscle fibers preventing effective action by an organ or muscle.

Fibrin (FĪ-brin) An insoluble protein that is essential to blood clotting; formed from fibrinogen by action of thrombin.

Fibrinogen (fī-BRIN-ō-jen) A high-molecular-weight protein in the blood plasma that by the action of thrombin is converted to fibrin.

Fibrinolysis (fī-brin-OL-i-sis) Dissolution of a blood clot by the action of a proteolytic enzyme that converts insoluble fibrin into a soluble substance.

Fibroblast (FĪ-brō-blast) A large, flat cell that forms collagenous and elastic fibers and the viscous ground substance of loose connective tissue.

Fibrocyte (FĪ-brō-sīt) A mature fibroblast that no longer produces fibers or matrix in connective tissue.

Fibromyositis (fī'-brō-mī-ō'-SĪ-tis) A group of symptoms including pain, tenderness, and stiffness of joints, muscles, or adjacent structures. Called **"charleyhorse"** when it involves the thigh.

Fibrosis (fī-BRŌ-sis) Abnormal formation of fibrous tissue.

Fibrous (FĪ-brus) **joint** A joint that allows little or no movement, such as a suture and syndesmosis.

Fibrous tunic (TOO-nik) The outer coat of the eyeball, made up of the posterior sclera and the anterior cornea.

Fight-or-flight response The effect of the stimulation of the sympathetic division of the autonomic nervous system.

Filiform papilla (FIL-i-form pa-PIL-a) One of the conical projections that are distributed in parallel rows over the anterior two-thirds of the tongue and contain no taste buds.

Filtration (fil-TRĀ-shun) The passage of a liquid through a filter or membrane that acts like a filter.

Filtration fraction The percentage of plasma entering the nephrons that actually becomes glomerular filtrate.

Filum terminale (FĪ-lum ter-mi-NAL-ē) Nonnervous fibrous tissue of the spinal cord that extends inferiorly from the conus medullaris to the coccyx.

Fimbriae (FIM-brē-ē) Fingerlike structures, especially the lateral ends of the uterine (Fallopian) tubes.

Fissure (FISH-ur) A groove, fold, or slit that may be normal or abnormal.

Fistula (FIS-choo-la) An abnormal passage between two organs or between an organ cavity and the outside.

Fixator A muscle that stabilizes the origin of the prime mover so that the prime mover can act more efficiently.

Fixed macrophage (MAK-rō-fāj) Stationary phagocytic cell found in the liver, lungs, brain, spleen, lymph nodes, subcutaneous tissue, and bone marrow. Also called a **histiocyte** (HIS-tē-ō-sīt).

Flaccid (FLAK-sid) Relaxed, flabby, or soft; lacking muscle tone.

Flagellum (fla-JEL-um) A hairlike, motile process on the extremity of a bacterium or protozoan. *Plural,* **flagella** (fla-JEL-a).

Flatfoot A condition in which the ligaments and tendons of the arches of the foot are weakened and the height of the longitudinal arch decreases.

Flatus (FLĀ-tus) Gas or air in the gastrointestinal tract; commonly used to denote passage of gas rectally.

Flexion (FLEK-shun) A folding movement in which there is a decrease in the angle between two bones anteriorly, except in flexion of the knee and toes, in which the bones are approximated posteriorly.

Flexor reflex A polysynaptic reflex arc that withdraws a part of the body from a harmful stimulus.

Fluoroscope (FLOOR-ō-skōp) An instrument for visual observation of the body by means of x ray.

Follicle-stimulating (FOL-i-kul) **hormone (FSH)** Hormone secreted by the adenohypophysis (anterior lobe) of the pituitary gland that initiates development of ova and stimulates the ovaries to secrete estrogens in females and initiates sperm production in males.

Fontanel (fon'-ta-NEL) A membrane-covered spot where bone formation is not yet complete, especially between the cranial bones of an infant's skull.

Foot The terminal part of the lower extremity.

Foramen (fo-RĀ-men) A passage or opening; a communication between two cavities of an organ or a hole in a bone for passage of vessels or nerves.

Foramen ovale (ō-VAL-ē) An opening in the fetal heart in the septum between the right and left atria. A hole in the greater wing of the sphenoid bone that transmits the mandibular branch of the trigeminal (V) nerve.

Forearm (FOR-arm) The part of the upper extremity between the elbow and the wrist.

Fornix (FOR-niks) An arch or fold; a tract in the brain made up of association fibers, connecting the hippocampus with the mammillary bodies; a recess around the cervix of the uterus where it protrudes into the vagina.

Fossa (FOS-a) A furrow or shallow depression.

Fourth ventricle (VEN-tri-kul) A cavity within the brain lying between the cerebellum and the medulla and pons.

Fracture (FRAK-chur) Any break in a bone.

Frontal plane A plane at a right angle to a midsagittal plane that divides the body or organs into anterior and posterior portions. Also called **coronal** (kō-RŌ-nal) **plane.**

Fulminate (FUL-mi-nāt') To occur suddenly with great intensity.

Functional residual (re-ZID-yoo-al) **volume** The sum of residual volume plus expiratory reserve volume; about 2,400 ml.

Fundus (FUN-dus) The part of a hollow organ farthest from the opening.

Fungiform papilla (FUN-ji-form pa-PIL-a) A mushroomlike elevation on the upper surface of the tongue appearing as a red dot; most contain taste buds.

Furuncle (FYOOR-ung-kul) A boil; painful nodule caused by bacterial infection and inflammation of a hair follicle or sebaceous (oil) gland.

Gallbladder A small pouch that stores bile, located under the liver, which is filled and emptied of bile via the cystic duct.

Gallstone A concretion, usually consisting of cholesterol, formed anywhere between bile canaliculi in the liver and the hepatopancreatic ampulla (ampulla of Vater), where bile enters the duodenum. Also called a **biliary calculus.**

Gamete (GAM-ēt) A male or female reproductive cell; the spermatozoon or ovum.

Gamete intrafallopian transfer (GIFT) Procedure in which aspirated ova are combined with a solution containing sperm outside the body and then the ova are inserted into the uterine (Fallopian) tubes.

Ganglion (GANG-glē-on) A group of nerve cell bodies that lie outside the central nervous system. *Plural,* **ganglia** (GANG-glē-a).

Gangrene (GANG-rēn) Death and rotting of a considerable mass of tissue that usually is caused by interruption of blood supply followed by bacterial invasion (*Clostridium*).

Gastroenterology (gas'-trō-en'-ter-OL-ō-jē) The medical specialty that deals with the structure, function, diagnosis, and treatment of diseases of the stomach and intestines.

Gastrointestinal (gas-trō-in-TES-ti-nal) **(GI) tract** A continuous tube running through the ventral body cavity extending from the mouth to the anus. Also called the **alimentary** (al'-i-MEN-tar-ē) **canal.**

Gavage (ga-VAZH) Feeding through a tube passed through the esophagus and into the stomach.

Gene (jēn) Biological unit of heredity; an ultramicroscopic, self-reproducing DNA particle located in a definite position on a particular chromosome.

General adaptation syndrome Wide-ranging set of bodily changes triggered by a stressor that gears the body to meet an emergency.

Generator potential The graded depolarization that occurs in the region of an afferent neuron adjacent to a receptor cell.

Genetics The study of heredity.

Genital herpes (JEN-i-tal HER-pēz) A sexually transmitted disease caused by type II herpes simplex virus.

Genitalia (jen'-i-TĀL-ya) Reproductive organs.

Genotype (JĒ-nō-tīp) The total hereditary information carried by an individual; the genetic makeup of an organism.

Geriatrics (jer'-ē-AT-riks) The branch of medicine devoted to the medical problems and care of elderly persons.

Germinal (JER-mi-nal) **epithelium** A layer of epithelial cells that covers the ovaries and lines the seminiferous tubules of the testes.

Germinativum (jer'-mi-na-TĒ-vum) Skin layers where new cells are germinated.

Gestation (jes-TĀ-shun) The period of intrauterine fetal development.

Giantism (GĪ-an-tizm) Condition caused by hypersecretion of growth hormone (GH) during childhood characterized by excessive bone growth and body size. Also called **gigantism.**

Gingivae (jin-JI-vē) Gums. They cover the alveolar processes of the mandible and maxilla and extend slightly into each socket.

Gingivitis (jin'-je-VĪ-tis) Inflammation of the gums.

Gland Single or group of specialized epithelial cells that secrete substances.

Glans penis (glanz PĒ-nis) The slightly enlarged region at the distal end of the penis.

Glaucoma (glaw-KŌ-ma) An eye disorder in which there is increased pressure due to an excess of fluid within the eye.

Gliding joint A synovial joint having articulating surfaces that are usually flat, permitting only side-to-side and back-and-forth movements, as between carpal bones, tarsal bones, and the scapula and clavicle. Also called an **arthrodial** (ar-THRŌ-dē-al) **joint.**

Glomerular (glō-MER-yoo-lar) **capsule** A double-walled globe at the proximal end of a nephron that encloses the glomerulus. Also called **Bowman's** (BŌ-manz) **capsule.**

Glomerular filtration The first step in urine formation in which substances in blood are filtered at the endothelial-capsular membrane and the filtrate enters the proximal convoluted tubule of a nephron.

Glomerular filtration rate (GFR) The total volume of fluid that enters all the glomerular (Bowman's) capsules of the kidneys in 1 min; about 125 ml/min.

Glomerulonephritis (glō-mer-yoo-lō-nef-RĪ-tis) Inflammation of the glomeruli of the kidney. Also called **Bright's disease.**

Glomerulus (glō-MER-yoo-lus) A rounded mass of nerves or blood vessels, especially the microscopic tuft of capillaries that is surrounded by the glomerular (Bowman's) capsule of each kidney tubule.

Glottis (GLOT-is) The air passageway between the vocal folds in the larynx.

Glucagon (GLOO-ka-gon) A hormone produced by the pancreas that increases the blood glucose level.

Glucocorticoids (gloo-kō-KOR-ti-koyds) A group of hormones of the adrenal cortex.

Gluconeogenesis (gloo'-kō-nē'-ō-JEN-e-sis) The conversion of a substance other than carbohydrate into glucose.

Glucose (GLOO-kōs) A six-carbon sugar, $C_6H_{12}O_6$; the major energy source for every cell type in the body. Its metabolism is possible by every known living cell for the production of ATP.

Glycogen (GLĪ-ko-jen) A highly branched polymer of glucose containing thousands of subunits; functions as a compact store of glucose molecules in liver and muscle cells.

Glycogenesis (glī'-kō-JEN-e-sis) The process by which many molecules of glucose combine to form a molecule called glycogen.

Glycogenolysis (glī'-kō-je-NOL-i-sis) The process of converting glycogen to glucose.

Glycolysis (glī-KOL-i-sis) A series of chemical reactions that break down glucose into pyruvic acid, with a net gain of two molecules of ATP.

Glycosuria (glī'-kō-SOO-rē-a) The presence of glucose in the urine.

Gnostic (NOS-tik) Pertaining to the faculties of perceiving and recognizing.

Gnostic area Sensory area of the cerebral cortex that receives and integrates sensory input from various parts of the brain so that a common thought can be formed.

Goblet cell A goblet-shaped unicellular gland that secretes mucus. Also called a **mucus cell.**

Goiter (GOY-ter) An enlargement of the thyroid gland.

Golgi (GOL-jē) **complex** An organelle in the cytoplasm of cells consisting of four to eight flattened channels, stacked upon one another, with expanded areas at their ends; functions in packaging secreted proteins, lipid secretion, and carbohydrate synthesis.

Golgi tendon organ *See* **Tendon organ.**

Gomphosis (gom-FŌ-sis) A fibrous joint in which a cone-shaped peg fits into a socket.

Gonad (GŌ-nad) A gland that produces gametes and hormones; the ovary in the female and the testis in the male.

Gonadocorticoids (gō-na-dō-KOR-ti-koydz) Sex hormones secreted by the adrenal cortex.

Gonadotropic (gō'-nad-ō-TRŌ-pik) **hormone** A hormone that regulates the functions of the gonads.

Gonorrhea (gon'-ō-RĒ-a) Infectious, sexually transmitted disease caused by the bacterium *Neisseria gonorrhoeae* and characterized by inflammation of the urogenital mucosa, discharge of pus, and painful urination.

Gout (gowt) Hereditary condition associated with excessive uric acid in the blood; the acid crystallizes and deposits in joints and the kidneys.

Graafian follicle *See* **Vesicular ovarian follicle.**

Gray commissure (KOM-i-shur) A narrow strip of gray matter connecting the two lateral gray masses within the spinal cord.

Gray matter Area in the central nervous system consisting of nonmyelinated nerve tissue.

Gray ramus communicans (RĀ-mus ko-MYOO-ni-kans) A short nerve containing postganglionic sympathetic fibers; the cell bodies of the fibers are in a sympathetic chain ganglion, and the nonmyelinated axons run by way of the gray ramus to a spinal nerve and then to the periphery to supply smooth muscle in blood vessels, arrector pili muscles, and sweat glands. *Plural,* **rami communicantes** (RĀ-mē kō-myoo-ni-KAN-tēz).

Greater omentum (ō-MEN-tum) A large fold in the serosa of the stomach that hangs down like an apron over the front of the intestines.

Greater vestibular (ves-TIB-yoo-lar) **glands** A pair of glands on either side of the vaginal orifice that open by a duct into the space between the hymen and the labia minora. Also called **Bartholin's** (BAR-to-linz) **glands.**

Groin (groyn) The depression between the thigh and the trunk; the inguinal region.

Gross anatomy The branch of anatomy that deals with structures that can be studied without using a microscope. Also called **macroscopic anatomy.**

Growth hormone (GH) Hormone secreted by the adenohy-pophysis (anterior lobe) of the pituitary gland that causes growth of body tissues, especially skeletal and muscular. Also known as **somatotropin** and **somatotropic hormone (STH).**

Gustatory (GUS-ta-tō'-rē) Pertaining to taste.

Gynecology (gī'-ne-KOL-ō-jē) The branch of medicine dealing with the study and treatment of disorders of the female reproductive system.

Gyrus (JĪ-rus) One of the convolutions of the cerebral cortex of the brain. *Plural,* gyri (JĪ-rī).

Hair A threadlike structure produced by hair follicles that develops in the dermis. Also called **pilus** (PI-lus).

Hair follicle (FOL-li-kul) Structure composed of epithelium surrounding the root of a hair from which hair develops.

Hallucination (ha-loo'-sī-NĀ-shun) A sensory perception of something that does not really exist in the world, that is, a sensory experience created from within the brain.

Hand The terminal portion of an upper extremity, including the carpus, metacarpus, and phalanges.

Haploid (HAP-loyd) Having half the number of chromosomes characteristically found in the somatic cells of an organism; characteristic of mature gametes.

Hard palate (PAL-at) The anterior portion of the roof of the mouth, formed by the maxillae and palatine bones and lined by mucous membrane.

Haustra (HAWS-tra) The sacculated elevations of the colon.

Haversian canal *See* **Central canal.**

Haversian system *See* **Osteon.**

Head The superior part of a human, cephalic to the neck. The superior or proximal part of a structure.

Heart A hollow muscular organ lying slightly to the left of the midline of the chest that pumps the blood through the cardiovascular system.

Heart block An arrhythmia (dysrhythmia) of the heart in which the atria and ventricles contract independently because of a blocking of electrical impulses through the heart at a critical point in the conduction system.

Heart-lung machine A device that pumps blood, functioning as a heart, and removes carbon dioxide from blood and oxygenates it, functioning as lungs; used during heart transplantation, open-heart surgery, and coronary artery bypass surgery.

Heat exhaustion Condition characterized by cool, clammy skin, profuse perspiration, and fluid and electrolyte (especially salt) loss that results in muscle cramps, dizziness, vomiting, and fainting. Also called **heat prostration.**

Heatstroke Condition produced when the body cannot easily lose heat and characterized by reduced perspiration and elevated body temperature. Also called **sunstroke.**

Heimlich maneuver *See* **Abdominal thrust maneuver.**

Hematocrit (hē-MAT-ō-krit) An expression of the percentage of the volume occupied by blood cells. Usually calculated by centrifuging a blood sample in a graduated tube and then reading off the volumes of cells and total blood.

Hematology (hē'-ma-TOL-ō-jē) The study of the blood.

Hematoma (hē'-ma-TŌ-ma) A tumor or swelling filled with blood.

Hematopoiesis (hem'-a-tō-poy-Ē-sis) Blood cell production occurring in the red marrow of bones. Also called **hemopoiesis** (hē-mō-poy-Ē-sis).

Hematuria (hē'-ma-TOOR-ē-a) Blood in the urine.

Hemiballismus (hem'-i-ba-LIZ-mus) Violent muscular restlessness of half of the body, especially of the upper extremity.

Hemiplegia (hem-i-PLĒ-jē-a) Paralysis of the upper extremity, trunk, and lower extremity on one side of the body.

Hemocytoblast (hē'-mō-SĪ-tō-blast) Immature stem cell in bone marrow that develops along different lines into all the different mature blood cells.

Hemodialysis (hē'-mō-dī-AL-i-sis)   Filtering of the blood by means of an artificial device so that certain substances are removed from the blood as a result of the difference in rates of their diffusion through a selectively permeable membrane while the blood is being circulated outside the body.

Hemodynamics (hē-mō-dī-NAM-iks)   The study of factors and forces that govern the flow of blood through blood vessels.

Hemoglobin (hē'-mō-GLŌ-bin) **(Hb)**   A substance in erythrocytes consisting of the protein globin and the iron-containing, red pigment heme and constituting about 33 percent of the cell volume; involved in the transport of oxygen and carbon dioxide.

Hemolysis (hē-MOL-i-sis)   The escape of hemoglobin from the interior of the red blood cell into the surrounding medium; results from disruption of the integrity of the cell membrane by toxins or drugs, freezing or thawing, or hypotonic solutions.

Hemolytic disease of the newborn   A hemolytic anemia of a newborn child that results from the destruction of the infant's rbc's by antibodies produced by the mother; usually the antibodies are due to an Rh blood type incompatibility. Also called **erythroblastosis fetalis** (e-rith'-rō-blas-TŌ-sis fe-TAL-us).

Hemophilia (hē'-mō-FĒL-ē-a)   A hereditary blood disorder where there is a deficient production of certain factors involved in blood clotting, resulting in excessive bleeding into joints, deep tissues, and elsewhere.

Hemorrhage (HEM-or-rij)   Bleeding; the escape of blood from blood vessels, especially when it is profuse.

Hemorrhoids (HEM-ō-royds)   Dilated or varicosed blood vessels (usually veins) in the anal region. Also called **piles.**

Hemostasis (hē-mō-STĀ-sis)   The stoppage of bleeding.

Hepatic (he-PAT-ik)   Refers to the liver.

Hepatic duct   A duct that receives bile from the bile capillaries. Small hepatic ducts merge to form the larger right and left hepatic ducts that unite to leave the liver as the common hepatic duct.

Hepatic portal circulation   The flow of blood from the digestive organs to the liver before returning to the heart.

Hepatitis (hep-a-TĪ-tis)   Inflammation of the liver due to a virus, drugs, or chemicals.

Hepatopancreatic (hep'-a-tō-pan'-krē-A-tik) **ampulla**   A small, raised area in the duodenum where the combined common bile duct and main pancreatic duct empty into the duodenum. Also called the **ampulla of Vater** (VA-ter).

Hering-Breuer reflex   *See* **Inflation reflex.**

Hernia (HER-nē-a)   The protrusion or projection of an organ or part of an organ through the wall of the cavity containing it.

Herniated (her'-nē-Ā-ted) **disc**   A rupture of an intervertebral disc so that the nucleus pulposus protrudes into the vertebral cavity. Also called a **slipped disc.**

Heterozygous (he-ter-ō-ZĪ-gus)   Possessing a pair of different genes on homologous chromosomes for a particular hereditary characteristic.

Hilus (HĪ-lus)   An area, depression, or pit where blood vessels and nerves enter or leave an organ. Also called a **hilum.**

Hinge joint   A synovial joint in which a convex surface of one bone fits into a concave surface of another bone, such as the elbow, knee, ankle, and interphalangeal joints. Also called a **ginglymus** (JIN-gli-mus) **joint.**

Hirsutism (HER-soot-izm)   An excessive growth of hair in females and children, with a distribution similar to that in adult males, due to the conversion of vellus hairs into large terminal hairs in response to higher-than-normal levels of androgens.

Histamine (HIS-ta-mēn)   Substance found in many cells, especially mast cells, basophils, and platelets, released when the cells are injured; results in vasodilation, increased permeability of blood vessels, and bronchiole constriction.

Histology (his-TOL-ō-jē)   Microscopic study of the structure of tissues.

Hodgkin's disease   A malignant disorder, usually arising in lymph nodes.

Holocrine (HŌL-ō-krin) **gland**   A type of gland in which the entire secreting cell, along with its accumulated secretions, makes up the secretory product of the gland, as in the sebaceous (oil) glands.

Holter monitor   Electrocardiograph worn by a person while going about everyday routines.

Homeostasis (hō'-mē-ō-STĀ-sis)   The condition in which the body's internal environment remains relatively constant, within limits.

Homologous (hō-MOL-ō-gus)   Similar in structure and origin, but not necessarily in function.

Homozygous (hō'-mō-ZĪ-gus)   Possessing a pair of identical genes on homologous chromosomes for a particular hereditary characteristic.

Horizontal plane   A plane that runs parallel to the ground and divides the body or organs into superior and inferior portions. Also called **transverse plane.**

Hormone (HOR-mōn)   A secretion of endocrine tissue that alters the physiological activity of target cells of the body.

Human chorionic gonadotropin (kō-rē-ON-ik gō-nad-ō-TRŌ-pin)   A hormone produced by the developing placenta that maintains the corpus luteum.

Human chorionic somatomammotropin (sō-mat-ō-mam-ō-TRŌ-pin) **(HCS)**   A hormone produced by the chorion of the placenta that may stimulate breast tissue for lactation, enhance body growth, and regulate metabolism.

Human leucocyte associated (HLA) antigens   Surface proteins on white blood cells and other nucleated cells that are unique for each person (except for identical twins) and are used to type tissues and help prevent rejection.

Humoral (YOO-mor-al) **immunity**   That component of immunity in which lymphocytes (B cells) develop into plasma cells that produce antibodies that destroy antigens. Also called **antibody-mediated immunity.**

Hunger center   A cluster of neurons in the lateral nuclei of the hypothalamus that, when stimulated, brings about feeding.

Hyaluronic (hī'-a-loo-RON-ik) **acid**   An amorphous matrix material found outside the cell.

Hyaluronidase (hī'-a-loo-RON-i-dās)   An enzyme that breaks down hyaluronic acid, increasing the permeability of connective tissues by dissolving the substances that hold body cells together.

Hydrocele (HĪ-drō-sēl)   A fluid-containing sac or tumor. Specifically, a collection of fluid formed in the space along the spermatic cord and in the scrotum.

Hydrocephalus (hī-drō-SEF-a-lus)   Abnormal accumulation of cerebrospinal fluid on the brain.

Hydrophobia (hī'-drō-FŌ-bē-a)   Rabies; a condition characterized by severe muscle spasms when attempting to drink water. Also, an abnormal fear of water.

Hymen (HĪ-men)   A thin fold of vascularized mucous membrane at the vaginal orifice.

Hyperbaric oxygenation (hī'-per-BA-rik ok'-sē-je-NĀ-shun) **(HBO)**   Use of pressure supplied by a hyperbaric chamber to cause more oxygen to dissolve in blood to treat patients infected with anaerobic bacteria (tetanus and gangrene bacteria). Also used to treat carbon monoxide poisoning, asphyxia, smoke inhalation, and certain heart disorders.

Hypercalcemia (hī'-per-kal-SĒ-mē-a) An excess of calcium in the blood.

Hypercapnia (hī'-per-KAP-nē-a) An abnormal increase in the amount of carbon dioxide in the blood.

Hyperemia (hī'-per-Ē-mē-a) An excess of blood in an area or part of the body.

Hyperextension (hī'-per-ek-STEN-shun) Continuation of extension beyond the anatomical position, as in bending the head backward.

Hyperglycemia (hī'-per-glī-SĒ-mē-a) An elevated blood sugar level.

Hypermetropia (hī'-per-mē-TRŌ-pē-a) A condition in which visual images are focused behind the retina with resulting defective vision of near objects; farsightedness.

Hyperplasia (hī'-per-PLĀ-zē-a) An abnormal increase in the number of normal cells in a tissue or organ, increasing its size.

Hyperpolarization (hī'-per-PŌL-a-ri-zā'-shun) Increase in the internal negativity across a cell membrane, thus increasing the voltage and moving it farther away from the threshold value.

Hypersecretion (hī'-per-se-KRĒ-shun) Overactivity of glands resulting in excessive secretion.

Hypersensitivity (hī'-per-sen-si-TI-vi-tē) Overreaction to an allergen that results in pathological changes in tissues. Also called **allergy.**

Hypertension (hī'-per-TEN-shun) High blood pressure.

Hyperthermia (hī'-per-THERM-ē-a) An elevated body temperature.

Hypertonic (hī'-per-TON-ik) Having an osmotic pressure greater than that of a solution with which it is compared.

Hypertrophy (hī-PER-trō-fē) An excessive enlargement or overgrowth of tissue without cell division.

Hyperventilation (hī'-per-ven-ti-LĀ-shun) A rate of respiration higher than that required to maintain a normal level of plasma PCO_2.

Hypervitaminosis (hī'-per-vī'-ta-min-Ō-sis) An excess of one or more vitamins.

Hypocalcemia (hī'-pō-kal-SĒ-mē-a) A below normal level of calcium in the blood.

Hypochloremia (hī'-pō-klō-RĒ-mē-a) Deficiency of chloride in the blood.

Hypoglycemia (hī'-pō-glī-SĒ-mē-a) An abnormally low concentration of glucose in the blood; can result from excess insulin (injected or secreted).

Hypokalemia (hī'-pō-kā-LĒ-mē-a) Deficiency of potassium in the blood.

Hypomagnesemia (hī'-pō-mag'-ne-SĒ-mē-a) Deficiency of magnesium in the blood.

Hyponatremia (hī'-pō-na-TRĒ-mē-a) Deficiency of sodium in the blood.

Hyponychium (hī'-pō-NIK-ē-um) Free edge of the fingernail.

Hypophyseal (hī'-pō-FIZ-ē-al) **pouch** An outgrowth of ectoderm from the roof of the stomodeum (mouth) from which the adenohypophysis (anterior lobe) of the pituitary gland develops. Also called **Rathke's** (rath-KEZ) **pouch.**

Hypophysis (hī-POF-i-sis) Pituitary gland

Hyposecretion (hī'-pō-se-KRĒ-shun) Underactivity of glands resulting in diminished secretion.

Hypothalamic-hypophyseal (hī'-pō-thal-AM-ik hī'-po-FIZ-ē-al) **tract** A bundle of nerve processes made up of fibers that have their cell bodies in the hypothalamus but release their neurosecretions in the posterior pituitary gland or neurohypophysis.

Hypothalamus (hī'-pō-THAL-a-mus) A portion of the diencephalon, lying beneath the thalamus and forming the floor and part of the wall of the third ventricle.

Hypothermia (hī-pō-THER-mē-a) Low body temperature; in surgical procedures, it refers to deliberate cooling of the body to slow down metabolism and reduce oxygen needs of tissues.

Hypotonic (hī-pō-TON-ik) Having an osmotic pressure lower than that of a solution with which it is compared.

Hypoventilation (hī-pō-ven-ti-LĀ-shun) A rate of respiration lower than that required to maintain a normal level of plasma PCO_2.

Hypoxia (hī-POKS-ē-a) Lack of adequate oxygen at the tissue level.

Hysterectomy (his-te-REK-to-mē) The surgical removal of the uterus.

Ileocecal (il'-ē-ō-SĒ-kal) **sphincter** A fold of mucous membrane that guards the opening from the ileum into the large intestine. Also called the **ileocecal valve.**

Ileum (IL-ē-um) The terminal portion of the small intestine.

Immunity (i-MYOON-i-tē) The state of being resistant to injury, particularly by poisons, foreign proteins, and invading parasites, due to the presence of antibodies.

Immunogenicity (im-yoo-nō-jen-IS-it-ē) Ability of an antigen to stimulate antibody production.

Immunoglobulin (im-yoo-nō-GLOB-yoo-lin) **(Ig)** An antibody synthesized by special lymphocytes (plasma cells) in response to the introduction of antigen. Immunoglobulins are divided into five kinds (IgG, IgM, IgA, IgD, IgE) based primarily on the larger protein component present in the immunoglobulin.

Immunologic (im'-yoo-nō-LOJ-ik) **tolerance** A state of specific nonresponsiveness to an antigen, such as a self antigen, following exposure to the antigen.

Immunology (im'-yoo-NOL-ō-jē) The branch of science that deals with the responses of the body when challenged by antigens.

Immunosuppression (im'-yoo-nō-su-PRESH-un) Inhibition of the immune response.

Impetigo (im'pe-TĪ-go) A contagious skin disorder characterized by pustular eruptions.

Implantation (im-plan-TĀ-shun) The insertion of a tissue or a part into the body. The attachment of the blastocyst to the lining of the uterus 7 to 8 days after fertilization.

Impotence (IM-pō-tens) Weakness; inability to copulate; failure to maintain an erection.

Incontinence (in-KON-ti-nens) Inability to retain urine, semen, or feces, through loss of sphincter control.

Infant respiratory distress syndrome (RDS) A disease of newborn infants, especially premature ones, in which insufficient amounts of surfactant are produced and breathing is labored. Also called **hyaline** (HĪ-a-lin) **membrane disease (HMD).**

Infarction (in-FARK-shun) The presence of a localized area of necrotic tissue, produced by inadequate oxygenation of the tissue.

Infectious mononucleosis (mon-ō-nook'-lē-Ō-sis) **(IM)** Contagious disease caused by the Epstein-Barr virus (EBV) and characterized by an elevated mononucleocyte and lymphocyte count, fever, sore throat, stiff neck, cough, and malaise.

Inferior (in-FĒR-ē-or) Away from the head or toward the lower part of a structure. Also called **caudad** (KAW-dad).

Inferior vena cava (VĒ-na CĀ-va) Large vein that collects blood from parts of the body inferior to the heart and returns it to the right atrium.

Infertility Inability to conceive or to cause conception. Also called **sterility.**

Inflammation (in'-fla-MĀ-shun) Localized, protective response to tissue injury designed to destroy, dilute, or wall off the infecting agent or injured tissue; characterized by redness, pain, heat, swelling, and sometimes loss of function.

Inflation reflex   Reflex that prevents overinflation of the lungs. Also called the **Hering-Breuer reflex.**

Infundibulum (in'-fun-DIB-yoo-lum)   The stalklike structure that attaches the pituitary gland (hypophysis) to the hypothalamus of the brain. The funnel-shaped, open, distal end of the uterine (Fallopian) tube.

Ingestion (in-JES-chun)   The taking in of food, liquids, or drugs, by mouth.

Inguinal canal   An oblique passageway in the anterior abdominal wall just superior and parallel to the medial half of the inguinal ligament that transmits the spermatic cord and ilioinguinal nerve in the male and round ligament of the uterus and ilioinguinal nerve in the female.

Inheritance   The acquisition of body characteristics and qualities by transmission of genetic information from parents to offspring.

Inhibin   A male sex hormone secreted by sustentacular (Sertoli) cells that inhibits FSH release by the adenohypophysis (anterior pituitary) and thus spermatogenesis.

Inhibitory postsynaptic potential (IPSP)   The increase in the internal negativity of the membrane potential so that the voltage moves further from the threshold value.

Insertion (in-SER-shun)   The manner or place of attachment of a muscle to the bone that it moves.

Insomnia (in-SOM-nē-a)   Difficulty in falling asleep and, usually, frequent awakening.

Inspiration (in-spi-RĀ-shun)   The act of drawing air into the lungs.

Inspiratory (in-SPĪ-ra-tō-rē) **capacity**   The total inspiratory ability of the lungs; the sum of tidal volume and inspiratory reserve; about 3,600 ml.

Inspiratory reserve volume   The volume of air in excess of tidal volume that can be inhaled by forced inspiration; about 3,100 ml.

Insula (IN-su-la)   A triangular area of cerebral cortex that lies deep within the lateral cerebral fissure, under the parietal, frontal, and temporal lobes, and cannot be seen in an external view of the brain. Also called the **island** or **isle of Reil** (RĪL).

Insulin (IN-su-lin)   A hormone produced by the pancreas that decreases the blood glucose level.

Integumentary (in-teg'-yoo-MEN-tar-ē)   Relating to the skin.

Intercalated (in-TER-ka-lāt-ed) **disc**   An irregular transverse thickening of sarcolemma that separates cardiac muscle fibers (cells) from each other.

Intercellular fluid   That portion of extracellular fluid that bathes the cells of the body; the internal environment of the body. Also called **interstitial** (in'-ter-STISH-al) **fluid.**

Interferon (in'-ter-FĒR-on)   Three principal types of protein (alpha, beta, gamma) naturally produced by virus-infected host cells that inhibit intracellular viral replication in uninfected host cells; artificially synthesized through recombinant DNA techniques.

Intermediate   Between two structures, one of which is medial and one of which is lateral.

Internal   Away from the surface of the body.

Internal capsule   A thick sheet of white matter made up of myelinated fibers connecting various parts of the cerebral cortex and lying between the thalamus and the caudate and lentiform nuclei of the basal ganglia.

Internal ear   The inner ear or labyrinth, lying inside the temporal bone, containing the organs of hearing and balance.

Internal nares (NA-rēz)   The two openings posterior to the nasal cavities opening into the nasopharynx. Also called the **choanae** (kō-A-nē).

Internal respiration   The exchange of respiratory gases between the blood and body cells.

Interphase (IN-ter-fāz)   The period during its life cycle when a cell is carrying on every life process except division; the stage between two mitotic divisions.

Interstitial cell of Leydig   *See* **Interstitial endocrinocyte.**

Interstitial (in'-ter-STISH-al) **endocrinocyte**   A cell located in the connective tissue between seminiferous tubules in a mature testis that secretes testosterone. Also called an **interstitial cell of Leydig** (LĪ-dig).

Interstitial fluid   The fluid filling the microscopic spaces between the cells of tissues.

Interstitial growth   Growth from within, as in the growth of cartilage. Also called **endogenous** (en-DOJ-e-nus) **growth.**

Intervertebral (in'-ter-VER-te-bral) **disc**   A pad of fibrocartilage located between the bodies of two vertebrae.

Intestinal gland   Simple tubular gland that opens onto the surface of the intestinal mucosa and secretes digestive enzymes. Also called a **crypt of Lieberkühn** (LE-ber-kyoon).

Intracellular (in'-tra-SEL-yoo-lar)   Within cells.

Intrafusal (in'-tra-FYOO-zal) **fibers**   Three to ten specialized muscle fibers (cells), partially enclosed in a connective tissue capsule that is filled with lymph; the fibers compose muscle spindles.

Intramembranous ossification (in'-tra-MEM-bra-nus os'-i-fi-KĀ-shun)   The method of bone formation in which the bone is formed directly in membranous tissue.

Intraocular (in-tra-OC-yoo-lar) **pressure**   Pressure in the eyeball, produced mainly by aqueous humor.

Intrapleural pressure   Air pressure between the two pleural layers of the lungs, usually subatmospheric. Also called **intrathoracic pressure.**

Intrapulmonic pressure   Air pressure within the lungs. Also called **intraalveolar pressure.**

Intrauterine device (IUD)   A small metal or plastic object inserted into the uterus for the purpose of preventing pregnancy.

Intrinsic clotting pathway   Sequence of reactions leading to blood clotting that is initiated by the release of a substance contained *within* blood itself.

Intrinsic factor   A glycoprotein synthesized and secreted by the gastric mucosa that facilitates vitamin B_{12} absorption.

Intubation (in'-too-BĀ-shun)   Insertion of a tube through the nose or mouth into the larynx and trachea for entrance of air or to dilate a stricture.

In utero (YOO-ter-ō)   Within the uterus.

Invagination (in-vaj'-i-NĀ-shun)   The pushing of the wall of a cavity into the cavity itself.

Inversion (in-VER-zhun)   The movement of the sole inward at the ankle joint.

In vitro (VĒ-trō)   Literally, in glass; outside the living body and in an artificial environment such as a laboratory test tube.

In vivo (VĒ-vō)   In the living body.

Ion (Ī-on)   Any charged particle or group of particles; usually formed when a substance, such as a salt, dissolves and dissociates.

Ipsilateral (ip'-si-LAT-er-al)   On the same side, affecting the same side of the body.

Ischemia (is-KĒ-mē-a)   A lack of sufficient blood to a part due to obstruction of circulation.

Island of Reil   *See* **Insula.**

Islet of Langerhans   *See* **Pancreatic islet.**

Isometric contraction   A muscle contraction in which tension on the muscle increases, but there is minimal muscle shortening so that no movement is produced.

Isotonic (ī'-sō-TON-ik) Having equal tension or tone. Having equal osmotic pressure between two different solutions or between two elements in a solution.

Isotonic contraction A muscle contraction in which tension remains constant, but the muscle shortens and pulls on another structure to produce movement.

Isotope (Ī-sō-tōpe') A chemical element that has the same atomic number as another but a different atomic weight. Radioactive isotopes change into other elements with the emission of certain radiations.

Isovolumetric (ī-sō-vol-yoo'-MET-rik) **contraction** The period of time, about 0.05 sec, between the start of ventricular systole and the opening of the semilunar valves; there is contraction of the ventricles, but no emptying, and there is a rapid rise in ventricular pressure.

Isovolumetric relaxation The period of time, about 0.05 sec, between the opening of the atrioventricular (AV) valves and the closing of the semilunar valves; there is a drastic decrease in ventricular pressure without a change in ventricular volume.

Isthmus (IS-mus) A narrow strip of tissue or narrow passage connecting two larger parts.

Jaundice (JAWN-dis) A condition characterized by yellowness of skin, white of eyes, mucous membranes, and body fluids.

Jejunum (jē-JOO-num) The middle portion of the small intestine.

Joint kinesthetic (kin'-es-THET-ik) **receptor** A proprioceptive receptor located in a joint, stimulated by joint movement.

Juxtaglomerular (juks-ta-glō-MER-yoo-lar) **apparatus** Consists of the macula densa (cells of the distal convoluted tubule adjacent to the afferent and efferent arteriole) and juxtaglomerular cells (modified cells of the afferent and sometimes efferent arteriole); secretes renin when blood pressure starts to fall.

Karyotype (KAR-ē-ō-tīp) Chromosome characteristics of an individual or a group of cells.

Keratin (KER-a-tin) An insoluble protein found in the hair, nails, and other keratinized tissues of the epidermis.

Keratinocyte (ker-A-tin'-ō-sīt) The most numerous of the epidermal cells that function in the production of keratin.

Keratohyalin (ker'-a-tō-HĪ-a-lin) A compound involved in the formation of keratin.

Ketone (KĒ-ton) **bodies** Substances produced primarily during excessive fat metabolism, such as acetone, acetoacetic acid, and β-hydroxybutyric acid.

Ketosis (kē-TŌ-sis) Abnormal condition marked by excessive production of ketone bodies.

Kidney (KID-nē) One of the paired reddish organs located in the lumbar region that regulates the composition and volume of blood and produces urine.

Kidney stone A concretion, usually consisting of calcium oxalate, uric acid, and calcium phosphate crystals that may form in any portion of the urinary tract. Also called a **renal calculus.**

Kilocalorie (KIL-ō-kal'-ō-rē) **(kcal)** The amount of heat required to raise the temperature of 1,000 g of water 1°C; the unit used to express the heating value of foods and to measure metabolic rate.

Kinesiology (ki-nē'-sē-OL-ō-jē) The study of the movement of body parts.

Kinesthesia (kin-is-THĒ-zē-a) Ability to perceive extent, direction, or weight of movement; muscle sense.

Korotkoff (kō-ROT-kof) **sounds** The various sounds that are heard while taking blood pressure.

Krebs cycle A series of energy-yielding chemical reactions in which carbon dioxide is formed and energy is transferred to carrier molecules for subsequent liberation. Also called the **citric acid cycle** and **tricarboxylic acid (TCA) cycle.**

Kupffer's cell See **Stellate reticuloendothelial cell.**

Kyphosis (kī-FŌ-sis) An exaggeration of the thoracic curve of the vertebral column, resulting in a "round-shouldered" or hunchback appearance.

Labial frenulum (LĀ-bē-al FREN-yoo-lum) A medial fold of mucous membrane between the inner surface of the lip and the gums.

Labia majora (LĀ-bē-a ma-JO-ra) Two longitudinal folds of skin extending downward and backward from the mons pubis of the female.

Labia minora (min-OR-a) Two small folds of mucous membrane lying medial to the labia majora of the female.

Labium (LĀ-bē-um) A lip. A liplike structure. *Plural,* **labia** (LĀ-bē-a).

Labyrinth (LAB-i-rinth) Intricate communicating passageway, especially in the internal ear.

Labyrinthine (lab-i-RIN-then) **disease** Malfunction of the internal ear characterized by deafness, tinnitus, vertigo, nausea, and vomiting.

Laceration (las'-er-Ā-shun) A torn, ragged, or mangled wound due to trauma.

Lacrimal (LAK-ri-mal) **canal** A duct, one on each eyelid, commencing at the punctum at the medial margin of an eyelid and conveying the tears medially into the nasolacrimal sac.

Lacrimal gland Secretory cells located at the superior lateral portion of each orbit that secrete tears into excretory lacrimal ducts that open onto the surface of the conjunctiva.

Lacrimal sac The superior expanded portion of the nasolacrimal duct that receives the tears from a lacrimal canal.

Lactation (lak-TĀ-shun) The secretion and ejection of milk by the mammary glands.

Lacteal (LAK-tē-al) One of many intestinal lymph vessels in villi that absorb fat from digested food.

Lacuna (la-KOO-na) A small, hollow space, such as that found in bones in which the osteoblasts lie. *Plural,* **lacunae** (la-KOO-nē).

Lambdoidal suture (lam-DOY-dal) The line of union in the skull between the parietal bones and the occipital bone.

Lamellae (la-MEL-ē) Concentric rings found in compact bone.

Lamellated corpuscle Oval pressure receptor located in subcutaneous tissue and consisting of concentric layers of connective tissue wrapped around an afferent nerve fiber. Also called a **Pacinian** (pa-SIN-ē-an) **corpuscle.**

Lamina (LAM-i-na) A thin, flat layer or membrane, as the flattened part of either side of the arch of a vertebra. *Plural,* **laminae** (LAM-i-nē).

Lamina propria (PRO-prē-a) The connective tissue layer of a mucous membrane.

Lanugo (lan-YOO-gō) Fine downy hairs that cover the fetus.

Large intestine The portion of the gastrointestinal tract extending from the ileum of the small intestine to the anus, divided into the cecum, colon, rectum, and anal canal.

Laryngitis (la-rin-JĪ-tis) Inflammation of the mucous membrane lining the larynx.

Laryngopharynx (la-rin'-gō-FAR-inks) The inferior portion of the pharynx, extending downward from the level of the hyoid bone to divide posteriorly into the esophagus and anteriorly into the larynx.

Laryngoscope (la-RING-ō-skōp) An instrument for examining the larynx.

Laryngotracheal (la-rin'-gō-TRA-kē-al) **bud** An outgrowth of endoderm of the foregut from which the respiratory system develops.

Larynx (LAR-inks) The voice box, a short passageway that connects the pharynx with the trachea.

Lateral (LAT-er-al) Farther from the midline of the body or a structure.

Lateral ventricle (VEN-tri-kul) A cavity within a cerebral hemisphere that communicates with the lateral ventricle in the other cerebral hemisphere and with the third ventricle by way of the interventricular foramen.

Leg The part of the lower extremity between the knee and the ankle.

Lens A transparent organ lying posterior to the pupil and iris of the eyeball and anterior to the vitreous humor.

Lesion (LĒ-zhun) Any localized, abnormal change in tissue formation.

Lesser omentum (ō-MEN-tum) A fold of the peritoneum that extends from the liver to the lesser curvature of the stomach and the commencement of the duodenum.

Lesser vestibular (ves-TIB-yoo-lar) **gland** One of the paired mucus-secreting glands that have ducts that open on either side of the urethral orifice in the vestibule of the female.

Leucocyte (LOO-kō-sīt) A white blood cell.

Leucocytosis (loo'-kō-sī-TŌ-sis) An increase in the number of white blood cells, characteristic of many infections and other disorders.

Leucopenia (loo-kō-PĒ-nē-a) A decrease of the number of white blood cells below 5,000/mm³.

Leukemia (loo-KĒ-mē-a) A malignant disease of the blood-forming tissues characterized by either uncontrolled production and accumulation of immature leukocytes in which many cells fail to reach maturity (acute), or an accumulation of mature leukocytes in the blood because they don't die at the end of their normal life span (chronic).

Leukoplakia (loo-kō-PLĀ-kē-a) A disorder in which there are white patches in the mucous membranes of the tongue, gums, and cheeks.

Libido (li-BĒ-dō) The sexual drive, conscious or unconscious.

Ligament (LIG-a-ment) Dense, regularly arranged connective tissue that attaches bone to bone.

Limbic system A portion of the forebrain, sometimes termed the visceral brain, concerned with various aspects of emotion and behavior, that includes the limbic lobe (hippocampus and associated areas of gray matter plus the cingulate gyrus), certain parts of the temporal and frontal cortex, some thalamic and hypothalamic nuclei, and parts of the basal ganglia.

Lingual frenulum (LIN-gwal FREN-yoo-lum) A fold of mucous membrane that connects the tongue to the floor of the mouth.

Lipase (LĪ-pās) A fat-splitting enzyme.

Lipid An organic compound composed of carbon, hydrogen, and oxygen that is usually insoluble in water, but soluble in alcohol, ether, and chloroform; examples include fats, phospholipids, steroids, and prostaglandins.

Lipogenesis (li-pō-GEN-e-sis) The synthesis of lipids from glucose or amino acids by liver cells.

Lipoma (li-PŌ-ma) A fatty tissue tumor, usually benign.

Liver Large gland under the diaphragm that occupies most of the right hypochondriac region and part of the epigastric region; functionally, it produces bile salts, heparin, and plasma proteins; converts one nutrient into another; detoxifies substances; stores glycogen, minerals, and vitamins; carries on phagocytosis of blood cells and bacteria; and helps to activate vitamin D.

Local Pertaining to or restricted to one spot or part.

Locus coeruleus (LŌ-kus sē-ROO-lē-us) A group of neurons in the brain stem where norepinephrine (NE) is concentrated.

Lordosis (lor-DŌ-sis) An exaggeration of the lumbar curve of the vertebral column.

Lou Gehrig's disease *See* **Amyotrophic lateral sclerosis.**

Lower extremity The appendage attached at the pelvic (hip) girdle, consisting of the thigh, knee, leg, ankle, foot, and toes.

Lumbar (LUM-bar) Region of the back and side between the ribs and pelvis; loins.

Lumbar plexus (PLEK-sus) A network formed by the anterior branches of spinal nerves L1 through L4.

Lumen (LOO-men) The space within an artery, vein, intestine, or tube.

Lung One of the two main organs of respiration, lying on either side of the heart in the thoracic cavity.

Lunula (LOO-nyoo-la) The moon-shaped white area at the base of a nail.

Luteinizing (LOO-tē-in'-īz-ing) **hormone (LH)** A hormone secreted by the adenohypophysis (anterior lobe) of the pituitary gland that stimulates ovulation, progesterone secretion by the corpus luteum, and readies the mammary glands for milk secretion in females and stimulates testosterone secretion by the testes in males.

Lymph (limf) Fluid confined in lymphatic vessels and flowing through the lymphatic system to be returned to the blood.

Lymphangiography (lim-fan'-jē-OG-ra-fē) A procedure by which lymphatic vessels and lymph organs are filled with a radiopaque substance in order to be x-rayed.

Lymphatic (lim-FAT-ik) Pertaining to lymph. A large vessel that collects lymph from lymph capillaries and converges with other lymphatics to form the thoracic and right lymphatic ducts.

Lymphatic tissue A specialized form of reticular tissue that contains large numbers of lymphocytes.

Lymph capillary Blind-ended microscopic lymph vessel that begins in spaces between cells and converges with other lymph capillaries to form lymphatics.

Lymph node An oval or bean-shaped structure located along the lymphatic vessels.

Lymphocyte (LIM-fō-sīt) A type of white blood cell, found in lymph nodes, associated with the immune system.

Lysosome (LĪ-sō-sōm) An organelle in the cytoplasm of a cell, enclosed by a double membrane and containing powerful digestive enzymes.

Lysozyme (LĪ-sō-zīm) A bactericidal enzyme found in tears, saliva, and perspiration.

Macrophage (MAK-rō-fāj) Phagocytic cell derived from a monocyte. May be fixed or wandering.

Macula (MAK-yoo-la) A discolored spot or a colored area. A small, flat region on the wall of the utricle and saccule that serves as a receptor for static equilibrium.

Macula lutea (LOO-tē-a) The yellow spot in the center of the retina.

Magnetic resonance imaging (MRI) A diagnostic procedure that focuses on the nuclei of atoms of a single element in a tissue, usually hydrogen, to determine if they behave normally in the presence of an external magnetic force; used to indicate the biochemical activity of a tissue. Formerly called **nuclear magnetic resonance (NMR).**

Malaise (ma-LĀYZ) Discomfort, uneasiness, and indisposition, often indicative of infection.

Malignant (ma-LIG-nant) Referring to diseases that tend to become worse and cause death; especially the invasion and spreading of cancer.

Mammary (MAM-ar-ē) **gland** Modified sweat gland of the female that secretes milk for the nourishment of the young.

Margination (mar'-ji-NĀ-shun) Accumulation and adhesion of neutrophils to the endothelium of blood vessels at the site of injury during the early stages of inflammation.

Marrow (MAR-ō) Soft, spongelike material in the cavities of bone. Red marrow produces blood cells; yellow marrow, formed mainly of fatty tissue, has no blood-producing function.

Mast cell   A cell found in loose connective tissue along blood vessels that produces heparin, an anticoagulant. The name given to a basophil after it has left the bloodstream and entered the tissues.

Mastectomy (mas-TEK-tō-mē)   Surgical removal of breast tissue.

Mastication (mas'-ti-KĀ-shun)   Chewing.

Meatus (mē-Ā-tus)   A passage or opening, especially the external portion of a canal.

Mechanoreceptor (MEK'-a-nō-rē'-sep'-tor)   Receptor that detects mechanical deformation of the receptor itself or adjacent cells; stimuli so detected include touch, pressure, vibration, proprioception, hearing, equilibrium, and blood pressure.

Medial (MĒ-dē-al)   Nearer the midline of the body or a structure.

Medial lemniscus (lem-NIS-kus)   A flat band of myelinated nerve fibers extending through the medulla, pons, and midbrain and terminating in the thalamus on the opposite side. Second-order sensory neurons in this tract transmit impulses for discriminating touch, pressure, and vibration sensations.

Median plane   A vertical plane dividing the body into right and left halves. Situated in the middle.

Mediastinum (mē'-dē-as-TĪ-num)   A broad, median partition, actually a mass of tissue found between the pleurae of the lungs that extends from the sternum to the vertebral column.

Medulla (me-DULL-la)   An inner layer of an organ, such as the medulla of the kidneys.

Medulla oblongata (ob'-long-GA-ta)   The most inferior part of the brain stem.

Medullary (MED-yoo-lar'-ē) **cavity**   The space within the diaphysis of a bone that contains yellow marrow.

Medullary rhythmicity (rith-MIS-i-tē) **area**   Portion of the respiratory center in the medulla that controls the basic rhythm of respiration.

Meibomian gland   *See* **Tarsal gland.**

Meiosis (mē-Ō-sis)   A type of cell division restricted to sex-cell production involving two successive nuclear divisions that result in daughter cells with the haploid number of chromosomes.

Meissner's corpuscle   *See* **Corpuscle of touch.**

Melanin (MEL-a-nin)   A dark black, brown, or yellow pigment found in some parts of the body such as the skin.

Melanocyte (MEL-a-nō-sīt')   A pigmented cell located between or beneath cells of the deepest layer of the epidermis that synthesizes melanin.

Melanocyte-stimulating hormone (MSH)   A hormone secreted by the adenohypophysis (anterior lobe) of the pituitary gland that stimulates the dispersion of melanin granules in melanocytes.

Melanoma (mel'-a-NŌ-ma)   A usually dark, malignant tumor of the skin containing melanin.

Melatonin (mel-a-TŌN-in)   A hormone secreted by the pineal gland that may inhibit reproductive activities.

Membrane   A thin, flexible sheet of tissue composed of an epithelial layer and an underlying connective tissue layer, as in an epithelial membrane, or of loose connective tissue only, as in a synovial membrane.

Membrane potential   The voltage present, at any instant, across the cell membrane; measured with microelectrodes inside and outside the cell, it usually registers "resting" values of about −70 mV. Also called a **resting potential.**

Membranous labyrinth (mem-BRA-nus LAB-i-rinth)   The portion of the labyrinth of the inner ear that is located inside the bony labyrinth and separated from it by the perilymph; made up of the membranous semicircular canals, the saccule and utricle, and the cochlear duct.

Memory   The ability to recall thoughts; commonly classified as activated (short-term) and long-term.

Menarche (me-NAR-kē)   Beginning of the menstrual function.

Ménière's (men-YAIRZ) **syndrome**   A type of labyrinthine disease characterized by fluctuating loss of hearing, vertigo, and tinnitus.

Meninges (me-NIN-jēz)   Three membranes covering the brain and spinal cord, called the dura mater, arachnoid, and pia mater. *Singular,* **meninx** (MEN-inks).

Meningitis (men-in-JĪ-tis)   Inflammation of the meninges, most commonly the pia mater and arachnoid.

Menopause (MEN-ō-pawz)   The termination of the menstrual cycles.

Menstrual (MEN-stroo-al) **cycle**   A series of changes in the endometrium of a nonpregnant female that prepares the lining of the uterus to receive a fertilized ovum.

Menstruation (men'-stroo-Ā-shun)   Periodic discharge of blood, tissue fluid, mucus, and epithelial cells that usually lasts for 5 days; caused by a sudden reduction in estrogens and progesterone. Also called the **menstrual phase** or **menses.**

Merocrine (MER-ō-krin) **gland**   A secretory cell that remains intact throughout the process of formation and discharge of the secretory product, as in the salivary and pancreatic glands.

Mesenchyme (MEZ-en-kīm)   An embryonic connective tissue from which all other connective tissues arise.

Mesentery (MEZ-en-ter'-ē)   A fold of peritoneum attaching the small intestine to the posterior abdominal wall.

Mesocolon (mez'-ō-KŌ-lon)   A fold of peritoneum attaching the colon to the posterior abdominal wall.

Mesoderm   The middle of the three primary germ layers that gives rise to connective tissues, blood and blood vessels, and muscles.

Mesothelium (mez'-ō-THĒ-lē-um)   The layer of simple squamous epithelium that lines serous cavities.

Mesovarium (mez'-ō-VAR-ēum)   A short fold of peritoneum that attaches an ovary to the broad ligament of the uterus.

Metabolism (me-TAB-ō-lizm)   The sum of all the biochemical reactions that occur within an organism, including the synthetic (anabolic) reactions and decomposition (catabolic) reactions.

Metacarpus (met-a-KAR-pus)   A collective term for the five bones that make up the palm of the hand.

Metaphase (MET-a-phāz)   The second stage of mitosis in which chromatid pairs line up on the equatorial plane of the cell.

Metaphysis (me-TAF-i-sis)   Growing portion of a bone.

Metarteriole (met'-ar-TĒ-rē-ōl)   A blood vessel that emerges from an arteriole, traverses a capillary network, and empties into a venule.

Metastasis (me-TAS-ta-sis)   The transfer of disease from one organ or part of the body to another.

Metatarsus (met'-a-TAR-sus)   A collective term for the five bones located in the foot between the tarsals and the phalanges.

Micelle (mī-SEL)   A spherical aggregate of bile salts that dissolves fatty acids and monoglycerides so that they can be transported into small intestinal epithelial cells.

Microcephalus (mi-kro-SEF-a-lus)   An abnormally small head; premature closing of the anterior fontanel so that the brain has insufficient room for growth, resulting in mental retardation.

Microfilament (mī-kro-FIL-a-ment)   Rodlike cytoplasmic structure, ranging in diameter from 30 nm to 120 nm; comprises contractile units in muscle cells and provides support, shape, and movement in nonmuscle cells.

Microglia (mi-krō-GLĒ-a)   Neuroglial cells that carry on phagocytosis. Also called **brain macrophages** (MAK-rō-fāj-ez).

Microphage (MĪK-rō-fāj) Granular leucocyte that carries on phagocytosis, especially neutrophils and eosinophils.

Microtomography (mī-krō-tō-MOG-rō-fē) A procedure that combines the principles of electron microscopy and computed tomography to produce high magnification, three-dimensional images of living cells.

Microtubule (mī-krō-TOOB-yool') Cylindrical cytoplasmic structure, ranging in diameter from 18 to 30 nm, consisting of the protein tubulin; provides support, structure, and transportation.

Microvilli (mī'-krō-VIL-ē) Microscopic, fingerlike projections of the cell membranes of small intestinal cells that increase surface area for absorption.

Micturition (mik'-too-RISH-un) The act of expelling urine from the urinary bladder. Also called **urination** (yoo-ri-NĀ-shun).

Midbrain The part of the brain between the pons and the diencephalon. Also called the **mesencephalon** (mes'-en-SEF-a-lon).

Middle ear A small, epithelial-lined cavity hollowed out of the temporal bone, separated from the external ear by the eardrum and from the internal ear by a thin bony partition containing the oval and round windows; extending across the middle ear are the three auditory ossicles. Also called the **tympanic** (tim-PAN-ik) **cavity.**

Midsagittal plane A vertical plane through the midline of the body that divides the body or organs into *equal* right and left sides. Also called a **median plane.**

Milking Contraction of skeletal muscle around veins causing venous valves to open and thus helping to drive blood back to the heart.

Milk letdown Contraction of alveolar cells to force milk into ducts of mammary glands, stimulated by oxytocin (OT), that is released from the posterior pituitary gland in response to suckling action.

Mineral Inorganic, homogeneous solid substance that may perform a function vital to life; examples include calcium, sodium, potassium, iron, phosphorus, and chlorine.

Mineralocorticoids (min'-er-al-ō-KOR-ti-koyds) A group of hormones of the adrenal cortex.

Minimal volume The volume of air in the lungs even after the thoracic cavity has been opened forcing out some of the residual volume.

Minute volume of respiration (MVR) Total volume of air taken into the lungs per minute; about 6,000/ml.

Mitochondrion (mī'-tō-KON-drē-on) A double-membraned organelle that plays a central role in the production of ATP; known as the "powerhouse" of the cell.

Mitosis (mī-TŌ-sis) The orderly division of the nucleus of a cell that ensures that each new daughter nucleus has the same number and kind of chromosomes as the original parent nucleus. The process includes the replication of chromosomes and the distribution of the two sets of chromosomes into two separate and equal nuclei.

Mitotic apparatus Collective term for continuous and chromosomal microtubules and centrioles; involved in cell division.

Mitotic spindle The combination of continuous and chromosomal microtubules, involved in chromosomal movement during mitosis.

Modality (mō-DAL-i-tē) Any of the specific sensory entities, such as vision, smell, or taste.

Modiolus (mō-DĪ-ō'-lus) The central pillar or column of the cochlea.

Mole The weight, in grams, of the combined atomic weights of the atoms that comprise a molecule of a substance.

Molecule (MOL-e-kyool) The chemical combination of two or more atoms.

Monoclonal antibody Antibody produced by in vitro clones of B cells hydridized with cancerous cells.

Monocyte (MON-ō-sīt') A type of white blood cell characterized by agranular cytoplasm; the largest of the leucocytes.

Mons pubis (monz PYOO-bis) The rounded, fatty prominence over the symphysis pubis, covered by coarse pubic hair.

Morbid (MOR-bid) Diseased; pertaining to disease.

Morula (MOR-yoo-la) A solid mass of cells produced by successive cleavages of a fertilized ovum a few days after fertilization.

Motor area The region of the cerebral cortex that governs muscular movement, particularly the precentral gyrus of the frontal lobe.

Motor unit A motor neuron together with the muscle cells it stimulates.

Mucous (MYOO-kus) **cell** A unicellular gland that secretes mucus. Also called a **goblet cell.**

Mucous membrane A membrane that lines a body cavity that opens to the exterior. Also called the **mucosa** (myoo-KŌ-sa).

Mucus The thick fluid secretion of the mucous glands and mucous membranes.

Multiple sclerosis (skler-Ō-sis) Progressive destruction of myelin sheaths of neurons in the central nervous system, short-circuiting conduction pathways.

Mumps Inflammation and enlargement of the parotid glands accompanied by fever and extreme pain during swallowing.

Murmur An unusual heart sound; may indicate a disorder such as a malfunctioning mitral valve or may have no clinical significance.

Muscarinic (mus'-ka-RIN-ik) **receptor** Receptor found on all effectors innervated by parasympathetic postganglionic axons and some effectors innervated by sympathetic postganglionic axons; so named because the actions of acetylcholine (ACh) on such receptors are similar to those produced by muscarine.

Muscle An organ composed of one of three types of muscle tissue (skeletal, cardiac, or visceral), specialized for contraction to produce voluntary or involuntary movement of parts of the body.

Muscle spindle An encapsulated receptor in a skeletal muscle, consisting of specialized muscle cells and nerve endings, stimulated by changes in length or tension of muscle cells; a proprioceptor. Also called a **neuromuscular** (noo-rō-MUS-kyoo-lar) **spindle.**

Muscle tissue A tissue specialized to produce motion in response to nerve impulses by its qualities of contractility, extensibility, elasticity, and excitability.

Muscle tone A sustained, partial contraction of portions of a skeletal muscle in response to activation of stretch receptors.

Muscular dystrophy (DIS-trō-fē') Inherited myopathy characterized by degeneration of muscle cells that leads to progressive atrophy.

Muscularis (MUS-kyoo-la'-ris) A muscular layer or tunic of an organ.

Muscularis mucosae (myoo-KŌ-sē) A thin layer of smooth muscle cells located in the outermost layer of the mucosa of the alimentary canal, underlying the lamina propria of the mucosa.

Musculi pectinati (MUS-kyoo-lī pek-ti-NA-tē) Projecting muscle bundles of the anterior atrial walls and the lining of the auricles.

Mutation (myoo-TĀ-shun) Any change in the sequence of bases in the DNA molecule resulting in a permanent alteration in some inheritable characteristic.

Myasthenia (mī-as-THĒ-nē-a) **gravis** Weakness of skeletal muscles caused by antibodies directed against acetylcholine receptors that inhibit muscle contraction.

Myelin (MĪ-e-lin) **sheath** A white, phospholipid, segmented covering, formed by neurolemmocytes (Schwann cells), around the axons and dendrites of many peripheral neurons.

Myenteric plexus   A network of nerve fibers from both autonomic divisions located in the muscularis coat or tunic of the small intestine. Also called the **plexus of Auerbach** (OW-er-bak).

Myocardial infarction (mī'-ō-KAR-dē-al in-FARK-shun)   Gross necrosis of myocardial tissue due to interrupted blood supply. Also called a **heart attack.**

Myocardium (mī'-ō-KAR-dē-um)   The middle layer of the heart wall, made up of cardiac muscle, comprising the bulk of the heart, and lying between the epicardium and the endocardium.

Myofibril (mī'-ō-FĪ-bril)   A threadlike structure, running longitudinally through a muscle cell, consisting mainly of thick myofilaments (myosin) and thin myofilaments (actin).

Myoglobin (mī-ō-GLŌ-bin)   The oxygen-binding, iron-containing conjugated protein complex present in the sarcoplasm of muscle fibers; contributes the red color to muscle.

Myogram (MĪ-ō-gram)   The record or tracing produced by the myograph, the apparatus that measures and records the effects of muscular contractions.

Myology (mī-OL-ō-jē)   The study of the muscles.

Myometrium (mī'-ō-MĒ-trē-um)   The smooth muscle coat or tunic of the uterus.

Myopia (mī-Ō-pē-a)   Defect in vision so that objects can be seen distinctly only when very close to the eyes; nearsightedness.

Myosin (MĪ-ō-sin)   The contractile protein that makes up the thick myofilaments of muscle cells.

Myotonia (mī-ō-TO-nē-a)   A continuous spasm of muscle; increased muscular irritability and tendency to contract, and less ability to relax.

Myxedema (mix-e-DĒ-ma)   Condition caused by hypothyroidism during the adult years characterized by swelling of facial tissues.

Nail   A hard plate, composed largely of keratin, that develops from the epidermis of the skin to form a protective covering on the dorsal surface of the distal phalanges of the fingers and toes.

Nail matrix (MĀ-triks)   The part of the nail beneath the body and root from which the nail is produced.

Nasal (NĀ-zal) **cavity**   A mucosa-lined cavity on either side of the nasal septum that opens onto the face at an external naris and into the nasopharynx at an internal naris.

Nasal septum (SEP-tum)   A vertical partition composed of bone and cartilage, covered with a mucous membrane, separating the nasal cavity into left and right sides.

Nasolacrimal (nā'-zō-LAK-ri-mal) **duct**   A canal that transports the lacrimal secretion from the nasolacrimal sac into the nose.

Nasopharynx (nā'-zō-FAR-inks)   The uppermost portion of the pharynx, lying posterior to the nose and extending down to the soft palate.

Nebulization (neb'-yoo-li-ZĀ-shun)   Treatment with medication by spray method.

Neck   The part of the body connecting the head and the trunk. A constricted portion of an organ such as the neck of a femur or the neck of the uterus.

Necrosis (ne-KRŌ-sis)   Death of a cell or group of cells as a result of disease or injury.

Negative feedback   The principle governing most control systems; a mechanism of response in which a stimulus initiates actions that reverse or reduce the stimulus.

Neonatal (nē'-ō-NĀ-tal)   Pertaining to the first 4 weeks after birth.

Neoplasm (NĒ-ō-plazm)   A mass of new, abnormal tissue; a tumor.

Nephritis (ne-FRĪT-is)   Inflammation of the kidney.

Nephron (NEF-ron)   The functional unit of the kidney.

Nephrosis (nef-RŌ-sis)   A degenerative disease of the kidney in which glomerular damage results in leakage of protein into urine.

Nerve   A cordlike bundle of nerve fibers and their associated connective tissue coursing together outside the central nervous system.

Nervous tissue   Tissue that initiates and transmits nerve impulses to coordinate homeostasis.

Neural plate   A thickening of ectoderm that forms early in the third week of development and represents the beginning of the development of the nervous system.

Neuritis (noo-RĪ-tis)   Inflammation of a nerve.

Neuroeffector (noo-rō-e-FEK-tor) **junction**   Collective term for neuromuscular and neuroglandular junctions.

Neurofibral node (noo-rō-FĪ-bral)   A space, along a myelinated nerve fiber, between the individual neurolemmocytes (Schwann cells) that form the myelin sheath and the neurolemma. Also called **node of Ranvier** (ron-VĒ-ā).

Neurofibril (noo-rō-FĪ-bril)   One of the delicate threads that forms a complicated network in the cytoplasm of the cell body and processes of a neuron.

Neuroglandular (noo-rō-GLAND-yoo-lar) **junction**   Area of contact between a motor neuron and a gland.

Neuroglia (nyoo-RŌ-glē-a)   Cells of the nervous system that are specialized to perform the functions of connective tissue. The neuroglia of the central nervous system are the astrocytes, oligodendrocytes, microglia, and ependyma; neuroglia of the peripheral nervous system include the neurolemmocytes (Schwann cells) and the ganglion satellite cells. Also called **glial** (GLĒ-al) **cells.**

Neurohypophyseal (noo'-rō-hī'-po-FIZ-ē-al) **bud**   An outgrowth of ectoderm located on the floor of the hypothalamus that gives rise to the neurohypophysis (posterior lobe) of the pituitary gland.

Neurohypophysis (noo-rō-hī-POF-i-sis)   The posterior lobe of the pituitary gland.

Neurolemma (noo-rō-LEM-ma)   The peripheral, nucleated cytoplasmic layer of the neurolemmocyte (Schwann cell). Also called **sheath of Schwann** (SCHVON).

Neurolemmocyte   A neuroglial cell of the peripheral nervous system that forms the myelin sheath and neurolemma of a nerve fiber by wrapping around a nerve fiber in a jelly-roll fashion. Also called a **Schwann** (SCHVON) **cell.**

Neurology (noo-ROL-ō-jē)   The branch of science that deals with the normal functioning and disorders of the nervous system.

Neuromuscular (noo-rō-MUS-kyoo-lar) **junction**   The area of contact between the axon terminal of a motor neuron and a portion of the sarcolemma of a muscle fiber. Also called a **myoneural** (mi-o-NOO-ral) **junction** or **motor end plate.**

Neuron (NOO-ron)   A nerve cell, consisting of a cell body, dendrites, and an axon.

Neuropeptide (noo-rō-PEP-tīd)   Chain of 2 to about 40 amino acids that occurs naturally in the brain that acts primarily to modulate the response of or the response to a neurotransmitter. Examples are enkephalins and endorphins.

Neurophysin (noo-rō-FĪ-sin)   Small protein that aids in the storage of oxytocin (OT) and antidiuretic hormone (ADH) in the neurohypophysis and their subsequent release.

Neurosecretory (noo-rō-SĒC-re-tō-rē) **cell**   A cell in a nucleus in the hypothalamus that produces oxytocin (OT) or antidiuretic hormone (ADH), hormones stored in the neurohypophysis of the pituitary gland.

Neurosyphilis (noo-rō-SIF-i-lis)   A form of the tertiary stage of syphilis in which various types of nervous tissue are attacked by the bacteria and degenerate.

Neurotransmitter One of a variety of molecules synthesized within the nerve axon terminals, released into the synaptic cleft in response to an action potential, and affecting the membrane potential of the postsynaptic neuron. Also called a **transmitter substance.**

Neutrophil (NOO-trō-fil) A type of white blood cell characterized by granular cytoplasm that stains as readily with acid or basic dyes. Also called a **polymorph** (POL-ē-morf).

Nicotinic (nik'-ō-TIN-ik) **receptor** Receptor found on both sympathetic and parasympathetic postganglionic neurons so named because the actions of acetylcholine (ACh) in such receptors are similar to those produced by nicotine.

Night blindness Poor or no vision in dim light or at night, although good vision present during bright illumination; frequently caused by a deficiency of vitamin A. Also referred to as **nyctalopia** (nik'-ta-LŌ-pē-a).

Nipple A pigmented, wrinkled projection on the surface of the mammary gland that is the location of the openings of the lactiferous ducts for milk release.

Nissl bodies *See* **Chromatophilic substance.**

Nociceptor (nō-sē-SEP-tor) A receptor that detects pain.

Node of Ranvier *See* **Neurofibral node.**

Nonessential amino acid An amino acid that can be synthesized by body cells through transamination, the transfer of an amino group from an amino acid to another substance.

Nongonococcal urethritis (NGU) A sexually transmitted disease characterized by inflammation of the urethra caused by a microorganism other than *Neisseria gonorrhoeae* or by nonmicrobial factors.

Norepinephrine (nor'-ep-ē-NEF-rin) **(NE)** A hormone secreted by the adrenal medulla that produces actions similar to those that result from sympathetic stimulation. Also called **noradrenaline** (nor-a-DREN-a-lin).

Notochord (NŌ-tō-cord) A flexible rod of embryonic tissue that lies where the future vertebral column will develop.

Nuclear medicine The branch of medicine concerned with the use of radioisotopes in the diagnosis of disease and therapy.

Nuclease (NOO-klē-ās) An enzyme that breaks nucleotides into pentoses and nitrogen bases; examples are ribonuclease and deoxyribonuclease.

Nucleic (noo-KLĒ-ic) **acid** An organic compound that is a long polymer of nucleotides, with each nucleotide containing a pentose sugar, a phosphate group, and one of four possible nitrogen bases (adenine, cytosine, guanine, and thymine or uracil).

Nucleolus (noo-KLĒ-ō-lus) Nonmembranous spherical body within the nucleus composed of protein, DNA, and RNA that functions in the synthesis and storage of ribosomal RNA.

Nucleosome (NOO-klē-ō-sōm) Elementary structural subunit of a chromosome consisting of histones and DNA.

Nucleus (NOO-klē-us) A spherical or oval organelle of a cell that contains the hereditary factors of the cell, called genes. A cluster of nerve cell bodies in the central nervous system. The central portion of an atom made up of protons and neutrons.

Nucleus cuneatus (kyoo-nē-Ā-tus) A group of nerve cells in the inferior portion of the medulla oblongata in which fibers of the fasciculus cuneatus terminate.

Nucleus gracilis (gra-SIL-is) A group of nerve cells in the medulla oblongata in which fibers of the fasciculus gracilis terminate.

Nucleus pulposus (pul-PŌ-sus) A soft, pulpy, highly elastic substance in the center of an intervertebral disc, a remnant of the notochord.

Nystagmus (nis-TAG-mus) Constant, involuntary, rhythmic movement of the eyeballs; horizontal, rotary, or vertical.

Obesity (ō-BĒS-i-tē) Body weight 10 to 20 percent over a desirable standard as a result of excessive accumulation of fat.

Obligatory water reabsorption The absorption of water from proximal convoluted tubules of nephrons as a function of osmosis.

Obstetrics (ob-STET-riks) The specialized branch of medicine that deals with pregnancy, labor, and the period of time immediately following delivery.

Occlusion (ō-KLOO-zhun) The act of closure or state of being closed.

Olfactory (ōl-FAK-tō-rē) Pertaining to smell.

Olfactory bulb A mass of gray matter at the termination of an olfactory (I) nerve, lying beneath the frontal lobe of the cerebrum on either side of the crista galli of the ethmoid bone.

Olfactory cell A bipolar neuron with its cell body lying between supporting cells located in the mucous membrane lining the upper portion of each nasal cavity.

Olfactory tract A bundle of axons that extends from the olfactory bulb posteriorly to the olfactory portion of the cortex.

Oligodendrocyte (o-lig-ō-DEN-drō-sīt) A neuroglial cell that supports neurons and produces a phospholipid myelin sheath around axons of neurons of the central nervous system.

Oligospermia (ol'-i-gō-SPER-mē-a) A deficiency of spermatozoa in the semen.

Olive A prominent oval mass on each lateral surface of the superior part of the medulla.

Oncogene (ONG-kō-jēn) Gene that has the ability to transform a normal cell into a cancerous cell.

Oncology (ong-KOL-ō-jē) The study of tumors.

Oogenesis (ō'-ō-JEN-e-sis) Formation and development of the ovum.

Oophorectomy (ō'-of-ō-REK-tō-mē) The surgical removal of the ovaries.

Ophthalmology (of'-thal-MOL-ō-jē) The study of the structure, function, and diseases of the eye.

Opsonization (op-sō-ni-ZĀ-shun) The action of some antibodies that renders bacteria and other foreign cells more susceptible to phagocytosis. Also called **immune adherence.**

Optic (OP-tik) Refers to the eye, vision, or properties of light.

Optic chiasma (kī-AZ-ma) A crossing point of the optic (II) nerves, anterior to the pituitary gland.

Optic disc A small area of the retina containing openings through which the fibers of the ganglion neurons emerge as the optic (II) nerve. Also called the **blind spot.**

Optic tract A bundle of axons that transmits impulses from the retina of the eye between the optic chiasma and the thalamus.

Oral contraceptive (OC) A hormonal compound that is swallowed and that prevents ovulation, and thus pregnancy. Also called **"the pill."**

Ora serrata (Ō-ra ser-RĀ-ta) The irregular margin of the retina lying internal and slightly posterior to the junction of the choroid and ciliary body.

Orbit (OR-bit) The bony, pyramid-shaped cavity of the skull that holds the eyeball.

Organ A structure of definite form and function composed of two or more different kinds of tissues.

Organelle (or-gan-EL) A permanent structure within a cell with characteristic morphology that is specialized to serve a specific function in cellular activities.

Organism (OR-ga-nizm) A total living form; one individual.

Orgasm (OR-gazm) Sensory and motor events involved in ejaculation for the male and involuntary contraction of the perineal muscles in the female at the climax of sexual intercourse.

Orifice (OR-i-fis)   Any aperture or opening.

Origin (OR-i-jin)   The place of attachment of a muscle to the more stationary bone, or the end opposite the insertion.

Oropharynx (or'-ō-FAR-inks)   The second portion of the pharynx, lying posterior to the mouth and extending from the soft palate down to the hyoid bone.

Orthopedics (or'-thō-PĒ-diks)   The branch of medicine that deals with the preservation and restoration of the skeletal system, articulations, and associated structures.

Osmosis (os-MŌ-sis)   The net movement of water molecules through a selectively permeable membrane from an area of high water concentration to an area of lower water concentration until an equilibrium is reached.

Osmotic pressure   The pressure required to prevent the movement of pure water into a solution containing solutes when the solutions are separated by a selectively permeable membrane.

Osseous (OS-ē-us)   Bony.

Ossicle (OS-si-kul)   Small bone, as in the middle ear (malleus, incus, stapes).

Ossification (os'-i-fi-KĀ-shun)   Formation of bone. Also called **osteogenesis.**

Osteoblast (OS-tē-ō-blast')   A bone-forming cell.

Osteoclast (OS-tē-ō-clast')   A large, multinuclear cell that destroys or resorbs bone tissue.

Osteocyte (OS-tē-ō-sīt')   A mature osteoblast that has lost its ability to produce new bone tissue.

Osteogenic layer (os'-tē-ō-JEN-ik)   The inner layer of the periosteum that contains cells responsible for forming new bone during growth and repair.

Osteology (os'-tē-OL-ō-jē)   The study of bones.

Osteomalacia (os'-tē-ō-ma-LĀ-shē-a)   A deficiency of vitamin D in adults causing demineralization and softening of bone.

Osteomyelitis (os'-tē-ō-mī-i-LĪ-tis)   Inflammation of bone marrow or of the bone and marrow.

Osteon   The basic unit of structure in adult compact bone, consisting of a central (Haversian) canal with its concentrically arranged lamellae, lacunae, osteocytes, and canaliculi. Also called a **Haversian** (ha-VER-shun) **system.**

Osteoporosis (os'-tē-ō-pō-RŌ-sis)   Age-related disorder characterized by decreased bone mass and increased susceptibility to fracture.

Otic (Ō-tik)   Pertaining to the ear.

Otitis media (ō-TĪ-tus MĒ-dē-a)   Acute infection of the middle ear cavity characterized by an inflamed tympanic membrane, subject to rupture.

Otolith (Ō-tō-lith)   A particle of calcium carbonate embedded in the otolithic membrane that functions in maintaining static equilibrium.

Otolithic (ō-tō-LITH-ik) **membrane**   Thick, gelatinous, glycoprotein layer located directly over hair cells of the macula in the saccule and utricle of the inner ear.

Otorhinolaryngology (ō'-tō-rī-nō-lar'-in-GOL-ō-jē)   The branch of medicine that deals with the diagnosis and treatment of diseases of the ears, nose, and throat.

Oval window   A small opening between the middle and inner ear into which the footplate of the stapes fits. Also called the **fenestra vestibuli** (fe-NES-tra ves-TIB-yoo-lē).

Ovarian (ō-VAR-ē-an) **cycle**   A monthly series of events in the ovary designed to produce a mature ovum.

Ovarian follicle (FOL-i-kul)   A general name for oocytes (immature ova) in any stage of development, along with their surrounding epithelial cells.

Ovarian ligament (LIG-a-ment)   A rounded cord of connective tissue that attaches the ovary to the uterus.

Ovary (Ō-var-ē)   Female gonad that produces ova and the hormones estrogens, progesterone, and relaxin.

Ovulation (ō-vyoo-LĀ-shun)   The rupture of a vesicular ovarian follicle (Graafian follicle) with discharge of an immature ovum into the pelvic cavity.

Ovum (Ō-vum)   The female reproductive or germ cell; an egg cell.

Oxidation (ok-si-DĀ-shun)   The removal of electrons and hydrogen ions (hydrogen atoms) from a molecule or, less commonly, the addition of oxygen to a molecule. The oxidation of glucose in the body is also called **cellular respiration.**

Oxygen debt   The volume of oxygen required to oxidize the lactic acid produced by muscular exercise.

Oxyhemoglobin (ok'-sē-HĒ-mō-glō-bin) **(HbO$_2$)**   Hemoglobin combined with oxygen.

Oxyphil cell   A cell found in the parathyroid gland that secretes parathyroid hormone (PTH).

Oxytocin (ok'-sē-TŌ-sin) **(OT)**   A hormone secreted by neurosecretory cells in the paraventricular nucleus of the hypothalamus that stimulates contraction of the smooth muscle cells in the pregnant uterus and contractile cells around the ducts of mammary glands.

Pacinian corpuscle   *See* **Lamellated corpuscle.**

Paget's (PAJ-ets) **disease**   A disorder characterized by a greatly accelerated remodeling process in which osteoclastic resorption is massive and new bone formation by osteoblasts is extensive. As a result, there is an irregular thickening and softening of the bones.

Palate (PAL-at)   The horizontal structure separating the oral and the nasal cavities; the roof of the mouth.

Palliative (PAL-ē-a-tiv)   Serving to relieve or alleviate without curing.

Palpate (PAL-pāt)   To examine by touch; to feel.

Pancreas (PAN-krē-as)   A soft, oblong organ lying along the greater curvature of the stomach and connected by usually two ducts to the duodenum. It is both exocrine (secreting pancreatic juice) and endocrine (secreting insulin, glucagon, and somatostatin).

Pancreatic (pan'-krē-AT-ik) **duct**   A single, large tube that unites with the common bile duct from the liver and gallbladder and drains pancreatic juice into the duodenum at the hepatopancreatic ampulla (ampulla of Vater). Also called the **duct of Wirsung.**

Pancreatic islet   A cluster of endocrine gland cells in the pancreas that secretes insulin, glucagon, and somatostatin. Also called an **islet of Langerhans** (LANG-er-hanz).

Papanicolaou (pap'-a-NIK-ō-la-oo) **test**   A cytological staining test for the detection and diagnosis of premalignant and malignant conditions of the female genital tract. Cells scraped from the genital epithelium are smeared, fixed, stained, and examined microscopically. Also called a **Pap smear.**

Paralysis (pa-RAL-a-sis)   Loss or impairment of motor function due to a lesion of nervous or muscular origin.

Paranasal sinus (par'-a-NĀ-zal SĪ-nus)   A mucus-lined air cavity in a skull bone that communicates with the nasal cavity. Paranasal sinuses are located in the frontal, maxillary, ethmoid, and sphenoid bones.

Paraplegia (par-a-PLĒ-jē-a)   Paralysis of both lower extremities.

Parasagittal plane   A vertical plane that does not pass through the midline and that divides the body or organs into *unequal* left and right portions.

Parasympathetic (par'-a-sim-pa-THET-ik) **division**   One of the two subdivisions of the autonomic nervous system, having cell bodies of preganglionic neurons in nuclei in the brain stem and in the lateral gray matter of the sacral portion of the spinal cord; primarily concerned with activities that restore and conserve body energy. Also called the **craniosacral** (krā'-nē-ō-SĀ-kral) **division.**

Parathyroid (par'-a-THĪ-royd) **gland** One of four small endocrine glands embedded on the posterior surfaces of the lateral lobes of the thyroid.

Parathyroid hormone (PTH) A hormone secreted by the parathyroid glands that decreases blood phosphate level and increases blood calcium level.

Paraurethral (par'-a-yoo-RĒ-thral) **gland** Gland embedded in the wall of the urethra whose duct opens on either side of the urethral orifice and secretes mucus. Also called **Skene's (SKĒNZ) gland.**

Parenchyma (par-EN-ki-ma) The functional parts of any organ, as opposed to tissue that forms its ground substance or framework.

Parenteral (par-EN-ter-al) Situated or occurring outside the intestines; referring to introduction of substances into the body other than by way of the intestines such as intradermal, subcutaneous, intramuscular, intravenous, or intraspinal.

Parietal (pa-RĪ-e-tal) Pertaining to or forming the outer wall of a body cavity.

Parietal cell The secreting cell of the gastric glands that produces hydrochloric acid and intrinsic factor. Also called an **oxyntic cell.**

Parietal pleura (PLOO-ra) The outer layer of the serous pleural membrane that encloses and protects the lungs; the layer that is attached to the wall of the pleural cavity.

Parkinson's disease Progressive degeneration of the basal ganglia and substantia nigra of the cerebrum resulting in decreased production of dopamine (DA) that leads to tremor, slowing of voluntary movements, and muscle weakness.

Parotid (pa-ROT-id) **gland** One of the paired salivary glands located inferior and anterior to the ears connected to the oral cavity via a duct that opens into the inside of the cheek opposite the upper second molar tooth.

Paroxysm (PAR-ok-sizm) A sudden periodic attack or recurrence of symptoms of a disease.

Pars intermedia A small avascular zone between the adenohypophysis and neurohypophysis of the pituitary gland.

Parturition (par'-too-RISH-un) Act of giving birth to young; childbirth, delivery.

Patellar (pa-TELL-ar) **reflex** Extension of the lower leg by contraction of the quadriceps femoris muscle in response to tapping the patellar ligament.

Patent ductus arteriosus Congenital anatomical heart defect in which the fetal connection between the aorta and pulmonary trunk remains open instead of closing completely after birth.

Pathogenesis (path'-ō-JEN-e-sis) The development of disease or a morbid or pathological state.

Pathogen (PATH-ō-jen) A disease-producing organism.

Pathological (path'-ō-LOJ-i-kal) **anatomy** The study of structural changes caused by disease.

Pectoral (PEK-tō-ral) Pertaining to the chest or breast.

Pediatrician (pē'-dē-a-TRISH-un) A physician who specializes in the care and treatment of children and their illnesses.

Pedicel (PED-i-sel) Footlike structure, as on podocytes of a glomerulus.

Pedicle (PED-i-kul) A short, thick process found on vertebrae.

Pelvic (PEL-vik) **cavity** Inferior portion of the abdominopelvic cavity that contains the urinary bladder, sigmoid colon, rectum, and internal female and male reproductive structures.

Pelvic inflammatory disease (PID) Collective term for any extensive bacterial infection of the pelvic organs, especially the uterus, uterine (Fallopian) tubes, and ovaries.

Pelvimetry (pel-VIM-e-trē) Measurement of the size of the inlet and outlet of the birth canal.

Pelvis The basinlike structure formed by the two pelvic (hip) bones, the sacrum, and the coccyx. The expanded, proximal portion of the ureter, lying within the kidney and into which the major calyces open.

Penis (PĒ-nis) The male copulary organ, used to introduce spermatozoa into the female vagina.

Pepsin Protein-digesting enzyme secreted by zymogenic (chief) cells of the stomach as the inactive form pepsinogen, which is converted to active pepsin by hydrochloric acid.

Peptic ulcer An ulcer that develops in areas of the gastrointestinal tract exposed to hydrochloric acid; classified as a gastric ulcer if in the lesser curvature of the stomach and as a duodenal ulcer if in the first part of the duodenum.

Perforating canal A minute passageway by means of which blood vessels and nerves from the periosteum penetrate into compact bone. Also called **Volkmann's (FŌLK-manz) canal.**

Pericardial (per'-i-KAR-dē-al) **cavity** Small potential space between the visceral and parietal layers of the serous pericardium.

Pericardium (per'-i-KAR-dē-um) A loose-fitting membrane that encloses the heart, consisting of an outer fibrous layer and an inner serous layer.

Perichondrium (per'-i-KON-drē-um) The membrane that covers cartilage.

Perikaryon (per'-i-KAR-ē-on) The nerve cell body that contains the nucleus and other organelles.

Perilymph (PER-i-lymf') The fluid contained between the bony and membranous labyrinths of the inner ear.

Perimetrium (per-i-MĒ-trē-um) The serosa of the uterus.

Perimysium (per'-i-MĪZ-ē-um) Invagination of the epimysium that divides muscles into bundles.

Perineum (per'-i-NĒ-um) The pelvic floor; the space between the anus and the scrotum in the male and between the anus and the vulva in the female.

Perineurium (per'-i-NYOO-rē-um) Connective tissue wrapping around fascicles in a nerve.

Periodontal (per-ē-ō-DON-tal) **disease** A collective term for conditions characterized by degeneration of gingivae, alveolar bone, periodontal ligament, and cementum.

Periodontal membrane The periosteum lining the alveoli (sockets) for the teeth in the alveolar processes of the mandible and maxillae.

Periosteum (per'-ē-OS-tē-um) The membrane that covers bone and is essential for bone growth, repair, and nutrition.

Peripheral (pe-RIF-er-al) Located on the outer part or a surface of the body.

Peripheral nervous system (PNS) The part of the nervous system that lies outside the central nervous system—nerves and ganglia.

Peripheral resistance The hindrance to blood flow due to friction between blood and blood vessels; related to viscosity of blood and diameter and length of blood vessels.

Peristalsis (per'-i-STAL-sis) Successive muscular contractions along the wall of a hollow muscular structure.

Peritoneum (per'-i-tō-NĒ-um) The largest serous membrane of the body that lines the abdominal cavity and covers the viscera.

Peritonitis (per'-i-tō-NĪ-tis) Inflammation of the peritoneum.

Peroxisome (pe-ROKS-ī-sōm) Organelle similar in structure to a lysosome that contains enzymes related to hydrogen peroxide metabolism; abundant in liver cells.

Perspiration Substance produced by sudoriferous (sweat) glands containing water, salts, urea, uric acid, amino acids, ammonia, sugar, lactic acid, and ascorbic acid; helps maintain body temperature and eliminate wastes.

Peyer's patches See **Aggregated lymphatic follicles.**

pH A symbol of the measure of the concentration of hydrogen ions in a solution. The pH scale extends from 0 to 14, with a value of 7 expressing neutrality, values lower than 7 expressing increasing acidity, and values higher than 7 expressing increased alkalinity.

Phagocytosis (fag'-ō-sī-TŌ-sis)   The process by which cells ingest particulate matter; especially the ingestion and destruction of microbes, cell debris, and other foreign matter.

Phalanx (FĀ-lanks)   The bone of a finger or toe. *Plural*, **phalanges** (fa-LAN-jēz).

Phantom pain   A sensation of pain as originating in a limb that has been amputated.

Pharmacology (far'-ma-KOL-ō-jē)   The science that deals with the effects and uses of drugs in the treatment of disease.

Pharynx (FAR-inks)   The throat; a tube that starts at the internal nares and runs partway down the neck where it opens into the esophagus posteriorly and into the larynx anteriorly.

Pheochromocytoma (fē-ō-krō'-mō-sī-TŌ-ma)   Tumor of the chromaffin cells of the adrenal medulla that results in hypersecretion of medullary hormones.

Phenotype (FĒ-nō-tīp)   The observable expression of genotype; physical characteristics of an organism determined by genetic makeup and influenced by interaction between genes and internal and external environmental factors.

Phenylketonuria (fen'-il-kē'-tō-NOO-rē-a) **(PKU)**   A disorder characterized by an elevation of the amino acid phenylalanine in the blood.

Phosphocreatine (fos'-fō-KRĒ-a-tin)   A high-energy molecule found in muscle fibers. It can decompose to creatine and phosphate, and the released energy is used to generate ATP from ADP.

Photopigment   A substance that can absorb light and undergo structural changes that can lead to the development of a generator potential. An example is rhodopsin.

Photoreceptor   Receptor that detects light on the retina of the eye. Also called **electromagnetic receptor.**

Physiology (fiz'-ē-OL-ō-jē)   Science that deals with the functions of an organism or its parts.

Pia mater (PĒ-a MĀ-ter)   The inner membrane (meninx) covering the brain and spinal cord.

Pilonidal (pī'-lō-NĪ-dal)   Containing hairs resembling a tuft inside a cyst or sinus.

Pineal (PIN-ē-al) **gland**   The cone-shaped gland located in the roof of the third ventricle. Also called the **epiphysis cerebri** (ē-PIF-i-sis se-RĒ-brē).

Pinealocyte (pin-ē-AL-ō-sīt)   Secretory cell of the pineal gland that produces hormones.

Pinna (PIN-na)   The projecting part of the external ear composed of elastic cartilage and covered by skin and shaped like the flared end of a trumpet. Also called the **auricle** (OR-i-kul).

Pinocytosis (pi'-nō-sī-TŌ-sis)   The process by which cells ingest liquid.

Pituicyte (pi-TOO-i-sīt)   Supporting cell of the posterior lobe of the pituitary gland.

Pituitary (pi-TOO-i-tar'-ē) **dwarfism**   Condition caused by hyposecretion of growth hormone (GH) during the growth years and characterized by childlike physical traits in an adult.

Pituitary gland   A small endocrine gland lying in the sella turcica of the sphenoid bone and attached to the hypothalamus by the infundibulum; nicknamed the "master gland." Also called the **hypophysis** (hī-POF-i-sis).

Pivot joint   A synovial joint in which a rounded, pointed, or conical surface of one bone articulates with a shallow depression of another bone, as in the joint between the atlas and axis and between the proximal ends of the radius and ulna. Also called a **trochoid** (TRŌ-koid) **joint.**

Placenta (pla-SEN-ta)   The special structure through which the exchange of materials between fetal and maternal circulations occurs. Also called the **afterbirth.**

Plantar flexion (PLAN-tar FLEK-shun)   Extension of the foot at the ankle joint.

Plaque (plak)   A cholesterol-containing mass in the tunica media of arteries. A mass of bacterial cells, dextran (polysaccharide), and other debris that adheres to teeth.

Plasma (PLAZ-ma)   The extracellular fluid found in blood vessels; blood minus the formed elements.

Plasma cell   Cell that produces antibodies and develops from a B cell (lymphocyte).

Plasma (cell) membrane   Outer, limiting membrane that separates the cell's internal parts from extracellular fluid and the external environment.

Platelet plug   Aggregation of thrombocytes at a damaged blood vessel to prevent blood loss.

Pleura (PLOOR-a)   The serous membrane that enfolds the lungs and lines the walls of the chest and diaphragm.

Pleural cavity   Small potential space between the visceral and parietal pleurae.

Plexus (PLEK-sus)   A network of nerves, veins, or lymphatic vessels.

Plexus of Auerbach   *See* **Myenteric plexus.**

Plexus of Meissner   *See* **Submucous plexus.**

Plicae circulares (PLĪ-kē SER-kyoo-lar-es)   Permanent, deep, transverse folds in the mucosa and submucosa of the small intestine that increase the surface area for absorption. Also called **circular folds.**

Pneumonia (noo-MŌ-nē-a)   Acute infection or inflammation of the alveoli of the lungs.

Pneumotaxic (noo-mō-TAK-sik) **area**   Portion of the respiratory center in the pons that continually sends inhibitory impulses to the inspiratory area that limit inspiration and facilitate expiration.

Podiatry (pō-DĪ-a-trē)   The diagnosis and treatment of foot disorders.

Polar body   The smaller cell resulting from the unequal division of cytoplasm during the meiotic divisions of an oocyte. The polar body has no function and is resorbed.

Polarized   A condition in which opposite effects or states exist at the same time. In electrical contexts, having one portion negative and another positive; for example, a polarized nerve cell membrane has the outer surface positively charged and the inner surface negatively charged.

Poliomyelitis (pō'-lē-ō-mī-e-LĪ-tis)   Viral infection marked by fever, headache, stiff neck and back, deep muscle pain and weakness, and loss of certain somatic reflexes; a serious form of the disease, **bulbar polio,** results in destruction of motor neurons in anterior horns of spinal nerves that leads to paralysis.

Polycythemia (pol'-ē-sī-THĒ-mē-a)   An abnormal increase in the number of red blood cells.

Polysaccharides (pol'-ē-SAK-a-rīds)   Three or more monosaccharides joined chemically.

Polyunsaturated fat   A fat that contains two or more double covalent bonds between its carbon atoms; examples are corn oil, safflower oil, and cottonseed oil.

Polyuria (pol'-ē-YOO-rē-a)   An excessive production of urine.

Pons (ponz)   The portion of the brain stem that forms a "bridge" between the medulla and the midbrain, anterior to the cerebellum.

Positron emission tomography (PET)   A type of radioactive scanning based on the release of gamma rays when positrons collide with negatively charged electrons in body tissues; it indicates where radioisotopes are used in the body.

Postabsorptive state   Metabolic state during which absorption is complete and energy needs of the body must be satisfied by nutrients already in the body. Also called the **fasting state.**

Postcentral gyrus A gyrus immediately posterior to the central sulcus that contains the general sensory area of the cerebral cortex.

Posterior (pos-TĒR-ē-or) Nearer to or at the back of the body. Also called **dorsal.**

Posterior root The structure composed of afferent (sensory) fibers lying between a spinal nerve and the dorsolateral aspect of the spinal cord. Also called the **dorsal root** or **sensory root.**

Posterior root ganglion A group of cell bodies of sensory (afferent) neurons and their supporting cells located along the posterior root of a spinal nerve. Also called a **dorsal** or **sensory root ganglion** (GANG-glē-on).

Postganglionic neuron (pōst'-gang-lē-ON-ik NOO-ron) The second visceral efferent neuron in an autonomic pathway, having its cell body and dendrites located in an autonomic ganglion and its unmyelinated axon ending at cardiac muscle, smooth muscle, or a gland.

Postpartum (pōst-PAR-tum) After parturition; occurring after the delivery of a baby.

Postsynaptic (pōst-sin-AP-tik) **neuron** The nerve cell that is activated by the release of a neurotransmitter substance from another neuron and carries action potentials away from the synapse.

Pouch of Douglas *See* **Rectouterine pouch.**

Precentral gyrus A gyrus immediately anterior to the central sulcus that contains the primary motor area of the cerebral cortex.

Preeclampsia (pre'-e-KLAMP-sē-a) A syndrome characterized by sudden hypertension, large amounts of protein in urine, and generalized edema; it might be related to an autoimmune or allergic reaction due to the presence of a fetus.

Preganglionic (prē'-gang-lē-ON-ik) **neuron** The first visceral efferent neuron in an autonomic pathway, with its cell body and dendrites in the brain or spinal cord and its myelinated axon ending at an autonomic ganglion, where it synapses with a postganglionic neuron.

Pregnancy Sequence of events including fertilization, implantation, embryonic growth, and fetal growth that normally terminates in birth.

Premenstrual syndrome (PMS) Severe physical and emotional stress occurring late in the postovulatory phase of the menstrual cycle and sometimes overlapping with menstruation.

Premonitory (prē-MON-i-tō-rē) Giving previous warning; as premonitory symptoms.

Prepuce (PRĒ-pyoos) The loose-fitting skin covering the glans of the penis and clitoris. Also called the **foreskin.**

Presbyopia (prez-bē-Ō-pē-a) A loss of elasticity of the lens of the eye due to advancing age with resulting inability to focus clearly on near objects.

Presynaptic (prē-sin-AP-tik) **neuron** A nerve cell that carries action potentials toward a synapse.

Prevertebral ganglion (prē-VERT-e-bral GANG-lē-on) A cluster of cell bodies of postganglionic sympathetic neurons anterior to the spinal column and close to large abdominal arteries. Also called a **collateral ganglion** (GANG-lē-on).

Primary germ layer One of three layers of embryonic tissue, called ectoderm, mesoderm, and endoderm, that give rise to all tissues and organs of the organism.

Primary motor area A region of the cerebral cortex in the precentral gyrus of the frontal lobe of the cerebrum that controls specific muscles or groups of muscles.

Primary somesthetic (sō-mes-THET-ik) **area** A region of the cerebral cortex posterior to the central sulcus in the postcentral gyrus of the parietal lobe of the cerebrum that localizes exactly the points of the body where sensations originate.

Prime mover The muscle directly engaged in contractions as distinguished from muscles that are relaxing at the same time. Also called an **agonist** (AG-ō-nist).

Primigravida (prī-mi-GRAV-i-da) A woman pregnant for the first time.

Primitive gut Embryonic structure composed of endoderm and mesoderm that gives rise to most of the gastrointestinal (GI) tract.

Primordial (prī-MOR-dē-al) Existing first; especially primordial egg cells in the ovary.

Principal cell Cell found in the parathyroid glands that secretes parathyroid hormone (PTH). Also called a **chief cell.**

Proctology (prok-TOL-ō-jē) The branch of medicine that treats the rectum and its disorders.

Progeny (PROJ-e-nē) Refers to offspring or descendants.

Progesterone (prō-JES-te-rōn) A female sex hormone produced by the ovaries that helps prepare the endometrium for implantation of a fertilized ovum and the mammary glands for milk secretion.

Prognosis (prog-NŌ-sis) A forecast of the probable results of a disorder; the outlook for recovery.

Projection (prō-JEK-shun) The process by which the brain refers sensations to their point of stimulation.

Prolactin (prō-LAK-tin) **(PRL)** A hormone secreted by the adenohypophysis (anterior lobe) of the pituitary gland that initiates and maintains milk secretion by the mammary glands.

Prolapse (PRŌ-laps) A dropping or falling down of an organ, especially the uterus or rectum.

Proliferation (prō-lif'-er-Ā-shun) Rapid and repeated reproduction of new parts, especially cells.

Pronation (prō-NĀ-shun) A movement of the flexed forearm in which the palm of the hand is turned posteriorly.

Properdin (prō-PER-din) A protein found in serum capable of destroying bacteria and viruses.

Prophase (PRŌ-fāz) The first stage in mitosis during which chromatid pairs are formed and aggregate around the equatorial plane region of the cell.

Proprioception (prō-prē-ō-SEP-shun) The receipt of information from muscles, tendons, and the labyrinth that enables the brain to determine movements and position of the body and its parts. Also called **kinesthesia** (kin'-es-THĒ-zē-a).

Proprioceptor (prō'-prē-ō-SEP-tor) A receptor located in muscles, tendons, or joints that provides information about body position and movements.

Prostaglandin (pros'-ta-GLAN-din) **(PG)** A membrane-associated lipid composed of 20-carbon fatty acids with 5 carbon atoms joined to form a cyclopentane ring; synthesized in small quantities and basically mimics hormones in activities.

Prostatectomy (pros'-ta-TEK-tō-mē) The surgical removal of part of or the entire prostate gland.

Prostate (PROS-tāt) **gland** A muscular, doughnut-shaped gland inferior to the urinary bladder that surrounds the superior portion of the male urethra and secretes an alkaline fluid.

Prosthesis (pros-THĒ-sis) An artificial device to replace a missing body part.

Protein An organic compound consisting of carbon, hydrogen, oxygen, nitrogen, and sometimes sulfur and phosphorus, and made up of amino acids linked by peptide bonds.

Prothrombin (prō-THROM-bin) An inactive protein synthesized by the liver, released into the blood, and converted to active thrombin in the process of blood clotting.

Prothrombin time A test to determine the amount of prothrombin in blood, based on the time required for a blood sample to clot after being treated with an anticoagulant and coagulation factors; normal time about 11 to 15 sec.

Protraction (prō-TRAK-shun) The movement of the mandible or clavicle forward on a plane parallel with the ground.

Proximal (PROK-si-mal) Nearer the attachment of an extremity to the trunk or a structure; nearer to the point of origin.

Pruritus (proo'-RĪ-tus) Itching.

Pseudopodia (soo'-dō-dō-PŌ-dē-a) Temporary, protruding projections of cytoplasm.

Psoriasis (sō-RĪ-a-sis) Chronic skin disease characterized by reddish plaques or papules covered with scales.

Psychosomatic (sī'-kō-sō-MAT-ik) Pertaining to the relationship between mind and body. Commonly used to refer to those physiological disorders thought to be caused entirely or partly by emotional disturbances.

Ptosis (TŌ-sis) Drooping, as of the eyelid or the kidney.

Puberty (PYOO-ber-tē) The time of life during which the secondary sex characteristics begin to appear and the capability for sexual reproduction is possible; usually between the ages of 10 and 15.

Pudendum (poo-DEN-dum) A collective designation for the external genitalia of the female.

Pulmonary (PUL-mo-ner'-ē) Concerning or affected by the lungs.

Pulmonary circulation The flow of deoxygenated blood from the right ventricle to the lungs and the return of oxygenated blood from the lungs to the left atrium.

Pulmonary edema (e-DĒ-ma) An abnormal accumulation of interstitial fluid in the tissue spaces and alveoli of the lungs due to increased pulmonary capillary permeability or increased pulmonary capillary pressure.

Pulmonary embolism (EM-bō-lizm) **(PE)** The presence of a blood clot or other foreign substance in a pulmonary arterial blood vessel that obstructs circulation to lung tissue.

Pulmonary ventilation The inflow (inspiration) and outflow (expiration) of air between the atmosphere and the lungs. Also called **breathing.**

Pulp cavity A cavity within the crown and neck of a tooth, filled with pulp, a connective tissue containing blood vessels, nerves, and lymphatics.

Pulsating electromagnetic fields (PEMFs) A procedure that uses electrotherapy to treat improperly healing fractures.

Pulse The rhythmic expansion and recoil of the elastic arteries caused by the ejection of blood from the left ventricle. Pulse rate corresponds to the heart rate.

Pulse pressure The difference between the maximum (systolic) and minimum (diastolic) pressures; normally a value of about 40 mm Hg.

Pupil The hole in the center of the iris, the area through which light enters the posterior cavity of the eyeball.

Purkinje fiber *See* **Conduction myofiber.**

Pus The liquid product of inflammation containing leucocytes or their remains and debris of dead cells.

P wave The deflection wave of an electrocardiogram that records atrial depolarization.

Pyemia (pī-Ē-mē-a) Infection of the blood, with multiple abscesses, caused by pus-forming microorganisms.

Pyloric (pī-LOR-ik) **sphincter** A thickened ring of smooth muscle through which the pylorus of the stomach communicates with the duodenum. Also called the **pyloric valve.**

Pyogenesis (pi'-ō-JEN-e-sis) Formation of pus.

Pyramid (PIR-a-mid) A pointed or cone-shaped structure; one of two roughly triangular structures on the ventral side of the medulla composed of the largest motor tracts that run from the cerebral cortex to the spinal cord; a triangular-shaped structure in the renal medulla composed of the straight segments of renal tubules.

Pyramidal (pi-RAM-i-dal) **pathways** Collections of motor nerve fibers arising in the brain and passing down through the spinal cord to motor cells in the anterior horns.

Pyrexia (pī-REK-sē-a) A condition in which the temperature is above normal.

Pyuria (pī-YOO-rē-a) The presence of leucocytes and other components of pus in urine.

QRS wave The deflection wave of an electrocardiogram that records ventricular depolarization.

Quadrant (KWAD-rant) One of four parts.

Quadriplegia (kwad'-ri-PLĒ-jē-a) Paralysis of the two upper and two lower extremities.

Radiographic (rā'-dē-ō-GRAF-ic) **anatomy** Diagnostic branch of anatomy that includes the use of x rays.

Rami communicantes (RĀ-mē ko-myoo-ni-KAN-tēz) Branches of a spinal nerve. *Singular,* **ramus communicans** (RĀ-mus ko-MYOO-ni-kans).

Rapid eye movement (REM) sleep A level of sleep characterized by symmetrical flutter of the eyes and eyelids and brain wave patterns similar to those of an awake person.

Rathke's pouch *See* **Hypophyseal pouch.**

Reactivity (rē-ak-TI-vi-tē) Ability of an antigen to react specifically with the antibody whose formation it induced.

Receptor A specialized cell or a nerve cell terminal modified to respond to some specific sensory modality, such as touch, pressure, cold, light, or sound. A specific molecule or arrangement of molecules organized to accept only molecules with a complementary shape.

Receptor-mediated endocytosis A highly selective process in which cells take up large molecules or particles (ligands). In the process, successive compartments called vesicles, endosomes, and CURLs form. Ligands are eventually broken down by enzymes in lysosomes.

Recessive gene A gene that is not expressed in the presence of a dominant gene on the homologous chromosome.

Reciprocal innervation (re-SIP-rō-kal in-ner-VĀ-shun) The phenomenon by which impulses stimulate contraction of one muscle and simultaneously inhibit contraction of antagonistic muscles.

Recombinant DNA Synthetic DNA, formed by joining a fragment of DNA from one source to a portion of DNA from another.

Recruitment (rē-KROOT-ment) The process of increasing the number of active motor units.

Rectouterine pouch A pocket formed by the parietal peritoneum as it moves posteriorly from the surface of the uterus and is reflected onto the rectum; the lowest point in the pelvic cavity. Also called the **pouch** or **cul de sac of Douglas.**

Rectum (REK-tum) The last 20 cm (8 inches) of the gastrointestinal tract, from the sigmoid colon to the anus.

Red nucleus A cluster of cell bodies in the midbrain, occupying a large portion of the tegumentum and sending fibers into the rubroreticular and rubrospinal tracts.

Reduction The addition of electrons and hydrogen ions (hydrogen atoms) to a molecule or, less commonly, the removal of oxygen from a molecule.

Referred pain Pain that is felt at a site remote from the place of origin.

Reflex Fast response to a change in the internal or external environment that attempts to restore homeostasis; passes over a reflex arc.

Reflex arc The most basic conduction pathway through the nervous system, connecting a receptor and an effector and consisting of a receptor, a sensory neuron, a center in the central nervous system for a synapse, a motor neuron, and an effector.

Refraction (rē-FRAK-shun)   The bending of light as it passes from one medium to another.

Refractory (re-FRAK-to-rē) **period**   A time during which an excitable cell cannot respond to a stimulus that is usually adequate to evoke an action potential.

Regimen (REJ-i-men)   A strictly regulated scheme of diet, exercise, or activity designed to achieve certain ends.

Regional anatomy   The division of anatomy dealing with a specific region of the body, such as the head, neck, chest, or abdomen.

Regulating factor   Chemical secretion of the hypothalamus that can either stimulate or inhibit secretion of hormones of the adenohypophysis (anterior pituitary).

Regurgitation (rē-gur'-ji-TA-shun)   Return of solids or fluids to the mouth from the stomach; flowing backward of blood through incompletely closed heart valves.

Relapse (rē-LAPS)   The return of a disease weeks or months after its apparent cessation.

Relaxin   A female hormone produced by the ovaries that relaxes the symphysis pubis, helps dilate the uterine cervix to facilitate delivery, and plays a role in increasing sperm motility.

Remodeling   Replacement of old bone by new bone tissue.

Renal (RĒ-nal)   Pertaining to the kidney.

Renal corpuscle (KOR-pus'-l)   A glomerular (Bowman's) capsule and its enclosed glomerulus.

Renal erythropoietic (ē-rith'-rō-poy-Ē-tik) **factor**   An enzyme released by the kidneys and liver in response to hypoxia that acts on a plasma protein to bring about the production of erythropoietin, which stimulates red blood cell production.

Renal failure   Inability of the kidneys to function properly, due to abrupt failure (acute) or progressive failure (chronic).

Renal pelvis   A cavity in the center of the kidney formed by the expanded, proximal portion of the ureter, lying within the kidney, and into which the major calyces open.

Renal pyramid   A triangular structure in the renal medulla composed of the straight segments of renal tubules.

Renin (RĒ-nin)   An enzyme released by the kidney into the plasma where it converts angiotensinogen into angiotensin I.

Renin-angiotensin (an'-jē-ō-TEN-sin) **pathway**   A mechanism for the control of aldosterone secretion by angiotensin II, initiated by the secretion of renin by the kidney in response to low blood pressure.

Residual (re-ZID-yoo-al) **volume**   The volume of air still contained in the lungs after a maximal expiration; about 1,200 ml.

Resistance   Ability to ward off disease. The hindrance encountered by an electrical charge as it moves through a substance from one point to another. The hindrance encountered by blood as it flows through the vascular system or by air through respiratory passageways.

Respiration (res-pi-RA-shun)   Overall exchange of gases between the atmosphere, blood, and body cells consisting of pulmonary ventilation, external respiration, and internal respiration.

Respiratory center   Neurons in the reticular formation of the brain stem that regulate the rate of respiration.

Respiratory distress syndrome (RDS) of the newborn   A disease of newborn infants, especially premature ones, in which insufficient amounts of surfactant are produced and breathing is labored. Also called **hyaline** (HĪ-a-lin) **membrane disease (HMD).**

Resting potential   The voltage that exists between the inside and outside of a cell membrane when the cell is not respond-

ing to a stimulus; about −70 to −90 mV, with the inside of the cell negative.

Resuscitation (rē-sus'-i-TA-shun)   Act of bringing a person back to full consciousness.

Retention (rē-TEN-shun)   A failure to void urine due to obstruction, nervous contraction of urethra, or absence of sensation of desire to urinate.

Rete (RĒ-tē) **testis**   The network of ducts in the testes.

Reticular (re-TIK-yoo-lar) **activating system (RAS)**   An extensive network of branched nerve cells running through the core of the brain stem. When these cells are activated, a generalized alert or arousal behavior results.

Reticular formation   A network of small groups of nerve cells scattered among bundles of fibers beginning in the medulla as a continuation of the spinal cord and extending upward through the central part of the brain stem.

Reticulocyte (rē-TIK-yoo-lō-sīt)   An immature red blood cell.

Retina (RET-i-na)   The inner coat of the eyeball, lying only in the posterior portion of the eye and consisting of nervous tissue and a pigmented layer comprised of epithelial cells lying in contact with the choroid. Also called the **nervous tunic** (TOO-nik).

Retinal (RE'-ti-nal)   The pigment portion of the photopigment rhodopsin. Also called **visual yellow.**

Retraction (rē-TRAK-shun)   The movement of a protracted part of the body backward on a plane parallel to the ground, as in pulling the lower jaw back in line with the upper jaw.

Retroflexion (re-trō-FLEK-shun)   A malposition of the uterus in which it is tilted posteriorly.

Retroperitoneal (re'-trō-per-i-tō-NĒ-al)   External to the peritoneal lining of the abdominal cavity.

Rheumatism (ROO-ma-tizm')   Any painful state of the supporting structures of the body—bones, ligaments, joints, tendons, or muscles.

Rh factor   An inherited agglutinogen on the surface of red blood cells.

Rhinology (rī-NOL-ō-jē)   The study of the nose and its disorders.

Rhodopsin (rō-DOP-sin)   A photopigment in rods of the retina, consisting of a protein scotopsin plus retinal, that is sensitive to low levels of illumination. Also called **visual purple.**

Ribonucleic (rī'-bō-nyoo-KLĒ-ik) **acid (RNA)**   A single-stranded nucleic acid constructed of nucleotides consisting of one of four possible nitrogen bases (adenine, cytosine, guanine, or uracil), ribose, and a phosphate group; three types are messenger RNA (mRNA), transfer RNA (tRNA), and ribosomal RNA (rRNA), each of which cooperates with DNA for protein synthesis.

Ribosome (RĪ-bō-sōm)   An organelle in the cytoplasm of cells, composed of ribosomal RNA and ribosomal proteins, that synthesizes proteins; nicknamed the "protein factory."

Rickets (RIK-ets)   Condition affecting children characterized by soft and deformed bones resulting from inadequate calcium metabolism due to a vitamin D deficiency.

Right heart (atrial) reflex   A reflex concerned with maintaining normal venous blood pressure.

Right lymphatic (lim-FAT-ik) **duct**   A vessel of the lymphatic system that drains lymph from the upper right side of the body and empties it into the right subclavian vein.

Rigor mortis   State of partial contraction of muscles following death due to lack of ATP that causes cross bridges of thick myofilaments to remain attached to thin myofilaments, thus preventing relaxation.

Rod   A visual receptor in the retina of the eye that is specialized for vision in dim light.

Roentgenogram (RENT-gen-ō-gram)   A photographic image.

Root canal A narrow extension of the pulp cavity lying within the root of a tooth.

Root hair plexus (PLEK-sus) A network of dendrites arranged around the root of a hair as free or naked nerve endings that are stimulated when a hair shaft is moved.

Root of penis Attached portion of penis that consists of the bulb and crura.

Rotation (rō-TĀ-shun) Moving a bone around its own axis, with no other movement.

Round ligament (LIG-a-ment) A band of fibrous connective tissue enclosed between the folds of the broad ligament of the uterus, emerging from a point on the uterus just below the uterine (Fallopian) tube, extending laterally along the pelvic wall, and penetrating the abdominal wall through the deep inguinal ring to end in the labia majora.

Round window A small opening between the middle and inner ear, directly below the oval window, covered by the secondary tympanic membrane. Also called the **fenestra cochlae** (fe-NES-tra KŌK-lē-ē).

Rugae (ROO-jē) Large folds in the mucosa of an empty hollow organ, such as the stomach and vagina.

Saccule (SAK-yool) The lower and smaller of the two chambers in the membranous labyrinth inside the vestibule of the inner ear containing a receptor organ for static equilibrium.

Sacral hiatus (hi-Ā-tus) Inferior entrance to the vertebral canal formed when the laminae of the fifth sacral vertebra (and sometimes fourth) fail to meet.

Sacral plexus (PLEK-sus) A network formed by the anterior branches of spinal nerves L4 through S3.

Sacral promontory (PROM-on-tor'-ē) The superior surface of the body of the first sacral vertebra that projects anteriorly into the pelvic cavity; a line from the sacral promontory to the superior border of the symphysis pubis divides the abdominal and pelvic cavities.

Saddle joint A synovial joint in which the articular surfaces of both of the bones are saddle-shaped or concave in one direction and convex in the other direction, as in the joint between the trapezium and the metacarpal of the thumb. Also called a **sellaris** (sel-LA-ris) **joint.**

Sagittal (SAJ-i-tal) **plane** A vertical plane that divides the body or organs into left and right portions. Such a plane may be **midsagittal (median)**, in which the divisions are equal, or **parasagittal**, in which the divisions are unequal.

Saliva (sa-LĪ-va) A clear, alkaline, somewhat viscous secretion produced by the three pairs of salivary glands; contains various salts, mucin, lysozyme, and salivary amylase.

Salivary amylase (SAL-i-ver-ē AM-i-lās) An enzyme in saliva that initiates the chemial breakdown of starch, mostly in the mouth.

Salivary gland One of three pairs of glands that lie outside the mouth and pour their secretory product (called saliva) into ducts that empty into the oral cavity; the parotid, submandibular, and sublingual glands.

Salpingitis (sal'-pin-JĪ-tis) Inflammation of the uterine (Fallopian) or auditory (Eustachian) tube.

Saltatory (sal-ta-TŌ-rē) **conduction** The propagation of an action potential (nerve impulse) along the exposed portions of a myelinated nerve fiber. The action potential appears at successive neurofibral nodes (nodes of Ranvier) and therefore seems to jump or leap from node to node.

Sarcolemma (sar'-kō-LEM-ma) The cell membrane of a muscle cell, especially of a skeletal muscle cell.

Sarcoma (sar-KŌ-ma) A connective tissue tumor, often highly malignant.

Sarcomere (SAR-kō-mēr) A contractile unit in a striated muscle cell extending from one Z line to the next Z line.

Sarcoplasm (SAR-kō-plazm) The cytoplasm of a muscle cell.

Sarcoplasmic reticulum (sar'-kō-PLAZ-mik re-TIK-yoo-lum) A network of saccules and tubes surrounding myofibrils of a muscle cell, comparable to endoplasmic reticulum; functions to reabsorb calcium ions during relaxation and to release them to cause contraction.

Satiety (sa-TĪ-e-tē) Fullness or gratification, as of hunger or thirst.

Satiety center A collection of nerve cells located in the ventromedial nuclei of the hypothalamus that, when stimulated, brings about the cessation of eating.

Saturated fat A fat that contains no double bonds between any of its carbon atoms, all are single bonds and all carbon atoms are bonded to the maximum number of hydrogen atoms; found naturally in animal foods such as meat, milk, milk products, and eggs.

Scala tympani (SKA-la TIM-pan-ē) The lower spiral-shaped channel of the bony cochlea, filled with perilymph.

Scala vestibuli (ves-TIB-yoo-lē) The upper spiral-shaped channel of the bony cochlea, filled with perilymph.

Schwann cell *See* **Neurolemmocyte.**

Sciatica (sī-AT-i-ka) Inflammation and pain along the sciatic nerve; felt at the back of the thigh running down the inside of the leg.

Sclera (SKLE-ra) The white coat of fibrous tissue that forms the outer protective covering over the eyeball except in the area of the anterior cornea; the posterior portion of the fibrous tunic.

Scleral venous sinus A circular venous sinus located at the junction of the sclera and the cornea through which aqueous humor drains from the anterior chamber of the eyeball into the blood. Also called the **canal of Schlemm** (SHLEM).

Sclerosis (skle-RŌ-sis) A hardening with loss of elasticity of the tissues.

Scoliosis (skō'-lē-Ō-sis) An abnormal lateral curvature from the normal vertical line of the spine.

Scotoma (skō-TŌ-ma) A blind spot or area of depressed vision within the visual field.

Scotopsin (skō-TOP-sin) The protein portion of the visual pigment rhodopsin found in rods of the retina.

Scrotum (SKRŌ-tum) A skin-covered pouch that contains the testes and their accessory structures.

Sebaceous (se-BĀ-shus) **gland** An exocrine gland in the dermis of the skin, almost always associated with a hair follicle, that secretes sebum. Also called an **oil gland.**

Sebum (SĒ-bum) Secretion of sebaceous (oil) glands.

Secondary sex characteristic A feature characteristic of the male or female body that develops at puberty under the stimulation of sex hormones, but is not directly involved in sexual reproduction, such as distribution of body hair, voice pitch, body shape, and muscle development.

Secretion (se-KRĒ-shun) Production and release from a gland cell of a fluid, especially a functionally useful product as opposed to a waste product.

Selectively permeable membrane A membrane that permits the passage of certain substances, but restricts the passage of others. Also called a **semipermeable** (sem'-ē-PER-mē-a-bl) **membrane.**

Sella turcica (SEL-a TUR-si-ka) A depression on the superior surface of the sphenoid bone that houses the pituitary gland.

Semen (SĒ-men) A fluid discharged at ejaculation by a male that consists of a mixture of spermatozoa and the secretions of the seminal vesicles, the prostate gland, and the bulbourethral (Cowper's) glands. Also called **seminal** (SEM-i-nal) **fluid.**

Semicircular canals   Three bony channels projecting superiorly and posteriorly from the vestibule of the inner ear, filled with perilymph, in which lie the membranous semicircular canals filled with endolymph. They contain receptors for equilibrium.

Semicircular ducts   The membranous semicircular canals filled with endolymph and floating in the perilymph of the bony semicircular canals. They contain cristae that are concerned with dynamic equilibrium.

Semilunar (sem'-ē-LOO-nar) **valve**   A valve guarding the entrance into the aorta or the pulmonary trunk from a ventricle of the heart.

Seminal vesicle (SEM-i-nal VES-i-kul)   One of a pair of convoluted, pouchlike structures, lying posterior and inferior to the urinary bladder and anterior to the rectum, that secrete a component of semen into the ejaculatory ducts.

Seminiferous tubule (sem'-i-NI-fer-us TOOB-yool)   A tightly coiled duct, located in a lobule of the testis, where spermatozoa are produced.

Senescence (se-NES-ens)   The process of growing old; the period of old age.

Senile macular (MAK-yoo-lar) **degeneration (SMD)**   A disease in which blood vessels grow over the macula lutea.

Senility (se-NIL-i-tē)   A loss of mental or physical ability due to old age.

Sensation   A state of awareness of external or internal conditions of the body.

Sensory area   A region of the cerebral cortex concerned with the interpretation of sensory impulses.

Septal defect   An opening in the septum (interatrial or interventricular) between the left and right sides of the heart.

Septum (SEP-tum)   A wall dividing two cavities.

Serosa (ser-Ō-sa)   Any serous membrane. The outermost layer or tunic of an organ formed by a serous membrane. The membrane that lines the pleural, pericardial, and peritoneal cavities.

Serous (SIR-us) **membrane**   A membrane that lines body cavities that does not open to the exterior. Also called the **serosa** (se-RŌ-sa).

Serum   Plasma minus its clotting proteins.

Sesamoid bones (SES-a-moyd)   Small bones usually found in tendons.

Sex chromosomes   The twenty-third pair of chromosomes, designated X and Y, which determine the genetic sex of an individual; in males, the pair is XY; in females, XX.

Sexual intercourse   The insertion of the erect penis of a male into the vagina of a female. Also called **coitus** (KŌ-i-tus).

Sexually transmitted disease (STD)   General term for any of a large number of diseases spread by sexual contact.

Sheath of Schwann   *See* **Neurolemma.**

Shingles   Acute infection of the peripheral nervous system caused by a virus.

Shinsplints   Soreness or pain along the tibia probably caused by inflammation of the periosteum brought on by repeated tugging of the muscles and tendons attached to the periosteum. Also called **tibia stress syndrome.**

Shoulder   A synovial or diarthrotic joint where the humerus joins the scapula.

Sigmoid colon (SIG-moyd KŌ-lon)   The S-shaped portion of the large intestine that begins at the level of the left iliac crest, projects inward to the midline, and terminates at the rectum at about the level of the third sacral vertebra.

Sign   Any objective evidence of disease such as a lesion, swelling, or fever.

Sinoatrial (si-nō-Ā-trē-al) **(SA) node**   A compact mass of cardiac muscle cells specialized for conduction, located in the right atrium beneath the opening of the superior vena cava. Also called the **sinuatrial node** or **pacemaker.**

Sinus (SĪ-nus)   A hollow in a bone (paranasal sinus) or other tissue; a channel for blood; any cavity having a narrow opening.

Sinusitis (sīn-yoo-SĪT-is)   Inflammation of the mucous membrane of a paranasal sinus.

Sinusoid (SĪN-yoo-soyd)   A microscopic space or passage for blood in certain organs such as the liver or spleen.

Skeletal muscle   An organ specialized for contraction, composed of striated muscle fibers (cells), supported by connective tissue, attached to a bone by a tendon or an aponeurosis, and stimulated by somatic efferent neurons.

Skene's gland   *See* **Paraurethral gland.**

Skull   The skeleton of the head consisting of the cranial and facial bones.

Sliding-filament theory   The most commonly accepted explanation for muscle contraction in which actin and myosin myofilaments move into interdigitation with each other, decreasing the length of the sarcomeres.

Small intestine   A long tube of the gastrointestinal tract that begins at the pyloric sphincter of the stomach, coils through the central and lower part of the abdominal cavity, and ends at the large intestine; divided into three segments: duodenum, jejunum, and ileum.

Smooth muscle   An organ specialized for contraction, composed of smooth muscle fibers (cells), located in the walls of hollow internal structures, and innervated by a visceral efferent neuron.

Sodium-potassium pump   An active transport system located in the cell membrane that transports sodium ions out of the cell and potassium ions into the cell at the expense of cellular ATP. It functions to keep the ionic concentrations of these elements at physiological levels.

Soft palate (PAL-at)   The posterior portion of the roof of the mouth, extending posteriorly from the palatine bones and ending at the uvula. It is a muscular partition lined with mucous membrane.

Solution   A homogeneous molecular or ionic dispersion of one or more substances (solutes) in a usually liquid dissolving medium (solvent).

Somatic (sō-MAT-ik) **nervous system (SNS)**   The portion of the peripheral nervous system made up of the somatic efferent fibers that run between the central nervous system and the skeletal muscles and skin.

Somesthetic (sō'-mes-THET-ik)   Pertaining to sensations and sensory structures of the body.

Somite (SŌ-mīt)   Block of mesodermal cells in a developing embryo that is distinguished into a myotome (which forms most of the skeletal muscles), dermatome (which forms connective tissues), and sclerotome (which forms the vertebrae).

Spasm (spazm)   An involuntary, convulsive, muscular contraction.

Spermatic (sper-MAT-ik) **cord**   A supporting structure of the male reproductive system, extending from a testis to the deep inguinal ring, that includes the ductus (vas) deferens, arteries, veins, lymphatics, nerves, cremaster muscle, and connective tissue.

Spermatogenesis (sper'-ma-tō-JEN-e-sis)   The formation and development of the spermatozoa.

Spermatozoon (sper'-ma-tō-ZŌ-on)   A mature sperm cell.

Spermicide (SPER-mi-sīd')   An agent that kills spermatozoa.

Spermiogenesis (sper'-mē-ō-JEN-e-sis)   The maturation of spermatids into spermatozoa.

Sphincter (SFINGK-ter)   A circular muscle constricting an orifice.

Sphincter of Oddi   *See* **Sphincter of the hepatopancreatic ampulla.**

Sphincter of the hepatopancreatic ampulla   A circular muscle at the opening of the common bile and main pancreatic ducts in the duodenum. Also called the **sphincter of Oddi** (OD-ē).

Sphygmomanometer (sfig'-mō-ma-NOM-e-ter) An instrument for measuring arterial blood pressure.

Spina bifida (SPĪ-na BIF-i-da) A congenital defect of the vertebral column in which the halves of the neural arch of a vertebra fail to fuse in midline.

Spinal (SPĪ-nal) **cord** A mass of nerve tissue located in the vertebral canal from which 31 pairs of spinal nerves originate.

Spinal (lumbar) puncture Withdrawal of some of the cerebrospinal fluid from the subarachnoid space in the lumbar region.

Spinal nerve One of the 31 pairs of nerves that originate on the spinal cord from posterior and anterior roots.

Spinal shock A period of time, from several days to several weeks, following transection of the spinal cord and characterized by the abolition of all reflex activity.

Spinous (SPĪ-nus) **process** A sharp or thornlike process or projection. Also called a **spine.** A sharp ridge running diagonally across the posterior surface of the scapula.

Spiral organ The organ of hearing, consisting of supporting cells and hair cells that rest on the basilar membrane and extend into the endolymph of the cochlear duct. Also called the **organ of Corti** (KOR-tē).

Spirometer (spī-ROM-e-ter) An apparatus used to measure air capacity of the lungs. Also called a **respirometer.**

Spleen (SPLĒN) Large mass of lymphatic tissue between the fundus of the stomach and the diaphragm that functions in phagocytosis, production of lymphocytes, and blood storage.

Sprain Forcible wrenching or twisting of a joint with partial rupture or other injury to its attachments without dislocation.

Sputum (SPYOO-tum) Substance ejected from the mouth containing saliva and mucus.

Starling's law of the capillaries The movement of fluid between plasma and interstitial fluid is in a state of near equilibrium at the arterial and venous ends of a capillary, that is, filtered fluid and absorbed fluid plus that returned to the lymphatic system are nearly equal.

Starling's law of the heart The force of muscular contraction is determined by the length of the cardiac muscle fibers; the greater the length of stretched fibers, the stronger the contraction.

Starvation (star-VĀ-shun) The loss of energy stores in the form of glycogen, fats, and proteins due to inadequate intake of nutrients or inability to digest, absorb, or metabolize ingested nutrients.

Stasis (STĀ-sis) Stagnation or halt of normal flow of fluids, as blood, urine, or of the intestinal mechanism.

Static equilibrium (ē-kwi-LIB-rē-um) The maintenance of posture in response to changes in the orientation of the body, mainly the head, relative to the ground.

Stellate reticuloendothelial (STEL-āte re-tik'-yoo-lō-en'-dō-THĒ-lē-al) **cell** Phagocytic cell that lines a sinusoid of the liver. Also called a **Kupffer's** (KOOP-ferz) **cell.**

Stenosis (sten-Ō-sis) An abnormal narrowing or constriction of a duct or opening.

Stereocilia (ste'-rē-ō-SIL-ē-a) Groups of extremely long, slender, nonmotile microvilli projecting from epithelial cells lining the epididymis.

Stereognosis (ste'-rē-og-NŌ-sis) The ability to recognize the size, shape, and texture of an object by touch.

Sterile (STE-rīl) Free from any living microorganisms. Unable to conceive or produce offspring.

Sterilization (ster'-i-li-ZĀ-shun) Elimination of all living microorganisms. The rendering of an individual incapable of reproduction (e.g., castration, vasectomy, hysterectomy).

Sternal puncture Introduction of a wide-bore needle into the marrow cavity of the sternum for aspiration of a sample of red bone marrow.

Stimulus Any change in the environment capable of altering the membrane potential.

Stomach The J-shaped enlargement of the gastrointestinal tract directly under the diaphragm in the epigastric, umbilical, and left hypochondriac regions of the abdomen, between the esophagus and small intestine.

Strabismus (stra-BIZ-mus) A condition in which the visual axes of the two eyes differ, so that they do not both fix on the same object.

Straight tubule (TOOB-yool) A duct in a testis leading from a convoluted seminiferous tubule to the rete testis.

Stratum basalis (STRĀ-tum ba-SAL-is) The outer layer of the endometrium, next to the myometrium, that is maintained during menstruation and gestation and produces a new fuctionalis following menstruation or parturition.

Stratum functionalis (funk'-shun-AL-is) The inner layer of the endometrium, the layer next to the uterine cavity, that is shed during menstruation and that forms the maternal portion of the placenta during gestation.

Stressor A stress that is extreme, unusual, or long-lasting and triggers the general adaptation syndrome.

Stretch receptor Receptor in the walls of bronchi, bronchioles, and lungs that sends impulses to the respiratory center that prevents overinflation of the lungs.

Stretch reflex A monosynaptic reflex triggered by a sudden stretch of a muscle and ending with a contraction of that same muscle.

Stricture (STRIK-cher) A local contraction of a tubular structure.

Stroke volume The volume of blood ejected by either ventricle in one systole; about 70 ml.

Stroma (STRŌ-ma) The tissue that forms the ground substance, foundation, or framework of an organ, as opposed to its functional parts.

Subarachnoid (sub'-a-RAK-noyd) **space** A space between the arachnoid and the pia mater that surrounds the brain and spinal cord and through which cerebrospinal fluid circulates.

Subcutaneous (sub'-kyoo-TĀ-nē-us) Beneath the skin. Also called **hypodermic** (hī-po-DER-mik).

Subcutaneous layer A continuous sheet of loose connective tissue and adipose tissue between the dermis of the skin and the deep fascia of the muscles. Also called the **superficial fascia** (FASH-ē-a).

Subdural space (sub-DOO-ral) A space between the dura mater and the arachnoid of the brain and spinal cord that contains a small amount of fluid.

Sublingual (sub-LING-gwal) **gland** One of a pair of salivary glands situated in the floor of the mouth under the mucous membrane and to the side of the lingual frenulum, with a duct that opens into the floor of the mouth.

Submandibular (sub'-man-DIB-yoo-lar) **gland** One of a pair of salivary glands found beneath the base of the tongue under the mucous membrane in the posterior part of the floor of the mouth, posterior to the sublingual glands, with a duct situated to the side of the lingual frenulum. Also called the **submaxillary** (sub'-MAK-si-ler-ē) **gland.**

Submucosa (sub-myoo-KŌ-sa) A layer of connective tissue located beneath a mucous membrane, as in the gastrointestinal tract or the urinary bladder, where a submucosa connects the mucosa to the muscularis tunic.

Submucous plexus A network of autonomic nerve fibers located in the outer portion of the submucous layer or tunic of the small intestine. Also called the **plexus of Meissner** (MĪS-ner).

Substrate A substance with which an enzyme reacts.

Subthreshold stimulus A stimulus of such weak intensity that it cannot initiate an action potential (nerve impulse). Also called a **subliminal stimulus.**

Sudoriferous (soo'-dor-IF-er-us) **gland** An apocrine or eccrine exocrine gland in the dermis or subcutaneous layer that produces perspiration. Also called a **sweat gland.**

Sulcus (SUL-kus) A groove or depression between parts, especially between the convolutions of the brain. *Plural,* **sulci** (SUL-sē).

Summation (sum-MĀ-shun) The algebraic addition of the excitatory and inhibitory effects of many stimuli applied to a nerve cell body. The increased strength of muscle contraction that results when stimuli follow in rapid succession.

Superficial (soo'-per-FISH-al) Located on or near the surface of the body.

Superficial fascia (FASH-ē-a) A continuous sheet of fibrous connective tissue between the dermis of the skin and the deep fascia of the muscles. Also called **subcutaneous** (sub'-kyoo-TĀ-nē-us) **layer.**

Superior (soo-PĒR-ē-or) Toward the head or upper part of a structure. Also called **cephalad** (SEF-a-lad) or **craniad.**

Superior vena cava (VĒ-na CĀ-va) Large vein that collects blood from parts of the body superior to the heart and returns it to the right atrium.

Supination (soo-pī-NĀ-shun) A movement of the forearm in which the palm of the hand is turned anteriorly.

Suppuration (sup'-yoo-RĀ-shun) Pus formation and discharge.

Surface anatomy The study of the structures that can be identified from the outside of the body.

Surfactant (sur-FAK-tant) A phospholipid substance produced by the lungs that decreases surface tension.

Susceptibility (sus-sep'-ti-BIL-i-tē) Lack of resistance of a body to the deleterious or other effects of an agent such as pathogenic microorganisms.

Suspensory ligament (sus-PEN-so-rē LIG-a-ment) A fold of peritoneum extending laterally from the surface of the ovary to the pelvic wall.

Sustentacular (sus'-ten-TAK-yoo-lar) **cell** A supporting cell of seminiferous tubules that produces secretions for supplying nutrients to spermatozoa and the hormone inhibin. Also called a **Sertoli** (ser-TŌ-lē) **cell.**

Sutural bone (SOO-cher-al) A small bone located within a suture of certain cranial bones. Also called **Wormian** (WER-mē-an) **bone.**

Suture (SOO-cher) A fibrous joint, especially in the skull, where bone surfaces are closely united.

Sympathetic (sim'-pa-THET-ik) **division** One of the two subdivisions of the autonomic nervous system, having cell bodies of preganglionic neurons in the lateral gray columns of the thoracic segment and first two or three lumbar segments of the spinal cord; primarily concerned with processes involving the expenditure of energy. Also called the **thoracolumbar** (thō'-ra-kō-LUM-bar) **division.**

Sympathomimetic (sim'-pa-thō-mi-MET-ik) Producing effects that mimic those brought about by the sympathetic division of the autonomic nervous system.

Symphysis (SIM-fi-sis) A line of union. A slightly movable cartilaginous joint such as the symphysis pubis between the anterior surfaces of the pubic bones.

Symphysis pubis (PYOO-bis) A slightly movable cartilaginous joint between the anterior surfaces of the pelvic bones.

Symptom (SIMP-tum) An observable abnormality that indicates the presence of a disease or disorder of the body.

Synapse (SIN-aps) The junction between the process of two adjacent neurons; the place where the activity of one neuron affects the activity of another.

Synaptic (sin-AP-tik) **cleft** The narrow gap that separates the axon terminal of one nerve cell from another nerve cell or muscle fiber (cell) and across which a neurotransmitter diffuses to affect the postsynaptic cell.

Synaptic delay The length of time between the arrival of the action potential at the axon terminal and the membrane potential (IPSP or EPSP) change on the postsynaptic membrane; usually about 0.5 msec.

Synaptic end bulb Expanded distal end of an axon terminal that contains synaptic vesicles. Also called a **synaptic knob** or **end foot.**

Synaptic gutter Invaginated portion of a sarcolemma under an axon terminal. Also called a **synaptic trough** (TROF).

Synaptic vesicle Membrane-enclosed sac in a synaptic end bulb that stores neurotransmitter substances.

Synarthrosis (sin'-ar-THRŌ-sis) An immovable joint.

Synchondrosis (sin'-kon-DRŌ-sis) A cartilaginous joint in which the connecting material is hyaline cartilage.

Syncope (SIN-kō-pē) A temporary cessation of consciousness, most commonly due to cerebral ischemia; a faint.

Syndesmosis (sin'-dez-MŌ-sis) A fibrous joint in which articulating bones are united by dense fibrous tissue.

Syndrome (SIN-drōm) A group of signs and symptoms that occur together in a pattern characteristic of a particular disease or abnormal condition.

Syneresis (si-NER-e-sis) The process of clot retraction.

Synergist (SIN-er-jist) A muscle that assists the prime mover by reducing undesired action or unnecessary movement.

Synostosis (sin'-os-TŌ-sis) A joint in which the dense fibrous connective tissue that unites bones at a suture has been replaced by bone, resulting in a complete fusion across the suture line.

Synovial (si-NŌ-vē-al) **cavity** The space between the articulating bones of a synovial or diarthrotic joint, filled with synovial fluid.

Synovial fluid Secretion of synovial membranes that lubricates joints and nourishes articular cartilage.

Synovial joint A fully movable or diarthrotic joint in which a joint or synovial cavity is present between the two articulating bones.

Synovial membrane The inner of the two layers of the articular capsule of a synovial joint, composed of loose connective tissue covered with epithelium that secretes synovial fluid into the joint cavity.

Syphilis (SIF-i-lis) A sexually transmitted disease caused by the bacterium *Treponema pallidum.*

System An association of organs that have a common function.

Systemic (sis-TEM-ik) Affecting the whole body; generalized.

Systemic anatomy The study of particular systems of the body, such as the skeletal, muscular, nervous, cardiovascular, or urinary systems.

Systemic circulation The routes through which oxygenated blood flows from the left ventricle through the aorta to all the organs of the body and deoxygenated blood returns to the right atrium.

Systemic lupus erythematosus (er-i-them-a-TŌ-sus) **(SLE)** An autoimmune, inflammatory disease that may affect every tissue of the body.

Systole (SIS-tō-lē) In the cardiac cycle, the phase of contraction of the heart muscle, especially of the ventricles.

Systolic (sis-TO-lik) **blood pressure** The force exerted by blood on arterial walls during ventricular contraction; the highest pressure measured in the large arteries, about 120 mm Hg under normal conditions for a young, adult male.

Tachycardia (tak'-i-KAR-dē-a) A rapid heartbeat or pulse rate.

Tactile (TAK-tīl) Pertaining to the sense of touch.

Tactile disc An encapsulated, cutaneous receptor for touch located in deeper layers of epidermal cells. Also called **Merkel's** (MER-kelz) **disc.**

Taenia coli (TĒ-nē-a KŌ-lī) One of three flat bands of thickened, longitudinal muscles running the length of the large intestine.

Target cell A cell whose activity is affected by a particular hormone.

Tarsal gland Sebaceous (oil) gland that opens on the edge of each eyelid. Also called a **Meibomian** (mī-BO-mē-an) **gland.**

Tarsus (TAR-sus) A collective term for the seven bones of the ankle.

Tay-Sachs disease Inherited, progressive neuronal degeneration of the central nervous system due to a deficient lysosomal enzyme that causes excessive accumulations of a lipid called ganglioside.

T cell A lymphocyte that can differentiate into one of six kinds of cells—killer, helper, suppressor, memory, amplifier, or delayed hypersensitivity—all of which function in cellular immunity.

Tectorial (tek-TŌ-rē-al) **membrane** A gelatinous membrane projecting over and in contact with the hair cells of the spiral organ (organ of Corti) in the cochlear duct.

Telophase (TEL-ō-fāz) The final stage of mitosis in which the daughter nuclei become established.

Tendinitis (ten'-din-Ī-tis) Inflammation of a tendon and synovial membrane at a joint. Also called **tenosynovitis** (ten'-o-sin'-ō-VĪ-tis).

Tendon (TEN-don) A white fibrous cord of dense, regularly arranged connective tissue that attaches muscle to bone.

Tendon organ A proprioceptive receptor, sensitive to changes in muscle tension, found chiefly near the junction of tendons and muscles. Also called **Golgi** (GOL-jē) **tendon organ.**

Tendon reflex A polysynaptic, ipsilateral reflex that is designed to protect tendons and their associated muscles from damage that might be brought about by excessive tension. The receptors involved are called tendon organs (Golgi tendon organs).

Tentorium cerebelli (ten-TŌ-rē-um ser'-e-BEL-ē) A transverse shelf of dura mater that forms a partition between the occipital part of the cerebral hemispheres and the cerebellum and that covers the cerebellum.

Teratogen (TER-a-tō-jen) Any agent or factor that causes physical defects in a developing embryo.

Testis (TES-tis) Male gonad that produces sperm and the hormones testosterone and inhibin.

Testosterone (tes-TOS-te-rōn) A male sex hormone (androgen) secreted by interstitial endocrinocytes (cells of Leydig) of a mature testis; controls the growth and development of male sex organs, secondary sex characteristics, spermatozoa, and body growth.

Tetanus (TET-a-nus) An infectious disease caused by the toxin of *Clostridium tetani,* characterized by tonic muscle spasms and exaggerated reflexes, lockjaw, and arching of the back. A smooth, sustained contraction produced by a series of very rapid stimuli to a muscle.

Tetany (TET-a-nē) A nervous condition caused by hypoparathyroidism and characterized by intermittent or continuous tonic muscular contractions of the extremities.

Tetralogy of Fallot (tet-RAL-ō-jē of fal-Ō) A combination of four congenital heart defects: (1) constricted pulmonary semilunar valve, (2) interventricular septal opening, (3) emergence of aorta from both ventricles instead of from the left only, and (4) enlarged right ventricle.

Thalamus (THAL-a-mus) A large, oval structure located above the midbrain, consisting of two masses of gray matter covered by a thin layer of white matter.

Thalassemia (thal'-a-SĒ-mē-a) A group of hereditary hemolytic anemias.

Therapy (THER-a-pē) The treatment of a disease or disorder.

Thermoreceptor (THER-mō-rē-sep-tor) Receptor that detects changes in temperature.

Thigh The portion of the lower extremity between the hip and the knee.

Third ventricle (VEN-tri-kul) A slitlike cavity between the right and left halves of the thalamus and between the lateral ventricles.

Thoracic (thō-RAS-ik) **cavity** Superior component of the ventral body cavity that contains two pleural cavities, the mediastinum, and the pericardial cavity.

Thoracic duct A lymphatic vessel that begins as a dilation called the cisterna chyli, receives lymph from the left side of the head, neck, and chest, the left arm, and the entire body below the ribs, and empties into the left subclavian vein. Also called the **left lymphatic** (lim-FAT-ik) **duct.**

Thoracolumbar (thō'-ra-kō-LUM-bar) **outflow** The fibers of the sympathetic preganglionic neurons, which have their cell bodies in the lateral gray columns of the thoracic segment and first two or three lumbar segments of the spinal cord.

Thorax (THŌ-raks) The chest.

Threshold potential The membrane voltage that must be reached in order to trigger an action potential (nerve impulse).

Threshold stimulus Any stimulus strong enough to initiate an action potential (nerve impulse). Also called a **liminal** (LIM-i-nal) **stimulus.**

Thrombin (THROM-bin) The active enzyme formed from prothrombin that acts to convert fibrinogen to fibrin.

Thrombocyte (THROM-bō-sīt) A fragment of cytoplasm enclosed in a cell membrane and lacking a nucleus; found in the circulating blood; plays a role in blood clotting. Also called a **platelet** (PLĀT-let).

Thrombophlebitis (throm'-bo-fle-BĪ-tis) A disorder in which inflammation of a vein wall is followed by the formation of a blood clot (thrombus).

Thromboplastin (throm-bō-PLAS-tin) A factor, or collection of factors, whose appearance initiates the blood clotting process.

Thrombosis (throm-BŌ-sis) The formation of a clot in an unbroken blood vessel.

Thrombus A clot formed in an unbroken blood vessel.

Thymus (THĪ-mus) **gland** A bilobed organ, located in the upper mediastinum posterior to the sternum and between the lungs, that plays a role in the immune mechanism of the body.

Thyroglobulin (thī-rō-GLŌ-byoo-lin) **(TGB)** A large glycoprotein molecule secreted by follicle cells of the thyroid gland in which iodine is combined with tyrosine to form thyroid hormones.

Thyroid cartilage (THĪ-royd KAR-ti-lij) The largest single cartilage of the larynx, consisting of two fused plates that form the anterior wall of the larynx. Also called the **Adam's apple.**

Thyroid colloid (KOL-loyd) A complex in thyroid follicles consisting of thyroglobulin and stored thyroid hormones.

Thyroid follicle (FOL-i-kul) Spherical sac that forms the parenchyma of the thyroid gland and consists of follicular cells that produce thyroxine (T_4) and triiodothyronine (T_3) and parafollicular cells that produce calcitonin (CT).

Thyroid gland An endocrine gland with right and left lateral lobes on either side of the trachea connected by an isthmus located in front of the trachea just below the cricoid cartilage.

Thyroid-stimulating hormone (TSH) A hormone secreted by the adenohypophysis (anterior lobe) of the pituitary gland that stimulates the synthesis and secretion of hormones produced by the thyroid gland.

Thyroxine (thī-ROK-sēn) **(T₄)**   A hormone secreted by the thyroid gland that regulates metabolism, growth and development, and the activity of the nervous system.

Tic   Spasmodic, involuntary twitching by muscles that are ordinarily under voluntary control.

Tidal volume   The volume of air breathed in and out in any one breath; about 500 ml in quiet, resting conditions.

Tinnitus (ti-NĪ-tus)   A ringing or tingling sound in the ears.

Tissue   A group of similar cells and their intercellular substance joined together to perform a specific function.

Tissue plasminogen activator (t-PA)   An enzyme that dissolves small blood clots by initiating a process that converts plasminogen to plasmin, which degrades the fibrin of a clot.

Tissue rejection   Phenomenon by which the body recognizes the proteins (HLA antigens) in transplanted tissues or organs as foreign and produces antibodies against them.

Tongue   A large skeletal muscle on the floor of the oral cavity.

Tonsil (TON-sil)   A multiple aggregation of large lymphatic nodules embedded in mucous membrane.

Torn cartilage   A tearing of an articular disc in the knee.

Total lung capacity   The sum of tidal volume, inspiratory reserve volume, expiratory reserve volume, and residual volume; about 6,000 ml.

Toxic (TOK-sik)   Pertaining to poison; poisonous.

Toxic shock syndrome (TSS)   A disease caused by the bacterium *Staphylococcus aureus,* occurring among menstruating females who use tampons and characterized by high fever, sore throat, headache, fatigue, irritability, and abdominal pain.

Trabecula (tra-BEK-yoo-la)   Irregular latticework of thin plate of spongy bone. Fibrous cord of connective tissue serving as supporting fiber by forming a septum extending into an organ from its wall or capsule. *Plural,* **trabeculae** (tra-BEK-yoo-lē).

Trabeculae carnae (tra-BEK-yoo-lē KAR-nē)   Ridges and folds of the myocardium in the ventricles.

Trachea (TRĀ-kē-a)   Tubular air passageway extending from the larynx to fifth thoracic vertebra. Also called the **windpipe.**

Tracheostomy (trā-kē-OS-tō-mē)   Creation of an opening into the trachea through the neck, with insertion of a tube to facilitate passage of air or evacuation of secretions.

Trachoma (tra-KŌ-ma)   A chronic infectious disease of the conjunctiva and cornea of the eye caused by the TRIC agent.

Tract   A bundle of nerve fibers in the central nervous system.

Transcription (trans-KRIP-shun)   The first step in the transfer of genetic information in which a single strand of a DNA molecule serves as a template for the formation of an RNA molecule.

Translation (trans-LĀ-shun)   The construction of a new protein on the ribosome of a cell as dictated by the sequence or codons in mRNA.

Transplantation (trans-plan-TĀ-shun)   The replacement of injured or diseased tissues or organs with natural ones.

Transvaginal oocyte retrieval   Procedure in which aspirated ova are combined with a solution containing sperm outside the body and then the fertilized ova are implanted in the uterus.

Transverse colon (trans-VERS KŌ-lon)   The portion of the large intestine extending across the abdomen from the right colic (hepatic) flexure to the left colic (splenic) flexure.

Transverse fissure (FISH-er)   The deep cleft that separates the cerebrum from the cerebellum.

Transverse tubules (TOOB-yools) **(T tubules)**   Minute, cylindrical invaginations of the muscle cell membrane that carry the surface action potentials deep into the muscle cell.

Trauma (TRAW-ma)   An injury, either a physical wound or

psychic disorder, caused by an external agent or force, such as a physical blow or emotional shock; the agent or force that causes the injury.

Treppe (TREP-ē)   The gradual increase in the amount of contraction by a muscle caused by rapid, repeated stimuli of the same strength.

Triad (TRĪ-ad)   A complex of three units in a muscle cell composed of a T tubule and the segments of sarcoplasmic reticulum on both sides of it.

Tricuspid (trī-KUS-pid) **valve**   Artrioventricular (AV) valve on the right side of the heart.

Trigeminal neuralgia (trī-JEM-i-nal noo-RAL-jē-a)   Pain in one or more of the branches of the trigeminal (V) nerve. Also called **tic douloureux** (doo-loo-ROO).

Trigone (TRĪ-gon)   A triangular area at the base of the urinary bladder.

Triiodothyronine (trī-ī-od-ō-THĪ-rō-nēn) **(T₃)**   A hormone produced by the thyroid gland that regulates metabolism, growth and development, and the activity of the nervous system.

Trophectoderm (trōf-EK-tō-derm)   The outer covering of cells of the blastocyst.

Tropic (TRŌ-pik) **hormone**   A hormone whose target is another endocrine gland.

Trunk   The part of the body to which the upper and lower extremities are attached.

Tubal ligation (lī-GĀ-shun)   A sterilization procedure in which the uterine (Fallopian) tubes are tied and cut.

Tuberculosis (too-berk-yoo-LŌ-sis)   An inflammation of the lungs and pleurae caused by *Mycobacterium tuberculosis* resulting in destruction of lung tissue and its replacement by fibrous connective tissue.

Tubular maximum (Tm)   The maximum amount of a substance that can be reabsorbed by renal tubules under any condition.

Tubular reabsorption   The movement of filtrate from renal tubules back into blood in response to the body's specific needs.

Tubular secretion   The movement of substances in blood back into filtrate in response to the body's specific needs.

Tumor (TOO-mor)   A growth of excess tissue due to an unusually rapid division of cells.

Tunica albuginea (TOO-ni-ka al'-byoo-JIN-ē-a)   A dense layer of white fibrous tissue covering a testis or deep to the surface of an ovary.

Tunica externa (eks-TER-na)   The outer coat of an artery or vein, composed mostly of elastic and collagenous fibers. Also called the **adventitia.**

Tunica interna (in-TER-na)   The inner coat of an artery or vein, consisting of a lining of endothelium and its supporting layer of connective tissue. Also called the **tunica intima** (IN-ti-ma).

Tunica media (MĒ-dē-a)   The middle coat of an artery or vein, composed of smooth muscle and elastic fibers.

T wave   The deflection wave of an electrocardiogram that records ventricular repolarization.

Twitch   Rapid, jerky contraction of a muscle in response to a single stimulus.

Tympanic (tim-PAN-ik) **membrane**   A thin, semitransparent partition of fibrous connective tissue between the external auditory meatus and the middle ear. Also called the **eardrum.**

Type II cutaneous mechanoreceptor   A receptor embedded deeply in the dermis and deeper tissues that detects heavy and continuous touch sensations. Also called **end organ of Ruffini.**

Ulcer (UL-ser)   An open lesion of the skin or a mucous membrane of the body with loss of substance and necrosis of the tissue.

Umbilical (um-BIL-i-kal) **cord**   The long, ropelike structure, containing the umbilical arteries and vein, which connects the fetus to the placenta.

Umbilicus (um-BIL-i-kus)   A small scar on the abdomen that marks the former attachment of the umbilical cord to the fetus. Also called the **navel.**

Unsaturated fat   A fat that contains one or more double covalent bonds between its carbon atoms; examples are olive oil and peanut oil.

Upper extremity   The appendage attached at the shoulder girdle, consisting of the arm, forearm, wrist, hand, and fingers.

Uremia (yoo-RĒ-mē-a)   Accumulation of toxic levels of urea and other nitrogenous waste products in the blood, usually resulting from severe kidney malfunction.

Ureter (YOO-re-ter)   One of two tubes that connect the kidney with the urinary bladder.

Urethra (yoo-RĒ-thra)   The duct from the urinary bladder to the exterior of the body that conveys urine in females and urine and semen in males.

Urinalysis   The physical, chemical, and microscopic analysis or examination of urine.

Urinary (YOO-ri-ner-ē) **bladder**   A hollow, muscular organ situated in the pelvic cavity posterior to the symphysis pubis.

Urinary tract infection (UTI)   An infection of a part of the urinary tract or the presence of large numbers of microbes in urine.

Urine   The fluid produced by the kidneys that contains wastes or excess materials and excreted from the body through the urethra.

Urobilinogenuria   The presence of urobilinogen in urine.

Urogenital (yoo'-rō-JEN-i-tal) **triangle**   The region of the pelvic floor below the symphysis pubis, bounded by the symphysis pubis and the ischial tuberosities and containing the external genitalia.

Urology (yoo-ROL-ō-jē)   The specialized branch of medicine that deals with the structure, function, and diseases of the male and female urinary systems and the male reproductive system.

Uterine (YOO-ter-in) **tube**   Duct that transports ova from the ovary to the uterus. Also called the **Fallopian** (fal-LŌ-pē-an) **tube** or **oviduct.**

Uterosacral ligament (yoo'-ter-ō-SĀ-kral LIG-a-ment)   A fibrous band of tissue extending from the cervix of the uterus laterally to attach to the sacrum.

Uterovesical pouch (yoo'-ter-ō-VES-ī-kal)   A shallow pouch formed by the reflection of the peritoneum from the anterior surface of the uterus, at the junction of the cervix and the body, to the posterior surface of the urinary bladder.

Uterus (YOO-te-rus)   The hollow, muscular organ in females that is the site of menstruation, implantation, development of the fetus, and labor. Also called the **womb.**

Utricle (YOO-tri-kul)   The larger of the two divisions of the membranous labyrinth located inside the vestibule of the inner ear, containing a receptor organ for static equilibrium.

Uvea (OO-vē-a)   The three structures that make up the vascular tunic of the eye.

Uvula (OO-vyoo-la)   A soft, fleshy mass, especially the V-shaped pendant part, descending from the soft palate.

Vagina (va-JĪ-na)   A muscular, tubular organ that leads from the uterus to the vestibule, situated between the urinary bladder and the rectum of the female.

Valvular stenosis (VAL-vyoo-lar STEN-Ō-sis)   A narrowing of a heart valve, usually the mitral valve.

Varicose (VAR-i-kōs)   Pertaining to an unnatural swelling, as in the case of a varicose vein.

Vasa vasorum (VĀ-sa va-SŌ-rum)   Blood vessels supplying nutrients to the larger arteries and veins.

Vascular (VAS-kyoo-lar)   Pertaining to or containing many blood vessels.

Vascular spasm   Contraction of the smooth muscle in the wall of a damaged blood vessel to prevent blood loss.

Vascular tunic (TOO-nik)   The middle layer of the eyeball, composed of the choroid, the ciliary body, and the iris.

Vascular (venous) sinus   A vein with a thin endothelial wall that lacks a tunica media and externa and is supported by surrounding tissue.

Vasectomy (va-SEK-tō-mē)   A means of sterilization of males in which a portion of each ductus (vas) deferens is removed.

Vasoconstriction (vāz-ō-kon-STRIK-shun)   A decrease in the size of the lumen of a blood vessel caused by contraction of the smooth muscle in the wall of the vessel.

Vasodilation (vās'-ō-DĪ-lā-shun)   An increase in the size of the lumen of a blood vessel caused by relaxation of the smooth muscle in the wall of the vessel.

Vasomotion (vāz-ō-MŌ-shun)   Intermittent contraction and relaxation of the smooth muscle of metarterioles and precapillary sphincters that result in an intermittent blood flow.

Vasomotor (vā-sō-MŌ-tor) **center**   A cluster of neurons in the medulla that controls the diameter of blood vessels, especially arteries.

Vein   A blood vessel that conveys blood from tissues back to the heart.

Vena cava (VĒ-na KĀ-va)   One of two large veins that open into the right atrium, returning to the heart all of the deoxygenated blood from the systemic circulation except from the coronary circulation.

Ventral (VEN-tral)   Pertaining to the anterior or front side of the body; opposite of dorsal.

Ventral body cavity   Cavity near the ventral aspect of the body that contains viscera and consists of a superior thoracic cavity and an inferior abdominopelvic cavity.

Ventricle (VEN-tri-kul)   A cavity in the brain or an inferior chamber of the heart.

Ventricular fibrillation (ven-TRIK-yoo-lar fib-ri-LĀ-shun)   Asynchronous ventricular contractions that result in circulatory failure.

Venule (VEN-yool)   A small vein that collects blood from capillaries and delivers it to a vein.

Vermiform appendix (VER-mi-form a-PEN-diks)   A twisted, coiled tube attached to the cecum.

Vermillion (ver-MIL-yon)   The area of the mouth where the skin on the outside meets the mucous membrane on the inside.

Vermis (VER-mis)   The central constricted area of the cerebellum that separates the two cerebellar hemispheres.

Vertebral (VER-te-bral) **canal**   A cavity within the vertebral column formed by the vertebral foramina of all the vertebrae and containing the spinal cord. Also called the **spinal canal.**

Vertebral column   The 26 vertebrae; encloses and protects the spinal cord and serves as a point of attachment for the ribs and back muscles. Also called the **spine, spinal column,** or **backbone.**

Vertigo (VER-ti-go)   Sensation of dizziness.

Vesicular ovarian follicle   A relatively large, fluid-filled follicle containing an immature ovum and its surrounding tissues that secretes estrogens. Also called a **Graafian** (GRAF-ē-an) **follicle.**

Vestibular (ves-TIB-yoo-lar) **membrane**   The membrane that separates the cochlear duct from the scala vestibuli.

Vestibule (VES-ti-byool)   A small space or cavity at the beginning of a canal, especially the inner ear, larynx, mouth, nose, and vagina.

Villus (VIL-lus)   A projection of the intestinal mucosal cells containing connective tissue, blood vessels, and a lymphatic vessel; functions in the absorption of food. *Plural,* **villi** (VIL-ē).

Viscera (VIS-er-a)   The organs inside the ventral body cavity. *Singular,* **viscus** (VIS-kus).

Visceral (VIS-er-al)   Pertaining to the organs or to the covering of an organ.

Visceral effector (e-FEK-tor)   Cardiac muscle, smooth muscle, and glandular epithelium.

Visceral muscle   An organ specialized for contraction, composed of smooth muscle cells, located in the walls of hollow internal structures, and stimulated by visceral efferent neurons.

Visceral pleura (PLOO-ra)   The inner layer of the serous membrane that covers the lungs.

Visceroceptor (vis'-er-ō-SEP-tor)   Receptor that provides information about the body's internal environment.

Viscosity (vis-KOS-i-tē)   The state of being sticky or thick.

Vital capacity   The sum of inspiratory reserve volume, tidal volume, and expiratory reserve volume; about 4,800 ml.

Vitamin   An organic molecule necessary in trace amounts that acts as a catalyst in normal metabolic processes to the body.

Vitiligo (vit-i-LĪ-gō)   Patchy, white spots on the skin due to partial or complete loss of melanocytes.

Vitreous humor (VIT-rē-us HYOO-mor)   A soft, jellylike substance that fills the posterior cavity of the eyeball, lying between the lens and the retina.

Vocal folds   Pair of mucous membrane folds below the ventricular folds that function in voice production. Also called **true vocal cords.**

Volkmann's canal   *See* **Perforating canal.**

Vomiting   Forcible expulsion of the contents of the upper gastrointestinal tract through the mouth.

Vulva (VUL-va)   Collective designation for the external genitalia of the female. Also called **pudendum** (poo-DEN-dum).

Wallerian (wal-LE-rē-an) **degeneration**   Degeneration of the distal portion of the axon and myelin sheath.

Wandering macrophage (MAK-rō-fāj)   Phagocytic cell that de-velops from a monocyte, leaves the blood, and migrates to infected tissues.

Wart   Generally benign tumor of epithelial skin cells caused by a virus.

White matter   Aggregations or bundles of myelinated axons located in the brain and spinal cord.

White ramus communicans (RĀ-mus co-MYOO-ni-kans)   The portion of a preganglionic sympathetic nerve fiber that branches away from the anterior ramus of a spinal nerve to enter the nearest sympathetic trunk ganglion.

Wormian bone   *See* **Sutural bone.**

Xiphoid (ZĪ-foyd)   Sword-shaped. The lowest portion of the sternum.

Yolk sac   An embryonic membrane that connects with the midgut during early embryonic development, but is nonfunctional in humans.

Zona fasciculata (ZŌ-na fa-sik'-yoo-LA-ta)   The middle zone of the adrenal cortex that consists of cells arranged in long, straight cords and that secretes glucocorticoid hormones.

Zona glomerulosa (glo-mer'-yoo-LŌ-sa)   The outer zone of the adrenal cortex, directly under the connective tissue covering, that consists of cells arranged in arched loops or round balls and that secretes mineralocorticoid hormones.

Zona pellucida (pe-LOO-si-da)   Gelantinous covering that surrounds an ovum.

Zona reticularis (ret-ik'-yoo-LAR-is)   The inner zone of the adrenal cortex, consisting of cords of branching cells that secrete sex hormones, chiefly androgens.

Zygote (ZĪ-gōt)   The single cell resulting from the union of a male and female gamete; the fertilized ovum.

Zymogenic (zī'-mō-JEN-ik) **cell**   One of the cells of a gastric gland that secretes the principal gastric enzyme precursor, pepsinogen. Also called a **peptic cell.**

Bibliography

UNIT I

Berkow, R. (Editor-in-Chief). *The Merck Manual,* 14th ed. Rahway, N.J.: Merck Sharp & Dohme, 1982.

Berridge, M. J., "The Molecular Basis of Communication Within the Cell," *Scientific American,* October 1985.

Bishop, M. J. "Oncogenes," *Scientific American,* March 1982.

Bretscher, M. S. "The Molecules of the Cell Membrane," *Scientific American,* October 1985.

Cairns, J. "The Treatment of Diseases and the War Against Cancer," *Scientific American,* November 1985.

Clemente, C. D. *Anatomy: A Regional Atlas of the Human Body,* 2d ed. Baltimore: Urban & Schwarzenberg, 1981.

Croce, C. M., and G. Klein. "Chromosome Translocations and Human Cancer," *Scientific American,* March 1985.

Dahl, M. V. "Acne: How It Happens and How It's Treated," *Modern Medicine,* September 1982.

Dautry-Varsat, A., and H. F. Lodish. "How Receptors Bring Proteins and Particles into Cells," *Scientific American,* May 1984.

de Duve, C. *A Guided Tour of the Living Cell,* New York: Scientific American Books, Inc., 1984.

————. "Microbodies in the Living Cell," *Scientific American,* May 1983.

Edelson, L. E., and J. M. Fink. "The Immunologic Function of Skin," *Scientific American,* June 1985.

Geller, S. A. "Autopsy," *Scientific American,* March 1983.

Hunter, T. "The Proteins of Oncogenes," *Scientific American,* August 1984.

Keiffer, S. A., and E. R. Heitzman. *An Atlas of Cross-Sectional Anatomy.* New York: Harper & Row, 1979.

Levine, N. "Sunburn: How to Stop the Pain, How to Prevent the Damage," *Modern Medicine,* July 1984.

Luterman, A., and P. W. Curreri. "Emergency Protocol for the Severely Burned Patient," *Hospital Medicine,* October 1984.

Netter, F. H., *CIBA Collection of Medical Illustrations.* Vols. 1–7. Summit, N.J.: CIBA, 1962–1983.

Porter, K. R., and J. B. Tucker. "The Ground Substance of the Living Cell," *Scientific American,* March 1981.

Rohen, J. W., and C. Yokochi, *Color Atlas of Anatomy.* Tokyo, New York: Igaku-Shoin, Ltd., 1983.

Ross, M. H., and E. J. Reith. *Histology: A Text and Atlas.* New York: Harper & Row, 1985.

Rothman, J. E. "The Compartmental Organization of the Golgi Apparatus," *Scientific American,* September 1985.

Silberner, J. "Cancer Virus Redux," *Science News,* 1 June 1985.

Synder, S. H. "The Molecular Basis of Communication Between Cells," *Scientific American,* October 1985.

Tonegawa, S. "The Molecules of the Immune System," *Scientific American,* October 1985.

Tortora, G. J. *Principles of Human Anatomy,* 4th ed. New York: Harper & Row, 1986.

Unwin, N., and R. Henderson. "The Structure of Proteins in Biological Membranes," *Scientific American,* February 1984.

Vance, J. C. "Skin Cancer: Making a Difference with Early Dx," *Modern Medicine,* August 1983.

Weinberg, R. A. "A Molecular Basis of Cancer," *Scientific American,* November 1983.

————. "The Molecules of Life," *Scientific American,* October 1985.

Yokochi, C., and J. W. Rohen. *Photographic Anatomy of the Human Body,* 2d ed. Tokyo, New York: Igaku-Shoin, Ltd., 1978.

UNIT II

Allman, W. F. "The Knee," *Science 83,* November 1983.

————. "Weight-Lifting: Inside the Pumphouse," *Science 84,* September 1984.

Arehart-Treichel, J. "Boning Up on Osteoporosis," *Science News,* 27 August 1983.

————. "The Joint Destroyers," *Science News,* September 1982.

Bassett, C. A., S. N. Mitchell, and S. R. Gaston. "Pulsing Electromagnetic Field Treatment in Ununited Fractures and Failed Arthrodeses," *J. Amer. Med. Assoc.,* 5 February 1982.

Bienenstock, H. "Diagnosis: Arthritis," *Hospital Medicine,* October 1984.

Cameron, N. U. "Electromagnetism—Quackery or Useful Therapy?," *Modern Medicine,* March 1984.

Johnson, G. T. (ed). "Arthroscopic Surgery on the Knee," *The Harvard Medical School Health Letter,* March 1982.

————. "Steroids: A Primer," *The Harvard Medical School Health Letter,* October 1983.

Lisak, R. P. "Myasthenia Gravis: Mechanisms and Management," *Hospital Practice,* March 1983.

Rosse, C., and D. K. Clawson. *The Musculoskeletal System in Health and Disease.* New York: Harper & Row, 1980.

Wilson, F. C. *The Musculoskeletal System: Basic Processes and Disorders,* 2d ed. Philadelphia: Lippincott, 1983.

UNIT III

Barr, M. L., and J. A. Kiernan. *The Human Nervous System,* 4th ed. New York: Harper & Row, 1983.

Begley, S., J. Carey, and R. Sawhill. "How the Brain Works," *Newsweek,* 7 February 1983.

Bloom, F. E. "Neuropeptides," *Scientific American,* October 1981.

"The Brain," *Scientific American,* September 1979 (entire issue devoted to the brain and the human nervous system).

Brown, M. R., and L. A. Fisher. "Brain Peptides as Intercellular Messengers," *J. Amer. Med. Assoc.,* 9 March 1984.

Carney, R. M. "Clinical Applications of Relaxation Training," *Hospital Practice,* July 1983.

Carpenter, M. B. *Human Neuroanatomy,* 7th ed. Baltimore: Williams & Wilkins, 1976.

Crapo, L. *Hormones: Messengers of Life.* New York: W. H. Freeman & Co., 1985.

Dunant, Y., and M. Israël. "The Release of Acetylcholine," *Scientific American,* March 1985.

Ezrin, C., J. O. Godden, and R. Volpe, *Systematic Endocrinology,* 2d ed. New York: Harper & Row, 1979.

Fellman, B. "A Clockwork Gland," *Science 85,* May 1985.

Gluhbegovic, N., and T. H. Williams. *The Human Brain: A Photographic Guide,* New York: Harper & Row, 1980.

Goldfinger, S. E. (ed). "The Diseases Called Sugar Diabetes," (Part 1), *The Harvard Medical School Health Letter,* February 1985. "Insulin and the Diabetic," *The Harvard Medical School Health Letter,* March 1985, "Diabetes: The Long Run," *The Harvard Medical School Health Letter,* April 1985.

———. "The Endorphins—The Body's Own Opiates," *The Harvard Medical School Health Letter,* January 1983.

———. "New Help for the Deaf," *The Harvard Medical School Health Letter,* August 1985.

———. "Parkinson's Disease: Less Mysterious, More Manageable," *The Harvard Medical School Health Letter,* November 1984.

———. "Shingles," *The Harvard Medical School Health Letter,* June 1984.

Guyton, A. C. *Human Physiology and Mechanisms of Disease,* 3d ed. Philadelphia: Saunders, 1982.

Hall, S. S. "Aplysia and Hermissenda," *Science 85,* May 1985.

Herbert, W. "Remembrance of Things Partly," *Science News,* 10 December 1983.

Hudspeth, A. J. "The Hair Cells of the Inner Ear," *Scientific American,* January 1983.

Johnson, G. T. (ed). "Is Spinal Anesthesia Best for You?" *Mayo Clinic Health Letter,* October 1984.

Llinas, R. R. "Calcium in Synaptic Transmission," *Scientific American,* October 1982.

Loeb, G. E. "The Functional Replacement of the Ear," *Scientific American,* February 1985.

Mathewson, J. "Alzheimer's Disease: Source Searching," *Science News,* 13 July 1985.

Melamed, M. A. "Cataracts: Recognition and Assessment," *Hospital Medicine,* July 1982.

Miller, J. A. "Cell Communication Equipment: Do-It-Yourself," *Science News,* 14 April 1984.

Morell, P., and W. T. Norton. "Myelin," *Scientific American,* May 1980.

Morrison, A. R. "A Window on the Sleeping Brain," *Scientific American,* April 1983.

Nauta, W. J. H., and M. Feirtag. *Fundamental Neuroanatomy.* New York; W. H. Freeman and Co., 1985.

Pulec, J. J. "Ménière's Syndrome," *Hospital Medicine,* June 1983.

Seyle, H. *Stress in Health and Disease.* London: Butterworths, 1976.

Stevens, C. F. "The Neuron," *Scientific American,* September 1979.

Thompson, R. F. *The Brain: An Introduction to Neuroscience.* New York: W. H. Freeman and Co., 1985.

Williams, R. *Textbook of Endocrinology.* 6th ed. Philadelphia: Saunders, 1981.

Wurtman, R. J. "Alzheimer's Disease," *Scientific American,* January 1985.

UNIT IV

Ackerman, S. "The Management of Obesity," *Hospital Practice,* March 1983.

Aledort, L. M. "Current Concepts in Diagnosis and Management of Hemophilia," *Hospital Practice,* October 1982.

Allman, W. F. "Scrambled News on Cholesterol," *Science 84,* April 1984.

Ash, S. R., et al. "Peritoneal Dialysis for Acute and Chronic Renal Failure: An Update," *Hospital Practice,* January 1983.

Brown, M. S., and J. L. Goldstein. "How LDL Receptors Influence Cholesterol and Atherosclerosis," *Scientific American,* November 1984.

Cantin, M., and J. Genest. "The Heart as an Endocrine Gland," *Scientific American,* February 1986.

Capra, J. D., and A. B. Edmundson. "The Antibody Combining Site," *Scientific American,* January 1977.

Carpenter, C. B., and T. B. Strom. "Transplantation: Immunogenetic and Clinical Aspects," *Hospital Practice,* December 1982.

Carr, D. T. "Malignant Lung Disease," *Hospital Practice,* January 1981.

Comroe, J. H. "Doctor, You Have Six Minutes," *Science 84,* January/February 1984.

Cowart, V. S. "Artificial Heart," *J. Amer. Med. Assoc.,* 17 May 1985.

Dixon, B. "Of Different Bloods," *Science 84,* November 1984.

Doolittle, R. F. "Fibrinogen and Fibrin," *Scientific American,* December 1981.

Dusheck, J. "Fish, Fatty Acids, and Physiology," *Science News,* 19 October 1985.

Erickson, R. A. "Peptic Ulcer Disease in 1985: A Diagnostic Update," *Modern Medicine,* February 1985.

Escher, D. J. W. "Use of Cardiac Pacemakers," *Hospital Practice,* September 1981.

Falk, J. S., B. Kaufman, and M. H. Weil. "Cardiopulmonary Resuscitation: An Update," *Hospital Medicine,* January 1984.

Goldfinger, S. E. (ed). "The AIDS Epidemic," *The Harvard Medical School Health Letter,* June 1983.

———. "AIDS: Update (Part 1)," *The Harvard Medical School Health Letter,* November 1985.

———. "Exercise and Well-Being," *The Harvard Medical School Health Letter,* February 1985.

———. "Keeping Up with Cholesterol," *The Harvard Medical School Health Letter,* June 1985.

———. "Putting the Squeeze on Angina," *The Harvard Medical School*

Health Letter, January 1984.

_____. "Sealing Out Tooth Decay," *The Harvard Medical School Health Letter,* February 1984.

Harrison, D. C. "Coronary Bypass: The First 10 Years," *Hospital Practice,* June 1981.

Heimlich, H. J. "A Life-Saving Maneuver to Prevent Food Choking," *J. Amer. Med. Assoc.,* 27 October 1975.

Jarvik, R. K. "The Total Artificial Heart," *Scientific American,* January 1981.

Jenkins, W. "Large Bowel Exam: When to Use What Test," *Modern Medicine,* March 1984.

Johnson, G. T. (ed). "The Tragedy of Anorexia Nervosa," *The Harvard Medical School Health Letter,* December 1981.

_____. "Heart Arrhythmias," *The Harvard Medical School Health Letter,* January 1982.

_____. "Heart Attacks and Counter Attacks," *The Harvard Medical School Health Letter,* February 1982.

_____. "The Ups and Downs of Blood Pressure Numbers," *The Harvard Medical School Health Letter,* November 1982.

_____. "Weight Control," *The Harvard Medical School Health Letter,* December 1980.

_____. "Some Passing Thoughts on Kidney Stones," *The Harvard Medical School Health Letter,* March 1983.

Kaufman, C. E. "Essential Hypertension: Essentials of Management," *Hospital Medicine,* May 1982.

Kelly, D. H., and D. C. Shannon, "Sudden Infant Death Syndrome—Parts I and II," *Hospital Medicine,* October

and November 1982.

Kiester, E. "A Little Fever Is Good for You." *Science 84,* November 1984.

Kiley, J. M. (ed). "Echocardiography," *Mayo Clinic Health Letter,* May 1985.

Kopp, U. C., and G. F. Di Bona. "Diagnosis: Edema," *Hospital Medicine,* February 1982.

Laurence, J. "The Immune System In AIDS," *Scientific American,* December 1985.

Leder, P. "The Genetics of Antibody Diversity," *Scientific American,* May 1982.

Leon, A. S. "Exercise and Coronary Heart Disease," *Hospital Medicine,* March 1983.

McIntyre, K. "Cardiac Arrest," *Hospital Medicine,* November 1984.

Marrach, P., and J. Kappler. "The T Cell and Its Receptor," *Scientific American,* February 1986.

Miller, J. A. "Antibodies for Sale," *Science News,* 7 May 1983.

Milstein, C. "Monoclonal Antibodies," *Scientific American,* October 1980.

Moog, F. "The Lining of the Small Intestine," *Scientific American,* November 1981.

Paris, A. C., and B. Safai. "Acquired Immune Deficiency Syndrome," *Hospital Medicine,* September 1984.

Rios, J. C. (ed). "Cholesterol Kills," *Cardiac Alert,* April 1984.

_____. "Exercise and the Heart," *Cardiac Alert,* February 1984.

_____. "Reviewing the Most Popular Diets," *Cardiac Alert,* February 1984.

Rogers, W. J. "Diagnosis: Coronary Artery Disease," *Hospital Medicine,* August 1984.

Sabetta, J. R. "Diagnosis: Infectious Mononucleosis," *Hospital Medicine,* March 1984.

Sawka, M. M., et al. "Influence of Hydration Level and Body Fluids on Exercise Performance in the Heart," *J. Amer. Med. Assoc.,* 7 September 1984.

Schimke, R. N. "Metabolic Diseases," *Hospital Practice,* January 1983.

Shabetai, R. "Answers to Questions on Cardiac Catheterization," *Hospital Medicine,* August 1982.

Silberner, J. "AIDS: Disease, Research Efforts Advance," *Science News,* 27 April 1985.

_____. "Anatomy of Atherosclerosis," *Science News,* 16 March 1985.

_____. "Clot Dissolver May Save Heart Tissue," *Science News,* 26 November 1983.

Smolens, P. "Acute Renal Failure," *Hospital Medicine,* August 1984.

Snider, G. "Chronic Obstructive Pulmonary Disease: Advice on Dx and Rx," *Modern Medicine,* December 1982.

Solomon, H. A. "Coronary Risk Factors," *Heart Health,* December 1983.

Steinberg, S. "Cancer and Cuisine," *Science News,* 1 October 1983.

Tonegawa, S. "The Molecules of the Immune System," *Scientific American,* October 1985.

Vichinsky, E. P., D. Hurst, and B. Lubin. "Sickle Cell Disease: Basic Concepts." *Hospital Medicine,* September 1983.

Wasserburg, J. D. "Running Does Not Prolong Life or Improve Health," *Modern Medicine,* February 1985.

Zucker, M. D. "The Functioning of Blood Platelets," *Scientific American,* June 1980.

UNIT V

Begley, S., and J. Carey. "How Human Life Begins," *Newsweek,* 11 January 1982.

Carlson, B. M. *Patten's Foundations of Human Embryology,* 4th ed. New York: McGraw-Hill Book Co., 1981.

Cohen, R. J. "Diagnosis: Breast Cancer," *Hospital Medicine,* July 1984.

Federman, D. D. "Impotence: Etiology and Management," *Hospital Practice,* March 1982.

Fuchs, F. "Genetic Amniocentesis," *Scientific American,* June 1980.

Gehring, W. J. "The Molecular Basis of Development," *Scientific American,* October 1985.

Gold, M. "The Baby Makers," *Science 85,* April 1985.

Goldfinger, S. E. (ed). "A Pill For Herpes?" *The Harvard Medical School Health Letter,* September 1984.

_____. "Lumpy Breasts," *The Harvard Medical School Health Letter,* November 1984.

_____. "Surgery for Breast Cancer: Preserve and Protect," *The Harvard Medical School Health Letter,* July 1985.

Goldman, L. "Premenstrual Syndrome: Is It Oversold or Overdiagnosed?" *Modern Medicine,* June 1985.

Grobstein, C. "External Human Fertilization," *Scientific American,* June 1979.

Hafez, E. S. E. *Human Reproduction,* 2d ed. New York: Harper & Row, 1980.

Handsfield, H. H. "Sexually Transmitted

Diseases," *Hospital Practice,* January 1982.

Henahan, J. F. "Fertilization, Embryo Transfer Procedures Raise Many Questions," *J. Amer. Med. Assoc.,* 17 August 1984.

Johnson, G. T. (ed). "Sexually Transmitted Diseases," *The Harvard Medical School Health Letter,* April 1981.

_____. "Cesarean Section," *The Harvard Medical School Health Letter,* June 1981.

Kaplan, A. L. "Carcinoma of the Uterine Corpus," *Hospital Medicine,* September 1984.

Langman, J. *Medical Embryology,* 4th ed. Baltimore: Williams & Wilkins, 1981.

Miller, J. A. "The Littlest Babies," *Sci-

ence News, 15 October 1983.

_____. "Window on the Womb," *Science News,* 2 February 1985.

Rugh, R., and L. B. Shettles. *From Conception to Birth.* New York: Harper & Row, 1971.

Singer, S. *Human Genetics.* New York: W. H. Freeman & Co., 1985.

Sweet, R. L. "Toxic Shock: Accurate Diagnosis and Aggressive Treatment," *Modern Medicine,* February 1983.

Temple, M. J. "Chromosomal Syndromes," *Hospital Practice,* February 1983.

Wilbanks, G. D. "Genital Herpes," *Modern Medicine,* March 1984.

Index

Page numbers followed by the letter *E* indicate terms to be found in exhibits.

A band, 198
A,B,C's of cardiopulmonary resuscitation (CPR), 586
Abdominal aorta, 463, 498, 500E, 508E
Abdominal cavity, 13
Abdominal reflex, 292
Abdominal thrust (Heimlich) maneuver, 586
Abdominopelvic cavity, 13
Abdominopelvic diaphragm, 13, 234E
Abdominopelvic quadrants
  left lower, 13
  left upper, 13
  right lower, 13
  right upper, 13
Abdominopelvic regions, 13
Abduction, 179
Abductor muscle, 219E
Abnormal uterine bleeding, 737
ABO blood grouping, 453–455
Abortion, 763
Abscess, 542
Absolute refractory period, 271
Absorption, 8, 79, 618
  carbohydrate, 618
  electrolyte, 621
  lipid, 621
  protein, 619
  vitamin, 621
  water, 621
Absorptive state, 646–649
Accessory duct, 610
Accessory sex glands, 706, 714, 715
Accessory structures of digestion, 590
  gallbladder, 614
  liver, 611
  pancreas, 609–610
  salivary glands, 596
  teeth, 598–599
  tongue, 596
Accessory structures of eye, 374, 376
Accommodation, 380
Acetabular labrum, 188
Acetabulum, 165
Acetylcholine (ACh), 200, 275, 325
Acetylcholinesterase (AChE), 202, 275
Acetyl coenzyme A (CoA), 638, 639, 645
Achalasia, 601
Achilles reflex, 292
Achilles tendon. *See* Calcaneal tendon
Achondroplasia, 130
Achromatopsia, 395
Acid, 34–37
Acid–base balance, 697–700
Acid–base imbalances, 700
Acidosis
  metabolic, 700
  respiratory, 700
  treatment of, 700
Acinar gland, 85
Acini, 610

Acne, 110
Acquired immune deficiency syndrome (AIDS), 551, 552
Acromegaly, 408
Acromion, 160
Acrosome, 711, 743
Actin, 59, 198
Actin binding site, 198
Action potential. *See* Nerve impulse
Activated memory, 351, 352
Activation energy, 635
Active processes, 48, 51–53
Active site, 635
Active transport, 51
Acupuncture
  definition of, 343
  and pain, 343
Acute prostatitis, 737
Adam's apple. *See* Thyroid cartilage
Adaptation, 339, 340
  dark, 384
  light, 384
  to smell, 371
  to taste, 373
Addison's disease, 422
Adduction, 181
Adductor muscle, 219E
Adenine, 41
Adenitis, 551
Adenohypophysis, 404–406
Adenosarcoma, 70
Adenosine diphosphate
  chemical composition of, 42
  and electron transport system, 640–642
  functions of, 42
  and glycolysis, 636–639
  and Krebs cycle, 639–640
  and muscle contraction, 202–203
Adenosine triphosphate
  chemical composition of, 42
  and electron transport system, 640–642
  function of, 42
  and glycolysis, 636–639
  and Krebs cycle, 639, 640
  and muscle contraction, 202, 203
  production in cellular respiration, 642
Adenylate cyclase, 274, 401
Adherence, phagocytic, 539
Adhesion, 97
Adipocyte, 93
Adipose capsule, 666
Adipose tissue, 93
ADP. *See* Adenosine diphosphate
Adrenal cortex, 419–423
Adrenal gland, 419–424
Adrenal medulla, 423–424
Adrenergic fiber, 363
Adrenocorticotropic hormone (ACTH), 408
Adrenogenital syndrome, 423
Adrenoglomerulotropin, 428
Adult connective tissue
  definition of, 86
  types of, 86, 88–91, 93

Adventitia, 601
Aerobic respiration, 638
Afferent lymphatic vessel, 532
Afferent nervous system, 262
Afferent neuron, 267
A fiber, 272
Afterbirth, 524, 751
Afterimage, 340
Agglutination, 454
Agglutinin, 454
Agglutinogen, 453
Aggregated lymphatic follicle, 616
Aging
  and cardiovascular system, 524
  and cells, 71, 72
  characteristics of, 71, 72
  definition of, 71
  and digestive system, 625
  and endocrine system, 430
  and integumentary system, 110
  and muscle tissue, 210
  and nervous system, 327
  and reproductive systems, 732–733
  and respiratory system, 582–583
  and skeletal system, 126
  theories of, 72
  and urinary system, 686
Agnosia, 335
Agranular leucocyte, 445
Agraphia, 322
Airway resistance, 569, 570
Ala, 556
Alarm reaction of stress, 430–432
Albinism, 104, 117
Albino, 104
Albumin, 448, 682
Albuminuria, 682
Aldosterone, 420
Aldosteronism, 421
Alimentary canal. *See* Gastrointestinal (GI) tract
Alkalosis,
  metabolic, 700
  respiratory, 700
  treatment of, 700
Allantois, 749, 750
Allergen, 550
Allergy. *See* Hypersensitivity
Allograft, 550
All-or-none principle, 203, 271
Alpha cell, 424
Alpha receptor, 364
Alpha wave, 322
Alveolar-capillary (respiratory) membrane, 564, 565
Alveolar duct, 564
Alveolar macrophage, 564
Alveolar sac, 564
Alveolus
  of lung, 564
  of mammary gland, 728
  of mandible, 145
  of maxilla, 143

Alzheimer's disease (AD), 334
Amaerine cell, 384
Amenorrhea, 737
Ametropia, 395
Amino acid, 40, 646
    essential, 646
    nonessential, 646
Amniocentesis, 737
Amnion, 749
Amniotic cavity, 747
Amniotic fluid, 749
Amphiarthrosis, 176
Amplifier T cells, 546
Ampulla
    of ductus (vas) deferens, 713
    of mammary glands, 728
    of semicircular canal, 390
    of uterine (Fallopian) tube, 720
    of Vater. See Hepatopancreatic ampulla
Ampulla of Vater. See Hepatopancreatic ampulla
Amygdaloid nucleus, 319
Amylase. See Pancreatic amylase; Salivary amylase
Anabolic steroid, 199
Anabolism
    of carbohydrates, 643
    definition of, 31, 634
    of lipids, 646
    of proteins, 646
Anaerobic process, 210
Anal canal, 622
Anal column, 622
Analgesia, 335
Anal triangle, 236E, 728
Anamnestic response, 548
Anaphase, 67
Anaphylaxis, 550
Anastomosis, 469, 489
Anatomical position, 9
Anatomy
    branches of, 6, 17, 18
    definition of, 6
Anchoring filament, 530
Androgen, 423
Anemia, 456
    aplastic, 456
    hemolytic, 456
    hemorrhagic, 456
    nutritional, 456
    pernicious, 456
    sickle-cell, 456, 457
    thalassemia, 456
Anesthetic, 335
Aneurysm, 525
Angina pectoris, 469
Angiocardiography, 526
Angiography, 482
Angiotensin, 326
Angiotensin I, 420
Angiotensin II, 420
Angiotensinogen, 420
Angular movement, 179
Anhidrosis, 113
Anion, 28
Ankle, 170
Ankyloglossia, 596
Ankylosis, 192
Annulus fibrosus, 146
Anopsia, 327E, 395
Anorexia nervosa, 628
Anosmia, 327E
ANS. See Autonomic nervous system
Antagonist, 218
Anteflexion, 721
Anterior cavity of eyeball, 379

Anterior chamber of eyeball, 379
Anterior corticospinal tract, 287E, 350
Anterior gray horn, 284
Anterior root of spinal nerve, 287
Anterior spinocerebellar tract, 286E, 348
Anterior spinothalamic pathway, 347
Anterior spinothalamic tract, 286E, 347
Anterior white column, 284
Anterior white commissure, 284
Antibody
    agglutinin, 454
    and allergy, 550
    anamnestic response and, 548
    and autoimmune disease, 110, 551
    cellular immunity and, 544–547
    classes of, 544
    definition of, 544
    formation of, 547
    humoral immunity and, 547, 548
    monoclonal, 549
    primary response and, 547
    structure of, 544
    and tissue rejection, 550
Antibody titer, 547, 548
Anticoagulant, 452
Anticodon, 62
Antidiuretic, 411
Antidiuretic hormone (ADH), 411, 412
Antigen
    agglutinogen, 453
    and allergy, 550
    and autoimmune disease, 110, 551
    and cellular immunity, 544–547
    characteristics of, 542–544
    complete, 542
    definition of, 542
    HLA, 445
    and humoral immunity, 547, 548
    partial, 543
    and tissue rejection, 550
    tumor-specific, 549
Antigen-antibody response, 446
Antigen, bivalent, 543
Antigenic determinant site, 542, 543
Antigenic valence, 542
Antihelix, 387
Antiperspirant, 113
Antisense strand, 62
Antitragus, 387
Anulus fibrosus, 146, 155
Anus, 622
Aorta, 463, 498, 500E
Aortic body, 494, 581
Aortic insufficiency, 484
Aortic reflex, 476
Aortic semilunar valve, 466
Apex
    of heart, 461
    of lung, 562
    of nose, 556
    of patella, 166
Aphasia, 322
Apheresis, 448
Apical foramen, 598
Apnea, 582
Apneustic area, 581
Apocrine gland, 85
Apocrine sweat gland, 108
Apoenzyme, 635
Aponeurosis
    definition of, 93, 196
    galea aponeurotica, 222E
Apothecary system of measurement, A-3, A-4
Appendicitis, 627, 628
Appendicular skeleton, 133, 159–173

Appendix. See Vermiform appendix
Appositional growth, 94
Apraxia, 335
Aqueous humor, 379
Arachnoid, 284, 307
Arachnoid villi, 308
Arbor vitae, 324
Arch
    of aorta, 463, 498, 500E, 504E
    longitudinal, of foot, 171
    palatoglossal, 596
    palatopharyngeal, 596
    transverse, of foot, 171
    vertebral, 150
Areflexia, 302
Areola, 728
Arm
    blood vessels of, 504E, 505
    bones of, 160, 162
    musculature of, 242E, 243
Arousal, 353
Arrector pili muscle, 107
Arrhythmia, 483
Arterial pressure, 472
Arteries
    abdominal aorta, 463, 498, 500E, 508E
    afferent arteriole, 668
    anastomosis, 469, 489
    anterior cerebral, 504E
    anterior communicating, 504E
    anterior interventricular, 468
    anterior tibial, 508E
    arch of aorta, 463, 498, 500E, 504E
    arcuate, 667, 668, 722
    ascending aorta, 463, 498, 500E
    axillary, 504E
    basilar, 504E
    brachial, 504E
    brachiocephalic, 504E
    bronchial, 506E
    celiac, 508E
    of cerebral arterial circle (circle of Willis), 504E
    circumflex, 468
    coats of, 488
    collateral circulation of, 489
    common carotid, 504E
    common hepatic, 508E
    common iliac, 498, 508E
    coronary, 463, 468, 469, 503E
    of coronary circulation, 468
    deep palmar, 504E
    definition of, 488, 489
    digital, 504E
    digital subtraction angiography (DSA) and, 489
    disorders of, 481–484
    dorsalis pedis, 508E
    ductus arteriosus, 523
    efferent arteriole, 668
    elastic, 489
    end, 489
    esophageal, 506E
    external carotid, 504E
    external iliac, 508E
    femoral, 508E
    of fetal circulation, 522–524
    gastric, 508E
    gonadals, 508E
    hepatic, 520
    of hepatic portal circulation, 520
    histology of, 488
    inferior hypophyseal, 410
    inferior mesenteric, 508E
    inferior phrenic, 508E
    interlobar, 667

interlobular, 668
internal carotid, 504E
internal iliac, 508E
lateral plantar, 508E
lumbar, 508E
marginal, 468
medial plantar, 508E
mediastinal, 506E
middle sacral, 508E
muscular, 489
ovarian, 508E
pericardial, 506E
peroneal, 508E
popliteal, 508E
posterior cerebral, 504E
posterior communicating, 504E
posterior intercostal, 506E
posterior interventricular, 468
posterior tibial, 508E
pulmonary, 520
of pulmonary circulation, 498, 520
pulmonary trunk, 520
radial, 504E, 722
renal, 508E, 667
spiral arteriole, 722
splenic, 508E
straight arteriole, 722
subclavian, 504E
subcostal, 506E
superficial palmar, 504E
superior hypophyseal, 404
superior mesenteric, 508E
superior phrenic, 506E
suprarenal, 508E
of systemic circulation, 500–510
testicular, 508E
thoracic aorta, 463, 498, 500E, 506E
ulnar, 504E
umbilical, 522
uterine, 722
vasoconstriction of, 489
vasodilation of, 489
vertebral, 504E
Arteriogram, 482
Arteriole, 489, 490
Arthritis
    definition of, 191
    gouty, 191
    osteoarthritis, 191
    rheumatoid (RA), 191
Arthrology, 176
Arthroscopy, 178
Arthrosis, 192
Articular capsule
    of coxal joint, 187
    definition of, 177
    of humeroscapular joint, 183
    of tibiofemoral joint, 188
Articular cartilage, 94, 120, 125, 177
Articular disc
    definition of, 178
    lateral, of knee, 190
    medial, of knee, 190
Articulations
    amphiarthrosis, 176, 183, 186E
    ball-and-socket, 182, 183, 186E
    cartilaginous, 176, 177, 183, 186E
    coxal (hip), 187, 188
    definition of, 176
    diarthrosis, 176, 183, 186E
    disorders of, 191, 192
    ellipsoidal, 182, 183, 186E
    fibrous, 176, 183, 186E
    as fulcrums, 216
    functional classification of, 176
    gliding, 181, 183, 186E

gomphosis, 176, 183, 186E
hinge, 181, 183, 186E
humeroscapular (shoulder), 183, 186, 187
medical terminology, 192
movements at, 178, 181
pivot, 182, 183, 186E
saddle, 182, 183, 186E
structural classification of, 176
suture, 176, 183, 186E
symphysis, 177, 183, 186E
symphysis pubis, 171
synarthrosis, 176, 183, 186E
synchondrosis, 176, 183, 184, 186E
syndesmosis, 176, 183, 186E
synovial, 176–183, 186E
tibiofemoral (knee), 188, 190
Artificial ear, 391
Artificial heart, 478
Artificial insemination, 730
Artificial kidney, 679, 680
Artificial pacemaker, 470, 472
Artificial sweetener, 37, 38
Arytenoid cartilage, 558
Ascending aorta, 463, 498, 500E
Ascending colon, 622
Ascending limb of loop of nephron (Henle),
    666
Ascending tract, 281, 284, 286E
Ascites, 592
Asphyxia, 586
Aspiration (of knee), 192
Assimilation, 8
Association areas of cerebrum, 322
Association fiber, 318
Association (interneuron) neuron, 267
Astereognosis, 395
Asthma, 584
Astigmatism, 382
Astrocyte, 264E
Ataxia, 325
Atelectasis, 569
Atherosclerosis, 333, 481, 482
Atherosclerotic plaque, 481
Athlete's foot, 113
Atlas, 150
Atmospheric pressure, 572
Atom, 25
Atomic number, 25
Atomic structure
    electron, 25
    neutron, 25
    nucleus, 25
    proton, 25
Atomic weight, 25
ATP. See Adenosine triphosphate
ATP-binding site, 198
Atresia, 725
Atrial fibrillation, 484
Atrial flutter, 484
Atrial natriuretic factor (ANF), 430
Atrial pressure, 472
Atrial systole, 472
Atrioventricular (AV) block, 483
Atrioventricular (AV) bundle, 470
Atrioventricular (AV) node, 470
Atrioventricular (AV) valve, 465, 466
Atrium, 462, 463, 480
Atrophy, 72, 205
Auditory association area, 322
Auditory ossicle, 388
Auditory sensation, 386–392
Auditory tube, 388
Auricle, of heart, 462
Auricle, of ear. See Pinna
Auricular surface, 152
Auscultation, 473

Autograft, 550
Autoimmune disease, 110, 551
Autologous transfusion, 457
Autolysis, 59
Autonomic control, of heart, 475–477
Autonomic ganglia
    prevertebral, 358
    sympathetic trunk, 358
    terminal, 358, 360
Autonomic nervous system, 262, 357–368
    activities of, 364–366
    and biofeedback, 367, 368
    comparison with somatic efferent, 357
    control of, 367, 368
    definition of, 262, 357
    ganglia of, 358, 360
    and heart control, 364–366, 475–477
    and meditation, 368
    neurotransmitters, 363
    parasympathetic division, 262, 263
    postganglionic neurons, 360, 361
    preganglionic neurons, 358
    reflex of, 366, 367
    structure of, 358–363
    sympathetic division, 361, 362
    visceral efferent neurons of, 358–361
Autonomic plexus, 360
Autopsy, 16
Autoregulation, 494
Autosome, 762
Avascular, 77
Avitaminosis, 652
Axial skeleton, 133–155
Axis, 150
Axoaxonic synapse, 272
Axodentritic synapse, 272
Axolemma, 264
Axon, 264, 266, 267
    postganglionic, 358
    preganglionic, 358
Axonal transport, 264
Axon collateral, 264
Axon hillock, 264
Axon reaction, 302, 303
Axon terminal, 264
Axoplasm, 264
Axoplasmic flow, 264
Axosomatic synapse, 272

Babinski sign, 292, 351
Back
    bones of, 146, 149–152
    musculature of, 246E, 247
Backbone. See Vertebral column
Bainbridge reflex, 477
Ball-and-socket joint, 182
Baroreceptor, 475, 494
Bartholin's gland. See Greater vestibular gland
Basal ganglion, 318, 319
Basal lamina, 77
Basal metabolic rate (BMR), 656
Base
    of heart, 461
    inorganic, 34–37
    of lung, 562
    of patella, 166
Basement membrane, 77
Basilar membrane, 390
Basophil, 445, 446
B cell, 545
Bedsore. See Decubitus ulcer
Bell's palsy, 328E
Belly, of skeletal muscle, 216
Benign tumor, 69, 70
Beri-beri, 653E
Beta blocker, 482

Beta cell, 424
Beta endorphin, 326
Beta oxidation, 645
Beta receptor, 364
Beta wave, 323
B fiber, 272
Biaxial movement, 182
Bicuspid (mitral) valve, 465, 466
Bilateral symmetry, 9
Bile, 612
Bile capillary, 611
Bile secretion, regulation of, 613, 614
Bilirubin, 612, 683
Bilirubinuria, 683
Billroth's cord. See Splenic cord
Biofeedback, 367, 368
Biopsy, 72
Bipennate fasciculus, 218
Bipolar neuron, 267, 379
Birth. See Parturition
Birth control
    castration, 731
    chemical methods, 732
    condom, 732
    contraceptive sponge, 732
    diaphragm, 732
    hysterectomy, 731
    intrauterine device (IUD), 732
    laparoscopy, 732
    mechanical method, 731, 732
    natural methods, 732
    oophorectomy, 732
    oral contraception, 732
    tubal ligation, 731
    vasectomy, 731
Black eye, 137
Blackhead. See Comedo
Blastocoel, 744
Blastocyst, 744
Blastomere, 744
Bleeding time, 453
Blepharitis, 395
Blind spot. See Optic disc
Blood
    coagulation, 449–453
    disorders of, 456, 457
    formed elements, 440–448
    functions of, 440
    groups, 453–456
    hemostasis of, 449–453
    and interstitial fluid and lymph, 456
    medical terminology, 457, 458
    origin, 441
    physical characteristics of, 440
    plasma, 448, 449
    tests, 453
Blood-brain barrier (BBB), 310
Blood-cerebrospinal fluid barrier, 308
Blood clot, 449, 450
Blood clotting. See Blood coagulation
Blood coagulation
    definition of, 449
    extrinsic pathway of, 450
    factors in, 449
    fibrinolysis and, 452
    intrinsic pathway of, 450–452
    prevention of, 452
    reactions of, 449–453
    retraction and, 452
    stages of, 449, 450
    tests, 453
Blood colloidal-osmotic pressure, 672
Blood flow, 492
Blood grouping (typing)
    ABO, 453–455
    Rh, 455, 456

Blood hydrostatic pressure (BHP), 495
Blood island, 524, 525
Blood osmotic pressure (BOP), 495
Blood plasma
    chemical composition of, 449E
    definition of, 448
Blood pressure (BP)
    control of, 493, 494
    definition of, 492
    factors affecting, 492
    homeostasis of, 20
    measurement of, 497, 498
Blood reservoir, 492
Blood substitute, 444
Blood sugar
    control of, 424–426
    homeostasis of, 20, 21
Blood urea nitrogen (BUN), 682
Blood vessels
    arteries, 488, 489
    arterioles, 489, 490
    as blood reservoirs, 492
    capillaries, 490
    of cerebral circulation, 504E
    of coronary (cardiac) circulation, 468
    definition of, 524, 525
    disorders of, 525, 526
    and exercise, 524
    of fetal circulation, 521–524
    of hepatic portal circulation, 520
    medical terminology, 526
    pressure within, 492–494
    of pulmonary circulation, 520
    of pulse, 496, 497
    sinusoid, 490
    of systemic circulation, 498–519
    vascular sinus, 492
    veins, 490, 491
    venules, 490
Blood volume, 492
Blue baby, 483
BMR. See Basal metabolic rate
Body
    of epididymis, 713
    of humerus, 162
    of pancreas, 609
    of penis, 716
    of pubis, 165
    of rib, 155
    of scapula, 160
    of sternum, 152
    of stomach, 603
    of uterus, 721
Body cavity
    abdominal, 13
    abdominopelvic, 13
    cranial, 13
    definition of, 13
    dorsal, 13
    nasal, 556
    oral, 595–600
    pelvic, 13
    pericardial, 13, 462
    peritoneal, 592
    pleural, 13, 562
    thoracic, 13
    ventral, 13
    vertebral, 13
Body heat
    loss of, 656, 657
    measurement of, 656
    production of, 655
    regulation of, 657, 658
Body sections
    cross, 13

frontal, 13
    midsagittal, 13
Body temperature, regulation by skin, 109, 657
Bohr effect, 576
Bolus, 596
Bone marrow
    red, 120
    yellow, 120
Bone marrow transplantation, 447
Bones
    atlas, 150
    auditory ossicles, 388
    axis, 150
    calcaneus, 170
    capitate, 164
    carpus, 164
    cervical vertebrae, 150
    clavicle, 160
    coccyx, 152
    coxal, 165, 166
    cranial, 136–143
    cuboid, 170
    ethmoid, 142
    facial, 143–146
    femur, 166
    fibula, 170
    first (medial) cuneiform, 170
    fontanels of, 136
    frontal, 136, 137
    hamate, 164
    humerus, 160, 162
    hyoid, 146
    incus, 388
    inferior nasal concha, 146
    lacrimal, 145
    as levers, 216
    lower extremity, 166–171
    lumbar vertebrae, 150
    lunate, 164
    malleus, 388
    mandible, 144, 145
    maxilla, 143, 144
    metacarpus, 164
    metatarsus, 170, 171
    nasal, 143
    navicular, of foot, 170
    occipital, 140
    organization into skeleton, 136E
    palatine, 145
    paranasal sinuses of, 144
    parietal, 137
    patella, 166
    pectoral (shoulder) girdle, 159, 160
    pelvic (hip) girdle, 165, 166
    phalanges, of fingers, 164
    phalanges, of toes, 171
    pisiform, 164
    radius, 164
    ribs, 154, 155
    sacrum, 150–152
    scaphoid, 164
    scapula, 160
    second (intermediate) cuneiform, 170
    skull, 133–146
    sphenoid, 141
    stapes, 388
    sternum, 152–154
    surface markings of, 134E
    sutures of, 134
    talus, 170
    tarsus, 170
    temporal, 137, 140
    third (lateral) cuneiform, 170
    thoracic vertebrae, 150
    thorax, 152–155
    tibia, 168

trapezium, 164
trapezoid, 164
triquetral, 164
types of, 133
ulna, 162, 164
upper extremity, 160–164
vertebral column, 146–152
vomer, 146
zygomatic, 144
Bone tissue. *See* Osseous tissue
Bony labyrinth, 389
Borders
    of heart, 461
    of scapula, 160
Botulism, 629
Bowman's capsule. *See* Glomerular capsule
Bowman's gland. *See* Olfactory gland
Boyle's law, 566
Brachial plexus, 296, 297, 298E
Bradycardia, 497
Bradykinesia, 332
Brain
    association areas, 322
    basal ganglia, 318, 319
    blood-brain barrier (BBB) of, 310
    blood supply, 310, 504E
    brain stem, 310–314
    brain waves, 323
    cerebellum, 323–325
    cerebrospinal fluid (CSF), 307–310
    cerebrum, 316–323
    cranial meninges, 307
    cranial nerves and, 327–329
    development of, 328–332
    diencephalon, 314–316
    disorders of, 322–335
    electrical activity mapping, 334
    hypothalamus, 314–316
    injuries to, 321
    lateralization and, 323
    limbic system, 319, 320
    lobes of, 318
    medical terminology, 335
    medulla oblongata, 310–313
    midbrain, 313, 314
    motor areas, 322
    neurotransmitters, 325–327
    parts of, 307
    pons, 313
    protection, 307
    reticular formation, 312
    sensory areas, 321, 322
    thalamus, 314
    ventricles of, 308
    white matter, 281, 318
Brain death, 323
Brain electrical activity mapping (BEAM), 334
Brain lateralization, 323
Brain sand, 428
Brain stem, 310–314
Brain tumor, 332
Brain wave, 323
Breast. *See* Mammary gland
Breast cancer, 729, 730, 738
Breast cancer screening indicator, 730
Breast feeding, advantages of, 760
Breathing. *See* Pulmonary ventilation
Bridge of nose, 556
Brim of pelvis, 165
Broad ligament, 721
Broca's area of cerebrum. *See* Motor speech
    area
Bronchi
    alveolar duct, 564
    bronchioles, 560
    development of, 583

primary, 560
    respiratory, 564
    secondary (lobar), 560
    terminal, 560
    tertiary (segmental), 560
Bronchial asthma, 584
Bronchial tree, 560
Bronchiole, 560
Bronchitis, 585
Bronchogenic carcinoma, 583, 584
Bronchogram, 560
Bronchography, 560
Bronchopulmonary segment, 564
Bronchoscopy, 560
Brunner's gland. *See* Duodenal gland
Buccal gland, 596
Buffer system, 36, 37, 698, 699
    carbonic acid-bicarbonate, 36, 37, 698
    hemoglobin-oxyhemoglobin, 698, 699
    phosphate, 698
    protein, 699
Bulb
    of hair, 106
    of penis, 717
Bulbar conjunctiva, 376
Bulbar polio, 332
Bulbourethral gland, 715
Bulbus cordis, 480
Bulimia, 629
Bullae, 111
Bundle branch, 470
Bundle of His. *See* Atrioventricular (AV) bundle
Bunion, 171
Burkitt's lymphoma, 71
Burn
    classification of, 111, 112
    definition of, 111
    Lund-Browder method, 111, 112
    systemic effects of, 111
    treatment of, 112
Bursae
    bursitis, 178, 191, 192
    definition of, 178
    infrapatellar, 190
    lateral, 190
    medial, 190
    prepatellar, 190
    subacromial, 186
    subcoracoid, 186
    subdeltoid, 186
    subscapular, 186
    suprapatellar, 190
Bursitis, 178, 191, 192
Buttocks
    intramuscular injection of, 255
    musculature of, 248E, 249

Caisson disease. *See* Decompression sickness
Calcaneal tendon, 252E, 253
Calcaneus, 170
Calcification, 123, 124
Calcitonin (CT), 416
Calcium
    distribution of, 696
    functions of, 696
    regulation of, 696
Calcium-ATPase, 202
Calculus
    biliary, 613, 628
    renal, 684
Calf, 252E, 253
Callus, 113, 130
Calmodulin, 403
Calorie, 655
Calorigenic effect, 414

Calorimeter, 655
Calsequestrin, 202
Calyx, 666
Canaliculi, 94, 121
Canal of Schlemm. *See* Scleral venous sinus
Cancellous bone tissue. *See* Spongy bone
Cancer
    of adrenal gland, 423
    antibodies and, 549
    antigens and, 549
    of brain, 332
    of breasts, 729, 730, 738
    Burkitt's lymphoma, 71
    causes of, 71
    definition of, 69, 70
    of GI tract, 628
    Laetrile and, 71
    of larynx, 559
    of lung, 574
    malignant tumor, 69, 70
    metastasis, 70
    nasopharyngeal, 71
    primary tumor, 70
    of prostate gland, 727
    of skin, 104, 111
    types of, 70, 71
    of uterus, 721, 738
Capacitation, 743
Capillaries, 490
    continuous, 490
    fenestrated, 490
    true, 490
Capitate, 164
Capitulum, 162
Capsular hydrostatic pressure, 672
Capsule
    adipose, 666
    articular. *See* Articular capsule
    fibrous, 177
    of kidney, 666
    of lymph node, 531
    of spleen, 534
    of thymus gland, 536
Carbaminohemoglobin, 578
Carbohydrate
    absorption of, 618
    definition of, 37
    digestion of, 618
    disaccharide, 37
    fate of, 636
    functions of, 37
    metabolism of, 636–643
    monosaccharide, 37
    polysaccharide, 38
    structure of, 37, 38
Carbon dioxide
    chloride shift and, 578
    and hemoglobin, 578
    and respiratory control, 581
    transport of, 578
Carbonic acid-bicarbonate buffer, 36, 37, 698
Carbonic anhydrase, 576
Carbon monoxide poisoning, 578
Carboxypolypeptidase, 611, 618
Carbuncle, 113
Carcinogen, 71
Carcinoma, 70
Cardiac arrest, 484
Cardiac catheterization, 479
Cardiac center, 475
Cardiac cycle, 472–474
Cardiac muscle tissue
    definition of, 195
    development of, 210, 211
    histology of, 206, 208
Cardiac notch, 564

Cardiac output (CO), 474, 475, 492
Cardiac reserve, 475
Cardiac tamponade, 462
Cardia of stomach, 603
Cardinal ligament, 721
Cardioaccelerator center (CAC), 475
Cardioinhibitory center (CIC), 475
Cardiology, 461
Cardiomegaly, 484
Cardiopulmonary resuscitation (CPR), 586
Cardiovascular system, 8E, 440–528
    adjustments of infant at birth, 759
    aging and, 524
    arteries, 488–510
    blood, 440–456
    blood pressure, 492–498
    capillaries, 490
    circulatory routes, 498–524
    coronary (cardiac) circulation, 468, 469, 498
    definition of, 440
    development of, 524, 525
    disorders of, 456, 457, 481–484, 525, 526
    exercise and, 524
    fetal circulation, 521–524
    formed elements, 440–448
    heart, 461–484
    hepatic portal circulation, 520
    interstitial fluid, lymph and, 456
    medical terminology, 457, 458, 484, 526
    physiology of circulation, 492–498
    plasma, 448, 449E
    pulmonary circulation, 520
    pulse, 496
    surgery and, 478, 479
    systemic circulation, 498, 519
    veins, 490, 510–519
Carina, 560
Carotene, 104
Carotid body, 494, 581
Carotid sinus, 475, 476
Carotid sinus reflex, 475, 476
Carpal tunnel syndrome, 297
Carpus, 164
Cartilage
    articular, 94, 120, 125, 177
    arytenoid, 558
    corniculate, 558
    costal, 94, 154
    cricoid, 558
    cuneiform, 558
    definition of, 93
    development of, 93, 94
    epiglottis, 538, 558
    growth of, 94
    thyroid, 558
    tracheal, 560
    types of, 93, 94
Cartilaginous joint
    definition of, 176
    types of, 176, 177
Cast, 684
Castration, 731
Catabolism
    of carbohydrates, 636–643
    definition of, 31, 634
    of lipids, 644–646
    of proteins, 646
Catalyst, 635
Cataract, 379, 394
Catheter, 479
Cation, 28
Cauda equina, 281
Caudate lobe of liver, 611
Caudate nucleus, 318
Cautery, 763

Cavitron ultrasonic surgical aspirator (CUSA), 313
Cecum, 622
Celiac disease, 659
Cell
    adipocyte, 93
    and aging, 71, 72
    alpha, 424
    alveolar macrophage, 564
    B, 545
    beta, 424
    and cancer, 69–71
    cytoplasm of, 53
    definition of, 6, 46
    delta, 424
    division of, 64–69
    enteroendocrine, 605
    extracellular materials of, 61, 62
    gametes, 67, 708–711, 720
    generalized animal, 46
    hepatic, 611
    inclusions of, 61
    interstitial endocrinocyte, 708
    medical terminology, 72
    mucous, 605
    organelles of, 53–61
    parietal, 605
    parts of, 46
    plasma membrane of, 46–53
    spermatozoon, 708, 711
    stellate reticuloendothelial, 490, 611
    sustentacular, 708
    T, 545–547
    zymogenic, 605
Cell body, of neuron, 264
Cell division
    abnormal, 69–71
    definition of, 46, 64
    somatic, 64–67
    types of, 64, 65
Cell inclusion
    definition of, 61
    types of, 61
Cell function, 77
Cell membrane. See Plasma membrane
Cellular immunity, 544–547
Cellular respiration, 634, 636
Cementum, 598
Center, of reflex arc, 287
Center of ossification, 123
Central canal
    of bone, 94, 120
    of spinal cord, 284
Central fovea, 379
Central nervous system, 281–337
    brain, 307–327
    cerebrospinal fluid of, 307–310
    cranial nerves, 327–329
    definition of, 262
    development of, 328–332
    dermatomes of, 302
    disorders of, 302–304, 332–335
    grouping of neural tissue, 281
    medical terminology, 335
    neurotransmitters, 325–327, 363
    plexuses of, 294–300
    protection and coverings, 281, 284, 307
    reflexes and, 287–292
    spinal cord, 281–304
    spinal nerves, 292–302
    spinal puncture, 284
    tracts of, 285, 286E, 287E
Central sulcus, 318
Central vein, 611
Centriole, 60
Centromere, 65

Centrosome, 60
Cephalgia
    extracranial, 334
    intracranial, 334
Cerebellar cortex, 323
Cerebellar hemisphere, 323
Cerebellar nucleus, 324
Cerebellar peduncle
    inferior, 324
    middle, 324
    superior, 324
Cerebellum, 323–325
Cerebral aqueduct, 308
Cerebral arterial circle, 504
Cerebral circulation, 498
Cerebral cortex, 316
Cerebral hemisphere, 318
Cerebral nucleus, 281, 318, 319
Cerebral palsy, 332
Cerebral peduncle, 313
Cerebral sulcus
    central, 318
    definition of, 318
    lateral, 318
    parietooccipital, 318
Cerebrospinal fluid (CSF), 307–310
Cerebrovascular accident (CVA), 333
Cerebrum
    association areas, 322
    basal ganglia, 318, 319
    brain lateralization, 323
    brain waves, 323
    convolutions, 316
    cortex of, 316
    fissures, 316
    lesions of, 319
    limbic system, 319, 320
    lobes, 318, 319
    motor areas, 322
    sensory areas, 321, 322
    sulci, 318
    white matter, 218, 318
Cerumen, 108, 387
Ceruminous gland, 108, 386, 387
Cervical canal, 721
Cervical cancer, 738
Cervical dysplasia, 738
Cervical enlargement, 281
Cervical plexus, 294, 295E
Cervical vertebrae, 150
Cervix, of uterus, 721
Cesarean section, 758
C fiber, 272
Chalazion, 376
Chambers of heart, 462, 463
Chancre, 736
Charleyhorse, 211
Cheek, 595
Chemical bond
    covalent, 28–30
    hydrogen, 30
    ionic, 27, 28
Chemical compound, 27
    inorganic, 33–37
    organic, 37–43
Chemical digestion, 590
Chemical element, 25, 34E
Chemical exfoliation, 104
Chemical method, of contraception, 732
Chemical reaction
    and covalent bonds, 28–30
    decomposition, 31
    and energy, 33
    exchange, 31
    and hydrogen bonds, 30
    and ionic bonds, 27, 28

reversible 31, 32
    synthesis, 31
Chemical symbol, 25
Chemical theory, of olfaction, 371
Chemical thermogenesis, 657
Chemonucleolysis, 155, 156
Chemoreceptor, 340, 494, 581
Chemosensitive area, 581
Chemotactic agent, 539
Chemotaxis, 445, 541
Chest. See Thorax
Cheyne-Stokes respiration, 586
Chill, 658
Chloride
    control of, 696
    deficiency of, 696
    distribution of, 696
    functions of, 696
Chloride shift, 578
Cholecystectomy, 628
Cholecystitis, 629
Cholecystokinin (CCK), 326, 327, 608E
Cholelithiasis, 629
Cholesterol, 481
Cholinergic fiber, 363
Chondriotin sulfate, 61
Chondroblast, 123, 126
Chondrocyte, 94
Chondrosarcoma, 71
Chordae tendineae, 465
Chorion, 749
Chorionic villi, 750
Chorionic villi sampling (CVS), 756
Choroid, 376, 377
Choroid plexus, 308
Chromaffin cell, 423
Chromatid, 65
Chromatin, 54, 65
Chromatolysis, 302
Chromatophilic substance, 264
Chromosomal microtubule, 66
Chromosome, 54, 65
Chromosome number, 708
Chronic prostatitis, 737
Chvostek sign, 417–419
Chylomicron, 621
Chyme, 605
Chymotrypsin, 618
Chymotrypsinogen, 618
Cigarette smoking, 583
Cilia, 61, 83, 538
Ciliary body, 377
Ciliary muscle, 377
Ciliary process, 377
Cine CT, 17
Circadian rhythm, 352
Circle of Willis. See Cerebral circulation
Circuits, 276–278
Circular fasciculus, 218
Circulation time, 496
Circulatory routes, 498–524
Circulatory shock, 477, 478
Circumcision, 717
Circumduction, 181
Circumvallate papilla, 372, 373, 596
Cirrhosis, 628
Cisterna
    of endoplasmic reticulum, 56
    of Golgi complex, 56
Cisterna chyli, 537, 550
Citric acid cycle. See Krebs cycle
Claustrum, 319
Clavicle, 160
Clawfoot, 171
Cleavage, 744
Cleavage furrow, 67

Cleft lip, 144
Cleft palate, 144
Climacteric, 726
Climax. See Orgasm
Clinical perineum, 728
Clitoris, 727
Cloaca, 687
Cloacal membrane, 625
Clot retraction, 452
Clotting time, 453
CNS. See Central nervous system
Coagulation factors, 449
Coarctation of aorta, 483
Coccyx, 152
Cochlea, 390
Cochlear duct, 390
Codon, 62
Coenzyme, 635
Coenzyme A (CoA), 638
Cofactor, 635
Coitus. See Sexual intercourse
Cold sore, 113
Colitis, 629
Collagen, 62, 88
Collagen implant, 104
Collagenous fiber, 62, 88
Collateral circulation, 489
Collecting duct, 667
Collision theory, 32, 33
Colon, 622
Colonoscopy, 628
Color blindness, 384, 762
Color vision, 384
Colostomy, 629
Colostrum, 760
Colposcopy, 722
Column
    anterior white, 284
    definition of, 281
    lateral white, 284
    posterior white, 284
Columnar cell, 78
Coma, 335
Comedo, 108, 113
Commissural fiber, 318
Commissure
    anterior white, 284
    gray, 284
Common baldness, 108
Common bile duct, 611
Common cold. See Coryza
Common hepatic duct, 611
Compact bone, 120–123
Compensation, 477, 478
Complement, 539
Complete antigen, 542
Complete blood count (CBC), 448
Complete tetanus, 204
Complete transection, 302
Compliance, 484, 569
Compound gland, 85
Computed tomography (CT), 17
Computed tomography (CT) scan, 17
Concentration gradient, 48
Concha
    inferior nasal, 146
    middle nasal, 142
    of pinna, 387
    superior nasal, 142
Concussion, 321
Condom, 732
Conduction, 268–275
Conduction, of heat, 656
Conduction myofiber, 470
Conduction pathway, 287
Conduction system, of heart, 469, 470

Conductivity, 9
Condyle
    definition of, 134E
    lateral, of femur, 166
    lateral, of tibia, 168
    medial, of femur, 166
    medial, of tibia, 168
    occipital, 140
Cone, 379
Cone biopsy, 721
Congenital heart defect, 482
Congestive heart failure (CHF), 484
Conjunctiva, 376
Conjunctivitis, 394
Connecting neuron. See Association neuron
Connective tissue
    classification of, 86
    definition of, 85, 86
    loose and dense, 86, 88, 91, 93
    types of, 86–94
Connective tissue proper, 86, 88–93
Consciousness, 353
Constipation, 625
Constriction of pupil, 382
Contact inhibition, 70, 105
Continuous ambulatory peritoneal dialysis
        (CAPD), 680
Continuous capillary, 490
Continuous conduction, 271
Continuous microtubule, 66
Contraception, 731, 732
Contraceptive diaphragm, 732
Contraceptive sponge, 732
Contraceptive techniques
    abstinence, 731
    condom, 732
    diaphragm, 732
    intrauterine device (IUD), 732
    oral contraception, 732
Contractility, 9, 195
Contraction, muscular, 199–206
Contraction period, 203
Contralateral reflex arc, 291
Contusion, 321
Conus medullaris, 281
Convection, 656
Convergence, 272, 277, 382
Convergent fasciculus, 218
Converging circuit, 277
Convolution, 316
Convulsion, 212
Cooper's ligament. See Suspensory ligament
        of the breast
Copulation. See Sexual intercourse
Cord, of nerve plexus, 297
Cordotomy, 342
Corn, 113
Cornea, 376
Corneal transplant, 376
Corniculate cartilage, 558
Corona of penis, 717
Corona radiata, 743
Coronary artery, 463, 468, 469, 503E
Coronary artery disease (CAD), 481, 525
Coronary artery spasm, 483
Coronary artery bypass surgery, 482
Coronary (cardiac) circulation, 468, 469, 498
Coronary sinus, 463, 469, 498, 510E
Coronary sulcus, 463
Corpora cavernosa penis, 716
Corpora quadrigemina, 313
Cor pulmonale, 484
Corpus albicans, 725
Corpus callosum, 318
Corpuscle of touch, 103, 341
Corpus hemorrhagicum, 725

Corpus luteum, 720
Corpus spongiosum penis, 716
Corpus striatum, 318
Cortex
    adrenal, 419–423
    of cerebellum, 323
    of cerebrum, 316
    of hair shaft, 106
    of kidney, 666
    of lymph node, 532
    of ovary, 719
    of thymus gland, 536
Cortical nephron, 666
Cortical sinus, 532
Corticobulbar tract, 350
Corticolipotroph cell, 404
Corticosterone, 421
Corticotropin releasing factor (CRF), 408, 422
Cortisol (hydrocortisone), 421
Cortisone, 421
Coryza, 585
Costal breathing, 567
Costal cartilage, 94, 154
Countercurrent mechanism, 679
Covalent bond, 28–30
Covering and lining epithelium, 77–84
Cowper's gland. See Bulbourethral gland
Coxal bone, 165, 166
Coxal (hip) joint, 187–188
Cramp, muscular, 212
Cranial bones, 133–142
Cranial cavity, 13
Cranial meninges, 307
Cranial nerve damage, 327–329
Cranial nerves, 327–329
Cranial parasympathetic outflow, 362
Craniosacral outflow, 358
Craniotomy, 130
Creatinine, 672
Cremaster muscle, 707
Crenation, 50
Crest
    definition of, 134E
    iliac, 165
    interarticular, 154
    lateral sacral, 152
    median sacral, 152
Cretinism, 415, 416
Cribriform plate, 142
Cricoid cartilage, 558
Crisis, 658
Crista
    of inner ear, 393
    of mitochondrion, 58
Crista galli, 142
Cross bridge, 198
Crossed extensor reflex, 291
Crossing-over, 69, 710
Crown, of tooth, 598
Crura, of penis, 717
Crutch palsy, 297
Crypt of Lieberkühn. See Intestinal gland
Cryptorchidism, 707
CT scan. See Computed tomography scan
Cuboid, 170
Cuboidal cell, 78
Culdoscopy, 763
Cuneiform bones, 170
Cuneiform cartilage, 558
Cupula, 393
Curvature, 156
Curves of vertebral column, 149
Cushing's syndrome, 422, 423
Cuspid (canine), 599
Cutaneous membrane, 96, 101–104
Cutaneous sensation
    definition of, 340

pain, 342, 343
pressure, 341, 342
thermoreceptive, 342
touch, 341
and two-point discrimination, 340
vibration, 342
Cuticle
    of hair follicle, 106
    of hair shaft, 106
    of nail, 109
Cutis. See Skin
Cyanosis, 483, 526
Cyclic adenosine-3',5'-monophosphate
    chemical composition of, 42
    and hormones, 401–403
    and impulse transmission, 274
Cyclic AMP. See Cyclic adenosine-3',5'-mono-
    phosphate
Cyst, 113
Cystic duct, 611
Cystic fibrosis, 659
Cystitis, 688
Cystoscopy, 685, 686
Cytochrome, 640, 641
Cytochrome system. See Electron transport
    chain
Cytokinesis, 67
Cytology, 6, 46
Cytoplasm
    composition of, 53
    definition of, 46, 53
    functions of, 53
Cytosine, 41
Cytoskeleton, 59, 60

Dalton's law, 572
Damping, 351
Dark adaptation, 384
Dartos, 706, 707
Dead air volume, 570
Deafness
    conduction, 394
    nerve, 394
Deamination, 646
Decarboxylation, 639
Decibel (dB), 390
Decidua, 750
Decidua basalis, 750
Decidua capsularis, 750
Decidua parietalis, 750
Deciduous teeth, 598, 599
Decompression sickness, 573
Decubitus ulcer, 111
Decussation of pyramids, 311
Deep fascia, 195
Deep inguinal ring, 714
Deep venous thrombosis (DVT), 526
Defecation, 624–625
Defensin, 445
Deflection waves, 470
Degeneration of nerve cells, 303
Deglutition, stages of, 600, 601
Deglutition center, 600
Dehydration, 693
Dehydration synthesis, 37
    of carbohydrate, 37
    of fat, 38
    of protein, 40
Delayed hypersensitivity T cell, 546
Delayed nerve grafting, 302
Delta cell, 424
Delta wave, 323
Demineralization, 128
Dendrite, 264
Dense body, 208
Dense bone tissue. See Compact bone
Dense (collagenous) connective tissue, 93

Dental caries, 627
Dental plaque, 627
Denticulate ligament, 284
Dentin, 598
Dentition, 598–600
Deodorant, 113
Deoxygenated blood, 573
Deoxyribonuclease, 611, 617, 618
Deoxyribonucleic acid
    in cell division, 65–67
    chemical composition of, 40, 41
    and chromatin, 65
    and chromosomes, 65
    functions of, 40, 41
    and protein synthesis, 62–64
    recombinant, 64
    repair of, 64
    structure of, 40, 41
Depilatory, 107
Depolarized membrane, 269, 270
Depression, 181
Depressor muscle, 219E
Dermabrasion, 113
Dermal papilla, 103
Dermatology, 101
Dermatome, 113, 302
Dermis
    definition of, 101, 103
    development of, 110
    papilla of, 103
    receptors of, 103
    regions of, 103
Descending aorta, 498
Descending colon, 622
Descending limb of loop of nephron (Henle),
    666
Descending tract, 281, 284, 287E
Detached retina, 379
Deterioration, 72
Detrusor muscle, 685
Development
    birth control, 731–733
    disorders of, 759, 760
    embryonic period, 747–751
    embryo transfer, 746
    external human fertilization, 746
    fertilization, 743–744
    fetal growth, 751–752
    gamete formation, 708–711, 720
    gamete intrafallopian transfer (GIFT), 746
    gestation, 753, 755
    hazards to embryo and fetus, 759–760
    hormones, 752–753
    implantation, 744, 746
    lactation, 760
    medical terminology, 763
    pregnancy, 743–753
    sexual intercourse, 730, 731
    transvaginal oocyte retrieval, 746
Developmental anatomy, 6, 743–755
    of adrenal cortex, 429
    of adrenal medulla, 429
    of anus, 625
    of blood cells, 525
    of blood vessels, 525
    of bone tissue, 127
    of brain, 328–332
    of bronchi, 583
    of cardiac muscle, 211
    of cardiovascular system, 480, 524–525
    of cartilage, 126
    of dermis, 110
    of digestive system, 625, 627
    of endocrine system, 428, 429
    of epidermis, 110
    of esophagus, 625
    of external genitals, 735

of extremities, 127, 128
of gallbladder, 627
of gonads, 733, 735
of hair follicles, 110
of heart, 480
of integumentary system, 110
of kidneys, 687
of large intestine, 627
of larynx, 583
of liver, 627
of lower extremities, 128
of lungs, 583
of lymphatic system, 549–550
of lymphatic vessels, 549–550
of lymph nodes, 549–550
of muscular system, 210–211
of nails, 110
of nerves, 328–332
of nervous system, 328–332
of oral cavity, 625
of ovaries, 735
of pancreas, 429, 627
of parathyroid glands, 429
of pharynx, 625
of pineal gland, 429
of pituitary gland, 428
of reproductive system, 733–735
of respiratory system, 583
of salivary glands, 627
of sebaceous glands, 110
of skeletal muscle, 210
of skeletal system, 127, 128
of small intestine, 627
of smooth muscle, 211
of stomach, 625
of sudoriferous glands, 110
of testes, 735
of thymus gland, 429
of thyroid gland, 429
of trachea, 583
of upper extremities, 127–128
of ureter, 688
of urethra, 688
of urinary bladder, 688
of urinary system, 686–688
of vertebral column, 128
Deviated nasal septum, 146
Dextran, 627
Diabetes insipidus, 412
Diabetes mellitus, 426–427
  type I, 427
  type II, 427
Diabetogenic effect, 405
Dialysis, 50
Diapedesis, 445, 541
Diaphragm
  abdominopelvic, 13, 234E
  contraceptive, 732
  pelvic, 235E
  urogenital, 236E
Diaphragmatic breathing, 567
Diaphragmatic surface, of spleen, 534
Diaphysis, 120
Diarrhea, 625
Diarthrosis, 176
Diastole, 472
Diastolic blood pressure, 498
Diencephalon, 314–316, 329
Dietary fiber, 629
Differential count, 447
Differentiation, 9
Diffuse lymphatic tissue, 530
Diffusion, 48, 49
Digestion
  definition of, 8, 37, 590
  in large intestine, 624
  in mouth, 599, 600

phases of, 590
  in small intestine, 617, 618
  in stomach, 605–607
Digestive system, 8E, 590–629
  absorption by, 618–621
  aging and, 625
  definition of, 590
  development of, 625–627
  disorders of, 627–629
  general histology, 591, 592
  medical terminology, 629
  organization of GI tract, 590, 591
  organs of, 595–625
  peritoneum of, 592–595
Digital substraction angiography (DSA), 489
Diphtheria, 586
Diplegia, 302
Diploid cell, 67
Diplopia, 327E
Directional terms, 9E, 10E
Direct transfusion, 457
Disaccharide, 37
Discharge zone, 276
Discriminative touch, 341
Dislocation, 192
  of hip, 188
  of jaw, 145
  of shoulder, 187
Disorders
  of articulations, 191, 192
  of brain, 332–335
  of the cardiovascular system, 456, 457, 481–
    484, 525, 526
  of the digestive system, 627–629
  of the endocrine system, 407, 408, 412, 415–
    417, 421–423, 426, 427
  of the integumentary system, 110–113
  of the lymphatic system and immunity, 550–
    552
  of metabolism, 658–660
  of muscle tissue, 211, 212
  of the reproductive system, 735–738
  of the respiratory system, 583–586
  of the skeletal system, 155, 156
  of skeletal tissue, 128–130
  of special senses, 394, 395
  of the spinal cord, 302–304
  of the urinary system, 688, 689
Dissociation. See Ionization
Distal convoluted tubule, 666
Diuresis, 681
Diuretic, 681
Divergence in synapses, 272, 277
Diverging circuits, 277
Diverticulitis, 628
Diverticulosis, 628
Dizygotic (fraternal) twins, 744
DNA. See Deoxyribonucleic acid
Dominant gene, 761
Dopamine (DA), 325
Doppler detector, 755
Dorsal body cavity, 13
Dorsal ramus, 294
Dorsiflexion, 181
Dorsum nasi, 556
Down's syndrome, 756
Dual innervation, 357
Duct
  accessory, 610
  cochlea, 390
  collecting, 667
  common bile, 611
  common hepatic, 611
  cystic, 611
  ductus (vas) deferens, 713, 714
  ductus epididymis, 713
  efferent, 713

ejaculatory, 714
excretory lacrimal, 376
hepatic, 611
lactiferous, 728
mammary, 728
nasolacrimal, 376
pancreatic, 610
papillary, 667
parotid, 596
rete testis, 713
straight tubules, 713
sublingual, 596
submandibular, 596
urethra, 684, 685
Ductless glands. See Endocrine system
Duct of Santorini. See Accessory duct
Duct of Wirsung. See Pancreatic duct
Ductus arteriosus, 523
Ductus (vas) deferens, 713–714
Ductus epididymis, 713
Ductus venosus, 523
Duodenal gland, 616
Duodenal papilla, 610
Duodenal ulcer, 627
Duodenum, 614
Dura mater, 281, 307
Dynamic equilibrium, 392, 393
Dynamic spatial reconstructor (DSR), 17
Dynorphin, 326
Dyskinesia, 395
Dyslexia, 334, 335
Dysmenorrhea, 737
Dysphagia, 629
Dysplasia, 72
Dyspnea, 586
Dystocia, 757
Dysuria, 689

Ear
  artificial, 391
  disorders of, 394, 395
  and equilibrium, 392, 393
  external, 386, 387
  and hearing, 390, 392
  internal, 389, 390
  middle, 388
  surface anatomy of, 387
Eardrum. See Tympanic membrane
Eccrine sweat gland, 108
ECF. See Extracellular fluid
ECG. See Electrocardiogram
Echocardiography, 479
Eclampsia, 763
Ectoderm
  derivatives of, 747E
  development of, 747
Ectopic focus, 484
Ectopic pregnancy (EP), 721
Eczema, 113
Edema, 537, 697
EEG. See Electroencephalogram
Effective filtration pressure (P_{eff}), 495, 673, 697
Effector, 288
Efferent duct, 713
Efferent lymphatic vessel, 532
Efferent nervous system, 262
Efferent neuron, 199, 267
Ejaculation, 730
Ejaculatory duct, 714
Ejection period, 472
Elastic artery, 489
Elastic cartilage, 94
Elastic connective tissue, 93
Elastic fiber, 62, 88, 89
Elasticity, 195
Elastin, 62
Elbow, 162, 163

Electrocardiogram (ECG or EKG)
    definition of, 470
    deflection waves of, 470
    interpretation of, 470
    use in diagnosis, 470
Electroconvulsive therapy (ECT), 335
Electroencephalogram (EEG), 323
Electrogenic pump, 269
Electrolysis, 107
Electrolyte, 34, 694–697
    absorption by small intestine, 621
    concentration of, 694
    distribution of, 694
    effects on osmosis, 694
    functions of, 694–697
Electromagnetic receptor, 340
Electromyography (EMG), 212
Electron micrograph, definition of, 46
Electron transport chain, 640–642
Eleidin, 102
Elephantiasis, 552
Elevation, 181
Ellipsoidal joint, 182
Embolus, 333, 452
Embryo
    definition of, 86, 747
    development of, 747–751
    fertilization of, 743, 744
    implantation of, 744, 746
    membranes of, 747, 749, 750
    placenta of, 750
    potential hazards to, 759, 760
    primary germ layers of, 747E
Embryology, 6, 747
Embryonic connective tissue, 86
Embryonic disc, 747
Embryonic membranes
    allantois, 749, 750
    amnion, 749
    chorion, 749
    yolk sac, 749
Embryonic period
    definition of, 747
    embryonic membranes, 747, 749, 750
    placenta and umbilical cord, 750, 751
    primary germ layers, 747E
Embryo transfer, 746
Emission, 730
Emmetropic eye, 382
Emphysema, 584
Emulsification, 612, 618
Enamel, 598
End artery, 489
End-diastolic volume (EDV), 474
Endocardium, 462
Endochondral ossification, 123–126
Endocrine gland, 84, 399, 404–428
Endocrine system, 8E, 398–433
    aging and, 430
    control of hormonal secretions, 403, 404
    development of, 428–430
    definition of, 8E, 399
    disorders of, 407, 408, 412, 415–417, 421–423, 426–427
    glands of, 404–428
    hormones of, 404–429
    mechanism of hormonal action, 399–403
    medical terminology, 433
    prostaglandins and, 403
    and stress, 430–433
Endocrinology, 398
Endocytosis, 51
    phagocytosis, 51
    pinocytosis, 51
    receptor-mediated, 51–53

Endoderm
    derivatives of, 747E
    development of, 747
Endodontics, 598
Endogenous pyrogen, 658
Endolymph, 389
Endometriosis, 738
Endometrium, 722
Endomysium, 196
Endoneurium, 293
Endoplasmic reticulum (ER)
    agranular, 56
    granular, 56
End organ of Ruffini. See Type II cutaneous
        mechanoreceptor
Endorphin, 325, 326
Endoscopy, 605
Endosome, 53
Endosteum, 120
Endothelial-capsular membrane, 666
Endothelial tube, 480
Endothelium, 78
End-systolic volume (ESV), 474, 475
Energy
    and chemical reactions, 33
    electrical, 33
    heat, 655
    kinetic, 33
    mechanical, 33
    potential, 33
    radiant, 33
Energy level, 26, 27
Engram, 351
Enkephalin, 325
Enlarged prostate, 737
Enteric gastrin, 608
Enteroceptor. See Visceroceptor
Enteroendocrine cell, 605
Enterogastric reflex, 608
Enterohepatic circulation, 619, 621
Enterokinase, 611
Enuresis, 689
Enzymes
    acetylcholinesterase (AChE), 202, 275
    active site of, 635
    adenylate cyclase, 43, 401
    carboxypolypeptidase, 611, 618
    catalase, 59
    cholinesterase, 202, 275
    chymotrypsin, 618
    definition of, 635
    dehydrogenase, 635
    deoxyribonuclease, 611, 617, 618
    enterokinase, 611
    enzyme-substrate complex, 635
    function of, 635
    gastric lipase, 606
    hyaluronidase, 88
    lactase, 617, 618
    maltase, 617, 618
    nuclease, 618
    pancreatic amylase, 611, 618
    pancreatic lipase, 611, 618
    pepsin, 605–618
    peptidase, 617, 618
    renal erythropoietic factor, 430
    rennin, 607
    ribonuclease, 611, 617, 618
    salivary amylase, 599, 618
    sucrase, 617, 618
    trypsin, 611, 618
    tyrosinase, 104
Enzyme-substrate complex, 635
Eosinophil, 445
Ependyma, 264E
Ependymal layer, of neural tube, 328

Epicardium, 462
Epicondyle
    definition of, 134E
    medial, of femur, 166
    medial, of humerus, 162
    lateral, of femur, 166
    lateral, of humerus, 162
Epidemiology, 72
Epidermal derivatives
    ceruminous glands, 108
    development of, 110
    hair, 106–108
    nails, 109
    sebaceous glands, 108
    sudoriferous glands, 108
Epidermal groove, 104
Epidermal ridge, 104
Epidermis
    definition of, 101
    development of, 110
    grooves of, 104
    layers of, 101–103
    pigments of, 104
    ridges of, 104
Epididymis, 713
Epidural space, 284
Epiglottis, 538, 558
Epilepsy, 333
Epimysium, 196
Epinephrine, 423
Epineurium, 293
Epiphyseal line, 125
Epiphyseal plate
    growth of, 125
    zones of, 125
Epiphysis, 120
Epiphysis cerebri. See Pineal gland
Epiploic appendage, 624
Episiotomy, 728
Epithelial membrane, 96
Epithelial tissue
    arrangement of layers, 77
    basement membrane of, 77
    cell shapes, 77, 78
    classification, 78
    covering and lining, 77–84
    definition of, 77
    glandular, 84, 85
    subtypes, 77–84
Epithelioreticular tissue, 536
Eponychium, 109
Epstein-Barr virus (EBV), 71, 457
Equatorial division, 69
Equilibrium, 48, 392, 393
ER. See Endoplasmic reticulum
Erection, 717, 730, 731
Erythema, 113
Erythroblastosis fetalis. See Hemolytic disease
        of newborn
Erythrocyte
    anemia and, 445
    functions of, 444
    hematocrit, 445
    lifespan, 444
    number per cubic millimeter, 444
    origin of, 444, 445
    production of, 444, 445
    reticulocyte count, 445
    structure of, 442, 444
Erythropoiesis, 444, 445
Erythropoietin, 430, 444
Esophageal hiatus, 601
Esophagus, 600–602
Estrogens, 427, 722, 752
Ethmoid bone, 142
Etiology, 128

Eupnea, 567
Eustachian tube. *See* Auditory tube
Evaporation, 656, 657
Eversion, 181
Exchange transfusion, 457
Excitability, 8, 195, 269–271
Excitatory postsynaptic potential (EPSP), 273, 274
Excitatory transmitter-receptor interaction, 273, 274
Excretory lacrimal duct, 376
Exercise and cardiovascular health, 524
Exhaustion reaction, of stress, 432, 433
Exocrine gland
    definition of, 84, 399
    functional classification of, 85
    structural classification of, 84, 85
Exocytosis, 51
Exogenous pyrogen, 658
Exophthalmic goiter, 416
Exophthalmos, 416
Expiration, 569
Expiratory reserve volume, 571
Extensibility, 195
Extension, 179
Extensor muscle, 219E
External anal sphincter, 622
External auditory canal, 386
External ear, 386, 387
External human fertilization, 746
External medullary lamina, 314
External nares, 556
External nose, 556
External os, 721
External respiration, 573–575
External root sheath, 106
External sphincter, of urinary bladder, 685
Exteroceptor, 340
Extracellular fluid (ECF), 17, 693
Extracellular materials
    definition of, 46, 61
    types of, 61, 62
Extracorporeal shock wave lithotripsy (ESWL), 684
Extraembryonic coelom, 747
Extrafusal fiber, 344
Extrapyramidal pathway, 350
Extrinsic blood clotting pathway, 450
Extrinsic muscle
    of eyeball, 380
    of tongue, 596
Exudate, 540
Eyeball
    accessory structures of, 374–376
    afferent pathway to brain, 384, 386
    disorders of, 394
    extrinsic muscles of, 380
    intrinsic muscles of, 380
    structure of, 376–380
    surface anatomy of, 375
    and vision, 380–386
Eyebrows, 374
Eyelashes, 376
Eyelids, 375

Face
    blood vessels of, 504E, 505
    bones of, 143–146
    musculature of, 222E, 223E
Facet
    articular, 150
    definition of, 134E
    demifacet on vertebrae, 150
    on rib, 150
    on vertebra, 150
Facial bones, 143–146

Facilitated diffusion, 49
Facilitated zone, 276
Facilitation, 273
Facultative reabsorption, 674, 675
FAD. *See* Flavin adenine dinucleotide
Falciform ligament, 593, 611
Fallopian tube. *See* Uterine tube
False labor, 756
Falx cerebelli, 323
Falx cerebri, 318
Fascia
    deep, 195
    definition of, 195
    superficial, 93, 101, 196
Fasciculus
    arrangement of, 216, 218
    of nerve, 293
    of skeletal muscle, 196
    of spinal cord. *See* Tract
Fasciculus cuneatus, 286E
Fasciculus gracilis, 286E
Fast-twitch red fiber, 206
Fast-twitch white fiber, 206
Fat. *See* Lipid
Fat-soluble vitamins, 652E, 653E
Fat storage, 644
Fatty acid, 38, 645
Fauces, 557, 595
Feces, 624
Feedback system
    and blood pressure, 20
    and blood sugar level, 20, 21
    and body temperature, 109, 110
    and hormones, 20, 21, 403, 404
    input, 20
    negative, 20, 403
    output, 20
    positive, 20
Feeding center (hunger), 316, 590
Female pronucleus, 743
Female reproductive system, 717–730
Female skeleton, 172, 173
Feminizing adenoma, 423
Femur, 166
Fenestra cochlea, 388
Fenestrated capillary, 490
Fenestra vestibuli, 388
Fertilization, of ovum, 743, 744
Fetal alcohol syndrome (FAS), 759
Fetal circulation, 498, 521–524
Fetal growth
    changes associated with, 751, 752
    definition of, 751
    and ultrasound, 730, 755
Fetal hemoglobin, 577
Fetal period, 751
Fetus
    definition of, 86, 747
    development of, 751, 752E
    potential hazards to, 759, 760
Fever, 542, 658
Fibrillation, 212
Fibrin, 542
Fibrinogen, 448, 450–452, 542
Fibrinolysis, 452
Fibroadenoma, 738
Fibroblast, 89
Fibrocartilage, 94
Fibrocyte, 89
Fibromyositis, 211
Fibrosis, 97, 211
Fibrositis, 211
Fibrous capsule, 177
Fibrous joints
    definition of, 176
    types of, 176

Fibrous pericardium, 462
Fibrous periosteum, 120
Fibrous tunic, 376
Fibula, 170
Fight-or-flight response, 366, 430–432
Filiform papilla, 373, 596
Film-screen mammography, 730
Filtrate, glomerular, 671–673
Filtration
    definition of, 50
    glomerular, 671–673
Filtration fraction, 673
Filtration slit, 666
Filum terminale, 281
Fimbriae, 720
Final common pathway, 351
Fingers
    blood vessels of, 504E, 505
    bones of, 164
    musculature of, 244E, 245
First-class lever, 216
First-degree burn, 111
First (medial) cuneiform, 170
First messenger, 401
First-order neuron, 346
First polar body, 720
Fissure
    anterior median, 281
    of bone, 134E
    of brain, 316, 318
    horizontal, 564
    inferior orbital, 144
    longitudinal, 316, 318
    oblique, 564
    of lung, 564
    palpebral, 275
    superior orbital, 141
    transverse, 318
Fixator, 218
Fixed macrophage, 539
Flaccid, 205
Flaccid paralysis, 351
Flagellum, 61
Flat bone, 133
Flatfoot, 171
Flatus, 629
Flavin adenine dinucleotide (FAD), 639
Flavoprotein, 639
Flexion, 179
Flexor muscle, 219E
Flexor reflex, 291
Flu. *See* Influenza
Fluid balance, 693
Fluid compartments, 693
Fluid intake, 693
Fluid mosaic model, 46
Fluid output, 693
Fluids, body
    aqueous humor, 379
    balance of, 693
    cerebrospinal, 307–310
    edema and, 537, 697
    extracellular, 693
    intake of, 693
    interstitial, 17, 440, 456
    intracellular, 693
    lymph, 530, 536, 537
    movement between interstitial and intracel-
        lular, 697
    movement between plasma and interstitial,
        697
    output of, 693
    pericardial, 462
    perilymph, 389
    peritoneal, 592

Fluids (*Continued*)
  plasma, 448
  synovial, 96, 177
Folia, 324
Follicle-stimulating hormone (FSH), 408
Follicular cell, 413
Fontanel
  anterior, 136
  anterolateral, 136
  definition of, 136
  posterior, 136
  posterolateral, 136
Food choking. *See* Abdominal thrust (Heimlich) maneuver
Food intake, regulation of, 590
Foot
  arches of, 171
  blood vessels of, 508E, 509
  bones of, 170, 171
  musculature of, 252E, 253, 254
  plate, 127
Foramen
  anterior sacral, 152
  apical, 598
  carotid, 147E
  definition of, 134E
  greater palatine, 147E
  hypoglossal, 147E
  incisive, 147E
  inferior orbital, 147E
  infraorbital, 147E
  interventricular, 308
  intervertebral, 150
  jugular, 147E
  lacerum, 147E
  lacrimal, 147E
  lesser palatine, 147E
  magnum, 147E
  mandibular, 147E
  mastoid, 147E
  mental, 147E
  obturator, 165
  olfactory, 147E
  optic, 147E
  ovale, 147E
  posterior sacral, 152
  rotundum, 147E
  spinosum, 147E
  stylomastoid, 147E
  superior orbital, 147E
  supraorbital, 137, 147E
  transverse, 150
  vertebral, 150
  zygomaticofacial, 147E
Foramen ovale, of heart, 480, 523
Foramen ovale, of skull, 147E
Forearm
  blood vessels of, 504E, 505
  bones of, 162, 164
  musculature of, 244E, 245
Forebrain vesicle. *See* Prosencephalon
Foregut, 429, 625
Forehead, 136, 137
Foreskin. *See* Prepuce
Formed elements, of blood, 440–448
Fornix, 726
Fossa
  coronoid, 162
  definition of, 134E
  glenoid, 160
  iliac, 165
  incisive, 144
  infraspinous, 160
  intercondylar, 166
  mandibular, 140
  olecranon, 162
  ovalis, 463, 524

  radial, 162
  subscapular, 160
  supraspinous, 160
  triangular, 387
Fossa ovalis, 463, 524
Fourth ventricle, 308
Fracture hematoma, 129
Fractures
  of clavicle, 160
  closed reduction, 128
  Colles', 129
  comminuted, 128
  complete, 128
  compound, 128
  definition of, 128
  displaced, 129
  greenstick, 128
  impacted, 129
  nondisplaced, 129
  open reduction, 128
  partial, 128
  pathologic, 129
  Pott's, 129
  repair of, 129, 130
  simple, 128
  spiral, 128
  stress, 129
  transverse, 129
  of vertebral column, 156
Freckles, 104
Free nerve ending, 341
Frontal bone, 136, 137
Frontal eye field area, 322
Frontal lobe of cerebrum, 318
Frontal plane, 12
Fulcrum, 216
Full-thickness burn. *See* Third-degree burn
Functional residual capacity, 572
Fundus
  of stomach, 603
  of uterus, 721
Fungiform papilla, 373, 596
Furuncle, 113
Fusiform fasciculus, 218

Galea aponeurotica, 222E
Gallbladder, 614
Gallstone. *See* Calculus, biliary
Gamete, 67
Gamete formation, 708
Gamete intrafallopian transfer (GIFT), 746
Gamma aminobutyric acid (GABA), 275, 325
Gamma globulin, 457
Ganglion
  autonomic, 358, 360
  celiac, 362
  ciliary, 362
  definition of, 281
  geniculate, 328E
  inferior cervical, 362
  inferior mesenteric, 362
  middle cervical, 362
  otic, 362
  posterior root, 285
  prevertebral, 385
  pterygopalatine, 362
  spiral, 328E
  submandibular, 362
  superior cervical, 361
  superior mesenteric, 362
  sympathetic trunk, 358
  terminal, 358, 360
  vestibular, 329E
Ganglion neuron, 379
Gangrene, 212
Gastric absorption, 609
Gastric emptying, 608, 609

Gastric gland, 605
Gastric inhibiting peptide (GIP), 608E
Gastric juice, 538, 605
Gastric lipase, 606
Gastric secretion, regulation of, 607, 608
Gastric ulcer, 627
Gastrin
  enteric, 608E
  stomach, 608E
Gastroenterology, 590
Gastroileal reflex, 624
Gastrointestinal (GI) tract
  disorders of, 627–629
  general histology of, 591, 592
  organs of, 595–625
  peritoneum of, 592–595
Gastroscopy, 605
Gastrulation, 747
Gene
  action of, 62–65
  activation of, 403
  definition of, 42, 64
  structure of, 62, 65
General adaptation syndrome, 430
Generalized animal cell, 46
General senses
  cutaneous, 340–343
  pain, 342, 343
  proprioceptive, 344, 345
General sensory area. *See* Primary somesthetic area
Generator potential, 339
Genetic engineering, 64
Genetics, 760
Genital herpes, 736
Genital tubercle, 735
Genotype, 760, 761
Geriatrics, 71
Germinal center, 532
Germinal epithelium, 718
Gestation, 753–755
Giantism, 407
Gingivae, 598
Gland
  acinar, 85
  adrenal (suprarenal), 419–424
  apocrine, 85
  apocrine sweat, 108
  buccal, 596
  bulbourethral, 715
  ceruminous, 108
  compound, 85
  definition, 84
  duodenal, 616
  eccrine, 108
  endocrine, 84, 399, 404–428
  exocrine, 84, 399
  functional classification of, 85
  gastric, 605
  greater vestibular, 727
  holocrine, 85
  intestinal, 616
  lacrimal, 376
  lesser vestibular, 727
  mammary, 728–730
  merocrine, 85
  multicellular, 85
  olfactory, 371
  ovaries, 717–720
  pancreas, 424–427, 609–611
  parathyroids, 416–419
  parotid, 596
  pineal, 428
  pituitary, 404–410
  prostate, 715
  salivary, 596
  sebaceous, 108

sebaceous ciliary, 376
seminal vesicles, 714, 715
simple, 85
structural classification of, 84, 85
sublingual, 596
submandibular, 596
sudoriferous, 108
tarsal, 375
testes, 707, 708
thymus, 428, 536
thyroid, 412–416
tubular, 85
tubuloacinar, 85
unicellular, 84, 85
Gland of Zeis. See Sebaceous ciliary gland
Glandular epithelium, 84, 85
Glans
    of clitoris, 727
    of penis, 717
Glaucoma, 380, 394
Glenoid cavity, 160
Glenoid labrum, 186
Glial cells. See Neuroglia
Gliding joint, 181, 183, 186E
Gliding movement, 178
Globulin, 448
Globus pallidus, 318
Glomerular blood hydrostatic pressure, 672
Glomerular capsule, 666
Glomerular filtration, 671–673
Glomerular filtration rate (GFR), 673
Glomerulonephritis, 688
Glomerulus, 666
Glottis, 558
Glucagon, 424
Glucocorticoid, 421–423
Gluconeogenesis, 421, 643
Glucose
    absorption of, 618
    alarm reactions and, 430–432
    anabolism of, 643
    and brain, 310
    catabolism of, 636–643
    in dehydration synthesis, 37
    and gluconeogenesis, 421, 643
    and glycogenesis, 643
    and glycogenolysis, 643
    in hydrolysis, 37
    movement into cells, 636
    storage of, 643
    structure of, 37
    transport of, 49
Glucostatic theory, 590
Glycerol, 38, 645
Glycogen, 61, 643
Glycogenesis, 643
Glycogenolysis, 643
Glycolysis, 636–638
Glycoprotein, 46
Glycosuria
    pathological, 682
    temporary (alimentary), 682
Gnostic area, 322
Goblet cell, 83
Goiter, 416
Golgi complex, 56–58
Golgi tendon organ. See Tendon organ
Gomphosis, 176
Gonad, 706
Gonadocorticoid, 423
Gonadotroph cell, 404
Gonadotropic hormone, 405
Gonadotropin releasing factor (GnRF), 408
Gonorrhea, 735, 736
"Goosebumps," 108
Gossypol, 732
Gout, 688

Graafian follicle. See Vesicular ovarian follicle
Grafting, skin, 112
Granstein cell, 101
Granular leucocytes, 445
Granulation tissue, 97, 105
Gravity inversion, 149
Gray commissure, 284
Gray matter, 281
Gray ramus communicans, 362
Greater curvature, 603
Greater omentum, 593
Greater (false) pelvis, 165
Greater trochanter, 166
Greater vestibular gland, 727
Groove, 134E
    bicipital, 160
    costal, 155
    intertubercular, 160
Gross anatomy, 6
Growth, 9
Growth hormone (GH), 405–408
Growth hormone inhibiting factor (GHIF).
        See Somatostatin
Growth hormone releasing factor (GHRF),
        405
GTP. See Guanosine triphosphate
Guanine, 41
Guanosine triphosphate (GTP), 642
Gums. See Gingivae
Gustation, mechanism of, 373
Gustatory cell, 372
Gustatory hair, 372
Gustatory sensation, 372, 373
Gynecology, 717
Gynecomastia, 423

Hair
    development, 110
    functions of, 106
    replacement of, 106
    structure of, 106, 107
Hair cell, 392, 393
Hair follicle, 106
Hair root, 106
Hair shaft, 106
Hamate, 164
Hand
    blood vessels of, 504E, 505
    bones of, 164
    musculature of, 244E, 245
Hand plate, 127
Haploid cell, 68, 708
Hard palate, 595
Haustra, 624
Haustral churning, 624
Haversian canal. See Central canal
Haversian system. See Osteon
Head
    blood vessels of, 504E, 505, 512E, 513
    bones of, 133–146
    bony process, 134E
    divisions of, 133–146
    of epididymis, 713
    of femur, 166
    of fibula, 170
    of humerus, 160
    musculature of, 222E, 223E
    of pancreas, 609
    of radius, 164
    of rib, 154
    of spermatozoon, 711
    of ulna, 164
Headache. See Cephalgia
Heart
    angina pectoris and, 469
    artificial, 478
    atria of, 462

autonomic control of, 475–477
blood supply of, 468, 469
borders of, 461, 462
cardiac cycle of, 472–474
cardiac output and, 474, 475
cardiac reserve and, 475
catheterization of, 479
chambers of, 462, 463
chemicals and, 477
circulatory shock and, 477, 478
conduction system of, 469, 470
development of, 480
disorders of, 481–484
electrocardiogram of, 470, 471
emotions and, 477
endocardium of, 462
epicardium of, 462
great vessels of, 463, 465
location, 461, 462
medical terminology, 484
murmur of, 474
myocardial infarction, 469
myocardium of, 462
pericardium, 96, 462
pressure changes in, 472
regulation of rate, 475–477
sex and age and, 477
sounds of, 473, 474
surface anatomy of, 466–468
surface projection of, 466–468
temperature and, 477
timing of, 472, 473
valves of, 465, 466
ventricles of, 462
wall of, 462
Heart attack, 479
Heartbeat, 472–474
Heart block, atrioventricular (AV), 483
Heartburn, 602
Heart-lung bypass, 479
Heart murmur, 474
Heart sounds, 473, 474
Heart surgery, 478, 479
Heart valve, 465, 466
Heat cramp, 658
Heat exhaustion, 658
Heat-losing center, 657
Heat-promoting center, 657
Heatstroke, 658
Heel, 170
Helix, of ear, 386
Helper T cell, 546
Hematocrit, 445
Hematology, 440
Hematuria, 682
Hemiballismus, 319
Hemiplegia, 302
Hemisection, 302
Hemisphere
    cerebellar, 323
    cerebral, 318
Hemocytoblast, 441
Hemodialysis, 679, 680
Hemoglobin
    and carbon dioxide, 578
    and DPG, 577
    and oxygen, 575, 576
    and pH, 576
    and PO_2, 575
    saturation of, 575, 576
    structure of, 442, 444
    and temperature, 577
Hemoglobin-oxyhemoglobin buffer, 698, 699
Hemolysis, 50, 454
Hemolytic disease of newborn, 455, 456
Hemophilia, 450–452, 763
Hemopoiesis, 120, 441

Hemorrhage, 458
Hemorrhoid, 623
Hemostasis, 449–453
Hemothorax, 562
Henle's layer. *See* Pallid layer
Henry's law, 573
Heparin, 452
Hepatic cell, 611
Hepatic duct, 611
Hepatic portal circulation, 520
Hepatitis, 628
Hepatitis B virus (HBV), 71
Hepatopancreatic ampulla, 610
Hereditary traits in humans, 760–763
Hermaphroditism, 763
Hernia, 629
Herniated (slipped) disc, 155
Hertz (Hz), 323
Heterozygous, 761
High altitude sickness, 575
High-density lipoprotein (HDL), 481, 482
Hilus
    of kidney, 664
    of lung, 562
    of lymph node, 531
    of ovary, 718
    of spleen, 534
Hindbrain vesicle. *See* Rhombencephalon
Hindgut, 625
Hinge joint, 181, 182
Hip, 165, 166
Hippocampus, 319
Hippuric acid, 683E
Histamine, 539, 541
Histology, 6, 77
Histone, 54
HLA antigen, 445
Hodgkin's disease, 71, 551
Holocrine gland, 85
Holoenzyme, 635
Holter monitor, 470
Homeostasis
    of blood pressure, 20
    of blood sugar level, 20, 21
    and body temperature, 109, 110
    of bone, 126
    and circulatory shock, 477, 478
    definition of, 17
    and enzymes, 635
    and feedback systems, 20, 21, 109, 110, 403, 404
    kidneys and, 680
    of muscle tissue, 210
    stress and, 18–20
    and tissue inflammation, 540–542
Homeotherm, 109
Homologous chromosome, 68, 709
Homozygous, 761
Horizontal cell, 384
Horizontal plane, 12
Hormonal action, mechanism of, 399–403
Hormones
    activation of genes, 403
    adrenocorticotropic (ACTH), 408
    adrenoglomerulotropin, 428
    aldosterone, 420
    androgens, 423
    antidiuretic (ADH), 411
    atrial natriuretic factor (ANF), 430
    and bone growth, 126
    calcitonin (CT), 416
    chemistry of, 399
    cholecystokinin (CCK), 608E
    corticosterone, 421
    cortisol (hydrocortisone), 421
    cortisone, 421

and cyclic AMP, 401–403
    definition of, 399
    enteric gastrin, 608E
    epinephrine, 423
    erythropoietin, 430
    estrogens, 427, 722, 752
    follicle-stimulating (FSH), 408
    gastric inhibiting peptide (GIP), 608E
    glucagon, 424
    glucocorticoids, 421, 422
    gonadocorticoids, 423
    gonadotropic, 405
    growth (GH), 405, 408
    human chorionic gonadotropin (HCG), 725, 726, 752
    human chorionic somatomammotropin (HCS), 753
    inhibin, 428, 712, 713
    insulin, 424–426
    interaction with intracellular receptors, 403
    interaction with plasma membrane receptors, 401–403
    luteinizing (LH), 408
    mechanism of action, 399–403
    melanocyte-stimulating (MSH), 409
    melatonin, 428
    mineralocorticoids, 420, 421
    and negative feedback control, 403–404
    norepinephrine (NE), 423
    oxytocin (OT), 410, 760
    parathyroid (PTH), 416–419
    of pregnancy, 752, 753
    progesterone, 427, 723, 752
    prolactin (PRL), 408, 409, 760
    and prostaglandin, 403
    relaxin, 427, 428, 723, 753
    secretin, 608E
    stomach gastrin, 608E
    testosterone, 428, 711, 712
    thymosin, 428
    thyroid, 412–414
    thyroid-stimulating (TSH), 408
    thyroxine (T$_4$), 413
    triiodothyronine (T$_3$), 413
    tropic, 405
Horn
    anterior gray, 284
    definition of, 281
    lateral gray, 284
    posterior gray, 284
Horner's syndrome, 366
Human chorionic gonadotropin (HCG), 725, 726, 752
Human chorionic somatomammotropin (HCS), 753
Humeroscapular joint, 183, 186, 187
Human T-cell leukemia lymphoma virus (HTLV-1), 71, 457
Humeral immunity, 547, 548
Humerus, 160, 162
Humoral immunity, 547, 548
Huntington's chorea, 335, 761, 762
Huxley's layer. *See* Granular layer
Hyaline cartilage, 94
Hyaline membrane disease. *See* Infant respiratory distress syndrome (RDS)
Hyaluronic acid, 61, 86
Hyaluronidase, 88
Hybridoma cell, 549
Hydrocephalus
    external, 310
    internal, 310
Hydrocortisone. *See* Cortisol
Hydrogen bond, 30
Hydrolysis, 37
Hydrostatic pressure, 672
Hydroxyapatites, 120

Hymen, 726
Hyoid bone, 146
Hyperacusia, 388
Hyperbaric oxygenation (HBO), 573
Hypercalcemia, 416
Hypercapnia, 581
Hyperemesis gravidarum, 746
Hyperextension, 179
Hyperglycemia, 405
Hyperinsulinism, 427
Hypermagnesemia, 696
Hypermetropia, 382
Hyperosmotic urine, 678
Hyperplasia, 72, 433
Hyperpolarization, 274
Hypersecretion, 407
Hypersensitivity, 550
Hypersomnia, 354
Hypertension
    primary, 525
    secondary, 525
Hypertonic solution, 50
Hypertrophy, 72, 205
Hyperventilation, 581
Hypervitamosis, 652
Hypocalcemia, 416
Hypochloremia, 696
Hypodermis, 93, 101, 113
Hypoglycemia, 406
Hypokalemia, 696
Hypokinesia, 332
Hypomagnesemia, 696
Hyponatremia, 696
Hyponychium, 109
Hypophyseal pouch, 429
Hypophysis. *See* Pituitary gland
Hypoplasia, 433
Hyposecretion, 407
Hyposmotic urine, 678
Hypotension, 526
Hypothalamic-hypophyseal tract, 410
Hypothalamic nuclei, 315–316
Hypothalamus, 314–316
Hypothermia, 479
Hypotonic solution, 50
Hypoventilation, 581
Hypoxia, 444, 445, 578
Hysterectomy, 731
H zone, 198

I band, 198
Ileocecal sphincter (valve), 614, 622
Ileum, 614
Iliac crest, 165
Ilium, 165
Immunity
    acquired immune deficiency syndrome (AIDS) and, 551–552
    allergy and, 550
    antibodies and, 544
    antigens and, 542–544
    autoimmune diseases and, 110, 551
    and cancer, 549
    cellular, 544–547
    definition of, 538, 542
    humoral, 544, 547, 548
    and monoclonal antibodies, 549
    and skin, 548
    tissue rejection and, 550
Immunogenicity, 542
Immunoglobulin. *See* Antibody
Immunologic competence, 545
Immunologic escape, 549
Immunologic surveillance, 549
Immunologic tolerance, 551
Immunology, 542

Immunosuppressive drug, 551
Impacted cerumen, 108
Imperforate hymen, 726
Impetigo, 113
Implantation, of fertilized egg, 744, 746
Impotence, 737
Incisor, 599
Incompetent valve, 484
Incomplete tetanus, 204
Incontinence, 686
Incus, 388
Indirect transfusion, 458
Infant, adjustments of, at birth, 758–759
Infarction, 469
Infectious mononucleosis (IM), 457
Inferior colliculi, 313
Inferior nasal concha, 146
Inferior vena cava, 463, 480, 498, 510E
Infertility, 737, 738
Inflammation. See Inflammatory response
Inflammatory bowel disease (IBD), 629
Inflammatory bowel syndrome, 629
Inflammatory response
    definition of, 540
    stages of, 540–542
    symptoms of, 540
Inflation (Hering-Breuer) reflex, 581
Influenza, 585
Infrapatellar fat pad, 190
Infundibulum
    of pituitary, 404
    of uterine (Fallopian) tube, 720
Ingestion, digestive, 8, 590
Ingestion, phagocytic, 540
Inguinal canal, 713, 714
Inguinal hernia, 714
Inheritance
    of color-blindness, 762
    genotype, 760
    phenotype, 760
    of PKU, 761
    of sex, 762
    X-linked, 763
Inhibin, 428, 712, 713
Inhibiting factor, 404
Inhibitory postsynaptic potential (IPSP), 274
Inhibitory transmitter-receptor interaction, 274
Initial heat, 210
Inner cell mass, 744
Innervation of skeletal muscles, 222E–252E
Inorganic compound, 33
Insertion of skeletal muscle, 216
Insidious, 72
Insomnia, 354
Inspiration, 565–569
Inspiratory capacity, 572
Inspiratory reserve volume, 571
Insula, 318
Insulin, 424–426
Insulin-like growth factor (IGF), 405
Integral protein, 46
Integrated response, 368
Integration at synapses, 274
Integrative functions of brain, 274, 275, 351–354
Integumentary system, 8E
    aging and, 110
    definition of, 8E, 101
    development of, 110
    disorders of, 110–112
    epidermal derivatives, 105–109
    medical terminology, 113
    and nonspecific resistance, 538
    skin, 101–105
Interatrial septal defect, 483

Interatrial septum, 463, 480
Intercalated disc, 206
Intercostal (thoracic) nerve, 300, 302
Intercostal space, 155
Interferon (IFN), 538
Interlukin 1, 545
Interlukin 2, 546
Intermediate filaments, 60, 208
Intermediate mass, 314
Internal anal sphincter, 622
Internal capsule, 314
Internal ear, 389, 390
Internal medullary lamina, 314
Internal nares, 556
Internal os, 721
Internal respiration, 575
Internal root sheath, 106
Internal sphincter, of urinary bladder, 685
Internuncial neuron. See Association neuron
Interphase, 65
Intersegmental reflex arc, 291
Interstitial cell of Leydig. See Interstitial endocrinocyte
Interstitial endocrinocyte, 708
Interstitial fluid, 17, 440, 456
Interstitial fluid hydrostatic pressure (IFHP), 495
Interstitial fluid osmotic pressure (IFOP), 495
Interstitial growth, 94
Interventricular foramen, 308
Interventricular septal defect, 482
Interventricular septum, 463, 480
Intervertebral disc, 146, 149
Intervillous space, 750
Intestinal gland, 616
Intestinal juice, 616, 617
Intestinal secretion, regulation of, 618
Intracellular fluid (ICF), 17, 693
Intracerebral hemorrhage, 333
Intrafusal fiber, 344
Intramembranous ossification, 123, 126
Intramuscular injection, 255
Intraocular pressure, 379
Intrapleural (intrathoracic) pressure, 568
Intrapulmonic (intraalveolar) pressure, 569
Intrauterine device (IUD), 732
Intravenous pyelogram (IVP), 689
Intrinsic blood clotting pathway, 450
Intrinsic factor, 445
Intrinsic muscle
    of eyeball, 380
    of tongue, 596
Intubation, 560
Inversion, 181
Inverted image formation, 382
Ion
    anion, 28
    cation, 28
Ionic bond, 27, 28
Ionization, 34E
Ipsilateral reflex arc, 289
Iris, 377
Irregular bone, 133
Irritable bowel syndrome, 629
Ischemia, 469
Ischium, 165
Island of Reil. See Insula
Islets of Langerhans. See Pancreatic islet
Isograft, 550
Isometric contraction, 204, 205
Isotonic contraction, 204, 205
Isotonic solution, 50
Isotope, 30
Isovolumetric contraction, 472
Isovolumetric relaxation, 472

Isthmus
    of thyroid gland, 412
    of uterine (Fallopian) tube, 720
    of uterus, 721

Jaundice, 613
    hemolytic, 613
    neonatal, 613
    obstructive, 613
Jejunum, 614
Joint kinesthetic receptor, 345
Joints. See Articulations
Jugular lymph sac, 549, 550
Jugular notch, 153
Juxtaglomerular apparatus, 669, 670
Juxtaglomerular cell, 669
Juxtamedullary nephron, 666

Kaposi's sarcoma, 551
Karyolymph, 54
Karyotype, 763
Keratin, 84, 101
Keratinization, 103
Keratinocyte, 101
Keratohyalin, 101
Keratoplasty. See Corneal transplant
Keratosis, 113
Ketogenesis, 645
Ketone bodies, 645
Ketosis, 427, 646, 682, 683
Kidney, 664–684
Kidney stone. See Renal calculus
Killer T cell, 545, 546
Kilocalorie, 655
Kinesthetic sense. See Proprioception
Kinetic energy, 33
Kinin, 541
Kinocilium, 392
Knee, 188, 190, 191
    dislocated, 191
    runner's, 190, 191
    swollen, 191
Knee injury, 190
Knee jerk. See Patellar reflex
Korotkoff sound, 498
Krebs cycle, 639, 640
Kupffer cell. See Stellate reticuloendothelial cell
Kwashiorkor, 660
Kyphosis, 156

Labial frenulum, 595
Labia majora, 727
Labia minora, 727
Labiosacral swelling, 735
Labor
    false, 756
    placental stage, 756
    stage of dilation, 756
    stage of expulsion, 756
    true, 756
Labyrinth. See Internal ear
Labyrinthine disease, 394
Laceration, 321
Lacrimal apparatus, 376, 538
Lacrimal bone, 145
Lacrimal canal, 376
Lacrimal caruncle, 375
Lacrimal gland, 376
Lacrimal sac, 376
Lacrimal secretion, 376
Lactase, 617, 618
Lactation, 729, 760
Lacteal, 616
Lactiferous duct, 728
Lactose intolerance, 618

Lactotroph cell, 404
Lacuna
   in bone, 94, 121
   in cartilage, 94
   definition of, 94
Laetrile, 71
Lamella, 94, 121, 122
Lamellated corpuscle, 103, 342
Lamina of vertebra, 150
Lamina propria, 96, 591
Langerhans' cell, 101
Langer's lines. See Lines of cleavage
Language area of cerebrum, 322
Lanugo, 110
Laparoscopy, 731
Large intestine, 622–625
Laryngitis, 559
Laryngopharynx, 557
Laryngotracheal bud, 583
Larynx, 558, 559
Latent period, in muscle contraction, 203
Lateral aperture, 308
Lateral cerebral sulcus, 318
Lateral commissure, 375
Lateral condyle
   of femur, 166
   of tibia, 168
Lateral corticospinal tract, 287E, 350
Lateral epicondyle
   of femur, 166
   of humerus, 162
Lateral gray horn, 284
Lateral malleolus, 170
Lateral mass, 142, 150
Lateral patellar retinacula, 190
Lateral spinothalamic pathway, 346, 347
Lateral spinothalamic tract, 286E, 347
Lateral umbilical ligament, 524
Lateral ventricle, 308
Lateral white column, 284
Learning, 351
Left colic flexure, 622
Left lobe of liver, 611
Leg
   blood vessels of, 508E, 516E, 519E
   bones of, 168, 170
   musculature of, 250E, 252E
Lens, 379
Lenses, corrective, 381, 382
Lentiform nucleus, 318
Lesser curvature, 603
Lesser omentum, 593
Lesser (true) pelvis, 165
Lesser trochanter, 166
Lesser vestibular gland, 727
Lethal gene, 763
Leucocyte
   agranular, 445
   and differential count, 447
   functions of, 445, 446
   granular, 445
   life span, 447
   number per cubic millimeter, 447
   origin of, 447
   structure of, 445
   types of, 445
Leucocytosis, 447
Leucocytosis-promoting factor, 542
Leucopenia, 447
Leukemia, 71, 456
Leukorrhea, 738
Levator muscle, 219E
Levels of structural organization, 6, 8
Lever
   classes of, 216
   definition of, 216

Leverage, 216
Life processes, 8, 9
Ligaments
   accessory, 178
   anterior cruciate, 190
   arcuate popliteal, 190
   artificial, 93
   coracohumeral, 183
   definition of, 93, 177
   denticulate, 284
   extracapsular, 178
   fibular collateral, 190
   glenohumeral, 186
   of head of femur, 188
   iliofemoral, 187
   intraarticular, 190
   intracapsular, 178
   ischiofemoral, 187
   oblique popliteal, 190
   patellar, 190, 250E
   posterior cruciate, 190
   pubofemoral, 187
   suspensory, 379, 718
   tibial collateral, 190
   transverse, of acetabulum, 188
   transverse humeral, 186
Ligamentum arteriosum, 465, 524
Ligamentum teres. See Round ligament
Ligamentum venosum, 524
Light adaptation, 384
Light rays, 380–382
Light touch, 341
Limb bud, 127
Limbic lobe, 319
Limbic system, 319, 320
Liminal stimulus. See Threshold stimulus
Line
   definition of, 134E
   linea aspera, 166
   transverse, 152
Linea alba, 233
Linea aspera, 166
Lines of cleavage, 103
Lingual frenulum, 596
Lingual tonsil, 533
Lipid
   absorption of, 619, 620
   anabolism of, 646
   catabolism of, 644, 645
   classification of, 39E
   definition of, 38
   digestion of, 618
   fate of, 644
   functions of, 38, 39, 644
   storage, 61, 644
   structure of, 38, 39
Lipofuscin, 264
Lipogenesis, 646
Lipoprotein, 481, 482
Lips, 556
Liver, 611–614
Lobe
   of cerebellum, 323
   of cerebrum, 318, 319
   of liver, 611
   of lungs, 564
   of mammary gland, 728
   of thymus gland, 536
   of thyroid gland, 412
Lobule
   of ear, 386
   of liver, 611
   of lung, 564
   of mammary gland, 728
   of testis, 707
   of thymus gland, 536

Local disease, 72
Lochia, 763
Locus coeruleus, 325
Long bone, 133
Longitudinal arch, of foot, 171
Longitudinal fissure, 316, 318
Long-term memory, 352
Loop of Henle. See Loop of the nephron
Loop of the nephron, 666
Loose connective tissue, 86, 88–91, 93
Lordosis, 156
Low-density lipoprotein (LDL), 481, 482
Lower esophageal sphincter, 601
Lower extremity
   blood vessels of, 508E, 519E
   bones of, 166–171
   musculature of, 250E, 252E
Lower motor neuron, 350, 351
Lubrication, during sexual intercourse, 730,
   731
Lumbago, 211
Lumbar enlargement, 281
Lumbar plexus, 297, 298, 301E
Lumbar vertebra, 150
Lumen, of blood vessel, 488
Lunate, 164
Lund-Browder method, 111, 112
Lung, 561–565
Lung bud, 583
Lunula, 109
Luteinizing hormone (LH), 408
Luxation. See Dislocation
Lymph, 530
   circulation of, 536, 537
Lymphadenopathy, 551
Lymphangiogram, 531
Lymphangiography, 531
Lymphatic, 530, 531
Lymphatic nodule, 532
Lymphatic organ, 530
Lymphatic system, 8E, 530–552
   and allergy, 550
   and autoimmune disease, 110, 551
   definition of, 530
   development of, 549, 550
   disorders of, 550–552
   and immunity, 542–548
   lymphatic tissue, 531–536
   lymphatic vessels, 530, 531
   lymph nodes, 531–533
   medical terminology, 552
   plan of lymph circulation, 536
   and tissue rejection, 550
Lymphatic tissue, 531–536
Lymphatic vessels
   afferent, 532
   bronchomediastinal trunk, 536, 537
   development of, 549, 550
   efferent, 532
   intestinal trunk, 536, 537
   jugular trunk, 536, 537
   lumbar trunk, 536, 537
   lymph capillaries, 530
   right lymphatic duct, 537
   subclavian trunk, 536, 537
   thoracic duct, 536, 537
Lymph capillary, 530, 531
Lymph circulation, 536, 537
Lymphedema, 551
Lymph node
   development of, 550
   histology of, 531, 532
   infection of, 532, 533
   lymph flow through, 532, 533
Lymph nodule, 530
Lymphoblast, 441

Lymphocyte, 445–447
Lymphokine, 546
Lymphoma, 71, 551
Lymphotoxin, 545
Lymph trunk, 536
Lysis, cell, 539
Lysosome, 58
Lysozyme, 538

Macrophage, 89, 539
    fixed, 539
    wandering, 539
Macrophage activating factor (MAF), 546
Macrophage chemotactic factor, 546
Macula, 392
Macula densa, 670
Macula lutea, 379
Magnesium
    control of, 697
    deficiency of, 696
    distribution of, 696
    excess of, 696
    functions of, 696
Magnetic resonance imaging (MRI), 25
Major calyx, 666
Male pronucleus, 743
Male reproductive system, 706–717
Male skeleton, 172, 173
Malignant tumor, 69
Malleolus
    lateral, 170
    medial, 168
Malleus, 388
Malpighian corpuscle. See Splenic nodule
Maltase, 617, 618
Mammary duct, 728
Mammary gland, 728–730
Mammillary body, 319
Mammography, 729, 730
Mandible, 144
Mandibular, 145
Mantle layer of neural tube, 328
Manubrium, 152
Marey's law of the heart, 476
Marginal layer of neural tube, 328
Margination, 541
Marrow biopsy, 154
Marrow transplantation, 447
Mass peristalsis, 624
Mast cell, 91
Mastectomy, 738
Mastication, 599
Mastoiditis, 140
Mastoid process, 140
Matrix
    of hair follicle, 106
    of mitochondrion, 58
    of nail, 109
Maxillae, 143, 144
Measurement, of body, 21
Meatus
    definition of, 134E
    external auditory, 140
    inferior, 556
    internal auditory, 140
    middle, 556
    superior, 556
Mechanical digestion, 590
Mechanical method, of contraception, 731, 732
Mechanoreceptor, 340
Medial commissure, 375
Medial condyle
    of femur, 166
    of tibia, 168

Medial epicondyle
    of femur, 166
    of humerus, 162
Medial lemniscus, 314
Medial malleolus, 168
Medial patellar retinacula, 190
Median aperture, 308
Mediastinum, 13
Medical terminology
    associated with articulations, 192
    associated with brain, 335
    associated with the cardiovascular system, 457, 458, 484, 526
    associated with cells, 72
    associated with development, 763
    associated with the digestive system, 629
    associated with the endocrine system, 433
    associated with the integumentary system, 113
    associated with the lymphatic system and immunity, 552
    associated with muscle tissue, 212
    associated with the reproductive systems, 738
    associated with the respiratory system, 586
    associated with skeletal tissue, 130
    associated with special senses, 395
    associated with the urinary system, 689
Meditation, 368
Medulla
    adrenal, 419
    of brain, 310–313
    of hair shaft, 106
    of kidney, 666
    of lymph node, 532
    of ovary, 719
    of thymus gland, 536
Medulla oblongata
    cranial nerves of, 313
    nuclei of, 311–313
    olive, 313
    pyramids of, 311
    reflex centers of, 312, 313
    reticular formation of, 312
Medullary cavity, 120
Medullary cord, 532
Medullary rhythmicity area, 579–581
Medullary sinus, 532
Megakaryoblast, 441
Meibomian gland. See Tarsal gland
Meiosis, 65, 68, 69
Meissner's corpuscle. See Corpuscle of touch
Melanin, 61, 104
Melanocyte, 91, 101, 104
Melanocyte-stimulating hormone (MSH), 409
Melanocyte-stimulating hormone inhibiting factor (MIF), 409
Melanocyte-stimulating hormone releasing factor (MRF), 409
Melanoma, 104, 111
Melatonin, 428
Membrane
    cutaneous, 96, 101–104
    definition of, 96
    epithelial, 96
    mucous, 96
    pleural, 96, 561, 562
    serous, 96
    synovial, 97
    types of, 96
Membrane capsule, 93
Membrane potential, 268, 269
Membranous labyrinth, 389
Membranous urethra, 714

Memory
    activated, 351
    consolidation, 352
    engram of, 351
    long-term, 352
    theories of, 352
Memory B cell, 547
Memory consolidation, 352
Memory T cell, 547
Menarche, 726
Ménière's syndrome, 394
Meningeal ramus, 294
Meninges
    cranial, 307
    spinal, 261, 264, 281, 284
Meningitis, 284
Meniscus. See Articular disc
Menopause, 726
Menstrual cycle
    hormonal control of, 722, 723
    menstrual phase, 723–725
    ovulation, 725
    postovulatory phase, 725
    preovulatory phase, 725, 726
Menstruation, 723, 724
Merkel's disc. See Tactile disc
Merocrine gland, 85
Mesencephalon. See Midbrain
Mesenchyme, 86
Mesentery, 593
Mesoappendix, 622
Mesocolon, 593, 622
Mesoderm
    derivatives of, 747E
    development of, 747
Mesonephric duct, 687
Mesonephros, 687
Mesothelium, 78
Mesovarium, 718
Messenger RNA, 62–64
Metabolic rate, 655, 656
Metabolism
    anabolism, 8, 32, 634, 643, 646
    basal metabolic rate, 656
    and calories, 655
    of carbohydrates, 636–643
    catabolism, 8, 634, 636–646
    definition of, 8, 634
    of lipids, 644–646
    and minerals, 649, 650, 651E, 652E
    and obesity, 658, 659
    of proteins, 646
    regulation of, 649
    and vitamins, 652E–655E
Metacarpus, 164
Metanephric duct, 688
Metanephros, 687
Metaphase, 67
Metaphysis, 120, 125
Metaplasia, 72
Metarteriole, 490
Metarubricyte, 444
Metastasis, 70, 72, 533
Metatarsus, 170, 171
Metencephalon, 329
Metric units
    of length, A-2
    of mass, A-3
    of volume, A-3
Micelle, 619
Microfilament, 59
Microglia, 264E
Microphage, 539
Microsurgery, 755
Microtomography, 46

Microtubule, 60
  chromosomal, 66
  continuous, 66
Microvillus, 82, 616
Micturition, 686
Micturition reflex, 686
Midbrain
  cerebral peduncles, 313
  corpora quadrigemina, 313
  cranial nerves of, 313, 314
  nuclei of, 313, 314
  parts of, 313, 314
  vesicle, 329
Midbrain vesicle, 329
Middle ear, 388
Midgut, 625
Midpiece, of spermatozoon, 711
Midsagittal plane, 12
Migration inhibition factor (MIF), 546
Milking, 496
Milk letdown, 760
Milk secretion. *See* Lactation
Milliequivalents per liter (meq/l), 694, 695E
Milliosmol, 676
Mineral, 649, 650, 651E, 652E
Mineralocorticoid, 420, 421
Minimal volume, 572
Minor calyx, 666
Minute volume of respiration (MVR), 570
Mitochondrion, 58, 59
Mitogenic factor, 546
Mitosis, 64–67
Mitotic apparatus, 66
Mitotic spindle, 66
Mitral valve. *See* Bicuspid valve
Mitral valve prolapse (MVP), 474
Mixed nerve, 293, 327
Mixing wave, 605
M line, 198
Modality, 340
Modified respiratory movements, 570
Modiolus, 390
Molar, 599
Mole, 35
Molecule, 27
Monaxial joint, 182
Monoblast, 441
Monoclonal antibody, 549
Monocyte, 445
Monoplegia, 302
Monosaccharide, 37
Monosynaptic reflex arc, 288
Monozygotic (identical) twins, 744
Mons pubis, 727
Morning sickness. *See* Hyperemesis gravidarum
Morula, 744
Motion sickness, 394
Motor areas of cerebrum, 322
Motor cortex, 349
Motor end plate. *See* Neuromuscular junction
Motor neuron. *See* Efferent neuron
Motor pathways
  extrapyramidal, 350, 351
  pyramidal, 349, 350
  and sensory input, 348
Motor speech area, 322
Motor unit, 209, 211
Mouth. *See* Oral cavity
mRNA. *See* Messenger RNA
Mucosa
  definition of, 96
  of ductus (vas) deferens, 713
  of esophagus, 601
  of gallbladder, 614
  of GI tract, 591, 592

  of large intestine, 623, 624
  of larynx, 558, 559
  of nasal cavity, 556
  and nonspecific resistance, 538
  of oral cavity, 595
  of pharynx, 557
  of small intestine, 612
  of stomach, 605
  of tongue, 596
  of trachea, 560
  of ureter, 684
  of urethra, 685
  of urinary bladder, 685
  of uterine (Fallopian) tube, 720
  of uterus, 722
  of vagina, 726
Mucous cell, 605
Mucous connective tissue, 86
Mucous membrane. *See* Mucosa
Mucus, 61, 538
Müller's duct. *See* Paramesonephric duct
Multicellular gland, 85
Multiple myeloma, 446
Multiple sclerosis (MS), 333
Multipolar neuron, 267
Multiunit smooth muscle tissue, 209
Multivalent antigen, 542
Mumps, 596
Muscarinic receptor, 364
Muscle contraction
  all-or-none principle, 203
  kinds of, 203–205
  motor unit, 200, 201
  muscle length and, 203
  neuromuscular junction, 199, 200
  physiology of, 201, 202
  role of ATP, 201–203
  role of calcium, 201, 202
  role of phosphocreatine, 202, 203
  sliding-filament theory, 199
  tone and, 205
Muscle fatigue, 210
Muscle fiber, 197, 205, 206
Muscles. *See* Skeletal muscle
Muscle spindle, 205, 344, 345
Muscle tissue
  aging and, 210
  all-or-none principle of, 203
  characteristics of, 77, 94, 95, 195
  connective tissue components of, 195
  contraction of, 199–206
  development of, 210, 211
  disorders of, 211, 212
  energy and, 202, 203
  fascia of, 195, 196
  fast (white), 206
  fatigue of, 210
  functions of, 195
  heat production by, 210
  histology of, 77, 95, 197–199
  homeostasis and, 210
  medical terminology, 212
  nerve and blood supply, 197
  oxygen debt and, 210
  physiology of, 201–205
  slow (red), 206
  types of, 195
Muscle tone, 205
Muscular artery, 489
Muscular dystrophy, 212
Muscular hypertrophy, 205
Muscularis
  definition of, 592
  of ductus (vas) deferens, 713
  of esophagus, 601
  of gallbladder, 614

  of GI tract, 592
  of large intestine, 624
  of pharynx, 557
  of small intestine, 616
  of stomach, 605
  of trachea, 560
  of ureter, 684
  of urethra, 685
  of urinary bladder, 685
  of uterine (Fallopian) tube, 720
  of uterus, 721, 722
  of vagina, 726
Muscularis mucosae, 591
Muscular system, 8E, 216–255
  actions of, 219E
  arrangement of fasciculi, 216, 218
  characteristics of, 195
  connective tissue components of, 195
  contraction and, 199–205
  definition of, 216
  development of, 210, 211
  disorders of, 211, 212
  fascia of, 195
  functions of, 195
  group actions of, 218
  histology of, 195–198, 206, 208, 209
  insertions of, 216
  intramuscular injections, 255
  leverage and, 216
  lever systems of, 216
  medical terminology, 212
  naming skeletal muscles, 218, 219
  nerve and blood supply of, 197
  origins of, 216
Musculi pectinati, 462
Mutation, 763
Myasthenia gravis (MG), 212
Myelencephalon, 329
Myelinated fiber, 266
Myelination, 266
Myelin sheath, 266
Myeloblast, 441
Myelography, 284
Myeloma, 70, 71
Myenteric plexus, 592
Myocardial infarction (MI), 469
Myocardium, 462
Myofibril, 198
Myofilament
  thick, 198
  thin, 198
Myoglobin, 205
Myogram, 203
Myology, 195
Myometrium, 721, 722
Myoneural junction. *See* Neuromuscular junction
Myopathy, 212
Myopia, 380
Myosin, 59, 198
Myosin-binding site, 198
Myositis, 212
Myotome, 210
Myotonia, 212
Myxedema, 416

NAD. *See* Nicotinamide adenine dinucleotide
Nail, 109
Narcolepsy, 354
Nasal bone, 143
Nasal cavity, 556
Nasal septum, 146, 556
Nasal visual field, 384
Nasofacial angle, 556
Nasolacrimal duct, 376
Nasopharyngeal carcinoma, 71

Nasopharynx, 557
Natural killer (NK) cells, 547
Natural method, of contraception, 731
Nausea, 629
Navel. *See* Umbilicus
Navicular, of foot, 170
Nebulization, 564
Neck
    blood vessels of, 504E, 505, 512E, 513
    bones of, 150
    of femur, 166
    of humerus, 160
    musculature of, 233E
    of rib, 154
    of tooth, 598
Necrosis, 72, 130
Negative feedback system, 20, 403, 404
Neoplasia, 738
Neoplasm, 72
Nephron, 666–679
Nephrosis, 688
Nerve
    abducens (VI), 328E
    accessory (XI), 328E
    ansa cervicalis, 295E
    axillary, 298E
    cardiac (accelerator), 475
    cochlear, 328E
    common peroneal, 301E
    damage to, 297, 300, 302
    deep peroneal, 301E
    definition of, 268, 281
    development, 328, 329
    dorsal scapular, 298E
    facial (VII), 329E
    femoral, 301E
    genitofemoral, 301E
    glossopharyngeal (IX), 329E
    greater auricular, 295E
    hypoglossal (XII), 329E
    iliohypogastric, 301E
    ilioinguinal, 301E
    inferior gluteal, 301E
    intercostal, 300, 302
    intercostobrachial, 298E
    lateral antebrachial cutaneous, 298E
    lateral femoral cutaneous, 301E
    lateral pectoral, 298E
    lateral plantar, 301E
    lesser occipital, 295E
    long thoracic, 298E
    lower lateral brachial cutaneous, 298E
    lower subscapular, 298E
    mandibular, 328E
    maxillary, 328E
    medial antebrachial cutaneous, 298E
    medial brachial cutaneous, 298E
    medial pectoral, 298E
    medial plantar, 301E
    median, 298E
    musculocutaneous, 298E
    obturator, 301E
    to obturator internus, 301E
    oculomotor (III), 327E
    olfactory (I), 327E
    ophthalmic, 328E
    optic (II), 327E, 371, 384
    pelvic splanchnic, 363
    perforating cutaneous, 301E
    phrenic, 295E
    to piriformis, 301E
    posterior antebrachial cutaneous, 298E
    posterior brachial cutaneous, 298E
    posterior cutaneous, 301E
    pudendal, 301E
    to quadratus femoris, 301E
    radial, 298E
    repair, 302, 303
    sciatic, 301E
    segmental, 295E
    splanchnic, 362
    to subclavius, 298E
    superficial peroneal, 301E
    superior gluteal, 301E
    supraclavicular, 295E
    suprascapular, 298E
    thoracodorsal, 298E
    tibial, 301E
    transverse cervical, 295E
    trigeminal (V), 328E
    trochlear (IV), 328E
    ulnar, 298E
    upper lateral brachial cutaneous, 298E
    upper subscapular, 298E
    vagus (X), 329E
    vestibular, 329E
    vestibulocochlear (VIII), 328E, 329E
Nerve damage, 297, 300, 302
Nerve fiber
    A, 272
    adrenergic, 363
    B, 272
    C, 272
    cholinergic, 363
    definition of, 266
    general somatic afferent, 268
    general somatic efferent, 268
    general visceral afferent, 268
    general visceral efferent, 268
    postganglionic, 358
    preganglionic, 358
Nerve impulse
    all-or-none principle, 271
    conduction across synapses, 272–275
    definition of, 270
    excitability, 269–271
    facilitation and, 273
    inhibitory transmission, 274
    integration at synapses, 274, 275
    membrane potential, 268, 269
    neurotransmitters, 275, 276
    one-way conduction, 272, 273
    organization of synapses, 276–278
    saltatory conduction, 271, 272
    speed of, 272
    summation and, 274
    synapse and, 273–275
Nervous system, 8E, 262–335, 357–368
    acupuncture and, 343
    afferent, 262
    aging and, 327
    autonomic, 262, 357–368
    basal ganglia, 318, 319
    and biofeedback, 367, 368
    blood supply of, 310
    brain, 307–327
    central, 262, 281–337
    cerebrospinal fluid (CSF), 307–310
    cranial nerves, 327–329
    definition of, 262
    dermatomes, 302
    development of, 328–332
    disorders of, 302–304, 332–335
    efferent, 262
    electroencephalogram (EEG), 323
    excitatory transmission of, 273, 274
    grouping of neural tissue, 281
    histology of, 263–268
    inhibitory transmission of, 274
    integrative functions of, 274, 275, 351–354
    medical terminology, 335
    meditation and, 368
    memory and, 351, 352
    meninges, 261, 264, 281, 284
    motor pathways of, 348–351
    nerve impulse of, 268–275
    neuroglia of, 263, 264E
    neurons of, 264–268
    neurotransmitters of, 199, 325–327, 363
    organization of, 262, 263
    peripheral, 262
    plexuses, 294–302
    reflexes and, 285, 287–292
    regeneration of, 202, 203, 276
    sensations and, 331–345
    sensory pathways of, 345–348
    sleep and wakefulness and, 352–354
    somatic, 262
    spinal cord, 281–329
    spinal nerves, 292–303
    synapses of, 272–275
Nervous tissue, 77, 95, 262–278
    all-or-none principle of, 203
    classification of, 263–268
    and conductivity, 268–276
    excitability of, 269–271
    excitatory transmission of, 273, 274
    grouping of, 281
    histology of, 77, 95, 263–268
    impulse speed of, 272
    inhibitory transmission of, 274
    and integration at synapses, 274, 275
    nerve impulse and, 268–276
    neuroglia of, 263–264E
    organization of, 262, 263
    regeneration of, 202, 203, 276
    saltatory conduction and, 271, 272
    structural variations of, 267
    synapses of, 273–275
Nervous tunic. *See* Retina
Neural crest, 328, 429
Neural fold, 328
Neural groove, 328
Neuralgia, 335
Neural plate, 328
Neural tube, 328
Neural pathway
    for auditory sensations, 390, 391
    for blood pressure regulation, 493, 494
    for defecation, 624, 625
    for deglutition, 600, 601
    for ejaculation, 730
    for equilibrium, 392, 393
    for erection, 730
    for gastric secretion, 607, 608
    for gustation, 373
    for heart rate, 475–477
    for liver secretion, 613
    for micturition, 686
    for muscle tone, 205, 347
    for olfaction, 371
    for pain, 342, 346, 347
    for pancreatic secretion, 611
    for posture, 347, 348
    for pressure, 346, 347
    for proprioception, 346, 347
    for respiration, 579–581
    for salivation, 598
    for skeletal muscles, 349–351
    for temperature, 346, 347
    for touch, 346, 347
    for vibration, 346, 347
    for vision, 384, 386
Neuritic plaque, 334
Neuritis, 303
Neuroblastoma, 433
Neuroeffector junction, 272, 363
Neurofibral node, 267

Neurofibril, 264
Neurofibrilary tangle, 334
Neuroglandular junction, 272
Neuroglia
    astrocytes, 264E
    ependyma, 264E
    microglia, 264E
    oligodendrocytes, 264E
Neurohypophyseal bud, 428
Neurohypophysis, 409–412
Neurolemma, 266
Neurolemmocyte, 266
Neurology, 262
Neuromuscular junction, 199, 200, 272
Neuronal pool, 276, 277
Neuron
    afferent, 262, 267
    all-or-none principle of, 271
    alpha efferent, 344, 345
    association, 267, 268
    bipolar, 266, 379
    classification of, 267, 268
    and conductivity, 268–275
    definition of, 95, 264
    efferent, 262, 267
    excitability, 269–271
    first-order, 346
    gamma efferent, 344
    ganglion, 379
    impulse speed of, 272
    inhibitory transmission of, 274
    and integration at synapses, 274, 275
    lower motor, 351
    membrane potential of, 268, 269
    multipolar, 267
    nerve impulse and, 268–275
    neurotransmitters of, 275, 276
    and one-way impulse conduction, 272, 273
    photoreceptor, 379
    physiology of, 268–275
    postganglionic, 360, 361
    postsynaptic, 272
    preganglionic, 358
    presynaptic, 272
    regeneration of, 276, 302, 303
    saltatory conduction and, 271, 272
    second-order, 346
    structural variations of, 267
    structure of, 264–268
    synapses of, 273–275
    third-order, 346
    unipolar, 267
    upper motor, 351
    visceral efferent postganglionic, 358–361
    visceral efferent preganglionic, 358
Neuropeptide, 325–327
Neurophysin, 410
Neurosecretory cell, 410
Neurosyphilis, 736
Neurotransmitter, 199, 273, 275, 325–327, 363
Neurotransmitter receptor, 274
Neutrophil, 445
Neurophysin, 410
Nevus, 113
Nicotinamide adenine dinucleotide, 639
Nicotinic receptor, 363
Night blindness, 384, 652E
Nipple, 728
Nissl body. See Chromatophilic substance
Nitrogen base, 41
Nitrogen narcosis, 573
Nitroglycerin, 482
Nociceptor, 340, 342
Node of Ranvier. See Neurofibral node
Nodule, 113
Nonaxial movement, 181

Nonelectrolyte, 694
Nongonococcal urethritis (NGU), 736
Nonimmune phagocytosis, 539, 540
Nonrapid eye movement (NREM) sleep, 353, 354
Nonspecific resistance
    antimicrobial substances, 538, 539
    chemical factors, 538
    fever, 542, 658
    inflammation, 540–542
    mechanical factors, 538
    phagocytosis, 539, 540
Nonspecific urethritis (NSU). See Nongono-
    coccal urethritis (NGU)
Norepinephrine (NE), 325, 423
Nose, 556, 557
Nostril. See External nares
Notch
    acetabular, 166
    clavicular, 153
    fibular, 168
    greater sciatic, 165
    jugular, 153
    lesser sciatic, 165
    mandibular, 145
    radial, 162
    trochlear, 162
    ulnar, 164
Notochord, 128
Nuclear bag fiber, 344
Nuclear cardiology, 482
Nuclear chain fiber, 344
Nuclear magnetic resonance (NMR). See
    Magnetic resonance imaging (MRI)
Nuclear medicine, 30
Nuclear membrane, 54
Nuclease, 618
Nucleic acid
    chemical composition of, 41
    definition of, 40
    digestion of, 618
    DNA, 40, 62–69
    functions of, 41
    RNA, 40, 62–64
    structure of, 40, 41
Nucleolus, 54
Nucleosome, 54
Nucleotide, 40
Nucleus
    amygdaloid, 319
    anterior, 314, 319
    caudate, 318
    of cell, 54
    cerebellar, 324
    cerebral, 281, 318, 319
    claustrum, 319
    corpus striatum, 318
    cuneatus, 311
    globus pallidus, 318
    gracilis, 311
    hypothalamic, 315, 316
    inferior salivatory, 598
    lateral geniculate, 314
    lentiform, 318
    medial geniculate, 314
    of nervous system, 281, 318, 319
    pulposus, 128, 146, 155
    putamen, 318
    raphe, 706
    red, 313
    reticular, 314
    substantia nigra, 313
    subthalamic, 319
    superior salivatory, 598
    ventral anterior, 314
    ventral lateral, 314

    ventral posterior, 314
    vestibular complex, 313
Nucleus cuneatus, 311
Nucleus gracilis, 311
Nucleus pulposus, 128, 146, 155
Nutrient
    absorption of, 618–621
    and calories, 655
    carbohydrates, 636–643
    definition of, 634
    digestion of, 617, 618
    and disease, 658–670
    lipids, 644–646
    and metabolic rate, 656
    metabolism of, 634–655
    minerals, 649, 650, 651E, 652E
    and obesity, 658, 659
    proteins, 646
    storage of, 643, 644, 646
    tissue repair and, 97
    vitamins, 652E–654E
    water, 621
Nystagmus, 395

Obesity
    causes, 659
    hypertrophic, 658
    hyperplastic, 658
    regulation of food intake, 590
    treatment, 659
Obligatory reabsorption, 674
Oblique fissure, 564
Obstetrics, 753
Occipital bone, 140
Occipital lobe, of cerebrum, 318
Occlusion, 526
Odor threshold, 371
Oil gland. See Sebaceous gland
Olecranon, 162
Olfaction, theories of, 371
Olfactory bulb, 371
Olfactory cell, 371
Olfactory gland, 371
Olfactory hair, 371
Olfactory region, 556
Olfactory sensation, 371
Olfactory threshold, 371
Olfactory tract, 371
Olfactory vesicle, 371
Oligodendrocyte, 264E
Olive, of medulla, 313
Oncogene, 71
Oncologist, 69
Oncology, 69
One-way impulse conduction, 272, 273
Oogenesis, 720
Oogonium, 720
Oophorectomy, 731
Ootid, 720
Ophthamology, 373
Opsonization, 539, 540
Optic chiasma, 384
Optic disc, 379
Optic tract, 384
Oral cavity, 595–600
Oral contraception, 732
Oral membrane, 625
Ora serrata, 377
Orbit, 136
Organ, 6, 101
Organelle
    centriole, 60
    centrosome, 60
    cilium, 61
    definition of, 46, 53
    endoplasmic reticulum, 56

flagellum, 61
Golgi complex, 56–58
intermediate filament, 60
lysosome, 59
microfilament, 59, 60
microtubule, 60
mitochondrion, 58, 59
nucleus, 54
peroxisome, 59
ribosome, 55
Organic compound, 33, 37
Organism, 6
Organ of Corti. *See* Spiral organ
Orgasm, 730, 731
Origin of skeletal muscle, 216
Oropharynx, 557
Orthopedics, 120
Orthopnea, 586
Osmosis, 49, 50
Osmotic pressure, 49, 50, 672, 673
Osseous tissue
    development of, 126–128
    disorders of, 128–130
    functions of, 120
    growth of, 125
    histology, 94, 120–123
    homeostasis of, 126
    medical terminology, 130
    ossification, 123–125
    remodeling, 126
Ossification
    definition of, 123
    endochondral, 123–125
    intramembranous, 123
Osteitis, 130
Osteitis fibrosa cystica, 418, 419
Osteoblast, 120, 127
Osteoblastoma, 130
Osteochondroma, 130
Osteoclast, 126
Osteocyte, 94, 120, 121
Osteogenic periosteum, 120
Osteogenic sarcoma, 70, 125
Osteoma, 128
Osteomalacia, 128
Osteomyelitis, 128
Osteon, 94, 120
Osteoporosis, 128
Osteosarcoma, 130
Otalgia, 395
Otitis media, 394
Otolith, 392
Otolithic membrane, 392
Otorhinolaryngology, 556
Otosclerosis, 395
Oval window. *See* Fenestra vestibuli
Ovarian cycle, 722–726
Ovarian cyst, 738
Ovarian follicle, 719
Ovarian ligament, 718
Ovary, 427, 428, 717–720
Overhydration, 697
Ovulation, 720, 725
Ovum, 720
Oxidation. *See* Cellular respiration
Oxygen
    and carbon monoxide poisoning, 578
    and DPG, 577
    and hemoglobin, 575, 576
    and pH, 576
    and temperature, 577
    transport of, 575–578
Oxygenated blood, 573
Oxygen debt, 210
Oxygen toxicity, 578
Oxyhemoglobin, 575

Oxyphil cell, 416
Oxytocin (OT), 410, 760

Pacemaker. *See* Sinoatrial node
Pacinian corpuscle. *See* Lamellated corpuscle
Paget's disease, 128
Pain
    and acupuncture, 343
    and endorphins, 325, 326
    phantom, 342
    referred, 342
    somatic, 342
    visceral, 342
Palatine bone, 145
Palatine tonsil, 533
Palatoglossal arch, 596
Palatopharyngeal arch, 596
Palm, 164
Palpebrae. *See* Eyelids
Palpebral conjunctiva, 376
Palpebral fissure, 375
Palpitation, 484
Pancreas, 424–427, 609–611
Pancreatic amylase, 611, 618
Pancreatic duct, 610
Pancreatic islet, 424, 610
Pancreatic juice, 610, 611
Pancreatic lipase, 611, 618
Pancreatic secretion, regulation of, 611
Papilla
    dermal, 103
    duodenal, 610
    of hair, 106
    of kidney, 666
    of tongue, 372, 373, 596
Papillary duct, 667
Papillary muscle, 465
Papillary region of dermis, 103
Papilloma virus, 71
"Pap" smear, 721
Papule, 113
Parafollicular cell, 413
Parallel after-discharge circuit, 277, 278
Parallel fasciculus, 216, 218
Paralysis, 212, 335
Paramesonephric duct, 733
Paranasal sinus
    definition of, 144
    ethmoidal, 142
    frontal, 137
    maxillary, 143
    sphenoidal, 141
Paraplegia, 302
Parasagittal plane, 12
Parasympathetic division of ANS, 262, 263
Parasympathetic nervous system, 262, 263
Parathyroid gland, 416–419
Parathyroid hormone (PTH), 416–419
Paraurethral gland, 727
Parenchymal repair, 97
Parenchymal tissue, 97
Parietal bone, 137
Parietal (oxyntic) cell, 605
Parietal lobe of cerebrum, 318
Parietal pericardium, 462
Parietal peritoneum, 592
Parietal pleura, 561
Parietooccipital sulcus, 318
Parkinson's disease, 332, 333
Parotid duct, 596
Parotid gland, 596
Paroxysmal tachycardia, 484
Pars intermedia, 404
Partial antigen, 543
Partial pressure, 572

Partial-thickness burn, 111
Parturition, 756, 757
Passive processes, 48–51
Patella, 166
Patellar ligament, 190, 250E
Patellar reflex, 289, 292
Patent ductus arteriosus, 483
Pathogen, 537
Pathological anatomy, 6
Pathological glycosuria, 682
Pectoral (shoulder) girdle, 159, 160
Pedicel, 666
Pedicle, 150
P_{eff}. *See* Effective filtration pressure
Pellegra, 654E
Pelvic cavity, 13
Pelvic diaphragm, 235E
Pelvic (hip) girdle, 165, 166
Pelvic inflammatory disease (PID), 738
Pelvic inlet, 165
Pelvic outlet, 165
Pelvimetry, 165
Pelvis, 165
Penis, 716, 717
Pennate fasciculus, 218
Pepsin, 605
Pepsinogen, 606
Peptidase, 617, 618
Peptide, 605, 606
Peptide bond, 40
Perception, 339
Percutaneous transluminal coronary angioplasty (PTCA), 482
Percutaneous ultrasonic lithotripsy (PUL), 684
Perforating canal, 120
Pericardial cavity, 13, 462
Pericardial fluid, 462
Pericarditis, 462
Pericardium, 96, 462
Perichondrium, 94, 123
Periderm, 110
Perikaryon. *See* Cell body, of neuron
Perilymph, 389
Perimetrium, 721
Perimysium, 196
Perineum, 236E, 728
Perineurium, 293, 728
Perinuclear cisterna, 54
Periodontal disease, 627
Periodontal ligament, 598
Periosteum, 120, 124
Peripheral nerve damage, 302
Peripheral nerve repair, 302, 303
Peripheral nervous system (PNS), 262
Peripheral protein, 46, 47
Peripheral resistance, 492
Peristalsis, 601, 617, 624
Peritoneal cavity, 592
Peritoneal fluid, 592
Peritoneum, 96, 592–595
Peritonitis, 627
Peritubular capillary, 668
Perivascular space, 310
Permanent dentition, 599
Peroxisome, 59
Perpendicular plate, 142
Perspiration, 108
Peyer's patch. *See* Aggregated lymphatic follicle
PG. *See* Prostaglandins
pH
    and buffer systems, 36, 37
    definition of, 35
    and kidney excretion, 675, 676
    maintenance of, 36, 37

pH (*Continued*)
  and respirations, 581, 582
  scale, 35
Phagocytic migration, 541, 542
Phagocytic vacuole, 540
Phagocytosis
  adherence phase of, 539, 540
  cells involved in, 539
  ingestion phase of, 540
  mechanism, 539, 540
Phagolysosome, 540
Phalanges
  of fingers, 164
  of toes, 171
Phantom pain, 342
Pharmacology, 72
Pharyngeal pouch, 429
Pharyngeal tonsil, 533
Pharynx, 556, 557
Phenotype, 760, 761
Phenylketonuria (PKU), 659, 761
Pheochromocytoma, 423, 525
Philtrum, 556
Phlebitis, 526
Phlebotomy, 458
Phosphate
  distribution of, 696
  functions of, 696
  regulation of, 696
Phosphate buffer system, 698
Phosphocreatine, 202
Phospholipid, 38, 39
Phospholipid bilayer, 46
Phosphorylation, 636
Photopigment, 383
Photopsin, 384
Photoreceptor neuron, 379
Physiology, 6
Pia mater, 284, 307
Pigmentation, 104
Pili. *See* Hair
Pineal gland, 428
Pinealocyte, 428
Pinna, 386
Pinocytosis, 51
Pisiform, 164
Pitch, 559
Pituicyte, 409
Pituitary dwarfism, 407
Pituitary gland
  adenohypophysis, 404, 405
  development of, 428, 429
  disorders of, 407, 408
  hormones of, 405–409, 410, 411
  neurohypophysis, 409, 410
Pivot joint, 182
Placenta, 521, 522, 750
Placental luteotropic releasing factor (pLRF),
    408, 753
Planes of body
  frontal, 12
  horizontal, 12
  midsagittal, 12
  parasagittal, 12
  sagittal, 12
Plantar flexion, 181
Plantar reflex, 292
Plaque
  atherosclerotic, 481
  dental, 627
Plasma cell, 91, 446, 547
Plasma coagulation factor, 449, 450E
Plasma membrane
  definition of, 46
  functions of, 47, 48
  movement of materials across, 48–53

  selective permeability of, 46–48
  structure of, 46, 47
Plasmapheresis, 212
Plasma protein, 448, 449E
Plasmin, 452
Plasminogen, 452
Platelet. *See* Thrombocyte
Platelet coagulation factor, 447
Platelet-derived growth factor (PDGF), 481
Platelet plug formation, 449
Pleural cavity, 562
Pleural membrane, 561, 562
Pleurisy, 562
Plexus
  autonomic, 363
  brachial, 296
  celiac, 362
  cervical, 294
  choroid, 308
  deep cardiac, 363
  definition of, 294
  hypogastric, 362
  inferior mesenteric, 362
  lumbar, 297
  myenteric, 592
  pulmonary, 363
  sacral, 298
  submucous, 592
  superficial cardiac, 363
  superior mesenteric, 362
Plexus of Auerbach. *See* Myenteric plexus
Plexus of Meissner. *See* Submucous plexus
Plicae circulares, 616
Pneumonia, 584
Pneumotaxic area, 581
Pneumothorax, 562
Podocyte, 666
Polarized membrane, 269
Poliomyelitis, 332
Polycystic disease, 688
Polycythemia, 457
Polyneuritis, 653E
Polyp, 113
Polyribosome, 62
Polysaccharide, 37
Polysomograph, 354
Polysynaptic reflex arc, 289
Polyunsaturated fat, 38
Polyuria, 462, 689
Pons
  cranial nerves of, 313
  nuclei of, 313
  reticular formation of, 313
  structure of, 313
Positive feedback system, 20
Positron emission tomography (PET), 20, 322
Postabsorptive (fasting) state, 646–649
Postcentral gyrus, 318
Posterior cavity of eyeball, 380
Posterior chamber of eyeball, 379
Posterior column pathway, 346, 347
Posterior gray horn, 284
Posterior lymph sac, 550
Posterior root, 285
Posterior root ganglion, 285
Posterior spinocerebellar tract, 286E, 347
Posterior white column, 284
Postganglionic fiber, 358
Postganglionic neuron, 358, 360, 361
Postovulatory phase, 725, 726
Postsynaptic neuron, 272
Postural hypotension, 526
Potassium
  control of, 696
  deficiency of, 696

  distribution of, 696
  functions of, 696
Potassium channel, 269
Potential difference of plasma membrane, 269
Potential hazards to embryo and fetus, 759,
    760
Pott's disease, 130
Pouch of Douglas. *See* Rectouterine pouch
Precapillary sphincter, 490
Precentral gyrus, 318
Preeclampsia, 763
Prefrontal lobotomy, 343
Preganglionic fiber, 358
Preganglionic neuron, 358
Pregnancy
  anatomical changes during, 753–755
  beginnings of organ systems during, 745–751
  birth control, 731–733
  definition of, 743
  development of blastocyst during, 744
  diagnosis during, 727, 752
  embryonic membranes during, 747, 749, 750
  embryonic period, 747–751
  external human fertilization, 746
  fertilization of ovum, 743, 744
  fetal growth, 751, 752
  formation of morula, 744
  gestation, 753–755
  hazards to embryo and fetus, 759, 760
  hormones of, 752, 753
  implantation during, 744, 746
  infant adjustments at birth, 758, 759
  lactation, 760
  parturition and labor, 756–759
  physiological changes during, 753–755
  placenta formation during, 750
  prenatal diagnostic techniques, 755, 756
  sexual intercourse and, 730, 731
  umbilical cord, 750
Pregnancy test, 752
Premature infant, 758
Premenstrual syndrome (PMS), 737
Premolar (bicuspid), 599
Premotor area, 322
Preoptic area, 657
Preovulatory phase, 725
Prepuce, 717, 727
Presbyopia, 382, 395
Pressure changes, of heart, 472
Pressure sensation, 341, 342
Presynaptic neuron, 272
Prevertebral ganglion, 358
Primary auditory area, 322
Primary bronchus, 560
Primary follicle, 724
Primary germ layers
  derivatives of, 747E
  development of, 747
Primary gustatory area, 322
Primary motor area, 322
Primary nail field, 110
Primary olfactory area, 322
Primary oocyte, 720
Primary ossification center, 124
Primary plexus, of adenohypophysis, 404
Primary plexus, of infundibulum, 404
Primary response, of antibody, 547, 548
Primary somesthetic area, 321
Primary spermatocyte, 709
Primary taste sensation, 373
Primary vesicle, 329
Primary visual area, 322
Prime mover (agonist), 218
Primitive gut, 625
Primitive heart tube, 480
Principal cell, 416

P-R interval, 470
Process
  alveolar, 143, 145
  condylar, 144
  coracoid, 160
  coronoid, 145, 162
  definition of, 134E
  inferior articular, 150
  mastoid, 140
  odontoid (dens), 150
  olecranon, 162
  palatine, 143
  pterygoid, 141
  spinous, 150
  styloid, of radius, 164
  styloid, of temporal bone, 140
  styloid, of ulna, 164
  superior articular, 150, 152
  temporal, 144
  transverse, 150, 152
  xiphoid, 152
  zygomatic, 137
Proctodeum, 625
Proctology, 623
Progesterone, 427, 723, 752
Projection, 339
Projection fiber, 318
Prolactin (PRL), 404, 408, 760
Prolactin inhibiting factor (PIF), 408
Prolactin releasing factor (PRF), 408–409
Pronation, 181
Pronator muscle, 219
Pronephric duct, 687
Pronephros, 687
Properdin, 539
Prophase, 65, 66
Proprioception, 344, 345
Proprioceptive sensation, 344, 345
Proprioceptor, 340, 344, 345
Prorubricyte, 444
Prosencephalon, 328
Prostaglandins (PG)
  chemical composition of, 38
  functions of, 38, 403
  and hormonal action, 403
  and inflammation, 541
Prostatectomy, 737
Prostate gland, 715
Prostatic urethra, 714
Prostatitis, 737
Protein
  absorption of, 619
  amino acids and, 40
  anabolism of, 646
  buffer, 699
  catabolism, 646
  classification of, 40E
  definition of, 39
  digestion of, 618
  fate of, 646
  functions of, 40E, 646, 647
  integral, 46
  levels of structural organization of, 40
  peptide bonds and, 40
  peripheral, 46, 47
  synthesis of, 62–64
Protein-bound iodine (PBI), 414
Protein buffer, 699
Protein kinases, 402
Protein synthesis, 62–64
Prothrombin, 449–452
Prothrombin activator, 449–542
Prothrombin time, 453
Proto-oncogene, 71
Protraction, 181
Provitamin, 652

Proximal convoluted tubule, 666
Pruritus, 113
Pseudopodia, 51, 540
Pseudostratified epithelium, 77, 84
Psoriasis, 111
Psychosomatic disorder, 316
Ptosis, 395, 666
Pubis, 165
Pudendal nerve block, 756, 757
Pudendum. See Vulva
Puerperal fever, 763
Pulled groin, 248E
Pulled hamstrings, 250E
Pulmonary artery, 520
Pulmonary capacity, 572
Pulmonary circulation, 498, 520
Pulmonary edema, 586
Pulmonary embolism (PE), 585, 586
Pulmonary semilunar valve, 466
Pulmonary trunk, 520
Pulmonary vein, 498, 520
Pulmonary ventilation, 565–570
Pulmonary volume, 570, 572
Pulp, 598
Pulp cavity, 598
Pulsating electromagnetic fields (PEMFs), 130
Pulse, 496
Pulse generator, 470, 472
Pulse pressure, 498
Punch biopsy, 721
Puncta lacrimalia, 376
Punnett square, 761
Pupil, 377
Purine, 41
Purkinje fiber. See Conduction myofiber
Pus, 542
Pustule, 113
Putamen, 318
P wave, 470
Pyelitis, 688
Pyelonephritis, 688
Pyloric sphincter (valve), 605
Pyloric stenosis, 605
Pylorospasm, 605
Pylorus, of stomach, 603
Pyogenesis, 542
Pyorrhea, 627
Pyramid, of medulla, 311
Pyramidal lobe, of thyroid gland, 412
Pyramidal pathway, 349, 350
Pyrimidine, 41
Pyrogen, 658
Pyruvic acid
  in glycolysis, 636–638
  in oxygen debt, 638
Pyuria, 682

QRS wave, 470
Quadrants, of abdominopelvic cavity, 13
Quadrate lobe of liver, 611
Quadriplegia, 302
Quantum, of acetylcholine, 275

Radial keratotomy, 376
Radiation, 656
Radiographic anatomy, 6, 17, 18
Radioisotope, 30
Radioisotopic scanning, 30
Radiology, 17
Radius, 164
Rales, 586
Rami communicantes, 294, 361, 362
Ramus
  of mandible, 144
  of pubis, 165
  of spinal nerves, 294

Raphe, 706
Rapid eye movement (REM) sleep, 353, 354
Rapidly adapting receptor, 339, 340
Rathke's pouch. See Hypophyseal pouch
Raynaud's disease, 526
Reactivity, 542
Receptor
  alpha, 364
  beta, 364
  definition of, 287, 339
  for discriminative touch, 341
  for equilibrium, 392, 393
  for gustation, 372, 373
  for hearing, 386, 392
  hormone, 400, 401
  for light touch, 341
  muscarinic, 363
  nicotinic, 363
  for olfaction, 371
  for pain, 342, 343
  for pressure, 341, 342
  for proprioception, 344, 345
  thermoreceptive, 342
  for vibration, 342
  for vision, 373–386
Receptor-mediated endocytosis, 51
Recessive gene, 761
Reciprocal innervation, 289
Reciprocal transfusion, 458
Recombinant DNA, 64
Recovery heat, 210
Recruitment, 200
Rectouterine pouch, 721
Rectum, 622
Red blood cell. See Erythrocyte
Red-green color blindness, 384, 762
Red marrow, 120
Red muscle fiber, 205, 206
Red nucleus, 313, 318
Red pulp, of spleen, 534
Reduction, chemical, 634
Reduction division, 69
Referred pain, 342
Reflex
  abdominal, 292
  Achilles, 292
  aortic, 476
  Babinski sign, 292
  Bainbridge, 477
  carotid sinus, 475, 476
  crossed extensor, 290–292
  deep tendon, 292
  defecation, 624, 625
  definition of, 288
  deglutition, 600
  and diagnosis, 292
  flexor, 290–292
  micturition, 686
  patellar, 292
  plantar, 292
  reinforcement of, 292
  right heart (atrial), 476, 477
  somatic, 288
  spinal, 288
  stretch, 288, 289
  superficial, 292
  tendon, 289, 290
  visceral autonomic, 366, 367
Reflex arc
  contralateral, 291
  definition of, 287
  intersegmental, 291
  ipsilateral, 289
  monosynaptic, 288
  polysynaptic, 289
Reflex center, 285, 286

Refraction, 380
Refractory period
    absolute, 271
    relative, 271
Regeneration, 276, 302, 303
Regional anatomy, 6
Regulating factor, 327, 404
Regulation of heart rate, 475–477
Reinforcement, 292
Rejection, of tissue, 550
Relative refractory period, 271
Relaxation period, 203, 473
Relaxin, 427, 428, 723, 753
Releasing factor, 404
Remodeling of bone, 126, 130
Renal calculus, 684
Renal capsule, 664
Renal clearance test, 682
Renal column, 666
Renal corpuscle, 666
Renal erythropoietic factor, 430
Renal failure
    acute, 688
    chronic, 688
Renal fascia, 666
Renal function, tests for, 682
Renal papilla, 666
Renal pelvis, 666
Renal plexus, 669
Renal pyramid, 666
Renal sinus, 664
Renal tubule, 666
Renin, 420, 674
Renin-angiotensin pathway, 420, 674
Rennin, 607
Repolarized membrane, 270, 271
Reproductive cell division, 67–69
Reproductive systems, 8E, 706–738
    aging and, 732, 733
    birth control and, 731, 732
    definition of, 706
    development of, 733, 735
    disorders of, 735–738
    estrogens and, 722, 752
    female organs of, 717–722
    gamete formation and, 708
    inhibin and, 712, 713
    male organs of, 706–717
    medical terminology, 738
    pregnancy and, 743–752
    progesterone and, 723, 752
    relaxin and, 723, 753
    sexual intercourse and, 730, 731
    testosterone and, 711, 712
Residual body, 540
Residual volume, 571
Resistance
    nonspecific, 537–542
    specific, 542–548
Resistance reaction, of stress, 743
Respiration, 542, 556
    air volumes exchanged, 572, 573
    control of, 579–582
    external, 573–575
    gas laws and, 572–573
    internal, 575
    modified movements, 570
    and nebulization, 564
    and pH, 699
    pulmonary ventilation, 565–570
    and smoking, 583
    transport of gases, 575–579
Respiration, control of
    chemical, 581, 582
    nervous, 579–581
Respirator, 586

Respiratory bronchiole, 564
Respiratory center, 579
Respiratory distress syndrome (RDS) of the
    newborn, 584, 585
Respiratory failure, 585
Respiratory gases
    exchange of, 572, 573
    gas laws and, 572, 573
    transport of, 575–579
Respiratory system, 555–586
    adjustments of infant at birth, 758, 759
    aging and, 582, 583
    air volumes exchanged, 559, 560
    control of, 579–582
    definition of, 555
    development of, 583
    disorders of, 583–586
    expiration, 569
    external respiration, 573–575
    gas laws and 572–573
    inspiration, 566–569
    internal respiration, 575
    medical terminology, 586
    modified movements, 570
    nebulization and, 564
    organs of, 556–565
    pulmonary ventilation, 565–570
    smoking and, 583
    transportation of gases, 575–579
Resting potential. See Membrane potential
Rest-repose system, 366
Retention, 686
Rete testis, 713
Reticular activating system (RAS), 353
Reticular connective tissue, 93
Reticular fiber, 62, 89
Reticular formation, 312
Reticular lamina, 77
Reticular nucleus, 314
Reticular region of dermis, 103
Reticulin, 62, 89
Reticulocyte count, 445
Retina, 378, 379
Retinal, 383
Retinal image formation, 380, 382
Retinoblastoma, 395
Retraction, 181
Retroflexion, 721
Retroperitoneal, 593, 664
Retroperitoneal lymph sac, 550
Reverberating (oscillatory) circuit, 277
Reye's syndrome (RS), 334
Rh blood grouping, 455, 456
Rheumatism, 191
Rhinitis, 586
Rhizotomy, 342, 343
Rhodopsin, 383
Rhombencephalon, 329
Rib
    false, 154
    floating, 154
    true, 154
Ribonuclease, 611, 617, 618
Ribonucleic acid
    chemical composition of, 42
    comparison with DNA, 42
    functions of, 42, 62–64
    messenger, 62–64
    and nucleoli, 62
    and protein synthesis, 62–64
    ribosomal, 62–64
    transfer, 62–64
Ribosomal RNA, 62–64
Ribosome
    definition of, 55
    and protein synthesis, 62–64

Rickets, 128
Right colic flexure, 622
Right heart (atrial) reflex, 476, 477
Right lobe of liver, 611
Right lymphatic duct, 537
Rigidity, 332
Rigor mortis, 200
Risk factors, heart disease, 479
RNA. See Ribonucleic acid
Rod, 379
Roentgenogram, 17
Root
    of hair, 106
    of lung, 564
    of nail, 109
    of nerve plexus, 294
    of nose, 556
    of penis, 717
    of tooth, 598
Root canal, 598
Root hair plexus, 108, 341
Rotation, 181
Rotator cuff, 187
Rotator muscle, 219E
Round ligament, 524, 611, 721
Round window. See Fenestra cochlea
rRNA. See Ribosomal RNA
Rubriblast, 441
Rubricyte, 444
Rubrospinal tract, 287E, 350
Rugae
    of gallbladder, 614
    of stomach, 605
    of urinary bladder, 685
    of vagina, 726
Ruptured tympanic membrane, 387

Saccule, 390
Sacral canal, 152
Sacral hiatus, 152
Sacral parasympathetic outflow, 362
Sacral plexus, 298, 300, 301E
Sacral promontory, 152
Sacrum, 150
Saddle joint, 182
Sagittal plane, 12
Saliva
    composition of, 596–598
    secretion of, 598
Salivary amylase, 599, 600
Salivary gland, 596
Salt, 27, 28
Saltatory conduction, 271
Sarcolemma, 197
Sarcoma
    definition of, 70
    types of, 70, 71
Sarcomere, 198
Sarcoplasm, 197
Sarcoplasmic reticulum, 197
Satiety center, 316, 590
Saturated fat, 38
Saturated hemoglobin, 575, 576
Scab, 97
Scala tympani, 390
Scala vestibuli, 390
Scaphoid, 164
Scapula, 160
Scar tissue, 97
Schwann cell. See Neurolemmocyte
Sciatica, 303
Sclera, 376
Scleral venous sinus, 376
Sclerotome, 210
Scoliosis, 155
Scotoma, 386

Scotopsin, 383
Scrotum, 706, 707
Sebaceous ciliary gland, 376
Sebaceous gland, 108
Sebum, 108
Secondary (lobar) bronchus, 560
Secondary follicle, 724
Secondary oocyte, 720
Secondary ossification center, 125
Secondary plexus, of adenohypophysis, 404
Secondary response, 548
Secondary sex characteristics, 712, 722, 723
Secondary spermatocyte, 710
Secondary tubule, 728
Secondary vesicle, 329
Second-class lever, 216
Second-degree burn, 111
Second (intermediate) cuneiform, 170
Second messenger, 401
Second-order neuron, 346
Second polar body, 720
Secretin, 608E
Secretion, 8, 79
Secretory granule, 58
Segmentation, 617
Segmentation nucleus, 743
Selective permeability
    definition of, 47
    factors determining, 47, 48
Self-excitability, 469
Sella turcica, 141
Semen, 715, 716
Semen analysis, 716
Semicircular canal, 390
Semicircular duct, 390
Semilunar valve, 466
Seminal fluid. See Semen
Seminal plasmin, 716
Seminal vesicle, 714, 715
Seminiferous tubule, 707, 708
Senescence, 72
Senile macular degeneration (SMD), 379
Sensations
    auditory, 386–392
    cerebellar tracts of, 347, 348
    characteristics of, 339, 340
    classifications, 340
    cutaneous, 340–344
    definition of, 339
    disorders of, 394, 395
    equilibrium, 392, 393
    general, 340–351
    gustatory, 372, 373
    levels of, 345
    medical terminology, 395
    motor pathways of, 348–351
    olfactory, 371
    pain, 342–344
    prerequisites for, 339
    pressure, 341, 342
    proprioceptive, 344, 345
    sensory pathways of, 345–348
    special, 371–393
    tactile, 341, 342
    thermoreceptive, 342
    touch, 341
    vibration, 342
    visual, 373–386
Sense strand, 62
Sensitization, 545
Sensory areas of cerebrum, 321, 322
Sensory neuron. See Afferent neuron
Sensory pathway
    anterior spinothalamic, 347
    lateral spinothalamic, 346, 347
    and motor responses, 348, 349

posterior column, 347
    spinothalamic, 346
Septal cell, 564
Septal defect, 483
Septicemia, 458
Serosa
    definition of, 592
    of gallbladder, 614
    of GI tract, 592
    of large intestine, 624
    of small intestine, 616
    of stomach, 605
    of ureter, 685
    of urinary bladder, 685
    of uterine (Fallopian) tube, 720
    of uterus, 721
Serotonin (5-HT), 325
Serous membrane
    definition of, 96
    parietal layer of, 96
    visceral layer of, 96
Serous pericardium, 462
Sertoli cell. See Sustentacular cell
Serum, 449
Serum creatine test, 682
Sesamoid bone, 133
Severe combined immunodeficiency (SCID), 551
Sex chromosomes, 762
Sexual intercourse, 730, 731
Sexually transmitted disease (STD), 735–736
Sheath of Schwann. See Neurolemma
Shingles, 303, 304
Shinsplints, 168, 170
Short bone, 133
Sigmoid colon, 622
Sigmoidoscope, 628
Sign of Babinski, 292, 351
Simple columnar epithelium, 82–84
Simple cuboidal epithelium, 78, 79
Simple epithelium
    definition of, 77
    types of, 77–84
Simple gland, 85
Simple goiter, 416
Simple series circuit, 276
Simple squamous epithelium, 78
Single binocular vision, 382
Sinoatrial (SA) node, 470
Sinusitis, 144
Sinusoid, 490
Sinus venosus, 480
Skeletal muscle
    actions of, 219E
    adductor brevis, 248E
    adductor longus, 248E
    adductor magnus, 248E
    anconeus, 242E
    arrangement of fasciculi, 216, 218
    biceps brachii, 242E
    biceps femoris, 250E
    brachialis, 242E
    brachioradialis, 242E
    buccinator, 223E
    bulbocavernosus, 236E
    coccygeus, 235E
    coracobrachialis, 242E
    corrugator supercilii, 223E
    cremaster, 707
    cricothyroid, 231E
    criteria for naming, 218, 219
    deep transverse perineus, 236E
    deltoid, 240E
    depressor labii inferioris, 222E
    diaphragm, 234E
    digastric, 230E

epicranius, 220E
erector spinae. See Skeletal muscle, sacrospinalis
extensor carpi radialis longus, 244E
extensor carpi ulnaris, 244E
extensor digitorum, 244E
extensor digitorum longus, 252E
extensor indicis, 244E
external anal sphincter, 236E
external intercostals, 234E
external oblique, 232E
extrinsic, of eye, 225E
extrinsic, of tongue, 227E
flexor carpi radialis, 244E
flexor carpi ulnaris, 244E
flexor digitorum longus, 252E
flexor digitorum profundus, 244E
flexor digitorum superficialis, 244E
frontalis, 222E
gastrocnemius, 252E
genioglossus, 227E
geniohyoid, 230E
gluteus maximus, 248E
gluteus medius, 248E
gluteus minimus, 248E
gracilis, 250E
group actions of, 218
hamstrings, 250E
hyoglossus, 227E
iliacus, 248E
iliococcygeus, 235E
iliocostalis cervicis, 246E
iliocostalis lumborum, 246E
iliocostalis thoracis, 246E
iliopsoas, 248E
inferior constrictor, 228E, 231E
inferior oblique, 225E
inferior rectus, 225E
infrahyoid (strap), 231E
infraspinatus, 240E
innervations of, 222E–252E
insertions of, 222E–252E
internal intercostals, 234E
internal oblique, 232E
intramuscular injections, 255
intrinsic, of tongue, 596
ischiocavernosus, 236E
lateral pterygoid, 224E
lateral rectus, 225E
latissimus dorsi, 240E
levator ani, 235E
levator labii superioris, 222E
levator palpebrae superioris, 223E, 375
levator scapulae, 238E
in lever systems, 216
longissimus capitis, 232E, 246E
longissimus cervicis, 246E
longissimus thoracis, 246E
longitudinalis inferior, 596
longitudinalis superior, 596
masseter, 224E
medial pterygoid, 224E
medial rectus, 225E
mentalis, 223E
middle constrictor, 228E, 231E
mylohyoid, 230E
naming of, 218, 219
obturator internus, 248E
occipitalis, 222E
omohyoid, 231E
orbicularis oculi, 223E
orbicularis oris, 222E
origins of, 222E–252E
palatoglossus, 227E
palatopharyngeus, 228E, 231E
palmaris longus, 244E

Skeletal muscle (*Continued*)
  pectineus, 248E
  pectoralis major, 240E
  pectoralis minor, 238E
  peroneus brevis, 252E
  peroneus longus, 252E
  peroneus tertius, 252E
  piriformis, 248E
  platysma, 223E
  pronator quadratus, 242E
  pronator teres, 242E
  psoas major, 248E
  pubococcygeus, 235E
  quadratus lumborum, 246E
  quadriceps femoris, 250E
  rectus abdominis, 232E, 246E
  rectus femoris, 250E
  rhomboideus major, 238E
  rhomboideus minor, 238E
  risorius, 223E
  sacrospinalis, 246E
  salpingopharyngeus, 228E
  sartorius, 250E
  semimembranosus, 250E
  semispinalis capitis, 232E
  semitendinosus, 250E
  serratus anterior, 238E
  soleus, 252E
  spinalis thoracis, 246E
  splenius capitis, 232E
  stapedius, 388
  sternocleidomastoid, 232E
  sternohyoid, 231E
  sternothyroid, 231E
  styloglossus, 227E
  stylohyoid, 230E
  stylopharyngeus, 228E, 231E
  subclavius, 238E
  subscapularis, 240E
  superficial transverse perineus, 236E
  superior constrictor, 228E
  superior oblique, 225E
  superior rectus, 225E
  supinator, 242E
  suprahyoid, 230E
  supraspinatus, 240E
  temporalis, 224E
  tensor fasciae latae, 248E
  tensor tympani, 388
  teres major, 240E
  teres minor, 240E
  thyrohyoid, 231E
  tibialis anterior, 252E
  tibialis posterior, 252E
  trachealis, 560
  transversus abdominis, 232E
  transversus linguae, 596
  trapezius, 238E
  triceps brachii, 242E
  urethral sphincter, 236E
  vastus intermedius, 250E
  vastus lateralis, 250E
  vastus medialis, 250E
  verticalis linguae, 596
  zygomaticus major, 222E
Skeletal muscle tissue
  all-or-none principle of, 203
  connective tissue components of, 195
  contraction of, 199–206
  definition of, 195
  development of, 210, 211
  energy and, 202, 203
  fascia of, 195
  fast and slow, 206
  flaccidity of, 205
  histology of, 77, 94, 95, 197–199

  hypertrophy of, 205
  motor unit of, 200, 201
  nerve and blood supply of, 197
  physiology of, 201–205
  sliding-filament theory of, 199
Skeletal system, 133–173
  appendicular skeleton of, 133, 159–173
  arches of foot, 171
  axial skeleton of, 133–155
  comparison of female and male, 172, 173
  definition of, 120, 133
  development of, 126
  disorders of, 128, 130, 156
  divisions of, 133
  fontanels, 136
  foramina of skull, 147E
  fracture repair, 129, 130
  fractures of, 128, 129
  functions of, 120
  histology of, 120–123
  homeostasis of, 126
  hyoid, 146
  lower extremities, 166–171
  medical terminology, 130
  ossification, 123–125
  paranasal sinuses of, 144
  pectoral (shoulder) girdles, 159, 160
  pelvic (hip) girdle, 165, 166
  remodeling of, 126
  skull, 133–146
  surface markings of, 133, 134E
  sutures, 134
  thorax, 152–155
  upper extremities, 160–164
  vertebral column, 149–152
Skene's gland. *See* Paraurethral gland
Skin
  acne of, 110
  aging and, 110
  and body temperature, 109, 657
  burns of, 111, 112
  cancer of, 111
  cleavage lines of, 103
  cold sores, 113
  color of, 104
  decubitus ulcer of, 111
  derivatives of, 105–109
  dermatomes of, 210
  dermis of, 103
  development of, 110
  disorders of, 110–112
  epidermal ridges and grooves, 104
  epidermis, 101–103
  functions of, 101
  and immunity, 548, 549
  impetigo, 113
  mechanical factors of, 538
  pruritus, 113
  psoriasis of, 111
  structure of, 101–103
  sunburn of, 111
  systemic lupus erythematosus (SLE), 110, 111
  temperature regulation and, 109, 110
  warts of, 113
Skin cancer, 111
Skull
  bones of, 133–146
  fontanels of, 136
  foramina of, 146, 147E
  paranasal sinuses of, 144
  sutures of, 134, 136
Sleep
  apnea, 570
  and arousal, 353
  and EEG, 353

  nonrapid eye movement (NREM), 353
  rapid eye movement (REM), 353, 354
  and reticular activating system (RAS), 353
Sliding-filament theory, 199
Slow-twitch red fiber, 206
Slowly adapting receptor, 340
Small intestinal absorption, 618–621
Small intestine, 614–621
Smegma, 738
Smoker's cough, 583
Smoking and respiration, 583
Smooth muscle tissue
  definition of, 195, 208
  development of, 211
  histology of, 208, 209
  kinds of, 209
Sodium
  control of, 696
  deficiency of, 696
  distribution of, 694
  functions of, 694, 696
Sodium channel, 269
Sodium-potassium pump, 51, 279
Soft palate, 595
Solenoid, 55
Solitary lymphatic nodule, 616
Solubility coefficient, 573
Solute, 33
Solution, 33
Solvent, 33
Somatic cell division, 64–67
Somatic nervous system (SNS), 262, 357
Somatic pain, 342
Somatomedin, 405
Somatosensory cortex, 345
Somatostatin, 405
Somatotroph cell, 404
Somesthetic association area, 322
Somite, 210
SOS response, 64
Sound wave, 390
Spasm, 212
Spastic, 335
Spasticity, 351
Spatial summation, 274
Special movements, 181
Special senses
  auditory, 386–392
  equilibrium, 392, 393
  gustatory, 372, 373
  olfactory, 371
  visual, 373–386
Specific gravity, of urine, 681
Spermatic cord, 713
Spermatid, 710
Spermatogenesis, 708–711
Spermatogonium, 708, 709
Spermatozoon, 708, 711
Spermiogenesis, 710, 711
Sphenoid bone, 141
Sphincter
  definition of, 219E
  external, of urinary bladder, 685
  external anal, 622
  of hepatopancreatic ampulla (Oddi), 614
  ileocecal, 614, 622
  internal, of urinary bladder, 685
  internal anal, 622
  lower esophageal, 601
  precapillary, 490
  pyloric, 597
  upper esophageal, 605
Sphincter muscle, 219E
Sphincter of hepatopancreatic ampulla, 614
Sphincter of Oddi. *See* Sphincter of hepatopancreatic ampulla

Sphygmomanometer, 497
Spina bifida, 156
Spinal column. *See* Vertebral column
Spinal cord
    column of, 281, 286, 287E
    cross-sectional structure, 284
    development of, 328, 329
    disorders of, 302–304
    external anatomy, 281
    functions of, 285–293
    general features, 281
    horns of, 281
    injury to, 302
    meninges, 281, 284
    nerves of, 292–302
    protection of, 281, 284
    reflex center, 285, 287–292
    spinal puncture, 284
    tracts of, 285, 286, 287E
Spinal cord injury, 302
Spinal meninges, 281, 284
Spinal nerves
    branches, 294
    cervical, 293
    coccygeal, 293
    connective tissue coverings of, 293
    definition of, 293
    development of, 328, 329
    lumbar, 293
    names of, 292, 293
    plexuses, 294–302
    sacral, 293
    thoracic, 300, 302
Spinal (lumbar) puncture, 284
Spinal segment, 281
Spinal shock, 302
Spine
    anterior superior iliac, 165
    definition of, 134E
    ischial, 165
    posterior superior iliac, 165
    of scapula, 160
    vertebral, 134E
Spinothalamic pathway, 346
Spinous process, 134E
Spiral organ, 390
Spirogram, 570
Spirometer (respirometer), 570
Spleen, 534, 535
    removal of, 536
    rupture of, 536
Splenectomy, 536
Splenic cord, 534
Splenic nodule, 534
Spongy bone, 120, 123
Spongy urethra, 714
Sprain, 192
Squamous cell, 77, 78
Squamous pulmonary epithelial cell, 564
Stapes, 388
Starling's law of the capillaries, 496
Starling's law of the heart, 475
Starvation, 659
Static equilibrium, 392
Stellate reticuloendothelial cell, 490, 611
Stensen's duct. *See* Parotid duct
Stereocilia, 392, 713
Stereognosis, 347
Sterilization, 731
Sternal puncture, 154
Sternum, 152
Steroid, 39E
Stimulus, 269, 339
Stokes-Adams syndrome, 484
Stomach, 603–609
Stomach gastrin, 608E

Stomodeum, 625
Strabismus, 395
Straight tubule, 713
Strain, 192
Stratified columnar epithelium, 84
Stratified cuboidal epithelium, 84
Stratified epithelium
    definition of, 77
    types of, 84
Stratified squamous epithelium, 84
Stratum basale, 101
Stratum basalis, 722
Stratum corneum, 102
Stratum functionalis, 722
Stratum germinativum, 101
Stratum granulosum, 101
Stratum lucidum, 102
Stratum spinosum, 101
Stratum zonale, 314
Stress
    alarm reaction and, 430–432
    and blood pressure, 20
    and blood sugar level, 20, 21
    definition of, 18, 430
    exhaustion reaction and, 432, 433
    general adaptation syndrome, 430
    and homeostasis, 18–20
    resistance reaction and, 432
    stressors and, 430
Stressor, 430
Stress-relaxation, 209
Stretch receptor, 581
Stretch reflex, 288, 289
Striae, 103
Stricture, 689
Stroke volume (SV), 474
Stroma, of ovary, 719
Stromal repair, 97
S-T segment, 470
Stupor, 335
Sty, 376
Styloid process
    of radius, 164
    of temporal bone, 140
    of ulna, 164
Subarachnoid space, 284
Subcutaneous layer, 93, 101, 195
Subdural space, 284
Sublingual duct, 596
Sublingual gland, 596
Subluxation, 192
Submandibular duct, 596
Submandibular gland, 596
Submucosa
    definition of, 592
    of GI tract, 592
    of large intestine, 624
    of small intestine, 616
    of stomach, 605
    of urethra, 685
    of urinary bladder, 685
Submucous plexus, 592
Subneural cleft, 200
Substance P, 326
Substantia nigra, 313, 319
Substrate, 635
Subthalamic nucleus, 319
Subthreshold stimulus, 203, 271
Sucrase, 617, 618
Sucrose, 37
Suction lipectomy, 93
Sudden infant death syndrome (SIDS), 583
Sudoriferous gland, 108
Sulcus
    anterior interventricular, 463
    of brain, 318

    central, 318
    coronary, 463
    intertubercular, 160
    lateral cerebral, 318
    parietooccipital, 318
    posterior interventricular, 463
    posterior median, 281
Summation
    spatial, 274
    temporal, 274
    of twitches, 203, 204
Summation of twitches, 203, 204
Sunburn
    definition of, 111
    effects of, 111
    protection against, 111
Superficial fascia. *See* Subcutaneous layer
Superficial inguinal ring, 714
Superior colliculi, 313
Superior vena cava, 463, 480, 498, 510E
Supination, 181
Supinator muscle, 219E
Suppressor T cell, 546
Suprarenal (adrenal) glands. *See* Adrenal
    gland
Surface anatomy
    definition of, 6
    ear, 387
    eye, 375
    heart, 466, 468
    nose, 556
Surface markings of bones, 133, 134E
Surface projection, of heart, 466, 468
Surfactant, 569, 584
Susceptibility, 537
Suspension, 33
Suspensory ligament, 379, 718
Suspensory ligament of the breast, 728
Sustentacular cell, 708
Sutural bone, 133
Suture
    coronal, 134
    definition of, 134
    lambdoidal, 134
    metopic, 136
    sagittal, 134
    squamosal, 134
    synostoses, 176
Swallowing. *See* Deglutition
Sweat gland. *See* Sudoriferous gland
Sympathectomy, 342
Sympathetic chain, 362
Sympathetic division of ANS, 361, 362
Sympathetic nervous system, 361, 362
Sympathetic trunk ganglia, 358
Sympathomimetic, 423
Symphysis joint, 177
Symphysis pubis, 177
Synapse
    classification of, 272
    convergence, 272
    definition of, 272
    divergence, 272
    excitatory transmitter-receptor interaction
      at, 273, 274
    facilitation at, 273
    factors affecting transmission at, 275, 276
    inhibitory transmitter-receptor interaction
      at, 274
    integration at, 274, 275
    neurotransmitters at, 275, 276
    one-way impulse conduction at, 272, 273
    organization at, 276–278
    summation at, 274
Synapsis, 69, 710
Synaptic cleft, 200, 272

Synaptic delay, 274
Synaptic end bulb, 199, 272
Synaptic gutter, 200
Synaptic vesicle, 199, 262, 263
Synarthrosis, 176
Synchondrosis, 176, 177
Syncope, 474
Syndesmosis, 176
Synergist, 218
Synovial cavity, 177
Synovial fluid, 96, 177
Synovial joint
    definition of, 177
    movements of, 178–181
    structure, 177, 178
    types of, 181–183, 186E
Synovial membrane, 97, 177
Synovitis, 192
Syphilis, 736
System
    cardiovascular, 8E, 440–528
    definition of, 6, 101
    digestive, 8E, 590–629
    endocrine, 8E, 398–432
    integumentary, 8E, 101–113
    lymphatic 8E, 530–552
    muscular, 8E, 216–255
    nervous, 8E, 262–368
    reproductive, 8E, 706–738
    respiratory, 8E, 555–586
    skeletal, 8E, 133–173
    urinary, 8E, 664–689
Systemic anatomy, 46
Systemic circulation, 498–520
Systemic disease, 72
Systemic lupus erythematosus (SLE), 110
Systole, 472
Systolic blood pressure, 498

Tachycardia, 497
Tactile disc, 101, 341
Tactile sensation, 341, 342
Taenia coli, 624
Tail
    of epididymis, 713
    of pancreas, 609
    of spermatozoon, 711
Talus, 170
Target cell, 399
Tarsal gland, 375
Tarsal plate, 375
Tarsus, 170
Taste bud, 372
Taste pore, 372
Taste threshold, 373
Taste zones of tongue, 374
Tay-Sachs disease, 334
T cell, 545–547
    amplifier, 546
    delayed hypersensitivity, 546
    helper, 546
    killer, 545, 546
    memory, 547
    suppressor, 546
Tectorial membrane, 390
Tectospinal tract, 287E, 350
Tectum of midbrain, 313
Teeth, 598–600
Telencephalon, 329
Telodendrium. See Axon terminal
Telophase, 67
Temperature, body
    abnormalities of, 658
    regulation of, 657, 658
Temporal bone, 137, 140
Temporal lobe of cerebrum, 318

Temporal summation, 274
Temporal visual field, 384, 386
Temporary (alimentary) glycosuria, 682
Tendinitis, 196
Tendon
    artificial, 93
    calcaneal, 252, 253
    definition, 93, 196
Tendon organ, 345
Tendon reflex, 289, 290
Tendon sheath, 196
Tension lines. See Lines of cleavage
Tensor muscle, 219E
Tentorium cerebelli, 323
Teratogen, 759
Terminal bronchiole, 560
Terminal ganglion, 358, 360
Termination codon, 62
Tertiary (segmental) bronchus, 560
Testes, 428, 707, 708
Testosterone, 428, 711, 712
Tetanus
    complete, 204
    incomplete, 204
Tetany, 417
Tetrad, 69, 710
Tetralogy of Fallot, 483
Thalamus, 314
Thallasemia, 456
Thallium scan, 482
Thermoreceptive sensation, 342
Thermoreceptor, 340
Theta wave, 323
Thigh
    blood vessels of, 508E, 519E
    bones of, 166
    musculature of, 250E
Third-class lever, 216
Third-degree burn, 111
Third (lateral) cuneiform, 170
Third-order neuron, 346
Third ventricle, 308
Thirst center, 316
Thoracic aorta, 463, 498, 500E, 506E
Thoracic cavity, 13
Thoracic duct, 536, 537
Thoracic vertebrae, 150
Thoracolumbar outflow, 358
Thorax
    blood vessels of, 506E, 516E
    bones of, 152–155
    musculature of, 238E
Thoroughfare channel, 490
Threshold level, 269, 270
Threshold stimulus, 203, 271
Thrombin, 449–452
Thrombocyte
    function of, 447
    number per cubic millimeter, 447
    origin of, 447
    structure of, 447
Thrombocytopenia, 458
Thrombophlebitis, 526
Thrombus, 452
Thymic corpuscle, 536
Thymine, 41
Thymosin, 428
Thymus gland, 428, 536
Thyroglobulin (TGB), 414
Thyroid cartilage, 558
Thyroid colloid, 414
Thyroid diverticulum, 429
Thyroid follicle, 412
Thyroid gland, 412–416
Thyroid hormones, 413
Thyroid-stimulating hormone (TSH), 408

Thyroid storm, 433
Thyrotroph cell, 404
Thyrotropin releasing factor (TRF), 408
Thyroxine (T$_4$), 413
Thyroxine-binding globulin (TBG), 414
Tibia, 168
Tibial tuberosity, 168
Tibiofemoral joint, 188–191
Tic, muscular, 212
Tidal volume, 570
Timing, of cardiac cycle, 472, 473
Tinnitus, 395
Tissue
    connective, 77, 85–94
    definition of, 6, 77
    epithelial, 77–85
    granulation, 97
    inflammation of, 540–542
    muscle, 77, 94, 95, 197–212
    nervous, 77, 95, 96, 262–278
    osseous, 94, 120–128
    rejection of, 550
    repair of, 97
    scar, 97
    types of, 77
    vascular, 94, 440–456
Tissue plasminogen activator (t-PA), 452
Tissue rejection, 550
Tissue repair
    conditions affecting, 97
    definition of, 97
    fibrosis and, 97
    mechanism of, 97
    parenchymal, 97
    stromal, 97
Tissue thromboplastin, 450
Toes, 171
Tongue, 596
Tongue-tied. See Ankyloglossia
Tonsils
    lingual, 533
    palatine, 533
    pharyngeal, 533
Tooth decay. See Dental caries
Torn cartilage, 178
Torpor, 335
Total lung capacity, 572
Touch sensation, 341
Toxic shock syndrome (TSS), 737–738
Trabecula
    of lymph node, 531
    of spongy bone, 123
    of thymus gland, 536
Trabeculae carneae, 465, 466
Trachea, 560
Tracheal cartilage, 560
Trachealis muscle, 560
Tracheostomy, 560
Trachoma, 394
Tract
    anterior corticospinal, 287E, 350
    anterior spinocerebellar, 286E, 348
    anterior spinothalamic, 286E, 377
    ascending, 281, 284, 286E
    corticobulbar, 350
    definition of, 281
    descending, 281, 284, 287E
    fasciculus cuneatus, 286E
    fasciculus gracilis, 286E
    hypothalamic-hypophyseal, 410
    lateral corticospinal, 287E, 350
    lateral spinothalamic, 286E, 347
    olfactory, 371
    optic, 384
    posterior spinocerebellar, 286E, 347
    rubrospinal, 287E, 350

tectospinal, 287E, 350
vestibulospinal, 287E, 350
Tragus, 387
Transamination, 646
Transcendental meditation (TM), 368
Transcription, 62
Transection, 302
Transfer factor, 546
Transfer RNA, 62–64
Transfusion, 458
   autologous, 457
   direct, 457
   exchange, 457
   indirect, 458
   reciprocal, 458
Transitional cell, 78
Transitional epithelium, 84
Translation, 62–64
Transmission of nerve impulses, 268–271
Transplantation
   definition of, 550
   immunosuppressive drugs and, 551
   types of, 550, 551
Transverse arch, of foot, 171
Transverse colon, 622
Transverse fissure, 318, 323
Transverse (T) tubule, 197
Trapezium, 164
Trapezoid, 164
Tremor, 332
Treppe, 204
Triad, 198
Triangular fossa, 387
Triaxial joint, 182
Trichinosis, 212
Trichomoniasis, 736
Tricuspid valve, 465, 466
Trigeminal neuralgia, 334
Triglyceride, 38
Trigone, 685
Triiodothyronine (T₃), 413
Triquetral, 164
tRNA. *See* Transfer RNA
Trochanter
   definition of, 134E
   greater, 166
   lesser, 166
Trochlea
   of humerus, 162
   of orbit, 225E, 226
Trophectoderm, 744
Tropic hormone, 405
Tropomyosin, 198
Tropomyosin-troponin complex, 198
Troponin, 198
Trousseau sign, 417–419
True capillary, 490
True labor
   placental stage, 756
   stage of dilation, 756
   stage of expulsion, 756
Truncus arteriosus, 480
Trunk of nerve plexus, 296
Trypsin, 611, 618
Trypsinogen, 611
T tubule. *See* Transverse tubule
Tubal ligation, 731
Tubercle
   articular, 140, 155
   conoid, 160
   definition of, 134E
   greater, 160
   lesser, 160
   of rib, 155
Tuberculosis, 584

Tuberosity
   costal, 160
   definition of, 134E
   deltoid, 162
   ischial, 165
   radial, 164
   tibial, 168
Tubular gland, 85
Tubular maximum (Tm), 673
Tubular reabsorption, 673–675
Tubular secretion, 675, 676
Tubuloacinar gland, 85
Tumor
   benign, 69, 70
   of brain, 332
   of breasts, 729, 730, 738
   of cervix, 738
   definition of, 69
   of GI tract, 628
   of larynx, 559
   of lung, 574
   malignant, 69
   nasopharyngeal, 71
   primary, 70
   of prostate, 727
   secondary, 70
   of skin, 104, 111
   of uterus, 721
Tumor-specific antigen, 549
Tunica albuginea, 707, 716, 718
Tunica externa, 488
Tunica interna, 488
Tunica media, 488
T wave, 470
Twitch contraction, 203
Two-point discrimination test, 340
Tympanic antrum, 388
Tympanic cavity. *See* Middle ear
Tympanic membrane, 387
Type II cutaneous mechanoreceptor, 341
Type 2 herpes simplex virus, 71
Tyrosinase, 104

Ulcer
   decubitus, 111
   duodenal, 627
   of inflamed tissue, 542
   gastric, 627
   peptic, 627
Ulna, 162, 164
Ultrasound, 730, 755
Umbilical cord, 751
Umbilicus, 751
Unicellular gland, 84, 85
Unipennate fasciculus, 218
Unipolar neuron, 267
Universal donor, 454
Universal recipient, 454
Unmyelinated fiber, 266
Unsaturated fat, 38
Upper esophageal sphincter, 601
Upper extremity
   blood vessels of, 504E, 515E
   bones of, 160–164
   divisions of, 160
   musculature of, 242E, 244E
Upper motor neuron, 350, 351
Uracil, 42
Urea, 672E
Uremia, 689
Ureter, 684, 685
Ureteric bud, 687
Urethra, 686, 714
Urethral fold, 735
Urethral groove, 735
Urethral orifice, 714, 727

Urethritis, 689
Uric acid, 672E
Urinalysis (UA), 682–684
Urinary bladder, 685–686
Urinary system, 664–689
   aging and, 686
   blood and nerve supply, 666–669
   definition of, 664
   development of, 686–688
   disorders of, 688, 689
   hemodialysis, 679, 680
   nephron physiology, 670–679
   urine, 680–684
Urinary tract infection (UTI), 689
Urine
   abnormal constituents, 682–684
   chemical composition, 683E
   concentration of, 678, 679
   dilution of, 676–678
   factors affecting volume, 680, 681
   physical characteristics, 681
Urobilinogen, 683
Urobilinogenuria, 683, 684
Urogenital diaphragm, 236E
Urogenital sinus, 688
Urogenital triangle, 236E, 728
Urology, 664
Uterine cavity, 721
Uterine tube, 720–721
Uterosacral ligament, 721
Uterus, 721, 722
Utricle, 390
Uvea, 376
Uvula, 595

Vagina, 726, 727
Vaginal orifice, 726, 727
Vaginitis, 738
Valence, 27
Valve, of heart, 465
Valvular stenosis, 483
Varicose vein, 491
Vasa recti, 668
Vasa vasorum, 488
Vascular sinus, 490
Vascular spasm, 449
Vascular tissue. *See* Blood
Vascular tunic, 376, 377
Vas deferens. *See* Ductus deferens
Vasectomy, 713, 731
Vasoconstriction, 489
Vasoconstrictor center. *See* Vasomotor center
Vasodilation, 489
Vasodilator substance, 494
Vasomotion, 495
Vasomotor center, 493
Vasomotor tone, 493
Veins
   accessory cephalic, 515E
   accessory hemiazygos, 516E
   anterior hypophyseal, 404
   anterior tibial, 519E
   arcuate, 669
   ascending lumbar, 516E
   axillary, 515E
   azygos, 516E
   basilic, 515E
   brachial, 515E
   brachiocephalic, 516E
   bronchial, 516E
   cardiac, 510E
   central, 611
   cephalic, 515E
   coats of, 490, 491
   common iliac, 516E
   coronary sinus, 463, 469, 498, 510E

Veins (*Continued*)
cystic, 520
deep, 515E
definition of, 488, 490
dorsalis pedis, 519E
dorsal metacarpal, 515E
dorsal venous arch, 515E, 519E
ductus venosus, 523
esophageal, 516E
external iliac, 516E
external jugular, 512E
femoral, 519E
gastric, 520
gastroepiploic, 520
gonadal, 516E
great cardiac, 469
great saphenous, 519E
hepatic, 516E, 520
hepatic portal, 520
of hepatic portal circulation, 520
hemiazygos, 516E
hypophyseal portal, 404
inferior mesenteric, 520
inferior phrenic, 516E
inferior sagittal sinus, 512E
inferior vena cava, 510E, 516E
intercostal, 516E
interlobar, 669
interlobular, 669
internal iliac, 516E
internal jugular, 512E
lateral plantar, 519E
lumbar, 516E
medial plantar, 519E
median antebrachial, 515E
median cubital, 515E
mediastinal, 516E
middle cardiac, 469
ovarian, 516E
palmar venous arch, 515E
pancreatic, 520
pericardial, 516E
peroneal, 519E
popliteal, 519E
posterior hypophyseal, 410
posterior tibial, 519E
pulmonary, 498, 520
of pulmonary circulation, 520
pyloric, 520
radial, 515E
renal, 516E, 669
sigmoid sinus, 512E
small saphenous, 519E
splenic, 520
straight sinus, 512E
subclavian, 515E
superficial, 515E
superior mesenteric, 520
superior sagittal sinus, 512E
superior vena cava, 510E
suprarenal, 516E
of systemic circulation, 510E–514E
testicular, 516E
transverse sinus, 512E
ulnar, 515E
umbilical, 522, 523
uterine, 722
varicose, 491
Velocity, of blood flow, 496
Venereal disease. *See* Sexually transmitted disease
Venesection, 458
Venous sinus, 535
Ventral body cavity, 13
Ventral ramus, 294

Ventricle
of brain, 308
of heart, 463, 480
Ventricular diastole, 472
Ventricular fibrillation, 484
Ventricular filling, 472
Ventricular fold, 558
Ventricular premature contraction (VPC), 484
Ventricular pressure, 472
Ventricular systole, 472
Venule, 488, 490
Vermiform appendix, 622
Vermilion, 595
Vermis, of cerebellum, 323
Vertebra
cervical, 150
coccyx, 152
lumbar, 150
parts of, 149, 150
sacrum, 150–152
thoracic, 150
typical, 149, 150
Vertebral arch, 150
Vertebral body cavity, 13
Vertebral canal, 150, 281
Vertebral column
cervical region, 150
coccygeal region, 152
curvatures, 156
curves, 149
fractures, 156
herniated (slipped) disc, 155
lumbar region, 150
sacral region, 150–152
spina bifida, 156
thoracic region, 150
Vertebral curve, 149
Vertebra prominens, 150
Vertigo, 394
Vesicouterine pouch, 721
Vesicular ovarian follicle, 720, 725
Vestibular membrane, 390
Vestibular nuclear complex, 313
Vestibule
of internal ear, 390
of nasal cavity, 556
of oral cavity, 595
of vulva, 727
Vestibulospinal tract, 287E, 350
Vibration sensation, 342
Villus, 616
Viral encephalitis, 335
Virilism, 423
Virilizing adenoma, 423
Viscera, definition of, 13
Visceral autonomic reflex, 366, 367
Visceral effector, 357
Visceral efferent neuron, 358–361
Visceral efferent pathway, 358–361
Visceral pain, 342
Visceral pericardium. *See* Epicardium
Visceral peritoneum, 592
Visceral pleura, 561
Visceral surface, of spleen, 534
Visceroceptor, 340
Viscosity, 440
Visual acuity, 379
Visual association area, 322
Visual field, 384–386
Visual sensation, 373–386
Vital capacity, 572
Vitamins
absorption by small intestine, 621
and bone growth, 126
deficiencies of, 652E–655E
fat-soluble, 652E, 653E

functions of, 652E–655E
megadoses of, 655E
sources of, 652E–655E
and tissue repair, 97
water-soluble, 653E, 654E
Vitiligo, 104
Vitreous humor, 380
Vocal fold, 559
Voice production, 588, 589
Volkmann's canal. *See* Perforating canal
Volkmann's contracture, 212
Vomer, 146
Vomiting, 609
Vulva, 727, 728

Waking state, 352, 353
Wallerian degeneration, 303
Wandering macrophage, 445
Warfarin, 452
Wart, 113
Water
absorption by small intestine, 621
and body fluids, 693
in chemical reactions, 33
and heat, 33
hydrostatic pressure and, 672
importance of, 33
and lubrication, 33
solvating property of, 33
Water intoxication, 697
Water-soluble vitamins, 653E, 654E
Wave summation, 204
Weight discrimination, 347
Wharton's duct. *See* Submandibular duct
Wharton's jelly. *See* Mucous connective tissue
Whiplash injury, 150
White blood cell. *See* Leucocyte
White matter, 281, 318
White muscle fiber, 205, 206
White pulp, of spleen, 534
White rami communicantes, 361
Whole blood, 458
Wisdom tooth, 599
Word blindness, 322
Word deafness, 322
Wormian bone. *See* Sutural bone
Wound healing
deep, 105
epidermal, 105
phases of, 105
Wrist, 164
Wryneck, 212

Xenograft, 550, 551
Xeroradiography, 730
Xiphoid process, 152
X-linked trait, 763

Yellow marrow, 120
Yoga, 368
Yolk sac, 749

Z line, 198
Zona fasciculata, 420
Zona glomerulosa, 419
Zona orbicularis, 187
Zona pellucida, 724, 743
Zona reticularis, 420
Zone of calcified matrix, 125
Zone of hypertrophic cartilage, 125
Zone of proliferating cartilage, 125
Zone of reserve cartilage, 125
Zygomatic arch, 137, 144
Zygomatic bone, 144
Zymogenic (peptic) cell, 605
Zygote, 744